Dear Larry,

Best wishes,

Jack

# PROSTATE CANCER

# PROSTATE CANCER

## SCIENCE AND CLINICAL PRACTICE

### SECOND EDITION

*Edited by*

JACK H. MYDLO

*Department of Urology, Temple University Hospital, Philadelphia, PA, USA*

CIRIL J. GODEC

*Department of Urology, SUNY Downstate University School of Medicine, Brooklyn, NY;*
*Long Island College Hospital, Long Island, NY, USA*

AMSTERDAM • BOSTON • HEIDELBERG • LONDON
NEW YORK • OXFORD • PARIS • SAN DIEGO
SAN FRANCISCO • SINGAPORE • SYDNEY • TOKYO

Academic Press is an Imprint of Elsevier

Academic Press is an imprint of Elsevier
125, London Wall, EC2Y 5AS, UK
525 B Street, Suite 1800, San Diego, CA 92101-4495, USA
225 Wyman Street, Waltham, MA 02451, USA
The Boulevard, Langford Lane, Kidlington, Oxford OX5 1GB, UK



**British Library Cataloguing-in-Publication Data**
A catalogue record for this book is available from the British Library

**Library of Congress Cataloging-in-Publication Data**
A catalog record for this book is available from the Library of Congress

ISBN: 978-0-12-800077-9

For information on all Academic Press publications
visit our website at http://store.elsevier.com/

Typeset by Thomson Digital

Publisher: Mica Haley
Acquisition Editor: Mara Conner
Editorial Project Manager: Jeffrey Rossetti
Production Project Manager: Lucía Pérez
Designer: Matthew Limbert

# Contents

## 18. Neuroendocrine Prostate Cancer

DANIEL C. PARKER, ALEXANDER KUTIKOV



## 19. Breast and Prostate Cancers: A Comparison of Two Endocrinologic Malignancies

MAAHUM HAIDER, PAUL H. LANGE



# IV

# PREVENTION OF PROSTATE CANCER

## 20. Heart Healthy = Prostate Healthy and S.A.M. are the Ideal "Natural" Recommendations for Prostate Cancer

MARK A. MOYAD



## 21. Effects of Smoking, Alcohol, and Exercise on Prostate Cancer

HADLEY WYRE, JAMES BRANTLEY THRASHER



## 22. Environmental and Occupational Exposures and Prostate Cancer

TULLIKA GARG



## 23. Level-1 Data From the REDUCE Study and the PCPT Data

GOUTHAM VEMANA, GERALD ANDRIOLE



# V

# CONSERVATIVE MANAGEMENT

## 24. Decision Support for Low-Risk Prostate Cancer

LEONARD P. BOKHORST, EWOUT W. STEYERBERG, MONIQUE J. ROOBOL



## 25. Active Surveillance: Rationale, Patient Selection, Follow-up, and Outcomes

LAURENCE KLOTZ

# VI

# SURGERY

## 26. Preoperative Risk Assessment

THOMAS J. GUZZO



## 27. Is Surgery Still Necessary for Prostate Cancer?

AHMED A. HUSSEIN, MATTHEW R. COOPERBERG



## 28. Indications for Pelvic Lymphadenectomy

JAY D. RAMAN, AWET GHEREZGHIHIR



## 29. The Surgical Anatomy of the Prostate

FAIRLEIGH REEVES, WOUTER EVERAERTS, DECLAN G. MURPHY, ANTHONY COSTELLO



## 30. Radical Retropubic Prostatectomy

LEONARD G. GOMELLA, CHANDAN KUNDAVARAM



## 31. Radiation-Resistant Prostate Cancer and Salvage Prostatectomy

YOUSEF AL-SHRAIDEH, SAMIR V. SEJPAL, JOSHUA J. MEEKS



## 32. Postradical Prostatectomy Incontinence

LISA PARRILLO, ALAN WEIN

## 33. Prognostic Significance of Positive Surgical Margins and Other Implications of Pathology Report

GEORGE R. SCHADE, JONATHAN L. WRIGHT, DANIEL W. LIN



## 34. Open Versus Robotic Prostatectomy

CESAR E. ERCOLE, ANDREW J. STEPHENSON



## 35. The Technique of Robotic Nerve-Sparing Prostatectomy

FIRAS ABDOLLAH, MANI MENON



## 36. Anterior Approach to Robotic Radical Prostatectomy

ANDREW C. HARBIN, DANIEL D. EUN



## 37. Posterior Approach to Robotic-Assisted Laparoscopic Radical Prostatectomy

MARK MANN, COSTAS D. LALLAS, EDOUARD J. TRABULSI



## 38. The Technique of Robotic Nerve Sparing Prostatectomy: Extraperitoneal Approach

CHRISTIAN P. PAVLOVICH



## 39. Clinical and Pathologic Staging of Prostate Cancer

ADAM C. REESE



## 40. Management of Bladder Neck Contracture in the Prostate Cancer Survivor

DANIEL C. PARKER, TIMOTHY J. TAUSCH, JAY SIMHAN



## 41. Reimbursement for Prostate Cancer Treatment

AHMED HADDAD, YAIR LOTAN



# VII

# RADIATION THERAPY

## 42. Fundamentals of Radiation Treatment for Prostate Carcinoma – Techniques, Radiation Biology, and Evidence Base

MOHAN P. ACHARY, CURTIS T. MIYAMOTO

## 43. Radiation with Hormonal Therapy

BROCK R. BAKER, JAHAN J. MOHIUDDIN, RONALD C. CHEN



## 44. Brachytherapy for Prostate Cancer: An Overview

NICHOLAS G. ZAORSKY, ERIC M. HORWITZ



## 45. Intensity Modulated Radiotherapy and Image Guidance

MELISSA R. YOUNG, JAMES B. YU



## 46. Proton Beam Therapy

CURTILAND DEVILLE



## 47. Radiotherapy After Radical Prostatectomy: Adjuvant Versus Salvage Approach

BARUCH MAYER GROB, TARYN G. TORRE, ALBERT PETROSSIAN



## 48. Emerging Modalities in Radiation Therapy for Prostate Cancer

CURTIS T. MIYAMOTO, MOHAN P. ACHARY



# VIII
# CLINICAL DILEMMAS

## 49. Management of PSA Recurrences After Radical Prostatectomy

JOHN B. EIFLER, JOSEPH A. SMITH, JR



## 50. Salvage Therapy for Locally Recurrent Prostate Cancer After External Beam Radiation Therapy

DAVID M. MARCUS, DAVID B. CAHN, DANIEL J. CANTER

# IX
# ADVANCED PROSTATE CANCER

## 65. Legal Implications of Prostate Cancer Screening

C.J. STIMSON



# XII

# NEW HORIZONS FOR PROSTATE CANCER

## 66. New Markers for Prostate Cancer Detection and Prognosis

E. DAVID CRAWFORD, KAREN H. VENTII, NEAL D. SHORE



## 67. Testosterone Therapy in Hypogonadal Men with Prostate Cancer

JOSHUA R. KAPLAN

# List of Contributors


**Philip H. Abbosh, MD, PhD** Department of Surgical Oncology, Fox Chase Cancer Center-Temple Health, Philadelphia, PA, USA

**Firas Abdollah, MD** Vattikuti Urology Institute & VUI Center for Outcomes Research Analytics and Evaluation, Henry Ford Hospital, Detroit, MI, USA

**Mohan P. Achary, PhD** Department of Radiation Oncology and Radiology, Temple University School of Medicine, Philadelphia, PA, USA

**Shaheen Alanee, MD, MPH** Division of Urology, Department of Surgery, Southern Illinois University School of Medicine, Springfield, IL, USA

**Peter C. Albertsen, MD, MS** Department of Surgery, University of Connecticut Health Center, Farmington, CT, USA

**Yousef Al-Shraideh, MD** Department of Urology, Northwestern University, Feinberg School of Medicine, Chicago, IL, USA

**Gerald Andriole, MD** Division of Urological Surgery, Washington University in St. Louis, St. Louis, MO, USA

**Janet E. Baack Kukreja, MD** Department of Urology, University of Rochester School of Medicine and Dentistry, Rochester, NY, USA

**Richard K. Babayan, MD** Boston University School of Medicine, Boston, MA, USA

**Brock R. Baker, BS** Department of Radiation Oncology, University of North Carolina, Chapel Hill, NC, USA

**Christopher E. Bayne, MD** Department of Urology, The George Washington University, Washington, DC, USA

**Marijo Bilusic, MD, PhD** Department of Medical Oncology, Fox Chase Cancer Center/Temple University, Philadelphia, PA, USA

**Leonard P. Bokhorst, MD** Department of Urology, Erasmus University Medical Center, Rotterdam, The Netherlands

**David B. Cahn, DO, MBS** Department of Urology, Einstein Healthcare Network, Philadelphia, PA, USA

**Daniel J. Canter, MD** Department of Urology, Fox Chase Cancer Center and Einstein Healthcare Network, Philadelphia, PA, USA

**David Y.T. Chen, MD, FACS** Department of Surgical Oncology, Fox Chase Cancer Center-Temple Health, Philadelphia, PA, USA

**Ronald C. Chen, MD, MPH** Department of Radiation Oncology, University of North Carolina, Chapel Hill, NC, USA

**Juan Chipollini, MD** Department of Urology, University of Miami Miller School of Medicine, Miami, FL, USA

**Peter L. Choyke, MD** Molecular Imaging Program, National Cancer Institute, National Institutes of Health, Bethesda, MD, USA

**Matthew R. Cooperberg, MD, MPH** Department of Urology and Helen Diller Family Comprehensive Cancer Center, University of California, San Francisco, CA; Department of Epidemiology and Biostatistics, University of California, San Francisco, CA, USA

**Anthony Costello, MD, FRACS, FRCSI** Departments of Urology and Surgery, Royal Melbourne Hospital and University of Melbourne, Melbourne; Epworth Prostate Centre, Epworth Healthcare, Melbourne, Australia

**E. David Crawford, MD** University of Colorado Health Science Center, Aurora, CO, USA

**Curtiland Deville, MD** Johns Hopkins University, The Sidney Kimmel Comprehensive Cancer Center, Sibley Memorial Hospital, Washington, DC, USA

**Essel Dulaimi, MD** Department of Pathology, Fox Chase Cancer Center, Philadelphia, PA, USA

**Danuta Dynda, MD** Division of Urology, Department of Surgery, Southern Illinois University School of Medicine, Springfield, IL, USA

**John B. Eifler, MD** Department of Urologic Surgery, Vanderbilt University Medical Center, Nashville, TN, USA

**Cesar E. Ercole, MD** Center for Urologic Oncology, Glickman Urological & Kidney Institute, Cleveland Clinic, Cleveland, OH, USA

**Daniel D. Eun, MD** Department of Urology, Temple University Hospital, Philadelphia, PA, USA

**Wouter Everaerts, MD, PhD** Division of Cancer Surgery, Peter MacCallum Cancer Centre, University of Melbourne, East Melbourne; Department of Urology, Royal Melbourne Hospital, Parkville; and Epworth Prostate Centre, Epworth Healthcare, Richmond, Victoria, Australia

**Izak Faiena, MD** Section of Urologic Oncology, Rutgers Cancer Institute of New Jersey and Division of Urology, Department of Surgery, Rutgers Robert Wood Johnson Medical School, New Brunswick, NJ, USA

**Michael A. Ferragamo, MD, FACS** Department of Urology, State University of New York, University Hospital, Stony Brook, NY, USA

**Chandra K. Flack, MD** Department of Urology, Indiana University School of Medicine, Indianapolis, IN, USA

**Tullika Garg, MD, MPH** Urology Department, Geisinger Health System, Danville, PA, USA

**Awet Gherezghihir, MD** Department of Surgery (Urology), The Penn State Milton S. Hershey Medical Center, Hershey, PA, USA

**Ciril J. Godec, MD, PhD** Department of Urology, SUNY Downstate University School of Medicine, Brooklyn, NY; Long Island College Hospital, Long Island, NY, USA

**Leonard G. Gomella, MD** Department of Urology, Kimmel Cancer Center, Thomas Jefferson University, Philadelphia, PA, USA

**Richard E. Greenberg, MD, FACS** Division of Urologic Oncology, Fox Chase Cancer Center, and Temple University School of Medicine, Philadelphia, PA, USA

**Baruch Mayer Grob, MD** Urology Section, Hunter Holmes McGuire VA Medical Center, and Division of Urology, Virginia Commonwealth University Health System, Richmond, VA, USA

**Giorgio Guazzoni, MD** Division of Oncology/Unit of Urology, URI, IRCCS Ospedale San Raffaele, Università Vita-Salute San Raffaele, Milan, Italy

**Thomas J. Guzzo, MD, MPH** Division of Urology, The Hospital of the University of Pennsylvania, Philadelphia, PA, USA

**Ahmed Haddad, MBChB, PhD** Department of Urology, University of Texas Southwestern Medical Center, Dallas, TX, USA

**Maahum Haider, MD** Department of Urology, University of Washington School of Medicine, Seattle, WA, USA

**Andrew C. Harbin, MD** Department of Urology, Temple University Hospital, Philadelphia, PA, USA

**Eric M. Horwitz, MD** Department of Radiation Oncology, Fox Chase Cancer Center, Philadelphia, PA, USA

**Ahmed A. Hussein, MD** Department of Urology and Helen Diller Family Comprehensive Cancer Center, University of California, San Francisco, CA, USA; Department of Urology, Cairo University, Oula, Giza, Egypt

**Timothy Ito, MD** Department of Surgical Oncology, Fox Chase Cancer Center, Philadelphia, PA, USA

**Thomas W. Jarrett, MD** Department of Urology, The George Washington University, Washington, DC, USA

**Lawrence C. Jenkins, MD, MBA** Sexual and Reproductive Medicine Program, Urology Service, Memorial Sloan-Kettering Cancer Center, New York, USA

**Joshua R. Kaplan, MD** Department of Urology, Temple University School of Medicine, Philadelphia, PA, USA

**Mark H. Katz, MD** Boston University School of Medicine, Boston, MA, USA

**Louis R. Kavoussi, MD, MBA** The Arthur Smith Institute for Urology, Hofstra North Shore LIJ School of Medicine, New Hyde Park, NY, USA

**Jonathan Kiechle, MD** Department of Urology, University Hospital Case Medical Center, Case Western Reserve University, Cleveland, OH, USA

**Simon P. Kim, MD, MPH** Department of Urology, University Hospital Case Medical Center, Case Western Reserve University, Cleveland, OH; Cancer Outcomes and Public Policy Effectiveness Research, COPPER Center, Yale University, New Haven, CT, USA

**Laurence Klotz, MD** Division of Urology, Sunnybrook Health Sciences Centre, University of Toronto, Toronto, ON, Canada

**Michael O. Koch, MD** Department of Urology, Indiana University School of Medicine, Indianapolis, IN, USA

**Chandan Kundavaram, MD** Department of Urology, Kimmel Cancer Center, Thomas Jefferson University, Philadelphia, PA, USA

**Alexander Kutikov, MD, FACS** Department of Urology, Fox Chase Cancer Center, Einstein Urologic Institute, Philadelphia, PA, USA

**Costas D. Lallas, MD, FACS** Department of Urology, Sidney Kimmel Cancer Center, Thomas Jefferson University, Philadelphia, PA, USA

**Paul H. Lange, MD, FACS** Department of Urology, University of Washington School of Medicine; Institute of Prostate Cancer Research, University of Washington and Fred Hutchinson Cancer Research Center, Seattle, WA, USA

**Massimo Lazzeri, MD, PhD** Division of Oncology/Unit of Urology, URI, IRCCS Ospedale San Raffaele, Università Vita-Salute San Raffaele, Milan, Italy

**Daniel W. Lin, MD** Department of Urology, University of Washington School of Medicine, Seattle; Division of Public Health Sciences, Fred Hutchinson Cancer Research Center, Seattle, WA, USA

**Yair Lotan, MD** Department of Urology, University of Texas Southwestern Medical Center, Dallas, TX, USA

**Casey Lythgoe, MD** Division of Urology, Department of Surgery, Southern Illinois University School of Medicine, Springfield, IL, USA

**Danil V. Makarov, MD, MHS** Department of Urology, Population Health, and Health Policy, NYU School of Medicine, New York, NY, USA

**Mark Mann, MD** Department of Urology, Sidney Kimmel Cancer Center, Thomas Jefferson University, Philadelphia, PA, USA

**David M. Marcus, MD** Department of Radiation Oncology, Winship Cancer Institute, Emory University, Atlanta, GA, USA

**Viraj A. Master, MD, PhD, FACS** Department of Urology, Emory University, Atlanta, GA, USA

**Joshua J. Meeks, MD, PhD** Department of Urology, Northwestern University, Feinberg School of Medicine, Chicago, IL, USA

**Neil Mendhiratta, BA** School of Medicine, New York University Langone Medical Center, New York, NY, USA

**Mani Menon, MD** Vattikuti Urology Institute & VUI Center for Outcomes Research Analytics and Evaluation, Henry Ford Hospital, Detroit, MI, USA

**Edward M. Messing, MD, FACS** Department of Urology, University of Rochester School of Medicine and Dentistry, Rochester, NY, USA

**Curtis T. Miyamoto, MD** Department of Radiation Oncology and Radiology, Temple University School of Medicine, Philadelphia, PA, USA

**Parth K. Modi, MD** Section of Urologic Oncology, Rutgers Cancer Institute of New Jersey and Division of Urology, Department of Surgery, Rutgers Robert Wood Johnson Medical School, New Brunswick, NJ, USA

**Jahan J. Mohiuddin, BS** Department of Radiation Oncology, University of North Carolina, Chapel Hill, NC, USA

**M. Francesca Monn, MD, MPH** Department of Urology, Indiana University School of Medicine, Indianapolis, IN, USA

**Francesco Montorsi, MD** Division of Oncology/Unit of Urology, URI, IRCCS Ospedale San Raffaele, Università Vita-Salute San Raffaele, Milan, Italy

**Daniel Moon, MBBS(Hon), FRACS** Division of Cancer Surgery, University of Melbourne, Peter MacCallum Cancer Centre, Melbourne; Robotic Surgery, Epworth Healthcare, Melbourne, Australia

**Kelvin A. Moses, MD, PhD** Department of Urologic Surgery, Vanderbilt University Medical Center, Nashville, TN, USA

**Judd W. Moul, MD, FACS** Department of Surgery and the Duke Cancer Institute, Division of Urology, Duke University School of Medicine, Durham, NC, USA

**Mark A. Moyad, MD, MPH** Department of Urology, University of Michigan Medical Center, Ann Arbor, MI, USA

**Phillip Mucksavage, MD** Perelman School of Medicine, University of Pennsylvania, Pennsylvania, USA

**John P. Mulhall, MD, MSc, FECSM, FACS** Sexual and Reproductive Medicine Program, Urology Service, Memorial Sloan-Kettering Cancer Center, New York, USA

**Declan G. Murphy, MB, FRCS Urol** Division of Cancer Surgery, Peter MacCallum Cancer Centre, University of Melbourne, East Melbourne; Department of Urology, Royal Melbourne Hospital, Parkville; and Epworth Prostate Centre, Epworth Healthcare, Richmond, Victoria, Australia

**Jack H. Mydlo, MD, FACS** Department of Urology, Temple University Hospital, Philadelphia, PA, USA

**Joel B. Nelson, MD** Department of Urology, University of Pittsburgh Medical Center, Pittsburgh, PA, USA

**Jaspreet Singh Parihar, MD** Section of Urologic Oncology, Rutgers Cancer Institute of New Jersey and Division of Urology, Department of Surgery, Rutgers Robert Wood Johnson Medical School, New Brunswick, NJ, USA

**Daniel C. Parker, MD** Department of Urology, Temple University Hospital; Department of Urology, Fox Chase Cancer Center, Einstein Urologic Institute, Philadelphia, PA, USA

**Lisa Parrillo, MD** Department of Urology, Perelman Center for Advanced Medicine, Philadelphia, PA, USA

**Neal Patel, MD** Section of Urologic Oncology, Rutgers Cancer Institute of New Jersey and Division of Urology, Department of Surgery, Rutgers Robert Wood Johnson Medical School, New Brunswick, NJ, USA

**Christian P. Pavlovich, MD** James Buchanan Brady Urological Institute, Johns Hopkins University School of Medicine, MD, USA

**Albert Petrossian, MD** Division of Urology, Virginia Commonwealth University Health System, Richmond, VA, USA

**Eugene Pietzak, MD** University of Pennsylvania Health Care System, Pennsylvania, USA

**Peter Pinto, MD** Urologic Oncology Branch, National Cancer Institute, National Institutes of Health, Bethesda, MD, USA

**Zachary Piotrowski, MD** Department of Urology, Fox Chase Cancer Center, and Temple University School of Medicine, Philadelphia, PA, USA

**Michel A. Pontari, MD** Department of Urology, Temple University School of Medicine, Philadelphia, PA, USA

**Sanoj Punnen, MD** Department of Urology, University of Miami Miller School of Medicine, Miami, FL, USA

**Jay D. Raman, MD** Department of Surgery (Urology), The Penn State Milton S. Hershey Medical Center, Hershey, PA, USA

**Adam C. Reese, MD** Department of Urology, Temple University School of Medicine, Philadelphia, PA, USA

**Fairleigh Reeves, MB, BS** Departments of Urology and Surgery, Royal Melbourne Hospital and University of Melbourne, Melbourne, Australia

**Simon Van Rij, MD, PhD** Division of Cancer Surgery, Peter MacCallum Cancer Centre, University of Melbourne, East Melbourne; Department of Urology, Royal Melbourne Hospital, Parkville; and Epworth Prostate Centre, Epworth Healthcare, Richmond, Victoria, Australia

**Benjamin T. Ristau, MD** Department of Urology, University of Pittsburgh Medical Center, Pittsburgh, PA, USA

**Monique J. Roobol, PhD** Department of Urology, Erasmus University Medical Center, Rotterdam, The Netherlands

**Simpa S. Salami, MD, MPH** The Arthur Smith Institute for Urology, Hofstra North Shore LIJ School of Medicine, New Hyde Park, NY, USA

**Amirali H. Salmasi, MD** Section of Urologic Oncology, Rutgers Cancer Institute of New Jersey and Division of Urology, Department of Surgery, Rutgers Robert Wood Johnson Medical School, New Brunswick, NJ, USA

**Sandeep Sankineni, MD** Molecular Imaging Program, National Cancer Institute, National Institutes of Health, Bethesda, MD, USA

**Kristen R. Scarpato, MD, MPH** Department of Urologic Surgery, Vanderbilt University Medical Center, Nashville, TN, USA

**George R. Schade, MD** Department of Urology, University of Washington School of Medicine, Seattle, WA, USA

**Matthew S. Schaff, MD** Department of Urology, Temple University Hospital, Philadelphia, PA, USA

**Samir V. Sejpal, MD, MPH** Department of Radiation Oncology, Northwestern University, Feinberg School of Medicine, Chicago, IL, USA

**Neal D. Shore, MD, FACS** Carolina Urologic Research Center, Atlantic Urology Clinics, Myrtle Beach, SC, USA

**Jay Simhan, MD** Department of Urology, Fox Chase Cancer Center, Einstein Urologic Institute, Philadelphia, PA, USA

**Susan F. Slovin, MD, PhD** Genitourinary Oncology Service, Sidney Kimmel Center for Prostate and Urologic Cancers, Memorial Sloan-Kettering Cancer Center, New York, NY; Department of Medicine, Weill-Cornell Medical College, USA

**Marc C. Smaldone, MD** Department of Surgical Oncology, Fox Chase Cancer Center, Philadelphia, PA, USA

**Joseph A. Smith, Jr, MD** William L. Bray Professor of Urologic Surgery, Department of Urologic Surgery, Vanderbilt University Medical Center, Nashville, TN, USA

**Andrew J. Stephenson, MD, FACS, FRCS(C)** Center for Urologic Oncology, Glickman Urological & Kidney Institute, Cleveland Clinic, Cleveland; Cleveland Clinic Lerner College of Medicine, Case Western Reserve University, Cleveland, OH, USA

**Ewout W. Steyerberg, PhD** Department of Public Health, Erasmus University Medical Center, Rotterdam, The Netherlands

**C.J. Stimson, MD, JD** Department of Urologic Surgery, Vanderbilt University Medical Center, Nashville, TN, USA

**Siobhan Sutcliffe, PhD** Department of Surgery, Division of Public Health Sciences, Washington University School of Medicine, Siteman Cancer Center, St. Louis, MO, USA

**Samir S. Taneja, MD** Department of Urology and Radiology, New York University Langone Medical Center, New York, NY, USA

**Vincent Tang, MBBS, MSc, DIC, FRCS Urol** Department of Urology, Royal Melbourne Hospital, Melbourne, Australia; Division of Cancer Surgery, University of Melbourne, Peter MacCallum Cancer Centre, Melbourne, Australia

**Timothy J. Tausch, MD** Department of Urology, UT Southwestern Medical Center, Dallas, TX, USA

**James Brantley Thrasher, MD** Department of Urology, University of Kansas Medical School, Kansas City, KS, USA

**Taryn G. Torre, MD** Division of Urology, Radiation Oncology Associates, a Division of Virginia Urology, Richmond, VA, USA

**Edouard J. Trabulsi, MD, FACS** Department of Urology, Sidney Kimmel Cancer Center, Thomas Jefferson University, Philadelphia, PA, USA

**Baris Turkbey, MD** Molecular Imaging Program, National Cancer Institute, National Institutes of Health, Bethesda, MD, USA

**Robert M. Turner II, MD** Department of Urology, University of Pittsburgh Medical Center, Pittsburgh, PA, USA

**Willie Underwood III, MD, MPH, MSci** Department of Urology, Roswell Park Cancer Institute, Buffalo, NY, USA

**Goutham Vemana, MD** Division of Urological Surgery, Washington University in St. Louis, St. Louis, MO, USA

**Shilpa Venkatachalam, PhD, MA** NYU Langone Medical Center, NYU School of Medicine, New York, NY, USA

**Karen H. Ventii, PhD** School of Medicine, Emory University, Atlanta, GA, USA

**Alan Wein, MD, PhD(Hon)** Department of Urology, Perelman Center for Advanced Medicine, Philadelphia, PA, USA

**Jonathan L. Wright, MD, MS** Department of Urology, University of Washington School of Medicine, Seattle; Division of Public Health Sciences, Fred Hutchinson Cancer Research Center, Seattle, WA, USA

**Hadley Wyre, MD** Department of Urology, University of Kansas Medical School, Kansas City, KS, USA

**Isaac Yi Kim, MD, PhD** Section of Urologic Oncology, Rutgers Cancer Institute of New Jersey and Division of Urology, Department of Surgery, Rutgers Robert Wood Johnson Medical School, New Brunswick, NJ, USA

**Melissa R. Young, MD, PhD** Department of Therapeutic Radiology, Yale School of Medicine, New Haven, CT, USA

**James B. Yu, MD** Department of Therapeutic Radiology, Yale School of Medicine, New Haven, CT, USA

**Nicholas G. Zaorsky, MD** Department of Radiation Oncology, Fox Chase Cancer Center, Philadelphia, PA, USA

# Preface

The field of urology has changed so much since the first edition of this book came out in 2003. Robotic surgery was still in its infancy, PSA screening was very active, and the economics of health care were different than they are today.

Many of the previous contributors of the first edition of this book have retired or passed, and therefore we have a new generation of innovative, talented, dedicated urologists who have made up the majority of this second edition. New chapters concerning testosterone replacement, legal ramifications of PSA screening, and billing codes have also been included.

The aim of the editors has been to get the latest, most comprehensive topics together in one book that would be of great value to the student, resident, researcher, scientist, and practicing urologist. The next decade will see another generation of new urologists and new therapies for urologic diseases. For now, we hope that this book covers everything you need to know about prostate cancer and its treatment.

*Jack H. Mydlo*
*Ciril J. Godec*

PART I

# ETIOLOGY, PATHOLOGY, AND TUMOR BIOLOGY

CHAPTER

# 1

# Population Screening for Prostate Cancer and Early Detection

*Judd W. Moul, MD, FACS*

**Department of Surgery and the Duke Cancer Institute, Division of Urology, Duke University School of Medicine, Durham, NC, USA**

## INTRODUCTION

Prostate cancer (PCa) is the most commonly diagnosed new solid cancer and the second most common cause of cancer-related deaths in men in the United States. The American Cancer Society (ACS) estimated that approximately 241,740 new cases and 28,170 PCa-related deaths would occur in the United States in 2014.[1] PCa is now the second-leading cause of cancer death in men, exceeded only by lung cancer. It accounts for 29% of all male cancers and 9% of male cancer-related deaths.

The prostate-specific antigen (PSA) concentration in the blood is a test approved by the US Food and Drug Administration as an aid to the early detection of PCa. Screening with PSA has been widely used to detect PCa for decades in the United States and many industrialized nations but it continues to be controversial.[2–5] Until a few years ago, the ACS and the American Urological Association (AUA) recommended PCa screening for men at average risk who are 50 years or older and have a life expectancy of at least 10 years, after the patient and physician discuss the risks and benefits of screening and intervention.[6–7] Screening was recommended for higher-risk groups, such as African-American (AA) men, beginning at age 40 years. There is much debate over the sensitivity of PSA tests and the ultimate effect on morbidity and mortality of detecting PCa at early, potentially clinically insignificant stages, so called "overdetection."[8–9]

In May 2012, the US Preventive Services Task Force (USPSTF) issued a new guideline against PSA testing and gave the test a "D-rating" indicating that it resulted in more harm than good.[10] Since this time, the debate has been heated and the AUA and other organizations have now come together to oppose population-based screening.[2–9] In this chapter, I will further explore this controversy and discuss the rationale and basis for using PSA in the early detection of PCa.

### The USPSTF Firestorm

In May 2012, the USPSTF issued an opinion opposing PCa screening using the PSA test.[10] The Task Force is an independent government-appointed panel of experts in prevention and evidence-based medicine composed of primary care providers (internists, pediatricians, family physicians, nurses, gynecologists/obstetricians, and health behavior specialists) who do not typically treat PCa patients. They receive funding from the US Government through the Department of Health and Human Services Agency for Health Care Research and Quality with the charge to conduct scientific evidence reviews of clinical preventive health care services and to develop recommendations for primary care clinicians and health systems.

The recommendations of the USPSTF ignited a firestorm of controversy. I was part of a group, composed of a number of recognized experts in the diagnosis and treatment of PCa, along with specialists in preventive medicine, oncology, and primary care, issued a brief commentary opposing their recommendations when they were initially published.[11] We also have articulated a critical analysis of the USPSTF and AUA recommendations.[12]

## SPECIFIC CRITICISMS OF THE USPSTF REPORT

In the initial presentation of their review and analysis, the USPSTF proposed to answer four main questions.[13] The first question asked if PSA-based screening decreases PCa-specific or all-cause mortality.

Prostate Cancer. http://dx.doi.org/10.1016/B978-0-12-800077-9.00001-3

The Task Force based their recommendations on a review of five studies of the potential benefits of PSA testing but gave significant weight to only two of these studies. The Prostate, Lung, Colorectal, and Ovarian (PLCO) Cancer Screening Trial was conducted in the United States from 1993 to 2001, randomly assigning 76,693 men aged 55–74 years at 10 US study centers to receive either screening, defined as annual PSA testing for 6 years and digital rectal examinations for 4 years or "usual care" as the control group.[14] After 7 years of follow-up, there were 2820 PCas diagnosed in the screened group and 2322 in the control group (RR, 1.22; 95% CI, 1.16–1.29). The incidence of death per 10,000 person-years was 2.0 in the screened group and 1.7 in the control group (rate ratio, 1.13; 95% CI, 0.75–1.70).

The European Randomized Study of Screening for Prostate Cancer (ERSPC), with screening in all centers between 1994 and 2006, and ongoing continued follow-up and screening in several of the major centers, involved 162,000 men, aged 55–69 years, randomly assigned either to a group offered PSA screening once every 4 years or to a control group that did not receive screening.[15] The cumulative incidence of PCa was 8.2% in the screened group and 4.8% in the control group, but the rate ratio for death from PCa in the screened group as compared with the control group was 0.80 (95% CI, 0.65–0.98; adjusted $p = 0.04$), resulting in decreased PCa-specific mortality. In the additional follow-up years 10 and 11, the rate ratio for death from PCa dropped to 0.62 (95% CI, 0.45–0.85; adjusted $p = 0.003$) for screened patients.[16] The USPSTF maintained that the cancer-specific mortality advantages in the two studies did not justify the "harms" of screening. Since 2012, the ERSPC has updated results with 13 years of follow-up as well as additional analysis that further support a beneficial impact on mortality.[17–22]

The critique is that the USPSTF did not acknowledge known methodological flaws in these two major studies. The best documented objection being "contamination" of the control populations by PSA screening performed outside the study protocols by subject choice. In the PLCO study, fully 44% of the control subjects had PSA tests performed prior to randomization, and 83% had at least one PSA test outside the study protocol at some point during the 6-year screening period of the trial. Out of these, the 17% had 1–2 tests, 42% had 3–4, and 24% had 5–6 PSA tests during the 6-year study period.[23–25] This "contamination" of the control group invalidated the original comparison hypotheses of the study. There was also a significant protocol violation as 60–70% of the men with an abnormal PSA screening test failed to get a timely prostate biopsy, although 64% did eventually undergo a biopsy within 3 years.[25,26] In the ERSPC trial, there was also nonprotocol screening, but far less than in the PLCO study. A reanalysis of the Rotterdam data set from the ERSPC trial, corrected for nonattendance and contamination, showed that PSA screening reduced the risk of dying of PCa by up to 31% as compared to the prior estimate of 20%.[27,28]

A further and perhaps even more serious problem with the USPSTF analysis of mortality was the inadequate follow-up time, median of about 6 years in the PLCO trial and 9 years in the ERSPC study. The time for the clinical evolution of PCa is much longer; in a Swedish study of an unscreened population, the lead time from onset of elevated PSA level to PCa diagnosis reached 10.7 years.[29] In a different Swedish study, a small cohort of 223 men with untreated localized PCa was followed for 32 years.[30] There were 90 (41.4%) local progression events and 41 (18.4%) cases of progression to distant metastasis. In total, 38 (17%) men died of PCa, but the relative rate of PCa-specific death increased from 4.9 at 5–9 years to 23.5 at 15–20 years, suggesting that cancer death accounts for a greater proportional mortality with advancing duration of the disease.

Interim follow-up results from the Göteborg randomized population-based Prostate-Cancer Screening Trial with a median follow up of 14 years, heighten the concern that the Task Force recommendation was based on inadequate duration of follow-up.[31] In this currently ongoing long-term study, 20,000 men aged 50–64 years were randomized either to a screening group to receive PSA testing every 2 years or to a nontested control group. There was a cumulative incidence of PCa of 12.7% in the screened group and 8.2% in the control group (hazard ratio, 1.64; 95% CI, 1.50–1.80; $p < 0.0001$). Yet despite a significantly higher incidence of PCa, the rate ratio for death from PCa for the screened group was lower at 0.56 (95% CI, 0.39–0.82; $p = 0.002$), a reduction of 44%.[31]

The second question that the USPSTF addressed was regarding the harms of PSA-based screening for PCa. The USPSTF considered false positive results (an elevated PSA level not leading to cancer diagnosis) as harmful. In the PLCO trial, the cumulative risk for at least one false-positive result was 13%, with a 5.5% risk for undergoing at least one unnecessary biopsy.[32] In the ERSPC trial, 76% of the prostate biopsies did not show PCa.[33] In a prior review in 2008, the USPSTF noted that false-positive PSA test results can cause adverse psychological effects.[34] In addition, there were biopsy complications such as infection, bleeding, and urinary difficulties in both trials: 68 events per 10,000 evaluations in the PLCO study; while the Rotterdam center of the ERSPC trial reported that among 5,802 prostate biopsies, 200 men (3.5%) developed a fever, 20 (0.4%) experienced urinary retention, and 27 (0.5%) required hospitalization due to prostatitis or urosepsis.[33]

The error in the Task Force argument is one of incomplete comparison, assuming that the absence of PSA screening completely avoids all diagnostic procedures. In fact, diagnostic procedures for PCa were frequently performed in the control groups of both studies. In the PLCO trial, the rate of diagnosis of PCa was only 20% less than in the screened group, possibly as a result of the very high level of nonprotocol screening that occurred. In the ERSPC trial, the PCa incidence rate was 8.2% in the screened group and 4.8% in the control group, but higher-grade cancer (Gleason score of 7 or more) was far more common in the control group than in the screened group (45.2% vs. 27.8%). The presence of metastatic disease was evaluated in a study conducted in four of the ERSPC centers where the necessary data were available. Metastatic disease was identified by imaging or by PSA values >100 ng/mL at diagnosis or during follow-up. After a median follow-up of 12 years, 666 men with metastatic PCa were detected; 256 in the screening arm and 410 in the control arm, resulting in a 42% ($p = 0.0001$) reduction in metastatic PCa for men who were actually screened.[35]

In a separate analysis of the Rotterdam ERSPC cohort, the investigators determined that, of 42,376 men randomized during the period of the first round of the trial (1993–1999), 1151 and 210 in the screening and control arms, respectively, were diagnosed with PCa.[36] Of these men, 420 (36.5%) screen-detected and 54 (25.7%) controls underwent radical prostatectomy with long-term follow-up data (median follow-up 9.9 years). Men from the screening arm had a significantly higher progression-free survival ($p = 0.003$), metastasis-free survival ($p < 0.001$), and cancer-specific survival ($p = 0.048$).[36] Yet, after adjusting for tumor volume in the surgical specimen, there was no longer a significant difference in biochemical recurrence rate between the screening and control arms. Thus, the key advantage of the PSA screened group appeared to be surgical intervention at an earlier stage of tumor growth. If one considers diagnostic procedures harmful, then it is undeniable that patients with high-grade or advanced PCa will eventually endure more invasive and harmful diagnostic or therapeutic procedures with correspondingly greater anxiety than those with localized cancer.

The third question that the USPSTF addressed was related to the benefits of treatment of early-stage or screening-detected PCa. The USPSTF acknowledged strong evidence that treatment of localized PCa reduced mortality as compared to observation alone, the so-called "watchful waiting." They endorsed a randomized controlled Scandinavian trial with 15 years of follow-up showing that radical prostatectomy resulted in a sustained decrease in PCa-specific mortality (15% vs. 21%; RR, 0.62; CI, 0.44–0.87) and all-cause mortality (RR, 0.75; CI, 0.61–0.92) while several additional studies of radical prostatectomy with durations of follow-up ranging from 4 years to 13 years, all showed a similar benefit.[37–42] The USPSTF also acknowledged an approximate 35% decrease in PCa-specific mortality with the alternative treatment of radiation therapy compared to observation alone.[43–45] However, they also argued against the apparent benefit of definitive treatment primarily on the basis of the results of the Prostate Intervention Versus Observation Trial (PIVOT) study, which followed a cohort of 731 men with initially localized PCa for 12 years in the Veterans Administration Hospital system assigned randomly either to radical prostatectomy or observation alone.[46] During the median follow-up of 10 years, 171 of 364 men (47%) assigned to radical prostatectomy died, as compared with 183 of 367 (49.9%) assigned to observation (hazard ratio, 0.88; 95% CI, 0.71–1.08; $p = 0.22$).

Most studies support the USPSTF analysis showing that both surgery and radiotherapy for early stage PCa are superior to watchful waiting in terms of eventual mortality. The PIVOT Trial results were marred by the failure to recruit the 2000 patients required in the original study design, leaving it severely underpowered statistically. Furthermore, the average age of their patients at baseline was 67 years, old enough to subject the cohort to significant mortality from causes other than cancer. Although "a life expectancy of at least 10 years" was an entry criterion, by 10 years almost half of the participants had died, leaving only 176 men in the surgery group and 187 observation men, and by 15 years only 30% were alive. According to the Social Security Administration Actuarial Office, median life expectancy in the United States for men aged 67 would be an additional 17.2 years. Thus, it appears that they recruited men with a limited life expectancy. There was also significant crossover and control group contamination even in the limited numbers of men recruited, with 23% of men assigned to the surgical arm not receiving radical prostatectomies while another 20% of the control group were treated with either surgery or radiation therapy.

The PIVOT Trial analysis also clearly illustrates the problem of relying on overall mortality to assess the benefit of screening. Disease-specific mortality is a more accurate measure, but can be confounded by reporting discrepancies and the difficulty in obtaining reliable cause of death statistics. The argument has been made that "excess mortality" of a cohort is a better measure of the effect of screening, but the emphasis on mortality as the ultimate comparison variable ignores the additional reality of the burden of advanced cancer in the living.[47] The USPSTF provided no assessment of comparative benefit in terms of potential avoidance of local and distant spread of not yet fatal PCa.

Finally, the USPSTF addressed the harms of treatment of early-stage or screening-detected PCa. The USPSTF cited a 30-day perioperative mortality rate postprostatectomy of about 0.5% with rates of serious cardiac events about 3% and vascular events (including pulmonary embolism and deep venous thrombosis) about

2%.[48–53] Due to the close proximity of the urinary tract and the penile nerves, radical prostatectomy carries a significant incidence of erectile dysfunction and other surgical complications. Surgical injuries to the rectum or the ureter range from 0.3% to 0.6%.[49,53] The rate of urinary incontinence in comparison to watchful waiting is from two- to fourfold higher.[53–57] Prostatectomy carries a 50–80% increased risk for erectile dysfunction as compared to watchful waiting.[53–59] Radiotherapy does not interrupt the normal anatomy and therefore does not damage urinary continence and erectile function as much as prostatectomy.[59–61] Due to the proximity of the rectum to the radiation fields, this form of therapy is sometimes associated with bowel complaints, primarily bowel urgency.[59–61]

It is undeniable that complications of surgery and radiotherapy are far lower at centers with specialization and experience in the treatment of PCa. For example, in the nationwide Scandinavian series, the 30-day mortality was only 0.11%.[62] Surprisingly, despite the many problems associated with surgery or radiotherapy, in multiple quality of life comparison studies, patients who undergo active treatment reported similar or better physical and/or emotional subscale forms on the SF-36 test as compared to watchful waiting patients.[54–60,62] The paradox of equivalent quality of life findings despite the sexual, urinary, and bowel issues may reflect a higher percentage of advanced cancer patients in the watchful waiting groups. Patients with advanced PCa report much lower quality of life scores than those with localized disease (48–51) and are susceptible to significant depression.[63–66] In a Swedish database of PCa patients, the standardized mortality (SMR) risk of suicide was increased for 22,929 men with locally advanced nonmetastatic tumors (SMR, 2.2; 95% CI, 1.6–2.9) and for 8,350 men with distant metastases (SMR, 2.1; 95% CI, 1.2–3.6).[67]

The USPSTF concluded that the complications of surgery and radiotherapy negatively impact the benefits of reduced mortality resulting from definitive treatment of PCa. The concern in their reasoning is failure to compare these complications with the harmful events associated with advanced PCa that occurs more frequently in unscreened and untreated populations. Manifestations of metastatic and advanced disease may include weight loss, pathologic fractures, spinal cord compression, hematuria, intractable pain, ureteral and/or bladder outlet obstruction, hydronephrosis, urinary retention, renal failure, and urinary incontinence.[68–70] Disseminated PCa is frequently characterized by painful bone metastases.[71–74] Furthermore, hormonal therapy used to suppress the growth of advanced PCa has many undesirable side effects including vasomotor flushing, loss of libido, erectile dysfunction, gynecomastia, cognitive decline, anemia, osteoporosis, and dyslipidemia resulting in reduced quality of life.[75–77]

In limiting their analysis to mortality alone, the USPSTF ignored the burden of suffering of individuals surviving with advanced cancer. In order to perform a valid comparison of the screened and control populations, it is not enough to assess mortality; one must systematically analyze the complications of treatment of local cancer as well as the adverse effects of preventable advanced cancer, more common in unscreened groups.[78] The ERSPC investigators have approached this issue quantitatively to determine the extent to which harms to quality of life resulting from "overdiagnosis" and treatment counterbalanced the lower mortality in their screened population.[79] They found that annual screening of 1000 men between the ages of 55 and 69 years would result in 9 fewer deaths from PCa (28% reduction), 14 fewer men receiving palliative therapy (35% reduction), and a total of 73 life-years gained (average, 8.4 years per PCa death avoided). Factoring in "harm" due to complications of diagnostic and treatment procedures, their analysis still showed 56 quality-of-life adjusted years gained by PSA screening.[79]

## SCREENING SPECIAL AT-RISK POPULATIONS

In the pure sense, population screening does not take into account patient history or special risk factors. Taking into account individual patient characteristics, risk factors, signs, and symptoms is "case finding" and not true screening. Nevertheless, the D-rating of the USPSTF implies that PSA and PCa early detection efforts should be discouraged whether it is true population screening, case finding or even in the setting of shared decision. The USPSTF blanket recommendation against PSA screening does not allow for special populations.

Hereditary PCa typically affects men at a younger age.[80–83] In a Scandinavian study of 44,788 twins, hereditary factors were found to account for 42% of the overall risk of development of PCa.[84] It is certainly reasonable to screen individuals with a strong family history of PCa earlier and more frequently than the general population.

Men of African ancestry and ethnicity are at 1.4 times higher risk of being diagnosed and 2–3 times higher risk of dying from PCa compared to European-American men.[85] Advanced PCa occurred at a 4:1 ratio of black to white men in the Detroit Surveillance, Epidemiology and End Results registry database.[86] Similar imbalances are observed in the United Kingdom where health care is free and does not depend on socioeconomic status; in the PROCESS study in the London–Bristol area, black men had a threefold higher risk of PCa compared to white men.[87–88]

In addition to the USPSTF, the current AUA Best Practice Policy on PSA screening released in May 2013 was weak on making special recommendations for high-risk groups.[89] These guidelines do endorse PSA testing every other year between the ages of 55 and 69 and are patterned after the Göteborg arm of the ERSPC. However, this is to be considered only after shared decision/informed consent. Critics have contended that the new guidelines are "soft" on high-risk groups, particularly AA men.[12]

## IMPACT OF AGE ON PROSTATE SCREENING DECISIONS

The 2012 USPSTF guidelines was opposed to PSA testing irrespective of age.[10] The expected lifespan in the United States for a male aged 75 is about 10 years.[90] In 2008, the USPSTF recommended against PSA screening of men over the age of 75.[34] It is plausible that the aging process in most 75-year-old men will progress to mortality prior to the advent of advanced PCa, but the current expected life span for men aged 45–50 years in the United States is about 30 years for whites and 27 years for blacks.[90] The recent blanket USPSTF recommendation may result in delayed diagnosis of potentially curable PCas in young men who will otherwise suffer advanced disease and death, and some otherwise healthy older men with high-grade PCa will also unnecessarily suffer advanced disease and cancer death.

## SCREENING/RISK ASSESSMENT IN YOUNG MEN

Whether PCa screening should include average-risk younger men is a subject of debate in the medical literature. The incidence of PCa in this age group is low, and the clinical significance of detected PCa in these men is unclear.[91] Recent data suggest that a PSA value above 0.7 ng/mL in young men is associated with greater risk for development of PCa.[92]

In 2006, the National Comprehensive Cancer Network (NCCN) adopted by "nonuniform consensus" a risk-stratification strategy for young men that includes a "baseline" PSA measurement at age 40 years and subsequent risk stratification based on the result of this initial test.[93] The NCCN early-detection protocol is supported by evidence that young men with PSA levels above the age-specific median are at greater risk for future PCa. In the Baltimore Longitudinal Study of Aging, the relative risk of PCa was 3.75 for men aged 40–49 years with a PSA level above the age-specific median of 0.6 ng/mL, whereas the risk was similar among men in their 50s with PSA levels above and below the age-specific median.[94] Similarly, among men aged 40–49 years who were at greater risk (either because of positive family history or AA race), a PSA level between the age-specific median (0.7 ng/mL) and 2.5 ng/mL was associated with a higher risk of PCa, with screening of 36 higher-risk men required to detect one additional case of cancer. The external validity of this finding is unclear; it is not known how many young men at average risk would need a PSA test to detect one additional case of cancer.[92]

Overdiagnosis of PCa using current detection strategies is not insignificant and leads to potentially avoidable harms in the course of treatment.[95–99] Etzioni et al. estimated an overdiagnosis rate of 29% among white patients aged <60 years, based on model comparisons to the National Cancer Institute's Surveillance, Epidemiology, and End Results (SEER) database.[96] Similarly, using data from the ERSPC, Draisma et al. found an overdiagnosis rate of 27% for men aged 55 years.[95] High overdiagnosis rates make the benefit of PCa screening unclear, but the adverse effects of treatment are clear and are similar for clinically detected and screening-related cancers.[99] If all of the 2.1 million 40-year-old men in the United States underwent PSA testing, this could result in up to 48,000 additional biopsies for elevated PSA levels alone, with up to 12,000 men diagnosed with PCa.[100–102] The incidence of treatment-related side effects (e.g., impotence and incontinence) are well known, and given current life expectancy among men at age 40 years, the effects of treatment could result in a significant decrement in quality-adjusted life-years. These costs should be balanced against the potential benefits of PCa screening in young men, which may include years of life saved and decreases in the costs of advanced disease management. Given the well-recognized limitations of retrospectively identified screening strategies and cost–benefit models, the issue of obtaining a baseline "risk-stratification" PSA remains controversial.[97,103]

The large majority of men in this age group who have a PSA test will not be diagnosed with PCa, although some may develop it later in life. Additional baseline testing of young men will likely have substantial costs, both direct and indirect.

Health care resource use associated with PSA testing could increase substantially under the NCCN protocol. Based on a prior study from my center, 22.5% of men aged 40–49 years have a PSA test annually.[104–106] In the United States, there are approximately 22.4 million men aged 40–49 years.[100] This translates into approximately 5 million annual PSA tests in this age group. If all men underwent PSA testing at age 40 years (current US population of 40-year-old men: 2.1 million), this would add up to 2 million annual tests.[100] For men with PSA levels above the age-specific median, annual PSA testing with digital rectal examination is recommended (e.g., up to 1 million annual PSA tests). Under this protocol, the number of PSA tests in men aged 40–49 years would more than double.

Despite the concerns, baseline PSA is a powerful predictor of future PCa and may be more powerful than family history or ethnicity in risk assessment. Our group from Duke University has examined this in a cohort of 3530 AA men and 6118 Caucasian-American (CA) men who were ≤50 years old at first PSA screening and the baseline PSA was ≤4 ng/mL.[107] Based on the initial PSA, patients were divided into 0.1–0.6, 0.7–1.4, 1.5–2.4, and 2.5–4.0 ng/mL. Chi-square test was used to compare the subsequent incidence of PCa among PSA groups. Univariate and age-adjusted multivariate logistic regression were used to estimate the relative risk of cancer in these PSA groups. The incidences of PCa during subsequent periods of follow-up were calculated. The median PSA was 0.7 ng/mL in both AA and CA groups. The subsequent incidences of PCa among PSA groups were significantly different both in AA and CA men. The incidence of PCa was not significantly different between initial PSA 0.7–1.4 and <0.6 ng/mL groups in both AA and CA men. Men with initial PSA at 1.5–2.4 ng/mL had 9.3- and 6.7-fold increase of age-adjusted relative risk of PCa detection in AA and CA, respectively. Initial PSA ≥1.5 ng/mL was associated with gradually increased detection rates during periods of follow-up in both AA and CA men. Using a PSA cutoff of 1.5 ng/mL may be better than median PSA (0.7 ng/mL) to determine the risk of subsequent PCa detection in both AA and CA men ≤50 years old. Furthermore, in a cohort of more than 4500 men followed for approximately 20 years at Duke Medical Center, men with a baseline PSA <4 ng/mL had a very low risk of death from PCa (0.6%), whereas those with a baseline PSA >4 ng/mL had a much higher risk of PCa death (5.9%).[108]

I also had the opportunity to work with Dr David Crawford et al. on another analysis of baseline PSA.[109] We conducted a retrospective review involving 21,502 men, selected from 800,000 patients enrolled in a large Midwestern health system. Men at least 40 years old with baseline PSA values between 0 ng/mL and 4.0 ng/mL and at least 4 years of follow-up after their initial PSA test were included. Adjustments were made for age and race to determine increased risk of PCa development over a 4-year period. The optimal PSA threshold and the predictive value of PSA values for development of PCa were calculated. PCa risk was about 15-fold higher in patients with PSA ≥1.5 ng/mL versus patients with PSA <1.5 ng/mL (7.85% vs. 0.51%). AA patients with baseline PSA <1.5 ng/mL faced PCa risks similar to that of the study population as a whole (0.54% vs. 0.51%, respectively). But AA patients with PSA 1.5–4.0 ng/mL faced a 19-fold increase in PCa risk versus AA patients with PSA values <1.5 ng/mL.

Finally, the recommendations of the USPSTF carry enormous weight since they have been given a very special status by the US Government. More than 50% of medical care reimbursement comes from the Department of Health and Human Services. The payment decisions made by the Centers for Medicare and Medicaid Services are invariably emulated by other third-party payers. Currently, as a result of the 2008 recommendations, PSA testing in men under 50 is frequently denied by third-party payers irrespective of family history or ethnicity.

## SCREENING PSA IN OLDER MEN

A cutpoint of age 75 for PCa screening as suggested in the 2008 USPSTF recommendation is supportable for most individuals if the advent of advanced PCa takes 10 years or more to manifest.[34] Conversely, the decision on screening or treatment of a malignancy should be made by informed individuals and their physicians rather than dictated by any government-appointed panel. In this 2008 guideline and in the updated 2012 guideline, the USPSTF did not separate screening/early detection from treatment. The development and use of active surveillance as a major treatment option, which minimizes definitive therapy for low-risk disease, was not even considered by the USPSTF. The issuance of a universal blanket prohibition of PSA screening for PCa based on chronologic age alone might be considered age discrimination.

Our group and others have examined PCa in older men that may have implications for screening decisions. A cohort of 4561 men who underwent radical prostatectomy between 1988 and 2008 at Duke University Medical Center were stratified into three age groups (<60, 60–70, and >70) and early and late PSA era groups based on the year of surgery (<2000 and ≥2000).[110] Race, body mass index, PSA, prostate weight, tumor volume, pathological Gleason sum, and pathological tumor stage were used in univariate and multivariate analyses. Survivals (PSA recurrence, distant metastasis, and disease specific death) were compared among the three age groups using univariate and multivariate methods. Compared with younger age groups (<60, 60–70), men aged >70 years had a higher proportion of pathological tumor stage 3/4 (33.0% vs. 44.3% vs. 52.1%, $p < 0.001$), pathological Gleason sum > 7 (9.5% vs. 13.4% vs. 17.2%, $p < 0.001$), and larger tumor volume (3.7% vs. 4.7% vs. 5.2%, $p < 0.001$). Pathological Gleason sum in men aged >70 years did not differ between the early and late PSA era groups ($p = 0.071$). Men aged >70 years showed a higher risk of PSA recurrence, distant metastasis, and disease-specific death in univariate ($p < 0.05$), but not multivariate analysis.

By eliminating PSA testing in all men above age 70, there will be a subset of older men who will be harmed.

I believe we need to take a more individualized approach in older men using PSA and PSA kinetics as these men approach an age when consideration should be given to discontinuation of testing. In one study from

Duke University, we evaluated PSA velocity to determine its value to stratify patients at risk of death from PCa and in aiding decision making regarding PSA screening in older men.[111] Our study cohort included 3525 patients aged ≥75 years with two or more PSA tests before a diagnosis of PCa. Cox proportional hazard model was used to evaluate which variables at time of last PSA measurement were associated with death from PCa. The rates of death from PCa after diagnosis in different PSA velocity groups were calculated. Kaplan–Meier and log rank test were used to assess the significant difference in death from PCa after diagnosis, stratified by PSA velocity cutoff. On multivariate analysis, men with a PSA velocity of ≥0.45 ng/mL/year had a 4.8-fold higher risk of death from PCa as compared to men with a PSA velocity of <0.45 ng/mL/year ($p = 0.013$). After a median 6.5 (up to 16.9) years of follow-up from diagnosis, 1.4% of the men with a PSA velocity <0.45 ng/mL/year had died of PCa as compared to 8.7% of those with a PSA velocity ≥0.45 ng/mL/year. The cumulative rate of death from PCa after diagnosis, stratified by a PSA velocity of 0.45 ng/mL/year, was statistically different (log rank test, $p < 0.001$). We concluded that men age ≥75 years old with a PSA velocity of <0.45 ng/mL/year are unlikely to die of PCa. It may be safe to discontinue PSA screening in these men. We have also examined total PSA and found that black men with a PSA level at <6.0 ng/mL and white men with a total PSA level of <3.0 mg/mL between 75 years and 80 years old had a very low risk of PCa mortality.[112]

## CONCLUSIONS

Population-based screening for PCa in the pure sense is not recommended by any professional organization and the USPSTF is generally against PSA testing. However, case finding using a risk stratified and shared decision approach is endorsed by the AUA, American Society of Clinical Oncology, ACS, and NCCN. The strongest current level I evidence from the Göteborg arm of the ERSPC supports the AUA guideline of every-other-year PSA testing for properly informed men between the ages of 55 and 69. Although baseline PSA is a strong predictor of future PCa in multiple studies, the recommendation for obtaining a risk assessment PSA for men in their 40s remains controversial being endorsed by NCCN but not by AUA. These guidelines are likely to evolve as studies continue to mature.

## References


1. DeStantis CE, Lin CC, Mariotto AB, et al. Cancer treatment and survivorship statistics, 2014. *CA Cancer J Clin* 2014;**64**(4):252–71.
2. Gomella LG, Liu XS, Trabulsi EJ, et al. Screening for prostate cancer: the current evidence and guidelines controversy. *Can J Urol* 2011;**18**(5):5875–83.
3. Borza T, Konijeti R, Kibel AS. Early detection, PSA screening, and management of overdiagnosis. *Hematol Oncol Clin North Am* 2013;**27**(6):1091–110.
4. Knight SJ. Decision making and prostate cancer screening. *Urol Clin North Am* 2014;**41**(2):257–66.
5. Van der Meer S, Löwik SA, Hirdes WH, et al. Prostate specific antigen testing policy worldwide varies greatly and seems not to be in accordance with guidelines: a systematic review. *BMC Fam Pract* 2012;**13**:100.
6. Smith RA, Manassaram-Baptisite D, Brooks D, et al. Cancer screening in the United States, 2014: a review of current American Cancer Society guidelines and current issues in cancer screening. *CA Cancer J Clin* 2014;**64**(1):30–51.
7. Carter HB. American Urological Association (AUA) guideline on prostate cancer detection: process and rationale. *BJU Int* 2013;**112**(5):543–7.
8. Murphy DG, Ahlering T, Catalona WJ, et al. The Melbourne Consensus Statement on the early detection of prostate cancer. *BJU Int* 2014;**113**(2):186–8.
9. Heidenreich A, Abrahamsson PA, Artibani W, et al. Early detection of prostate cancer: European Association of Urology recommendation. *Eur Urol* 2013;**64**(3):347–54.
10. Moyer VA. US Preventive Services Task Force. Screening for prostate cancer: U.S. Preventive Services Task Force recommendation statement. *Ann Intern Med* 2012;**157**(2):120–34.
11. Catalona WJ, D'Amico AV, Fitzgibbons WF, et al. What the U.S. Preventive Services Task Force missed in its prostate cancer screening recommendation. *Ann Intern Med* 2012;**157**(2):137–8.
12. Moul JW, Walsh PC, Rendell MS, et al. Re: Early detection of prostate cancer: AUA guideline: H.B. Carter, P.C. Albertsen, M.J. Barry, R. Etzioni, S.J. Freedland, K.L. Greene, L. Holmberg, P. Kantoff, B.R. Kontey, M.H. Murad, D.F. Penson and A.L. Zietman. J Urol 2013;190:419–426. *J Urol* 2013 Sep;**190**(3):1134–7.
13. Chou R, Croswell JM, Dana T, et al. Screening for prostate cancer: a review of the evidence for the U.S. Preventive Services Task Force. *Ann Intern Med* 2011;**155**(11):762–71.
14. Andriole GL, Crawford ED, Grubb III RL, et al. Mortality results from a randomized prostate-cancer screening trail. *N Engl J Med* 2009;**360**(13):1310–9.
15. Schröder FH, Hugosson J, Roobol MJ, et al. Screening and prostate-cancer mortality in a randomized European study. *N Engl J Med* 2009;**360**(260):1320–8.
16. Schröder FH, Hugosson J, Roobol MJ, et al. Prostate-cancer mortality at 11 years of follow-up. *N Engl J Med* 2012;**336**(11):981–90.
17. Schröder FH, Hugosson J, Roobol MJ, et al. Screening and prostate cancer mortality: results of the European Randomized Study of Screening for Prostate Cancer (ERSPC) at 13 years of follow-up. *Lancet* 2014;**384**:2027–35.
18. Luján M, Páez A, Angulo JC, et al. Prostate cancer incidence and mortality in the Spanish section of the European Randomized Study of Screening for Prostate Cancer (ERSPC). *Prostate Cancer Prostatic Dis* 2014;**17**(2):187–91.
19. Schröder FH. ERSPC, PLCO studies and critique of Cochrane review 2013. *Recent Results Cancer Res* 2014;**202**:59–63.
20. Schröder FH. Screening for prostate cancer: current status of ERSPC and screening-related issues. *Recent Results Cancer Res* 2014;**202**:47–51.
21. Zappa M, Puliti D, Hugosson J, et al. A different method of evaluation of the ERSPC trial confirms that prostate-specific antigen testing has a significant impact on prostate cancer mortality. *Eur Urol* 2014;**66**:401–3.
22. Bokhorst LP, Bangma CH, van Leenders GJ, et al. Prostate-specific antigen-based prostate cancer screening: reduction of prostate cancer mortality after correction for nonattendance and contamination in the Rotterdam section of the European Randomized Study of Screening for Prostate Cancer. *Eur Urol* 2014;**65**(2):329–36.

23. D'Amico AV. Prostate-cancer mortality after PSA screening. *N Engl J Med* 2012;**336**:2229.
24. Grubb III RL, Pinsky PF, Greenlee RT, et al. Prostate cancer screening in the Prostate, Lung, Colorectal and Ovarian cancer screening trial: update on findings from the initial four rounds of screening in randomized trial. *BJU Int* 2008;**102**(11):1524–30.
25. Pinsky PF, Blacka A, Kramer BS, et al. Assessing contamination and compliance in the prostate component of the Prostate, Lung, Colorectal, and Ovarian (PLCO) Cancer Screening Trial. *Clin Trials* 2010;**7**(4):303–11.
26. Pinsky PF, Andriole GL, Kramer BS, et al. Prostate biopsy following a positive screen in the prostate, lung, colorectal and ovarian cancer screening trial. *J Urol* 2005;**173**(3):746–50.
27. Roobol MJ, Kerkhof M, Schröder FH, et al. Prostate cancer mortality reduction by prostate-specific antigen-based screening adjusted for nonattendance and contamination in the European Randomized Study of Screening for Prostate Cancer (ERSPC). *Eur Urol* 2009;**56**(4):584–91.
28. Kerkhos M, Roobol MJ, Cuzick J, et al. Effect of the correction for noncompliance and contamination on the estimated reduction of metastatic prostate cancer within a randomized screening trail (ERSPC section Rotterdam). *Int J Cancer* 2010;**127**(11):2639–44.
29. Törnblom MT, Eriksson H, Franzen S, et al. Lead time associate with screening for prostate cancer. *Int J Cancer* 2004;**108**:122–9.
30. Popiolek M, Rider JR, Andrén O, et al. Natural history of early, localized prostate cancer: a final report from three decades of follow-up. *Eur Urol* 2012;**12**:1221–3.
31. Hugosson J, Carlsson S, Aus G, et al. Mortality results from the Göteborg randomized population-based prostate-cancer screening trial. *Lancet Oncol* 2010;**11**:725–32.
32. Croswell JM, Kramer BS, Kreimer AR, et al. Cumulative incidence of false-positive results in repeated, multimodal cancer screening. *Ann Fam Med* 2009;**7**(3):212–22.
33. Raaijmakers R, Kirkels WJ, Roobol MJ, et al. Complication rates and risk factors of 5802 transrectal ultrasound-guided sextant biopsies of the prostate within a population-based screening program. *Urology* 2002;**60**(5):826–30.
34. US Preventive Services Task Force. Screening for Prostate Cancer: U.S. Preventive Services Task Force Recommendation Statement. *Ann Intern Med* 2008;**149**:185–91.
35. Schröder FH, Hugosson J, Carlsson S, et al. Screening for prostate cancer decreases the risk of developing metastatic disease: findings from the European Randomized Study of Screening for Prostate Cancer (ERSPC). *Eur Urol* 2012;**62**(5):745–52.
36. Loeb S, Zhu Z, Schröder FH, et al. Long-term radical prostatectomy outcomes among participants from the European Randomized Study of Screening for Prostate Cancer (ERSPC) Rotterdam. *BJU Int* 2012;**110**(11):1678–83.
37. Bill-Axelson A, Holmberg L, Ruutu M, et al. Radical prostatectomy versus watchful waiting in early prostate cancer. *N Engl J Med* 2011;**364**(18):1708–17.
38. Albertsen PC, Hanley JA, Penson DF, et al. 13-year outcomes following treatment for clinically localized prostate cancer in a population based cohort. *J Urol* 2007;**177**:932–6.
39. Ladjevardi S, Sandblom G, Berglund A, et al. Tumor grade, treatment, and relative survival in a population-based cohort of men with potentially curable prostate cancer. *Eur Urol* 2010;**57**:631–8.
40. Merglen A, Schmidlin F, Fioretta G, et al. Short-and long-term mortality with localized prostate cancer. *Arch Intern Med* 2007;**167**:1944–50.
41. Schymura MJ, Kahn AR, German RR, et al. Factors associated with initial treatment and survival for clinically localized prostate cancer: results from the CDC-NPCR Patterns of Care Study (PoC1). *BMC Cancer* 2010;**10**:152.
42. Stattin P, Holmberg E, Johansson JE, et al. Outcomes in localized prostate cancer: National Prostate Cancer Register of Sweden follow-up study. *J Natl Cancer Inst* 2010;**102**:950–8.
43. Tewari A, Divine G, Chang P, et al. Long-term survival in men with high grade prostate cancer: a comparison between conservative treatment, radiation therapy and radical prostatectomy – a propensity scoring approach. *J Urol* 2007;**177**:911–5.
44. Wong YN, Mitra N, Hudes G, et al. Survival associated with treatment vs observation of localized prostate cancer in elderly men. *JAMA* 2006;**296**:2683–93.
45. Zhou EH, Ellis RJ, Cherullo E, et al. Radiotherapy and survival in prostate cancer patients: a population-based study. *Int J Radiat Oncol Biol Phys* 2009;**73**:15–23.
46. Wilt TJ, Brawer MK, Jones KM, et al. Radical prostatectomy versus observation for localized prostate cancer. *N Engl J Med* 2012;**367**(3):203–13.
47. van Leeuwen PJ, Kranse R, Hakulinen T, et al. Disease-specific mortality may underestimate the total effect of prostate cancer screening. *J Med Screen* 2010;**17**(4):204–10.
48. Alibhai SM, Leach M, Tomlinson G, et al. 30-day mortality and major complications after radical prostatectomy: influence of age and comorbidity. *J Natl Cancer Inst* 2005;**97**:1525–32.
49. Yao SL, Lu-Yao G. Population-based study of relationships between hospital volume of prostatectomies, patient outcomes, and length of hospital stay. *J Natl Cancer Inst* 1999;**91**:1950–6.
50. Begg CB, Riedel ER, Bach PB, et al. Variations in morbidity after radical prostatectomy. *N Engl J Med* 2002;**346**:1138–44.
51. Walz J, Montorsi F, Jeldres C, et al. The effect of surgical volume, age and comorbidities on 30-day mortality after radical prostatectomy: a population-based analysis of 9208 consecutive cases. *BJU Int* 2008;**101**:826–32.
52. Augustin H, Hammerer P, Graefen M, et al. Intraoperative and perioperative morbidity of contemporary radical retropubic prostatectomy in a consecutive series of 1243 patients: results of a single center between 1999 and 2002. *Eur Urol* 2003;**43**:113–8.
53. Johansson E, Bill-Axelson A, Holmberg L, et al. Time, symptom burden, androgen deprivation, and self-assessed quality of life after radical prostatectomy or watchful waiting: the Randomized Scandinavian Prostate Cancer Group Study Number 4 (SPCG-4) clinical trial. *Eur Urol* 2009;**55**:422–30.
54. Hoffman RM, Hunt WC, Gilliland FD, et al. Patient satisfaction with treatment decisions for clinically localized prostate carcinoma. Results from the Prostate Cancer Outcomes Study. *Cancer* 2003;**97**:1653–62.
55. Litwin MS, Hays RD, Fink A, et al. Quality-of-life outcomes in men treated for localized prostate cancer. *JAMA* 1995;**273**(2):129–35.
56. Schapira MN, Lawrence WF, Katz DA, et al. Effect of treatment on quality of life among men with clinically localized prostate cancer. *Med Care* 2001;**39**(3):243–53.
57. Smith DP, King MT, Egger S, et al. Quality of life three years after diagnosis of localized prostate cancer: population based cohort study. *BMJ* 2009;**339**:b4817.
58. Siegel T, Moul JW, Spevak M, et al. The development of erectile dysfunction in men treated for prostate cancer. *J Urol* 2001;**165**(2):430–5.
59. Fransson P, Damber JE, Tomic R, et al. Quality of life and symptoms in randomized trial of radiotherapy versus deferred treatment of localized prostate carcinoma. *Cancer* 2001;**92**:3111–9.
60. Thong MS, Mols F, Kil PJ, et al. Prostate cancer survivors who would be eligible for active surveillance but were either treated with radiotherapy or managed expectantly: comparisons on long-term quality of life and symptom burden. *BJU Int* 2010;**105**:652–8.
61. Bacom CG, Giovannucci E, Testa M, et al. The impact of cancer treatment of quality of life outcomes for patients with localized prostate cancer. *J Urol* 2001;**166**(5):1804–10.

62. Carlsson S, Adolfsson J, Bratt O, et al. Nationwide population-based study on 30-day mortality after radical prostatectomy in Sweden. *Scand J Urol Nephrol* 2009;**43**(5):350–6.
63. Galbraith ME, Ramirez JM, Pedro LW. Quality of life, health outcomes, and identity for patients with prostate cancer in five different treatment groups. *Oncol Nurs Forum* 2001;**28**(3):551–60.
64. Lubeack DP, Litwin MS, Henning JM, et al. Changes in health-related quality of life in the first year after treatment for prostate cancer: results from CaPSURE. *Urology* 1999;**53**:180–6.
65. Smith DS, Carvalhal GF, Schneider K, et al. Quality-of-life outcomes for men with prostate carcinoma detected by screening. *Cancer* 2000;**88**:1454–63.
66. Talcott JA, Manola J, Clark JA, et al. Time course and predictors of symptoms after primary prostate cancer therapy. *J Clin Oncol* 2003;**21**:3979–86.
67. Bill-Axelson A, Garmo H, Lambe M, et al. Suicide risk in men with prostate-specific antigen-detected early prostate cancer: a nationwide population-based cohort study from PCBaSe Sweden. *Eur Urol* 2010;**57**(3):390–5.
68. Bruchovsky N, Klotz L, Crook J, et al. Quality of life, morbidity, and mortality results of a prospective phase II study of intermittent androgen suppression for men with evidence of prostate-specific antigen relapse after radiation therapy for locally advanced prostate cancer. *Clin Genitourin Cancer* 2008;**6**(1):46–52.
69. Salvo N, Zeng L, Zhang L, et al. Frequency of reporting and predictive factors anxiety and depression in patients with advanced cancer. *Clin Oncol (R Coll Radiol)* 2012;**24**(2):139–48.
70. Khafagy R, Shackley D, Samuel J, et al. Complications arising in the final year of life in men dying from advanced prostate cancer. *J Palliat Med* 2007;**10**(3):705–11.
71. Warde P, Mason M, Ding K, et al. Combined androgen deprivation therapy and radiation therapy for locally advanced prostate cancer: a randomised, phase 3 trial. *Lancet* 2011;**378**(9809):2104–11.
72. Stephenson AJ, Bolla M, Briganti A, et al. Postoperative radiation therapy for pathologically advanced prostate cancer after radical prostatectomy. *Eur Urol* 2011;**61**(3):443–51.
73. Luz MA, Aprikian AG. Preventing bone complications in advanced prostate cancer current. *Oncology* 2010;**17**:S65–71.
74. Sturge J, Caley MP, Waxman J. Bone metastasis in prostate cancer: emerging therapeutic strategies. *Nat Rev Clin Oncol* 2011;**8**(6): 357–68.
75. Ahmadi H, Daneshmand S. Androgen deprivation therapy for prostate cancer: long-term safety and patient outcomes. *Patient Relat Outcome Meas* 2014;**5**:63–70.
76. Trost LW, Serefoglu E, Gokce A, et al. Androgen deprivation therapy impact on quality of life and cardiovascular health, monitoring therapeutic replacement. *J Sex Med* 2013;**10**(Suppl. 1):84–101.
77. Ahmadi H, Daneshmand S. Androgen deprivation therapy: evidence-based management of side effects. *BJU Int* 2013;**111**(4): 543–8.
78. Fowler Jr FJ, McNaughton Collins M, Walker Corkery E, et al. The impact of androgen deprivation on quality of life after radical prostatectomy for prostate carcinoma. *Cancer* 2002;**95**:287–95.
79. Hartzband P, Groopman J. There is more to life than death. *N Engl J Med* 2012;**367**:987–9.
80. Heijnsdijk EA, Wever EM, Auvinen A, et al. Quality-of-life effects of prostate-specific antigen screening. *N Engl J Med* 2012;**367**(7):595–605.
81. Stanford JL, Ostrander EA. Familial prostate cancer. *Epidemiol Rev* 2001;**23**(1):19–23.
82. Madersbacher S, Alcaraz A, Emberton M, et al. The influence of family history on prostate cancer risk: implications for clinical management. *BJU Int* 2011;**107**(5):716–21.
83. Chen YC, Page JH, Chen R, et al. Family history of prostate and breast cancer and the risk of prostate cancer in the PSA era. *Prostate* 2008;**68**:1582–91.
84. Carter BS, Bova GS, Beaty TH, et al. Hereditary prostate cancer: epidemiologic and clinical features. *J Urol* 1993;**150**(3):797–802.
85. Lichtenstein P, Holm NV, Verkasalo PK, et al. Environmental and heritable factors in the causation of cancer – analyses of cohorts of twins from Sweden, Denmark, and Finland. *N Engl J Med* 2000;**343**(2):78–85.
86. Chornokur G, Dalton K, Borysova ME, et al. Disparities at presentation, diagnosis, treatment, and survival in African American men, affected by prostate cancer. *Prostate* 2011;**71**(9):985–97.
87. Powell IJ, Bock CH, Ruterbusch JJ, et al. Evidence supports a faster growth rate and/or earlier transformation to clinically significant prostate cancer in black than in white American men, and influences racial progression and mortality disparity. *J Urol* 2010;**183**(5):1792–6.
88. Ben-Shlomo Y, Evans S, Ibrahim F, et al. The risk of prostate cancer amongst black men in the United Kingdom: the PROCESS cohort study. *Eur Urol* 2008;**53**(1):99–105.
89. Carter HB, Albertsen PC, Barry MJ, et al. Early detection of prostate cancer: AUA Guideline. *J Urol* 2013;**190**(2):419–26.
90. Arias E. United States life tables, 2003. *National Vital Statistics Reports*, vol. 58, no. 21, June 28, 2010.
91. Lin DW, Porter M, Montgomery B. Treatment and survival outcomes in young men diagnosed with prostate cancer: a population based cohort study. *Cancer* 2009;**115**(13):2863–71.
92. Loeb S, Roehl KA, Antenor JA, et al. Baseline prostate-specific antigen compared with median prostate-specific antigen for age group as predictor of prostate cancer risk in men younger than 60 years old. *Urology* 2006;**67**:316–20.
93. Smith CV, Bauer JJ, Connelly RR, et al. Prostate cancer in men age 50 years or younger: a review of the Department of Defense Center for Prostate Disease Research multicenter prostate cancer database. *J Urol* 2000;**164**:1964–7.
94. Fang J, Metter EJ, Landis P, et al. Low levels of prostate-specific antigen predict long-term risk of prostate cancer: results from the Baltimore Longitudinal Study of Aging. *Urology* 2001;**58**: 411–6.
95. Draisma G, Boer R, Otto SJ, et al. Lead times and overdetection due to prostate-specific antigen screening: estimates from the European Randomized Study of Screening for Prostate Cancer. *J Natl Cancer Inst* 2003;**95**:868–78.
96. Etzioni R, Penson DF, Legler JM, et al. Overdiagnosis due to prostate-specific antigen screening: lessons from U.S. prostate cancer incidence trends. *J Natl Cancer Inst* 2002;**94**:981–90.
97. Black WC. Overdiagnosis: an underrecognized cause of confusion and harm in cancer screening. *J Natl Cancer Inst* 2000;**92**:1280–2.
98. Ganz PA, Litwin MS. Prostate cancer: the price of early detection. *J Clin Oncol* 2001;**19**:1587–8.
99. Madalinska JB, Essink-Bot ML, de Koning HJ, et al. Health-related quality-of-life effects of radical prostatectomy and primary radiotherapy for screen-detected or clinically diagnosed localized prostate cancer. *J Clin Oncol* 2001;**19**:1619–28.
100. *US Census Bureau. Monthly Population Estimates.* http://www.census.gov/popest/estimates.php [accessed 10.07.2007].
101. Thompson IM, Pauler DK, Goodman PJ, et al. Prevalence of prostate cancer among men with a prostate-specific antigen level ≤4.0 ng per milliliter. *N Engl J Med* 2004;**350**:2239–46.
102. Welch HG, Schwartz LM, Woloshin S. Prostate-specific antigen levels in the United States: implications of various definitions for abnormal. *J Natl Cancer Inst* 2005;**97**:1132–7.
103. Etzioni RD, Ankerst DP, Thompson IM. Re: detection of life-threatening prostate cancer with prostate-specific antigen velocity during a window of curability. *J Natl Cancer Inst* 2007;**99**: 489–90.
104. Scales Jr CD, Curtis LH, Norris RD, et al. Prostate specific antigen testing in men older than 75 years in the United States. *J Urol* 2006;**176**:511–4.

105. Scales Jr CD, Curtis LH, Norris RD, et al. Relationship between body mass index and prostate cancer screening in the United States. *J Urol* 2007;**177**:493–8.
106. Scales Jr CD, Antonelli J, Curtis LH, et al. Prostate-specific antigen screening among young men in the United States. *Cancer* 2008;**113**(6):1315–23.
107. Tang P, Sun L, Uhlman MA, et al. Initial prostate specific antigen 1.5 ng/mL or greater in men 50 years old or younger predicts higher prostate cancer risk. *J Urol* 2010;**183**(3): 946–50.
108. Tang P, Sun L, Uhlman MA, et al. Baseline PSA as a predictor of prostate cancer-specific mortality over the past 2 decades: Duke University experience. *Cancer* 2010;**116**(20):4711–7.
109. Crawford ED, Moul JW, Rove KO, et al. Prostate-specific antigen 1.5–4.0 ng/mL: a diagnostic challenge and danger zone. *BJU Int* 2011;**108**(11):1743–9.
110. Sun L, Caire AA, Robertson CN, et al. Men older than 70 years have higher risk prostate cancer and poorer survival in the early and late prostate specific antigen eras. *J Urol* 2009;**182**(5):2242–8.
111. Tang P, Sun L, Uhlman MA, et al. Prostate-specific antigen-based risk-adapted discontinuation of prostate cancer screening in elderly men. *BJU Int* 2011;**108**(1):44–8.
112. Tang P, Sun L, Uhlman MA, et al. Prostate-specific antigen-based risk-adapted discontinuation of prostate cancer screening in elderly African American and Caucasian American men. *Urology* 2010;**76**(5):1058–62.

CHAPTER

# 2

# Inflammation and Infection in the Etiology of Prostate Cancer

*Siobhan Sutcliffe, PhD*, Michel A. Pontari, MD***

*Department of Surgery, Division of Public Health Sciences, Washington University School of Medicine, Siteman Cancer Center, St. Louis, MO, USA

**Department of Urology, Temple University School of Medicine, Philadelphia, PA, USA

## INTRODUCTION

The idea of a link between inflammation and cancer is not unique to prostate cancer (PCa). The idea of an association between inflammation and cancer dates back at least 150 years. In 1863, Virchow proposed that cancer originates at sites of chronic inflammation.[1] Chronic inflammation has been estimated to contribute to 15–20% of cancers and has been linked specifically to stomach cancer from *Helicobacter pylori* infection, liver cancer after infection with hepatitis B and C, colon cancer in association with inflammatory bowel disease and pertaining to urology, and bladder cancer after infection with schistosomiasis.[2]

Inflammation is commonly found in association with PCa, on the order of 60%.[3] There is evidence for an association for inflammation with PCa, but some reports do not confirm this. We will review the evidence for histologic inflammation with PCa, the data for certain types of inflammatory lesions, for clinical prostatitis, biomarkers, infection, genetics of inflammatory markers, diet, and finally the data on chemoprevention by inhibiting inflammation.

## HISTOLOGIC PROSTATIC INFLAMMATION AND PROSTATE CANCER

Studies have explored the inflammation/PCa hypothesis by examining possible associations between histologic prostatic inflammation and PCa. While these studies provide some direct evidence toward this hypothesis, they are complicated by difficulties in obtaining "normal" prostate control tissue to which to compare tissue from PCa cases. Therefore, most studies to date have used more readily available prostate tissue, such as from men undergoing prostate biopsy for indication because of either abnormal DRE findings or, more commonly because of elevated prostate-specific antigen (PSA) levels. In general, these studies have tended to observe null or even protective findings,[4–15] which are opposite to what would be expected based on an inflammation/PCa hypothesis. However, as PSA levels may rise for many different reasons, including both prostate inflammation and PCa,[16,17] these null to protective findings may reflect the dual influence of inflammation and PCa on PSA – that is, in men diagnosed with PCa, the predominant contribution to PSA may be cancer or initially undiagnosed cancer, whereas in men who remain undiagnosed, one of the predominant contributions to PSA may be inflammation, thereby creating a protective association. In contrast to these findings, in the only studies, to our knowledge, to compare tissue from PCa cases to controls without regard to indication for biopsy, the opposite findings were generally observed. In a study conducted in the Prostate Cancer Prevention Trial (PCPT), in which a large proportion of men not diagnosed with PCa during the trial underwent a prostate biopsy at the end of the trial irrespective of indication, a positive association was observed between histologic prostatic inflammation and high-grade PCa.[18] Although a null association was observed in another small autopsy study that compared histologic inflammation among men with and without histologic PCa, very few men had high-grade disease in this study.[19] Finally, positive associations have also been observed between tumor-associated histologic inflammation at PCa diagnosis, accompanied sometimes by postatrophic hyperplasia, and subsequent PCa recurrence or death.[20–23]

Prostate Cancer. http://dx.doi.org/10.1016/B978-0-12-800077-9.00002-5

## PROLIFERATIVE INFLAMMATORY ATROPHY

The lesion that links inflammation with PCa is called proliferative inflammatory atrophy or PIA. Whereas most areas of prostatic atrophy are quiescent, areas of atrophy next to inflammation have been found to proliferate. These areas, called PIA, have been suggested as precursors to PCa either directly or indirectly by progression to PIN.[24,25] Morphologic studies have reported transition from PIA to PIN and from PIA to cancer.[26] Proliferation in PAH lesions is significantly greater than in BPH or SA, but less than in PIN and PCa.[27]

There are biochemical changes in PIA lesions that are also found in PIN and PCa[28]:

- Glutathione-*S*-transferases (GST) is a cellular enzyme that neutralizes reactive oxygen species (ROS). Methylation inactivates this enzyme, increasing the chance of oxidative damage. This is found in areas of PIA, PIN, and PCa, but not found in normal prostate cells.
- Amplification of chromosome 8, reported as a marker for poor prognosis in prostate adenocarcinoma.
- Protein products of three prostate tumor suppressor genes, NKX3.1, CDKN1B, which encodes p27 and PTEN (phosphatase and tensin homolog), are all downregulated in focal atrophy lesions – present in normal prostate, and frequently decreased or absent in PIN and cancer.
- There is decreased apoptosis attributed to increased Bcl-2 expression.[29]

PIA also appears to go along with prognosis. Davidsson et al.[24] looked at TURP specimens in which stage T1a-b PCa was diagnosed. They compared men who died of PCa with those who lived 10 years without metastasis. Chronic inflammation itself had no correlation with survival but when chronic inflammation was associated with areas of atrophy, there was a fivefold increase in risk of dying of PCa.

## MECHANISMS OF INFLAMMATORY CARCINOGENESIS

One of the hypotheses linking inflammation to PCa is that chronic inflammation leads to release of inflammatory mediators and free radicals that damage the cells. The cells then attempt to regenerate in this environment to repair themselves. The mediators are also genotoxic and the end result is an increased risk of mutation and malignant transformation. Cytokines from macrophages, T lymphocytes or even tumor cells themselves may contribute to malignant progression.

Several cytokines and inflammatory mediators have been implicated in the development of PCa.

### Tumor Necrosis Alpha (TNF-α)

TNF-α is produced by macrophage/monocytes after exposure to noxious stimuli. TNF-α induces other inflammatory responses. High levels of TNF are toxic to tumors by destruction of tumor vascularity and apoptosis, but in small doses chronically, may be a tumor promoter.[1] It can lead to induction of Cox-2, matrix metalloproteases and chemokines, which are procarcinogenic.[30] TNF-α also stimulates tumor cell proliferation directly by activation of nuclear factor kappa (NF-κB).[31] This family of transcription factors key regulator of cell growth and inhibitors of apoptosis, thus inflammation can lead to decreased cell death.[32] In prostate cells, TNF also suppresses androgen receptor expression and can lead to loss of androgen sensitivity.[33]

### Reactive Species

ROS include the highly reactive molecules superoxide and hydrogen peroxide, which are produced by activated macrophages and by other cell types as by-products of oxidative metabolism. ROS are unstable and reactive as they contain unpaired electrons. They are neutralized by intracellular antioxidants such as vitamin E and beta carotenes and cellular enzymes like glutathione peroxides and GST.[34] Reactive nitrogen species (RNS) include nitric oxide (NO) and its reactive intermediates. NO is a toxic gas with free radical properties produced by inflammatory cells and most other cell types by iNOS following stimulation by inflammatory cytokines such as TNF-α and IL-1.[35] ROS and RNS both damage cellular lipids, proteins, and DNA, which can ultimately lead to carcinogenesis. They also inactivate DNA repair enzymes, producing a net imbalance between rate of mutagenesis and ability to repair mutations without error, also contributing to carcinogenesis.[31]

### Cyclo-oxygenase 2 (Cox-2)

Cox-2 is the inducible form of cyclo-oxygenase and catalyzes the conversion of arachidonic acid to prostaglandins. Cox-2 is expressed by inflammatory cells, such as macrophages, and can be induced by TNF and EGF.[36] The prostaglandins and eicosanoids produced can have a major role in the development of human cancers but the role is not as firmly established in PCa.[37] There is over-expression of Cox-2 in PCa and PIN compared to normal or hyperplastic prostate.[38,39] *In vitro* Cox-2 inhibitors inhibit PCA growth and increase apoptosis in PCa cell lines.[40] Cox-2 may play a role in more advanced T3 or T4 cancers.[41]

### Vascular Endothelial Growth Factor(VEGF)

VEGF is a mitogen for endothelial cells, induces angiogenesis and regulates vascular permeability. VEGF is induced by PGE2 and TNF-α.[31] Normally, only expressed in basal cells of the prostate, it is also expressed in PCA and PIN, with increased staining with increasing lack of differentiation.[42] Serum levels of VEGF are higher in men with PCa than in controls with a low risk of PCa, but not in controls at risk of PCa.[43]

## CLINICAL PROSTATITIS AND PROSTATE CANCER

The role of inflammation and prostate carcinogenesis has also been examined in epidemiologic studies of clinical prostatitis and PCa. In a meta-analysis of 11 studies conducted through 2000, a suggestive positive association was observed between a history of this condition and PCa (odds ratio (OR) = 1.6, 95% confidence interval (CI): 1.0–2.4),[44] as was also observed in several subsequent or additional studies.[45–52] However, most of these positive findings were limited to studies with methodologic considerations that may have led to either an underestimated prevalence of clinical prostatitis in controls or an inflated prevalence of prostatitis in cases. These considerations include: (1) exclusion of controls with a history of genitourinary conditions (e.g., prostatitis) or markers of these conditions (e.g., higher PSA) from the study population; (2) performance of the study in Asia, where former cultural practices included informing patients with cancer that they had benign (e.g., prostatitis) rather than malignant conditions (e.g., PCa); and (3) including recent diagnoses of clinical prostatitis in the exposure definition or not taking into consideration the frequency of PCa screening in the analysis.[45–48,51–57] This last consideration may have led to an inflated prevalence of clinical prostatitis among cases because men with prostatitis likely visit urologists more often than those without prostatitis, providing greater opportunity for PCa screening; and because prostatitis can lead to elevated levels of PSA, and thus prostate biopsy and incidental PCa detection. Considering studies that took detection bias into consideration in their analysis, null findings were largely observed,[58–62] with the exception of suggestive positive case-control findings for acute and chronic bacterial prostatitis, but not chronic nonbacterial prostatitis,[63] and modest positive cohort findings for a self-reported history of any prostatitis.[50] This largely null literature does not, however, rule out a role for prostatic inflammation in prostate carcinogenesis because clinical "prostatitis," as defined in most epidemiologic studies, does not necessarily imply prostate inflammation. In fact, only approximately half of "prostatitis" diagnoses involve inflammation, 10% as acute or chronic bacterial prostatitis (NIH prostatitis categories I and II), and 50% as inflammatory chronic prostatitis/chronic pelvic pain syndrome (category IIIa).[64] The remainder is not believed to be related to inflammation. Therefore, studies of "clinical prostatitis" may be less informative for the inflammation/PCa hypothesis than those examining histologic inflammation.

## INFECTION

Infectious agents are a frequently cited source of prostatic inflammation, as many agents are capable of infecting and eliciting inflammation within the prostate. These microorganisms include both sexually transmitted agents, such as *Neisseria gonorrhoeae*, as well as other urogenital organisms, such as those responsible for bacterial prostatitis (e.g., *Escherichia coli*).[65] With respect to sexually transmitted infections (STIs), numerous epidemiologic studies have investigated STIs over the past few decades with combined findings suggestive of a positive relation between self-reported STIs (typically symptomatic STIs, such as gonorrhea and syphilis) and PCa (relative risk (RR) = 1.49, 95% CI: 1.19–1.92).[66] However, as many of these studies were case-control in design and assessed STI history by self-report, recall bias may potentially explain these pooled findings (i.e., PCa cases may have been more likely to reply truthfully to STI questions than controls). This possible explanation is supported by null findings in several recent cohort studies,[50,61,67] with the exception of those from a recent Taiwanese study of older STI patients who were found to have a higher risk of PCa than the general population.[68] This unique finding may point toward a role for prolonged, even decades-long, STI exposure in prostate carcinogenesis, as STIs are typically acquired when men are younger (e.g., in their 20s).[69] Positive findings for STIs that tend to be asymptomatic and thus that may persist for longer periods of time in the urogenital tract without detection or treatment, such as *Trichomonas vaginalis* and mycoplasma infections, are also consistent with this hypothesis.[70–73] Other infectious agents of interest include *E. coli*, *Propionibacterium acnes*,[74,75] and possibly intracellular fungi, as variation in the *MSMB* gene, which encodes a fungicidal protein, has been found to be associated with PCa risk.[76] Finally, given that many infectious agents have the potential to infect and elicit prostate inflammation, it is also possible that no single infectious agent is solely responsible for PCa risk, but that many may contribute to risk, depending on their propensity to establish chronic prostate infections[70] or possibly to trigger long-term chronic inflammation even once they are cleared.[75] To date, no infectious agent has been clearly associated with the etiology or pathogenesis of PCa.

## PROSTATIC CALCULI AND OTHER PHYSICAL OR CHEMICAL IRRITANTS

In addition to infectious agents, noninfectious physical irritants, such as intraluminal corpora amylacea (small inclusion bodies) and prostatic calculi, have also been proposed to contribute to prostate carcinogenesis, most likely through inflammation and prostate cell damage.[77] Interestingly, this hypothesis may not be entirely distinct from an infectious hypothesis because infectious agents have been cultured from corpora amylacea and calculi, and because these foreign bodies have been found to be composed, at least in part, by acute inflammatory proteins and the remnants or imprints of microorganisms, such as *E. coli*.[75] To our knowledge, only a few studies have examined corpora amylacea or prostatic calculi in relation to PCa to date. Three of these studies observed positive or suggestive positive associations between corpora amylacea or calculi and PCa,[77–79] while the remaining five studies observed no association,[80–84] although many of these studies compared PCa cases to controls with BPH, which might also be influenced by corpora amylacea and calculi.

In addition to these small foreign bodies, other potential physical irritants, such as spermatozoa, as well as chemical irritants, such as dietary mutagens or refluxed urine, have also been proposed to contribute to prostate inflammation and cancer.[28] However, no studies, to our knowledge, have explored histologic evidence of these possible irritants in relation to PCa to date. Instead, studies have examined an easier measure to collect – self-reported frequency of ejaculation – in relation to PCa as a surrogate measure of the length of time that physical and chemical irritants contact prostate epithelial cells, as well as the rate of formation of prostatic calculi.[85] In general, these studies have observed protective associations between increased self-reported frequency of ejaculation earlier in life and PCa,[86–89] providing some support for the hypothesis that increased "flushing" of the intraluminal contents of the prostate may reduce possible damage caused by physical or chemical irritants and the risk of PCa development.

## CHANGES IN GENES ASSOCIATED WITH INFECTION/INFLAMMATION THAT INCREASE THE RISK OF PCA

Several genes associated with inflammation and/or infection have been linked to PCa. One is the RNAse L gene. RNAse L, a ribonuclease that is activated in viral infections, cleaves ribosomal RNA and leads to apoptosis. Inactivating mutations (E265X and M11) segregate with PCa in two PCa families: E265X with one of European descent and M11 with one of African descent. Lymphoblasts from carriers of either mutation were found to be deficient in enzyme RNASE activity, but no other phenotype besides PCa were obvious.[90] When *RNASEL* is activated in cells by its cognate interferon inducible ligands, 2′5′ linked oligoadenylates, mRNA species are consistently induced, one of which is encoded by the MIC1 gene (macrophage inhibitory cytokine), another PCa susceptibility locus. The RNASEL SNP rs12757998 has also been associated with a prolonged time to biochemical occurrence or death from PCa in men who underwent radiation, but not radical prostatectomy.[91] Macrophage scavenger receptor 1 or *MSR1* gene codes for a scavenger receptor on macrophages that binds many ligands including gram-negative and gram-positive bacteria. Mutant alleles code for proteins that can no longer bind. Two different alleles, alleles R293X and H441R, are found in several PCa families.[92]

Toll-like receptor 4 (TLR4) is involved in innate immune response to gram-negative bacteria. Carriers of a single nucleotide polymorphism (SNP) in 3″UTR region of TLR4 (11381G/C) were found to be associated with PCa risk. Carriers of GC or CC instead of wild type GG had 26% increased risk of PCa, and a 39% risk of early onset PCa (before age 65). In a North American population, homozygosity for variant alleles of 8 SNPs in TLR4 were associated with lower risk of PCa; the site in Sweden did not confer increased risk.[93] There may be a dose-dependent effect of genetic risk. A case control study of over 1300 cases with PCa and 1300 controls compared over 142 SNPs in genes associated with inflammation. A group of 10 inflammatory SNPs were associated with PCA, and the effect was seen in a dose-dependent fashion[94]; the more SNPs present, the greater the risk of PCa.

## DIET AND PROSTATE CANCER RISK

Several dietary items have been described as either increasing or decreasing the risk of PCa. Green tea is derived from the plant *Camellia sinensis*, and contains polyphenolic compounds, which are reported to decrease risk and slow progression of PCa. The most abundant and best studied is EGCG (−)-epigallocatechin-3-gallate, an antioxidant 25–100 times more potent than vitamins C and E.[95] In a study of 49,920 men in Japan aged 40–69,[96] there was a decrease in risk of advanced PCa in men who consumed >5 cups per day (RR 0.52). There was no data to support benefit in those who are already diagnosed. Soy products are common in the Far East. Phytoestrogens are plant compounds with a chemical structure similar to estradiol, of which isoflavones are most important.[95] Isoflavones alter expression of numerous genes associated with PCa, mainly through Estrogen 2 receptors, suppress proliferation, but also downregulate IL-8 genes.[97] Meta analyses show an inverse relationship between soy consumption and risk of PCa with an OR, 0.7; CI, 0.57–0.84.[98]

Tomatoes and lycopenes may have an ameliorating effect on the development of PCa. Lycopene is a red carotenoid pigment in tomatoes, watermelon, and grapefruit. *In vitro* has antiproliferative effect by inhibiting cell cycle at G0/G1 phase.[99] It also produces an increase in number of IGF-1 binding proteins, resulting in net decrease of insulin-like growth factor-1, IGF-1, which has been associated with increased risk of PCa.[100] Meta-analysis of 21 studies concluded that there was an RR of 0.89 with eating lots of raw tomato and cooked tomato products.[101] However, a recent FDA panel concluded only limited evidence that lycopenes and tomato consumption reduces risk of PCa based on evaluation of 13 observational studies.[102] Similar to lycopenes, there was an initial interest in vitamin E after an initial report from the ATBC study (α-tocopherol, β-carotene cancer prevention study) looked at vitamin A (β-carotene) or vitamin E (α-tocopherol) supplementation in lung cancer. After 6 years, there was 34% reduction in incidence of PCa in those randomized to vitamin E.[94,103] However, large follow-up studies have not reproduced this benefit including in the SELECT trial.[104] Longer-term follow-up in the SELECT cohort showed that vitamin E supplementation actually increased the risk of PCa,[104] while selenium had no effect.

Polyunsaturated fats omega 3 and 6 are essential fatty acids derived entirely from the diet, as the body is unable to synthesize these *de novo*. *In vitro* and human studies have found ω-6 and trans-fatty acids (TFAs) to be proinflammatory and ω-3 to be anti-inflammatory.[95] One would think that the anti-inflammatory lipids may protect against PCa. However, in the PCPT and the SELECT trial, blood levels of phospholipid fatty acids have reported increased risk of high-grade PCa with high levels of ω-3 fatty acids.[105,106] The risk of lethal PCa is increased with several SNPs of fatty acid synthase, and the risk is associated with BMI. In obese but not lean men, the polymorphism rs1127678 was associated with a risk of advanced PCa of 2.49 and a risk of mortality from PCa of 2.04.[107]

Eating large amounts of red meat and certain processed meats appears to increase the risk of PCa. In a review of a large study of men aged 50–71, red meat increased the risk of PCa by 12% and red processed meat increased the risk by 7%. Grilling meat resulted in an 11% increased risk of total PCa and a 36% increased risk of advanced PCa. The risk was associated with levels of nitrites and nitrates in the meat.[108] In a study of 7949 African-American men, there was no association with red, white, or processed meat but there was a 20% greater risk for nonadvanced PCa in those consuming red meats cooked at high temperature.[109]

Obesity itself may be a risk for PCa. Adipose cells are a source of inflammation as well as macrophages in adipose, which release inflammatory mediators.[110] In the PCPT, there was a protective effect of 18% for men with a BMI >30 for low-grade cancer (Gleason < 7), but these same men had a 29% increased risk of high-grade PCa (Gleason ≥ 7).[111] Other studies have also reported a significantly positive association between PCa aggressiveness and obesity.[112] One of the difficulties in interpreting results for obese men is that they have greater circulating plasma volume and therefore a "hemodilution" of PSA.[113] However, the clinical impact of this finding on predicting tumor volume and stage is not certain, as some studies have found no need to adjust for BMI.[114]

## BIOMARKERS FOR INFLAMMATION AND PROSTATE CANCER

C reactive protein (CRP) was initially reported to be positively associated with PCa in small studies, but a more recent meta-analysis of five studies did not show an association of CRP and PCa with a risk estimate of 1.[115,116] A study of over 17,000 men showed no association of neither CRP nor other available markers thought to indicate inflammation – albumin, haptoglobin, hemoglobin, and leukocytes – with risk of PCa.[117] A recent study however did show a correlation between CRP and albumin, called the modified Glasgow predictive score (mGPS). In almost 900 patients with PCa, the mGPS predicted poorer overall 5-year survival and relative survival independent of age and disease grade.[118] IL-6 is elevated in men with untreated metastatic or castration-resistant PCa, and is associated with an aggressive phenotype of cancer.[119] IL-6 may play a role in the development of castration-resistant PCa by activation of the androgen receptor.[120] A polymorphism for IL-6 gene is also reported to be associated with more aggressive PCa.[121]

## CHEMOPREVENTION WITH ASA OR NSAIDs

Animal studies have suggested that using aspirin or nonsteroidal anti-inflammatory drugs (NSAIDs) can reduce the growth of PCa.[56,122] Large studies in humans have had mixed results. Recent large studies include a meta-analysis of 24,000 patients, which showed that aspirin use reduced the risk of overall PCa by 17%, and 19% for advanced cancer. Nonaspirin NSAIDs were less consistent in effect but still suggestive of a benefit.[123] Using the Saskatchewan database and examining over 9000 men with PCa, there was a class effect seen. Propionates (Ibuprofen and Naprosyn) were most effective with the most risk reduction and OR 0.90, while other types of NSAIDs were not associated with a PCa risk reduction, nor was aspirin. The anti-inflammatories were most effective if started 11–15 years prior to the study date.[124]

## CONCLUSIONS

There is not a clear, linear relationship between inflammation, infection, and PCa. The evidence points to an interaction of both environmental factors, in this case infection, diet, BMI, and probably anti-inflammatory medications with an individual's genetic predisposition. Inflammation appears to play a role in some men with PCa. This association presents multiple possible targets for intervention and prevention.

## References


1. Balkwill F, Mantovani A. Inflammation and cancer: back to Virchow? *Lancet* 2001;**357**(9255):539–45.
2. Coussens LM, Werb Z. Inflammation and cancer. *Nature* 2002;**420**(6917):860–7.
3. Gerstenbluth RE, Seftel AD, MacLennan GT, et al. Distribution of chronic prostatitis in radical prostatectomy specimens with up-regulation of bcl-2 in areas of inflammation. *J Urol* 2002;**167**(5): 2267–70.
4. Wolters T, Roobol MJ, Schroder FH, et al. Can non-malignant biopsy features identify men at increased risk of biopsy-detectable prostate cancer at re-screening after 4 years? *BJU Int* 2008;**101**(3):283–8.
5. Sajadi KP, Kim T, Terris MK, et al. High yield of saturation prostate biopsy for patients with previous negative biopsies and small prostates. *Urology* 2007;**70**(4):691–5.
6. Karakiewicz PI, Benayoun S, Begin LR, et al. Chronic inflammation is negatively associated with prostate cancer and high-grade prostatic intraepithelial neoplasia on needle biopsy. *Int J Clin Prac* 2007;**61**(3):425–30.
7. Terakawa T, Miyake H, Kanomata N, et al. Inverse association between histologic inflammation in needle biopsy specimens and prostate cancer in men with serum PSA of 10–50 ng/mL. *Urology* 2008;**72**(6):1194–7.
8. Bassett WW, Bettendorf DM, Lewis JM, et al. Chronic periglandular inflammation on prostate needle biopsy does not increase the likelihood of cancer on subsequent biopsy. *Urology* 2009;**73**(4):845–9.
9. Gann PH, Fought A, Deaton R, et al. Risk factors for prostate cancer detection after a negative biopsy: a novel multivariable longitudinal approach. *J Clin Oncol* 2010;**28**(10):1714–20.
10. Krystyna A, Safi T, Briggs WM, et al. Correlation of hepatitis C and prostate cancer, inverse correlation of basal cell hyperplasia or prostatitis and epidemic syphilis of unknown duration. *Int Braz J Urol* 2011;**37**(2):223–9 discussion 30.
11. Fujita K, Hosomi M, Tanigawa G, et al. Prostatic inflammation detected in initial biopsy specimens and urinary pyuria are predictors of negative repeat prostate biopsy. *J Urol* 2011;**185**(5):1722–7.
12. Kryvenko ON, Jankowski M, Chitale DA, et al. Inflammation and preneoplastic lesions in benign prostate as risk factors for prostate cancer. *Mod Pathol* 2012;**25**(7):1023–32.
13. Zhao C, Venigalla S, Miyamoto H. Chronic inflammation on initial benign prostate biopsy is a negative predictor of subsequent cancer detection. *Pathol Int* 2012;**62**(11):774–6.
14. Yli-Hemminki TH, Laurila M, Auvinen A, et al. Histological inflammation and risk of subsequent prostate cancer among men with initially elevated serum prostate-specific antigen (PSA) concentration in the Finnish prostate cancer screening trial. *BJU Int* 2013;**112**(6):735–41.
15. Moreira DM, Nickel JC, Gerber L, et al. Baseline prostate inflammation is associated with a reduced risk of prostate cancer in men undergoing repeat prostate biopsy: results from the REDUCE study. *Cancer* 2014;**120**(2):190–6.
16. Oesterling JE. Prostate specific antigen: a critical assessment of the most useful tumor marker for adenocarcinoma of the prostate. *J Urol* 1991;**145**(5):907–23.
17. Sandhu JS. Prostate cancer and chronic prostatitis. *Curr Urol Rep* 2008;**9**(4):328–32.
18. Gurel B, Lucia MS, Thompson Jr IM, et al. Chronic inflammation in benign prostate tissue is associated with high-grade prostate cancer in the placebo arm of the prostate cancer prevention trial. *Cancer Epidemiol Biomarkers Prev* 2014;**23**(5):847–56.
19. Delongchamps NB, de la Roza G, Chandan V, et al. Evaluation of prostatitis in autopsied prostates – is chronic inflammation more associated with benign prostatic hyperplasia or cancer? *J Urol* 2008;**179**(5):1736–40.
20. Irani J, Goujon JM, Ragni E, et al. High-grade inflammation in prostate cancer as a prognostic factor for biochemical recurrence after radical prostatectomy. *Urology* 1999;**54**(3):467–72.
21. Karja V, Aaltomaa S, Lipponen P, et al. Tumour-infiltrating lymphocytes: a prognostic factor of PSA-free survival in patients with local prostate carcinoma treated by radical prostatectomy. *Anticancer Res* 2005;**25**(6C):4435–8.
22. Nonomura N, Takayama H, Nishimura K, et al. Decreased number of mast cells infiltrating into needle biopsy specimens leads to a better prognosis of prostate cancer. *Br J Cancer* 2007;**97**(7): 952–6.
23. Nonomura N, Takayama H, Nakayama M, et al. Infiltration of tumour-associated macrophages in prostate biopsy specimens is predictive of disease progression after hormonal therapy for prostate cancer. *BJU Int* 2011;**107**(12):1918–22.
24. Davidsson S, Fiorentino M, Andren O, et al. Inflammation, focal atrophic lesions, and prostatic intraepithelial neoplasia with respect to risk of lethal prostate cancer. *Cancer Epidemiol Biomarkers Prev* 2011;**20**(10):2280–7.
25. Platz EA, De Marzo AM. Epidemiology of inflammation and prostate cancer. *J Urol* 2004;**171**(2 Pt. 2):S36–40.
26. Wang W, Bergh A, Damber JE. Morphological transition of proliferative inflammatory atrophy to high-grade intraepithelial neoplasia and cancer in human prostate. *Prostate* 2009;**69**(13): 1378–86.
27. Shah R, Mucci NR, Amin A, et al. Postatrophic hyperplasia of the prostate gland: neoplastic precursor or innocent bystander? *Am J Pathol* 2001;**158**(5):1767–73.
28. De Marzo AM, Platz EA, Sutcliffe S, et al. Inflammation in prostate carcinogenesis. *Nat Rev Cancer* 2007;**7**(4):256–69.
29. De Marzo AM, Marchi VL, Epstein JI, et al. Proliferative inflammatory atrophy of the prostate: implications for prostatic carcinogenesis. *Am J Pathol* 1999;**155**(6):1985–92.
30. Balkwill F. Tumor necrosis factor or tumor promoting factor? *Cytokine Growth Factor Rev* 2002;**13**(2):135–41.
31. Lucia MS, Torkko KC. Inflammation as a target for prostate cancer chemoprevention: pathological and laboratory rationale. *J Urol* 2004;**171**(2 Pt. 2):S30–4 discussion S5.
32. Sumitomo M, Tachibana M, Nakashima J, et al. An essential role for nuclear factor kappa B in preventing TNF-alpha-induced cell death in prostate cancer cells. *J Urol* 1999;**161**(2):674–9.
33. Mizokami A, Gotoh A, Yamada H, et al. Tumor necrosis factor-alpha represses androgen sensitivity in the LNCaP prostate cancer cell line. *J Urol* 2000;**164**(3 Pt. 1):800–5.
34. Kang DH. Oxidative stress, DNA damage, and breast cancer. *AACN Clin Issues* 2002;**13**(4):540–9.
35. Lala PK, Chakraborty C. Role of nitric oxide in carcinogenesis and tumour progression. *Lancet Oncol* 2001;**2**(3):149–56.
36. Williams CS, Mann M, DuBois RN. The role of cyclooxygenases in inflammation, cancer, and development. *Oncogene* 1999;**18**(55): 7908–16.
37. Cao Y, Prescott SM. Many actions of cyclooxygenase-2 in cellular dynamics and in cancer. *J Cell Physiol* 2002;**190**(3):279–86.

38. Pruthi RS, Derksen E, Gaston K. Cyclooxygenase-2 as a potential target in the prevention and treatment of genitourinary tumors: a review. *J Urol* 2003;**169**(6):2352–9.
39. Aparicio Gallego G, Diaz Prado S, Jimenez Fonseca P, et al. Cyclooxygenase-2 (COX-2): a molecular target in prostate cancer. *Clin Transl Oncol* 2007;**9**(11):694–702.
40. Kirschenbaum A, Liu X, Yao S, et al. The role of cyclooxygenase-2 in prostate cancer. *Urology* 2001;**58**(2 Suppl. 1):127–31.
41. Shao N, Feng N, Wang Y, et al. Systematic review and meta-analysis of COX-2 expression and polymorphisms in prostate cancer. *Mol Biol Rep* 2012;**39**(12):10997–1004.
42. Kollermann J, Helpap B. Expression of vascular endothelial growth factor (VEGF) and VEGF receptor Flk-1 in benign, premalignant, and malignant prostate tissue. *Am J Clin Pathol* 2001;**116**(1): 115–21.
43. Botelho F, Pina F, Lunet N. VEGF and prostatic cancer: a systematic review. *Eur J Cancer Prev* 2010;**19**(5):385–92.
44. Dennis LK, Lynch CF, Torner JC. Epidemiologic association between prostatitis and prostate cancer. *Urology* 2002;**60**(1):78–83.
45. Hsing AW, Wang RT, Gu FL, et al. Vasectomy and prostate cancer risk in China. *Cancer Epidemiol Biomarkers Prev* 1994;**3**(4): 285–8.
46. Ritchie JM, Vial SL, Fuortes LJ, et al. Organochlorines and risk of prostate cancer. *J Occup Environ Med* 2003;**45**(7):692–702.
47. Patel DA, Bock CH, Schwartz K, et al. Sexually transmitted diseases and other urogenital conditions as risk factors for prostate cancer: a case-control study in Wayne County, Michigan. *Cancer Causes Control* 2005;**16**(3):263–73.
48. Daniels NA, Ewing SK, Zmuda JM, et al. Correlates and prevalence of prostatitis in a large community-based cohort of older men. *Urology* 2005;**66**(5):964–70.
49. Sarma AV, McLaughlin JC, Wallner LP, et al. Sexual behavior, sexually transmitted diseases and prostatitis: the risk of prostate cancer in black men. *J Urol* 2006;**176**(3):1108–13.
50. Cheng I, Witte JS, Jacobsen SJ, et al. Prostatitis, sexually transmitted diseases, and prostate cancer: the California Men's Health Study. *PloS ONE* 2010;**5**(1):e8736.
51. Hosseini M, SeyedAlinaghi S, Mahmoudi M, et al. A case-control study of risk factors for prostate cancer in Iran. *Acta Medica Iranica* 2010;**48**(1):61–6.
52. Hennis AJ, Wu SY, Nemesure B, et al. Urologic characteristics and sexual behaviors associated with prostate cancer in an African-Caribbean population in Barbados, West Indies. *Prostate Cancer* 2013;**2013**:682750.
53. Wynder EL, Mabuchi K, Whitmore Jr WF. Epidemiology of cancer of the prostate. *Cancer* 1971;**28**(2):344–60.
54. Mishina T, Watanabe H, Araki H, et al. Epidemiological study of prostatic cancer by matched-pair analysis. *Prostate* 1985;**6**(4): 423–36.
55. Honda GD, Bernstein L, Ross RK, et al. Vasectomy, cigarette smoking, and age at first sexual intercourse as risk factors for prostate cancer in middle-aged men. *Br J Cancer* 1988;**57**(3):326–31.
56. Nakata S, Imai K, Yamanaka H. Study of risk factors for prostatic cancer. *Hinyokika Kiyo Acta Urologica Japonica* 1993;**39**(11):1017–24 discussion 24–25.
57. Wei Q, Tang X, Yang Y, et al. Risk factors of prostate cancer – a matched case-control study. *J West China Univ Med Sci* 1994;**25**(1): 87–90.
58. Zhu K, Stanford JL, Daling JR, et al. Vasectomy and prostate cancer: a case-control study in a health maintenance organization. *Am J Epidemiol* 1996;**144**(8):717–22.
59. Rosenblatt KA, Wicklund KG, Stanford JL. Sexual factors and the risk of prostate cancer. *Am J Epidemiol* 2001;**153**(12):1152–8.
60. Lightfoot N, Conlon M, Kreiger N, et al. Medical history, sexual, and maturational factors and prostate cancer risk. *Ann Epidemiol* 2004;**14**(9):655–62.
61. Sutcliffe S, Giovannucci E, De Marzo AM, et al. Gonorrhea, syphilis, clinical prostatitis, and the risk of prostate cancer. *Cancer Epidemiol Biomarkers Prev* 2006;**15**(11):2160–6.
62. Weinmann S, Shapiro JA, Rybicki BA, et al. Medical history, body size, and cigarette smoking in relation to fatal prostate cancer. *Cancer Causes Control* 2010;**21**(1):117–25.
63. Roberts RO, Bergstralh EJ, Bass SE, et al. Prostatitis as a risk factor for prostate cancer. *Epidemiology* 2004;**15**(1):93–9.
64. Krieger JN, Nyberg Jr L, Nickel JC. NIH consensus definition and classification of prostatitis. *JAMA* 1999;**282**(3):236–7.
65. Sutcliffe S, Platz EA. Inflammation in the etiology of prostate cancer: an epidemiologic perspective. *Urol Oncol* 2007;**25**(3):242–9.
66. Caini S, Gandini S, Dudas M, et al. Sexually transmitted infections and prostate cancer risk: a systematic review and meta-analysis. *Cancer Epidemiol* 2014;**38**(4):329–38.
67. Huang WY, Hayes R, Pfeiffer R, et al. Sexually transmissible infections and prostate cancer risk. *Cancer Epidemiol Biomarkers Prev* 2008;**17**(9):2374–81.
68. Chung SD, Lin YK, Huang CC, et al. Increased risk of prostate cancer following sexually transmitted infection in an Asian population. *Epidemiol Infect* 2013;**141**(12):2663–70.
69. Centers for Disease Control and Prevention. *Sexually transmitted disease surveillance 2010*. Atlanta, GA: US Department of Health and Human Services; 2011.
70. Sutcliffe S. Sexually transmitted infections and risk of prostate cancer: review of historical and emerging hypotheses. *Future Oncol* 2010;**6**(8):1289–311.
71. Urbanek C, Goodison S, Chang M, et al. Detection of antibodies directed at *M. hyorhinis* p37 in the serum of men with newly diagnosed prostate cancer. *BMC Cancer* 2011;**11**:233.
72. Barykova YA, Logunov DY, Shmarov MM, et al. Association of *Mycoplasma hominis* infection with prostate cancer. *Oncotarget* 2011;**2**(4):289–97.
73. Hrbacek J, Urban M, Hamsikova E, et al. Serum antibodies against genitourinary infectious agents in prostate cancer and benign prostate hyperplasia patients: a case-control study. *BMC Cancer* 2011;**11**:53.
74. Sfanos KS, De Marzo AM. Prostate cancer and inflammation: the evidence. *Histopathology* 2012;**60**(1):199–215.
75. Sfanos KS, Isaacs WB, De Marzo AM. Infections and inflammation in prostate cancer. *Am J Clin Exp Urol* 2013;**1**(1):3–11.
76. Sutcliffe S, De Marzo AM, Sfanos KS, et al. MSMB variation and prostate cancer risk: clues towards a possible fungal etiology. *Prostate* 2014;**74**(6):569–78.
77. Ioachim H. La lithiase prostatique peut-elle etre consideree comme une facteur cancerogene. *Urologia* 1961;**28**(1):1–11.
78. Kovi J, Rao MS, Heshmat MY, et al. Incidence of prostatic calcification in blacks in Washington, D.C., and selected African cities. Correlation of specimen roentgenographs and pathologic findings. Cooperative Prostatic Research Group. *Urology* 1979;**14**(4): 363–9.
79. Hwang EC, Choi HS, Im CM, et al. Prostate calculi in cancer and BPH in a cohort of Korean men: presence of calculi did not correlate with cancer risk. *Asian J Androl* 2010;**12**(2):215–20.
80. Cristol DS, Emmett MD. The incidence of coincident prostatic calculi, prostatic hyperplasia and carcinoma of the prostate gland. *JAMA* 1944;**124**(10):646.
81. Golden M, Abeshouse BS. Prostatic calculi and neoplasm. *Sinai Hosp J* 1952;**1**(1):20–8.
82. Niijima T, Koiso K. Incidence of prostatic cancer in Japan and Asia. *Scand J Urol Nephrol Suppl* 1980;**55**:17–21.
83. Sondergaard G, Vetner M, Christensen PO. Prostatic calculi. *Acta Pathol Microbiol Immunol Scand A* 1987;**95**(3):141–5.
84. Checkoway H, DiFerdinando G, Hulka BS, et al. Medical, lifestyle, and occupational risk factors for prostate cancer. *Prostate* 1987;**10**(1):79–88.

85. Isaacs JT. Prostatic structure and function in relation to the etiology of prostatic cancer. *Prostate* 1983;**4**(4):351–66.
86. Banerjee AK. Carcinoma of prostate and sexual activity. *Urology* 1986;**28**(2):159.
87. Hsieh CC, Thanos A, Mitropoulos D, et al. Risk factors for prostate cancer: a case-control study in Greece. *Int J Cancer* 1999;**80**(5): 699–703.
88. Giles GG, Severi G, English DR, et al. Sexual factors and prostate cancer. *BJU Int* 2003;**92**(3):211–6.
89. Leitzmann MF, Platz EA, Stampfer MJ, et al. Ejaculation frequency and subsequent risk of prostate cancer. *JAMA* 2004;**291**(13): 1578–86.
90. Silverman RH. Implications for RNase L in prostate cancer biology. *Biochemistry* 2003;**42**(7):1805–12.
91. Schoenfeld JD, Margalit DN, Kasperzyk JL, et al. A single nucleotide polymorphism in inflammatory gene *RNASEL* predicts outcome after radiation therapy for localized prostate cancer. *Clin Cancer Res* 2013;**19**(6):1612–9.
92. Sun J, Hsu FC, Turner AR, et al. Meta-analysis of association of rare mutations and common sequence variants in the MSR1 gene and prostate cancer risk. *Prostate* 2006;**66**(7):728–37.
93. Zheng SL, Liu W, Wiklund F, et al. A comprehensive association study for genes in inflammation pathway provides support for their roles in prostate cancer risk in the CAPS study. *Prostate* 2006;**66**(14):1556–64.
94. Kwon EM, Salinas CA, Kolb S, et al. Genetic polymorphisms in inflammation pathway genes and prostate cancer risk. *Cancer Epidemiol Biomarkers Prev* 2011;**20**(5):923–33.
95. Hori S, Butler E, McLoughlin J. Prostate cancer and diet: food for thought? *BJU Int* 2011;**107**(9):1348–59.
96. Takachi R, Inoue M, Ishihara J, et al. Fruit and vegetable intake and risk of total cancer and cardiovascular disease: Japan Public Health Center-Based Prospective Study. *Am J Epidemiol* 2008;**167**(1): 59–70.
97. Handayani R, Rice L, Cui Y, et al. Soy isoflavones alter expression of genes associated with cancer progression, including interleukin-8, in androgen-independent PC-3 human prostate cancer cells. *J Nutr* 2006;**136**(1):75–82.
98. Hwang YW, Kim SY, Jee SH, et al. Soy food consumption and risk of prostate cancer: a meta-analysis of observational studies. *Nutr Cancer* 2009;**61**(5):598–606.
99. Amir H, Karas M, Giat J, et al. Lycopene and 1,25-dihydroxyvitamin D3 cooperate in the inhibition of cell cycle progression and induction of differentiation in HL-60 leukemic cells. *Nutr Cancer* 1999;**33**(1):105–12.
100. Karas M, Amir H, Fishman D, et al. Lycopene interferes with cell cycle progression and insulin-like growth factor I signaling in mammary cancer cells. *Nutr Cancer* 2000;**36**(1): 101–11.
101. Etminan M, Takkouche B, Caamano-Isorna F. The role of tomato products and lycopene in the prevention of prostate cancer: a meta-analysis of observational studies. *Cancer Epidemiol Biomarkers Prev* 2004;**13**(3):340–5.
102. Kavanaugh CJ, Trumbo PR, Ellwood KC. The U.S. Food and Drug Administration's evidence-based review for qualified health claims: tomatoes, lycopene, and cancer. *J Natl Cancer Inst* 2007;**99**(14): 1074–85.
103. The Alpha-Tocopherol, Beta Carotene Cancer Prevention Study Group. The effect of vitamin E and beta carotene on the incidence of lung cancer and other cancers in male smokers. *New Engl J Med* 1994;**330**(15):1029–35.
104. Lippman SM, Klein EA, Goodman PJ, et al. Effect of selenium and vitamin E on risk of prostate cancer and other cancers: the Selenium and Vitamin E Cancer Prevention Trial (SELECT). *JAMA* 2009;**301**(1):39–51.
105. Brasky TM, Darke AK, Song X, et al. Plasma phospholipid fatty acids and prostate cancer risk in the SELECT trial. *J Natl Cancer Inst* 2013;**105**(15):1132–41.
106. Brasky TM, Till C, White E, et al. Serum phospholipid fatty acids and prostate cancer risk: results from the prostate cancer prevention trial. *Am J Epidemiol* 2011;**173**(12):1429–39.
107. Nguyen PL, Ma J, Chavarro JE, et al. Fatty acid synthase polymorphisms, tumor expression, body mass index, prostate cancer risk, and survival. *J Clin Oncol* 2010;**28**(25):3958–64.
108. Sinha R, Park Y, Graubard BI, et al. Meat and meat-related compounds and risk of prostate cancer in a large prospective cohort study in the United States. *Am J Epidemiol* 2009;**170**(9):1165–77.
109. Major JM, Cross AJ, Watters JL, et al. Patterns of meat intake and risk of prostate cancer among African-Americans in a large prospective study. *Cancer Causes Control* 2011;**22**(12):1691–8.
110. Fuentes E, Fuentes F, Vilahur G, et al. Mechanisms of chronic state of inflammation as mediators that link obese adipose tissue and metabolic syndrome. *Mediators Inflamm* 2013;**2013**:136584.
111. Gong Z, Neuhouser ML, Goodman PJ, et al. Obesity, diabetes, and risk of prostate cancer: results from the prostate cancer prevention trial. *Cancer Epidemiol Biomarkers Prev* 2006;**15**(10): 1977–83.
112. Su LJ, Arab L, Steck SE, et al. Obesity and prostate cancer aggressiveness among African and Caucasian Americans in a population-based study. *Cancer Epidemiol Biomarkers Prev* 2011;**20**(5):844–53.
113. Banez LL, Hamilton RJ, Partin AW, et al. Obesity-related plasma hemodilution and PSA concentration among men with prostate cancer. *JAMA* 2007;**298**(19):2275–80.
114. Mitchell CR, Umbreit EC, Rangel LJ, et al. Does body mass index "dilute" the predictive property of prostate-specific antigen for tumor volume at radical prostatectomy? *Urology* 2011;**78**(4): 868–72.
115. Heikkila K, Ebrahim S, Lawlor DA. A systematic review of the association between circulating concentrations of C reactive protein and cancer. *J Epidemiol Community Health* 2007;**61**(9):824–33.
116. Heikkila K, Harris R, Lowe G, et al. Associations of circulating C-reactive protein and interleukin-6 with cancer risk: findings from two prospective cohorts and a meta-analysis. *Cancer Causes Control* 2009;**20**(1):15–26.
117. Van Hemelrijck M, Jungner I, Walldius G, et al. Risk of prostate cancer is not associated with levels of C-reactive protein and other commonly used markers of inflammation. *Int J Cancer* 2011;**129**(6):1485–92.
118. Shafique K, Proctor MJ, McMillan DC, et al. Systemic inflammation and survival of patients with prostate cancer: evidence from the Glasgow Inflammation Outcome Study. *Prostate Cancer Prostatic Dis* 2012;**15**(2):195–201.
119. Nguyen DP, Li J, Tewari AK. Inflammation and prostate cancer: the role of interleukin 6 (IL-6). *BJU Int* 2014;**113**(6):986–92.
120. Lee SO, Lou W, Johnson CS, et al. Interleukin-6 protects LNCaP cells from apoptosis induced by androgen deprivation through the Stat3 pathway. *Prostate* 2004;**60**(3):178–86.
121. Tan D, Wu X, Hou M, et al. Interleukin-6 polymorphism is associated with more aggressive prostate cancer. *J Urol* 2005;**174**(2): 753–6.
122. Bardia A, Platz EA, Yegnasubramanian S, et al. Anti-inflammatory drugs, antioxidants, and prostate cancer prevention. *Curr Opin Pharmacol* 2009;**9**(4):419–26.
123. Mahmud SM, Franco EL, Aprikian AG. Use of nonsteroidal anti-inflammatory drugs and prostate cancer risk: a meta-analysis. *Int J Cancer* 2010;**127**(7):1680–91.
124. Mahmud SM, Franco EL, Turner D, et al. Use of non-steroidal anti-inflammatory drugs and prostate cancer risk: a population-based nested case-control study. *PloS ONE* 2011;**6**(1):e16412.

CHAPTER

# 3

# Androgen Receptor

*Parth K. Modi, MD, Izak Faiena, MD, Isaac Yi Kim, MD, PhD*

**Section of Urologic Oncology, Rutgers Cancer Institute of New Jersey and Division of Urology, Department of Surgery, Rutgers Robert Wood Johnson Medical School, New Brunswick, NJ, USA**

## INTRODUCTION

The androgen receptor (AR) is part of the steroid hormone receptor family of molecules, which also includes the progesterone, estrogen, mineralocorticoid, retinoic acid, thyroid, and vitamin D receptors.[1–3] The AR is a transcription factor, primarily responsible for mediating the physiologic effects of androgens through binding of the androgen–AR complex to specific DNA target sequences and inducing or suppressing the transcription of target genes.[4]

## ANDROGEN RECEPTOR STRUCTURE AND FUNCTION

The AR gene is a single-copy gene located on chromosome Xq11.2-q12. The gene spans approximately 90 kb of DNA, and is oriented with the 5′ end toward the centromere.[4] The coding sequence is made up of eight exons that code for a 10.6 kb mRNA forming a 2757-base pair open reading frame.[4] The AR (Figure 3.1), like all steroid receptors, contains three regions: (1) a ligand-binding domain (LBD), (2) a DNA-binding domain (DBD), and (3) a transactivation domain.[5] The first exon codes for the entire N-terminal domain (NTD), which is the transcriptional regulatory region of the protein. This transactivation domain contains a polyglutamine sequence of variable length, which is encoded by a CAG triplet repeat sequence.[5] This CAG sequence begins at codon 58 and extends for 11–31 (average 21) repeats in normal men.[4,5] The variable nature of this polyglutamine sequence is thought to be due to slipping of the DNA polymerase during DNA replication.[4] While some studies suggest that CAG repeat length is inversely associated with prostate cancer (PCa) risk, this concept still remains controversial.[5] Long CAG repeat length (40–62 repeats) is seen in Kennedy's disease, an inherited neuromuscular degenerative disease.[4]

The second and third exons encode the DBD, which is a highly conserved region among steroid hormone receptors.[4] The DBD is composed of two zinc finger motifs. The first zinc finger is encoded by exon 2 and confers receptor-specificity for androgen-response elements (AREs); the second zinc finger is encoded by exon 3 and is involved in the dimerization of the receptor and its binding to the major groove of DNA.[4] Finally, exons 4–8 code for the C-terminal LBD, which is significantly variable among steroid hormone receptors. The LBD folds into a ligand-binding pocket composed of 12 helices, which encloses bound ligand and induces structural change in the receptor allowing for coactivator binding.[4] A hinge region connects the DBD and LBD, and contains a nuclear localization signal.

### Mechanism of Action

The inactive AR resides in the cytoplasm, bound to chaperones.[6] AR activation proceeds in multiple steps beginning with initial complex formation with certain chaperonins, binding of ligand, posttranslational modifications, and dimerization (Figure 3.2). The receptor complex translocates into the nucleus and forms transcriptional coactivator complexes that remodel chromatin in order to access the target initiation site. Finally, the complex acts to stabilize the RNA polymerase II machinery for repeated rounds of transcription.

### DNA-Binding Domain (DBD)

The DBD is a highly conserved region that contains 70 amino acids. In the DNA-binding region there are two zinc fingers complexed with four cysteine residues each. This region allows for recognition of specific AREs. The

Prostate Cancer. http://dx.doi.org/10.1016/B978-0-12-800077-9.00003-7

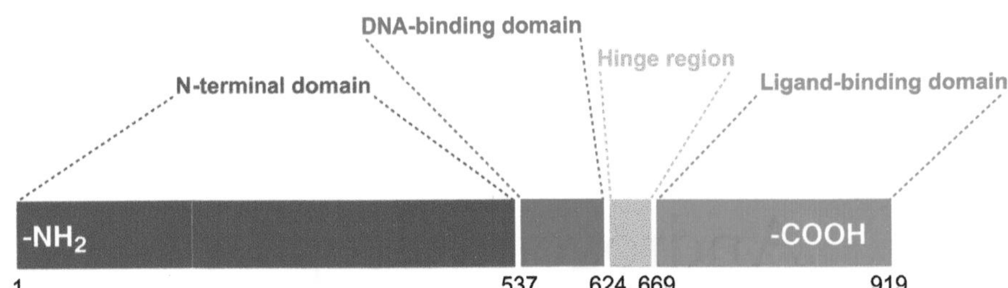


FIGURE 3.1 **The AR is a 110 kDa Protein with N-Terminal, DBD, and LBD.** *Image courtesy of Brian Eisinger, PhD.*

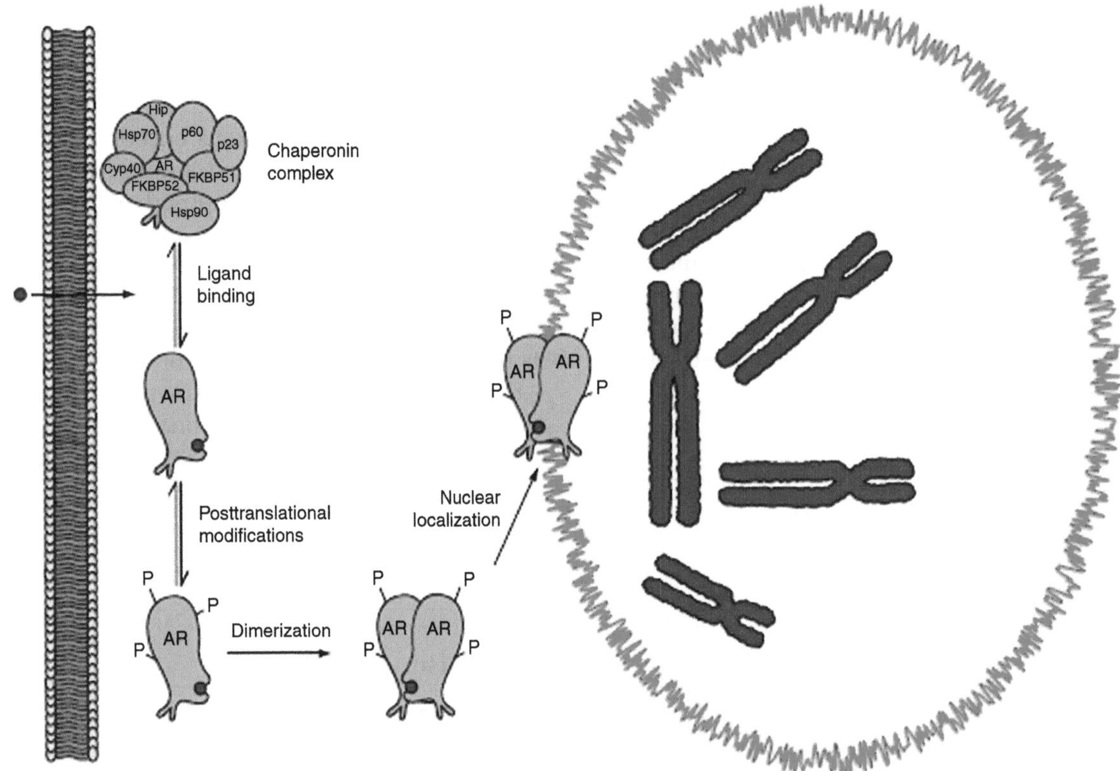


FIGURE 3.2 **The Steps in AR Activation by Ligand.** *Image courtesy of Berman et al.*[7]

first zinc finger is positioned in the major groove where several amino acids make base-specific contacts, which is the basis for the sequence-specific recognition of the ARE.[4] The second zinc finger stabilizes the binding complex by hydrophobic interactions with the first finger and contributes to the specificity of receptor-DNA binding.[8] The DBD binds to AREs, which are characterized by two hexameric half-sites: 5′-AGAACA-3′ or 5′-AGGTCA-3′.[9] Other nuclear receptors, such as the mineralocorticoid, glucocorticoid, and progesterone receptors, also bind to these sequences.

## Ligand-Binding Domain (LBD)

The three-dimensional structure of the LBD is similar to that of other hormone receptors, and is characterized by a 12-helix structure that folds to form a ligand-binding pocket.[4] When ligands, such as DHT or testosterone, are bound, helix 12 folds to enclose the ligand pocket and exposes a groove that binds a region of the NTD, which is the main site of coactivator binding.[10] The bound receptor initiates downstream effects of ligand-receptor dimerization, posttranslational modification, nuclear translocation, and finally, target-gene activation.[7] When unbound, helix

12 folds in a way that interferes with binding of coactivators or promotes corepressor binding.[11]

## Chaparonin Binding and Coregulators

In the cytoplasm, the AR is sequestered by multiple chaparonins, such as the heat shock proteins[12] (Hsp90, Hsp70, Hip, p60, p23, FKBP51, FKBP52, and Cyp40) and cytoskeletal proteins.[13] Evidence suggests that the heat shock proteins tether the AR to cytoskeletal proteins such as filamin-A and interact with the negative regulatory domain in the AR hinge region.[14] Once bound to its ligand, such as testosterone or dihydrotestosterone (DHT), a conformational change in the AR causes the heat shock proteins to dissociate and allows the AR to rapidly translocate into the nucleus.[4]

Coregulatory proteins act upon the AR with and without bound ligand in a variety of cellular locations. The coregulators are either coactivators or corepressors and primarily function as transactivation chaperones. They influence DNA binding, nuclear translocation, chromatin remodeling, binding interruption of other coregulators, AR stability, and bridging AR with the cell's transcriptional machinery.[15]

Coactivators function primarily to enhance AR transactivation, and regulate histone modification, proteosomal degradation, chaperones, sumoylation, chromatin remodeling, and the cytoskeleton.[16] Many coactivators have been identified; however, their relative importance remains unclear. Well-known coactivators include the steroid receptor coactivator family (SRC/p160) and those containing histone acetyltransferase (HAT) activity, such as cAMP response element binding protein (CREB) binding protein (CBP)/p300 and p300/CBP-associated factor (p/CAF). SRC/p160 coactivators are characterized by three LxxLL motifs contained in the center of their peptide sequence and a C-terminal glutamine-rich region that stabilizes ligand-bound AR thus enhancing transactivation.[17] Furthermore, SRC family of proteins are able to recruit transcriptional factors as well as HAT containing coregulators.[18] CREB-binding protein (CBP)/p300 and p/CAF contain HAT activity and interact with the AR to induce chromatin remodeling and allow for transcription.[17] Another novel coactivator, Vav3, was found to be overexpressed in a significant number of human PCa and upregulated following progression to castration resistance in preclinical models.[19]

Corepressors, on the other hand, act to inhibit transcription initiation of androgen responsive genes. Some of the more well-known corepressors are the nuclear receptor corepressor (NCoR) and silencing mediator for retinoid and thyroid hormone receptors (SMRT). SMRT is able to inhibit transcription by interacting with the NTD and the LBD and disrupt the N-terminal/C-terminal (N/C) interaction and/or compete with the SRC/p160 coactivators.[20] This can occur in the presence or absence of ligand, whereas NCoR only functions in the presence of an agonist.[17] Furthermore, SMRT is able to recruit histone deacetylase (HDAC), which induces DNA packaging into nucleosomes and prevents access to the AR gene promoter regions, thereby inhibiting transcription.[20]

## Posttranslational Modifications

Once the AR dissociates from its chaperones, it is susceptible to posttranslational modifications such as phosphorylation, acetylation, sumoylation, ubiquitination, or methylation that may allow for cross-talk interaction through other signaling pathways[21] and alter AR function in significant ways.[22–24] For example, AR is a substrate and downstream target of receptor-tyrosine kinase (RTK) and G-protein coupled receptor (GPCR) signaling, which can activate AR independently of androgen[15,25] while protecting the AR from proteolytic degradation.[26] Furthermore, phosphorylation by several protein kinases also acts to enhance AR response to low levels of androgens, estrogens and antiandrogens and aids in recruitment of nuclear coactivators required for chromatin remodeling.[27] The function of acetylation is known from studies of mutated acetyl receptor sites where signaling is no longer able to activate the AR, and coactivators such as SRC1 are inhibited, resulting in reduced binding to AREs. This shifts the AR equilibrium toward a repressed form, which has increased binding to corepressor proteins such as NCoR.[28,29] The various types of posttranslational modifications and their interactions that have been described highlight the importance of AR regulation. As these posttranslational modifications are better characterized, the complex regulation of AR activity may begin to be better understood.

## Dimerization

AR dimerization appears to play an integral role in the process of ligand-dependent activation by influencing nuclear localization, cofactor binding, DNA binding, and transactivation potential. There are three forms of AR dimerization: (1) DNA-dependent dimerization mediated through the DBD, (2) the androgen-induced N/C interaction occurring between specific surfaces in the AR-NTD and AR-LBD, and (3) LBD–LBD dimerization.[30] DBD-mediated dimerization through coordination of the zinc fingers is the most crucial step in AR signaling. The Pbox amino-acid region of the first zinc finger of each AR monomer binds to one half-site of the ARE and D-box amino-acid region of the second zinc finger binds to the partner monomer.[30–32] Dimerization can also occur via N/C interaction that can only be induced when bound to an agonist. When an agonist binds, the AR-LBD undergoes a conformational rearrangement that forms the surface known as activation function 2 (AF-2).[33] The AF-2 domain preferentially interacts with FxxLF-like (FQNLF) motifs contained in the AR-NTD with the coactivator-binding groove of the LBD,[34,35] and will initially form an intramolecular N/C interaction, but gradually this conformation will change into an intermolecular N/C dimerization in the nucleus.[36] Finally, LBD–LBD dimerization, which occurs prior to DNA binding, is thought to restrict DBD-mediated interactions to those receptors that have already formed dimers through their LBD.[30,37] The functional significance of and the specific motifs involved in this interaction have not been fully elucidated.[30]

### Nuclear Localization

After the AR undergoes ligand-binding, it translocates into the nucleus through the nuclear pore complex. This nuclear transport is mediated by a superfamily of transport receptors known collectively as karyopherins (importins and exportins). Karyopherins bind to specific nuclear localization signals (NLS) that facilitate nuclear translocation. NLSs have been identified in many proteins and share similarities that provide specificity through interaction with import receptor proteins.[38] The most well characterized import signals are the monopartite NLS in simian virus 40 (SV40) (PKKKRRV) and the bipartite NLS in nucleoplasmin (KRPAATKKAGQAKKKK).[39] There are two receptor regions that are involved in nuclear transport that share a homology among other steroid receptors. The first region is the second DNA-binding zinc finger region as well as the flanking hinge region (NL1) consisting of a bipartite signal including flanking leucines and the core signal 628RKLKKLGN.[40] Nuclear translocation occurs in multiple steps including receptor recognition of the NLS through binding of the basic amino-acid nuclear localization signal to importins α and β; translocation through the nuclear pore complex; and dissociation of the receptor-NLS complex via RanGTP–mediated release.[41]

## ANDROGEN RECEPTOR AND THE PROSTATE

The AR is part of the nuclear receptor family and functions as a transcription factor by binding to AREs in the regulatory regions of target genes. These actions are mediated by AR ligands, primarily testosterone and its metabolites, 5α-DHT, and estradiol.[42] Reproductive tissues are more sensitive to DHT, while muscle responds to testosterone and adipose tissue aromatizes testosterone to estradiol.[43] Reproductive tissues require androgen for differentiation and development in the prenatal and pubertal periods.[43] Furthermore, DHT is required for the development and maintenance of prostate tissue while castration causes its regression.[43] Studies of AR binding sites using chromatin immuno-precipitation (ChIP) sequencing have identified over 1200 genes actively regulated by the AR.[44] These include pathways regulating cell cycle progression, steroid biosynthesis, and metabolism.

Early in embryonic development, the endodermal urogenital sinus and its surrounding mesenchyme respond to fetal androgens and develop into the rudimentary prostate gland.[45] Early development of the prostate relies on the AR found in the urogenital mesenchyme and its interaction with the prostatic epithelium.[45] The adult prostatic epithelium is dependent on the presence of androgen to inhibit apoptosis and maintain a growth-quiescent state.[45] Tissue-specific knockout of AR in prostatic epithelium has been shown to result in decreased epithelial height, decreased expression of epithelial markers, and increased epithelial sloughing.[42] Knockdown of AR in prostatic stromal (smooth muscle and fibroblast) tissue, on the other hand, resulted in less invasive PCa in cell culture studies.[46] These findings are supported by the more recent studies that suggests AR activity promotes cell growth in terminally differentiated luminal epithelial cells and stromal cells, while suppressing intermediate, basal, and stem/progenitor cells[47] These studies collectively provide evidence that AR signaling has varied effects in prostatic tissue and that prostatic stromal–epithelial interaction may be essential for PCa growth and progression.[47,48]

The cell cycle of androgen-dependent prostate cells is also regulated by AR signaling. In androgen-deprived conditions, these cells exit the cell cycle.[49] This effect is thought to be mediated by inhibition of cyclin D1 and D3 expression in the absence of AR signaling and resultant inactivation of cdk4 and cdk6.[49] Cyclin D1, in addition to being upregulated by AR signaling, appears to inhibit AR transcriptional activity by directly interacting with the AR N/C dimerization and suppressing transcriptional activity.[49] This negative feedback loop, which allows for AR control over cell-cycle progression, is dysregulated in PCa.[49] AR also regulates the transcription of Cdc6, a regulator of DNA replication and AR silencing can trigger cell-cycle arrest in PCa cells.[50]

The AR plays a role in regulating cell death in the normal prostate and this regulatory function is thought to be aberrant in PCa. Multiple apoptotic pathways, including TNF-α/FAS, Bcl-2, p53, and caspase-mediated, are suppressed by AR signaling.[47,51] The AR also promotes prostate cell survival by directly promoting the transcription of DNA repair genes, which protect the cell from DNA damage caused by ionizing radiation.[52,53] This provides a mechanistic explanation for the sensitizing effect of androgen deprivation therapy (ADT) to radiation therapy for PCa.

## ANDROGEN RECEPTOR ACTIVITY IN CASTRATION-RESISTANT PROSTATE CANCER

Since the seminal discovery of the androgen sensitivity of PCa by Huggins and Hodges,[54] the androgen axis has been the primary target of systemic therapy for advanced and metastatic PCa. Despite the widespread initial response to androgen deprivation, PCa progression inevitably occurs. This state was initially labeled "androgen-independent PCa." Further research, however, demonstrated that this refractory PCa continued to depend on AR signaling, leading to its designation as castration-resistant prostate cancer (CRPC).

Evidence for the dependence of CRPC on the AR axis is now abundant. CRPC continues to express the AR gene,[55] AR targets,[56] as well as PSA. Furthermore, CRPC cells produce androgen-synthesizing enzymes, which maintain an elevated intratumoral androgen concentration.[57] These findings are indicative of a persistently functional AR axis in CRPC, which has prompted drug discovery targeting the AR. In this section, various mechanisms by which PCa achieves castrate resistance and potential targets for therapy will be examined.

Multiple mechanisms have been proposed for the activation of the AR signaling axis in castrate conditions. Broadly, these can be classified into three groups (modified from Chen et al.[58]): (1) CRPC with normal AR function, (2) CRPC with an abnormal AR, and (3) CRPC that activates downstream AR targets but bypasses the AR itself. The first category includes mechanisms in which CRPC cells have a normal AR. Despite a structurally and functionally normal AR, resistance to hormonal therapy may develop due to intracrine androgen production, increased expression of the AR, or an abnormal balance of AR coregulators. The discovery of elevated intratumoral androgens in CRPC led to the development and approval of abiraterone acetate, an inhibitor of testosterone biosynthesis with significant activity against metastatic CRPC.[59] Overexpression of the AR gene and overproduction of the AR protein have also been implicated as mechanisms of castrate resistance.[4,60] Increased levels of AR can compensate for castrate androgen levels and continue AR signaling and downstream function.

The AR, as a transcription factor, recruits multiple coregulators (coactivators and corepressors) to its binding sites.[61] These coregulator proteins participate in the interaction between the AR, the cell's transcriptional machinery and the bound DNA, and either enhance or inhibit the transcriptional activity of the AR by a variety of mechanisms.[62] Over 170 such coregulators have been identified and are being studied to identify their contributions to the development of castrate resistance.[63] ARA70 is a coregulator of the AR that exists in two isoforms. Expression of the ARA70α isoform is reduced in PCa and its overexpression inhibits proliferation in PCa cells *in vitro*.[64] ARA70β, on the other hand, is overexpressed in PCa and promotes cell growth and invasion, even in androgen-independent conditions.[64] The SRC (steroid receptor coactivator) family of proteins has been shown to bind the AR and be increased in PCa.[64] One member of this family, SRC-1, has been shown to potentiate the activity of the AR in the setting of low hormone levels and to be present in higher levels in CRPC.[65] Another study has demonstrated increased transcription of the SRC-2 gene in primary and metastatic PCa tissue.[66] Additional coregulators of AR function have been identified and are being studied. An imbalance in the activity of these coactivators and corepressors may contribute to the development of CRPC.[63]

A second category of postulated mechanisms for castrate resistance involves abnormalities of the AR, specifically those that result in constitutive AR activity or activation by noncanonical ligand binding. One such mutation, known as T877A, was found in the commonly studied PCa cell line, LNCaP.[67] This point mutation in the LBD weakens ligand-specificity of the AR and allows activation by estrogens, progestin, corticosteroids, and even antiandrogens.[63,67] Multiple similar mutations have been found in the LBD of the AR, resulting in broad ligand specificity.[68] These findings paved the way for the recognition of the AR LBD as a promising target for trials in CRPC. Eventually, this line of research led to the development and approval of enzalutamide, a second-generation antiandrogen, which effectively antagonizes the AR with improved binding affinity, and without partial agonism.[69] Recent studies examining mechanisms of resistance to second-generation antiandrogens, however, have identified another mutation of the AR LBD, F876L, which confers resistance to enzalutamide and ARN-509.[70,71] This mutation was found to induce partial agonist activity by enzalutamide and ARN-509 *in vitro* and in tumor xenograft models.[70,71] Such mutations, thought to be induced by the selective pressure of AR-targeted therapy, may continue to be found in the ongoing search for future-generation antiandrogens.

In addition to broadened ligand-binding, some CRPC demonstrates constitutive AR activity. In studies of CRPC cell lines, novel AR variants (AR-Vs) were isolated, and found to be lacking the c-terminal LBD.[72] These AR variants are a result of the splicing of a cryptic exon of the AR gene, resulting in constitutively active AR forms that are independent of ligand-binding, resistant to conventional androgen deprivation, and able to activate AR target genes.[72] Studies in human PCa specimens confirmed the presence of one major AR splice variant (AR-V7) and its overexpression in CRPC and ability to direct the transcription of a unique set of target genes.[73,74] Further research demonstrated that AR-V7 can sustain castration-resistant PCa growth and may mediate resistance to enzalutamide[75] and abiraterone.[76] Studies examining the mechanism of AR-V generation suggest that AR-V expression is regulated by the full-length AR.[76–78] Interestingly, AR-V is rapidly expressed in response to castration,[76,78] and suppression of full-length AR signaling leads to increased AR-V expression.[77] These data suggest that AR-Vs are an important mechanism of castrate resistance and, given their lack of a functional LBD, may confer resistance to new LBD-targeted agents (such as enzalutamide and ARN-509) in CRPC.

A third category of mechanisms for AR activity in CRPC includes processes that bypass the AR itself while maintaining the downstream effects of AR activation. One such mechanism recently reported is the induction of glucocorticoid receptor (GR) expression, which was shown to mimic AR function by activating a similar set of genes.[79] CRPC tissue from mice and a cell-line designed to be resistant to enzalutamide were found to have increased GR expression.[79] These cells were able to establish tumors when injected into castrated mice on antiandrogen therapy while injection of PCa cells without GR expression were not.[79] GR expression was shown to be reduced with the withdrawal of antiandrogen, and regained with the re-exposure of cells to antiandrogen, providing strong evidence that GR expression is controlled by AR-mediated negative feedback.[79] These data conflict with current treatment paradigms as corticosteroids are routinely administered with ketoconazole and abiraterone, and thought to provide a benefit by suppressing ACTH and adrenal androgen production. While blockade of the AR was a relatively well-tolerated mechanism to attack PCa, suppressing GR function would likely have significant clinical side-effects and may potentially limit the use of this strategy in treating CRPC.

Another mechanism by which CRPC may be able to bypass conventional AR signaling is by interaction with an alternate signaling pathway. One example is the phosphatidyl inositol 3-kinase (PI3K) pathway, which is a signaling pathway that has multiple effects in cell growth, survival, and metabolism.[80] Abnormalities in this pathway are frequently seen in various human cancers. The most common abnormality of PI3K signaling is inactivation of the phosphatase and tensin homolog deleted on chromosome 10 (PTEN) tumor suppressor.[80] In PCa, PTEN mutation has been associated with higher Gleason score, higher rate of metastasis, and poorer prognosis.[81] Loss of PTEN has also been shown to promote androgen-independent growth of PCa *in vitro* and *in vivo*.[82] More recent research has suggested that the PI3K and AR pathways regulate each other via reciprocal negative feedback.[83] This hypothesis suggests that inhibition of either pathway may activate the other and allow for continued survival of PCa. Early preclinical research has demonstrated that inhibition of AR and PI3K pathways produces significant tumor regression in cell culture and murine models.[83] Studies investigating the inhibition of the PI3K pathway with and without AR inhibition in the treatment of PCa are ongoing.

## CONCLUSIONS

The AR plays a central role in the normal growth and function of the prostate as well as the pathologic development and progression of PCa. Advances in our understanding of the AR in the last decade have led to numerous discoveries in PCa with ongoing research promising many more in the decades to come.

### References

1. Chang C, Saltzman A, Yeh S, et al. Androgen receptor: an overview. *Crit Rev Eukaryot Gene Expr* 1995;**5**(2):97–125.
2. Chang CS, Kokontis J, Liao ST. Molecular cloning of human and rat complementary DNA encoding androgen receptors. *Science* 1988;**240**(4850):324–6.
3. Chang CS, Kokontis J, Liao ST. Structural analysis of complementary DNA and amino acid sequences of human and rat androgen receptors. *Proc Natl Acad Sci USA* 1988;**85**(19):7211–5.
4. Gelmann EP. Molecular biology of the androgen receptor. *J Clin Oncol* 2002;**20**(13):3001–15.
5. Sartor O, Zheng Q, Eastham JA. Androgen receptor gene CAG repeat length varies in a race-specific fashion in men without prostate cancer. *Urology* 1999;**53**(2):378–80.
6. Black BE, Paschal BM. Intranuclear organization and function of the androgen receptor. *Trends Endocrinol Metab* 2004;**15**(9):411–7.
7. Berman DM, Rodriguez R, Veltri RW. Development, molecular biology, and physiology of the prostate. *Campbell-Walsh Urol.* 2012; 2533-69.
8. Schoenmakers E, Alen P, Verrijdt G, et al. Differential DNA binding by the androgen and glucocorticoid receptors involves the second Zn-finger and a C-terminal extension of the DNA-binding domains. *Biochem J* 1999;**341**(Pt 3):515–21.
9. Shaffer PL, Jivan A, Dollins DE, et al. Structural basis of androgen receptor binding to selective androgen response elements. *Proc Natl Acad Sci USA* 2004;**101**(14):4758–63.
10. Alen P, Claessens F, Verhoeven G, et al. The androgen receptor amino-terminal domain plays a key role in p160 coactivator-stimulated gene transcription. *Mol Cell Biol* 1999;**19**(9):6085–97.
11. Brzozowski AM, Pike AC, Dauter Z, et al. Molecular basis of agonism and antagonism in the oestrogen receptor. *Nature* 1997;**389**(6652):753–8.
12. Veldscholte J, Berrevoets CA, Zegers ND, et al. Hormone-induced dissociation of the androgen receptor-heat-shock protein complex: use of a new monoclonal antibody to distinguish transformed from nontransformed receptors. *Biochemistry* 1992;**31**(32):7422–30.
13. Ozanne DM, Brady ME, Cook S, et al. Androgen receptor nuclear translocation is facilitated by the f-actin cross-linking protein filamin. *Mol Endocrinol* 2000;**14**(10):1618–26.
14. Loy CJ, Sim KS, Yong EL. Filamin-A fragment localizes to the nucleus to regulate androgen receptor and coactivator functions. *Proc Natl Acad Sci USA* 2003;**100**(8):4562–7.
15. Bennett NC, Gardiner RA, Hooper JD, et al. Molecular cell biology of androgen receptor signalling. *Int J Biochem Cell Biol* 2010;**42**(6):813–27.
16. Culig Z, Santer FR. Androgen receptor co-activators in the regulation of cellular events in prostate cancer. *World J Urol* 2012;**30**(3):297–302.

17. Heinlein CA, Chang C. Androgen receptor (AR) coregulators: an overview. *Endocr Rev* 2002;**23**(2):175–200.
18. Lemon B, Tjian R. Orchestrated response: a symphony of transcription factors for gene control. *Genes Dev* 2000;**14**(20):2551–69.
19. Wu F, Peacock SO, Rao S, et al. Novel interaction between the co-chaperone Cdc37 and Rho GTPase exchange factor Vav3 promotes androgen receptor activity and prostate cancer growth. *J Biol Chem* 2013;**288**(8):5463–74.
20. Liao G, Chen LY, Zhang A, et al. Regulation of androgen receptor activity by the nuclear receptor corepressor SMRT. *J Biol Chem* 2003;**278**(7):5052–61.
21. Gioeli D, Paschal BM. Post-translational modification of the androgen receptor. *Mol Cell Endocrinol* 2012;**352**(1-2):70–8.
22. Fu M, Rao M, Wu K, et al. The androgen receptor acetylation site regulates cAMP and AKT but not ERK-induced activity. *J Biol Chem* 2004;**279**(28):29436–49.
23. Goueli SA, Holtzman JL, Ahmed K. Phosphorylation of the androgen receptor by a nuclear cAMP-independent protein kinase. *Biochem Biophy Res Commun* 1984;**123**(2):778–84.
24. Coffey K, Robson CN. Regulation of the androgen receptor by post-translational modifications. *J Endocrinol* 2012;**215**(2):221–37.
25. Cao X, Qin J, Xie Y, et al. Regulator of G-protein signaling 2 (RGS2) inhibits androgen-independent activation of androgen receptor in prostate cancer cells. *Oncogene* 2006;**25**(26):3719–34.
26. Blok LJ, de Ruiter PE, Brinkmann AO. Forskolin-induced dephosphorylation of the androgen receptor impairs ligand binding. *Biochemistry* 1998;**37**(11):3850–7.
27. Rochette-Egly C. Nuclear receptors: integration of multiple signalling pathways through phosphorylation. *Cell Signal* 2003;**15**(4): 355–66.
28. Xu K, Shimelis H, Linn DE, et al. Regulation of androgen receptor transcriptional activity and specificity by RNF6-induced ubiquitination. *Cancer Cell* 2009;**15**(4):270–82.
29. Fu M, Rao M, Wang C, et al. Acetylation of androgen receptor enhances coactivator binding and promotes prostate cancer cell growth. *Mol Cell Biol* 2003;**23**(23):8563–75.
30. Centenera MM, Harris JM, Tilley WD, et al. The contribution of different androgen receptor domains to receptor dimerization and signaling. *Mol Endocrinol* 2008;**22**(11):2373–82.
31. Dahlman-Wright K, Wright A, Gustafsson JA, et al. Interaction of the glucocorticoid receptor DNA-binding domain with DNA as a dimer is mediated by a short segment of five amino acids. *J Biol Chem* 1991;**266**(5):3107–12.
32. Luisi BF, Xu WX, Otwinowski Z, et al. Crystallographic analysis of the interaction of the glucocorticoid receptor with DNA. *Nature* 1991;**352**(6335):497–505.
33. Sack JS, Kish KF, Wang C, et al. Crystallographic structures of the ligand-binding domains of the androgen receptor and its T877A mutant complexed with the natural agonist dihydrotestosterone. *Proc Natl Acad Sci USA* 2001;**98**(9):4904–9.
34. He B, Lee LW, Minges JT, et al. Dependence of selective gene activation on the androgen receptor $NH_2$- and COOH-terminal interaction. *J Biol Chem* 2002;**277**(28):25631–9.
35. Hur E, Pfaff SJ, Payne ES, et al. Recognition and accommodation at the androgen receptor coactivator binding interface. *PLoS Biol* 2004;**2**(9):E274.
36. Dubbink HJ, Hersmus R, Verma CS, et al. Distinct recognition modes of FXXLF and LXXLL motifs by the androgen receptor. *Mol Endocrinol* 2004;**18**(9):2132–50.
37. Savory JG, Prefontaine GG, Lamprecht C, et al. Glucocorticoid receptor homodimers and glucocorticoid-mineralocorticoid receptor heterodimers form in the cytoplasm through alternative dimerization interfaces. *Mol Cell Biol* 2001;**21**(3):781–93.
38. Ni L, Llewellyn R, Kesler CT, et al. Androgen induces a switch from cytoplasmic retention to nuclear import of the androgen receptor. *Mol Cell Biol* 2013;**33**(24):4766–78.
39. Lange A, Mills RE, Lange CJ, et al. Classical nuclear localization signals: definition, function, and interaction with importin alpha. *J Biol Chem* 2007;**282**(8):5101–5.
40. Poukka H, Karvonen U, Yoshikawa N, et al. The RING finger protein SNURF modulates nuclear trafficking of the androgen receptor. *J Cell Sci* 2000;**113**(Pt 17):2991–3001.
41. Pemberton LF, Paschal BM. Mechanisms of receptor-mediated nuclear import and nuclear export. *Traffic* 2005;**6**(3):187–98.
42. De Gendt K, Verhoeven G. Tissue- and cell-specific functions of the androgen receptor revealed through conditional knockout models in mice. *Mol Cell Endocrinol* 2012;**352**(1-2):13–25.
43. Mooradian AD, Morley JE, Korenman SG. Biological actions of androgens. *Endocr Rev* 1987;**8**(1):1–28.
44. Massie CE, Lynch A, Ramos-Montoya A, et al. The androgen receptor fuels prostate cancer by regulating central metabolism and biosynthesis. *EMBO J* 2011;**30**(13):2719–33.
45. Cunha GR, Ricke W, Thomson A, et al. Hormonal, cellular, and molecular regulation of normal and neoplastic prostatic development. *J Steroid Biochem Mol Biol* 2004;**92**(4):221–36.
46. Niu Y, Chang TM, Yeh S, et al. Differential androgen receptor signals in different cells explain why androgen-deprivation therapy of prostate cancer fails. *Oncogene* 2010;**29**(25):3593–604.
47. Wen S, Niu Y, Lee SO, et al. Androgen receptor (AR) positive vs negative roles in prostate cancer cell deaths including apoptosis, anoikis, entosis, necrosis and autophagic cell death. *Cancer Treat Rev* 2014;**40**(1):31–40.
48. Bhowmick NA, Moses HL. Tumor-stroma interactions. *Curr Opin Genet Dev* 2005;**15**(1):97–101.
49. Schiewer MJ, Augello MA, Knudsen KE. The AR dependent cell cycle: mechanisms and cancer relevance. *Mol Cell Endocrinol* 2012;**352**(1-2):34–45.
50. Jin F, Fondell JD. A novel androgen receptor-binding element modulates Cdc6 transcription in prostate cancer cells during cell-cycle progression. *Nucl Acids Res* 2009;**37**(14):4826–38.
51. Lu S, Liu M, Epner DE, et al. Androgen regulation of the cyclin-dependent kinase inhibitor p21 gene through an androgen response element in the proximal promoter. *Mol Endocrinol* 1999;**13**(3):376–84.
52. Polkinghorn WR, Parker JS, Lee MX, et al. Androgen receptor signaling regulates DNA repair in prostate cancers. *Cancer Discov* 2013;**3**(11):1245–53.
53. Goodwin JF, Schiewer MJ, Dean JL, et al. A hormone-DNA repair circuit governs the response to genotoxic insult. *Cancer Discov* 2013;**3**(11):1254–71.
54. Huggins C, Hodges CV. Studies on prostatic cancer. I. The effect of castration, of estrogen and androgen injection on serum phosphatases in metastatic carcinoma of the prostate. *CA Cancer J Clin* 1972;**22**(4):232–40.
55. de Vere White R, Meyers F, Chi SG, et al. Human androgen receptor expression in prostate cancer following androgen ablation. *Eur Urol* 1997;**31**(1):1–6.
56. Holzbeierlein J, Lal P, LaTulippe E, et al. Gene expression analysis of human prostate carcinoma during hormonal therapy identifies androgen-responsive genes and mechanisms of therapy resistance. *Am J Pathol* 2004;**164**(1):217–27.
57. Montgomery RB, Mostaghel EA, Vessella R, et al. Maintenance of intratumoral androgens in metastatic prostate cancer: a mechanism for castration-resistant tumor growth. *Cancer Res* 2008;**68**(11):4447–54.
58. Chen CD, Welsbie DS, Tran C, et al. Molecular determinants of resistance to antiandrogen therapy. *Nature Med* 2004;**10**(1): 33–9.
59. Attard G, Reid AH, Yap TA, et al. Phase I clinical trial of a selective inhibitor of CYP17, abiraterone acetate, confirms that castration-resistant prostate cancer commonly remains hormone driven. *J Clin Oncol* 2008;**26**(28):4563–71.

60. Shiota M, Yokomizo A, Naito S. Increased androgen receptor transcription: a cause of castration-resistant prostate cancer and a possible therapeutic target. *J Mol Endocrinol* 2011;**47**(1):R25–41.
61. Sung YY, Cheung E. Androgen receptor co-regulatory networks in castration-resistant prostate cancer. *Endocr Relat Cancer* 2014;**21**(1):R1–R11.
62. Heemers HV, Tindall DJ. Androgen receptor (AR) coregulators: a diversity of functions converging on and regulating the AR transcriptional complex. *Endocr Rev* 2007;**28**(7):778–808.
63. Egan A, Dong Y, Zhang H, et al. Castration-resistant prostate cancer: adaptive responses in the androgen axis. *Cancer Treat Rev* 2014;**40**(3):426–33.
64. Peng Y, Li CX, Chen F, et al. Stimulation of prostate cancer cellular proliferation and invasion by the androgen receptor co-activator ARA70. *Am J Pathol* 2008;**172**(1):225–35.
65. Agoulnik IU, Vaid A, Bingman 3rd WE, et al. Role of SRC-1 in the promotion of prostate cancer cell growth and tumor progression. *Cancer Res* 2005;**65**(17):7959–67.
66. Taylor BS, Schultz N, Hieronymus H, et al. Integrative genomic profiling of human prostate cancer. *Cancer Cell* 2010;**18**(1):11–22.
67. Veldscholte J, Berrevoets CA, Ris-Stalpers C, et al. The androgen receptor in LNCaP cells contains a mutation in the ligand binding domain which affects steroid binding characteristics and response to antiandrogens. *J Steroid Biochem Mol Biol* 1992;**41**(3-8):665–9.
68. Steketee K, Timmerman L, Ziel-van der Made AC, et al. Broadened ligand responsiveness of androgen receptor mutants obtained by random amino acid substitution of H874 and mutation hot spot T877 in prostate cancer. *Int J Cancer* 2002;**100**(3):309–17.
69. Scher HI, Beer TM, Higano CS, et al. Antitumour activity of MDV3100 in castration-resistant prostate cancer: a phase 1-2 study. *Lancet* 2010;**375**(9724):1437–46.
70. Joseph JD, Lu N, Qian J, et al. A clinically relevant androgen receptor mutation confers resistance to second-generation antiandrogens enzalutamide and ARN-509. *Cancer Discov* 2013;**3**(9):1020–9.
71. Korpal M, Korn JM, Gao X, et al. An F876L mutation in androgen receptor confers genetic and phenotypic resistance to MDV3100 (enzalutamide). *Cancer Discov* 2013;**3**(9):1030–43.
72. Dehm SM, Schmidt LJ, Heemers HV, et al. Splicing of a novel androgen receptor exon generates a constitutively active androgen receptor that mediates prostate cancer therapy resistance. *Cancer Res* 2008;**68**(13):5469–77.
73. Guo Z, Yang X, Sun F, et al. A novel androgen receptor splice variant is up-regulated during prostate cancer progression and promotes androgen depletion-resistant growth. *Cancer Res* 2009;**69**(6):2305–13.
74. Hu R, Dunn TA, Wei S, et al. Ligand-independent androgen receptor variants derived from splicing of cryptic exons signify hormone-refractory prostate cancer. *Cancer Res* 2009;**69**(1):16–22.
75. Li Y, Chan SC, Brand LJ, et al. Androgen receptor splice variants mediate enzalutamide resistance in castration-resistant prostate cancer cell lines. *Cancer Res* 2013;**73**(2):483–9.
76. Yu Z, Chen S, Sowalsky AG, et al. Rapid induction of androgen receptor splice variants by androgen deprivation in prostate cancer. *Clin Cancer Res* 2014;**20**(6):1590–600.
77. Hu R, Lu C, Mostaghel EA, et al. Distinct transcriptional programs mediated by the ligand-dependent full-length androgen receptor and its splice variants in castration-resistant prostate cancer. *Cancer Res* 2012;**72**(14):3457–62.
78. Watson PA, Chen YF, Balbas MD, et al. Constitutively active androgen receptor splice variants expressed in castration-resistant prostate cancer require full-length androgen receptor. *Proc Natl Acad Sci USA* 2010;**107**(39):16759–65.
79. Arora Vivek K, Schenkein E, Murali R, et al. Glucocorticoid receptor confers resistance to antiandrogens by bypassing androgen receptor blockade. *Cell* 2013;**155**(6):1309–22.
80. Courtney KD, Corcoran RB, Engelman JA. The PI3K pathway as drug target in human cancer. *J Clin Oncol* 2010;**28**(6):1075–83.
81. Pourmand G, Ziaee AA, Abedi AR, et al. Role of PTEN gene in progression of prostate cancer. *Urol J* 2007;**4**(2):95–100.
82. Jiao J, Wang S, Qiao R, et al. Murine cell lines derived from Pten null prostate cancer show the critical role of *PTEN* in hormone refractory prostate cancer development. *Cancer Res* 2007;**67**(13):6083–91.
83. Carver BS, Chapinski C, Wongvipat J, et al. Reciprocal feedback regulation of PI3K and androgen receptor signaling in PTEN-deficient prostate cancer. *Cancer Cell* 2011;**19**(5):575–86.

CHAPTER

# 4

# Novel Research on Fusion Genes and Next-Generation Sequencing

*Philip H. Abbosh, MD, PhD, David Y.T. Chen, MD, FACS*
Department of Surgical Oncology, Fox Chase Cancer Center-Temple Health, Philadelphia, PA, USA

## RECURRENT TRANSLOCATIONS IN PROSTATE CANCER

### Discovery of Recurrent Translocations in Prostate Cancer

As of 2005, recurrent translocations were only associated with sporadic solid tumors in sarcoma, some thyroid cancers, and pediatric malignancies. Using gene expression arrays and a bioinformatics filtering algorithm called cancer outlier profile analysis (COPA), Chinnaiyan and coworkers first identified recurrent translocations occurring in prostate cancer, finding juxtaposition of the 5′ untranslated region of the prostate-specific gene *TMPRSS2* with the proximal coding region of either *ERG* or *ETV1*,[1] members of the ETS family of transcription factors. In general, carcinogenic translocations occur as one of two classes: (1) Juxtaposition of two disparate coding regions that in combination create a chimeric fusion protein having novel function, such as with breakpoint-cluster region (*BCR*)-Abelson kinase (*ABL*) in CML; or (2) juxtaposition of a proto-oncogene with the promoter of a highly expressed gene, which is the nature of recurrent translocations in prostate cancer. The COPA algorithm detects the second class of translocation by virtue of high expression of oncogenes in a particular subset of cancers (outliers). *TMPRSS2* expression is driven by the androgen receptor (AR), so translocation results in high expression of any coding sequences fused distal to the *TMPRSS2* breakpoint. *ERG* and *ETV1* had been previously identified as elements of recurrent translocations in Ewing sarcomas,[2,3] and so a similar carcinogenic mechanism was implicated in prostate cancer. The proteins encoded by ETS gene family members are transcription factors with homology to proto-oncogenic sequences within the E26 virus. They have properties similar to classical proto-oncogenes such as *Src* kinase,[4] in that they have been shown to have transforming capacity when expressed in normal cells. The human genome encodes 27 ETS family members that all bind to a consensus sequence GGA(A/T).[5] The gene family includes *ERG* and *ETV1*, which contain the aforementioned 3′ translocations in Ewing sarcoma and have been identified in about half of prostate cancers. Although this family of transcription factors have promiscuous binding to regulatory elements (by virtue of their short DNA response elements), interactions with other transcription-regulating proteins elicits a tissue-specific and biology-specific response.[6]

The ETS transcription factor family members are therefore considered *bona fide* oncogenes. However, the translocations identified by Tomlins et al. involving *ETV1* were identified at a lower-than expected frequency than that predicted by COPA, suggesting unrecognized ETS rearrangements. Subsequent studies identified alternative 5′ translocation partners, in place of *TMPRSS2*, which more frequently juxtaposed *ETV1*.[7] These translocations had a similar mechanism: the *ETV1* locus was fused downstream of one of four genes strongly expressed in prostate. Additionally, rare 3′ translocation partners were identified, which involved other ETS family proto-oncogenes, *ETV4*,[8] *FLI1*,[9] and *ETV5*.[10] In summary, this category of recurrent translocations juxtapose the regulatory element of one of several 5′ partners that are strongly expressed in prostate or prostate cancer, linking them to the majority of the coding sequence of one of at least six ETS transcription factor proto-oncogenes (see Figure 4.1). When expressed, the proteins are often truncated at the N-terminus. Tomlins et al. have suggested that ETS rearrangements can be subdivided into one of five classes based on upstream regulatory sequence.[7] The immediate consequence of these translocations is the overexpression of ERG, ETV1, ETV4, ETV5, ELK4,

Prostate Cancer. http://dx.doi.org/10.1016/B978-0-12-800077-9.00004-9

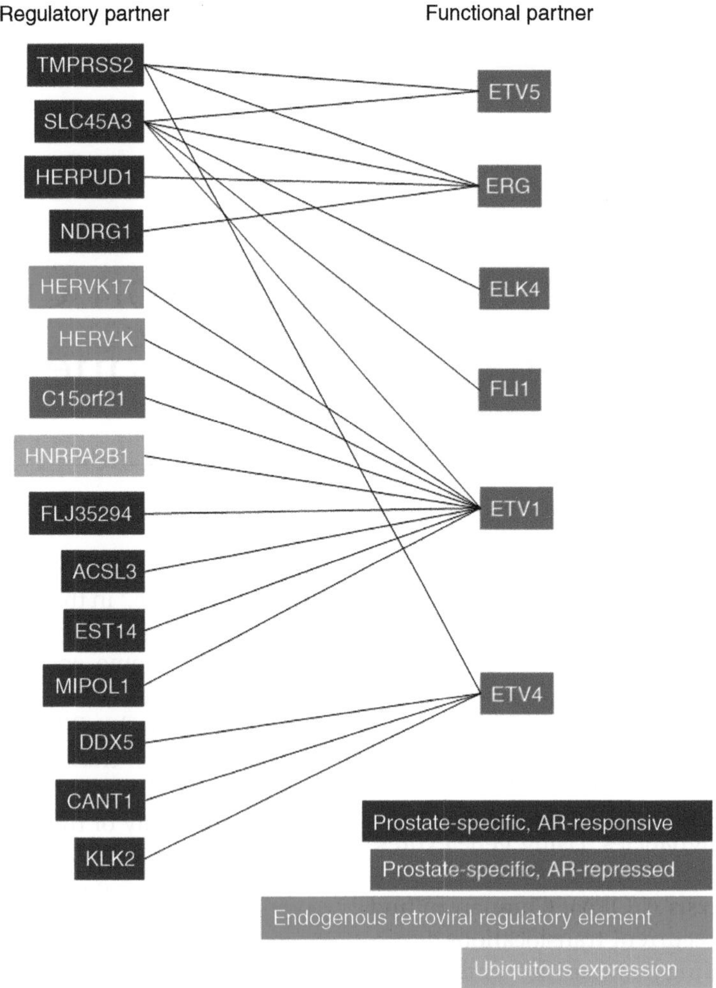


**FIGURE 4.1 Overview of ETS family rearrangements in prostate cancer.** Regulatory sequences from genes that are expressed in prostate cancer are rearranged upstream to the majority of the coding sequence for one of six ETS family transcription factors. Rearrangement partners are promiscuous in that one regulatory partner may have several functional partners and vice versa. Regulatory partners may be prostate-specific, prostate-cancer-specific, or ubiquitously expressed in all tissues. Corf21 is paradoxically highly expressed in prostate cancer but repressed by androgens in model systems. *Adapted from* Journal of Clinical Oncology *29:3659.*

or FLI1 protein, resulting in downstream transcriptional profile changes.

## Function of Recurrent Prostate Cancer Translocations

The role of any oncogenic transcription factor will presumably result from alterations in downstream gene expression, so research has focused on determining the targets of the ETS family. Because *ERG* is the most common *TMPRSS2* translocation partner, ERG function in prostate cancer has been the most intensely studied. In particular, identification of ERG transcriptional targets might explain its role in carcinogenesis and suggest a focus for therapeutic intervention. The first effect attributed to overexpression of ETS family genes was increased invasiveness when expressed in RWPE cells,[7,11] which are immortalized normal prostate epithelial cells. This invasive capacity is mediated by an invasion-associated gene expression program,[11] and is also associated with epithelial–mesenchymal transition.[12] Although historically proto-oncogenes have been shown to transform model systems, surprisingly ETS overexpression did not lead to transformation in this study.[11] *ERG* or *ETV1* translocations seem to be necessary but not independently sufficient for development of prostate cancer, and it is suspected that additional genetic lesions are required. Supporting this belief, when transgenic mice harboring the prostate-specific overexpression of ERG also have heterozygous loss of the *PTEN* tumor suppressor, they develop prostate cancer.[13–15] This is clinically relevant as *PTEN* loss is frequent in prostate cancer and seems to coincide with *ETS* gene rearrangements.[13,14] ETS family overexpression in combination with *PTEN* loss is therefore a clinically meaningful model of prostate cancer, but knowledge of the oncogenic mechanism is still evolving (explained in more detail later).

Rather than identifying targets by examining gene expression changes after introduction of ERG, the ETS cistrome (the gene regulatory regions to which ETS directly binds) can be directly characterized using chromatin immunoprecipitation (ChIP).[16,17] ChIP was used to show that ETS family binding sites often overlap AR transcriptional targets.[18] Unexpectedly, sequence analysis of AR targets identified using ChIP often did not contain canonical 15-base pair androgen response elements (AREs). Instead, AR targets contained a partial six-base pair ARE, which was also associated with an ETS response element in 70% of the targets, suggesting coregulation of gene expression.[18] Additionally, androgen treatment of LNCaP prostate cancer cells induced redistribution of *ETS1*, as well as increased *ETS1* binding to the AR cistrome. AR and ETS1 interact in immunoprecipitation experiments, and so although *ETS1* has not been formally shown to be involved in recurrent translocations in prostate cancer, the potential that other ETS family members associated with translocations could substitute for *ETS1* is intriguing.

The finding of ETS transcription factors and AR target co-occupancy was confirmed and further explored using next-generation sequencing (NGS) of ChIP targets.[19] The ETS response element was the second most enriched motif after the ARE in the AR cistrome: 40% of the AR binding sites had an ARE, 29% had an ETS response element, and this overlapping pattern was recapitulated across cell lines and found in a human prostate cancer sample.[19] Furthermore, ERG and AR binding sites also overlap in a mouse model of prostate-specific ERG overexpression in a *PTEN* heterozygous context.[15] ERG and AR proteins directly interact, but unexpectedly, ERG binds the *AR* promoter to downregulate AR expression *in vitro*; this leads to low AR-induced gene expression in tumors harboring ERG translocations,[19] which seems counterintuitive. Lastly, ERG-mediated AR target repression seems to be mediated by methylation of histone H3-K27, an enzymatic reaction carried out by EZH2 and Polycomb Repressive Complex 2 (PRC2).[19]

## Epidemiology/Biomarker Studies

When initially described, the *TMPRSS2–ERG* and *TMPRSS2–ETV1* fusions were identified in 23 of 29 samples (79%)[1] analyzed independent of any knowledge regarding the more general involvement of ETS family gene expression or translocations. However, this value underestimates the rate of translocations involving the ETS family in prostate cancer and overestimates the frequency of *TMPRSS2-ERG* translocations. Because multiple 5′ and 3′ translocations have been identified, which result in overexpression of an ETS family member, any study looking at just one pair of translocation partners will underestimate the overall rate of translocation. *TMPRSS2-ERG*, however, are the most common translocations and account for 90% of all prostate cancer oncogenic translocations and occur in about 50% of all prostate cancer specimens examined to date,[20] although this was mostly evaluated in prostate-specific antigen (PSA)-screened populations.

Because of the ubiquity of these translocations and the evolving debate around the benefit of PSA-based prostate cancer screening,[21] ETS-associated translocations have been proposed as a molecular marker for prostate cancer. The most compelling data for *TMPRSS2-ERG* as a biomarker examined urine from >1300 men with prostate cancer and correlated it with pathological indicators of disease aggressiveness.[22] Subanalysis of >1100 men undergoing prostate biopsy had urinary *TMPRSS2-ERG* measurement prior to prostate biopsy. Presence of a positive signal correlated strongly with risk of cancer diagnosis, number of positive cores, total cancer involvement of all cores, and clinically significant cancer as defined by Epstein criteria.[23] The relationship of fusion transcript presence and pathologic surrogates of disease aggressiveness was further examined in a cohort of patients who underwent radical prostatectomy, and the presence of *TMPRSS2-ERG* correlated with clinically significant disease (defined as tumor size >1 $cm^3$, Gleason score >6, or nonorgan confined disease). This is provocative preliminary work, and requires greater delineation of thresholds and prospective validation before it can be considered for routine application.

## Correlation to Oncologic Features and Outcomes

Prostate cancers associated with an ETS family translocation may have specific clinical features due to their common genetic basis. Mosquera et al. first described such features in 253 prostate cancer cases, ~90% of which were radical prostatectomy specimens.[24] The main findings of the study were that features, such as cribriform growth pattern (by definition, Gleason pattern 4 disease), signet ring cells, blue-tinged mucin, marconuclei, and intraductal tumor spread, were more common in fusion-containing tumors than noncontaining tumors, and that the presence of three or more of these features was 93% specific for identifying cases as translocation-positive. A second study confirmed the association of mucin-producing tumors, cribriform growth, signet ring cells, and *TMPRSS2–ERG* fusions; however, only mucin production showed a statistically significant correlation.[25] This result was based on only

1/3 the number of cases as the study of Mosquera et al., so it is likely it was underpowered to recognize other meaningful variables.

Correlation of presence of *ETS* fusion genes to metrics of oncologic outcome, which require follow up (as opposed to pathologic parameters), is less clear. Multiple studies are unable to show correlation between ETS-family gene status and biochemical recurrence (BCR), development of metastases (DM), or prostate-cancer-specific mortality (PCSM),[26–33] while others in contrast have found a correlation between these outcomes and ETS-family gene translocations.[34–37] This may be in part explained by (1) variable approaches by which groups define translocation (immunohistochemical staining for ERG, FISH, sequencing, mRNA expression, assays for different parts of the involved genes, etc.); (2) geographic, chronologic, and screening differences among independent patient cohorts; (3) heterogeneous nature of prostate cancer; (4) not accounting for all ETS-family gene fusion partners; (5) small cohorts, small subgroups, or small event frequency in many cohorts; and (6) inconsistent definition of outcome metrics (BCR, DM, and PCSM).

Robust studies have been completed and show no correlation between *ERG* status and outcome. A study of 481 patients from the European Randomized Study of Screening for Prostate Cancer (ERSPC) showed no association between ERG expression and recognized pathological risk factors for BCR or local recurrence.[31] Because ERG is not usually expressed in prostate cancer except by means of translocation,[1] protein expression correlates with translocation presence,[38] but translocations involving other ETS family members (though infrequent) may have been missed. Another study evaluating 1180 men from the Physicians' Health and Health Professionals Follow-Up Studies also showed no correlation between *ERG* status and BCR or PCSM.[33] While a meta-analysis of >5000 men followed for BCR and >2000 men followed for PCSM did show an association between *ERG* status and prostate cancer stage at diagnosis, there was no association between *ERG* status and subsequent BCR or PCSM.[33]

In contrast, study of the well-annotated Örebero watchful waiting cohort[39] of men treated only having symptomatic localized prostate cancer or clinically evident metastases showed that *TMPRSS2–ERG* fusion was associated with DM.[36] Fusion was enriched in men with metastases, with 29% of men with DM having fusion compared to 10% of men who were nonmetastatic. Furthermore, 53% of all fusion-positive cases showed DM whereas only 23% of fusion-negative men had metastasis. In a separate study of 26 men with Gleason score 7 prostate cancer, *TMPRSS2–ERG* fusion was the only significant predictor of BCR at a median of 12 months.[37] However, a cohort this small may be subject to bias; notably, established predictive factors of BCR, like pathologic stage, grade, and margin status, did not show an association with BCR in this cohort. Another source of bias is that only half of the cohort was analyzed for *ERG* translocation or overexpression, and surprisingly only 15% of the patients in the cohort had evidence of fusion, compared to 50% in contemporary series. This may suggest that *ERG* translocation is actually selected against during metastasis. Simultaneous *TMRPSS2-ERG* translocation, amplification of the *ERG* translocation arm, and deletion of the intervening chromosome 21 sequence (both genes are on the same arm of chromosome 21) within the same tumor cell was identified in 2008.[35] Deletion of the intervening sequence alone associated with a 2.9-fold hazard ratio for PCSM, but deletion of that sequence combined with amplification of the translocation increased the hazard ratio for PCSM to 6.1-fold. These genetic aberrations are uncommon, as 14 and 6% of the 445 patient cohort had deletion of the intervening sequence or deletion plus amplification, respectively. Although this rare subgroup appears to be most at risk for adverse outcome, the correlation failed validation in a later study where *ERG* amplification but not intervening sequence deletion with translocation was associated with BCR.[29]

Finally, genetic substratification of prostate cancers using *ERG* and *PTEN* correlates with long-term outcome in combinatorial analysis. The first suggestion that a "subset of a subset" identified genetically based on ETS family rearrangement showed that men with prostate cancer without *TMPRSS2-ERG* translocation and with normal *PTEN* status had the most favorable rate of BCR, compared to men with tumors harboring either a translocation or homozygous deletion of *PTEN*, who showed intermediate rates of BCR, and tumors having both translocation and homozygous *PTEN* loss demonstrated the highest rate of BCR[40] (see Figure 4.2). Again, presence of *TMPRSS2–ERG* fusion alone did not correlate with BCR, but amplification of the fused partners independently predicted for BCR. The finding that *ERG*-nonrearranged tumors with normal *PTEN* status are a subgroup with favorable prognosis was corroborated in a separate analysis. In that cohort, ERG expression (and by proxy, *ERG* translocation) did not enrich with stage,[41] arguing against a more aggressive phenotype of these tumors when translocation alone is assessed as a prognostic marker. A third cohort of 308 men without therapy diagnosed with prostate cancer from transurethral resection showed a somewhat similar genetic *PTEN/ERG* substratification profile, which predicted for PCSM.[42] Given that these results mirror transgenic mouse models where overexpression of ETS family members in the setting of *PTEN* loss causes prostate cancer,[13–15] a rationale for 2 × 2 substratification of prostate cancer patients based on these genetic lesions emerges.

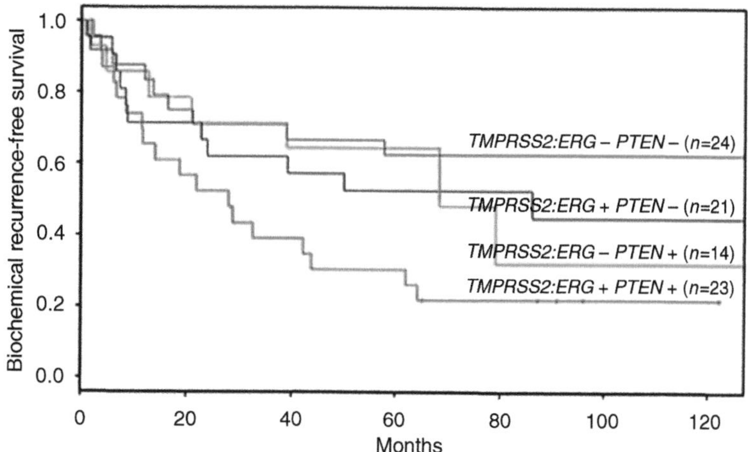


**FIGURE 4.2 Substratification by genetic lesions predicts biochemical recurrence.** Tumors (*n*=125) containing all combinations of *TMPRSS22:ERG* rearrangements (*TMPRSS:ERG* +) or evidence of *PTEN* deletion (*PTEN* +) were subjected to Kaplan–Meier analysis with BCR after radical prostatectomy. *Reprinted with permission from Yoshimoto et al.*[40]

# SPECTRUM OF POINT MUTATION AND COPY NUMBER ALTERATIONS IN PROSTATE CANCER

## Mutated Gene Spectrum in Prostate Cancer is Very Different Than Other Common Cancers

Prostate cancer has a unique spectrum of genomic alterations. Mutations in genes that are frequently altered in multiple types of other cancers – *TP53*, the RAS family, *EGFR*, *PIK3CA*, *RB1*, and so on – together are only mutated in much less than half of current prostate cancer cases with sequencing and copy number alteration (CNA) data publicly available[43] (see Table 4.1). The exception to this rule is metastatic cases (the Michigan cohort), where such mutations are much more frequent. *PTEN* is probably the most common prostate cancer gene that is also commonly mutated in other cancers (i.e., altered in >10%), and it is altered in some way (point mutation, or CNA) in about 20% of prostate cancers overall. Even obvious candidates such as *AR* are so infrequently mutated in the bulk of prostate cancers, that *AR* mutation was only readily identified in highly selected populations (shown later). For this reason, identifying novel prostate-cancer-specific mutations was difficult until the advent of NGS technologies and refinement of bioinformatics algorithms to detect them. This underscores the significance of translocations involving ETS family genes as early events, and the importance of finding initiating or driver prostate cancer genes that might be better prevention, early detection, and treatment targets.

Another commonly altered gene in prostate cancer is *NKX3.1*, a homeodomain-containing master regulator of prostate development,[44] which is often the target of CNA loss in prostate cancer precursor tissue, as well as up to 90% of prostate cancer specimens.[45–47] *NKX3.1* loss may cooperate with *TMPRSS2–ERG* fusion because it physically binds to the *TMRPRSS2* promoter to promote its downregulation,[48] and reciprocally ERG mediates downregulation of *NKX3.1*.[49] Finally, double heterozygote *Nkx3.1;Pten* mice develop prostate cancer,[50] and those surviving past 1 year of age develop lymph node metastases.[51]

## Mutations of AR

Although AR seems central to prostate carcinogenesis, mutations, splice variants, and gene amplification have curiously only been identified in men almost exclusively after androgen deprivation therapy (ADT), suggesting that hormone ablation is a positive selection pressure for such mutants. This emphasizes the central role that AR plays in prostate cancer, as cancers are required to circumvent therapeutic approaches in order to maintain signaling through AR and survive, akin to mutations in the *ABL* kinase in imatinib resistance.[52] Prostate cancer can then be considered a disease where "oncogene addiction"[53] to AR is the essential pathway for growth and survival. The continued central role for AR in antiandrogen resistance was elegantly shown by genetic disruption of the AR mRNA using a hammerhead ribozyme, which cleaved AR in an "androgen-independent" cell line.[54] Although *androgen-independent*, the cells were still *AR-dependent* for proliferation. This critical understanding has refocused the central dogma in prostate cancer therapy, shifting thinking away from "androgen-independence" to "castration-resistance" and led to a renaissance in AR-targeted therapies.

Some of the earliest point mutations identified in prostate cancer were in the AR.[55–58] Interestingly, these mutations were shown to impart promiscuity on the part of the mutated receptor as such AR mutants are transactivated by estrogens and progesterones,[56,57] glucocorticoids,[55] and

**TABLE 4.1** Percentage of Tumors with Genetic Alterations in Publicly Available NGS Datasets

| | TCGA ($n$ = 246) | MSKCC ($n$ = 103) | Michigan ($n$ = 61) | Broad Institute 2013 ($n$ = 56) | Broad Institute 2012 ($n$ = 109) |
|---|---|---|---|---|---|
| ***PERCENTAGE OF TUMORS WITH POINT MUTATION OR CNA IN COMMONLY ALTERED CANCER GENES*** | | | | | |
| *TP53* | 11 | 5 | 52 | 11 | 6 |
| *RB1* | 11 | 5 | 28 | 11 | 3 |
| *PIK3CA* | 7 | 3 | 10 | 7 | 5 |
| *PTEN* | 22 | 14 | 49 | 18 | 7 |
| *NRAS* | 0 | 1 | 0 | 0 | 0 |
| *HRAS* | 1 | 0 | 2 | 0 | 1 |
| *KRAS* | 0 | 3 | 7 | 5 | 0 |
| *EGFR* | 2 | 6 | 2 | 0 | 4 |
| *ERBB2* | 1 | 2 | 2 | 0 | 4 |
| | **TCGA ($n$ = 261)** | **MSKCC ($n$ = 103)** | **Michigan ($n$ = 61)** | **Broad Institute 2013 ($n$ = 57)** | **Broad Institute 2012 ($n$ = 112)** |
| ***PERCENTAGE OF TUMORS WITH POINT MUTATION ONLY IN COMMONLY ALTERED CANCER GENES*** | | | | | |
| *TP53* | 9 | 3 | 41 | 11 | 6 |
| *RB1* | 1 | 0 | 10 | 0 | 0 |
| *PIK3CA* | 3 | 3 | 0 | 4 | 4 |
| *PTEN* | 5 | 2 | 10 | 4 | 4 |
| *NRAS* | 0 | 0 | 0 | 0 | 0 |
| *HRAS* | 1 | 0 | 0 | 0 | 1 |
| *KRAS* | 0 | 1 | 2 | 0 | 0 |
| *EGFR* | 1 | 2 | 0 | 0 | 1 |
| *ERBB2* | 0 | 2 | 2 | 0 | 1 |

*Accessed from http://cbioportal.org.*

disturbingly, even by antiandrogens.[58] Receptor-ligand promiscuity provided a tenable mechanism for castration resistance, as well as for clinical observations such as the success of antiandrogen withdrawal, and explains why switching ADT modality would sometimes result in a clinical response and decreased PSA. Furthermore, these observations directly led to the development of MDV-3100 (enzalutamide),[59] a second-generation specific AR antagonist having stronger binding affinity to AR.

More recently, splice variants have been implicated in AR signaling. An AR splice variant that excludes exons 5, 6, and 7 (causing loss of the ligand binding domain) was shown to behave in a constitutively activated manner and was present in >40% of tumors sampled at the time of rapid autopsy.[60] This may be related to constitutive nuclear translocation in such splice variants.[61] Such splice variants are also constitutively resistant to enzalutamide.[62] Clinical application of these variants was used retrospectively to identify a mechanism of enzalutamide and/or abiraterone resistance.[63] Circulating tumor cells from patients with metastatic prostate cancer enrolled in trials evaluating these two drugs were isolated, and the level of AR variant mRNA that lacks the ligand binding domain was quantitated. No patients expressing the variant had a PSA response to either drug, while 53 and 68% of patients with full length AR mRNA had a PSA response after enzalutamide and abiraterone, respectively. This further implicates splice variants in advanced stages of prostate cancer and provides another mechanism cancers use to attain castration resistance. Targeting AR variants will require completely novel approaches because such variants do not bind androgens or antiandrogens. Perhaps drugs that target AR coactivators or other AR-interacting proteins might prove useful.

Another mechanism for antiandrogen resistance has been shown to involve *AR* amplification. This was initially shown to occur in up to 30% of tumors with rising PSA after ADT[64] and has since been recapitulated in independent cohorts. *AR* amplifications have been found in exceedingly rare frequencies in hormone-naïve settings,[65–70] being found in only 9 of 989 samples, strongly suggesting that the more frequent amplification seen post-ADT is a therapy-resistance mechanism and results from selective evolutionary pressure.

## Prostate Cancer Whole Genome Sequencing Studies

NGS technology has been applied to several prostate cancer specimen cohorts to describe in fine detail the landscape of the prostate cancer genome.[71–76] Although all used similar technical approaches, only a few of the studies show overlap in the mutations identified. PTEN/PI3K pathway alterations, ETS-family translocations, *CHD1* deletion, and mutation of chromatin-modifying enzymes have been identified in independent reports. Prostate cancer is one subtype, adenocarcinoma, in the overwhelming majority of cases, which is unlike lung or kidney cancers, where histopathologic subtyping gives the clinician some clue of the expected clinical outcome. Annotation of specimens according to mutually exclusive mutation patterns is leading to molecular subclassification of prostate cancer[77,78] akin to breast cancer subtyping,[79] with the intent that future precision therapies will be assigned in a subtype-driven manner. Historically, AR has been the primary target of prevention and treatment strategies, and this focus will likely continue, as many if not all mutations seem to impinge on AR as a final common pathway. Indeed, point mutations[55–58] and splice variants[60–62] in AR and its interactome have also been identified and shown to or been hypothesized to contribute to prostate cancer carcinogenesis.

## Prostate Cancer Chromoplexy

A major breakthrough in the understanding of prostate cancer initiation occurred in 2011 with whole genome sequencing and annotation of seven prostate cancer genomes and the corresponding normal tissues. Berger et al. identified numerous copy-neutral translocations in three specimens, which also harbored ETS family translocations.[74] Their model explains how multiple prostate cancer genes can be disrupted simultaneously: several double-strand breaks are repaired by "braiding" chromosomes back together (see Figure 4.3). This closed chain pattern of breakage/repair in tumors with ETS fusions frequently occurred at areas of open chromatin, and RNA polymerase II and AR were also frequently bound to translocation break points, suggesting that transcription-coupled repair contributes to the process of breakage/fusion. There were between 43 and 213 translocations per tumor, even in tumors without ETS-family rearrangements, and that translocations frequently occurred in cancer-related genes such as *TP53,*

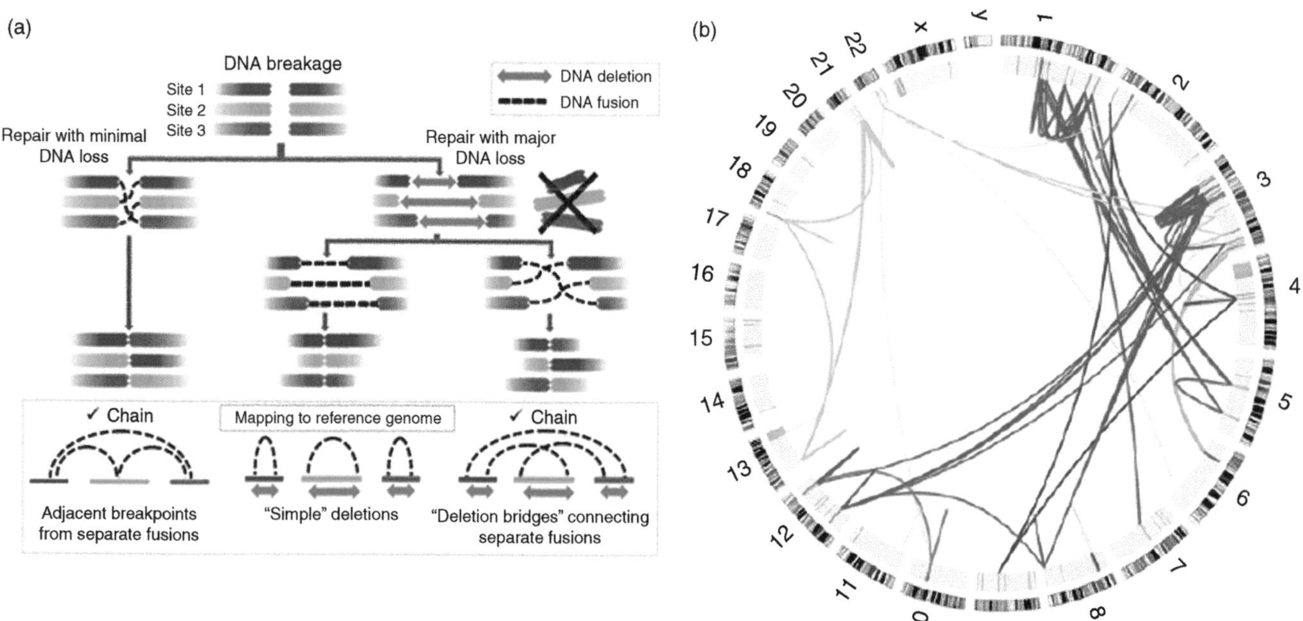


**FIGURE 4.3 Chromoplexy in prostate cancer.** (a) Three models of DNA repair that might explain whole-genome sequencing findings of Baca et al.:[76] (1) rearrangement with minimal loss of breakpoint DNA, (2) simple deletion without rearrangement, and (3) rearrangement with major loss of breakpoint DNA. Only deletion (minor or major) in combination with rearrangement can explain the identified chromosome chains (reprinted from Baca et al.).[76] (b) Circos plot depicting the complexity of chain rearrangements in a single patient. Chromosomes are arranged on the outer circle with each chain in the breakage-fusion cycle assigned to a different color. Gray rearrangements are "simple" and not part of a chain. There are eight chains identified in this patient, each with multiple chain links. The inner circle shows CNA gains and losses in blue and red, respectively (reprinted from Baca et al.).[76]

*ABL1*, *PTEN*, and *MAGI2* (a PTEN-interacting protein). This provides a mechanism whereby many prostate cancer genes could be silenced simultaneously.

This group subsequently coined this breakage/fusion process chromoplexy – catastrophic, simultaneous breakage/fusion cycles, which are also associated with deletion of intervening sequences with end-joining to new partners.[76] These events are interdependent and occur simultaneously, resulting in the observed partnering of breakpoints. Chromoplexy was observed in 50 of 57 (88%) of tumors, but was much more frequent in tumors harboring ETS gene rearrangements, and *ERG* rearrangements themselves were frequently the result of chromoplexy-associated translocations. In tumors without ETS family mutations, a related but separate process occurs, which is akin to chromothripsis,[80,81] and results in more numerous genetic events. Chromothripsis results from "shattering" of one or two chromosomal regions and isolation of chromosomes into micronuclei. As strictly defined, chromothripsis is a single event in the life cycle of a cell, whereas chromoplexy seems to be processive or iterative. Baca et al. showed that chromoplexy events occur throughout the life cycle of prostate cancers, equating NGS data to the fossil record as reinvented by Eldridge and Gould to suggest that prostate cancer genomes also undergo "punctuated equilibria" during their evolution. The genomic insult in ETS-negative tumors is frequently associated with functional genetic disruption of *CHD1*, a gene whose protein seems to regulate genome integrity, and *CHD1* loss is associated with homozygous deletion of several chromosomal regions.[82]

## SPOP Point Mutations

Whole exome sequencing of 112 prostate cancers and normal tissue controls has revealed mutations in several genes, one of which was the E3 ubiquitin ligase *SPOP*.[73] This gene has been subsequently confirmed to be commonly mutated in prostate cancer cohorts[83–86] as well as in other cancer types.[87,88] *SPOP* mutations cluster to a substrate binding cleft within the MATH domain of the protein suggesting that the mutations modify its interaction with other proteins without disrupting its enzymatic activity. *SPOP* mutations were mutually exclusive from ETS family translocations in these 112 tumors, suggesting these mutations occur early and lead to distinct prostate cancer subtypes. Indeed, in a large cohort, *SPOP* mutation was shown to associate with *CHD1* deletions,[83] implicating *SPOP* mutations in chromothripsis. There is a large body of literature implicating ubiquitin signaling in DNA repair,[89] so it is foreseeable that SPOP may be important in DNA repair and genome integrity.

The wild type SPOP protein has been shown to bind to and mediate degradation of AR by the ubiquitin/proteasome pathway whereas mutant *SPOP* cannot.[90] Furthermore, *AR* which is mutated at the hinge region (which is a region lacking in most AR splice variants described previously), is also unable to bind SPOP and be degraded, unifying the consequences of two cancer-associated genetic changes. A second target of SPOP-mediated degradation is SRC3,[91] an AR coactivator[92] whose overexpression is associated with failure of ADT.[93] Similarly, wild type SPOP targets SRC3 for ubiquitination/degradation, but *SPOP* mutants found in prostate cancer cannot. These findings corroborate the notion that *SPOP* mutations serve to alter the proteins that it can target for degradation, and that their build-up is oncogenic. However, these targets were identified and studied in LNCaP cells, which have an ETS fusion, and thus may evolve through a different genetic pathway. It is possible that *SPOP*-derived prostate cancers develop through mechanisms independent of androgen signaling, rendering these SPOP targets irrelevant for *SPOP*-subtype prostate cancer development.

## Prostate Cancer Subtyping

A major goal of genomic landscaping, identification of specific mutants, and molecular subclassification of prostate cancer (or other cancers) is to lead to personalized or precision medicine. In this manner, treatment would be tailored to each patient's tumor based on specific biomarker assignment. Several studies have started to recognize subgroups within prostate cancer cohorts; the earliest study substratified on the basis of gene expression profiling.[94] Lapointe et al. showed that three distinct subtypes segregate by unsupervised hierarchical clustering and correlate with clinical variables and outcome. Subtype I had an expression profile including normally expressed genes and predicted for the most favorable prognosis, whereas subtype III had the worst oncologic outcome and contained most tumors with lymph node metastases as well as localized tumors with a lymph node-metastatic profile. Subtype II was the largest subtype and shared features of the subtype III profile. Two proteins, MUC1 and AZGP1, could correctly assign subgrouping to subtype II and III versus subtype I, and correlated to BCR when assessed individually and in combination. Furthermore, unsupervised hierarchical clustering of CNAs resolved three subtypes with breakdown of the same cases into each group as identified previously by gene expression profiling.[95] Subtype II was associated with ETS family gene fusions, whereas subtype III tumors had CNA increases of regions involving *MYC* and deletion of the region coding for *PTEN*.

Other groups have focused on ETS-rearranged tumors to identify commonalities. Demichelis et al. showed that specific CNAs associate translocation status; in rearranged cases 7q is deleted and 16q is amplified, whereas 6q is deleted in tumors without ETS translocations.[96] The most comprehensive subtype assignment has been generated based on CNA analysis, expression analysis, and focused exon sequencing in 218 prostate tumors.[71] This analysis identified two subgroups based on CNA analysis: one with minimal CNAs, which split into four clusters, and another with frequent CNA, which split into two clusters and contained most of the patients with metastasis. In addition to different risk of metastasis, these subgroups correlated to different risk of BCR.

Finally, with the evolution of sequencing technology, molecular subtypes can be identified on the basis of NGS annotation. The two main prostate cancer subtypes identified as a result of NGS technologies revolve around genome structure and global genomic changes. These subtypes juxtapose chromoplexy, the erroneous repair of a small number of chromosome breaks at sites of active transcription that result in chain rearrangements and translocations and are epitomized by ETS translocations, and a chromothripsis-like process that cosegregates with deletion of the genome integrity maintenance gene *CHD1*.[74,76] Although this model does not incorporate gene expression profiling into classification of subtype, it does account for several phenomena that are mutually exclusive events in prostate carcinogenesis, including the types and frequency of CNAs and the genesis of ETS rearrangements, and it suggests a role for androgen signaling in the formation of translocations by chromoplexy.[97] Supporting this model, Burkhardt et al. illustrate the central role of *CHD1* loss in determining *ERG* translocation status; *ERG* translocations occur 16 times less frequently *in vitro* in a setting of lost CHD1 function, and *CHD1* homozygous deletion occurs in 0.2% of *ERG* rearranged tumors but 4.1% of *ERG* nonrearranged tumors in a cohort of >2000 specimens.[98]

## CONCLUSIONS

In reviewing the history of our understanding of the prostate cancer genome, the pace of discovery is striking and has primarily occurred only over the last decade. Great advances have been made in explaining the major components of prostate carcinogenesis, although many details are still missing. Why is the prostate prone to chromoplexy or a chromothripsis-like process? Can these processes be targeted in prevention or treatment strategies, and how? Can these newly identified alterations be used in place of or in addition to PSA or other screening modalities? We have progressed from broad genomic expression profiles to more analysis of CNA gains and losses and ultimately will reach "base-pair resolution" in determining the significance of genetic changes and their relative roles in the development and biology of prostate cancer. These changes will be evaluated at the individual patient level, providing more refined subtyping with the goal of providing precision therapies that can be appropriately assigned to patients of each subtype. Additionally, a common paradigm in cancer drug development is to identify drugs that target mutated genes and apply them in a subtype-specific manner. This approach has been successful in targeting ABL kinase[99] and EGFR inhibitors,[100] for example, and could similarly be applied in targeted therapy for prostate cancer.

## References


1. Tomlins SA, Rhodes DR, Perner S, et al. Recurrent fusion of TMPRSS2 and ETS transcription factor genes in prostate cancer. *Science* 2005;**310**:644–8.
2. Jeon IS, Davis JN, Braun BS, et al. A variant Ewing's sarcoma translocation (7;22) fuses the EWS gene to the ETS gene *ETV1*. *Oncogene* 1995;**10**:1229–34.
3. Sorensen PH, Lessnick SL, Lopez-Terrada D, et al. A second Ewing's sarcoma translocation, t(21;22), fuses the *EWS* gene to another ETS-family transcription factor ERG. *Nat Genet* 1994;**6**:146–51.
4. Stehelin D, Varmus HE, Bishop JM, et al. DNA related to the transforming gene(s) of avian sarcoma viruses is present in normal avian DNA. *Nature* 1976;**260**:170–3.
5. Gutierrez-Hartmann A, Duval DL, Bradford AP. ETS transcription factors in endocrine systems. *Trends Endocrinol Metab* 2007;**18**:150–8.
6. Sharrocks AD. The ETS-domain transcription factor family. *Nat Rev Mol Cell Biol* 2001;**2**:827–37.
7. Tomlins SA, Laxman B, Dhanasekaran SM, et al. Distinct classes of chromosomal rearrangements create oncogenic ETS gene fusions in prostate cancer. *Nature* 2007;**448**:595–9.
8. Tomlins SA, Mehra R, Rhodes DR, et al. *TMPRSS2:ETV4* gene fusions define a third molecular subtype of prostate cancer. *Cancer Res* 2006;**66**:3396–400.
9. Paulo P, Barros-Silva JD, Ribeiro FR, et al. *FLI1* is a novel ETS transcription factor involved in gene fusions in prostate cancer. *Genes Chromosomes Cancer* 2012;**51**:240–9.
10. Helgeson BE, Tomlins SA, Shah N, et al. Characterization of *TMPRSS2:ETV5* and *SLC45A3:ETV5* gene fusions in prostate cancer. *Cancer Res* 2008;**68**:73–80.
11. Tomlins SA, Laxman B, Varambally S, et al. Role of the *TMPRSS2-ERG* gene fusion in prostate cancer. *Neoplasia* 2008;**10**:177–88.
12. Leshem O, Madar S, Kogan-Sakin I, et al. *TMPRSS2/ERG* promotes epithelial to mesenchymal transition through the ZEB1/ZEB2 axis in a prostate cancer model. *PLoS ONE* 2011;**6**:e21650.
13. King JC, Xu J, Wongvipat J, et al. Cooperativity of *TMPRSS2-ERG* with PI3-kinase pathway activation in prostate oncogenesis. *Nat Genet* 2009;**41**:524–6.
14. Carver BS, Tran J, Gopalan A, et al. Aberrant *ERG* expression cooperates with loss of *PTEN* to promote cancer progression in the prostate. *Nat Genet* 2009;**41**:619–24.
15. Chen Y, Chi P, Rockowitz S, et al. ETS factors reprogram the androgen receptor cistrome and prime prostate tumorigenesis in response to *PTEN* loss. *Nat Med* 2013;**19**:1023–9.
16. Collas P. The current state of chromatin immunoprecipitation. *Mol Biotechnol* 2010;**45**:87–100.
17. Jackson V. Studies on histone organization in the nucleosome using formaldehyde as a reversible cross-linking agent. *Cell* 1978;**15**:945–54.
18. Massie CE, Adryan B, Barbosa-Morais NL, et al. New androgen receptor genomic targets show an interaction with the *ETS1* transcription factor. *EMBO Rep* 2007;**8**:871–8.
19. Yu J, Yu J, Mani RS, et al. An integrated network of androgen receptor, polycomb, and *TMPRSS2-ERG* gene fusions in prostate cancer progression. *Cancer Cell* 2010;**17**:443–54.

20. Kumar-Sinha C, Tomlins SA, Chinnaiyan AM. Recurrent gene fusions in prostate cancer. *Nat Rev Cancer* 2008;**8**:497–511.
21. Moyer VA. USPST Force. Screening for prostate cancer: U.S. Preventive Services Task Force recommendation statement. *Ann Intern Med* 2012;**157**:120–34.
22. Tomlins SA, Aubin SM, Siddiqui J, et al. Urine *TMPRSS2:ERG* fusion transcript stratifies prostate cancer risk in men with elevated serum PSA. *Sci Transl Med* 2011;**3**:94ra72.
23. Epstein JI, Walsh PC, Carmichael M, et al. Pathologic and clinical findings to predict tumor extent of nonpalpable (stage T1c) prostate cancer. *JAMA* 1994;**271**:368–74.
24. Mosquera JM, Perner S, Demichelis F, et al. Morphological features of *TMPRSS2-ERG* gene fusion prostate cancer. *J Pathol* 2007;**212**:91–101.
25. Tu JJ, Rohan S, Kao J, et al. Gene fusions between TMPRSS2 and ETS family genes in prostate cancer: frequency and transcript variant analysis by RT-PCR and FISH on paraffin-embedded tissues. *Mod Pathol* 2007;**20**:921–8.
26. Gopalan A, Leversha MA, Satagopan JM, et al. *TMPRSS2-ERG* gene fusion is not associated with outcome in patients treated by prostatectomy. *Cancer Res* 2009;**69**:1400–6.
27. Eguchi FC, Faria EF, Neto CS, et al. The role of *TMPRSS2:ERG* in molecular stratification of PCa and its association with tumor aggressiveness: a study in Brazilian patients. *Sci Rep* 2014;**4**:5640.
28. Xu B, Chevarie-Davis M, Chevalier S, et al. The prognostic role of *ERG* immunopositivity in prostatic acinar adenocarcinoma: a study including 454 cases and review of the literature. *Hum Pathol* 2014;**45**:488–97.
29. Toubaji A, Albadine R, Meeker AK, et al. Increased gene copy number of ERG on chromosome 21 but not *TMPRSS2-ERG* fusion predicts outcome in prostatic adenocarcinomas. *Mod Pathol* 2011;**24**:1511–20.
30. Minner S, Enodien M, Sirma H, et al. *ERG* status is unrelated to PSA recurrence in radically operated prostate cancer in the absence of antihormonal therapy. *Clin Cancer Res* 2011;**17**:5878–88.
31. Hoogland AM, Jenster G, van Weerden WM, et al. *ERG* immunohistochemistry is not predictive for PSA recurrence, local recurrence or overall survival after radical prostatectomy for prostate cancer. *Mod Pathol* 2012;**25**:471–9.
32. Sabaliauskaite R, Jarmalaite S, Petroska D, et al. Combined analysis of *TMPRSS2-ERG* and *TERT* for improved prognosis of biochemical recurrence in prostate cancer. *Genes Chromosomes Cancer* 2012;**51**:781–91.
33. Pettersson A, Graff RE, Bauer SR, et al. The *TMPRSS2:ERG* rearrangement, *ERG* expression, and prostate cancer outcomes: a cohort study and meta-analysis. *Cancer Epidemiol Biomarkers Prev* 2012;**21**:1497–509.
34. Barwick BG, Abramovitz M, Kodani M, et al. Prostate cancer genes associated with *TMPRSS2-ERG* gene fusion and prognostic of biochemical recurrence in multiple cohorts. *Br J Cancer* 2010;**102**:570–6.
35. Attard G, Clark J, Ambroisine L, et al. Duplication of the fusion of TMPRSS2 to ERG sequences identifies fatal human prostate cancer. *Oncogene* 2008;**27**:253–63.
36. Demichelis F, Fall K, Perner S, et al. *TMPRSS2:ERG* gene fusion associated with lethal prostate cancer in a watchful waiting cohort. *Oncogene* 2007;**26**:4596–9.
37. Nam RK, Sugar L, Wang Z, et al. Expression of *TMPRSS2:ERG* gene fusion in prostate cancer cells is an important prognostic factor for cancer progression. *Cancer Biol Ther* 2007;**6**:40–5.
38. Park K, Tomlins SA, Mudaliar KM, et al. Antibody-based detection of *ERG* rearrangement-positive prostate cancer. *Neoplasia* 2010;**12**:590–8.
39. Johansson JE, Andren O, Andersson SO, et al. Natural history of early, localized prostate cancer. *JAMA* 2004;**291**:2713–9.
40. Yoshimoto M, Joshua AM, Cunha IW, et al. Absence of *TMPRSS2:ERG* fusions and *PTEN* losses in prostate cancer is associated with a favorable outcome. *Mod Pathol* 2008;**21**:1451–60.
41. Reid AH, Attard G, Ambroisine L, et al. Molecular characterisation of *ERG*, *ETV1* and *PTEN* gene loci identifies patients at low and high risk of death from prostate cancer. *Br J Cancer* 2010;**102**:678–84.
42. Leinonen KA, Saramaki OR, Furusato B, et al. Loss of *PTEN* is associated with aggressive behavior in *ERG*-positive prostate cancer. *Cancer Epidemiol Biomarkers Prev* 2013;**22**:2333–44.
43. Cerami E, Gao J, Dogrusoz U, et al. The cBio cancer genomics portal: an open platform for exploring multidimensional cancer genomics data. *Cancer Discov* 2012;**2**:401–4.
44. Bhatia-Gaur R, Donjacour AA, Sciavolino PJ, et al. Roles for Nkx3.1 in prostate development and cancer. *Genes Dev* 1999;**13**:966–77.
45. Emmert-Buck MR, Vocke CD, Pozzatti RO, et al. Allelic loss on chromosome 8p12-21 in microdissected prostatic intraepithelial neoplasia. *Cancer Res* 1995;**55**:2959–62.
46. Bergerheim US, Kunimi K, Collins VP, et al. Deletion mapping of chromosomes 8, 10, and 16 in human prostatic carcinoma. *Genes Chromosomes Cancer* 1991;**3**:215–20.
47. Haggman MJ, Wojno KJ, Pearsall CP, et al. Allelic loss of 8p sequences in prostatic intraepithelial neoplasia and carcinoma. *Urology* 1997;**50**:643–7.
48. Thangapazham R, Saenz F, Katta S, et al. Loss of the *NKX3.1* tumorsuppressor promotes the *TMPRSS2-ERG* fusion gene expression in prostate cancer. *BMC Cancer* 2014;**14**:16.
49. Kunderfranco P, Mello-Grand M, Cangemi R, et al. ETS transcription factors control transcription of EZH2 and epigenetic silencing of the tumor suppressor gene Nkx3.1 in prostate cancer. *PLoS ONE* 2010;**5**:e10547.
50. Kim MJ, Cardiff RD, Desai N, et al. Cooperativity of Nkx3.1 and Pten loss of function in a mouse model of prostate carcinogenesis. *Proc Natl Acad Sci USA* 2002;**99**:2884–9.
51. Abate-Shen C, Banach-Petrosky WA, Sun X, et al. Nkx3.1; Pten mutant mice develop invasive prostate adenocarcinoma and lymph node metastases. *Cancer Res* 2003;**63**:3886–90.
52. Gorre ME, Mohammed M, Ellwood K, et al. Clinical resistance to STI-571 cancer therapy caused by *BCR-ABL* gene mutation or amplification. *Science* 2001;**293**:876–80.
53. Weinstein IB, Joe A. Oncogene addiction. *Cancer Res* 2008;**68**:3077–80 discussion 80.
54. Zegarra-Moro OL, Schmidt LJ, Huang H, et al. Disruption of androgen receptor function inhibits proliferation of androgen-refractory prostate cancer cells. *Cancer Res* 2002;**62**:1008–13.
55. Zhao XY, Malloy PJ, Krishnan AV, et al. Glucocorticoids can promote androgen-independent growth of prostate cancer cells through a mutated androgen receptor. *Nat Med* 2000;**6**:703–6.
56. Taplin ME, Bubley GJ, Shuster TD, et al. Mutation of the androgen-receptor gene in metastatic androgen-independent prostate cancer. *N Engl J Med* 1995;**332**:1393–8.
57. Gaddipati JP, McLeod DG, Heidenberg HB, et al. Frequent detection of codon 877 mutation in the androgen receptor gene in advanced prostate cancers. *Cancer Res* 1994;**54**:2861–4.
58. Taplin ME, Bubley GJ, Ko YJ, et al. Selection for androgen receptor mutations in prostate cancers treated with androgen antagonist. *Cancer Res* 1999;**59**:2511–5.
59. Tran C, Ouk S, Clegg NJ, et al. Development of a second-generation antiandrogen for treatment of advanced prostate cancer. *Science* 2009;**324**:787–90.
60. Sun S, Sprenger CC, Vessella RL, et al. Castration resistance in human prostate cancer is conferred by a frequently occurring androgen receptor splice variant. *J Clin Invest* 2010;**120**:2715–30.
61. Chan SC, Li Y, Dehm SM. Androgen receptor splice variants activate androgen receptor target genes and support aberrant prostate cancer cell growth independent of canonical androgen receptor nuclear localization signal. *J Biol Chem* 2012;**287**:19736–49.
62. Li Y, Chan SC, Brand LJ, et al. Androgen receptor splice variants mediate enzalutamide resistance in castration-resistant prostate cancer cell lines. *Cancer Res* 2013;**73**:483–9.

63. Antonarakis ES, Lu C, Wang H, et al. AR-V7 and resistance to enzalutamide and abiraterone in prostate cancer. *N Engl J Med* 2014;**371**:1028–38.
64. Visakorpi T, Hyytinen E, Koivisto P, et al. *In vivo* amplification of the androgen receptor gene and progression of human prostate cancer. *Nat Genet* 1995;**9**:401–6.
65. Edwards J, Krishna NS, Grigor KM, et al. Androgen receptor gene amplification and protein expression in hormone refractory prostate cancer. *Br J Cancer* 2003;**89**:552–6.
66. Brown RS, Edwards J, Dogan A, et al. Amplification of the androgen receptor gene in bone metastases from hormone-refractory prostate cancer. *J Pathol* 2002;**198**:237–44.
67. Bubendorf L, Kononen J, Koivisto P, et al. Survey of gene amplifications during prostate cancer progression by high-throughout fluorescence *in situ* hybridization on tissue microarrays. *Cancer Res* 1999;**59**:803–6.
68. Palmberg C, Koivisto P, Hyytinen E, et al. Androgen receptor gene amplification in a recurrent prostate cancer after monotherapy with the nonsteroidal potent antiandrogen Casodex (bicalutamide) with a subsequent favorable response to maximal androgen blockade. *Eur Urol* 1997;**31**:216–9.
69. Koivisto P, Kononen J, Palmberg C, et al. Androgen receptor gene amplification: a possible molecular mechanism for androgen deprivation therapy failure in prostate cancer. *Cancer Res* 1997;**57**:314–9.
70. Merson S, Yang ZH, Brewer D, et al. Focal amplification of the androgen receptor gene in hormone-naive human prostate cancer. *Br J Cancer* 2014;**110**:1655–62.
71. Taylor BS, Schultz N, Hieronymus H, et al. Integrative genomic profiling of human prostate cancer. *Cancer Cell* 2010;**18**:11–22.
72. Kumar A, White TA, MacKenzie AP, et al. Exome sequencing identifies a spectrum of mutation frequencies in advanced and lethal prostate cancers. *Proc Natl Acad Sci USA* 2011;**108**:17087–92.
73. Barbieri CE, Baca SC, Lawrence MS, et al. Exome sequencing identifies recurrent *SPOP*, *FOXA1* and *MED12* mutations in prostate cancer. *Nat Genet* 2012;**44**:685–9.
74. Berger MF, Lawrence MS, Demichelis F, et al. The genomic complexity of primary human prostate cancer. *Nature* 2011;**470**:214–20.
75. Grasso CS, Wu YM, Robinson DR, et al. The mutational landscape of lethal castration-resistant prostate cancer. *Nature* 2012;**487**:239–43.
76. Baca SC, Prandi D, Lawrence MS, et al. Punctuated evolution of prostate cancer genomes. *Cell* 2013;**153**:666–77.
77. Brenner JC, Chinnaiyan AM. Disruptive events in the life of prostate cancer. *Cancer Cell* 2011;**19**:301–3.
78. Barbieri CE, Tomlins SA. The prostate cancer genome: perspectives and potential. *Urol Oncol* 2014;**32**:53.e15–22.
79. Perou CM, Sorlie T, Eisen MB, et al. Molecular portraits of human breast tumours. *Nature* 2000;**406**:747–52.
80. Stephens PJ, Greenman CD, Fu B, et al. Massive genomic rearrangement acquired in a single catastrophic event during cancer development. *Cell* 2011;**144**:27–40.
81. Rausch T, Jones DT, Zapatka M, et al. Genome sequencing of pediatric medulloblastoma links catastrophic DNA rearrangements with TP53 mutations. *Cell* 2012;**148**:59–71.
82. Liu W, Lindberg J, Sui G, et al. Identification of novel *CHD1*-associated collaborative alterations of genomic structure and functional assessment of *CHD1* in prostate cancer. *Oncogene* 2012;**31**:3939–48.
83. Blattner M, Lee DJ, O'Reilly C, et al. *SPOP* mutations in prostate cancer across demographically diverse patient cohorts. *Neoplasia* 2014;**16**:14–20.
84. Kim MS, Je EM, Oh JE, et al. Mutational and expressional analyses of *SPOP*, a candidate tumor suppressor gene, in prostate, gastric and colorectal cancers. *APMIS* 2013;**121**:626–33.
85. Zuhlke KA, Johnson AM, Tomlins SA, et al. Identification of a novel germline *SPOP* mutation in a family with hereditary prostate cancer. *Prostate* 2014;**74**:983–90.
86. Buckles E, Qian C, Tadros A, et al. Identification of speckle-type POZ protein somatic mutations in African American prostate cancer. *Asian J Androl* 2014;**16**(6):829–32.
87. Le Gallo M, O'Hara AJ, Rudd ML, et al. Exome sequencing of serous endometrial tumors identifies recurrent somatic mutations in chromatin-remodeling and ubiquitin ligase complex genes. *Nat Genet* 2012;**44**:1310–5.
88. Jia D, Dong R, Jing Y, et al. Exome sequencing of hepatoblastoma reveals novel mutations and cancer genes in the Wnt pathway and ubiquitin ligase complex. *Hepatology* 2014;**60**(5):1686–96.
89. Messick TE, Greenberg RA. The ubiquitin landscape at DNA double-strand breaks. *J Cell Biol* 2009;**187**:319–26.
90. An J, Wang C, Deng Y, et al. Destruction of full-length androgen receptor by wild-type *SPOP*, but not prostate-cancer-associated mutants. *Cell Rep* 2014;**6**:657–69.
91. Geng C, He B, Xu L, et al. Prostate cancer-associated mutations in speckle-type POZ protein (SPOP) regulate steroid receptor coactivator 3 protein turnover. *Proc Natl Acad Sci USA* 2013;**110**:6997–7002.
92. Zhou XE, Suino-Powell KM, Li J, et al. Identification of SRC3/AIB1 as a preferred coactivator for hormone-activated androgen receptor. *J Biol Chem* 2010;**285**:9161–71.
93. Agoulnik IU, Vaid A, Nakka M, et al. Androgens modulate expression of transcription intermediary factor 2, an androgen receptor coactivator whose expression level correlates with early biochemical recurrence in prostate cancer. *Cancer Res* 2006;**66**:10594–602.
94. Lapointe J, Li C, Higgins JP, et al. Gene expression profiling identifies clinically relevant subtypes of prostate cancer. *Proc Natl Acad Sci USA* 2004;**101**:811–6.
95. Lapointe J, Li C, Giacomini CP, et al. Genomic profiling reveals alternative genetic pathways of prostate tumorigenesis. *Cancer Res* 2007;**67**:8504–10.
96. Demichelis F, Setlur SR, Beroukhim R, et al. Distinct genomic aberrations associated with *ERG* rearranged prostate cancer. *Genes Chromosomes Cancer* 2009;**48**:366–80.
97. Haffner MC, Aryee MJ, Toubaji A, et al. Androgen-induced TOP2B-mediated double-strand breaks and prostate cancer gene rearrangements. *Nat Genet* 2010;**42**:668–75.
98. Burkhardt L, Fuchs S, Krohn A, et al. *CHD1* is a 5q21 tumor suppressor required for *ERG* rearrangement in prostate cancer. *Cancer Res* 2013;**73**:2795–805.
99. Druker BJ, Sawyers CL, Kantarjian H, et al. Activity of a specific inhibitor of the BCR-ABL tyrosine kinase in the blast crisis of chronic myeloid leukemia and acute lymphoblastic leukemia with the Philadelphia chromosome. *N Engl J Med* 2001;**344**:1038–42.
100. Lynch TJ, Bell DW, Sordella R, et al. Activating mutations in the epidermal growth factor receptor underlying responsiveness of non-small-cell lung cancer to gefitinib. *N Engl J Med* 2004;**350**:2129–39.

CHAPTER

# 5

# Should Gleason Score 6 Still Be Called Cancer?

*Robert M. Turner II, MD, Benjamin T. Ristau, MD, Joel B. Nelson, MD*

**Department of Urology, University of Pittsburgh Medical Center, Pittsburgh, PA, USA**

## INTRODUCTION

While the lifetime risk of a prostate cancer diagnosis for a man living the United States is estimated to be 1 in 7, the incidence to mortality ratio is nearly 8:1.[1] Thus, most men diagnosed with prostate cancer will not die of the disease. The era of prostate-specific antigen (PSA) screening has resulted in the diagnosis of lower grade and volume cancers. While PSA screening has reduced the incidence of advanced disease and mortality, the trade-offs have included over-diagnosis and overtreatment. These issues have led to an effort to identify and surveil men with "clinically insignificant" disease. Some have argued that Gleason 6 prostate cancer is not in and of itself a lethal cancer and have advocated for reclassification of this pattern as a nonmalignant precursor lesion.[2,3] This chapter outlines differences in the molecular and clinical characteristics of Gleason 6 prostate cancer compared with higher grade disease as well as the issues of diagnostic uncertainty that have prevented acceptance of a shift in nomenclature.

## MOLECULAR CHARACTERISTIC OF GLEASON PATTERN 3 VERSUS 4

In a framework provided by Hanahan and Weinberg,[4] neoplastic cells acquire six biological capabilities during the multistep progression to a neoplastic state: sustaining proliferative signaling, evading growth suppressors, resisting cell death, enabling replicative immortality, inducing angiogenesis, and activating invasion and metastasis. In an elegant review, Ahmed et al.[3] present a convincing argument that pure Gleason pattern 3 prostate cancers lack the molecular characteristics of each of these hallmarks (Figure 5.1).

### Self-Sufficiency in Growth Signals

The EGFR pathway has been proposed to play a prominent role in prostate cancer progression to castration-resistant disease. Ross et al.[5] demonstrated differential in expression *EGF* and *EGFR* in cancer cells isolated by laser capture microdissection from low-grade (Gleason 3 + 3 = 6) or high-grade (Gleason 4 + 4 = 8) tumors of men with localized disease. Both *EGF* and *EGFR* regulate cell growth, differentiation, motility, and adhesion by signaling through several signal transduction mechanisms.[6] Also, amplification of *HER2/neu*, a member of the *EGFR* family, is associated with high tumor volume of greater than 2.0 $cm^3$. There is greater degree of *HER2/neu* amplification in Gleason pattern 4 lesions when compared to Gleason pattern 3 lesions.[7]

### Insensitivity to Antigrowth Signals

The D-type cyclins (cyclins D1, D2, and D3) are involved in the regulation of transition from $G_1$ to S during the cell cycle. Inactivation of cyclin D2 by promoter hypermethylation has been reported in breast and prostate cancer, supporting the tumor suppressor role in these tumor types. Tumors with a Gleason sum ≥7 have been shown to have significantly greater methylation frequency than tumors with a Gleason sum ≤6.[8] Guo et al.[9] showed progressive loss of cell cycle inhibitor $p27^{Kip1}$ with increasing Gleason grade by immunohistochemistry of 101 radical prostatectomy specimens.

Prostate Cancer. http://dx.doi.org/10.1016/B978-0-12-800077-9.00005-0

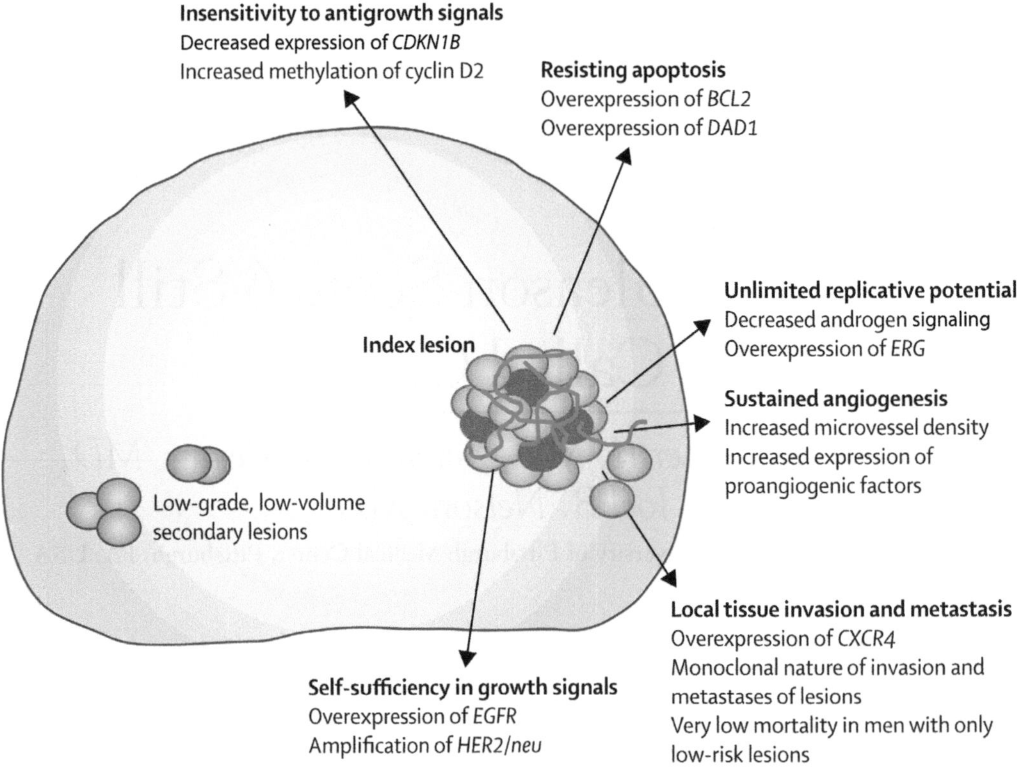


**FIGURE 5.1** Hallmarks of cancer as applied to the index and secondary lesions in prostate cancer. *Reproduced with permission from Ahmed et al.*[3]

## Resisting Cell Death

DAD1, a downstream target of the NFκB survival pathway, exhibits an antiapoptotic function and has been associated with perineural invasion in *in vitro* studies.[10] True et al.[11] evaluated the expression of DAD1 protein by immunohistochemistry and RT-PCR in prostatectomy cores and demonstrated higher expression in Gleason pattern 4 and 5 cancers than low-grade cancers of pattern 3. *BCL2*, another antiapoptotic gene, has likewise been demonstrated to be upregulated in Gleason pattern 4 and 5 lesions compared with those that were only pattern 3.[12]

## Unlimited Replicative Potential

Fusion of the PSA-regulated *TMPRSS2* gene to ETS family oncogenic transcription factors, commonly ERG, has been identified as a common molecular alteration in prostate cancer.[13] *TMPRSS2-ERG* fusion, for example, has been identified in approximately 40–60% of tumors in surgical series.[14] While the association of *ERG* gene rearrangement and aggressive prostate cancer phenotype remains controversial, overexpression of ERG has been associated with Gleason score, tumor volume, and cancer-specific mortality. In a cross-sectional study in which individual cancers from radical prostatectomy specimens were analyzed, there was higher ERG protein expression Gleason 7 compared with Gleason 6 tumors.[14]

## Sustained Angiogenesis

Vascular endothelial growth factor (VEGF) expression has been associated with PSA, grade, and poor prognosis in prostate cancer.[15] Similarly, endocrine-gland-derived vascular endothelial growth factor/prokineticins (EG-VEGF/PK1 and PK2) expression is greater in high-grade lesions compared with Gleason pattern 3 lesions.[16] The effects of neoangiogenesis have been associated with poor prognosis. Mucci et al.[17] demonstrated that patients with higher-grade tumors exhibited greater microvessel density, greater irregularity of the vessel lumen, and small vessels. Men with tumors exhibiting the smallest vessel diameter and those with the most irregularly shaped vessels were significantly more likely to develop lethal disease.

## Tissue Invasion and Metastasis

The molecular characteristics that permit cell migration, local invasion, and metastatic potential seem to be comparatively overexpressed in higher-grade lesions. The chemokine receptor CXCR4, which permits directional migration of cells down a concentration gradient of its ligand CSCL12, is upregulated in Gleason pattern 4 lesions compared to Gleason pattern 3 lesions. CXCR4 upregulation has been associated with lymph node and bone metastasis.[3] Although by definition, Gleason 3 + 3 = 6 prostate cancer has invaded the surrounding benign prostatic tissue, it is rarely associated with extraprostatic extension.

The sum of these biochemical findings support a growing consensus that pure Gleason pattern 6 disease is not as aggressive at a molecular level when compared with higher-grade disease. While these data may be interesting, the indolent clinical characteristics of Gleason 6 are more compelling.

# NATURAL HISTORY OF TREATED GLEASON 6 CANCER

Studies of pathologically confirmed Gleason 6 cancer show that the rate of metastasis and cancer-specific death is essentially zero. In the study by Hernandez and coworkers,[18] in 6081 men with pathologically organ-confined, Gleason score 6 or less disease after radical prostatectomy, the 15-year actuarial probability of biochemical recurrence was 1.3%. A total of five men (0.2%) had a local recurrence, four of whom received salvage radiotherapy and subsequently had an undetectable PSA level. No patients developed distant metastases or died of prostate cancer.

In a similar study of 14,124 patients with pathologically confirmed Gleason 6 prostate cancer following radical prostatectomy, 22 patients were initially reported to have lymph node metastases. All 22 of these patients were upgraded on subsequent review; thus, there were no cases of confirmed Gleason 6 nodal metastases in this large series.[19]

Eggner et al.[20] reported a 20-year prostate-cancer-specific mortality of 0.2% after radical prostatectomy in 1200 men with Gleason 6 disease between the ages of 60 and 69 years. The authors speculated that they would likely have observed even fewer cancer specific deaths had all surgical specimens been subjected to a contemporary pathologic review.

In the author's experience, of 434 consecutive radical prostatectomies from 2005 until the present (median follow-up 54 months) with Gleason 3 + 3 = 6 on final analysis, there have been only three patients who developed a biochemical recurrence, for a rate of 0.7%.

While the treatment effect of radical prostatectomy cannot be ignored in these series, the data certainly support the benign behavior of contemporarily diagnosed Gleason 6 disease. There is certainly very little evidence of metastatic disease upon presentation or local recurrence that subsequently progressed to metastases. These series question the need for treatment of patients with Gleason 6 disease and evoke the frequently quoted statement of Dr Willet Whitmore, Jr., "Is cure possible? Is cure necessary? Is cure possible only when it is not necessary?"[21] Unfortunately, it is the biopsy Gleason sum, not the radical prostatectomy Gleason score, available at the time patients and clinicians make treatment-related decisions regarding a newly diagnosed prostate cancer.

# NATURAL HISTORY OF UNTREATED GLEASON 6 CANCER

## Historical Cohort

Our understanding of the natural history of untreated Gleason 6 prostate cancer is largely based on data from the pre-PSA era. To determine the natural history of clinically localized prostate cancer, Albertsen et al.[22] performed a retrospective population-based cohort study of 767 men diagnosed between 1971 and 1984. Most recently, they reported the 20-year outcome of patients treated with observation or androgen deprivation alone. Men with Gleason scores of 6 and 7 had mortality rates of 30 and 65 deaths per 1000 person-years, respectively. When stratified by age, 20–30% of Gleason 6 patients died of prostate cancer.

The Albertsen cohort does not represent a modern cohort of men with clinically localized prostate cancer in several respects. PSA screening has introduced lead-time bias; screen-detected men are diagnosed approximately 10 years earlier than clinically detected men.[23] Also, the classical Gleason scoring system used in the Albertsen cohort was modified in 2005 by the International Society of Urological Pathology (ISUP).[24] This has resulted in an accepted upward grading shift, in which many patients formerly classified as Gleason 6 have been reclassified at Gleason 7. One example is that of cribiform pattern, which was previously permissible in Gleason pattern 3, is now classified as Gleason pattern 4. Also, for needle biopsy tissue where there are more than two patterns present, and the worst grade is neither the predominant nor the secondary grade, the predominant and highest grade are now selected to arrive at a score. It has been estimated that one-third of men with Gleason 6 disease in the Albertsen study would be upgraded according to current grading criteria.[25,26] Thus, it was concluded that if one-third of the Gleason 6 tumors in the Albertsen series behaved like Gleason 7, with a 45% prostate cancer mortality at 20 years, this would account for 17% of the deaths. If another one-third were graded correctly but actually harbored higher-grade cancer, that would account for virtually all of the prostate cancer deaths in that group.[26]

Despite the historical biases in evaluating the natural history of Gleason 6 prostate cancer in the pre-PSA era, it is clear that the rate of metastasis and mortality in this "low-risk" group is low.

## Modern Cohort

Contemporary active surveillance cohorts provide a means to understand the natural history of Gleason 6 prostate cancer as diagnosed by prostate needle biopsy in the modern era of PSA screening. These series have lower prostate-cancer-specific mortality when compared with an observational historical cohort;[27] however, the occurrence of prostate-cancer-specific death in these patients highlights the fact that these "low-risk" patients are a heterogeneous group that are at risk for undergrading and progression.

In the University of Toronto experience,[28] 450 men were observed on active surveillance with a mean follow up of 6.8 years. While the majority of men included in this series had Gleason 3 + 3 = 6 disease, 18% had Gleason 3 + 4 = 7 disease. During surveillance, 30% of men were reclassified as higher risk and were offered definitive therapy. Of 117 patients treated radically, the PSA failure rate was 50%. The 10-year prostate cancer actuarial survival was 97.2%. It is certainly concerning that four of five patients who died of prostate cancer within 4–10 years of diagnosis had pure Gleason 6 on initial prostate biopsy.

In the Johns Hopkins experience,[29,30] 769 men with very-low-risk prostate cancer (all had biopsy Gleason score 6 or less) were followed prospectively. The median survival free of intervention was 6.5 years after diagnosis. The proportions of men remaining free of intervention after 2, 3, and 10 years of follow-up were 81, 59, and 41%, respectively. Overall, 33% of men underwent intervention at a mean of 2.2 years after diagnosis, and the rate of PSA failure following delayed radical therapy was 9.3%. While there were no prostate cancer deaths, this series had a limited median follow-up of 2.7 years.

In a multi-institutional retrospective study of 262 men on active surveillance,[31] the 2- and 5-year probabilities of remaining on active surveillance were 91 and 75%, respectively. All men had a biopsy Gleason sum of 6 or less. Of 43 patients who underwent delayed treatment, 4 (5%) had PSA failure at a median of 23 months follow-up. Skeletal metastases developed in one patient 38 months after starting active surveillance.

While these series demonstrate that the majority of patients with biopsy Gleason 6 disease can be safely followed with active surveillance, they also highlight the risks of progression and cancer-specific death in patients with biopsy Gleason 6 disease. While treatment effect certainly cannot be overlooked, the differences in outcomes in patients on active surveillance for biopsy Gleason 6 disease compared to those with Gleason 6 disease on the radical prostatectomy specimen may in part be explained by undersampling inherent to biopsy techniques.

## UNDERGRADING

The conclusion that pure Gleason pattern 6 lacks metastatic potential implies that treatment is unnecessary if one is confident that a patient has only Gleason 6 cancer. Unfortunately, current diagnosis via transrectal ultrasound-guided prostate needle biopsy is imperfect. A substantial number of men diagnosed with Gleason 6 prostate cancer at biopsy and undergo prostatectomy are upgraded on their final pathology specimen. Reported rates of upgrading range from 25% to 50% (Table 5.1). In the author's experience, 771/3021 (25.5%) patients were upgraded from their biopsy Gleason score to final pathology. Reported significant predictors of upgrading include older age, higher preoperative PSA, higher clinical stage, smaller prostate size, and higher percentage of tumor in prostate biopsy cores. Almost universally, patients who are upgraded at the time of radical prostatectomy have adverse characteristics relative to nonupgraded patients including higher rates of extraprostatic extension, seminal vesicle invasion, positive surgical margins, lymphovascular invasion, and biochemical recurrence.[32,33]

**TABLE 5.1** Select Series Demonstrating Gleason Score Upgrading from Biopsy to Radical Prostatectomy

| Author(s) | Patients (*n*) | Upgraded tumors (%) | Predictors for upgrading |
|---|---|---|---|
| Chun et al.[34] | 4789 | 28.2 | Higher clinical stage, higher PSA level |
| Pinthus et al.[35] | 875 | 34.1 | Greater percentage cancer in biopsy, higher PSA level |
| Freedland et al.[36] | 1113 | 27 | Higher PSA level, greater percentage cores with cancer, obesity |
| Dong et al.[37] | 268 | 50 | Higher PSA, higher cancer volume in prostate, larger prostate weight |
| Tilki et al.[32] | 684 | 29.7 | Higher PSA, smaller prostate volume, greater PSA density |
| Davies et al.[38] | 1251 | 31 | Smaller prostate size |
| Gershman et al.[39] | 1836 | 29.6 | Higher PSA, older age, smaller prostate weight |
| Epstein et al.[40] | 7643 | 36.3 | Older age, higher PSA, smaller prostate weight, higher percentage of cancer per core |
| Nelson (2014, unpublished raw data) | 3021 | 25.5 | |

Some institutions have attempted to mitigate this issue by utilizing repeat biopsy in patients with low-risk prostate cancer eligible for active surveillance. Choo and coworkers showed that Gleason score was unchanged in 31%, upgraded in 35%, and downgraded in 32%.[41] Similarly, Berglund et al.[42] demonstrated that 27% of patients were upgraded following repeat biopsy (and therefore deemed ineligible for active surveillance). Unfortunately "restaging" prostate needle biopsy fails to entirely eliminate the issue of undersampling. Of patients who were not upgraded or upstaged and went on to radical prostatectomy, 42% were upgraded to Gleason 7 on final pathology.[42]

In the ideal world, we would have a diagnostic method that would tell us the Gleason grade of a given patient's prostate cancer with 100% accuracy without the necessity of removing the gland. Alternate imaging modalities (e.g., MRI, MRI-TRUS fusion) promise improvements in our ability to accurately diagnose prostate cancer grade and minimize sampling bias. However, we will only be able to consider not actively following this low risk patient population when we are certain that a biopsy demonstrating pure Gleason 6 is a true representation of what one would find on a radical prostatectomy specimen.

## NOMENCLATURE

Cancer is a term that engenders considerable fear, apprehension, and anxiety in patients. However, there is some evidence that such emotions can be tempered with the proper counseling. Several studies have looked at anxiety in the management of low-risk prostate cancer by active surveillance.[43–45] In general, most patients on active surveillance report low levels of anxiety and high health-related quality of life.[46,47] In fact, only about 5–10% of patients on active surveillance choose to have treatment without an absolute indication to do so.[28] Even though these numbers are relatively small, there is clearly a negative emotional response to being told that one has cancer.

Given that Gleason pattern 3 + 3 = 6 disease does not express the hallmarks of cancer, there is reason to consider renaming the disease, and there have been several proposed benefits changing the nomenclature of indolent lesions. The first is decreased patient anxiety and an overall reduction in public confusion about the significance of a cancer diagnosis.[2] Some have also argued that changing the nomenclature might result in a reduction of overtreatment and its inherent side effects.[48] Additionally, recognizing pure Gleason pattern 3 as a nonmalignant entity would promote the search for tests to more accurately identify clinically significant disease. Finally, better characterization of aggressive versus clinically insignificant lesions could enhance acceptance of minimally invasive/focal therapies for borderline disease.[49]

Renaming this entity would not be a novel concept. In the bladder cancer spectrum, the term papillary urothelial neoplasm of low malignant potential (PUNLMP) is used to describe bladder lesions falling between a benign papilloma and a low-grade papillary urothelial carcinoma. The clinical implication of a PUNLMP diagnosis is that these lesions need to be followed, but they are not associated with invasion or metastasis.[50]

In 2009, Esserman et al.[51] proposed the term *indolent lesions of epithelial origin* (IDLE) to describe minimal risk lesions for which the reduction or elimination of therapeutic intervention while maintaining good outcome could be achieved. In prostate cancer, this would apply to low-volume lesions with low Gleason scores – precisely those that are included in most active surveillance protocols. This concept gained further traction with the 2011 NIH state-of-the-science conference statement on active surveillance in prostate cancer, which concluded the following: "because of the very favorable prognosis of low-risk prostate cancer, strong consideration should be given to modifying the anxiety-provoking term 'cancer' for this condition."[52] Despite these suggestions, no specific term has garnered widespread acceptance. Much of this lies in the inability to diagnose this state of clinically insignificant disease with certainty. As we gain more sophisticated techniques to better characterize indolent lesions, confidence in the new nomenclature will grow.

## CONCLUSIONS

Pure Gleason 3 + 3 = 6 prostate cancer lacks the established hallmarks of malignancy at the molecular level. In patients undergoing radical prostatectomy with Gleason 6 disease on final pathology, there is little evidence of metastatic disease upon presentation or local recurrence that subsequently progressed to metastases. Contemporary active surveillance series demonstrate an indolent course for patients with Gleason 6 disease, but also highlight the undersampling and undergrading associated with current biopsy techniques. While a change in terminology may decrease the anxiety, overdiagnosis, and overtreatment of clinically insignificant disease, these diagnostics shortcomings preclude a shift in nomenclature at this time.

## References


1. Siegel R, Ma J, Zou Z, et al. Cancer statistics, 2014. *CA Cancer J Clin* 2014;**64**(1):9–29.
2. Nickel JC, Speakman M. Should we really consider Gleason 6 prostate cancer? *BJU Int* 2012;**109**(5):645–6.
3. Ahmed HU, Arya M, Freeman A, et al. Do low-grade and low-volume prostate cancers bear the hallmarks of malignancy? *Lancet Oncol* 2012;**13**(11):e509–17.

4. Hanahan D, Weinberg RA. The hallmarks of cancer. *Cell* 2000;**100**(1):57–70.
5. Ross AE, Marchionni L, Vuica-Ross M, et al. Gene expression pathways of high grade localized prostate cancer. *Prostate* 2011;**71**(14):1568–77.
6. Gan Y, Shi C, Inge L, et al. Differential roles of ERK and Akt pathways in regulation of EGFR-mediated signaling and motility in prostate cancer cells. *Oncogene* 2010;**29**(35):4947–58.
7. Skacel M, Ormsby AH, Pettay JD, et al. Aneusomy of chromosomes 7, 8, and 17 and amplification of *HER-2/neu* and epidermal growth factor receptor in Gleason score 7 prostate carcinoma: a differential fluorescent in situ hybridization study of Gleason pattern 3 and 4 using tissue microarray. *Hum Pathol* 2001;**32**(12):1392–7.
8. Padar A, Sathyanarayana UG, Suzuki M, et al. Inactivation of cyclin D2 gene in prostate cancers by aberrant promoter methylation. *Clin Cancer Res* 2003;**9**(13):4730–4.
9. Guo Y, Sklar GN, Borkowski A, et al. Loss of the cyclin-dependent kinase inhibitor p27(Kip1) protein in human prostate cancer correlates with tumor grade. *Clin Cancer Res* 1997;**3**(12 Pt 1):2269–74.
10. Ayala GE, Dai H, Ittmann M, et al. Growth and survival mechanisms associated with perineural invasion in prostate cancer. *Cancer Res* 2004;**64**(17):6082–90.
11. True L, Coleman I, Hawley S, et al. A molecular correlate to the Gleason grading system for prostate adenocarcinoma. *Proc Natl Acad Sci USA* 2006;**103**(29):10991–6.
12. Fleischmann A, Huland H, Mirlacher M, et al. Prognostic relevance of Bcl-2 overexpression in surgically treated prostate cancer is not caused by increased copy number or translocation of the gene. *Prostate* 2012;**72**(9):991–7.
13. Tomlins SA, Rhodes DR, Perner S, et al. Recurrent fusion of TMPRSS2 and ETS transcription factor genes in prostate cancer. *Science* 2005;**310**(5748):644–8.
14. Bismar TA, Dolph M, Teng LH, et al. ERG protein expression reflects hormonal treatment response and is associated with Gleason score and prostate cancer specific mortality. *Eur J Cancer* 2012;**48**(4):538–46.
15. West AF, O'Donnell M, Charlton RG, et al. Correlation of vascular endothelial growth factor expression with fibroblast growth factor-8 expression and clinico-pathologic parameters in human prostate cancer. *Br J Cancer* 2001;**85**(4):576–83.
16. Pasquali D, Rossi V, Staibano S, et al. The endocrine-gland-derived vascular endothelial growth factor (EG-VEGF)/prokineticin 1 and 2 and receptor expression in human prostate: up-regulation of EG-VEGF/prokineticin 1 with malignancy. *Endocrinology* 2006;**147**(9):4245–51.
17. Mucci LA, Powolny A, Giovannucci E, et al. Prospective study of prostate tumor angiogenesis and cancer-specific mortality in the health professionals follow-up study. *J Clin Oncol* 2009;**27**(33):5627–33.
18. Hernandez DJ, Nielsen ME, Han M, et al. Natural history of pathologically organ-confined (pT2), Gleason score 6 or less, prostate cancer after radical prostatectomy. *Urology* 2008;**72**(1):172–6.
19. Ross HM, Kryvenko ON, Cowan JE, et al. Do adenocarcinomas of the prostate with Gleason score (GS) </=6 have the potential to metastasize to lymph nodes? *Am J Surg Pathol* 2012;**36**(9):1346–52.
20. Eggener SE, Scardino PT, Walsh PC, et al. Predicting 15-year prostate cancer specific mortality after radical prostatectomy. *J Urol* 2011;**185**(3):869–75.
21. Montie JE, Smith JA. Whitmoreisms: memorable quotes from Willet F. Whitmore, Jr, M.D. *Urology* 2004;**63**(1):207–9.
22. Albertsen PC, Hanley JA, Fine J. 20-Year outcomes following conservative management of clinically localized prostate cancer. *JAMA* 2005;**293**(17):2095–101.
23. Draisma G, Boer R, Otto SJ, et al. Lead times and overdetection due to prostate-specific antigen screening: estimates from the European Randomized Study of Screening for Prostate Cancer. *J Natl Cancer Inst* 2003;**95**(12):868–78.
24. Epstein JI, Allsbrook Jr WC, Amin MB, et al. The 2005 International Society of Urological Pathology (ISUP) Consensus Conference on Gleason Grading of Prostatic Carcinoma. *Am J Surg Pathol* 2005;**29**(9):1228–42.
25. Albertsen PC, Hanley JA, Barrows GH, et al. Prostate cancer and the Will Rogers phenomenon. *J Natl Cancer Inst* 2005;**97**(17):1248–53.
26. Klotz L. Active surveillance: patient selection. *Curr Opin Urol* 2013;**23**(3):239–44.
27. Klotz L, Thompson I. Early prostate cancer – treat or watch? *New Engl J Med* 2011;**365**(6):569.
28. Klotz L, Zhang L, Lam A, et al. Clinical results of long-term follow-up of a large, active surveillance cohort with localized prostate cancer. *J Clin Oncol* 2010;**28**(1):126–31.
29. Duffield AS, Lee TK, Miyamoto H, et al. Radical prostatectomy findings in patients in whom active surveillance of prostate cancer fails. *J Urol* 2009;**182**(5):2274–8.
30. Tosoian JJ, Trock BJ, Landis P, et al. Active surveillance program for prostate cancer: an update of the Johns Hopkins experience. *J Clin Oncol* 2011;**29**(16):2185–90.
31. Eggener SE, Mueller A, Berglund RK, et al. A multi-institutional evaluation of active surveillance for low risk prostate cancer. *J Urol* 2013;**189**(1 Suppl):S19–25.
32. Tilki D, Schlenker B, John M, et al. Clinical and pathologic predictors of Gleason sum upgrading in patients after radical prostatectomy: results from a single institution series. *Urol Oncol* 2011;**29**(5):508–14.
33. Corcoran NM, Hong MK, Casey RG, et al. Upgrade in Gleason score between prostate biopsies and pathology following radical prostatectomy significantly impacts upon the risk of biochemical recurrence. *BJU Int* 2011;**108**(8 Pt 2):E202–10.
34. Chun FK, Briganti A, Shariat SF, Graefen M, Montorsi F, Erbersdobler A, et al. Significant upgrading affects a third of men diagnosed with prostate cancer: predictive nomogram and internal validation. *BJU Int* 2006;**98**(2):329–34.
35. Pinthus JH, Witkos M, Fleshner NE, Sweet J, Evans A, Jewett MA, et al. Prostate cancers scored as Gleason 6 on prostate biopsy are frequently Gleason 7 tumors at radical prostatectomy: implication on outcome. *J Urol* 2006;**176**(3):979–84.
36. Freedland SJ, Kane CJ, Amling CL, et al. Upgrading and downgrading of prostate needle biopsy specimens: risk factors and clinical implications. *Urology* 2007;**69**(3):495–9.
37. Dong F, Jones JS, Stephenson AJ, Magi-Galluzzi C, Reuther AM, Klein EA. Prostate cancer volume at biopsy predicts clinically significant upgrading. *J Urol* 2008;**179**(3):896–900.
38. Davies JD, Aghazadeh MA, Phillips S, et al. Prostate size as a predictor of Gleason score upgrading in patients with low risk prostate cancer. *J Urol* 2011;**186**(6):2221–7.
39. Gershman B, Dahl DM, Olumi AF, et al. Smaller prostate gland size and older age predict Gleason score upgrading. *Urol Oncol* 2013;**31**(7):1033–7.
40. Epstein JI, Feng Z, Trock BJ, et al. Upgrading and downgrading of prostate cancer from biopsy to radical prostatectomy: incidence and predictive factors using the modified Gleason grading system and factoring in tertiary grades. *Eur Urol* 2012;**61**(5):1019–24.
41. Choo R, Danjoux C, Morton G, et al. How much does Gleason grade of follow-up biopsy differ from that of initial biopsy in untreated, Gleason score 4-7, clinically localized prostate cancer? *Prostate* 2007;**67**(15):1614–20.
42. Berglund RK, Masterson TA, Vora KC, et al. Pathological upgrading and up staging with immediate repeat biopsy in patients eligible for active surveillance. *J Urol* 2008;**180**(5):1964–7 discussion 7–8.

43. Venderbos LD, van den Bergh RC, Roobol MJ, et al. A longitudinal study on the impact of active surveillance for prostate cancer on anxiety and distress levels. *Psychooncology* 2014;**24**(3):348–54.
44. Liu D, Lehmann HP, Frick KD, et al. Active surveillance versus surgery for low risk prostate cancer: a clinical decision analysis. *J Urol* 2012;**187**(4):1241–6.
45. Hayes JH, Ollendorf DA, Pearson SD, et al. Active surveillance compared with initial treatment for men with low-risk prostate cancer: a decision analysis. *JAMA* 2010;**304**(21):2373–80.
46. Anderson J, Burney S, Brooker JE, et al. Anxiety in the management of localised prostate cancer by active surveillance. *BJU Int* 2014;**114**(Suppl 1):55–61.
47. Wilcox CB, Gilbourd D, Louie-Johnsun M. Anxiety and health-related quality of life (HRQL) in patients undergoing active surveillance of prostate cancer in an Australian centre. *BJU Int* 2014;**113**(Suppl. 2):64–8.
48. Klotz L. Prostate cancer overdiagnosis and overtreatment. *Curr Opin Endocrinol Diab Obes* 2013;**20**(3):204–9.
49. Valerio M, Ahmed HU, Emberton M, et al. The role of focal therapy in the management of localised prostate cancer: a systematic review. *Eur Urol* 2013;**66**(4):732–51.
50. Epstein JI, Amin MB, Reuter VR, et al. The World Health Organization/International Society of Urological Pathology consensus classification of urothelial (transitional cell) neoplasms of the urinary bladder. Bladder Consensus Conference Committee. *Am J Surg Pathol* 1998;**22**(12):1435–48.
51. Esserman L, Shieh Y, Thompson I. Rethinking screening for breast cancer and prostate cancer. *JAMA* 2009;**302**(15):1685–92.
52. Ganz PA, Barry JM, Burke W, et al. National Institutes of Health State-of-the-Science Conference: role of active surveillance in the management of men with localized prostate cancer. *Ann Inter Med* 2012;**156**(8):591–5.

CHAPTER

# 6

# High Grade Prostatic Intraepithelial Neoplasia and Atypical Glands

*Timothy Ito, MD*, Essel Dulaimi, MD**, Marc C. Smaldone, MD**

*Department of Surgical Oncology, Fox Chase Cancer Center, Philadelphia, PA, USA

**Department of Pathology, Fox Chase Cancer Center, Philadelphia, PA, USA

## INTRODUCTION

High-grade prostatic intraepithelial neoplasia (HGPIN) and atypical small acinar proliferation (ASAP) are intermediaries to distinctly benign or malignant diagnoses on prostate biopsy. HGPIN is a true pathologic entity and the only accepted precursor to prostate adenocarcinoma. ASAP on the other hand is not a defined pathologic entity, but instead represents diagnostic ambiguity with atypical findings suspicious but not sufficient for a diagnosis of malignancy. Given this distinction in pathologic classification, the management for ASAP has remained consistent over time while the management of HGPIN has changed over the past decade and a half due to evolving practice patterns with regards to PSA (prostate-specific antigen) screening and biopsy strategies, each of which influence the detection rates of HGPIN and prostate cancer. The following chapter reviews these two entities – their pathologic features, their prediction of subsequent cancer, and their current recommended management.

## HIGH-GRADE PROSTATIC INTRAEPITHELIAL NEOPLASIA

HGPIN is currently the only accepted precursor to prostate adenocarcinoma. It was initially described by John McNeal in 1969 by the term "intraductal dysplasia of the prostate," but did not become widely accepted until its morphologic criteria were more precisely characterized by McNeal and David Bostwick in 1986.[1–3] Initial PIN classification occurred on a scale of I–III; however, this was simplified to differentiation into low (I) and high (II, III) grades only. Low-grade PIN (LGPIN) is no longer reported by pathologists because of a lack of interobserver reproducibility due to its similarity to benign prostate, as well as because of its lack of clinical significance.[4–6] In contrast, moderate to good interobserver agreement for the diagnosis of HGPIN has been demonstrated, particularly among genitourinary pathologists.[4,7]

Historically, isolated HGPIN on sextant prostate biopsy was associated with high rates of concurrent prostate cancer and thus the recommendation for patients with this finding was for immediate repeat biopsy.[8] Controversy regarding this guideline has evolved over the past two decades as initial biopsy strategies have increased sampling of the prostate through extended core biopsy schemes. The resulting identification of smaller foci of cancer on initial biopsy likely has contributed to decreased rates of cancer detection on repeat biopsy after HGPIN.[4]

### Incidence

Autopsy and cystoprostatectomy series best estimate true HGPIN incidence, and in these studies the incidence ranges from 45% to 85%.[2,9–13] Incidence varies based on age and ethnicity. PIN incidence increases with age. Sakr et al. reported a 9 and 20% incidence of LGPIN in the third and fourth decades of life, with HGPIN first found in the fifth decade.[14] PIN in this study preceded the development of cancer by approximately 10 years. The frequency of HGPIN diagnosis by decade has been estimated to be 15, 24, 47, 58, and 70% in the fifth through ninth decades, respectively.[5] When considering race, African-American men have been shown to have an increased frequency of HGPIN decade for decade in comparison to Caucasian men, similar to trends seen with prostate

Prostate Cancer. http://dx.doi.org/10.1016/B978-0-12-800077-9.00006-2

adenocarcinoma. Sakr et al. in a separate study examining 370 prostates from autopsy, 60% of which were from African-Americans, found HGPIN in the African-American cohort in 18, 31, 69, 78, and 86% in the fourth through eighth decades, respectively, compared to 14, 21, 38, 50, and 63% in prostates from Caucasian patients.[15] This study also found that in patients under 60 years old, a higher proportion of African-American men were found to have extensive HGPIN on radical prostatectomy specimen when compared to Caucasian men (57% vs. 33%).

The incidence of HGPIN found on needle biopsy is more clinically relevant information but it underestimates the true incidence given sampling error. The incidence of HGPIN on needle biopsy reported in the literature is highly variable, ranging from 1.7% to 45% (Table 6.1).[16–37] Reported mean incidence is between 4% and 9%.[3–5,38,39] Multiple potential explanations for this wide variation exist. Biopsy strategy may influence the frequency of a diagnosis of HGPIN, with saturation biopsies associated with higher rates of detection. Schoenfield et al. demonstrated a 22% incidence of HGPIN with initial 24-core saturation biopsy.[20] Lane et al. found that with an initial saturation biopsy with a mean of 24 cores (range 20–33) the incidence of HGPIN was 45%.[34] Differences

**TABLE 6.1** Summary of HGPIN and ASAP Incidence Rates Reported in Recent Studies

| Study | Year | N | Core number | HGPIN incidence (%) | ASAP incidence (%) |
|---|---|---|---|---|---|
| Kamoi et al.[33] | 2000 | 611 | Not reported | 10.0 | 1.0 |
| O'dowd et al.[41] | 2000 | 132,426 | Not reported | 3.7 | 2.5 |
| Borboroglu et al.[24] | 2001 | 1,391 | Mean 7.7 ± 2.2 | 6.1 | 4.4 |
| Ouyang et al.[42] | 2001 | 331 | Not reported | – | 6.3 |
| Iczkowski et al.[43] | 2002 | 7,081 | Not reported | – | 2.6 |
| Lefkowitz et al.[36] | 2002 | 1,223 | 12 | 9.7 | – |
| Abdel-Khalek et al.[31] | 2004 | 3,081 | 6 | 2.7 | – |
| Brausi et al.[44] | 2004 | 1,327 | Mean 9.5 (range 8–14) | | 5.4 |
| Naya et al.[25] | 2004 | 1,086 | Not reported | 8.7 | 1.9 |
| Moore et al.[37] | 2005 | 1,188 | ≥10 | 2.5 | 6.0 |
| Scattoni et al.[45] | 2005 | 3,350 | 10–12 | – | 3.8 |
| Tan et al.[23] | 2006 | 1,219 | 10 (range 6–18) | 4.6 | – |
| Amin et al.[32] | 2007 | 2,265 | Median 10 (range 6–14) | 23.7 | 2.5 |
| Lopez[19] | 2007 | 4,770 | 6–8 | 2.6 | 0.9 |
| Mancuso et al.[16] | 2007 | 1,632 | Not reported | 6.2 | 1.9 |
| Schoenfield et al.[20] | 2007 | 100 | 24 | 22.0 | 5.0 |
| Gallo et al.[30] | 2008 | 1,223 | Mean 16.3 (range 12–20) | 11.4 | – |
| Lane et al.[34] | 2008 | 257 | Median 24 (range 20–33) | 45.0 | 15.0 |
| De Nunzio et al.[28] | 2009 | 650 | 12 | 22.0 | – |
| Merrimen et al.[26] | 2009 | 12,304 | Not reported | 10.4 | 2.7 |
| Ploussard et al.[21] | 2009 | 2,006 | 21 | 1.7 | 1.1 |
| Singh et al.[22] | 2009 | 2,087 | Median 12 cores (range 8–14) | 4.2 | – |
| Laurila et al.[18] | 2010 | 17,884 (HGPIN); 21,043 (ASAP) | Mostly 6, Finland 10–12 for >2002 | 2.8 | 2.4 |
| Lee et al.[35] | 2010 | not reported | Not reported | 18.2 | – |
| Ryu et al.[46] | 2010 | 3,130 | 6 cores if volume <30 bcc; 12 cores if ≥30 $cm^3$ | – | 7.8 |
| Koca et al.[17] | 2011 | 2,433 | Mean 10.8 (range 8–12) | 4.7 | 2.4 |
| Garcia-Cruz et al.[29] | 2012 | 1,000 | Not reported | 8.2 | – |
| El Shafei et al.[27] | 2013 | 682 | 8–14 | 43.6 | 13.2 |

in the composition of the study cohort also results in variation of HGPIN incidence. Bostwick and Cheng observed that lower likelihood of HGPIN diagnosis was seen in PSA screening populations when compared with studies reported from a urology practice.[40] Finally, differences in pathologic processing and interpretation may also account for some of the variability. Use of non-standard fixatives may enhance nuclear and nucleolar features resulting in higher rates of diagnosis whereas suboptimal thick sectioning may obscure nuclear detail and result in the opposite. A lack of standardized criteria regarding degree and extent of nucleolar enhancement and enlargement leaves room for subjective interpretation. While interobserver agreement is acceptably high among urologic pathologists, there is some decline in reproducibility with general pathologists.[4,7] Laurila et al. in examining the diagnosis of HGPIN in the ERSPC cohort found significant country-to-country variability (incidence 0.8–7.6%) which the authors suggested was a result of interobserver variation given the lack of centralized pathologic review.[18]

## Histology

HGPIN is characterized by epithelial cell proliferation within architecturally normal prostate glands and ducts, accompanied by cellular changes that mimic cancer. These cytologic changes are primarily nuclear and include nuclear enlargement and hyperchromasia, as well as prominent nucleoli typically visible at 20× magnification.[3,5,6,39,47] Similar to other types of malignancies, reversal of proliferative orientation is seen in HGPIN as growth at the luminal surface predominates, in contrast to proliferation in normal glands, which occurs at the level of the basal cells.[5] As a premalignant lesion, HGPIN lacks invasion of the stroma, maintaining a basal cell layer, which may be continuous or disjointed. Using basal-cell-specific stains detecting the presence of P63, or high molecular weight cytokeratin via the 34βE12 antibody allows for differentiation of HGPIN from adenocarcinoma, which lacks a basal cell layer (Figure 6.1).[48,49]

There are four main histologic subtypes of HGPIN: tufting, micropapillary, cribriform, and flat (Figure 6.2).[50] Typically multiple subtypes are present in a single case. The tufting pattern is the most common (in up to 97% of cases), resulting from variable degrees of hyperplasia within prostatic acini. Elongation of hyperplastic columns results in the micropapillary pattern. Flat HGPIN consists of a single layer of atypical cells.[3] The cribriform variant must be distinguished from intraductal carcinoma of the prostate, which may be similar in appearance but is associated with high-grade, high-volume, infiltrative tumors.[4,51] Rarer HGPIN variants described in the literature include signet ring cell, small cell neuroendocrine, mucinous, foamy gland (microvacuolated),

(a)

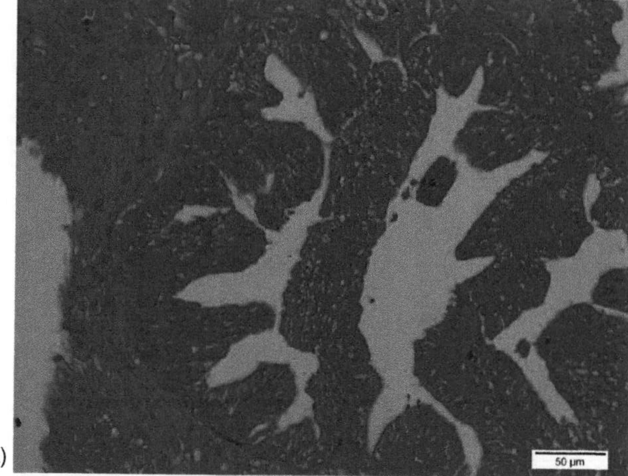


(b)

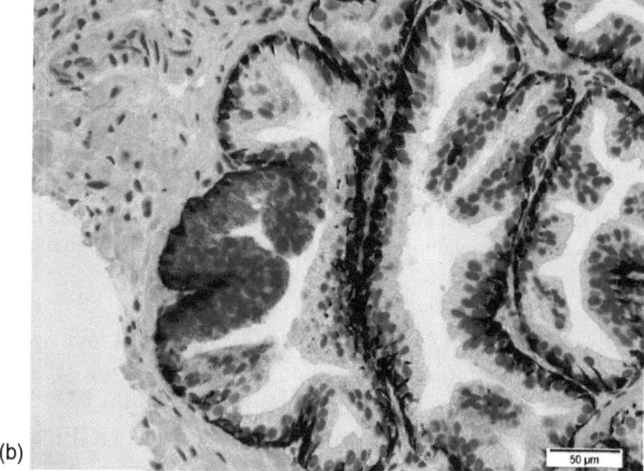


**FIGURE 6.1 High grade prostatic intraepithelial neoplasia.** (a) H&E stain; (b) AMACR stain delineates a continuous basal cell layer.

hobnail (inverted), squamous, and desquamating apoptotic patterns [3,40,52–54] Multiple studies have demonstrated a lack of variation in the degree of risk of subsequent diagnosis of prostate cancer based on the different histologic subtypes.[23,50,55,56] Only one study has suggested that HGPIN variants may be used for risk stratification. Kronz et al. demonstrated that in patients requiring multiple biopsies for the diagnosis of cancer after an initial HGPIN diagnosis, predominant micropapillary or cribriform histology were associated with a higher rate of cancer on follow-up than predominant flat or tufting HGPIN (58.3% vs. 16.7%, $p = 0.002$).[57] This study included a small number of patients however ($n = 144$) and only 17% of them had predominant micropapillary/cribriform HGPIN.

Histologic evidence for HGPIN representing a premalignant state arises from its similarities to, and close association with, prostate cancer. HGPIN is frequently multifocal, predominantly located in the peripheral zone (75–80%), and less often found in the transitional zone (10–15%) and central zone (<5%), mirroring prostate cancer.[6] Increasing grade of HGPIN has been shown to be associated

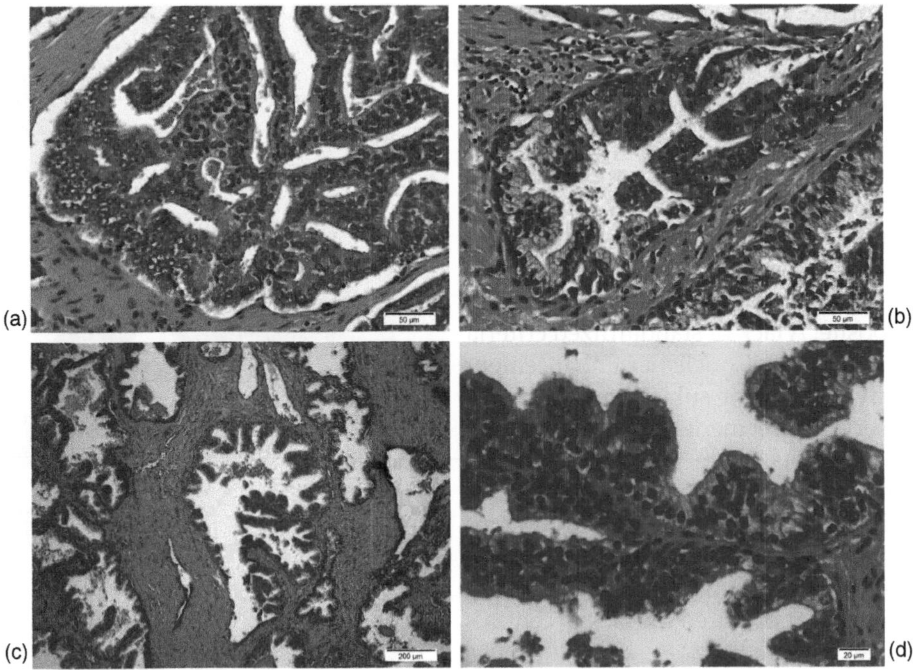

**FIGURE 6.2 High grade prostatic intraepithelial neoplasia histologic variants.** (a) Cribriform HGPIN, (b) tufted variant HGPIN, (c) papillary HGPIN low magnification, and (d) papillary HGPIN high magnification.

with increasing fragmentation of the basal cell layer, suggesting a mechanism for early carcinogenesis via stromal invasion through basal cell layer disruptions.[5,50,58,59] In prostatectomy and cystoprostatectomy series, HGPIN has been found in continuity with adenocarcinoma 60–100% of the time and its volume has been shown to increase with increasing Gleason grade and pathologic stage of disease.[11–13,60] Interestingly, HGPIN volume has been shown in some studies to demonstrate an inverse relationship to tumor volume (i.e., glands with large tumors are associated with small amounts of HGPIN), which suggests that areas previously occupied by HGPIN may be replaced with adenocarcinoma.[61] Direct conclusions based solely on histologic evaluation, however, are limited, given the static nature of the technique.

## Genetics and Molecular Biology

Further evidence for the clonal progression of HGPIN to prostate cancer is provided by the genetic and molecular similarities found between the two. Cytogenetic anomalies shared by HGPIN and prostate cancer include telomere shortening,[62,63] rate of allelic imbalances,[64] allelic loss of 8p12-21,[65] c-myc amplification,[66,67] loss of heterozygosity at chromosomes 6 and 8, and gain of chromosomes 7, 8, 10, and 12.[68] Similar epigenetic changes between HGPIN and prostate cancer include hypermethylation of tumor suppressor genes *APC* and *RARB2*,[69,70] and higher global methylation rates when compared to benign prostatic tissue.[70,71] Additional select shared molecular abnormalities include overexpression of p16,[72] c-Met,[73] and alpha-methylacyl-CoA racemase,[74] as well as reduction of annexin I,[75] and the presence of the *TMPRSS2–ERG* gene fusion.[76–79] Together, these similarities in molecular and genetic derangements represent shared pathways leading to proliferative deregulation and development of an invasive phenotype via genetic instability supporting the model of progression from HGPIN to prostate adenocarcinoma.

## Risk of Subsequent Cancer

Given HGPIN is a premalignant lesion, its ability to predict the presence of cancer as well as the potential for the development of cancer is of paramount clinical interest. The predictive value of isolated HGPIN is influenced by the extent of sampling on initial biopsy and the period of time to repeat biopsy.[38] Cancer detection on immediate repeat biopsy likely indicates the presence of occult cancer missed at the time of the initial biopsy as opposed to development of cancer in the short interim between biopsies given on autopsy studies prostate cancer development occurs 5–10 years after initial diagnosis of PIN.[14] A summary of recent studies examining cancer detection rates after an initial biopsy with HGPIN is provided in Table 6.2. While sextant biopsy strategies were primarily being employed, increased sampling error resulted in prostate cancer detection rates of 22–79% on immediate repeat biopsy, with mean incidence around 30%, and thus the recommendation was for short interval repeat

**TABLE 6.2** Summary of Recent Studies Examining Prostate Cancer Detection Rate on Repeat Biopsy Following an Initial Diagnosis of HGPIN

| Study | Year | *N* | Cores on initial biopsy | Cores on repeat biopsy | Time to rebiopsy | Overall CaP (prostate cancer) rate (%) | CaP first rebiopsy (%) | Time to second biopsy | CaP second rebiopsy (%) |
|---|---|---|---|---|---|---|---|---|---|
| Kamoi et al.[33] | 1999 | 45 | 6 | Not reported | Within 1 year | 22.0 | 22.0 | – | – |
| O'dowd et al.[41] | 2000 | 1306 | For all initial bx, ≤8 in 99.5% | For all repeat bx, ≤8 in 99.1% | <12 month | 22.6 | 22.6 | – | – |
| Borboroglu et al.[24] | 2001 | 45 | Mean 7.7 ± 2.2 | Not reported | Mean 3.9 month | 44.0 | 44.0 | – | – |
| Kronz et al.[57] | 2001 | 245 | Not reported | Not reported | Median 5.3 month | 32.2 | 24.5 | – | – |
| Lefkowitz et al.[80] | 2001 | 43 | 12 | 12 | Within 1 year | 2.3 | 2.3 | – | – |
| Stewart et al.[123] | 2001 | 64 | 6 | Mean 23 (range 14–45) | Median 2.4 year from initial to saturation bx | 30.7 | – | – | – |
| Lefkowitz et al.[36] | 2002 | 31 | 12 | Not reported | 3 year | 25.8 | 25.8 | – | – |
| Abdel-Khalek et al.[31] | 2004 | 83 | 6 | 11 | Not reported | 43.4 | 36.0 | Not reported | 19.0 |
| Bishara et al.[55] | 2004 | 132 | 6 in 60% (more in rest) | 6 in 61% (more in rest) | Mean 7 month (range 1–33) | 28.8 | 20.5 | Not reported | 17.9 |
| Naya et al.[25] | 2004 | 47 | Not reported | Not reported | Not reported | 10.6 | 10.6 | – | – |
| Gokden et al.[90] | 2005 | 190 | Mean 6.6 (range 5–14) | Not reported | Not reported | 30.5 | 13.2 | Not reported | 13.7 |
| Moore et al.[37] | 2005 | 22 | ≥10 | ≥10 | 15 weeks | 4.5 | 4.5 | Not reported | 0.0 |
| Schlesinger et al.[82] | 2005 | 204 | At least 6, most 8–10 | Mean 9.9 | 6 month | 23.0 | 23.0 | – | – |
| Herawi et al.[81] | 2006 | 791 | 332 with 6, 323 with ≥8 | 345 with 6–7, 426 with ≥8 | Mean 4.6 month | 17.6 | 17.6 | – | – |
| Netto et al.[83] | 2006 | 41 | 10.6 (range 5–16) | 10.4 (range 6–16) | 11 month (range 1–41 month) | 39.0 | 39.0 | – | – |
| Tan et al.[23] | 2006 | 29 | Median 10 (range 6–18) | Not reported | Mean 9 month (range 1–60 month) | 24.1 | 13.8 | – | – |
| Amin et al.[32] | 2007 | 201 | 10 | Median 10 | Not reported | 21.9 | 15.9 | Not reported | 13.6 |
| Loeb et al.[97] | 2007 | 96 | ≥6 in 91% | Not reported | 8 month (range 0–51 month) | 57.3 | 48.0 | – | – |
| Lopez JI[19] | 2007 | 125 | 6–8 | 10–12 | Not reported | 16.8 | Not reported | – | – |
| Gallo et al.[30] | 2008 | 65 | 12–20 | 12–20 | 3–12 month | 21.5 | 1.6 | 13–24 month | 11.6 |
| Merrimen et al.[26] | 2009 | 564 | Not reported | Not reported | 0.91 year | 27.5 | – | – | – |
| Ploussard et al.[21] | 2009 | 2006 | 21 | 21 | Mean 12.4 month (range 2–66) | 19.0 | 19.0 | – | – |
| Singh et al.[22] | 2009 | 67 | Median 12 (range 8–14) | Mean 11 | Median 13.5 month | 41.8 | 35.8 | – | – |
| Laurilaet al.[18] | 2010 | 626 | Mostly 6, Finland 10–12 far >2002 | Mostly 6, Finland 10–12 for >2002 | Within 6 weeks | 12.9 | Not reported | – | – |

*(Continued)*

**TABLE 6.2** Summary of Recent Studies Examining Prostate Cancer Detection Rate on Repeat Biopsy Following an Initial Diagnosis of HGPIN *(cont.)*

| Study | Year | *N* | Cores on initial biopsy | Cores on repeat biopsy | Time to rebiopsy | Overall CaP (prostate cancer) rate (%) | CaP first rebiopsy (%) | Time to second biopsy | CaP second rebiopsy (%) |
|---|---|---|---|---|---|---|---|---|---|
| Lee et al.[35] | 2010 | 328 | Mean 11.9 (range 6–26) | Mean 15.1 (range 6–32) | 1.31 year (±1.25 year) | 35.7 | 35.7 | – | – |
| Mitterberger et al.[93] | 2010 | 104 | Not reported | Systematic 10 cores + targeted cores to hypervascular region (up to 5) | Within 6 months | 25.0 | 25.0 | – | – |
| Antonelli et al.[86] | 2011 | 546 | 10.8 (range 6–23) | Not reported | Mean 7.8 month | 31.8 | 21.2 | – | – |
| Fleshner et al.[87] | 2011 | 310 | Not reported | Not reported | 6, 12, 24, 36 month | 26.4 | – | – | – |
| Godoy et al.[88] | 2011 | 112 | Not reported | Not reported | Median 34.4 month | 32.1 | 22.3 | 66.2 month from initial | 23.4 |
| Hailemariam et al.[56] | 2011 | 66 | Not reported | Median 6 | Median 3 month (range 0.5–68) | 33.3 | 30.0 | 4 month from second bx (range 1–32) | 20.0 |
| Koca et al.[17] | 2011 | 40 | Mean 10.8 (range 8–12) | Mean 10.8 (range 8–12) | Not reported | 0.0 | 0.0 | Not reported | 0.0 |
| Marshall et al.[111] | 2011 | 269 | 10.5 placebo, 10.7 selenium | 9.7 placebo, 10.6 selenium | Not reported | 36.1 | – | – | – |
| Garcia-Cruz et al.[29] | 2012 | 45 | 10 | 10 | 3–6 month | 22.2 | 22.2 | – | – |
| He et al.[103] | 2012 | 94 | Mean 12 (range 6–20) | Not reported | Not reported | 38.3 | – | – | – |
| Roscigno et al.[84] | 2012 | 262 | Median 12 (range 6–24) | Mean 22 (range 20–26) | Median 12 month (range 3–30) | 36.3 | 31.7 | Mean 14.5 month from last (range 3–29) | 12.9 |
| De Nunzio et al.[85] | 2013 | 117 | 12 | 12 | 6 month | 31.6 | 18.8 | 6 months from last biopsy | 35.0 |
| El Shafei et al.[27] | 2013 | 297 | 8–14 | Mean 15.7 (std deviation 4.6) | Mean 1.92 year | 32.7 | 32.7 | – | – |

biopsy in patients with isolated HGPIN.[38] With increasing utilization of extended core biopsies and the detection of smaller foci of prostate cancer initially, the rate of cancer detection on immediate repeat biopsy appears to have significantly decreased. Lefkowitz et al. demonstrated that after an extended 12-core biopsy the risk of cancer detection on biopsy done within a year occurred at a rate of 2.3%.[80] Similarly, Moore et al. found that prostate cancer risk was just 4.5% when repeat biopsy was performed within 15 weeks of an initial biopsy involving greater than or equal to 10 cores.[37] A study by Herawi et al. further highlights the effect of core number on the risk of prostate cancer within a year of HGPIN diagnosis. When six-core biopsy was used initially, the risk of cancer on repeat biopsy was significantly higher than eight or more cores that were obtained initially (20.8% vs. 13.3%, $p = 0.01$). In patients with an initial six-core biopsy, cancer detection rates after repeat biopsy with six cores and eight or more cores was 14.1 and 31.9%, respectively. In contrast, extended biopsy followed by a repeat extended biopsy resulted in only a 14.6% cancer detection rate within 1 year.[81]

When examining all studies with mean time to repeat biopsy of less than a year after an initial extended core biopsy (≥10 cores), observed cancer detection rates range from 2.3% to 39.0%.[23,29,30,37,80,82–86] The majority of these studies demonstrated cancer detection rates of less than 23%, which is comparable to the average rate of 19% after benign biopsy reported by Epstein and Herawi.[4] The two studies with over 30% cancer detection included patients who had repeat biopsies up to 3–4 years later,[83,84] and one study included patients with widespread HGPIN only,[83] and thus these studies likely overestimate actual short-term cancer risk.

While mandated short-term interval biopsy for isolated HGPIN after an extended core biopsy appears to be unnecessary based on these results, optimal long-term biopsy strategy is yet to be determined due to a dearth of data examining long-term risk. Overall incidence of prostate cancer in studies examining biopsies performed at a mean of greater than 1 year from initial biopsy ranges from 19% to 41.8%.[21,22,27,30,35,36,86–88] Lefkowitz et al. found that on interval biopsy at 3 years after initial diagnosis of HGPIN, 25.8% of men had prostate cancer.[36] A follow up study by the same group demonstrated a persistently elevated risk of prostate cancer 6 years after initial biopsy in those with a negative 3-year biopsy (positive in 23.4%), without significant differences found in PSA levels between benign and malignant biopsies. This led the authors to recommend that empiric biopsy be performed every 2–3 years after diagnosis of HGPIN.[88] However, these findings are not universal. In a study by Gallo et al., overall risk of prostate cancer after a diagnosis of HGPIN on extended core biopsy (12–20 cores) was 21.5%, and diagnosis at less than 1, 1–2, 2–3, and 3–4 years after initial biopsy was only 1.6, 11.6, 5.7, and 6.1%, respectively.[30]

Another consideration with regards to the utility and timing of long-term biopsy strategies after HGPIN diagnosis is the characteristics of cancer that is diagnosed. While bias exists given the increased screening these men undergo, it is important to note that for the most part cancer diagnosed after an initial biopsy with HGPIN demonstrates favorable characteristics.[88–90] Based on radical prostatectomy specimens, Al-Hussain and Epstein demonstrated that patients with prior HGPIN diagnosis had small tumors (mean 0.3 mL), which were more often organ-confined in comparison to those diagnosed without a prior diagnosis of HGPIN (84% vs. 65%, $p = 0.0007$).[89] Eminaga et al. found on examination of radical prostatectomy specimens that cancer associated with HGPIN was smaller, lower grade, and less advanced stage-wise than cancer not associated with HGPIN.[91] Balaji et al. found no association between HGPIN diagnosis on biopsy and subsequent biochemical recurrence after radical prostatectomy.[92]

## Risk Factors and Biomarkers Associated With Subsequent Cancer Diagnosis

Given transrectal prostate biopsy is not without morbidity, the identification of tools to risk-stratify patients with the diagnosis of isolated HGPIN is desirable to minimize unnecessary biopsies in patients least likely to benefit. Many risk factors for cancer have been examined in patients with HGPIN but few are accepted as reliable prognosticators.

### *Multifocal HGPIN*

The extent of HGPIN on biopsy has been examined for its predictive ability. Multiple studies have demonstrated increased risk of subsequent prostate cancer diagnosis in patients with multiple cores positive for HGPIN.[26,31,32,35,55,57,84,85,93] Netto and Epstein found that patients with "widespread HGPIN," defined as four or more cores of HGPIN, had a 39% risk of cancer on subsequent biopsy.[83] Lee et al. reported that 1-, 3-, and 5-year cancer detection rates for multifocal and bilateral HGPIN were 9, 29, and 48% and 12, 38, and 58%, respectively. On multivariate analysis controlling for age, PSA, DRE, and number of cores, multifocal and bilateral HGPIN were significant predictors of subsequent biopsy (HR 2.56 and 2.20, respectively, for both $p < 0.0001$).[35] In the largest study examining the effect of HGPIN multifocality, Merrimen et al. retrospectively examined 564 patients with HGPIN on initial biopsy, and reported that the OR for diagnosis of prostate cancer increased in a step-wise fashion with each additional site involved with HGPIN. Unifocal HGPIN demonstrated a risk similar to those patients without HGPIN (OR 1.02, $p = 0.9$) while patients with 2, 3, and 4 sites involved had significantly increased risk of subsequent cancer diagnosis (OR 1.55, 1.99, and 2.66, respectively, all $p < 0.03$). In addition, this study found a higher volume of Gleason 4 or 5 in patients with prior multifocal HGPIN when compared to unifocal HGPIN (11.2% vs. 2.9%, $p = 0.01$). This body of data is consistent enough to conclude that patients with multiple cores positive for HGPIN should be monitored more closely and should undergo repeat biopsy at a shorter interval.

### *PSA*

It is generally accepted that HGPIN does not significantly elevate PSA levels.[13,94] Given this fact, one would expect that the presence of prostate cancer in the setting of HGPIN would result in an increase of PSA relative to patients with HGPIN in the absence of cancer; however, most studies demonstrate no difference in PSA levels between these two groups.[25,28,30,31,57,85,88] Several older studies have found a relationship between higher PSA level and ultimate prostate cancer diagnosis on repeat biopsy; however, these studies used limited biopsy schemes primarily and thus applicability to current practice is questionable.[41,95,96] Only three studies have demonstrated a significant relation between PSA and cancer detection with initial extended biopsies. Singh et al. found that after an initial biopsy with a median of 12 cores, patients later diagnosed with prostate cancer had a higher baseline PSA (17.5 ng/mL vs. 8.4 ng/mL, $p < 0.005$) and mean change in PSA (4.8 ng/mL vs. 0.5 ng/mL, $p < 0.05$).[22] Roscigno et al. similarly found a higher PSA level associated with cancer detection with repeat saturation biopsy (HR 1.08, $p = 0.04$).[84] Antonelli et al. reported a slightly higher PSA (8.6 ng/mL vs. 7.2 ng/mL, $p = 0.004$) in patients diagnosed with cancer on first rebiopsy after an initial diagnosis of HGPIN; however, this difference was no longer seen on subsequent rebiopsy (7.4 ng/mL vs. 7.0 ng/mL, $p = 0.24$).[86]

Limited data exist regarding the utility of PSA density (PSAD), free-to-total PSA ratio, and PSA velocity (PSAV). De Nunzio et al. found no significant difference in PSAD or free-to-total PSA ratio in patients with or without cancer on repeat biopsy.[85] In contrast Abdel-Khalek et al. found that elevated PSAD was a significant predictor of cancer detection utilizing a threshold of PSAD ≥0.15.[31] Regarding PSAV, Loeb et al. reported that it could be used to follow men with HGPIN as increased PSAV in their cohort was a significant predictor of ultimate cancer diagnosis when controlling for age and total PSA (OR 4.56, $p = 0.004$). This group recommended the use of a threshold PSAV 0.75 ng/mL/year to identify men who should undergo repeat biopsy.[97]

### *Other Biomarkers*

Several other biomarkers have been examined including α-methylacyl-CoA-racemase (AMACR), and the TMPRSS2:ERG fusion protein. However, data for each of these markers is preliminary and at this time their clinical utility has yet to be validated.

AMACR staining is useful for distinguishing adenocarcinoma from benign glands. HGPIN exhibits high rates of AMACR expression.[98] Wu et al. found in an examination of 140 radical prostatectomy specimens that HGPIN glands immediately adjacent to adenocarcinoma was significantly more likely to stain for AMACR than those located farther away (56% vs. 14%, $p < 0.0001$) suggesting that staining of AMACR on biopsy may identify patients most likely to harbor concurrent cancer.[99] Furthermore, Stewart et al. reported that patients with at least one AMACR positive HGPIN on needle biopsy had a 5.2-fold increased risk of cancer detection on repeat biopsy ($p = 0.004$).[100] Contrary to these findings, however, Hailemariam et al. found that AMACR staining was the same in HGPIN patients with and without cancer on subsequent biopsy (77% vs. 78%, respectively). This study also looked at Ki-67, annexin II, and Bcl-2 and none of these demonstrated significant differences either.[56]

*TMPRSS2:ERG* gene fusion is prevalent in approximately 50% of prostate cancers and has been shown to be present in around 20% of HGPIN as well.[76,101] This gene rearrangement is frequently accompanied by loss of tumor suppressor *PTEN*, which together contributes to the progression of HGPIN to prostate cancer.[102] Studies examining the ability of *TMPRSS2:ERG* staining to identify patients at risk for subsequent cancer have demonstrated mixed results. He et al. found no difference in cancer detection between patients whose HGPIN biopsies were found to have ERG expression on immunohistochemistry versus those who did not. The number of ERG positive biopsies in this study were low however ($n = 5$).[103] In contrast, Gao et al. found an ERG rearrangement rate of ≥1.6% on FISH was an independent predictor of subsequent cancer diagnosis (OR 2.45, $p < 0.001$).[104] Park et al. similarly saw increased cancer detection in HGPIN patients demonstrating ERG expression on biopsy cores compared to ERG-negative patients (53% vs. 35%, $p = 0.01$).[105] Further studies are required to further evaluate the utility of this potential marker, as well as AMACR, in the future.

## Recommended Management of Isolated HGPIN

Current NCCN guidelines recommend a repeat biopsy within 6 months for patients with multifocal HGPIN defined as more than two sites. In patients with unifocal HGPIN or a negative repeat biopsy after multifocal HGPIN, follow up in 1 year with PSA and DRE is recommended.[106] The EAU guidelines state that repeat biopsy is warranted in patients with multifocal HGPIN; however, they do not comment on the timing of the repeat biopsy, merely stating that the optimal interval to biopsy is yet to be determined.[107]

## Chemoprevention

As a premalignant lesion, HGPIN represents an ideal marker for patients most likely to benefit from chemopreventative strategies to prevent or delay progression to frank adenocarcinoma. To date no agent is FDA-approved for use in this role though multiple randomized controlled trials have been conducted.

Two large-scale randomized controlled trials have examined the role for 5-alpha reductase inhibitors in the chemoprevention of prostate cancer: the Prostate Cancer Prevention Trial (PCPT) and the Reduction by Dutasteride of Prostate Cancer Events (REDUCE) trial. In the PCPT, daily finasteride administration resulted in a 24.8% relative risk reduction in prostate cancer overall, primarily low-grade disease. A significant increase in high-grade prostate was seen.[108] In a secondary analysis of the effect of finasteride on the development of HGPIN, Thompson et al. found that finasteride significantly decreased the risk of HGPIN diagnosis when compared to placebo (11.7% vs. 9.2%, $p < 0.001$).[109] The REDUCE trial results paralleled that of the PCPT with a 22.8% relative risk reduction of prostate cancer with the administration of dutasteride, along with an increase in the incidence of high-grade cancer in Year 4 of the study. This study found that dutasteride decreased diagnosis of HGPIN by 39.2% (3.7% vs. 6%, $p < 0.001$).[110] Neither of these drugs were approved for the chemoprevention of prostate cancer due to the questionable clinical benefit of prevention of low-grade prostate cancer along with concerns for the induction of higher-grade disease in a small number.

Based on promising preclinical data, nutritional supplements have also been tested for chemoprevention of prostate cancer. The SWOG S9917 phase III randomized placebo-controlled trial cohort consisted of patients with HGPIN only and demonstrated no difference between daily selenium and placebo with respect to cancer detection rate at 3 years (35.6% vs. 36.6%, $p = 0.73$).[111] Another trial examining selenium, reported by Fleshner et al. also yielded a negative result, with a rate of diagnosis of cancer after administration of soy, selenium, and vitamin E similar to that after placebo (HR 1.03, $p = 0.88$).[87]

Finally, toremifene, a selective estrogen receptor modulator used in the treatment of advanced breast cancer, was tested on cohort of patients with HGPIN with or without ASAP in a phase III randomized controlled trial completed in 2013. Despite phase IIb data showing a significant decrease in cumulative prostate cancer incidence with toremifene at 12 months (24.4% vs. 31.2%, $p = 0.048$), the phase III trial failed to demonstrate any chemopreventative effect. Cancer was detected in 34.7% of men in the toremifene group and 32.3% of men receiving placebo.[112,113]

While chemoprevention to decrease the burden of increased prostate cancer diagnosis remains a desirable goal, significant strides will need to be made before it is a reality.

## ATYPICAL SMALL ACINAR PROLIFERATION

### Introduction

ASAP is a term utilized to describe pathologic features of prostate biopsies that are suspicious for malignancy but cannot be definitively described as such. Given concerns over the overdiagnosis of prostate cancer – particularly given the significant potential morbidity and patient burden that accompanies the diagnosis – in the absence of absolute evidence for cancer, this diagnostic risk category exists to identify patients that require close follow up and short interval biopsy due to high rates of concurrent prostate cancer.

### Diagnosis

A diagnosis of ASAP encompasses a wide variety of pathologic findings including inadequately sampled adenocarcinoma and benign mimickers of malignant proliferation. Typically, this results from a presence of conflicting morphologic or histochemical findings compounded by the small size of the focus of concern.[114] Confusing morphologic findings may include insufficient nuclear atypia, exaggerated nuclear hyperchromasia obscuring nucleolar findings, or clustered growth resembling benign processes like atypical adenomatous hyperplasia. Immunohistochemical findings in ASAP may include the absence of AMACR staining or focally positive basal cell staining in otherwise overtly malignant appearing glands. Iczkowski et al. reported the most frequent histologic feature resulting in an ASAP diagnosis; however, it has limited size and number of acini involved (70% of cases).[115] This may be a result of sampling or a byproduct of suboptimal sample handling and processing.[116] Reyes and Humphrey found that gaining additional information via examining additional sections of the biopsy of concern can lead to a change in diagnosis from benign to malignant in a minority of patients.[117]

Immunohistochemical staining may also aid to clarify a diagnosis. The absence of a basal cell layer is a hallmark of cancer and thus basal cell staining for p63 or HMWK using 34βE12 antibodies can rule in malignancy for those glands that do not stain (Figure 6.3). The possibility of false negatives exists however. The use of both basal cell stains has been shown to increase the sensitivity for basal cells and may help to decrease uncertainty in these cases.[118,119] AMACR staining may also help reclassify ASAP into benign or malignant diagnoses as moderate to strong staining has been shown to be associated with a 97% sensitivity and 100% specificity for adenocarcinoma.[120] Zhou et al. found that AMACR staining converted the diagnosis from ASAP to cancer in about 10% on "expert review" and when used in combination with basal cell markers further discriminative ability has been demonstrated.[45,121,122]

### Incidence

ASAP incidence ranges from 0.9% to 15% in contemporary biopsy series, with the majority of studies demonstrating incidences less than 5% (Table 6.1).[16–21,24,25,27,32–34,37,41–46] Some of the variation in incidence is a result of interobserver variation. Van der Kwast et al. reported

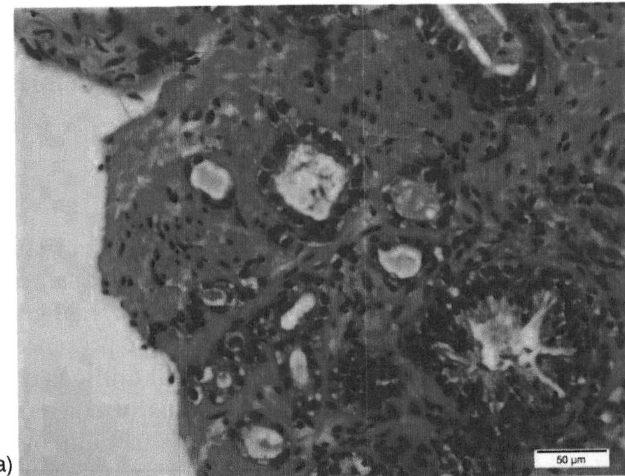


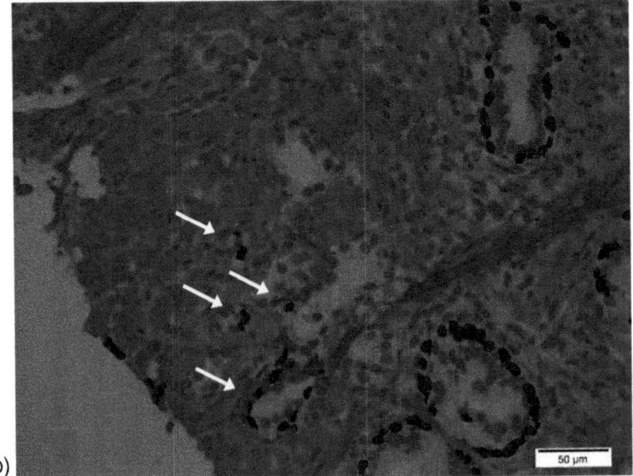


**FIGURE 6.3 Atypical small acinar proliferation.** (a) H&E stain; (b) p63 stain demonstrates scant basal cells (white arrows).

on the pathologic review of 20 biopsies with small atypical foci by 5 urologic pathologists and 7 nonspecialist pathologists and found higher levels of agreement in diagnosis between urologic pathologists than the general pathologists. Significant variation in diagnosis was seen in 5 and 10 cases, respectively, particularly in cases where atypical foci comprised fewer than 6 glands. The authors also reported that experts diagnosed carcinoma significantly more frequently than the general pathologists (49% vs. 32%, $p < 0.001$). Based on these results the authors recommended specialist consultation in difficult cases, particularly when a scarcity of atypical glands was present.

## Risk for Subsequent Cancer

In contrast to HGPIN, the risk of subsequent cancer on repeat biopsy after a diagnosis of ASAP has remained fairly constant over time, ranging from 26.7% to 59.1% (Table 6.3).[16–19,21,24,27,32,37,41–43,45,46,82,123–125] This high risk of cancer on repeat biopsy reflects the rate of undersampling of cancer that the typical ASAP diagnosis represents. Kim et al. examined ASAP biopsy sites on radical prostatectomy specimens and found cancer at 36 of 46 sites (78%).[126] Similarly, Brausi et al. found that in 25 patients who underwent immediate radical prostatectomy after ASAP diagnosis rather than repeat biopsy, all 25 had prostate cancer on final pathology.[44]

There are few reported predictors of cancer on follow up biopsy after a diagnosis of ASAP. Stratification of ASAP based on level of suspicion has not been shown to be a significant predictor of subsequent cancer.[114] Patients with ASAP and HGPIN concurrently on biopsy have been shown to have rates of cancer detection on repeat biopsy similar to patients with ASAP alone and thus these patients should be treated with similarly high levels of suspicion. Merrimen et al. found that increasing amounts of concurrent HGPIN with ASAP actually resulted in a higher rate of positive repeat biopsies when compared to patients with ASAP alone (71% vs. 43%).[125] Given the lack of reliable predictive variables and the high rate of cancer detection, it is currently recommended that all patients with a diagnosis of ASAP should undergo rebiopsy at a short interval.[106]

## Recommended Management

Current NCCN guidelines reflect the findings of these studies. After a biopsy diagnosis of ASAP, the current recommendation is for an extended core repeat biopsy within 6 months, with increased sampling at the location of the biopsy positive for ASAP and adjacent areas.[106] Borboroglu et al. found only a 47% correlation between the initial ASAP biopsy sextant and eventual biopsy site positive for adenocarcinoma, and so targeted biopsies to the affected site alone are not sufficient.[24] If repeat biopsy is negative, the NCCN guidelines recommend close follow up with continued screening with PSA and DRE at an initial interval of 1 year. EAU guidelines similarly recommend repeat biopsy though no time interval is specified. The EAU guidelines do not recommend an empiric third biopsy unless clinically indicated for an individual patient. The use of multiparametric MRI is suggested in patients with persistently negative biopsies despite continued high clinical suspicion.[107]

# CONCLUSIONS

HGPIN and ASAP are prostate biopsy findings in between benign and malignant diagnoses, which indicate the need for close follow-up. While the risk for

**TABLE 6.3** Summary of Recent Studies Examining Prostate Cancer Detection Rate on Repeat Biopsy Following an Initial Diagnosis of ASAP

| Study | Year | *N* | Cores on initial biopsy | Cores on repeat biopsy | CaP (prostate cancer) (%) |
|---|---|---|---|---|---|
| O'dowd et al.[41] | 2000 | 1321 | For all initial bx, ≤8 in 99.5% | For all repeat bx, ≤8 in 99.1% | 40.0 |
| Borboroglu et al.[24] | 2001 | 48 | Mean 7.7 ± 2.2 | Not reported | 48.0 |
| Ouyang et al.[42] | 2001 | 17 | Not reported | Not reported | 52.9 |
| Stewart et al.[123] | 2001 | 27 | 6 | Mean 23 (range 14–45) | 48.2 |
| Iczkowski et al.[43] | 2002 | 129 | Not reported | Not reported | 39.5 |
| Brausi et al.[44] | 2004 | 48 | Mean 9.5 (range 8–14) | Mean 13 (range 12–14) | 70.8 |
| Moore et al.[37] | 2005 | 72 | ≥10 | ≥10 | 41.5 |
| Scattoni et al.[45] | 2005 | 84 | 10–12 | Not reported | 35.0 |
| Schlesinger et al.[82] | 2005 | 78 | Not reported | Not reported | 37.2 |
| Amin et al.[32] | 2007 | 22 | Mean 8.3 (no range provided) | Median 10 (no range provided) | 59.1 |
| Lopez[19] | 2007 | 45 | 6–8 | 10–12 | 26.7 |
| Mancuso et al.[16] | 2007 | 31 | Mean 10.1 (range 6–20) | Mean 15.5 (range 10–20) | 54.8 |
| Ploussard et al.[21] | 2009 | 17 | 21 | 21 | 41.2 |
| Laurila et al.[18] | 2010 | 582 | Mostly 6, Finland 10–12 for >2002 | Mostly 6, Finland 10–12 for >2002 | 33.8 |
| Ryu et al.[46] | 2010 | 170 | 6 cores if volume <30 $cm^3$; 12 cores if ≥30 $cm^3$ | 6 cores if volume <30 $cm^3$; 12 cores if ≥30 $cm^3$ | 33.5 |
| Koca et al.[17] | 2011 | 97 | Mean 10.8 (range 8–12) | Mean 10.8 (range 8–12) | 39.2 |
| Merrimen et al.[125] | 2011 | 189 | Mean 7.6 (range not reported) | Not reported | 48.1 |
| El Shafei et al.[27] | 2013 | 90 | 8–14 | Mean 15.7 (std deviation 4.6) | 38.9 |
| Zhang et al.[124] | 2013 | 179 | Mean 16 (range 6–64) | Not reported | 36.3 |

adenocarcinoma after an initial extended core biopsy discovers HGPIN is similar to patients with a benign baseline biopsy, patients with multifocal HGPIN remain at a higher risk for subsequent cancer detection and warrant repeat prostate biopsy. ASAP on prostate biopsy represents undersampled prostate cancer 40–50% of the time and thus patients with this finding on biopsy require repeat biopsy at a short interval. Further studies identifying risk factors for subsequent cancer may help to curb unnecessary biopsies in the future. To date no agents for chemoprevention in the setting of HGPIN have proven effective. This remains an additional area for future research.

## References


1. McNeal JE. Origin and development of carcinoma in the prostate. *Cancer* 1969;**23**:24–34.
2. McNeal JE, Bostwick DG. Intraductal dysplasia: a premalignant lesion of the prostate. *Hum Pathol* 1986;**17**:64–71.
3. Epstein JI. Precursor lesions to prostatic adenocarcinoma. *Virchows Arch* 2009;**454**:1–16.
4. Epstein JI, Herawi M. Prostate needle biopsies containing prostatic intraepithelial neoplasia or atypical foci suspicious for carcinoma: implications for patient care. *J Urol* 2006;**175**:820–34.
5. Bostwick DG, Qian J. High-grade prostatic intraepithelial neoplasia. *Mod Pathol* 2004;**17**:360–79.
6. Ayala AG, Ro JY. Prostatic intraepithelial neoplasia: recent advances. *Arch Pathol Lab Med* 2007;**131**:1257–66.
7. Allam CK, Bostwick DG, Hayes JA, et al. Interobserver variability in the diagnosis of high-grade prostatic intraepithelial neoplasia and adenocarcinoma. *Mod Pathol* 1996;**9**:742–51.
8. Bostwick DG. Prostatic intraepithelial neoplasia is a risk factor for cancer. *Semin Urol Oncol* 1999;**17**:187–98.
9. Oyasu R, Bahnson RR, Nowels K, et al. Cytological atypia in the prostate gland: frequency, distribution and possible relevance to carcinoma. *J Urol* 1986;**135**:959–62.
10. Sakr WA. Prostatic intraepithelial neoplasia: a marker for high-risk groups and a potential target for chemoprevention. *Eur Urol* 1999;**35**:474–8.
11. Silvestri F, Bussani R, Pavletic N, et al. Neoplastic and borderline lesions of the prostate: autopsy study and epidemiological data. *Pathol Res Pract* 1995;**191**:908–16.
12. Bruins HM, Djaladat H, Ahmadi H, et al. Incidental prostate cancer in patients with bladder urothelial carcinoma: comprehensive analysis of 1,476 radical cystoprostatectomy specimens. *J Urol* 2013;**190**:1704–9.
13. Kim HL, Yang XJ. Prevalence of high-grade prostatic intraepithelial neoplasia and its relationship to serum prostate specific antigen. *Int Braz J Urol* 2002;**28**:413–6 discussion 7.
14. Sakr WA, Haas GP, Cassin BF, et al. The frequency of carcinoma and intraepithelial neoplasia of the prostate in young male patients. *J Urol* 1993;**150**:379–85.

15. Sakr WA, Grignon DJ, Haas GP, et al. Epidemiology of high grade prostatic intraepithelial neoplasia. *Pathol Res Pract* 1995;**191**:838–41.
16. Mancuso PA, Chabert C, Chin P, et al. Prostate cancer detection in men with an initial diagnosis of atypical small acinar proliferation. *BJU Int* 2007;**99**:49–52.
17. Koca O, Caliskan S, Ozturk MI, et al. Significance of atypical small acinar proliferation and high-grade prostatic intraepithelial neoplasia in prostate biopsy. *Korean J Urol* 2011;**52**:736–40.
18. Laurila M, van der Kwast T, Bubendorf L, et al. Detection rates of cancer, high grade PIN and atypical lesions suspicious for cancer in the European Randomized Study of Screening for Prostate Cancer. *Eur J Cancer* 2010;**46**:3068–72.
19. Lopez JI. Prostate adenocarcinoma detected after high-grade prostatic intraepithelial neoplasia or atypical small acinar proliferation. *BJU Int* 2007;**100**:1272–6.
20. Schoenfield L, Jones JS, Zippe CD, et al. The incidence of high-grade prostatic intraepithelial neoplasia and atypical glands suspicious for carcinoma on first-time saturation needle biopsy, and the subsequent risk of cancer. *BJU Int* 2007;**99**:770–4.
21. Ploussard G, Plennevaux G, Allory Y, et al. High-grade prostatic intraepithelial neoplasia and atypical small acinar proliferation on initial 21-core extended biopsy scheme: incidence and implications for patient care and surveillance. *World J Urol* 2009;**27**:587–92.
22. Singh PB, Nicholson CM, Ragavan N, et al. Risk of prostate cancer after detection of isolated high-grade prostatic intraepithelial neoplasia (HGPIN) on extended core needle biopsy: a UK hospital experience. *BMC Urol* 2009;**9**:3.
23. Tan PH, Tan HW, Tan Y, et al. Is high-grade prostatic intraepithelial neoplasia on needle biopsy different in an Asian population: a clinicopathologic study performed in Singapore. *Urology* 2006;**68**:800–3.
24. Borboroglu PG, Sur RL, Roberts JL, et al. Repeat biopsy strategy in patients with atypical small acinar proliferation or high grade prostatic intraepithelial neoplasia on initial prostate needle biopsy. *J Urol* 2001;**166**:866–70.
25. Naya Y, Ayala AG, Tamboli P, et al. Can the number of cores with high-grade prostate intraepithelial neoplasia predict cancer in men who undergo repeat biopsy? *Urology* 2004;**63**:503–8.
26. Merrimen JL, Jones G, Walker D, et al. Multifocal high grade prostatic intraepithelial neoplasia is a significant risk factor for prostatic adenocarcinoma. *J Urol* 2009;**182**:485–90 discussion 90.
27. Elshafei A, Li YH, Hatem A, et al. The utility of PSA velocity in prediction of prostate cancer and high grade cancer after an initially negative prostate biopsy. *Prostate* 2013;**73**:1796–802.
28. De Nunzio C, Trucchi A, Miano R, et al. The number of cores positive for high grade prostatic intraepithelial neoplasia on initial biopsy is associated with prostate cancer on second biopsy. *J Urol* 2009;**181**:1069–74 discussion 74–75.
29. Garcia-Cruz E, Piqueras M, Ribal MJ, et al. Low testosterone level predicts prostate cancer in re-biopsy in patients with high grade prostatic intraepithelial neoplasia. *BJU Int* 2012;**110**:E199–202.
30. Gallo F, Chiono L, Gastaldi E, et al. Prognostic significance of high-grade prostatic intraepithelial neoplasia (HGPIN): risk of prostatic cancer on repeat biopsies. *Urology* 2008;**72**:628–32.
31. Abdel-Khalek M, El-Baz M, Ibrahiem el H. Predictors of prostate cancer on extended biopsy in patients with high-grade prostatic intraepithelial neoplasia: a multivariate analysis model. *BJU Int* 2004;**94**:528–33.
32. Amin MM, Jeyaganth S, Fahmy N, et al. Subsequent prostate cancer detection in patients with prostatic intraepithelial neoplasia or atypical small acinar proliferation. *Can Urol Assoc J* 2007;**1**:245–9.
33. Kamoi K, Troncoso P, Babaian RJ. Strategy for repeat biopsy in patients with high grade prostatic intraepithelial neoplasia. *J Urol* 2000;**163**:819–23.
34. Lane BR, Zippe CD, Abouassaly R, et al. Saturation technique does not decrease cancer detection during followup after initial prostate biopsy. *J Urol* 2008;**179**:1746–50 discussion 50.
35. Lee MC, Moussa AS, Yu C, et al. Multifocal high grade prostatic intraepithelial neoplasia is a risk factor for subsequent prostate cancer. *J Urol* 2010;**184**:1958–62.
36. Lefkowitz GK, Taneja SS, Brown J, et al. Followup interval prostate biopsy 3 years after diagnosis of high grade prostatic intraepithelial neoplasia is associated with high likelihood of prostate cancer, independent of change in prostate specific antigen levels. *J Urol* 2002;**168**:1415–8.
37. Moore CK, Karikehalli S, Nazeer T, et al. Prognostic significance of high grade prostatic intraepithelial neoplasia and atypical small acinar proliferation in the contemporary era. *J Urol* 2005;**173**:70–2.
38. Godoy G, Taneja SS. Contemporary clinical management of isolated high-grade prostatic intraepithelial neoplasia. *Prostate Cancer Prostatic Dis* 2008;**11**:20–31.
39. Zynger DL, Yang X. High-grade prostatic intraepithelial neoplasia of the prostate: the precursor lesion of prostate cancer. *Int J Clin Exp Pathol* 2009;**2**:327–38.
40. Bostwick DG, Cheng L. Precursors of prostate cancer. *Histopathology* 2012;**60**:4–27.
41. O'dowd GJ, Miller MC, Orozco R, et al. Analysis of repeated biopsy results within 1 year after a noncancer diagnosis. *Urology* 2000;**55**:553–9.
42. Ouyang RC, Kenwright DN, Nacey JN, et al. The presence of atypical small acinar proliferation in prostate needle biopsy is predictive of carcinoma on subsequent biopsy. *BJU Int* 2001;**87**:70–4.
43. Iczkowski KA, Chen HM, Yang XJ, et al. Prostate cancer diagnosed after initial biopsy with atypical small acinar proliferation suspicious for malignancy is similar to cancer found on initial biopsy. *Urology* 2002;**60**:851–4.
44. Brausi M, Castagnetti G, Dotti A, et al. Immediate radical prostatectomy in patients with atypical small acinar proliferation. Over treatment? *J Urol* 2004;**172**:906–8 discussion 8–9.
45. Scattoni V, Roscigno M, Freschi M, et al. Predictors of prostate cancer after initial diagnosis of atypical small acinar proliferation at 10 to 12 core biopsies. *Urology* 2005;**66**:1043–7.
46. Ryu JH, Kim YB, Lee JK, et al. Predictive factors of prostate cancer at repeat biopsy in patients with an initial diagnosis of atypical small acinar proliferation of the prostate. *Korean J Urol* 2010;**51**:752–6.
47. Iczkowski KA. Current prostate biopsy interpretation: criteria for cancer, atypical small acinar proliferation, high-grade prostatic intraepithelial neoplasia, and use of immunostains. *Arch Pathol Lab Med* 2006;**130**:835–43.
48. Shah RB, Zhou M, LeBlanc M, et al. Comparison of the basal cell-specific markers, 34betaE12 and p63, in the diagnosis of prostate cancer. *Am J Surg Pathol* 2002;**26**:1161–8.
49. Wojno KJ, Epstein JI. The utility of basal cell-specific anti-cytokeratin antibody (34 beta E12) in the diagnosis of prostate cancer. A review of 228 cases. *Am J Surg Pathol* 1995;**19**:251–60.
50. Bostwick DG, Amin MB, Dundore P, et al. Architectural patterns of high-grade prostatic intraepithelial neoplasia. *Hum Pathol* 1993;**24**:298–310.
51. Shah RB, Magi-Galluzzi C, Han B, et al. Atypical cribriform lesions of the prostate: relationship to prostatic carcinoma and implication for diagnosis in prostate biopsies. *Am J Surg Pathol* 2010;**34**:470–7.
52. Reyes AO, Swanson PE, Carbone JM, et al. Unusual histologic types of high-grade prostatic intraepithelial neoplasia. *Am J Surg Pathol* 1997;**21**:1215–22.
53. Cohen RJ, O'Brien BA, Wheeler TM. Desquamating apoptotic variant of high-grade prostatic intraepithelial neoplasia: a possible precursor of intraductal prostatic carcinoma. *Hum Pathol* 2011;**42**:892–5.
54. Argani P, Epstein JI. Inverted (Hobnail) high-grade prostatic intraepithelial neoplasia (PIN): report of 15 cases of a previously undescribed pattern of high-grade PIN. *Am J Surg Pathol* 2001;**25**:1534–9.

55. Bishara T, Ramnani DM, Epstein JI. High-grade prostatic intraepithelial neoplasia on needle biopsy: risk of cancer on repeat biopsy related to number of involved cores and morphologic pattern. *Am J Surg Pathol* 2004;**28**:629–33.
56. Hailemariam S, Vosbeck J, Cathomas G, et al. Can molecular markers stratify the diagnostic value of high-grade prostatic intraepithelial neoplasia? *Hum Pathol* 2011;**42**:702–9.
57. Kronz JD, Allan CH, Shaikh AA, et al. Predicting cancer following a diagnosis of high-grade prostatic intraepithelial neoplasia on needle biopsy: data on men with more than one follow-up biopsy. *Am J Surg Pathol* 2001;**25**:1079–85.
58. de la Torre M, Haggman M, Brandstedt S, et al. Prostatic intraepithelial neoplasia and invasive carcinoma in total prostatectomy specimens: distribution, volumes and DNA ploidy. *Br J Urol* 1993;**72**:207–13.
59. Bostwick DG, Brawer MK. Prostatic intra-epithelial neoplasia and early invasion in prostate cancer. *Cancer* 1987;**59**:788–94.
60. Qian J, Wollan P, Bostwick DG. The extent and multicentricity of high-grade prostatic intraepithelial neoplasia in clinically localized prostatic adenocarcinoma. *Hum Pathol* 1997;**28**:143–8.
61. Sakr WA, Grignon DJ. Prostatic intraepithelial neoplasia and atypical adenomatous hyperplasia. Relationship to pathologic parameters, volume and spatial distribution of carcinoma of the prostate. *Anal Quant Cytol Histol* 1998;**20**:417–23.
62. Meeker AK, Hicks JL, Platz EA, et al. Telomere shortening is an early somatic DNA alteration in human prostate tumorigenesis. *Cancer Res* 2002;**62**:6405–9.
63. Vukovic B, Park PC, Al-Maghrabi J, et al. Evidence of multifocality of telomere erosion in high-grade prostatic intraepithelial neoplasia (HPIN) and concurrent carcinoma. *Oncogene* 2003;**22**:1978–87.
64. Bostwick DG, Shan A, Qian J, et al. Independent origin of multiple foci of prostatic intraepithelial neoplasia: comparison with matched foci of prostate carcinoma. *Cancer* 1998;**83**:1995–2002.
65. Emmert-Buck MR, Vocke CD, Pozzatti RO, et al. Allelic loss on chromosome 8p12-21 in microdissected prostatic intraepithelial neoplasia. *Cancer Res* 1995;**55**:2959–62.
66. Jenkins RB, Qian J, Lieber MM, et al. Detection of c-myc oncogene amplification and chromosomal anomalies in metastatic prostatic carcinoma by fluorescence *in situ* hybridization. *Cancer Res* 1997;**57**:524–31.
67. Qian J, Jenkins RB, Bostwick DG. Detection of chromosomal anomalies and c-myc gene amplification in the cribriform pattern of prostatic intraepithelial neoplasia and carcinoma by fluorescence *in situ* hybridization. *Mod Pathol* 1997;**10**:1113–9.
68. Qian J, Bostwick DG, Takahashi S, et al. Chromosomal anomalies in prostatic intraepithelial neoplasia and carcinoma detected by fluorescence *in situ* hybridization. *Cancer Res* 1995;**55**:5408–14.
69. Jeronimo C, Henrique R, Hoque MO, et al. Quantitative RARbeta2 hypermethylation: a promising prostate cancer marker. *Clin Cancer Res* 2004;**10**:4010–4.
70. Henrique R, Jeronimo C, Teixeira MR, et al. Epigenetic heterogeneity of high-grade prostatic intraepithelial neoplasia: clues for clonal progression in prostate carcinogenesis. *Mol Cancer Res* 2006;**4**:1–8.
71. Yang B, Sun H, Lin W, et al. Evaluation of global DNA hypomethylation in human prostate cancer and prostatic intraepithelial neoplasm tissues by immunohistochemistry. *Urol Oncol* 2013;**31**:628–34.
72. Henshall SM, Quinn DI, Lee CS, et al. Overexpression of the cell cycle inhibitor p16INK4A in high-grade prostatic intraepithelial neoplasia predicts early relapse in prostate cancer patients. *Clin Cancer Res* 2001;**7**:544–50.
73. Nakashiro K, Hayashi Y, Oyasu R. Immunohistochemical expression of hepatocyte growth factor and c-Met/HGF receptor in benign and malignant human prostate tissue. *Oncol Rep* 2003;**10**:1149–53.
74. Luo J, Zha S, Gage WR, et al. Alpha-methylacyl-CoA racemase: a new molecular marker for prostate cancer. *Cancer Res* 2002;**62**:2220–6.
75. Kang JS, Calvo BF, Maygarden SJ, et al. Dysregulation of annexin I protein expression in high-grade prostatic intraepithelial neoplasia and prostate cancer. *Clin Cancer Res* 2002;**8**:117–23.
76. Mosquera JM, Perner S, Genega EM, et al. Characterization of *TMPRSS2-ERG* fusion high-grade prostatic intraepithelial neoplasia and potential clinical implications. *Clin Cancer Res* 2008;**14**:3380–5.
77. Cerveira N, Ribeiro FR, Peixoto A, et al. *TMPRSS2-ERG* gene fusion causing ERG overexpression precedes chromosome copy number changes in prostate carcinomas and paired HGPIN lesions. *Neoplasia (New York, NY)* 2006;**8**:826–32.
78. Furusato B, Gao CL, Ravindranath L, et al. Mapping of *TMPRSS2-ERG* fusions in the context of multi-focal prostate cancer. *Mod Pathol* 2008;**21**:67–75.
79. Tomlins SA, Laxman B, Varambally S, et al. Role of the *TMPRSS2-ERG* gene fusion in prostate cancer. *Neoplasia (New York, NY)* 2008;**10**:177–88.
80. Lefkowitz GK, Sidhu GS, Torre P, et al. Is repeat prostate biopsy for high-grade prostatic intraepithelial neoplasia necessary after routine 12-core sampling? *Urology* 2001;**58**:999–1003.
81. Herawi M, Kahane H, Cavallo C, et al. Risk of prostate cancer on first re-biopsy within 1 year following a diagnosis of high grade prostatic intraepithelial neoplasia is related to the number of cores sampled. *J Urol* 2006;**175**:121–4.
82. Schlesinger C, Bostwick DG, Iczkowski KA. High-grade prostatic intraepithelial neoplasia and atypical small acinar proliferation: predictive value for cancer in current practice. *Am J Surg Pathol* 2005;**29**:1201–7.
83. Netto GJ, Epstein JI. Widespread high-grade prostatic intraepithelial neoplasia on prostatic needle biopsy: a significant likelihood of subsequently diagnosed adenocarcinoma. *Am J Surg Pathol* 2006;**30**:1184–8.
84. Roscigno M, Scattoni V, Freschi M, et al. Diagnosis of isolated high-grade prostatic intra-epithelial neoplasia: proposal of a nomogram for the prediction of cancer detection at saturation re-biopsy. *BJU Int* 2012;**109**:1329–34.
85. De Nunzio C, Albisinni S, Cicione A, et al. Widespread high grade prostatic intraepithelial neoplasia on biopsy predicts the risk of prostate cancer: a 12 months analysis after three consecutive prostate biopsies. *Arch Ital Urol Androl* 2013;**85**:59–64.
86. Antonelli A, Tardanico R, Giovanessi L, et al. Predicting prostate cancer at rebiopsies in patients with high-grade prostatic intraepithelial neoplasia: a study on 546 patients. *Prostate Cancer Prostatic Dis* 2011;**14**:173–6.
87. Fleshner NE, Kapusta L, Donnelly B, et al. Progression from high-grade prostatic intraepithelial neoplasia to cancer: a randomized trial of combination vitamin-E, soy, and selenium. *J Clin Oncol* 2011;**29**:2386–90.
88. Godoy G, Huang GJ, Patel T, et al. Long-term follow-up of men with isolated high-grade prostatic intra-epithelial neoplasia followed by serial delayed interval biopsy. *Urology* 2011;**77**:669–74.
89. Al-Hussain TO, Epstein JI. Initial high-grade prostatic intraepithelial neoplasia with carcinoma on subsequent prostate needle biopsy: findings at radical prostatectomy. *Am J Surg Pathol* 2011;**35**:1165–7.
90. Gokden N, Roehl KA, Catalona WJ, et al. High-grade prostatic intraepithelial neoplasia in needle biopsy as risk factor for detection of adenocarcinoma: current level of risk in screening population. *Urology* 2005;**65**:538–42.
91. Eminaga O, Hinkelammert R, Abbas M, et al. High-grade prostatic intraepithelial neoplasia (HGPIN) and topographical distribution in 1,374 prostatectomy specimens: existence of HGPIN near prostate cancer. *Prostate* 2013;**73**:1115–22.
92. Balaji KC, Rabbani F, Tsai H, et al. Effect of neoadjuvant hormonal therapy on prostatic intraepithelial neoplasia and its prognostic significance. *J Urol* 1999;**162**:753–7.

93. Mitterberger M, Horninger W, Aigner F, et al. Contrast-enhanced colour Doppler-targeted vs a 10-core systematic repeat biopsy strategy in patients with previous high-grade prostatic intraepithelial neoplasia. *BJU Int* 2010;**105**:1660–2.
94. Alexander EE, Qian J, Wollan PC, et al. Prostatic intraepithelial neoplasia does not appear to raise serum prostate-specific antigen concentration. *Urology* 1996;**47**:693–8.
95. Davidson D, Bostwick DG, Qian J, et al. Prostatic intraepithelial neoplasia is a risk factor for adenocarcinoma: predictive accuracy in needle biopsies. *J Urol* 1995;**154**:1295–9.
96. Raviv G, Janssen T, Zlotta AR, et al. Prostatic intraepithelial neoplasia: influence of clinical and pathological data on the detection of prostate cancer. *J Urol* 1996;**156**:1050–4 discussion 4–5.
97. Loeb S, Roehl KA, Yu X, et al. Use of prostate-specific antigen velocity to follow up patients with isolated high-grade prostatic intraepithelial neoplasia on prostate biopsy. *Urology* 2007;**69**:108–12.
98. Kunju LP, Chinnaiyan AM, Shah RB. Comparison of monoclonal antibody (P504S) and polyclonal antibody to alpha methylacyl-CoA racemase (AMACR) in the work-up of prostate cancer. *Histopathology* 2005;**47**:587–96.
99. Wu CL, Yang XJ, Tretiakova M, et al. Analysis of alpha-methylacyl-CoA racemase (P504S) expression in high-grade prostatic intraepithelial neoplasia. *Hum Pathol* 2004;**35**:1008–13.
100. Stewart J, Fleshner N, Cole H, et al. Prognostic significance of alpha-methylacyl-coA racemase among men with high grade prostatic intraepithelial neoplasia in prostate biopsies. *J Urol* 2008;**179**:1751–5 discussion 5.
101. Zhang S, Pavlovitz B, Tull J, et al. Detection of *TMPRSS2* gene deletions and translocations in carcinoma, intraepithelial neoplasia, and normal epithelium of the prostate by direct fluorescence *in situ* hybridization. *Diagn Mol Pathol* 2010;**19**:151–6.
102. Carver BS, Tran J, Gopalan A, et al. Aberrant ERG expression cooperates with loss of PTEN to promote cancer progression in the prostate. *Nat Genet* 2009;**41**:619–24.
103. He H, Osunkoya AO, Carver P, et al. Expression of ERG protein, a prostate cancer specific marker, in high grade prostatic intraepithelial neoplasia (HGPIN): lack of utility to stratify cancer risks associated with HGPIN. *BJU Int* 2012;**110**:E751–5.
104. Gao X, Li LY, Zhou FJ, et al. ERG rearrangement for predicting subsequent cancer diagnosis in high-grade prostatic intraepithelial neoplasia and lymph node metastasis. *Clin Cancer Res* 2012;**18**:4163–72.
105. Park K, Dalton JT, Narayanan R, et al. *TMPRSS2:ERG* gene fusion predicts subsequent detection of prostate cancer in patients with high-grade prostatic intraepithelial neoplasia. *J Clin Oncol* 2014;**32**:206–11.
106. Prostate Cancer Early Detection Version 1.2014. NCCN Guidelines.
107. Heidenreich A, Bastian PJ, Bellmunt J, et al. EAU guidelines on prostate cancer. Part 1: screening, diagnosis, and local treatment with curative intent – update 2013. *Eur Urol* 2014;**65**:124–37.
108. Thompson IM, Goodman PJ, Tangen CM, et al. The influence of finasteride on the development of prostate cancer. *N Engl J Med* 2003;**349**:215–24.
109. Thompson IM, Lucia MS, Redman MW, et al. Finasteride decreases the risk of prostatic intraepithelial neoplasia. *J Urol* 2007;**178**:107–9 discussion 10.
110. Andriole GL, Bostwick DG, Brawley OW, et al. Effect of dutasteride on the risk of prostate cancer. *N Engl J Med* 2010;**362**:1192–202.
111. Marshall JR, Tangen CM, Sakr WA, et al. Phase III trial of selenium to prevent prostate cancer in men with high-grade prostatic intraepithelial neoplasia: SWOG S9917. *Cancer Prev Res (Philadelphia, PA)* 2011;**4**:1761–9.
112. Taneja SS, Morton R, Barnette G, et al. Prostate cancer diagnosis among men with isolated high-grade intraepithelial neoplasia enrolled onto a 3-year prospective phase III clinical trial of oral toremifene. *J Clin Oncol* 2013;**31**:523–9.
113. Price D, Stein B, Sieber P, et al. Toremifene for the prevention of prostate cancer in men with high grade prostatic intraepithelial neoplasia: results of a double-blind, placebo controlled, phase IIB clinical trial. *J Urol* 2006;**176**:965–70 discussion 70–71.
114. Bostwick DG, Meiers I. Atypical small acinar proliferation in the prostate: clinical significance in 2006. *Arch Pathol Lab Med* 2006;**130**:952–7.
115. Iczkowski KA, MacLennan GT, Bostwick DG. Atypical small acinar proliferation suspicious for malignancy in prostate needle biopsies: clinical significance in 33 cases. *Am J Surg Pathol* 1997;**21**:1489–95.
116. Montironi R, Scattoni V, Mazzucchelli R, et al. Atypical foci suspicious but not diagnostic of malignancy in prostate needle biopsies (also referred to as "atypical small acinar proliferation suspicious for but not diagnostic of malignancy"). *Eur Urol* 2006;**50**:666–74.
117. Reyes AO, Humphrey PA. Diagnostic effect of complete histologic sampling of prostate needle biopsy specimens. *Am J Clin Pathol* 1998;**109**:416–22.
118. Zhou M, Shah R, Shen R, et al. Basal cell cocktail (34betaE12 + p63) improves the detection of prostate basal cells. *Am J Surg Pathol* 2003;**27**:365–71.
119. Shah RB, Kunju LP, Shen R, et al. Usefulness of basal cell cocktail (34betaE12 + p63) in the diagnosis of atypical prostate glandular proliferations. *Am J Clin Pathol* 2004;**122**:517–23.
120. Rubin MA, Zhou M, Dhanasekaran SM, et al. alpha-Methylacyl coenzyme A racemase as a tissue biomarker for prostate cancer. *JAMA* 2002;**287**:1662–70.
121. Zhou M, Aydin H, Kanane H, et al. How often does alpha-methylacyl-CoA-racemase contribute to resolving an atypical diagnosis on prostate needle biopsy beyond that provided by basal cell markers? *Am J Surg Pathol* 2004;**28**:239–43.
122. Hameed O, Sublett J, Humphrey PA. Immunohistochemical stains for p63 and alpha-methylacyl-CoA racemase, versus a cocktail comprising both, in the diagnosis of prostatic carcinoma: a comparison of the immunohistochemical staining of 430 foci in radical prostatectomy and needle biopsy tissues. *Am J Surg Pathol* 2005;**29**:579–87.
123. Stewart CS, Leibovich BC, Weaver AL. Prostate cancer diagnosis using a saturation needle biopsy technique after previous negative sextant biopsies. *J Urol* 2001;**166**:86–91 discussion 91–92.
124. Zhang M, Amberson JB, Epstein JI. Two sequential diagnoses of atypical foci suspicious for carcinoma on prostate biopsy: a follow-up study of 179 cases. *Urology* 2013;**82**:861–4.
125. Merrimen JL, Jones G, Hussein SA, et al. A model to predict prostate cancer after atypical findings in initial prostate needle biopsy. *J Urol* 2011;**185**:1240–5.
126. Kim KH, Kim YB, Lee JK, et al. Pathologic results of radical prostatectomies in patients with simultaneous atypical small acinar proliferation and prostate cancer. *Korean J Urol* 2010;**51**:398–402.

CHAPTER

# 7

# Prostate Cancer in the Elderly

*Janet E. Baack Kukreja, MD, Edward M. Messing, MD, FACS*

**Department of Urology, University of Rochester School of Medicine and Dentistry, Rochester, NY, USA**

## INTRODUCTION

Prostate cancer is a disease of the elderly, with the peak incidence of diagnosed disease at 80 years of age.[1] While many view prostate cancer as an indolent disease, it still remains the second leading cause of cancer death among men in the United States, second only to lung cancer.[2] Men aged ≥75 at diagnosis account for over 50% of prostate cancer deaths and almost 25% of the incidence; however, this group represents fewer than 10% of all men in the United States ≥40 years of age.[3]

In 2010, the US census estimated the number of men ≥75 years old at about 7,250,000.[4] Of those who are aged ≥75 years old, 1,136,000 were living with prostate cancer diagnosed sometime in the past 20 years. Thus about 15% of men aged ≥75 have prostate cancer or were treated for prostate cancer in the past.[5] As the male population ages and life expectancy increases, the prevalence of prostate cancer in the ≥75 age group will increase. By 2021 males born in the Baby Boom era (1946–1964) will begin to mature into the age group of men ≥75 years old. The actuarial predictions in Table 7.1 illustrate life expectancy for healthy elderly men.

## THE MAGNITUDE OF PROSTATE CANCER IN ELDERLY MEN

The impact prostate cancer will have on elderly individuals is largely determined by the grade and stage of their disease at diagnosis in the context of their health status. The elderly often have high-risk prostate cancer at presentation. Of those with clinical stage T1c (no palpable abnormalities on digital rectal exam [DRE], with an elevated prostate-specific antigen [PSA] serum level), M0 tumors, and age ≥75, greater than 25% have high-risk disease (defined as PSA ≥20 ng/mL and/or Gleason score ≥8).[8] In contrast, high-risk disease accounts for less than 16% of all men with stage T1c prostate cancer at diagnosis.[8]

Not only do the elderly commonly have high-risk prostate cancer at diagnosis, the frequency of distant metastases (M1) at diagnosis among men increases with age.[9] In a recent SEER study, metastatic disease was present in 8, 13, and 17% of men in age groups 80–84, 85–89, and >90, respectively.[9] Indeed almost half of all men with M1 disease at diagnosis are ≥75 years old when diagnosed.[9] Similarly, the 5-year prostate-specific mortality was 13, 20, and 30% for men diagnosed with prostate cancer at ages 80–84, 85–89, and >90, respectively.[9]

Higher-stage disease at diagnosis leads to a higher chance of disease-specific mortality. However, this could be explained by selection bias. Bias may stem from urologists only performing biopsies on older men with rapidly rising or markedly elevated PSAs, or with large prostatic nodules/masses on DRE. Such practice would result in a higher proportion of elderly men diagnosed with prostate cancer having very aggressive disease compared with younger men; however, it could not explain how, of all men diagnosed with stage T1cNxM0 disease, PSA < 10, Gleason score ≥8, almost 40% are aged 75 and older at diagnosis.[8] These findings along with those from autopsy studies from the United States and overseas, give compelling support to the contention that the elderly do have more aggressive prostate cancer.

## PATHOLOGIC CHARACTERISTICS

In an autopsy study, men ≥70 years old were not only more likely to have any prostate cancer, but also more likely to have larger and higher-grade cancers than younger men.[10] There are many hypothesized reasons for the greater incidence of prostate cancer with higher Gleason scores (on average) among the elderly. These include relatively impaired immune responses in the elderly, which reduce host resistance to malignancies

**Prostate Cancer. http://dx.doi.org/10.1016/B978-0-12-800077-9.00007-4**

TABLE 7.1 Life Expectancy for a Healthy Male

| References | Albert et al. (2010)[6] | Walter and Covinsky (2001)[7] |
|---|---|---|
| Age | Years of life left | |
| 75 | 14.8 | 14.3 |
| 80 | 11.5 | 10.8 |
| 85 | 8.3 | 7.9 |
| 90 | 5.6 | 5.8 |

developing.[11] Additionally, epigenetic changes and cumulative exposure to carcinogens with aging result in greater DNA abnormalities, the development of more prostate as well as other causes, and less differentiated cancers.[12]

## PSA SCREENING AND DIAGNOSIS WITH PROSTATE BIOPSY

Life expectancy should be considered in each individual prior to PSA testing, prostate biopsy, and prostate cancer treatment. The use of a comorbidity-adjusted life-expectancy calculator or adjusted life tables can help clinicians objectively determine life years left for elderly men.[13–15] Many of these tools are new, hence, validation in a prospective study is lacking. However, for estimation purposes, these are deemed better than clinician intuition.

PSA testing has recently come under great scrutiny by the United States Preventative Services Task Force (USPSTF) for all age groups, especially the elderly (≥75 years old).[16] Importantly, many screening trials have excluded men >75 under the assumption that men in this age group have too few remaining years of life to realize any benefit from early detection of prostate cancer.[16,17] However, according to Table 7.1, those who are 75 years old and healthy will live 14 more years, on average. This represents roughly 33% of the male population more than 75 years of age.[6]

The International Society of Geriatric Oncology recommends that the decision to test for prostate cancer be made on health status rather than age alone.[18] PSA testing should be individualized, even for men having life expectancies of greater than 10 years. Men 70–79 years old should be considered for early diagnosis, and even those patients who are 80–84, if fit, should be included in this group.[18] Given the increased incidence of aggressive prostate cancer in the elderly, patients may wish to be tested.[18] Discussion with patients prior to PSA testing should inform them of the risks and benefits of such testing. Routine PSA testing may lower the likelihood of being diagnosed with metastatic prostate cancer, and improve overall and prostate cancer-specific survival rates.[19] In the pre-PSA era, the rates of metastatic prostate cancer at diagnosis were three times higher than the actual number observed in the 2008 US population, and most of this difference was noted among the elderly.[20] Thus, PSA testing should be considered in the healthier segment of the population ≥75 years of age. Predictors of elderly men undergoing a PSA test, billed as a screening test, are often based not on comorbidities of a patient, but rather on age, marital status, and education.[21]

Current data on the benefits of PSA testing in the elderly are limited. Moreover, greater research must be done on the results of newer ways to identify prostate cancers (e.g., PHI,[22] PCA-3,[23] Mi-Prostate Score,[24] OPKO 4K,[25] and MRI[26]) to determine what role these tests have in diagnosing prostate cancer in the elderly. While there is no specific reason to believe that these tests will perform differently in the young and old, they have primarily been tested in men who have undergone radical prostatectomy, which is rarely performed on men of advanced age. Specifically, test results found useful in younger men may not be so in the elderly.

When patients are found to have elevated PSAs for their age, biopsy should be considered, specifically if they have more than 10 years of remaining life.[27] Some patients may want their PSAs checked, and if it is elevated but they have less than 10 years of life left, the decision to biopsy may prove difficult. Urologists who were surveyed overwhelmingly answered they would not biopsy a man >75 years old for a PSA between 4 ng/mL and 10 ng/mL, although information from large national databases indicates this is often not adhered to.[8,19,28] Accordingly, some advocate that those over the age of 80 do not need a prostate biopsy to determine the presence of prostate cancer, as the biopsy results do not often alter cancer care.[29] For special cases, treatment of elderly without a prostate biopsy may be considered, especially in patients with dementia, spinal cord compression, or inability to come off anticoagulation. While individual cases may vary, a candidate for empiric treatment generally would need to have significant symptoms of local or metastatic disease not responding to low morbidity therapies as well as a total serum PSA >50 ng/mL (with a 98.5% positive predictive value) and/or a lower PSA with a very abnormal DRE.[30]

A negative prostate biopsy before age 75 results in a lower chance of prostate cancer diagnosis on future biopsies.[31] Among the elderly with previously negative biopsies, the likelihood of prostate cancer detection on future biopsies is just about 10%.[31]

Earlier reports have cited that complications from ultrasound-guided prostate needle biopsy increases with age.[32] However, more recent research has found that neither age nor underlying comorbidities are associated with increased complications from ultrasound-guided prostate biopsy.[33] While it is beyond the scope of this chapter to discuss the merits of prostatic cancer screening, we strongly believe that the elderly with comorbidity-adjusted life expectancy of ≥10 years, especially those with other prostate cancer risk factors (family history, race, etc.), merit strong consideration for PSA testing and thereafter biopsy, if treatment would be recommended for those with high-risk and possibly intermediate-risk disease. This becomes a common clinical quandary for the healthy elderly man who has lower urinary tract voiding symptoms (LUTS) and is being considered for treatment beyond alpha adrenergic blocker therapy. In such individuals, the use of 5-alpha reductase inhibitors (which lower PSA) or more invasive interventions, particularly those that do not provide tissue for histologic diagnosis, should be withheld until you have confirmed whether it is cancer or benign prostatic hyperplasia that you are treating. In such individuals, measuring PSA, examining the prostate, and performing a biopsy if either yields suspicious results, seems appropriate.

In elderly men with age-specific elevated PSAs, it is important to differentiate prostate cancer into risk classifications. Categorization is often done with the D'Amico classification (low, intermediate, and high risk), which can only be done with a prostate biopsy.[34] Survival advantages are seen as early as 5 years after treatment, if administered with curative intent for those with high-risk prostate cancer.[27]

## TREATMENT

Knowledge of the disease's aggressiveness is often needed to determine the necessity or appropriateness of prostate cancer treatment. A healthy elderly male with high-risk disease and a life expectancy of more than 10 years should be considered for treatment. On the other hand, a healthy elderly male with low-risk disease is not expected to benefit from any form of therapy.[35]

## PATIENT PREFERENCES AND TREATMENT PRACTICE PATTERNS IN THE ELDERLY WITH PROSTATE CANCER

Elderly patients with prostate cancer are not offered curative treatment as often as their younger counterparts are.[36–38] Elderly or even younger prostate cancer patients with serious comorbidities and life expectancies of ≤10 years are less likely to benefit from curative treatment of localized prostate cancer; however, in certain circumstances, particularly high-risk disease and a life expectancy of >5 years, treatment may be appropriate.[27,35,37–41] Still, it has been estimated that up to 15% of healthy ≥75 year-old men have been undertreated for local and regional prostate cancer.[39] More importantly, the undertreatment occurs not just in those with a long life expectancy, but also in those with high-grade prostate cancer, a group of men who would likely benefit from treatment.[42] Review of the Ontario Cancer Registry confirmed that patients with high-risk cancer are more often not offered curative treatment, in otherwise healthy elderly men.[40]

Chronological age by itself should not preclude curative treatment. Compared to other factors, age is often given more weight when considering prostate cancer treatment options, resulting in an age bias for localized prostate cancer treatment.[35] Key treatment considerations in all prostate cancer patients, especially the elderly, include other medical conditions, functional status, life expectancy, and patient preferences.[18] Additionally, high-risk disease should be the focus of most treatment efforts made in the elderly.[41] Lastly, when making treatment decisions, the risk of developing prostate cancer and treatment-related complications which might interfere with quality of life should be considered, particularly in light of preexisting comorbidities.[41,43]

All prostate cancer patients and their families should be involved in informed discussion of treatment options, regardless of age. It is also necessary to incorporate patient preferences into treatment decision making.[43] Elderly patients with aggressive prostate cancer reported a satisfactory quality of life after treatment with curative intent, specifically with either radical prostatectomy or external beam radiation therapy.[44]

## FITNESS FOR PROSTATE CANCER TREATMENT IN THE ELDERLY

One important facet for individualizing prostate cancer management in the elderly is the geriatric assessment. The geriatric assessment is a battery of tests usually administered by a medical oncologist or geriatrician in patients ≥70 years old who have physical and/or cognitive impairments.[45] It helps find treatable conditions that may improve treatment tolerability, and predict cancer survival probability.[45,46] The assessment primarily works to determine the patient's overall fitness to undergo treatment and the treatment's potential tolerability when weighed against the risk posed by the malignancy. Furthermore, in determining prostate cancer treatment, social and practical barriers (e.g., how well an elderly man who does not drive can get to daily radiation therapy treatments for several weeks) must be identified during evaluation.[46] For prostate cancer, geriatric assessments must also determine the ability to endure therapy with endocrine treatments and chemotherapy.[46]

"Fit" is a term used to describe elderly patients who are functionally independent without relevant comorbidities, and who may be candidates to receive the same treatment as younger men.[1] The "unfit" are a diverse population who are considered "vulnerable" and/or "frail." Using the cumulative illness ratings scale along with evaluation of activities of daily living, these patients can be placed in two categories.[1] The first category, "vulnerable" patients, are functionally independent in basic activities of daily living (e.g., dressing, bathing, and eating), but not in "instrumental" activities (e.g., driving, cooking, and remembering medications).[1] The vulnerable patient has less than two Grade 3 diseases on the cumulative illness ratings scale (Table 7.2).[48] The vulnerable patient usually will benefit from geriatric interventions.[1] The second category, "frail" patients, on the other hand, have higher 2-year mortality rates, usually >40%.[1] The frail are typically over 85 years of age and have geriatric syndromes, at least three Grade 3 comorbidities or 1 comorbidity that is a Grade 4, and dependence on others for activities of daily living.[1] These patients are unlikely to return to their previous state of independence and health after a significant health event.[1]

## TREATMENT OF PROSTATE CANCER IN ELDERLY PATIENTS

In unfit patients with localized prostate cancer, watchful waiting probably represents the best strategy, especially when they are unlikely to live 10 years.[49] Watchful waiting is a type of expectant management, but differs from active surveillance. In watchful waiting, the primary aim is to avoid curative treatment, and instead offer ways relieve symptoms from the disease. In active surveillance, one continues to monitor for prostate cancer progression, and curative treatment is usually offered if there is disease progression.[50]

An individualized approach based on a patient's specific factors should be implemented.[51] In an asymptomatic elderly patient with locally advanced disease, it is reasonable to consider the treatment risks and benefits prior to proceeding with any treatment. For example, androgen deprivation therapy (ADT) may be given alone or with radiation therapy for intermediate or particularly high-risk disease. However, ADT may accelerate frailty in older men by causing difficulty with mobility, balance, cognition, physical endurance, and nutrition.[51]

Treatment of the frail elderly man should take into account the stage-specific recommendations during informed decision-making conversations. Those who are frail, asymptomatic, and are diagnosed with incidental low- or intermediate-risk prostate cancer will most certainly not benefit from upfront treatment.[52] Watchful waiting is highly recommended for frail elderly patients diagnosed with stage T1a to T2b prostate cancer.[27,52] ADT may be considered as a treatment, but baseline cognitive function should be determined as this can decline even further in this patient population.[53,54]

Prostatectomy or radiation therapy may be a reasonable treatment for high-risk elderly patients if their life expectancy is longer than 10 years.[55,56] Additionally, if life expectancy is greater than 5 years, radiation treatment with ADT may be reasonable for high-risk disease patients, including those with clinical T3–T4 disease (even if PSA is <20 and Gleason score is <8).[57,58] ADT alone may be considered in elderly men "unfit" for other treatments, when they are not able to receive curative treatment.[57]

In elderly prostate cancer patients with a biochemical recurrence after definitive therapy, watchful waiting may also be considered due to their susceptibility to side effects from ADT, especially among those who are frail.[51,59] "ADT light" may be considered in men who are unlikely to tolerate traditional ADT.[60] An antiandrogen (e.g., biclutamide) alone or in combination with a 5-alpha reductase inhibitor (e.g., finasteride) is considered ADT light or "peripheral ADT."[61] While claiming less side effects, the ADT light regimen may prove less effective. Since gynecomastia and breast tenderness is a common sequella of this therapy, prophylactic breast radiation should probably be administered before ADT light treatment is started.[61] Finally, if ADT is contemplated, it should be noted that intermittent ADT has been found to be as efficacious as continuous ADT, and is less morbid and costly than continuous ADT for men with rising PSAs after local therapy.[61] This is important particularly for elderly men with >10 years life expectancy.

Metastatic disease is more likely to develop in advanced age, particularly as life expectancy continues to increase.[1] A geriatric assessment can also prove useful in counseling this age group on treatment options.[62] A fit senior must be offered the same treatments extended to younger patients.[3,63] Today most trials for metastatic and metastatic castrate-resistant prostate cancer (mCRPC) exclude patients who fall into the unfit category.[63] Table 7.3 summarizes available treatments for advanced prostate cancer, including the number of patients who were ≥75 years of age in the pivotal trials that led to FDA approval. In these studies, even if advanced age was not an exclusion criterion, given that the purpose was to see if the therapy had a beneficial effect on prostate cancer progression, patients with comorbidities, which reduced study treatment tolerance or posed an imminent threat to survival, were almost always excluded from participation. These factors would make it impossible to assess the effects of the treatment itself on prostate cancer, and require larger sample sizes and longer follow-up experiences for the studies' endpoints to be reached. Thus most participants would be in the "fit" category. Additionally, depending on the study, many of the patients had to be fit enough to have previously undergone docetaxel

**TABLE 7.2** Cumulative Illness Ratings Scale[47]

| Organ system | Grade of disease 0 | 1 | 2 | 3 | 4 |
|---|---|---|---|---|---|
| General disease severity | None | Current mild problem or past significant problem | Moderate disability or morbidity OR requires "first line" therapy | Severe/constant significant disability OR "uncontrollable" chronic problem | Extremely severe OR immediate treatment required OR end organ failure OR severe impairment in function |
| Heart | None | Remote MI (>5 years previous) OR occasional angina treated with prn meds | CHF compensated with meds OR daily antiangina meds OR left ventricular hypertrophy OR atrial fibrillation OR bundle branch block OR daily antiarrhythimic drugs | Previous MI within 5 years OR abnormal stress test OR status post percutaneous coronary angioplasty or coronary artery bypass graft surgery | Marked activity restriction secondary to cardiac status |
| Vascular | None | HTN compensated with salt restriction and weight loss/serum cholesterol >200 mg/dL serum cholesterol above normal | Daily anti-HTN meds OR one symptom of atherosclerotic disease OR aortic aneurysm <4 cm | Two or more symptoms of atherosclerosis | Previous surgery for vascular problem/aortic aneurysm >4 cm |
| Hematology | None | Hgb: females > 10 < 12, males > 12 < 14 OR anemia of chronic disease | Hgb: females > 8 < 10, males > 10 < 12 OR anemia secondary to iron, vitamin B12, or folate deficiency or chronic renal failure OR total WBC >2000 or <4000 | Hgb: females < 8, males < 10 OR total WBC < 2000 | Any leukemia, any lymphoma |
| Respiratory | None | Recurrent episodes of acute bronchitis/currently treated asthma with prn inhalers OR cigarette smoker >10 but <20 pack years | X-ray evidence of COPD OR requires daily theophylline or inhalers OR treated for pneumonia two or more times in the past 5 years OR smoked 20–40 pack years | Limited ambulation secondary to limited respiratory capacity OR requires oral steroids for lung disease OR smoked >40 pack years | Requires supplemental oxygen OR at least one episode of respiratory failure requiring assisted ventilation OR any lung cancer |
| Sight, ears, nose, and throat | None | Corrected vision 20/40 OR chronic sinusitis OR mild hearing loss | Corrected vision 20/60 or reads newsprint with difficulty OR requires hearing aid OR chronic sinus or nasal complaints requiring medication OR requires medication for vertigo | Partially blind OR unable to read newsprint OR conversational hearing still impaired with hearing aid | Functional blindness OR functional deafness OR laryngectomy OR requires surgical intervention for vertigo |
| Upper GI | None | Hiatal hernia OR heartburn complaints treated with prn meds | Needs daily H2 blocker or antacid OR documented gastric or duodenal ulcer within 5 years | Active ulcer OR guiac positive stools OR any swallowing disorder or dysphagia | Gastric cancer OR history of perforated ulcer OR melena or hematochezia from UGI source |
| Lower GI | None | Constipation managed with prn meds OR active hemorrhoids OR status posthernia repair | Requires daily bulk laxatives or stool softeners OR diverticulosis OR untreated hernia | Bowel impaction in the past year OR daily use of stimulant laxatives or enemas | Hematochezia from lower GI source, currently impacted, diverticulitis flare up OR status postbowel obstruction OR bowel carcinoma |
| Liver | None | History of hepatitis >5 years ago OR cholecystectomy | Mildly elevated LFTs (>150% of normal) OR hepatitis within 5 years OR cholelithiasis OR daily or heavy alcohol use within 5 years | Elevated bilirubin (total >2) OR marked elevation of LFTs (>150% of normal) OR requires supplemental pancreatic enzymes for digestion | Biliary obstruction OR any biliary tree carcinoma OR cholecystitis OR pancreatitis OR active hepatitis |

*(Continued)*

TABLE 7.2 Cumulative Illness Ratings Scale[47] *(cont.)*

| Organ system | Grade of disease | | | | |
|---|---|---|---|---|---|
| | 0 | 1 | 2 | 3 | 4 |
| Renal | None | s/p Kidney stone passage <10 years or asymptomatic kidney stone OR pyelonephritis < 5 year | Serum creatinine >1.5 but <3.0 without diuretic or antihypertensive medication | Serum creatinine >3.0 or serum creatinine >1.5 in conjunction with diuretic, antihypertensive, or bicarbonate therapy OR current pyelonephritis | Requires dialysis OR renal carcinoma |
| Genitourinary | None | Stress incontinence OR hysterectomy OR BPH without urinary symptoms | Abnormal pap smear OR frequent UTIs (≥3 in past year) OR urinary incontinence (nonstress) in females OR BPH with hesitancy or frequency OR current UTI OR any urinary diversion procedure OR status post-TURP | Vaginal bleeding OR cervical carcinoma *in situ* OR hematuria OR urosepsis in <1 year | Acute urinary retention |
| Muscle and skin | None | Uses prn meds for arthritis or has mildly limited ADLs from joint pathology OR excised nonmelanoma skin cancers OR skin infections requiring antibiotics within a year | Daily arthritis meds or use of assistive devices or moderate limitation in ADLs OR daily meds for chronic skin conditions OR melanoma without metastasis | Severely impaired ADLs secondary to arthritis OR requires steroids for arthritis condition OR vertebral compression fractures from osteoporosis | Wheelchair-bound OR severe joint deformity or severely impaired usage OR osteomyelitis OR any bone or muscle carcinoma OR metastatic melanoma |
| Neurological | None | Frequent headaches requiring prn meds without interference with daily activities OR a history of TIA phenomena (at least one) | Requires daily meds for chronic headaches or headaches that regularly interfere with daily activities OR S/P CVA without significant residual OR neurodegenerative disease-mild severity | S/P CVA with mild residual dysfunction OR any CNS neurological procedure OR neurodegenerative disease-moderate severity | S/P CVA with residual functional hemiparesis or aphasia OR neurodegenerative disease-severe |
| Endocrine | None | DM compensated with diet OR obesity: BMI > 30 requires thyroid hormone replacement | DM requiring insulin or oral agents OR fibrocystic breast disease | Any electrolyte disturbance requiring hospital treatment OR morbid obesity BMI >45 | Brittle or poorly controlled DM or diabetic coma in the past year OR requires adrenal hormone replacement OR adrenal, thyroid, or breast carcinoma |
| Psychiatric | None | Minor psychiatric condition or history of previous outpatient mental health treatment during a crisis OR outpatient treatment for depression >10 years ago OR current usage of minor tranquilizers for episodic anxiety (occasional) OR mild early dementia | A history of major depression (<10 years) OR mild dementia OR any previous psychiatric hospitalization OR any psychotic episode OR substance abuse history >10 years ago | Currently meets criteria for major depression or 2 or more episodes of major depression <10 years OR moderate dementia OR current usage of daily antianxiety medication OR currently meets criteria for substance abuse or dependence OR requires daily antipsychotic medication | Current mental illness requiring psychiatric hospitalization, institutionalization, or intensive outpatient management, patients with severe or suicidal depression, acute psychosis or psychotic decompensation, severe agitation from dementia, severe substance abuse OR severe dementia |
| Oncology | None | Cancer diagnosed in the past without evidence of recurrence or consequence in <10 years | Cancer diagnosed in the past without evidence of recurrence or consequence in <5 years | Required chemotherapy, radiation, hormonal therapy or surgical procedure in <5 years | Recurrent malignancy of life-threatening potential OR failed containment of the primary OR palliative treatment stage |

MI, myocardial infarction; CHF, congestive heart failure; HTN, hypertension; Hgb, hemoglobin; GI, gastrointestinal; LFT, liver function test; BPH, benign prostatic hyperplasia; ADL, adult daily activities of living; TIA, transient ischemic attack; CVA, cerebrovascular accident; CNS, central nervous system; DM, diabetes mellitus; and BMI, body mass index.
*Table created from Miller and Towers.*[47]

**TABLE 7.3** Treatments for Advanced Prostate Cancer and Elderly Inclusion Criteria

| Generic | Brand | Treatment indication | FDA approval date | Type of study | Age range | Patients aged ≥75 in treatment group (%) | Median survival in months (alternative treatment) | Performance status criteria | Important limitations | References |
|---|---|---|---|---|---|---|---|---|---|---|
| Docetaxel + prednisone | Taxotere Docefrez | mCRPC | 5/2004 | Phase III, open label, RCT. | 36–92 | 185 (34%) | 18.9 versus 16.5 (mitoxantrone) | Karnofsky PS >60% | 87% Karnofsky PS >70% | [64] |
| | | | | Retrospective | ≥75 | 175 (100%) | 15 | | 82% of patients were ECOG ≤1 | [3] |
| Degarelix | Firmagon | Advanced prostate cancer | 12/2008 | Open label RCT-noninferiority | | 172 (42%) | | | Previous docetaxel | [65] |
| Sipuleucel-T | Provenge | Asymptomatic or minimally symptomatic mCRPC | 4/2010 | Phase III-Double blind RCT with placebo | 40–89 | 438 ≥ 65 (72.9%) | 25.8 versus 21.9 (placebo) | ECOG ≤1 | Initial enrollment of Gleason <7 (75% of pts), no previous long bone fractures | [66,67] |
| Cablitaxel + prednisone | Jevtana | mCRPC | 6/2010 | Open label RCT | | 70 (18.9%) | 15.1 versus 12.7 (mitoxantrone) | ECOG ≤2 | Previous docetaxel; ≥65 more likely to experience fatal outcomes | [66,68] |
| Abiraterone + prednisone | Zytiga | mCRPC | 4/2011 | Phase III-Double blind RCT with placebo | 44–95 | 185 (34%) | 16.5 versus 8.3 (prednisone only) | ECOG ≤1 | No previous chemotherapy | [66,69,70] |
| | | | | Phase III-Double blind RCT with placebo | 39–95 | 220 (28%) | 14.8 versus 10.9 (placebo) | ECOG ≤2 | Previous docetaxel | [71] |
| Enzalutamide | Xtandi | mCRPC | 8/2012 | Phase III-Double blind RCT | 42–93 | 240 (25%); 274 (31.4%) | 18.4 versus 13.6 (placebo) | ECOG ≤1 | Previous docetaxel | [66,72] |
| Radium 223 dichloride | Xofigo | mCRPC to bone (symptomatic) | 5/2013 | Phase III-Double blind RCT with placebo | 44–94 | 198 (33%) | 14.9 versus 11.3 (placebo) | ECOG ≤2 | No visceral metastasis | [66,73] |

mCRPC, metastatic castrate resistant prostate cancer; ECOG, Eastern Cooperative Group; PS, performance status; and RCT, randomized control trial.

chemotherapy and still have an Eastern Cooperative Group (ECOG) performance of <1 or <2. Comorbidities are present in many elderly men, and we do not know if study findings apply to elderly men in the same way they do to healthier men and/or younger men.[6]

## CONCLUSIONS

Men over 75 years old represent less than 10% of the male population in the United States and about 25% of all men diagnosed with prostate cancer. Moreover, almost 50% of men with distant metastasis at diagnosis and over 50% who die from prostate cancer are over age 75 at diagnosis. Because these are actual numbers of patients and not just the proportion of men in the age group with these features, one must conclude that this disease is more aggressive in the elderly, a conclusion substantiated by autopsy studies and how often high-grade cancer is found even in stage T1c, PSA < 10 disease. Older men thus experience a disproportionate suffering and mortality from prostate cancer.

The elderly are a group of patients with unique challenges for prostate cancer diagnosis and treatment. It is anticipated that this group will continue to grow in size as men are living longer. Developing strategies to address these challenges will continue to be important areas of research, both for diagnosis and treatment of all stages of prostate cancer.

Elderly fit or healthy men may benefit from PSA testing if they have a life expectancy of >10 years, and should not be excluded from considering such testing based on advanced age. The same applies for considerations of prostate cancer treatment in high-risk disease. Numerous studies reveal that this age group is actually being undertreated for high-risk prostate cancer.

However, for less fit elderly men, life expectancy calculations should be performed prior to diagnostic testing. Moreover, even for high-risk disease, geriatric assessments, life expectancy calculations as well as consideration of personal and family preferences must be incorporated into the decision making. Using this approach is critical to optimize management for every elderly man at risk for or with prostate cancer.

## References


1. Falci C, Morello E, Pierre Droz J. Treatment of prostate cancer in unfit senior adult patients. *Cancer Treat Rev* 2009;**35**:522–7.
2. US Cancer Statistics Working Group. *United States cancer statistics: 1999–2010 incidence and mortality web-based report*. Atlanta: US Department of Health and Human Services, Centers for Disease Control and Prevention and National Cancer Institute; 2013. Available from: www.cdc.gov/uscs.
3. Italiano A, Ortholan C, Oudard S, et al. Docetaxel-based chemotherapy in elderly patients (age 75 and older) with castration-resistant prostate cancer. *Eur Urol* 2009;**55**:1368–76.
4. Werner CA. The older population: 2010. 2010 census briefs. United States Census Bureau; 2011. Available at: http://www.census.gov/2010census/data/.
5. National Cancer Institute. SEERStat Software. Available from: http://seer.cancer.gov/seerstat. [accessed 10.08.2014].
6. Albert FS, Bragg JM, Brooks JC. Health expectancy. The Actuarial Practice Forum. February 2010.
7. Walter LC, Covinsky KE. Cancer screening in elderly patients: a framework for individualized decision making. *JAMA* 2001;**285**(21):2750–6.
8. Zhang H, Messing EM, Travis LB, et al. Age and racial differences among PSA-detected (AJCC stage T1cN0M0) prostate cancer in the U.S.: a population-based study of 70,345 men. *Front Oncol* 2013;**3**(312):1–20.
9. Scosyrev E, Messing EM, Mohile S, et al. Prostate cancer in the elderly: frequency of advanced disease at presentation and disease-specific mortality. *Cancer* 2012;**118**:3062–670.
10. Delongchamps NB, Wang CY, Chandan V, et al. Pathological characteristics of prostate cancer in elderly men. *J Urol* 2009;**182**:927–30.
11. Leibovitz A, Baumoehl Y, Segal R. Increased incidence of pathological and clinical prostate cancer with age: age related alterations in local immune surveillance. *J Urol* 2004;**172**:435–7.
12. Damaschke NA, Yang B, Bhusari S, et al. Epigenetic susceptibility factors for prostate cancer with aging. *Prostate* 2013;**73**:1721–30.
13. Cho H, Klabunde CN, Yabroff R, et al. Comorbidity-adjusted life expectancy: a new tool to inform recommendations for optimal screening strategies. *Ann Intern Med* 2013;**159**(10):667–77.
14. Mariotto AB, Wang Z, Klabunde CN, et al. Life tables adjusted for co-morbidity more accurately estimate non-cancer survival for recently diagnosed cancer patients. *J Clin Epidemiol* 2013;**66**:1376–85.
15. Repetto L, Comandini D, Mammoliti S. Life expectancy, comorbidity and quality of life: the treatment equation in the older cancer patients. *Crit Rev Oncol Hematol* 2001;**37**:147–52.
16. US Preventive Services Task Force. Screening for prostate cancer: U.S. Preventive Services Task Force recommendation statement. *Ann Intern Med* 2008;**149**:185–91.
17. Kawachi MH, Babaian RJ, Bahnson RR, editors. *National comprehensive cancer network clinical practice guidelines in oncology: prostate cancer early detection*. Jenkintown, PA: National Comprehensive Cancer Network; 2010 Version 2.
18. Droz JP, Aapro M, Balducci L, et al. Management of prostate cancer in older patients: updated recommendations of a working group of the International Society of Geriatric Oncology. *Lancet* 2014;**15**:e404–14.
19. Scosyrev E, Wu G, Golijanin D, et al. Prostate-specific antigen testing in older men in the United States: data from the behavioral risk factor surveillance system. *BJU Int* 2012;**110**(10):1485–90.
20. Hu JC, Williams SB, Carter SC, et al. Population-based assessment of prostate-specific antigen testing for prostate cancer in the elderly. *Urol Oncol* 2015;**33**:69.e29–34.
21. Scosyrev E, Wu G, Mohile S, et al. Prostate-specific antigen screening for prostate cancer and the risk of overt metastatic disease at presentation: analysis of trends over time. *Cancer* 2012;**118**:5768–76.
22. Lazzeri M, Haese A, Abrate A, et al. Clinical performance of serum prostate-specific antigen isoform [−2]proPSA (p2PSA) and its derivatives, %p2PSA and the prostate health index (PHI), in men with a family history of prostate cancer: results from a multicentre European study, the PROMEtheuS project. *BJU Int* 2013;**112**:313–21.
23. Crawford ED, Rove KO, Trabulsi EJ, et al. Diagnostic performance of PCA3 to detect prostate cancer in men with increased prostate specific antigen: a prospective study of 1,962 cases. *J Urol* 2012;**188**:1726–31.
24. Salami SS, Schmidt F, Laxman B, et al. Combining urinary detection of TMPRSS2:ERG and PCA3 with serum PSA to predict diagnosis of prostate cancer. *Urol Oncol* 2013;**31**:566–71.

25. Carlsson S, Maschino A, Schröder F, et al. Predictive value of four kallikrein markers for pathologically insignificant compared with aggressive prostate cancer in radical prostatectomy specimens: results from the European Randomized Study of Screening for Prostate Cancer section Rotterdam. *Eur Urol* 2013;**64**:693–9.
26. Phillips R. MRI reduces low-risk prostate cancer diagnosis and biopsy. *Nat Rev Urol* 2014;**11**(5):245.
27. Mistry S, Mayer W, Khavari R, et al. Who's too old to screen? Prostate cancer in elderly men. *Can Urol Assoc J* 2009;**3**(3):205–10.
28. Fowler Jr FJ, Bin L, Collins MM, et al. Prostate cancer screening and beliefs about treatment efficacy: a national survey of primary care physicians and urologists. *Am J Med* 1998;**104**:526–32.
29. Bott SRJ, Foley CL, Bull MD, et al. Are prostatic biopsies necessary in men aged ≥80 years? *BrJUrol Int* 2006;**99**:335–8.
30. Gerstenbluth RE, Seftel AD, Hampel N, et al. The accuracy of the increased prostate specific antigen level (greater than or equal to 20 ng/ml) in predicting prostate cancer: is a biopsy always necessary? *J Urol* 2002;**168**:1990–3.
31. Hoffman RM, Denberg T, Hunt WC, et al. Prostate cancer testing following a negative prostate biopsy: over testing the elderly. *J Gen Intern Med* 2007;**22**(8):1129–43.
32. Roberts RO, Bergstralh EJ, Besse JA, et al. Trends and risk factors for prostate biopsy complications in the pre-PSA and PSA eras, 1980 to 1997. *Urology* 2002;**59**:79.
33. Chiang N, Chang SJ, Pu YS, et al. Major complications and associated risk factors of transrectal ultrasound guided prostate needle biopsy: a retrospective study of cases in Taiwan. *J Formos Med Assoc* 2007;**106**(11):929–34.
34. D'Amico AV, Whittington R, Malkowicz SB, et al. Biochemical outcome after radical prostatectomy, external beam radiation therapy, or interstitial radiation therapy for clinically localized prostate cancer. *JAMA* 1998;**280**:969–74.
35. Bian SX, Hoffman KE. Management of prostate cancer in elderly men. *Semin Radiat Oncol* 2013;**23**:198–205.
36. Bubolz T, Wasson JH, Lu-Yao G, et al. Treatments for prostate cancer in older men: 1984–1997. *Urology* 2001;**58**:977–82.
37. Houterman S, Janssen-Heijnen ML, Verheij CD, et al. Greater influence of age than comorbidity on primary treatment and complications of prostate cancer patients: an in-depth population-based study. *Prostate Cancer Prostatic Dis* 2006;**9**:179–84.
38. Fitzpatrick JM. Management of localized prostate cancer in senior adults: the crucial role of comorbidity. *BJU Int* 2008;**101**(Suppl. 2):16–22.
39. Alibhai SMH, Krahn MD, Cohen MM, et al. Is there an age bias in the treatment of localized prostate carcinoma? *Cancer* 2004;**100**(1):72–81.
40. Schwartz KL, Alibhai SM, Tomlinson G, et al. Continued undertreatment of older men with localized prostate cancer. *Urology* 2003;**62**:860–5.
41. Chen RC, Carpenter WR, Hendrix LH, et al. Receipt of guideline-concordant treatment in elderly prostate cancer patients. *Int J Radiat Oncol Biol Phys* 2014;**88**(2):332–8.
42. Llyod A, Penson D, Dewilde S, et al. Eliciting patient preferences for hormonal therapy options in the treatment of metastatic prostate cancer. *Prostate Cancer Prostatic Dis* 2008;**11**:153–9.
43. Namiki S, Ishidoya S, Kawamura S, et al. Quality of life among elderly men treated for prostate cancer with either radical prostatectomy or external beam radiation therapy. *J Cancer Res Clin Oncol* 2010;**136**:379–86.
44. Terret C, Albrand G, Droz JP. Geriatric assessment in elderly patients with prostate cancer. *Crit Rev Oncol Hematol* 2005;**55**(3): 241–52.
45. Extermann M, Aapro M, Bernabei R, et al. Use of comprehensive geriatric assessment in older cancer patients: recommendations from the task force on CGA of the International Society of Geriatric Oncology (SIOG). *Crit Rev Oncol Hematol* 2005;**55**:241–52.
46. Mohile SG, Bylow K, Dale W, et al. A pilot study of the Vulnerable Elders Survey-13 compared with the comprehensive geriatric assessment for identifying disability in older patients with prostate cancer who receive androgen ablation. *Cancer* 2007;**109**:802–10.
47. Miller MD, Towers A. *Manual of guidelines for scoring the cumulative illness rating scale for geriatrics (CIRS-G)*. Pittsburg, PA: University of Pittsburgh; 1991.
48. D'Amico AV, Moul J, Carroll PR, et al. Cancer-specific mortality after surgery or radiation for patients with clinically localized prostate cancer managed during the prostate-specific antigen era. *J Clin Oncol* 2003;**11**:2163–72.
49. Bill-Axelson A, Holmberg L, Ruutu M, et al. Radical prostatectomy versus watchful waiting in early prostate cancer. *N Engl J Med* 2011;**364**:1708–17.
50. Bylow K, Mohile SG, Stadler WM, et al. Does androgen-deprivation therapy accelerate the development of frailty in older men with prostate cancer? *Cancer* 2007;**15**:2604–13.
51. Logothetis CJ, Hoosein NM, Hsieh JT. The clinical and biological study of androgen independent prostate cancer (AI PCa). *Semin Oncol* 1994;**21**(5):620–9.
52. Heidenreich A, Aus G, Bolla M, et al. European Association of Urology guidelines on prostate cancer. *Eur Urol* 2008;**53**:68–80.
53. Holmes L, Chan W, Jiang Z, et al. Effectiveness of androgen deprivation therapy in prolonging survival of older men treated for locoregional prostate cancer. *Prostate Cancer Prostatic Dis* 2007;**10**:388–95.
54. Mohile SG, Lacy M, Rodin M, et al. Cognitive effects of androgen deprivation therapy in an older cohort of men with prostate cancer. *Crit Rev Oncol Hematol* 2010;**75**:152–9.
55. Alibhai SMH, Leach M, Tomlinson G, et al. 30-Days mortality and major complications after radical prostatectomy: influence of age and comorbidity. *J Natl Cancer Inst* 2005;**97**(20):1525–32.
56. Rogers CG, Sammon JD, Sukumar S, et al. Robot assisted radical prostatectomy for elderly patients with high risk prostate cancer. *Urol Oncol* 2013;**31**:193–7.
57. Antonarakis ES, Eisenberger MA. Does short-term ADT before and during radiation therapy improve outcomes in locally advanced prostate cancer? *Nat Clin Pract Urol* 2008;**5**(9):480–1.
58. Kirk D. Prostate cancer in elderly. *Eur J Surg Oncol* 1998;**24**:379–83.
59. Fung C, Dale W, Mohile SG. Prostate cancer in the elderly patient. *J Clin Oncol* 2014;**32**:1–8.
60. Monk JP, Halabi S, Picus J, et al. Efficacy of peripheral androgen blockade in prostate cancer patients with biochemical failure after definitive local therapy: results of cancer and leukemia group B (CALBG) 9782. *Cancer* 2012;**118**:4139–47.
61. Droz JP, Chaladaj A. Management of metastatic prostate cancer: the crucial role of geriatric assessment. *BJU Int* 2008;**101**(Suppl. 2):23–9.
62. Ohlmann CH, Engelmann UH, Heidenreich A. Second-line chemotherapy with docetaxel for prostate-specific antigen (PSA) relapse in men with hormone-refractory prostate cancer (HRPC) previously treated with docetaxel-based chemotherapy. *J Clin Oncol* 2005;**23**:4682.
63. Tannock IF, de Wit R, Berry WR, et al. Docetaxel plus prednisone or mitoxantrone plus prednisone for advanced prostate cancer. *N Engl J Med* 2004;**351**:1502–12.
64. Pazdur R. Latest update of FDA prostate cancer drug information. http://www.cancer.gov/cancertopics/druginfo/fda-degarelix; 2013 [accessed 08.09.2014].
65. Mukherji D, Pezaro CJ, Shamseddine A, et al. New treatment developments applied to elderly patients with advanced prostate cancer. *Cancer Treat Rev* 2013;**39**:578–83.
66. Kantoff PW, Higano CS, Shore ND, et al. Sipuleucel-T immunotherapy for castration-resistant prostate cancer. *N Engl J Med* 2010;**363**(5):411–22.
67. de Bono JS, Oudard S, Ozguroglu M, et al. Prednisone plus cabazitaxel or mitoxantrone for metastatic castration-resistant prostate

cancer progressing after docetaxel treatment: a randomized open label trial. *Lancet* 2010;**376**(9747):1147–54.

68. Ryan CJ, Smith MR, de Bono JS, et al. Abiraterone in metastatic prostate cancer without previous chemotherapy. *N Engl J Med* 2013;**368**(2):138–48.
69. Mulders PF, Molina A, Marberger M, et al. Efficacy and safety of abiraterone acetate in an elderly patient subgroup (aged 75 and older) with metastatic castration-resistant prostate cancer after docetaxel-based chemotherapy. *Eur Urol* 2014;**65**:875–83.
70. de Bono JS, Logothetis CJ, Molina A, et al. Abiraterone and increased survival in metastatic prostate cancer. *N Engl J Med* 2011;**364**(21):1995–2005.
71. Beer TM, Armstrong AJ, Rathkopf DE, et al. Enzalutamide in metastatic prostate cancer before chemotherapy. *N Engl J Med* 2014;**371**(5):424–33.
72. Scher HI, Fizazi K, Saad F, et al. Increased survival with Enzalutamide in prostate cancer after chemotherapy. *N Engl J Med* 2012;**367**(13):1187–97.
73. Parker C. Overall survival benefit of radium-223 chloride (Alpharadin™) in the treatment of patients with symptomatic bone metastases in castration-resistant prostate cancer (CRPC): a phase III randomized trial (ALSYMPCA). In: *European Multidisciplinary Cancer Congress 2011*. Stockholm; 2011.

CHAPTER

# 8

# Prostate Cancer and Other Primary Malignancies

*Matthew S. Schaff, MD, Jack H. Mydlo, MD, FACS*

**Department of Urology, Temple University Hospital, Philadelphia, PA, USA**

## BACKGROUND

As the detection and treatment of cancer improve, the prevalence of multiple primary malignancies has increased and will undoubtedly continue to increase. One out of every two men will receive a cancer diagnosis in their lifetime, with prostate cancer comprising 44% of these diagnoses.[1] According to the Surveillance, Epidemiology, and End Results (SEER) Program database, as of 2013, 16% of new cancer diagnoses represented a second- or higher-order malignancy. Prostate cancer is no exception. In a review of the Minneapolis Veterans Affairs tumor registry, Powell et al. concluded that one out of every nine cancer diagnoses represented second-order malignancies, and prostate cancer was the most common primary cancer diagnosis and the second most common secondary cancer diagnosis.[2] Approximately 17% of all prostate cancer diagnoses occur in combination with another primary malignancy diagnosed before or after the time of diagnosis of prostate cancer.[3] The overall incidence of multiple primary tumors in prostate cancer patients ranges from as low as 11.5%[4] to as high as 20%.[5] Evaluation of autopsy specimens by Hajdu and Hajdu reported an incidence of multiple malignancies in prostate cancer patients as high as 27%[6]; however, current data correlate with previously stated range.

When considering multiple malignancies of the same patient, it is important to consider the timing of diagnosis and distinguish between metachronous and synchronous malignancies. Metachronous malignancies are likely a result of the improvements in treatment modalities and the increased cancer-specific survivorship. As treatment strategies improve and patients are surviving longer with the diagnosis of prostate cancer, the probability of developing a second malignancy increases. Synchronous malignancies are likely the result of epidemiologic risk factors in a given population. Family history is one the greatest risk factors in the development of prostate cancer and many other malignancies. It can therefore be inferred that patients who develop one malignancy may have a genetic predisposition to developing malignancies at other anatomic locations. The advent of screening protocols for prostate cancer and other malignancies such as colon cancer has allowed for the diagnosis of lower-grade and lower-stage cancers that previously would have been clinically undiagnosed. The most common secondary cancer diagnoses made after a primary diagnosis of prostate cancer are colorectal, lung, and bladder.[2,3] Renal cell cancer comprises approximately 5% of cancer diagnosis made in patients following a diagnosis of prostate cancer.[7]

## PROSTATE CANCER AND BLADDER CANCER

After colorectal and lung cancer, bladder cancer is the third most common malignancy diagnosed in patients with a previous diagnosis of prostate cancer. It stands to reason that other urologic malignancies would more likely be diagnosed in prostate cancer patients, as there is an inherent bias in evaluation by the urologist. Van Hemelrijck et al. reported an increased diagnosis of bladder and renal cell cancers in patients with prostate cancer both before and after the diagnosis of prostate cancer, thus suggesting a detection bias.[3] Liskow et al. noted an increased detection of secondary bladder cancers in prostate cancer patients, which is likely related to the routine use of cystoscopy in the workup of their prostate cancer patients.[4] Local radiotherapy for the treatment of prostate cancer has been implicated as an etiologic agent for the development of bladder cancer.

**Prostate Cancer. http://dx.doi.org/10.1016/B978-0-12-800077-9.00008-6**

Brenner et al. showed a small but increased risk of solid tumor development with local radiotherapy when compared to surgery alone.[8] This risk becomes clinically relevant as nearly 25% of patients with prostate cancer undergo some form of radiotherapy as initial treatment. Boorjian et al. reported a 1.5% overall incidence of bladder cancer in patients with prostate cancer. Smoking and radiotherapy independently increased the risk of bladder cancer in these patients two- and threefold, respectively. Patients who smoked and received some form of radiotherapy had a nearly fourfold increased risk of developing subsequent bladder cancer.[9] Local radiotherapy not only increases the risk of developing bladder cancer but also leads to more aggressive cancer at the time of diagnosis. In a study published by Yee et al., patients who had received radiotherapy presented with higher-grade tumors and progressed to higher-stage disease when compared to patients who had not received radiotherapy. The incidence of muscle invasive disease was much higher in the irradiated compared to nonirradiated patients (70% vs. 43%).[10]

Previously undiagnosed prostate cancer is discovered on nearly 20% of cystoprostatectomy specimens. Incidental prostate cancer tends to be organ-confined. Pignot et al. reported an overall incidence of 21.4% of incidental prostate cancer in their cystoprostatectomy patients.[11] Review of data from multiple centers showed incidence of prostate cancer on cystoprostatectomy specimens ranging from 10.9% to 32.5%. Pignot et al. noted a significant variation in incidence related to age at time of surgery. Prostate cancer was diagnosed in 5.2% of patients under the age of 50 and diagnosed in 30.5% of patients over the age of 75.[11] These data correlate with the well-established age-related incidence of prostate cancer. Along with the increased incidence of incidental prostate cancer in older patients undergoing radical cystoprostatectomy, older patients were noted to have more advanced disease. Higher Gleason score and higher TNM stage were noted in patients aged 75 or older compared to patients aged 50 or younger.[11] As treatment of invasive bladder cancer becomes more advanced with the advent of orthotopic bladder substitutes, prostate-sparing cystectomy has been described as a means to improve urinary and sexual function in this patient population. However, the incidence of prostate cancer as previously described has kept this from becoming a front-line treatment option. Apex-sparing surgery has recently been described as a potential compromising solution, although it still carries significant risk of subsequent prostate cancer development. Gakis et al. reviewed 95% of patients who underwent cystoprostatectomy and evaluated the pathology of the prostate with specific attention to the apical tissue. Of the 95 patients, 26 (27%) had incidental prostate cancer, with seven (7.3%) having cancer in the apical tissue.[12] These data support and stress the need for careful dissection of the prostate at the time of cystoprostatectomy.

## PROSTATE CANCER AND KIDNEY CANCER

Though the incidence of kidney cancer in patients with prostate cancer is not as high as that of bladder cancer, it is of particular importance and interest to urologists. Van Hemelrjick et al. report that renal cell cancer comprises 5% of malignancies diagnosed in patients carrying a diagnosis of prostate cancer.[7] Greenberg et al. reported an increased incidence of renal cell carcinoma in patients with prostate cancer when compared to incidence reports of the general population.[13] Greenberg suggested an underlying common etiologic factor to account for the increased incidence of renal cell carcinoma in prostate cancer patients. This conjecture has been supported by subsequent studies. In a review of the SEER database from 1973 to 1996, Barocas and Rabbani reported an increased incidence of renal cell carcinoma in prostate cancer patients when compared to the general population. However, there was no increased incidence of renal cell carcinoma among patients with prostate cancer when compared to patients with one of six other index malignancies.[14] These data suggest an underlying etiologic factor in the development of multiple malignancies at different anatomical locations. Conversely, they reported an increased incidence of prostate cancer in men with renal cell carcinoma when compared to the general population, which was not seen in patients with one of the six other index malignancies.[14] Barocas offers detection bias as an explanation for this latter finding, suggesting that patients undergoing treatment for renal cell carcinoma by an urologist are more likely to receive rigorous screening for prostate cancer. In a review of the SEER database from 1992 to 2010, Davis et al. also reported an increased incidence of renal cell carcinoma in patients with prostate cancer when compared to the general population. They noted that the incidence was most increased in the African-American population. They also observed a markedly increased incidence in renal cell carcinoma diagnosed in the 12 months immediately following diagnosis of their prostate cancer, thus suggesting a detection bias.[15]

## PROSTATE CANCER AND COLORECTAL CANCER

Colorectal cancers (e.g., colon, rectal, and anal cancers) are the most common cancers diagnosed in men carrying a diagnosis of prostate cancer. Colon, rectal, and anal cancers comprise 17.1% of cancer diagnoses in men with prostate cancer.[7] The correlation between prostate cancer and colorectal is multifaceted. Detection bias has been proposed as colorectal cancer, like prostate cancer, has a well-established screening protocol. The workup and surgical planning for the treatment of prostate cancer may also lead to increased detection of colorectal malignancies. Heme-positive stools have been discovered during digital rectal examination on patients referred for workup of elevated PSA.[16] Similarly, abdominal imaging is a common step in the surgical planning of malignancies and can lead to the diagnosis of otherwise clinically unapparent colorectal lesions. Computed tomography is also used as a tool in treatment planning for the radiation oncologist treating low-risk prostate cancers and may result in the detection of concurrent malignancies.[17]

The treatment of prostate cancer has also been implicated as an etiologic factor in the development of colon, rectal, and anal cancers. Local radiotherapy has been shown to increase the risk of developing rectal cancer. Nieder et al. reviewed 243,082 patients in the SEER database who underwent radical prostatectomy or radiotherapy for prostate cancer between 1988 and 2003. They found an increased risk of rectal cancer in patients who received some form of radiotherapy compared to those who received surgery alone. The highest incidence was seen in patients who received external beam radiotherapy with a standardized incidence ratio of 1.26.[18] Interestingly, the converse relationship has not been born out by the data. Smith-Gagen et al. reviewed the SEER database and identified 29,230 patients with stage II or III rectal cancer who received some form of radiotherapy. A total of 2424 patients developed a secondary malignancy. The incidence or prostate cancer however was lower in patients who had received radiotherapy when compared to a similar cohort that had not received radiotherapy.[19] Prostate cancer's sensitivity to radiotherapy has been postulated as an explanation for this finding. Systemic therapies have also been shown to increase the risk of colorectal cancer. Van Hemelrijck et al. found that the highest incidence proportion of colon cancer after diagnosis of prostate cancer was seen in patients undergoing endocrine therapy.[3] These data are supported by Gillessen et al. who reviewed 107,859 patients over the age of 67 diagnosed with prostate cancer from 1993 through 2002, and compared rates of colorectal cancer among patients receiving androgen deprivation either by orchiectomy or GnRH agonists with patients who did not receive androgen deprivation. They found an incidence rate of 6.3 colon cancers per 1000 person-years in patients who underwent orchiectomy, 4.4 colon cancers per 1000 person-years in patients taking GnRH agonists, and 3.7 cancers per 1000 person-years in patients who did not receive androgen deprivation. They found that the risk of colon cancer also increased with duration of GnRH agonist therapy.[20]

Lifestyle risk factors are also common between prostate cancer and colorectal cancer. Smoking has been shown in exhausting detail to increase the risk of multiple cancers including prostate and colorectal cancers. Obesity has long been shown to increase the risk of multiple cancers including prostate and colon cancer.[21] High-carbohydrate and high-fat western diets have been shown to increase concentrations of insulin and several circulating growth factors. High-circulating concentrations of insulin-like growth factor 1 (IGF-1) have been shown to increase the risk of both prostate cancer and colon cancer. The correlation between high levels of IGF-1 and colon cancer has been well established in the colorectal literature.[22] Roddam et al. compared circulating concentrations of multiple IGFs and IGF binding proteins in patients with prostate cancer and case-matched control. They found an increased risk of prostate cancer associated with increased concentrations of IGF-1.[23]

Prostate cancer risk has also been shown to be increased in familial syndromes associated with colon cancer. Lynch syndrome, caused by a mutation in mismatch repair genes, is associated with a three- to eightfold increased incidence of prostate cancer. Haraldsdottir et al. reviewed genetic data from 188 patients with known Lynch syndrome and reported a fivefold increased incidence of prostate cancer when compared to population-based control. They also noted a significantly increased risk of prostate cancer with the MSH2 mutation compared to the MLH1, MSH6, or PMS2 mutations.[24] Recent data published by Rosty et al. argues that prostate cancer should be included as part of the Lynch syndrome. They report that prostate cancer was the first or only diagnosed tumor in 37% of Lynch syndrome carriers.[25]

## TREATMENT OF MULTIPLE MALIGNANCIES

Diagnosis of a second malignancy in the face of prostate cancer or the diagnosis of prostate cancer in the setting of another primary malignancy presents a dilemma in treatment. In the case of synchronous lesions, the question in management is simultaneous versus staged resection. In a report by Mydlo and Gerstein, several urologic oncologists were questioned as to how to handle multiple lesions. Most of the urologists polled noted that the secondary lesion is usually asymptomatic and that surgical management of one tumor does not significantly alter the natural history of the other. Despite this fact, most of the urologists polled advocated for simultaneous resections unless one tumor was noted to be more aggressive than the other.[16] As mentioned earlier, as many as 20% of patients undergoing radical cystoprostatectomy for muscle invasive bladder cancer are found to have incidental prostate cancer on the surgical specimen.

The vast majority of these lesions is organ-confined and requires no further treatment.

Metachronous lesions present a separate therapeutic dilemma. Surgical intervention is rarely questioned if the second lesion is amenable to such an approach. However, systemic and nonsurgical local therapies require closer attention. As discussed earlier, local radiotherapy for the treatment of prostate cancer has been implicated in the development and progression of local epithelial malignancies including bladder and rectal cancer. Systemic hormone therapy has also been shown to increase the risk of colon cancer. It is therefore of utmost importance to counsel patients accordingly. A multidisciplinary approach should be carried out for the care of these patients with a team comprised of the surgeons specializing in the resection of these malignancies, a medical and/or radiation oncologist, and the patient's primary care physician.

In conclusion, there may be a small cohort of urologic cancer patients with other primary malignancies because of the increased aging population, increased exposure to causative agents, improved treatments, and advances in cancer detection. The need for the specialist to focus not just on their organ of interest cannot be overemphasized. In many reports of multiple tumors, the second lesion was discovered during the workup of the first. Vigilance for these multiple malignancies should rest initially with the primary care physician. However, the subspecialist should be on the alert. As understanding of cancer biology, screening protocols, and treatment options continues to improve, the incidence of multiple primary malignancies will undoubtedly increase, prompting the need for a multidisciplinary approach to patient care.

## References


1. Hayat MJ, Howlader N, Reichman ME, et al. Cancer statistics, trends, and multiple primary cancer analyses from the Surveillance, Epidemiology, and End Results (SEER) Program. *Oncologist* 2007;**12**:20–37.
2. Powell S, Tarchand G, Rector T, et al. Synchronous and metachronous malignancies: analysis of the Minneapolis Veterans Affairs (VA) tumor registry. *Cancer Cause Control* 2013;**24**:1565–73.
3. Van Hemelrijck M, Drevin L, Holmberg L, et al. Primary cancers before and after prostate cancer diagnosis. *Cancer* 2012;**118**:6207–16.
4. Liskow AS, Neugut AI, Benson M, et al. Multiple primary neoplasms in association with prostate cancer in black and white patients. *Cancer* 1987;**59**:380–4.
5. Lynch H, Larsen A, Magnuson C, et al. Prostate carcinoma and multiple primary malignancies: study of a family and 109 consecutive prostate cancer patients. *Cancer* 1966;**19**:1891–7.
6. Hajdu SI, Hajdu EO. Multiple primary malignant tumors. *J Am Geriatr Soc* 1968;**16**:16–26.
7. Van Hemelrijck M, Feller A, Garmo H, et al. Incidence of second malignancies for prostate cancer. *PLoS ONE* 2014;**9**:e102596.
8. Brenner DJ, Curtis RE, Hall EJ, et al. Second malignancies in prostate carcinoma patients after radiotherapy compared with surgery. *Cancer* 2000;**88**:398–406.
9. Boorjian S, Cowan JE, Konety BR, et al. Bladder cancer incidence and risk factors in men with prostate cancer: results from Cancer of the Prostate Strategic Urologic Research Endeavor. *J Urol* 2007;**177**:883–7 discussion 887–888.
10. Yee DS, Shariat SF, Lowrance WT, et al. Impact of previous radiotherapy for prostate cancer on clinical outcomes of patients with bladder cancer. *J Urol* 2010;**183**:1751–6.
11. Pignot G, Salomon L, Lebacle C, et al. Prostate cancer incidence on cystoprostatectomy specimens is directly linked to age: results from a multicentre study. *BJU Int* 2015;**115**:87–93.
12. Gakis G, Schilling D, Bedke J, et al. Incidental prostate cancer at radical cystoprostatectomy: implications for apex-sparing surgery. *BJU Int* 2010;**105**:468–71.
13. Greenberg RS, Rustin ED, Clark WS. Risk of genitourinary malignancies after cancer of the prostate. *Cancer* 1988;**61**:396–401.
14. Barocas D, Rabbani F. A population-based study of renal cell carcinoma and prostate cancer in the same patients. *BJU Int* 2006;**97**:30–6.
15. Davis EJ, Beebe-Dimmer JL, Yee CL, et al. Risk of second primary tumors in men diagnosed with prostate cancer: a population-based cohort study. *Cancer* 2014;**120**:1–7.
16. Mydlo J, Gerstein M. Patients with urologic cancer and other nonurologic malignancies: analysis of a sample and review of the literature. *Urology* 2001;**58**(6):864–9 http://www.sciencedirect.com/science/article/pii/S0090429501013942 [accessed 22.06.2014].
17. Miller JS, Puckett ML, Johnstone PAS. Frequency of coexistent disease at CT in patients with prostate carcinoma selected for definitive radiation therapy: is limited treatment-planning CT adequate? *Radiology* 2000;**215**:41–4.
18. Nieder AM, Porter MP, Soloway MS. Radiation therapy for prostate cancer increases subsequent risk of bladder and rectal cancer: a population-based cohort study. *J Urol* 2008;**180**:2005–9 discussion 20092010.
19. Smith-Gagen J, Goodwin GA, Tay J. Multiple primary tumors following stage II and III rectal cancer in patients receiving radiotherapy, 1998–2010. *J Cancer Res Clin Oncol* 2014;**140**:949–55.
20. Gillessen S, Templeton A, Marra G, et al. Risk of colorectal cancer in men on long-term androgen deprivation therapy for prostate cancer. *J Natl Cancer Inst* 2010;**102**:1760–70.
21. Wolin KY, Carson K, Colditz GA. Obesity and cancer. *Oncologist* 2010;**15**:556–65.
22. Giovannucci E. Insulin, insulin-like growth factors and colon cancer: a review of the evidence 1. *J Nutr* 2001;**131**:3109–20.
23. Roddam A, Allen N, Appleby P, et al. Insulin-like growth factors, their binding proteins, and prostate cancer risk: analysis of individual patient data from 12 prospective studies. *Ann Intern Med* 2008;**149**:83–8.
24. Haraldsdottir S, Hampel H, Wei L, et al. Prostate cancer incidence in males with Lynch syndrome. *Genet Med* 2014;**16**:553–7.
25. Rosty C, Walsh MD, Lindor NM, et al. High prevalence of mismatch repair deficiency in prostate cancers diagnosed in mismatch repair gene mutation carriers from the colon cancer family registry. *Fam Cancer* 2014;**13**:573–82.

CHAPTER

# 9

# Biopsy Prophylaxis, Technique, Complications, and Repeat Biopsies

*Richard K. Babayan, MD, Mark H. Katz, MD*

**Boston University School of Medicine, Boston, MA, USA**

## BIOPSY PROPHYLAXIS

The majority of prostate biopsies are performed with transrectal ultrasound (TRUS) guidance and tissue sampling, with only a small proportion being done via a transperineal approach. The bacterial colonization of the rectum with gram-negative organisms poses a significant and *potentially* devastating risk of infection and sepsis during TRUS biopsy. Rates of sepsis after prostate biopsy have been traditionally reported to be 1–3%.

Risk factors for postbiopsy sepsis include antibiotic therapy within the past 6 months, history of prostatitis, and, most importantly, noncompliance with biopsy antibiotic prophylaxis recommendations. Those men with chronic indwelling urethral catheters, neurogenic bladder with elevated postvoid residuals, and history of exposure to multidrug resistant organisms (e.g., hospital employee) also deserve special attention to minimize risk of infection.

The American Urological Association (AUA) recommends antibiotic prophylaxis for all men undergoing TRUS biopsy of the prostate. This recommendation is based on level-1 evidence from randomized clinical trials (RCTs). The best-practice statement was last updated on January 1, 2014.[1] First line agents for prophylaxis include oral quinolones, trimethoprim-sulfamethoxazole (tmp-smx), and oral cephalosporins (first, second, and third generation). Alternative regimens include intravenous (IV)/intramuscular (IM) aminoglycosides or Aztreonam. The recommended duration of therapy is 24 h since several RCTs have demonstrated equivalent efficacy of 24-h dosing compared with 3-day regimens.[2–4] These recommendations are for empiric therapy, assuming preprocedure urine cultures are negative. When organisms have been isolated from rectal swab cultures, antimicrobial therapy should be selected based in culture data. For those clinicians working in the hospital setting, it is prudent to review institutional data on organism susceptibilities to various antimicrobials. For example, the rate of quinolone- and tmp-smx-resistance may vary widely between and within institutions over time. Updated hospital sensitivity data should be consulted when selecting the best prophylaxis regimen.

Minimizing the risk of postbiopsy infection has been extensively investigated, and in additional to adhering to the previous recommendations, isolated studies have demonstrated adjunctive maneuvers to further decrease the rate of infection. The use of broad-spectrum IV antibiotics prior to biopsy has demonstrated a decreased rate of infection compared with standard oral regimens. In a prospective study ($n = 170$), the incidence of sepsis after prostate biopsy was compared in men receiving IV ertapenem plus oral agents versus oral agents alone. The group receiving the additional IV ertapenem demonstrated a significantly decreased incidence of sepsis (0% vs. 6.7%, $p = 0.03$). The benefit of decreased sepsis rates with IV antibiotics must be weighed against the additional cost and inconvenience with this strategy.[5]

Fluoroquinolone resistance rates over 20% have been demonstrated in the rectal carriage of men undergoing prostate biopsy.[6] Indeed, rectal swab culture taken in advance of prostate biopsy is one strategy to minimize the risk of postprocedure infection. Taylor et al. compared targeted prophylaxis (culture swab data) and empiric prophylaxis in a study of 457 men and showed a significant decrease in infectious complications using the targeted approach. Importantly, the targeted therapy approach was significantly more cost-effective than empiric therapy due to the fact that managing infectious complications is extremely expensive.[7] The use of betadine-impregnated enemas prior to biopsy has yielded decreased bacterial colony counts in the rectum as

Prostate Cancer. http://dx.doi.org/10.1016/B978-0-12-800077-9.00009-8

well as fewer infectious complications.[8] Disinfection of the needle tip with formalin after each biopsy core taken has also shown a decreased risk of sepsis in a retrospective study.[9]

The authors recommend following the Best Practice Guidelines of the AUA and tailoring prophylaxis choices to local institution bacterial sensitivities. If, after review of internal data, infection rates are still higher than expected (1–2%), then consider employing one or more of the previously mentioned strategies to minimize infection.

Historically, all antiplatelet and anticoagulant agents such as aspirin and warfarin were discontinued prior to prostate biopsy for fear of increased bleeding (hematuria, rectal bleeding). However, several recent studies show that continuing these medications does not increase the risk of postbiopsy hemorrhage.[10,11] This finding is particularly important in patients with strong indications for antiplatelet agents (e.g., recent coronary stent) and anticoagulants (e.g., atrial fibrillation, embolic stroke). It is our recommendation to continue these agents prior to prostate biopsy for the majority of men, especially those with significant cardiovascular and cerebrovascular disease.

## BIOPSY TECHNIQUE AND SPECIMEN PROCESSING

Prostate biopsy technique has dramatically evolved over the past several decades. The days of finger-guided biopsies in the preultrasound era and the classic 6-core "sextant" biopsy from the 1980s and 1990s have made way for the modern day 12-core prostate biopsy. Biopsy technique has continued to evolve with the goals of maximizing clinically significant cancer-detection rates (CDR), avoidance of false negative biopsies (better NPV), minimizing the identification of indolent cancers, providing good concordance with surgical specimen pathology to reliably assess risk and guide treatment options, and containing costs. *The modern day TRUS-guided prostate biopsy can be described as a 12-core systematic sampling, including the far lateral and apical regions of the prostate.* Newer techniques involving magnetic-resonance imaging (MRI)-guided biopsy and MRI–Ultrasound fusion biopsy are beyond the scope of this chapter but will be addressed elsewhere in this text.

### Transrectal Ultrasound-Guided Prostate Biopsy – Technique

After adherence to antibiotic prophylaxis is confirmed and informed consent obtained, patients are placed in the left lateral decubitus position with the knees flexed upward toward the chest. The well-lubricated, TRUS probe is gently placed and images of the prostate in the longitudinal and sagittal plane are taken for measurement of volume as well as other anatomical considerations (e.g., median lobe, hypoechoic zones, suspected invasion into seminal vesicles, or extracapsular extension). An ultrasound-guided, periprostatic anesthetic block with 2% lidocaine is performed bilaterally at the base of the prostate. Multiple studies have demonstrated improved pain control with periprostatic block when compared with placebo or rectal insertion of lidocaine jelly. The 12 cores are then taken, 2 each from the base, midzone, and apex of the prostate bilaterally. The second core from each zone is taken in a more lateral location. Additional suspicious areas (e.g., large hypoechoic zones, seminal vesicle) are subsequently sampled if indicated. Biopsies can be taken either in the longitudinal or sagittal plane, depending on physician or ultrasonographer preference. The probe is then removed and specimens are labeled and placed in formalin for processing. Additional details about labeling and processing are delineated in succeeding sections. TRUS biopsy can be reliably done well by the physician alone, but a two-person team approach (sonographer and physician or technician and physician) significantly speeds up the procedure. Since much of the patient discomfort stems from the rectal probe stretching the anal sphincter, a faster procedure translates into less pain and more satisfaction. After the probe is removed, gentle pressure is placed against the anal sphincter for 2–3 min to compress any venous bleeding. Alternatively, digital pressure can be placed inside the rectum, but this is more uncomfortable and usually unnecessary. Placing the patient supine after the biopsy, for several minutes, will also serve to compress the perirectal area and reduce bleeding. It is of utmost importance to remind patients of postbiopsy warning signs that require urgent attention (e.g., fever, chills, urinary retention, unremitting gross hematuria, or rectal bleeding). Potential sepsis can be life-threatening and must be identified and treated as soon as possible.

### Specimen Storage – Number of Containers

Urologists must decide how to label and separate biopsy specimens for pathological analysis. Number of cores per specimen jar and labeling of the containers are the two variables, which can be readily modified. The optimal specimen storage scheme yields the highest CDRs, fewest indeterminate results, detailed anatomical information to help guide therapeutic decisions, and is as cost-effective as possible.

There are several options for the number of cores placed in each formalin jar. The most cost-effective is stratifying the biopsy cores into two formalin jars based on laterality (left/right). Thus, for a 12-core biopsy, 6 cores are placed in each container. Some studies, however, have demonstrated more diagnostic accuracy when fewer cores are placed per specimen container. Although the optimal number of cores per container is unknown, placing two to three specimens per container has clearly demonstrated improved CDR. The etiologies for improved diagnostic accuracy with fewer cores per container include less specimen fragmentation and tangling as well as more precise sectioning of the cores. There has been some evidence to suggest that urologists with a financial interest in the pathologic processing and analysis of specimens tend to use more specimen jars.

## Specimen Storage – Biopsy Core Labeling

Labeling the biopsy cores with their corresponding anatomical location provides additional information necessary to guide therapeutic decisions. For the most common treatments involving the entire gland such as surgery, radiation, and cryoablation, knowledge of which locations contain cancer can affect treatment strategies. This is especially true for 12-core biopsies, which have shown a strong concordance to radical prostatectomy specimens (shown later). For example, a surgeon may take a wider margin near a location with positive biopsies. Radiation and cryoablation templates can also be modified based on positive biopsy locations. Because apical and far lateral areas containing prostate cancer have demonstrated increased prognostic significance (e.g., likelihood for extracapsular disease), treatment modifications can be made if these areas are positive. Moreover, for patients placed on active surveillance, special attention to the anatomical zone(s) positive for cancer can be taken during subsequent follow-up biopsy or imaging (e.g., endorectal coil MRI). Finally, for the increasingly popular focal ablative therapies, anatomical details about disease location within the gland are paramount to successful treatment.

Urologists at Boston University utilize a cost-containing, single-container storage strategy that does not compromise the knowledge of anatomical zones sampled. Biopsies taken from a given anatomical zone are initially placed on small, 1 × 1 in. paper squares labeled with the anatomical zone. The paper squares with the tissue cores are then placed in small pouches. All the pouches are subsequently placed in a single formalin jar and sent to pathology. The pathologist can then identify and process the cores in each pouch, providing the detailed anatomical information usually obtained with multiple containers (Figures 9.1–9.3).

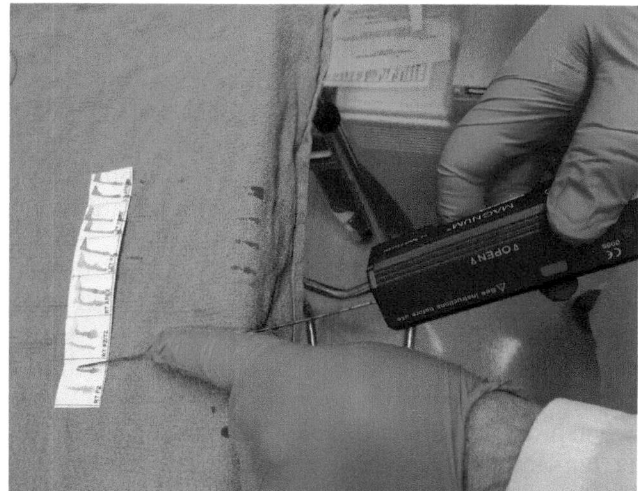

FIGURE 9.1 **Prostate biopsy core placed onto labeled paper template with anatomical zones demarcated.**

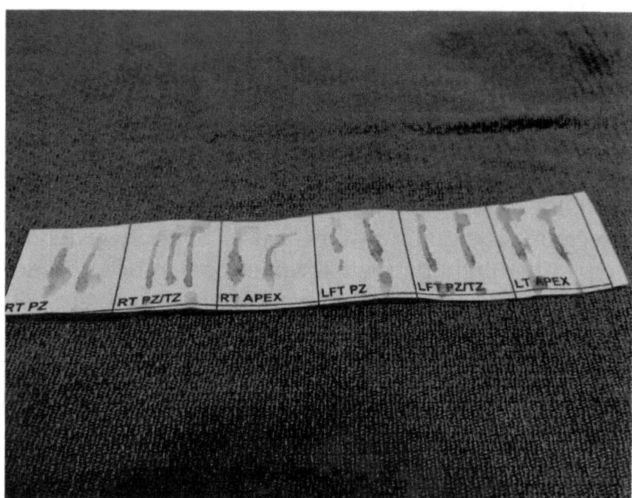


FIGURE 9.2 **Each prostate biopsy core placed on labeled paper template based on anatomical zone sampled.**

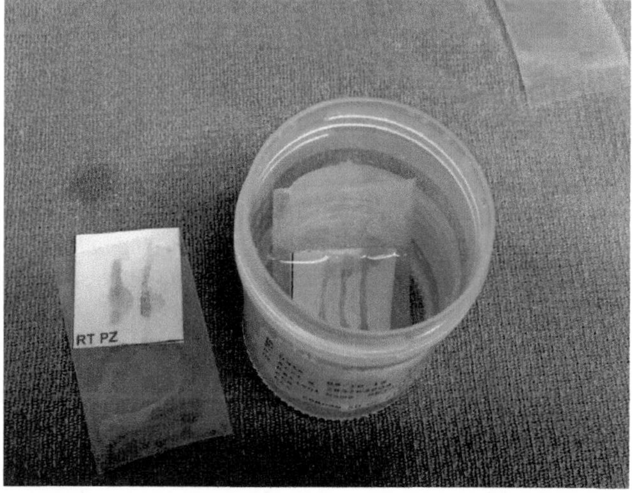


FIGURE 9.3 **Prostate biopsy cores from each anatomical zone placed in individual pouches and all the pouches placed in one formalin specimen container.**

## Number of Cores Sampled

Several studies have proven improved CDR with a 12-core extended prostate biopsy compared with sextant sampling. CDRs increase by 20–33% with the 12-core approach. A large meta-analysis of 87 studies including more than 20,000 patients confirmed that an increased number of cores taken resulted in higher cancer yield.[12] The 12-core biopsy schemes that included laterally directed cores detected, on average, 31% more cancers than sextant biopsies. Furthermore, complication rates for sextant and 12-core schemes were similar.

Evidence also suggests that increasing the number of cores taken beyond 12 has only a marginal, clinically insignificant, increase in CDR. However, when stratifying results based on prostate size, one study identified higher CDRs with an 18-core biopsy versus 12 cores when prostate volume was 55 $cm^3$ or greater, suggesting better sampling of larger glands. The meta-analysis by Eichler et al. failed to demonstrate improved CDR with 18–24-core biopsy schemes. There is a paucity of data comparing complication rates between 12-core and 18–24-core biopsies.

### Concordance with Prostatectomy Specimens

Compared with sextant biopsy, extended 12-core sampling improves concordance with surgical specimen pathology. Gleason score upgrading at the time of prostatectomy is significantly decreased with an extended biopsy, highlighting the likelihood of significant undersampling and underrepresentation of the prostate with sextant biopsy. This is especially true for large-volume prostates or those with anterior tumors. As a result, more accurate and informative shared decision-making between patient and physician can take place after 12-core biopsy. Results from saturation biopsies (24 or more cores) suggest an even greater concordance with prostatectomy specimens.

### Detection of Clinically Insignificant Cancers

One theoretical fear of increasing the number of biopsy cores from 6 to 12 or more is the increased identification of small, clinically indolent cancers. This could result in unnecessary treatment, subjecting men to unwanted side effects, or increased anxiety, doctors visits, serum PSA monitoring, imaging, and subsequent biopsies while on active surveillance. *Clearly, the optimal biopsy scheme strikes a fine balance between thorough sampling of the entire gland (avoid missing significant cancer-increased sensitivity) and minimizing the identification of clinically insignificant cancers.*

As delineated in the preceding paragraph, compared with sextant biopsy, the modern-day 12-core biopsy improves CDR. However, is this at the expense of detecting insignificant cancers that do not require therapy? The data suggests the answer to this question is no. The 12-core sampling does not overdiagnose indolent prostate cancer.

## COMPLICATIONS

TRUS-guided prostate biopsy is a well-tolerated, office-based procedure in most instances. The incidence of major complications is very low; however, minor, self-limiting complications are quite common and often unavoidable. The most important aspect of managing TRUS biopsy complications is informed consent. A detailed discussion of expectations during and after the procedure helps patients identify severe complications that require urgent intervention, and also allay the fears of minor complications that are self-limited. A discussion of the procedure and possible complications should be undertaken, and the initial office visit and important points reiterated on the day of the biopsy. In addition, providing literature summarizing what to expect after the biopsy and when to call the physician or seek emergent care can be helpful as a reference.

The most-feared complication after TRUS biopsy is sepsis due to urinary tract infection (UTI). This can be life-threatening and may require hospital level of care with IV antibiotics, fluid resuscitation, and possible intensive care monitoring for shock. When sepsis criteria are met, urine and blood cultures should be sent immediately and broad-spectrum IV antibiotics initiated. Antibiotic coverage can subsequently be tailored based on culture results. The most common pathogen responsible for postbiopsy sepsis is quinolone-resistant *Escherichia coli.* Several large retrospective series report an incidence of 0.5–2.2% for sepsis after biopsy. As detailed earlier in this chapter, biopsy prophylaxis and knowledge of local bacterial sensitivity profiles help to minimize the risk of sepsis. If quinolone resistance is prohibitively high, other antibiotic choices should be initiated. Special precautions may be taken for men with indwelling urinary catheters, immune-compromised patients, men with a prior history of prostatitis, and those who work in a medical facility since there is a predilection for increased bacterial resistance. In some cases, preprocedure stool culture and/or anal swab culture can be obtained.

Aside from sepsis, the other potential complications after prostate biopsy can be considered minor and, typically, self-limiting. Localized UTI and/or prostatitis have an incidence of up to 10% (1–10%) in the literature, with an increased incidence in patients with recent exposure to quinolones. In patients with localized voiding symptoms (e.g., dysuria) after prostate biopsy without signs of sepsis, a urine culture should be obtained with the suspicion of quinolone resistance. Empiric treatment can be initiated with an oral third-generation cephalosporin until culture results are available.

In men with large prostate glands, urinary retention after prostate biopsy is a rare, transient complication. Atkas et al. reviewed 92 men who underwent TRUS biopsy, of which 6 patients developed acute urinary retention (AUR).[13] Five of the six men had a prostate volume >38 cm$^3$, suggesting larger glands were a risk factor for postprocedure urinary retention. The authors also found that deterioration in voiding parameters and quality of life one week after prostate biopsy were more prevalent in men with larger glands. In a prospective study of 66 men undergoing prostate biopsy, 4 patients developed AUR.[14] The cohort that received tamsulosin for 30 days, starting 1 day before the biopsy, had a lower incidence of urinary retention and less deterioration in voiding symptoms. These studies highlight the importance of counseling men, especially those with large prostate volumes, on the small, risk of AUR after biopsy.

Bleeding after prostate biopsy is a common and usually self-limiting complication. Gross hematuria, hematospermia, and rectal bleeding can all occur. Typically, hematuria and rectal bleeding will self-resolve after a few days, whereas hematospermia may persist for several weeks to months. Proper preprocedure counseling of the risks of self-limited bleeding is paramount to allay anxiety and to manage proper expectations. Patients should always be instructed to call the physician if bleeding of any origin is worrisome. Men taking anticoagulants and/or antiplatelet agents have a theoretical increased risk of bleeding after prostate biopsy, and these agents should ideally be held for one week prior to the procedure. Consider a discussion with the primary care physician or cardiologist about holding these agents prior to biopsy, especially in those men with significant cardiovascular disease (e.g., coronary artery disease, atrial fibrillation, and cerebrovascular accident). Some men cannot safely stop the previously mentioned agents due to competing risks, and a recent study suggests that prostate biopsy can still be safely performed in this setting. Chowdhury et al. studied 902 men who underwent an 8–10-core TRUS prostate biopsy, of which 68 were taking warfarin, 216 low-dose aspirin (1 both), and 617 neither agent.[10] The incidence of rectal bleeding and hematuria was similar in all three groups, suggesting that biopsy could be safely performed in men taking these agents. Approximately 30 and 15% of men in each group reported minor hematuria and rectal bleeding, respectively. There were no severe bleeding complications that required intervention. A smaller study by Raheem et al. also demonstrated no increased risk of bleeding after biopsy when anticoagulants were continued, and, importantly, this study included patients taking clopidogrel, a potent antiplatelet agent often prescribed after coronary stenting.[11] Thus, bleeding appears to be a minor, albeit relatively frequent, self-limiting complication after prostate biopsy. Anticoagulation and/or antiplatelet agents should be held before the procedure if possible, but limited data suggest it is safe to proceed even with the continued use of these agents.

The vast majority of TRUS-guided prostate biopsies should take place in the office setting under local anesthesia (periprostatic infiltration with local anesthetic). The procedure is typically well tolerated with minimal discomfort. On rare occasions, patients will not tolerate the procedure while awake, usually due to pain from the TRUS probe insertion. In these instances, the procedure can be done under IV sedation in the operating room or procedure suite.

## REPEAT BIOPSIES

Repeat prostate biopsies may be indicated for several reasons including persistently elevated or rising PSA after negative biopsy, atypical small acinar proliferation (ASAP) on initial biopsy, scheduled active surveillance rebiopsy, and targeted biopsy based on imaging (MRI) findings.

Perhaps the most common indication for repeat biopsy is a persistently elevated or rising PSA after negative initial biopsy. In these cases, the PSA may be abnormal for nonmalignant causes (e.g., BPH, inflammation, and infection) or the prostate was undersampled at initial biopsy and the malignant focus was missed. Typically, the repeat biopsy is done with the standard 12-core approach, and additional samples are taken based on palpable or imaging abnormalities, if present. Endorectal coil MRI has been increasingly used as an adjunct in this scenario. The MRI can be done in advance and abnormal zones targeted on subsequent TRUS biopsy (cognitive fusion). The MRI can be "fused" with the real-time ultrasound, allowing for the ultrasound to be superimposed over MRI images. Finally, the MRI can be used to target perineal prostate biopsies without the use of ultrasound technology. Further discussion of these MRI-assisted biopsy techniques is beyond the scope of this chapter.

The recommendation for ASAP has traditionally been a repeat prostate biopsy between 3 months and 12 months. A repeat 12-core biopsy is the gold standard but, again, with the growing popularity of MRI, targeted repeat biopsies are becoming a popular approach in this setting.

Active surveillance is becoming increasingly recommended for clinically low-risk prostate cancer. Repeat biopsy schedules are variable, but most agree that a repeat biopsy between 6 months and 12 months after initial diagnosis is recommended. MRI can be used as part of the active surveillance regimen to guide the repeat biopsies as described earlier.

# CONCLUSIONS

Despite advances in imaging and molecular technologies, the foundation of prostate cancer diagnosis will always be biopsy-proven pathology. The modern-day prostate biopsy has evolved significantly over the past several decades to maximize diagnostic accuracy, minimize complications, and improve patient comfort and safety. By far the most devastating potential complication of TRUS-guided prostate biopsy is bacteremia and sepsis. The cornerstone of prevention is antibiotic prophylaxis prior to biopsy. However, an emerging increased prevalence of quinolone-resistant gram-negative organisms has made prophylaxis more challenging. Clinicians must be aware of local antibiotic resistance data and change their practice accordingly. High-risk populations (e.g., recent antibiotic therapy, history of prostatitis, chronic indwelling catheters, and neurogenic bladder) deserve special attention and may benefit from broader antibiotic coverage. A periprostatic lidocaine block is now the standard of care, and measures to expedite the procedure should be taken since most of the discomfort stems from the ultrasound probe in the rectum. The number of core samples should be at least 12 (base, midgland, apex), with additional directed biopsies of suspicious areas. The second core from each zone should be laterally directed. The 12-core sampling maximizes CDR while minimizing overdiagnosis of indolent cancers. Biopsy cores should be labeled and separated based on anatomical location. Separate specimen jars or pouches for each anatomical zone sampled may increase diagnostic accuracy and offer information critical for treatment planning. Endorectal prostate MRI, with and without TRUS fusion, is being increasingly utilized for directed biopsies and may emerge as a future standard.

## References

1. Taneja SS, Bjurlin MA, Carter HB, et al. White paper: AUA /optimal techniques of prostate biopsy and specimen handling. January 2014.
2. Aron M, Rajeev TP, Gupta NP. Antibiotic prophylaxis for transrectal needle biopsy of the prostate: a randomized controlled study. *BJU Int* 2000;**85**:682–5.
3. Sabbagh R, McCormack M, Peloquin F, et al. A prospective randomized trial of 1-day versus 3-day antibiotic prophylaxis for transrectal ultrasound guided prostate biopsy. *Can J Urol* 2004;**11**:2216–9.
4. Shigemura K, Tanaka K, Yasuda M, et al. Efficacy of 1-day prophylaxis medication with fluoroquinolone for prostate biopsy. *World J Urol* 2005;**23**:356–60.
5. Losco G, Studd R, Blackmore T. Ertapenem prophylaxis reduces sepsis after transrectal biopsy of the prostate. *BJU Int* 2014;**113**(Suppl. 2):69–72.
6. Cohen JE, Landis P, Trock BJ, et al. Fluoroquinolone resistance in the rectal carriage of men in an active surveillance cohort: longitudinal analysis. *J Urol* 2015;**193**:552–6.
7. Taylor AK, Zembower TR, Nadler RB, et al. Targeted antimicrobial prophylaxis using rectal swab cultures in men undergoing transrectal ultrasound guided prostate biopsy is associated with reduced incidence of postoperative infectious complications and cost of care. *J Urol* 2012;**187**:1275–9.
8. Park DS, Oh JJ, Lee JH, et al. Simple use of the suppository type povidone-iodine can prevent infectious complications in transrectal ultrasound-guided prostate biopsy. *Adv Urol* 2009;doi: 10.1155/2009/750598.
9. Issa MM, Al-Qassab UA, Hall J, et al. Discontinuation of anticoagulant or antiplatelet therapy for transrectal ultrasound-guided prostate biopsies: a single-center experience. *J Urol* 2013;**190**:1769–75.
10. Chowdhury R, Abbas A, Idriz S, et al. Should warfarin or aspirin be stopped prior to prostate biopsy? An analysis of bleeding complications related to increasing sample number regimes. *Clin Radiol* 2012;**67**:64–70.
11. Raheem OA, Casey RG, Galvin DJ, et al. Discontinuation of anticoagulant or antiplatelet therapy for transrectal ultrasound-guided prostate biopsies: a single-center experience. *Korean J Urol* 2012;**53**:234–9.
12. Eichler K, Hempel S, Wilby J, et al. Diagnostic value of systematic biopsy methods in the investigation of prostate cancer: a systematic review. *J Urol* 2006;**175**:1605–12.
13. Atkas BK, Bulut S, Gokkaya CS, et al. Association of prostate volume with voiding impairment and deterioration in quality of life after prostate biopsy. *Urology* 2014;**83**:617–21.
14. Bozlu M, Ulusoy E, Doruk E, et al. Voiding impairment after prostate biopsy: does tamsulosin treatment before biopsy decrease this morbidity? *Urology* 2003;**62**:1050–3.

CHAPTER

# 10

# Total and Free PSA, PCA3, PSA Density and Velocity

*Massimo Lazzeri, MD, PhD, Giorgio Guazzoni, MD, Francesco Montorsi, MD*

**Division of Oncology/Unit of Urology, URI, IRCCS Ospedale San Raffaele, Università Vita-Salute San Raffaele, Milan, Italy**

## INTRODUCTION

Since the late 1980s, the diagnosis and follow-up of prostate cancer (PCa) has relied on the use of prostate-specific antigen (PSA), a blood laboratory measurement that was shown to be associated with pathological diagnosis of cancer and had both diagnostic and prognostic clinical validity and utility. In 1986 the Food and Drug Administration approved the test to monitor those men already diagnosed with cancer, and in 1994 it went further, authorizing the test to help detect cancer in men aged 50 and older.[1] Through the years, PSA has provided significant advancements in diagnosis and prognosis of PCa, although it was counterbalanced by its low sensitivity and specificity.[2] PSA clinical availability triggered a frenzied hunt for the tumor, but its indiscriminate use let critics of the testing, once regarded as heretics, gain credibility. In 2004 the World Health Organization arranged an international consultation to assess new markers recognizing the limitation of PSA testing.[3] Recently, PSA has been thrust into the public spotlight after several publications showed the risk of overdiagnosis and overtreatment of low-risk PCa in particular, which showed that nonperformance of PSA testing would not have affected the longevity or the quality of life.[4] Such shortcoming led urologists to optimized the use of PSA (PSA density and velocity), to investigate some isoforms of PSA (free PSA, [−2]proPSA) and to develop novel molecular markers (PCA3 or molecular markers, i.e., cell cycling processing genes).

In this chapter we will summarize the current use of PSA for PCa detection, integrating it with an insight into the new biomarkers using blood, urine, and tissue.

## PSA

### Biology

PSA is an androgen-regulated chymotrypsin-like serine protease, part of the family of proteases known as kallikreins encoded by a cluster of genes located on human chromosome 19q13.4,[5,6] and are thus also called human kallikrein (hK) 3. It is produced in high levels within the prostatic ductal and acinar epithelium, with a 17-amino acid leader sequence (preproPSA) that is cleaved co-translationally to generate an inactive 244-amino acid precursor protein (proPSA), with seven additional amino acids compared to mature PSA.[7,8] Generally, proPSA is normally secreted from the prostate luminal epithelial cells, and after its release into the lumen, the proleader part is removed and converted to its active form by the effect of hK-2 and hK-4, which have a trypsin-like activity and are expressed predominantly by prostate secretory epithelium.[9] Other kallikreins, localized in the prostate, such as hK-216 or prostin 17, are involved in the conversion and activation of proPSA. Cleavage of the N-terminal seven amino acids from proPSA generates the active enzyme (PSA), which has a mass of 33 kDa. In the seminal fluid, PSA cleaves the seminogelin I and II, thus promoting the semen liquefaction.[10,11] The enzymatically active PSA is normally confined within the prostate gland by a tight and orderly prostatic glandular architecture. Consequently, only a few PSA leaks into the circulation, and its serum concentration ($<4$ μg/L) is a million-fold lower than that in the seminal plasma (0.5–5 g/L).[12] The measurable serum total PSA (tPSA) comprises either a complexed form (cPSA, 70–90%), bound by protease inhibitors (primarily α1-antichymotrypsin)

**Prostate Cancer. http://dx.doi.org/10.1016/B978-0-12-800077-9.00010-4**

and a noncomplexed form (free PSA, fPSA).[13] fPSA has recently been discovered to exist in at least three molecular forms: proPSA, benign PSA (BPSA), and inactive intact PSA (iPSA), covering approximately 33, 28, and 39% of fPSA, respectively.[14] BPSA is a degraded form of PSA that is identical to the native, mature PSA with 237 amino acids but contains two internal peptide bond cleavages at Lys182 and Lys145. Immunohistochemical studies have shown that BPSA is expressed preferentially in the transitional zone of the prostate and is associated with pathological benign prostate hyperplasia (BPH).[15] iPSA is similar to native PSA but is inactive due to structural or conformational changes. The partial removal of the leader sequence of the preproPSA leads to other truncated forms of proPSA. Hence, theoretically, seven isoforms of proPSA should exist, although only [−1], [−2], [−4], [−5], [−7]proPSA were found, while there is still no evidence of [−3], [−6]proPSA. All these forms of proPSA are enzymatically inactive, but they might play a role in cancer detection, especially [−2]proPSA.

## PSA Cutoff: Sensitivity and Specificity

Before 1986, when the FDA officially approved the use of PSA for follow-up in patients with a diagnosis of PCa, the only serum marker for PCa was human prostatic acid phosphatase (PAP). Patients with metastatic disease to bone have increased PAP activity at the site of metastasis as well as elevated serum levels of the protein.[16] Serum PAP was used in the past as a biomarker for pathological prediction, cancer progression, and response to androgen deprivation therapy.[17] However, with the introduction of PSA, serum PAP was rendered almost obsolete, as it was inferior to PSA for PCa detection, staging, and prognostication.[18] In 1991, Catalona et al. reported a large-scale population-based study examining the use of PSA and digital rectal examination (DRE) for PCa screening.[19] This study demonstrated that serum PSA measurements provided a useful adjunct to DRE for enhanced PCa detection. The FDA officially approved PSA for PCa detection in 1994, defining 4.0 ng/mL as the upper limit of normal. After its clinical introduction, it was soon discovered that around 20% of patients with PCa had tPSA value below 4 ng/mL.[2] PSA is indeed an organ-specific marker and not disease-specific. Numerous nonmalignant processes, including BPH and prostatitis which occur in many men, frequently lead to serum PSA elevations, limiting the specificity of PSA elevation for cancer detection; even prostatic manipulation (DRE), sexual activity, and medical interventions such as cystoscopy can confound elevated PSA measurements.[20] Furthermore, in a pooled meta-analysis, the positive predictive value of a PSA >4.0 ng/mL, found in 10.1% of the population, was only 25% (low specificity).[21] The PSA also has a low sensitivity as confirmed by studies using a lower cutoff value of abnormal PSA levels.[2] Tornblom et al. and Lodding et al., in two different studies, showed that of a total population with prostate carcinoma, 14% have a PSA level of less than 3 ng/mL, 23–24% have a PSA level of 3–4 ng/mL, and 62% have a PSA level of more than 4 ng/mL.[22,23] Thompson et al. showed that many cancers are missed with a cutoff of 4 ng/mL and that most patients with high values have no cancer, thus confirming that the current reference cutoff fails to demonstrate both a high sensitivity and specificity.[2] There is rather a continuum of PCa risk at all PSA values.[24]

## Age-Adjusted PSA

The current international standard normal tPSA between 0.0 ng/mL and 4.0 ng/mL does not account for age-related PSA changes due to prostate volume changes related to the development of BPH. Early in the history of PSA, investigators introduced the concept of age-adjusted tPSA. Oesterling et al. suggested considering age-related reference ranges in order to improve cancer detection rates in young men.[25] The increase of specificity for age-adjusted PSA ranges was confirmed by Partin in the early era of PSA. He found that the potential detection of PCa increased by 18% in younger men and decreased by 22% in older men.[26] Several studies have reported the association between baseline PSA measurements at a young age and PCa risk. Antenor et al. showed that men with an initial PSA value above the median age-adjusted PSA level of 0.7–0.9 in younger men (in compliance with The Global Age Watch Index 2013[27]) face an increased risk of PCa.[28] Using frozen sera samples from the Baltimore Longitudinal Study of Aging (BLSA), investigators found that low levels of PSA in young men (between 40 years and 50 years of age) were directly associated with the development of PCa decades after the performance of the baseline PSA test.[29] A greater baseline PSA is also associated with more aggressive tumor features and a greater biochemical progression rate following treatment. Loeb et al. examined the usefulness of baseline PSA testing in 13,943 men younger than 60 years from a large prospective PCa screening trial in the United States, in which biopsy was performed for a PSA >2.5 ng/mL or suspicious findings on DRE.[30] High-risk men in their 40s with a baseline PSA level between 0.7 ng/mL (age-specific median) and 2.5 ng/mL had a 14.6 times higher risk of PCa compared with those with a baseline PSA below the age-specific median. Similarly, men in their 50s with a PSA from 0.9 ng/mL (age-specific median) to 2.5 ng/mL had a 7.6 times higher risk of PCa, compared with men in their 50s with a PSA below the age-specific median. Vickers et al. found that the tPSA value at age 60 years not only predicts a lifetime risk of clinically detected PCa, but also metastasis and death from the disease.[31] This suggests that men aged

60 years with a PSA level below the median of 1 ng/mL might harbor PCa; however, it is unlikely to become life-threatening. In a recent review, Loeb et al. confirmed that baseline PSA measurements at a young age are stronger predictors of PCa risk than race and family history. Furthermore, they found that baseline PSA measurements at a young age are robust predictors of aggressive PCa, metastasis, and disease-specific mortality many years later.[32] Baseline PSA testing might therefore be useful for risk stratification and for individualizing screening protocols. Unfortunately, not all urologists agree with such a conclusion. The American Urological Association guidelines recommended against routine PSA testing for average-risk asymptomatic men younger than 55 years of age.[33] For others at higher risk (e.g., those with a family history or of African-American race), the panel recommended individualizing decisions – especially for those with a family history of early-onset PCa (before the age of 55 years) and/or PCa in multiple first-degree relatives or multiple generations. While some men in their 40s and early 50s will benefit from PSA testing, the evidence suggests that harms will likely outweigh benefits for most men at average risk.

## PSA and PCa Screening

Any form of screening aims to reduce disease-specific and overall mortality, and to improve a patient's future quality of life. PSA-based screening for PCa has generated considerable debate within the medical and broader community, as demonstrated by the varying recommendations made by medical organizations and governed by national policies. In 2009 the efficacy of PSA as a widespread screening tool was assessed by two large population-based trials, which reached two opposite conclusions. In the US Prostate, Lung, Colorectal, and Ovarian (PLCO) Cancer Screening Trial, Andriole et al. report no mortality benefit from combined screening with PSA testing and DRE during a median follow-up of 11 years. In the European Randomized Study of Screening for Prostate Cancer (ERSPC) trial, Schröder et al. report that PSA screening without DRE was associated with a 20% relative reduction in the death rate from PCa at a median follow-up of 9 years, with an absolute reduction of about 7 PCa deaths per 10,000 men screened.[34,35] It must be stressed that the PLCO trial had an extremely high contamination rate of PSA testing upward of 70% in the control arm. This led to the conclusion that annual screening is not better than ad hoc screening, although it does not shed light on the question of screening versus no screening. Updated results from the European trial using longer follow-up data were published in 2012.[36] They showed an increase in relative risk with a reduction in mortality from 20% to 29% ; the need to be invited to undergo screening (NNI) and the need to be detected in order to prevent one death (NND) from PCa decreased from 1410 to 936 and from 48 to 33, respectively (absolute benefit). On the basis of ERSPC follow-up data, Heijnsdijk et al. used a simulation screening analysis to predict the number of PCas, treatments, deaths, and quality-adjusted life-years (QALYs) gained after the introduction of PSA screening.[37] The base model predicted a gain of 56 QALYs (range, −21–97) over an overall unadjusted 73 life-years gained for men between the ages of 55 and 69 years. It means that 23% [73 − 56 = 17 (23%)] of the unadjusted life-years gained by screening would be counterbalanced by a loss in quality of life. Authors concluded that the predicted difference between life-years and QALYs gained is caused by overdiagnosis and overtreatment induced by PSA screening. Based on those results, the US Preventive Services Task Force (USPSTF) published in 2012 a review of the evidence for PSA screening for PCa and made a clear recommendation against screening.[38] The agency concluded that "there is moderate or high certainty that this service (PSA screening) has no net benefit or that the harms outweigh the benefits." However, the USPSTF report contained a number of important errors of fact, interpretation, and statistics which cast doubts about the statement.[39] The Cochrane collaborative group recently performed an update of its previous review in 2006.[40] In conclusion, the meta-analysis, which was focused on the last RCT, showed that PSA screening did not significantly decrease prostate-cancer-specific mortality in a combined meta-analysis of five RCTs. Only one study (ERSPC[36]) reported a 21% significant reduction in PCa-specific mortality in a prespecified subgroup of men aged 55–69 years. Pooled data currently demonstrates no significant reduction in prostate-cancer-specific and overall mortality. Harms associated with PSA-based screening and subsequent diagnostic evaluations are frequent, and moderate in severity. Overdiagnosis and overtreatment are common and are associated with treatment-related harms. Men should be informed of this and the demonstrated adverse effects when they are deciding whether or not to undertake screening for PCa. Any reduction in prostate-cancer-specific mortality may take up to 10 years to accrue; therefore, men who have a life expectancy less than 10–15 years should be informed that screening for PCa is unlikely to be beneficial. No studies examined the independent role of screening by DRE.

## Free PSA and %fPSA

Some circulating PSA do not form complexes with serum antiproteases such as serine protease inhibitors [α1-antichymotrypsin, protein C inhibitor (PCI), α1-antitrypsin (API)], and other classes of antiproteases, including α2-macroglobulin (AMG) and pregnancy-zone protein (PZP).[41] This form of PSA was defined as free PSA (fPSA). The clinical importance of fPSA was investigated as proportionally related to tPSA. It was found that men with PCa have a lower ratio of free to total PSA (%fPSA) than those without. Catalona et al. first demonstrated its usefulness in 773 men with PSA values of 4–10 ng/mL and normal DRE.[42] In their study, 773 men with a PSA value ranging between 4 ng/mL and 10 ng/mL and a normal DRE underwent measurement of free and total PSA prior to prostate biopsy. Using a cutoff of <25% yielded a sensitivity of 95% while reducing unnecessary biopsies by 20%; its association was independent of total PSA. After this preliminary data, several studies confirmed the correlation of lower percentage-free PSA with higher probability of PCa on biopsy and paved the way for FDA approval of %fPSA for use in the PSA 4–10 ng/mL range.[43,44] After the preliminary evidence, other investigators evaluated the clinical performance of %f/tPSA for other ranges. Partin et al. found that %fPSA performed well for different ranges. A %fPSA cutoff of <20% had a sensitivity of 95% in a cohort of 219 men with PSA levels of 2–20 ng/mL.[45] Other studies established the utility of %fPSA in the PSA 2–4 ng/mL range. Catalona et al. found that at a threshold of <27%, the sensitivity was at 90%, and an 18% reduction in unnecessary biopsies could be avoided.[46] Carter et al. showed that %fPSA could be used to predict a future diagnosis of aggressive cancer.[47] Indeed, they found that men with adverse pathological outcomes (clinical stage T3, lymph node or bone metastasis, Gleason score >7) had a significantly lower %fPSA up to 10 years prior to diagnosis than those with more favorable pathology ($p = 0.008$), whereas total PSA did not differentiate between the groups. More recent studies about the clinical utility of %fPSA have been less impressive than the initial studies. Haese et al. reported that the use of an 18–20 %fPSA cutoff in the PSA 2–4 ng/mL range would only slightly increase the number of biopsies necessary to detect one significant cancer (biopsy-to-cancer ratio of 3:1 to 4:1), compared to men with PSA levels of 4–10 ng/mL.[48] Finally, according to some authors, %fPSA could help urologists in the decision-making process after a negative initial biopsy as %fPSA performed best at second and third or more repeat biopsy.[49,50]

Some caveats were discovered. As for tPSA, %fPSA increases with patient age and prostate volume, and it decreases as total PSA increases. Serum fPSA is not very stable compared with complexed PSA as it showed a greater analytic variability. This requires strict sample handling and processing within a few hours of collection; otherwise, it should be kept frozen to provide optimal analysis.[51] Furthermore, free PSA measurements of the same specimen using kits from different manufacturers are not accurately reproducible. Regardless of the extent to which these issues explain the inconsistent performance of percentage-free PSA, this marker to date has not become widely used as a primary screening tool. Finally, prostatic manipulation (DRE, biopsy, and urethral instrumentation) increases the fPSA component of total PSA[52] and, as with total PSA, interassay variability is a potential problem when the diagnostic performance of fPSA is compared across studies.[53]

## PSA DENSITY

The concept of PSA density (PSAD) was introduced by Benson et al. As PSA is influenced by prostate volume, they introduced the concept to correlate PSA and prostate volume.[54] PSAD is defined as the total serum PSA divided by prostate volume, as determined by transrectal ultrasound (TRUS) measurement. The idea was that theoretically, PSAD could help the decision-making process (biopsy or not) in distinguishing between PCa and BPH in men whose PSA ranged from 4 ng/mL to 10 ng/mL. It was noted that increasing PSAD is associated with a greater risk of PCa.[55] Today one of the most used cutoffs is 0.15. In the early era of PSA, Seaman et al. reported that the value of PSAD could improve the detection rate of cancer, making it useful in selecting patients for radical prostatectomy, and may be capable of identifying patients with micrometastatic disease at that cutoff value.[56] Brawer et al. failed to support the proposed cutoff. They studied 107 men with PSA levels in the 4–10 ng/mL range and found no statistical difference between those with positive and those with negative biopsy findings using the 0.15 cutoff.[57] After that, several concerns emerged about the clinical utility of PSAD. The value of PSAD may be limited as it is strongly dependent on the individual performing the prostate volume measurement during the urltrasonography. Furthermore, the prostate volume in men with BPH does not always correlate with serum PSA values as variations are reported between individuals in their epithelial-to-stromal ratios. Doubts about its clinical utility were raised in recent years. Catalona et al. were concerned about the ideal cutoff of 0.15. They reported that nearly 50% of cancers would be missed using the 0.15 cutoff.[58] Beyond its use in screening, PSAD is also predictive of PCa aggressiveness. For example, Kundu et al. showed that increasing PSAD was associated with a greater risk of non-organ-confined disease, Gleason score ≥7, and increasing tumor volume after radical prostatectomy.[59] PCa patients with a low PSAD (<0.15), no Gleason pattern 4 or 5, fewer than three positive biopsy cores, and <50% involvement of any biopsy core were more likely to have pathologically insignificant cancer.[60] Koie et al. retrospectively examined the ability of preoperative PSAD to predict pathological and oncological outcomes in high-risk PCa patients who underwent radical prostatectomy (RP), and they found that PSAD might be an independent predictor of advanced pathological features and biochemical recurrence in high-risk PCa patients treated with RP alone. PSAD may be used for further risk stratification of high-risk PCa patients.[61]

## PSA VELOCITY

Changes in serum PSA levels over time have been proposed to increase the accuracy of tPSA in detecting PCa and improve its testing or screening. Those changes were defined as PSA velocity (PSAV). PSAV consists of the change in PSA per year (ng/mL/year). In 1992, Carter et al. estimated serial PSA measurements in men with and without prostate disease.[62] They found that in men with total PSA levels of 4–10, a PSAV >0.75 mg/mL/year was significantly associated with PCa. After such report, several other studies confirmed the relationship between a PSAV >0.75 ng/mL/year and PCa in the same tPSA range.[63,64] In men with a tPSA lower than 4 ng/mL, a PSAV >0.4 ng/mL/year was deemed more accurate for detecting PCa in such population.[65,66] Further studies have shown a significant relationship between PSAV and PCa aggressiveness. D'Amico et al. reported that men with a preoperative PSAV >2 ng/mL/year had a ninefold increased risk of PCa death after RP on multivariate analysis controlled for PSA and other prognostic variables.[67] D'Amico et al. expanded their observation to men who underwent radiotherapy, finding out that a PSAV >2 ng/mL/year was associated with a higher cancer-specific mortality rate.[68] Similar prognostic data were reported after brachytherapy.[69]

In contrast to the previously quoted evidence, which generally cast the value of PSAV in a positive light, new studies published in the last few years have questioned its value for clinical decision-making. Vikers et al. very critically reviewed the literature covering the topic and drew significant take-home message.[70] They concluded that high PSAV could not be considered an indication for biopsy. However, for those men with a PSA below biopsy thresholds (generally 4 ng/mL) but a high PSAV (>0.75 mg/mL/year), consideration should be given to having PSA taken at a shorter interval. In men with a first-round negative biopsy, PSAV could not determine the need for repeat biopsy, although PSA over time can aid decision-making on biopsy, as informed by the clinical context. Other doubts were cast on the use of PSAV for active treatment in men on active surveillance as it is an "unreliable trigger" for definitive treatment.[71]

## PCA3

The prostate cancer antigen 3 gene (PCA3) is a PCa-specific gene located on chromosome 9q21–22. It is a long noncoding RNA. In 1999 researchers reported that PCA3 was highly overexpressed in PCa tissue compared to normal prostate or BPH tissue.[72] After the release of such data, PCA3 tests measuring mRNA expression in urine were developed and evaluated for pre- and analytical validity before the assessment of clinical validity.[73] This led to the development of a commercial molecular urinary assay in which the first 30 mL of urine are collected after a DRE. Preliminary results for PCa detection were encouraging, showing a balance between sensitivity and specificity of 69 and 79%, respectively, and outperforming the tPSA.[74–76] It was then noted that PCA3 RNA levels are independent of prostate volume and serum PSA.[77,78]

In 529 men scheduled for prostate biopsy, urine PCA3 score was correlated with prostate volume determined by TRUS at the time of biopsy. Prostate volume was divided into three categories: <30 cm$^3$, 30–50 cm$^3$, and >50 cm$^3$. In contrast to serum PSA, PCA3 scores did not increase with prostate volume. The mean PCA3 scores for the three groups were 45, 38, and 43, respectively. These encouraging results suggest that age- and volume-related effects that complicate application of serum PSA in PCa screening, particularly affecting specificity of mild PSA elevations, will not similarly be encountered with PCA3 testing.[79] Furthermore, PCA3 score was also independent of the number of previous biopsies.[80] The most studied role of PCA3 has been in men who maintained a risk of bearing PCa but with a prior negative biopsy. Considering the high rate of negative biopsies after the first round, the need to reduce the number of unnecessary biopsies was one of the main targets of this urine molecular biomarker. In the United States, Marks et al. first evaluated the use of PCA3 in 226 consecutive patients undergoing repeat biopsy, showing a higher accuracy of PCA3 versus tPSA (AUC: 0.68 vs. 0.52; $p = 0.008$, respectively) in predicting PCa.[81] At a cutoff of PCA3, the sensitivity and specificity were at 58 and 72%, respectively. A progressive increase in the number of men with cancer was found when correlated with PCA3 score: men with a PCA3 score of >100 had a 50% probability of positive biopsy compared with a probability of only 12% for men with scores of <5. Subsequently, Ankerst et al., Deras et al., van Gils et al., and Wu et al., all confirmed that PCA3 has a diagnostic accuracy that is superior to that of PSA in multivariable regression models.[82–85]

In February 2012 the FDA approved the PROGENSA® PCA3 assay to aid decision-making on biopsy in men age 50 years or older who had one or more negative biopsies (no clearance was done about initial biopsy as data were not submitted).[86]

An ideal clinical biomarker for PCa should be able to distinguish men with cancer from men without, as well as indolent cancer from clinically significant cancer. Different authors thus analyzed the relationships between PCA3 score and the different characteristics of tumor aggressiveness. Haese reported that the PCA3 score was significantly higher in men with HGPIN versus those without HGPIN, clinical stage T2 versus T1, Gleason score ≥7, and "significant" versus "indolent" (clinical stage T1c, PSAD, 0.15 ng/mL, Gleason score in biopsy ≤ 6, and percentage positive cores ≤ 33%) PCa.[80] Other distinct papers supported Hease's findings.[87,88] Similar results were not confirmed by others. Augustin et al. assessed that in 127 patients treated with radical prostatectomy for clinically localized PCa, the PCA3 score relate to the cancer aggressiveness.[89] They found that the PCA3 score had no impact on the prediction of aggressive PCas.

Bradley et al. recently performed a comparative effectiveness review of PCA3's role in the diagnosis and management of PCa.[90] The authors investigated the performance of PCA3 for two intended uses: the prediction of positive biopsy in men at risk for PCa to inform decision-making on initial or repeat biopsy; and the distinction of men with aggressive disease. Three important conclusions were drawn from their analyses. First, the PCA3 scores had greater diagnostic accuracy for positive biopsy than tPSA increases. Second, the relative performance of PCA3 versus tPSA did not appear to be dependent on whether it was a repeat biopsy. Finally, the information provided by PCA3 score was essentially independent from that provided by tPSA increases. Unfortunately, although PCA3 had a higher diagnostic accuracy than tPSA, the strength of evidence was low. The study showed that the evidence was insufficient to conclude that PCA3 testing leads to improved health outcomes (clinical utility) for all other outcomes of interest.

## References


1. http://www.fda.gov/
2. Thompson IM, Pauler DK, Goodman PJ, et al. Prevalence of prostate cancer among men with a prostate-specific antigen level < or = 4.0 ng per milliliter. *N Engl J Med* 2004;**350**(22):2239–46.
3. Stenman UH, Abrahamsson PA, Aus G, et al. Prognostic value of serum markers for prostate cancer. *Scand J Urol Nephrol* 2005;**39**:64–81.
4. Prostate cancer: help or harm. The Economist. March 8, 2014: 69–70.
5. Kumar A, Mikolajczyk SD, Goel AS, et al. Expression of pro form of prostate-specific antigen by mammalian cells and its conversion to mature, active form by human kallikrein 2. *Cancer Res* 1997;**57**:3111–4.
6. Clements J, Hooper J, Dong Y, et al. The expanded human kallikrein (KLK) gene family: genomic organisation, tissue-specific expression and potential functions. *Biol Chem* 2001;**382**:5–14.
7. Lovgren J, Rajakoski K, Karp M, et al. Activation of the zymogen form of prostate-specific antigen by human glandular kallikrein 2. *Biochem Biophys Res Commun* 1997;**238**:549–55.
8. Takayama TK, Fujikawa K, Davie EW. Characterization of the precursor of prostate-specific antigen. Activation by trypsin and by human glandular kallikrein. *J Biol Chem* 1997;**272**:21582–8.
9. Takayama TK, McMullen BA, Nelson PS, et al. Characterization of hK4 (prostase), a prostate-specific serine protease: activation of the precursor of prostate specific antigen (pro-PSA) and single-chain urokinase-type plasminogen activator and degradation of prostatic acid phosphatase. *Biochemistry* 2001;**40**:15341–8.
10. Lilja H, Oldbring J, Rannevik G, et al. Seminal vesicle-secreted proteins and their reactions during gelation and liquefaction of human semen. *J Clin Invest* 1987;**80**:281–5.
11. Stephan C, Jung K, Lein M, et al. PSA and other tissue kallikreins for prostate cancer detection. *Eur J Cancer* 2007;**43**:1918–26.
12. Lilja H, Christensson A, Dahlen U, et al. Prostate-specific antigen in serum occurs predominantly in complex with alpha 1-antichymotrypsin. *Clin Chem* 1991;**37**:1618–25.
13. Stenman UH, Leinonen J, Alfthan H, et al. A complex between prostate-specific antigen and alpha 1-antichymotrypsin is the major form of prostate-specific antigen in serum of patients with prostatic cancer: assay of the complex improves clinical sensitivity for cancer. *Cancer Res* 1991;**51**:222–6.
14. Mikolajczyk SD, Marks LS, Partin AW, et al. Free prostate-specific antigen in serum is becoming more complex. *Urology* 2002;**59**:797–802; Mikolajczyk SD, Rittenhouse HG. Pro PSA: a more cancer specific form of prostate specific antigen for the early detection of prostate cancer. *Keio J Med* 2003;**52**:86–91.
15. Mikolajczyk SD, Millar LS, Wang TJ, et al. "BPSA," a specific molecular form of free prostate-specific antigen, is found predominantly in the transition zone of patients with nodular benign prostatic hyperplasia. *Urology* 2000;**55**:41–5.
16. Veeramani S, Yuan TC, Chen SJ, et al. Cellular prostatic acid phosphatase: a protein tyrosine phosphatase involved in androgen-independent proliferation of prostate cancer. *Endocr Relat Cancer* 2005;**12**(4):805–22.
17. Whitesel JA, Donohue RE, Mani JH, et al. Acid phosphatase: its influence on the management of carcinoma of the prostate. *J Urol* 1984;**131**:70–2.
18. Lowe FC, Trauzzi SJ. Prostatic acid phosphatase in 1993. Its limited clinical utility. *Urol Clin North Am* 1993;**20**:589–95.
19. Catalona WJ, Smith DS, Ratliff TL, et al. Measurement of prostate-specific antigen in serum as a screening test for prostate cancer. *N Engl J Med* 1991;**324**(17):1156–61.
20. Nadler RB, Humphrey PA, Smith DS, et al. Effect of inflammation and benign prostatic hyperplasia on elevated serum prostate specific antigen levels. *J Urol* 1995;**154**:407–13.
21. Mistry K, Cable G. Meta-analysis of prostate-specific antigen and digital rectal examination as screening tests for prostate carcinoma. *J Am Board Fam Pract* 2003;**16**:95–101.
22. Tornblom M, Norming U, Adolfsson J, et al. Diagnostic value of percent free prostate-specific antigen: retrospective analysis of a population-based screening study with emphasis on men with PSA levels less than 3 ng/mL. *Urology* 1999;**53**:945–50.
23. Lodding P, Aus G, Bergdahl S, et al. Characteristics of screening detected prostate cancer in men 50–66 years old with 3–4 ng/mL prostate-specific antigen. *J Urol* 1998;**159**:899–903.
24. Thompson I, Ankerst D, Chi C, et al. Operating characteristics of prostate-specific antigen in men with an initial PSA level of 3.0 ng/mL or lower. *JAMA* 2005;**294**:66–70.
25. Oesterling JE, Rice DC, Glenski WJ, et al. Effect of cystoscopy, prostate biopsy, and transurethral resection of prostate on serum prostate-specific antigen concentration. *Urology* 1993;**42**:276–82.
26. Partin AW, Pearson JD, Landis PK, et al. Evaluation of serum prostate-specific antigen velocity after radical prostatectomy to distinguish local recurrence from distant metastases. *Urology* 1994;**43**:649–59.
27. http://www.helpage.org/download/52440f486ab91/
28. Antenor J, Han M, Roehl K, et al. Relationship between initial prostate-specific antigen level and subsequent prostate cancer detection in a longitudinal screening study. *J Urol* 2004;**172**:90–3.
29. Fang J, Metter EJ, Landis P, et al. Low levels of prostate-specific antigen predict long-term risk of prostate cancer: results from the Baltimore Longitudinal Study of Aging. *Urology* 2001;**58**:411–6.
30. Loeb S, Roehl KA, Antenor JAV, et al. Baseline prostate-specific antigen compared with median prostate-specific antigen for age

group as predictor of prostate cancer risk in men younger than 60 years old. *Urology* 2006;**67**:316–20.
31. Vickers A, Cronin AM, Björk T, et al. Prostate-specific antigen concentration at age 60 and death or metastasis from prostate cancer: case-control study. *BMJ* 2010;**341**:c4521.
32. Loeb S, Carter HB, Catalona WJ, et al. Baseline prostate-specific antigen testing at a young age. *Eur Urol* 2012;**61**:1–7.
33. Carter HB, Albertsen PC, Barry MJ, et al. Early detection of prostate cancer: AUA Guideline. *J Urol* 2013;**190**:419–26.
34. Andriole G, Crawford ED, Grubb III RL, et al. Mortality results from a randomized prostate-cancer screening trial. *N Engl J Med* 2009;**360**:1310–9.
35. Schröder F, Hugosson J, Roobol MJ, et al. Screening and prostate-cancer mortality in a randomized European study. *N Engl J Med* 2009;**360**:1320–8.
36. Schröder FH, Hugosson J, Roobol MJ, et al. Prostate-cancer mortality at 11 years of follow-up. *N Engl J Med* 2012;**366**:981–90.
37. Heijnsdijk EA, Wever EM, Auvinen A, et al. Quality-of-life effects of prostate-specific antigen screening. *N Engl J Med* 2012;**367**(7): 595–605.
38. Moyer VA. Screening for prostate cancer: U.S. Preventive Services Task Force recommendation statement. *Ann Intern Med* 2012;**157**(2):120–34.
39. Carlsson S, Vickers AJ, Roobol M, et al. Prostate cancer screening: facts, statistics, and interpretation in response to the US Preventive Services Task Force Review. *J Clin Oncol* 2012;**30**(21): 2581–4.
40. Ilic D, Neuberger MM, Djulbegovic M, et al. Screening for prostate cancer. *Cochrane Database Syst Rev* 2013;**1**:CD004720.
41. Christensson A, Laurell CB, Lilja H. Enzymatic activity of prostate-specific antigen and its reactions with extracellular serine proteinase inhibitors. *Eur J Biochem* 1990;**194**:755–63.
42. Catalona WJ, Partin AW, Slawin KM, et al. Use of the percentage of free prostate-specific antigen to enhance differentiation of prostate cancer from benign prostatic disease: a prospective multicenter clinical trial. *JAMA* 1998;**279**:1542–7.
43. Djavan B, Zlotta A, Kratzik C, et al. PSA, PSA density, PSA density of transition zone, free/total PSA ratio, and PSA velocity for early detection of prostate cancer in men with serum PSA 2.5–4.0 ng/mL. *Urology* 1999;**54**:517–22.
44. Djavan B, Zlotta A, Remzi M, et al. Optimal predictors of prostate cancer on repeat prostate biopsy: a prospective study of 1,051 men. *J Urol* 2000;**163**:1144–9.
45. Partin AW, Brawer MK, Subong EN, et al. Prospective evaluation of percent free-PSA and complexed-PSA for early detection of prostate cancer. *Prostate Cancer Prostatic Dis* 1998;**1**:197–203.
46. Catalona WJ, Smith DS, Ornstein DK. Prostate cancer detection in men with serum PSA concentrations of 2.6–4.0 ng/mL and benign prostate examination. Enhancement of specificity with free PSA measurements. *JAMA* 1997;**277**:1452–5.
47. Carter HB, Partin AW, Luderer AA, et al. Percentage of free prostate-specific antigen *in sera* predicts aggressiveness of prostate cancer a decade before diagnosis. *Urology* 1997;**49**:379–84.
48. Haese A, Dworschack RT, Partin AW. Percent free prostate specific antigen in the total prostate specific antigen 2–4 ng/mL range does not substantially increase the number of biopsies needed to detect clinically significant prostate cancer compared to the 4 to 10 ng/mL range. *J Urol* 2002;**168**:504–8.
49. Lee BH, Hernandez AV, Zaytoun O, et al. Utility of percent free prostate-specific antigen in repeat prostate biopsy. *Urology* 2011;**78**:386–91.
50. Auprich M, Augustin H, Budäus L, et al. A comparative performance analysis of total prostate-specific antigen, percentage free prostate-specific antigen, prostate-specific antigen velocity and urinary prostate cancer gene 3 in the first, second and third repeat prostate biopsy. *BJU Int* 2012;**109**:1627–35.
51. Piironen T, Pettersson K, Suonpää M, et al. *In vitro* stability of free prostate-specific antigen (PSA) and prostate-specific antigen (PSA) complexed to alpha 1-antichymotrypsin in blood samples. *Urology* 1996;**48**(6A Suppl.):81–7.
52. Lilja H, Haese A, Björk T, et al. Significance and metabolism of complexed and noncomplexed prostate specific antigen forms, and human glandular kallikrein 2 in clinically localized prostate cancer before and after radical prostatectomy. *J Urol* 1999;**162**:2029–34.
53. Nixon RG, Wener MH, Smith KMB. Biological variation of prostate specific antigen levels in serum: an evaluation of day-to-day physiological fluctuations in a well-defined cohort of 24 patients. *J Urol* 1997;**157**:2183–90.
54. Benson MC, Whang IS, Pantuck A, et al. Prostate specific antigen density: a means of distinguishing benign prostatic hypertrophy and prostate cancer. *J Urol* 1992;**147**:815–6.
55. Sfoungaristos S, Katafigiotis I, Perimenis P. The role of PSA density to predict a pathological tumour upgrade between needle biopsy and radical prostatectomy for low risk clinical prostate cancer in the modified Gleason system era. *Can Urol Assoc J* 2013;**7**:E722–7.
56. Seaman EK, Whang IS, Cooner W, et al. Predictive value of prostate-specific antigen density for the presence of micrometastatic carcinoma of the prostate. *Urology* 1994;**43**:645–8.
57. Brawer MK, Aramburu EA, Chen GL, et al. The inability of prostate specific antigen index to enhance the predictive the value of prostate specific antigen in the diagnosis of prostatic carcinoma. *J Urol* 1993;**150**:369–73.
58. Catalona WJ, Southwick PC, Slawin KM, et al. Comparison of percent free PSA, PSA density, and age-specific PSA cutoffs for prostate cancer detection and staging. *Urology* 2000;**56**:255–60.
59. Kundu SD, Roehl KA, Yu X, et al. Prostate specific antigen density correlates with features of prostate cancer aggressiveness. *J Urol* 2007;**177**:505–9.
60. Carter HB, Sauvageot J, Walsh PC, et al. Prospective evaluation of men with stage T1C adenocarcinoma of the prostate. *J Urol* 1997;**157**:2206–9.
61. Koie T, Mitsuzuka K, Yoneyama T, et al. Prostate-specific antigen density predicts extracapsular extension and increased risk of biochemical recurrence in patients with high-risk prostate cancer who underwent radical prostatectomy. *Int J Clin Oncol* 2014;**20**(1):176–81.
62. Carter HB, Morrell CH, Pearson JD, et al. Estimation of prostatic growth using serial prostate-specific antigen measurements in men with and without prostate disease. *Cancer Res* 1992;**52**:3323–8.
63. Smith DS, Catalona WJ. Rate of change in serum prostate specific antigen levels as a method for prostate cancer detection. *J Urol* 1994;**52**:1163–7.
64. Bjurlin MA, Loeb S. PSA velocity in risk stratification of prostate cancer. *Rev Urol* 2013;**15**:204–6.
65. Moul JW, Sun L, Hotaling JM, et al. Age adjusted prostate specific antigen and prostate specific antigen velocity cut points in prostate cancer screening. *J Urol* 2007;**177**:499–503.
66. Loeb S, Roehl KA, Yu X, et al. Use of prostate-specific antigen velocity to follow up patients with isolated high-grade prostatic intraepithelial neoplasia on prostate biopsy. *Urology* 2007;**69**:108–12.
67. D' Amico AV, Chen MH, Roehl KA, et al. Preoperative PSA velocity and the risk of death from prostate cancer after radical prostatectomy. *N Engl J Med* 2004;**351**:125–35.
68. D' Amico AV, Renshaw AA, Sussman B, et al. Pretreatment PSA velocity and risk of death from prostate cancer following external beam radiation therapy. *JAMA* 2005;**294**:440–4.
69. Eggener SE, Roehl KA, Yossepowitch O, et al. Prediagnosis prostate specific antigen velocity is associated with risk of prostate cancer progression following brachytherapy and external beam radiation therapy. *J Urol* 2006;**176**:1399–403.
70. Vickers AJ, Thompson IM, Klein E, et al. A commentary on PSA velocity and doubling time for clinical decisions in prostate cancer. *Urology* 2014;**83**:592–8.

71. Ross AE, Loeb S, Landis P, et al. Prostate-specific antigen kinetics during follow-up are an unreliable trigger for intervention in a prostate cancer surveillance program. *J Clin Oncol* 2010;**28**:2810–6.
72. Bussemakers MJ, van Bokhoven A, Verhaegh GW, et al. DD3: a new prostate-specific gene, highly overexpressed in prostate cancer. *Cancer Res* 1999;**59**:5975–9.
73. Sokoll LJ, Ellis W, Lange P, et al. A multicenter evaluation of the PCA3 molecular urine test: pre-analytical effects, analytical performance, and diagnostic accuracy. *Clin Chim Acta* 2008;**389**:1–6.
74. Hessels D, Klein Gunnewiek J, van Oort I, et al. DD3 (PCA3)-based molecular urine analysis for the diagnosis of prostate cancer. *Eur Urol* 2003;**44**:8–16.
75. Tinzl M, Marberger M, Horvath S, et al. DD3 RNA analysis in urine? A new perspective for detecting prostate cancer. *Eur Urol* 2004;**46**:182–7.
76. Groskopf J, Aubin SM, Deras IL, et al. APTIMA PCA3 molecular urine test: development of a method to aid in the diagnosis of prostate cancer. *Clin Chem* 2006;**52**:1089–95.
77. Haese A, de la Taille A, Van Poppel H. Clinical utility of the PCA3 urine assay in European men scheduled for repeat biopsy. *Eur Urol* 2008;**54**:1081–8.
78. Nakanishi H, Groskopf J, Fritsche HA, et al. PCA3 molecular urine assay correlates with prostate cancer tumor volume: implication in selecting candidates for active surveillance. *J Urol* 2008;**179**:1804–10.
79. Groskopf J, Blase A, Koo S. The PCA3 score is independent of prostate gland volume, and can synergize with other patient information for predicting biopsy outcome. *Eur Urol* 2007;**6**:48.
80. Haese A, de la Taille A, van Poppel H, et al. Clinical utility of the PCA3 urine assay in European men scheduled for repeat biopsy. *Eur Urol* 2008;**54**:1081–8.
81. Marks L, Fradet Y, Deras IL, et al. PCA3 molecular urine assay for prostate cancer in men undergoing repeat biopsy. *Urology* 2007;**69**:532–5.
82. Ankerst D, Groskopf J, Day JR, et al. Predicting prostate cancer risk through incorporation of prostate cancer gene 3. *J Urol* 2008;**180**:1303–8.
83. Deras I, Aubin SM, Blase A, et al. PCA3: A molecular urine assay for predicting prostate biopsy outcome. *J Urol* 2008;**179**:1587–92.
84. van Gils M, Hessels D, van Hooij O, et al. The time-resolved fluorescence-based PCA3 test on urinary sediments after digital rectal examination; a Dutch multicenter validation of the diagnostic performance. *Clin Cancer Res* 2007;**13**:939–43.
85. Wu A, Reese A, Cooperberg M, et al. Utility of PCA3 in patients undergoing repeat biopsy for prostate cancer. *Prostate Cancer Prostatic Dis* 2012;**15**:100–5.
86. US Food and Drug Administration: Medical Devices: PROGENSA PCA3 Assay, 2012. Available from: www.fda.gov/MedicalDevices/Productsand/MedicalProcedures/DeviceApprovalsandClearances/Recently-ApprovedDevices/ucm294907.htm. [accessed 11.01.2013].
87. Durand X, Xylinas E, Radulescu C, et al. The value of urinary prostate cancer gene 3 (PCA3) scores in predicting pathological features at radical prostatectomy. *BJU Int* 2012;**110**:43–9.
88. van Poppel H, Haese A, Graefen M, et al. The relationship between Prostate CAncer gene 3 (PCA3) and prostate cancer significance. *BJU Int* 2012;**109**:360–6.
89. Augustin H, Mayrhofer K, Pummer K, et al. Relationship between prostate cancer gene 3 (PCA3) and characteristics of tumor aggressiveness. *Prostate* 2013;**73**:203–10.
90. Bradley LA, Palomaki GE, Gutman S, et al. Comparative effectiveness review: prostate cancer antigen 3 testing for the diagnosis and management of prostate cancer. *J Urol* 2013;**190**:389–98.

C H A P T E R

# 11

# Imaging in Localized Prostate Cancer

*Sandeep Sankineni, MD*, Peter L. Choyke, MD*, Peter Pinto, MD**, Baris Turkbey, MD**

*Molecular Imaging Program, National Cancer Institute, National Institutes of Health, Bethesda, MD, USA
**Urologic Oncology Branch, National Cancer Institute, National Institutes of Health, Bethesda, MD, USA

## INTRODUCTION

Prostate cancer affects one in seven men in the United States, and it is estimated that 12% of affected men will die from their disease.[1] Prostate cancer represents a spectrum of disease ranging from indolent with a low risk of mortality to very aggressive with a high risk of metastases leading to death. For most patients, prostate cancer will not be the cause of death but may nonetheless be a source of great anxiety and may lead to treatment-related adverse events. The paradox of prostate cancer is that while it is mostly a slow-growing, nonlife-threatening diagnosis, it can also be an aggressive cancer in which metastases portend a median 5-year survival of only 28%.[2] Motivated by the hope that early detection could lead to improved outcomes, population screening with prostate-specific antigen (PSA) was initiated beginning in the late 1980s.[3] Now that over a quarter of a decade has elapsed, it is clear that PSA screening has also led to overdiagnosis and overtreatment of low-grade prostate cancer since PSA cannot discriminate between low- and high-risk cancers with reasonable specificity. Moreover, many men underwent workups for a variety of pathologies including prostatitis and benign prostatic hyperplasia (BPH), which cause false-positive increases in PSA.[4] The results of two recently completed large trials of screening, including the Prostate, Lung, Colorectal, and Ovarian (PLCO) Cancer Screening Trial, and the European Randomized Study of Screening for Prostate Cancer showed a high rate (17–50%) of overdiagnosis of low-grade prostate cancer based on routine PSA screening with no perceptible benefit to the patient.[5] Up to the time of this report, many men with low-risk disease were being aggressively treated with surgery or radiation therapy for low-risk disease. While this treatment no doubt cured patients of their prostate cancer, it also left them with decades of decreased quality of life. As a result of these two studies and several other smaller studies that reached the same conclusion, the US Preventative Screening Task Force (USPSTF) in 2012 issued a grade of "D" for PSA screening using existing criteria. This has been widely interpreted to mean PSA screening is of no value to patients.[6] However, there is still no suitable alternative and no better screening method currently available. In 2015, a 3221-patient study in Toronto, Canada looked at the impact of the new PSA screening guidelines set forth by the USPSTF. As a result of curtailment of PSA screening by family practitioners, there was a 42.8% decrease in per-month detection of clinically significant prostate cancer (Gleason >7) in the year following the release of the updated PSA screening recommendation.[7] Thus, while reducing the use of PSA screening certainly reduces the overdiagnosis of low-grade disease, it also reduces the potentially life-saving diagnosis of higher-grade disease. It is therefore important that a new balance be struck between the extremes of screening everyone and screening no one.

There is some hope that imaging, particularly prostate MRI, might prove a useful adjunct to PSA screening. It is often lost in the discussion that PSA is only partly to blame for the current dilemma. Elevated PSA leads to a systematic but undirected prostate biopsy where it is quite likely that small incidental islands of low-grade disease could be discovered. Thus, this "random" biopsy is equally to blame for the unintended diagnosis of many low-grade tumor islets. Meanwhile, larger and more consequential lesions outside of the normal biopsy template are likely to be missed. Therefore, the addition of MRI could be helpful in detecting lesions within the prostate more likely to be clinically significant and directing biopsies into those lesions while avoiding biopsies of normal-appearing tissue, which is nonetheless likely to contain

Prostate Cancer. http://dx.doi.org/10.1016/B978-0-12-800077-9.00011-6

indolent islands of low-grade cancer. Over the past decade, there has been rapid improvement in image-guided biopsy technology.[8] Most importantly, the acceptance and use of multiparametric prostate MRI in combination with transrectal ultrasound (TRUS) biopsy has increased clinicians' ability to detect clinically significant disease while reducing the detection of low-grade disease, and is now directly impacting the clinical management of patients with prostate cancer. While it is still too early to definitively state that this approach will result in a more efficacious screening method for prostate cancer, it is nonetheless very promising and readily applied today whereas new biomarkers are still years away from the clinic. This chapter will discuss the current state of imaging in prostate cancer detection, including ultrasound, multiparametric MRI (mpMRI), and PET/CT.

## ULTRASOUND

TRUS is a widely available, portable, readily repeatable, and relatively inexpensive imaging technique. Ultrasound enables real-time visualization of the prostate, thus allowing the clinician to determine the gland volume and the distinction between the peripheral zone (PZ) and transition zone (TZ). Prostate cancers usually appear as hypoechoic regions on TRUS; however, this pattern can be mimicked by a variety of other pathologies such as BPH and inflammatory process. Moreover, not every prostate cancer is hypoechoic, and lesions commonly appear isoechoic, which reduces the diagnostic utility of TRUS. ECE (extracapsular extension) detection by TRUS is also limited except in very extensive cases. Color or power Doppler modes in TRUS can improve the tumor detection rate, but this is dependent upon the extent of angiogenesis. In smaller or less aggressive cancers, this affect may be minimal and lead to lower sensitivity rates. Today TRUS is mainly used for the purpose of guiding biopsies during systematic, blind sampling of the prostate.

More recently, contrast-enhanced TRUS with *microbubbles* has been reported to improve the sensitivity for tumor detection. Microbubbles are 5–10 μm gas-filled bubbles that can be seen on ultrasound. However, microbubbles tend to act like blood pool agents because of their large size, and often, only the vessels themselves are visualized. A study on contrast-enhanced TRUS demonstrated recently the ability to differentiate prostate cancer from normal tissue. It reported a sensitivity of 100%, but a specificity of only 48% in patients having previous negative biopsies but rising PSA values.[9] While the addition of contrast-enhancement ultrasound (CEUS) is a promising contribution to conventional TRUS biopsy, the value of CEUS is controversial, with experts achieving excellent results but several multicenter trials achieving mediocre results. As with all facets of ultrasound, the method is still highly operator-dependent and results will vary according to skill and experience. More research is warranted before this can be considered a routine option in clinical practice.

## MAGNETIC RESONANCE IMAGING

mpMRI is now considered to be the most powerful noninvasive diagnostic method for the detection of prostate cancer. While needle biopsy and histopathology remain the gold standard of diagnosis, mpMRI offers a high level of confidence in diagnosing clinically significant cancers.[10] A study of 1003 men by the National Cancer Institute showed the significant improvement in prostate cancer detection with the use of mpMRI followed by targeted MRI-TRUS fusion-guided biopsy.[11]

Prostate MRI can be acquired at 1.5T or 3T, with or without the use of an endorectal coil (ERC). There is an understanding that while the lower field strength of 1.5T is sufficient for evaluation, imaging at 3T is likely to be of higher quality. This is because 3T scanners exhibit a higher signal-to-noise ratio (SNR). However, multiple studies have shown minimal difference in outcomes between prostate MRIs conducted at 1.5T with an ERC versus 3T using only phased-array surface coils.[12,13] The routine use of the ERC is still debated, particularly at 3T. Newer MRI units are capable of performing excellent prostate MRIs without the ERC, thus reducing cost, time, and patient discomfort. Nonetheless, several studies have determined that there is benefit to using ERC at 3T. Heijmink et al. found mpMRI with ERC to have better accuracy for tumor localization when compared to mpMRI with body coil.[14] Turkbey et al. recently showed increased positive predictive value of 80% versus 64% for mpMRI with and without an ERC, respectively, in 20 patients who underwent radical prostatectomy.[15] However, the clinical consequence of this slightly increased sensitivity with the ERC has not yet been determined. Thus, in general, better-quality scans can be obtained with ERC, and this translates to higher sensitivity and specificity; however, it is unclear whether the use of ERC is justified in terms of better patient outcomes.

Eventually, it is expected that routine use of ERC will no longer be used for the reasons just mentioned. With time and technological advancement in MR technology, mpMRI with ERC may be reserved for very specific purposes – such as in local staging after histopathological diagnosis and especially in the postprostatectomy follow-up after biochemical recurrence.

### Anatomic MR Imaging (T1W, T2W MRI)

Anatomic MR imaging includes T1W and T2W MRI. T1W images are not used for diagnostic purposes because zonal anatomy is difficult to identify and tumors are typically not well seen. However, T1W images are helpful in determining if there is residual hemorrhage due to prior biopsy, which appears hyperintense on T1W images.[16] It is important to identify hemorrhage prior to further sequence evaluation as hemorrhage can result in false-positive diagnoses on mpMRI. It is widely agreed that it is best to wait at least 8–10 weeks postbiopsy to reduce the chance of hemorrhagic artifact related to biopsy.[17]

T2W imaging offers high-resolution and excellent anatomical detail (Figure 11.1a, b). T2W imaging is acquired in axial, sagittal, and coronal views. The high clarity offered by T2W imaging allows the clinician to determine zonal anatomy. The normal PZ tends to have a higher signal, whereas the TZ has an intermediate-lower signal intensity pattern mainly due to benign hyperplasia. Cancers appear as hypointense foci within PZ and TZ, but may be more difficult to detect in the latter due to the heterogeneous background in the TZ. T2W MRI for detection of cancer is known to have a wide range of sensitivities and specificities, 27–100%, and 32–99%, respectively, but this reflects a broad range of experience, patient selection, MR quality, and diagnostic criteria.[18–20]

A major concern for clinicians and a motivation behind the acquisition of the T2W sequence in routine diagnosis of prostate cancer is to detect ECE (Figure 11.2). ECE is seen as a direct extension of the tumor into the periprostatic fat.[21] This is important for preoperative staging since the presence of ECE would upstage a patient to Stage T3A. This could redirect a patient's clinical management approach including altering the surgical approach or even ruling out the possibility of surgery. On MRI, indirect features of ECE include capsular bulge, broad capsular base, and border irregularity, obliteration of rectoprostatic angle, and/or capsular retraction.[22] Despite the superior spatial resolution of T2W MRI, it is still limited for the evaluation of microscopic ECE with sensitivity and specificity ranges of 14.4–100% and 67–100%, respectively.[23,24] Seminal vesicles, which are located superiorly at the base of the prostate, are another important structure to evaluate for accurate staging. On T2W MRI, they normally appear as hyperintense saccules due to their high fluid content. When there is seminal vesicle invasion, they often appear hypointense and may enhance after contrast media.[25] MRI with an ERC has a high reported accuracy (91%) for detection of seminal vesicle invasion.[26]

## Functional MR Imaging

### *Diffusion Weighted MRI and Apparent Diffusion Coefficient Maps (DWI and ADC)*

DWI is based on the Brownian motion of free water within tissues. Cellular density increases in malignancies when compared with normal tissue. Due to the increased number of cell membranes, water diffusion is more restricted in the tumor site. This restriction or impedance can be noninvasively detected on DWI and can be quantified by calculating the apparent diffusion coefficient (ADC). DWI does not require contrast injection, and can be done readily in almost all clinical scanners along with T2W MRI. A meta-analysis of 5892 patients showed that DWI alone yielded higher sensitivity, specificity, and area under the ROC curve than T2W alone.[27] On ADC maps, prostate cancer shows decreased signal intensity relative to normal tissue (Figure 11.1c). DWI and ADC maps demonstrate a wide range of sensitivities and specificities, 57–93.3% and 57–100%, respectively, for tumor detection.[28–32] Additionally, it has been shown that the use of high b value imaging, (e.g., 1000–2000 s/mm$^2$) also improves performance for lesion detection (Figure 11.1d).[33,34] DWI is a major component of the mpMRI and has great utility for improving prostate cancer detection and management. However, it has some challenges to be addressed, as it has a lower spatial resolution and is highly susceptible to bulk motion.[35] Also, air within the rectum may distort the DWI image and cause susceptibility artifacts if an ERC filled with a perfluorocarbon is not used. Improvements in technology (parallel imaging and field strength) may mitigate these technical shortcomings.

### Dynamic Contrast Enhanced (DCE) MRI

DCE MRI uses a gadolinium-chelate-based agent injected as bolus and offers information on the vascularity of a lesion.[36] Blood vessel proliferation increases in response to oxygen and nutrient deprivation as tumors surpass a few millimeters in diameter. These new vessels are tortuous and permeable in comparison to normal vessels. DCE MRI consists of a T1W gradient echo imaging prior to, during, and after intravenous bolus injection of a low molecular weight gadolinium chelate agent. The prostate imaging is acquired continuously for several minutes, and the interval between the beginning and end of each component scan is known as the temporal resolution, which typically varies from 5 s to 30 s. Shorter temporal resolution is preferred in order to image the enhancement peak after injection as early enhancement can be an indicator of a positive study. On DCE MRI, malignant lesions appear with early focal enhancement and rapid washout [37] (Figure 11.1e); however, BPH and prostatitis may also show early enhancement leading to false positives. Analysis of DCE MRI can be done qualitatively, semiquantitatively, or quantitatively. Qualitative evaluation simply includes visual assessment of serially acquired images in a video mode, and tumoral enhancement is usually characterized by rapid and early opacification (1–3 phases after the contrast arrives in the major arteries such as femoral artery) compared to its surrounding. Semiquantitative analysis mostly relies on evaluating the DCE by matching the time-signal curve to

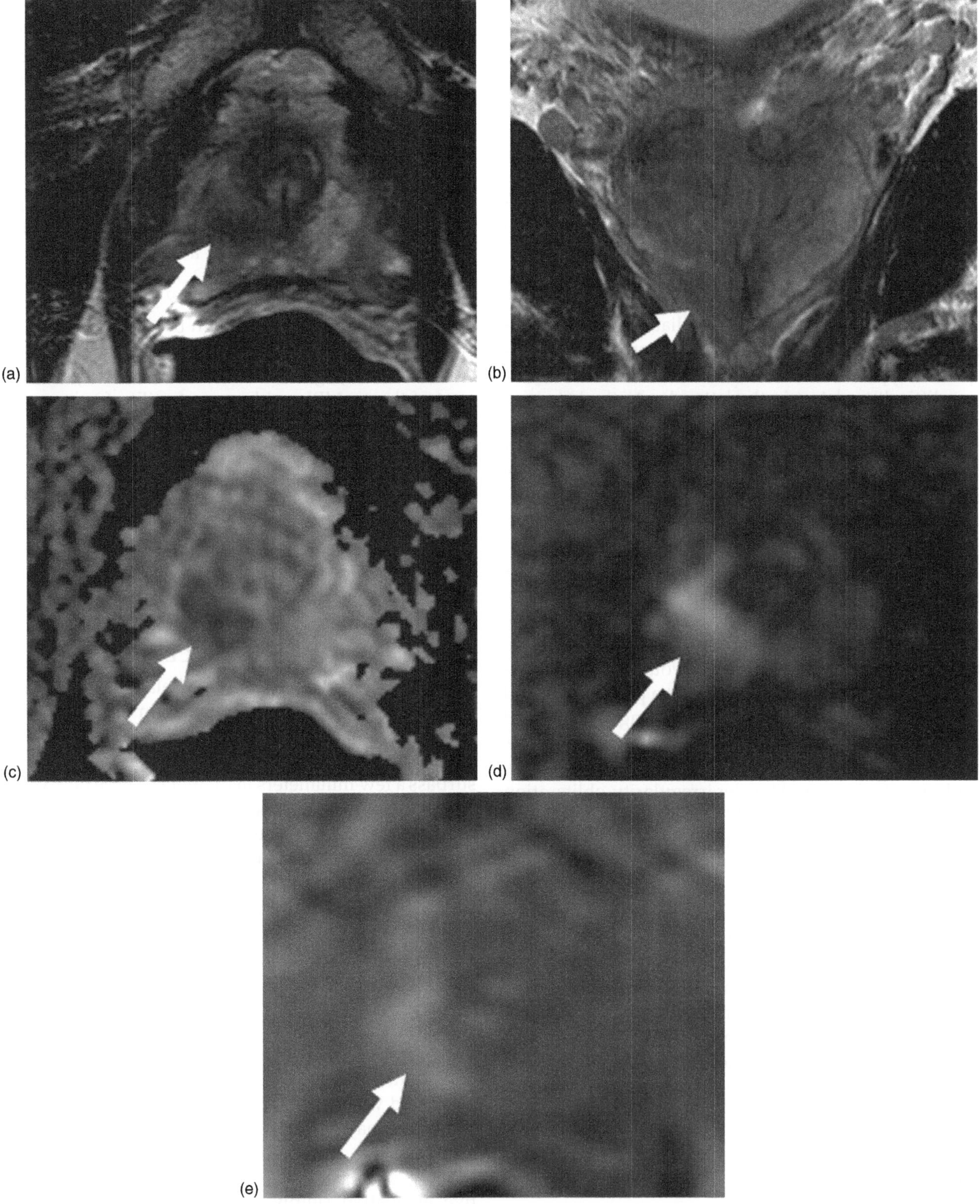

**FIGURE 11.1 An mpMRI was acquired at 3T with an ERC in a 58-year-old male with serum PSA = 4.46 ng/mL, DRE negative, with no prior prostate biopsy.** (a, b) Axial and coronal T2W imaging shows a lesion in the right apical peripheral zone, (c) ADC maps of DWI identify the same right apical peripheral zone region with hypo-intense features, (d) b2000 shows a focal hyperenhancing region corresponding to the right apical PZ, and (e) DCE shows early focal enhancement of the right apical PZ. The lesion, located in the peripheral zone, is scored with PI-RADS Version 2.0 based on the diffusion imaging, and therefore receives a score of 4, with moderately-high level of suspicion for clinically significant prostate cancer; it is therefore recommended that the patient undergo a targeted biopsy. Following the targeted MRI/TRUS fusion-guided biopsy, histopathology confirms a Gleason 3 + 4 (60%) with perineural invasion.

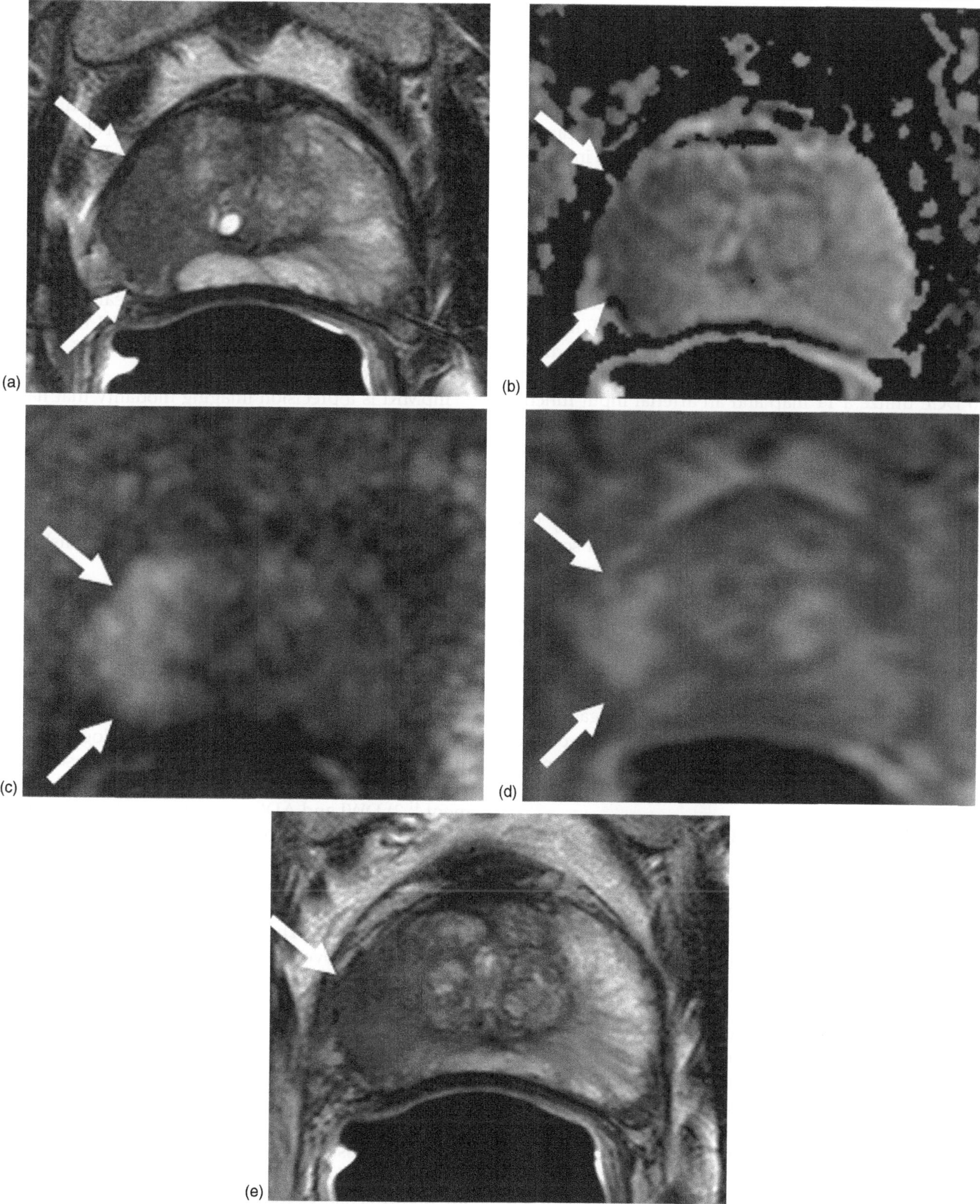


**FIGURE 11.2 An mpMRI was acquired at 3T with an ERC in a 66-year-old male with serum PSA = 12.1 ng/mL, DRE negative, with one prior positive conventional prostate biopsy (Gleason 3 + 3 in the right mid peripheral zone), who was on active surveillance at the time of his mpMRI.** (a) Axial T2W imaging shows a large lesion in the right midbase peripheral zone with possible ECE, (b) ADC maps of DWI show a larger hypointense area corresponding with the T2W imaging, (c) b2000 shows a focal enhancement of the suspicious region, (d) DCE shows early focal enhancement of the suspected lesion, (e) axial T2W imaging shows a possible ECE of the lesion. The lesion, located in the peripheral zone, is scored with PI-RADS Version 2.0 based on its diffusion imaging, and therefore receives a score of 5. There is a high level of suspicion for clinically significant prostate cancer and the patient requires a targeted biopsy for this reason. Following the targeted MRI/TRUS fusion-guided biopsy, histopathology confirms a Gleason 4 + 4, in 95% of the targeted core from the right midbase peripheral zone.

one of three different curve types. Type 1 curve refers to benign enhancement pattern where there is continuous gradual enhancement, whereas Type 2 curves, wherein the signal initially increases and then plateaus, can represent inflammatory processes. Finally, Type 3 includes a rapid wash in and out of contrast material and is highly predictive of a malignant lesion. However, almost all of the curve types can be identified in any given lesion within the prostate, regardless of whether they are benign or malignant. Thus, one typically selects the "worst" curve (i.e., Type 3) to represent the DCE of the malignant lesion. This can reduce the effectiveness of DCE in distinguishing prostate cancers from other benign conditions such as inflammation or BPH. Quantitative approaches to DCE utilize a T1 map and an arterial input function (either personalized or population-based), which are integrated into a two-compartment model to calculate several quantitative parameters, such as forward and reverse contrast flow rates (Ktrans and kep, respectively). Qualitative evaluation is the most commonly used technique because it does not require special software. Compared to T2W and DWI, DCE MRI plays a relatively modest role in prostate cancer diagnosis. However, it may be beneficial in identifying recurrences in the post-treatment setting. Despite its utility, some challenges exist for DCE MRI including standardization of image acquisition protocols and analysis techniques. Patients with severe renal failure cannot receive gadolinium contrast material; hence, the use of gadolinium-contrast agents should be carefully evaluated in such patients.

## MR SPECTROSCOPIC IMAGING

MR spectroscopy is another functional MRI technique that aims to display certain compounds inside the prostate, such as choline and citrate, by representing the amount of signal from particular proton resonance frequencies that correspond to those compounds. MR spectroscopic imaging (MRSI) thus produces a spectrum of peaks corresponding to relevant and nonrelevant molecules. While it truly provides molecular information about the chemicals within the prostate, this technique is largely limited due to its long acquisition times, necessity for expert personnel (e.g., MR physicist), and complex processing methods. Because it utilizes a much lower spatial resolution than the other sequences previously discussed, it has a lower sensitivity due to significant partial volume effects. Therefore, it is not recommended in routine clinical practice; however, it is still used for research purposes in a limited number of centers.

mpMRI is therefore a powerful imaging technique for detection and staging of prostate cancer. The most underappreciated benefit of prostate mpMRI is its strong negative predictive value. Studies have typically reported negative predictive values of mpMRI ranging from 70% to 80%.[38] The ability to rule out clinically significant disease is one of its most important features. In patients with BPH or prostatitis, who often have an elevated PSA (>4.0), PSAD (>0.15), and/or abnormal DRE, the failure to identify a prostate lesion can be highly useful in eliminating significant cancers. Without an mpMRI and a targeted biopsy approach, a patient may go on to have numerous negative biopsies. Negative MRIs give physicians confidence to minimize the number of biopsy sessions a patient undergoes over several years. In a recent study by Gupta et al., the NPV of mpMRI in detecting prostate cancer has been reported as high as 73.1% for all prostate cancer and 89.7% for ECE.[39]

mpMRI is a precision imaging modality that has the future potential to be tailored for personalized imaging. In case of suspicion for prostate cancer, a patient-specific mpMRI may be performed. The individual mpMRI sequences can be acquired for screening, diagnosis, staging, and follow-up. Further trials and studies will need to be performed before guidelines are established for such a practice.

One important aspect of mpMRI is the adoption of a standardized reporting system. mpMRI as defined by the 2014 American College of Radiology's Prostate Imaging Reporting and Data System (PI-RADS version 2.0) consists of evaluation of numerous MR sequences to determine a reproducible scoring system which may be readily communicated among radiologists, urologists, and their patients.[40] The sequences used are 1) T2-weighted (T2W) pulse sequence for anatomic assessment,[41] 2) DWI and ADC maps with high-b value imaging, and 3) dynamic contrast enhanced (DCE) imaging for functional assessment.[42] In PI-RADS, DWI is used primarily for the scoring of the PZ, but T2W is primarily used in the TZ due to overlap between cancer lesions and BPH nodules. The use of PI-RADS is therefore expected to improve education and communication and thereafter improve the overall utility of mpMRI in patients with suspected cancer.

## PET/CT

Conventional CT and mpMRI are unable to detect changes in tissue at the molecular level. These changes can be imaged by using targeted radionuclides, which are highly sensitive for detecting malignancies. Imaging of nodal and bony metastatic spread may benefit from such studies, whereas they are likely too expensive for routine use in localized prostate cancer.

The current standard of care for bone metastasis is the 99mTc methyldiphosphonate (MDP) bone scan. Radiolabeled MDP agent is taken up in osteoblastic lesions seen in metastatic prostate cancer.[43] Unfortunately, because it is a single photon emitter, sensitivity is less than optimal, and false-positive findings in nonmalignant conditions such as benign neoplasms, trauma, and degenerative joint disease are commonplace. Because bone scans are typically not acquired with a CT scan, it can be difficult to resolve whether a particular focus of uptake is associated with a benign condition on CT. For this reason, there has been interest in the development of $^{18}$F-sodium fluoride ($^{18}$F-NaF) PET/CT as a viable alternative for detection of bone metastases. A recent study evaluated the efficacy of $^{18}$F-NaF versus that of 99mTc-bone scintigraphy in detecting bone metastases in patients with lung, breast, and prostate cancer.[44] $^{18}$F-sodium fluoride outperformed the bone scan in all components and had a sensitivity and NPV of 100%. However, due to a high rate of false positives, these scans must be correlated with the coregistered CT. This can be a time-consuming process, but new automated methods of detecting and eliminating benign lesions from further consideration are under development.

More recently, there has been a tremendous growth in research in radiotracers for PET/CT diagnostics in localized prostate cancer. Results of studies have been published on several of these agents.

$^{18}$F-DCFBC (PSMA target), developed at Johns Hopkins University (Baltimore, MD), is one such promising radiotracer. A preliminary study in five patients identified 32 PET-positive suspected metastatic sites. Out of these, 21 (65%) of the sites were positive on both PET and conventional imaging suggestive of metastatic disease, while the remaining 11 PET positive-only lesions (34%) were indeterminant but were likely to be metastatic.[45] A Phase 2 clinical trial, with an enrollment of 45 patients, is now underway at the National Institutes of Health (NIH, Bethesda, MD) to determine the efficacy of $^{18}$F-DCFBC.

Another related agent is $^{68}$Ga-PSMA (gallium labeled PSMA ligand). In early 2014 a comparison study from Germany reported results on $^{68}$Ga-PSMA versus $^{18}$F-choline PET/CTs. A total of 78 lesions were detected in 32 patients using $^{68}$Ga-PSMA, and 56 lesions were detected in 26 patients using $^{18}$F-choline. The higher detection rate in $^{68}$Ga-PSMA PET/CT was statistically significant ($p = 0.04$).[46,47] Another radiotracer unrelated to PSMA was developed at Emory University (Atlanta, GA). This agent, $^{18}$F-FACBC (radiolabeled leucine analog targeting protein metabolism), has now also been studied at multiple sites. A sector-based comparison with histopathologic analysis performed at NCI revealed sensitivity and specificity of 67 and 66%, respectively, for $^{18}$F-FACBC in patients with localized prostate cancer although the performance was inferior to T2W MRI (sensitivity and specificity at 73 and 79%, respectively). Yet the combined positive predictive value (FACBC + T2W MRI) was shown to be higher, at 82%.[48,49]

$^{18}$F-FDHT, targeting the androgen receptor, was created in 2004 at Memorial Sloan Kettering (New York, NY). An initial trial by Larson and coworkers compared $^{18}$F-FDHT with $^{18}$F-FDG PET. In the study, 59 lesions were found by conventional imaging. $^{18}$F-FDG PET was positive in 57 of 59 lesions (97%), while $^{18}$F-FDHT PET was positive in 46 of 59 lesions (78%).[50]

$^{11}$C-acetate (targeting fatty acid metabolism) for prostate cancer, first used in Japan, now also has published results from comparison studies. Mena et al. performed a sector-based comparison with histopathology, which determined a sensitivity and specificity of 61.6 and 80.0%, respectively, for $^{11}$C-acetate, and 82.3 and 95.1%, respectively, for MRI for localized disease.[51,52]

Finally, $^{18}$F-choline and $^{11}$C-choline (targeting increased membrane turnover) have been used in numerous studies of prostate cancer. Umbehr et al. reported a systematic meta-analysis of 10 studies for staging 637 patients with proven but untreated prostate cancer, and showed a pooled sensitivity and specificity of 84 and 79%, respectively.[53] Thus, $^{18}$F-choline and $^{11}$C-choline are promising PET/CT agents for staging prostate cancer. It should be noted that $^{11}$C-choline and $^{11}$C-acetate have a half-life of only 20 min so they must be produced on site, whereas, the $^{18}$F labeled compounds have the potential to be centrally produced and shipped to various medical sites for injection.

While nearly all of the aforementioned agents have shown promising early results, the question currently remains regarding which one is the best and even what metric should be utilized to better assess these agents against one another. There are numerous clinical trials currently underway to better understand the capabilities of these agents, the results of which are awaited with much anticipation.[54]

# CONCLUSIONS

The development and acceptance of precise and powerful MRI techniques has altered the diagnosis and management of localized prostate cancer. MRI is currently the most valuable imaging modality available for localized disease. mpMRI employs a combination of T2W, DWI, and DCE. T2W provides the most detailed anatomic information while DWI is the most effective functional technique and has good specificity. DCE MRI is sensitive for prostate cancer but plays a secondary role in diagnosis and characterization. Using these sequences in combination may improve initial detection and staging of clinically significant disease, and also improve targeted biopsy techniques. These advances will continue to improve both the decision-making and utility of active surveillance and focal therapies. Novel imaging techniques such as PET/CT radiotracer agents are emerging to improve the detection and staging of prostate cancer.

## References


1. Siegel R, Ma J, Zou Z, et al. Cancer statistics, 2014. *CA Cancer J Clin* 2014;**64**:9–29.
2. The American Cancer Society. *Cancer Facts & Figures* 2015. http://www.cancer.org/research/cancerfactsstatistics/cancerfactsfigures2015/; 2015[accessed 10.01.15].
3. Stamey TA, Yang N, Hay AR, et al. Prostate-specific antigen as a serum marker for adenocarcinoma of the prostate. *N Engl J Med* 1987;**317**:909.
4. Schröder FH, Carter HB, Wolters T, et al. Early detection of prostate cancer in 2007. Part 1: PSA and PSA kinetics. *Eur Urol* 2008;**53**: 468–77.
5. Miller AB. New data on prostate cancer mortality after PSA screening. *N Engl J Med* 2012;**366**:1047–8.
6. Moyer VA, LeFevre ML, Siu AL, et al. Screening for prostate cancer: U.S. Preventive Services Task Force recommendation statement. *Ann Intern Med* 2012;**157**(2):120–34.
7. Bhindi B, Mamdani M, Kulkarni GS, et al. Impact of the U.S. Preventive Services Task Force recommendations against PSA screening on prostate biopsy and cancer detection rates. *J Urol* 2014.
8. Turkbey B, Choyke PL. Decade in review-imaging: a decade in image-guided prostate biopsy. *Nat Rev Urol* 2014;**11**(11):611–2.
9. Taymoorian K, Thomas A, Slowinski T, et al. Transrectal broadband-Doppler sonography with intravenous contrast medium administration for prostate imaging and biopsy in men with an elevated PSA value and previous negative biopsies. *Anticancer Res* 2007;**27**:4315–20.
10. Muller BG, Shih J, Sankineni S, et al. Prostate Cancer: Interobserver Agreement and Accuracy with the Revised Prostate Imaging Reporting and Data System at Multiparametric MR Imaging. *Radiology* 2015;142818 [Epub ahead of print].
11. Siddiqui MM, Rais-Bahrami S, Turkbey B, et al. Comparison of MR/ultrasound fusion-guided biopsy with ultrasound-guided biopsy for the diagnosis of prostate cancer. *JAMA* 2015;**313**(4): 390–7.
12. Sosna J, Pedrosa I, Dewolf WC, et al. MR imaging of the prostate at 3 Tesla: comparison of an external phased-array coil to imaging with an endorectal coil at 1.5 Tesla. *Acad Radiol* 2004;**11**:857–862; Beyersdorff D, Taymoorian K, Knösel T, et al. MRI of prostate cancer at 1.5 and 3.0 T: comparison of image quality in tumor detection and staging. *Am J Roentgenol* 2005;**185**:1214–1220.
13. Park BK, Kim B, Kim CK, et al. Comparison of phased-array 3.0-T and endorectal 1.5-T magnetic resonance imaging in the evaluation of local staging accuracy for prostate cancer. *J Comput Assist Tomogr* 2007;**31**:534–8.
14. Heijmink SWTPJ, Futterer JJ, Hambrock T, et al. Prostate cancer: body-array versus endorectal coil MR imaging at 3 T – comparison of image quality, localization, and staging performance. *Radiology* 2007;**244**:184–95.
15. Turkbey B, Merino MJ, Gallardo EC, et al. Comparison of endorectal coil and nonendorectal coil T2W and diffusion-weighted MRI at 3 Tesla for localizing prostate cancer: correlation with whole-mount histopathology. *J Magn Reson Imaging* 2014;**39**:1443–8.
16. Barrett T, Vargas HA, Akin O, et al. Value of the hemorrhage exclusion sign on T1-weighted prostate MR images for the detection of prostate cancer. *Radiology* 2012;**263**:751–7.
17. Qayyam A, Coakley FV, Lu Y, et al. Organ-confined prostate cancer: effect of prior transrectal biopsy on endorectal MRI and MR spectroscopic imaging. *Am J Roentgenol* 2004;**183**:1079–83.
18. Aydin H, Kizilgoz V, Tatar IG, et al. Detection of prostate cancer with magnetic resonance imaging: optimization of T1-weighted, T2-weighted, dynamic-enhanced T1-weighted, diffusion-weighted imaging apparent diffusion coefficient mapping sequences and MR spectroscopy, correlated with biopsy and histopathological findings. *J Comput Assist Tomogr* 2012;**36**(1):30–45.
19. Chen M, Dang HD, Wang JY, et al. Prostate cancer detection: comparison of T2-weighted imaging, diffusion-weighted imaging, proton magnetic resonance spectroscopic imaging, and the three techniques combined. *Acta Radiol* 2008;**49**(5):602–10.
20. Tan CH, Wei W, Johnson V, et al. Diffusion-weighted MRI in the detection of prostate cancer: meta-analysis. *Am J Roentgenol* 2012;**199**(4):822–9.
21. Wang L, Mullerad M, Chen HN, et al. Prostate cancer: incremental value of endorectal MR imaging findings for prediction of extracapsular extension. *Radiology* 2004;**232**:133–9.
22. Claus FG, Hricak H, Hattery RR. Pretreatment evaluation of prostate cancer: role of MR imaging and H-1 MR spectroscopy. *Radiographics* 2004;**232**:133–9.
23. Nakashima J, Tanimoto A, Imai Y, et al. Endorectal MRI for prediction of tumor site, tumor size, and local extension of prostate cancer. *Urology* 2004;**64**:101–5.
24. Bloch BN, Furman-Haran E, Helbich TH, et al. Prostate cancer: accurate determination of extracapsular extension with high-spatial-resolution dynamic contrast-enhanced and T2-weighted MR imaging – initial results. *Radiology* 2007;**245**:176–85.
25. Raskolnikov D, George AK, Rais-Bahrami S, et al. Multiparametric magnetic resonance imaging and image-guided biopsy to detect seminal vesicle invasion by prostate cancer. *J Endourol* 2014;**28**(11):1283–9.
26. Sala E, Akin O, Moskowitz CS, et al. Endorectal MR imaging in the evaluation of seminal vesicle invasion: diagnostic accuracy and multivariate feature analysis. *Radiology* 2006;**238**(3):929–37.
27. Tan CH, Wei W, Johnson V, et al. Diffusion-weighted MRI in the detection of prostate cancer: meta-analysis. *Am J Roentgenol* 2012;**199**:822–9.
28. Kim CK, Park BK, Lee HM, et al. Value of diffusion-weighted imaging for the prediction of prostate cancer location at 3T using a phased-array coil: preliminary results. *Invest Radiol* 2007;**42**:842–7.
29. Miao H, Fukatsu H, Ishigaki T. Prostate cancer detection with 3-T MRI: comparison of diffusion-weighted and T2-weighted imaging. *Eur J Radiol* 2007;**61**:297–302.
30. Haider MA, van der Kwast TH, Tanguay J, et al. Combined T2-weighted and diffusion-weighted MRI for localization of prostate cancer. *Am J Roentgenol* 2007;**189**:323–8.
31. Kozlowski P, Chang SD, Jones EC, et al. Combined diffusion-weighted and dynamic contrast-enhanced MRI for prostate cancer diagnosis – correlation with biopsy and histopathology. *J Magn Reson Imaging* 2006;**24**:108–13.
32. Woodfield CA, Tung GA, Grand DJ, et al. Diffusion-weighted MRI of peripheral zone prostate cancer: comparison of tumor apparent diffusion coefficient with Gleason score and percentage of tumor on core biopsy. *Am J Roentgenol* 2010;**194**:316–22.
33. Tamada T, Kanomata N, Sone T, et al. High b value (2000 s/mm$^2$) diffusion-weighted magnetic resonance imaging in prostate cancer at 3 Tesla: comparison with 1000 s/mm$^2$ for tumor conspicuity and discrimination of aggressiveness. *PLoS ONE* 2014;**9**:e96619.
34. Grant KB, Agarwal HK, Shih JH, et al. Comparison of calculated and acquired high b value diffusion-weighted imaging in prostate cancer. *Abdom Imaging* 2015;**40**:578–86.
35. Hoeks CMA, Berentsz JO, Hambrock T, et al. Prostate cancer: multiparametric MR imaging for detection, localization, and staging. *Radiology* 2011;**261**:46–66.
36. Alonzi R, Padhani AR, Allen C. Dynamic contrast enhanced MRI in prostate cancer. *Eur J Radiol* 2007;**63**:335–50.
37. Bonekamp D, Macura KJ. Dynamic contrast-enhanced magnetic resonance imaging in the evaluation of the prostate. *Top Magn Reson Imaging* 2008;**19**:273–84.
38. Panebianco V, Barchetti F, Sciarra A. Multiparametric magnetic resonance imaging vs. standard care in men being evaluated for prostate cancer: a randomized study. *Urol Oncol* 2015;**33**(1):17.e1–e7.

39. Gupta RT, Faridi KF, Singh AA, et al. Comparing 3-T multiparametric MRI and the Partin Tables to predict organ-confined prostate cancer after radical prostatectomy. *Urol Oncol* 2014;**32**(8): 1292–9.
40. American College of Radiology. *PI-RADS v2 - Prostate Imaging and Reporting and Data System: Version 2*. http://www.acr.org/Quality-Safety/Resources/PIRADS/; 2015[accessed 10.01.15].
41. Bhavsar A, Verma S. Anatomic imaging of the prostate. *Biomed Res Int* 2014;**2014**:728539.
42. Sankineni S, Osman M, Choyke PL. Functional MRI in prostate cancer detection. *Biomed Res Int* 2014;**2014**:590638.
43. Bouchelouche K, Tagawa ST, Goldsmith SJ, et al. PET/CT imaging and radioimmunotherapy of prostate cancer. *Semin Nucl Med* 2011;**41**:29–44.
44. Damle NA, Bal C, Bandopadhyaya GP, et al. The role of 18F-fluoride PET-CT in the detection of bone metastases in patients with breast, lung and prostate carcinoma: a comparison with FDG PET/CT and 99mTc-MDP bone scan. *Jpn J Radiol* 2013;**31**:262–9.
45. Cho SY, Gage KL, Mease RC, et al. Biodistribution, tumor detection, and radiation dosimetry of 18F-DCFBC, a low-molecular-weight inhibitor of prostate-specific membrane antigen, in patients with metastatic prostate cancer. *J Nucl Med* 2012;**53**(12):1883–91.
46. Afshar-Oromieh A, Zechmann CM, Malcher A, et al. Comparison of PET imaging with a (68)Ga-labelled PSMA ligand and (18)F-choline-based PET/CT for the diagnosis of recurrent prostate cancer. *Eur J Nucl Med Mol Imaging* 2014;**41**(1):11–20.
47. Afshar-Oromieh A, Malcher A, Eder M, et al. PET imaging with a [68Ga]gallium-labelled PSMA ligand for the diagnosis of prostate cancer: biodistribution in humans and first evaluation of tumour lesions. *Eur J Nucl Med Mol Imaging* 2013;**40**(4):486–95.
48. Turkbey B, Mena E, Shih J, et al. Localized prostate cancer detection with 18F-FACBC PET/CT: comparison with MR imaging and histopathologic analysis. *Radiology* 2014;**270**(3):849–56.
49. Schuster DM, Votaw JR, Nieh PT, et al. Initial experience with the radiotracer anti-1-amino-3-18F-fluorocyclobutane-1-carboxylic acid with PET/CT in prostate carcinoma. *J Nucl Med* 2007;**48**(1): 56–63.
50. Larson SM, Morris M, Gunther I, et al. Tumor localization of 16beta-18F-fluoro-5alpha-dihydrotestosterone versus 18F-FDG in patients with progressive, metastatic prostate cancer. *J Nucl Med* 2004;**45**(3):366–73.
51. Mena E, Turkbey B, Mani H, et al. 11C-acetate PET/CT in localized prostate cancer: a study with MRI and histopathologic correlation. *J Nucl Med* 2012;**53**(4):538–45.
52. Oyama N, Akino H, Kanamaru H, et al. 11C-acetate PET imaging of prostate cancer. *J Nucl Med* 2002;**43**(2):181–6.
53. Umbehr MH, Müntener M, Hany T, et al. The role of 11C-choline and 18F-fluorocholine positron emission tomography (PET) and PET/CT in prostate cancer: a systematic review and meta-analysis. *Eur Urol* 2013;**64**(1):106–17.
54. US National Institutes of Health. *Prostate PET/CT*. https://ClinicalTrials.gov; 2015[accessed 5.02.15].

PART II

# GENETIC SUSCEPTIBILITY AND HEREDITARY PREDISPOSITION, SCREENING, AND COUNSELING

CHAPTER

# 12

# Prostate Cancer Prevention: Strategies and Realities

*Jonathan Kiechle, MD**, *Simon P. Kim, MD, MPH**,**

*Department of Urology, University Hospital Case Medical Center, Case Western Reserve University, Cleveland, OH, USA

**Cancer Outcomes and Public Policy Effectiveness Research, COPPER Center, Yale University, New Haven, CT, USA

## INTRODUCTION

Prostate cancer is the most commonly diagnosed male malignancy in the United States with approximately 230,000 patients diagnosed each year.[1] Worldwide, about 900,000 men are diagnosed annually, making prostate cancer a global public health problem.[2] In the era of prostate-specific antigen (PSA) screening, most men are now diagnosed with clinically localized prostate cancer and face difficult treatment decisions due to the protracted history of the disease as well as uncertainty about optimal treatment and treatment-related morbidity.[3] Furthermore, the widespread adoption of PSA screening has led to a dramatic increase in the lifetime risk of prostate cancer diagnosis from 9% in the prescreening era to 17% at present.[4] Indeed, overdiagnosis and overtreatment of prostate cancer has become an area of significant concern due to limited mortality benefits found in screening trials.[5]

While controversies surround prostate cancer screening and treatment, particularly for low-risk disease, prevention of this common malignancy represents an attractive solution to lower the cancer burden for patients. Reducing the incidence of prostate cancer could decrease the number of men who die of prostate cancer, require prostate biopsy, or undergo primary treatment of prostate cancer which often have significant quality of life implications. Furthermore, while healthcare currently consumes almost 20% of the United States' Gross Domestic Product, primary prevention also offers the opportunity to lower the cost attributed to the treatment and management of prostate cancer. As a consequence, it is essential to understand the evidence regarding prostate cancer prevention and the current clinical role of prevention in prostate cancer. In this context, the aim of this chapter is to comprehensively review different primary prevention strategies for prostate cancer in regards to chemoprevention, lifestyle modifications, and the future direction of cancer prevention research.

## CHEMOPREVENTION

### 5-α Reductase Inhibitors

It is widely accepted that the development of prostate cancer depends on androgens. Androgens, specifically testosterone and dihydrotestosterone, are essential for growth and development of the prostate gland. When an androgen binds to the androgen receptor, several cellular processes occur, ultimately resulting in the recruitment and activation of gene products. This activation then leads to androgen-receptor-dependent gene transcription and translation which promotes both prostate growth and carcinogenesis.[6] Androgen blockade through medical or surgical castration has been shown to induce apoptosis and slow cancer progression, and has become the standard of care in treatment-naive metastatic prostate cancer.[7–9]

5-α Reductase inhibitors (5-ARIs) are a class of medications that block the activity of the enzyme 5-α reductase. The conversion of testosterone to dihydrotestosterone (DHT) through 5-α reductase is crucial in accelerating prostate growth, given the greater effects and potency of DHT.[10] 5-α Reductase exists as two different isoenzymes, type I and type II, and the latter has been shown to have high activity in the prostate.[11]

Prostate Cancer. http://dx.doi.org/10.1016/B978-0-12-800077-9.00012-8

Several landmark clinical trials have critically evaluated the effects of 5-ARIs in treating benign prostatic hyperplasia (BPH). The Finasteride Study Group was a multicentered, double-blinded study where patients with BPH were randomly assigned to receive finasteride (5 mg or 1 mg) or placebo.[12] The primary outcomes in this study were urinary symptoms, urinary flow and volume, and serum levels of DHT. Among patients randomized to 5 mg of finasteride, there was a significant decrease in prostate volume of nearly 20% by the end of 1 year. The Medical Therapy of Prostatic Symptoms (MTOPS) study was another large randomized trial designed to determine whether monotherapy with finasteride or doxazosin (an alpha receptor blocker), or combined medical therapy with both medications was more effective in managing patients with BPH.[13] Similarly, patients who were randomized to receive finasteride had a 20% reduction in prostate volume. These patients also experienced significant improvements in urinary flow, lower rates of urinary retention, and a decreased need for invasive secondary treatments.[13]

Against this backdrop, it is conceivable that 5-ARIs would be successful agents for chemoprevention of prostate cancer, given their biologic role in reducing DHT and their clinical effect on reducing prostate volume. Two large, multicentered clinical trials of 5-ARIs were conducted to critically evaluate the key question – do 5-ARIs reduce the risk of prostate cancer diagnosis and mortality? The Prostate Cancer Prevention Trial (PCPT) was a large, multicentered randomized clinical trial to test whether finasteride, a type II 5-ARI, reduced the risk of prostate cancer among healthy men. Men were eligible for inclusion if they were <55 years, with a normal digital rectal exam (DRE) and with an American Urological Association Symptom Score of <20 at the time of accrual. All patients recruited and enrolled underwent annual PSA testing and DRE for a total of 7 years. PSAs for men randomized to finasteride were adjusted by a factor of 2.3, while PSAs for men on placebo were not adjusted. Indications for transrectal ultrasound (TRUS)-guided prostate biopsy were abnormal DRE or a PSA ≥4 ng/mL. The primary outcome of this clinical trial was the incidence of newly diagnosed men with prostate cancer during the study. In this landmark trial, 18,882 patients were randomized to receive 5 mg finasteride daily or placebo for a period of 7 years. In 2003, Thompson et al. published the initial results from the PCPT. Among patients included in the final analysis, 18.4% of men randomized to finasteride were diagnosed with prostate cancer, while 24.4% of men randomized to placebo were diagnosed with prostate cancer (relative risk or RR 24.8% [95% CI: 18.6–30.6%] $p < 0.001$).[14] Although the initial results from the PCPT were impressive in terms of decreasing the overall rate of prostate cancer diagnosis, a greater percentage of men randomized to finasteride developed high-grade prostate cancer (Gleason ≥7) compared with those on placebo (6.4% vs. 5.1%; RR: 1.67 [95% CI: 1.44–1.93%] $p = 0.005$). Furthermore, men on finasteride were more likely to experience erectile dysfunction (67.4% vs. 61.5%; $p < 0.001$), loss of libido (65.4% vs. 59.6%; $p < 0.001$), and gynecomastia (4.5% vs. 2.8%; $p < 0.001$) compared with placebo. Finally, and possibly related to these adverse events, the rate of nonadherence to finasteride was significantly higher than nonadherence to placebo (36.8% vs. 28.9%; $p < 0.001$).[14]

In a recent update of the PCPT study after 17 years of followup,[15] randomization to finasteride was associated with lower rates of prostate cancer diagnosis compared with placebo (10.5% vs. 14.9%; RR: 0.70 [95% CI: 0.65 to 0.76] $p < 0.001$). However, patients taking finasteride continued to have a higher rate of high-grade (Gleason 7–10) prostate cancer diagnosis compared with those receiving placebo (3.5% vs. 3.0%; RR: 1.17 [95% CI: 1.00–1.37] $p = 0.05$). Furthermore, the 15-year survival rates across both arms were nearly identical (78.0% vs. 78.2%; adjusted hazard ratio or HR: 1.03 [95% CI: 0.98–1.09] $p = 0.26$). Following a diagnosis of prostate cancer, the long-term overall survival remained similar without any significant differences between groups.[15]

The Reduction by Dutasteride of Prostate Cancer Events (REDUCE) trial was another double-blinded, placebo-controlled, multicenter study that followed patients for 4 years.[16] The investigators chose dutasteride for this study as it inhibits both the type I and II isoforms of 5-α reductase. The study enrolled men thought to be at higher risk of prostate cancer; specifically men aged 50–75 years, with a PSA of 2.5–10.0 ng/mL, and a history of one prior negative prostate biopsy within 6 months of study enrollment. Patients were randomized to dutasteride (0.5 mg/daily) or placebo and both groups underwent prostate biopsy at 2 and 4 years. The primary outcome was detection of prostate cancer after 2 and 4 years of treatment or placebo.[16] In the final analysis, 8231 patients were enrolled, with 4105 randomized to dutasteride and 4126 randomized to placebo.[17] Of these patients, 6729 underwent biopsy during the course of the study and were included in the final analysis. Overall, 19.9% of men randomized to dutasteride and 25.1% of men randomized to placebo were diagnosed with prostate cancer. Treatment with dutasteride conferred a relative risk reduction of 22.8% (95% CI: 15.2–29.8%; $p < 0.001$) compared with placebo. While there were no significant differences in the number of patients diagnosed with high-grade prostate cancer (Gleason ≥7) during the study, a greater number of patients in the treatment group were diagnosed with high-grade cancer in the final 2 years of the study (12 vs. 1; $p = 0.0003$). There were no deaths attributed to prostate cancer in either group during the course of the study. Randomization to dutasteride was also associated with lower risks of developing sequelae of BPH compared with placebo, including acute urinary retention (1.6 vs. 6.7%; $p < 0.001$), urinary tract infections (5.3% vs. 8.8%; $p < 0.001$), and BPH-related surgeries (1.4% vs. 5.1%; $p < 0.001$). However, patients treated with dutasteride also demonstrated greater adverse events from the hormonal effects of 5-ARI including decreased libido (3.3% vs. 1.6%; $p < 0.001$), erectile dysfunction (9.0% vs. 5.7%; $p < 0.001$), and gynecomastia (1.9% vs. 1.0%; $p = 0.002$).[17]

Data from both the PCPT and REDUCE trials clearly show a benefit in reducing the incidence of low-grade prostate cancer. However, in the PCPT trial, this decrease in the incidence of low-grade cancer came at the expense of an increased rate of diagnosis of high-grade disease. Similarly, the REDUCE investigators found an increase in the rate of high-grade diagnosis in the treatment group in the last 2 years of their trial. Although both finasteride and dutasteride led to a decreased risk of prostate cancer detection overall, the findings that 5-ARIs were associated with a greater risk of high-grade prostate cancer has led to significant controversy about the use of these medications as preventive agents.[18] Furthermore, even with 17 years of follow-up, treatment with finasteride has not conferred a survival benefit for patients in the PCPT,[15] and several studies have suggested that 5-ARIs have long-term adverse effects, especially loss of libido and erectile dysfunction.[19–22] Hence, to allow patients to make informed decisions, 5-ARIs should only be used for primary prevention in prostate cancer after thorough discussions about the risks and benefits of treatment. In fact, the American Society of Clinical Oncology (ASCO) and the American Urological Association (AUA) recently removed from their websites their joint guideline on the use of 5-ARIs as chemopreventive agents since the FDA has not approved dutasteride for prostate cancer prevention , while 5-ARI labels now include information about the increased risk of developing high-grade prostate cancer while taking these medications.[23]

## Selenium and Vitamin E

Essential elements and vitamins have also been investigated for their potential role as chemopreventive agents for prostate cancer. Selenium is critical to the enzymatic pathway for glutathione peroxidase and can act as an anti-oxidant.[24–26] Similarly, it has been postulated that vitamin E (alpha-tocopherol) can help protect against carcinogenesis and other chronic diseases by reducing oxidative stress.[27] In a clinical setting, secondary analyses of randomized controlled trials had shown a potential protective benefit from selenium and vitamin E on prostate cancer incidence, thus leading to the design of a clinical trial to specifically address the issue.[28–30]

The Selenium and Vitamin E Cancer Prevention Trial (SELECT) was a large, multicenter, placebo-controlled, randomized, double-blinded study designed to test whether selenium, alpha-tocopherol, or the combination would confer a protective benefit in lowering the incidence of prostate cancer.[31] Secondary endpoints included overall cancer diagnosis, all-cause mortality, and cardiovascular events. Inclusion criteria included patient age ≥55 years of age (≥50 years for African-Americans), no history of prostate cancer, PSA <4 ng/mL, and a normal DRE. Of note, men were not required to undergo annual PSA screening or DRE, and there were no strict criteria to undergo prostate biopsy. Over 7 years, the SELECT trial enrolled 35,533 men from 427 participating sites. Patients were randomly assigned to study groups including placebo, vitamin E (400 IU/day), selenium (200 μg/day), or both.[31]

In 2008, with a median follow-up of 5.46 years, the independent data safety and review committee ended the trial early since the data demonstrated no differences in any of the primary or secondary endpoints. Compared with patients receiving placebo, the incidence of prostate cancer was similar for those patients randomized to vitamin E (HR: 1.13 [95% CI: 0.95–1.35]), selenium (HR: 1.04 [95% CI: 0.87–1.25]), and both (HR: 1.05 [95% CI: 0.88–1.25]). Patients randomized to vitamin E actually trended to a higher incidence of prostate cancer ($p = 0.06$).[31] The SELECT trial clearly demonstrated that among healthy men, supplementation with selenium or vitamin E does not help reduce the incidence of prostate cancer.

## Fish Oil

Eicosapentaenoic acid (EPA) and docosahexaenoic acid (DHA) are the two most common long chain *n*-3 polyunsaturated fatty acids (PUFAs) in fish oil. In experimental studies, marine fatty acids have been shown to have antitumor effects on prostate cancer cells.[32] Using data from the Health Professionals Follow-up Study, Augustsson et al. published data exploring the relationship between seafood intake and prostate cancer.[28] The Health Professionals Follow-up Study was a prospective cohort study that enrolled 47,882 men, 40–75 years of age. During 12 years of follow-up, 2400 cases of prostate cancer were diagnosed. Dietary habits were determined with a validated 131-item semiquantitative food frequency questionnaire. The consumption of fish more than three times per week was associated with an overall decreased risk of prostate cancer (RR: 0.56 [95% CI: 0.37–0.86]) compared with men eating fish less than twice monthly. Interestingly, use of fish oil supplements, as compared with no use, was not associated with a decreased risk of prostate cancer. However, only 4% of men in the study used fish oil supplements, hence, data regarding dose, frequency, and duration of supplementation was lacking.

## Aspirin

Aspirin is a commonly used nonsteroidal anti-inflammatory drug (NSAID), and has been shown to have protective effects in colorectal cancer by inhibition of cyclooxygenase 2 (COX-2). COX-2, is overexpressed in prostate cancer tissue and plays a role in prostate cancer cell growth, with high expression being associated with a poor prognosis.[29,30] Thus, NSAIDs, which inhibit the COX pathway, have been investigated as a method for chemoprevention. Unfortunately, although many studies and several meta-analyses have been published exploring the potential role of NSAIDs in chemoprevention, the data remains relatively contradictory, making conclusions about effectiveness challenging.

Huang et al.[33] conducted a meta-analysis of 24 studies examining the relationship between aspirin use and prostate cancer. Regular aspirin use was found to be associated with a reduction in both overall risk of prostate cancer diagnosis (pooled RR: 0.86 [95% CI: 0.81–0.92]) and risk of advanced prostate cancer (pooled RR: 0.83 [95% CI: 0.75–0.91]). The authors performed a further analysis, including only studies that provided data for long-term aspirin users (≥4 years of use), and found that the association with decreased risk became stronger with longer-term aspirin therapy.

Another meta-analysis by Liu et al.,[34] included 31 studies relating NSAID use to prostate cancer incidence. The authors performed separate analyses with NSAIDs, aspirin, and nonaspirin NSAIDs. The pooled effects for NSAIDs as a group and nonaspirin NSAIDs did not significantly impact prostate cancer incidence. However, the authors found that aspirin use was associated with a significant decrease in incidence risk both for overall prostate cancer (overall risk or OR: 0.92 [95% CI: 0.87–0.97]) and for advanced disease (OR: 0.81 [95% CI: 0.73–0.89]). Long-term aspirin use (≥4 years) was again associated with a reduced incidence of prostate cancer (OR: 0.88 [95% CI: 0.79–0.99]). Finally, when analyzing eight studies that contained survival data, compared with nonuse, use of aspirin resulted in a 14% decrease in prostate-cancer-specific mortality.

Dhillon et al.[35] analyzed data from the Health Professionals Follow-up Study to determine the effect of aspirin usage on prostate cancer incidence and mortality. The authors found that men taking ≥6 adult-strength aspirin tablets weekly had a reduced risk of developing high-grade (Gleason score 8–10) and "lethal" prostate cancer (M1, bone metastases, or prostate cancer death) (HR: 0.72 [95% CI: 0.54–0.96]). Interestingly, no significant associations were found between aspirin use and the development of regionally advanced cancer (T3b-T4 or N1).

In conclusion, while aspirin may lower the risk of prostate cancer, questions remain about the dose and duration of treatment necessary to help prevent prostate cancer. Furthermore, the cellular mechanism underlying aspirin's success as a chemopreventive agent remains unclear as similar effects have not been seen with non-aspirin NSAIDs.

## Statin Drug Use

Statins, classes of drugs for cholesterol reduction, are used by 25% of adults over the age of 45.[36] *In vitro* studies have shown that statins cause prostate cancer cell apoptosis, cell cycle arrest, autophagy, and degradation of the androgen receptor. Statins are also theorized to have a chemopreventive role in cancer by inhibiting angiogenesis and inflammation.[37] However, statins are known to reduce serum PSA, an effect that could confound prostate cancer diagnosis.[38]

Using the SEER database, Geybels et al.[39] did a retrospective study with 1001 patients and noted that prostate-cancer-specific mortality (PCSM) was lower with statin use. There was a 1% risk of PCSM in patients taking statins compared with a 5% risk of PCSM in nonusers. Furthermore, Nielsen et al.[40] published a nationwide study based on the Danish Cancer Registry that found a HR of 0.81 (95% CI: 0.75–0.88, $p < 0.001$) for PCSM for patients taking statins. Murtola et al.[41] reported results from another European cohort study of over 23,000 men in the screening arm of the Finnish prostate cancer screening trial. Among men taking statins, the overall incidence of prostate cancer was decreased compared with men not taking statins (HR: 0.75 [95% CI: 0.63–0.89]). This decrease in incidence was dose-dependent, with lower risk associated with higher cumulative doses of statins. Furthermore, the incidence of prostate cancer was not significantly different in men taking nonstatin cholesterol medications. Finally, a meta-analysis that included 27 nonrandomized studies found that statin use was associated with a 7% relative risk reduction in overall risk of prostate cancer, with a 20% relative risk reduction in advanced prostate cancer.[42]

In contrast, several large studies have been published showing no protective benefit from statins on the development of prostate cancer. A prospective cohort study of over 5000 elderly men by Chan et al.[43] showed no association between statin use and prostate cancer incidence, grade, or stage. The PCPT placebo cohort of 9457 men has also been analyzed to determine the role of statins in chemoprevention.[44] With 7 years of follow-up data, statin use during the PCPT was not associated with a change in the risk of prostate cancer diagnosis (HR: 1.03 [95% CI: 0.82–1.30]), nor grade of prostate cancer diagnosed.

Unfortunately, the data regarding the effect of statins on prostate cancer incidence remains contradictory, and no data from randomized, controlled trials designed to specifically look at the effect of statins on the incidence of prostate cancer exists. However, several randomized controlled trials are currently recruiting patients to assess the effects of statin use on prostate cancer. It is hoped that the results of these trials will help to further elucidate the role of statins in the prevention and treatment of prostate cancer.

## LIFESTYLE FACTORS IN PROSTATE CANCER PREVENTION

### Obesity and Physical Activity

There have been several studies assessing the relationship between obesity, weight loss, physical activity, and prostate cancer incidence and specific mortality with conflicting results. As part of the National Institutes of Health-AARP Diet and Health Study, data from 287,760 men aged 50–71 years of age was examined to evaluate the effect of body mass index (BMI) and weight change on prostate cancer incidence and mortality.[45] During 5 years of follow-up, 9986 men were diagnosed with prostate cancer and 173 men died of prostate cancer during 6 years of follow-up. Men with BMI ≥40 had a significantly reduced risk of being diagnosed with prostate cancer compared to men with BMI ≤25. On the other hand, the risk of dying of prostate cancer increased with increasing BMI. Finally, the authors identified a positive association between weight gain and prostate cancer mortality though this trend did not reach significance.

A 2008 meta-analysis reviewing 141 articles exploring the association between BMI and developing a variety of different cancers as well as 27 studies covering over 3 million patients were conducted specifically to study prostate cancer incidence as a function of BMI.[46] The meta-analysis found no significant relationship between BMI and the incidence of prostate cancer (RR: 1.03 [95% CI: 1.00–1.07] $p = 0.11$).

Most recently, researchers in the United Kingdom published a prospective cohort study including over 5 million individuals looking for relationships between BMI and the diagnosis of common cancers.[47] Almost 25,000 prostate cancers were diagnosed in the cohort and increasing BMI was found to be significantly associated with a decreased risk of prostate cancer (HR: 0.98 [95% CI: 0.95–1.00] $p = 0.0042$). The association between BMI and prostate cancer was also found to be nonlinear, with the highest risk of diagnosis noted for patients with a BMI of 27. It is possible that the lower rates of prostate cancer diagnosed in obese men are secondary to detection bias. Obesity has also been linked to lower PSA levels, thought to be secondary to hemodilution.[48] Hence, obese men are likely to get fewer PSA-directed biopsies and are therefore diagnosed at more advanced stages of disease. DREs can also be more difficult to perform in obese patients, leading to the diagnosis of fewer clinically palpable tumors. This detection bias could also help explain the association between obesity and higher risk of prostate cancer mortality in some studies.

Although these factors may play a role in the clinical delay in diagnosis of prostate cancer in obese men, there are also molecular mechanisms that are seen to lead to a higher incidence of advanced prostate cancer in obese patients. While outside the scope of this chapter, these mechanisms include the insulin/insulin like growth factor-1 axis, sex hormones, leptin, and adipokine signaling.[49] Increased circulating IGF-1 has been linked to increased prostate cancer incidence.[50] Furthermore, leptin, a hormone that is elevated in obese individuals, has been shown to induce cell proliferation and inhibit apoptosis in *in vitro* studies of prostate cancer cell lines and leptin.[51,52]

### Alcohol Intake and Prostate Cancer

A study by Fowke et al.[53] using data from the REDUCE trial assessed the role of alcohol intake in prostate cancer. As part of the original study, participants completed a questionnaire regarding alcohol consumption. Of the 75% of patients who reported alcohol intake, 26% were heavy drinkers (more than seven drinks per week). In the placebo group there was no association between alcohol intake and the risk of low- or high-grade prostate cancer. However, in the dutasteride group, a significant 86% increase ($p = 0.01$) in the risk of high-grade cancer was seen in patients reporting heavy alcohol use. Similarly, the protective effect of dutasteride was lost in men reporting heavy alcohol use (OR: 0.99 [95% CI: 0.67–1.45]).[53]

Data from the PCPT has also been examined to determine the effect of alcohol in men taking 5-ARIs. While there was no difference in the risk of developing high-grade disease between the placebo or treatment arms, men who were heavy drinkers (≥4 daily drinks on ≥5 days per week) or reported heavy alcohol consumption (≥50 g alcohol daily) had an increased risk of developing high- grade prostate cancer compared to men with lower alcohol consumption (RR: 2.17 [95% CI: 1.42–3.30] and RR: 2.01 [95% CI: 1.33–3.05]). Furthermore, heavy alcohol consumption was associated with an increased risk of low-grade disease in the finasteride group as heavy alcohol consumption eliminated the protective effect of finasteride on low-grade prostate cancer (RR: 1.89 [95% CI: 1.39–2.56]).[54]

These studies suggest that heavy alcohol use may lead to an increased risk of developing prostate cancer, specifically in patients taking 5-ARIs. Of note, the PCPT data also showed an increased risk of high-grade cancer in heavy consumers of alcohol. Thus, it should be recommended that men limit their alcohol intake, especially if starting 5-ARIs for chemoprevention.

## THE FUTURE OF PROSTATE CANCER PREVENTION

Given the prevalence of the disease and significant morbidity associated with treatment, prostate cancer prevention will undoubtedly remain a topic of significant clinical research for many years. Unfortunately, while large-scale, randomized trials of 5-ARIs have shown a decrease in the diagnosis of low-grade prostate cancer, this has come at the expense of diagnosing more high-grade prostate cancers. Furthermore, these studies failed to demonstrate a significant decrease in mortality from primary prevention efforts with 5-ARIs.

These studies also illustrate the hurdles that future prevention studies will have to overcome. Prostate cancer has a long natural history from diagnosis to mortality. Therefore, to detect small survival differences in prevention studies, an extremely large number of patients would need to be recruited and maintained on placebo or prevention protocols for many years, all without the assurance that the prevention strategy would be successful. None of the previously discussed trials included prostate cancer survival as a primary outcome due to the inherent difficulties associated with designing and financing such a study. Nonetheless, such a study would be the best way to determine the true benefit of a prostate cancer prevention regimen. Unfortunately, such a study would likely entail excessive costs and would require enroling thousands of patients to achieve adequate power to detect small differences in incidence of clinically significant prostate cancer and cancer-specific mortality.

Nonetheless, given current clinical evidence, it is reasonable to discuss the risks and benefits of 5-ARIs with patients, and point out these medications' potential to prevent low-grade prostate cancer while potentially increasing the risk of developing high-grade cancer. Other potential prevention strategies such as taking aspirin or dietary supplementation with vitamin E, selenium, fish oil, isoflavones, or other dietary supplements are not currently supported by well-designed clinical trials. The future of prostate cancer prevention research will clearly require significant basic science and translational research to further elucidate the potential biologic mechanisms behind some of these supplements and compounds and ultimately enable clinical trial dollars to be spent in the most cost-effective way possible. For now, while a panacea for preventing prostate cancer remains undiscovered, research will continue to attempt to limit the burden of this common malignancy.

## Acknowledgment

Dr Simon P. Kim is supported through a career development award by the Conquer Cancer Foundation of the American Society of Clinical Oncology.

## References

1. Siegel R, Ma J, Zou Z, et al. Cancer statistics, 2014. *CA Cancer J Clin* 2014;**64**(1):9–29.
2. Ferlay J, Shin HR, Bray F, et al. Estimates of worldwide burden of cancer in 2008: GLOBOCAN 2008. *Int J Cancer* 2010;**127**(12):2893–917.
3. Wilt TJ, MacDonald R, Rutks I, et al. Systematic review: comparative effectiveness and harms of treatments for clinically localized prostate cancer. *Ann Intern Med* 2008;**148**(6):435–48.
4. Welch HG, Albertsen PC. Prostate cancer diagnosis and treatment after the introduction of prostate-specific antigen screening: 1986–2005. *J Natl Cancer Inst* 2009;**101**(19):1325–9.
5. Welch HG, Black WC. Overdiagnosis in cancer. *J Natl Cancer Inst* 2010;**102**(9):605–13.
6. Ta HQ, Gioeli D. The convergence of DNA damage checkpoint pathways and androgen receptor signaling in prostate cancer. *Endocr Relat Cancer* 2014;**21**(5):395–407.
7. Huggins C. Androgen and anaplasia. *Yale J Biol Med* 1947;**19**(3):319–30.
8. Huggins C, Hodges CV. Studies on prostatic cancer: I. The effect of castration, of estrogen, and of androgen injection on serum phosphatases in metastatic carcinoma of the prostate. 1941. *J Urol* 2002;**168**(1):9–12.
9. Jenster G. The role of the androgen receptor in the development and progression of prostate cancer. *Semin Oncol* 1999;**26**(4):407–21.
10. Jung C, Park Y, Kim YR, et al. Five-alpha reductase inhibitor influences expression of androgen receptor and HOXB13 in human hyperplastic prostate tissue. *Int Braz J Urol* 2013;**39**(6):875–83.
11. Steers WD. 5alpha-reductase activity in the prostate. *Urology* 2001;**58**(6 Suppl 1):17–24 discussion 24.
12. Gormley GJ, Stoner E, Bruskewitz RC, et al. The effect of finasteride in men with benign prostatic hyperplasia. The Finasteride Study Group. *N Engl J Med* 1992;**327**(17):1185–91.
13. McConnell JD, Roehrborn CG, Bautista OM, et al. The long-term effect of doxazosin, finasteride, and combination therapy on the clinical progression of benign prostatic hyperplasia. *N Engl J Med* 2003;**349**(25):2387–98.
14. Thompson IM, Goodman PJ, Tangen CM, et al. The influence of finasteride on the development of prostate cancer. *N Engl J Med* 2003;**349**(3):215–24.
15. Thompson Jr IM, Goodman PJ, Tangen CM, et al. Long-term survival of participants in the prostate cancer prevention trial. *N Engl J Med* 2013;**369**(7):603–10.
16. Andriole G, Bostwick D, Brawley O, et al. Chemoprevention of prostate cancer in men at high risk: rationale and design of the reduction by dutasteride of prostate cancer events (REDUCE) trial. *J Urol* 2004;**172**(4 Pt 1):1314–7.
17. Andriole GL, Bostwick DG, Brawley OW, et al. Effect of dutasteride on the risk of prostate cancer. *N Engl J Med* 2010;**362**(13):1192–202.
18. Walsh PC. Survival in the prostate cancer prevention trial. *N Engl J Med* 2013;**369**(20):1967.
19. Fwu CW, Eggers PW, Kaplan SA, et al. Long-term effects of doxazosin, finasteride and combination therapy on quality of life in men with benign prostatic hyperplasia. *J Urol* 2013;**190**(1):187–93.
20. Fwu CW, Eggers PW, Kirkali Z, et al. Change in sexual function in men with lower urinary tract symptoms/benign prostatic hyperplasia associated with long-term treatment with doxazosin, finasteride and combined therapy. *J Urol* 2013;**191**(6):1828–34.
21. Irwig MS. Persistent sexual side effects of finasteride: could they be permanent? *J Sex Med* 2012;**9**(11):2927–32.
22. Gur S, Kadowitz PJ, Hellstrom WJ. Effects of 5-alpha reductase inhibitors on erectile function, sexual desire and ejaculation. *Expert Opin Drug Saf* 2013;**12**(1):81–90.

23. *American Society of Clinical Oncology. Use of 5-alpha reductase inhibitors for prostate cancer chemoprevention: ASCO-AUA 2008 Clinical Practice Guideline.* http://www.asco.org/quality-guidelines/use-5-alpha-reductase-inhibitors-prostate-cancer-chemoprevention-asco-aua-2008 [accessed 10.09.2014].
24. Chow CK. Nutritional influence on cellular antioxidant defense systems. *Am J Clin Nutr* 1979;**32**(5):1066–81.
25. Combs Jr GF, Clark LC. Can dietary selenium modify cancer risk? *Nutr Rev* 1985;**43**(11):325–31.
26. Mercurio SD, Combs Jr GF. Drug-induced changes in selenium-dependent glutathione peroxidase activity in the chick. *J Nutr* 1985;**115**(11):1459–70.
27. Ladas EJ, Jacobson JS, Kennedy DD, et al. Antioxidants and cancer therapy: a systematic review. *J Clin Oncol* 2004;**22**(3):517–28.
28. Augustsson K, Michaud DS, Rimm EB, et al. A prospective study of intake of fish and marine fatty acids and prostate cancer. *Cancer Epidemiol Biomarkers Prev* 2003;**12**(1):64–7.
29. Gupta S, Srivastava M, Ahmad N, et al. Over-expression of cyclooxygenase-2 in human prostate adenocarcinoma. *Prostate* 2000;**42**(1):73–8.
30. Yoshimura R, Sano H, Masuda C, et al. Expression of cyclooxygenase-2 in prostate carcinoma. *Cancer* 1 2000;**89**(3): 589–96.
31. Lippman SM, Klein EA, Goodman PJ, et al. Effect of selenium and vitamin E on risk of prostate cancer and other cancers: the Selenium and Vitamin E Cancer Prevention Trial (SELECT). *JAMA* 2009;**301**(1):39–51.
32. Li CC, Hou YC, Yeh CL, et al. Effects of eicosapentaenoic acid and docosahexaenoic acid on prostate cancer cell migration and invasion induced by tumor-associated macrophages. *PLoS ONE* 2014;**9**(6):e99630.
33. Huang TB, Yan Y, Guo ZF, et al. Aspirin use and the risk of prostate cancer: a meta-analysis of 24 epidemiologic studies. *Int Urol Nephrol* 2014;**46**:1715–28.
34. Liu Y, Chen JQ, Xie L, et al. Effect of aspirin and other non-steroidal anti-inflammatory drugs on prostate cancer incidence and mortality: a systematic review and meta-analysis. *BMC Med* 2014;**12**:55.
35. Dhillon PK, Kenfield SA, Stampfer MJ, et al. Long-term aspirin use and the risk of total, high-grade, regionally advanced, and lethal prostate cancer in a prospective cohort of health professionals, 1988–2006. *Int J Cancer* 2011;**128**(10):2444–52.
36. CDC/NCHS. Health, United States. 2010.
37. Papadopoulos G, Delakas D, Nakopoulou L, et al. Statins and prostate cancer: molecular and clinical aspects. *Eur J Cancer* 2011;**47**(6): 819–30.
38. Hamilton RJ, Goldberg K, Platz EA, et al. The influence of statin medications on prostate-specific antigen levels. *J Natl Cancer Inst* 2008;**1001**(21):1511–8.
39. Geybels MS, Wright JL, Holt SK, et al. Statin use in relation to prostate cancer outcomes in a population-based patient cohort study. *Prostate* 2013;**73**(11):1214–22.
40. Nielsen SF, Nordestgaard BG, Bojesen SE. Statin use and reduced cancer-related mortality. *N Engl J Med* 2012;**367**(19):1792–802.
41. Murtola TJ, Tammela TL, Maattanen L, et al. Prostate cancer and PSA among statin users in the Finnish prostate cancer screening trial. *Int J Cancer* 2010;**127**(7):1650–9.
42. Bansal D, Undela K, D'Cruz S, et al. Statin use and risk of prostate cancer: a meta-analysis of observational studies. *PLoS ONE* 2012;**7**(10):e46691.
43. Chan JM, Litwack-Harrison S, Bauer SR, et al. Statin use and risk of prostate cancer in the prospective Osteoporotic Fractures in Men (MrOS) Study. *Cancer Epidemiol Biomarkers Prev* 2012;**21**(10):1886–8.
44. Platz EA, Tangen CM, Goodman PJ, et al. Statin drug use is not associated with prostate cancer risk in men who are regularly screened. *J Urol* 2014;**192**(2):379–84.
45. Wright ME, Chang SC, Schatzkin A, et al. Prospective study of adiposity and weight change in relation to prostate cancer incidence and mortality. *Cancer* 2007;**109**(4):675–84.
46. Renehan AG, Tyson M, Egger M, et al. Body-mass index and incidence of cancer: a systematic review and meta-analysis of prospective observational studies. *Lancet* 2008;**371**(9612):569–78.
47. Bhaskaran K, Douglas I, Forbes H, et al. Body-mass index and risk of 22 specific cancers: a population-based cohort study of 5.24 million UK adults. *Lancet* 2014;**384**(9945):755–65.
48. Banez LL, Hamilton RJ, Partin AW, et al. Obesity-related plasma hemodilution and PSA concentration among men with prostate cancer. *JAMA* 2007;**298**(19):2275–80.
49. Roddam AW, Allen NE, Endogenous Hormones and Prostate Cancer Collaborative Group. et al. Endogenous sex hormones and prostate cancer: a collaborative analysis of 18 prospective studies. *J Natl Cancer Inst* 2008;**100**(3):170–83.
50. Price AJ, Allen NE, Appleby PN, et al. Insulin-like growth factor-I concentration and risk of prostate cancer: results from the European Prospective Investigation into Cancer and Nutrition. *Cancer Epidemiol Biomarkers Prev* 2012;**21**(9):1531–41.
51. Hoda MR, Popken G. Mitogenic and anti-apoptotic actions of adipocyte-derived hormone leptin in prostate cancer cells. *BJU Int* 2008;**102**(3):383–8.
52. Huang CY, Yu HS, Lai TY, et al. Leptin increases motility and integrin up-regulation in human prostate cancer cells. *J Cell Physiol* 2011;**226**(5):1274–82.
53. Fowke JH, Howard L, Andriole GL, et al. Alcohol intake increases high-grade prostate cancer risk among men taking dutasteride in the REDUCE trial. *Eur Urol* 2014;**66**(6):1133–8.
54. Gong Z, Kristal AR, Schenk JM, et al. Alcohol consumption, finasteride, and prostate cancer risk: results from the Prostate Cancer Prevention Trial. *Cancer* 2009;**115**(16):3661–9.

CHAPTER

# 13

# Prostate-Specific Antigen Screening Guidelines

*Kristen R. Scarpato, MD, MPH*, Peter C. Albertsen, MD, MS***

*Department of Urologic Surgery, Vanderbilt University Medical Center, Nashville, TN, USA

**Department of Surgery, University of Connecticut Health Center, Farmington, CT, USA

## WHAT IS PSA?

Prostate-specific antigen (PSA) is a serum protease inhibitor produced only by prostate tissue following androgen stimulation. PSA typically forms a complex with α1-antichymotrypsin and plays a role in semen liquefaction. Serum PSA values usually correlate with prostate volume. Several conditions elevate PSA including benign prostatic hypertrophy (BPH), prostate cancer, prostate infection, prostate manipulation, and ejaculation. Androgen deprivation or castration, certain medications including 5α-reductase inhibitors, and prostate excision or ablation decrease PSA.

Testing for PSA began in the late 1980s in the United States and has dramatically increased the incidence of this disease. In 2013, over 80% of the prostate cancers identified by PSA were organ-confined.[1] Researchers have developed assays for various forms of PSA and have constructed statistical techniques to improve the detection of prostate cancer including the measurement of PSA velocity, PSA density, and percent free PSA.[2] Although several isoforms of PSA have been identified and utilized, specifically free and complex PSA, they have not improved the performance of PSA as a screening test. PSA measurements vary depending upon the laboratory and assay used, thus limiting the precision of any individual PSA test. Mildly elevated PSA values should prompt repeat serum PSA testing before patients proceed to prostate biopsy, as many men will have normal PSA values on subsequent serum studies.[3]

Draisma et al. estimate that PSA testing has advanced the date of diagnosis by approximately 12.3 years for men aged 55 years and by 6 years for men aged 75 years.[4] Unfortunately, there is no definitive cut point that separates men harboring clinically significant prostate cancer from those that do not. Men younger than age 50 usually have a PSA <2.5 ng/mL while men older than age 65 frequently have values as high as 6.5 ng/mL or greater. Not all prostate cancers produce PSA, hence, there is no value below which prostate cancer can be definitively excluded.

Early detection and treatment are common goals for the management of most cancers. Over a century ago, Hugh Hampton Young advocated screening for prostate cancer using a digital rectal examination (DRE) and treatment with radical surgery.[5] During the past quarter century, PSA testing has become the gold standard for the early detection of this disease. Unfortunately, early detection is a necessary but not sufficient condition to confirm the validity of a screening test.

## WHAT IS THE NATURAL HISTORY OF PROSTATE CANCER?

The natural history of prostate cancer is extraordinarily variable, ranging from indolent to highly aggressive. Autopsy studies have revealed a high prevalence of prostate cancer in men of all ages, including men under age 50. Estimates suggest that 14–17% of men in their 60s and 31–83% of men in their 70s have pathologic evidence of prostate cancer.[6] Why most of these tumors remain dormant continues to be a mystery. The Prostate Cancer Prevention Trial (PCPT) followed 18,882 men for 7 years and demonstrated that the prevalence of prostate cancer ranged from 18.4% to 24.4% in the general population.[7] These rates are three to four times higher than previously estimated. Most cancers detected in this study were low-grade Gleason score 3 + 3 tumors and would likely have never been detected in the absence of PSA testing.

Prostate Cancer. http://dx.doi.org/10.1016/B978-0-12-800077-9.00013-X

Several studies have described the natural history of prostate cancer. In 1998 and 2005 Albertsen et al. reported the long-term outcomes of a competing risk analysis of 767 men diagnosed between 1971 and 1984 who were managed expectantly for clinically localized disease.[8,9] In these studies, only 4–7% of men with Gleason 2–4 tumors had progression leading to death within 20 years of diagnosis while men with Gleason 5 and 6 tumors progressed at a rate of 6–11 and 18–30%, respectively. Men with low-grade prostate cancers are less likely to benefit from early detection because of the slow progression of disease and the relatively high death rate from competing medical conditions over time. Men with Gleason grade 7 and higher cancers harbor aggressive cancers that are much more likely to metastasize during their lifetime. These men have a much higher potential to benefit from prostate cancer screening if they receive an effective treatment.

The Gleason scoring system, derived by analyzing prostate glandular architecture under low-power magnification, has undergone several changes since it was introduced. The original system identified nine unique growth patterns that were subsequently stratified into the five Gleason patterns. Over the past decade, some features have been transferred from pattern 3 to pattern 4, and patterns 1 and 2 have been essentially eliminated.[10,11] This reclassification has resulted in an upgrading of many samples such that tumors once classified as Gleason score 5 or less are now graded as Gleason 6, and previous Gleason 6 tumors are now often classified as Gleason 7 tumors. These revisions have altered our understanding of the natural history of this disease, resulting in dramatically improved survival of contemporary patients when compared to the historical series.[12] This reclassification bias is often described as the Will Rogers effect and has been documented to occur in prostate cancer.[13] The recently published 18-year follow-up of the Scandinavian Prostate Cancer Group 4 study provides relatively contemporary data on the natural history of this disease.[14]

Prostate cancer mortality has decreased during the past two decades in both the United States and the United Kingdom. The cause of this decline is controversial. In the United States great emphasis has been placed on screening and early treatment in contrast to the United Kingdom where screening is relatively uncommon and patients are treated only when disease becomes clinically apparent. Both environmental and genetic factors appear to have had an impact. Therefore, the observed decline in prostate cancer mortality appears to be a combination of earlier detection and treatment, more effective therapies, alterations in exposure to risk factors, and changes in the attribution of cause of death.[15,16]

## ARE PROSTATE CANCER TREATMENTS EFFECTIVE?

Screening aims to detect cancer before it becomes clinically apparent with the assumption that early intervention will improve outcomes. For screening to be effective, readily accessible, and available, treatments must improve outcomes. Surgery and radiation are the two primary treatments employed to manage prostate cancer.

Two studies from Sweden have demonstrated that these therapies can effectively treat some men with localized prostate cancer.[14,17,18] The Scandinavian Prostate Cancer Group-4 (SPC-4) is a randomized control trial comparing surgery against watchful waiting and was recently updated in 2014. Nearly 700 men were recruited over a 10-year period from 1989 to 1999 and were randomized to undergo radical prostatectomy or observation. Importantly, only 12% of the participants were identified on the basis of PSA testing. The majority of the patients had localized low-grade disease, while 5% of patients had Gleason 8 or higher disease, and 12% had an unknown grade. After 18 years of follow-up, the disease-specific mortality was 29% for the watchful waiting group versus 18% in the prostatectomy group, indicating a relative risk reduction of 56% and an absolute risk reduction of 11% for men undergoing surgery. This benefit was only seen in men younger than 65 years of age and was most pronounced for men with Gleason 7 disease. Prostate cancer progressed slowly and infrequently for men with low-grade disease, resulting in a minimal survival difference between the two groups. Men with high-grade disease, unfortunately, were cured infrequently with surgery and therefore progressed despite surgical intervention. How these results should advise a population of men with screen-detected disease is unclear. PSA testing introduces a significant lead time resulting in earlier detection of disease. Whether men diagnosed earlier would accrue the same benefit from treatment must be confirmed independently.

In 2009, Widmark et al. published findings from the Scandinavian Prostate Cancer Group 7 (SPG-7) trial that evaluated the impact of radiation among a group of men with locally advanced prostate cancer who were treated with endocrine therapy.[18] Between 1996 and 2002, 875 patients with T3 prostate cancers were randomized to receive endocrine therapy that included continuous flutamide with or without pelvic irradiation. Only 2% of the patient population was identified on the basis of PSA testing. The 10-year prostate cancer mortality rate was 23.9% in the endocrine-alone group and 11.9% in the endocrine-plus-radiation group, indicating a relative risk reduction of 44% and an absolute risk reduction of 12–44% for combined therapy. Whether these results apply to men with screen- detected T1c disease can be debated. Because of the lead time associated with testing and the identification of disease at an earlier stage, benefits from radiation are likely to accrue to a screen-detected population many years later than was seen in the SPG-7 study.

More recently, Wilt et al. published findings from the Prostate Cancer Intervention versus Observation Trial (PIVOT).[19] In this study, 731 men with localized prostate were randomly assigned to radical prostatectomy or observation and followed for 8 years. The primary outcome metrics were all cause and prostate-cancer-specific mortality. Most of these men had disease identified by PSA testing; the median value was 7.8 ng/mL. Men qualified for the study if they were medically fit for radical prostatectomy, had histologically confined localized disease, a negative bone scan, a life expectancy of more than 10 years, and a PSA <50 ng/mL. After 12 years of follow-up, 47% of the surgery group versus 50% of the observation group had died, yielding an absolute risk reduction of 2.9%. Prostate-cancer-specific mortality was not significantly different between the two groups. Interestingly, results suggested that men with a PSA >10 ng/mL and men with more advanced disease benefited from surgical intervention. Although the study had relatively low power, it has raised concerns about the relative efficacy of surgery especially among men with low-grade disease. These men appear to receive no benefit from surgery but are exposed to all of the harms.

Other contemporary randomized trials include the ProtecT trial that is scheduled to report in 2016.[20] This trial has enroled over 1500 men in a three-armed study comparing surgery, radiation, and conservative management in men with screen-detected prostate cancer. Study outcomes include all-cause mortality, prostate cancer mortality, and several measures evaluating quality of life and cost effectiveness.

## IS PROSTATE-SPECIFIC ANTIGEN AN EFFECTIVE SCREENING TEST?

The dramatic rise in the incidence of prostate cancer observed wherever PSA testing has been introduced provides strong evidence that PSA testing can identify prostate cancer. Whether early detection lowers prostate cancer mortality rates, however, requires evaluation by a randomized trial.

Two studies provide the primary evidence evaluating the efficacy of PSA screening in lowering prostate cancer mortality. They are the European Randomized Study of Screening for Prostate Cancer (ERSPC) and the Prostate, Lung, Colorectal, and Ovarian (PLCO) trial.[21,22] The ERSPC trial is comprised of seven separate PSA screening trials conducted in the Netherlands, Belgium, Sweden, Finland, Italy, Spain, and Switzerland. The combined trial enroled over 182,000 men aged 55–69 years and showed an absolute risk reduction of 0.71 deaths per 1000 men and a relative risk reduction of 20% after 9 years of follow-up. Interpreting these findings has proved difficult because the trial did not use the same protocol at each site. Recruitment and randomization procedures differed among countries, as did the screening protocols. For instance, in the early 1990s, Dutch and Belgium sites relied on DRE, transrectal ultrasound (TRUS), and PSA tests, while later in the study, most centers relied solely on serum PSA values. PSA cut-points differed by site and changed during the study from 4.0 ng/mL to 3.0 ng/mL as the study progressed. The screening interval also varied by site. Most men were screened every 4 years except in Sweden where the interval was every 2 years. Finally, the number of biopsies varied by site, with many centers using sextant biopsies while others performed 10–12 core biopsies.

The findings of the ERSPC study were published as if the trial were conducted as a single study rather than as seven separate trials. This is relevant because only two of the seven trials yielded statistically significant results: Sweden and the Netherlands. Finland contributed the largest number of patients, but the trial there showed no significant difference in outcomes between the screened and unscreened populations. The other four trials were underpowered and did not contribute to the overall findings of the study.

The PLCO study was initiated in 2003 in the United States and enroled 76,000 men. This study showed no significant difference in prostate cancer mortality after 7–10 years of follow-up between screened and unscreened populations. Men were randomly assigned to receive either annual PSA tests for 6 years and DRE for 4 years or "usual care." Primary caregivers were notified when a study participant was found to have a PSA at or above the threshold of 4.0 ng/mL. Unfortunately, the trial suffered from a high rate of PSA testing prior to enrollment and frequent PSA testing in the control arm such that 50% of men in the control arm underwent at least one PSA test. This severely compromised the study's power to detect a difference in prostate cancer mortality.

The PLCO trial was designed to test the public health impact of annual PSA testing as compared to the standard of care in practice at that time. Ultimately, this trial compared more thoroughly screened men with less thoroughly screened men in an environment that allowed men to choose how they would deal with a positive study. Although significantly more cancers were found in the screening arm, after 7 years of follow-up, only 50 deaths had occurred in the screening arm and 44 deaths in the control arm.

The Swedish component of the ERSPC trial has received considerable independent attention following publication of findings in 2010.[23] The study randomly sampled 20,000 men from the population registry of Goteborg, Sweden, who were then randomly assigned to a screening group or a control group. The screened group was invited for PSA testing every 2 years, and men with an elevated PSA were offered further testing with DRE and TRUS biopsy. During a median follow-up of 14 years, the cumulative prostate cancer incidence was 12.7% in the screening group and 8.2% in the control group with a relative risk reduction of prostate cancer mortality of 40%. This is the strongest evidence to date of the benefit of PSA testing. Whether results can be generalized to the United States, however, is uncertain because the prevalence of clinically significant prostate cancer is much higher in Sweden. As a consequence, PSA testing is likely to be more effective in Sweden than in the United States.

## ARE THERE RISKS ASSOCIATED WITH PSA SCREENING?

The assessment of any medical intervention requires an evaluation of both the potential benefits and harms. Screening for prostate cancer with PSA testing is no exception. Several researchers have evaluated the harms associated with PSA testing and their findings were summarized in the USPSTF report.[24]

Men found to have an elevated PSA often proceed to undergo TRUS and prostate biopsy. This invasive test has been associated with physical harms including hematuria, dysuria, urinary retention, pain, prostatitis, urosepsis, and a 2% risk of hospitalization. Men with newly diagnosed disease reported increased anxiety and were found to have increased rates of suicide and cardiovascular events. The ERSPC found that 76% of prostate biopsies for an elevated PSA identified no cancer while the PLCO trial found a cumulative false-positive rate of 13% using a PSA cutoff of 4.0 ng/mL.[21,25]

Recent epidemiologic studies have confirmed that PSA testing leads to substantial overdiagnosis and overtreatment of this disease. Draisma et al. estimated that overdiagnosis could be as high as 50% for men aged 75 years because most men identified through screening are not destined to die from prostate cancer.[4] About 90% of screen-detected cancers are indolent localized lesions that do not require treatment. Unfortunately, no test exists that can distinguish clinically significant cancers from indolent tumors.[26,27] This places patients in an uncomfortable dilemma of choosing between active surveillance programs or aggressive interventions, knowing that each approach is associated with potential morbidity.

More than 75% of men diagnosed with localized prostate cancer elect aggressive treatment with either surgery or radiation and often encounter significant treatment-related side effects including urinary incontinence, erectile dysfunction, bowel problems, and decreased quality of life.[26,28–31] Depending upon the definition of incontinence and the age of the patient, long-term incontinence persists in 2–22% of patients following surgery.[32] In the SPCG-4 clinical trial, there was a 28% absolute increase in the risk of urinary incontinence following prostatectomy.[28] One small randomized trial comparing radiation and watchful waiting also found an increased risk of urinary incontinence.[33]

Erectile dysfunction affecting 20–90% of men is a common outcome following prostatectomy and radiation. The SPCG-4 trial found erectile dysfunction increased by 36% while several cohort studies found a median increase of 14% following radiation.[28–31,34–36] Bowel dysfunction was less common following prostatectomy and watchful waiting, but occurred more frequently following radiation therapy.[28–31,33,37–39] Miller et al. found that men undergoing external beam radiation or brachytherapy had significantly worse overall health-related quality of life in at least one of four domains compared to control groups.[40]

The use of androgen deprivation therapy (ADT) has also increased as a consequence of PSA screening. Bolla et al. have clearly demonstrated a role for androgen deprivation therapy in conjunction with radiation therapy in men with advanced localized disease, but the role of ADT as primary treatment for localized disease remains controversial.[41,42] No clinical trial supports the use of ADT as primary treatment for localized disease. Prolonged treatment increases the incidence of cardiovascular disease, osteoporosis, diabetes, cognitive impairment, and abnormal lipid profiles.

## WHAT ADVICE SHOULD PHYSICIANS GIVE THEIR PATIENTS?

The initial enthusiasm that accompanied the introduction of PSA screening has begun to give way to concerns that repeated PSA testing results in more harm than benefit. This position was well articulated in the USPSTF report. The ERSPC trial and the recent update from the SPCG-4 randomized trial suggest that there is a subset of men that may benefit from PSA testing. Patients considering a PSA test should be aware that in the absence of screening, they face a 1 in 200 chance of dying from prostate cancer. With testing this chance is lowered to 1 in 250. The men most likely to benefit from PSA testing are aged 55–65 years. There is no evidence that testing lowers prostate cancer mortality in older men.

Men who test positive for PSA must undergo a TRUS and biopsy. This carries a 2% risk of sepsis requiring hospitalization.[24] If the biopsy shows prostate cancer, men with Gleason 7 disease are the ones most likely to benefit from treatment. For men under the age of 65, radical prostatectomy can lower the death rate from prostate cancer by 11%. Men with low-grade disease, however, may want to consider active surveillance. Data from the SPCG-4 study suggest that outcomes from surgery and watchful waiting for these men are comparable. Men found to have high-grade disease are likely to have disease progression. They should consider multimodality therapy. Unfortunately, little data are available to quantify the potential benefit of treatment.

With this information in mind, the American Urological Association recently revised its guidelines for the early detection of prostate cancer and updated its Best Practice Statement regarding PSA testing for the pretreatment staging and posttreatment management of prostate cancer.[43,44] Their recommendations are as follows: (1) Men under the age of 40 should not undergo PSA testing; (2) Men between the ages 40 and 54 years at average risk for prostate cancer should not undergo testing; (3) Men between ages 55 and 69 years should weigh the potential harms associated with screening and treatment against the potential benefit of lowering the chance of dying from prostate cancer. This should be a shared decision between the patient and the physician; (4) To reduce the potential harms of PSA testing, the interval for routine screening can safely be increased to every 2 years; and (5) Men aged 70 years and older and any man with a life expectancy of less than 10–15 years should not undergo PSA testing.

## FUTURE DIRECTIONS AND RESEARCH NEEDS

The relatively poor ability of PSA testing to identify men with clinically significant prostate cancer highlights the need for a more targeted screening approach. Current practices have resulted in excessive overdiagnosis and overtreatment of this disease. Numerous researchers are exploring genomic-based testing to address this issue. Unfortunately, current assays based on PSA isoforms, urinary biomarkers, and DNA single nucleotide peptide alternations have provided little additional information.

New imaging techniques offer an alternative screening algorithm. Pelvic MRI imaging may provide a mechanism to identify those men with clinically significant cancers. Current procedures need to be refined and tested. Finally, the value of screening depends upon the effectiveness of treatment. Improvement in the treatment risk/benefit ratio requires more detailed evidence concerning how much men benefit from specific treatments. This highlights the need for more randomized clinical trials. Until we are able to identify which men harbor clinically significant disease and have treatments that alter the natural history of their disease, prostate cancer screening will remain mired in controversy.

## References


1. Siegel R, Ma J, Zou Z, et al. Cancer statistics, 2014. *CA Cancer J Clin* 2014;**64**:9–29.
2. Catalona WJ, Partin AW, Slawin KM. Use of percentage of free prostate-specific antigen to enhance differentiation of prostate cancer from benign prostatic disease. *JAMA* 1998;**279**:1542–7.
3. Eastham JA, Riedel E, Scardino PT, et al. Variation of serum prostate-specific antigen levels: an evaluation of year-to-year fluctuations. *JAMA* 2003;**289**:2695–700.
4. Draisma G, Boer R, Otto SJ, et al. Lead times and overdetection due to prostate-specific antigen screening: estimates from the European Randomized Study of Screening for Prostate Cancer. *J Natl Cancer Inst* 2003;**95**:868–78.
5. Young HH. Early diagnosis and radical cure of carcinoma of the prostate. *Bull Johns Hopkins Hosp* 1905;**16**:314–21.
6. Sakr WA, Haas GP, Cassin BF, et al. The frequency of carcinoma and intraepithelial neoplasia of the prostate in young male patients. *J Urol* 1993;**150**:379–85.
7. Thompson IM, Goodman PJ, Tangen CM, et al. The influence of finasteride on the development of prostate cancer. *N Engl J Med* 2003;**349**:215–24.
8. Albertsen PC, Hanley JA, Gleason DF, et al. Competing risk analysis of men aged 55 to 74 years at diagnosis managed conservatively for clinically localized prostate cancer. *JAMA* 1998;**280**: 975–80.
9. Albertsen PC, Hanley JA, Fine J. 20 year outcomes following conservative management of clinically localized prostate cancer. *JAMA* 2005;**293**:2095–101.
10. Gleason DF. Mellinger G.T. for the Veterans Administration Cooperative Urological Research Group. Prediction of prognosis for prostatic adenocarcinoma by combined histologic grading and clinical staging. *J Urol* 1974;**111**:58–64.
11. Epstein JI. Gleason score 2–4 adenocarcinoma of the prostate on needle biopsy. *Am J Surg Pathol* 2000;**24**:477–8.
12. Lu-Yao GL, Albertsen PC, Moore DF, et al. Outcomes of localized prostate cancer following conservative management. *JAMA* 2009;**302**:1202–9.
13. Albertsen PC, Hanley JA, Barrows GH, et al. Prostate cancer and the Will Rogers phenomenon. *J Natl Cancer Inst* 2005;**97**:1248–53.
14. Collin SM, Martin RM, Metcalfe C, et al. Prostate cancer mortality in the USA and UK in 1974–2004: an ecological study. *Lancet Oncol* 2008;**9**:445–52.
15. Murtola TJ, Tammela TL, Maattanen L, et al. Prostate cancer and PSA among statin users in the Finnish prostate cancer screening trial. *Int J Cancer* 2010;**127**:1650–9.
16. Bill-Axelson A, Holmberg L, Garmo H, et al. Radical prostatectomy or watchful waiting in early prostate cancer. *N Engl J Med* 2014;**370**:932–42.
17. Bill-Axelson A, Holmberg L, Filén F, et al. Radical prostatectomy versus watchful waiting in localized prostate cancer: the Scandinavian prostate cancer group-4 trial. *J Natl Cancer Inst* 2008; **100**:1144–54.

18. Widmark A, Klepp O, Solberg A, et al. Endocrine treatment, with or without radiotherapy, in locally advanced prostate cancer (SPCG-7/SFUO-3): an open randomized phase III trial. *Lancet* 2009;**373**:301–8.
19. Wilt TJ, Brawer MK, Jones KM, et al. Radical prostatectomy versus observation for localized prostate cancer. *N Engl J Med* 2012;**367**:203–13.
20. Lane JA, Hamdy FC, Martin RM, et al. Latest results from the UK trials evaluating prostate cancer screening and treatment: the CAP and ProtecT studies. *Eur J Cancer* 2010;**46**:3095–101.
21. Schroder FH, Hugosson J, Roobol MJ, et al. Screening and prostate-cancer mortality in a randomized European study. *N Engl J Med* 2009;**360**:1320–8.
22. Andriole GL, Crawford ED, Grubb RL, et al. Mortality results from a randomized prostate-cancer screening trial. *N Engl J Med* 2009;**360**:1310–9.
23. Hugosson J, Carlsson S, Aus G, et al. Mortality results from the Goteborg randomised population-based prostate-cancer screening trial. *Lancet Oncol* 2010;**11**:725–32.
24. Moyer VA. Screening for prostate cancer: U.S. Preventive Services Task Force recommendation statement. *Ann Intern Med* 2012;**157**:120–44.
25. Croswell JM, Kramer BS, Kreimer AR, et al. Cumulative incidence of false-positive results in repeated, multimodal cancer screening. *Ann Fam Med* 2009;**7**:212–22.
26. Cooperberg MR, Broering JM, Carroll PR. Time trends and local variation in primary treatment of localized prostate cancer. *J Clin Oncol* 2010;**28**:1117–23.
27. Welch HG, Albertsen PC. Prostate cancer diagnosis and treatment after the introduction of prostate-specific antigen screening: 1986–2005. *J Natl Cancer Inst* 2009;**101**:1325–9.
28. Johansson E, Bill-Axelson A, Homberg L, et al. Time, symptom burden, androgen deprivation, and self-assessed quality of life after radical prostatectomy or watchful waiting: the randomized Scandinavian Prostate Cancer Group Study Number 4 (SPCG-4) clinical trial. *Eur Urol* 2009;**55**:422–30.
29. Hoffman RM, Hunt WC, Gilliland FD, et al. Patient satisfaction with treatment decisions for clinically localized prostate carcinoma. Results from the Prostate Cancer Outcomes Study. *Cancer* 2003;**97**:1653–62.
30. Litwin MS, Hays RD, Fink A, et al. Quality-of-life outcomes in men treated for localized prostate cancer. *JAMA* 1995;**273**:129–35.
31. Smith DP, King MT, Egger S, et al. Quality of life three years after diagnosis of localised prostate cancer: population based cohort study. *BMJ* 2009;**339**:b4817.
32. Sanda MG, Dunn RL, Michalski J, et al. Quality of life and satisfaction with outcome among prostate-cancer survivors. *N Engl J Med* 2008;**358**:1250–61.
33. Fransson P, Damber JE, Tomic R, et al. Quality of life and symptoms in a randomized trial of radiotherapy versus deferred treatment of localized prostate carcinoma. *Cancer* 2001;**92**:3111–9.
34. Schapira MM, Lawrence WF, Katz DA, et al. Effect of treatment on quality of life among men with clinically localized prostate cancer. *Med Care* 2001;**39**:243–53.
35. Siegal T, Moul JW, Spevak M, et al. The development of erectile dysfunction in men treated for prostate cancer. *J Urol* 2001;**165**:430–5.
36. Thong MS, Mols F, Kil PJ, et al. Prostate cancer survivors who would be eligible for active surveillance but were either treated with radiotherapy or managed expectantly: comparisons on long-term quality of life and symptom burden. *BJU Int* 2010;**105**:652–8.
37. Steineck G, Helgesen F, Adolfsson J, et al. Quality of life after radical prostatectomy or watchful waiting. *N Engl J Med* 2002;**347**:790–6.
38. Litwin MS, Sadetsky N, Pasta D, et al. Bowel function and bother after treatment for early stage prostate cancer: a longitudinal quality of life analysis from CaPSURE. *J Urol* 2004;**172**:515–9.
39. Talcott JA, Rossi C, Shipley WU, et al. Patient-reported long-term outcomes after conventional and high-dose combined proton and photon radiation for early prostate cancer. *JAMA* 2010;**303**:1046–53.
40. Miller DC, Sanda MG, Dunn RL, et al. Long-term outcomes among localized prostate cancer survivors: health-related quality of life changes after radical prostatectomy, external radiation, and brachytherapy. *J Clin Oncol* 2005;**23**:2772–80.
41. Bolla M, deReijke TM, Van Tienhoven GV, et al. Duration of androgen suppression in the treatment of prostate cancer. *N Engl J Med* 2009;**360**:2516–27.
42. Lu-Yao GL, Albertsen PC, Moore DF, et al. Survival following primary androgen deprivation therapy among men with localized prostate cancer. *JAMA* 2008;**300**:173–81.
43. Carter HB, Albertsen PC, Barry MJ, et al. Early detection of prostate cancer: AUA guideline. *J Urol* 2013;**190**:419–26.
44. Carroll P, Albertsen PC, Greene K. PSA testing for the pretreatment staging and post-treatment management of prostate cancer: 2013 revision of 2009 Best Practice Statement. Available from: www.AUAnet.org/education/guidelines/prostate-specific-antigen.cfm.

## PART III

# EPIDEMIOLOGY

CHAPTER

# 14

# Cancer of the Prostate: Incidence in the USA

*Christopher E. Bayne, MD, Thomas W. Jarrett, MD*

Department of Urology, The George Washington University, Washington, DC, USA

## INTRODUCTION

Prostate cancer is the most common noncutaneous cancer diagnosed in the United States. In 2014 it was estimated that 233,000 men will be newly diagnosed with prostate cancer.[1] Prostate cancer represents 14% of all new cancer diagnoses in the United States during 2014. Only skin cancer affects more men than prostate cancer in the United States.

Prostate cancer is the second leading cancer killer of men in the United States behind lung and bronchus cancer. The disease was projected to kill 29,480 men in the United States in 2014.[1] In comparison, 232,670 and 40,000 women were expected to be diagnosed with and killed by breast cancer, respectively, in 2014.[1]

Men living in the United States have a 15.33% (1 in 7 men) lifetime risk of developing prostate cancer. Lifetime probability of dying from prostate cancer in the United States is 2.71% (one in 37 men).[2] Current 5-year relative survival rates for local and regional prostate cancer approach 100%. Men with distant prostate cancer at diagnosis have a 27.9% probability of 5-year survival.[2]

Prostate cancer incidence varies throughout the continental US. The District of Columbia (194.4), Delaware (177.3), and Louisiana (169.3) have the highest incidence of prostate cancer diagnoses per 100,000 men.[1] The District of Columbia also has the highest prostate cancer mortality per 100,000 men (38.3), with Alabama (28.9) and Georgia (27.1) having the second- and third-highest, respectively.

## MODERN TRENDS IN PROSTATE CANCER DIAGNOSIS IN THE UNITED STATES

The discovery and adoption of prostate-specific antigen (PSA) testing has led to a tremendous increase in the diagnosis and treatment of prostate cancer. PSA is a protease derived from prostate tissue and acts to cleave gel-forming proteins from the seminal vesicles, thereby liquefying ejaculate and increasing sperm mobilization for improved fertilization.[3] Though PSA is now routinely used in the detection of prostate cancer and surveillance of prostate cancer treatment, the enzyme is not a cancer-specific marker and is measurable in patients without prostate pathology as well as benign disease. Independent of the presence of cancer, serum PSA increases with age and prostate volume.

PSA was not initially valued as a prostate cancer screening marker. Even after PSA was discovered, its lack of specificity for malignant disease resulted in a slow adoption.[3] In 1986 the Food and Drug Administration (FDA) approved PSA testing to monitor relapse after curative prostate cancer treatment. The availability of FDA-approved PSA kits, along with concurrent refinement of the sextant transrectal-ultrasound-guided prostate biopsy and improvements in local prostate cancer treatment led by Walsh's development of the nerve-sparing retropubic radical prostatectomy, paved the way for a dramatic paradigm shift in the early detection of prostate cancer.[3]

In 1991, in the *New England Journal of Medicine*, Catalona et al. reported the use of PSA test as part of first-line

Prostate Cancer. http://dx.doi.org/10.1016/B978-0-12-800077-9.00014-1

prostate cancer screening that detected nearly 32% of prostate cancers that would have been missed by digital rectal examination (DRE) alone.[4] Then again in 1994, Catalona et al. published a landmark multisite, prospective trial evaluating the use of PSA as a reliable indicator for the detection of prostate cancer via sextant prostate biopsy.[5] The study noted that for approximately every 100 men aged 50–75 years screened with PSA, approximately 15 would have a serum PSA greater than 4.0 ng/mL, and approximately four to five of these men would have clinically significant prostate cancer detected on biopsy.[3,5] That same year, the FDA approved PSA as the first serum tumor marker in the early detection of cancer.[3]

After the seminal paper published in 1987 in which Stamey et al. identified PSA as a serum marker for prostate cancer,[6] the incidence of prostate cancer in the United States rose approximately 12% each year until it peaked in 1992 (at 237.2 per 100,000 men).[7] Welch and Albertsen reported data from the National Cancer Institute's Surveillance, Epidemiology, and End Results (SEER) program, that showed that overall incidence of prostate cancer in the United States rose 26% (from 119 to 150.5 per 100,000 men) between 1986 and 2005.[7] PSA testing was adopted faster in the United States than in European countries, which may explain the increased incidence of prostate cancer in the United States compared to many European countries over the same time period.[8] Figure 14.1 shows age-adjusted prostate cancer incidence in the United States from 1975 to 2011.

Between 2001 and 2010, the incidence of prostate cancer in the United States decreased by approximately 2.2% annually.[10] During this time, prostate cancer mortality also decreased to an average of 3.4% annually. African-American men experienced a twofold higher death rate from prostate cancer compared to Caucasian men (50.9 vs. 21.2 per 100,000 men) during this time, yet African-American men only experienced a mild improvement in death rate (−3.8% annually) compared to Caucasian men (−3.1% annually) over the same time period.

The median age of men at the time of prostate cancer diagnosis was 72.2 years in 1988–1989 before PSA. The median age dropped to 67.2 years between 2004 and 2005.[11] The increase in prostate cancer diagnoses between 1986 and 2005 was greatest in younger men: the incidence increased over sevenfold among men younger than 50 years (1.3–9.4 per 100,000 men), threefold for men aged 50–59 years (58.4–212.7 per 100,000 men), and twofold for men aged 60–69 years (349.4–666.9 per 100,000 men). For older men aged 70–79 years, there was little change (819.2–896.8 per 100,000 men) over the same time period. For men older than 80 years, prostate cancer incidence declined dramatically (1146.5–637.4 per 100,000 men).[11]

The modern era of prostate cancer detection has led to a stage migration at time of prostate cancer diagnosis. In 1986, fewer than one-third of men diagnosed with prostate cancer had organ-confined disease. Today well over two-thirds of diagnosed men have organ-confined disease.[3] Between 1984 and 1988, 34% of prostate cancer tumors were classified as "well differentiated" compared to only 4% of patients from 1999 to 2003.[12] This is likely due to changes in pathological grading rather than disease characteristics.[13] Over nearly the same time period, the incidence of stage IV prostate cancer decreased approximately 6.4% each year, from 28.1 per 100,000 men in 1988 to 12.3 per 100,000 men in 2003.[14] Five-year survival for stage IV prostate cancer without distant metastasis improved over the time span, from 41.6% in 1988 to 62.3% in 2001.[14]

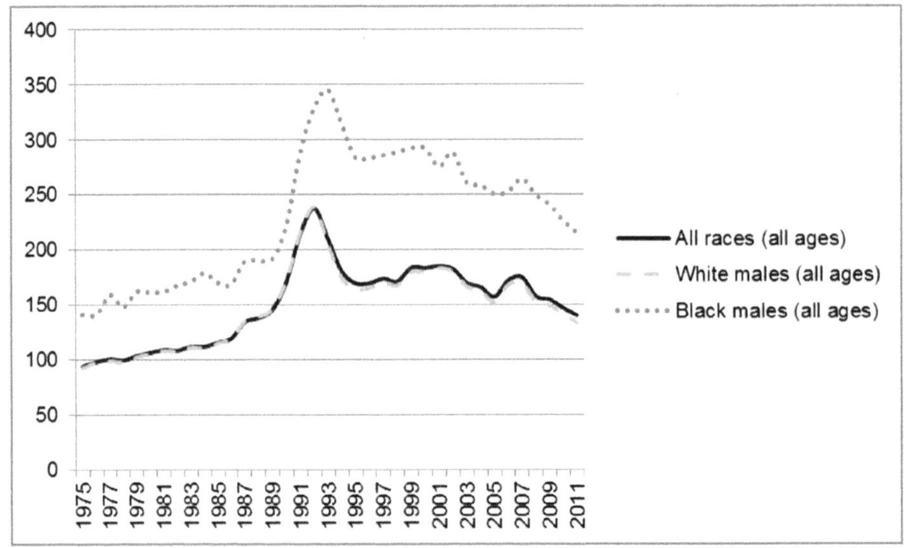


**FIGURE 14.1 Prostate cancer incidence in the United States from 1975–2011.** *Data adapted from the National Cancer Institute Surveillance Epidemiology and End Results program and age-adjusted per 100,000 men to the 2000 US standard population.*[9]

## RISK FACTORS FOR THE DEVELOPMENT OF PROSTATE CANCER

Age, family history, race/ethnicity, *BRCA1/2* mutations, and Lynch syndrome are the only well-established risk factors for the diagnosis of prostate cancer. Several other factors have been espoused in recent years, particularly in the lay press, to contribute to prostate cancer risk. As will be explored here, much of the associations between these factors and prostate cancer lack definitive scientific evidence to extract causation.

### Age

Prostate cancer is a disease of older age. The probability of developing prostate cancer increases significantly with age. The overall lifetime risk of developing prostate cancer in the United States is 15.3% (one in seven men).[1] The risk of a man developing prostate cancer between birth and 49 years of age is 0.3% (1 in 298 men).[1] The risk increases over sevenfold in the next decade of life (50–59 years) to 2.3% (1 in 43 men). Men 60–69 years old have a 6.4% risk (1 in 16 men), and men older than 70 years have a 11.2% risk (one in nine men).[1]

### Race/Ethnicity

The incidence of prostate cancers across racial lines has been well established in large population-based studies. In the United States (as well as globally), the incidence of prostate cancer is highest in African-American men (220.0 per 100,000 men), followed by non-Hispanic Caucasians (138.6 per 100,000 men), Hispanics (124.2 per 100,000 men), and Asian-Americans (75.0 per 100,000 men).[1] American-Indians have a prostate cancer incidence (104.1 per 100,000 men) between Caucasians and Hispanics.[1] African-American men are also at a higher risk for aggressive prostate cancer compared to other races. The disparity between prostate cancer incidence and mortality across racial lines continues to be a major health concern and the subject of intense research, which will be explored elsewhere in this textbook.

### Socioeconomic Status

The difference in prostate cancer incidence across racial lines is often attributed to genetic, environmental, and possibly socioeconomic status (SES). Generally accepted is the relationship between higher SES and increased prostate cancer incidence and decreased prostate cancer mortality. This relationship is likely related to increased healthcare access to early prostate cancer screening and follow-up after prostate cancer treatment. Indeed, among SEER data, the lowest SES status is associated with an increased risk of distant-stage prostate cancer at initial diagnosis.[15]

SES alone cannot account for the differences in prostate cancer incidence or mortality across racial lines. Cheng et al. examined SES and prostate cancer incidence and mortality in California according to the California Cancer Registry and SES-index using the 2000 US Census and SEER data.[16] Uniformly across all races, higher SES was associated with increased incidence of prostate cancer. Younger (age 45–64 years) and older (65–74 years) African-American men had a higher incidence of prostate cancer across all SES levels. Interestingly, among the oldest men (older than 75 years), Hispanic men had the highest incidence of prostate cancer. Across all SES levels, African-American men had a two- to fivefold increased risk of prostate cancer mortality.[16]

### Family History of Prostate Cancer

Family history of prostate cancer increases a man's risk of prostate cancer incidence and death. Broadly speaking of general consensus, a man with a father or brother with prostate cancer has more than a twofold risk of developing the disease over that of the general population. The risk is greater when the affected family member is a brother versus a father, and the more family members affected, the greater the risk.

The Swedish Family-Cancer Database contains the largest prospectively collected population database examining family history and prostate cancer incidence.[17] Table 14.1 shows the age-specific hazard ratios of prostate cancer incidence according to affected family member(s). In men under the age of 75 years, having a father with prostate cancer portended over a twofold increased risk of prostate cancer over the general population.[17] If a brother has been diagnosed, a man's risk jumps to nearly threefold that of the general population. A man with three brothers diagnosed with prostate cancer has a 17-fold risk of being diagnosed himself.[17]

There are obvious limitations in trying to adapt the Swedish data to the US population. Perhaps the biggest problem is the diversity in the US population compared to that in Sweden. The Swedish data has been similarly replicated in smaller US studies. Furthermore, Mordukhovich et al. reported that prostate cancer family history affected African-American and Caucasian men similarly across four US studies examining racial differences in prostate cancer.[18] Unfortunately, large US databases, such as SEER, as well as large US studies on prostate cancer, such as Prostate, Lung, Colorectal, and Ovarian Cancer Screening Trial, Reduction by Dutasteride of Prostate Cancer Events (REDUCE), and Prostate Cancer Prevention Trial (PCPT), do not have relevant data necessary to draw conclusions on family history and prostate cancer.

One of the difficulties in assessing familial risk of prostate cancer is an inherent screening selection bias:

TABLE 14.1 Age-Specific Hazard Ratios for the Diagnosis of Prostate Cancer in Men with First-Degree Relatives Diagnosed with the Disease

| | 0–54 years | 55–64 years | 65–74 years | All ages |
|---|---|---|---|---|
| Father only | 2.93 | 2.22 | 1.78 | 2.12 |
| Brother only | 4.41 | 3.15 | 2.56 | 2.96 |
| Father and one brother | 11.32 | 6.48 | 3.46 | 5.51 |
| Two brothers | 5.90 | 8.93 | 6.49 | 7.71 |
| Father and two brothers | 8.06 | 11.03 | 4.91 | 8.51 |
| Three brothers | 23.26 | 23.77 | 9.03 | 17.74 |

*Data adapted from the Swedish Family-Cancer Database.*[17]

men with a family history of prostate cancer are more likely to be screened for prostate cancer. Until genomic studies identify specific patterns in familial lineages, this selection bias must be accepted as a limitation to all population-based studies.

## Diet and Obesity

Western diet and prostate cancer risk have collectively garnered attention since earlier epidemiological studies showed increased prostate cancer incidence in Eastern populations adopting a Westernized diet and Eastern immigrants to the United States.[19,20] It is difficult to draw concrete conclusions regarding prostate cancer incidence and the Western diet given the complexities and spectrum of the diet itself. Large systematic reviews have found trends between consumption of a diet high in carbohydrates, red meat, dairy, and fats, and prostate cancer risk and progression.[21,22] In reality, a link between a Westernized diet and prostate cancer probably rests in the overall health status rather than singular dietary components.

Obesity lowers serum PSA levels,[23] and this likely has implications on screening. There is increasing evidence linking obesity with aggressive prostate cancer, increased rates of progression of low-risk prostate cancer managed on active surveillance protocols, increased rates of biochemical failure after definitive local tumor and hormonal treatment, and increased prostate-cancer-specific mortality (CSM).[24,25] With a sophisticated projection method, Fesinmeyer et al. calculated that an increase in US obesity prevalence since 1980 corresponds with an age-adjusted 15.5% increase in high-risk prostate cancer and 7–23% increase in prostate cancer mortality over the same time period.[26] The authors noted a slight, insignificant decrease in low-grade prostate cancer incidence over the time period.[26]

How obesity may contribute to prostate carcinogenesis, in general, remains unclear and is certain to become an increasingly important research focus given current obesity trends in the United States.

## Nutrition Supplements

### *Folic Acid*

Folate is a naturally occurring vitamin B. Folate deficiency is known to increase the risk of fetal neural tube defects. To help prevent this birth defect, certain US foods are fortified with folic acid, the synthetic, more stable version of the vitamin. Two large meta-analyses have shown statistically insignificant trends toward folic acid supplementation and the incidence of prostate cancer.[27,28] Interestingly, it is suggested that folate may have opposing roles, if any, in prostate carcinogenesis: an initial protective effect against DNA damage, then a promoter of cellular proliferation after neoplastic transformation.[29] A recent study on folic acid supplementation in patients receiving definitive local therapy for prostate cancer found no effect between supplementation and biochemical recurrence.[30] At present, there is insufficient quality evidence to draw definitive conclusions on the possible role of folate in the development or progression of prostate carcinogenesis.

### *Fish Oil*

Discussion in the lay press surrounding fish oil supplements and a potential link to prostate cancer climaxed in 2013 after Brasky et al. published a case-cohort design study showing that increased serum levels of long chain omega-3 polyunsaturated fatty acids (LCω-3PUFA) predicted a higher overall risk for prostate cancer.[31] The study examined serum LCω-3PUFA concentrations from men in the Selenium and Vitamin E Cancer Prevention Trial (SELECT). Compared to men in the lowest quartile of serum LCω-3PUFA concentrations, men in the highest quartile of serum LCω-3PUFA concentrations had an overall higher risk of low-grade, high-grade, and overall prostate cancer.[31] There are prominent limitations to the Brasky et al. study that should be noted.[32,33] First, the study was not a randomized control trial comparing men taking fish oil supplements to men abstaining. Second, it is not known whether men in the study took fish oil supplements, and if so, how much. While the result of the Brasky et al. study is intriguing, its data do not allow us to draw conclusions between fish oil supplementation and prostate cancer risk, and clinical decisions should not be based on the study's conclusions.

## Inherited Conditions

There is strong evidence for the inheritability of prostate cancer. Several prostate-cancer-specific genes have been identified.[34] An in-depth discussion of the genomics of prostate cancer is beyond the scope of this chapter and will be discussed elsewhere in this textbook. However, two inheritable conditions have garnered attention for their associated increased incidence of prostate cancer and are worth specific mention here.

### Lynch Syndrome

Lynch syndrome is an inheritable cancer syndrome associated with mutations in mismatch repair (MMR) genes. The syndrome leads to a substantially increased risk of colorectal and other cancers including prostate cancer. Investigation of two US familial cancer registries has shown men with mutations in Lynch syndrome MMR genes have an overall twofold greater risk of prostate cancer than the general population.[35] This risk is nearly 2.5-fold for men younger than 60 years.[35]

### BRCA Genes

Tumor suppressor genes *BRCA1* and *BRCA2* have long been implicated in breast and ovarian cancer, but there is strong evidence that mutations in the genes predispose to prostate cancer as well. Carriers of a *BRCA2* mutation have up to a fivefold higher risk of prostate cancer.[34] The incidence increases to over sevenfold in carriers with a family history of early-onset prostate cancer (<65 years). Furthermore, men with a *BRCA2* mutation tend to harbor more aggressive prostate cancers relative to the general population. *BRCA1* mutation carriers under 65 years have a 3.8-fold increased risk of prostate cancer.[34]

Given the increased incidence of prostate cancer in hereditary syndromes, particularly in men carrying *BRCA1/2* mutations, cancer screening in these populations is of particular interest. The IMPACT (Identification of Men with a genetic predisposition to ProstAte Cancer: Targeted screening in *BRCA1/2* mutation carriers and controls) study is an international, multicenter consortium evaluating targeted prostate cancer screening in men with *BRCA1/2* mutations.[36] The first-year screening results were recently published. Using a PSA cutoff of >3 ng/mL to trigger biopsy, the trial reported a trend toward improved positive predictive value of biopsy in *BRCA1* and *BRCA2* carriers compared to controls. Furthermore, a significantly higher number of intermediate- and high-risk prostate cancers were detected in *BRCA2* carriers.[36] Additional years of results need to be reported before drawing hard conclusions.

## Occupational Exposure

No specific occupational exposures have been definitively linked to prostate cancer. Of all the potential exposures, pesticides, specifically Agent Orange, have the strongest evidence. Several studies have shown higher incidences of prostate cancer among those with the highest exposure to Agent Orange in the Vietnam jungle compared to controls and the general population in Southeast Asia.[37] Causative associations between prostate cancer and other pesticides have not been established.[38] Cadmium and the nitrosamines in the rubber industry have also been implicated to increase prostate cancer risk but lack strong evidence.[37] No association between lead and zinc and prostate cancer has been identified.[37]

## Pharmaceuticals

### 5-ARIs

The use of 5α-reductase inhibitors (5-ARIs) in the prevention of prostate cancer was the central focus of two large, multicenter, placebo-controlled trials in the United States. The REDUCE and PCPT trials showed an approximately 25% reduction in the overall incidence of prostate cancer at 4 and 7 years, respectively.[39,40] The decreased incidence of prostate cancer was limited to Gleason 6 or less disease, and both trials showed an overall increased incidence of high-grade prostate cancer in the 5-ARI arms. There has been much debate in literature over whether 5-ARIs caused the increase in high-risk prostate cancer in these trials or if the incidence of high-risk cancer was an artifact of other factors, including study design, tumor grading, and prostate volume reduction with 5-ARI use. The consensus in the urological literature, including the assessment of a FDA advisory meeting in December 2010, is that while a direct link between 5-ARI use and high-risk prostate cancer incidence can be refuted on many levels, there is overall insufficient evidence to ignore the finding altogether.[41] 5-ARIs are not routinely used for the chemoprevention of prostate cancer in the United States.[41]

### Statins

Probably secondary to the large proportion of US men on medications for hyperlipidemia, statin medications have been the subject of intense investigation in the

United States. Farwell et al. reported a significantly decreased risk of prostate cancer incidence in military veterans taking statin medications compared to those taking antihypertensive medications.[42] This study is notable in that not only did it report that veterans taking statins were 14% less likely to be diagnosed with low-grade prostate cancer than veterans on antihypertensives, but the veterans on statin medications experienced a 60% decreased incidence of high-grade prostate cancer.[42] These results have been replicated, though to a less significant degree, in a Cleveland Clinic cohort and a large meta-analysis of observational studies.[43,44] Whether or not a decreased risk in all or high-risk prostate cancer incidence translates into clinical significance in the face of regular prostate cancer screening remains to be determined. Platz et al. reported no difference in all, low-grade, and high-grade prostate cancer incidence at 7-year follow-up of the placebo arm of the PCPT cohort.[45]

A separate issue remains in how statin use may impact prostate cancer progression and cancer-specific survival. A large United Kingdom study reported decreased prostate CSM in patients taking statins, with a more pronounced CSM decrease in patients taking statins before prostate cancer diagnosis.[46] A small cohort study from Washington State reported similar findings.[47] Given the confounds inherent in observational studies, randomized trials are needed before drawing definitive conclusions on statin use and their potential benefit in prostate carcinogenesis.

### *Testosterone Supplementation*

Testosterone therapy (TT) has many potential benefits in the treatment of hypogonadism, yet a barrier to TT use in hypogonadal men has been the widespread belief that high testosterone levels fuel the growth of prostate cancer cells.[48] This belief has largely given way to the saturation model in which only a relatively low level of androgen is necessary to activate androgen receptors in prostate cancer cells. Maximum androgen-receptor binding is achieved at testosterone levels well below the physiologic range.[49] Numerous longitudinal population-based studies as well as placebo-controlled TT studies have shown that men with higher endogenous levels of testosterone and men receiving exogenous TT are at no higher risk for developing prostate cancer than the general population.[50] Most recently, a median 5 years of follow-up was reported from three prospective registries of hypogonadal men.[51] The centers found no increased risk of prostate cancer in men receiving TT for hypogonadism.

One unanticipated result from pooling of data on testosterone and prostate cancer is that perhaps low endogenous levels of testosterone predispose not only to a higher risk of prostate cancer but also to more aggressive cancers.[50] This will certainly be an intense area of research in years to come.

## FUTURE OF PROSTATE CANCER INCIDENCE IN THE UNITED STATES

The incidence of prostate cancer in the United States surged in the late 1980s with the widespread adoption of PSA testing. The incidence has expectedly decreased steadily since the early 1990s. Prostate cancer incidence is likely to continue to decrease since the United States Preventative Services Task Force released their Grade D recommendation against the routine use of PSA screening for prostate cancer in 2012.[52] The key to future prostate cancer screening will be more intelligent use of PSA, development and adoption of prostate cancer biomarkers, and a more thorough understanding of prostate cancer genetics. In the future, overall prostate cancer incidence will be less important than a dichotomized focus on early identification of aggressive prostate cancers as well as accurate surveillance of low-risk prostate cancers.

## References

1. American Cancer Society. *Cancer facts & figures 2014*. Atlanta, GA: American Cancer Society; 2014.
2. Howlader N, Noone AM, Krapcho M, et al. Editors. *SEER Cancer Statistics Review, 1975–2010*. Bethesda, MD: National Cancer Institute. http://seer.cancer.gov/csr/1975_2010/, based on November 2012 SEER data submission, posted on the SEER web site, April 2013.
3. De Angelis G, Rittenhouse HG, Mikolajczyk SD, et al. Twenty years of PSA: from prostate antigen to tumor marker. *Rev Urol* 2007;**9**(3):113–23.
4. Catalona WJ, Smith DS, Ratliff TL, et al. Measurement of prostate-specific antigen in serum as a screening test for prostate cancer. *N Engl J Med* 1991;**324**(17):1156–61.
5. Catalona WJ, Hudson MA, Scardino PT, et al. Selection of optimal prostate-specific antigen cutoffs for early detection of prostate cancer: receiver operating characteristic curves. *J Urol* 1994;**152**(6 pt 1): 2037–42.
6. Stamey TA, Yang N, Hay AR, et al. Prostate-specific antigen as a serum marker for adenocarcinoma of the prostate. *N Engl J Med* 1987;**317**(15):909–16.
7. Welch HG, Albertsen PC. Prostate cancer diagnosis and treatment after the introduction of prostate-specific antigen screening: 1986–2005. *J Natl Cancer Inst* 2009;**101**:1325–9.
8. Neppl-Huber C, Zappa M, Coebergh JW, et al. Changes in incidence, survival and mortality of prostate cancer in Europe and the United States in the PSA era: additional diagnoses and avoided deaths. *Ann Oncol* 2012;**23**(5):1325–34.
9. Howlader N, Noone AM, Krapcho M, et al. Editors. *SEER Cancer Statistics Review, 1975–2011*. Bethesda, MD: National Cancer Institute. http://seer.cancer.gov/csr/1975_2011/, based on November 2013 SEER data submission, posted to on the SEER web site, April 2014.
10. Edwards BK, Noone AM, Mariotto AB, et al. Annual report to the nation on the status of cancer, 1975–2010, featuring prevalence of comorbidity and impact on survival among persons with lung, colorectal, breast, or prostate cancer. *Cancer* 2014;**120**(9): 1290–314.
11. Li J, Djenaba JA, Soman A, et al. Recent trends in prostate cancer incidence by age, cancer stage, and grade, the United States, 2001–2007. *Prostate Cancer* 2012;**2012**:691380.

12. Jani AB, Master VA, Rossi PJ, et al. Grade migration in prostate cancer: an analysis using the Surveillance, Epidemiology, and End Results registry. *Prostate Cancer Prostatic Dis* 2007;**10**(4):347–51.
13. Brawley OW. Trends in prostate cancer in the United States. *J Natl Cancer Inst Monogr* 2012;**2012**(45):152–6.
14. Cetin K, Beebe-Dimmer JL, Fryzek JP, et al. Recent time trends in the epidemiology of stage IV prostate cancer in the United States: analysis of data from the Surveillance, Epidemiology, and End Results Program. *Urology* 2010;**75**(6):1396–404.
15. Clegg LX, Reichman ME, Miller BA, et al. Impact of socioeconomic status on cancer incidence and stage at diagnosis: selected findings from the Surveillance, Epidemiology, and End results: National Longitudinal Mortality Study. *Cancer Causes Control* 2009;**20**(4):417–35.
16. Cheng I, Witte JS, McClure LA, et al. Socioeconomic status and prostate cancer incidence and mortality rates among the diverse population of California. *Cancer Causes Control* 2009;**20**(8):1431–40.
17. Brandt A, Bermejo JL, Sundquist J, et al. Age-specific risk of incident prostate cancer and risk of death from prostate cancer defined by the number of affected family members. *Eur Urol* 2010;**58**(2):275–80.
18. Mordukhovich I, Reiter PL, Backes DM, et al. A review of African American-white differences in risk factors for cancer: prostate cancer. *Cancer Causes Control* 2011;**22**(3):341–57.
19. Shimizu H, Ross RK, Bernstein L, et al. Cancers of the prostate and breast among Japanese and white immigrants in Los Angeles County. *Br J Cancer* 1991;**63**(6):963–6.
20. Park SK, Sakoda LC, Kang D, et al. Rising prostate cancer rates in South Korea. *Prostate* 2006;**66**:1285–91.
21. Mandair D, Rossi RE, Pericleous M, et al. Prostate cancer and the influence of dietary factors and supplements: a systematic review. *Nutr Metab (Lond)* 2014;**11**:30.
22. Masko EM, Allott EH, Freedland SJ. The relationship between nutrition and prostate cancer: is more always better? *Eur Urol* 2013;**63**(5):810–20.
23. Bañez LL, Hamilton RJ, Partin AW, et al. Obesity-related plasma hemodilution and PSA concentration among men with prostate cancer. *JAMA* 2007;**298**(19):2275–80.
24. Allott EH, Masko EM, Freedland SJ. Obesity and prostate cancer: weighing the evidence. *Eur Urol* 2013;**63**(5):800–9.
25. Bhindi B, Kulkarni GS, Finelli A, et al. Obesity is associated with risk of progression for low-risk prostate cancers managed expectantly. *Eur Urol* 2014;**66**:841–8.
26. Fesinmeyer MD, Gulati R, Zeliadt S, et al. Effect of population trends in body mass index on prostate cancer incidence and mortality in the United States. *Cancer Epidemiol Biomarkers Prev* 2009;**18**(3):808–15.
27. Vollset SE, Clarke R, Lewington S, et al. Effects of folic acid supplementation on overall and site-specific cancer incidence during the randomised trials: meta-analyses of data on 50,000 individuals. *Lancet* 2013;**381**(9871):1029–36.
28. Wien TN, Pike E, Wisløff T, et al. Cancer risk with folic acid supplements: a systematic review and meta-analysis. *BMJ Open* 2012;**2**(1):e000653.
29. Rycyna KJ, Bacich DJ, O'Keefe DS. Opposing roles of folate in prostate cancer. *Urology* 2013;**82**(6):1197–203.
30. Tomaszewski JJ, Richman EL, Sadetsky N, et al. Impact of folate intake on prostate cancer recurrence following definitive therapy: data from CaPSURE™. *J Urol* 2014;**191**(4):971–6.
31. Brasky TM, Darke AK, Song X, et al. Plasma phospholipid fatty acids and prostate cancer risk in the SELECT trial. *J Natl Cancer Inst* 2013;**105**(15):1132–41.
32. Chow K, Murphy DG. Words of wisdom. Re: plasma phospholipid fatty acids and prostate cancer risk in the SELECT trial. *Eur Urol* 2013;**64**(6):1015–6.
33. Taneja SS. Re: plasma phospholipid fatty acids and prostate cancer risk in the SELECT trial. *J Urol* 2014;**191**(3):658.
34. Eeles R, Goh C, Castro E, et al. The genetic epidemiology of prostate cancer and its clinical implications. *Nat Rev Urol* 2014;**11**(1): 18–31.
35. Raymond VM, Mukherjee B, Wang F, et al. Elevated risk of prostate cancer among men with Lynch syndrome. *J Clin Oncol* 2013;**31**(14):1713–8.
36. Bancroft EK, Page EC, Castro E, et al. Targeted prostate cancer screening in *BRCA1* and *BRCA2* mutation carriers: results from the initial screening round of the IMPACT study. *Eur Urol* 2014;**66**: 489–99.
37. Mullins JK, Loeb S. Environmental exposures and prostate cancer. *Urol Oncol* 2012;**30**(2):216–9.
38. Mink PJ, Adami HO, Trichopoulos D, et al. Pesticides and prostate cancer: a review of epidemiologic studies with specific agricultural exposure information. *Eur J Cancer Prev* 2008;**17**(2):97–110.
39. Andriole GL, Bostwick DG, Brawley OW, et al. Effect of dutasteride on the risk of prostate cancer. *N Engl J Med* 2010;**362**(13): 1192–202.
40. Thompson IM, Goodman PJ, Tangen CM, et al. The influence of finasteride on the development of prostate cancer. *N Engl J Med* 2003;**349**(3):215–24.
41. Theoret MR, Ning YM, Zhang JJ, et al. The risks and benefits of 5α-reductase inhibitors for prostate-cancer prevention. *N Engl J Med* 2011;**365**(2):97–9.
42. Farwell WR, D'Avolio LW, Scranton RE, et al. Statins and prostate cancer diagnosis and grade in a veterans population. *J Natl Cancer Inst* 2011;**103**(11):885–92.
43. Tan N, Klein EA, Li J, et al. Statin use and risk of prostate cancer in a population of men who underwent biopsy. *J Urol* 2011;**186**(1): 86–90.
44. Bansal D, Undela K, D'Cruz S, et al. Statin use and risk of prostate cancer: a meta-analysis of observational studies. *PLoS ONE* 2012;**7**(10):e46691.
45. Platz EA, Tangen CM, Goodman PJ, et al. Statin drug use is not associated with prostate cancer risk in men who are regularly screened. *J Urol* 2014;**192**(2):379–84.
46. Yu O, Eberg M, Benayoun S, et al. Use of statins and the risk of death in patients with prostate cancer. *J Clin Oncol* 2014;**32**(1): 5–11.
47. Geybels MS, Wright JL, Holt SK, et al. Statin use in relation to prostate cancer outcomes in a population-based patient cohort study. *Prostate* 2013;**73**(11):1214–22.
48. Morgentaler A. Testosterone and prostate cancer: an historical perspective on a modern myth. *Eur Urol* 2006;**50**(5):935–9.
49. Morgentaler A, Traish AM. Shifting the paradigm of testosterone and prostate cancer: the saturation model and the limits of androgen-dependent growth. *Eur Urol* 2009;**55**(2):310–20.
50. Morgentaler A. Testosterone and prostate cancer: what are the risks for middle-aged men? *Urol Clin North Am* 2011;**38**(2): 119–24.
51. Haider A, Zitzmann M, Doros G, et al. Incidence of prostate cancer in hypogonadal men receiving testosterone therapy: observations from five year-median follow-up of three registries. *J Urol* 2015;**193**:80–6.
52. Moyer VA. Screening for prostate cancer: U.S. Preventive Services Task Force recommendation statement. *Ann Intern Med* 2012;**157**:120.

CHAPTER

# 15

# International Trends in Prostate Cancer

*Simon Van Rij, MD, PhD, Wouter Everaerts, MD, PhD, Declan G. Murphy, MB, FRCS Urol*

**Division of Cancer Surgery, Peter MacCallum Cancer Centre, University of Melbourne, East Melbourne; Department of Urology, Royal Melbourne Hospital, Parkville; and Epworth Prostate Centre, Epworth Healthcare, Richmond, Victoria, Australia**

The world population climbed over 7 billion in 2011. Over the next 40 years, the predicted growth in population will almost all come from the less-developed world. This is due to its higher birth rate and younger current population compared to those in the more developed world. The world population will continue to change at an exponential rate; in particular, the more developed world will account for a decreasing percentage of the total population.[1] These facts are particularly important to keep in mind when investigating global trends in prostate cancer. In 2012 there were 1,111,689 new diagnoses of prostate cancer worldwide as well as 307,471 deaths.[2] There is a tendency to forget that almost all literature on prostate cancer comes from the developed world, which can skew our perception of the disease. Prostate cancer is often described as a disease of the developed world, a fact confirmed by its higher incidence rates. However, of the 307,471 deaths from prostate cancer reported in 2012, over 50% occurred outside of the United States and Europe. These numbers will only get higher, hence, it is extremely important for us to further understand the population trends in prostate cancer not just the United States but throughout the world. This chapter aims to provide up-to-date information on prostate cancer incidence and mortality along with epidemiological changes in the disease, specifically from countries outside the United States.

Gathering global statistics on prostate cancer is not without difficulty. Different countries report data in different forms using different types of methodology. Databases are only as viable as the funding that supported them; therefore, the comparative accuracy of the statistics of more developed nations and those of their less developed counterparts needs to be taken into account.

The World Health Organization's International Agency for Research on Cancer (IACR) publishes the incidence, mortality, and prevalence rates for all cancers. Known as GLOBOCAN, the IACR project publishes results on a periodic basis, with the latest update released in 2012. A total of 184 countries throughout the world have data collected, and for statistical purposes, a distinction is made between "more developed" and "less developed" regions. The more developed regions include all regions of Europe plus Northern America, Australia/New Zealand, and Japan. The less developed regions consist of all regions of Africa, Asia (excluding Japan), Latin America and the Caribbean, Melanesia, Micronesia, and Polynesia. This is the source of data used for comparison of rates between countries. It is however very difficult to compare global rates over time against each other; one of the disclaimers of GLOBOCAN is that comparison should not be performed between years as significant changes may be more due to methodology change rather than actual trends.

## INCIDENCE RATES OF PROSTATE CANCER

Globally, there were 1,111,689 new cases of prostate cancer diagnosed in 2012. Table 15.1 outlines these rates broken down by economic and geographic groupings. The most striking statistic is the massive difference between countries' incidence rates. Of particular interest is the nearly five-times-higher incidence rates in the more developed regions over those in the less developed ones. However, even among less developed regions, there is a stark difference in incidence rates. In Southeast Asia

**Prostate Cancer. http://dx.doi.org/10.1016/B978-0-12-800077-9.00015-3**

TABLE 15.1 Incidence Rates of Prostate Cancer Broken Down by Geographic and Socioeconomic Regions

| Population | Numbers | Crude rate | ASR (W) | Cumulative risk |
|---|---|---|---|---|
| World | 1,111,689 | 31.2 | 31.1 | 3.84 |
| More developed regions | 758,739 | 125.2 | 69.5 | 8.84 |
| Less developed regions | 352,950 | 12 | 14.5 | 1.65 |
| Very high human development | 750,896 | 131.9 | 73.7 | 9.34 |
| High human development | 195,839 | 38.2 | 37.5 | 4.55 |
| Medium human development | 115,942 | 6.4 | 7 | 0.72 |
| Low human development | 47,809 | 7.3 | 14.9 | 1.84 |
| WHO Africa region (AFRO) | 51,689 | 11.8 | 26.8 | 3.24 |
| WHO Americas region (PAHO) | 412,739 | 87.6 | 75 | 9.35 |
| WHO East Mediterranean region (EMRO) | 18,585 | 5.8 | 9.7 | 1.13 |
| WHO Europe region (EURO) | 436,688 | 99.9 | 60.9 | 7.82 |
| WHO South-East Asia region (SEARO) | 38,515 | 4.1 | 5.5 | 0.62 |
| WHO Western Pacific region (WPRO) | 153,167 | 16.2 | 12.6 | 1.37 |
| IARC membership (24 countries) | 807,515 | 61.5 | 51.3 | 6.49 |
| Middle-East and Northern Africa (MENA) | 29,377 | 12.9 | 19.7 | 2.38 |
| Africa | 59,493 | 11.1 | 23.2 | 2.81 |
| Sub-Saharan Africa | 51,945 | 12 | 27.9 | 3.37 |
| Eastern Africa | 17,187 | 9.8 | 23.3 | 2.82 |

*Source: Ferlay et al.*[2]

the incidence rate of 5.5/100,000 is ten times lower than South America's rate of 75/100,000. Figure 15.1 shows pictorially the incidence rates for countries measured by age-adjusted standardized rates per 100,000 population. Age-adjusted standardized rates supposedly attempt to limit the confounding effect of an aging population, thus increasing the amount of prostate cancer incidence. The reason for the marked difference in incidence rates between countries is likely to be multifactorial. These factors are likely to include genetic and lifestyle differences between populations along with differences in prostate cancer screening and treatment standards.

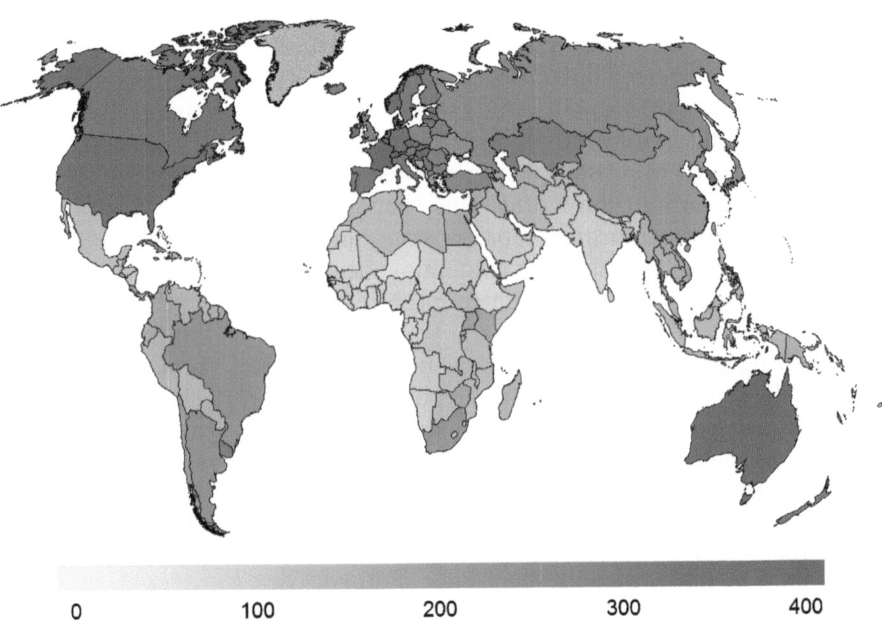


**FIGURE 15.1 Estimated prostate cancer incidence worldwide in 2012 using age-standardized rates per 100,000.** *Source: Ferlay et al.*[2]

The comparison between nations is important, but just as crucial is the change in incidence rates over time within populations. This on a global scale is difficult to assess due to changes potentially being caused by methodology/reporting changes rather than by actual disease changes. Therefore, only robust databases that have not altered over time can be used for this analysis. Center et al. analyzed 40 different countries worldwide and compared prostate cancer incidence rates over a 5-year period.[3] Results showed that the incidence rate in 32 of the 40 countries had increased in the last 5 years. In particular, large increases were noted in Lithuania, China, and South Korea. This has been linked directly to an increase in the amount of PSA screening in these populations. The countries with stable incidence rates of prostate cancer are from the more developed world where large percentages of the population have already had PSA screening for a longer time period; such nations include Australia, New Zealand, and the United States.

## PSA SCREENING

The incidence of prostate cancer is obviously linked to detection rate. Prostate cancer is often detected through screening of asymptomatic men with PSA testing. Screening aims to detect the cancer early on to avoid mortality and morbidity associated with prostate cancer. A downside of screening is overdiagnosis of clinically insignificant cancers. Internationally, efforts have been made to investigate whether or not the use of PSA testing to screen for prostate cancer is beneficial. The most important of these is the European Randomized study of Screening for Prostate Cancer (ERSPC) conducted in eight European countries and enroling 162,388 men. The 2014 update of these results showed that at 13 years of follow-up, the rate ratio of death from prostate cancer in those in the PSA screening arm as compared to the control group who did not receive screening was 0.79 (0.69–0.91). When we look specifically at the incidence rates between the screened and the nonscreened arms, we see a significantly higher incidence rate in the screened arm.[4]

The ERSPC results contradict those of the Prostate Cancer Screening in the Randomized Prostate, Lung, Colorectal, and Ovarian Cancer Screening Trial (PLCO) performed in the United States. The PLCO trial randomized 76,685 men to PSA screening versus no screening. The results at 13 years of follow-up show no difference in prostate cancer mortality between those who were assigned to an organized screening program and those who were not.[5] The methodology and clinical significance of these trials have been discussed in much detail in previous publications.[6] However, from an international perspective, it highlights the different population groups that were studied between the two trials as well as the probable reasons behind the difference in results. The US population enrolled for this trial already had a high rate of PSA screening at baseline which affected the outcome. Almost half of the control arm in this study actually underwent PSA testing, making a difference in prostate-cancer-specific mortality rates almost impossible to demonstrate. Caution should therefore be used when applying the results of the PLCO trial to other populations that do not have the same high baseline PSA testing rates noted in the United States. These populations may in fact realize a benefit in terms of mortality reduction from introducing PSA screening.

Internationally, efforts have been made outside of clinical trials to introduce screening initiatives into individual countries, ultimately to gain insight into the real-world benefit of PSA testing. In Austria in 1993, the district of Tyrol was chosen to allow men to receive free PSA screening for prostate cancer. The aim of the exercise was to compare outcomes with the rest of the Austrian population who did not undergo screening. Published after 2008, results showed a significant reduction (risk ratio 0.7) in prostate cancer mortality in Tyrol compared to the rest of the population. The authors concluded that the decrease in mortality was due to the aggressive downstaging of prostate cancer along with successful treatment.[7]

More recently in 2006, Lithuania introduced the Early Prostate Cancer Detection Program, providing free access to PSA screening for all men over the age of 50 (or younger if there is family history of prostate cancer). Recent results show a dramatic rise in the incidence of prostate cancer in Lithuania in the years following the introduction of the program. However, it is still too early to report information as to whether this program has altered mortality rates in the country.[8]

Outside of Europe, the Japanese Prospective Cohort Study of Screening for Prostate Cancer (JPSPC) is looking at whether or not PSA screening is effective in altering mortality rates. A total of 100,000 men aged 50–79 in the screening arm are being compared to a matched cohort group. The results of the study are expected to be released in 2015.[9]

From the data, it appears that the amount of PSA testing in a population reflects the incidence rate of prostate cancer. Further research has attempted to clarify the exact amount of testing that is performed in different countries. Using laboratory data and population records, it has been shown that even though no international society recommends mass screening of all asymptomatic men, large numbers of men undergo PSA testing. In New Zealand, which has one of the highest incidence rates of prostate cancer in the world, approximately 55% of men over the age of 50 have had a PSA test within a 12-month period. The rates are similar in other developed countries such as France, Australia, and the United States.[10] Unfortunately, there are no papers published from the less developed world to show the exact rate of PSA testing. This detail would be interesting to compare between countries to determine whether a population's screening rate and the rate of prostate cancer diagnosis are correlated.

## GENETIC LINKS

We now see that a key determining factor in the incidence rate of prostate cancer is the amount of screening performed in a population. However, screening cannot be the sole reason for differences in prostate cancer incidence rates between populations. We know genetically that prostate cancer is a polygenic disease with low-penetrance alleles.[11] Our understanding of the genetics of prostate cancer has improved over time, allowing us to identify inherited genetic traits, the most commonly reported of which are those in African-American and Caribbean populations. Compared to Caucasian males, African- American men have a fourfold increased risk of prostate cancer. Genetic changes in the 8q24 loci have been identified as contributing to this increased risk.[12] However, it is interesting to note that the prostate cancer incidence rate in African males is very low compared to that of African-American males, a difference that is not to be expected between populations coming from a similar genetic line. One theory to explain this difference is the genetic drift caused by historic colonization patterns of Northern European nations.[13] This colonization caused an admixture of the Northern European genome into the populations that it settled such as the African-Americans.[14] Northern European nations have some of the highest rates of cancer in the world. Even when compared to nations that are geographically close with similar screening rates, Northern European nations still report higher incidence rates of prostate cancer. There is likely a genetic predisposition in this population that increases their risk of prostate cancer.[15] Gunderson et al. acquired epidemiological evidence from tracking the colonization pathways of Northern Europeans and comparing this to areas that were not colonized by Northern Europeans or other European nations. Such data showed that those areas historically settled by Northern Europeans, including the United States and the Caribbean, have some of the highest incidence rates of prostate cancer in the world. In Africa, for example, we note that the countries that were populated by Northern Europeans, such as South Africa, have considerably higher rates of prostate cancer compared to nations that have had minimal European settlement, such as Ethiopia.[16] What is the significance of these findings? Though the exact genetic traits have not been identified from the Northern European lineage, this information can be used to more selectively target those at increased risk of prostate cancer. They may also account for the rising incidence of prostate cancer throughout the world. The admixture of genomes harboring prostate cancer risk is likely to have increased with globalization and greater migration of population groups to new nations.

Given lower rates of prostate cancer in Asia, it would appear that there may be a decreased genetic susceptibility to prostate cancer in these populations. Even when these populations emigrate to developed nations such as the United States, the incidence rates of prostate cancer in Asian-Americans are still lower than the rest of the US population.[17] However, that does not mean that there are no increased genetic risks in Asian populations. Researchers from Japan have identified genetic risk factors for familial prostate cancer while aggressive diseases have been identified within the 8q24 chromosomal region in Asian individuals.[18]

Certain genetic variants have already been identified as increasing a person's risk of prostate cancer. There are also a number of lifestyle factors and dietary habits that have been shown to impact the risk of developing prostate cancer. Green tea is a food group that has been touted to prevent prostate cancer. A meta-analysis of three studies showed that Asian men who drank green tea more frequently had lower rates of prostate cancer. Notably, this effect was not seen in the US men who were studied.[19] Similarly, studies looking into the benefit of isoflavones have shown reduced risk of cancer in Asian men with higher consumption levels; however, this benefit has not been shown in western populations.[20] A possible reason for this difference in effect between nations may be a required genetic predisposition that varies between populations. Another explanation may be that a dietary factor must be heavily ingrained in a population for a long time or from a young age for it to yield actual benefits. Methodological differences between studies may also be an underlying reason.

One of the other key factors that have been shown to increase the risk of prostate cancer is the current western diet, one that has high levels of saturated fats and processed meats, and lesser amounts of vegetables and fruit.[21] Migration studies comparing native Asian men to those who have migrated to the Western world may help shed light on the matter. Native Asian men have lower rates of prostate cancer than those who have emigrated such as Asian-Americans. As discussed throughout this section, a multifocal cause combining genetic changes, lifestyle factors, and screening habits most likely impacts the risk of cancer.[22]

## MORTALITY

The mismatch in international prostate cancer statistics is that the countries with the highest incidence rates of prostate cancer do not have the highest mortality rates

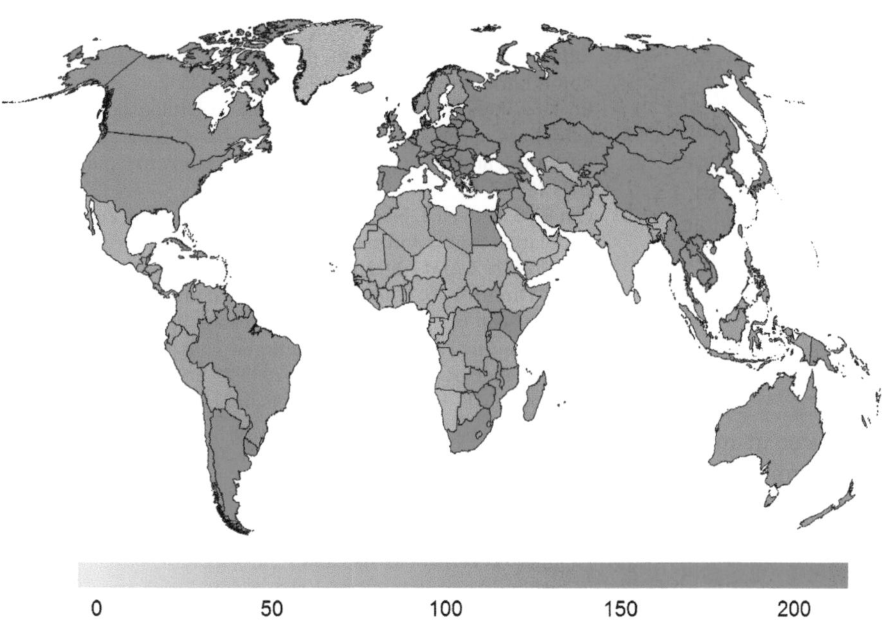


**FIGURE 15.2** Estimated prostate cancer mortality worldwide in 2012 using age-standardized rates per 100,000. *Source: Ferlay et al.*[2]

from the disease. Figure 15.2 is a pictorial graph showing the mortality rates for countries throughout the world. Focusing initially on the more developed world, a database analysis of mortality rates shows a progressive drop over the last 15 years. The GLOBOCAN mortality rate for the developed world in 2012 was 10/100,000. Within these nations, as prostate cancer incidence rates have stabilized, there is clear evidence of a decrease in mortality.[23] This is a far cry from the situation in certain countries within the less developed world where prostate cancer mortality rates remain high, more specifically in countries in South America, the Caribbean, and sub-Saharan Africa where rates are as high as 29.3/100,000. The only countries to report an increase in prostate cancer mortality rates are those from the less developed world, particularly from Eastern Europe, Africa, and Asia.[24] The reason for this may be related to the westernization of diets and lifestyles within these nations; however, it may also be a false effect due to better reporting of data.

Overall mortality rates in Asian countries are generally low compared to those in the Western world and have been calculated at 3.8/100,000. An attempt to determine the effect screening and treatment have on the outcomes of prostate cancer can be made by comparing the mortality-rate-to-incidence-rate ratio for prostate cancer in different countries. The lower the number, the fewer patients diagnosed with prostate cancer die from the disease. Using population data from California as a standard, the ratio of population diagnosed with cancer to mortality was between 0–1 and 1–0.18. Comparing Asian countries, the mortality-to-incidence ratio is low in the more developed nations such as Japan and Korea (0.18–0.22) but high in the less developed nations such as Thailand, the Philippines, and Malaysia (0.5–0.6).[25] Therefore, although cancer mortality rates in many Asian countries are low, it would appear from reduced levels of screening or access to treatment that a higher proportion of those with the disease die of the disease. This highlights the fact that equal access to prostate cancer treatment is important throughout the world, even in those countries that supposedly do not have high incidence rates of prostate cancer. Data revealed that the countries with the highest incidence rates have the lowest mortality/incidence ratios. This could be attributed to screening to identify cancers early on and curb morbidity and mortality. However, the clinical significance of these findings is difficult to assess as diagnosis of insignificant cancers could also be a reason for the lower ratio.

How do we explain the wide difference in prostate cancer mortality rates between nations? This may be difficult to answer due to the long latent period that many patients with prostate cancer have. First, lower mortality rates are the effect of more widely available prostate cancer screening in wealthier nations. Death rates have declined since the introduction of readily available PSA testing in the 1990s. The screening trials discussed earlier point to an advantage but not as great as the difference in mortality rates between the likes of the United States and sub-Saharan Africa. Potentially, the declining mortality rates may be more due to the availability of treatment for prostate cancer and the improvement in care seen in more developed nations. Such new treatment options for advanced disease are less likely to be available in the less developed world. Furthermore, certain genetic traits that may predispose to prostate cancer may be more prevalent in certain populations. Given these multiple factors, one cannot just deduce from the mortality data whether certain populations have more aggressive forms of prostate cancer. All these however emphasize that future trials investigating prostate cancer must take into account the population that is being targeted as well as potential results confounders such as rate of PSA screening within the population, lifestyle factors, genetic variants, and access to health care. Doing so will ensure the applicability of the results to other countries' populations.

## CONCLUSIONS

Prostate cancer is the most common cancer in males worldwide. The incidence and mortality rates for prostate cancer vary widely between population groups around the world. This is due to differences in screening, provision of health services, genetic predispositions, and lifestyle factors. A clear understanding of these factors is important for future research to ensure that all nations benefit from improved survival from prostate cancer. Specifically, as the population of the less developed world grows, so would the demand for better prostate cancer management in the countries involved.

## References

1. United Nations. *World Population Prospect: The 2012 Revision. Data Sources and Meta Information*. New York: Department of Economic and Social Affairs, Population Division; 2013. Available from: http://esa.un.org/unpd/wpp/sources/country.aspx and http://esa.un.org/unpd/wpp/Excel-Data/WPP2012_F02_METAINFO.xls.
2. Ferlay J, Soerjomataram I, Ervik M, et al. GLOBOCAN 2012 v1.0, Cancer Incidence and Mortality Worldwide: IARC Cancer Base No. 11 [Internet]. Lyon, France: International Agency for Research on Cancer; 2013. Available from: http://globocan.iarc.fr.
3. Center MM, Jemal A, Lortet-Tieulent J, et al. International variation in prostate cancer incidence and mortality rates. *Eur Urol* 2012;**61**(6):1079–92.
4. Schröder FH, Hugosson J, Roobol MJ, et al. Screening and prostate cancer mortality: results of the European Randomised Study of Screening for Prostate Cancer (ERSPC) at 13 years of follow-up. *Lancet* 2014;**384**:2027–35.
5. Andriole GL, Crawford ED, Grubb III RL, et al. Prostate cancer screening in the randomized Prostate, Lung, Colorectal, and Ovarian Cancer Screening Trial: mortality results after 13 years of follow-up. *J Natl Cancer Inst* 2012;**104**(2):125–32.
6. Nam RK, Klotz LH. Does screening for prostate cancer reduce prostate cancer mortality? *Can Urol Assoc J* 2009;**3**(3):187–8.
7. Oberaigner W, Siebert U, Horninger W, et al. Prostate-specific antigen testing in Tyrol, Austria: prostate cancer mortality reduction was supported by an update with mortality data up to 2008. *Int J Public Health* 2012;**57**(1):57–62.
8. Smailyte G, Aleknaviciene B. Incidence of prostate cancer in Lithuania after introduction of the Early Prostate Cancer Detection Programme. *Public Health* 2012;**126**:1075–7.
9. Ito K, Kakehi Y, Naito S, et al. Japanese Urological Association guidelines on prostate-specific antigen-based screening for prostate cancer and the ongoing cluster cohort study in Japan. *Int J Urol* 2008;**15**:763–8.
10. van Rij S, Dowell T, Nacey J. PSA screening in New Zealand: total population results and general practitioners' current attitudes and practices. *N Z Med J* 2013;**126**(1381):27–36.
11. Xu J, Dimitrov L, Chang BL, et al. A combined genomewide linkage scan of 1,233 families for prostate cancer-susceptibility genes conducted by the international consortium for prostate cancer genetics. *Am J Hum Genet* 2005;**77**:219–29.
12. Freedman ML, Haiman CA, Patterson N, et al. Admixture mapping identifies 8q24 as a prostate cancer risk locus in African-American men. *Proc Natl Acad Sci USA* 2006;**103**:14068–73.
13. Gunderson K, Wang CY, Wang R. Global prostate cancer incidence and the migration, settlement, and admixture history of the Northern Europeans. *Cancer Epidemiol* 2011;**35**(4):320–7.
14. Higgins PB, Fernandez JR, Goran MI, et al. Early ethnic difference in insulin-like growth factor-1 is associated with African genetic admixture. *Pediatr Res* 2005;**58**:850–4.
15. Gunderson K, Wang CY, Wang R. Global prostate cancer incidence and the migration, settlement, and admixture history of the Northern Europeans. *Cancer Epidemiol* 2011;**35**(4):320–7.
16. Ferlay J, Soerjomataram I, Ervik M, et al. GLOBOCAN 2012 v1.0, Cancer Incidence and Mortality Worldwide: IARC CancerBase No. 11 [Internet]. Lyon, France: International Agency for Research on Cancer; 2013. Available from: http://globocan.iarc.fr.
17. Ito K. Prostate cancer in Asian men. *Nat Rev Urol* 2014;**11**(4):197–212.
18. Suzuki K, Matsui H, Ohtake N, et al. A p53 codon 72 polymorphism associated with prostate cancer development and progression in Japanese. *J Biomed Sci* 2003;**10**:430–5.
19. Zheng J, Yang B, Huang T, et al. Green tea and black tea consumption and prostate cancer risk: an exploratory meta-analysis of observational studies. *Nutr Cancer* 2011;**63**:663–72.
20. Kurahashi N, Iwasaki M, Inoue M, et al. Plasma isoflavones and subsequent risk of prostate cancer in a nested case-control study: the Japan Public Health Center. *J Clin Oncol* 2008;**26**:5923–9.
21. World Cancer Research Fund and American Institute for Cancer Research. *Food, nutrition, physical activity, and the prevention of cancer: a global perspective*. American Institute for Cancer Research; 2007. p. 305–309.
22. Ito K. Prostate cancer in Asian men. *Nat Rev Urol* 2014;**11**(4):197–212.
23. Center MM, Jemal A, Lortet-Tieulent J, et al. International variation in prostate cancer incidence and mortality rates. *Eur Urol* 2012;**61**(6):1079–92.
24. Ferlay J, Soerjomataram I, Ervik M, et al. GLOBOCAN 2012 v1.0, Cancer Incidence and Mortality Worldwide: IARC CancerBase No. 11 [Internet]. Lyon, France: International Agency for Research on Cancer; 2013. Available from: http://globocan.iarc.fr.
25. McCracken M, Olsen M, Chen Jr MS, et al. Cancer incidence, mortality, and associated risk factors among Asian Americans of Chinese, Filipino, Vietnamese, Korean, and Japanese ethnicities. *CA Cancer J Clin* 2007;**57**(4):190–205.

CHAPTER

# 16

# Race, Ethnicity, Marital Status, Literacy, and Prostate Cancer Outcomes in the United States

*Kelvin A. Moses, MD, PhD*, Viraj A. Master, MD, PhD, FACS**, Willie Underwood, III, MD, MPH, MSci*[†]

*Department of Urologic Surgery, Vanderbilt University Medical Center, Nashville, TN, USA
**Department of Urology, Emory University, Atlanta, GA, USA
[†]Department of Urology, Roswell Park Cancer Institute, Buffalo, NY, USA

## EPIDEMIOLOGY

African-Americans (AA) have a 25% greater cancer mortality rate compared to Caucasians, and prostate cancer is a significant contributor to the overall cancer mortality gap in men.[1] There will be an estimated 220,800 new cases of prostate cancer in 2015, with AA men having the highest incidence of prostate cancer among men in the United States.[1,2] Asian-American/Pacific Islanders have the lowest incidence, whereas Hispanic men have an incidence rate slightly lower than non-Hispanic Caucasians (124.2 vs. 138.6 per 100,000, respectively).[2] AA men had an almost 2.4 × rate of mortality from prostate cancer compared to Caucasian men from 2006 to 2010, a gap that has not narrowed significantly within the last 20 years.[2,3] In order to elucidate the source of this glaring disparity in AA men, it is important to examine variance in stage at diagnosis and treatment, as well as the potential impact of genetic differences and social factors.

## PROSTATE CANCER DISPARITY IN AFRICAN-AMERICAN MEN

AA men are diagnosed with prostate cancer at an earlier age compared to Caucasian men. Data from the Surveillance, Epidemiology, and End Results (SEER) Program 1996–2002 show that AA men are approximately 3 years younger at the time of prostate cancer diagnosis and are significantly at higher risk (relative risk 1.93) of prostate cancer diagnosis prior to age 45.[4] These data are in line with the findings of Powell et al., who examined autopsy specimens in men who died from nonprostate cancer causes, and compared them with prostatectomy specimens from their institution and Detroit SEER data. Their results suggest that while subclinical prostate cancer develops at a similar age among young AA and Caucasian men, the higher rate of metastatic disease and greater tumor volume among AA in the same geographical area are the result of a faster growth rate and/or earlier transformation to aggressive disease.[5] Data from SEER and the Cancer of the Prostate Strategic Urologic Research Endeavor (CaPSURE™) show that AA men present with higher-grade and higher-stage disease at the time of diagnosis.[6,7] Differences in screening patterns likely contribute to differences in stage at diagnosis, both of which are significant contributors to higher prostate cancer-specific mortality in men.[8]

### Treatment Disparity

There is no consensus regarding the optimal treatment for prostate cancer, regardless of the risk classification at diagnosis. However, among AA men, there has been a consistent difference in both the actual receipt of treatment as well as the type of treatment received compared to Caucasian men. AA men, and to a lesser degree Hispanic men, were more likely to receive radiation therapy (RT), androgen deprivation therapy (ADT), or watchful waiting (WW), and were less likely to receive

Prostate Cancer. http://dx.doi.org/10.1016/B978-0-12-800077-9.00016-5

a radical prostatectomy (RP).[6,7,9–11] Even among AA men who receive RP, there are some data that indicate that AA men have worse survival rates compared to Caucasian men.[12] One potential explanation for this is that when AA men do receive RP, they are less likely than Caucasian, Hispanic, and Asian men to receive pelvic lymph node dissection, especially when diagnosed with Gleason 8–10 prostate cancer.[13] In regards to WW, Hispanic and AA men, in general, receive less medical monitoring and have longer median times from diagnosis to receipt of a medical monitoring visit or procedure compared to Caucasian men.[14]

There have been multiple studies utilizing administrative datasets spanning several decades that demonstrate this trend in disparate treatment. SEER data from 1988 to 1989 showed that AA men were less likely to receive aggressive therapy (defined as RP or RT) than Caucasian men, and were less likely to receive RP if they received definitive treatment.[9] Zeliadt et al. analyzed the receipt of aggressive treatment versus WW in men ≥65 years of age with nonmetastatic prostate cancer utilizing linked SEER-Medicare data from 1991 to 1999.[15] Adjusting for age, socioeconomic status, comorbidity, and Gleason grade, AA men were 26% less likely to receive aggressive therapy (RP, RT, or brachytherapy) than Caucasian men, and were significantly more likely to receive conservative therapy. Using SEER data from 1992 to 1999, Underwood et al. reported the trends in treatment by ethnicity to determine the impact of brachytherapy use.[16] Caucasian and Hispanic men were more likely to undergo RP, whereas AA men were more likely to receive ADT/WW. Compared to Caucasian men, Hispanic and AA men were significantly more likely to receive ADT/WW. AA men were more likely than Caucasian and Hispanic men to receive RT. Harlan et al. examined data from the Prostate Cancer Outcomes Study (PCOS) from 1994 to 1995 detailing treatment trends among men with localized disease.[10] After adjusting for age, clinical stage, baseline PSA, and Gleason grade, they noted that AA men aged 60 and older were less likely to receive aggressive treatment compared to Caucasian and Hispanic men. Data from CaPSURE™ from 1995 to 2008, when adjusted for D'Amico risk level, age, health perception, number of comorbidities, education level, and insurance status, showed that Caucasian men were significantly less likely than AA men to receive ADT over RP.[7] There was a trend toward Caucasian men being less likely to receive RT instead of RP compared to AA men. The authors also showed that Caucasians were much less likely to receive ADT than AA men (OR 0.29) for low-risk disease. Presley et al. examined SEER-Medicare data among men 67–84 years of age diagnosed with localized prostate cancer from 1998 to 2007.[17] They defined the group with highest benefit as those men with moderate-risk tumors (Gleason 7–10 and cT2b-T2c) and ≥10 years of life expectancy, with patients in low- or intermediate-benefit groups having either lower-grade tumors and/or lower life expectancy. AA men were less likely to receive any therapy within all the predefined benefit groups, with the largest disparity in the highest benefit group. Specifically, AA men with moderate-risk tumors and normal life expectancy were 43% less likely to receive treatment than Caucasian men in the same category. Using SEER data, Underwood et al. reported that among men diagnosed with localized/regional prostate cancer from 1992 to 1999, AA men were less likely to receive definitive therapy.[16] They also noted that AA men with high-grade (Gleason 8–10) prostate cancer were less likely to receive definitive therapy than Caucasian men with moderate (Gleason 5–7) prostate cancer.

These results are disconcerting in light of several reports showing a significant benefit to surgical approach in AA men. In a study of 4926 Caucasian men and 362 AA men who underwent RP at Johns Hopkins from 1988 to 2004, there was no difference in the rate of high-grade disease (Gleason >4 + 3 = 7), positive surgical margins, extraprostatic extension, or seminal vesicle invasion, despite higher preoperative PSA, BMI, and high-grade disease in AA men.[18] Tewari et al. reported risk factors for PSA recurrence in 646 Caucasian men and 402 AA men who underwent RP for clinically localized prostate cancer.[19] Despite higher preoperative PSA and shorter preoperative PSA-DT, AA race was not an independent factor for PSA recurrence with mean follow-up of greater than 6 years. Using the Shared Equal Access Regional Cancer Hospital (SEARCH) database, Freedland et al. analyzed 1547 patients who underwent RP from 1988 to 2001.[20] There were significant differences between racial groups for several presurgical factors such as age, clinical stage, PSA, and grade; however, race was not an independent predictor of biochemical recurrence.

The importance of addressing prostate cancer treatment disparity in AA and Hispanic men, and the consistent utilization of nonsurgical treatment in men with potentially curable disease, cannot be overstated. Data from the Detroit SEER registry from 1988 to 1992 showed that men of low socioeconomic status and those who received nonsurgical treatment were significantly at greater risk of death, whereas when controlling for age or tumor grade, there was little effect on survival differences.[21] Bach et al. performed a MEDLINE review of the English-language literature regarding overall survival for AA and Caucasian patients with cancer, including prostate cancer. They found that AA and Caucasians who received similar treatment for cancer had nonsignificant differences in survival rates, whereas treatment differences, stage at diagnosis, and comorbidity were significant drivers of survival.[22] This indicates that in areas where there are large disparities (prostate cancer screening, treatment, etc.), there remains an urgent need to improve access to care, and to increase awareness among patients and physicians of the need for aggressive treatment in appropriate circumstances.

### Genetic Alterations Associated With Prostate Cancer

Because faster growth/tumor transformation is postulated in AA men and other high-risk populations, some investigators have initiated a search for genetic variants or other molecular markers for aggressive disease in men of different ethnic backgrounds. Recent studies have yielded some clues to a biological basis of aggressive tumor features, some of which may obviate the social construct of race and place risk of aggressive disease on specific cross-cultural mutations. Haiman et al. performed a genome-wide study in men of African ancestry and found a significant association with a variant of chromosome 17q21 in prostate cancer cases rarely seen in men of other populations.[23] Other potential genetic targets, such as 11q22 and Xq21, have been identified by genome-wide linkage analysis in the African-American Hereditary Prostate Cancer study.[24] Specific alterations in chromosomes 8p, 8q, and 10q have yielded potential associations to aggressive prostate cancer in Caucasian, AA, Hispanic, and Asian men.[25–29] Other genetic mutations that have been associated with prostate cancer risk or risk of aggressive cancer include the G84E mutation in *HOXB13*,[30] loss of *PTEN*,[31] and *TMPRSS2-ERG* fusion,[32] all of which have varying penetrance in certain ethnic groups but are not exclusively expressed by any one group. These studies, while still in the early stages of discovery and elucidation of significance, are very promising in that clinicians may be able to provide personalized, disease-specific treatments regardless of ethnic background.

## MARITAL STATUS AND PROSTATE CANCER OUTCOMES

There are numerous studies evaluating prostate cancer outcomes in relation to marital status. One of the best-studied outcomes has been survival. Uniformly, patients who are married fare better. In a recent, large single-institution study of 869 patients at an academic safety net hospital, Rand et al. found that married patients were almost twice as likely to survive after treatment with diverse forms of therapy for prostate cancer compared to single patients.[33] SEER registry data have also been analyzed to establish the association of marital status and prostate cancer outcomes. In a study of approximately 26,000 patients, single patients were 1.37 times more likely to die of prostate cancer, compared to married patients.[34] While the majority of large studies do show that marital status is important, a few studies have contradicted this finding, but these tend to be retrospective studies of rather limited numbers of patients. In summary, being married is protective of both overall survival and prostate-cancer-specific survival.

Most prostate cancer patients tend to be long-term survivors, especially as most prostate cancers are indolent, suggesting that long-term cancer survivors need survivorship care plans. Marital status is independently associated with follow-up care use in a large cohort of patients with several malignancies. Very interestingly, the lack of tangible and emotional support was most strongly associated with prostate cancer although this did not achieve statistical significance.[35] Many of these studies do look at the subtypes of being single. For example, a man may be single because he was never married or was never in a relationship, or is divorced, widowed, or separated. There may be outcome differences for those patients with prostate cancer. Sprehn et al. found that among unmarried patients, those patients who were separated at the time of cancer diagnosis had the lowest survival, followed sequentially by those who were widowed, divorced, and never married.[36] The 5-year and 10-year relative survival rates of separated patients were only 72% and 64%, respectively, compared to those of married patients.

Some of the best data on this subject come from large registry studies in European countries. Burns et al. were interested in factors driving inequality in prostate cancer survival and studied over 20,000 patients with prostate cancer in an Irish registry. This group found that compared to men who were either single, divorced, or widowed, married men were 20% more likely to survive, even after adjusting for cofounding variables, including stage and Gleason grade.[37] Interestingly, marital status may not predict for discontinuing active surveillance in other countries. Loeb et al. examined over 11,000 patients in the national prostate cancer registry of Sweden and found that marital status did not predict for discontinuation.[38]

Clearly, the observation that marriage is generally considered protective in the diagnosis, treatment, and survival of patients with all types cancer is not new. The seminal work of Goodwin et al. in 1987 is generally thought of as an early study that specified explanations for survival.[39] This group used a large dataset of almost 28,000 patients and showed that unmarried persons with cancer had a 1.23-fold higher risk of overall mortality. Unmarried persons tended to be diagnosed at a later stage and were found less likely to be treated for cancer after diagnosis. Interestingly, even after adjusting for stage and despite the presence of treatment, unmarried patients still had inferior survival. Of interest, for most ages, both men and women benefit in terms of survival by being married, although men seemed to derive more benefit than women, as they got older, for certain cancers.[40]

## THE IMPACT OF LITERACY AND NUMERACY

Health literacy, defined as "the degree to which people have the capacity to obtain, process, and understand basic health information and services needed to make appropriate health decisions,"[41] has received increasing attention as a problem in the United States, as low literacy has a significant impact on the ability of patients to navigate the increasingly complex medical system. The 2003 National Assessment of Adult Literacy (NAAL) concluded that nearly 30 million Americans had below basic health literacy.[42] This means millions of patients cannot read a prescription label that instructs patients to take 1 tablet twice a day. Furthermore, the median reading level of Americans was the seventh grade. A more recent study is the 23-country Program for the International Assessment of Adult Competencies (PIAAC), known as the International Survey of Adult Skills (ISAS) in the United States. Data from 2013 indicates that one in six Americans, or 36 million adults, lack basic skills.[43] Perhaps most significantly, the majority of those who perform poorly on literacy assessments were the young adults.

Numeracy is a separate concept defined as the ability to access, process, interpret, communicate, and act on numerical, quantitative, graphical, biostatistical, and probabilistic health information needed to make effective health decisions.[44,45] Therefore, innumerate patients are more likely to make poorly informed treatment decisions, participate less in their care, or misinterpret risk of disease.[46,47] There are no robust data regarding the prevalence of innumeracy, but the relationship of numeracy with literacy likely indicates that this is a relatively common phenomenon.

Two different forces currently affecting the US healthcare system magnify the problem of widespread low health literacy and numeracy, particularly in prostate cancer. First, because of declining patient reimbursements, many physicians and other healthcare providers see more patients in a shorter period of time, thus limiting the amount of time to assess just how much the patient has understood about their prostate cancer diagnosis. Second, there is emphasis on shared decision-making between the doctor and the patient, moving away from a model where the physician would make treatment choices for the patient. For patients with prostate cancer who are faced with a multitude of management options, a high degree of sophistication is required to understand the nuanced details of differing options. There is a surprising paucity of data surrounding literacy and prostate cancer. For example, in a PubMed search of, "prostate AND cancer AND literacy," only 62 articles were found. Even fewer were found with numeracy as a search term.

In one of the earliest large studies examining the association of patient educational level involving 3484 patients, Kane et al. concluded that education was an important factor predictive of treatment selection. For example, men above the age of 75 with a higher educational level were more likely to choose RT and less likely to choose other forms of therapy, including WW. The choice of surgery for prostate cancer was magnified especially among AA patients, with less than 40% choosing RP if they had less than a high school education, as compared with nearly 80% of AA men choosing RP if they had postcollegiate education.[48] Tobias-Machado et al. examined the association between three factors: literacy, compliance with prostate cancer screening, and cancer aggressiveness. Of the over 17,000 men studied, illiterate men were less likely to come for further evaluation after having been told they had an abnormal digital rectal examination (DRE), which may have significant bearing on future health. An associated finding was that illiterate men presented with higher PSA levels and higher stage, and were even 67% more likely to have Gleason 8–10 disease.[49]

Finally, it is vital to reassess both sides of the physician and patient understanding of literacy and numeracy. It is assumed that modern training results in sequentially more numerate physicians; however, this may not be true. In a multi-institutional study of over 300 medical students exposed to classes in statistics, one in three were not numerate, a number that has generally been observed among other medical trainees as well.[50,51] When faced with complex decision-making, in this case choosing between nine different Medicare Part D plans, highly educated physicians fared no better than their older patients.[52]

It is likely that discussions of document literacy and numeracy may seem disconnected from actual patient care. However, recent data has spotlighted the actual terminology used frequently by physicians when discussing prostate cancer. Terms that make up the urinary, sexual, and bowel domains are frequently used with the implicit sense that patients completely understand what these terms mean. Kilbridge et al. first examined this question, finding that the vast majority of prostate cancer terms used by healthcare providers were not understood by a group of low-literacy patients.[53] For example, only 5% of patients understood the term, "incontinence." Subsequently, Wang et al. used the identical methodology on a cohort of more educated patients and found similar findings.[54] In this study, only 15% understood the term "incontinence." Compound terminology such as "urinary urgency" was particularly difficult for patients. If the patient cannot understand the very words used during conversations about prostate cancer, it may be hard for the patient to fully participate in informed and shared decision-making.

The future of the very real barriers to understanding the nuances of prostate cancer may be found in technology, particularly technology which patients and their families or caregivers may be able to access outside of in-room time with a physician. Many investigators have looked at this issue, and found that media approaches that involve more than just written words are the most effective. Kim et al. studied the effect of giving patients a CD-ROM shared decision-making program, and found that more than three-quarters of patients in their Veterans Administration clinic population were very satisfied with this tool.[55] However, importantly, prostate cancer knowledge scores varied among participants after participation in the CD-ROM intervention program, causing the investigators to wonder whether patients were adequately informed to make the best choices regarding their treatment. It appeared that low literacy hindered patient understanding of the shared decision-making program, as they found that lower prostate cancer knowledge scores corresponded to lower literacy scores. Volk used the concept of entertainment education with the use of video "novellas" to explain concepts such as prostate cancer screening to patients, with excellent response rates.[56] For many studies, both in prostate cancer as well as other disease entities, a significant barrier has been the tradeoff between trying to give patients the fullest understanding of the disease process and treatments, and establishing patients' ability to take in all of the information. In a recent study using tablets to deliver combined audio, video, and written description of prostate cancer terms, it was apparent that patient fatigue with lengthy programs impeded understanding.[57] As many patients use the Internet to access sources of information regarding urological conditions and care, several authors have conducted an assessment of educational materials by various urological departments. Essentially, most online urological educational material is written at or above the eleventh grade level.[58] As technology improves, barriers to implementing smartphone/tablet/computer decision-support multimedia tools will decrease, with the possibility of making quick changes as new knowledge is obtained.

## STRATEGIES TO REDUCE DISPARITY IN HIGH-RISK POPULATIONS

While it remains critical to identify disparities in prostate cancer among ethnically and economically disadvantaged populations, it is just as important to address barriers to quality care and improve cultural competence within the healthcare system. Betancourt et al. conducted a systematic review of the literature and identified three areas where cultural competence could be improved at the organizational, structural, and clinical level.[59] This review highlighted the fact that there is a lack of minority representation in leadership positions within medical and public health school faculties and healthcare management organizations, and an under-representation of minority physicians within the healthcare workforce. The lack of minority representation at leadership levels can contribute to an underprioritization of programs necessary to address the cultural needs of minority communities.[60] Efforts to combine the resources of large academic and publicly funded institutions with community-based organizations may prove to be an ideal conduit to improve access and care for underserved populations.

It has been demonstrated that AA and Hispanic physicians are more likely to care for AA and Hispanic patients in medically underserved areas, provide more effective care due to lack of cultural or language barriers, and more likely to care for patients with Medicaid or no insurance.[61,62] However, efforts to increase minority representation in medical and public health schools so that they reflect the general population have been reduced due to legal challenges to previously successful affirmative action programs.[63] The resulting decrease in recruitment of qualified providers from diverse backgrounds has had a direct, negative impact on the most economically vulnerable populations who will continue to have poor outcomes unless institutional diversity is addressed in health policy, health delivery, and medical education systems.[64] Despite the obstacles to diversify the medical trainee population, there are some programs that have been successfully implemented to increase the number of graduates who go on to serve medically underserved populations.[65,66] Efforts to recruit a diverse cadre of physicians interested in providing healthcare to economically disadvantaged populations should be supported.

Consistent access to subspecialty care and information about prostate cancer risk and treatment options are also important areas that need to be addressed, particularly for men in lower sociodemographic strata. Physician distrust is an important factor in patient decision-making and potentially can alter a patient's receipt of care or even follow-up for treatment.[67,68] In a study from the North Carolina Cancer Registry, AA and Caucasian men recently diagnosed with prostate cancer were surveyed regarding attitudes and beliefs regarding prostate cancer screening and treatment, access to care, and physician trust.[68] The authors showed that AA men were more likely to be uninsured and not possess Medicare supplementary insurance, and more likely to utilize a public clinic or hospital emergency room for routine care. This reduction in access to care resulted in lower receipt of timely prostate cancer screening tests, self-reported delay in care, and physician mistrust. These findings were not necessarily due to AA men not acknowledging their risk of prostate cancer, as most of them admitted that they knew AA men are at higher risk of prostate cancer diagnosis and were more likely to agree that traditional treatment was better than alternative therapies. The issue of physician trust is very important, as AA men are significantly less likely to receive information about treatment options and more likely to experience treatment regret.[69] AA men who trust their physician are more likely to undergo prostate cancer screening; and, with a combination of more information about prostate cancer and involvement of family and friends, men can make more informed screening and treatment decisions.[70] Improving access to quality care for economically disadvantaged men with prostate cancer can be accomplished, as has been demonstrated by the IMProving Access, Counseling and Treatment for Californians with Prostate Cancer (IMPACT) program, and further efforts both on the state and national levels should be encouraged.

## References


1. Siegel R, Ward E, Brawley O, et al. Cancer statistics, 2011: the impact of eliminating socioeconomic and racial disparities on premature cancer deaths. *CA Cancer J Clin* 2011;**61**(4):212–36.
2. Siegel RL, Miller KD, Jemal A. Cancer statistics. *CA Cancer J Clin* 2015;**65**(1):5–29.
3. Aizer AA, Wilhite TJ, Chen MH, et al. Lack of reduction in racial disparities in cancer-specific mortality over a 20-year period. *Cancer* 2014;**120**(10):1532–9.
4. Karami S, Young HA, Henson DE. Earlier age at diagnosis: another dimension in cancer disparity? *Cancer Detect Prev* 2007;**31**(1): 29–34.
5. Powell IJ, Bock CH, Ruterbusch JJ, et al. Evidence supports a faster growth rate and/or earlier transformation to clinically significant prostate cancer in black than in white American men, and influences racial progression and mortality disparity. *J Urol* 2010;**183**(5):1792–6.
6. Underwood III W, Jackson J, Wei JT, et al. Racial treatment trends in localized/regional prostate carcinoma: 1992–1999. *Cancer* 2005;**103**(3):538–45.
7. Moses KA, Paciorek AT, Penson DF, et al. Impact of ethnicity on primary treatment choice and mortality in men with prostate cancer: data from CaPSURE. *J Clin Oncol* 2010;**28**(6):1069–74.
8. Taksler GB, Keating NL, Cutler DM. Explaining racial differences in prostate cancer mortality. *Cancer* 2012;**118**(17):4280–9.
9. Schapira MM, McAuliffe TL, Nattinger AB. Treatment of localized prostate cancer in African-American compared with Caucasian men. Less use of aggressive therapy for comparable disease. *Med Care* 1995;**33**(11):1079–88.
10. Harlan LC, Potosky A, Gilliland FD, et al. Factors associated with initial therapy for clinically localized prostate cancer: prostate cancer outcomes study. *J Natl Cancer Inst* 2001;**93**(24):1864–71.
11. Shavers VL, Brown ML, Potosky AL, et al. Race/ethnicity and the receipt of watchful waiting for the initial management of prostate cancer. *J Gen Intern Med* 2004;**19**(2):146–55.
12. Godley PA, Schenck AP, Amamoo MA, et al. Racial differences in mortality among Medicare recipients after treatment for localized prostate cancer. *J Natl Cancer Inst* 2003;**95**(22):1702–10.
13. Hayn MH, Orom H, Shavers VL, et al. Racial/ethnic differences in receipt of pelvic lymph node dissection among men with localized/regional prostate cancer. *Cancer* 2011;**117**(20):4651–8.
14. Shavers VL, Brown M, Klabunde CN, et al. Race/ethnicity and the intensity of medical monitoring under 'watchful waiting' for prostate cancer. *Med Care* 2004;**42**(3):239–50.
15. Zeliadt SB, Potosky AL, Etzioni R, et al. Racial disparity in primary and adjuvant treatment for nonmetastatic prostate cancer: SEER-Medicare trends 1991–1999. *Urology* 2004;**64**(6):1171–6.
16. Underwood W, De Monner S, Ubel P, et al. Racial/ethnic disparities in the treatment of localized/regional prostate cancer. *J Urol* 2004;**171**(4):1504–7.
17. Presley CJ, Raldow AC, Cramer LD, et al. A new approach to understanding racial disparities in prostate cancer treatment. *J Geriatr Oncol* 2013;**4**(1):1–8.
18. Nielsen ME, Han M, Mangold L, et al. Black race does not independently predict adverse outcome following radical retropubic prostatectomy at a tertiary referral center. *J Urol* 2006;**176**(2):515–9.
19. Tewari A, Horninger W, Badani KK, et al. Racial differences in serum prostate-specific antigen (PSA) doubling time, histopathological variables and long-term PSA recurrence between African-American and white American men undergoing radical prostatectomy for clinically localized prostate cancer. *BJU Int* 2005;**96**(1):29–33.
20. Freedland SJ, Amling CL, Dorey F, et al. Race as an outcome predictor after radical prostatectomy: results from the Shared Equal Access Regional Cancer Hospital (SEARCH) database. *Urology* 2002;**60**(4):670–4.
21. Schwartz K, Powell IJ, Underwood III W, et al. Interplay of race, socioeconomic status, and treatment on survival of patients with prostate cancer. *Urology* 2009;**74**(6):1296–302.
22. Bach PB, Schrag D, Brawley OW, et al. Survival of blacks and whites after a cancer diagnosis. *JAMA* 2002;**287**(16):2106–13.
23. Haiman CA, Chen GK, Blot WJ, et al. Genome-wide association study of prostate cancer in men of African ancestry identifies a susceptibility locus at 17q21. *Nat Genet* 2011;**43**(6):570–3.
24. Baffoe-Bonnie AB, Kittles RA, Gillanders E, et al. Genome-wide linkage of 77 families from the African American Hereditary Prostate Cancer study (AAHPC). *Prostate* 2007;**67**(1):22–31.
25. Chang BL, Hughes L, Chen DY, et al. Validation of association of genetic variants at 10q with prostate-specific antigen (PSA) levels in men at high risk for prostate cancer. *BJU Int* 2014;**113**(5b): E150–6.
26. Barros-Silva JD, Ribeiro FR, Rodrigues A, et al. Relative 8q gain predicts disease-specific survival irrespective of the *TMPRSS2-ERG* fusion status in diagnostic biopsies of prostate cancer. *Genes Chromosomes Cancer* 2011;**50**(8):662–71.
27. Beuten J, Gelfond JA, Martinez-Fierro ML, et al. Association of chromosome 8q variants with prostate cancer risk in Caucasian and Hispanic men. *Carcinogenesis* 2009;**30**(8):1372–9.
28. Hui J, Xu Y, Yang K, et al. Study of genetic variants of 8q21 and 8q24 associated with prostate cancer in Jing-Jin residents in northern China. *Clin Lab* 2014;**60**(4):645–52.
29. Zeegers MP, Nekeman D, Khan HS, et al. Prostate cancer susceptibility genes on 8p21-23 in a Dutch population. *Prostate Cancer Prostatic Dis* 2013;**16**(3):248–53.
30. Xu J, Lange EM, Lu L, et al. *HOXB13* is a susceptibility gene for prostate cancer: results from the International Consortium for Prostate Cancer Genetics (ICPCG). *Hum Genet* 2013;**132**(1):5–14.
31. Lotan TL, Carvalho FL, Peskoe SB, et al. *PTEN* loss is associated with upgrading of prostate cancer from biopsy to radical prostatectomy. *Mod Pathol* 2014;**28**(1):128–37.
32. Magi-Galluzzi C, Tsusuki T, Elson P, et al. *TMPRSS2-ERG* gene fusion prevalence and class are significantly different in prostate cancer of Caucasian, African-American and Japanese patients. *Prostate* 2011;**71**(5):489–97.
33. Rand AE, Agarwal A, Ahuja D, et al. Patient demographic characteristics and disease stage as drivers of disparities in mortality in

prostate cancer patients who receive care at a safety net academic medical center. *Clin Genitourin Cancer* 2014;**12**(6):455–60.

34. Rusthoven CG, Waxweiler TV, DeWitt PE, et al. Gleason stratifications prognostic for survival in men receiving definitive external beam radiation therapy for localized prostate cancer. *Urol Oncol* 2015;**33**(2):e11–9.
35. Forsythe LP, Alfano CM, Kent EE, et al. Social support, self-efficacy for decision-making, and follow-up care use in long-term cancer survivors. *Psychooncology* 2014;**23**(7):788–96.
36. Sprehn GC, Chambers JE, Saykin AJ, et al. Decreased cancer survival in individuals separated at time of diagnosis: critical period for cancer pathophysiology? *Cancer* 2009;**115**(21):5108–16.
37. Burns RM, Sharp L, Sullivan FJ, et al. Factors driving inequality in prostate cancer survival: a population-based study. *PLoS ONE* 2014;**9**(9):e106456.
38. Loeb S, Folkvaljon Y, Makarov DV, et al. Five-year nationwide follow-up study of active surveillance for prostate cancer. *Eur Urol* 2015;**67**(2):233–8.
39. Goodwin JS, Hunt WC, Key CR, et al. The effect of marital status on stage, treatment, and survival of cancer patients. *JAMA* 1987;**258**(21):3125–30.
40. Wang L, Wilson SE, Stewart DB, et al. Marital status and colon cancer outcomes in US Surveillance, Epidemiology and End Results registries: does marriage affect cancer survival by gender and stage? *Cancer Epidemiol* 2011;**35**(5):417–22.
41. Parker RM, Ratzan SC, Lurie N. Health literacy: a policy challenge for advancing high-quality health care. *Health Aff (Millwood)* 2003;**22**(4):147–53.
42. NCES.http://nces.ed.gov/surveys/piaac/index.asp; 2014 [accessed 9.2.2014].
43. OECD. http://www.oecd.org/site/piaac/surveyofadultskills.htm; 2014 [accessed 9.1.2014].
44. Donelle L, Hoffman-Goetz L, Arocha JF. Assessing health numeracy among community-dwelling older adults. *J Health Commun* 2007;**12**(7):651–65.
45. Golbeck AL, Ahlers-Schmidt CR, Paschal AM, et al. A definition and operational framework for health numeracy. *Am J Prev Med* 2005;**29**(4):375–6.
46. Ciampa PJ, Osborn CY, Peterson NB, et al. Patient numeracy, perceptions of provider communication, and colorectal cancer screening utilization. *J Health Commun* 2010;**15**(Suppl 3):157–68.
47. Reyna VF, Nelson WL, Han PK, et al. How numeracy influences risk comprehension and medical decision making. *Psychol Bull* 2009;**135**(6):943–73.
48. Kane CJ, Lubeck DP, Knight SJ, et al. Impact of patient educational level on treatment for patients with prostate cancer: data from CaPSURE. *Urology* 2003;**62**(6):1035–9.
49. Tobias-Machado M, Carvalhal GF, Freitas Jr CH, et al. Association between literacy, compliance with prostate cancer screening, and cancer aggressiveness: results from a Brazilian screening study. *Int Braz J Urol* 2013;**39**(3):328–34.
50. Johnson TV, Abbasi A, Schoenberg ED, et al. Numeracy among trainees: are we preparing physicians for evidence-based medicine? *J Surg Educ* 2014;**71**(2):211–5.
51. Sheridan SL, Pignone M. Numeracy and the medical student's ability to interpret data. *Eff Clin Pract* 2002;**5**(1):35–40.
52. Barnes AJ, Hanoch Y, Martynenko M, et al. Physician trainees' decision making and information processing: choice size and Medicare Part D. *PLoS ONE* 2013;**8**(10):e77096.
53. Kilbridge KL, Fraser G, Krahn M, et al. Lack of comprehension of common prostate cancer terms in an underserved population. *J Clin Oncol* 2009;**27**(12):2015–21.
54. Wang DS, Jani AB, Tai CG, et al. Severe lack of comprehension of common prostate health terms among low-income inner-city men. *Cancer* 2013;**119**(17):3204–11.
55. Kim SP, Knight SJ, Tomori C, et al. Health literacy and shared decision making for prostate cancer patients with low socioeconomic status. *Cancer Invest* 2001;**19**(7):684–91.
56. Volk RJ, Jibaja-Weiss ML, Hawley ST, et al. Entertainment education for prostate cancer screening: a randomized trial among primary care patients with low health literacy. *Patient Educ Couns* 2008;**73**(3):482–9.
57. Wang DS. Video-based educational tool improves patient comprehension of common prostate health terminology. *Cancer* 2015;**121**(3):733–40.
58. Colaco M, Svider PF, Agarwal N, et al. Readability assessment of online urology patient education materials. *J Urol* 2013;**189**(3):1048–52.
59. Betancourt JR, Green AR, Carrillo JE, et al. Defining cultural competence: a practical framework for addressing racial/ethnic disparities in health and health care. *Public Health Rep* 2003;**118**(4):293–302.
60. Nickens HW. The rationale for minority-targeted programs in medicine in the 1990s. *JAMA* 1992;**267**(17):2390 5.
61. Komaromy M, Grumbach K, Drake M, et al. The role of black and Hispanic physicians in providing health care for underserved populations. *N Engl J Med* 1996;**334**(20):1305–10.
62. Walker KO, Moreno G, Grumbach K. The association among specialty, race, ethnicity, and practice location among California physicians in diverse specialties. *J Natl Med Assoc* 2012;**104**(1-2):46–52.
63. Saha S, Taggart SH, Komaromy M, et al. Do patients choose physicians of their own race? *Health Aff (Millwood)* 2000;**19**(4):76–83.
64. Byrd WM. Race, biology, and health care: reassessing a relationship. *J Health Care Poor Underserved* 1990;**1**(3):278–96.
65. Ko M, Heslin KC, Edelstein RA, et al. The role of medical education in reducing health care disparities: the first ten years of the UCLA/Drew Medical Education Program. *J Gen Intern Med* 2007;**22**(5):625–31.
66. Rabinowitz HK, Diamond JJ, Markham FW, et al. Critical factors for designing programs to increase the supply and retention of rural primary care physicians. *JAMA* 2001;**286**(9):1041–8.
67. Xu J, Dailey RK, Eggly S, et al. Men's perspectives on selecting their prostate cancer treatment. *J Natl Med Assoc* 2011;**103**(6):468–78.
68. Talcott JA, Spain P, Clark JA, et al. Hidden barriers between knowledge and behavior: the North Carolina prostate cancer screening and treatment experience. *Cancer* 2007;**109**(8):1599–606.
69. Hu JC, Kwan L, Krupski TL, et al. Determinants of treatment regret in low-income, uninsured men with prostate cancer. *Urology* 2008;**72**(6):1274–9.
70. Jones RA, Steeves R, Williams I. How African American men decide whether or not to get prostate cancer screening. *Cancer Nurs* 2009;**32**(2):166–72.

CHAPTER

# 17

# Hereditary Prostate Cancer

*Casey Lythgoe, MD, Danuta Dynda, MD, Shaheen Alanee, MD, MPH*

Division of Urology, Department of Surgery, Southern Illinois University School of Medicine, Springfield, IL, USA

## INTRODUCTION

The McKusick catalogue (1990) lists 66 Mendelian disorders that involve cancer or a predisposition for cancer as a component of the phenotype.[1,2] The genetic loci of 21 of these 66 traits have been mapped by genetic epidemiological investigations.[1] Prostate cancer has a well-established familial clustering, including both dominant and recessive syndromes similar to those found in breast and ovarian cancer syndrome families. However, extensive research in families affected by prostate cancer has so far failed to identify major genes as risk factors for prostate cancer. More recently, research has been directed to a "common disease common variant" approach employing a very large number of cases and controls in genome-wide association studies (GWASs). These association studies produced a large number of genetic variants that are common in the population, and are associated with prostate cancer, but their effect on prostate cancer risk was only modest. The most recent development in the field is the increasing utilization of high-throughput sequencing technology that is providing us with an unprecedented opportunity to thoroughly investigate the genome at single base resolution, and thus opening the door to understanding the role of genetic variants, common and rare, in the development of prostate cancer.

In this chapter, we summarize the evidence from family-based and GWASs. We also briefly discuss the clinical implications of our developing knowledge of the genetic basis of prostate cancer, and possible future directions for inherited prostate cancer research.

## EPIDEMIOLOGIC STUDIES OF FAMILY HISTORY AND PROSTATE CANCER RISK

Studies of familial clustering of cancer provided the rudimentary basis for the investigation of a heritable component to prostate cancer. Morganti and Woolf were the first to demonstrate an increased risk of prostate cancer and prostate cancer death in relatives as compared to the general population.[3,4] These findings were corroborated by later studies using a database of Utah Mormons.[5] Steinbert and Walsh reported a case-control study to estimate the relative risk of developing prostate cancer in men with a positive family history. They found a 2-, 5-, or 11-fold increase in men with 1, 2, or 3 affected first-degree relatives. This implied that an increased number of affected relatives increased the genetic disposition for prostate cancer.[6] These findings were substantiated in a more contemporary study by Lesko et al., who studied a case-control cohort of 1266 men from Massachusetts in the 1990s. The authors found an odds ratio (OR) of 2.3 in men who had a brother or a father with prostate cancer. The OR was 2.2 for men with a single relative with prostate cancer and 3.9 for men with two or more relatives with prostate cancer.[7]

The relationship between family history of prostate cancer and age of onset was also examined by Lesko et al. who found that the association between family history and the risk of developing prostate cancer was greatest when prostate cancer was diagnosed in a family member younger than age 65 with an OR of 4.1. The OR for that association dropped to 0.76 if the affected family member was 75 years or older at the time of diagnosis

Prostate Cancer. http://dx.doi.org/10.1016/B978-0-12-800077-9.00017-7

establishing an inverse relationship between risk of prostate cancer and age of affected relative at time of diagnosis.[7]

Many studies have attempted to determine if familial prostate cancer is related to biologic aggressiveness or risk of recurrence with mixed results. Kupelian et al. showed 3-year freedom from biochemical recurrence rates of 52.5% versus 72% for patients with and without family history, respectively. The 5-year rates showed a similar trend with biochemical freedom from recurrence of 29% versus 52%. When the authors stratified the cohort into two groups based on PSA <10, Gleason score <7, and clinical stage T1-2, they found that patients with a family history of prostate cancer had a 5-year biochemical recurrence free survival rate of 49% versus 80% for those with no family history. Patients showed a similar difference in biochemical recurrence free survival of 20% versus 35% (with and without family history, respectively).[8]

The results from Kupelian's study hinted to the possibility of increased aggressiveness of familial prostate cancer. Bova et al. attempted to characterize familial prostate cancer aggressiveness by matching a cohort of men who met the criteria for HPC to others by Gleason score, pathologic stage, and time interval of prostatectomy. The cohort was followed for 5 years and did not show a significant difference in the likelihood of freedom from biochemical recurrence.[9]

In an effort to quantify the survival of patients with familial prostate cancer, Siddiqui et al. examined a cohort of 3560 patients who underwent prostatectomy for cancer from 1987 to 1997. A total of 865 patients were found to have either familial or hereditary prostate cancer. The authors found that although preoperative PSA was higher in patients with hereditary prostate cancer, no significant difference was found in the pathological stage, Gleason score, seminal vesicle involvement, margin positivity, or DNA ploidy. In addition, postoperative adjuvant therapy was equivalent in all groups. Importantly, the 10-year data from this cohort showed no difference in biochemical recurrence free survival, systemic progression free survival, or cancer-specific survival.[10]

One major limitation of familial studies in prostate cancer is the potential for detection bias. Almost all of the authorities on screening for prostate cancer, including the American Urologic Association, recommend early screening for prostate cancer in males with family history of the disease. In addition, men with family history of prostate cancer have been shown in previous studies to be more likely to ask for prostate cancer screening.[11] This higher prevalence of screening in males with family history of prostate cancer would obviously lead to a higher detection bias. A familial history of prostate cancer could also indicate a shared environmental exposure among members of the same family. However, twin studies were able to isolate genetic from environmental risk factors by comparing the risk of prostate cancer between monozygotic (MZ) and dizygotic (DZ) twins. Grönberg et al. studied 4840 male twin pairs, in which 458 prostate cancers were identified between 1959 and 1989. Among these, 16 MZ and 6 DZ twin pairs were concordant for prostate cancer. The rate of concordance for MZ versus DZ, respectively was 0.192 and 0.043, and the correlation of liability for the twin pairs was found to be 0.40 and −0.05. The authors concluded that the differences in concordance rates between the two groups were notable and implicate genetic factors in the development of prostate cancer.[12] Along the same lines, a study of World War II veteran twins examined concordance rates to estimate the risk of a twin developing cancer given that the other twin had cancer. This study found that MZ versus DZ twins had a concordance rate of 27.1% versus 7.1%, which calculated to a risk ratio of 3.83. They concluded that concordance rates were higher in MZ pairs and estimated 50% of the variability in liability in this prostate cancer cohort was due to genetics.[13]

## SEGREGATION AND LINKAGE STUDIES

Twin and familial studies of prostate cancer were the impetus for further investigation into the possibility of a genetic cause of prostate cancer. Once a genetic cause is suspected, a segregation analysis can be performed to determine if a Mendelian mode of inheritance exists. Mendelian inheritance denotes the possibility of a specific gene involved in prostate cancer, a gene that could be theoretically shared by familial and sporadic forms of prostate cancer. While some of the early segregation studies have suggested a possible autosomal dominant transmission of a major rare risk factor for prostate cancer,[14–19] other studies suggest more of a recessive, or polygenic, genetic basis for this disease.[20]

With the distinct possibility of a genetic component indicated by segregation analyses, the next step in further qualifying the genetic causes for prostate cancer was to perform linkage analyses to specific genetic landmarks, and measure the extent of the linkage. The results from these studies were given in terms of a logarithm of odds (LOD) score (logarithm of the OR in favor of linkage vs. no linkage) as a metric for the likelihood of linkage.[21–23] Table 17.1 demonstrates many of the contemporary studies attempting to establish genetic linkages to prostate cancer in different populations.

While many putative prostate cancer genes have been found and examined, evidence for the elusive high-risk allele is still inconclusive. The absence of a major finding from these linkage studies is strong evidence to the fact that prostate cancer risk is determined by a group of genes with small or modest risk effects. Having said that,

**TABLE 17.1** Published Results of Linkage Analysis Studies of Heritable Prostate Cancer Risk

| Chromosomes | LOD | References |
|---|---|---|
| 1q24-25, 1p36, 1p | 3.1–4.74 | [24–29] |
| 3p24, 3p25-26, 3q26, 3q21 | 2.37–3.39 | [25,30–32] |
| 4q12, 4q35 | 2.28, 2.56 | [25,33] |
| 5q34, 5q35, 5q11.2, 5q22, 5p13, 5q23 | 1.96–5.94 | [25,26,29,34–36] |
| 6p22.3, 6q21 | 2.51, 1.87 | [37,38] |
| 7q21, 7p22 | 1.88, 2.47 | [25,26,38] |
| 8q24, 8q34, 8q22 | 2.2–5.66 | [25,32,35] |
| 10q13, 10q14 | 3.36, 6.12 | [30,35] |
| 11q14, 11q24 | 1.42, 2.27–5.69 | [25,31,35] |
| 12q11, 12q13-14 | 2.1, 1.99 | [25,38] |
| 13q12 | 4.76 | [39] |
| 15q11, 15q12, 15q13, 15q23 | 1.88–5.57 | [25,33,39,40] |
| 16q21, 16q22, 16p13.3 | 1.99–3.16 | [25,33,34] |
| 17, 17q22 | 3.16–4.77 | [33,41] |
| 19q12, 19p13, 19q13 | 1.95–5.15 | [25,35,42] |
| 20q, 20q12 | 2.3, 2.42 | [25,29] |

the research for a major gene in prostate cancer should continue in families and homogeneous populations as previous research in similar groups have provided us with keys to understanding the biology of important cancers (Von Hippel–Lindau in kidney cancer, and the *MSH2* gene in colon cancer).

## GENOME-WIDE ASSOCIATION STUDIES IN PROSTATE CANCER

The development of commercially available and inexpensive arrays capable of genotyping up to one million single nucleotide polymorphisms (SNPs) in parallel, in combination with the cataloging human SNPs through the Human Genome Project and the International HapMap Project, has made it possible for prostate cancer research to move from familial and linkage studies to investigating the entire human genome for genetic variants associated with prostate cancer risk. These GWASs are based on the concept of linkage disequilibrium, which occurs when a certain allele combination (haplotype) is preserved over subsequent generations. What follows is that the likelihood of recombination between two alleles is a function of the genetic distance between them. Thus, tagging SNPs that are correlated with neighboring SNPs can be used to infer the human haplotype. This linkage phenomenon would therefore allow us to examine genetic variants in the entire genome without the need to tag every single SNP. These haplotypes are then compared between cases and controls, and are proving to be a very powerful tool to identify genetic variants associated with different diseases.

The large number of GWAS performed in prostate cancer research since 2007 is a testimony to the high success rates of such studies in discovering genetic variants associated with prostate cancer.[43] The first GWAS in prostate cancer revealed a risk locus at 8q24, rs6983267 to be associated with prostate cancer.[44] Ever since, GWAS have identified nearly 100 common variants associated with susceptibility to prostate cancer.[45] Not only are these GWAS carried out to investigate the association of genetic changes with prostate cancer, but they are also helping to understand the genetic basis for disease aggressiveness, treatment complications, and outcomes. Herein, we summarize the findings of GWAS and examine the functional relevance of some of the SNPs that have been reported in multiple studies. This discussion is not all-inclusive of the studied SNPs, and the interested reader should review the original studies dealing with the subject.

### 8q24

Multiple GWAS and replication studies have confirmed the association of the chromosome 8q24 region with prostate cancer.[46–52] Two SNPs, rs1447295 and rs6983267 are the most frequently replicated SNP on 8q24 to be associated with prostate cancer risk.[46–49,51] This region of the genome, 8q24, is a 600-kbp region that was originally thought to be a gene desert. However, there is now some evidence that *POU5F1P1*, originally thought to be a pseudogene within 8q24, can encode for a weak transcriptional activator that plays a role in carcinogenesis.[53] Recent studies have also shown 8q24 to possibly encode for enhancers of *MYC*,[54,55] and an 8q24 SNP (rs6983267), which lies within the region of the *MYC* enhancer, was found to be associated with increased risk of prostate cancer metastasis.[56]

### HNF1B

The region of 17q12 has two SNPs (rs4430796 and rs7501939) that were reported in GWAS,[46,47,49–51] and confirmed in replication studies[57,58] to be associated with prostate cancer risk. Both of these SNPs are located in the intronic regions of *HNF1B*, which encodes for a transcription factor (TCF2), and whose role in prostate cancer pathways is yet to be determined.

### JAZF1

One SNP rs10486567 within intron 2 of *JAZF1* (juxtaposed with another zinc finger 1) was reported to be associated with prostate cancer risk and aggressiveness in preliminary studies.[48] However, this SNP was confirmed to be associated with risk, but not aggressiveness, in follow-up replication studies in males of European and African descent.[57,59] *JAZF1* variants were also shown to be associated with diabetes and human height suggesting that this gene may explain some of the epidemiologic association of prostate cancer with metabolic syndrome.[57]

### KLK2–3

One SNP (rs2735839), located between *KLK2* and *KLK3*, was found to be associated with prostate cancer in a GWAS by Eeles et al. in 2008[47] and was confirmed in replication studies.[60,61] *KLK3* encodes PSA protein, and SNPs in proximity to this gene have been associated with aggressive prostate cancer risk.[62] *KLK2* encodes kallikrein-related peptidase, another biomarker that is associated with prostate cancer risk.[63]

### LILRA3

One SNP (rs103294) on 19q13.4 was found to be associated with prostate cancer risk in Han Chinese subjects.[64] This SNP is in strong linkage equilibrium with a 6.7-kb germline deletion that removes the first six of seven exons in leukocyte immunoglobulin-like receptor A3 (*LILRA3*), which is a gene that regulates inflammatory response. With the increased evidence on the role of inflammation in prostate cancer, further research is needed to elucidate the role of *LILRA3* in prostate cancer.

### MSMB

There is one SNP (rs10993994) in the promoter of *MSMB* (β-microseminoprotein) that has been shown to be associated with prostate cancer in multiple GWAS.[47,48,50,51] *MSMB* encodes a protein (PSP94) whose expression has been shown to be affected by the variant allele of rs10993994. This same SNP has also been suggested to affect the mRNA expression levels of the nearby gene, NCOA4, whose product is a coactivator of the androgen receptor transcriptional activity.[65,66] Further studies of the prostate cancer associated T allele of rs10993994 found it to be associated with the increased risk of metastatic prostate cancer.[56]

### TMPRSS2

In a recent meta-analysis of 87,040 individuals that identified multiple new loci associated with prostate cancer, one new SNP, rs1041449 on chromosome 21q22, was determined to be a significant predictor of the risk of this disease.[45] This SNP is situated 20 kb 5′ to the *TMPRSS2* gene. Expression of *TMPRSS2* is highly specific to prostate tissue, and chromosomal translocation resulting in fusion of the *TMPRSS2* promoter–enhancer region with the ETS transcription factors *ERG* and *ETV1* are frequently observed in prostate cancer.[67] The authors of the meta-analysis found no association between the risk allele of rs1041449 and *TMPRSS2–ERG* fusion status when analyzing data of 552 tumors characterized for such fusion. They also found little evidence that this variant influences *TMPRSS2* expression in prostate tumors or in normal prostate tissue.[45]

### SNPs Predicting Aggressiveness of Prostate Cancer

Multiple SNPs (rs10788165, rs10749408, and rs10788165 on 10q26; rs4775302 and rs1994198 on 15q21; rs4054823; and rs11672691) were recently found to predict aggressive prostate cancer.[68–70] While none of these SNPs is functionally annotated, they represent an interesting cohort of genetic markers for further research since more emphasis is directed toward treating only aggressive forms of prostate cancer and watching low-risk indolent ones.

## NEXT-GENERATION SEQUENCING IN PROSTATE CANCER

As previously mentioned, familial and linkage studies have provided inconsistent results. In contrast, GWAS was successful in the identification of approximately 100 SNPs that are associated with prostate cancer. However, the magnitude of risk attributed to each variant is low, and these SNPs, in aggregate, account for only 33% of the familial risk.[45] Next-generation sequencing technologies have provided scientists with the tools to examine large genomic segments in a rapid and comprehensive manner, and families with strong history of prostate cancer are probably enriched in such variants, and have been shown to be very valuable in fine mapping studies. The bioinformatics limitations in dealing with 3 billion variants are immense, but they are slowly being resolved to unlock the great potential of next-generation sequencing in the field of cancer genetics. One recent example on the success of next-generation sequencing is the identification of HOXB13 G84E mutation as a risk factor for prostate cancer. In a recent study, Ewing et al. sequenced 202 genes in the 17q21-22 region, implicated in prostate cancer susceptibility, in germline DNA of 94 unrelated familial prostate cancer patients selected for linkage to the candidate region.[71] Probands from four families were heterozygous for a rare missense mutation (G84E) in *HOXB13*. In the replication phase, this mutation was found in 72 (1.4%) of the 5083 men with prostate cancer but only 1 (0.1%) of the 1401 control subjects (0.1%) ($P = 8.5 \times 10^{-7}$). Furthermore, the mutation was significantly more prevalent in men with early-onset, familial prostate cancer (3.1%) than in those with late-onset disease and no family history of prostate cancer (0.6%). Subsequent studies by other groups have also reported the prevalence of G84E mutation in familial prostate cancer patients. In a study cohort consisting of 928 familial prostate cancer patients and 930 controls, Breyer et al. found the G84E mutation in 1.9% of probands, with a 2.7% occurrence in patients with more than three affected family members.[72] Akbari et al. performed a case-control study, sequencing germline DNA from peripheral leukocytes of 1843 men diagnosed with prostate cancer and 2225 controls to identify possible mutations in *HOXB13*. The mutation was more prevalent in Caucasian men than in ethnically matched controls (0.7% vs. 0.1%, $P = 0.01$).[73] Finally, Karlsson et al. genotyped samples from two population-based, Swedish, case-control studies and found the prevalence of the G84E mutation to be more than 1% in the general Swedish population, which is higher than we and others have reported for controls (around 0.1%).[74] As of now, there is no conclusive evidence to the association of *HOXB13* mutations with more aggressive prostate cancer.

## DIRECT TO CONSUMER GENETIC TESTING

The identification of genetic markers associated with prostate cancer risk has made it possible to provide direct consumer genetic testing. Multiple online websites are now available where men with a family history of prostate cancer can upload their information and mail-in saliva to be tested for germline variants associated with prostate cancer risk. While the overall results of such commercial testing are usually accurate, the clinical significance remains questionable for many reasons. To start, there is no effective chemopreventive strategy in prostate cancer. The prostate cancer prevention trial did show reduction in prostate cancer risk in patients treated with 5-α reductase inhibitors, but there was an increased risk of more aggressive prostate cancer in the treatment group.[75] Similarly, the Selenium and Vitamine E Cancer Prevention Trial showed possible harmful effect from the prevention intervention, and was closed early because of futility.[76] There is also an absence of safe prophylactic treatments in the world of prostate cancer. Surgery, radiation therapy, and hormonal ablation have well-known side effects that could permanently alter the patient's quality of life for an uncertain benefit. However, the door remains open for more direct intervention as we continue to study the genetics of prostate cancer, and may find genetic variants that are strongly associated with aggressive forms of the disease similar to breast cancer associated-*BRCA* genetic variants.

## CONCLUSIONS

A major risk gene for prostate cancer remains elusive. Familial and linkage studies have so far provided inconclusive results. GWAS have uncovered a large number of low- to modest-effect genetic variants, but their association with prostate cancer aggressiveness is unproven, which limits their clinical utility. More advanced techniques, like next-generation sequencing, hold the promise for better understanding of the genetic bases of prostate cancer. These techniques, with epigenetic studies and better understanding of the gene-environment interactions, are hot topics for future prostate cancer research.

## References


1. McKusick VA. *Mendelian inheritance in man: catalogs of autosomal dominant, autosomal recessive, and x-linked phenotypes*. 9th ed. Baltimore: Johns Hopkins University Press; 1990.
2. Carter BS, Bova GS, Beaty TH, et al. Hereditary prostate cancer: epidemiologic and clinical features. *J Urol* 1993;**150**(3):797–802.
3. Morganti G, Gianferrari L, Cresseri A, et al. Clinico-statistical and genetic research on neoplasms of the prostate. *Acta Genet Stat Med* 1956;**6**(2):304–5.
4. Woolf CM. An investigation of the familial aspects of carcinoma of the prostate. *Cancer* 1960;**13**:739–44.
5. Bishop DT, Skolnick MH. Genetic epidemiology of cancer in Utah genealogies: a prelude to the molecular genetics of common cancers. *J Cell Physiol Suppl* 1984;**3**:63–77.
6. Steinberg GD, Carter BS, Beaty TH, et al. Family history and the risk of prostate cancer. *Prostate* 1990;**17**(4):337–47.
7. Lesko SM, Rosenberg L, Shapiro S. Family history and prostate cancer risk. *Am J Epidemiol* 1996;**144**(11):1041–7.
8. Kupelian PA, Kupelian VA, Witte JS, et al. Family history of prostate cancer in patients with localized prostate cancer: an independent predictor of treatment outcome. *J Clin Oncol* 1997;**15**(4):1478–80.
9. Bova GS, Partin AW, Isaacs SD, et al. Biological aggressiveness of hereditary prostate cancer: long-term evaluation following radical prostatectomy. *J Urol* 1998;**160**(3 Pt. 1):660–3.
10. Siddiqui SA, Sengupta S, Slezak JM, et al. Impact of familial and hereditary prostate cancer on cancer specific survival after radical retropubic prostatectomy. *J Urol* 2006;**176**(3):1118–21.
11. Wallner LP, Sarma AV, Lieber MM, et al. Psychosocial factors associated with an increased frequency of prostate cancer screening in men ages 40–79 years: the Olmsted County study. *Cancer Epidemiol Biomarkers Prev* 2008;**17**(12):3588–92.
12. Grönberg H, Damber L, Damber JE. Studies of genetic factors in prostate cancer in a twin population. *J Urol* 1994;**152**(5 Pt. 1):1484–7 discussion 1487–1489.
13. Page WF, Braun MM, Partin AW, et al. Heredity and prostate cancer: a study of World War II veteran twins. *Prostate* 1997;**33**(4):240–5.
14. Grönberg H, Damber L, Damber JE, et al. Segregation analysis of prostate cancer in Sweden: support for dominant inheritance. *Am J Epidemiol* 1997;**146**(7):552–7.
15. Schaid DJ, McDonnell SK, Blute ML, et al. Evidence for autosomal dominant inheritance of prostate cancer. *Am J Hum Genet* 1998;**62**(6):1425–38.
16. Cui J, Staples MP, Hopper JL, et al. Segregation analyses of 1,476 population-based Australian families affected by prostate cancer. *Am J Hum Genet* 2001;**68**(5):1207–18.

17. Gong G, Oakley-Girvan I, Wu AH, et al. Segregation analysis of prostate cancer in 1,719 white, African-American and Asian-American families in the United States and Canada. *Cancer Cause Control* 2002;**13**(5):471–82.
18. Baffoe-Bonnie AB, Kiemeney LA, Beaty TH, et al. Segregation analysis of 389 Icelandic pedigrees with breast and prostate cancer. *Genet Epidemiol* 2002;**23**(4):349–63.
19. Valeri A, Briollais L, Azzouzi R, et al. Segregation analysis of prostate cancer in France: evidence for autosomal dominant inheritance and residual brother–brother dependence. *Ann Hum Genet* 2003;**67**(Pt. 2):125–37.
20. Pakkanen S, Baffoe-Bonnie AB, Matikainen MP, et al. Segregation analysis of 1,546 prostate cancer families in Finland shows recessive inheritance. *Hum Genet* 2007;**121**(2):257–67.
21. Suarez BK, Lin J, Witte JS, et al. Replication linkage study for prostate cancer susceptibility genes. *Prostate* 2000;**45**(2):106–14.
22. Grönberg H, Smith J, Emanuelsson M, et al. In Swedish families with hereditary prostate cancer, linkage to the HPC1 locus on chromosome 1q24-25 is restricted to families with early-onset prostate cancer. *Am J Hum Genet* 1999;**65**(1):134–40.
23. Cooney KA, McCarthy JD, Lange E, et al. Prostate cancer susceptibility locus on chromosome 1q: a confirmatory study. *J Natl Cancer Inst* 1997;**89**(13):955–9.
24. Smith JR, Freije D, Carpten JD, et al. Major susceptibility locus for prostate cancer on chromosome 1 suggested by a genome-wide search. *Science* 1996;**274**(5291):1371–4.
25. Xu J, Dimitrov L, Chang BL, et al. A combined genome-wide linkage scan of 1,233 families for prostate cancer-susceptibility genes conducted by the international consortium for prostate cancer genetics. *Am J Hum Genet* 2005;**77**(2):219–29.
26. Maier C, Herkommer K, Hoegel J, et al. A genome-wide linkage analysis for prostate cancer susceptibility genes in families from Germany. *Eur J Hum Genet* 2005;**13**(3):352–60.
27. Gibbs M, Stanford JL, McIndoe RA, et al. Evidence for a rare prostate cancer-susceptibility locus at chromosome 1p36. *Am J Hum Genet* 1999;**64**(3):776–87.
28. Matsui H, Suzuki K, Ohtake N, et al. Genome-wide linkage analysis of familial prostate cancer in the Japanese population. *J Hum Genet* 2004;**49**(1):9–15.
29. Camp NJ, Farnham JM, Cannon Albright LA. Genomic search for prostate cancer predisposition loci in Utah pedigrees. *Prostate* 2005;**65**(4):365–74.
30. Schleutker J, Baffoe-Bonnie AB, Gillanders E, et al. Genome-wide scan for linkage in Finnish hereditary prostate cancer (HPC) families identifies novel susceptibility loci at 11q14 and 3p25-26. *Prostate* 2003;**57**(4):280–9.
31. Rökman A, Baffoe-Bonnie AB, Gillanders E, et al. Hereditary prostate cancer in Finland: fine-mapping validates 3p26 as a major predisposition locus. *Hum Genet* 2005;**116**(1–2):43–50.
32. Xu J, Gillanders EM, Isaacs SD, et al. Genome-wide scan for prostate cancer susceptibility genes in the Johns Hopkins hereditary prostate cancer families. *Prostate* 2003;**57**(4):320–5.
33. Gillanders EM, Xu J, Chang BL, et al. Combined genome-wide scan for prostate cancer susceptibility genes. *J Natl Cancer Inst* 2004;**96**(16):1240–7.
34. Wiklund F, Gillanders EM, Albertus J, et al. Genome-wide scan of Swedish families with hereditary prostate cancer: suggestive evidence of linkage at 5q11.2 and 19p13.3. *Prostate* 2003;**57**(4):290–7.
35. Chang BL, Lange EM, Dimitrov L, et al. Two-locus genome-wide linkage scan for prostate cancer susceptibility genes with an interaction effect. *Hum Genet* 2006;**118**(6):716–24.
36. Slager SL, Zarfas KE, Brown WM, et al. Genome-wide linkage scan for prostate cancer aggressiveness loci using families from the University of Michigan Prostate Cancer Genetics Project. *Prostate* 2006;**66**(2):173–9.
37. Janer M, Friedrichsen DM, Stanford JL, et al. Genomic scan of 254 hereditary prostate cancer families. *Prostate* 2003;**57**(4):309–19.
38. Stanford JL, FitzGerald LM, McDonnell SK, et al. Dense genome-wide SNP linkage scan in 301 hereditary prostate cancer families identifies multiple regions with suggestive evidence for linkage. *Hum Mol Genet* 2009;**18**(10):1839–48.
39. Lange EM, Ho LA, Beebe-Dimmer JL, et al. Genome-wide linkage scan for prostate cancer susceptibility genes in men with aggressive disease: significant evidence for linkage at chromosome 15q12. *Hum Genet* 2006;**119**(4):400–7.
40. Lange EM, Beebe-Dimmer JL, Ray AM, et al. Genome-wide linkage scan for prostate cancer susceptibility from the University of Michigan Prostate Cancer Genetics Project: suggestive evidence for linkage at 16q23. *Prostate* 2009;**69**(4):385–91.
41. Cunningham JM, McDonnell SK, Marks A, et al. Genome linkage screen for prostate cancer susceptibility loci: results from the Mayo Clinic Familial Prostate Cancer Study. *Prostate* 2003;**57**(4):335–46.
42. Schaid DJ, Stanford JL, McDonnell SK, et al. Genome-wide linkage scan of prostate cancer Gleason score and confirmation of chromosome 19q. *Hum Genet* 2007;**121**(6):729–35.
43. Juran BD, Lazaridis KN. Genomics in the post-GWAS era. *Semin Liver Dis* 2011;**31**(2):215–22.
44. Yeager M, Orr N, Hayes RB, et al. Genome-wide association study of prostate cancer identifies a second risk locus at 8q24. *Nat Genet* 2007;**39**(5):645–9.
45. Al Olama AA, Kote-Jarai Z, Berndt SI, et al. A meta-analysis of 87,040 individuals identifies 23 new susceptibility loci for prostate cancer. *Nat Genet* 2014;**46**(10):1103–9.
46. Gudmundsson J, Sulem P, Manolescu A, et al. Genome-wide association study identifies a second prostate cancer susceptibility variant at 8q24. *Nat Genet* 2007;**39**(5):631–7.
47. Eeles RA, Kote-Jarai Z, Giles GG, et al. Multiple newly identified loci associated with prostate cancer susceptibility. *Nat Genet* 2008;**40**(3):316–21.
48. Thomas G, Jacobs KB, Yeager M, et al. Multiple loci identified in a genome-wide association study of prostate cancer. *Nat Genet* 2008;**40**(3):310–5.
49. Yeager M, Chatterjee N, Ciampa J, et al. Identification of a new prostate cancer susceptibility locus on chromosome 8q24. *Nat Genet* 2009;**41**(10):1055–7.
50. Takata R, Akamatsu S, Kubo M, et al. Genome-wide association study identifies five new susceptibility loci for prostate cancer in the Japanese population. *Nat Genet* 2010;**42**(9):751–4.
51. Schumacher FR, Berndt SI, Siddiq A, et al. Genome-wide association study identifies new prostate cancer susceptibility loci. *Hum Mol Genet* 2011;**20**(19):3867–75.
52. Cheng I, Chen GK, Nakagawa H, et al. Evaluating genetic risk for prostate cancer among Japanese and Latinos. *Cancer Epidemiol Biomarkers Prev* 2012;**21**(11):2048–58.
53. Panagopoulos I, Möller E, Collin A, et al. The *POU5F1P1* pseudogene encodes a putative protein similar to *POU5F1* isoform 1. *Oncol Rep* 2008;**20**(5):1029–33.
54. Sotelo J, Esposito D, Duhagon MA, et al. Long-range enhancers on 8q24 regulate c-Myc. *Proc Natl Acad Sci USA* 2010;**107**(7):3001–5.
55. Wasserman NF, Aneas I, Nobrega MA. An 8q24 gene desert variant associated with prostate cancer risk confers differential *in vivo* activity to a MYC enhancer. *Genome Res* 2010;**20**(9):1191–7.
56. Ahn J, Kibel AS, Park JY, et al. Prostate cancer predisposition loci and risk of metastatic disease and prostate cancer recurrence. *Clin Cancer Res* 2011;**17**(5):1075–81.
57. Stevens VL, Ahn J, Sun J, et al. *HNF1B* and *JAZF1* genes, diabetes, and prostate cancer risk. *Prostate* 2010;**70**(6):601–7.
58. Kim HJ, Bae JS, Lee J, et al. *HNF1B* polymorphism associated with development of prostate cancer in Korean patients. *Urology* 2011;**78**(4):969.e1–6.

59. Chang BL, Spangler E, Gallagher S, et al. Validation of genome-wide prostate cancer associations in men of African descent. *Cancer Epidemiol Biomarkers Prev* 2011;**20**(1):23–32.
60. Hooker S, Hernandez W, Chen H, et al. Replication of prostate cancer risk loci on 8q24, 11q13, 17q12, 19q33, and Xp11 in African Americans. *Prostate* 2010;**70**(3):270–5.
61. Bensen JT, Xu Z, Smith GJ, et al. Genetic polymorphism and prostate cancer aggressiveness: a case-only study of 1,536 GWAS and candidate SNPs in African-Americans and European-Americans. *Prostate* 2013;**73**(1):11–22.
62. He Y, Gu J, Strom S, et al. The prostate cancer susceptibility variant rs2735839 near *KLK3* gene is associated with aggressive prostate cancer and can stratify Gleason score 7 patients. *Clin Cancer Res* 2014;**20**(19):5133–9.
63. Klein RJ, Halldén C, Cronin AM, et al. Blood biomarker levels to aid discovery of cancer-related single-nucleotide polymorphisms: kallikreins and prostate cancer. *Cancer Prev Res (Philadelphia)* 2010;**3**(5):611–9.
64. Xu J, Mo Z, Ye D, et al. Genome-wide association study in Chinese men identifies two new prostate cancer risk loci at 9q31.2 and 19q13.4. *Nat Genet* 2012;**44**(11):1231–5.
65. Lou H, Li H, Yeager M, et al. Promoter variants in the *MSMB* gene associated with prostate cancer regulate *MSMB/NCOA4* fusion transcripts. *Hum Genet* 2012;**131**(9):1453–66.
66. FitzGerald LM, Zhang X, Kolb S, et al. Investigation of the relationship between prostate cancer and *MSMB* and *NCOA4* genetic variants and protein expression. *Hum Mutat* 2013;**34**(1):149–56.
67. Morris DS, Tomlins SA, Montie JE, et al. The discovery and application of gene fusions in prostate cancer. *BJU Int* 2008;**102**(3):276–82.
68. Nam RK, Zhang W, Siminovitch K, et al. New variants at 10q26 and 15q21 are associated with aggressive prostate cancer in a genome-wide association study from a prostate biopsy screening cohort. *Cancer Biol Ther* 2011;**12**(11):997–1004.
69. Xu J, Zheng SL, Isaacs SD, et al. Inherited genetic variant predisposes to aggressive but not indolent prostate cancer. *Proc Natl Acad Sci USA* 2010;**107**(5):2136–40.
70. Amin Al Olama A, Kote-Jarai Z, Schumacher FR, et al. A meta-analysis of genome-wide association studies to identify prostate cancer susceptibility loci associated with aggressive and non-aggressive disease. *Hum Mol Genet* 2013;**22**(2):408–15.
71. Ewing CM, Ray AM, Lange EM, et al. Germline mutations in *HOXB13* and prostate-cancer risk. *N Engl J Med* 2012;**366**(2):141–9.
72. Breyer JP, Avritt TG, McReynolds KM, et al. Confirmation of the *HOXB13* G84E germline mutation in familial prostate cancer. *Cancer Epidemiol Biomarkers Prev* 2012;**21**(8):1348–53.
73. Akbari MR, Trachtenberg J, Lee J, et al. Association between germline *HOXB13* G84E mutation and risk of prostate cancer. *J Natl Cancer Inst* 2012;**104**(16):1260–2.
74. Karlsson R, Aly M, Clements M, et al. A population-based assessment of germline *HOXB13* G84E mutation and prostate cancer risk. *Eur Urol* 2014;**65**:169–76.
75. Goodman PJ, Thompson Jr IM, Tangen CM, et al. The prostate cancer prevention trial: design, biases and interpretation of study results. *J Urol* 2006;**175**(6):2234–42.
76. Lippman SM, Klein EA, Goodman PJ, et al. Effect of selenium and vitamin E on risk of prostate cancer and other cancers: the Selenium and Vitamin E Cancer Prevention Trial (SELECT). *JAMA* 2009;**301**(1):39–51.

CHAPTER

# 18

# Neuroendocrine Prostate Cancer

*Daniel C. Parker, MD*,**, Alexander Kutikov, MD, FACS***

*Department of Urology, Temple University Hospital, Philadelphia, PA, USA

**Department of Urology, Fox Chase Cancer Center, Einstein Urologic Institute, Philadelphia, PA, USA

## INTRODUCTION

Carcinoma of the prostate represents the most common cancer diagnosis in American men and is the second leading cause of deaths from cancer. The large majority of prostate cancers are histologically adenocarcinomas. The development and implementation of serum prostate-specific antigen (PSA) screening among at-risk men lead to the high incidence of prostate cancers discovered in the late twentieth and early twenty-first centuries. While far more prostate carcinomas are being diagnosed at an early/localized stage in the PSA era, there is still a significant population of American men with prostate cancer who are diagnosed with advanced-stage disease or who progress following treatment.[1]

Historically, neuroendocrine differentiation (NED) of prostate carcinoma is considered an important and largely lethal step of carcinogenesis in patients with advanced hormone refractory prostate cancer (HRPC). Indeed, NED is strongly associated with antiandrogen treatment failure, disease progression, and worsening prognosis independent of tumor stage and grade.[2] Neuroendocrine (NE) cells appear to be a ubiquitous component of normal prostate tissue, serving a regulatory function via exocrine, endocrine, paracrine, and autocrine pathways. In its purest form, NED manifests in prostate cancer as small-cell carcinoma, a rare variant associated with extremely poor prognosis, representing at most 2% of diagnosed prostate cancers.[3] Nevertheless, it is important to understand that to varying degrees NED is present in almost all adenocarcinomas. Decades of scientific investigation have focused on mechanisms that underpin the association of NED and prostate cancer aggressiveness, attempting to identify serum markers to detect excessive levels of NED and to pinpoint therapeutic opportunities against the disease.[4–6] As such, this chapter will introduce the reader to the physiologic role of the neuroendocrine system in the development of normal prostate tissue, explain historical and current theories about NED's part in carcinogenesis of prostate malignancy and describe therapeutic options currently available to patients with advanced prostate cancer that exhibits NED features.

## NEUROENDOCRINE CELLS IN THE HEALTHY PROSTATE

### Embryologic Origin of Prostatic Neuroendocrine Cells

Neuroendocrine cells of the prostate constitute a small part of an otherwise dispersed network of regulatory endocrine cells first described by Feyrter in 1938.[7] Anthony Pearse later assigned the functional moniker *amine precursor uptake and decarboxylation* (APUD) to the system, which is now referred to more simply as the *neuroendocrine system.*[8] Nests of NE cells are relatively abundant in the respiratory and gastrointestinal epithelium as well as C-cells of the thyroid and Islets of Langerhans in the pancreas. Their presence in the prostate, however, was not recognized until 1944.[9]

Tracking the embryological origin of prostatic NE cells has been challenging. By the mid-1980s, the prevailing theory that the neural crest gives rise to a population of migratory cells that proliferate and differentiate in the mesenchyme of the urogenital sinus had been disproven.[10] The ensuing decades saw numerous studies examining the developmental origin of prostatic NE cells. A "common precursor" theory, which purports that the luminal secretory, basal, and NE cells of the prostate are all derived from a single cell line in the developing endoderm, was supported by findings that all three cell types express PSA.[11] Conversely, evidence exists supporting the origin of prostatic NE cells as a distinct neurogenic lineage, independent of the associated basal and

Prostate Cancer. http://dx.doi.org/10.1016/B978-0-12-800077-9.00018-9

secretory cells that arise in the urogenital sinus. Immunohistochemical (IHC) staining of prostatic tissue from fetuses ranging from 10 weeks' gestation to term demonstrated that the prostatic epithelium was devoid of endocrine cells in fetuses younger than 12 weeks' gestation. Staining of adjacent paraganglia for chromogranin A (CgA), however, demonstrated positivity prior to the tenth week, with subsequent penetration of CgA positive cells into the prostatic mesenchyme after the tenth week.[12]

Ultimately, NE cells arrive primarily within the ducts and acini of the prostatic and urethral epithelium. Cohen et al. were the first to describe a temporal and spatial relationship in the developing prostate using neuron-specific enolase (NsE) and CgA as IHC markers. In the first 3 months of life, they identified NE cells dispersed throughout the prostate, including the peripheral zone, which disappeared between 3 months of age and the onset of puberty. Only the periurethral acini and ducts were positive for NsE and CgA staining in this age group. After puberty, however, stain positivity returned to the peripheral zone of the prostate. Indeed, across all ages, the highest density of NsE and CgA positivity in the prostate was found in the ducts of the prostate and the prostatic urethra.[13] Men between the ages of 25–54 have the highest concentration of prostatic NE cells.[14]

## Physiology of Prostatic Neuroendocrine Cells

The exact function of the normal NE cell in the prostate is unknown; however, the cells likely influence the growth of surrounding prostatic epithelium and regulate prostatic secretions. Morphologically, prostatic NE cells are either an "open type," featuring apical processes extending toward the lumen, or those of the "closed type" with cytoplasmic projections that resemble dendritic processes of neuronal cells (Figure 18.1).[15] This difference in morphology suggests variation in function among NE cells within the prostate where open cells function primarily in an exocrine fashion,[17] while the closed-type cells participate in paracrine and/or neuroendocrine interactions. Data suggest that both cell types may have the capacity for autocrine and endocrine function.[18] The concept of endocrine–paracrine heterogeneity is further supported by the disparities in cytoplasmic granular content among NE cells (Table 18.1). The cytoplasmic granules of prostatic NE cells contain peptide hormones and prohormones that are exocytosed via fusion with the cellular membrane in response to a number of potential stimuli including direct neural stimulation.[15] Indeed, neuroendocrine differentiation in prostate adenocarcinoma was established by use of immunohistochemistry to demonstrate the presence of serotonin,[19] chromogranin A,[20] and neuron-specific enolase[21] in NE cells. Other known cytoplasmic contents of NE cells include compounds that are

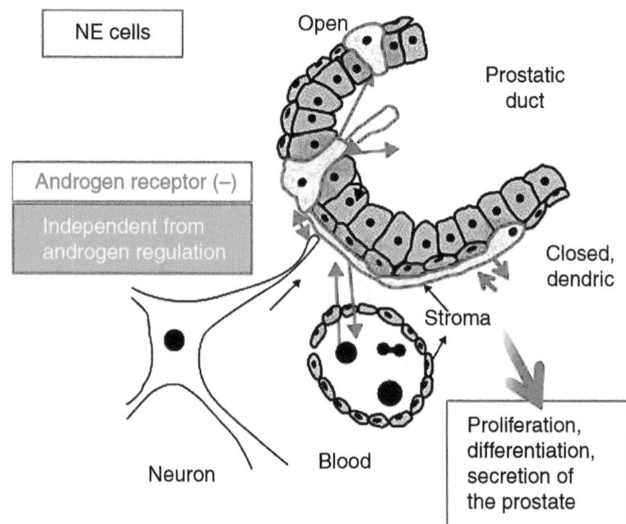


**FIGURE 18.1 Neuroendocrine cells in the normal prostate.** *Reproduced with permission from Komiya et al.*[16]

**TABLE 18.1** General Characteristics, Functional Rules, Products, and Receptors of the Neuroendocrine Cell

| |
|---|
| General characteristics |
| Androgen-receptor negative |
| Nonproliferating |
| PSA-negative |
| Bcl-2-negative |
| Express intermediate and luminal cytokeratins |
| Functional roles |
| Regulation of cell growth and differentiation |
| Regulation of homeostasis |
| Regulation of prostatic secretion |
| Products |
| Calcitonin gene family |
| Chromogranin A |
| Chromogranin B |
| Cholecystokinin (CCK) |
| Gastrin-releasing peptide |
| Histamine |
| Neuron-specific enolase |
| Neuropeptide Y |
| Parathyroid hormone-related protein |
| Proadrenomedullin N-terminal peptide |
| Serotonin |
| Somatostatin |
| TSH-like peptide |
| Vascular endothelial growth factor |
| Receptors |
| Gastrin releasing peptide (GRPR) |
| Serotonin (5HTR1A, B) |
| Somatostatin (SSTR 1–5) |
| Calcitonin (hCTR-2) |
| Cholecystokinin |
| Neuropeptide Y |
| Vasoactive intestinal peptide |
| PTHrP receptor (highly expressed in bone metastases from prostate) |

*Reproduced with permission from Vashchenko and Abrahamsson.*[50]

members of the chromogranin family (chromogranin A, B and secretogranin), calcitonin family (calcitonin, calcitonin gene-related peptide, katacalcin),[22] bombesin,[23] alpha-human chorionic gonadotropin,[24] somatostatin,[25] and thyroid stimulating hormone-like peptide.[18] Finally, the presence of parathyroid hormone-related protein has been implicated as an important factor in the paracrine and autocrine signaling of NE cells.[26]

NE cells are devoid of surface markers of cellular proliferation, namely, the antigens MIB-1 and Ki-67, suggesting that the normal NE cell in the prostate is fully differentiated and postmitotic.[27] In addition, prostatic NE cells are androgen-insensitive owing to their lack of expressed surface androgen receptor (AR),[19] which may have implications for treatment of hormone-refractory prostate cancer and overall disease prognosis. Likely direct regulators of prostatic NE cells, in addition to paracrine and autocrine regulatory effects mentioned earlier, include epidermal growth factor (EGF)[17] and transforming growth factor-alpha.[28] As such, the morphology of the neuroendocrine cell as well as its integral role in the regulation of normal prostate tissue places it at the center of critical checkpoints in carcinogenesis and prostate cancer progression.

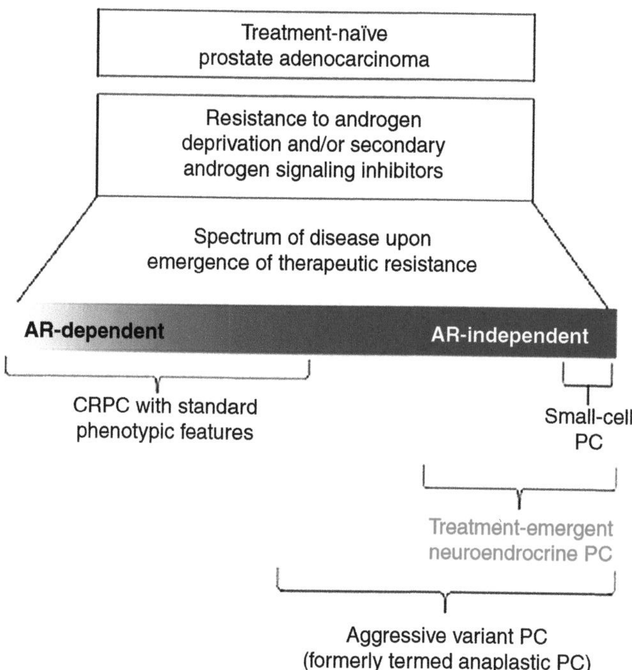


**FIGURE 18.2 Conceptual construct outlining clinical phenotypes of prostate cancer.** *Figure adapted from Aggarwal et al.*[29]

## NEUROENDOCRINE DIFFERENTIATION IN PROSTATE CANCER

Recently, two conceptual constructs have been proposed to outline how NED fits into various clinical phenotypes of prostate cancer. The first divides patients into two clinical cohorts based on their androgen-independent disease phenotype: patients with pure small-cell prostate cancer, and those who's NED emerges after standard treatments aimed at achieving castrate serum androgen levels (Figure 18.2).[29] The term "anaplastic prostate cancer" previously referred to the latter population; however, more recently some authors suggested replacing the term with "aggressive variant," because many NE prostate cancers do not display anaplastic pathologic features and some anaplastic prostate cancers do not exhibit neuroendocrine differentiation. The second construct is histologic and subdivides NE prostate cancer histology into the following categories: (1) *Small-cell carcinoma*, which appears as bland sheets of cells, frequently infiltrating and without defined margins; (2) *Large cell prostate cancer*, which is more rare and demonstrates uniform cells with bulky nuclei and cytoplasm that stains readily with eosin[30]; (3) *Paneth cell containing adenocarcinoma* identified by cells with abundant intracytoplasmic granules[31]; and (4) *carcinoid tumors*, which tend to be more well-differentiated and lower grade[32]; and (5) *typical acinar adenocarcinoma mixed with foci of neuroendocrine disease.*[33] It has been observed that as the subtype progresses toward a small cell appearance, so too does the tumor behave more aggressively and require earlier administration of chemotherapy for treatment response. As the modalities used to detect NED have become more sensitive since the 1970s with the advent of immunohistochemistry, so too has the incidence of NED in prostate cancer risen. Early studies reported approximately 10% of tumors staining positive for neuroendocrine markers[34]; however, using contemporary techniques, some authors have suggested that virtually all prostate cancers will demonstrate some degree of NED.[35]

*In vivo* studies have helped to better understand NED in prostatic tissues. The first human-derived androgen-sensitive prostate cancer cell line became available to researchers in the late 1970s. These LnCaP cells were found to exhibit a neuroendocrine phenotype *in vitro* under specific conditions. Interestingly, PSA production and sensitivity to androgen stimulation significantly decreased upon NED.[36] Further evidence suggesting that IL-6 induced NED in prostate cancer contributed to the current understanding that at least in part castration resistance is modulated through NE differentiation.[37] These initial *in vitro* findings have been supported by data from *in vivo* animal studies.[38,39]

Molecular mechanisms behind prostate cancer NED are still ill-defined. Several studies have supported the concept of transdifferentiation, whereby neoplastic prostatic cells attain the phenotype of NE cells. For instance, frequency of *ERG* gene rearrangements are known to be common in conventional acinar prostate cancer but rare in small-cell cancers. Lotan et al., studying *ERG* gene rearrangement, described a finding suggesting a common precursor cell to non-neuroendocrine and neuroendocrine tissue elements.[40] A biochemical mechanism was proposed involving the serine/threonine kinase AKT pathway. Wu et al. demonstrated that activating AKT via the phosphatidylinositol 3-kinase-AKT-mamallian target of rapamycin (PI3K-AKT-mTOR) pathway was essential to produce NED in androgen-deprived LNCaP cells. Meanwhile, inactivation of AKT inhibited NED.[41] Androgen deprivation alone resulted in increased cAMP levels and NED in LNCaP cells, suggesting a causal relationship between NED and castration-resistant prostate cancer.[42] When transdifferentiated NE cells are cultured and cloned from the LNCaP cell line, the absence of AR and PSA expression with concomitant production of various neuroendocrine markers can be observed. Importantly, it appears that transdifferentiation is an irreversible phenomenon.[43]

Specific genetic loci have been implicated in playing a key role in NED. For instance, higher levels of protocadherin-PC gene (PCDH-PC) gene are expressed in castration-resistant prostate cancer cells,[44] and specifically in androgen-deprived LNCaP cells in which neuroendocrine transdifferentiation and apoptosis resistance also occurred.[45] NED can also be induced by blocking expression of the TMPRSS2-ERG fusion protein via administration of enzalutamide and subsequent blockade of androgen receptor signaling.[46] Other transcription factor candidates recognized as playing a genetic role in the transdifferentiation of prostate cancer into a neuroendocrine phenotype include those derived from Forkhead box a (foxa) and effectors of the HOX gene family.[47]

NED has long been thought to have an effect on overall prostate cancer prognosis, especially in hormone-refractory disease; however, this remains controversial. Studies correlating tissue markers of NED with prognosis are conflicting. Jiborn et al. showed that tumors with larger numbers of NE cells were associated with advanced grade and stage of prostate cancer. The association was stronger after androgen deprivation.[48] Conversely, tissue levels of CgA positivity in pathology specimens were not correlated with prognosis in 90 patients after radical prostatectomy in which long-term follow-up data were available.[49] Variations in the distribution of NE cells within tissue specimens, patient populations, and methodology have all been proposed as reasons why the literature remains discordant.[50] Downregulation of the *regulator of neuronal gene expression (REST)* protein was seen in approximately half of 218 prostate tumors with NED and has been proposed as an emerging tissue marker.[51] Similarly, enhanced expression of *amplification of Aurora kinase A (AURKA)* protein was evident in all studied NE prostate cancers by Beltran et al. In 4/7 of the tumors, the parent gene of AURKA was also amplified.[33]

In contrast to tissue markers, data for use of serum markers of NED appear more promising. Serum levels of chromogranin A (CgA)[52] and neuron-specific enolase (NSE)[53] have been shown to correlate with NED, castration resistance, and disease prognosis. The amount of CgA positivity in prostate tissues is reliably reflected by serum CgA levels.[54] Serum measurement of CgA may, in fact, be superior to NSE in assessing prognosis in patients with advanced disease when used alongside PSA and this is the basis for why serum CgA measurement is considered useful for prognosis in some clinical settings.

The mechanism by which NED affects prognosis may be through enhancement of the proliferative capacity of the cancer's nonneuroendocrine component. For instance, assessment of Ki-67 expression as an index of cell proliferation in the setting of NED demonstrated that more robust proliferation was noted in tumors with strong NED, compared to those cells weakly positive for NED or none at all.[55] NED may influence the invasiveness of prostate cancer, as well. Nagakawa et al. first demonstrated that the C-terminal fragment of CgA escalated migration of prostate cancer cells in association with production of urokinase-type plasminogen activator.[56] The gene products of NE cells, including calcitonin gene-related peptide and parathyroid hormone-related peptide were also linked to aggressive prostate cancer phenotype.[57] Finally, NED may increase the ability or likelihood of prostate cancer cells to metastasize. Gelsolin, a key regulator of actin assembly and disassembly, was upregulated by the secretory products of NE cells resulting in augmented pulmonary metastasis of LNCaP cells in a mouse model.[58]

## CLINICAL CHALLENGES AND OPPORTUNITIES

Treatment of prostate carcinoma that exhibits NED presents significant clinical diagnostic and therapeutic challenges. Pure small-cell carcinoma of the prostate is often primarily encountered at an advanced stage characterized by loco-regional invasion, bone involvement, and visceral metastasis. Commonly seeded organs include the liver, lung, and brain.[3] CNS imaging with MRI and whole-body FDG-PET/CT scans are indicated for those instances in which it is appropriate to first rule out widespread disease prior to local treatment.[59] A high clinical suspicion for NE prostate cancer is warranted for any patient who exhibits an aggressive cancer phenotype after being treated with ADT or newer antiandrogens such as abiraterone or enzalutamide. Indeed, recent data suggest that widespread use of these agents is associated with developing a NE prostate cancer phenotype.[60] Aggarwal et al. suggested that clinicians should assess the risk of a patient harboring a NE prostate cancer phenotype before initiating the latest-generation antiandrogen. Clinical features, such as rapid (<6 months) disease progression on ADT, a PSA nadir of >4 ng/mL after ADT, a low PSA-to-disease burden ratio, and high levels of carcinoembryonic antigen (CEA) or lactate dehydrogenase (LDH), should raise alarm regarding a patient potentially harboring NED prostate cancer. Additionally, the presence of visceral metastasis or involvement of the CNS with or without lytic bone involvement is strongly associated with NED. These key opinion leaders thus recommend biopsy for the purposes of detecting tissue markers in cases where the clinical suspicion for NE prostate cancer is supported by serum markers and the aforementioned features. Histopathologic examination of the biopsy tissue for small-cell characteristics and staining with CgA and NSE should be routine in these instances.[29]

As with small-cell carcinoma of the lung, therapeutic guidelines for pure small-cell carcinoma of the prostate advocate for radiation therapy in instances where the disease is detected while still locally advanced. Simultaneous chemotherapy with platinum and etoposide are ideal in this scenario, but still the treatment failure rates are high. Nevertheless, most small-cell prostate carcinomas present with dispersed metastatic disease for which radiation is unreasonable. Due to its high proliferative index, widely advanced small-cell carcinoma is treated primarily with cytotoxic systemic therapy. A 62% treatment response rate was seen in 21 patients undergoing platinum plus etoposide combination therapy and thus this regimen has become standard.[61]

There are no such established treatment algorithms for those cancers that exhibit a heterogeneous acinar and NE phenotype; however, several trials have been reported.[6,62,63] Carboplatin with or without docetaxel has become a common first line choice based on a phase II trial demonstrating up to a 65% treatment response despite a median survival of less than a year. Patients who progressed after this regimen showed positive yet limited progression-free survival when continued on platinum-based therapy with etoposide added in combination.

Opportunities for targeted therapy have been proposed and include the employment of AURKA inhibitors. In preclinical trials, AURKA inhibitors were shown to suppress neuroendocrine marker expression in NE prostate cancer. The clinical trial NCT01799278 is examining the effect of an oral AURKA inhibitor on patients with confirmed or suspected metastatic NE prostate cancer.

## CONCLUSIONS

Neuroendocrine differentiation is an important factor influencing the development of prostate cancer toward a particularly androgen-independent, lethal phenotype. Small-cell carcinoma represents the most uniform example of NED, but many prostate cancers are composed of a heterogeneous mix of typical acinar and neuroendocrine cells. Indeed, the presence of NED is an independent negative prognostic indicator in patients with prostate carcinoma. Recent advances have improved our understanding of the pathophysiology of NED but better clinical markers of early-stage NE prostate cancer are still needed. Current guidelines advocate platinum-based systemic cytotoxic therapy for metastatic prostate cancer with small-cell features; however, molecular targets have been identified and are currently being explored as therapeutic opportunities in clinical trials.

## References


1. di Sant'Agnese PA. Neuroendocrine differentiation in prostatic carcinoma: an update on recent developments. *Ann Oncol* 2001;**12**(2):6.
2. Weinstein MH, Partin AW, Veltri RW, et al. Neuroendocrine differentiation in prostate cancer: enhanced prediction of progression after radical prostatectomy. *Hum Pathol* 1996;**27**(7):683–7.
3. Marcus DM, Goodman M, Jani AB, et al. A comprehensive review of incidence and survival in patients with rare histological variants of prostate cancer in the United States from 1973 to 2008. *Prostate Cancer Prostatic Dis* 2012;**15**(3):283–8.
4. Hirano D, Okada Y, Minei S, et al. Neuroendocrine differentiation in hormone refractory prostate cancer following androgen deprivation therapy. *Eur Urol* 2004;**45**(5):586–92 discussion 592.
5. Berruti A, Bollito E, Cracco CM, et al. The prognostic role of immunohistochemical chromogranin a expression in prostate cancer patients is significantly modified by androgen-deprivation therapy. *Prostate* 2010;**70**(7):718–26.
6. Aparicio AM, Harzstark AL, Corn PG, et al. Platinum-based chemotherapy for variant castrate-resistant prostate cancer. *Clin Cancer Res* 2013;**19**(13):3621–30.
7. Feyrter F. *Uber diffuse endocrine epithelial organe*. Leipzig: JA Barth; 1938.
8. Pearse A. The cytochemistry and ultrastructure of polypeptide hormone-producing cells of the APUD series and the embryologic, physiologic, and pathologic implications of the concept. *J Histochem Cytochem* 1969;**17**:303–13.
9. Pretl K. Zur frage der endokrinie der menschlichem vorsteherdruse. *Virchows Arch* 1944;**312**:392–408.
10. Andrew A, Kramer B, Rawdon BB. Gut and pancreatic amine precursor uptake and decarboxylation cells are not neural crest derivatives. *Gastroenterology* 1983;**84**(2):429–31.
11. Aprikian AG, Cordon-Cardo C, Fair WR, et al. Characterization of neuroendocrine differentiation in human benign prostate and prostatic adenocarcinoma. *Cancer* 1993;**71**(12):3952–65.
12. Aumuller G, Leonhardt M, Janssen M, et al. Neurogenic origin of human prostate endocrine cells. *Urology* 1999;**53**(5):1041–8.
13. Cohen RJ, Glezerson G, Taylor LF, et al. The neuroendocrine cell population of the human prostate gland. *J Urol* 1993;**150**(2 Pt 1): 365–8.
14. Battaglia S, Casali AM, Botticelli AR. Age-related distribution of endocrine cells in the human prostate: a quantitative study. *Virchows Arch* 1994;**424**(2):165–8.
15. di Sant'Agnese PA. Neuroendocrine differentiation in carcinoma of the prostate: diagnostic, prognostic, and therapeutic implications. *Cancer* 1992;**70**(1):15.

16. Komiya A, Suzuki H, Imamoto T, et al. Neuroendocrine differentiation in the progression of prostate cancer. *Int J Urol* 2009;**16**(1):37–44.
17. Iwamura M, Gershagen S, Lapets O, et al. Immunohistochemical localization of parathyroid hormone-related protein in prostatic intraepithelial neoplasia. *Hum Pathol* 1995;**26**(7):797–801.
18. Abrahamsson PA, Wadström LB, Alumets J, et al. Peptide-hormone- and serotonin-immunoreactive cells in normal and hyperplastic prostate glands. *Pathol Res Pract* 1986;**181**(6):675–83.
19. Sciarra A, Mariotti G, Gentile V, et al. Neuroendocrine differentiation in human prostate tissue: is it detectable and treatable? *BJU Int* 2003;**91**(5):438–45.
20. Isshiki S, Akakura K, Komiya A, et al. Chromogranin a concentration as a serum marker to predict prognosis after endocrine therapy for prostate cancer. *J Urol* 2002;**167**(2 Pt 1):512–5.
21. Kamiya N, Akakura K, Suzuki H, et al. Pretreatment serum level of neuron specific enolase (NSE) as a prognostic factor in metastatic prostate cancer patients treated with endocrine therapy. *Eur Urol* 2003;**44**(3):309–14 discussion 314.
22. Fetissof F, Bruandet P, Arbeille B, et al. Calcitonin-secreting carcinomas of the prostate. An immunohistochemical and ultrastructural analysis. *Am J Surg Pathol* 1986;**10**(10):702–10.
23. Sunday ME, Kaplan LM, Motoyama E, et al. Gastrin-releasing peptide (mammalian bombesin) gene expression in health and disease. *Lab Invest* 1988;**59**(1):5–24.
24. Fetissof F, Arbeille B, Guilloteau D, et al. Glycoprotein hormone alpha-chain-immunoreactive endocrine cells in prostate and cloacal-derived tissues. *Arch Pathol Lab Med* 1987;**111**(9):836–40.
25. di Sant'Agnese PA, de Mesy Jensen KL. Somatostatin and/or somatostatin-like immunoreactive endocrine-paracrine cells in the human prostate gland. *Arch Pathol Lab Med* 1984;**108**(9):693–6.
26. di Sant'Agnese PA. Neuroendocrine differentiation in prostatic carcinoma: recent findings and new concepts. *Cancer* 1995;**75**(7):10.
27. Bonkhoff H, Stein U, Remberger K. Endocrine-paracrine cell types in the prostate and prostatic adenocarcinoma are postmitotic cells. *Hum Pathol* 1995;**26**(2):167–70.
28. Nilsson O, Wängberg B, Kölby L, et al. Expression of transforming growth factor alpha and its receptor in human neuroendocrine tumours. *Int J Cancer* 1995;**60**(5):645–51.
29. Aggarwal R, Zhang T, Small EJ, et al. Neuroendocrine prostate cancer: subtypes, biology, and clinical outcomes. *J Natl Compr Canc Netw* 2014;**12**(5):719–26.
30. Evans AJ, et al. Large cell neuroendocrine carcinoma of prostate: a clinicopathologic summary of 7 cases of a rare manifestation of advanced prostate cancer. *Am J Surg Pathol* 2006;**30**(6):684–93.
31. Adlakha H, Bostwick DG. Paneth cell-like change in prostatic adenocarcinoma represents neuroendocrine differentiation: report of 30 cases. *Hum Pathol* 1994;**25**(2):135–9.
32. Almagro UA. Argyrophilic prostatic carcinoma. Case report with literature review on prostatic carcinoid and "carcinoid-like" prostatic carcinoma. *Cancer* 1985;**55**(3):608–14.
33. Beltran H, Rickman DS, Park K, et al. Molecular characterization of neuroendocrine prostate cancer and identification of new drug targets. *Cancer Discov* 2011;**1**(6):487–95.
34. Azzopardi JG, Evans DJ. Argentaffin cells in prostatic carcinoma: differentiation from lipofuscin and melanin in prostatic epithelium. *J Pathol* 1971;**104**:5.
35. Abrahamsson PA. Neuroendocrine cells in tumour growth of the prostate. *Endocr Relat Cancer* 1999;**6**(4):503–19.
36. Shen R, Dorai T, Szaboles M, et al. Transdifferentiation of cultured human prostate cancer cells to a neuroendocrine cell phenotype in a hormone-depleted medium. *Urol Oncol* 1997;**3**(2):67–75.
37. Mori S, Murakami-Mori K, Bonavida B. Interleukin-6 induces G1 arrest through induction of p27(Kip1), a cyclin-dependent kinase inhibitor, and neuron-like morphology in LNCaP prostate tumor cells. *Biochem Biophys Res Commun* 1999;**257**(2):609–14.
38. Noordzij MA, van Weerden WM, de Ridder CM, et al. Neuroendocrine differentiation in human prostatic tumor models. *Am J Pathol* 1996;**149**(3):859–71.
39. Jongsma J, Oomen MH, Noordzij MA, et al. Kinetics of neuroendocrine differentiation in an androgen-dependent human prostate xenograft model. *Am J Pathol* 1999;**154**(2):543–51.
40. Lotan TL, Gupta NS, Wang W, et al. ERG gene rearrangements are common in prostatic small cell carcinomas. *Mod Pathol* 2011;**24**(6):820–8.
41. Wu C, Huang J. Phosphatidylinositol 3-kinase-AKT-mammalian target of rapamycin pathway is essential for neuroendocrine differentiation of prostate cancer. *J Biol Chem* 2007;**282**(6):3571–83.
42. Burchardt T, Burchardt M, Chen MW, et al. Transdifferentiation of prostate cancer cells to a neuroendocrine cell phenotype *in vitro* and *in vivo*. *J Urol* 1999;**162**(5):1800–5.
43. Yuan X, Li T, Wang H, et al. Androgen receptor remains critical for cell-cycle progression in androgen-independent CWR22 prostate cancer cells. *Am J Pathol* 2006;**169**(2):682–96.
44. Terry S, Queires L, Gil-Diez-de-Medina S, et al. Protocadherin-PC promotes androgen-independent prostate cancer cell growth. *Prostate* 2006;**66**(10):1100–13.
45. Chen MW, Vacherot F, De La Taille A, et al. The emergence of protocadherin-PC expression during the acquisition of apoptosis-resistance by prostate cancer cells. *Oncogene* 2002;**21**(51):7861–71.
46. Pagliarini R. ERG influences cell fate decisions in prostate cancer. *Presented at the Twentieth Annual Prostate Cancer Foundation Scientific Retreat*. National Harbor, Maryland; 2013.
47. Cindolo L, Cantile M, Vacherot F, et al. Neuroendocrine differentiation in prostate cancer: from lab to bedside. *Urol Int* 2007;**79**(4):287–96.
48. Jiborn T, Bjartell A, Abrahamsson PA. Neuroendocrine differentiation in prostatic carcinoma during hormonal treatment. *Urology* 1998;**51**(4):585–9.
49. Noordzij MA, van der Kwast TH, van Steenbrugge GJ, et al. The prognostic influence of neuroendocrine cells in prostate cancer: results of a long-term follow-up study with patients treated by radical prostatectomy. *Int J Cancer* 1995;**62**(3):252–8.
50. Vashchenko N, Abrahamsson PA. Neuroendocrine differentiation in prostate cancer: implications for new treatment modalities. *Eur Urol* 2005;**47**(2):147–55.
51. Lapuk AV, Wu C, Wyatt AW, et al. From sequence to molecular pathology, and a mechanism driving the neuroendocrine phenotype in prostate cancer. *J Pathol* 2012;**227**(3):286–97.
52. Kadmon D, Thompson TC, Lynch GR, et al. Elevated plasma chromogranin-A concentrations in prostatic carcinoma. *J Urol* 1991;**146**(2):358–61.
53. Tarle M, Rados N. Investigation on serum neurone-specific enolase in prostate cancer diagnosis and monitoring: comparative study of a multiple tumor marker assay. *Prostate* 1991;**19**(1): 23–33.
54. Angelsen A, Syversen U, Haugen OA, et al. Neuroendocrine differentiation in carcinomas of the prostate: do neuroendocrine serum markers reflect immunohistochemical findings? *Prostate* 1997;**30**(1):1–6.
55. Grobholz R, Griebe M, Sauer CG, et al. Influence of neuroendocrine tumor cells on proliferation in prostatic carcinoma. *Hum Pathol* 2005;**36**(5):562–70.
56. Nagakawa O, Murakami K, Ogasawara M, et al. Effect of chromogranin A (pancreastatin) fragment on invasion of prostate cancer cells. *Cancer Lett* 1999;**147**(1-2):207–13.
57. Nagakawa O, Ogasawara M, Murata J, et al. Effect of prostatic neuropeptides on migration of prostate cancer cell lines. *Int J Urol* 2001;**8**(2):65–70.
58. Uchida K, Masumori N, Takahashi A, et al. Characterization of prostatic neuroendocrine cell line established from neuroendocrine carcinoma of transgenic mouse allograft model. *Prostate* 2005;**62**(1):40–8.

59. Furtado P, Lima MV, Nogueira C, et al. Review of small cell carcinomas of the prostate. *Prostate Cancer* 2011;**2011**:543272.
60. Yuan TC, Veeramani S, Lin MF. Neuroendocrine-like prostate cancer cells: neuroendocrine transdifferentiation of prostate adenocarcinoma cells. *Endocr Relat Cancer* 2007;**14**(3):531–47.
61. Kalemkerian GP, Akerley W, Bogner P, et al. *NCCN clinical practice guidelines in oncology: small cell lung cancer. Version 2.2014.* [26.08.2014].
62. Mohler JL, Kuet KP, Armstrong AJ, et al. NCCN clinical practice guidelines in oncology: prostate cancer. Version 2.2014.[cited 26.08.2014].
63. Papandreou CN, Daliani DD, Thall PF, et al. Results of a phase II study with doxorubicin, etoposide, and cisplatin in patients with fully characterized small-cell carcinoma of the prostate. *J Clin Oncol* 2002;**20**(14):3072–80.

CHAPTER

# 19

# Breast and Prostate Cancers: A Comparison of Two Endocrinologic Malignancies

*Maahum Haider, MD**, *Paul H. Lange, MD, FACS**,**

*Department of Urology, University of Washington School of Medicine, Seattle, WA, USA

**Institute of Prostate Cancer Research, University of Washington and Fred Hutchinson Cancer Research Center, Seattle, WA, USA

## INTRODUCTION

Prostate and breast cancers are two common malignancies that have some striking and surprising similarities. As we know, they are both tumors of accessory sex organs and both are characterized primarily by being epithelial, hormone-driven malignancies, which respond to so-called "endocrine therapy." They constitute the second leading cause of cancer-related mortality in the United States for men and women, respectively.[1] In both cancers, there is often a latency period spanning many years from the time of diagnosis of the primary tumor to the development of metastasis. Recent research has revealed that both breast and prostate cancer are complex and heterogeneous disease processes that encompass multiple pathologies with variable treatment options, as opposed to one distinct disease with a uniform treatment. A great deal of experimental and clinical research has been undertaken for each of these cancers resulting in many new strategies of care. Until recently, this has been truer for breast cancer than for prostate cancer. This is partly due to the fact that breast cancer was recognized centuries before prostate cancer, and that it is a more accessible organ. More important reasons are the early vigorous advocacy of breast cancer survivors for research support and the misconception that prostate cancer was an "old man's disease" meriting little concern. More recently, support for and progress in prostate cancer research has greatly expanded.

Therefore, it seems appropriate now to further explore some of the interesting parallels between these two cancers in addition to some of the differences and areas of ignorance. It is hoped that this comparison will highlight important areas for further inquiry.

## HISTORY

The earliest descriptions of breast cancer can be traced back to ancient Egypt around 1500 BCE.[2] At that time, it was treated locally with cauterization techniques, but there was no concept of metastatic disease and prognosis was obviously poor. The first mastectomy was performed as early as 548 AD and the first radical mastectomy was performed in the seventeenth century in France, after physicians developed a better understanding of the human anatomy.[2] In 1882, William S. Halstead performed the first radical mastectomy in the United States, and this became the standard of care for all suspicious breast lumps until the mid-1970s. Although the surgery was morbid and later deemed excessive, it improved both staging and long-term survival significantly. With a better understanding of the endocrinological background of breast cancer came the idea of castration (i.e., removing estrogen). In 1889, a German physician named Albert Schinzinger proposed oophorectomy as a treatment for breast cancer.[3] Hypophysectomy was also used for a time to target the hypothalamo-pituitary axis. These endocrine approaches resulted in dramatic, though mostly temporary, remissions.

Prostate cancer was recognized much later and was first described in 1853. Removal of the entire gland and

Prostate Cancer. http://dx.doi.org/10.1016/B978-0-12-800077-9.00019-0

seminal vesicles to treat cancer was not performed until 1904 by Hugh Hampton Young, who used a perineal approach.[4] Gradually, the retropubic approach for radical prostatectomy became safer, and thanks in large part to the work of Patrick Walsh, beginning around 1983, it became the predominant approach for this surgery.[5] A major advantage of this approach was that the pelvic lymph nodes could simultaneously be removed for staging and possibly improved survival. Charles B. Huggins introduced the idea of castration (i.e., removing testosterone) to treat metastatic prostate cancer; a discovery that would earn him the Nobel Prize in 1966.[6] Castration was accomplished either by orchiectomy or the use of estrogen to oppose testosterone. Here again, the responses to castration were dramatic though often temporary. Parenthetically, Huggins et al. also popularized the use of oophorectomy (and adrenalectomy) for breast cancer.[3] The gonadotropin-releasing hormone receptor was later discovered by Andrew Schally, and then agonists and later antagonists were developed and are now the predominant form of castration for prostate cancer. These discoveries earned Andrew Schally the Nobel Prize in 1977.[7]

## EPIDEMIOLOGY

Prostate and breast cancers have remarkable epidemiologic parallels, which may hint at an underlying genetic or environmental link. Thanks to the widespread adoption of screening, they are the two most commonly diagnosed cancers in the United States with an estimated 233,000 and 232,670 incident cases of prostate cancer and breast cancer (in women), respectively, in 2014 alone.[1] These two cancers are also the second leading cause of cancer-related deaths in the United States, with 40,000 deaths attributable to breast cancer in women and 29,480 to prostate cancer in 2014.[1] According to statistics from the American Cancer Society, between 2008 and 2010, US men had a one in seven chance of developing prostate cancer during their lifetime, while women had a one in eight chance of developing breast cancer.[1] The reasons for this current incidence and the controversies surrounding it will be discussed in the subsequent section on screening.

Important epidemiological data have been gathered from autopsy studies. Such studies were considerably more difficult to perform in the case of breast cancer, given the larger volume and fatty composition of the organ (averaging hundreds of paraffin blocks per specimen).[8] A compilation of the most important studies from the 1980s revealed the prevalence of asymptomatic invasive and *in situ* disease to be 1.3 and 8.9%, respectively.[9] Autopsy studies of the prostate revealed a much higher occurrence of the disease in aging men.[10,11] Almost 60% of men aged 90 or greater were found to have incidental prostate cancer on autopsy.[10] These significant differences found on autopsy might be partially explained by a less detailed pathological analysis of breast tissue and a greater number of autopsies being performed on postmenopausal women, thereby possibly causing some cancers to regress. These provisos notwithstanding, the prevailing evidence suggests that autopsy cancer in breast is much less common than prostate cancer. Why this should be the case deserves further investigation.

In terms of race, prostate and breast cancers have higher mortality rates among African-Americans.[1] African-American men are more likely to be diagnosed with and to die of their prostate cancer while African-American women are less likely to be diagnosed with breast cancer than their Caucasian counterparts, but are more likely to die of their disease.[1] Asians have the lowest mortality rates for both cancers (Table 19.1), but these rates do increase after relocation to Western countries.[12]

**TABLE 19.1** Similarities and Differences Between Breast and Prostate Cancers

| | Breast cancer | Prostate cancer |
|---|---|---|
| Epidemiology | 232,670 cases/year<br>More common in Caucasians<br>40,000 cancer-related deaths/year | 233,000 cases/year<br>More common in African- Americans<br>29,480 cancer-related deaths/year |
| Risk factors | Family history<br>*BRCA1/2* genes<br>Exposure to estrogen<br>Race<br>Dietary fat | Family history<br>*BRCA2* gene<br>Exposure to androgen<br>Race<br>Dietary fat |
| Prevention | Tamoxifen<br>Low-fat diet | Finasteride<br>Low-fat diet |
| Adjuvant therapies | Well-established | Vary by practice |
| Metastatic profile | Bone metastasis common (osteoclastic) | Bone metastasis common (osteoblastic) |

## RISK FACTORS

Virtually all experts believe that breast and prostate cancers occur because of the interaction of a variety of environmental factors and genetic susceptibility. In breast cancer, the most important risk factor is probably family history and thus genetic predisposition. So-called hereditary breast cancer accounts for 5–10% of all breast cancers and the relative risk increases with the number of first-degree relatives with the disease.[13]

The best known genetic susceptibility factors are mutations in the tumor suppressor genes *BRCA1* and *BRCA2*, which are responsible for as much as 80% of cases of inherited breast cancer.[14] These mutations confer a lifetime risk of breast cancer of 60–85%, and interestingly, a lifetime risk of ovarian cancer of 15–40%.[15] Mutations in the BRCA2 gene have been found in several nonbreast cancers, including cancer of the prostate.[13] In fact, in 1999 a study by the Breast Cancer Linkage Consortium found a relative risk of 4.65 for prostate cancer in families with *BRCA2* mutations.[16]

Another important risk factor for breast cancer is prolonged and unopposed exposure to estrogen (early menarche, nulliparity, first live birth after age 30, and delayed menopause).[14] Other risk factors include age, race, smoking, previous breast disease, breast tissue density, radiation exposure, weight, exercise, and alcohol consumption.[15]

Prostate cancer has many of the same risk factors as breast cancer. Like breast cancer, a positive family history is an important risk factor. The relative risk for men with a single first-degree relative with prostate cancer increases by a factor of 2.1–2.8, and having a first-degree and a second-degree relative with prostate cancer can increase the risk by up to four- to sixfold when compared to the general population.[14] Unfortunately, intensive studies have not yet revealed any genes that have the predictive power of *BRCA1* and *BRCA2*. However, some experts believe that in subtle, yet-to-be-discovered ways, heredity genes influence both breast and prostate cancer development and progression in as much as 40% of cases.[17]

Also, like breast cancer, it seems that prolonged exposure to steroid hormones, specifically in this case androgens, is an important risk factor for prostate cancer. This fact is evidenced by the elimination of prostate cancer risk in men who undergo early castration or have genetic defects blocking androgen synthesis.[14] Other risk factors for prostate cancer include age, high-fat and low-fiber diet, obesity, and prostatic inflammation.

## SCREENING

Screening for both breast and prostate cancers greatly changed the incidence of the disease. The adoption of breast cancer screening preceded that of prostate cancer by two decades. It came to involve ever more accurate X-ray mammography and needle biopsy of detected lesions.[18] First screening mammography is reported to have a positive predictive value of 22.2% for women aged 50–59, and this value increases with increasing age.[19] This approach was widely recommended on a yearly basis for mostly all women initially. This resulted in a large increase in breast cancer incidence so that, as mentioned previously, today one in seven women are diagnosed with the cancer. A great many women advocacy groups vigorously promoted breast cancer screening and facilitated delivery of screening services to women with little to no access to healthcare.

In the last 5–6 years, the thinking about breast cancer screening has shifted considerably, generating great emotion among involved clinicians and patient support groups. Investigators and clinicians came to realize that while the mortality rates had dropped 20–30% in many randomized studies, the rise in incidence was far greater and many women diagnosed with this cancer had asymptomatic, nonaggressive disease that would not kill them and if treated, could cause more harm than good.[20] Nevertheless, based on these facts and the mortality rates among women who had been screened and those who had not, the United States Preventive Services Task Force (USPSTF) released recommendations in 2009 for breast cancer screening that advised against the routine teaching of breast self-examination and recommended biennial mammographic screening for women aged 50–74 but not for women under 50.[21] Even more recently, other groups have been actually advocating the abolition of breast cancer screening much like the case for prostate cancer (*vide infra*).[22]

Prostate cancer screening is also highly controversial. Screening began in earnest with the introduction of the prostate-specific antigen (PSA) blood test in 1992. What is often overlooked is that the surge in screening was also due in part to better biopsy techniques and the perfection of transrectal ultrasonography. Thereafter, prostate cancer incidence soared, with mortality rates seemingly halved.[23] A variety of randomized clinical trials assessing screening and surgery versus observation were conducted, but the details are outside the scope of this chapter. In general, these studies mostly showed that mortality was decreased by 20–30% with early detection and intervention.[22] However, it certainly did lead to overdiagnosis and overtreatment of men with low-risk disease, who may never have been symptomatic from their cancer. Furthermore, some men suffer from complications of the prostate needle biopsy, from undue anxiety over a cancer that requires no intervention, and most significantly from complications of treatment. Over time, as the indolent nature of most prostate cancers became clear and the risks of prostate biopsy and cancer treatment became more familiar, the utility of yearly PSA screening came into question. In 2012, the USPSTF released the very controversial guidelines regarding prostate cancer screening, which essentially recommended against routine PSA testing (grade D recommendation).[24] A large number of clinicians and investigators (e.g., the American Urologic Association) vigorously reject the USPSTF's analysis of the data and continue to advocate early diagnosis for men with a life expectancy of greater than 10–15 years, with the provisory that careful patient counseling be conducted. Unlike breast cancer however, most experts now agree that for low-risk cancers, a program of active surveillance is appropriate and desireable.[25]

One important concept that may explain why screening does not affect mortality more significantly is the real possibility that the fatal cancers are already systemic even when the most optimal screening approaches are utilized. Certainly in the case of breast cancer, many experts, most prominently Bernard Fisher, posited that the fatal breast cancers are already systemic at diagnosis.[26] Indeed, as we will discuss later, this hypothesis is strengthened by data showing lumpectomy with appropriate radiation therapy and systemic therapy is often the best management approach. Certainly, there is evidence that many early-diagnosis prostate cancers are already systemic when diagnosed but can still be cured or at least advantageously affected by aggressive local or regional therapy.

## PREVENTION

Preventative options for high-risk women range from chemopreventative measures to aggressive prophylactic surgery for those women with the BRCA genes. A number of large studies have investigated the use of tamoxifen, a selective estrogen receptor modulator, in the prevention of breast cancer. In breast tissue, it competitively binds the estrogen receptor, thereby blocking the effect of estrogen on proliferative activity. In the endometrium, it can act as an agonist and has been shown to be associated with endometrial cancer. Thus, the Breast Cancer Prevention Trial (BCPT) was conducted in the mid-1990s and showed that tamoxifen significantly decreased the risk of ER-positive breast cancer in women who were considered high-risk at the time of enrollment. However, women in the tamoxifen group also had increased rates of endometrial cancer and thromboembolic events.[27] The Study of Tamoxifen and Raloxifene for the Prevention of Breast Cancer (STAR trial) found that both drugs reduced the risk of breast cancer in postmenopausal women by about 50%, but that raloxifene posed a lower risk of thromboembolic events and cataracts.[28] The USPSTF released recommendations in 2013 advocating for shared decision-making between clinicians and women at increased risk for breast cancer regarding the use of these medications for chemoprevention (Grade B recommendation). Importantly, they recommend against the use of these agents in women who are not at increased risk for the disease (Grade D recommendation).[29]

While surgical prophylaxis in the form of bilateral mastectomy and/or oophorectomy is highly effective in reducing the risk of breast cancer and ovarian cancer, respectively, these procedures are obviously associated with many physiological and psychological ramifications.[30] As a result, surgical prophylaxis is only recommended for women with a genetic predisposition, or those with a strong calculated breast cancer risk similar to that of women with a genetic predisposition.[30]

Prostate-specific medications and nutritional supplements have garnered a considerable amount of interest for their potential role in the prevention of prostate cancer. The Prostate Cancer Prevention Trial (PCPT) conducted from 1993 to 2003 was designed to assess the ability of a 5-$\alpha$ reductase inhibitor (5-ARI), finasteride, to prevent prostate cancer in men aged 55 and older. The study was terminated early due to an interim analysis, which showed that finasteride reduced the risk of developing prostate cancer by 25%. However, while the overall cancer rate was lower in the finasteride group, the proportion of men with high-grade prostate cancer was slightly higher.[31] While some experts felt that the increased high-grade cancer was real and worrisome, others thought it was merely due to sampling error. To back up that reassurance claim, an analysis on long-term data from the initial study cohort showed that finasteride decreased the risk of prostate cancer by 30%, and while high-grade prostate cancer was slightly more common in the finasteride group than the placebo group, after 18 years of follow up, there was no significant difference in overall survival between groups.[32] In another trial (REDUCE), the value of dutasteride, another 5-$\alpha$ reductase inhibitor that blocks both subtypes of the enzyme was studied in men with one pretrial negative biopsy. This study also found about a 20% lower risk of cancer in the treated arm, but, again, the risk of high-grade cancer was slightly higher.[33] After much debate, the FDA discouraged the use of 5-$\alpha$ reductase inhibitors for prostate cancer prevention.[34] Finally, the Selenium and Vitamin E Cancer Prevention Trial (SELECT) was conducted from 2001 to 2008. The study was stopped early for several reasons including the finding that there was a significant increase in prostate cancer risk with vitamin E.[35,36] Other smaller studies on vitamins, betacarotenoids, and lycopene have shown variable effects on prostate cancer, and none have been statistically significant.[37] Thus, in prostate cancer, no prevention strategies are yet officially recommended.

## ENDOCRINOLOGY

Since both breast and prostate cancers are endocrinologically driven, many of the nonsurgical treatment options target their respective hormonal axes. A complete review of each cancer is beyond the scope of this chapter, but it is worth reviewing the basic endocrinological basis with respect to treatment targets for each.

Breast cancer, like prostate cancer, is a heterogeneous disease and is classified based on the expression of three receptors: estrogen receptor-α (ERα), progesterone receptor (PR), and the erythroblastosis oncogene-B2 (*ErbB2, HER2/neu*). The expression of these markers is predictive of clinical response and prognosis.[38] Estrogens play a major role in promoting the proliferation of both normal and neoplastic breast epithelium. The biologic activities of estrogens are mediated by the nuclear estrogen receptor (ER) which, on activation by ligands, forms a homodimer with another ER–ligand complex and activates transcription of specific genes containing the estrogen response elements.[39]

Progesterone is another major, although controversial, player in mammary gland biology. In conjunction with estrogen, it acts through its specific receptor (PgR) in the normal epithelium for regulating breast development.[40] The content of ERα and PgR in the lobular structures of the breast is directly proportional to the rate of cell proliferation.[40]

The risk of developing breast cancer has been traditionally linked to exposure to estrogen, mainly because a majority of breast cancers contain receptors for this hormone. The ERα content of a tumor is considered to be a parameter of prognostic significance and guides treatment.[41] First-line therapy for receptor-positive tumors therefore consists of tamoxifen with or without an aromatase inhibitor.[42] These treatments in effect block the hypothalamo-pituitary-gonadal axis. This is similar to what is done as first-line androgen deprivation therapy (ADT) for men with metastatic prostate cancer, specifically using luteinizing hormone-releasing hormone (LHRH) antagonists to block the axis. Like prostate cancer however, breast cancer can also develop hormone-independence.[42]

Just as intratumoral androgens have been targeted in advanced prostate cancer, neoadjuvant aromatase inhibitor treatment has been shown to lower the levels of intratumoral estrogen in postmenopausal women.[43] Again as in the case of prostate cancer, the role of hormones and receptors are becoming more complex and its elucidation promises the development of more effective therapies.

Prostate carcinogenesis is similarly deeply intertwined with the male sexual hormonal axis. Androgens influence the development, maturation, and maintenance of the prostate, affecting both proliferation and differentiation of the luminal epithelium. Testosterone is irreversibly converted to dihydrotestosterone (DHT), which is the more potent androgen, and this in turn binds to the intracytoplasmic androgen receptor (AR), enhancing translocation of the steroid–receptor complex into the nucleus and activating androgen response elements.[44] Insufficient exposure of the prostate to DHT appears to protect against the development of prostate cancer. While androgen exposure seems to be a key factor in the later development of prostate cancer, the duration and magnitude of androgen exposure needed to promote carcinogenesis is unknown.[44] The treatment of advanced prostate cancer is thus centered on androgen deprivation in the serum with the use of drugs that block the hypothalamo-pituitary testicular axis, and in the tissue with AR blockers. Recently, it was discovered that malignant tissue, particularly in castration-resistant prostate cancer, produces its own androgens and upregulates the AR, thus providing a hormonally driven escape mechanism for what was heretofore called "castration-resistant" disease. This realization has led to a host of new drugs, which are now approved, (enzalutamide and abiraterone) or being developed to extend life in advanced disease. While these developments have generated great excitement, problems remain. We know that eventually, the patient becomes refractory to these new hormonal drugs from a variety of known and postulated reasons. Very briefly, they include the following: (1) the appearance of different splice variants of the AR, which may bind less specifically to ligands or be constitutively active, leading to proliferation even in the absence of androgens; (2) the development of a neuroendocrine phenotype; or (3) the development of other non-AR growth factors.[45]

Estrogens were traditionally considered to be protective against prostate cancer and have been used as treatment for advanced disease. This treatment effect is primarily through a negative feedback on the hypothalamo-pituitary-gonadal axis, and also through a direct inhibitory effect of estrogens on prostate epithelial cell growth.[44] However, there is increasing evidence across multiple species that estrogens acting through stromal ERα may contribute to prostate carcinogenesis and cancer progression.[46] Prostate epithelial ERβ may play an important role in initiation of prostate cancer, with loss of ERβ potentially contributing to disease progression in organ-confined disease.[46] In addition, the re-emergence of ERβ expression in metastatic prostate cancer also suggests a potential role in progression to castrate-resistant disease.[44]

Interestingly, based on the findings of a European cohort study, men with Klinefelter's syndrome (47 XXY) have a statistically significant increased risk of breast cancer and a statistically significant lower risk of prostate cancer.[47,48] This is thought to be due in part to the higher circulating levels of estradiol and the higher ratio of estradiol to testosterone in the serum of these men.[48]

## MOLECULAR CROSS TALK

Several studies have demonstrated that the AR and ERα can interact with each other directly to inhibit each other's transcriptional activity and cell-signaling pathways.[49] Another study showed that ERα activity was sufficiently inhibited in breast cancer cells transfected with the *AR* DNA-binding domain. This would suggest that direct competition for binding sites is also a mechanism of cross talk between the pathways. Lastly, AR and ERα are known to be influenced by common cofactors of transcription.[50]

Although ERα plays a pivotal role in driving breast cancer growth, the most commonly expressed hormone receptor in *in situ*, invasive, and metastatic breast cancers is the AR.[51] The AR is present in up to 90% of primary tumors and 25% of metastases, and appears to play various roles in different subtypes of the disease.[52–54] A growing body of evidence indicates the importance of AR activity in ER-breast cancer. In fact, a recently conducted phase II trial showed clinical benefit in 19% of metastatic breast cancer patients treated with the antiandrogen bicalutamide.[54]

## MANAGEMENT

The treatment of local disease for both breast and prostate cancers is complicated and can only be summarized briefly here. The local treatment of breast cancer has greatly changed in the last 50 years. Previously, women were subjected to radical mastectomy with various extensive lymphadenectomies. This caused significant morbidity and anxiety, and the cure rate was disappointing even after extensive screening approaches were applied. Gradually, it was hypothesized that a less-extensive surgical lumpectomy and more limited lymphadenectomy followed in most cases, and that radiotherapy was equally effective with significantly less morbidity. Later, chemotherapy with or without hormone therapy was added before and/or after surgery and dictated by the subtype, receptor, and Her2 status. More specifically, lobular carcinoma *in situ* (LCIS) is now considered to have a benign course and there is controversy over the need for open surgical excision. The National Comprehensive Cancer Network recommends only imaging follow-up for LCIS of the usual type found as a result of routine screening without imaging discordance. The mainstay of treatment for ductal carcinoma *in situ* and invasive breast cancers is surgical excision with radiation, in conjunction with systemic therapy in an adjuvant or neoadjuvant setting.[38] The choice of adjuvant and neoadjuvant therapy (radiation, hormonal therapy, or chemotherapy) is individualized to each woman based on molecular profiling of the tumor.[27] More recently, the role of lymphadenectomy has also changed. For some time, sentinel node biopsy was used to decide if and how a lymphadenectomy was to be performed. Yet recently, randomized controlled trials suggest that even in patients with positive sentinel nodes, the treatment should be systemic without further lymphadenectomy.[55] All of these developments have supported the idea that serious breast cancer is primarily a systemic disease. Thus, while local therapy helps prevent local recurrence, it does not greatly influence the clinical development and progression of systemic disease.

The local treatment of prostate cancer by surgery or radiation has also evolved over the last few decades. In surgery, the emphasis has been on better methods of total removal of the prostate and seminal vesicles (i.e., radical prostatectomy). Radiation has also concentrated on local and locoregional treatment of the whole prostate with better focus of the radiation beam and thus increasing dosages (i.e., intensity-modulated radiation therapy or IMRT). These improvements lessened its morbidity, especially with regards to impotence and incontinence, and improved cure rates. Since PSA is a supreme serum marker for persistent disease after locoregional treatment, good-, intermediate-, and poor-risk groups have been developed for pathological outcomes and PSA recurrence for both surgery and radiation based on presurgery parameters. Thanks to the revelations of screening, very-low-risk disease is now typically managed with active surveillance. The surgical treatment for the other risk groups is radical prostatectomy. Lymphadenectomy was previously a part of all prostatectomies but is now reserved for the high-risk groups and done in a more extensive fashion. Radiation is given in an adjuvant fashion for adverse pathology or as salvage therapy for PSA recurrence. There is no standardized neoadjuvant treatment for surgery.

Radiation treatment given in the form of brachytherapy or external beam is an alternative primary treatment strategy. For intermediate- and high-risk disease, temporary hormone therapy is also given. Newer, more minimally invasive options for total or even focal treatment of the prostate, such as cryotherapy and high-intensity focused ultrasound, are gaining popularity but are lacking in long-term data.[56]

In contrasting breast cancer with prostate cancer, the differences in local treatments are striking. In prostate cancer, the hope for cure still concentrates on destruction or removal of the whole gland and regional lymph nodes. In part, this is because breast cancer is probably more focal. But more importantly, the concept of the cancer as a systemic disease initially is more accepted in breast cancer. Hormonal assessment of biopsied or excised tumor is not part of prostate cancer management but is central to breast cancer management. Molecular profiling is established management for breast cancer but is just beginning in prostate cancer. Systemic neoadjuvant and adjuvant therapies are more well-established in breast cancer treatment than they are in prostate cancer management. Undoubtedly, many of these differences are due to the much greater clinical trial activity in breast cancer.

## ADVANCED DISEASE

The approach to metastatic carcinoma has changed drastically in the last decade. Both breast and prostate cancers readily metastasize to bone, leading to a host of related complications. According to one autopsy series, 64% of patients with breast cancer and 66% of patients with prostate cancer were found to have bone metastases.[58] Unlike metastatic breast cancer however, metastatic prostate cancer can be managed for years with ADT, chemotherapy, and a host of other drugs aimed at targeting different steps in the androgen receptor pathway.

The options for treating metastatic breast cancer are somewhat limited as the disease has a more rapid course than prostate cancer once metastasis is discovered. Women with visceral or life-threatening disease are typically treated with chemotherapy regardless of hormone receptor status, for faster palliation of symptoms.[58] Those without life-threatening or symptomatic visceral disease whose tumors are ER and PR positive, receive hormonal therapy first.[44] Chemotherapy is eventually initiated if the tumor becomes unresponsive to hormonal therapy. For those with Her-2/neu+ disease, trastuzumab-based therapy is the standard of care.[58]

Advanced or recurrent prostate cancer is treated with androgen deprivation, most commonly in the form of a luteinizing hormone-releasing hormone agonist, with or without an antiandrogen. As there is no molecular profiling of the tumors, this therapy is standard for all prostate adenocarcinomas. Over time, most tumors develop resistance to androgen blockade, but ADT is continued alongside the addition of various other drugs in hopes of targeting the population of cells that may still be hormone-responsive. Once metastatic lesions are confirmed, the treatment options broaden and choice of therapy becomes dependent on the patient's performance status, disease-related symptoms, and previous treatment with docetaxel. In addition to taxane-based chemotherapy, other agents used to treat metastatic prostate cancer include second-generation antiandrogens, CYP 17,20-lyase inhibitors, dendtritic cell-based immunotherapy, and radium-122.[56] While the development of novel agents in the treatment of advanced disease is promising, there is still much to be desired in the way of individualizing treatment to the prostate cancer patient. Doing so could lead to honing in on the most effective drugs for each patient early on in their treatment course and minimizing the lost time and adverse side effects of ineffective treatments.

## DORMANCY

Dormancy is a process observed frequently in certain solid tumors, characterized by a long disease-free interval between treatment of the primary tumor and relapse. Both breast and prostate cancers have the potential for long dormancy periods before disease recurrence ultimately takes the patient's life. The exact mechanisms surrounding induction of dormancy and eventual disease recurrence are not known, but molecular characterization of circulating tumor cells (CTC) and disseminated tumor cells (DTC) is gradually improving our understanding of this process. In advanced breast and prostate cancers, several studies have shown that the presence of CTCs at the time of treatment and first follow-up after primary treatment is directly and independently related to disease-specific and overall survival.[59–62] CTCs are usually rare in patients with local (i.e., early) disease but DTCs (i.e., cells detected in the bone marrow) are sometimes found in breast cancer; DTCs are found in about 15–20% of women with apparently local disease and their presence is a prognostic factor for adverse outcomes.[63] In prostate cancer, DTCs are more common. In our series, 60–70% of men have DTCs in their bone marrow prior to surgery. More importantly, almost 40% of men who appear to be NED (no evidence of disease) at 5 years continue to have DTCs, and their presence is a necessary but not sufficient condition for recurrence. Many men therefore seem to have persistent DTCs and yet will never have clinical recurrence. Thus, a hypothesis is that treating the local disease does not prevent tumor cells from escaping, but rather stops the cells that have always been escaping from turning into cells with less dormancy potential or from releasing factors that convert the escaped dormant cells to actively growing cells.[64,65] Obviously, more investigations into determining the differences in DTC, including single-cell analysis, are key to developing a better understanding.[66]

## DISCUSSION

The well-known similarities between breast and prostate cancers are increasing as we learn more about these diseases. Witness the change in data and attitudes about screening or the increasing knowledge about the molecular biology of hormone therapy. Much of the obvious differences between these cancers may be more apparent than real. For example, the much greater number of clinical trials in breast cancer have resulted in much more early systemic therapies. Also, women have greater variations in organ-stimulating hormone levels when they are young, and with advancing age, all women eventually undergo castration with menopause. It may well be that breast and prostate cancers are much more similar when these differences in normal stimulation are taken into account. What is certain is that deeper understanding of both cancers by aficionados of each disease will accelerate progress.

## References


1. *Cancer Facts & Figures 2014*. Atlanta, GA: American Cancer Society; 2014. http://www.cancer.org/research/cancerfactsstatistics/cancerfactsfigures2014 [accessed 26.09.2014].
2. Olson JS. *Bathseba's breast: women, cancer and history*. Baltimore, Maryland, USA: Johns Hopkins University Press; 2005.
3. Love RR, Philips J. Oophorectomy for breast cancer: history revisited. *J Natl Cancer Inst* 2002;**94**(19):1433–4.
4. Young HH. The early diagnosis and radical cure of carcinoma of the prostate. Being a study of 40 cases and presentation of a radical operation which was carried out in four cases. 1905. *J Urol* 2002;**168**(3):914–21.
5. Walsh PC, Lepor H, Eggleston JC. Radical prostatectomy with preservation of sexual function: anatomical and pathological considerations. *Prostate* 1983;**4**(5):473–85.
6. Huggins C. Two principles in endocrine therapy of cancers: hormone deprival and hormone interference. *Cancer Res* 1965;**25**(7):1163–7.
7. Schally AV, Kastin AJ, Arimura A. Hypothalamic follicle-stimulating hormone (FSH) and luteinizing hormone (LH)-regulating hormone: structure, physiology, and clinical studies. *Fertil Steril* 1971;**22**(11):703–21.
8. Nielsen M, Thomsen JL, Primdahl S, et al. Breast cancer and atypia among young and middle-aged women: a study of 110 medicolegal autopsies. *Br J Cancer* 1987;**56**(6):814–9.
9. Welch H, Black W. Using autopsy series to estimate the disease "reservoir" for ductal carcinoma *in situ* of the breast: how much more breast cancer can we find? *Ann Inter Med* 1997;**127**(11):1023–8.
10. Takahashi S, Shirai T, Hasegawa R, et al. Latent prostatic carcinomas found at autopsy in men over 90 years old. *Jpn J Clin Oncol* 1992;**22**(2):117–21.
11. Tulinius H. Latent malignancies at autopsy: a little used source of information on cancer biology. *IARC Sci Pub* 1991;(112):253–61.
12. Shimizu H, Ross RK, Bernstein L, et al. Cancers of the prostate and breast among Japanese and white immigrants in Los Angeles County. *Br J Cancer* 1991;**63**(6):963–6.
13. Kaylene J, Ready BKA. *Breast cancer*. 2nd ed. New York, NY: Springer Science + Business Media, LLC; 2008.
14. Yonover EG, Steven PM, Campbell SC. Breast and prostate cancer: a comparison of two common endocrinologic malignancies. In: Mydlo JH, Godec CJ, editors. *Prostate cancer science and clinical practice*. 1st ed. San Diego, California, USA: Academic Press; 2003. p. 164–9.
15. Korde LA, Calzone KA, Zujewski J. Assessing breast cancer risk: genetic factors are not the whole story. *Postgrad Med* 2004;**116**(4):6–8 chapters 11–14, 19–20.
16. Breast Cancer Linkage Consortium. Cancer risks in BRCA2 mutation carriers. *J Natl Cancer Inst* 1999;**91**(15):1310–6.
17. Hjelmborg JB, Scheike T, Holst K, et al. The heritability of prostate cancer in the Nordic Twin Study of Cancer. *Cancer Epidemiol Biomarkers Prev* 2014;**23**(11):2303–10.
18. Lerner BH. "To see today with the eyes of tomorrow": a history of screening mammography. *Can Bull Med Hist* 2003;**20**(2):299–321.
19. Kerlikowske K, Grady D, Barclay J, et al. Positive predictive value of screening mammography by age and family history of breast cancer. *JAMA* 1993;**270**(20):2444–50.
20. Catsburg C, Miller AB, Rohan TE. Adherence to cancer prevention guidelines and risk of breast cancer. *Int J Cancer* 2014;**135**(10):2444–52.
21. Karimi P, Shahrokni A, Moradi S. Evidence for U.S. Preventive Services Task Force (USPSTF) recommendations against routine mammography for females between 40–49 years of age. *Asian Pac J Cancer Prev* 2013;**14**(3):2137–9.
22. Biller-Andorno N, Juni P. Abolishing mammography screening programs? A view from the Swiss Medical Board. *N Engl J Med* 2014;**370**(21):1965–7.
23. Gomella LG, Liu XS, Trabulsi EJ, et al. Screening for prostate cancer: the current evidence and guidelines controversy. *Can J Urol* 2011;**18**(5):5875–83.
24. Moyer VA. Screening for prostate cancer: U.S. Preventive Services Task Force recommendation statement. *Ann Intern Med* 2012;**157**(2):120–34.
25. *Early detection of prostate cancer: AUA Guidelines*. https://www.auanet.org/education/guidelines/prostate cancer-detection.cfm; 2014 [accessed 26.09.2014].
26. Fisher B, Jeong JH, Anderson S, et al. Twenty-five-year follow-up of a randomized trial comparing radical mastectomy, total mastectomy, and total mastectomy followed by irradiation. *N Engl J Med* 2002;**347**(8):567–75.
27. Fisher B, Costantino JP, Wickerham DL, et al. Tamoxifen for prevention of breast cancer: report of the National Surgical Adjuvant Breast and Bowel Project P-1 Study. *J Natl Cancer Inst* 1998;**90**(18):1371–88.
28. Vogel VG, Costantino JP, Wickerham DL, et al. Effects of tamoxifen vs raloxifene on the risk of developing invasive breast cancer and other disease outcomes: the NSABP Study of Tamoxifen and Raloxifene (STAR) P-2 trial. *JAMA* 2006;**295**(23):2727–41.
29. Moyer VA. Medications to decrease the risk for breast cancer in women: recommendations from the U.S. Preventive Services Task Force recommendation statement. *Ann Intern Med* 2013;**159**(10):698–708.
30. Bevers TB. Primary prevention of breast cancer, screening for early detection of breast cancer, and diagnostic evaluation of clinical and mammographic breast abnormalities. In: Hunt KK, Robb GL, Strom EA, Ueno NT, editors. *Breast cancer*. 2nd ed. New York, USA: Springer Science + Business Media, LLC; 2003. p. 27–55.
31. Thompson IM, Tangen C, Goodman P. The Prostate Cancer Prevention Trial: design, status, and promise. *World J Urol* 2003;**21**(1):28–30.
32. Thompson Jr IM, Goodman PJ, Tangen CM, et al. Long-term survival of participants in the prostate cancer prevention trial. *N Engl J Med* 2013;**369**(7):603–10.
33. Andriole GL, Bostwick DG, Brawley OW, et al. Effect of dutasteride on the risk of prostate cancer. *N Engl J Med* 2010;**362**(13):1192–202.
34. *FDA drug safety communication: 5-alpha reductase inhibitors (5-ARIs) may increase the risk of a more serious form of prostate cancer*. http://www.fda.gov/Drugs//Drug Safety//ucm258314.htm; 2014 [accessed 26.09.2014].
35. Lippman SM, Klein EA, Goodman PJ, et al. Effect of selenium and vitamin E on risk of prostate cancer and other cancers: the Selenium and Vitamin E Cancer Prevention Trial (SELECT). *JAMA* 2009;**301**(1):39–51.

36. Klein EA, Thompson Jr IM, Tangen CM, et al. Vitamin E and the risk of prostate cancer: the Selenium and Vitamin E Cancer Prevention Trial (SELECT). *JAMA* 2011;**306**(14):1549–56.
37. Vance TM, Su J, Fontham ET, et al. Dietary antioxidants and prostate cancer: a review. *Nutr Cancer* 2013;**65**(6):793–801.
38. Fioretti FM, Sita-Lumsden A, Bevan CL, et al. Revising the role of the androgen receptor in breast cancer. *J Mol Endocrinol* 2014;**52**(3):R257–65.
39. Tsai MJ, O'Malley BW. Molecular mechanisms of action of steroid/thyroid receptor superfamily members. *Annu Rev Biochem* 1994;**63**:451–86.
40. Russo J, Hu YF, Yang X, et al. Developmental, cellular, and molecular basis of human breast cancer. *J Natl Cancer Inst* 2000;(27):17–37.
41. Khan SA, Rogers MA, Khurana KK, et al. Estrogen receptor expression in benign breast epithelium and breast cancer risk. *J Natl Cancer Inst* 1998;**90**(1):37–42.
42. Mary Pinder AB. Endocrine therapy for breast cancer. In: Hunt KK, Robb GL, Strom EA, Ueno NT, editors. *Breast cancer*. 2nd ed. New York, USA: Springer Science + Business Media; 2008. p. 411–34.
43. Takagi K, Ishida T, Miki Y, et al. Intratumoral concentration of estrogens and clinicopathological changes in ductal carcinoma *in situ* following aromatase inhibitor letrozole treatment. *Br J Cancer* 2013;**109**(1):100–8.
44. Robert Abouassaly IMTJ, Platz EA, Klein EA. Epidemiology, etiology and prevention of prostate cancer. 10th ed. Alan Wein LK, Novick A, Partin A, Peters C, editors. *Campbell-Walsh Urology*, 3. Philadelphia, PA: Elsevier Saunders; 2012. p. 2712–5.
45. Mostaghel EA. Beyond T and DHT – novel steroid derivatives capable of wild type androgen receptor activation. *Int J Biol Sci* 2014;**10**(6):602–13.
46. Prins GS, Korach KS. The role of estrogens and estrogen receptors in normal prostate growth and disease. *Steroids* 2008;**73**(3):233–44.
47. Swerdlow AJ, Schoemaker MJ, Higgins CD, et al. Cancer incidence and mortality in men with Klinefelter syndrome: a cohort study. *J Natl Cancer Inst* 2005;**97**(16):1204–10.
48. De Sanctis V, Fiscina B, Soliman A, et al. Klinefelter syndrome and cancer: from childhood to adulthood. *Pediatr Endocrinol Rev* 2013;**11**(1):44–50.
49. Panet-Raymond V, Gottlieb B, Beitel LK, et al. Interactions between androgen and estrogen receptors and the effects on their transactivational properties. *Mol Cell Endocrinol* 2000;**167**(1-2):139–50.
50. Risbridger GP, Davis ID, Birrell SN, et al. Breast and prostate cancer: more similar than different. *Nat Rev Cancer* 2010;**10**(3):205–12.
51. Fioretti FM, Sita-Lumsden A, Bevan CL, et al. Revising the role of the androgen receptor in breast cancer. *J Mol Endocrinol* 2014;**52**(3):R257–65.
52. Soreide JA, Lea OA, Varhaug JE, et al. Androgen receptors in operable breast cancer: relation to other steroid hormone receptors, correlations to prognostic factors and predictive value for effect of adjuvant tamoxifen treatment. *Eur J Surg Oncol* 1992;**18**(2):112–8.
53. Park S, Koo J, Park HS, et al. Expression of androgen receptors in primary breast cancer. *Ann Oncol* 2010;**21**(3):488–92.
54. Gucalp A, Tolaney S, Isakoff SJ, et al. Phase II trial of bicalutamide in patients with androgen receptor-positive, estrogen receptor-negative metastatic breast cancer. *Clin Cancer Res* 2013;**19**(19):5505–12.
55. Lyman GH, Temin S, Edge SB, et al. Sentinel lymph node biopsy for patients with early-stage breast cancer: American Society of Clinical Oncology clinical practice guideline update. *J Clin Oncol* 2014;**32**(13):1365–83.
56. *NCCN Guidelines: Prostate cancer*. http://www.nccn.org/professionals/physician_gls/pdf/
57. Yoneda T. Cellular and molecular mechanisms of breast and prostate cancer metastasis to bone. *Eur J Cancer* 1998;**34**(2):240–5.
58. Green MC, Hortobagyi GN. Chemotherapy for metastatic breast cancer. In: Hunt KK, Robb GL, Strom EA, Ueno NT, editors. *M. D. Anderson cancer care series breast cancer*. 2nd ed. New York, NY, USA: Springer Science + Business Media, LLC; 2008. p. 364–79.
59. Cristofanilli M, Budd GT, Ellis MJ, et al. Circulating tumor cells, disease progression, and survival in metastatic breast cancer. *N Engl J Med* 2004;**351**(8):781–91.
60. Saloustros E, Perraki M, Apostolaki S, et al. Cytokeratin-19 mRNA-positive circulating tumor cells during follow-up of patients with operable breast cancer: prognostic relevance for late relapse. *Breast Cancer Res* 2011;**13**(3):R60.
61. Ma X, Xiao Z, Li X, et al. Prognostic role of circulating tumor cells and disseminated tumor cells in patients with prostate cancer: a systematic review and meta-analysis. *Tumour Biol* 2014;**35**(6):5551–60.
62. Cristofanilli M, Hayes DF, Budd GT, et al. Circulating tumor cells: a novel prognostic factor for newly diagnosed metastatic breast cancer. *J Clin Oncol* 2005;**23**(7):1420–30.
63. Janni W, Vogl FD, Wiedswang G, et al. Persistence of disseminated tumor cells in the bone marrow of breast cancer patients predicts increased risk for relapse – a European pooled analysis. *Clin Cancer Res* 2011;**17**(9):2967–76.
64. Morgan TM, Lange H, Porter MP, et al. Disseminated tumor cells in prostate cancer patients after radical prostatectomy and without evidence of disease predicts biochemical recurrence. *Clin Cancer Res* 2009;**15**(2):677–83.
65. Ruppender NS, Morrissey C, Lange PH, et al. Dormancy in solid tumors: implications for prostate cancer. *Cancer Metastasis Rev* 2013;**32**(3-4):501–9.
66. Welty CJ, Coleman I, Coleman RET-AL>. Single cell transcriptomic analysis of prostate cancer cells. *BMC Mol Biol* 2013;**14**(6):1–11.

P A R T IV

# PREVENTION OF PROSTATE CANCER

CHAPTER

# 20

# Heart Healthy = Prostate Healthy and S.A.M. are the Ideal "Natural" Recommendations for Prostate Cancer

*Mark A. Moyad, MD, MPH*

**Department of Urology, University of Michigan Medical Center, Ann Arbor, MI, USA**

## KEY POINTS

Cardiovascular disease (CVD) preventive strategies via lifestyle, dietary supplement, and pharmacologic agents appear to have the best current evidence for prostate cancer prevention, especially aggressive disease.

Obesity, dyslipidemia, glucose intolerance, lack of exercise, caloric excess, and even metabolic syndrome should be considered prostate cancer risk factors.

Dietary supplements in high-doses that have not been found to prevent CVD and could increase the risk of CVD and could increase the risk of prostate cancer.

Statins, aspirin, and metformin (S.A.M.) are heart healthy, have unique mechanisms of action and appear to be the best potential interventional nontraditional agents for prostate cancer when considering the benefit to risk ratio.

Heart Healthy = Prostate Healthy and Heart Unhealthy = Prostate Unhealthy along with S.A.M. should be a simplistic teaching tool utilized by clinicians to practically and logically reinforce simplistic antiprostate cancer strategies.

## INTRODUCTION

Men's health issues should be triaged before recommending and construing any ideal prostate cancer prevention or treatment program. Emphasizing the primary causes of past and current morbidity and mortality allows for an easier understanding of lifestyle, supplement, or pharmacologic additions or deletions. This advice needs to be quick but pithy, simple, logical, and practical for the patient as well as the clinician. When overall health concerns are triaged, it will be easier to construe and advocate for the ideal prostate cancer patient program supported by the phrase heart healthy = prostate healthy.[1]

Cardiovascular disease (CVD) is the number one overall cause of mortality in the United States and in other industrialized countries, and it is currently the number one cause of death worldwide.[2,3] Cancer is the second leading cause of death in the United States and in most developed countries, and it is expected to equal or surpass the number of deaths from CVD in the future. CVD has been the number one cause of male deaths in the United States for approximately 114 out of the last 115 years, only surpassed one year by the influenza pandemic in 1918 (classified at that time as "diseases of the heart," CVD ranked number two that year). If a college football team was number one in the United States for virtually 115 years, whether or not one enjoys watching this sport would not be the issue because there would arguably be such dominant awareness and attention paid to this streak because it would be so remarkable and unprecedented. How aware would such a dynasty be to the public and clinicians if it involved more than just football, but individual risk of morbidity and mortality?

If cancer becomes the primary cause of death in the near future, the majority of what is known concerning lifestyle and dietary change for CVD prevention appears to directly apply to cancer prevention and most other prevalent chronic urologic and nonurologic diseases.[4,5] For example, one of the most significant reductions in early morbidity and mortality rates in US history for CVD and cancer was via a common behavioral/lifestyle change – smoking cessation.[6–8] The opposite side of this pendulum is also true; for example, the global prevalence of tobacco use remains high and is still the largest

Prostate Cancer. http://dx.doi.org/10.1016/B978-0-12-800077-9.00020-7

preventable cause of death from CVD and cancer.[9] Another example, obesity, which not only continues to increase the risk of CVD, but numerous cancers, diabetes, early morbidity, and mortality, may erase the advances in the declines previously observed in early CVD and cancer mortality from smoking cessation.[10–12] Over the past 30 plus years not a single country has reported success in combating the obesity epidemic.[13]

Men have a consistently lower life expectancy in most countries, and have a higher morbidity and mortality from heart disease, hypertension, cancer, and diabetes.[14] Yet, heart healthy changes are tantamount to overall men's health improvements regardless of the part of the human anatomy that is receiving attention, including the prostate and the penis.[1,15] Heart healthy changes should be advocated to men concerned about prostate cancer because it places research and probability into perspective, which can impact all-cause mortality as well as potentially prevent or slow the progression of prostate cancer.

## PRIMARY PREVENTION TRIALS UTILIZING A PHARMACOLOGIC AGENT – THE UNTOLD STORY

A fascinating feature of large randomized clinical trials, especially for primary prevention, is that they appear to mirror the current health status and risk issues of not only the subjects being tested but perhaps the general population. For example, results of the PCPT seem to have garnered attention and controversy regarding the use of finasteride daily versus placebo to reduce the risk of prostate cancer.[16–19] However, another observation from this trial has not received adequate exposure in the medical literature. Over 18,000 healthy men were included in this randomized trial, and five men died from prostate cancer in the finasteride arm, while five men died of prostate cancer in the placebo arm. However, 1123 men in total died during this primary prevention trial. Prostate cancer was responsible for approximately less than 1% of the deaths, while the majority of the overall causes of mortality were from CVD and other notable causes.[20,21] The mean BMI, systolic blood pressure, and total and HDL cholesterol were the following: 27–28 (50% overweight, and approximately 25% obese), 138–140 mmHg (prehypertensive), 212 mg/dL and 42–43 mg/dL (dyslipidemia or at-risk). Despite 85% of men with no history of CVD, approximately 50% of the men reported some level of erectile dysfunction (ED).[20]

The more recent international dutasteride prevention trial known as REDUCE (Reduction by Dutasteride of Prostate Cancer Events) had somewhat similar issues to the North American PCPT in terms of newsworthy controversies,[22–24] but what was not discussed was that the BMI and several other abnormal CVD parameter issues mentioned earlier were similar in the two trials. On average, men in REDUCE were overweight (BMI of 27–28).[22] There were 8231 men randomized and after the 4-year trial in this group of high-risk men there were 147 total deaths, primarily from cardiovascular events and none from prostate cancer. Of further note, men in the placebo arm of PCPT with low cholesterol (<200 mg/dL) had a 59% ($p = 0.02$) apparent reduction in risk of being diagnosed with aggressive prostate cancer (Gleason 8–10) compared to men with high cholesterol ($\geq$200 mg/dL),[25] and men with coronary heart disease (CHD) at baseline in REDUCE were found to have a significantly higher risk of a prostate cancer diagnosis, and this included low-grade (OR = 1.34, $p = 0.02$) and high-grade cancer (OR = 1.34, $p = 0.09$).[26] These observations place the overall risk of morbidity and mortality in a more proper perspective. Patients inquiring about the advantages and disadvantages of 5-alpha reductase inhibitors for prostate cancer prevention need to be reminded that the number one risk to them in general is CVD, and in both clinical trials the researchers found that heart health was tantamount to prostate health (Table 20.1).[1]

## NOTABLE DIETARY SUPPLEMENT CANCER PREVENTION TRIALS – THE UNTOLD STORY

The Selenium and Vitamin E Cancer Prevention Trial (SELECT) was the largest dietary supplement trial ever completed in prostate cancer.[27] It randomized over 35,000 men into four groups: high-dose vitamin E (400 IU/day), high-dose selenium (200 μg/day), vitamin E and selenium, or placebo. The trial was terminated early after a median of 5.5 years due to a lack of efficacy, although at the time a nonsignificant ($p = 0.06$) increase risk of prostate cancer in the vitamin E arm, and type 2 diabetes in the selenium group ($p = 0.16$) were observed.

Still, participant follow-up continued (54,464 added person-years), which provided more clarity of the further health impacts after the cessation of these agents.[28] A significant ($p = 0.008$; HR = 1.17) increased risk of prostate cancer was observed in the vitamin E group, and the increased risk with this individual supplement began to emerge after only 3 years, and was found to be consistent for low- and high-grade disease types. The risk of Gleason 7 or higher disease was greater for the three intervention arms compared to placebo, but did not reach statistical significance. The HR and $p$-value for Gleason 7 and higher disease compared to placebo was 1.16 ($p = 0.20$), 1.21 ($p = 0.11$), and 1.23 ($p = 0.08$) for vitamin E, selenium, and the combination, respectively.

TABLE 20.1 Heart Healthy, Heart Unhealthy, and Other CVD Parameters and Interventions and Their Potential Correlation With the Risk and Progression of Prostate Cancer

| CVD parameter/interventions | Correlation with prostate cancer risk |
|---|---|
| Aspirin (low dose, originally derived from willow bark) | Determine if a man concerned about prostate cancer qualifies for aspirin based on CVD risk working with his physicians and utilizing risk scores such as Framingham or Reynolds Risk Score. Men with intermediate to high risk of a CVD risk may qualify especially if they have a low risk of an ulcer or bleeding event. |
| Caloric control or reduction/diet | Reducing caloric intake to assist in preventing weight gain may reduce prostate cancer risk and PSA velocity, which could reduce unneeded biopsies and maintain focus on heart health. |
| Diabetes/glucose intolerance | Associated with a higher risk of aggressive prostate cancer. |
| Dieting | A variety of "Fad" diets have the ability to be heart healthy. Patients should be monitored for changes in cholesterol, glucose, blood pressure, weight/waist, and mood during a specific diet to determine if it is heart healthy. Also, following one low-cost inflammatory marker such as hs-CRP is also advised along with PSA and testosterone. |
| Dyslipidemia | Associated with a higher risk of aggressive prostate cancer. |
| Exercise | Associated with a lower risk of prostate cancer and possibly aggressive disease. |
| Folic acid supplements/high-dose B-vitamin supplements | Folic acid in excessive dosages has been associated with a higher risk of prostate cancer. |
| Heart Healthy = Prostate Healthy and S.A.M. | The patient and clinician mantra in terms of what lifestyle changes to follow and pills to consider before and after a prostate cancer diagnosis. |
| Hypertension | Part of the spectrum of metabolic syndrome that could increase the risk of prostate cancer. |
| Metformin | Reduces IGF-1, diabetes risk, weight, and gluconeogenesis, which together may reduce the risk of prostate cancer and reduce the side effects of LHRH treatment. |
| Multivitamin | One pill of day of a low-dose multivitamin or children's multivitamin in adults is safe and may lower overall cancer risk, but more than 1 multivitamin a day may increase risk of aggressive prostate cancer. Centrum Silver is the only multivitamin utilized in a phase-3 like trial. |
| Obesity | Associated with an increased risk of aggressive and fatal prostate cancer, and hemodilution of the PSA test (false negatives). |
| Selenium dietary supplements | High-dose selenium supplements may increase the risk of diabetes and could significantly increase the risk of aggressive prostate cancer in men already replete with selenium from dietary sources. |
| Smoking/tobacco | Increases the risk of aggressive and fatal prostate cancer, and reduces the blood level of numerous nutrients that otherwise are protective for overall and prostate health. |
| Statins | Associated with a lower risk of aggressive and advanced prostate cancer. |
| Vitamin D supplements | Potentially a U- or J-shaped curve with higher blood levels showing an increased risk and normalization of deficient or insufficient levels showing a reduction in risk. Caution should be advised with supplementation and the daily recommendation is 600–800 IU per day. |
| Vitamin E supplements | High-dose vitamin E supplements (400 IU) significantly increase the risk of prostate cancer and could increase the risk of aggressive disease. |
| Zinc supplements | High-dose individual zinc supplements have not been shown to reduce risk and may actually increase risk of other urologic conditions. |

The negative findings from SELECT cannot be construed by increased biopsy rates or bias, but rather the high-dose dietary supplements themselves were responsible, and the confidence intervals to support this thought have continuously narrowed over time.[27,28] Other observations from secondary endpoint analyses included other cancers and cardiovascular events, but no statistical differences were found compared to placebo. Yet, what should receive more attention was the observation that CVD events and deaths represented the primary cause of morbidity and mortality overall in this trial in all four treatment arms. There were over 4200 cardiovascular events and over 500 CVD deaths that occurred compared to 1750 prostate cancers diagnosed and one death from prostate cancer. There were 3363 cancers diagnosed overall (including prostate) and 476 deaths from cancer, which again emphasizes the need for future agents or lifestyle interventions to demonstrate activity against CVD and cancer, because the global burden of cancer is beginning to compete with CVD,[29] again as reflected in the SELECT trial.

The results of SELECT could have been even more negative over time if the interventions were continued. Regardless, even if any of these agents would have prevented prostate cancer it is questionable whether they would have provided a tangible overall clinical advance. The controversy that plagued high-dose vitamin E and selenium supplements from past clinical trials was the lack of evidence or at times negative impact these supplements had on overall mortality and on CVD.[30–35]

Neither the form of vitamin E (synthetic or natural source – a seemingly constant debate), or even frequency of utilization of this supplement in larger dosages would have arguably provided a difference in the SELECT trial, especially with regard to CVD and cancer outcomes.[31,33,36–39] One noteworthy trial (HOPE TOO) actually found a significantly higher rate of heart failure with a naturally derived vitamin E supplement.[33] Another large randomized trial of vitamin E and prostate cancer risk in healthy males, the Physicians Health Study II (PHSII), found no impact of 400 IU of vitamin E every other day compared to placebo,[36] but a significant increased risk of hemorrhagic stroke was documented.[37]

Additionally, the Alpha-Tocopherol, Beta Carotene (ATBC) trial of over 29,000 men demonstrated a 35% risk reduction of prostate cancer risk with a vitamin E supplement from a secondary endpoint, and provided some verve for the design and initiation of SELECT.[38] However, the dosage utilized in the ATBC was only 50 IU, approximately eight times lower compared to SELECT, and a higher rate of hemorrhagic stroke was also found in ATBC. The number one and two overall cause of death during ATBC and at postintervention follow-up was ischemic heart disease and lung cancer.[38,39] Men in ATBC were chronic 36-year on average smokers, and tobacco users are at a higher risk of diverse nutrient deficiencies including vitamin E.[40] Less than 10% of SELECT participants were current smokers,[27] and one can only ponder the outcome of this trial or others had a lower dose been utilized in a more representative population of generally healthy men. Healthy and primarily nonsmoking or former smoking men (85% of the participants) from a unique randomized trial (SUVIMAX) utilizing far lower doses of vitamin E (30 IU) and several other supplement ingredients demonstrated the potential for significant overall benefit and prostate cancer prevention, but also harm for men with higher baseline PSA levels.[41,42] Less may be more in otherwise healthy men.

What about selenium dietary supplements? What should patients know? Again, the impact of high-dose selenium supplements on heart and overall health from past studies were arguably as troubling as past vitamin E clinical research especially in those replete with this nutrient,[43,44] and included a potential significant increased risk of type-2 diabetes and nonmelanoma skin cancer recurrence.[45,46] Interestingly, the increased risk of skin cancer recurrence was the final conclusion of the primary endpoint of the randomized selenium supplement US trial (Nutritional Prevention of Cancer or NPC) initiated in the 1980s and completed in the 1990s.[45,47] It was the NPC trial secondary endpoint results, for example, the lower rate of prostate cancer, that were an impetus for the design and initiation of the SELECT trial.

It is also plausible that the SELECT researchers or even future investigators testing individual supplements for cancer or CVD prevention will not be capable of initiating a nutritionally uncontaminated trial. This could be due to a novel situation occurring with the ongoing US and global popularity of functional foods and supplements,[48,49] and I have referred to it as the "over anti-oxidation of the population."[50] If any nutrient appears to impact some common medical condition without adequate long-term research, no entity exists to discourage the ability of nutritional commercial products in the United States to add more of these nutrients to multivitamins, protein bars, energy drinks, water, and so on! For example, baseline serum selenium status in SELECT was actually 22 points higher (135 ng/mL vs. 113 ng/mL) compared to notable NPC trial completed in the 1990s in the United States. NPC participants who were selenium deficient eventually experienced a potential reduced prostate cancer risk, but a higher rate of cancer occurred in a small group of individuals with repletion of baseline selenium levels.[27,51,52] Most SELECT participants were already selenium sufficient at baseline and were recruited from all over the United States including some of the same geographical areas as the NPC trial participants only more than a decade later. I believe the increased addition of selenium (and vitamin E) in foods, beverages, and supplements, increased overall consumption of these functional foods and calories, and the reduction in smoking since the 1990s all greatly assisted in the normalization of selenium. For example, locating a multivitamin with selenium in the 1980s or 1990s was difficult and today finding any multivitamin without selenium is almost impossible. Approximately 30% of NPC versus 8% of SELECT participants were current smokers (smokers have lower selenium levels). Finally, as a striking example of this overall concern, during the time this chapter was being written, members of the SELECT research team reported a significant increased risk of aggressive prostate cancer in men replete with selenium before utilizing a selenium supplement.[53] This continues to raise concerns over adding antioxidants in large quantities for healthy individuals already receiving sufficient levels of these compounds (as argued previously).

Some have claimed that the reason numerous antioxidant trials have been neutral or negative overall in medicine is because nutritionally sufficient rather than insufficient or deficient individuals are subjects of these studies.[54] However, multiple years are required to initiate any large-scale clinical trials. Thus, the initially depleted participants being tested will eventually be replete with the interventions being utilized before the trial officially commences. This will represent a challenge to any further nutrient trial in industrialized countries, and perhaps this is why other supplements such as omega-3 or vitamin D supplements for example have not been found to have dramatic impacts in other areas of medicine from recent trials or reviews.[55,56]

Utilizing high-dose supplements in an already replete population could result in the nutrient in question to function as a pro-oxidant or disease initiator and promoter rather than an antioxidant. This is what could have occurred in the case of vitamin E,[57] and with selenium,[58] or with another nutrient such as folic acid or other B-vitamins, which has already been observed in multiple randomized clinical trials to be a potential prostate and potentially other cancer risk factor in excessive dosages.[59,60]

I would also argue that the lack of attention toward supplements that are overall or heart healthy or first could "do no harm" is another reason for disappointing results from past clinical supplement trials. And, there is an immediate need to garner preliminary direct data initially from phase-1 and -2 trials (similar to drug development) before moving straight toward a phase-3 trial (as occurred with SELECT and other supplement studies).

## MULTIVITAMINS AND OTHER DIETARY SUPPLEMENTS – LESS IS MORE

Dietary supplement research within cancer should arguably revolve around testing lower dosages to ensure safety first and potential efficacy, or simply test supplements for specific conditions/side effects of conventional treatment rather than prevention itself. The result of the first major phase-3-like randomized trial of multivitamins versus placebo was published and the primary endpoints were total cancer incidence and cardiovascular events. The Physicians Health Study II (PHS2) found a significant ($p = 0.04$) 8% cancer reduction in total cancer incidence compared to placebo in a healthy group of subjects 50 years or older ($n$ = 14,641, 11.2 years of follow-up).[61] A larger nonsignificant 18% reduction was found for men aged 70 and over at baseline and those with a history of cancer (−27%), but no benefit was found for those with a parental history of cancer. Current smokers (less than 4% of the participants) appeared to receive a large benefit (−28%) compared to former and never smokers. There was minimal impact on prostate cancer, but it should be of interest that low-dose multivitamin with a similar side effect to placebo significantly and modestly reduced total cancer incidence in a group of primarily healthy men. Further subgroup analysis found that men consuming seven or more fruits and vegetables per day benefitted as much as those that consumed less than four servings a day, and those with a normal BMI appeared to benefit as much as overweight or obese men. There was no increased or decreased risk of this multivitamin on cardiovascular events, which is reassuring. Although, fatal myocardial infarction (a secondary endpoint) was reduced by 39% ($p = 0.05$) in the multivitamin group, but especially in those men without a baseline history of CVD (−44%, $p = 0.03$).[62] Additionally, cataract incidence (another secondary endpoint) was modestly but significantly reduced (HR = 0.91; $p = 0.04$) in PHS2,[63] and a recent meta-analysis has suggested a similar overall observation in the literature with the use of multivitamins from past studies.[64] No impact of multivitamins was observed for age-related macular degeneration or cognition.[65]

Healthcare professionals and patients often ask if I recommend a multivitamin for healthy individuals or those with a history of prostate cancer. Pundits appear to enjoy taking opinions on two ends of the spectrum: either do not waste your money or take another more expensive product to see a better result. My argument is simple from the PHS2: one of the cheapest multivitamins (Centrum® Silver) had the same safety as placebo over 11 years, modestly reduced the risk of cataracts and cancer (two massive burdens on the healthcare system) in healthy men, only requires one pill a day, and appears to be heart healthy so this is what I use and recommend (FYI, no conflict of interest). When this trial was designed, I believed that minimal to nothing would occur in these men because of their baseline health, which was again outstanding. I was incorrect and have been surprised by the positive data and so this is another reason I advocate the use of this product in those with and without a personal history of cancer.

It is also interesting that the original Centrum Silver utilized during this trial from 1997 to 2011 is not the exact OTC product offered to consumers currently because over time these nutritional formulations appear to change based on some science and marketing.

Therefore, if one is impressed by these data and my argument then in reality a single children's multivitamin could be recommended for an adult because this dosage appears to be similar to an older Centrum Silver, or the patient should just consume the newest Centrum Silver or something close to the formula that is detailed in the clinical trial publication.[61]

More has not shown to be better in terms of multivitamins. Some of the largest past prospective epidemiologic studies are suggesting a greater rate of aggressive and fatal prostate cancer when consuming more than one multivitamin a day, and even further increasing risk when other high-dose individual supplements are also utilized (selenium, vitamin E, and zinc).[66] Men with a family history of prostate cancer appeared to experience the largest and most significant elevated risks of this condition. Other large male observational studies have found somewhat similar results with multivitamins and some individual supplements.[67–69] Multivitamins are also replete in my experience with higher doses of B-vitamins such as B12 and folic acid, which have also recently been found to potentially have no impact on health or increase the risk of prostate cancer from the largest and most recent meta-analysis of clinical trials.[59,70,71] There is no consistent suggestion of benefit with a greater intake of multivitamins or any other vitamin or mineral in supplement form in healthy individuals, but there is a suggestion of either no impact or serious harm; it would be prudent to "first do no harm" and wait for more clarity from additional clinical data.[72]

Most other dietary supplements are replete with their own issues in terms of prostate cancer prevention or treatment. For example, vitamin D in mega-doses may have some similar issues to vitamin E or selenium. The tendency for some clinicians to recommend more vitamin D and patients to ingest more of this supplement is concerning. In the area of prostate cancer prevention vitamin D has not been impressive. Several epidemiologic studies have found either no impact or a potential increased risk of aggressive prostate cancer or total cancer at higher 25-OH vitamin D blood levels.[73,74] Vitamin D is important for bone health, but the amount required has been embellished. Vitamin D tends to function more like a hormone, which is why caution should be followed because the potential for a U- or even J-shaped risk curve does exist for male health in general.[75] One of the largest and longest randomized trials in elderly women found that excessively high blood levels of vitamin D from high-dose supplementation compared to placebo was actually associated with an increased risk of falls and fractures.[76] The normal level of vitamin D (25-OH) could be 30–40 ng/mL based on benefit versus risk philosophy and expert opinion from a review of past clinical trials accessing multiple outcomes.[77] Still, even vitamin D blood tests have a history of uncertainty based on the assay utilized.[78,79] And the cost of this nonvalidated test is also an issue. Monitoring vitamin D in men, especially higher-risk bone loss patients with vitamin D deficiency, for example, men on androgen deprivation therapy (ADT) for prostate cancer, may be more appropriate.[80] For prostate cancer prevention the vitamin D test may provide more harm than good until more clinical endpoints are followed in healthy individuals.[79] The latest Institute of Medicine (IOM) should also be a reminder that despite the perception, the recommended intakes of vitamin D have only increased by 200 IU (5 μg) in most groups and vitamin D supplements have the potential to increase the risk of hypercalcemia and nephrolithiasis.[81]

Vitamin D blood levels may also simply be a marker of healthy behavior.[1] Higher levels of systemic inflammation and greater burden of many diseases can inhibit vitamin D synthesis rendering supplementation futile and meaningless in some cases. Additionally, a lean man with a low cholesterol, who consumes fish and exercises regularly, is more likely to have a higher blood level of vitamin D compared to a physically inactive overweight or obese man with a high cholesterol level and other heart unhealthy parameters.[82,83] Patients should be reminded that improvement in heart healthy parameters could increase vitamin D levels without or with additional smaller increments in supplementation. This moment represents an opportunity to emphasize heart healthy lifestyle changes first before increasing pill counts.

## LIFESTYLE MATTERS (HEART HEALTHY = PROSTATE HEALTHY)

Any lifestyle change that mitigates the risk of heart disease has ample evidence it could reduce the risk of prostate cancer, and parameters that increase the risk of heart disease increase the risk of prostate cancer. Belaboring this point or reviewing mechanisms of action or lifestyle changes in extensive detail is not the purpose of this manuscript, and this detailed information is found in multiple past-written resources.[1,80,84,85]

Patients should be encouraged to do whatever is practical and plausible to reduce their risk of CVD to as close to zero as possible. This should provide the greatest potential to not only reduce the risk of prostate cancer, but other disease morbidity and even impact all-cause mortality. Most major behavioral risk factors for CVD morbidity and mortality appear to be correlated with a higher risk of aggressive prostate cancer and/or fatal prostate cancer. For example, smoking is the single largest preventable cause of death and disease in the United States with approximately 443,000 deaths occurring per year from tobacco-related disease, and approximately 20% of adults smoke, which is a number that has remained constant the past several years.[86,87] Smoking has been associated with a higher risk of being diagnosed with prostate cancer in past meta-analyses,[88] a higher risk of aggressive prostate cancer, and dying from prostate cancer.[89,90] Obesity is also associated with a higher risk of aggressive and fatal prostate cancer,[91] and a higher risk of recurrence post-treatment.[92] It is also plausible that obesity is associated with a lower risk of localized prostate cancer, and a higher risk of advanced disease due to the artificial lowering of PSA or hemodilution.[93,94]

Ancillary pertinent issues abound when dealing with obesity in patients. For example, ongoing evidence suggests an increased risk of cancer with insulin resistance, and this may include aggressive prostate cancer.[95–98] Higher concentrations of growth factors occur with increased insulin levels, but long-term diabetes may result in insulin, IGF, and androgen reduction, which may be correlated with a lower prostate cancer risk in the short-term ("diabetes paradox").[98] The profound increase in the diabetes epidemic,[99] along with the known 2–4 times increased risk of CVD events in diabetics over nondiabetics,[100] should make type II diabetes prevention strategies a priority for simultaneous prostate cancer prevention. In the United States, the impact is tangible; only 15 years ago three states in the United States, had a diabetes prevalence of 6% or higher, but now all 50 states in the United States, have a rate of at least 6% or higher.[99] Six states have rates of 10% or more along with Puerto Rico, and currently approximately 20 million people in the United States have diabetes and 7 million are undiagnosed. Perhaps, cancer prevention strategies can help to modestly curb this epidemic. Exercise (aerobic and resistance), dietary (caloric reduction), and other moderate to aggressive lifestyle changes have been shown to significantly prevent diabetes and metabolic syndrome in normal and high-risk individuals in many cases better than pharmacologic therapy.[101–105]

Regular vigorous exercise (approximately 3 h or more per week) is a profound potential strategy to significantly reduce prostate cancer death after diagnosis, and simultaneously reduce all-cause mortality to a similar degree (50–60%) in these same patients compared to men that perform only 1 h or less exercise per week.[106] A review of past studies including a recent summary of 22 studies published over the past 12 years supports the modest prostate cancer prevention effects of exercise.[107–109] Noteworthy reductions in blood pressure, diabetes, depression, dyslipidemia, cancer, CVD, fatigue, obesity, and multiple other conditions would arguably be enough to earn exercise a Nobel Prize if it was a pharmacologic agent.[1]

A primary risk factor for CVD and stroke is hypertension and almost a third of the US adult population have this condition.[110] Hypertension increases with age to approximately 70% of individuals 65 years and older. Hypertension is a contributing factor in one out of every seven deaths, and 70% of individuals who have a first heart attack or stroke have hypertension.[111] Treating hypertension has been correlated with dramatic reductions in the incidence of stroke (40%), heart attacks (25%), and heart failure (>50%).[112] The correlation between prostate cancer risk and hypertension and/or antihypertensive medications are currently weak.[113] However, high blood pressure as part of a continuum of unhealthy parameters, such as observed with metabolic syndrome (central obesity, dyslipidemia, and insulin resistance), is becoming a potential risk factor for prostate diseases including cancer.[114] Alpha-blockers, originally discovered for blood pressure control, are now one of those most effective treatments for men with prostate issues (BPH) and lower urinary tract symptoms (LUTS) despite not having consistent positive or negative impacts on prostate cancer risk.[115,116] Preventing or controlling hypertension can improve prostate health. And, perhaps the reason that blood pressure medications such as beta-blockers do not have data in the area of prostate cancer could be due to the lack of extensive research or inability of past studies to correct for pertinent confounding variables in order to resolve this question. Some recent clinical studies are suggesting a survival benefit in men on these medications that should continue to sustain the debate.[117] Dietary and lifestyle changes have provided similar effects to most low doses of standard antihypertensive pharmacologic agents in some of the largest clinical trials to address this issue. The D.A.S.H. study was arguably the most noteworthy and patients should be reminded that the synergistic impact of exercise and diet demonstrated blood pressure lowering effects seen with most antihypertensive drug classes.[118] Caloric reduction in overweight and obese patients with hypertension may also provide similar benefits.[119]

## THE IMPORTANCE OF S.A.M. (STATINS, ASPIRIN, AND/OR METFORMIN) AND PROSTATE CANCER

Often, I am asked about the primary supplements patients should consume along with their conventional treatment for prostate cancer or for prevention. Earlier, the discussion of "less is more," multivitamins, vitamin D, and others were covered. An ancillary pill in prostate cancer should be proven to be heart healthy, cost-effective, benefits need to outweigh the risks for patients, and there should be minimal to no chance that this agent interferes with proven conventional treatment. Thus, the list is short but three such interventions exist and arguably should be discussed and also referred to the patient's primary care doctor. The acronym S.A.M. (statins, aspirin, and metformin) should arguably receive the most initial attention (everything else is secondary in my discipline). All three of these agents are "natural" (statins were derived from yeast, aspirin from willow bark, and metformin from the French Lilac), generic, heart healthy, low cost, and if a patient qualifies (benefit outweighs the risk) then there are arguably no other ancillary pills/supplements that have as much overall positive data. Some supplements may be able to mimic the activity of these agents and could serve as an alternative for drug intolerant patients (for e.g., red yeast rice supplements instead of a statin).[1]

Approximately 35% of Americans have dyslipidemia,[120] and lower lipid levels have been associated with a lower risk of aggressive prostate cancer,[1] and increases in HDL ("good cholesterol") may also be protective.[121] Heart disease may increase the risk of prostate cancer from observations derived from two major pharmacologic studies of prostate cancer prevention,[25,26] and a variety of epidemiologic investigations.[1] Some of the largest observational studies have suggested a lower risk of aggressive prostate cancer with cholesterol lowering interventions even when controlling for multiple confounding variables.[122] Statins should be investigated as a prostate cancer adjuvant interventional agent, and there is a trial that is currently recruiting to determine the role of lipid lowering in the active surveillance prostate cancer population and its impact on the progression of this disease.[123] Some might argue that it is currently too difficult to conduct such a trial when a large proportion of men are already utilizing these agents. This is not accurate. The JUPITER trial is the most recent example of the dramatic potential impact on cardiovascular health when aggressive lipid lowering is accomplished in individuals who are in no apparent need of a statin based on their baseline LDL levels (approximately 100 mg/dL or mmol/L).[124,125] The low-cost, CVD impact, overall benefit-to-risk ratio in a healthy population of men, potential prostate cancer impact, and a plethora of the basic science and clinical evidence suggest that like the drug metformin, statins should be a priority intervention in clinical trials for the potential prevention of total and aggressive prostate cancer.[1,126–128] Some positive data existed over a decade ago to potentially study this class of agents in a large trial for prostate cancer prevention.[123,126–128]

Aspirin has been garnering an impressive amount of preliminary data as a potential anticancer agent, especially in the area of colorectal cancer prevention.[129,130] However, the overall data also suggest a reduced risk of prostate cancer or aggressive disease with this agent.[131–134] The ability of aspirin to further reduce prostate-cancer-specific mortality in men receiving radiation or surgery has also garnered some preliminary research.[135] If a benefit appears to be derived from aspirin it appears, as with statins, that men with high-risk disease features or a more aggressive molecular profile consuming these products for longer periods of time (5+ years) are the ones most likely to benefit.[136,137] Interestingly, the first large meta-analysis and systematic review of 39 studies (20 case-control and 19 cohort) of aspirin and prostate cancer was recently published and concluded with the following statement: "The present meta-analysis provides support for the hypothesis that aspirin use is inversely related to prostate cancer incidence and PCa-specific mortality."[138] Thus, as with statins, clinicians will need to address the positives and negatives of aspirin use in prostate cancer and refer the ultimate decision to utilize this agent based on a discussion with their primary care provider. If the patient qualifies for aspirin based on his total cardiovascular risk, then it should be encouraged.[139,140] I often ask patients to utilize the web site www.reynoldsriskscore.org to access their risk-versus-benefit ratio (about 30 s involving several questions and numbers) and discuss their results with their primary care doctor.[1]

Metformin significantly reduces diabetes risk long-term and is cost-effective with a low rate of adverse events, and has the ability to also reduce the risk of CVD events and impact all-cause mortality.[141–144] Metformin is also beginning to demonstrate evidence as a cancer prevention or recurrence inhibition agent in those with and without diabetes and is currently in a phase-3 trial in breast cancer patients with survival as the primary endpoint.[145] A preliminary small phase-2 trial has suggested it may have the potential to slow the progression of even advanced prostate cancer.[146] Additionally, a recent clinical trial of patients with prostate cancer on ADT for 6 months utilizing 850 mg twice a day of metformin with a suggested low glycemic diet (caloric restriction) were able to significantly reduce weight gain, BMI, waist circumference, and systolic blood pressure compared to the control group.[147] Men on metformin were also able to control glucose and HgbA1c levels. Metformin is cost effective, safe, reduces weight gain, diabetes risk, and arguably CVD and perhaps total cancer and prostate cancer risk,[1,148] and many laboratory analyses suggest it may be synergistic with a variety of standard and non-standard agents (including statins).[149,150] Such a combination fits my criteria for an ideal prostate cancer ancillary interventional agent.[151]

## CONCLUSIONS – SOLVING THE DIET AND SUPPLEMENT DEBATE

CVD is the number one cause of global mortality, resulting in over 17 million deaths per year, which is a number expected to rise to over 23.5 million by 2030.[152] CVD is the number one cause of death in the United States, and approximately 1 million heart attacks and over 700,000 strokes occur every year.[153] More than 2200 Americans die of CVD each day, over 800,000 per year, and 150,000 of these individuals are less than 65 years of age,[153] which is still lower than the average age of a prostate cancer diagnosis.[154] Therefore, while the debate over PSA screening continues,[155,156] so will the urgent need to place risk in perspective and highlight less-recognized observations from these same pieces of controversial data. For example, the notable PLCO US PSA screening trial, which was the major impetus for the US Preventive Services Task Force to recently discourage PSA screening,[155,156] followed an impressive 76,693 men in 10 US study centers.[157] After 10 years of follow-up there were 174 deaths from prostate cancer, 1834 total cancer deaths, 1700 deaths from ischemic heart disease, and 3323 deaths from CVD. The debate over who might benefit or not from PSA screening may continue for some time,[158] as will the positives and negatives of past failed interventions utilized for prostate cancer prevention and beyond.[159] However, this debate should be enjoying a more halcyon era if clinicians, patients, and the overall public become more aware and deferential of the number one cause of death for men for almost 115 years, and the simultaneous impact that risk factors for CVD have on the risk of prostate cancer itself and vice versa.

A simplistic shift in thinking is needed for clinical and research milieu changes. One could argue that it has been the lack of recognition or motion toward the simplistic that has caused such deviation from the forest and such gravitation toward the individual tree. One could even argue that hundreds of millions of dollars would not have been spent over the past two decades on adverse pharmacologic and dietary supplement interventions had more deference been paid to more simplistic correlations or innuendo between the CVD and prostate cancer nexus.

An individual chronic prevalent disease is usually not found to be isolated or insular in incidence or prevalence. Areas of the world or specific populations with some of the lowest rates of death from CVD simultaneously enjoy lower rates of mortality from a multitude of life-threatening diseases and this is what appears to increase their overall life expectancy.[160–167] Yet, if CVD risk increases at a later point in these same areas, then cancer and all-cause mortality also begin to increase.

Perhaps the moment has arrived that most cancer ancillary treatment strategies should be solidly embedded in CVD risk reduction strategies to maximize health benefits and longevity. It is time to prioritize and simplify health recommendations for men, women, and children especially at a time where just 1–2% of Americans are following multiple proven heart healthy lifestyle changes and parameters that could immediately impact disease prevalence and life expectancy.[168–172] The public must be regularly distracted and fatigued from a perceived infinity of incoming behavioral recommendations from countless health awareness campaigns and agendas via multimedia sources that are now open all day and night 365 days a year. How else does one construe the obsession clinicians witness regularly over medical minutiae such as the latest antiaging supplement or drug that apparently prevents most diseases compared to long-term evidence-based heart healthy interventions.

What if heart healthy interventions or lifestyle changes ultimately do not impact prostate cancer from some notable future randomized trial? Attempting to compress the impact of the number one cause of morbidity and mortality in the worst-case scenario is still at least a worst-case scenario with a positive outcome for patients.

In a never-ending desire or obsession with technology it is easy to occasionally lose sight of simplicity. Yet, it is the simplicity, which continues to garner the data in terms of efficacy. This can also occur with nutritional and supplement advice where an obsession for higher cost and caloric-laden exotic juices and pills garner some questionable evidence against prostate cancer,[173,174] and the resulting excitement by clinicians and patients is palpable. Yet, a meta-analysis of something so simple as carrot consumption and its potential antiprostate cancer effects does not make the headline news or a basic clinical conversation.[175] Unless the simple is embraced by the clinician and patient the perceived complex will dominate our collective attention and behavioral patterns. This is within your power to change right now.

## References


1. Moyad MA. *Complementary and alternative medicine for prostate and urologic health*. New York: Springer Publishing; 2014.
2. Lloyd-Jones D, Adams RJ, Brown TM, American Heart Association Statistics Committee and Stroke Statistics Subcommittee. et al. Heart disease and stroke statistics – 2010 update: a report from the American Heart Association. *Circulation* 2010;**121**:e46–e215.
3. Moran AE, Forouzanfar MH, Roth GA, et al. Temporal trends in ischemic heart disease mortality in 21 world regions, 1980 to 2010: the Global Burden of Disease 2010 study. *Circulation* 2014;**129**:1483–92.
4. Eyre H, Kahn R, Robertson RM, American Cancer Society, American Diabetes Association, and the American Heart Association. et al. Preventing cancer, cardiovascular disease, and diabetes: a common agenda for the American Cancer Society, the American Diabetes Association, and the American Heart Association. *Circulation* 2004;**109**:3244–55.
5. Moyad MA. Lifestyle changes to prevent BPH: heart healthy=prostate healthy. *Urol Nurs* 2003;**23**:439–41.
6. US Department of Health and Human Services. *The health benefits of smoking cessation: a report of the Surgeon General*. Bethesda, MD: US Public Health Service, Office on Smoking and Health; 1990.
7. Anthonisen NR, Skeans MA, Wise RA, et al. The effects of a smoking cessation intervention on 14.5-year mortality: a randomized clinical trial. *Ann Intern Med* 2005;**142**:233–9.
8. Wu J, Sin DD. Improved patient outcome with smoking cessation: when is it too late? *Int J Chron Obstruct Pulmon Dis* 2011;**6**:259–67.
9. Zheng W, McLerran DF, Rolland BA, et al. Burden of total and cause-specific mortality related to tobacco smoking among adults aged ≥ 45 years in Asia: a pooled analysis of 21 cohorts. *PLoS Med* 2014;**11**:e1001631.
10. Peeters A, Barendregt JJ, Willekens F, et al. Obesity in adulthood and its consequences for life expectancy: a life-table analysis. *Ann Intern Med* 2003;**138**:24–32.
11. Nagai M, Kuriyama S, Kakizaki M, et al. Impact of obesity, overweight and underweight on life expectancy and lifetime medical expenditures: the Ohsaki Cohort Study. *BMJ Open* 2012;**2**(3):e000940.
12. Walls HL, Backholer K, Proietto J, et al. Obesity and trends in life expectancy. *J Obes* 2012;**2012**:107989.
13. Ng M, Fleming T, Robinson M, et al. Global, regional, and national prevalence of overweight and obesity in children and adults during 1980–2013: a systematic analysis for the Global Burden of Disease Study 2013. *Lancet* 2014;**384**:766–81.

14. Pinkhasov RM, Shteynshlyuger A, Hakimian P, et al. Are men shortchanged on health? Perspective on life expectancy, morbidity, and mortality in men and women in the United States. *Int J Clin Pract* 2010;**64**:465–74.
15. Moyad MA. The optimal male health diet and dietary supplement program. *Urol Clin North Am* 2012;**39**:89–107.
16. Thompson IM, Goodman PJ, Tangen CM, et al. The influence of finasteride on the development of prostate cancer. *N Engl J Med* 2003;**349**:215–24.
17. Scardino PT. The prevention of prostate cancer – The dilemma continues. *N Engl J Med* 2003;**349**:297–9.
18. Lebdai S, Bigot P, Azzouzi AR. High-grade prostate cancer and finasteride. *BJU Int* 2010;**105**:456–9.
19. Svatek RS, Lee JJ, Roehrborn CG, et al. Cost-effectiveness of prostate cancer chemoprevention: a quality of life-years analysis. *Cancer* 2008;**112**:1058–65.
20. Thompson IM, Tangen CM, Goodman PJ, et al. Erectile dysfunction and subsequent cardiovascular disease. *JAMA* 2005;**294**:2996–3002.
21. Kristal AR, Arnold KB, Schenk JM, et al. Dietary patterns, supplement use, and the risk of symptomatic benign prostatic hyperplasia: results from the prostate cancer prevention trial. *Am J Epidemiol* 2008;**167**:925–34.
22. Andriole GL, Bostwick DG, Brawley OW, For the REDUCE Study Group. et al. Effect of dutasteride on the risk of prostate cancer. *N Engl J Med* 2010;**362**:1192–202.
23. Azzouni F, Mohler J. Role of 5-alpha-reductase inhibitors in prostate cancer prevention and treatment. *Urology* 2012;**79**:1197–205.
24. Walsh PC. Chemoprevention of prostate cancer. *N Engl J Med* 2010;**362**:1237–8.
25. Platz EA, Till C, Goodman PJ, et al. Men with low serum cholesterol have a lower risk of high-grade prostate cancer in the placebo arm of the prostate cancer prevention trial. *Cancer Epidemiol Biomarkers Prev* 2009;**18**:2807–13.
26. Thomas II JA, Gerber L, Banez LL, et al. Prostate cancer risk in men with baseline history of coronary artery disease: results from the REDUCE Study. *Cancer Epidemiol Biomarkers Prev* 2012;**21**:576–81.
27. Lippman SM, Klein EA, Goodman PJ, et al. Effect of selenium and vitamin E on risk of prostate cancer and other cancers: the Selenium and Vitamin E Cancer Prevention Trial (SELECT). *JAMA* 2009;**301**:39–51.
28. Klein EA, Thompson Jr IM, Tangen CM, et al. Vitamin E and the risk of prostate cancer: the Selenium and Vitamin E cancer prevention trial (SELECT). *JAMA* 2011;**306**:1549–56.
29. Beaglehole R, Bonita R, Maghusson R. Global cancer prevention: an important pathway to global health and development. *Public Health* 2011;**125**:821–31.
30. Eidelman RS, Hollar D, Hebert PR, et al. Randomized trials of vitamin E in the treatment and prevention of cardiovascular disease. *Arch Intern Med* 2004;**164**:1552–6.
31. Lee IM, Cook NR, Gaziano JM, et al. Vitamin E in the primary prevention of cardiovascular disease and cancer: the Women's Health Study: a randomized controlled trial. *JAMA* 2005;**294**:56–65.
32. Miller III ER, Pastor-Barriuso R, Dalal D, et al. Meta-analysis: high-dosage vitamin E supplementation may increase all-cause mortality. *Ann Intern Med* 2005;**142**:37–46.
33. Lonn E, Bosch J, Yusuf S, et al. Effects of long-term vitamin E supplementation on cardiovascular events and cancer: a randomized controlled trial. *JAMA* 2005;**293**:1338–47.
34. Stranges S, Marshall JR, Trevisan M, et al. Effects of selenium supplementation on cardiovascular disease incidence and mortality: secondary analyses in a randomized clinical trial. *Am J Epidemiol* 2006;**163**:694–9.
35. Neve J. Selenium as a risk factor for cardiovascular diseases. *J Cardiovasc Risk* 1996;**3**:42–7.
36. Gaziano JM, Glynn RJ, Christen WG, et al. Vitamins E and C in the prevention of prostate and total cancer in men: the Physicians' Health Study II randomized controlled trial. *JAMA* 2009;**301**:52–62.
37. Sesso HD, Buring JE, Christen WG, et al. Vitamins E and C in the prevention of cardiovascular disease in men: the Physicians' Health Study II randomized controlled trial. *JAMA* 2008;**300**:2123–33.
38. The Alpha-Tocopherol Beta Carotene Cancer Prevention Study Group. The effect of vitamin E and beta carotene on the incidence of lung cancer and other cancers in male smokers. *N Engl J Med* 1994;**330**:1029–35.
39. The ATBC Study Group. Incidence of cancer and mortality following alpha-tocopherol and beta-carotene supplementation. *JAMA* 2003;**290**:476–85.
40. Lodge JK. Vitamin E bioavailability in humans. *J Plant Physiol* 2005;**162**:790–6.
41. Hercberg S, Galan P, Preziosi P, et al. The SU.VI.MAX Study: a randomized, placebo-controlled trial of the health effects of antioxidant vitamins and minerals. *Arch Intern Med* 2004;**164**:2335–42.
42. Meyer F, Galan P, Douville P, et al. Antioxidant vitamin and mineral supplementation and prostate cancer prevention in the SU.VI.MAX trial. *Int J Cancer* 2005;**116**:182–6.
43. Navas-Acien A, Bleys J, Guallar E. Selenium intake and cardiovascular risk: what is new? *Curr Opin Lipidol* 2008;**19**:43–9.
44. Bleys J, Navas-Aclen A, Guallar E. Serum selenium levels and all-cause, cancer, and cardiovascular mortality among US adults. *Arch Intern Med* 2008;**168**:404–10.
45. Duffield-Lillico AJ, Slate EH, Reid ME, Nutritional Prevention of Cancer Study Group. et al. Selenium supplementation and secondary prevention of nonmelanoma skin cancer in a randomized trial. *J Natl Cancer Inst* 2003;**95**:1477–81.
46. Stranges S, Marshall JR, Natarajan R, et al. Effects of long-term selenium supplementation on the incidence of type 2 diabetes: a randomized trial. *Ann Intern Med* 2007;**147**:217–23.
47. Clark LC, Combs Jr GF, Turnbull BW, et al. Effects of selenium supplementation for cancer prevention in patients with carcinoma of the skin: a randomized controlled trial. *JAMA* 1996;**276**:1957–63.
48. Ozen AE, Pons A, Tur JA. Worldwide consumption of functional foods: a systematic review. *Nutr Rev* 2012;**70**:472–81.
49. Nahin RL, Barnes PM, Stussman BJ, et al. Costs of complementary and alternative medicine (CAM) and frequency of visits to CAM practitioners: United States, 2007. *Natl Health Stat Report* 2009;**18**:1–14.
50. Moyad MA. Heart healthy=prostate healthy: SELECT, the symbolic end of preventing prostate cancer via heart unhealthy and over anti-oxidation mechanisms? *Asian J Androl* 2012;**14**:243–4.
51. Duffield-Lillico AJ, Dalkin BL, Reid ME, et al. Selenium supplementation, baseline plasma selenium status and incidence of prostate cancer: an analysis of the complete treatment period of the Nutritional Prevention of Cancer Trial. *BJU Int* 2003;**91**:608–12.
52. Duffield-Lillico AJ, Reid ME, Turnbull BW, et al. Baseline characteristics and the effect of selenium supplementation on cancer incidence in a randomized clinical trial: a summary report of the Nutritional Prevention of Cancer Trial. *Cancer Epidemiol Biomarkers Prev* 2002;**11**:630–9.
53. Kristal AR, Darke AK, Morris JS, et al. Baseline selenium status and effects of selenium and vitamin E supplementation on prostate cancer risk. *J Natl Cancer Inst* 2014;**106**(3):djt456.
54. Morris MC, Tangney CC. A potential design flaw of randomized trials of vitamin supplements. *JAMA* 2011;**305**:1348–9.
55. Rizos EC, Ntzani EE, Bika E, et al. Association between omega-3 fatty acid supplementation and risk of major cardiovascular disease events: a systematic review and meta-analysis. *JAMA* 2012;**308**:1024–33.

56. Murdoch DR, Slow S, Chambers ST, et al. Effect of vitamin D3 supplementation on upper respiratory tract infections in healthy adults: the VIDARIS randomized controlled trial. *JAMA* 2012;**308**:1333–9.
57. Pearson P, Lewis SA, Britton J, et al. The pro-oxidant activity of high-dose vitamin E supplements *in vivo*. *BioDrugs* 2006;**20**:271–3.
58. Rayman MP. Selenium and human health. *Lancet* 2012;**379**:1256–68.
59. Wien TN, Pike E, Wisloff T, et al. Cancer risk with folic acid supplements: a systematic review and meta-analysis. *BMJ Open* 2012;**2**(1):e000653.
60. van Wijngaarden JP, Swart KM, Enneman AW, et al. Effect of daily vitamin B-12 and folic acid supplementation on fracture incidence in elderly individuals with an elevated plasma homocysteine concentration: B-PROOF, a randomized controlled trial. *Am J Clin Nutr* 2014;**100**:1578–86.
61. Gaziano JM, Sesso HD, Christen WG, et al. Multivitamins in the prevention of cancer in men: the Physicians' Health Study II randomized controlled trial. *JAMA* 2012;**308**:1871–80.
62. Sesso HD, Christen WG, Bubes V, et al. Multivitamins in the prevention of cardiovascular disease in men: the Physicians' Health Study II randomized controlled trial. *JAMA* 2012;**308**:1751–60.
63. Christen WG, Glynn RJ, Manson JE, et al. Effect of multivitamin supplement on cataract and age-related macular degeneration in a randomized trial of male physicians. *Ophthalmology* 2014;**121**:525–34.
64. Zhao LQ, Li LM, Zhu H, et al. The effect of multivitamin/mineral supplements on age-related cataracts: a systematic review and meta-analysis. *Nutrients* 2014;**6**:931–49.
65. Grodstein F, O'Brien J, Kang JH, et al. Long-term multivitamin supplementation and cognitive function in men: a randomized trial. *Ann Intern Med* 2013;**159**:806–14.
66. Lawson KA, Wright ME, Subar A, et al. Multivitamin use and risk of prostate cancer in the National Institutes of Health-AARP Diet and Health Study. *J Natl Cancer Inst* 2007;**99**:754–64.
67. Stevens VL, McCullough ML, Diver WR, et al. Use of multivitamins and prostate cancer mortality in a large cohort of US men. *Cancer Causes Control* 2005;**16**:643–50.
68. Neuhouser ML, Barnett MJ, Kristal AR, et al. Dietary supplement use and prostate cancer risk in the Carotene and Retinol Efficacy Trial. *Cancer Epidemiol Biomarkers Prev* 2009;**18**:2202–6.
69. Leitzmann MF, Stampfer MJ, Wu K, et al. Zinc supplement use and risk of prostate cancer. *J Natl Cancer Inst* 2003;**96**:1004–7.
70. Clarke R, Halsey J, Lewington S, et al. Effects of lowering homocysteine levels with B vitamins on cardiovascular disease, cancer, and cause-specific mortality. *Arch Intern Med* 2010;**170**:1622–31.
71. Collin SM, Metcalfe C, Refsum H, et al. Circulating folate, vitamin B12, homocysteine, vitamin B12 transport proteins, and risk of prostate cancer: a case-control study, systematic review, and meta-analysis. *Cancer Epidemiol Biomarkers Prev* 2010;**19**:1632–42.
72. Giovannucci E, Chan AT. Role of vitamin and mineral supplementation and aspirin use in cancer survivors. *J Clin Oncol* 2010;**28**:4081–5.
73. Barnett CM, Nielson CM, Shannon J, et al. Serum 25-OH vitamin D levels and risk of developing prostate cancer in older men. *Cancer Causes Control* 2010;**21**:1297–303.
74. Chung M, Lee J, Terasawa T, et al. Vitamin D with or without calcium supplementation for prevention of cancer and fractures: an updated meta-analysis for the U.S. Preventive Services Task Force. *Ann Intern Med* 2011;**155**:827–938.
75. Michaelsson K, Baron JA, Snellman G, et al. Plasma vitamin D and mortality in older men: a community-based prospective cohort study. *Am J Clin Nutr* 2010;**92**:841–8.
76. Sanders KM, Stuart AL, Williamson EJ, et al. Annual high-dose oral vitamin D and falls and fractures in older women. *JAMA* 2010;**303**:1815–22.
77. Bischoff-Ferrari HA, Giovannucci E, Willett WC, et al. Estimation of optimal serum concentrations of 25-hydroxyvitamin D for multiple health outcomes. *Am J Clin Nutr* 2006;**84**:18–28.
78. Zerwekh JE. Blood biomarkers of vitamin D status. *Am J Clin Nutr* 2008;**87**(Suppl.):1087S–91S.
79. Isenor JE, Ensom MH. Is there a role for therapeutic drug monitoring of vitamin D level as a surrogate marker for fracture risk. *Pharmacotherapy* 2010;**30**:254–64.
80. Moyad MA. *Promoting wellness for prostate cancer patients*. 4th ed. Ann Arbor, MI: Spry Publishing; 2013.
81. Ross AC, Taylor CL, Yaktine AL, Del Valle HB, editors. *Institute of Medicine of the National Academies. Dietary reference intakes for calcium and vitamin D*. Washington, DC: National Academies Press; 2011.
82. Ardawi MS, Siblany AM, Bakhsh TM, et al. High prevalence of vitamin D deficiency among healthy Saudi Arabian men: relationship to bone mineral density, parathyroid hormone, bone turnover markers, and lifestyle factors. *Osteoporos Int* 2012;**23**:675–86.
83. Jaaskelainen T, Knekt P, Marniemi J, et al. Vitamin D status is associated with sociodemographic factors, lifestyle and metabolic health. *Eur J Nutr* 2013;**52**(2):513–25.
84. Moyad MA. Heart health=urologic health and heart unhealthy=urologic unhealthy: rapid review of lifestyle changes and dietary supplements. *Urol Clin N Am* 2011;**38**:359–67.
85. Moyad MA, Park K. What do most erectile dysfunction guidelines have in common? No evidence-based discussion or recommendation of heart-healthy lifestyle changes and/or Panax ginseng. *Asian J Androl* 2012;**14**:830–41.
86. *US Department of Health and Human Services. How tobacco smoke causes disease: the biology and behavioral basis for smoking-attributable disease: a report of the Surgeon General, GA: US Department of Health and Human Services, CDC*. Available from: http://www.cdc.gov/tobacoo/data_statistics/sgr/2010/index.htm; 2010 [accessed 01.11.2012].
87. Centers for Disease Control and Prevention (CDC). Current cigarette smoking among adults – United States, 2011. *MMWR Morb Mortal Wkly Rep* 2012;**61**:889–94.
88. Huncharek M, Haddock KS, Reid R, et al. Smoking as a risk factor for prostate cancer: a meta-analysis of 24 prospective cohort studies. *Am J Public Health* 2010;**100**:693–701.
89. Zu K, Giovannucci E. Smoking and aggressive prostate cancer: a review of the epidemiologic evidence. *Cancer Causes Control* 2009;**20**:1799–810.
90. Kenfield SA, Stampfer MJ, Chan JM, et al. Smoking and prostate cancer survival and recurrence. *JAMA* 2011;**305**:2548–55.
91. Cao Y, Ma J. Body mass index, prostate cancer-specific mortality, and biochemical recurrence: a systematic review and meta-analysis. *Cancer Prev Res* 2011;**4**:486–501.
92. Joshu CE, Mondul AM, Menke A, et al. Weight gain is associated with an increased risk of prostate cancer recurrence after prostatectomy in the PSA era. *Cancer Prev Res (Phila)* 2011;**4**:544–51.
93. Discacciatl A, Orsini N, Wolk A. Body mass index and incidence of localized and advanced prostate cancer-a dose response meta-analysis of prospective studies. *Ann Oncol* 2012;**23**:1665–71.
94. Woodward G, Ahmed S, Podelski V, et al. Effect of Roux-en-Y gastric bypass on testosterone and prostate-specific antigen. *Br J Surg* 2012;**99**:693–8.
95. Tsugane S, Inoue M. Insulin resistance and cancer: epidemiological evidence. *Cancer Sci* 2010;**101**:1073–9.
96. Albanes D, Weinstein SJ, Wright ME, et al. Serum insulin, glucose, indices of insulin resistance, and risk of prostate cancer. *J Natl Cancer Inst* 2009;**101**:1272–9.
97. Yun SJ, Min BD, Kang HW, et al. Elevated insulin and insulin resistance are associated with the advanced pathological stage of prostate cancer in Korean population. *J Korean Med Sci* 2012;**27**:1079–84.

98. Grossmann M, Wittert G. Androgens, diabetes and prostate cancer. *Endocr Relat Cancer* 2012;**19**:F47–62.
99. Centers for Disease Control and Prevention (CDC). Increasing prevalence of diagnosed diabetes-United States and Puerto Rico, 1995–2010. *MMWR Morb Mortal Wkly Rep* 2012;**61**:918–21.
100. Snell-Bergeon JK, Wadwa RP. Hypoglycemia, diabetes, and cardiovascular disease. *Diabetes Technol Ther* 2012;**14**(Suppl. 1):S51–8.
101. Grontved A, Rimm EB, Willett WC, et al. A prospective study of weight training and risk of type 2 diabetes mellitus in men. *Ann Intern Med* 2012;**6**:1–7.
102. Knowler WC, Barrett-Connor E, Fowler SE, Diabetes Prevention Program Research Group. et al. Reduction in the incidence of type 2 diabetes with lifestyle intervention or metformin. *N Engl J Med* 2002;**346**:393–403.
103. Orchard TJ, Temprosa M, Goldberg R, Diabetes Prevention Program randomized trial. et al. The effect of metformin and intensive lifestyle intervention on the metabolic syndrome: the Diabetes Prevention Program randomized trial. *Ann Intern Med* 2005;**142**:611–9.
104. Diabetes Prevention Program Research Group. 10-year follow-up of diabetes incidence and weight loss in the Diabetes Prevention Program Outcomes Study. *Lancet* 2009;**374**:1677–86.
105. Perreault L, Pan Q, Mather KJ, Diabetes Prevention Program Research Group. et al. Effect of regression from prediabetes to normal glucose regulation on long-term reduction in diabetes risk: results from the Diabetes Prevention Program Outcomes Study. *Lancet* 2012;**379**:2243–51.
106. Kenfield SA, Stampfer MJ, Giovannucci E, et al. Physical activity and survival after prostate cancer diagnosis in the health professionals follow-up study. *J Clin Oncol* 2011;**29**:726–32.
107. Oliveria SA, Lee IM. Is exercise beneficial in the prevention of prostate cancer? *Sports Med* 1997;**23**:271–8.
108. Torti DC, Matheson GO. Exercise and prostate cancer. *Sports Med* 2004;**34**:363–9.
109. Young-McCaughan S. Potential for prostate cancer prevention through physical activity. *World J Urol* 2012;**30**:167–79.
110. CDC. Vital signs: prevalence, treatment, and control of hypertension – United States, 1992–2002 and 2005–2008. *MMWR Morb Mortal Wkly Rep* 2011;**60**:103–8.
111. Rogers VL, Go AS, Lloyd-Jones DM, et al. Heart disease and stroke statistics – 2011 update: a report from the American Heart Association. *Circulation* 2011;**123**:e18–e209.
112. Neal B, McMahon S, Chapman N. Effects of ACE inhibitors, calcium antagonists, and other blood pressure-lowering drugs. *Lancet* 2000;**356**:1955–64.
113. Bhaskaran K, Douglas I, Evans S, et al. Angiotensin receptor blockers and risk of cancer: cohort study among people receiving antihypertensive drugs in UK General Practice Research Database. *BMJ* 2012;**344**:e2697.
114. De Nunzio C, Aronson W, Freedland SJ, et al. The correlation between metabolic syndrome and prostatic diseases. *Eur Urol* 2012;**61**:560–70.
115. Loeb S, Gupta A, Losonczy L, et al. Does benign prostatic hyperplasia treatment with alpha-blockers affect prostate cancer risk? *Curr Opin Urol* 2012;**23**(1):2–4.
116. Roehrborn CG. BPH progression: concept and key learning from MTOPS, ALTESS, COMBAT, and ALF-ONE. *BJU Int* 2008;**101**(Suppl. 3):17–21.
117. Grytli HH, Fagerland MW, Fossa SD, et al. Association between use of beta-blockers and prostate cancer-specific survival: a cohort study of 3561 prostate cancer patients with high-risk or metastatic disease. *Eur Urol* 2014;**65**:635–41.
118. Siervo M, Lara J, Chowdhury S, et al. Effects of the Dietary Approach to Stop Hpyertension (DASH) diet on cardiovascular risk factors: a systematic review and meta-analysis. *Br J Nutr* 2014;**113**:1–15 epub ahead of print.
119. Saneel P, Salehi-Abargouei A, Esmallzadeh A, et al. Influence of Dietary Approaches to Stop Hypertension (DASH) diet on blood pressure: a systematic review and meta-analysis on randomized trials. *Nutr Metab Cardiovasc Dis* 2014;**24**:1253–61.
120. Centers for Disease Control and Prevention (CDC). Prevalence of cholesterol screening and high blood cholesterol among adults-United States, 2005, 2007, and 2009. *MMWR Morb Mortal Wkly Rep* 2012;**61**:697–702.
121. Kotani K, Sekine Y, Ishikawa S, et al. High-density lipoprotein and prostate cancer: an overview. *J Epidemiol* 2013;**23**:313–9.
122. Murtola TJ, Visakorpi T, Lahtela J, et al. Statins and prostate cancer prevention: where we are now and future directions. *Nat Clin Pract Urol* 2008;**5**:376–87.
123. Moyad MA, Klotz LH. Statin clinical trial (REALITY) for prostate cancer: an over 15-year wait is finally over thanks to a dietary supplement. *Urol Clin North Am* 2011;**38**:325–31.
124. Ridker PM, Danielson E, Fonseca FA, JUPITER Study Group. et al. Rosuvastatin to prevent vascular events in men and women with elevated C-reactive protein. *N Engl J Med* 2008;**359**:2195–207.
125. Everett BM, Glynn RJ, MacFadyen JG, et al. Rosuvastatin in the prevention of stroke among men and women with elevated levels of C-reactive protein: Justification for the Use of Statins in Prevention: an Intervention Trial Evaluating Rosuvastatin (JUPITER). *Circulation* 2010;**121**:143–50.
126. Solomon KR, Freeman MR. The complex interplay between cholesterol and prostate malignancy. *Urol Clin North Am* 2011;**38**:243–59.
127. Moyad MA. Why a statin and/or another proven heart healthy agent should be utilized in the next major cancer chemoprevention trial: part I. *Urol Oncol* 2004;**22**:466–71.
128. Moyad MA. Why a statin and/or another proven heart healthy agent should be utilized in the next major cancer chemoprevention trial: part II. *Urol Oncol* 2004;**22**:472–7.
129. Crosara Teixeira M, Braghiroli MI, Sabbaga J, et al. Primary prevention of colorectal cancer: myth or reality? *World J Gastroenterol* 2014;**20**:15060–9.
130. Sostres C, Gargalio CJ, Lanas A. Aspirin, cyclooxygenase inhibition and colorectal cancer. *World J Gastrointest Pharmacol Ther* 2014;**5**:40–9.
131. Shebl FM, Sakoda LC, Black A, et al. Aspirin but not ibuprofen use is associated with reduced risk of prostate cancer: a PLCO study. *Br J Cancer* 2012;**107**:207–14.
132. Bosetti C, Rosato V, Gallus S, et al. Aspirin and cancer risk: a quantitative review to 2011. *Ann Oncol* 2012;**23**:1403–15.
133. Rothwell PM, Fowkes FG, Belch JK, et al. Effect of daily aspirin on long-term risk of death due to cancer: analysis of individual patient data from randomized trials. *Lancet* 2011;**377**:31–41.
134. Mahmud S, Franco E, Aprikian A. Prostate cancer and use of nonsteroidal anti-inflammatory drugs: systematic review and meta-analysis. *Br J Cancer* 2004;**90**:93–9.
135. Choe KS, Cowan JE, Chan JM, et al. Aspirin use and the risk of prostate cancer mortality in men treated with prostatectomy or radiotherapy. *J Clin Oncol* 2012;**30**:3540–5.
136. Jacobs EJ, Newton CC, Stevens VL, et al. Daily aspirin use and prostate cancer-specific mortality in a large cohort of men with nonmetastatic prostate cancer. *J Clin Oncol* 2014;**32**:3716–22.
137. Bosetti C, Rosato V, Gallus S, et al. Aspirin and prostate cancer prevention. *Recent Results Cancer Res* 2014;**202**:93–100.
138. Liu Y, Chen JQ, Xie L, et al. Effect of aspirin and other nonsteroidal anti-inflammatory drugs on prostate cancer incidence and mortality: a systematic review and meta-analysis. *BMC Med* 2014;**12**:55.
139. Hennekens CH, Dalen JE. Aspirin in the primary prevention of cardiovascular disease: current knowledge and future research needs. *Trends Cardiovasc Med* 2014;**24**:360–6.

140. Halvorsen S, Andreotti F, ten Berg JM, et al. Aspirin therapy in primary cardiovascular disease prevention: a position paper of the European Society of Cardiology working group on thrombosis. *J Am Coll Cardiol* 2014;**64**:319–27.
141. Anfossi G, Russo I, Bonomo K, et al. The cardiovascular effects of metformin: further reasons to consider an old drug as a cornerstone in the therapy of type 2 diabetes mellitus. *Curr Vasc Pharmacol* 2010;**8**:327–37.
142. Ekstrom N, Scholer L, Svensson AM, et al. Effectiveness and safety of metformin in 51,675 patients with type 2 diabetes and different levels of renal function: a cohort study from the Swedish National Diabetes Register. *BMJ Open* 2012;**13**:2.
143. Roumie CL, Hung AM, Greevy RA, et al. Comparative effectiveness of sulfonylurea and metformin monotherapy on cardiovascular events in type 2 diabetes mellitus: a cohort study. *Ann Intern Med* 2012;**157**:601–10.
144. Klachko D, Whaley-Connell A. Use of metformin in patients with kidney and cardiovascular diseases. *Cardiorenal Med* 2011;**1**:87–95.
145. Jalving M, Gietema JA, Lefrandt JD, et al. Metformin: taking away the candy for cancer? *Eur J Cancer* 2010;**46**:2369–80.
146. Rothermundt C, Havoz S, Templeton AJ, et al. Metformin in chemotherapy-naive castration-resistant prostate cancer : a multicenter phase 2 trial (SAKK 08/09). *Eur Urol* 2014;**66**:468–74.
147. Nobes JP, Langley SE, Klopper T, et al. A prospective, randomized pilot study evaluating the effects of metformin and lifestyle intervention on patients with prostate cancer receiving androgen deprivation therapy. *BJU Int* 2012;**109**:1495–502.
148. Clements A, Gao B, Yeap SH, et al. Metformin in prostate cancer: two for the price of one. *Ann Oncol* 2011;**22**:2556–60.
149. Colquhoun AJ, Venier NA, Vandersluis AD, et al. Metformin enhances the antiproliferative and apoptotic effect of bicalutamide in prostate cancer. *Prostate Cancer Prostatic Dis* 2012;**15**:346–52.
150. Babcock MA, Sramkoski RM, Fujioka H, et al. Combination simvastatin and metformin induces G1-phase cell cycle arrest and Ripk1- and Ripk3-dependent necrosis in C4-2B osseous metastatic castration-resistant prostate cancer cells. *Cell Death Dis* 2014;**5**:e1536.
151. Moyad MA. Re: A prospective, randomized pilot study evaluating the effects of metformin and lifestyle intervention on patients with prostate cancer receiving androgen deprivation therapy. *Eur Urol* 2012;**61**:623–4.
152. Mitka M. New basic care goals seek to rein in global rise in cardiovascular disease. *JAMA* 2012;**308**:1725–6.
153. Rogers VL, Go AS, Lloyd-Jones DM, et al. Heart disease and stroke statistics – 2012 update: a report from the American Heart Association. *Circulation* 2012;**125**:e2–e220.
154. Cross DS, Ritter M, Reding DJ. Historical prostate cancer screening and treatment outcomes from a single institution. *Clin Med Res* 2012;**10**:97–105.
155. Payton S. Prostate cancer: PSA screening – more data, more debate. *Nat Rev Urol* 2012;**9**:59.
156. Gomella LG, Llu XS, Trabulsi EJ, et al. Screening for prostate cancer: the current evidence and guidelines controversy. *Can J Urol* 2011;**18**:5875–83.
157. Andriole GL, Crawford ED, Grubb III RL, PLCO Project Team. et al. Mortality results from a randomized prostate-cancer screening trial. *N Engl J Med* 2009;**360**:1310–9.
158. Crawford ED, Grubb III R, Black A, et al. Comorbidity and mortality results from a randomized prostate cancer screening trial. *J Clin Oncol* 2011;**29**:355–61.
159. Theoret MR, Ning YM, Zhang JJ, et al. The risks and benefits of 5alpha-reductase inhibitors for prostate-cancer prevention. *N Engl J Med* 2011;**365**:97–9.
160. Perez-Lopez FR, Chedraui P, Haya J, et al. Effects of the Mediterranean diet on longevity and age-related morbid conditions. *Maturitas* 2009;**64**:67–79.
161. Vasto S, Scapagnini G, Rizzo C, et al. Mediterranean diet and longevity in Sicily: survey in a Sicani Mountains population. *Rejuvenation Res* 2012;**15**:184–8.
162. Vasto S, Rizzo C, Caruso C. Centenarians and diet: what they eat in the Western part of Sicily. *Immun Ageing* 2012;**23**:10.
163. Pauwels EK. The protective effect of the Mediterranean diet: focus on cancer and cardiovascular risk. *Med Princ Pract* 2011;**20**:103–11.
164. Fraser GE. Associations between diet and cancer, ischemic heart disease, and all-cause mortality in non-Hispanic white California Seventh-day Adventists. *Am J Clin Nutr* 1999;**70**(3 Suppl.):532S–8S.
165. Miyagi S, Iwama N, Kawabata T, et al. Longevity and diet in Okinawa, Japan: the past, present and future. *Asia Pac J Public Health* 2003;**15**(Suppl.):S3–9.
166. Willcox DC, Willcox BJ, Todoriki H, et al. The Okinawan diet: health implications of a low-calorie, nutrient-dense, antioxidant-rich dietary pattern low in glycemic load. *J Am Coll Nutr* 2009;**28**(Suppl.):500S–16S.
167. Ishijima H, Nagai M, Shibazaki S, et al. Age and cause of death contributing to reduction of disparity in age-adjusted overall mortality between males in Okinawa and mainland Japan. *Nihon Koshu Eisei Zasshi* 2007;**54**:695–703.
168. Knoops KT, de Groot LC, Kromhout D, et al. Mediterranean diet, lifestyle factors, and 10-year mortality in elderly European men and women: the HALE project. *JAMA* 2004;**292**:1433–9.
169. Wilcox BJ, He Q, Chen R, et al. Midlife risk factors and healthy survival in men. *JAMA* 2006;**296**:2343–50.
170. Yates LB, Djousse L, Kurth T, et al. Exceptional longevity in men: modifiable factors associated with survival and function to age 90 years. *Ann Intern Med* 2008;**168**:284–90.
171. McCullough ML, Patel AV, Kushi LH, et al. Following cancer prevention guidelines reduces risk of cancer, cardiovascular disease, and all-cause mortality. *Cancer Epidemiol Biomarkers Prev* 2011;**20**:1089–97.
172. Yang Q, Cogswell ME, Flanders WD, et al. Trends in cardiovascular health metrics and associations with all-cause and CVD mortality among US adults. *JAMA* 2012;**307**:1273–83.
173. Pantuck AJ, Leppert JT, Zomorodian N, et al. Phase II study of pomegranate juice for men with rising prostate-specific antigen following surgery or radiation for prostate cancer. *Clin Cancer Res* 2006;**12**:4018–26.
174. Paller CJ, Ye X, Wozniak PJ, et al. A randomized phase II study of pomegranate extract for men with a rising PSA following initial therapy for localized prostate cancer. *Prostate Cancer Prostatic Dis* 2013;**16**:50–5.
175. Xu X, Cheng Y, Li S, et al. Dietary carrot consumption and the risk of prostate cancer. *Eur J Nutr* 2014;**53**:1615–23.

CHAPTER

# 21

# Effects of Smoking, Alcohol, and Exercise on Prostate Cancer

*Hadley Wyre, MD, James Brantley Thrasher, MD*

**Department of Urology, University of Kansas Medical School, Kansas City, KS, USA**

## INTRODUCTION

Multiple studies have been performed in an effort to determine the cause of cancer. Certainly, both genetic and environmental factors play a role in the pathogenesis of malignancies. Patients often wish to know what lifestyle behaviors can be modified to decrease their risk of developing cancer. One of the most common questions asked by patients with a new diagnosis of cancer is, "What caused this cancer?" Often unaffected family members want to know what, if anything, can be done to prevent their own diagnosis of cancer. Smoking and alcohol have been studied extensively as potential risk factors for cancers. Physical activity has also recently been determined to be protective against a number of different malignancies. In this chapter, we examine the association between the risk of prostate cancer and the modifiable lifestyle behaviors of smoking, alcohol, and physical activity.

## CIGARETTE SMOKING

Smoking is a major risk factor for the development of multiple malignancies, lung cancer being the most widely studied and recognized. The urologist is no stranger to cigarette-induced malignancies, with smoking being the leading risk factor for bladder cancer in the United States.[1] Renal cancer also shows a link with cigarette use, albeit to a lesser extent than that of bladder cancer.[2] Certainly the logical progression would be to examine the effect of cigarette smoke on the development of prostate cancer. Many studies, including several meta-analyses, have examined the effect of cigarette smoking on the risk of developing prostate cancer as well as the risk of developing advanced or fatal prostate cancer.

### Risk of Developing Prostate Cancer

Multiple studies, both case-control and cohort in design, have attempted to examine the effect of smoking on developing prostate cancer. However, the data from these studies are conflicting, with some showing an increased risk while others showing no change in the risk of prostate cancer.

The majority of cohort studies were unable to identify an increased risk of prostate cancer among smokers. These studies included large population-based cohorts from multiple ethnicities and countries, including Korea,[3] the United States,[4,5] Europe,[6] and Singapore.[7] Only one cohort study conducted in Finland showed a slight increase in the incidence of prostate cancer among smokers.[8] One study even showed an inverse relationship of smoking with prostate cancer in Japanese men.[9]

Several recently performed case control series also showed mixed results. Plaskon et al. showed in a study using the Seattle-Puget Sound Cancer Registry that current smokers had an increased risk of prostate cancer, with an odds ratio of 1.4. They also found a dose–response relationship between number of pack-years smoked and prostate cancer risk.[10] Another study in the United States showed an increased risk of prostate cancer in heavy smoking African-American men, with an odds ratio of 2.57.[11] Contrary to these results, Darlington et al. were unable to show an effect in their case control study from Ontario.[12]

Huncharek et al. performed a meta-analysis on 24 cohort studies, which included over 21,000 prostate cancer case participants. In their pooled data, they found no increased risk of prostate cancer in current smokers. However, when the studies contained data where participants were stratified by amount smoked, there was an increased risk of prostate cancer among the highest smokers, although the increase was small, with an

Prostate Cancer. http://dx.doi.org/10.1016/B978-0-12-800077-9.00021-9

RR = 1.11. In this same study, former smokers also had a very slight increase in the incidence of prostate cancer with a RR of 1.09.[13] It appears from the mentioned studies that the overall risk of prostate cancer is increased only slightly, if at all, from cigarette smoking.

### Risk of Developing Advanced/Fatal Prostate Cancer

While the data relating to the incidence of prostate cancer is mixed, there is more consistent data that smoking increases the risk of developing advanced or fatal prostate cancer.

Several studies attempted to examine the risk of developing advanced or high-grade cancer in smokers or former smokers. In their cohort study of Japanese men, Sawada et al. examined the effect of smoking on the development of advanced cancers that were detected secondary to symptoms. They were able to show that, while smoking was inversely associated with incident prostate cancer, it tended to increase the risk of developing advanced prostate cancer.[9] Similarly, Murphy et al. attempted to study the effect of smoking on high-grade cancers in a case control study consisting of patients in urology clinics in two major US cities. They showed that in African-American men, who made up a large percentage of the cases and controls, heavy smoking increased the risk of developing high-grade cancer when compared to never or light smokers, with an odds ratio of 1.89.[11]

In a large cohort study conducted on 26,810 men in Maryland, Rohrmann et al. showed that current smokers of 20 or more cigarettes per day and former smokers had a greater risk of death from prostate cancer, with relative risks of 2.38 and 2.75, respectively.[4] Similarly, another large cohort study conducted on American men showed that current smokers were 1.7 times more likely to develop fatal prostate cancer, even though they were not more likely to develop prostate cancer overall. This increased risk of fatal cancer was not identified in former smokers in the same study.[5] In their large meta-analysis, Huncharek et al. also found that current smokers had an increased risk of fatal prostate cancer, and also showed that the heaviest smokers had a 24–30% greater risk of death from prostate cancer than did nonsmokers.[13] Similarly, Rohrmann et al. showed in a separate study of over 145,000 European men, the heaviest smokers (25+ cigarettes per day) or those who had smoked for an extended period of time (40 plus years) had a higher risk of prostate cancer death, with a relative risk of 1.81 and 1.38, respectively, both of which were statistically significant.[6] In their review of the current literature, Zu and Giovannucci go so far as to conclude that the majority of prospective studies showed that current smoking is associated with an increase of 30% in fatal prostate cancer risk when compared to never/nonsmokers.[14]

### Summary

The data from several cohort and case-control studies is mixed in regard to smoking leading to an increased risk of prostate cancer. However, several studies did show that current and former smokers were more likely to develop fatal or advanced prostate cancer and these data appear to be more consistent across multiple studies. While the data are not definitive on the effect of smoking, all physicians can surely encourage their patients to discontinue smoking owing to the myriad of health benefits from smoking cessation.

## ALCOHOL

Alcohol has also been shown to have an effect on the development of multiple cancers. Table 21.1 shows the malignancies that have an established or suspected association with alcohol intake.[15] Many studies have examined the effect of alcohol intake on the risk of developing prostate cancer and the risk of developing advanced or fatal prostate cancer. Several of these studies also attempted to determine if the type of alcohol consumed affected the prostate cancer risk. Interestingly, subgroup analysis of the Prostate Cancer Prevention Trial (PCPT) and the REDUCE trial were performed looking at the effect alcohol had on the use of the 5-alpha reductase inhibitors finasteride and dutasteride in the prevention of prostate cancer.

### Risk of Developing Prostate Cancer

Several cohort and case-control studies have attempted to answer the question of alcohol's effect on the risk of developing prostate cancer. Five recent cohort studies have been performed in different populations and again the results are mixed. Three studies were unable to identify an association in European,[16] American,[17] and Australian cohorts.[18] Contrary to this, Watters et al. found that the risk of prostate cancer was 25% higher for men consuming greater than or equal to six drinks

**TABLE 21.1** Cancers Associated with Alcohol Use

| |
|---|
| Causal association has been established |
| • Oral cavity |
| • Pharynx |
| • Larynx |
| • Esophagus |
| • Colon |
| • Rectum |
| • Breast |
| Causal association is suspected |
| • Pancreas |
| • Lung |

per day, 19% higher for men consuming three to six drinks per day, and 6% higher for men consuming up to three drinks per day when compared to nondrinkers.[19] Velicer et al. also found a small increased risk of prostate cancer in men who consumed one drink per month, with a relative risk of 1.2.[20] The results from the case-control studies were also mixed, with two studies showing no increased risk of prostate cancer.[21,22] One study however showed an increased risk of prostate cancer, with a 1.78 relative risk for nonaggressive tumors in the highest intake quartile in Canadian men, and another study based out of Sweden found that prostate cancer cases were more likely than control groups to be current or former drinkers.[23,24]

Two recent meta-analyses have been performed to try to examine the relationship between alcohol and prostate cancer. In their meta-analysis of over 70 studies including nearly 53,000 prostate cancer cases, Rota et al. were unable to provide evidence of a material association between alcohol drinking and prostate cancer.[25] Contradictory to this, Middleton Fillmore et al. performed a meta-analysis and concluded that there was a linear association with heavier alcohol use and prostate cancer incidence, albeit with only a slight increase in relative risk of 1.16. They felt that the use of population-based cohort studies is the reason that their data contradicted other meta-analyses performed on the subject.[26]

Several studies have also attempted to examine if the type of alcohol, that is, red wine, modified the risk of developing prostate cancer. Chao et al. examined data collected from the California Men's Heath Study and wished to specifically examine the effect of red wine on prostate cancer risk. They were unable to find a clear association between red wine and prostate cancer. Furthermore, they were not able to observe an association when they subclassified for other types of alcohol, such as beer, white wine, liquor, or combined intake.[17] Sutcliffe et al. were also unable to identify a protective or causative effect of red wine in their examination of the Health Professionals Follow-up Study.[27] Additionally, in an Australian cohort, Baglietto et al. did not find an association with type of alcoholic beverage or pattern of drinking and prostate cancer risk.[18]

In contrast to these studies, two studies did find a modified risk of developing prostate cancer when examining the type of alcohol consumed. Schoonen et al. found that while overall prostate cancer risk was not associated with alcohol intake, there appeared to be a protective effect of red wine consumption. They found that each additional glass of red wine per week showed a 6% decrease in risk of prostate cancer. There were no association with beer and liquor use.[22] Velicer et al. found that white wine intake increases your risk of prostate cancer, with a relative risk of 1.27 when comparing any white wine to no white wine consumption.[20]

## Risk of Developing Advanced/Fatal Prostate Cancer

In comparison to the previously mentioned studies that examined overall incidence of prostate cancer, several studies attempted to look at the effect alcohol had on developing high-grade or advanced prostate cancer. Again, the results of the studies looking at high-grade and advanced prostate cancer were mixed. Two studies showed that the risk of aggressive prostate cancer was not increased by alcohol.[18,19] Conversely, Sawada et al. were able to identify a dose-dependent association between alcohol and the risk of developing advanced prostate cancer, with a hazard ratio of 1.41 in the highest quartile of alcohol intake, while McGregor et al. showed a relative risk of 2.0 in the highest quartile of alcohol intake.[9,23] Interestingly, one case control study from Britain found a decreased risk of low Gleason-grade prostate cancer (RR = 0.96) but a very slight increase in high-grade prostate cancer (RR = 1.04) per 10 units/week increase in alcohol consumption.[28]

## Interaction of Alcohol with Finasteride and Dutasteride Chemoprevention of Prostate Cancer

The PCPT and the REDUCE trial were performed to examine the use of finasteride and dutasteride, respectively, as primary chemoprevention of prostate cancer. As most know, both trials showed a decrease in overall malignancy but an increased incidence of higher-grade cancers.[29,30] Because of the latter finding, the FDA did not award a secondary indication for their use as chemoprevention for prostate cancer.[31] Interestingly, both studies collected data on patient alcohol use and subgroup analyses were performed to study the interaction between alcohol use and the chemoprevention of prostate cancer by the 5-alpha reductase inhibitors.

Gong et al. examined data on alcohol use and prostate cancer prevention in over 10,000 men enrolled in PCPT. They found that heavy alcohol consumption and regular drinking were both associated with an increased risk of high-grade cancers, with relative risks of 2.01 and 2.17, respectively; this association did not differ between the treatment arms. In contrast, they found that the placebo arm and the treatment arm differed significantly with respect to the association of alcohol and low-grade cancers. There was an increased risk of low-grade cancers in heavy drinkers taking finasteride when compared to nondrinkers and light drinkers taking Finasteride. They therefore concluded that heavy drinking made finasteride ineffective at preventing low-grade cancers.[32]

Similarly, Fowke et al. looked at the effect alcohol use had on prostate cancer risk in men taking dutasteride in the REDUCE trial. They found that alcohol did not affect the risk of low-grade cancers in either arm. They also showed that heavy alcohol use did not increase high-grade cancers in the placebo arm. However, they did show that dutasteride significantly decreased the risk of high-grade cancers in alcohol abstainers but this reduction in high-grade cancers was not seen in heavy alcohol users. From this, they concluded that heavy alcohol consumption negated the protective association between dutasteride and high-grade prostate cancer.[33]

### Summary

As with smoking, the data on alcohol intake and prostate cancer risk are mixed. There is clearly not enough evidence to state definitively if alcohol intake is a modifiable risk factor for prostate cancer. There are also no clear findings as to state whether the type of alcohol affects subsequent prostate cancer risk. Interestingly though, two studies did find that alcohol intake may affect the chemoprevention properties of both 5-alpha reductase inhibitors currently on the market.

## PHYSICAL ACTIVITY

Physical activity has been extensively studied in the prevention of cancer. In fact, in one review of physical activity as a modifiable risk factor for multiple different types of malignancies, Friedenreich et al. estimated that 9–19% of all cancers in Europe could be attributed to lack of physical activity.[34] One can only deduce that this number would be higher in the United States as the lack of physical activity has reached epidemic proportions. Many studies have been undertaken to determine if increased physical activity can decrease the development of prostate cancer. This phenomenon can be examined in multiple ways, including level of overall or recreational physical activity, occupational physical activity, or level of physical activity and its interaction with obesity.

### Risk of Prostate Cancer

Many recent studies, including cohort, case-control, and meta-analyses, have been performed to examine the effect of physical activity, both recreational and occupational, on prostate cancer risk. As we will see, the majority of the studies have found that increased physical activity leads to a decreased risk of developing prostate malignancy.

Several cohort studies attempted to find a link between prostate cancer risk and total or recreational physical activity. In a large cohort study of nearly 46,000 men with over 2700 prostate cancer cases, Orsini et al. found that prostate cancer incidence was decreased by 16% in the top quartile of lifetime physical activity when compared to the lowest quartile.[35] Moore et al. examined the data from the National-Institutes of Health-AARP Diet and Health Study and found that the interaction between physical activity and prostate cancer risk may also be influenced by race. They were unable to find an association in white men between physical activity and prostate cancer. However, they found that black males engaging in greater than or equal to 4 h of moderate/vigorous physical activity during ages 19–29 years had a 35% reduction in prostate cancer when compared to inactive men.[36] Giovannucci examined the data from the Health Professionals Follow-up Study, which contained data from over 47,000 American men. They were unable to find an association for total, vigorous, and nonvigorous physical activity and prostate cancer.[37] Similarly, two large cohort studies from Europe, one from Norway with over 29,000 men and another from the Netherlands with over 58,279 men, were able to link physical activity with a decreased prostate cancer incidence.[38,39]

While the results of the cohort studies were mixed, the case-control studies more consistently found that increased physical activity was associated with a decreased risk of overall prostate cancer incidence. Only one case-control study from Sweden was unable to identify a link between lifetime physical activity and prostate cancer risk.[40] Only one other case-control study showed mixed results. Friedenreich et al. found no association for lifetime physical activity and prostate cancer risk. However, when activity was examined by intensity (low, moderate, or vigorous) they found that vigorous activity was associated with decreased prostate cancer risk.[37] The remainder of the case-control studies found that increased physical activity appeared to be protective against prostate cancer. Darlington et al. found that strenuous activity by men in their early 50s was associated with a reduced risk of subsequent prostate cancer, with an odds ratio of 0.8.[12] Showing similar results for the importance of physical activity in mid-life, a small study of Malaysian men found that past history of not engaging in any physical activity between the ages of 45–54 increased risk of prostate cancer by almost threefold.[41] Another group from Italy showed similar findings. When comparing active versus inactive men, they found an odds ratio of 0.5 for prostate cancer risk.[42] Lastly, Jian et al. examined how moderate physical activity affected prostate cancer risk in Chinese men. They again found that moderate physical activity was inversely related to prostate cancer risk, with and adjusted odds ratio of 0.2 when comparing the upper versus the lower quartiles of weekly metabolic equivalent task-hours. They also found that the dose–response relationship was also significant.[43]

While the previously mentioned studies looked at total physical activity or recreational physical activity, several groups attempted to look at how occupational physical activity interacted with prostate cancer risk. In their case-control study from Italy, Pierotti et al. found that the highest level of occupational physical activity at age 30–39 and 50–59 were associated with a decrease risk of prostate cancer, with odds ratios of 0.78 and 0.75, respectively.[44] Similarly, Lund Haheim et al. showed that sedentary versus intermediate physical activity at work was a significant predictor of prostate cancer and this, in combination with other factors of the metabolic syndrome, worsened the risk of prostate cancer even further.[45] A small case-control series among workers at a nuclear and rocket engine testing facility in Southern California found that high activity levels at work was associated with a significant decrease in prostate cancer risk among aerospace workers, with an odds ratio of 0.55.[46] Two other case-control studies, both from Canada, also showed that high occupational work activity was associated with decreased odds ratio for prostate cancer, 0.54 in one study and 0.90 in the other.[34,47] Orsini et al. also found that compared to men who mostly sit at work, men who sit half of the time at work had a 20% lower risk of prostate cancer.

Lastly, a meta-analysis was performed by Liu et al. and included over 88,000 cases from 19 eligible cohort and 24 eligible case-control studies. They found that total physical activity was significantly associated with a decreased risk of prostate cancer (RR = 0.90). When separated into occupational activity and recreational activity, both showed a decrease in prostate cancer risk, which was significant, with relative risks of 0.81 and 0.95, respectively.[48]

Interestingly, one study looked to see if men with increased physical activity were less likely to have a positive biopsy when compared to men with less physical activity. Antonelli et al. found that men who reported 9 or more metabolic equivalent task hours per week were less likely to have cancer on their biopsy, with an odds ratio of 0.35, and had a lower risk of high-grade disease (Gleason 7 or greater) with an odds ratio of 0.14.[49]

## Risk of Advanced Prostate Cancer

While the previously mentioned studies looked at overall risk of developing cancer, several studies examined the effect of physical activity on developing advanced or fatal prostate cancer. Five recent studies, all cohort in design, examined this question, and all five studies were able to show a decreased risk of advanced or fatal cancer with increased physical activity. Nilsen et al. examined a cohort of over 29,100 men from the Norwegian Cancer registry. They showed that the highest category of physical activity, when compared to no physical activity, was associated with relative risks of 0.64 for advanced prostate cancer and 0.67 for prostate cancer death.[38] Another large cohort study from Europe included nearly 128,000 men from eight European countries, found that higher level of occupational physical activity was associated with a significantly lower risk of advanced prostate cancer.[50] Similar results were found by Orsini et al. who found that the rate ratio linearly decreased by 12% for advanced prostate cancer for every 30 min per day increment of lifetime walking or bicycling.[35] Giovannucci et al. also found a decreased risk of both advanced and fatal prostate cancer in men 65 years or older in the highest category of vigorous activity, with relative risk of 0.33 and 0.26, respectively. They also showed that men with high levels of physical activity were less likely to be diagnosed with a Gleason Grade greater than or equal to 7.[37] Lastly, Patel et al. found a decreased risk of aggressive prostate cancer in men participating in greater than 35 metabolic equivalent hours per week.[51]

## Physical Activity and Obesity

The effect of obesity and the development of prostate cancer are discussed elsewhere in this text. However, two studies specifically looked at the interaction between obesity, physical activity, and prostate cancer. Littman et al. found that physical activity (greater than or equal to 10.5 MET-h per week) was associated with a decreased risk of prostate cancer in normal-weight individuals, with a hazard ratio of 0.69. However, greater physical activity was associated with an increased risk of prostate cancer in obese males, HR equal to 1.5.[52] Zeegers et al. found similar results in their examination of the Netherlands Cohort Study. They were unable to identify a protective effect of physical activity for prostate cancer. However, they too found an increased risk of prostate cancer for men with a BMI greater than 30 and for men with high baseline energy intake who were physically active for greater than 1 h per day.[39] Both of these studies point to a possible interaction between the increased risk of prostate cancer in obese males with a protective effect of physical activity.

## Summary

The majority of studies show that the risk of prostate cancer is decreased as physical activity increases, whether as recreational, occupational, or a combination of both. These findings are important for physicians in counseling patients and their family members in terms of modifiable lifestyle risk factors for prostate cancer.

## CONCLUSIONS

During patient counseling, patients and family members often want to know what lifestyle changes they can make to help prevent malignancies. Studies have been extensively carried out on the modifiable risk factors for malignancies, and more specifically for prostate cancer. The data for smoking and alcohol and the association with prostate cancer risk are mixed, with some studies indicating that smoking and alcohol use increase the risk of prostate cancer. However, even if the effect on prostate cancer is minimal, physicians can certainly counsel patients on smoking cessation and curbing heavy alcohol use, as adjusting these lifestyle behaviors is associated with multiple other health benefits besides chemoprevention of prostate malignancy. Conversely, the majority of the data with physical activity strongly indicate that increased exercise and physical activity, either recreational or occupational, decreases the subsequent risk of developing prostate cancer. Physicians can counsel patients to increase physical activity, as this likely decreases subsequent prostate risk, but also has multiple other positive health and psychological benefits. One in three men in the United States will die from cardiovascular disease. In this light, even if the benefits of physical activity on prostate cancer are slight and the increased risk of prostate cancer with smoking is not clearly defined, there are a myriad of health benefits from quitting smoking and regular exercise, such as weight loss, decreased risk of lung cancer, and improved cardiovascular health. Certainly any member of the medical community can encourage their patients to make these lifestyle changes for these clearly defined improvements in health, even if the effects on prostate cancer are minimal.

## References


1. Montie JE, Strope SA. The causal role of cigarette smoking in bladder cancer initiation and progression, and the role of urologists in smoking cessation. *J Urol* 2008;**180**(1):31–7.
2. Hunt JD, van der Hel OL, McMillan GP, et al. Renal cell carcinoma in relation to cigarette smoking: meta-analysis of 24 studies. *Int J Cancer* 2005;**114**(1):101–8.
3. Bae JM, Li ZM, Shin MH, et al. Cigarette smoking and prostate cancer risk: negative results of the Seoul Male Cancer Cohort Study. *Asian Pac J Cancer Prev* 2013;**14**(8):4667–9.
4. Rohrmann S, Genkinger JM, Burke A, et al. Smoking and risk of fatal prostate cancer in a prospective U.S. study. *Urology* 2007;**69**(4): 721–5.
5. Watters JL, Park Y, Hollenbeck A, et al. Cigarette smoking and prostate cancer in a prospective US cohort study. *Cancer Epidemiol Biomarkers Prev* 2009;**18**(9):2427–35.
6. Rohrmann S, Linseisen J, Allen N, et al. Smoking and the risk of prostate cancer in the European Prospective Investigation into Cancer and Nutrition. *Br J Cancer* 2013;**108**(3):708–14.
7. Butler LM, Wang R, Wong AS, et al. Cigarette smoking and risk of prostate cancer among Singapore Chinese. *Cancer Causes Control* 2009;**20**(10):1967–74.
8. Malila N, Virtanen MJ, Virtamo J, et al. Cancer incidence in a cohort of Finnish male smokers. *Eur J Cancer Prev* 2006;**15**(2):103–7.
9. Sawada N, Inoue M, Iwasaki M, et al. Alcohol and smoking and subsequent risk of prostate cancer in Japanese men: the Japan Public Health Center-based prospective study. *Int J Cancer* 2014;**134**(4): 971–8.
10. Plakson LA, Penson DF, Vaughan TL, et al. Cigarette smoking and risk of prostate cancer in middle-aged men. *Cancer Epidemiol Biomarkers Prev* 2003;**12**(7):604–9.
11. Murphy AB, Akereyeni F, Nyame YA, et al. Smoking and prostate cancer in a multi-ethnic cohort. *Prostate* 2013;**73**(14):1518–28.
12. Darlington GA, Kreiger N, Lightfoot N, et al. Prostate cancer risk and diet, recreational physical activity and cigarette smoking. *Chronic Dis Can* 2007;**27**(4):145–53.
13. Huncharek M, Haddock KS, Reid R, et al. Smoking as a risk factor for prostate cancer: a meta-analysis of 24 prospective cohort studies. *Am J Public Health* 2010;**100**(4):697–701.
14. Zu K, Giovannucci E. Smoking and aggressive prostate cancer: a review of the epidemiologic evidence. *Cancer Causes Control* 2009;**20**(10):1799–810.
15. Baffetta P, Hashibe M. Alcohol and cancer. *Lancet Oncol* 2006; **7**(2):149–56.
16. Rohrmann S, Linseisen J, Key TJ, et al. Alcohol consumption and the risk for prostate cancer in the European Prospective Investigation into Cancer and Nutrition. *Cancer Epidemiol Biomarkers Prev* 2008;**17**(5):1282–7.
17. Chao C, Haque R, Van Den Eeden SK, et al. Red wine consumption and risk of prostate cancer: the California Men's Health study. *Int J Cancer* 2010;**126**(1):171–9.
18. Baglietto L, Severi G, English DR, et al. Alcohol consumption and prostate cancer risk: results from the Melbourne collaborative cohort study. *Int J Cancer* 2006;**119**(6):1501–4.
19. Watters JL, Park Y, Hollenbeck A, et al. Alcoholic beverages and prostate cancer in a prospective US cohort study. *Am J Epidemiol* 2010;**172**(7):773–80.
20. Velicer CM, Kristal A, White E. Alcohol use and the risk of prostate cancer: results from the VITAL cohort study. *Nutr Cancer* 2006;**56**(1):50–6.
21. Crispo A, Talamini R, Gallus S, et al. Alcohol and the risk of prostate cancer and benign prostatic hyperplasia. *Urology* 2004;**64**(4): 717–22.
22. Schoonen WM, Salinas CA, Kiemeney LA, et al. Alcohol consumption and risk of prostate cancer in middle-aged men. *Int J Cancer* 2005;**113**(1):133–40.
23. McGregor SE, Coumeya KS, Kopciuk KA, et al. Case-control study of lifetime alcohol intake and prostate cancer risk. *Cancer Causes Control* 2013;**24**(3):451–61.
24. Chang ET, Hedelin M, Adami HO, et al. Alcohol drinking and risk of localized versus advanced and sporadic versus familial prostate cancer in Sweden. *Cancer Causes Control* 2005;**16**(3): 275–84.
25. Rota M, Scotti L, Turati F, et al. Alcohol consumption and prostate cancer risk: a meta-analysis of the dose-risk relation. *Eur J Cancer Prev* 2012;**21**(4):350–9.
26. Middleton Fillmore K, Chikritzhs T, Stockwell T, et al. Alcohol use and prostate cancer: a meta-analysis. *Mol Nutr Food Res* 2009;**53**(2):240–55.
27. Sutcliffe S, Giovannucci E, Leitzmann MF, et al. A prospective cohort study of red wine consumption and risk of prostate cancer. *Int J Cancer* 2007;**120**(7):1529–35.
28. Zuccolo L, Lewis SJ, Donovan JL, et al. Alcohol consumption and PSA-detected prostate cancer risk – a case-control nested in the ProtecT study. *Int J Cancer* 2013;**132**(9):2176–85.
29. Thompson IM, Goodman PJ, Tangen CM, et al. The influence of finasteride on the development of prostate cancer. *N Engl J Med* 2003;**349**(3):215–24.

30. Andriole GL, Bostwick DG, Brawley OW, REDUCE study group. et al. Effect of dutasteride on the risk of prostate cancer. *N Engl J Med* 2010;**362**(13):1192–202.
31. Eastman P. ODAC recommends against Finasteride and Dutasteride for prostate cancer chemoprevention. *Oncol Times* 2011;**33**(1): 22–4.
32. Gong Z, Kristal AR, Schenk JM, et al. Alcohol consumption, finasteride, and prostate cancer risk: results from the Prostate Cancer Prevention Trial. *Cancer* 2009;**115**(16):3661–9.
33. Fowke JH, Howard L, Andriole GL, et al. Alcohol intake increases high-grade prostate cancer risk among men taking dutasteride in the REDUCE trial. *Eur Urol* 2014;**66**(6):1133–8.
34. Friedenreich CM, Neilson HK, Lynch BM. State of the epidemiological evidence on physical activity and cancer prevention. *Eur J Cancer* 2010;**2010**(46):14.
35. Orsini N, Bellocco R, Bottai M, et al. A prospective study of lifetime physical activity and prostate cancer incidence and mortality. *Br J Cancer* 2009;**101**(11):1932–8.
36. Moore SC, Peters TM, Ahn J, et al. Age-specific physical activity and prostate cancer risk among white men and black men. *Cancer* 2009;**115**(21):5060–70.
37. Giovannucci EL, Liu Y, Leitzmann MF, et al. A prospective study of physical activity and incident and fatal prostate cancer. *Arch Intern Med* 2005;**165**(9):1005–10.
38. Nilsen TI, Romundstad PR, Vatten LJ. Recreational physical activity and risk of prostate cancer: a prospective population-based study in Norway (the Hunt Study). *Int J Cancer* 2006;**119**(12):2943–7.
39. Zeegers MP, Dirx MJ, van den Brandt PA. Physical activity and the risk of prostate cancer in the Netherlands cohort study, results after 9.3 years of follow-up. *Cancer Epidemiol Biomarkers Prev* 2005;**14**(6):1490–5.
40. Wiklund F, Lageros YT, Chang E, et al. Lifetime physical activity and prostate cancer risk: a population-based case-control study in Swede. *Eur J Epidemiol* 2008;**23**(11):739–46.
41. Shahar S, Shafurah S, Hasan Shaari NS. Roles of diet, lifetime physical activity and oxidative DNA damage in the occurrence of prostate cancer among men in Klang Valley, Malaysia. *Asian Pac J Cancer Prev* 2011;**12**(3):605–11.
42. Gallus S, Foschi R, Talamini R, et al. Risk factors for prostate cancer in men aged less than 60 years: a case-control study from Italy. *Urology* 2007;**70**(6):1121–6.
43. Jian L, Shen ZJ, Lee AH, et al. Moderate physical activity and prostate cancer risk: a case-control study in China. *Eur J Epidemiol* 2005;**20**(2):155–60.
44. Pierotti B, Altieri A, Talamini R, et al. Lifetime physical activity and prostate cancer. *Int J Cancer* 2005;**114**(4):630–42.
45. Lund Haheim L, Wisløff TF, Holme I, et al. Metabolic syndrome predicts prostate cancer in a cohort of middle-aged Norwegian men followed for 27 years. *Am J Epidemiol* 2006;**164**(8):769–74.
46. Krishnadasan A, Kennedy N, Zhao Y, et al. Nested case-control study of occupational physical activity and prostate cancer among workers using a job exposure matrix. *Cancer Causes Control* 2008; **19**(1):107–14.
47. Parent ME, Rousseau MC, El-Zein M, et al. Occupational and recreation physical activity during adult life and risk of cancer among men. *Cancer Epidemiol* 2011;**35**(2):151–9.
48. Liu Y, Hu F, Wang F, et al. Does physical activity reduce the risk of prostate cancer? A systematic review and meta-analysis. *Eur Urol* 2011;**60**(5):1029–44.
49. Antonelli JA, Jones LW, Bañez LL, et al. Exercise and prostate cancer risk in a cohort of veterans undergoing prostate needle biopsy. *J Urol* 2009;**182**(5):2226–31.
50. Johnsen NF, Tjønneland A, Thomsen BL, et al. Physical activity and risk of prostate cancer in the European Prospective Investigation into Cancer and Nutrition (EPIC) cohort. *Int J Cancer* 2009;**125**(4):902–8.
51. Patel AV, Rodriguez C, Jacobs EJ, et al. Recreational physical activity and risk of prostate cancer in a large cohort of US men. *Cancer Epidemiol Biomarkers Prev* 2005;**14**(1):275–9.
52. Littman A, Kristal AR, White E. Recreational physical activity and prostate cancer risk (United States). *Cancer Causes Control* 2006;**17**(6):831–41.

CHAPTER

# 22

# Environmental and Occupational Exposures and Prostate Cancer

*Tullika Garg, MD, MPH*

Urology Department, Geisinger Health System, Danville, PA, USA

## INTRODUCTION

Prostate cancer is the most common solid organ malignancy diagnosed in men and the second leading cause of cancer deaths in American men.[1] The etiology and risk factors for prostate cancer are largely unknown. Some theories suggest that environmental carcinogens or occupation-related exposures may be associated with prostate cancer. In particular, research efforts have focused on the interplay between environmental exposures and hormonal aspects of prostate growth and carcinogenesis. Chemicals, such as endocrine disruptors, have garnered attention in recent years.[2] While no definitive associations between environmental or occupational exposures have been identified, this continues to be an active research area.

This review of the current literature of environmental/occupational exposures and prostate cancer will cover topics including Agent Orange (AO), endocrine disruptors, pesticides and farming, metal exposures, rubber manufacturing, whole body vibration, and air pollution.

## AGENT ORANGE

AO, named for the color of its storage containers, was used extensively as a defoliant in the Vietnam War. It was composed of a mixture of 2,4-dichloro-phenoxyacetic acid (2,4-D) and 2,4,5-trichloro-phenoxyacetic acid (2,4,5-T). The mixture was contaminated with 2,3,7,8-tetrachlorodibenzo-*p*-dioxin, a by-product and known carcinogen. Over the course of the conflict, eleven million gallons of AO were sprayed in southeast Asia.[3] In 1998, the National Academy of Science reviewed the association between AO and prostate cancer and the committee found "limited/suggestive evidence."[4]

Various studies have examined the relationship between AO and prostate cancer with conflicting results. A study of cancers in veterans who participated in "Operation Ranch Hand," an Air Force initiative for aerial AO spraying, found an increased incidence of prostate cancer in exposed veterans.[3]

A case-control study looked at whether AO exposure is associated with prostate cancer diagnosis. The authors identified 47 prostate cancer cases and 142 randomly selected controls from the Ann Arbor Veterans Affairs (VA). After controlling for age and race, veterans reporting AO exposure were more likely to have a diagnosis of prostate cancer, though this result did not reach statistical significance (OR 2.06, 95% CI 0.81–5.23).[5]

Chamie et al. assessed the incidence and characteristics of prostate cancer in AO-exposed veterans in the Northern California VA Health System. The cohort included a total of 13,144 veterans, 6214 exposed and 6930 unexposed based on the standard VA protocol for AO exposure determination. AO exposed veterans with prostate cancer were younger, had high-grade disease, and were more likely to present with metastases. After adjusting for other risk factors, the odds ratio for developing any prostate cancer after AO exposure was 4.83 (95% CI 3.42–6.81). High-grade cancer was independently associated with AO (OR 2.59, 95% CI 1.3–5.13). The authors conclude that AO exposure was the most important factor for predicting high-grade prostate cancer.[6]

Another study corroborated these results in a 2720 veteran cohort. In multivariable logistic regression analysis, AO exposure was associated with higher odds of prostate cancer on biopsy (OR 1.52, 95% CI 1.07–2.13), high-grade prostate cancer (OR 1.74, 95% CI 1.14–2.63), and Gleason 8 or greater prostate cancer (OR 2.1, 95% CI 1.22–3.62). Exposed veterans in this

Prostate Cancer. http://dx.doi.org/10.1016/B978-0-12-800077-9.00022-0

study were diagnosed with prostate cancer 5 years earlier than their unexposed counterparts.[7]

AO exposure may impact outcomes after prostate cancer treatment. A retrospective review of 1592 veterans from the West Los Angeles, Palo Alto, Augusta, and Durham VA hospitals found a higher risk of biochemical progression and shorter PSA doubling time after radical prostatectomy with AO exposure.[8]

Despite conflicting evidence, the Department of Veterans Affairs recognizes an association between AO and prostate cancer and provides compensation and healthcare to veterans with documented exposure.[9]

## PESTICIDES AND FARMING

Farming is associated with an increased risk of cancer, possibly due to on-the-job exposures to chemicals such as pesticides.[10] In the Agricultural Health Study there was a slightly higher standardized incidence ratio (SIR) of cancer in private farmers as compared to controls (1.19, 95% CI 1.14–1.25). The SIR was even higher for commercial pesticide applicators in the study (1.28, 95% CI 1–1.61).[11] A meta-analysis of 12 studies including 3978 prostate cancer cases and 7393 controls found an increased likelihood that cases were farmers by occupation when compared to controls (meta OR 1.38, 95% CI 1.16–1.64); however, there was an inverse relationship between pesticide exposure and prostate cancer (meta OR 0.68, 95% CI 0.49–0.96).[12]

Other conflicting studies include the Netherlands Cohort Study that after 9.3 years of follow-up and controlling for multiple variables found no association between occupational exposure to pesticides and prostate cancer.[13] Pesticide use is dominated by the United States and Europe who together consume 25–30% of pesticides worldwide.[14]

A meta-analysis of 22 studies from the United States, Canada, and Europe found an elevated meta-rate ratio of 1.13 (95% CI 1.04–1.22) for prostate cancer in farmers and pesticide applicators.[15] An expanded meta-analysis by the same authors included, in addition to farmers, occupations such as agricultural pesticide applicators, farmers licensed to use pesticides, farmers reporting exposures to pesticides, nursery and greenhouse workers, and employees of pesticide spraying companies. In the 22 studies analyzed, the meta-rate ratio for pesticide exposure and prostate cancer was 1.24 (95% CI 1.06–1.45), slightly higher than the smaller meta-analysis.[16] When limited to pesticide manufacturing workers, they found a meta-rate ratio of 1.28 (95% CI 1.05–1.58). The main chemical class exposures associated with prostate cancer were herbicides contaminated with dioxin and/or furan.[17]

Other studies have looked at specific agricultural chemicals and risk of prostate cancer. A case-control study based in the California Central Valley from 2005 to 2006 used the geographical information system (GIS) approach to quantify pesticide exposure and type and in 173 cancer registry cases and 162 control subjects. Methyl bromide and organochlorines were associated with an increased risk of prostate cancer.[18] Similarly, a systematic review of three epidemiologic studies found that prostate cancer risk increased in a dose-dependent pattern to methyl bromide exposure. This finding was pronounced for men with a family history of prostate cancer (OR 3.47, 95% CI 1.37–8.76).[19]

Methyl bromide binds covalently to DNA and creates DNA adducts that lead to sister chromatid exchange.[11] The purported mechanism of prostate cancer carcinogenesis occurs via inactivation of two gene products (pi-class glutathione S-transferases (GSTP1) and glutathione S-transferase *theta*) that protect cells from cytotoxic and carcinogenic agents. Methyl bromide has been used as a fumigant to strip soil off pathogens, though it has also been used to disinfect furniture, wood, barges, warehouses, buildings, cargo ships, and freight containers. Exposure is primarily through inhalation or direct skin contact. It was first recognized as a potential carcinogen in a cohort study of chemical plant workers who had long-term exposures and had higher mortality from testicular cancer than the general population.[20]

The largest study of pesticides and a variety of diseases including prostate cancer is the Agricultural Health Study (AHS), a prospective cohort study of nearly 89,000 licensed pesticide applicators from North Carolina and Iowa.[21] The study enrolled 82% of individuals applying for licenses for restricted use pesticides between December 1993 and December 1997. Participants completed detailed pesticide exposure, diet, and medical history questionnaires and cancer cases were identified from cancer registries. Follow-up of the cohort continues today.

An update of the AHS prostate cancer cohort identified 1962 cancers of which 919 were defined as aggressive based on Gleason score greater than 7, distant metastases or underlying cause of death being prostate cancer. Three organophosphate insecticides (fonofos, malathion, terbufos) and one organochlorine insecticide (aldrin) were associated with development of aggressive prostate cancer, especially in those with a family history (Table 22.1).[22] Organophosphate pesticides are metabolized to highly toxic intermediaries called oxons, which create reactive oxygen species that damage DNA.[23–25] Organochlorines are endocrine disruptors and accumulate in fat to have a potentially continuous effect.

The hormonal and endocrine disrupting effects of pesticides may cause prostate carcinogenesis. Vinclozolin, a fungicide for crops, has antiandrogen activity and interferes with androgen receptor activity. Maternal exposures may cause epigenetic changes that result

TABLE 22.1 Relative Risk of Aggressive Prostate Cancer Associated With the Highest Quartiles of Exposure to Insecticides in the Agricultural Health Study

| Insecticide | Class | Relative risk | 95% Confidence interval | *P* value |
|---|---|---|---|---|
| Fonofos | Organophosphate | 1.63 | 1.22–2.17 | <0.001 |
| Malathion | Organophosphate | 1.43 | 1.08–1.88 | 0.04 |
| Terbufos | Organophosphate | 1.29 | 1.02–1.64 | 0.03 |
| Aldrin | Organochlorine | 1.49 | 1.03–2.18 | 0.02 |

in premature acinar atrophy and aging-associated prostatitis. DDT and DDE are pesticides that function as 5-alpha-reductase inhibitors. In the Agricultural Health Study, chlorpyrifos, fonofos, and phorate inhibit P450 enzymes that metabolize sex hormones such as estradiol, estrone, and testosterone. These enzymes are found in the liver and the prostate.[26]

Certain individuals may have a genetic predisposition to prostate cancer after pesticide exposure, explaining the link between family history of prostate cancer and pesticides. The AHS prostate cancer cohort has been further subdivided into nested case-control studies consisting of 776 prostate cancer cases and 1444 controls designed to examine genetic susceptibilities. One recent study identified an interaction between the base excision repair (BER) gene *NEIL3*, increasing exposure to fonofos and prostate cancer. Men with a mutation-high fonofos use were 3.25 times more likely to develop prostate cancer than men reporting no use.[27] Derangement in BER genes leads to deficits in oxidative damage repair and has been associated with carcinogenesis.

A genome-wide association study from the same nested cohort found that men with prostate cancer single nucleotide polymorphisms (SNPs) who were exposed to pesticides were more susceptible to developing prostate cancer. Men with two T alleles at the rs2710647 SNP in the *EHBP1* gene and high exposure to the pesticide malathion had a 3.43 times higher risk of developing prostate cancer (95% CI 1.44–8.15). *EHBP1* is located at chromosome 17q24 and participates in signal transduction and other cell membrane functions. Similarly, men with two A alleles at the rs7679673 SNP in the *TET2* gene and high exposure to the pesticide aldrin had 3.67 times the risk of prostate cancer as unexposed men (95% CI 1.43–9.41). *TET2* is a tumor suppressor gene located on chromosome 4 and involved in androgen regulation in prostate cancer cell lines.[28]

## ENDOCRINE DISRUPTORS

Endocrine disruptors are defined as "a class of environmental toxicants that interfere with endocrine signaling pathways."[29] Throughout the mid-twentieth century, endocrine disruptors, such as polychlorinated biphenyls (PCBs) and polyhalogenated aromatic hydrocarbons including bisphenol A (BPA), dioxin, and dibenzofurans, have been used widely. These chemicals are lipophilic compounds that inhibit estrogen sulfotransferase, which results in increased levels of bioavailable estrogen. The majority of the general population has measurable serum levels of endocrine disruptors. A recent analysis showed that Swedish men with prostate cancer had higher concentrations of endocrine disruptors, specifically PCG153 and *trans*-chlordane, in fat.[30]

In recent years, there have been significant increases in the incidence of hormonally active cancers and conditions such as cancers of the breast, prostate and testis, and hypospadias and cryptorchidism.[31–33] Concurrently, there has also been a marked decrease worldwide in sperm quantity and quality. Some studies have shown significant relationships between exposures to estrogen-like substances and development of breast cancer. Similarly, mothers with high levels of urine phthalates, an endocrine disruptor found in plastics, are at increased risk of having boys with cryptorchidism. Perhaps the most commonly known endocrine disruptor is BPA.

BPA is a monomer used in manufacturing polycarbonate plastics and epoxy resins for many products including milk and food containers, baby formula bottles, water carboys, canned food liners, and dental resins. BPA is ubiquitous in the environment and is found in aerosols, dust particles, and drinking water. It enters cells through binding to membrane-bound and nuclear estrogen receptors alpha and beta.[34]

Laboratory data suggest that exposure to BPA has multiple hormonally mediated effects. For example, male mice exposed to BPA *in utero* had increased anogenital distances, increased prostate size and decreased epididymal weight. Male mice fetuses had increased prostatic glandular buds. In adult animals, BPA exposure leads to decreased daily sperm production and enlarged prostate size. Exposed mouse prostatic stroma demonstrated increased androgen-receptor expression, and in rats, altered prostate stroma differentiation and urethral malformations.

Early life exposure to low levels of BPA in neonate rats induces a higher incidence and grade of prostatic intraepithelial neoplasia (PIN). In a study of prostatospheres and side population analysis using flow cytometry, prostate cancer cells exposed to BPA or dioxin showed proliferation of progenitor cells, similar to what happens to cells exposed to estradiol.[29] BPA may also induce carcinogenesis through centrosome dysfunction. In a lab study using six prostate cancer cell lines, cells that were treated with BPA had a two- to eightfold increase in the number of cells with three or more centrosomes.[35] In other malignancies, centrosome dysfunction has been associated with the development of cancer and may also cause progression of the disease.

The same authors conducted a clinical study looking at the association between urine BPA concentrations and prostate cancer. In comparing 27 prostate cancer cases and 33 controls, patients with prostate cancer had higher urinary BPA levels and younger prostate cancer patients had the highest levels. The overall geometric mean BPA in this cohort was 2–2.5 times higher than baseline US levels in the National Health and Nutrition Examination Survey (NHANES) study.[29]

BPA interacts with mutant androgen receptors on prostate cancer cells. At nanomolar doses in androgen-dependent and androgen-independent prostate cancer cell lines, BPA activates multiple mutant androgen receptors, especially in the presence of dihydrotestosterone (DHT).[36] In prostate cancer cells with the codon 877 mutation, one of the most common androgen receptor mutations, BPA induced proliferation of cells.[37] These are concerning findings as BPA may bypass androgen deprivation therapy by activating mutant androgen receptors.

BPA interaction with mutant androgen receptors may also influence prostate cancer outcomes. In mice, low doses of BPA increased prostate tumor size and caused earlier biochemical failure.[38] A comparison of BPA-exposed and unexposed mice showed that BPA-exposed mice had higher PSA, larger tumor volumes, increased tumor cell proliferation, and shorter time for failure of therapy. BPA attachment to mutated androgen receptors leads to downstream signaling. The authors suggest that androgen receptor blockers, such as bicalutamide, may block BPA by binding to the androgen receptors first.[39]

Another class of endocrine disruptors that have been studied with prostate cancer are persistent organic pollutants (POPs). POPs consist of two classes: intentionally produced industrial chemicals or unintentional byproducts of industrial production.[40] Some of the most well-known POPs include dioxin and dichlorodiphenyltricholorethane (DDT). Dioxin exposure is associated with liver, thyroid, lung, skin, oral cavity, and ovarian cancers, among others.[41,42] Dioxin is a halogenated aromatic hydrocarbon that binds to the aryl hydrocarbon receptor (AhR), which cross talks with estrogen receptor-alpha. A meta-analysis of 17 cohort studies with a total of 40,286 participants with an average follow up of 20 years showed an elevated standardized mortality ratio for prostate cancer in subjects exposed to dioxin (meta-SMR 1.26, 95% CI 1.00–1.57).[43]

The relationship between other POPs and prostate cancer has been examined in small pilot studies. In one study of 58 prostate cancer cases and 23 controls with benign prostatic hyperplasia (BPH), analysis of POPs in abdominal fat biopsies showed higher odds of prostate cancer in the presence of polychlorinated biphenyls (PCB) congener 153 and *trans*-chlordane.[44] Another series of 58 prostate cancer cases and 99 controls showed that oxychlordane and PCB 180 were associated with increased prostate cancer risk after adjusting for age, BMI, and history of prostatitis.[45]

## VITAMIN D AND SUNLIGHT

Low vitamin D levels and lack of sunlight exposure may be involved in a variety of nonskin cancers. Vitamin D comes in several forms, but the biologically active version is 1,25-dihydroxyvitamin D, or calcitriol. Vitamin $D_3$ and vitamin $D_2$ (ergocalciferol) are ingested from dietary sources. $D_3$ can also be obtained from the conversion of 7-dehydrocholesterol in the skin via ultraviolet (UV) B light and undergoes two hydroxylations, in the liver and kidney, into the active calcitriol.[46] In the United States, sunlight is the biggest source of vitamin D as dietary sources are minimal.[47] A single, whole body dose of solar radiation makes 250 μg of vitamin D. Calcitriol is synthesized in various tissues including prostate cells.[48] Vitamin D influences at least 60 genes that control cell-cycle arrest, apoptosis, and prostate-cell differentiation. Active vitamin D binds to a vitamin D receptor (VDR), which leads to a complex with the retinoid X receptor (RXR), which then binds and influences specific areas on the promoter regions of genes that respond to vitamin D.[49]

African-Americans and northern Europeans have low vitamin D levels, due to reduced ultraviolet light absorption and low sunlight exposure, respectively. These two groups also have the highest age-adjusted rates of prostate cancer in the world.[50] Conversely, due to a diet rich in tuna and skipjack fish, the Japanese have the highest vitamin D levels in the world and the lowest rates of prostate cancer.[51,52] In a geographic map trend surface study of UV radiation in 3073 counties in the United States and prostate cancer mortality, Hanchette and Schwartz found an inverse relationship between the UV radiation map and the prostate cancer mortality map suggesting that areas with higher sunlight exposure had lower prostate cancer mortality.[53]

Epidemiologic studies of vitamin D levels have conflicting results. A review of three small serologic nested case-control studies showed null results.[54] Large prospective studies, however, have shown that low calcitriol levels are associated with increased prostate cancer risk. In the Physicians' Health Study, the relative risk for prostate cancer in men with vitamin D levels below the median and with VDR polymorphisms was 0.43 (95% CI 0.19–0.98).[55] In comparing 153 prostate cancer cases and 3414 controls from the NHANES I cohort, subjects that lived in southern states the longest and had the highest rates of solar radiation had a decreased relative risk of prostate cancer.[56] A Norwegian study found that patients diagnosed in the summer and fall months had lower case fatality rates when vitamin D levels were at their peak.[57]

A large study of 35,620 nonmelanoma cancer patients from national cancer registries in sunny (Australia, Singapore, and Spain) and less sunny countries (Canada, Denmark, Finland, Iceland, Norway, Scotland, Slovenia, and Sweden) found that standardized incidence ratios for solid primary cancers were lower in sunny (1.03, 95% CI 0.99–1.08) versus nonsunny countries (1.14, 95% CI 1.11–1.17). Among genitourinary malignancies, this relationship was strongest for prostate, bladder, and kidney cancers.[58] A longitudinal, nested case-control study of Norwegian subjects by the same authors found a U-shaped relationship between vitamin D levels and prostate cancer risk. They found that prostate cancer risk was highest in men with low (less than 19 nmol/L) and high (greater than 80 nmol/L) vitamin D levels and lowest in men with normal vitamin D levels.[59]

## METALS

The relationship between inorganic arsenic and prostate cancer is probably the most established of the metal exposures. The Environmental Protection Agency and World Health Organization set the upper limit for arsenic in drinking water at 10 parts per million.[60] Inorganic arsenic is found in contaminated drinking water from underground wells, atmospheric emissions, vegetables, and rice, and specific occupations such as mining, smelting wood preservation, and electronics. The association with prostate cancer was first identified in the 1980s in Taiwan. The age-standardized mortality from prostate cancer in groups exposed to the highest levels of arsenic was six times that of Taiwan's general population in a dose-dependent manner.[61] While there were limitations to this initial study, a follow-up study showed that age adjusted mortality for prostate cancer was 10 times higher in the group (9.18 deaths per 100,000) exposed to the highest levels of arsenic as compared to those exposed to the lowest levels (0.95 deaths per 100,000). The highest arsenic levels were greater than 0.60 parts per million.[62] More recent studies have found higher standardized mortality ratios for higher exposure levels in Utah and in Australia.[63,64]

It is unclear how arsenic mediates carcinogenesis in the prostate, but there is some suggestion that it binds to estrogen receptors, activates estrogen related genes, and pushes prostate cancer cells toward androgen independence.[29]

Other potential metal exposures include metalworking fluids, cadmium, and zinc. Metalworking fluids are used in machining and grinding processes. According to the International Agency for Research on Cancer (IARC), metalworking fluids contain PAH, alkanolamines, and nitrosamines, all of which are possible carcinogens. Two nested case-control studies of the United Autoworkers union have shown an association between prostate cancer and exposures to metalworking fluids during late puberty and 25 years prior to risk age.[65,66]

Cadmium is classified as a carcinogen by IARC and the National Toxicology Program. Cadmium accrues in the prostate leading to cumulative effects. It binds to the estrogen receptor as an estrogen mimic and causes proliferation of human prostate cells. Cadmium has been associated with castration-resistant prostate cancer.[26]

Similarly, zinc accumulates in the prostate and causes apoptosis via citrate oxidation. Low prostate zinc levels have been associated with prostate tumor growth. One GIS study looked at soil–zinc concentrations and prostate cancer rates in African-Americans in South Carolina. Census tracts with high groundwater use and low soil–zinc concentrations had higher rates of prostate cancer.[67] Low levels of zinc and other metals may influence prostate cancer outcomes. A nested case-control study from the Cooperative Prostate Cancer Tissue Resource showed that patients with biochemical recurrence after prostatectomy had 12% lower median iron levels and 21% lower zinc levels in normal tissue from prostatectomy specimens.[68]

## RUBBER MANUFACTURING

There are conflicting reports on the risk of prostate cancer resulting from work in the rubber industry. Rubber workers are exposed to nitrosamines, naturally occurring chemicals that are highly concentrated in the rubber industry. Exposures can occur in food, cosmetics, drugs, passenger areas of new cars, and tobacco, both smokeless and burned. In one review, a total of 4917 male rubber factory workers were followed through January 1983. Standardized mortality ratios were highest for multiple malignancies including prostate cancer.[69] A cohort study of 8933 workers from five rubber tire factories in Germany also found a nonsignificant trend of increasing mortality from prostate cancer with increasing exposure to nitrosamines.[70] A study of 11,660 men and 6087 women employed at a rubber tire plant between 1950 and 1995 showed more deaths than expected from prostate cancer based on standardized mortality ratios.[71]

A conflicting meta-analysis of 12 studies showed minimal to no association between work in rubber and tire manufacturing and prostate cancer (pooled risk estimate 1.03, 95% CI 0.96–1.11).[72] A study of 17,924 subjects employed in the North American synthetic rubber industry found lower than expected overall deaths in this population; however, there were greater than expected deaths from prostate cancer in particular subgroups, though the authors suggest that the deaths may not be attributable to occupational exposures.[73]

## WHOLE BODY VIBRATION

Up to 7% of all employees are exposed to whole body vibration (WBV), which occurs when mechanical energy is transmitted via a surface to the body.[74] WBV may cause disorders of the musculoskeletal, gastrointestinal, and genitourinary systems. In the genitourinary system, whole body vibration has been associated with prostatitis and higher testosterone levels.[75–77] ISO 26310-1 1997 standards recommend keeping WBV dose values below 8.5 $m/s^2$.[78]

A population-based case-control questionnaire study of occupational WBV exposure and prostate cancer in Canada identified 447 prostate cancer cases and 532 controls and found a significant risk of prostate cancer in transport equipment operators (OR 1.9, 95% CI 1.07–3.39); however, age-adjusted multivariable modeling found no association between occupation-based WBV and prostate cancer (OR 1.29, 95% CI 0.9–1.84).[79]

A meta-analysis on the topic identified eight studies of driving occupations including railway transport workers, truck drivers, heavy-equipment operators, and bus drivers. The pooled relative risk for prostate cancer trended toward significance (RR 1.14, 95% CI 1.06–1.22). The authors identified multiple problems with studies on this topic including misclassification bias as exposure was based on job title, few studies, and lack of standardized exposure measurement.[80]

## OTHER EXPOSURES

Industrial pollution and work in the aerospace and radiation industries have been evaluated as possible risk factors for prostate cancer. A study of 791,519 residents of a highly industrialized region of Spain found an increased risk of death from prostate cancer in subjects living near metal and chemical factories.[81] A nested case-control study of 362 prostate cancer cases and 1805 matched controls derived from a cohort of aerospace and radiation workers in Southern California found that increasing on-the-job exposure to trichloroethylene was associated with increasing risk of prostate cancer.[82] The authors also found a strong linear relationship between exposure to mineral oil from puberty to the age of 23 and subsequent development of prostate cancer (RR 2.38, 95% CI 1.31–4.33).

There has also been interest in the carcinogenic effects of air pollution. The World Trade Center attacks in 2001 released a large bolus of particulate matter into the atmosphere and survivors have higher rates of prostate and thyroid cancers, as well as multiple myeloma. There was no statistically significant increase in the cancers as related to level of exposure, and the follow-up is still relatively short, particularly for prostate cancer.[83]

## CONCLUSIONS

Despite extensive research into possible environmental and occupational causes of prostate cancer, definitive associations remain elusive. Many studies discussed here are small with short follow-up and have the risk of misclassification bias. Additionally, both studies and cohorts often predate the PSA testing era and may not be reflective of contemporary cases and management. Further study is needed to fully elucidate the weak associations described here.

### References

1. Siegel R, Ma J, Zou Z, et al. Cancer statistics. *CA Cancer J Clin* 2014;**64**: 9–29.
2. Soto AM, Sonnenschein C. Environmental causes of cancer: endocrine disruptors as carcinogens. *Nat Rev Endocrinol* 2010;**6**: 363–70.
3. Akhtar FZ, Garabrant DH, Ketchum NS, et al. Cancer in US Air Force veterans of the Vietnam War. *J Occup Environ Med* 2004; **46**:123–36.
4. Sun XL, Kido T, Okamoto R, et al. The relationship between Agent Orange and PSA: a comparison of a hotspot and non-sprayed area in Vietnam. *Environ Health Prev Med* 2013;**18**:356–60.
5. Giri VN, Cassidy AE, Beebe-Dimmer J, et al. Association between Agent Orange and prostate cancer: a pilot case-control study. *Urology* 2004;**63**:757–61.
6. Chamie K, DeVere White RW, Lee D, et al. Agent Orange exposure, Vietnam War veterans and the risk of prostate cancer. *Cancer* 2008;**113**:2464–70.
7. Ansbaugh N, Shannon J, Mori M, et al. Agent Orange exposure as a risk-factor for high-grade prostate cancer. *Cancer* 2013;**119**: 2399–404.
8. Shah SR, Freedland SJ, Aronson WJ, et al. Exposure to Agent Orange is a significant predictor of prostate-specific antigen (PSA)-based recurrence and a rapid PSA doubling time after radical prostatectomy. *BJU Int* 2009;**103**:1168–72.
9. *U.S. Department of Veterans Affairs*. http://www.benefits.va.gov/compensation/claims-postservice-agent_orange.asp [accessed 10.11.2014].

10. Acquavella J, Olson G, Cole P, et al. Cancer among farmers: a meta-analysis. *Ann Epidemiol* 1998;**8**:64–74.
11. Koutros S, Alavanja MC, Lubin JH, et al. An update of cancer incidence in the Agricultural Health Study. *J Occup Environ Med* 2010;**52**:1098–105.
12. Ragin C, Davis-Reyes B, Tadesse H, et al. Farming, reported pesticide use, and prostate cancer. *Am J Mens Health* 2012;**7**:102–9.
13. Boers D, Zeegers MPA, Swaen GM, et al. The influence of occupational exposure to pesticides, polycyclic aromatic hydrocarbons, diesel exhaust, metal dust, metal fumes, and mineral oil on prostate cancer: a prospective cohort study. *Occup Environ Med* 2005;**62**:531–7.
14. Dich J, Zahm SH, Hanberg A, et al. Pesticides and cancer. *Cancer Causes Control* 1997;**8**:420–43.
15. Van Maele-Fabry G, Willems JL. Occupation related pesticide exposure and cancer of the prostate: a meta-analysis. *Occup Environ Med* 2003;**60**:634–42.
16. Van Maele-Fabry G, Willems JL. Prostate cancer among pesticide applicators: a meta-analysis. *Int Arch Occup Environ Health* 2004;**77**:559–70.
17. Van Maele-Fabry G, Libotte V, Willems J, et al. Review and meta-analysis of risk estimates for prostate cancer in pesticide manufacturing workers. *Cancer Causes Control* 2006;**17**:353–73.
18. Cockburn M, Mills P, Zhang X, et al. Prostate cancer and ambient pesticide exposure in agriculturally intensive areas in California. *Am J Epidemiol* 2011;**173**:1280–8.
19. Budnik LT, Kloth S, Velasco-Garrido M, et al. Prostate cancer and toxicity from critical use exemptions of methyl bromide: environmental protection helps protect against human health risks. *Environ Health* 2012;**11**:1–12.
20. FAO: Food and Agriculture Organization of the United Nations International standards for phytosanitary measures. Manual on Pesticides. 2003.
21. *Agricultural Health Study*. http://aghealth.nih.gov/ [accessed 10.11.2014].
22. Koutros S, Beane Freeman LE, Lubin JH. Risk of total and aggressive prostate cancer and pesticide use in the Agricultural Health Study. *Am J Epidemiol* 2013;**177**:59–74.
23. Hodgson E, Rose RL. Organophosphorus chemicals: potent inhibitors of the human metabolism of steroid hormones, xenobiotics. *Drug Metab Rev* 2006;**38**:149–62.
24. Tang J, Cao Y, Rose RL, et al. Metabolism of chlorpyrifos by human cytochrome P450 isoforms and human, mouse, and rat liver microsomes. *Drug Metab Dispos* 2001;**29**:1201–4.
25. Wu JC, Hseu YC, Tsai JS, et al. Fenthion and terbufos induce DNA damage, the expression of tumor-related genes and apoptosis in HEPG2 cells. *Environ Mol Mutagen* 2011;**52**:529–37.
26. Prins GS. Endocrine disruptors and prostate cancer risk. *Endocr Relat Cancer* 2008;**15**:649–56.
27. Barry KH, Koutros S, Berndt SI, et al. Genetic variation in base excision repair pathway genes, pesticide exposure, and prostate cancer risk. *Environ Health Perspect* 2011;**119**:1726–32.
28. Koutros S, Berndt SI, Barry KH, et al. Genetic susceptibility loci, pesticide exposure and prostate cancer risk. *PloS ONE* 2013;**8**: e58195.
29. Hu WY, Shi GB, Hu DP, et al. Actions of estrogens and endocrine disrupting chemicals on human prostate stem/progenitor cells and prostate cancer risk. *Mol Cell Endocrinol* 2012;**354**: 63–73.
30. Hardell L, Andersson SO, Carlberg M, et al. Adipose tissue concentrations of persistent organic pollutants and the risk of prostate cancer. *J Occup Environ Med* 2006;**48**:700–7.
31. Ries LAG, Eisner MP, Kosary CL, Hankey BF, Miller BA, Clegg L, Edwards BK, editors. *SEER cancer statistics review*. Bethesda, MD: National Cancer Institute; 2002 1973–1999.
32. Carlsen E, Giwercman A, Keiding N, et al. Evidence for the decreasing quality of semen during the past 50 years. *BMJ* 1992;**305**: 609–12.
33. Giwercman A, Carlsen E, Keiding N, et al. Evidence for increasing incidence of abnormalities of the human testis: a review. *Environ Health Perspect* 1993;**101**:65–71.
34. Maffini MV, Rubin BS, Sonnenschein C, et al. Endocrine disruptors and reproductive health: the case of bisphenol A. *Mol Cell Endocrinol* 2006;**254–255**:179–86.
35. Tarapore P, Ying J, Ouyang B, et al. Exposure to BPA correlates with early-onset prostate cancer and promotes centrosome amplification and anchorage-independent growth *in vitro*. *PLoS ONE* 2014;**9**:e90332.
36. Wetherill YB, Petre CE, Monk KR, et al. The xenoestrogen bisphenol A induces inappropriate androgen receptor activation and mitogenesis in prostatic adenocarcinoma cells. *Mol Cancer Ther* 2002;**1**:515–24.
37. Wetherill YB, Fisher NL, Staubach A, et al. Xenoestrogen action in prostate cancer: pleiotropic effects dependent on androgen receptor status. *Cancer Res* 2005;**65**:54–65.
38. Wetherill YB, Hess-Wilson JK, Comstock CE, et al. Bisphenol A facilitates bypass of androgen ablation therapy in prostate cancer. *Mol Cancer Ther* 2006;**5**:3181–90.
39. Hess-Wilson JK. BPA may reduce the efficacy of androgen deprivation therapy in prostate cancer. *Cancer Causes Control* 2009;**20**: 1029–37.
40. *Environmental Protection Agency*. http://www2.epa.gov/international-cooperation/persistent-organic-pollutants-global-issue-global-response#pops [accessed 30.10.2014].
41. Knerr S, Schrenk D. Carcinogenicity of 2,3,7,8-tetrachlorodibenzo-*p*-dioxin in experimental models. *Mol Nutr Food Res* 2006;**50**: 897–907.
42. Davis BJ, McCurdy EA, Miller BD, et al. Ovarian tumors in rats induced by chronic 2,3,7,8-tetrachlorodibenzo-*p*-dioxin treatment. *Cancer Res* 2000;**60**:5414–9.
43. Leng L, Chen X, Li CP, et al. 2,3,7,8-Tetrachlorodibenzo-*p*-dioxin exposure and prostate cancer: a meta-analysis of cohort studies. *Public Health* 2014;**128**:207–13.
44. Hardell L, Adnersson SO, Carlberg M, et al. Adipose tissue concentrations of persistent organic pollutants and the risk of prostate cancer. *J Occup Environ Med* 2006;**48**:700–7.
45. Ritchie JM, Vial SL, Fuortes LK, et al. Organochlorines and risk of prostate cancer. *J Occup Environ Med* 2003;**45**:692–702.
46. John EM, Dreon DM, Koo J, et al. Residential sunlight exposure is associated with a decreased risk of prostate cancer. *J Steroid Biochem Mol Biol* 2004;**89–90**:549–52.
47. Vieth R. Vitamin D supplementation, 25-hydroxyvitamin D concentrations, and safety. *Am J Clin Nutr* 1999;**69**:842–56.
48. Schwartz GG, Whitlatch LW, Chen TC, et al. Human prostate cells synthesize 1,25-dihydroxyvitamin $D_3$ from 25-hydroxyvitamin $D_3$. *Cancer Epidemiol Biomarkers Prev* 1998;**7**:391–5.
49. Christako S, Dhawan P, Liu Y, et al. New insights into the mechanisms of vitamin D action. *J Cell Biochem* 2003;**88**:695–705.
50. Kurihara M, Aoki K, Hismamichi S, editors. *Cancer mortality statistics in the world, 1950–1985*. Nagoya, Japan: University of Nagoya Press; 1989.
51. Nakamura K, Nashimoto M, Hori Y, et al. Serum 25-hydroxyvitamin D concentrations and related dietary factors in peri- and postmenopausal Japanese women. *Am J Clin Nutr* 2000;**71**: 1161–5.
52. Nakamura K, Nashimoto M, Okuda Y, et al. Fish as a major source of vitamin D in the Japanese diet. *Nutrition* 2002;**18**:415–6.
53. Hanchette CL, Schwartz GG. Geographic patterns of prostate cancer mortality. Evidence for a protective effect of UV radiation. *Cancer* 1992;**70**:2861.

54. Schwartz GG. Vitamin D and the epidemiology of prostate cancer. *Semin Dial* 2005;**18**:276–89.
55. Ma J, Stampfer MJ, Gann PH, et al. Vitamin D receptor polymorphisms, circulating vitamin D metabolites, and risk of prostate cancer in United States physicians. *Cancer Epidemiol Biomarkers Prev* 1998;**7**:385–90.
56. John EM, Dreon DM, Koo J, et al. Residential sunlight exposure is associated with a decreased risk of prostate cancer. *J Steroid Biochem Mol Biol* 2004;**89–90**:549–52.
57. Robsahm TE, Tretli S, Dahlback A, et al. Vitamin D3 from sunlight may improve the prognosis of breast-, colon- and prostate cancer. *Cancer Causes Control* 2004;**15**:149–58.
58. Tuohimaa P, et al. Does solar exposure as indicated by the non-melanoma skin cancers, protect from solid cancers: vitamin D as a possible explanation. *Eur J Cancer* 2007;**43**:1701.
59. Tuohimaa P, et al. Both high and low levels of blood vitamin D are associated with a higher prostate cancer risk: a longitudinal, nested case-control study in the Nordic countries. *Int J Cancer* 2004;**108**:104.
60. *Harvard Men's Health Watch,* August 2009.
61. Chen CJ, Kuo TL, Wu MM. Arsenic and cancers. *Lancet* 1988;**1**:414–5.
62. Wu MM, Kuo TL, Hwang YH, et al. Dose–response relation between arsenic concentration in well water and mortality from cancers and vascular diseases. *Am J Epidemiol* 1989;**130**:1123–32.
63. Lewis DR, Southwick JW, Ouellet-Hellstrom R, et al. Drinking water arsenic in Utah: a cohort mortality study. *Environ Health Perspect* 1999;**107**:359–65.
64. Hinwood AL, Jolley DJ, Sim MR. Cancer incidence and high environmental arsenic concentrations in rural populations: results of an ecological study. *Int J Environ Health Res* 1999;**9**:131–41.
65. Agalliu I, Eisen EA, Kriebel D, et al. A biological approach to characterizing exposure to metalworking fluids and risk of prostate cancer. *Cancer Causes Control* 2005;**16**:323–31.
66. Agalliu I, Kriebel D, Quinn MM, et al. Prostate cancer incidence in relation to time windows of exposure to metalworking fluids in the auto industry. *Epidemiology* 2005;**16**:664–71.
67. Wagner SE, Burch JB, Hussey J, et al. Soil zinc content, groundwater usage, and prostate cancer incidence in South Carolina. *Cancer Causes Control* 2009;**20**:345–53.
68. Sarafanov AG, Todorov TI, Centeno JA, et al. Prostate cancer outcome and tissue levels of metal ions. *Prostate* 2011;**71**:1231–8.
69. Bernardinelli L, de Marco R, Tinelli C. Cancer mortality in an Italian rubber factory. *Br J Ind Med* 1987;**44**:187–91.
70. Straif K, Weiland SK, Bungers M, et al. Exposure to high concentrations of nitrosamines and cancer mortality among a cohort of rubber workers. *Occup Environ Med* 2000;**57**:180–7.
71. Wilczynska U, Szadkowska-Stanczyk I, Szeszenia-Dabrowska N, et al. Cancer mortality in rubber tire workers in Poland. *Int J Occup Med Environ Health* 2001;**14**:115–25.
72. Stewart RE, Dennis LK, Dawson DV, et al. A meta-analysis of risk estimates for prostate cancer related to tire and rubber manufacturing operations. *J Occup Environ Med* 1999;**41**:1079–84.
73. Delzell E, Sathiakumar N, Graff J, et al. An updated study of mortality among North American rubber industry workers. *Res Rep Health Eff Inst* 2006;**132**:1–63.
74. Bovenzi M, Hulshof CT. An updated review of epidemiologic studies on the relationship between exposure to whole-body vibration and low back pain (1986–1997). *Int Arch Occup Environ Health* 1999;**72**:351–65.
75. NIOSH. Relationship between whole-body vibration and morbidity patterns among heavy equipment operators. Cincinnati, OH: National Institute of Occupational Safety and Health, DHHS (NIOSH) Publication # 74–131, NIOSH; 1974.
76. Bosco C, Iacovelli M, Tsarpela O, et al. Hormonal responses to whole-body vibration in men. *Eur J Appl Physiol* 2000;**81**:449–54.
77. Sass-Kortsak AM, Purdham JT, Kreiger N, et al. Occupational risk factors for prostate cancer. *Am J Ind Med* 2007;**50**:568–76.
78. Cann AP, Salmoni AW, Vi P, et al. An exploratory study of whole-body vibration exposure and dose while operating heavy equipment in the construction industry. *Appl Occup Environ Hyg* 2003;**18**:999–1005.
79. Nadalin V, Krelger N, Parent ME, et al. Prostate cancer and occupational whole-body vibration exposure. *Ann Occup Hyg* 2012;**56**:968–74.
80. Young E, et al. Prostate cancer and driving occupations: could whole body vibration play a role? *Int Arch Occup Environ Health* 2009;**82**:551–6.
81. Ramis R, Diggle P, Cambra K, et al. Prostate cancer and industrial pollution: risk around putative focus in a multi-source scenario. *Environ Int* 2011;**37**:577–85.
82. Krishnadasan A, Kennedy N, Zhao Y, et al. Nested case-control study of occupational chemical exposures and prostate cancer in aerospace and radiation workers. *Am J Ind Med* 2007;**50**:383–90.
83. Li J, Cone JE, Kahn AR, et al. Association between World Trade Center exposure and excess cancer risk. *JAMA* 2012;**308**:2479–88.

CHAPTER

# 23

# Level-1 Data From the REDUCE Study and the PCPT Data

*Goutham Vemana, MD, Gerald Andriole, MD*

**Division of Urological Surgery, Washington University in St. Louis, St. Louis, MO, USA**

## INTRODUCTION

Alternatives to waiting for the clinical signs of a disease to become apparent are screening and prevention. These approaches may make treatment more effective and safer. Prostate cancer is seemingly an ideal disease for these approaches as it has a long latent period driven not just by mutation but also by many epigenetic variables, particularly those relating to the action of cell-signaling regulatory molecules, which can also be important determinants during the latent period before invasion and metastasis occur.[1] Pharmacologic modulation of these regulatory pathways, over and above the effective use of drugs and micronutrients that block mutagenic damage to DNA, thus offers great potential for prevention of prostate cancer.[1]

Chemoprevention reduces disease diagnosis by the use of chemical agents, drugs, or food nutrients. With respect to prostate cancer, risk reduction by 5-alpha-reductase (5 AR) inhibitors has been studied in two settings. The first assessed healthy men at low risk of the disease (Prostate Cancer Prevention Trial, PCPT) and the second assessed men at higher risk for the disease because of elevated prostate-specific antigen (PSA) levels. Two large randomized trials have attempted to address the goal of chemoprevention of prostate cancer by pharmacologic intervention in these two populations.

## INCIDENCE

In 2013, it was estimated that prostate cancer accounted for 27% (233,000) of incident cancer cases in men[2]. Additionally, prostate cancer also has rising incidence with age.[2] Nonetheless, though prostate cancer was typically a disease of the older man, the era of PSA-based screening has resulted in a trend toward diagnosis in younger men. As such, the 2005 incidence rate of prostate cancer in men younger than 50 is 7.23 times the 1986 incidence rate.[3]

## FACTORS FOR INCREASED RISK OF PROSTATE CANCER

### Inflammatory Influence

Recent literature has supported the idea that inflammation is a key component of tumor progression. Inflammatory cells have protumor actions such as releasing growth and survival factors, promoting angiogenesis and lymph angiogenesis, stimulating DNA damage, remodeling the extracellular matrix to facilitate invasion, coating tumor cells to make available receptors for disseminating cells via lymphatics and capillaries, and evading host defense mechanisms.[4] A review of the relationship between inflammation and prostate cancer suggested a causal relationship in men with underlying genetic predisposition.[5]

### Metabolic Influence

Adverse metabolic features, such as insulin resistance, hypertension, obesity, and dyslipidemia, may predispose a patient to develop prostate cancer.[6] Epidemiologic data from the PCPT demonstrated that obesity (BMI $> 30$) and diabetes were both associated with increased risk of developing higher-grade prostate cancer (Gleason $\geq 7$).[7] Conversely, a review of the literature by Hsing et al. demonstrated mixed or inconclusive results with regard to prostate cancer and obesity and diabetes and further research will be required to clarify the associations.[8]

Prostate Cancer. http://dx.doi.org/10.1016/B978-0-12-800077-9.00023-2

### Hormonal Influence

The development and growth of benign prostatic tissue as well as prostate cancer cells is mediated by androgens. Specifically, dihydrotestosterone (DHT) is the primary androgen; it is derived from testosterone by the enzyme 5 AR, which is located within the stroma and basal cells of the prostate.[9] There are two main isoforms of the 5 AR enzyme. Type I is expressed predominantly in the liver and skin and prostate while type II is expressed mostly in prostatic epithelium and genital tissues.[10] In addition to DHT promoting prostatic growth, it is also responsible for facial hair, acne, and male pattern baldness.[11]

### Genetic Factors

Some processes that may induce genetic changes for cancer growth and cell proliferation include oxidative stress and DNA damage, shortening of telomeres, and cellular senescence.[12] Despite ongoing research, the genetic factors of prostate cancer have yet to be clearly identified for routine clinical use.

## 5-ALPHA-REDUCTASE INHIBITORS

Males with congenital 5 AR deficiency have underdeveloped external genitalia and prostate.[11] These observations led to the initial findings that prostate differentiation and growth are significantly mediated by DHT. Additionally, it was observed that these individuals with congenital 5 AR deficiency did not develop benign prostatic hyperplasia (BPH) or prostate cancer.

Using this knowledge of the effects of DHT, researchers explored the idea of inhibiting 5 AR in order to prevent the conversion of testosterone to DHT among males whose external genitalia were fully developed and matured. They theorized that inhibiting 5 AR would result in a safe medication that could improve lower urinary tract symptoms by reducing BPH, treat male pattern baldness, and improve acne.[13] This led to the development and eventual US Food and Drug Administration (FDA) approval for finasteride (for use of BPH and male hair loss) in 1992 followed by dutasteride (for use of symptomatic BPH only) in 2002. These 5 AR inhibitors have been shown in multiple studies to be effective for the treatment and prevention of BPH.[14–17]

## THE PROSTATE CANCER PREVENTION TRIAL (PCPT)

Based on the findings of earlier studies suggesting the role of androgens in the development of prostate cancer,[18–20] the PCPT was developed to evaluate the ability of finasteride, a selective inhibitor of type II isoform of 5 AR, to prevent prostate cancer in men at low risk for the disease. The PCPT was a prospective randomized double-blinded study that evaluated the use of finasteride for 7 years versus placebo.[21] The primary objective of the study was to determine whether the use of finasteride would be associated with a risk reduction in the development of prostate cancer.

Men in PCPT were at least 55 years old, had a normal digital rectal examination, PSA level less than or equal to 3.0 ng/mL, minimal coexisting medical conditions, and without severe lower urinary tract symptoms (as defined by the American Urologic Symptom Score[22] of less than 20). This study randomly assigned 18,882 men who met the enrollment criteria to treatment with finasteride (5 mg per day) or placebo over the course of 7 years.[21] A prostate biopsy was recommended in the placebo group if a patient's annual PSA level exceeded 4.0 ng/mL or if digital rectal examination was abnormal.[21] In the finasteride group, PSA values were adjusted (multiplied by 2–2.7 over the course of the study) so that the biopsy rates were similar among finasteride and placebo treated men.

The findings of PCPT were not only significant but also controversial. Of the patients in the finasteride treatment arm, 18.4% had prostate cancer diagnosed as opposed to 24.4% in the placebo arm (a 24.8% relative risk reduction). Curiously, the proportion of patients diagnosed with prostate cancer in the placebo group after 7 years was much higher than expected (based on a projected lifetime risk of 17%). The patients in the finasteride arm had statistically significantly higher risk of sexual side effects (reduced ejaculate volume, erectile dysfunction, loss of libido, and gynecomastia) than the placebo group. However, they had statistically significant improvement in their urinary symptoms (urinary urgency, urinary frequency, prostatitis, urinary tract infection, and urinary retention).

The major controversy from PCPT was the finding of a statistically significantly increased prevalence of higher-grade prostate cancer (as defined by Gleason grade ≥ 7) in the treatment arm versus placebo (6.4% vs. 5.1%). Potential explanations for this finding included finasteride's effects on tumor morphology, PSA sensitivity, and prostatic size at the time of biopsy. A subsequent assessment showed that finasteride did not change prostate cancer morphology but its effects on PSA sensitivity and volume of the prostate made it more likely that a higher-grade tumor would be detected on biopsy.[23] Several post hoc logistic regression analysis supported the claims that the volume-shrinking effect of finasteride on the prostate, along with increased PSA sensitivity could explain the prevalence of higher-grade tumors.[24–27]

A long-term follow-up of men in PCPT[28] was recently reported. The primary objectives of the study were to estimate the time to metastases and the difference in all-cause and prostate-cancer mortality among men in the finasteride group, as compared with those in the placebo group. Survival information was prospectively collected on enrolled patients in the PCPT for 7 additional years. Additionally, survival data through 2011 was queried via the Social Security Death Index. Of the men evaluated, 3.5% in the finasteride group versus 3% in the placebo group were diagnosed with higher-grade prostate cancer. At 10 years, the rate of overall survival among patients with any grade of prostate cancer was 79.3% in the finasteride group and 79.5% in the placebo group. The overall survival of patients with low-grade prostate cancer was 83% in the finasteride group and 80.9% in the placebo group. The overall survival of patients with high-grade cancer was 73% in the finasteride group and 73.6% in the placebo group. Limiting the survival analysis to only the patients diagnosed with prostate cancer did not affect the results. Therefore, despite the increased prevalence of higher-grade prostate cancer in the treatment group, the overall survival was no different from the placebo group.

## REDUCTION BY DUTASTERIDE OF PROSTATE CANCER EVENTS (REDUCE)

While finasteride is a selective inhibitor of the type II isoform of 5 AR only, dutasteride is a dual inhibitor of types I and II isoforms of 5 AR. Type II 5 AR is predominant in prostatic and genital tissue and type I is overexpressed in some prostate cancers.[29–31] Furthermore, dutasteride achieves almost total elimination of serum DHT when compared to finasteride,[32] which reduces serum DHT by approximately 70%.

The rates of prostate cancer development were examined in three large randomized placebo-controlled BPH trials[33] of dutasteride. The combined results demonstrated that the incidence of prostate cancer diagnosis was 1.2% in the dutasteride group compared to 2.5% in the placebo group (a 51% reduction in incidence from dutasteride use). It was this observation that spurred the development of the Reduction by Dutasteride of Prostate Cancer Events (REDUCE) trial to clarify the efficacy of dutasteride for the prevention of prostate cancer.

The REDUCE trial[34] was a 4-year, international, multicenter, randomized, double-blinded, placebo-controlled trial. It was designed to assess the effects of dutasteride (0.5 mg daily dose) versus placebo on the development of prostate cancer among a cohort of men considered to be at higher risk of the disease. This cohort creation differed from PCPT, which looked at a group of men at lower risk for the development of prostate cancer.

REDUCE defined higher risk based on age, having an elevated PSA, as well as having a previously negative prostate biopsy that was performed for suspicion of harboring prostate cancer. This study enrolled 8231 men aged 50–75 years old with a baseline PSA of 2.5–10 ng/mL. All men included in the study had a previous negative biopsy within 6 months before enrollment to exclude relatively large volume cancer at baseline. This was deemed necessary as REDUCE had a planned biopsy 2 years after randomization.

Per the study design, 10-core prostate biopsies were performed at 2 and 4 years after randomization. However, cause biopsies (for example in men with a rising PSA) could be performed whenever clinically indicated. Of the enrolled patients in the study, 81.6% of men in the dutasteride group and 84.1% of men in the placebo group underwent at least one study biopsy. The vast majority (93.6%) of biopsies were driven by the study protocol rather than being for cause.

After the 4-year study period, 19.9% of men in the dutasteride group and 25.1% of men in the placebo group were diagnosed with prostate cancer. This finding demonstrated a statistically significant absolute risk reduction in the dutasteride group of 5.1%. The use of dutasteride was also associated with a statistically significant reduction in the diagnosis of Gleason 5 and 6 prostate cancers (13.2% in the dutasteride group vs. 18.1% in the placebo group). However, in contrast to the increased incidence of high-grade prostate cancer in the PCPT, over the entire 4-year study period, REDUCE showed no statistically significant difference in the two groups for diagnoses of Gleason 7–10 tumors (6.7% in the dutasteride group vs. 6.8% in the placebo group). However, there was a numerical increase of 10 more cases of Gleason 8–10 cancer in the dutasteride group. This increased number of Gleason 8–10 cancers occurred during years 3 and 4 of the trial and has been explained by smaller prostate volume in the treated men and to the REDUCE study design wherein fewer men in the placebo group underwent a biopsy during years 3 and 4, since many more such men were discovered to have cancer on the first round of biopsies during years 1 and 2[35] and were no longer in the study. It remains to be seen if long-term outcomes of these patients demonstrate any survival differences in the treatment groups, but using data from the PLCO cancer trial and modeling outcomes, no difference is anticipated.[36]

Dutasteride had many positive effects on urinary related symptoms. Prostate size for patients on dutasteride had a mean overall decrease by 17.5%. Additionally, patients treated with dutasteride had a relative risk reduction of acute urinary retention, BPH-related surgery, and urinary tract infection of 77.3, 73, and 40.7%, respectively.

There were some adverse events noted with the use of dutasteride. First, there were increased sexual side effects (decreased or loss of libido, erectile dysfunction, decreased semen volume, and gynecomastia). Second, there was an unexpected finding of increased cardiac events not seen in prior studies with dutasteride. These events were congestive heart failure, cardiac failure, acute cardiac failure, ventricular failure, cardiopulmonary failure, or congestive cardiomyopathy.

## THE CONTROVERSY

The FDA, in response to the findings of PCPT and REDUCE, issued a black box warning against the use of finasteride and dutasteride for the use of prostate cancer chemoprevention[37]. This recommendation was made as result of the observation of increased incidence of Gleason 7–10 prostate cancer in PCPT and the numerical increase in Gleason 8–20 in REDUCE. However, there does not appear to be an increased incidence of high-grade prostate cancer in men treated for BPH in Sweden[38] and the United States[39]. The FDA endorses that both finasteride and dutasteride may still be used for their primary indication for controlling symptoms of BPH. They have yet to comment on the risks and benefits of their use for male pattern baldness.

## CONCLUSIONS

PCPT and REDUCE have demonstrated level-one evidence in the ability to reduce the risk of low-risk prostate cancer. However, concerns about the increased incidence of higher-grade prostate cancer with 5 AR inhibitors use have halted their adoption for prostate cancer chemoprevention. However, further research as well as additional long-term follow-up data will generate increased understanding of the overall and cancer-specific survival of patients being treated with 5 AR inhibitors.

## References

1. Sporn MB, Suh N. Chemoprevention of cancer. *Carcinogenesis* 2000;**21**(3):525–30.
2. Siegel R, Desantis C, Jemal A. Colorectal cancer statistics. *CA Cancer J Clin* 2014;**64**(2):104–17.
3. Brawley OW. Prostate cancer epidemiology in the United States. *World J Urol* 2012;**30**(2):195–200.
4. Coussens LM, Werb Z. Inflammation and cancer. *Nature* 2002;**420**(6917):860–7.
5. De Marzo AM, Platz EA, Sutcliffe S, et al. Inflammation in prostate carcinogenesis. *Nat Rev Cancer* 2007;**7**(4):256–69.
6. De Nunzio C, Aronson W, Freedland SJ, et al. The correlation between metabolic syndrome and prostatic diseases. *Eur Urol* 2012;**61**(3):560–70.
7. Gong Z, Neuhouser ML, Goodman PJ, et al. Obesity, diabetes, and risk of prostate cancer: results from the prostate cancer prevention trial. *Cancer Epidemiol Biomarkers Prev* 2006;**15**(10):1977–83.
8. Hsing AW, Sakoda LC, Chua Jr S. Obesity, metabolic syndrome, and prostate cancer. *Am J Clin Nutr* 2007;**86**(3):s843–57.
9. Steers WD. 5alpha-reductase activity in the prostate. *Urology* 2001;**58**(6 Suppl. 1):17–24 discussion 24.
10. Imperato-McGinley J, Zhu YS. Androgens and male physiology the syndrome of 5alpha-reductase-2 deficiency. *Mol Cell Endocrinol* 2002;**198**(1–2):51–9.
11. Imperato-McGinley J, Guerrero L, Gautier T, Peterson RE. Steroid 5alpha-reductase deficiency in man: an inherited form of male pseudohermaphroditism. *Science* 1974;**186**(4170):1213–5.
12. Shen MM, Abate-Shen C. Molecular genetics of prostate cancer: new prospects for old challenges. *Genes Dev* 2010;**24**(18):1967–2000.
13. Marks LS. 5alpha-reductase: history and clinical importance. *Rev Urol* 2004;**6**(Suppl. 9):S11–21.
14. Andriole GL, Guess HA, Epstein JI, et al. Treatment with finasteride preserves usefulness of prostate-specific antigen in the detection of prostate cancer: results of a randomized, double-blind, placebo-controlled clinical trial. PLESS Study Group. Proscar Long-term Efficacy and Safety Study. *Urology* 1998;**52**(2):195–201 discussion 201–2.
15. McConnell JD, Roehrborn CG, Bautista OM, et al. The long-term effect of doxazosin, finasteride, and combination therapy on the clinical progression of benign prostatic hyperplasia. *N Engl J Med* 2003;**349**(25):2387–98.
16. Roehrborn CG, Siami P, Barkin J, et al. The effects of combination therapy with dutasteride and tamsulosin on clinical outcomes in men with symptomatic benign prostatic hyperplasia: 4-year results from the CombAT study. *Eur Urol* 2010;**57**(1):123–31.
17. McConnell JD, Bruskewitz R, Walsh P, et al. The effect of finasteride on the risk of acute urinary retention and the need for surgical treatment among men with benign prostatic hyperplasia. Finasteride Long-Term Efficacy and Safety Study Group. *N Engl J Med* 1998;**338**(9):557–63.
18. Hsing AW, Reichardt JK, Stanczyk FZ. Hormones and prostate cancer: current perspectives and future directions. *Prostate* 2002;**52**(3):213–35.
19. Ross RK, Bernstein L, Lobo RA, et al. 5-alpha-reductase activity and risk of prostate cancer among Japanese and US white and black males. *Lancet* 1992;**339**(8798):887–9.
20. Giovannucci E, Stampfer MJ, Krithivas K, et al. The CAG repeat within the androgen receptor gene and its relationship to prostate cancer. *Proc Natl Acad Sci USA* 1997;**94**(7):3320–3.
21. Thompson IM, Goodman PJ, Tangen CM, et al. The influence of finasteride on the development of prostate cancer. *N Engl J Med* 2003;**349**(3):215–24.
22. Barry MJ, Fowler Jr FJ, O'Leary MP, et al. The American Urological Association symptom index for benign prostatic hyperplasia. The Measurement Committee of the American Urological Association. *J Urol* 1992;**148**(5):1549–57 discussion 64.
23. Lucia MS, Epstein JI, Goodman PJ, et al. Finasteride and high-grade prostate cancer in the Prostate Cancer Prevention Trial. *J Natl Cancer Inst* 2007;**99**(18):1375–83.
24. Cohen YC, Liu KS, Heyden NL, et al. Detection bias due to the effect of finasteride on prostate volume: a modeling approach for analysis of the Prostate Cancer Prevention Trial. *J Natl Cancer Inst* 2007;**99**(18):1366–74.
25. Pinsky P, Parnes H, Ford L. Estimating rates of true high-grade disease in the prostate cancer prevention trial. *Cancer Prev Res* 2008;**1**(3):182–6.
26. Redman MW, Tangen CM, Goodman PJ, et al. Finasteride does not increase the risk of high-grade prostate cancer: a bias-adjusted modeling approach. *Cancer Prev Res* 2008;**1**(3):174–81.
27. Kaplan SA, Roehrborn CG, Meehan AG, et al. PCPT: evidence that finasteride reduces risk of most frequently detected intermediate- and high-grade (Gleason score 6 and 7) cancer. *Urology* 2009;**73**(5):935–9.
28. Thompson Jr IM, Goodman PJ, Tangen CM, et al. Long-term survival of participants in the prostate cancer prevention trial. *N Engl J Med* 2013;**369**(7):603–10.

29. Bruchovsky N, Sadar MD, Akakura K, et al. Characterization of 5alpha-reductase gene expression in stroma and epithelium of human prostate. *J Steroid Biochem Mol Biol* 1996;**59**(5–6):397–404.
30. Thigpen AE, Silver RI, Guileyardo JM, et al. Tissue distribution and ontogeny of steroid 5 alpha-reductase isozyme expression. *J Clin Invest* 1993;**92**(2):903–10.
31. Thomas LN, Douglas RC, Vessey JP, et al. 5alpha-reductase type 1 immunostaining is enhanced in some prostate cancers compared with benign prostatic hyperplasia epithelium. *J Urol* 2003;**170**(5):2019–25.
32. Clark RV, Hermann DJ, Cunningham GR, et al. Marked suppression of dihydrotestosterone in men with benign prostatic hyperplasia by dutasteride, a dual 5alpha-reductase inhibitor. *J Clin Endocrinol Metab* 2004;**89**(5):2179–84.
33. Andriole GL, Roehrborn C, Schulman C, et al. Effect of dutasteride on the detection of prostate cancer in men with benign prostatic hyperplasia. *Urology* 2004;**64**(3):537–41 discussion 542–3.
34. Andriole GL, Bostwick DG, Brawley OW, et al. Effect of dutasteride on the risk of prostate cancer. *N Engl J Med* 2010;**362**(13):1192–202.
35. Andriole GL, Bostwick DG, Gomella LG, et al. Modeling and analysis of Gleason score 8–10 prostate cancers in the REDUCE study. *Urology* 2014;**84**(2):393–9.
36. Pinsky PF, Black A, Grubb R, et al. Projecting prostate cancer mortality in the PCPT and REDUCE chemoprevention trials. *Cancer* 2013;**119**(3):593–601.
37. *Administration FaD. Briefing information and slides presented at the December 1, 2010, meeting of the Oncologic Drugs Advisory Committee.* http://www.fda.gov/AdvisoryCommittees/CommitteesMeetingMaterials/Drugs/OncologicDrugsAdvisoryCommittee/ucm195226.htm; 2010 [accessed 31.05.2014].
38. Robinson D, Garmo H, Bill-Axelson A, et al. Use of 5alpha-reductase inhibitors for lower urinary tract symptoms and risk of prostate cancer in Swedish men: nationwide, population based case-control study. *BMJ* 2013;**346**:f3406.
39. Preston MA, Wilson KM, Markt SC, et al. 5α-Reductase inhibitors and risk of high-grade or lethal prostate cancer. *JAMA Intern Med* 2014;**174**(8):1301–7.

P A R T V

# CONSERVATIVE MANAGEMENT

CHAPTER

# 24

# Decision Support for Low-Risk Prostate Cancer

*Leonard P. Bokhorst, MD*, Ewout W. Steyerberg, PhD**, Monique J. Roobol, PhD**

***Department of Urology, Erasmus University Medical Center, Rotterdam, The Netherlands**
****Department of Public Health, Erasmus University Medical Center, Rotterdam, The Netherlands**

## INTRODUCTION

Over the past decades, increasing numbers of men have been faced with the diagnosis of prostate cancer.[1,2] Conversely, prostate cancer mortality rates have decreased.[3] In addition, men are more often diagnosed at an earlier stage of the disease. The obvious reason for these profound changes is the introduction of the prostate-specific antigen (PSA) test for the early detection of prostate cancer. Early detection of prostate cancer at a more favorable stage allows the cancer to be treated with curative, rather than palliative, intend. Screening for prostate cancer using PSA was indeed shown to reduce the prostate-cancer-specific mortality rate by 21% after 13 years of follow-up in the largest randomized prostate cancer screening trial.[4] A major disadvantage is the detection of many additional prostate cancers, with an increase in prostate cancer incidence by over 50%.[4] Many of these additionally identified cancers are likely not to result in any symptoms or death during a man's lifetime. In fact, it is estimated that 50% of men diagnosed through screening have these so-called "overdiagnosed" prostate cancer.[5] The presence of many nonaggressive tumors has already been observed in autopsy studies, which showed that prostate cancer was present in a large percentage (10–30% depending on age) of men who died of other causes.[6] Because of the introduction of the PSA test resulting in subsequent prostate biopsies, this large reservoir of prostate cancers is now detected. Hence, radical treatment of PSA-detected tumors will for most men not result in any survival benefit.[7] It may however lead to side-effects such as impotency or incontinence, decreasing quality of life in a substantial number of men.[8] A way of avoiding the side-effects of radical treatment is to offer these patients active surveillance. This treatment option aims to avoid or delay treatment for most, while by monitoring tumor progression be able to offer radical treatment for those who benefit. Men diagnosed today with early detected prostate cancer are thus facing the difficult choice of radical treatment with its likely low benefit, but substantial risks of side-effects, versus conservative treatment, which might risk losing the benefit obtained by early detection. In this chapter, we will discuss some of the tools that are available to help men to choose between active treatment of their prostate cancer or to opt for the more conservative approach of active surveillance.

## DEFINING INDOLENT PROSTATE CANCER

Indolent prostate cancer is defined as a tumor that will not result in symptoms or death during a man's lifetime if left untreated. The dynamic aspect of this definition makes it difficult to operationalize. A tumor that might be moderately progressive over time could be no threat for a man with a short life expectancy, but might become problematic (e.g., metastasize or cause symptoms) for a man with a longer life. Furthermore, no single parameter yet will provide definitive information on future tumor development. Last, as tumor progression in a low-risk prostate cancer group is relatively rare and takes usually at least a decade to develop, statistical evaluation of this end-point is difficult. Therefore, several other definitions were proposed to indicate tumors that are latent and have low probabilities of developing

**Prostate Cancer. http://dx.doi.org/10.1016/B978-0-12-800077-9.00024-4**

TABLE 24.1 Definitions Used in the Context of Indolent/Low-Risk Prostate Cancer

| | Indolent disease over time | Indolent disease at radical prostatectomy | Low-risk disease |
|---|---|---|---|
| Definition | A tumor that will not result in symptoms or death during a man's lifetime if left untreated | Organ confined, no Gleason grade 4 or 5, and a tumor volume ≤0.5 cm$^3$ | A tumor with a high probability of being indolent at radical prostatectomy and/or over time, based on its clinical characteristics. |
| Advantage | Most optimal definition of indolent disease | Does not require follow-up | Does not require surgical excision |
| Disadvantage | • Can only be determined in retrospect<br>• Requires long follow-up<br>• Dependent on a man's life expectancy | • Requires pathological examination<br>• Could potentially, even with radical treatment, still progress and give rise to symptoms<br>• Definition might be too restrictive for men with a short life expectancy<br>• A tumor volume ≤0.5 cm$^3$ might be too restrictive[9] | • Uncertainty of prediction, caused by underestimation of Gleason score, tumor volume or T-stage<br>• Tumor is not removed and could progress over time, re-evaluation is therefore necessary |

symptoms, based on parameters that do not require follow-up in time. The most common definition of indolent disease is an organ-confined tumor, ≤0.5 cm$^3$, with no Gleason grade 4 or 5. This definition requires the removal of the prostate for pathological evaluation. In a conservative treatment strategy low-risk tumors are therefore defined as tumors with a high probability of being indolent on radical prostatectomy. Because the tumor is not removed and progression could thus occur, reevaluation of its low-risk character through time will be necessary. A summary of the different definitions used in the context of indolent/low-risk prostate cancer is given in Table 24.1.

Several clinical definitions were proposed to select men with high likelihood of indolent prostate cancer on radical prostatectomy. These men are best suitable for a conservative treatment approach. Most definitions use a combination of low tumor grade (Gleason ≤6), localized disease (cT1c-cT2c), low PSA or PSA density (PSA ≤10–15, PSA-density ≤0.15–0.2), and small tumor volume (≤2 cores positive on prostate biopsy, ≤50% tumor involvement per core). Different definitions used for the selection of men on active surveillance studies are shown in Table 24.2. Although commonly used in clinical practice, these rule-based definitions have some obvious disadvantages. The most important limitation of these criteria is that much of the predictive value of individual prognostic factors is lost. For example, a man with a PSA of 2 ng/mL and a man with a PSA of 10 ng/mL would both meet the same definition if the criterion is: PSA ≤10 ng/mL. Using the definition would assume these men are similar in risk of having indolent disease, while clearly the man with the lower PSA will have a higher probability of indolent disease. Moreover, using a rule-based definition of indolent disease exerts equal value to all individual risk factors. Both a man with a Gleason score of 7 and a man with a PSA of 10.5 ng/mL would not fit the rule-based definition. The first might have a truly higher chance of having aggressive disease and therefore not be suitable for conservative management, while the second may only marginally differ in risk of being indolent and could therefore still consider a nonradical approach. Combining risk factors into an individual risk estimation (risk-based), instead of "eligible" or "noneligible" (rule-based), may better inform

TABLE 24.2 Inclusion Criteria for Different Active Surveillance Studies

| Active surveillance study | Criteria for inclusion |
|---|---|
| Royal Marsden[10] | Gleason ≤3+4 (primary Gleason grade ≤3); PSA ≤15 ng/mL; cT1c-2a; ≤50% of cores positive |
| University of Miami[11] | Gleason ≤6; PSA ≤15 ng/mL; cT1c-2c; ≤2 cores positive; ≤20% of any core positive |
| Johns Hopkins[12] | Gleason ≤3+3; PSA density ≤0.15 ng/mL; cT1c; ≤2 cores positive; ≤50% of any core positive |
| University of California San Francisco[13] | Gleason ≤3+3; PSA ≤10 ng/mL; cT1c-2c; ≤33% of cores positive; ≤20% of any core positive |
| University of Toronto[14] | Gleason ≤6; PSA ≤10 ng/mL (until January 2000, for men age >70 years: Gleason ≤3+4; PSA ≤15 ng/mL) |
| Prostate cancer Research International Active Surveillance (PRIAS)[15] | Gleason ≤3+3; PSA ≤10 ng/mL; PSAD ≤0.2 ng/mL; cT1c-2c; ≤2 cores positive (age >70 years: Gleason ≤3+4, maximum 10% tumor per cores) |

the patient and his physician and help to make a more conscious decision on treatment choice.

## INDIVIDUALIZED PREDICTIONS

Individual risk estimation using a risk-calculator, or nomogram, is frequently applied to other areas of prostate cancer care, for instance, in the decision to perform a prostate biopsy. Multiple risk-calculators were developed by several study groups to predict the chance of having a positive prostate biopsy, as it was realized that performing a prostate biopsy only based on a single PSA cut-off was suboptimal.[16] These risk-calculators use different risk factors such as the age, PSA value, digital rectal examination, transrectal ultrasound findings, prostate volume, and prior biopsy status to calculate an individual risk of having a positive prostate biopsy.[17] It was estimated that using these risk-calculators to guide biopsy decisions could potentially reduce the number of unnecessary biopsy by 30% without missing important prostate cancers (defined as Gleason grade >3, PSA >20 ng/mL, T-stage 3 or 4, >50% positive cores, >20 mm cancer (indicates the millimeter of cancer) in the biopsy core(s), or <40 mm benign tissue in all cores) as compared to a single PSA cut-off approach.[18] Risk-calculators could even be used to calculate the risk of having a positive biopsy up to 8 years in the future.[19] After being validated in other cohorts,[20–24] the effect of using risk-calculators in clinical practice has been assessed.[25] In total, 83% of patients complied with the recommendation provided by the risk-calculator. If a biopsy was recommended, 96% complied with the recommendation versus 64% of men with a recommendation against prostate biopsy. Of men not complying with a negative biopsy recommendation only 3% were found to have a relevant tumor (Gleason >6).[26] The main reason for not complying with a negative biopsy recommendation were a PSA ≥3 ng/mL for urologists or wanting certainty for patients.[25] In the decision to perform a prostate biopsy, risk-calculators seem easy tools (most risk-calculators today can be found and used online as a web-based tool, for example, www.prostatecancer-riskcalculator.com, or even downloaded as an app for your mobile phone, for example, Rotterdam prostate cancer risk calculator (Google play store and Apple app store)) to increase patient participation and reduce unnecessary examinations.

Several risk-calculators were developed to predict the risk of having an indolent prostate cancer as defined by radical prostatectomy characteristics (Table 24.3). Kattan et al. developed a prediction model based on a clinical cohort of 409 men with cT1c or cT2a Gleason ≤6 prostate cancer who received radical prostatectomy.[27] In total, 20% of men had indolent disease (defined as organ-confined, Gleason score ≤6, prostate cancer with a tumor volume ≤0.5). The model, including PSA, primary and secondary Gleason grade, clinical stage, ultrasound prostate volume, and mm cancer and mm noncancerous tissue (indicates the millimeter of benign tissue) in the biopsy core(s)), could reasonably predict indolent disease with a receiver operating characteristics (ROC) area under the curve (AUC) of 0.79. As already noted by the authors of this nomogram the percentage of men with indolent disease increased over time in their population. As the percentage of indolent prostate cancers in the population increases the nomogram predictions might underestimate the chance of having an indolent prostate cancer. In 2007, the nomogram was therefore updated on their website (www.nomograms.org) to better fit a more contemporary population.

Because the underlying prevalence of indolent disease in the population the model is developed on makes a substantial difference in predicted risks, a model was developed to better apply to a more intensively PSA

**TABLE 24.3** Overview of Risk-Prediction Tools for Indolent Prostate Cancer

| | Indolent prostate cancer (Organ confined, Gleason score 6, tumor volume ≤0.5 cm$^3$) at radical prostatectomy | | | | | 10-year disease free survival |
|---|---|---|---|---|---|---|
| | Kattan et al.[27] | Steyerberg et al.[28] | Nakanishi et al.[29] | Chun et al.[30] | O'Brien et al.[31] | Kattan et al.[32] |
| Cohort origin | Clinical | Screening | Clinical | Clinical | Clinical | Clinical |
| Number of men used for nomogram development | 409 | 247 | 254 | 1132 | 2525 | 1310 |
| Percentage with indolent disease | 20% | 49% | 52% | 6% | 6% | – |
| Example of risk prediction (Man, age 65 year, PSA 5 ng/mL, prostate volume 50 cm$^3$, cT1c, Gleason 3+3, 1 of 12 cores positive, 5 mm cancerous tissue, 200 mm benign tissue, no early hormonal therapy) | 48% | 79% | 37% | 4–10%* | 1–5%* | 96–90%* |

** Risk predictions were based on graphical devices and therefore presented as an interval.*

screened population.[28] Steyerberg et al. adapted the model developed by Kattan et al. based on 278 men detected in the screening arm of the European Randomized study of Screening for Prostate Cancer (ERSPC) Rotterdam. At radical prostatectomy 49% of men had indolent disease. The new model uses the same predictors as the previous model (PSA, primary and secondary Gleason grade, clinical stage, ultrasound prostate volume, and mm cancer and mm noncancerous tissue) and is again able to predict indolent disease moderately well with an AUC of 0.76 (online available at www.prostatecancer-riskcalculator.com).[28] Based on this nomogram it was estimated that in a PSA screening setting, 30% of men would have indolent disease and could be suitable for conservative management.[33] Because the model was developed based on men diagnosed with sextant prostate biopsy, length of prostate cancer and length of noncancerous tissue might not be accurate for men diagnosed with more contemporary extended biopsy core schemes. Correction factors were therefore calculated to be able to accurately predict the risk of men diagnosed with 12- or 18-core biopsies (mm cancer should be divided by 2.03 or 2.72 and mm noncancerous tissue by 2 or 3 for 12- or 18-core biopsies, respectively, when applying the nomogram).[34]

Additional nomograms to predict indolent disease at radical prostatectomy (again defined as organ-confined, Gleason score ≤6, prostate cancer with a tumor volume ≤0.5) were developed by Nakanishi et al., Chun et al., and O'Brien et al.[29–31] The first was developed on a cohort of 258 men with only one positive biopsy core on an extended biopsy scheme (10–13 cores).[29] Because of this strict selection, the percentage of men with indolent disease at radical prostatectomy in the study by Nakanishi et al. was higher than previous studies at 52%. Using age, PSA density, and mm cancer tissue the model could moderately well predict indolent disease with an AUC of 0.73. Chun et al. developed a nomogram on a European cohort of men of all Gleason scores that were most likely not extensively PSA screened.[30] This was underlined by the very low rate of indolent disease at radical prostatectomy of only 6%. A model including PSA, biopsy Gleason sum score, length of cancer tissue, and percentage of positive biopsy cores was developed, which had an AUC of 0.90. In the same analysis the Kattan et al. nomogram was validated, which had an AUC of 0.81 in this cohort.[30] The nomogram developed by O'Brien et al. was developed in an Australian cohort of men with all Gleason scores and again a very low rate of indolent disease of 6%.[31] AUC was again high at 0.93. In addition, the nomogram slightly outperformed two rule-based inclusion criteria for active surveillance programs. Although the Chun et al. and the O'Brien et al. nomograms may have very high predictive capabilities, results need to be interpreted with caution. Both cohorts used for model development included a substantial part of men with biopsy Gleason scores >6 (41 and 68% for Chun et al. and O'Brien et al., respectively). These men almost per definition do not have indolent disease at radical prostatectomy (in the study of O'Brien et al. 0.5% of men with Gleason score >6 on prostate biopsy had indolent disease due to Gleason score downgrading at radical prostatectomy[31]). Inclusion of these men for model development tends to inflate the predictive capability of the model as indicated by the AUC. In addition, adding these men might alter the estimated prognostic effect of individual parameters for men with Gleason scores of 6.

All previously described nomograms are designed to predict the presence of indolent disease at immediate radical prostatectomy. However, as previously described, this is used as a surrogate for a tumor that would not cause any symptoms or death during a man's life. The last nomogram that will be discussed takes a different approach and aims to predict prostate-cancer-specific survival after 10 years if conservative treatment is opted for (i.e., watchful waiting, which differs from active surveillance in that it does not attempt to offer curative treatment, but only palliative treatment if symptoms occur).[32] The model was based on 1310 men diagnosed with prostate cancer either by biopsy or transurethral resection of the prostate between 1990 and 1996. Cox regression analysis was used to predict 10 year prostate cancer mortality rates using clinical stage, method of diagnosis, percentage of cancer tissue, PSA, age, Gleason sum score, and the use of early hormonal treatment (within 6 months). The concordance index (similar to the AUC, but for censored data) was moderate (0.73).[32] Low 10 year disease specific mortality rates obtained from this model, combined with a short to intermediate life expectancy, could be used to select men for watchful waiting.

## WHICH PREDICTION MODEL TO USE?

Clinicians and patients thus have several different nomograms at their disposal to help differentiate between indolent disease, most likely suitable for conservative management, or less indolent disease, which might require more aggressive treatment. But which of these nomograms should be used and which one is most suitable? Several aspects need to be addressed. First, a nomogram developed on a specific cohort might perform well on that cohort, but have limited predictive capabilities outside this setting.[35] External validation of a model is therefore essential.[36,37]

The Kattan et al. nomogram and the Steyerberg et al. nomogram were validated in an external population of 296 men with Gleason score 6, localized disease.[38] At radical prostatectomy 27% had indolent prostate cancer. Both models performed equally well in predicting indolent disease with an AUC of 0.77, which is similar to the predictive accuracy of the development cohort, indicating good generalizability. A second validation was done of all five nomograms described earlier in a contemporary cohort of 370 men with Gleason 6 disease on transrectal prostate biopsy.[39] In 38% of patients indolent disease was present on radical prostatectomy. The result indicated that the Kattan et al. and the Steyerberg et al. nomograms significantly outperformed the Nakanishi et al. nomogram, which in turn outperformed the Chun et al. and the O'Brien et al. nomograms. Predictive capabilities were again moderately well for the Kattan et al. and the Steyerberg et al. nomograms with an AUC of 0.77.[39] These two nomograms also showed good calibration and the highest net benefit.[39] It was noted that all models were most accurate at low predictive capabilities, indicating that these models are best at excluding indolent disease rather than accurately identifying it. One of the reasons for the lower performance at higher predictive values was the presence of anterior and apical tumors.[39] Both located at areas not frequently sampled with standard transrectal biopsy schemes. The nomogram predicting 10 year disease-free survival has not yet been externally validated.[32]

As is shown, the specific population at external validation may affect results. This is illustrated in an example of predicted probabilities of indolent disease (Table 24.3). If the chance of having indolent disease for a 65-year-old man with 5 mm prostate cancer in one biopsy core is calculated with all nomograms, predictions range from 1% to 79%. Ten-year disease-free survival is calculated at 90–96%. Predicted probabilities seem very dependent on the percentage of men with indolent disease in the development cohort. The Chun et al. and the O'Brien et al. nomograms, which were developed in a group of men often having Gleason scores >6, seem not well able to identify indolent disease in men with lower Gleason scores. Other important differences include the development in a clinical cohort or a screening cohort. The latter having more men with indolent disease. Furthermore, most cohorts were developed in white European or American men. Applicability to Asian or African men might be limited. Clinicians should be aware of these differences and try to select a nomogram best suitable for their patient. Overall, the Kattan et al. and Steyerberg et al. nomograms seem to be the most widely applicable, and outperform other nomograms.[38,39] In addition, both nomograms can be easily applied using an online tool (www.prostatecancer-riskcalculator.com and www.nomograms.org), which simply calculates the predicted probability after providing parameter data, greatly enhancing clinical usability.

## CLINICAL APPLICABILITY

Although many nomograms are developed to help clinicians and patients in treatment decisions, very few make it into clinical practice. Physicians may be reluctant to trust nomogram predictions and rather choose to follow well-established preconceptions. An example is provided by a study on the implementation of a risk-calculator to aid in the decision of prostate biopsy.[25] Although the risk-calculator (i.e., the ERSPC risk-calculator[17,18]) was proven superior over a single PSA cut-off, 36% of men were biopsied against a negative biopsy recommendation provided by the risk-calculator. When asked for the reason of ignoring the advice, 78% of times a PSA ≥3.0 ng/mL was replied. Of men ignoring the negative biopsy recommendation only 3% showed aggressive prostate cancer on biopsy.[26] The remaining 97% received unnecessary biopsies because of the prejudice that a PSA value ≥3.0 ng/mL should trigger further investigation. That said, it seems vital to not only conduct studies on how to improve selection, but also to better implement successful tools in clinical practice. This can be done by simply presenting the risk provided by a nomogram, but can also be more elaborate, for example, by combining individual risk scores with information on its meaning, prognosis, and the advantages and disadvantages of different treatment options into a (personalized) decision aid.

One study aimed to investigate the impact of using a nomogram to advise men on active treatment or active surveillance.[40] The Steyerberg et al. nomogram was used to predict the presence of indolent disease in 240 men diagnosed with prostate cancer in five Dutch hospitals. As a rule of thumb, a probability cut-off of ≥70% was chosen to advice men on active surveillance. With this cut-off 82% of patients adhered to the recommendation to choose active surveillance. Surprisingly, 29% of men with a probability <70% of indolent disease also choose an initial active surveillance strategy. The main reason being the patients' preference to delay the physical side effects of active treatment.[40] Measurements of the decisional conflict scale were low in this study, indicating that patients felt well informed by the nomogram and certain in their choice of treatment. Two other randomized trials have shown that using decision aids (not including individual risk assessments) for treatment choice in localized prostate cancer not only helped patients make more informed decisions on treatment, but also increase satisfaction with the decision made.[41,42] In addition, the better information provided and higher patient participation might have an effect on the treatment that is selected.[42,43] Decision aids, including personalized risk assessments, seem good tools not only to improve the selection of men with indolent disease, but will also increase patient understanding, participation, and satisfaction in the treatment chosen for the management of their prostate cancer.

## FUTURE PERSPECTIVES

Although the use of nomograms to help treatment decisions in the increasing number of men diagnosed with low-risk localized prostate cancer seems preferable over rule-based decision supports to reduce overtreatment, there are some limitations. As with all rule-based criteria, none of the presented nomograms is able to perfectly predict the presence of indolent disease. In fact the nomograms seem more suitable for excluding the presence of significant disease. Most likely the restrictions of currently used blind biopsy sampling, often missing anterior and apical tumors, contribute to this. Improvement of current nomograms is therefore essential. Promising and most likely to be quickly incorporated is the use of MRI. MRI seems especially useful in visualizing higher-grade prostate cancers.[44,45] MRI visualized lesions could trigger targeted biopsies, which might better represent tumor grade and volume. Nomogram predictions are therefore likely to improve if data on targeted biopsies could be added. In addition to the information provided by targeted biopsies, the MR images itself could provide new parameters on tumor characteristics. These not only include tumor volume as can be measured on MRI, but also water diffusion coefficients, which seem to correlate with tumor aggressiveness.[46] MRI, using spectroscopic imaging, could also be used to obtain information on a molecular level, which again might help to predict tumor aggressiveness.[47,48] Next to imaging, genomic and histological information could potentially provide better information on tumor behavior and help to decide on the most appropriate treatment strategy. A recent study genotyped 242,221 single nucleotide polymorphisms (SNPs) in blood DNA of men with Gleason 6 prostate cancer.[49] Fifteen SNPs were found to be able to predict Gleason score upgrading on radical prostatectomy; however, only one SNP remained predictive if other clinical information was added. The addition of the SNP to a clinical model significantly improved the predictive accuracy.[49] Future studies should validate if these findings remain significant and could improve the prediction of indolent disease.

## CONCLUSIONS

Several nomograms may aid men in assessing their risk of having an indolent tumor. These decision aids, although not perfect in their prediction of indolent disease, are preferable over commonly used rule-based selection criteria for active surveillance, because they provide a more individual risk-assessment, which helps to better inform men facing treatment decisions. For clinicians it is important to choose a nomogram that is most accurate, externally validated, and best fits the patient's characteristics. Well-validated nomograms with reasonable accuracy can be found online for a clinical population (www.nomograms.org) or for a more intensive PSA-screened population (www.prostatecancer-riskcalculator.com). Future developments, such as MRI and new genetic markers, will likely improve current nomograms. Implementation into clinical practice has, however, already shown to be valuable. The time has therefore arrived to start using these prediction tools in clinical practice to provide the best of care for the large number of men diagnosed with low-risk prostate cancer today.

### References

1. Siegel R, Ma J, Zou Z, et al. Cancer statistics. *CA Cancer J Clin* 2014;**64**(1):9–29.
2. Ferlay J, Steliarova-Foucher E, Lortet-Tieulent J, et al. Cancer incidence and mortality patterns in Europe: estimates for 40 countries in 2012. *Eur J Cancer* 2013;**49**(6):1374–403.
3. Center MM, Jemal A, Lortet-Tieulent J, et al. International variation in prostate cancer incidence and mortality rates. *Eur Urol* 2012;**61**(6):1079–92.
4. Schroder FH, Hugosson J, Roobol MJ, et al. Screening and prostate cancer mortality: results of the European Randomised Study of Screening for Prostate Cancer (ERSPC) at 13 years of follow-up. *Lancet* 2014;**384**(9959):2027–35.
5. Draisma G, Boer R, Otto SJ, et al. Lead times and overdetection due to prostate-specific antigen screening: estimates from the European Randomized Study of Screening for Prostate Cancer. *J Natl Cancer Inst* 2003;**95**(12):868–78.
6. Rich AR. On the frequency of occurrence of occult carcinoma. *J Urol* 1935;**33**:215–23.
7. Wilt TJ, Brawer MK, Jones KM, et al. Radical prostatectomy versus observation for localized prostate cancer. *N Engl J Med* 2012;**367**(3):203–13.
8. Sanda MG, Dunn RL, Michalski J, et al. Quality of life and satisfaction with outcome among prostate-cancer survivors. *N Engl J Med* 2008;**358**(12):1250–61.
9. Wolters T, Roobol MJ, van Leeuwen PJ, et al. A critical analysis of the tumor volume threshold for clinically insignificant prostate cancer using a data set of a randomized screening trial. *J Urol* 2011;**185**(1):121–5.
10. van As NJ, Norman AR, Thomas K, et al. Predicting the probability of deferred radical treatment for localised prostate cancer managed by active surveillance. *Eur Urol* 2008;**54**(6):1297–305.
11. Soloway MS, Soloway CT, Williams S, et al. Active surveillance; a reasonable management alternative for patients with prostate cancer: the Miami experience. *BJU Int* 2008;**101**(2):165–9.
12. Carter HB, Kettermann A, Warlick C, et al. Expectant management of prostate cancer with curative intent: an update of the Johns Hopkins experience. *J Urol* 2007;**178**(6):2359–64 discussion 64–65.
13. Dall'Era MA, Konety BR, Cowan JE, et al. Active surveillance for the management of prostate cancer in a contemporary cohort. *Cancer* 2008;**112**(12):2664–70.
14. Klotz L, Zhang L, Lam A, et al. Clinical results of long-term follow-up of a large, active surveillance cohort with localized prostate cancer. *J Clin Oncol* 2010;**28**(1):126–31.

15. van den Bergh RC, Vasarainen H, van der Poel HG, et al. Short-term outcomes of the prospective multicentre 'Prostate Cancer Research International: Active Surveillance' study. *BJU Int* 2010; **105**(7):956–62.
16. Zhu X, Albertsen PC, Andriole GL, et al. Risk-based prostate cancer screening. *Eur Urol* 2011;**61**(4):652–61.
17. Kranse R, Roobol M, Schroder FH. A graphical device to represent the outcomes of a logistic regression analysis. *Prostate* 2008;**68**(15):1674–80.
18. Roobol MJ, Steyerberg EW, Kranse R, et al. A risk-based strategy improves prostate-specific antigen-driven detection of prostate cancer. *Eur Urol* 2010;**57**(1):79–85.
19. Roobol MJ, Zhu X, Schroder FH, et al. A calculator for prostate cancer risk 4 years after an initially negative screen: findings from ERSPC Rotterdam. *Eur Urol* 2013;**63**(4):627–33.
20. Oliveira M, Marques V, Carvalho AP, et al. Head-to-head comparison of two online nomograms for prostate biopsy outcome prediction. *BJU Int* 2011;**107**(11):1780–3.
21. Ouzaid I, Yates DR, Hupertan V, et al. A direct comparison of the diagnostic accuracy of three prostate cancer nomograms designed to predict the likelihood of a positive initial transrectal biopsy. *Prostate* 2012;**72**(11):1200–6.
22. Roobol MJ, Schroder FH, Hugosson J, et al. Importance of prostate volume in the European Randomised Study of Screening for Prostate Cancer (ERSPC) risk calculators: results from the prostate biopsy collaborative group. *World J Urol* 2012;**30**(2):149–55.
23. Trottier G, Roobol MJ, Lawrentschuk N, et al. Comparison of risk calculators from the Prostate Cancer Prevention Trial and the European Randomized Study of Screening for Prostate Cancer in a contemporary Canadian cohort. *BJU Int* 2011;**108**(8 Pt 2):E237–44.
24. van Vugt HA, Roobol MJ, Kranse R, et al. Prediction of prostate cancer in unscreened men: external validation of a risk calculator. *Eur J Cancer* 2011;**47**(6):903–9.
25. van Vugt HA, Roobol MJ, Busstra M, et al. Compliance with biopsy recommendations of a prostate cancer risk calculator. *BJU Int* 2012;**109**(10):1480–8.
26. van Vugt HA, Kranse R, Steyerberg EW, et al. Prospective validation of a risk calculator which calculates the probability of a positive prostate biopsy in a contemporary clinical cohort. *Eur J Cancer* 2012;**48**(12):1809–15.
27. Kattan MW, Eastham JA, Wheeler TM, et al. Counseling men with prostate cancer: a nomogram for predicting the presence of small, moderately differentiated, confined tumors. *J Urol* 2003; **170**(5):1792–7.
28. Steyerberg EW, Roobol MJ, Kattan MW, et al. Prediction of indolent prostate cancer: validation and updating of a prognostic nomogram. *J Urol* 2007;**177**(1):107–12; discussion 112.
29. Nakanishi H, Wang X, Ochiai A, et al. A nomogram for predicting low-volume/low-grade prostate cancer: a tool in selecting patients for active surveillance. *Cancer* 2007;**110**(11):2441–7.
30. Chun FK, Haese A, Ahyai SA, et al. Critical assessment of tools to predict clinically insignificant prostate cancer at radical prostatectomy in contemporary men. *Cancer* 2008;**113**(4):701–9.
31. O'Brien BA, Cohen RJ, Ryan A, et al. A new preoperative nomogram to predict minimal prostate cancer: accuracy and error rates compared to other tools to select patients for active surveillance. *J Urol* 2011;**186**(5):1811–7.
32. Kattan MW, Cuzick J, Fisher G, et al. Nomogram incorporating PSA level to predict cancer-specific survival for men with clinically localized prostate cancer managed without curative intent. *Cancer* 2008;**112**(1):69–74.
33. Roemeling S, Roobol MJ, Kattan MW, et al. Nomogram use for the prediction of indolent prostate cancer: impact on screen-detected populations. *Cancer* 2007;**110**(10):2218–21.
34. Bul M, Delongchamps NB, Steyerberg EW, et al. Updating the prostate cancer risk indicator for contemporary biopsy schemes. *Can J Urol* 2011;**18**(2):5625–9.
35. Steyerberg EW. *Clinical prediction models*. Springer; 2009.
36. Bleeker SE, Moll HA, Steyerberg EW, et al. External validation is necessary in prediction research: a clinical example. *J Clin Epidemiol* 2003;**56**(9):826–32.
37. Steyerberg EW, Bleeker SE, Moll HA, et al. Internal and external validation of predictive models: a simulation study of bias and precision in small samples. *J Clin Epidemiol* 2003;**56**(5):441–7.
38. Dong F, Kattan MW, Steyerberg EW, et al. Validation of pretreatment nomograms for predicting indolent prostate cancer: efficacy in contemporary urological practice. *J Urol* 2008;**180**(1):150–4 discussion 154.
39. Iremashvili V, Soloway MS, Pelaez L, et al. Comparative validation of nomograms predicting clinically insignificant prostate cancer. *Urology* 2013;**81**(6):1202–8.
40. van Vugt HA, Roobol MJ, van der Poel HG, et al. Selecting men diagnosed with prostate cancer for active surveillance using a risk calculator: a prospective impact study. *BJU Int* 2012;**110**(2):180–7.
41. Chabrera C, Zabalegui A, Bonet M, et al. A decision aid to support informed choices for patients recently diagnosed with prostate cancer: a randomized controlled trial. *Cancer Nurs* 2014;**38**(3):E42–50.
42. Berry DL, Halpenny B, Hong F, et al. The Personal Patient Profile-Prostate decision support for men with localized prostate cancer: a multi-center randomized trial. *Urol Oncol* 2013;**31**(7):1012–21.
43. Auvinen A, Hakama M, Ala-Opas M, et al. A randomized trial of choice of treatment in prostate cancer: the effect of intervention on the treatment chosen. *BJU Int* 2004;**93**(1):52–6 discussion 56.
44. Pokorny MR, de Rooij M, Duncan E, et al. Prospective study of diagnostic accuracy comparing prostate cancer detection by transrectal ultrasound-guided biopsy versus magnetic resonance (MR) imaging with subsequent MR-guided biopsy in men without previous prostate biopsies. *Eur Urol* 2014;**66**(1):22–9.
45. Siddiqui MM, Rais-Bahrami S, Truong H, et al. Magnetic resonance imaging/ultrasound-fusion biopsy significantly upgrades prostate cancer versus systematic 12-core transrectal ultrasound biopsy. *Eur Urol* 2013;**64**(5):713–9.
46. Hambrock T, Hoeks C, Hulsbergen-van de Kaa C, et al. Prospective assessment of prostate cancer aggressiveness using 3-T diffusion-weighted magnetic resonance imaging-guided biopsies versus a systematic 10-core transrectal ultrasound prostate biopsy cohort. *Eur Urol* 2012;**61**(1):177–84.
47. Thormer G, Otto J, Horn LC, et al. Non-invasive estimation of prostate cancer aggressiveness using diffusion-weighted MRI and 3D proton MR spectroscopy at 3.0 T. *Acta Radiol* 2014;**56**(1): 121–8.
48. Kobus T, Vos PC, Hambrock T, et al. Prostate cancer aggressiveness: *in vivo* assessment of MR spectroscopy and diffusion-weighted imaging at 3 T. *Radiology* 2012;**265**(2):457–67.
49. Oh JJ, Park S, Lee SE, et al. The use of exome genotyping to predict pathological Gleason score upgrade after radical prostatectomy in low-risk prostate cancer patients. *PloS ONE* 2014;**9**(8):e104146.

CHAPTER

# 25

# Active Surveillance: Rationale, Patient Selection, Follow-up, and Outcomes

*Laurence Klotz, MD*

Division of Urology, Sunnybrook Health Sciences Centre, University of Toronto, Toronto, ON, Canada

## INTRODUCTION

The last century has witnessed a shift from invasive to minimally invasive surgery, and it has been proposed that the next century may bring the elimination of invasion altogether.[1] For the management of localized prostate cancer, this is a very apt expectation. Active surveillance represents a major advance at the noninvasive end of the treatment spectrum and offers the critically important promise of reducing the number of patients required to be treated to reduce prostate cancer mortality effectively. This approach is reviewed in this chapter.

## BACKGROUND

Prostate-specific antigen (PSA) testing reduces prostate cancer mortality.[2] However, in 2012 the US Preventive Services Task Force (USPSTF) published a Level D recommendation against PSA screening,[3] based on concerns about overdiagnosis and overtreatment. Subsequently, a number of other national health policy organizations published recommendations that reflected ambivalence about the net value of screening.[4] The result has been a steady decline in the rate of PSA testing and referral for biopsy over the last few years.

Histologic prostate cancer occurs normally with age, and the likelihood of harboring microfocal prostate cancer is approximately one's age as a percentage; 50% of men over 50, for example.[5] The effect of screening is to identify a large number of men with clinically insignificant prostate cancer who otherwise would not be diagnosed during their natural lifetime.

Prior to the USPSTF recommendation against PSA screening, almost all cases of localized prostate cancer were treated by either radical prostatectomy or high-dose radiation treatment.[6] The task-force recommendation, bolstered by substantial evidence regarding the indolent nature of low-grade disease and the favorable outcome with conservative management, has resulted in an international consensus that these therapies represent an unacceptable level of overtreatment for low-risk prostate cancer, and the most widely accepted alternative to radical treatment is active surveillance.[7,8]

## THE RATIONALE FOR SURVEILLANCE: THE NATURAL HISTORY AND MOLECULAR BIOLOGY OF LOW-GRADE PROSTATE CANCER

In all men, regardless of ethnic and racial background, microfocal low-grade prostate cancer occurs normally with age.[5] Most of these cancers are microfoci only (<1 $mm^3$) and low grade. Furthermore, a recent autopsy study in Japanese and Russian men who died of other causes showed that 50% of the cancers in Japanese men aged >70 were Gleason score 7 or above.[9] This finding suggests that, particularly in men over 70, microfocal Gleason 3 + 4 might also be a form of "autopsy" cancer, and not require treatment.

## GENETIC FEATURES OF LOW-GRADE PROSTATE CANCER

Genetic analyses comparing Gleason 3 and 4 patterns, the two most common histologic patterns of prostate cancer, have found that their molecular phenotypes differ profoundly. The hallmarks of cancer, described by Hanahan and Weinberg, provide a framework for comparing the degree of malignancy of these subtypes of prostate cancer.[10,11]

Prostate Cancer. http://dx.doi.org/10.1016/B978-0-12-800077-9.00025-6

**TABLE 25.1** Gleason 3 Lacks the Hallmarks of Cancer

| Characteristics of cancer | Gleason 3 | Gleason 4 |
|---|---|---|
| Absence of senescence[13,14] | Normal | Increased |
| Insensitivity to antigrowth signals such as cyclin D2 methylation, CKDN1β[15–17] | Expressed | Absent |
| *HER2/neu*[18] | Not present | Amplified |
| Expression of pro-proliferation embryonic and hematopoietic stem cell genes, EGF, and EGFR[19] | Not present | Overexpressed |
| Resistance to apoptosis: BCL2[20] | Negative | Strong expression |
| AKT pathway[21] | Not present | Aberrant |
| Tissue invasion and metastasis markers (CXCR4, others)[22] | Normal | Overexpressed |
| Sustained angiogenesis: VEGF[23] | Expression low | Increased |
| Other proangiogenic factors and microvessel density[23,24] | Normal | Increased |
| TMPRSS2-ERG translocation[25–28] | Present 45% | Present 50–60% |
| *PTEN*[29] | Present (7% deleted) | Deleted |
| Clinical evidence of metastasis mortality[30,31] | Virtually absent | Present |

The six hallmarks of cancer described originally include unlimited replicative potential, sustained angiogenesis, local tissue invasion, insensitivity to antigrowth signals, metastasis, and replicative self-sufficiency. An update in 2011 added two more: deregulating cellular energetics and evasion of immune destruction. The genetic pathways responsible for these hallmarks of malignancy have been described in detail over the last decade (Table 25.1). Considering that the Gleason score is based on low-power histology, it has a remarkable ability to segregate prostate cancer between genetically normal and abnormal cells.[12] There are many examples of this distinction. Proliferation pathway associated genes, including AkT and HER2neu, are expressed normally in Gleason 3 and abnormally in Gleason 4 (Table 25.1). Genetic pathways mediating apoptosis resistance, angiogenesis, and the development of other proangiogenic factors, genes involved in regulating cellular metabolomics, and metastasis and invasion processes, are similarly overexpressed in Gleason 4 and normal in 3.[13–25] There are exceptions; in particular, *pTEN*[26,27] and *TMPRSS2-ERG*,[28] commonly upregulated and present respectively in most Gleason 4s, have been reported to be altered in a proportion of Gleason 3. Given the limits of histology, this is not surprising. However, these isolated genetic alterations, particularly *TMPRSS2-ERG*, do not appear to translate into an aggressive metastatic phenotype.

## METASTATIC POTENTIAL

The biological behavior of prostate cancer is highly variable, from indolent to very aggressive. At the favorable end of this spectrum of disease, some cancers may even spontaneously involute and disappear.[29] In several large clinical series of the results of radical prostatectomy for localized disease, in which pathological grading of the entire prostate removes the possibility of occult higher-grade cancer, the metastasis rate for those men with only Gleason 6 is close to zero. One multicenter study of 24,000 men with long-term follow-up after surgery included 12,000 with surgically confirmed Gleason 6 cancer.[30] The 20 year prostate cancer mortality was 0.2%. About 4000 of these were treated at MSKCC; of these, one died of prostate cancer; a pathological review of this patient revealed Gleason 4 + 3 disease (Scott Eggener, personal communication). Thus, the prostate cancer mortality in this cohort of proven Gleason 6 disease was zero. A second study of 14,000 men with surgically confirmed Gleason 6 disease found only 22 with lymph node metastases; a review of these cases showed that all had higher-grade cancer in the primary tumor. The rate of node positive disease in the patients with no Gleason 4 or 5 disease in their prostates was also zero.[31] While these studies have limitations, the message is clear.

Of course, this exceedingly low rate of metastasis and death may simply reflect the excellent results of surgery. In the absence of surgery, however, a confounder is the presence of coexistent high-grade cancer missed on the systematic biopsy due to sampling. Higher-grade cancer is present in 25–30% of men initially diagnosed with Gleason 6 on biopsy. These lesions are responsible for most of the prostate cancer deaths reported in Gleason 6 patients managed conservatively. In the Albertsen and coworkers series of men managed with "watchful waiting," for example, the 20 year prostate cancer mortality among Gleason 6 was 22%, remarkably close to this proportion.[32]

Occasionally, genetic mutations may confer an aggressive phenotype that is prehistologic, that is, not reflected in the appearance of the cancer cells. A recent genetic analysis of multiple metastatic sites from a patient who had extensive Gleason 4 + 3 pT3a N1 disease resected at age 47, and died 17 years later of metastatic CRPC, reported that the metastatic lesions appeared to derive from a microfocus of Gleason pattern 3 disease, rather than, as expected, from the high-grade cancers elsewhere in the prostate.[33] This case report is a challenge to the view that Gleason pattern 3 does not metastasize. It has been proposed that in this case the Gleason 3 cancer evolved as a differentiated progeny of a higher-grade cancer that had metastasized, resulting in a shared genetic phenotype.[34] That Gleason 4–5 cancers can redifferentiate to pattern 3 is clearly evident from the common observation of prostate cancer metastases, where pattern 3 glands are often admixed with higher-grade glands.[35,36] Because only cancers with pattern 4 or 5 can metastasize, clearly a single cancer cell can produce low-grade and higher-grade progeny.

Regardless, the case emphasizes that biology is complex, and not 100% predictable. It is possible that histological Gleason pattern 3, particularly when it coexists with higher-grade cancer, can harbor prehistological genetic alterations that confer a more-aggressive phenotype.

This is the conceptual basis for genetically based predictive assays that disaggregate low-grade cancer into low- and higher-risk groups. This single case should be balanced against the extensive clinical evidence supporting the absence of metastatic potential in the vast majority of pure Gleason pattern 3 cancers.

Understanding that Gleason pattern 3 has little or no metastatic phenotype has altered the approach to patients with this cancer. Terms like "pseudo-cancer," "pseudo-disease," "part of the aging process," and "precancer" may be utilized in counseling these men. Changing the terminology from the emotionally loaded term "cancer" can significantly reassure the patient and derail the headlong rush into aggressive treatment. Several alternate nomenclatures have been proposed, including "IDLE" tumors (indolent epithelial neoplasms)[37] or "prostatic neoplasm of low malignant potential" (PNLMP).

Given the consensus on the problem of overdiagnosis and overtreatment, the evidence for the indolent nature of low-risk disease, and the concern about occult higher-risk disease in some patients, an approach of initial expectant management with selective delayed intervention for the subset who are reclassified as higher risk over time is sensible. This is consistent with the current Zeitgeist of "personalized medicine."

Active surveillance not only offers the prospect of reduced morbidity and improved quality of life, but an improvement in survival. How could less treatment improve cancer survival? PSA screening has been discarded by policy makers such as the USPSTF because of concerns about overtreatment and a high number needed to treat (NNT) for each death avoided. Selective treatment employing active surveillance would result in a decrease in the NNT for each death avoided. If widely adopted, active surveillance would eventually result in a reappraisal of the benefits of PSA screening, and a greater acceptance of its value by policy makers such as the USPSTF. The result will be "rehabilitation" of PSA screening, earlier identification of those with aggressive disease, lives saved, and an overall reduction in prostate cancer mortality (compared to no screening resulting from the perceived hazards of overtreatment).

## WHO IS A CANDIDATE?

The 2005 reclassification of the Gleason scoring system resulted in Gleason 2–5 being removed from needle-biopsy grading. Low-risk disease based on biopsy is widely defined as Gleason 6 and PSA <10 ng/mL. Patients with T stage > T2a are excluded, but in fact, most such patients are T1c. This group includes around 45% of newly diagnosed patients in the United States and Canada; approximately 150,000 men per year. (T1c, Gleason 6 cancer is the most common single grade and stage combination of disease categories in all of oncology.) Low-risk disease has been disaggregated between very low and low risk based on the number of cores, extent of core involvement, and PSA density. The Epstein criteria for very low-risk cancer are those with only one or two cores positive (regardless of how many cores were taken), no core with more than 50% involvement and PSA density <0.15. The Epstein criteria were based on those sextant biopsy criteria, which predicted for the Stamey definition of clinically insignificant disease (<0.5 $cm^3$ of Gleason 6 prostate cancer). Stamey's definition was based on the characteristics of unsuspected prostate cancer found in 149 cystoprostatectomy specimens from the pre-PSA era.[38] This definition of clinically insignificant cancer is too stringent, and would exclude many patients with low-risk disease who would otherwise be excellent candidates for conservative management.

An attempt to provide a more realistic definition of "clinically insignificant" cancer is based on the patients diagnosed with prostate cancer in ERSPC. Based on the finding that 50% of men in the screening arm of ERSPC were overdiagnosed, the threshold for clinically significant Gleason 6 disease was a cancer volume of >1.3 $cm^3$.[39] This is an important refinement of the traditional Stamey definition.

Based on the genetic characterization of Gleason pattern 3 and the clinical experience with Gleason 6, patients with only Gleason 6 cancer are at little or no risk of metastasis. The main significance of higher-volume disease is as a predictor of occult coexistent higher-grade cancer. In the absence of higher-grade cancer, metastasis is exceedingly unlikely. Thus, patients with high-volume Gleason 6 require close scrutiny to exclude coexistent higher-grade disease with as much certainty as possible, but do not in most cases require treatment (unless higher-grade cancer is identified).

Among Gleason 6 patients who are subsequently upgraded, most harbor higher-grade cancer at the time of diagnosis. True grade progression (Gleason 3 cells giving rise to Gleason 4 or 5 progeny) occurs, but is uncommon. In the Toronto surveillance cohort, we observed that the likelihood of grade progression increased approximately 1% per year from the time of the original biopsy.[40] This is a likely estimate of the frequency of biologic grade progression. The implication is that patients on surveillance require long-term follow-up, although in most cases the Gleason grade remains stable.

Low prostate volume, and a high PSA density (PSA: prostate volume ratio) has been demonstrated in many studies to be a predictor for risk progression. A high PSA density in some surveillance candidates reflects PSA arising from a large occult cancer. Increased caution is warranted in these cases.

Young age is not a contraindication to conservative management. The benefits of avoiding treatment with respect to maintenance of erectile function and continence are greater in young men, and the risks of second malignancies as sequelae of radiation are also greater in men with a long life expectancy. Microfocal low-grade cancer is present in 30–40% of men in their 40s.[5] Diagnosing this on a TRUS-guided biopsy does not mean that disease progression is inevitable. Men with high-volume Gleason pattern 3 do have a considerably higher risk of harboring higher-grade cancer. The reported "high volume" of Gleason 3 on biopsy at which point higher-grade cancer is more likely to be present is variable. A threshold effect of more than 8 mm of total cancer on systematic biopsy has recently been described.[41]

A special situation is that of young men (age <50 years) who have extensive Gleason 6 cancer on biopsy; this is uncommon. Given the uncertainty in these men about the risk of tumor progression over time, as well as the risk of harboring occult high-grade disease, it is reasonable to offer them treatment. Where exactly to draw the line in terms of age and cancer volume is a matter of clinical judgment.

African-Americans on AS have a higher rate of risk reclassification, and PSA failure when treated than Caucasian men.[42] Black men who are surveillance candidates also have a higher rate of large anterior cancers than Caucasians.[43] Japanese men younger than 60 have a lower rate of histological "autopsy" cancer than Caucasian men. Thus, the finding of low-grade prostate cancer in young Asian men is perhaps less likely to represent overdiagnosis. However, African-American and Asian patients diagnosed with low-grade prostate cancer include many men who have little or no probability of a prostate cancer related death during their remaining lives, and active surveillance is still an appealing option for those who have been appropriately risk-stratified.

Over time, about one-third of patients will be reclassified as higher risk for progression and offered treatment. This will depend on the inclusion criteria used for eligibility for surveillance. An inclusive approach, offering surveillance to all patients with Gleason 6 and PSA <15, for example, will include more patients with occult high-grade disease than a narrower approach, restricting surveillance to those who meet Epstein criteria (≤2 positive cores, <50% involvement of any one core, and PSA density <0.15). However, the more stringent eligibility denies the benefits of AS to many men with indolent disease who do not fit the Epstein criteria and thus are discouraged from choosing AS.

Several new biomarkers have recently been approved by the FDA based on their ability to predict progression in low-grade prostate cancer patients. These include the Prolaris assay[44] (Myriad Genetics), which looks for abnormal expression of cell-cycle-related genes, and the Oncotype DX assay (Genome Health), which identifies a panel of genes linked to a more aggressive phenotype.[45] The Mitomics assay, which identifies the presence of a functional mitochondrial DNA deletion associated with aggressive prostate cancer,[46] is not yet FDA approved. The Decipher test predicts for the likelihood of metastasis by characterizing the genetic phenotype of cells.[47] These tests hold the promise of interrogating the microfocus of Gleason 6 found on biopsy to identify the higher-grade cancer elsewhere in the prostate. That the biomarkers can achieve this confirms the interrelationship of heterogeneous multifocal cancers.

These tests, performed on biopsy tissue, are a proxy for predicting future biological behavior based on genetic alterations in low-grade cancer cells. A patient with low-grade prostate cancer and a "high risk" Oncotype DX or Prolaris score should have an MRI and be treated according to the best judgment of his disease characteristics. A further area for research is to better understand how to integrate the results of genetic biomarker tests and MRI. For example, optimal management of the patient in whom results are discrepant (i.e., genetic test indicates high risk but MRI is negative) is currently unknown. False positive and false negative results undoubtedly occur with both diagnostic approaches, but how commonly this occurs is unknown. Conceptually, the MRI is a snapshot of current disease extent and characteristics, while the genetic biomarkers provide information about future behavior (these are obviously related). While the molecular biomarkers address the unmet need of better risk assignment, further validation of their performance is needed before they are widely adopted in the surveillance scenario.

The utility of surveillance compared to surgery and radiation has been modeled by several groups. One propensity score analysis compared 452 men from the Toronto surveillance cohort to 6485 men having RP, 2264 treated with external beam, and 1680 with brachytherapy. There was no difference in prostate cancer mortality, and an improved overall survival in the surveillance group (due to an increase in other-cause mortality in the radiation patients).[48] A decision analysis of surveillance compared to initial treatment showed that surveillance had the highest QALE even if the relative risk of prostate-cancer-specific death for initial treatment versus active surveillance was as low as 0.6.[49] (In fact, it is almost certainly greater than 0.95 at 15 years.)

While surveillance has become more widely accepted over the last decade, the modification of the Gleason system in 2005 has, ironically, resulted in a decrease in the number of newly diagnosed Gleason 6 compared to 7, and therefore a smaller proportion of prostate cancer patients are eligible for surveillance. A key component of this modification was that small amounts of higher-grade cancer were incorporated into the final score. Thus, a 3 + 3 +tertiary 4 case is now 3 + 4. Clearly, therefore, many historic 3 + 3 cases would now be called Gleason 7 (3 + 4). Patients with Gleason 3 + 4 = 7, where the component of pattern 4 is small (<10%), have a very similar natural history to those with Gleason 3 + 3, perhaps reflecting this grade migration phenomenon.[50]

## SURVEILLANCE FOLLOW-UP PROTOCOLS

The clinical management of men on AS has evolved over the last 15 years. Currently, most experienced clinicians use the following approach or a variation of it: following the initial diagnosis of Gleason 6 prostate cancer on 10 or more core systematic biopsy, PSA is performed every 3–6 months for the first 2 years, and then every 6 months. A confirmatory biopsy must be carried out within 3–12 months of the initial diagnostic biopsy on which cancer was identified. This confirmatory biopsy should target the areas that are typically undersampled on the initial diagnostic biopsy. These include the anterior prostate, and the prostatic apex and base. The anterior prostate, in particular, has been shown to be the most common site of occult clinically significant disease. If the confirmatory biopsy is either negative or confirms microfocal Gleason 3 + 3 disease, subsequent biopsies are performed every 3–5 years (median of 4 years) until the patient reaches the age of 80, or has a life expectancy <5 years because of comorbidity. Multiparametric MRI should be performed on those patients whose PSA kinetics suggest more aggressive disease (usually defined as a PSA DT <3 years), whose biopsies show substantial Gleason 6 volume increase, or who are upgraded to Gleason 3 + 4 and surveillance is still desired as a management option. Identification of an MRI target suspicious for high-grade disease should warrant a targeted biopsy; or, if the lesion is large and unequivocal, intervention.

An alternative to the incorporation of MRI is the use of template biopsies using a transperineal approach and a brachytherapy needle placement guide. This technique, based on 1–2 cores/cm$^3$ of prostate, has been shown to provide reliable information about the extent of the disease. The drawbacks mainly relate to the resource issues, since general anesthesia, an hour of OR time, and a significantly greater amount of pathologist's time are required. This has limited the uptake of template biopsies in the surveillance cohorts. Comparative studies of MR versus template biopsies have shown that neither approach is 100% accurate; in one recent study, transperineal template biopsy missed 21% of Gleason 7 cancers, and MR targeted biopsy missed 12%.[51] Furthermore, the number of cores required to diagnose each significant cancer is three- to fourfold higher with template biopsies than MR-targeted biopsies.

Upgrading on the confirmatory biopsy is due to resampling (about 25% of patients). Eighty-five percent of patients who are upgraded have Gleason 3 + 4, and many of these cases may still be indolent.[52] The commonest cause of death in men on AS is cardiovascular disease. Death from prostate cancer is uncommon. In the most mature surveillance cohort,[53] with a follow-up ranging from 2 years to 20 years and median follow-up of 8 years, the cumulative hazard ratio (or relative risk) of nonprostate-cancer death was 10 times that for prostate cancer. To date, the published literature on surveillance includes 13 prospective studies, encompassing about 5000 men.[53–64] Most of these studies have a duration of follow-up that is insufficient to identify an increased risk of prostate cancer mortality as a result of surveillance. For example, a pivotal Swedish study reported that the risk of prostate cancer mortality in patients managed by watchful waiting was low for many years, but tripled after 15 years of follow-up.[65] ("Watchful waiting" meant no opportunity for selective delayed intervention, whereas about 35% of patients in the surveillance series have had radical treatment.) In the Toronto experience, 70 patients have been followed for 14 years; about 1.5% have had late disease progression, but there is no evidence of a sharp increase in mortality to date.[66] Thus, a critical question in this field is what the long-term prostate cancer mortality will be beyond 15 years. It will be 5–7 years before the most mature cohorts have a substantial cohort with >15 years of

TABLE 25.2 Outcomes of AS in Large Prospective Series

| References | n | Median follow-up (months) | Treated overall (%); treatment free (%) | Overall/disease-specific survival (%) | BCR postdeferred treatment (%) |
|---|---|---|---|---|---|
| Klotz et al.,[53,66] University of Toronto | 993 | 92 | 30; 72 at 5 year | 79/95 at 15 year | 25 (6 overall) |
| Bul et al.,[7] Multicentre, Europe | 2500 | 47 | 32; 43 at 10 year | 77/100 at 10 year | 20 |
| Dall'Era et al.,[54] UCSF | 328 | 43 | 24; 67 at 5 year | 100/100 at 5 year | NR |
| Kakehi et al.,[55] Multicentre, Japan | 118 | 36 | 51; 49 at 3 year | NR | NR |
| Tosoian et al.,[56] Johns Hopkins, USA | 407 | NR | 36 NR | NR | NR (50 "incurable" based on RP pathology) |
| Roemeling et al.,[57] Rotterdam Netherlands | 273 | 41 | 29; 71 at 5 year | 89/100 at 5 year | NR (31 of 13 RP positive margins) |
| Soloway et al.,[58] Miami, USA | 99 | 35 | 8; 85 at 5 year | NR | NR |
| Patel et al.,[59] Memorial Sloan Kettering, USA | 88 | 35 | 35; 58 at 5 year | NR | NR |
| Barayan et al.,[60] McGill, Canada | 155 | 65 | 20 | NR | NR |
| Rubio-Briones et al.,[61] Spain | 232 | 36 | 27 | 93 at 5 year/99.5 | |
| Godtman et al.,[62] | 439 | | 63 | 81/99.8 | 14 |
| Thomsen et al.,[63] Denmark | 167 | 40 | 35; 60 at 5 year | | |
| Selvadurai et al.,[64] UK | 471 | 67 | 30 | 98/99.7 | 12 |

follow-up. Table 25.2 summarizes the results of the 13 prospective series. The key outcome measures include the proportion of patients treated, overall, and cause-specific survival. Overall, about one-third of patients are treated; most series have few or no prostate cancer deaths. In the Toronto series, the actuarial prostate cancer mortality at 15 years is 5%. Few of the other publications have significant numbers of patients followed for more than 10 years. Some variation exists with respect to eligibility criteria and triggers for intervention, but there are consistent themes.

All groups have relied on systematic TRUS-guided biopsies performed serially, at intervals of 1–5 years. This technique has significant limitations. Most importantly, TRUS-guided biopsy tends to undersample the anterior prostate, apex, and anterolateral horn.

MRI has two potential benefits: reassurance that no higher-risk disease is present in those with no visualized disease; and earlier identification of higher-grade cancer. With respect to the former, the key metric is the NPV. This has been reported to be 97% for a cohort of 300 surveillance candidates at MSKCC.[67] Similarly, an MRI abnormality with a Prostate Imaging Reporting and Data System (PiRADS) score of 4 or 5/5 had a 90% positive predictive value for high-grade cancer. A PiRADS 4 or 5 lesion should lead at least to a targeted biopsy, or perhaps definitive intervention. A equivocal lesion (PiRADS 3/5) should trigger a targeted biopsy.

If the results of single center cohorts are validated, this performance of MRI as a diagnostic test would permit a level of confidence in a negative MRI that would allow it to replace the biopsy. This would decrease the number requiring biopsies (a major unmet need) and facilitate early identification of clinically significant disease. A limitation of multiparametric MRI is that the skill set for accurate interpretation is demanding, and not yet widely prevalent. This deficit is improving rapidly, however, due to the aggressive diffusion of the skill sets involved among radiologists.

PSA kinetics is currently used as a guide to identify patients at higher risk, but not to drive the decision to treat. This represents a shift in practice. Until multiparametric MRI became available, men on AS with poor PSA kinetics (doubling time < 3 years) were offered treatment. In the PRIAS multi-institutional AS registry, 20% of men being treated had intervention based on a PSA doubling time <3 years.[7] PSA kinetics does have sensitivity. In a report of the five men dying of metastatic prostate cancer in the Toronto cohort, all had a PSA doubling time <2 years.[68] The limitation of PSA kinetics is its lack of *specificity*. Vickers, in an overview of all of the studies of more than 200 patients examining the predictive value of PSA kinetics in localized prostate cancer, concluded that kinetics had no independent predictive value beyond the absolute value of PSA.[69] In a study of PSA kinetics in a large surveillance cohort, false positive PSA triggers (doubling time <3 years, or PSA velocity >2 ng/year) occurred in 50% of stable untreated patients, none of whom went on to progress, require treatment, or die of prostate cancer.[70] Put another way, the lability of PSA and the many factors that can increase PSA spuriously means that, in a healthy population of men followed over time, a sharp transient rise in PSA is common. This limits its use as a trigger for treatment.

## CONCLUSIONS

Active surveillance is an appealing approach for low-risk patients, and an important solution to the widely recognized problem of overtreatment of screen-detected prostate cancer. Widespread adoption of surveillance would result in a reduction in the number needed to treat for each death avoided without significantly increasing the risk of prostate cancer mortality. An objective and unbiased reassessment of PSA screening based on these improved metrics should lead to a reconsideration of the value of prostate cancer screening by organizations such as the USPSTF. Furthermore, ongoing improvements in diagnostic accuracy based on multiparametric MRI and genetic biomarkers hold the promise of reducing the need for systematic biopsies, improve the early identification of occult higher-risk disease, and enhance the ability to detect patients destined to have grade progression over time. A minimum management standard currently is a confirmatory biopsy targeting the anterolateral horn and anterior prostate within 6–12 months. PSA should be performed every 6 months and subsequent biopsies every 3–5 years until the patient is no longer a candidate for definitive therapy. MRI is indicated for men with a grade or volume increase, or adverse PSA kinetics. Treatment should be offered for most patients with upgraded disease. Selected patients with Gleason 3 + 4 disease, particularly if the percentage of Gleason 4 is small, may also be offered surveillance with the expectation of a low risk of disease progression. It is important to convey the message to patients that many so-called cancers pose little or no threat to their life, and likely do not require treatment. At the same time, there is no risk-free strategy. Rare patients (1–2%) may miss an opportunity for cure because of the delay in intervention, and progress to preventable metastatic disease. It is likely that the incorporation of MRI and/or biomarkers into the algorithm of management will reduce this low percentage further. The alternative is many patients suffering erectile dysfunction, incontinence, or other treatment-related quality of life effects for a condition that posed no threat to their quantity or quality of life.

## References


1. Gawende A. Two hundred years of surgery. *N Engl J Med* 2012;**366**:1716–23.
2. Schröder FH, Hugosson J, Roobol MJ, et al. Screening and prostate cancer mortality: results of the European Randomised Study of Screening for Prostate Cancer (ERSPC) at 13 years of follow-up. *Lancet* 2014;**384**(9959):2027–35.
3. http://ww.uspreventiveservicestaskforce.org/uspstf12/prostate/prostateart.htm
4. https://www.auanet.org/education/guidelines/prostate-cancer-detection.cfm
5. Sakr WA, Grignon DJ, Crissman JD, et al. High grade prostatic intraepithelial neoplasia (HGPIN) and prostatic adenocarcinoma between the ages of 20–69: an autopsy study of 249 cases. *In Vivo* 1994;**8**(3):439–43.
6. Cooperberg MR, Broering JM, Carroll PR. Time trends and local variation in primary treatment of localized prostate cancer. *J Clin Oncol* 2010;**28**(7):1117–23.
7. Bul M, Zhu X, Valdagni R, et al. Active surveillance for low-risk prostate cancer worldwide: the PRIAS study. *Eur Urol* 2013;**63**:597–603.
8. Amin MB, Lin DW, Gore JL, et al. The critical role of the pathologist in determining eligibility for active surveillance as a management option in patients with prostate cancer: consensus statement with recommendations supported by the College of American Pathologists, International Society of Urological Pathology, Association of Directors of Anatomic and Surgical Pathology, the New Zealand Society of Pathologists and the Prostate Cancer Foundation. *Arch Pathol Lab Med* 2014;**138**(10):1387–405.
9. Zlotta AR, Egawa S, Pushkar D, et al. Prevalence of prostate cancer on autopsy: cross-sectional study on unscreened Caucasian and Asian men. *J Natl Cancer Inst* 2013;**105**(14):1050–8.
10. Hanahan D, Weinberg RA. The hallmarks of cancer. *Cell* 2000; **100**:57–70.
11. Hanahan D, Weinberg RA. Hallmarks of cancer: the next generation. *Cell* 2011;**144**(5):646–74.
12. Ahmed H, Emberton M. Do low-grade and low-volume prostate cancers bear the hallmarks of malignancy? *Lancet Oncol* 2012;**13**(11):e509–17.
13. Hendriksen PJ, Dits NF, Kokame K, et al. Evolution of the androgen receptor pathway during progression of prostate cancer. *Cancer Res* 2006;**66**:5012–20.
14. Serrano M. Cancer: a lower bar for senescence. *Nature* 2010; **464**(7287):363–4.
15. Padar A, Sathyanarayana UG, Suzuki M, et al. Inactivation of cyclin D2 gene in prostate cancers by aberrant promoter methylation. *Clin Cancer Res* 2003;**9**:4730–4.
16. Susaki E, Nakayama KI. Multiple mechanisms for p27(Kip1) translocation and degradation. *Cell Cycle* 2007;**6**:3015–20.
17. Guo Y, Sklar GN, Borkowski A, et al. Loss of the cyclin-dependent kinase inhibitor p27(Kip1) protein in human prostate cancer correlates with tumor grade. *Clin Cancer Res* 1997;**3**:2269–74.
18. Skacel M, Ormsby AH, Pettay JD, et al. Aneusomy of chromosomes 7, 8, and 17 and amplification of *HER-2/neu* and epidermal growth factor receptor in Gleason score 7 prostate carcinoma: a differential fluorescent *in situ* hybridization study of Gleason pattern 3 and 4 using tissue microarray. *Hum Pathol* 2001;**32**:1392–7.
19. Ross AE, Marchionni L, Vuica-Ross M, et al. Gene expression pathways of high grade localized prostate cancer. *Prostate* 2011;**71**: 1568–77.
20. Fleischmann A, Huland H, Mirlacher M, et al. Prognostic relevance of Bcl-2 overexpression in surgically treated prostate cancer

is not caused by increased copy number or translocation of the gene. *Prostate* 2012;**72**:991–7.
21. True L, Coleman I, Hawley S, et al. A molecular correlate to the Gleason grading system for prostate adenocarcinoma. *Proc Natl Acad Sci USA* 2006;**103**:10991–6.
22. Tomlins SA, Mehra R, Rhodes DR, et al. Integrative molecular concept modeling of prostate cancer progression. *Nat Genet* 2007;**39**:41–51.
23. West AF, O'Donnell M, Charlton RG, et al. Correlation of vascular endothelial growth factor expression with fibroblast growth factor-8 expression and clinico-pathologic parameters in human prostate cancer. *Br J Cancer* 2001;**85**:576–83.
24. Sowalsky AG, Ye H, Bubley GJ, et al. Clonal progression of prostate cancers from Gleason grade 3 to grade 4. *Cancer Res* 2013;**73**(3):1050–5.
25. Bismar TA, Dolph M, Teng LH, et al. ERG protein expression reflects hormonal treatment response and is associated with Gleason score and prostate cancer specific mortality. *Eur J Cancer* 2012;**48**: 538–46.
26. Furusato B, Gao CL, Ravindranath L, et al. Mapping of *TMPRSS2-ERG* fusions in the context of multi-focal prostate cancer. *Mod Pathol* 2008;**21**:67–75.
27. Wang J, Cai Y, Ren C, et al. Expression of variant *TMPRSS2/ERG* fusion messenger RNAs is associated with aggressive prostate cancer. *Cancer Res* 2006;**66**:8347–51.
28. Berg KD, Vainer B, Thomsen FB, et al. ERG protein expression in diagnostic specimens is associated with increased risk of progression during active surveillance for prostate cancer. *Eur Urol* 2014;**66**(5):851–60.
29. Lotan TL, Carvalho FL, Peskoe SB, et al. *PTEN* loss is associated with upgrading of prostate cancer from biopsy to radical prostatectomy. *Mod Pathol* 2014;**28**(1):128–37.
30. Eggener S, Scardino P, Walsh P, et al. 20 Year prostate cancer specific mortality after radical prostatectomy. *J Urol* 2011;**185**(3): 869–75.
31. Ross HM, Kryvenko ON, Cowan JE, et al. Do adenocarcinomas of the prostate with Gleason score (GS) ≤6 have the potential to metastasize to lymph nodes? *Am J Surg Pathol* 2012;**36**(9): 1346–52.
32. Lu-Yao GL, Albertsen PC, Moore DF, et al. Outcomes of localized prostate cancer following conservative management. *JAMA* 2009;**302**(11):1202–9.
33. Haffner M, Yegnasubramanian S. The clonal origin of lethal prostate cancer. *J Clin Invest* 2013;**123**(11):4918–22 http://www.ncbi.nlm.nih.gov/pubmed/24135135.
34. Barbieri CE, Demichelis F, Rubin MA. The lethal clone in prostate cancer: redefining the index. *Eur Urol* 2014;**66**(3):307–95.
35. Berman DM, Epstein JI. When is prostate cancer really cancer? *Urol Clin North Am* 2014;**41**(2):339–46.
36. Cheng L, Slezak J, Bergstralh EJ, et al. Dedifferentiation in the metastatic progression of prostate carcinoma. *Cancer* 1999;**86**:657–63.
37. Esserman LJ, Thompson IM, Reid B, et al. Addressing overdiagnosis and overtreatment in cancer: a prescription for change. *Lancet Oncol* 2014;**15**(6):e234–42.
38. Stamey TA, Freiha FS, McNeal JE, et al. Relationship of tumor volume to clinical significance for treatment in prostate cancer. *Cancer* 1993;**71**:933–40.
39. Wolters T, Roobol M, Schröder F, et al. A critical analysis of the tumor volume threshold for clinically insignificant prostate cancer using a data set of a randomized screening trial. *J Urol* 2011;**185**:121–5.
40. Jain S, Loblaw A, Vesprini D, Zhang L, Kattan MW, Mamedov A, et al. Gleason Upgrading with Time in a Large Prostate Cancer Active Surveillance Cohort. *J Urol* 2015;**194**(1):79–84.
41. Bratt O, Folkvaljon Y, Loeb S, Klotz L, Egevad L, Stattin P. Optimizing the definition of very low risk prostate cancer. *BJU Int* 2014;Jul 23. doi: 10.1111/bju.12874; presented at EAU Stockholm March 2014.
42. Sundi D, Ross AE, Humphreys EB, et al. African American men with very low-risk prostate cancer exhibit adverse oncologic outcomes after radical prostatectomy: should active surveillance still be an option for them? *J Clin Oncol* 2013;**31**(24):2991–7.
43. Sundi D, Ross AE, Humphreys EB, et al. African American men with very low-risk prostate cancer exhibit adverse oncologic outcomes after radical prostatectomy: should active surveillance still be an option for them? *J Clin Oncol* 2013;**31**(24):2991–7.
44. Cuzick J, Berney DM, Fisher G, et al. Prognostic value of a cell cycle progression signature for prostate cancer death on conservatively managed needle biopsy cohort. *Br J Cancer* 2012;**106**: 1095–9.
45. Knezevic D, Goddard AD, Natraj N, et al. Analytical validation of the Oncotype DX prostate cancer assay – a clinical RT-PCR assay optimized for prostate needle biopsies. *BMC Genomics* 2013; **14**:690.
46. Robinson K, Creed J, Reguly B, et al. Accurate prediction of repeat prostate biopsy outcomes by a mitochondrial DNA deletion assay. *Prostate Cancer Prostatic Dis* 2013;**16**(4):398.
47. Ross AE, Feng FY, Ghadessi M, et al. A genomic classifier predicting metastatic disease progression in men with biochemical recurrence after prostatectomy. *Prostate Cancer Prostatic Dis* 2014;**17**(1):64–9.
48. Stephenson A, Klotz L. Comparative propensity analysis of active surveillance vs initial treatment. *AUA* 2013.
49. Hayes JH, Ollendorf DA, Pearson SD, et al. Active surveillance compared with initial treatment for men with low-risk prostate cancer: a decision analysis. *JAMA* 2010;**304**(21):2373–80.
50. Reese AC, Cowan JE, Brajtbord JS, et al. The quantitative Gleason score improves prostate cancer risk assessment. *Cancer* 2012;**118**(24):6046–54.
51. Radtke JP, Kuru TH, Boxler S, et al. Comparative analysis of transperineal template-saturation prostate biopsy versus MRI-targeted biopsy with MRI-US fusion-guidance. *J Urol* 2014;**193**(1):87–94.
52. Porten SP, Whitson JM, Cowan JE, et al. Changes in prostate cancer grade on serial biopsy in men undergoing active surveillance. *J Clin Oncol* 2011;**29**(20):2795–800.
53. Klotz L, Zhang L, Lam A, et al. Clinical results of long-term follow-up of a large, active surveillance cohort with localized prostate cancer. *J Clin Oncol* 2010;**28**(1):126–31.
54. Dall'Era MA, Konety BR, Cowan JE, et al. Active surveillance for the management of prostate cancer in a contemporary cohort. *Cancer* 2008;**112**(12):2664–70.
55. Kakehi Y, Kamoto T, Shiraishi T, et al. Prospective evaluation of selection criteria for active surveillance in Japanese patients with stage T1cN0M0 prostate cancer. *Jpn J Clin Oncol* 2008;**38**(2): 122–8.
56. Tosoian JJ, Trock BJ, Landis P, et al. Active surveillance program for prostate cancer: an update of the Johns Hopkins experience. *J Clin Oncol* 2011;**29**:2185–90.
57. Roemeling S, Roobol MJ, de Vries SH, et al. Active surveillance for prostate cancers detected in three subsequent rounds of a screening trial: characteristics, PSA doubling times, and outcome. *Eur Urol* 2007;**51**(5):1244–50.
58. Soloway MS, Soloway CT, Eldefrawy A, et al. Careful selection and close monitoring of low-risk prostate cancer patients on active surveillance minimizes the need for treatment. *Eur Urol* 2010;**58**:831–5.
59. Patel MI, DeConcini DT, Lopez-Corona E, et al. An analysis of men with clinically localized prostate cancer who deferred definitive therapy. *J Urol* 2004;**171**(4):1520–4.
60. Barayan GA, Brimo F, Bégin LR, et al. Factors influencing disease progression of prostate cancer under active surveillance: a McGill University Health Center cohort. *BJU Int* 2014;**114**(6b):E99–E104.

61. Rubio-Briones J, Iborra I, Ramírez M, et al. Obligatory information that a patient diagnosed of prostate cancer and candidate for an active surveillance protocol must know. *Actas Urol Esp* 2014;**38**(9):559–65.
62. Godtman RA, Holmberg E, Khatami A, et al. Outcome following active surveillance of men with screen-detected prostate cancer. Results from the Göteborg randomised population-based prostate cancer screening trial. *Eur Urol* 2013;**63**(1):101–7.
63. Thomsen FB, Røder MA, Hvarness H, et al. Active surveillance can reduce overtreatment in patients with low-risk prostate cancer. *Dan Med J* 2013;**60**(2):A4575.
64. Selvadurai ED, Singhera M, Thomas K, et al. Medium-term outcomes of active surveillance for localised prostate cancer. *Eur Urol* 2013;**64**:981–7.
65. Popiolek M, Rider JR, Andrén O, et al. Natural history of early, localized prostate cancer: a final report from three decades of follow-up. *Eur Urol* 2013;**63**(3):428–35.
66. Klotz L, Vesprini D, Loblaw A. Long term results of a large active surveillance cohort. AUA, Orlando Fl, 2014.
67. Vargas HA, Akin O, Afaq A, et al. Magnetic resonance imaging for predicting prostate biopsy findings in patients considered for active surveillance of clinically low risk prostate cancer. *J Urol* 2012;**188**(5):1732–8.
68. Krakowsky Y, Loblaw A, Klotz L. Prostate cancer death of men treated with initial active surveillance: clinical and biochemical characteristics. *J Urol* 2010;**184**(1):131–5.
69. Vickers A. Systematic review of pretreatment PSA velocity and doubling time as PCA predictors. *J Clin Oncol* 2008;**27**: 398–403.
70. Loblaw A, Zhang L, Lam A, et al. Comparing prostate specific antigen triggers for intervention in men with stable prostate cancer on active surveillance. *J Urol* 2010;**184**(5):1942–6.

P A R T VI

# SURGERY

CHAPTER

# 26

# Preoperative Risk Assessment

*Thomas J. Guzzo, MD, MPH*

Division of Urology, The Hospital of the University of Pennsylvania, Philadelphia, PA, USA

## INTRODUCTION

There was an estimated 233,000 cases of prostate cancer diagnosed in the United States in 2014.[1] There are a variety of therapeutic options for men with clinically localized prostate cancer including radical prostatectomy, radiation therapy, and active surveillance. Multiple variables ultimately factor into a patient's decision on which therapeutic avenue to pursue including overall health status and life expectancy, tolerance of side effects, and oncologic efficacy. Further complicating the decision-making process is the fact that a significant number of screen-detected prostate cancers have a low potential to ultimately result in patient mortality.[2]

Pretreatment decision-making for any disease is a complex process and interpretation of a multitude of factors, patient and disease based, must be considered. Given the multitude of treatment options, heterogeneity of tumor biology, and quality of life implications associated with treatment, the decision to move forward with treatment and which treatment can be agonizing decisions for patients and their clinicians. There are a variety of predictive tools that have been developed to aid physicians and patients in prostate cancer decision-making. Pretreatment predictive tools have been reported in various forms including risk stratification groupings, regression tree analysis, artificial neural networks, and nomograms.[3] Nomograms, defined as a graphic representation of a mathematical formula that incorporates several predictive variables into a model, have been increasingly popular in prostate cancer decision-making.

Incorporation of pretreatment decision tools is necessary as individual clinical judgment has been shown to be quite subjective and often inaccurate.[4,5] Pretreatment tools have been used with a variety of different goals including predicting pathologic outcomes, non-prostate-cancer related mortality, and biochemical and prostate-cancer-specific survival.[6] To this end, there now exists an extensive list of predictive tools that can be used as decision aids after an initial diagnosis of prostate cancer. The focus of this chapter will be the most commonly used and studied pretreatment decision tools.

## LIFE EXPECTANCY

No discussion on prostate cancer treatment decision-making would be complete without framing it within the context of life expectancy. Outcomes for prostate cancer treatment are measured in years and decades and therefore a thoughtful analysis of a patient's medical comorbidities and life expectancy is necessary for any prostate cancer treatment conversation. The national Comprehensive Cancer Network Guidelines do not recommend definite treatment for patients with low or intermediate risk prostate cancer whose life expectancy is <10 years.[7] While on the surface this is a very rational approach, it is often difficult for physicians to accurately assess comorbidity and life expectancy. In fact, the accuracy of clinicians to predict 10-year life expectancy based on age and comorbidity has been reported to be as low as 69%.[8,9]

The most basic life expectancy predictor is a life table, of which the most commonly used in clinical practice are the United States Life Tables prepared by the US Department of Health and Human Services. Using census data, life tables estimate life expectancy for a given age based on gender and race. A significant limitation of life tables is that they do not factor additional clinical information into the predictive model. For example, Walz et al. reported that life tables are only 60% accurate in predicting life expectancy in prostate cancer patients who were treated with radiation therapy.[10] In an attempt to better individualize life expectancy, a multitude of predictive tools incorporating patient comorbidity have been reported. One of the most commonly used is the Charlson Comorbidity Index (CCI), which predicts the 10-year mortality for a

Prostate Cancer. http://dx.doi.org/10.1016/B978-0-12-800077-9.00026-8

patient based on 22 comorbid conditions.[11] The CCI has also been applied to a number of malignancies to predict mortality.[12,13] Specifically in prostate cancer, a CCI of $\geq 2$ has been shown to be a predictor of non-prostate-cancer mortality after radical prostatectomy.[14]

Investigators have also incorporated comorbidity and cancer-specific variables to predict survival following prostate cancer treatment. Using four variables including age, CCI, biopsy Gleason score, and PSA, Tewari et al. retrospectively evaluated 1611 men with clinically localized prostate cancer and 4538 age-matched controls. The authors reported a validation C-index of 0.69 for overall survival using this four-variable model.[15] Similarly, Walz et al. evaluated 9131 men treated with either radical prostatectomy or external beam radiation using age and CCI to predict 10-year life expectancy.[16] Their nomogram for predicting 10-year life expectancy after either surgery or radiation was 84.3% accurate.

In summary, it is important for clinicians to take overall comorbidity and life expectancy into account when discussing various treatment alternatives with prostate cancer patients. While subjective physician assessment of life expectancy is inaccurate, several nomograms do exist that were developed specifically in prostate cancer patients, which can better discriminate life expectancy in men with prostate cancer.

## PREDICTORS OF FINAL PATHOLOGY

A major handicap in pretreatment decision-making is the discordance between clinical and pathologic stage. Even in patients with clinical low-risk disease, upgrading and upstaging can be significant. For example, El Hajj et al. evaluated 626 patients who met active surveillance criteria based on the Prostate Cancer Research International: Active Surveillance (PRIAS) criteria and noted an upgrading to Gleason score 7 or higher in 45% and upstaging (>pT2) in 21% of patients.[17] Pathologic upstaging for those with low-risk prostate cancer is currently a major limitation for patients seeking active surveillance protocols. Multiple predictive tools have been developed in an attempt to better predict final pathology based on preoperative clinical characteristics. Most of these tools have been developed based on large populations of radical prostatectomy cohorts. Each predictive tool is designed to predict various pathologic outcomes such as Gleason score upgrading, extra-capsular extension, seminal vesicle invasion, and lymph node metastasis.

## PARTIN TABLES

The Partin Tables, developed at Johns Hopkins, is one of the earliest predictive nomograms to be used in prostate cancer.[18–22] They incorporate three clinical pretreatment variables (PSA, biopsy Gleason score, and clinical stage) to predict final pathologic tumor stage. The initial iteration of the Partin Tables used regression analysis combining the three clinical variables from 703 radical prostatectomy patients to construct nomograms.[18] The combined nomogram was able to more accurately predict final pathology compared to any single clinical variable. Pathologic predictive variables included in the Partin Tables include the probability of organ-confined disease, extracapsular extension, seminal vesicle invasion, and lymph node metastasis. A major limitation of the original Partin Tables was that they were derived using data that largely included early or pre-PSA era patients. The Partin Tables have undergone several updates over the last two decades, most recently in 2012.[22] In their most recent update, based on 5629 radical prostatectomy patients from 2006 to 2011, the Partin Tables demonstrated good discrimination in predicting postoperative pathology. It is important to note that 92% of patients in the most recent analysis had a PSA of <10 and 85% had biopsy Gleason scores of 3 + 3 or 3 + 4. With the majority of patients having low-risk disease in this cohort, the predictive accuracy in higher-risk patients is limited. The Partin Tables have also been externally validated in several large patient cohorts.[23–25] It is important to consider the population in which a particular nomogram was studied as it may not be generalizable to all patient populations. For example, the Partin Tables have reported worse predictive ability in European men highlighting the importance of external validation of any nomogram.[26]

In addition to the Partin Tables, several prediction tools have been published to predict seminal vesicle invasion or lymph node metastasis at the time of radical prostatectomy. The largest such study attempting to predict seminal vesicle involvement was reported by Baccala et al., which comprised 6740 patients from three institutions.[27] There were 566 (8.4%) patients that had seminal vesicle involvement. The variables included in this nomogram included age, PSA, biopsy Gleason score, and clinical stage. The nomogram predicted seminal vesicle invasion with reasonable accuracy with an area under the curve of 0.80. Nomograms predicting the presence of seminal vesicle invasion may be useful when surgeons are considering seminal vesicle sparing surgery.[28] Cagiannos et al. reported on a nomogram for predicting pelvic lymph node metastasis using data from 5510 patients from six institutions.[29] Preoperative variables included in the nomogram included PSA, clinical stage, and biopsy Gleason score. Only 3.7% of patients in the cohort ultimately had lymph node metastasis on final pathology. The predictive accuracy of this model was 0.78. The negative predictive value was 0.99 to predict a 3% or less chance of positive lymph nodes and therefore this may be a valuable tool when determining when to perform a lymph node dissection at the time of radical prostatectomy.

In addition to pathologic stage prediction, several nomograms have been developed to predict Gleason score upgrading. One of the earliest studies analyzing Gleason upgrading was published by D'Amico et al. in 1999 attempting to predict upgrading to high-grade histology.[30] The ability to predict Gleason score upgrading has taken on greater importance over the last several years with our increased understanding of indolent prostate cancer and popularization of active surveillance protocols. To this end, Kulkarni et al. reported on 175 patients with low-risk prostate cancer, of which 34% were upgraded at the time of prostatectomy.[31] Using preoperative variables, including age, PSA, level of pathologist expertise, DRE results, prostate volume, TRUS results, and percentage of positive cores, the authors developed a nomogram with good predictive accuracy for upgrading (C-index = 0.71). A limitation of this nomogram is that it includes extended and sextant biopsy patients. Similarly, Capitanio et al. evaluated 301 low-risk prostate cancer patients who underwent a minimum of 10-core biopsy.[32] The nomogram variables included Gleason sum, total number of cores, and number of positive cores. Fewer total cores (10–12 cores vs. >18 cores) was found to be an independent predictor of Gleason score upgrading in this model. Adding the number of cores to their predictive model increased the accuracy for predicting Gleason score upgrading by 9%.

In summary, multiple predictive models exist to aid clinicians in predicting Gleason upgrading and pathologic upstaging based on easy to obtain pretreatment variables. These tools are useful not only to risk stratify patients prior to therapy but also can be helpful in optimizing therapeutic decision-making. Additionally, in the era of increased emphasis on active surveillance strategies for low-risk prostate cancer, the ability to predict which patients truly have low-risk organ-confined disease has taken on even greater precedence.

## PREOPERATIVE PREDICTORS OF BIOCHEMICAL RECURRENCE

Depending on the patient population and definition, biochemical recurrence rates following radical prostatectomy have been reported to be 15–40%.[33,34] While many patients with biochemical recurrence will not ultimately die of prostate cancer, a significant proportion will undergo secondary treatments for prostate cancer making it an important predictive end point. Several predictive models, based on a large series of patients, have gained fairly widespread clinical penetrance as tools in predicting biochemical outcome prior to definitive therapy.

## D'AMICO CRITERIA

In 1998, D'Amico reported a three-tiered preoperative risk stratification that is simple and easy to use.[35] Patients were categorized into low-risk (Gleason score ≤6, PSA up to 10 ng/mL and clinical stage cT1c-T2a), intermediate-risk (Gleason score 7, PSA 10–20 ng/mL and clinical stage T2b), and high-risk (Gleason score 8–10, PSA >20 ng/mL and clinical stage T2c). The authors applied these criteria to 1872 prostate cancer patients treated with either radical prostatectomy, brachytherapy, or external beam radiation therapy at a single institution. Increasing D'Amico risk category was associated with a higher likelihood of biochemical recurrence. Since its original report, a myriad of studies have investigated the utility of the D'Amico criteria in varying patient populations not only with regard to biochemical outcome but also disease progression and survival. Hernandez et al. evaluated the predictive performance of the D'Amico criteria in 6652 radical prostatectomy patients from a single institution.[36] The authors reported significant differences in biochemical failure rates between the three D'Amico classifications with 94.5, 76.6, and 54.6% 5 year biochemical recurrence-free survival rates for the low-, intermediate-, and high-risk cohorts, respectively. Boorjian et al. at the Mayo Clinic evaluated 7591 consecutive patients who underwent a radical prostatectomy at their institution.[37] The risk of prostate cancer mortality was significantly higher in intermediate- (HR 6.3, 95% CI 3.3–12.3) and high-risk (HR 11.5, 95% CI 5.9–22.3) patients compared to that of low-risk patients validating the predictive utility of D'Amico stratification.

## CAPRA SCORE

The Cancer of the Prostate Risk Assessment (CAPRA) score was first reported by Cooperberg et al. in 2005 as a simple method to predict the probability of biochemical recurrence and recurrence-free survival in men following radical prostatectomy.[38] The CAPRA score is derived by assigning weighted values to five easy-to-obtain preoperative variables including PSA, Gleason score, clinical stage, percent positive biopsy, and age. The original study population consisted of 1439 patients from the Cancer of the Prostate Strategic Urologic Research Endeavor (CaPSURE) database.[38] There were 88% of patients that had CAPRA scores in the 1–4 range, while only 2% had scores >6. In the initial publication, patients with a score of 0 or 1 had 3- and 5-year recurrence-free survival rates of 91 and 85%, respectively. For those with scores of 7 or more the rates dropped significantly (24 and 8%, respectively). An increase in CAPRA score by increments of two resulted in approximately double the risk of recurrence.

Cooperberg et al. also assessed the ability of the CAPRA score to predict outcome in patients who underwent nonsurgical treatment for prostate cancer (external beam radiation, brachytherapy, androgen deprivation monotherapy, and watchful waiting or active surveillance).[39] Similar to their previous study, for each point increase in the CAPRA score, there was an increased risk of bone metastases and cancer-specific mortality. The C-index for predicting metastases and cancer-specific mortality was 0.78 and 0.80, respectively.

Recently, the predictive capability of the CAPRA score has been analyzed in a large meta-analysis.[40] Using data from six validation studies ($n$ = 6081), CAPRA score correctly predicted 3-year recurrence-free survival in all three risk groups (low risk, RR 0.98, 95% CI 0.95–1.00; intermediate risk, RR 1.03, 95% CI 0.99–1.08; high risk, RR 0.87, 95% CI 0.73–1.05). However, the CAPRA score significantly under-predicted 5-year recurrence-free survival across the same risk strata.

## STEPHENSON NOMOGRAM

In 1998, the group at Memorial Sloan Kettering Cancer Center published a nomogram to predict the risk of biochemical failure at 5 years following radical prostatectomy.[41] The original nomogram was based on 983 patients who underwent a radical prostatectomy by a single surgeon. The variables in the nomogram include PSA, clinical stage, and biopsy Gleason score, and the discrimination of the nomogram was 0.76. It has been extensively studied since its original publication and independently validated in several patient populations.[42–44] The same group built upon this nomogram in later publications by adding systematic biopsy results (number of positive and negative cores) and accounting for stage migration by adjusting for year of surgery.[45] Using the updated nomogram, the 10-year progression-free probability could be predicted with a C-index of 0.78. The Stephenson nomogram is available online and easy to use in the clinical setting.

The three predictive models discussed earlier (D'Amico, CAPRA, and Stephenson) are the most widely used clinical tools to predict biochemical recurrence following radical prostatectomy. All three have been externally validated but there are few studies that have directly compared their predictive capabilities. Recently, Lughezzani et al. have evaluated all three models in a European cohort (1976 patients) of radical prostatectomy patients.[46] The C-index for 3-year biochemical recurrence-free survival for the D'Amico, CAPRA, and Stephenson models were 0.70, 0.74, and 0.75, respectively. The C-index for 5-year biochemical recurrence-free survival for the D'Amico, CAPRA, and Stephenson models were 0.67, 0.73, and 0.74, respectively.

The predictive tools described in this section are readily available and predict biochemical recurrence with reasonable discrimination in the pretreatment setting. Although biochemical recurrence is not an absolute surrogate for prostate cancer mortality it is a meaningful endpoint and can aid patients and clinicians in pretreatment decision-making.

## PREOPERATIVE PREDICTORS OF MORTALITY

PSA recurrences following radical prostatectomy occur in approximately 15–40% of patients; however, many will not ultimately die of prostate cancer.[33,34] Therefore, models that use biochemical recurrence as a predictive endpoint are inherently imperfect as they may not accurately reflect the most important end point (prostate cancer mortality). Many of the same authors and institutions have used similar variations of the preoperative models discussed in the previous section to predict prostate cancer mortality prior to treatment. The D'Amico criteria were evaluated using a multi-institutional database of 7316 patients treated with either radical prostatectomy or radiation therapy from 1988 to 2002.[47] Pretreatment D'Amico stratification was predictive of prostate cancer mortality in radical prostatectomy patients with a relative risk of 4.9 (95% CI 1.7–8.1) and 14.2 (95% CI 5.0–23.4) for intermediate- and high-risk categories, respectively. In the intermediate-risk prostatectomy cohort specifically, having all three factors also was associated with shorter interval to prostate-cancer-specific mortality compared to having any one or two risk factors alone.

Similarly, Cooperberg et al. have reported the ability of the CAPRA score to predict bone metastasis and cancer-specific mortality within the CaPSURE database.[48] The CAPRA score was predictive of the development of bone metastasis and cancer-specific mortality both when it was used as a continuous variable and as a three-tiered risk strata. Specifically, each point increase in CAPRA score was associated with an increased risk of bone metastases (HR 1.47, 95% CI 1.39–1.56) and cancer-specific mortality (HR 1.39, 95% CI 1.31–1.48). Prostate cancer survival at 10 years was 97.1, 91.6, and 79.1% for low-, intermediate-, and high-risk CAPRA scores in the CaPSURE cohort. The C-index for predicting cancer-specific mortality was 0.80.

Finally, Stephenson et al. reported on their nomogram including biopsy Gleason score, PSA, and clinical stage to predict 10 and 15 year prostate cancer-specific mortality.[49] As the in modeling cohort or the nomogram, 6398 patients from Memorial Sloan Kettering Cancer Center and Baylor were used. The nomogram was also externally validated in 4103 patients treated at the Cleveland Clinic and 2176 patients from the University of Michigan over the same time period. The externally validated C-index for 15-year prostate-cancer-specific mortality was 0.82. Only 4% of contemporary patients in this study had a predicted 15-year prostate cancer mortality rate of over 5%.

The predictive tools that have been used to predict biochemical failure based on pretreatment variables have also demonstrated good discrimination for predicting risk of metastasis and ultimately death from prostate cancer. These tools are simple, straightforward, and easy to use in the clinical setting and can also be found online.

## THE USE OF PRETREATMENT IMAGING TO PREDICT OUTCOMES

Diagnostic imaging, including pelvic cross-sectional imaging and bone scan, has long been a critical component in prostate cancer staging. Traditionally, these modalities have been used to rule out metastatic disease rather than predict local pathologic staging; however, prostate MRI has been increasingly studied over the last several years for its utility to predict biologic aggressiveness, pathologic staging, and outcome. Multiparametric MRI has been shown in several studies to have an improved ability to detect higher-grade disease compared to standard transrectal ultrasound-guided biopsy alone.[50,51] Additionally, multiparametric MRI has been investigated in patients who would otherwise qualify for active surveillance in an attempt to better identify patients with true low-risk disease. Stamatakis et al. evaluated 85 patients who would have otherwise qualified for active surveillance with a multiparametric MRI and fusion-guided prostate biopsy and reclassified 29% as not suitable for active surveillance.[52] Similarly, Shukla-Dave et al. evaluated 181 low-risk prostate cancer patients with MRI and MRI spectroscopy prior to radical prostatectomy.[53] Four nomograms for predicting insignificant disease at final pathology were evaluated, two of which incorporated MRI results. For the 181 patients included in this study, the nomograms incorporating MRI performed significantly better than standard clinical models as well as more comprehensive clinical models. While further studies in larger populations are needed to confirm these results, it appears that multiparametric MRI may play a role in the future in optimally identifying active surveillance patients.

Multiparametric MRI has also been recently investigated for its ability to predict biochemical failure following radical prostatectomy. In a study of 282 patients, the presence of a tumor on multiparametric-MRI was associated with an increased risk of biochemical recurrence (HR 2.38, $p = 0.047$).[54] While MRI appears to be promising in identifying high-risk disease and predicting pathologic stage it is subject to interreader variability and it is currently unclear whether studies reported at centers with significant MRI experience will be generalizable to the overall population. It does seem likely that with increased experience and advancing technology that diagnostic imaging will play a greater role in prostate decision-making in the future.

## CONCLUSIONS

The decision as to when to move forward with definitive treatment, and what treatment, is a complex one for clinicians and prostate cancer patients. Pretreatment models are available to help clinicians council patients regarding their potential prognosis prior to treatment, which can help patients make informed decisions. Simple, easy-to-use nomograms should be incorporated into clinical decision-making when counseling patients regarding treatment options.

### References


1. DeSantis CE, Lin CC, Mariotto AB, et al. Cancer treatment and survivorship statistics. *CA Cancer J Clin* 2014;**64**:252–71.
2. Klotz L. Prostate cancer overdiagnosis and overtreatment. *Curr Opin Endocrinol Diabetes Obes* 2013;**20**(3):204–9.
3. Lughezzani G, Briganti A, Karakiewicz PI, et al. Predictive and prognostic models in radical prostatectomy candidates: a critical analysis of the literature. *Eur Urol* 2010;**58**(5):687–700.
4. Elstein AS. Heuristics and biases: selected errors in clinical reasoning. *Acad Med* 1999;**74**:791–4.
5. Ross PL, Gerigk C, Gonen M, et al. Comparisons of nomograms and urologists predictions in prostate cancer. *Semin Urol Oncol* 2002;**20**:82–8.
6. Shariat SF, Kattan MW, Vickers AJ, et al. Critical review of prostate cancer predictive tools. *Future Oncol* 2009;**5**(10):1555–84.
7. National Comprehensive Cancer Network Clinical Practice Guidelines in Oncology. Prostate Cancer Early Detection. Version.1.2015. Available from: www.nccn.org.
8. Walz J, Gallina A, Perrotte P, et al. Clinicians are poor raters of life-expectancy before radical prostatectomy or definitive radiotherapy for localized prostate cancer. *BJU Int* 2007;**100**:1254–8.
9. Jeldres C, Latouff JB, Saad F. Predicting life expectancy in prostate cancer patients. *Curr Opin Support Palliat Care* 2009;**3**(3):166–9.
10. Walz J, Gallina A, Hutterer G, et al. Accuracy of life tables in predicting overall survival in candidates for radiotherapy for prostate cancer. *Int J Radiat Oncol Biol Phys* 2007;**69**:88–94.
11. Charlson ME, Pompei P, Ales KL, et al. A new method of classifying prognostic comorbidity in longitudinal studies: development and validation. *J Chronic Dis* 1987;**40**:373–83.
12. Lawindy SM, Kurian T, Kim T, et al. Important surgical considerations in the management of renal cell carcinoma with inferior vena cava tumor thrombus. *BJU Int* 2012;**110**(7):926–39.
13. Berger I, Martini T, Wehrberger C, et al. Perioperative complications and 90-day mortality of radical cystectomy in the elderly (75+): a retrospective, multicenter study. *Urol Int* 2014;**93**(3):296–302.
14. Guzzo TJ, Dluzniewski P, Orosco R, et al. Prediction of mortality after radical prostatectomy by Charlson comorbidity index. *Urology* 2010;**76**(3):553–7.

15. Tewari A, Johnson CC, Divine G, et al. Long-term survival probability in men with clinically localized prostate cancer: a case-control, propensity modeling study stratified by race, age, treatment and comorbidities. *J Urol* 2004;**171**(4):1513–9.
16. Walz J, Gallina A, Saad F, et al. A nomogram for predicting 10-year life expectancy in candidates for radical prostatectomy or radiotherapy for prostate cancer. *J Clin Oncol* 2007;**25**:3576–81.
17. El Hajj A, Ploussard G, de la Taille A, et al. Analysis of outcomes after radical prostatectomy in patients eligible for active surveillance (PRIAS). *BJU Int* 2013;**111**(1):53–9.
18. Partin AW, Yoo J, Carter HB, et al. The use of prostate specific antigen, clinical stage and Gleason score to predict pathological stage in men with localized prostate cancer. *J Urol* 1993;**150**:110–4.
19. Partin AW, Kattan MW, Subong EN, et al. Combination of prostate-specific antigen, clinical stage, and Gleason score to predict pathological stage of localized prostate cancer. A multi-institutional update. *JAMA* 1997;**277**:1445–51.
20. Partin AW, Mangold LA, Lamm DM, et al. Contemporary update of prostate cancer staging nomograms (Partin Tables) for the new millennium. *Urology* 2001;**58**:843–8.
21. Makarov DV, Trock BJ, Humphreys EB, et al. Updated nomogram to predict pathologic stage of prostate cancer given prostate-specific antigen level, clinical stage, and biopsy Gleason score (Partin Tables) based on cases from 2000 to 2005. *Urology* 2007;**69**:1095–101.
22. Eifler JB, Feng Z, Lin BM, et al. An updated prostate cancer staging nomogram (Partin Tables) based on cases from 2006 to 2011. *BJU Int* 2013;**111**(1):22–9.
23. Blute ML, Bergstralh EJ, Partin AW, et al. Validation of Partin Tables for predicting pathological stage of clinically localized prostate cancer. *J Urol* 2000;**164**:1591–5.
24. Karakiewicz PI, Bhojani N, Capitanio U, et al. External validation of the updated Partin Tables in a cohort of North American men. *J Urol* 2008;**180**:898–902.
25. Bhojani N, Nalomon L, Capitanio U, et al. External validation of the updated Partin Tables in a cohort of French and Italian men. *Int J Radiat Oncol Biol Phys* 2009;**73**:347–52.
26. Bhojani N, Ahyai S, Graefen M, et al. Partin tables cannot accurately predict the pathological stage at radical prostatectomy. *Eur J Surg Oncol* 2009;**35**:123–8.
27. Baccala Jr A, Reuther AM, Bianco Jr FJ, et al. Complete resection of seminal vesicles at radical prostatectomy results in substantial long-term disease-free survival: multi-institutional study of 6740 patients. *Urology* 2007;**69**(3):536–40.
28. Guzzo TJ, Vira M, Wang Y, et al. Preoperative parameters, including percent positive biopsy, in predicting seminal vesicle involvement in patients with prostate cancer. *J Urol* 2006;**175**(2):518–21.
29. Cagiannos I, Karakiewicz P, Eastham JA, et al. A preoperative nomogram identifying decreased risk of positive pelvic lymph nodes in patients with prostate cancer. *J Urol* 2003;**170**:1798–803.
30. D'Amico AV, Renshaw AA, Arsenault L, et al. Clinical predictors of upgrading to Gleason grade 4 or 5 disease at radical prostatectomy: potential implications for patient selection for radiation and androgen suppression therapy. *Int J Radiat Oncol Biol Phys* 1999;**145**(4):841–6.
31. Kulkarni GS, Lockwood G, Evans A, et al. Clinical predictors of Gleason score upgrading: implications for patients considering watchful waiting, active surveillance or brachytherapy. *Cancer* 2007;**109**(12):2432–8.
32. Capitanio U, Karakiewicz PI, Valiquette L, et al. Biopsy core number represents one of foremost predictors of clinically significant gleason sum upgrading in patients with low-risk prostate cancer. *Urology* 2009;**73**(5):1087–91.
33. Han M, Partin AW, Pound CR, et al. Long-term biochemical disease-free survival and cancer-specific survival following anatomic radical retropubic prostatectomy. The 15-year Johns Hopkins experience. *Urol Clin North Am* 2001;**28**(3):555–65.
34. Ward JF, Moul JW. Rising prostate-specific antigen after primary prostate cancer therapy. *Nat Clin Pract Urol* 2005;**2**(4):174–82.
35. D'Amico AV, Whittington R, Malkowicz SB, et al. Biochemical outcome after radical prostatectomy, external beam radiation therapy, or interstitial radiation therapy for clinically localized prostate cancer. *JAMA* 1998;**280**(11):969–74.
36. Hernandez DJ, Nielsen ME, Han M, et al. Contemporary evaluation of the D'Amico risk classification of prostate cancer. *Urology* 2007;**70**(5):931–5.
37. Boorjian SA, Karnes RJ, Rangel LJ, et al. Mayo Clinic validation of the D'amico risk group classification for predicting survival following radical prostatectomy. *J Urol* 2008;**179**(4):1354–60.
38. Cooperberg MR, Pasta DJ, Elkin EP, et al. The University of California, San Francisco Cancer of the Prostate Risk Assessment score: a straightforward and reliable preoperative predictor of disease recurrence after radical prostatectomy. *J Urol* 2005;**173**:1938–42.
39. Cooperberg MR, Broering JM, Carroll PR. Risk assessment for prostate cancer metastasis and mortality at the time of diagnosis. *J Natl Cancer Inst* 2009;**101**(12):878–87.
40. Meurs P, Galvin R, Fanning DM, et al. Prognostic value of the CAPRA clinical prediction rule: a systematic review and meta-analysis. *BJU Int* 2013;**111**(3):427–36.
41. Kattan MW, Eastham JA, Stapleton AM, et al. A preoperative nomogram for disease recurrence following radical prostatectomy for prostate cancer. *J Natl Cancer Inst* 1998;**90**(10):766–71.
42. Graefen M, Karakiewicz PI, Cagiannos I, et al. International validation of a preoperative nomogram for prostate cancer recurrence after radical prostatectomy. *J Clin Oncol* 2002;**20**(15):3206–12.
43. Bianco Jr FJ, Kattan MW, Scardino PT, et al. Radical prostatectomy nomograms in black American men: accuracy and applicability. *J Urol* 2003;**170**(1):73–6.
44. Greene KL, Meng MV, Elkin EP, et al. Validation of the Kattan preoperative nomogram for prostate cancer recurrence using a community based cohort: results from cancer of the prostate strategic urological research endeavor (capsure). *J Urol* 2004;**171**:2255–9.
45. Stephenson AJ, Scardino PT, Eastham JA, et al. Preoperative nomogram predicting the 10-year probability of prostate cancer recurrence after radical prostatectomy. *J Natl Cancer Inst* 2006;**98**:715–7.
46. Lughezzani G, Budaus L, Isbarn H, et al. Head-to-head comparison of the three most commonly used preoperative models for prediction of biochemical recurrence after radical prostatectomy. *Eur Urol* 2010;**57**(4):562–8.
47. D'Amico AV, Moul JW, Carroll PR, et al. Cancer-specific mortality after surgery or radiation for patients with clinically localized prostate cancer managed during the prostate-specific antigen era. *J Clin Oncol* 2003;**21**(11):2163–72.
48. Cooperberg MR, Broering JM, Carroll PR. Risk assessment for prostate cancer metastasis and mortality at the time of diagnosis. *J Natl Cancer Inst* 2009;**101**(12):878–87.
49. Stephenson AJ, Kattan MW, Eastham JA, et al. Prostate cancer-specific mortality after-radical prostatectomy for patients treated in the prostate-specific antigen era. *J Clin Oncol* 2009;**27**(26):4300–5.
50. Abd-Alazeez M, Ahmed HU, Arya M, et al. Can multiparametric magnetic resonance imaging predict upgrading of transrectal ultrasound biopsy results at more definitive histology? *Urol Oncol* 2014;**32**(6):741–7.
51. Siddiqui MM, Rais-Bahrami S, Truong H, et al. Magnetic resonance imaging/ultrasound-fusion biopsy significantly upgrades prostate cancer versus systematic 12-core transrectal ultrasound biopsy. *Eur Urol* 2013;**64**(5):713–9.

52. Stamatakis L, Siddiqui MM, Nix JW, et al. Accuracy of multiparametric magnetic resonance imaging in confirming eligibility for active surveillance for men with prostate cancer. *Cancer* 2013;**119**(18):3359–66.
53. Shukla-Dave A, Hricak H, Akin O, et al. Preoperative nomograms incorporating magnetic resonance imaging and spectroscopy for prediction of insignificant prostate cancer. *BJU Int* 2012;**109**(9):1315–22.
54. Park JJ, Kim CK, Park SY, et al. Prostate cancer: role of pretreatment multiparametric 3-T MRI in predicting biochemical recurrence after radical prostatectomy. *AJR Am J Roentgenol* 2014;**202**(5):459–65.

CHAPTER

# 27

# Is Surgery Still Necessary for Prostate Cancer?

*Ahmed A. Hussein, MD**,***,
*Matthew R. Cooperberg, MD, MPH**,†

*Department of Urology and Helen Diller Family Comprehensive Cancer Center, University of California, San Francisco, CA, USA
**Department of Urology, Cairo University, Oula, Giza, Egypt
†Department of Epidemiology and Biostatistics, University of California, San Francisco, CA, USA

## STAGE MIGRATION: INCREASED INCIDENCE OF LOCALIZED PROSTATE CANCER

In the prostate-specific antigen (PSA)-based early detection era, approximately 81% of men with prostate cancer (PCa) have disease considered confined to the prostate gland (clinically localized PCa), and only 4% present with distant disease.[1] Survival has improved and disease-specific mortality has declined with overall 5-year survival rates now at 99%.[2] With better understanding of the biological behavior of PCa, most clinically localized tumors are now understood to be indolent, with a relatively benign course even without treatment, whereas a minority is more aggressive and potentially lethal.[3] This subset still account for more cancer deaths among US men than any malignancy except lung cancer.[1]

Standard management options include observation/conservative strategies, including watchful waiting (WW) and active surveillance (AS), while active treatment options include radical prostatectomy (RP) (open, laparoscopic (LRP) or robotic assisted (RARP) approaches), and radiation therapies (RT) (various forms of external beam radiation therapy (EBRT) and brachytherapy (BT)). Primary androgen deprivation therapy (PADT), while not a standard option for localized PCa, is still adopted by some urologists for patients not fit for or refusing RP and RT.[4,5] Other less common treatments include cryotherapy, and outside the United States, high-intensity focused ultrasound (HIFU) therapy.

Treatment should be tailored to each patient, taking in consideration the patient's overall health, life expectancy, and the disease risk, including the PSA, tumor extent, and Gleason grade. Other factors that should further bear on the decision include the clinician's skill and experience, and the patients' preference to trade off potential benefits, side-effects, and complications (such as urinary leakage, sexual, and bowel dysfunction).[6,7] In an era of tightly constrained healthcare resources, costs must also be considered at the societal level, if not necessarily in the setting of individual patient discussions. In this chapter, we sought to discuss the role of surgery, and its comparative effectiveness with other strategies in the treatment of PCa.

## COMPARATIVE EFFECTIVENESS RESEARCH (CER) AMONG DIFFERENT APPROACHES

Comparative effectiveness research (CER) and related cost effectiveness studies are critical to help guide patients and physicians regarding the most appropriate modality and timing for each individual case.[8] In fact, the relative effectiveness of the management strategies for localized PCa was considered among the highest initial national priorities for CER by the Institute of Medicine (IOM) in 2009.[9] Several related factors must be addressed when comparing different treatments: oncological efficacy, measured in terms of recurrence, cancer specific, and/or overall survival rates; short- and long-term side

Prostate Cancer. http://dx.doi.org/10.1016/B978-0-12-800077-9.00027-X

effects and their effects on health-related quality of life (HRQOL); and the cost of the initial and subsequent treatments and their sequelae.

## CHALLENGES TO CER AMONG TREATMENTS FOR LOCALIZED PCa

CER among treatments for localized PCa is not an easy task given several factors, including a notable paucity of randomized controlled trials (RCTs) comparing the outcomes of different treatments. However, recently the use of observational cohorts from large disease registries accruing long-term follow-up has allowed for the comparison of different therapeutic strategies, in some cases based on data on "real-world" outcomes.[8] The long natural history of PCa creates further challenges in comparing long-term outcomes, and increases the cost and complexity of adequate follow-up.[10] Addressing the effects of different treatments on HRQOL – reported by the patients – is of utmost importance; it should be emphasized that physician-reported HRQOL assessments generally underestimate the adverse impacts of treatments when compared to patients' reports,[11,12] and HRQOL outcomes derived from administrative (billing/coding) data are entirely unreliable.

The various treatments have notably different effects on PSA and PSA kinetics, which severely complicate comparison among different treatment modalities for localized PCa based on biochemical recurrence rates.[13] The definition of biochemical recurrence varies from one treatment to another, and measured within-modality disease progression rates may vary up to 35% depending on the definition used.[13–15] For RP, the most widely accepted definition for biochemical recurrence is that of the American Urological Association (AUA), which is defined as serum PSA ≥0.2 ng/mL following surgery, with confirmation by a repeat test.[7] For patients who undergo RT, the Phoenix consensus criteria were adopted by the American Society for Radiation Oncology (ASTRO), specifying that a PSA rise of ≥2 ng/mL above the nadir PSA is considered as biochemical failure after RT, regardless of whether or not a patient receives ADT.[16]

These definitions diverge in their express intent: the RP definition is intended to identify persistent or recurrent disease early, whereas the Phoenix definition is intended to predict long-term clinical outcomes.[16] For this reason – and given fundamental differences in the time course of tumor ablation via surgery versus radiation – biochemical recurrence is wholly invalid as a means of comparing the oncological efficacy of surgery versus radiation therapy. A useful illustrative example was provided by Nielsen et al., who found that compared to the standard surgical definition, applying the Phoenix definition to cohort of RP patients caused overestimation of the biochemical recurrence-free survival, even after stratifying patients into the standard prognostic risk groups. Among those patients recurring, the apparent time to biochemical recurrence would be shifted forward *more than 5 years*, from 2.8 years to 7.9 years.[17] Unfortunately, papers continue to be published claiming oncological superiority of radiation over surgery based on biochemical outcomes.[18] Without exception, such analyses will be biased artifactually and irrecoverably in favor of radiation and against surgery. Thus, while biochemical recurrence might be used to compare external-beam with interstitial radiation, for example, *only* clinical endpoints (metastasis, cancer-specific mortality, overall mortality) can be used legitimately for comparison studies between surgery and radiation.

## RISK STRATIFICATION OF LOCALIZED PCa

Localized PCa exhibits extremely heterogeneous biology and aggressiveness, and therefore must be risk-stratified according to the risk of disease progression. Basic risk stratification uses the clinical stage of disease, baseline PSA, and the Gleason score to divide patients into prognostic categories; some instruments further incorporate measures of the extent of biopsy involvement. The AUA/D'Amico classification, the most widely used risk stratification, stratifies patients into low-risk patients (PSA ≤10 ng/mL, Gleason score ≤6, and stage T1-2a), intermediate-risk (PSA 10–20 ng/mL, Gleason score 7, or stage T2b), and high-risk patients (PSA >20, Gleason score ≥8, or stage T2c / T3a).[19] This venerable system is substantially limited for contemporary application by its lack of distinction between Gleason 3 + 4 and 4 + 3, its overemphasis of T-stage, and most importantly its inability to reflect the impact of multiple adverse risk factors.[20]

Many multivariable instruments, represented as nomograms, risk tables, or scoring systems, are available for PCa risk assessment; a minority of these have been well-validated. The Kattan preoperative nomogram predicts the 5-year probability of treatment failure among men with clinically localized PCa treated with RP,[21] and has been validated in multiple studies.[22–25] Stephenson et al. extended the predictions to 10 years and addressed the limitations of the Kattan nomogram by incorporating the systematic biopsy results.[26] Other nomograms were developed for estimating the probability of successful treatment after brachytherapy and EBRT for clinically localized PCa.[27–29] The University of California, San Francisco Cancer of the Prostate Risk Assessment score (UCSF-CAPRA) is likewise based on pretreatment data and predicts pathologic status, disease recurrence, and mortality after RP, RT, and primary ADT.[30,31] The CAPRA score has been successfully validated in other disease registries in multiple countries, and in some analyses outperforms competing instruments.[32–34]

## COMPARING ONCOLOGICAL OUTCOMES: RCTs AND OBSERVATIONAL STUDIES

While there are limitations to the observational data obtained from registries, in the setting of PCa, observational studies obtained from well-designed large-volume patient registries may provide valuable knowledge regarding oncologic and HRQOL outcomes among treatments, complementing clinical trials. RCTs in the setting of localized prostate are difficult due to the long natural history of the disease, high costs associated with long follow-up, and challenges to accrual.[35] Only few RCTs have been reported to date comparing different outcomes of various treatments.

Two important RCTs have compared surgery with observation. The Prostate Cancer Intervention Versus Observation Trial (PIVOT) showed that among men with clinically localized PCa, RP did not cause any significant reduction in mortality for low-risk disease when compared with observation, through a median of 10 years of follow-up. Further analyses suggested no significant difference according to age, race, comorbidities, or performance status. On the other hand, PIVOT also clearly showed that RP reduced mortality by 60% among men with high-risk disease, despite being notably underpowered in this subgroup.[36]

Extended follow-up of the Scandinavian Prostate Cancer Group Study 4 (SPCG-4) – which likewise randomized men between surgery and watchful waiting, but in the pre-PSA screening era – favored RP over observation, where RP showed an absolute 13% lower overall mortality (RR of death 0.71, (95% CI, 0.59–0.86; $P < 0.001$)); 11% lower PCa specific mortality (RR of death 0.56 (95% CI, 0.41–0.77; $P = 0.001$)); and 12% lower risk of metastasis (RR of distant metastases 0.57 (95% CI, 0.44–0.75; $P < 0.001$)). On further analysis, maximum benefit was demonstrated in men younger than 65 years and those with intermediate risk tumors, who showed significant reduction in all three end points. On the other hand, older men showed significant reduction in the risk of metastases but not mortality.[37] However, no contemporary trial has been reported comparing RP and any type of RT. The Prostate testing for cancer and Treatment (ProtecT) study is the only ongoing RCT including more than one active treatment arm (RP, EBRT, and observation),[38] with initial results expected in 2016. The other contemporary attempted trial, the Surgical Prostatectomy Versus Interstitial Radiation Intervention Trial (SPIRIT) was able to enroll only 56 patients from 31 centers over 2 years to compare RP versus BT and was prematurely terminated.[35]

Given the lack of RCT data comparing surgery and radiation, well-designed prospective registries and other robust cohorts provide valuable data to compare different treatments using metastasis, cancer-specific mortality, and overall mortality outcomes and to assess the effectiveness of care and patient satisfaction. Disease registries provide high data quality with numerous variables to obtain valid conclusions. Various methods may be used to minimize selection bias and the effects of possible measured and unmeasured confounders inherent to observational research, including risk adjustment (through regression or analysis of variance methods) and/or the use of propensity analysis or instrumental variables.[39] Additionally, population-based studies allow generalizability and reflect actual practice patterns. They also allow analysis of a large number of patients, minimizing the effect of unobserved differences, allowing for sufficient sample sizes, adequate power analyses, and significant statistical power – though in some cases at the cost of clinical detail. In any retrospective analysis, it is critical to ensure uniform data collection, via the use of well-validated objective assessment tools, allowing proper measurement of variables and confounders. It is also essential to maintain complete follow-up with as little loss to follow-up as possible, continued long enough to describe the study objectives[40] – which can be well over a decade at least in the case of mortality endpoints for localized prostate cancer.

A study from the community-based Cancer of the Prostate Strategic Urologic Research Endeavor (CaPSURE), in which patients were categorized according to Kattan and CAPRA scoring systems, found that men with localized PCa treated with RP have better cancer-specific and overall survival when compared to those treated with RT and PADT, and that either local treatment was superior to PADT alone. The greatest incremental benefits for surgery accrued to men with the highest-risk tumors on average.[41] Another study at two academic centers showed that treatment with RP was associated with a threefold reduced risk of metastasis at 8 years when compared with EBRT, again especially for high-risk disease. In this study, patients were stratified according to Kattan preoperative nomogram, and all radiation was given to at least 81 Gy.[42]

Similar results were obtained from Prostate Cancer Outcomes Study (PCOS) study after 15 years of follow-up, in which RP was associated with lower overall and cancer-specific mortality than EBRT.[43] Another study compared RP with EBRT, and notably included BT as well. The adjusted 10-year overall survival was 88.9, 82.6, and 81.7%, respectively, and the adjusted 10-year PCa-specific mortality was 1.8, 2.9, and 2.3%, respectively. RT was associated with decreased overall survival and increased PCa-specific mortality compared to RP.[44] A large Swedish observational study involving more than

**TABLE 27.1** Outcomes of Observational Studies Comparing the Outcomes of Radical Prostatectomy and External Beam Radiation Therapy for Localized Prostate Cancer

| Study/year | Average follow-up (years) | RP patients (*n*) | EBRT patients (*n*) | Outcome (HR, 95% CI) |
|---|---|---|---|---|
| Tewari et al.[47] | 4.6 | 119 | 137 | CSM HR 0.5 (0.3–1.0) favoring RP over EBRT |
| Albertsen et al.[48] | 13.3 | 802 | 702 | CSM RR 2.5 (1.7–3.5) favoring RP over EBRT |
| Cooperberg et al.[41] | 4.2 | 5,543 | 1,294 | CSM HR 2.2 (1.5–3.2) favoring RP over EBRT |
| Zelefsky et al.[42] | 5 | 1,318 | 1,062 | CSM HR 2.9 (1.3–7.7) favoring RP over EBRT |
| Stattin et al.[49] | 8.2 | 3,399 | 1,429 | CSM RR 1.4 (0.9–2.0) favoring RP over EBRT |
| Boorjian et al.[50] | 10.2 | 1,238 | 609 | CSM HR 1.1 (0.7–1.9) NS for RP versus EBRT |
| Kibel et al.[44] | 5.5 | 6,485 | 2,264 | CSM HR 1.5 (1.0–2.3) favoring RP over EBRT |
| Hoffman et al.[43] | 15 | 1,164 | 491 | CSM HR 2.9 (2.0–3.8) favoring RP over EBRT |
| Sun et al.[46] | – | 33,613 | 15,532 | CSM HR 2.5 (1.5–4.2) favoring RP over EBRT |
| Sooriakumaran et al.[45] | 15 | 21,533 | 12,982 | CSM HR 1.8 (1.5–2.1) favoring RP over EBRT |

*n*, number; RP, radical prostatectomy; EBRT, external beam radiation; CSM, cancer specific mortality; RR, relative risk; HR, hazards ratio; CI, confidence interval.

35,000 men with 15 years follow-up likewise found that RP was associated with better survival than RT, especially for younger men and those with less comorbidities, with intermediate or high-risk localized PCa.[45] Another study added that the survival benefit for RP is reserved for patients with more than 10 years life expectancy, regardless of the disease stage.[46] Table 27.1 summarizes the observational studies comparing RP with EBRT.

While there remain no published RCTs comparing the different techniques for RP, the open approach remains the standard for the surgical management of localized PCa by which newer technologies must be evaluated. Minimally invasive techniques reduce blood loss and in some cases hospital stays.[51] RARP in addition offers a three-dimensional view of the operative field and wristed laparoscopic instruments whose ergonomics closely mimic human wrist and hand movements, notably facilitating laparoscopic dissection.[52] Establishing the benefits to patients of RARP versus open surgery remains somewhat controversial. What seems clear from the emerging literature is that RARP is associated with fewer short-term complications than open surgery,[53] and in experienced hands with equivalent oncological outcomes.[53,54] Some studies have shown better urinary and/or sexual quality of life following RARP,[55,56] but achieving good outcomes in these domains requires high surgical experience and expertise, regardless of the surgical approach.

## COMPARING HRQOL

The development and publication of large studies incorporating patient-reported outcomes using validated questionnaires has allowed for substantial improvements in HRQOL research, allowing for high-quality comparisons between different treatments. All treatment options may result in side-effects (primarily urinary, bowel, and sexual), although the severity and frequency may vary from one modality to another and across practice settings.

An updated CaPSURE study compared HRQOL outcomes among RP, EBRT, WW, and PADT for the first 2 years following treatment. RP patients had low immediate postoperative scores particularly for urinary continence, but improvement in all domains at 1 year after surgery and sexual function continued to improve at 2 years. On the other hand, patients treated with EBRT, WW, and PADT had relatively stable scores except for sexual function, which decreased in all treatment groups with time, and differences between modalities attenuate with time up to 10 years after treatment.[57] Data from PCOS showed that men who elected for RP were more likely to have incontinence and erectile dysfunction than patients who received RT.[58] However, at 15 years, both showed declined outcomes in all functional domains. RT patients reported in addition more bowel dysfunction.[59] The CaPSURE and PCOS studies used urinary assessments that were sensitive to incontinence, but relatively insensitive to irritative symptoms more typically associated with radiation.

The Prostate Cancer Outcomes and Satisfaction with Treatment Quality Assessment study (PROST-QA) further described the HRQOL impact pattern for each treatment modality. RP patients experienced the greatest initial declines in sexual and urinary functions at 2 months with continued improvement at 2 years, with no change in their bowel symptoms. Patients who had RT experienced a gradual decline in sexual function over 2 years. Both forms of radiation were associated with significant declines in urinary irritation functional scores followed by recovery at 6 and 12 months. Moreover, significant decline in bowel function was observed at 2 years. Treatment-related to both that was associated with sexual function peaked at 1 year for all patients.[60]

The Spanish Multicentric Study of Clinically Localized Prostate Cancer compared different treatments for localized PCa after 3 years of follow-up, with similar findings. Men treated with RP suffered urinary incontinence and sexual dysfunction more than those treated with either forms of RT, but improved urinary irritative and obstructive scores when compared to baseline. The difference in sexual dysfunction scores between RP and BT decreased over time. EBRT patients showed better urinary scores than BT patients, but they tend to have similar scores over time. Adverse bowel events were only encountered in men who received EBRT. These differences between the three treatment groups remained significant even at 3 years.[61]

Another CaPSURE study that addressed the possible effects of multimodal therapy on HRQOL showed that combining ADT with RP or RT resulted in temporary loss of sexual function, which improved at 9 months after initiation of the treatment. On the other hand, when EBRT was combined with BT, decline in urinary functions and bother continued for 21 months. These findings may inform better patient counseling prior to initiation of a multimodal therapeutic approach.[62] Ultimately, there is relatively little controversy regarding the effects of treatments: surgery causes more incontinence, and radiation causes more urinary and bowel irritation symptoms. Both can affect sexual function, though the impact of surgery is typically felt earlier.

## COMPARING COSTS: WHO PAYS?

The economic burden of PCa management on the healthcare system is growing.[63] This may be attributed to many factors, including an aging population, increased detection and treatment of localized disease, and the advent of newer and more costly treatments. More studies are needed to elucidate the benefit of the newer treatment modalities (such as minimally invasive RP and IMRT) and to rationalize the associated increase in the cost. A recent review analyzed some of the reports addressing this subject, finding that direct costs varied within and between treatments among different studies. This variation may be a result of different patient characteristics, health care systems, years, and the method of cost evaluation among different countries.[64]

One study found that, for low-risk disease, AS can result in a net savings of $12,194 per patient at 5 years, with the main costs being those associated with annual biopsy.[65] Another study compared utilization of new technologies between 2002 and 2005 in the United States. Among patients who underwent RP, the use of minimally invasive RP increased from 1.5% in 2002 to 28.7% in 2005 while for RT, the use of IMRT increased from 29% to 82%. This resulted in substantial increase in expenditure, accounting for $282 million for IMRT, $59 million for BT plus IMRT, and $4 million for minimally invasive RP, compared to less costly alternatives for men diagnosed in 2005.[66] These discrepant figures reflect the oddities of the US reimbursement system that compensates highly for newer radiation technologies but not for surgical technologies.

Another recent study, using a complex Markov model including primary, secondary, and salvage procedures along with all their sequelae, HRQOL impacts, and costs, compared the cost effectiveness of different primary treatment modalities for localized PCa for each disease-risk group. For all risk groups, surgical options were equally or more effective as – and less expensive than – radiation options. Furthermore, all surgical modalities had similar outcomes and costs across all risk strata (approximately $20,000, $28,500, and $35,500, respectively, for low-, intermediate-, and high-risk patients).

Regarding radiation therapies, BT was the most effective radiation modality among low-risk patients, while a combination of EBRT + BT was the most effective for intermediate- and high-risk patients. In contrast to surgical options, costs for radiation modalities differed within risk strata. BT was the least expensive among all risk strata ($25,067, $32,553, and 43,952 for low-, intermediate-, and high-risk patients, respectively). IMRT was considered a dominated strategy, meaning it was not superior to RP or BT in low- and intermediate-risk men, but was more expensive.[67] It follows clearly that proton beam therapy is never cost effective for any patient with PCa based on the current costs and the state of evidence.[66,68]

## HIGH-RISK DISEASE AND SURGERY AS PART OF A MULTIMODAL APPROACH

Despite the widespread use of PSA screening, 15–26% of patients with PCa will present with high-risk disease.[69,70] There is no consensus yet about the optimal management of high-risk disease. RP in this setting was once believed to be associated with increased risk of PSA failure, systemic progression, and worse outcomes.[71–73] However, the current evidence shows that RP in select men may have an excellent oncological prognosis.[74,75] Previous studies showed that more than one-third of men diagnosed with presumed high-risk disease experience downgrading and/or downstaging at RP.[76,77] These men were shown to have survival outcomes that were comparable to men with lower-risk disease.[77,78] Therefore,

RP in this context remains a valuable option, maintaining the opportunity for cure while avoiding adjuvant treatment. Approximately 20–50% of men with high-risk disease will be cured with RP alone.[70,79]

In the context of high-risk disease, several studies have shown that RP may be superior to RT or ADT. One study (see Table 27.1) showed that RP was associated with a nearly 8–10% lower risk of progression to metastasis and a lower cancer-specific mortality in comparison to RT.[42] Another study added that men who had RT were 3.5 times more likely to receive adjuvant ADT when compared to men who had RP.[78] Additionally, RP compared with RT and ADT provides definitive pathological staging of the disease. Postoperative PSA provides an indicator of control and provides a valuable tool for early recognition of recurrent disease, and prompt salvage RT if necessary.[80] Adjuvant or salvage RT cures up to half of recurrences without significant morbidity.[81]

The role of adjuvant therapy in comparison with salvage therapy is yet to be defined. Adjuvant ADT is most relevant in the setting of positive lymph nodes, where it improved PSA recurrence, cancer-specific and overall survival.[82] Another study showed that adjuvant ADT was associated with significant improvements in PSA and local recurrence, but not in the rate of distant metastasis or cancer-specific survival.[83] Similarly, adjuvant RT remains controversial. The AUA/ASTRO joint guidelines included that adjuvant RT should be discussed with patients with adverse pathologic findings at RP (e.g., pT3, positive surgical margins), and salvage RT to patients with PSA or local recurrence. The decision should be made by the patient and multidisciplinary treatment team, considering the risk–benefit ratio for each patient.[84] RCTs of adjuvant RT in high-risk patients with adverse pathology showed improvements in PSA recurrence-free survival and overall survival compared with delayed treatment, but trials of adjuvant versus *early* salvage radiation therapy are still ongoing.[85–88]

In the setting of positive nodal disease, RP may decrease the risk of overall mortality in well-selected men when combined with adjuvant or salvage treatment and showed better outcomes than those treated with RT alone.[89,90] For distant metastatic prostate cancer, however, it is still unclear whether cytoreductive prostatectomy is of benefit as other malignancies or not. Men with metastatic PCa who had RP or RT had better survival and improved response to systemic therapy, although further evidence is warranted.[91]

## SALVAGE RP

Patients who experience local recurrence after RT are at a higher risk of metastasis and cancer-specific mortality.[92–94] Currently, salvage RP represents the best-established curative treatment option in this setting, provided it is done early in the course of the disease recurrence. Candidates for local salvage therapy should be patients with a life expectancy of 10 years or more, have a cancer that was initially and is now potentially curable with local therapy, and have no evidence of severe radiation cystitis or proctitis. In addition, lymph nodes should be negative if a pelvic lymph node dissection was done prior to radiotherapy.[95]

The treatment should be initiated as early as possible; for patients whose preoperative PSA level was 10 ng/mL or less, two-thirds had organ-confined disease and up to 70% progression-free survival rates,[95–98] this increased to 86% in patients whose serum PSA levels are less than 4 ng/mL. Conversely, the 5-year progression-free survival rates dropped to below 30% for patients with PSA values higher than 10 ng/mL.[95] However, the incidence of complications, as rectal injury, anastomotic stricture, urinary incontinence, and erectile dysfunction is higher.[95] This may be attributed to extensive fibrosis induced by irradiation resulting in obliteration of the surgical planes. The higher incontinence rates may be a result of radiation-induced sphincteric dysfunction. Anastomotic strictures are commonly resistant and require repeated interventions.[95]

Salvage RP was less utilized due to poor outcomes and high morbidity reported in the earlier studies. However, in most of these studies, patients were referred for RP long after PSA recurrence, and many of them had advanced disease prior to irradiation. In more contemporary series, the associated morbidity has decreased substantially due to improvements in radiation and surgical techniques, better patient selection, and early referral of patients who fail radiation therapy for salvage RP.[95] There is increasing evidence about the safety and efficacy of robotic salvage RP. Salvage RARP may provide a safe alternative to open salvage, while offering the advantages of enhanced visualization, lower blood loss and complication rates, and shorter hospital stays.[99]

## CONCLUSIONS

The PSA era has witnessed a stage migration, with more detection of clinically localized PCa, and as yet there remains no consensus about the most cost-beneficial approach to its management. Although RCTs are considered the gold standard for evidence-based practices, there are few comparing active treatment modalities. Multiple, well-conducted retrospective studies from large disease registries suggest better survival for RP compared to EBRT, especially for higher-risk disease, and RP should be considered, particularly in the context of multimodal treatment strategies, for men with high-risk disease. Minimally invasive approaches, especially RARP, have gained much popularity among urologists and patients, providing similar oncological outcomes but less pain, blood loss, and shorter operating times and hospital stay than the conventional open approach. HRQOL does not clearly favor any one modality over the other; rather patients should be carefully counseled as to the benefits and adverse effects of each alternative. In the United States, RT, especially IMRT and proton-beam therapy, are substantially more expensive than surgery. More studies are needed to evaluate the cumulative costs and benefits of the newer approaches to treatment of localized PCa, and whether increased expenditures are warranted. More studies are still needed to elucidate the cost effectiveness of minimally invasive surgical techniques. Salvage RP may be useful for well-selected patients for the management of local recurrence following RT. The most important principle is that treatments need to be targeted to the patients most likely to benefit, and the way forward likely will involve more use of AS for low-risk disease and multimodal treatment for high-risk disease.

## References

1. Siegel R, Ma J, Zou Z, et al. Cancer statistics. *CA Cancer J Clin* 2014;**64**(1):9–29.
2. National Cancer Institute. SEER Stat Fact Sheets: Prostate Cancer. Available from: http://seer.cancer.gov/statfacts/html/prost.html.
3. Ganz PA, Barry JM, Burke W, et al. NIH State-of-the-Science Conference Statement: role of active surveillance in the management of men with localized prostate cancer. *NIH Consens State Sci Statements* 2011;**28**(1):1–27.
4. Mottet N, Bellmunt J, Bolla M, et al. EAU guidelines on prostate cancer. Part II: treatment of advanced, relapsing, and castration-resistant prostate cancer. *Eur Urol* 2011;**59**(4):572–83.
5. Mohler J, Bahnson RR, Boston B, et al. NCCN clinical practice guidelines in oncology: prostate cancer. *J Natl Compr Canc Netw* 2010;**8**(2):162–200.
6. *Guideline for the Management of Clinically Localized Prostate Cancer.* Available from: http://www.auanet.org/education/guidelines/prostate-cancer.cfm; 2007 [accessed 30.04.2014].
7. Cookson MS, Aus G, Burnett AL, et al. Variation in the definition of biochemical recurrence in patients treated for localized prostate cancer: the American Urological Association Prostate Guidelines for Localized Prostate Cancer Update Panel report and recommendations for a standard in the reporting of surgical outcomes. *J Urol* 2007;**177**(2):540–5.
8. Hoffman RM, Penson DF, Zietman AL, et al. Comparative effectiveness research in localized prostate cancer treatment. *J Comp Eff Res* 2013;**2**(6):583–93.
9. *100 Initial priority topics for comparative effectiveness research.* http://www.iom.edu/~/media/Files/Report%20Files/2009/ComparativeEffectivenessResearchPriorities/Stand%20Alone%20List%20of%20100%20CER%20Priorities%20-%20for%20web.ashx; 2014 [accessed on 7.02.2014].
10. Schneider Chafen J, Newberry S, Maglione M, et al. Comparative effectiveness of therapies for clinically localized prostate cancer: surveillance report. Rockville, MD: Agency for Healthcare Research and Quality; 2012. Available from: www.effectivehealthcare.ahrq.gov/ehc/products/9/80/TX-for-Localized-Prostate-Cancer_SurveillanceAssesment_20120614.pdf.
11. Litwin MS, Lubeck DP, Henning JM, et al. Differences in urologist and patient assessments of health related quality of life in men with prostate cancer: results of the CaPSURE database. *J Urol* 1998;**159**(6):1988–92.
12. Sonn GA, Sadetsky N, Presti JC, et al. Differing perceptions of quality of life in patients with prostate cancer and their doctors. *J Urol* 2013;**189**(1 Suppl):S59–65 discussion S65.
13. Gretzer MB, Trock BJ, Han M, et al. A critical analysis of the interpretation of biochemical failure in surgically treated patients using the American Society for Therapeutic Radiation and Oncology criteria. *J Urol* 2002;**168**(4 Pt 1):1419–22.
14. Amling CL, Bergstralh EJ, Blute ML, et al. Defining prostate specific antigen progression after radical prostatectomy: what is the most appropriate cut point? *J Urol* 2001;**165**(4):1146–51.
15. Kuban DA, Thames HD, Shipley WU. Defining recurrence after radiation for prostate cancer. *J Urol* 2005;**173**(6):1871–8.
16. Roach III M, Hanks G, Thames Jr H, et al. Defining biochemical failure following radiotherapy with or without hormonal therapy in men with clinically localized prostate cancer: recommendations of the RTOG-ASTRO Phoenix Consensus Conference. *Int J Radiat Oncol Biol Phys* 2006;**65**(4):965–74.
17. Nielsen ME, Makarov DV, Humphreys E, et al. Is it possible to compare PSA recurrence-free survival after surgery and radiotherapy using revised ASTRO criterion – "nadir + 2"? *Urology* 2008;**72**(2):389–93 discussion 394–5.
18. Grimm P, Billiet I, Bostwick D, et al. Comparative analysis of prostate-specific antigen free survival outcomes for patients with low, intermediate and high risk prostate cancer treatment by radical therapy. Results from the Prostate Cancer Results Study Group. *BJU Int* 2012;**109**(Suppl 1):22–9.
19. D'Amico AV, Whittington R, Malkowicz SB, et al. Biochemical outcome after radical prostatectomy, external beam radiation therapy, or interstitial radiation therapy for clinically localized prostate cancer. *JAMA* 1998;**280**(11):969–74.
20. Mitchell JA, Cooperberg MR, Elkin EP, et al. Ability of 2 pretreatment risk assessment methods to predict prostate cancer recurrence after radical prostatectomy: data from CaPSURE. *J Urol* 2005;**173**(4):1126–31.
21. Kattan MW, Eastham JA, Stapleton AM, et al. A preoperative nomogram for disease recurrence following radical prostatectomy for prostate cancer. *J Natl Cancer Inst* 1998;**90**(10):766–71.
22. Graefen M, Karakiewicz PI, Cagiannos I, et al. A validation of two preoperative nomograms predicting recurrence following radical prostatectomy in a cohort of European men. *Urol Oncol* 2002;**7**(4):141–6.
23. Graefen M, Karakiewicz PI, Cagiannos I, et al. International validation of a preoperative nomogram for prostate cancer recurrence after radical prostatectomy. *J Clin Oncol* 2002;**20**(15):3206–12.
24. Greene KL, Meng MV, Elkin EP, et al. Validation of the Kattan preoperative nomogram for prostate cancer recurrence using a community based cohort: results from cancer of the prostate strategic urological research endeavor (CaPSURE). *J Urol* 2004;**171**(6 Pt 1):2255–9.
25. Bianco Jr FJ, Wood Jr DP, Cher ML, et al. Ten-year survival after radical prostatectomy: specimen Gleason score is the predictor in organ-confined prostate cancer. *Clin Prostate Cancer* 2003;**1**(4):242–7.
26. Stephenson AJ, Scardino PT, Eastham JA, et al. Preoperative nomogram predicting the 10-year probability of prostate cancer recurrence after radical prostatectomy. *J Natl Cancer Inst* 2006;**98**(10):715–7.
27. Kattan MW, Potters L, Blasko JC, et al. Pretreatment nomogram for predicting freedom from recurrence after permanent prostate brachytherapy in prostate cancer. *Urology* 2001;**58**(3):393–9.
28. Potters L, Roach III M, Davis BJ, et al. Postoperative nomogram predicting the 9-year probability of prostate cancer recurrence after permanent prostate brachytherapy using radiation dose as a prognostic variable. *Int J Radiat Oncol Biol Phys* 2010;**76**(4):1061–5.

29. Kattan MW, Zelefsky MJ, Kupelian PA, et al. Pretreatment nomogram that predicts 5-year probability of metastasis following three-dimensional conformal radiation therapy for localized prostate cancer. *J Clin Oncol* 2003;**21**(24):4568–71.
30. Cooperberg MR, Pasta DJ, Elkin EP, et al. The University of California, San Francisco Cancer of the Prostate Risk Assessment score: a straightforward and reliable preoperative predictor of disease recurrence after radical prostatectomy. *J Urol* 2005;**173**(6):1938–42.
31. Cooperberg MR, Broering JM, Carroll PR. Risk assessment for prostate cancer metastasis and mortality at the time of diagnosis. *J Natl Cancer Inst* 2009;**101**(12):878–87.
32. May M, Knoll N, Siegsmund M, et al. Validity of the CAPRA score to predict biochemical recurrence-free survival after radical prostatectomy. Results from a European multicenter survey of 1,296 patients. *J Urol* 2007;**178**(5):1957–62 discussion 1962.
33. Lughezzani G, Budaus L, Isbarn H, et al. Head-to-head comparison of the three most commonly used preoperative models for prediction of biochemical recurrence after radical prostatectomy. *Eur Urol* 2010;**57**(4):562–8.
34. Zhao KH, Hernandez DJ, Han M, et al. External validation of University of California, San Francisco, Cancer of the Prostate Risk Assessment score. *Urology* 2008;**72**(2):396–400.
35. Wallace K, Fleshner N, Jewett M, et al. Impact of a multi-disciplinary patient education session on accrual to a difficult clinical trial: the Toronto experience with the surgical prostatectomy versus interstitial radiation intervention trial. *J Clin Oncol* 2006;**24**(25):4158–62.
36. Wilt TJ, Brawer MK, Jones KM, et al. Radical prostatectomy versus observation for localized prostate cancer. *N Engl J Med* 2012;**367**(3):203–13.
37. Bill-Axelson A, Holmberg L, Garmo H, et al. Radical prostatectomy or watchful waiting in early prostate cancer. *N Engl J Med* 2014;**370**(10):932–42.
38. Donovan JL, Lane JA, Peters TJ, et al. Development of a complex intervention improved randomization and informed consent in a randomized controlled trial. *J Clin Epidemiol* 2009;**62**(1):29–36.
39. D'Agostino Jr RB, D'Agostino Sr RB. Estimating treatment effects using observational data. *JAMA* 2007;**297**(3):314–6.
40. Hannan EL. Randomized clinical trials and observational studies: guidelines for assessing respective strengths and limitations. *JACC Cardiovasc Interv* 2008;**1**(3):211–7.
41. Cooperberg MR, Vickers AJ, Broering JM, et al. Comparative risk-adjusted mortality outcomes after primary surgery, radiotherapy, or androgen-deprivation therapy for localized prostate cancer. *Cancer* 2010;**116**(22):5226–34.
42. Zelefsky MJ, Eastham JA, Cronin AM, et al. Metastasis after radical prostatectomy or external beam radiotherapy for patients with clinically localized prostate cancer: a comparison of clinical cohorts adjusted for case mix. *J Clin Oncol* 2010;**28**(9):1508–13.
43. Hoffman RM, Koyama T, Fan KH, et al. Mortality after radical prostatectomy or external beam radiotherapy for localized prostate cancer. *J Natl Cancer Inst* 2013;**105**(10):711–8.
44. Kibel AS, Ciezki JP, Klein EA, et al. Survival among men with clinically localized prostate cancer treated with radical prostatectomy or radiation therapy in the prostate specific antigen era. *J Urol* 2012;**187**(4):1259–65.
45. Sooriakumaran P, Nyberg T, Akre O, et al. Comparative effectiveness of radical prostatectomy and radiotherapy in prostate cancer: observational study of mortality outcomes. *BMJ* 2014;**348**:g1502.
46. Sun M, Sammon JD, Becker A, et al. Radical prostatectomy vs radiotherapy vs observation among older patients with clinically localized prostate cancer: a comparative effectiveness evaluation. *BJU Int* 2014;**113**(2):200–8.
47. Tewari A, Divine G, Chang P, et al. Long-term survival in men with high grade prostate cancer: a comparison between conservative treatment, radiation therapy and radical prostatectomy – a propensity scoring approach. *J Urol* 2007;**177**(3):911–5.
48. Albertsen PC, Hanley JA, Penson DF, et al. 13-year outcomes following treatment for clinically localized prostate cancer in a population based cohort. *J Urol* 2007;**177**(3):932–6.
49. Stattin P, Holmberg E, Johansson JE, et al. Outcomes in localized prostate cancer: National Prostate Cancer Register of Sweden follow-up study. *J Natl Cancer Inst* 2010;**102**(13):950–8.
50. Boorjian SA, Karnes RJ, Viterbo R, et al. Long-term survival after radical prostatectomy versus external-beam radiotherapy for patients with high-risk prostate cancer. *Cancer* 2011;**117**(13):2883–91.
51. Hu JC, Gu X, Lipsitz SR, et al. Comparative effectiveness of minimally invasive vs open radical prostatectomy. *JAMA* 2009;**302**(14):1557–64.
52. Finkelstein J, Eckersberger E, Sadri H, et al. Open versus laparoscopic versus robot-assisted laparoscopic prostatectomy: the European and US experience. *Rev Urol* 2010;**12**(1):35–43.
53. Tewari A, Sooriakumaran P, Bloch DA, et al. Positive surgical margin and perioperative complication rates of primary surgical treatments for prostate cancer: a systematic review and meta-analysis comparing retropubic, laparoscopic, and robotic prostatectomy. *Eur Urol* 2012;**62**(1):1–15.
54. Barocas DA, Salem S, Kordan Y, et al. Robotic assisted laparoscopic prostatectomy versus radical retropubic prostatectomy for clinically localized prostate cancer: comparison of short-term biochemical recurrence-free survival. *J Urol* 2010;**183**(3):990–6.
55. Novara G, Ficarra V, Mocellin S, et al. Systematic review and meta-analysis of studies reporting oncologic outcome after robot-assisted radical prostatectomy. *Eur Urol* 2012;**62**(3):382–404.
56. Novara G, Ficarra V, Rosen RC, et al. Systematic review and meta-analysis of perioperative outcomes and complications after robot-assisted radical prostatectomy. *Eur Urol* 2012;**62**(3):431–52.
57. Punnen S, Cowan JE, Chan JM, et al. Long-term health-related quality of life after primary treatment for localized prostate cancer: results from the CaPSURE registry. *Eur Urol* 2014;.
58. Potosky AL, Legler J, Albertsen PC, et al. Health outcomes after prostatectomy or radiotherapy for prostate cancer: results from the Prostate Cancer Outcomes Study. *J Natl Cancer Inst* 2000;**92**(19):1582–92.
59. Resnick MJ, Koyama T, Fan KH, et al. Long-term functional outcomes after treatment for localized prostate cancer. *N Engl J Med* 2013;**368**(5):436–45.
60. Sanda MG, Dunn RL, Michalski J, et al. Quality of life and satisfaction with outcome among prostate-cancer survivors. *N Engl J Med* 2008;**358**(12):1250–61.
61. Pardo Y, Guedea F, Aguilo F, et al. Quality-of-life impact of primary treatments for localized prostate cancer in patients without hormonal treatment. *J Clin Oncol* 2010;**28**(31):4687–96.
62. Wu AK, Cooperberg MR, Sadetsky N, et al. Health related quality of life in patients treated with multimodal therapy for prostate cancer. *J Urol* 2008;**180**(6):2415–22 discussion 22.
63. Roehrborn CG, Black LK. The economic burden of prostate cancer. *BJU Int* 2011;**108**(6):806–13.
64. Sanyal C, Aprikian AG, Chevalier S, et al. Direct cost for initial management of prostate cancer: a systematic review. *Curr Oncol* 2013;**20**(6):e522–31.
65. Dall'Era MA. The economics of active surveillance for prostate cancer. *Curr Opin Urol* 2013;**23**(3):278–82.
66. Nguyen PL, Gu X, Lipsitz SR, et al. Cost implications of the rapid adoption of newer technologies for treating prostate cancer. *J Clin Oncol* 2011;**29**(12):1517–24.
67. Cooperberg MR, Ramakrishna NR, Duff SB, et al. Primary treatments for clinically localised prostate cancer: a comprehensive lifetime cost-utility analysis. *BJU Int* 2013;**111**(3):437–50.

68. Konski A, Speier W, Hanlon A, et al. Is proton beam therapy cost effective in the treatment of adenocarcinoma of the prostate? *J Clin Oncol* 2007;**25**(24):3603–8.
69. Cooperberg MR, Lubeck DP, Mehta SS, CaPsure. et al. Time trends in clinical risk stratification for prostate cancer: implications for outcomes (data from CaPSURE). *J Urol* 2003;**170**(6 Pt 2):S21–5 discussion S26–7.
70. Cooperberg MR, Cowan J, Broering JM, et al. High-risk prostate cancer in the United States, 1990–2007. *World J Urol* 2008;**26**(3):211–8.
71. Chung BH. The role of radical prostatectomy in high-risk prostate cancer. *Prostate Int* 2013;**1**(3):95–101.
72. Grossfeld GD, Latini DM, Lubeck DP, et al. Predicting recurrence after radical prostatectomy for patients with high risk prostate cancer. *J Urol* 2003;**169**(1):157–63.
73. Walz J, Joniau S, Chun FK, et al. Pathological results and rates of treatment failure in high-risk prostate cancer patients after radical prostatectomy. *BJU Int* 2011;**107**(5):765–70.
74. Heidenreich A, Bellmunt J, Bolla M, et al. EAU guidelines on prostate cancer. Part 1: screening, diagnosis, and treatment of clinically localised disease. *Eur Urol* 2011;**59**(1):61–71.
75. Mohler JL, Armstrong AJ, Bahnson RR, et al. Prostate cancer, Version 3.2012: featured updates to the NCCN guidelines. *J Natl Compr Canc Netw* 2012;**10**(9):1081–7.
76. Reese AC, Sadetsky N, Carroll PR, et al. Inaccuracies in assignment of clinical stage for localized prostate cancer. *Cancer* 2011;**117**(2):283–9.
77. Boorjian SA, Karnes RJ, Crispen PL, et al. The impact of discordance between biopsy and pathological Gleason scores on survival after radical prostatectomy. *J Urol* 2009;**181**(1):95–104 discussion 104.
78. Meng MV, Elkin EP, Latini DM, et al. Treatment of patients with high risk localized prostate cancer: results from cancer of the prostate strategic urological research endeavor (CaPSURE). *J Urol* 2005;**173**(5):1557–61.
79. Freedland SJ, Partin AW, Humphreys EB, et al. Radical prostatectomy for clinical stage T3a disease. *Cancer* 2007;**109**(7):1273–8.
80. Kang HW, Lee JY, Kwon JK, et al. Current status of radical prostatectomy for high-risk prostate cancer. *Korean J Urol* 2014;**55**(10):629–35.
81. Trock BJ, Han M, Freedland SJ, et al. Prostate cancer-specific survival following salvage radiotherapy vs observation in men with biochemical recurrence after radical prostatectomy. *JAMA* 2008;**299**(23):2760–9.
82. Messing EM, Manola J, Yao J, et al. Immediate versus deferred androgen deprivation treatment in patients with node-positive prostate cancer after radical prostatectomy and pelvic lymphadenectomy. *Lancet Oncol* 2006;**7**(6):472–9.
83. Boorjian SA, Thompson RH, Siddiqui S, et al. Long-term outcome after radical prostatectomy for patients with lymph node positive prostate cancer in the prostate specific antigen era. *J Urol* 2007;**178**(3 Pt 1):864–70 discussion 870–1.
84. Freedland SJ, Rumble RB, Finelli A, et al. Adjuvant and salvage radiotherapy after prostatectomy: American Society of Clinical Oncology clinical practice guideline endorsement. *J Clin Oncol* 2014;**32**(34):3892–8.
85. Thompson IM, Tangen CM, Paradelo J, et al. Adjuvant radiotherapy for pathological T3N0M0 prostate cancer significantly reduces risk of metastases and improves survival: long-term followup of a randomized clinical trial. *J Urol* 2009;**181**(3):956–62.
86. Wiegel T, Bottke D, Steiner U, et al. Phase III postoperative adjuvant radiotherapy after radical prostatectomy compared with radical prostatectomy alone in pT3 prostate cancer with postoperative undetectable prostate-specific antigen: ARO 96-02/AUO AP 09/95. *J Clin Oncol* 2009;**27**(18):2924–30.
87. Thompson Jr IM, Tangen CM, Paradelo J, et al. Adjuvant radiotherapy for pathologically advanced prostate cancer: a randomized clinical trial. *JAMA* 2006;**296**(19):2329–35.
88. Bolla M, van Poppel H, Tombal B, et al. Postoperative radiotherapy after radical prostatectomy for high-risk prostate cancer: long-term results of a randomised controlled trial (EORTC trial 22911). *Lancet* 2012;**380**(9858):2018–27.
89. Engel J, Bastian PJ, Baur H, et al. Survival benefit of radical prostatectomy in lymph node-positive patients with prostate cancer. *Eur Urol* 2010;**57**(5):754–61.
90. Van Poppel H, Joniau S. An analysis of radical prostatectomy in advanced stage and high-grade prostate cancer. *Eur Urol* 2008;**53**(2):253–9.
91. Faiena I, Singer EA, Pumill C, et al. Cytoreductive prostatectomy: evidence in support of a new surgical paradigm (Review). *Int J Oncol* 2014;**45**(6):2193–8.
92. Lee WR, Hanks GE, Hanlon A. Increasing prostate-specific antigen profile following definitive radiation therapy for localized prostate cancer: clinical observations. *J Clin Oncol* 1997;**15**(1):230–8.
93. D'Amico AV, Cote K, Loffredo M, et al. Determinants of prostate cancer-specific survival after radiation therapy for patients with clinically localized prostate cancer. *J Clin Oncol* 2002;**20**(23):4567–73.
94. Pinover WH, Horwitz EM, Hanlon AL, et al. Validation of a treatment policy for patients with prostate specific antigen failure after three-dimensional conformal prostate radiation therapy. *Cancer* 2003;**97**(4):1127–33.
95. Stephenson AJ, Eastham JA. Role of salvage radical prostatectomy for recurrent prostate cancer after radiation therapy. *J Clin Oncol* 2005;**23**(32):8198–203.
96. Bianco Jr FJ, Scardino PT, Stephenson AJ, et al. Long-term oncologic results of salvage radical prostatectomy for locally recurrent prostate cancer after radiotherapy. *Int J Radiat Oncol Biol Phys* 2005;**62**(2):448–53.
97. Amling CL, Lerner SE, Martin SK, et al. Deoxyribonucleic acid ploidy and serum prostate specific antigen predict outcome following salvage prostatectomy for radiation refractory prostate cancer. *J Urol* 1999;**161**(3):857–62 discussion 862–3.
98. Gheiler EL, Tefilli MV, Tiguert R, et al. Predictors for maximal outcome in patients undergoing salvage surgery for radio-recurrent prostate cancer. *Urology* 1998;**51**(5):789–95.
99. Kaffenberger SD, Keegan KA, Bansal NK, et al. Salvage robotic assisted laparoscopic radical prostatectomy: a single institution, 5-year experience. *J Urol* 2013;**189**(2):507–13.

CHAPTER

# 28

# Indications for Pelvic Lymphadenectomy

*Jay D. Raman, MD, Awet Gherezghihir, MD*

Department of Surgery (Urology), The Penn State Milton S. Hershey Medical Center, Hershey, PA, USA

## INTRODUCTION

The introduction of prostate-specific antigen (PSA) screening during the 1990s has decreased the incidence of prostate cancer lymph node metastasis in contemporary surgical series.[1] This observed stage migration coupled with improvements in axial imaging has largely obviated the need for pelvic lymph node dissection (PLND) as an independent staging procedure. As a consequence, current practice patterns largely integrate PLND simultaneously with open or minimally invasive prostatectomy when clinically indicated. High-level scientific evidence in the form of randomized studies is lacking thereby predicating interpretation of the diagnostic and therapeutic benefit of LND from retrospective analyses.[2,3] In this chapter, we discuss the indications, role of predictive models, significance of anatomic boundaries, and therapeutic yield associated with PLND for the management of prostate cancer.

## RISK STRATIFICATION FOR PERFORMING PELVIC LYMPHADENECTOMY

PLND is an effective and reliable diagnostic modality for the staging of prostate cancer and discovery of metastatic disease.[4–6] Predictive models have also demonstrated that lymph node yield is a significant predictor of detecting positive lymph nodes with a linear increase in probability with each node removed.[7,8] Furthermore, some data implicate the potential for eradication of micrometastatic disease thereby contributing a therapeutic benefit realized via a survival advantage.[9] Such observations, however, are largely predicated on the likelihood of detecting positive lymph nodes (LN) at PLND. In particular, Klein et al. explored the concept of number needed to treat (NNT) to cure one patient with biologically or histologically positive LN at PLND and suggested that (not surprisingly) the NNT rose progressively at lower likelihoods of positive nodes.[10]

There are clearly some patients undergoing prostatectomy in whom PLND is unnecessary because of a low risk of nodal metastasis. With the increased use of PSA screening and the decreased macroscopic nodal burden at time of initial diagnosis, a trend toward less aggressive disease with significantly less nodal metastases has been clearly demonstrated at time of surgery.[3,9,11,12] Such observations have contributed to population registries noting decreased utilization of PLND. Analysis of the CAPSURE database, for example, highlighted a trend to decreased PLND for low- and intermediate-risk patients undergoing prostatectomy. Specifically, for low- to intermediate-risk patients, while 94% underwent PLND in 1992, this percentage decreased to 80% in 2004.[13]

A formal decision tree analysis has been used to evaluate the risk-to-benefit ratio of PLND at the time of prostatectomy. The analysis was based on two assumptions: (1) prostatectomy would be aborted if frozen section noted positive lymph nodes and (2) lymphadenectomy had no therapeutic benefit. With such considerations, the study suggested that lymph node dissection can be omitted in patients whose risk of lymph node involvement is less than 18%.[14] Collectively, such observations highlight the need for refined risk stratification to appropriately select patients suitable for PLND prior to surgery.

### Predictive Models and Nomograms

Current predictive models and nomograms seek to identify patients at greater risk of LN metastasis although they lack the ability to determine the therapeutic benefit of PLND. In 1995, Bishoff et al. used logistic regression analysis to generate probability curves allowing the estimation of LN positive disease.[15] Using variables including serum PSA, clinical stage, and biopsy Gleason score, and by comparing to the final pathologic stage, the authors found that in a cohort of 481 men PLND could be eliminated in up to 50% of patients. Bhatta-Dhar et al. explored this similar concept by retrospectively reviewing a cohort of over 800

Prostate Cancer. http://dx.doi.org/10.1016/B978-0-12-800077-9.00028-1

patients with favorable tumor characteristics undergoing RP with or without a limited PLND.[16] In this series, favorable characteristics were defined by a serum PSA ≤10 ng/mL, Gleason score ≤6, and clinical stage T1 or T2 not receiving adjuvant or neoadjuvant treatment. This study demonstrated 6-year biochemical recurrence-free rates of 86 and 88% in the PLND versus no-PLND groups, respectively ($p = 0.28$). Interval follow-up at 10-years from this same group additionally failed to demonstrate survival differences (83.8% vs. 87.9%, $p = 0.33$).[17] Such data underscored that low-risk patients are unlikely to have LN metastases and therefore may not benefit from PLND.

Nomograms have served as a more refined means to identify patients at high risk for lymphatic involvement. Most nomograms incorporate PSA at diagnosis, clinical T stage, and Gleason sum score at biopsy with fairly high predictive accuracies. The Partin Tables were one of the first such nomograms created from clinical and pathologic data of 703 men undergoing radical retropubic prostatectomy at a single institution.[18] Subsequent multi-institutional modeling comprised of over 4000 patients treated at three academic centers of excellence highlighted that in the validation analyses the nomograms were 72% accurate in correctly predicting the probability of a pathological stage to within 10%.[19] Further validation cohorts have demonstrated that the Partin nomograms are an accurate predictor of organ-confined disease.[20,21]

Refinements in predictive models have improved the accuracy of more contemporary nomograms to over 75%. For example, an update of the Partin Tables was published by Makarov et al. in 2007 to correct the effect of stage migration via a revised variable categorization and more contemporary patient cohort.[1] This analysis of 5730 men undergoing prostatectomy between 2000 and 2005 reported an 88% accuracy to predict lymph node involvement. Similarly, Cagiannos et al. developed a nomogram to assess lymph node invasion with a retrospective, nonrandomized analysis of 7014 patients using pretreatment PSA, clinical stage, and biopsy Gleason sum score.[22] Their bootstrap methodology corrected accuracy was reported to be 76%. Additionally, Briganti et al. studied 781 consecutive patients with clinically localized prostate cancer who underwent RRP and extended PLND with the intent of developing a multivariate nomogram to predict the rate of lymph node invasion based on preoperative PSA level, biopsy Gleason sum, clinical stage, and number of lymph nodes removed.[7] This was the first nomogram available that analyzed the extent of the pelvic lymphadenectomy as a means to determine likelihood of lymph node invasion. Their incremental accuracy using this additional factor was 1.8% compared to predictive tools that did not include the extent of lymphadenectomy. This nomogram was further updated utilizing the percentage of positive biopsy cores with univariate analysis suggesting this variable to be the most accurate in predicting lymph node invasion. This update from Briganti et al. reported a predictive accuracy for predicting LN involvement as 87.6% and these authors suggested to omit extended PLND in patients with an under 5% nomogram prediction for lymph node involvement.[23]

A summary of some predictive models and nomograms for lymph node involvement are highlighted in Table 28.1.[1,7,15,20,22–27]

**TABLE 28.1** Summary of Some Predictive Models and Nomograms Predicting Likelihood of Lymph Node Involvement for Patients with Clinically Localized Prostate Cancer

| Authors (year) | Patients | Studied variables | LN metastases (%) | Predictive accuracy (%) |
|---|---|---|---|---|
| Bishoff et al.[15] | 481 | PSA, clinical stage, biopsy Gleason sum, final pathologic stage | 7.7 | – |
| Kattan et al.[20] | 697 | PSA, clinical stage, biopsy Gleason sum | 8 | 76.8 |
| Conrad et al.[24] | 345 | PSA, clinical stage, biopsy Gleason sum, # of positive biopsies containing any Gleason grade 4 or 5 cancer | 8.1 | – |
| Cagiannos et al.[22] | 7014 | PSA, clinical stage, biopsy Gleason sum | 3.7 | 76 |
| Briganti et al.[7] | 781 | PSA, clinical stage, biopsy Gleason sum | 9.1 | 76 |
| Wang et al.[25] | 411 | PSA, biopsy Gleason sum, endorectal MRI | 5 | 89.2 |
| Briganti et al.[23] | 588 | PSA, clinical stage, biopsy Gleason sum, percentage of positive cores | 8.3 | 87.6 |
| Makarov et al.[1] | 5730 | PSA, clinical stage, biopsy Gleason sum | 1 | 88 |
| Karam et al.[26] | 425 | PSA, clinical stage, biopsy Gleason sum, preoperative plasma endoglin level | 3.3 | 97.8 |
| Bhojani et al.[27] | 839<br>235<br>(2 cohorts) | PSA, clinical stage, Gleason sum | 2<br>5 | 82<br>75 |

## Guidelines From National Organizations

Information as shown earlier has appropriately prompted national and international organizations to create guidelines to create better uniformity among practicing urologists. The 2007 American Urologic Association (AUA) prostate cancer guideline states that "Pelvic lymphadenectomy can be performed concurrently with radical prostatectomy and is generally reserved for patients with higher risk of nodal involvement."[28] The subsequently released Best Practice Statement for PSA Testing in the pretreatment staging and posttreatment management of prostate cancer revision refers to surgical staging (via lymphadenectomy) as "Generally unnecessary in low risk patients as defined by PSA ≤10 ng/mL and cT1/T2a disease and no pattern 4 or 5 disease."[29]

The EAU guidelines refer to PLND as the only reliable staging method when considering that current imaging modalities may not suffice to detect small metastases (<5 mm). These guidelines suggest that lymph node dissection can be omitted in patients with stage cT2 or less, PSA <10 ng/mL, a Gleason score ≤6, and <50% positive biopsy cores given that such patients have an under 10% likelihood of nodal metastases.[30] Finally, the NCCN Guidelines state that "A pelvic lymph node dissection can be excluded in patients with <2% predicated probability of nodal metastases by nomograms, although some patients with lymph node metastases will be missed."[31]

# ANATOMIC EXTENT OF PELVIC LYMPHADENECTOMY

Discussions regarding the therapeutic benefit of PLND for prostate cancer would be remiss without first considering lymphatic drainage and differences in the anatomic extent of dissection. In 2001, Brossner et al. mapped the lymphatic drainage after injecting oily contrast medium into the prostate of men undergoing needle biopsy followed by imaging with plain radiography and CT.[32] This group suggested that the lymphatic drainage was no different between the transition and peripheral zones and comprised three major routes: (1) the lymph nodes along the lateral bony wall of the pelvis to the angle of internal/external iliac lymph nodes to the common iliac lymph nodes; (2) perineal floor to the pudendal internal lymph nodes to the angle of the internal/external iliac lymph nodes to the common iliac lymph nodes; and (3) sacral lymph nodes. Subsequent work by Mattei et al. in patients with known prostate cancer undergoing prostatectomy involved mapping LN drainage using SPECT/CT or SPECT/MRI following Tc-99m nanocolloid injection to the prostate.[33] These authors highlighted that using this methodology 75% of prostate cancer harboring nodes involved not only the external and obturator regions but also lymphatics near the internal iliac and common iliac vessels extending up to the ureteric crossing. Several other groups have suggested that in order to achieve adequate staging, it is necessary to include the lymph nodes along the common iliac artery as upward of 50% of LN involvement has been demonstrated to involve this packet.[4,6,30,34,35]

In general, a standard PLND has been defined by the following borders: bladder (medially), pelvic side wall (laterally), external iliac vein (superiorly), obturator nerve (inferiorly), bifurcation of the common iliac vessels (proximally), and femoral canal (distally).[36] The boundaries of an extended PLND differ somewhat between institutions but can include the area outlined by the standard template as well as tissues in the hypogastric, common iliac, lateral to external iliac, and presacral regions.[2,36,37] The current NCCN guideline similarly suggests that "An extended PLND includes removal of all node-bearing tissue from an area bounded by the external iliac vein anteriorly, the pelvic side wall laterally, the bladder wall medially, the floor of the pelvis posteriorly, Cooper's ligament distally, and the internal iliac artery proximally."[6,31] A summary of standard and extended templates are demonstrated in Figures 28.1 and 28.2 and Table 28.2.[2,33,37] Despite these published templates, clear variability exists in practice patterns when considering extent of PLND. A recent survey study by Touijer et al. querying urologic oncologists found that 15% perform a PLND limited to the external iliac, while 30% include the external iliac, obturator fossa, and hypogastric lymph nodes.[3] Among surgeons using open and robotic approaches, 19% reported that the indication for and extent of lymphadenectomy performed differ based on the surgical approach used. The authors concluded that collaborative efforts were requisite to develop guidelines and better ensure standardization of technique.

Numerous groups have clearly demonstrated that a more extended lymph node dissection harvests a greater number of lymph nodes with an increased likelihood of detecting metastasis. In 2002, Heidenreich et al. reported on 103 patients who underwent extended PLND for prostate cancer in comparison to 100 patients who had a standard lymphadenectomy.[38] These authors noted positive lymph nodes in 26% of patients with the extended technique (including the internal iliac and presacral regions) compared to 12% using a standard/limited approach. Similar findings were noted by Allaf et al. when reviewing the Johns Hopkins experience comparing two surgeons performing either an extended or standard lymphadenectomy.[39] In this series, extended lymphadenectomy was associated with a greater LN yield (11.6 vs. 8.9, $p < 0.0001$) as well as LN positivity rate (3.3% vs. 1.2%, $p < 0.0001$). In contrast to these observations, however, Clark et al. prospectively randomized 123 patients undergoing radical prostatectomy to an extended PLND

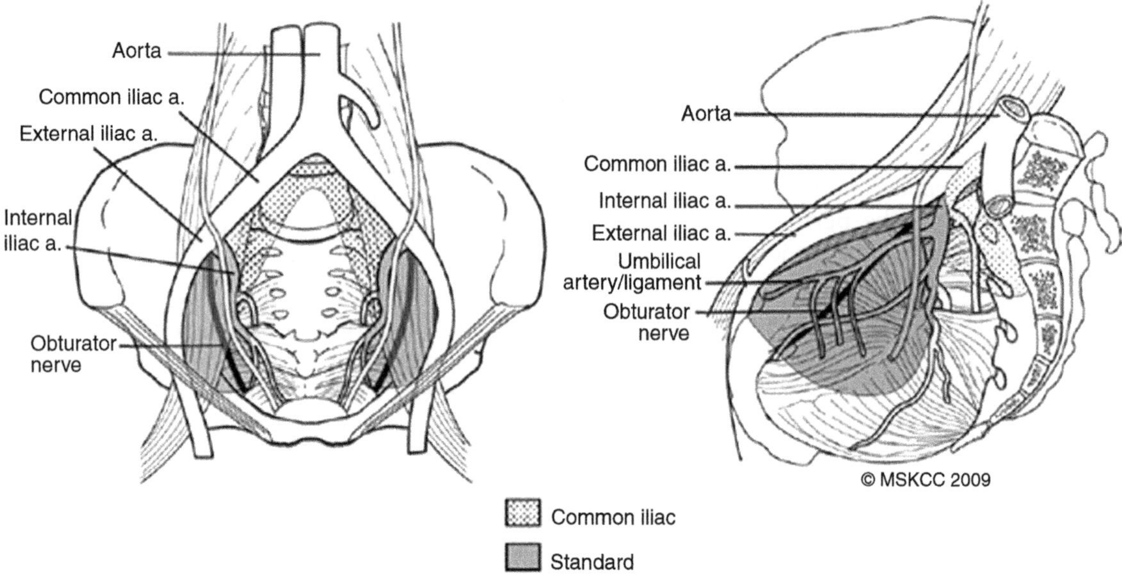


**FIGURE 28.1 Standard and extended (common iliac) pelvic lymph node dissection templates from the Memorial Sloan Kettering Cancer Center.** *Reprinted with permission from Yee et al.*[37]

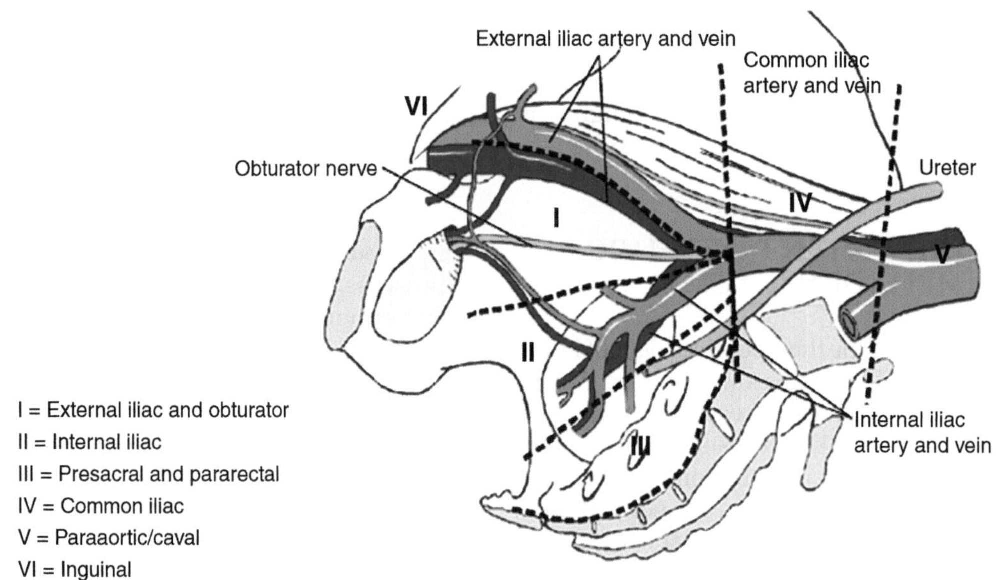


**FIGURE 28.2 Boundaries of pelvic lymph node dissection (PLND) subdivided into different regions.** "Limited" PLND removes tissue along with the external iliac vein and from the obturator fossa corresponding to region I. "Extended" template PLND removes tissue along the major pelvic vessels (external iliac vein, obturator fossa, and internal iliac artery and vein) corresponding to regions I and II. *Reprinted with permission from Mattei et al.*[33]

on the right to standard PLND on the left.[40] This study failed to demonstrate a significant difference in pelvic LN metastasis between the two templates although limitations of these data included inclusion of a large cohort of low-risk patients and arbitrary selection of side of PLND irrespective of disease location in prostate.

With such information, national guidelines do appear to support extended PLND when the decision is made to perform a lymph node dissection at time of prostatectomy. In particular, the NCCN guidelines specifically state that "An extended PLND will discover metastases approximately twice as often as a limited PLND. Extended PLND provides more complete staging and may cure some men with microscopic metastases; therefore, an extended PLND is preferred when PLND is performed."[6,31]

**TABLE 28.2** Lymphatic Tissue Removed During Pelvic Lymphadenectomy

| | Standard template | Extended template |
|---|---|---|
| Medial to external iliac vein | Yes | Yes |
| Obdurator | Yes | Yes |
| Hypogastric | No | Yes |
| Presacral | No | Yes |
| Common iliac | No | Yes |
| Lateral to external iliac vein | No | Yes |

*Reprinted with permission from Palapattu et al.*[2]

## DOES PELVIC LYMPH NODE DISSECTION CONFER A THERAPEUTIC BENEFIT (I.E., A SURVIVAL ADVANTAGE)

The effect of pelvic lymphadenectomy on long-term survival outcomes for prostate cancer has been the subject of multiple studies and debates over the years. Due to the lack of prospective randomized trials, it remains a challenging (if not impossible) question to answer. Furthermore, with variable lymph node dissection templates (i.e., standard vs. extended), additional confounding factors exist that can limit interpretation of existing datasets even from high volume centers of excellence. In summary, contemporary data remain conflicting with regard to survival benefit. Some of the more salient studies are highlighted later in the chapter.

Using an undetectable PSA as a surrogate for disease-free status after prostatectomy, Han et al. from Johns Hopkins have demonstrated excellent outcomes in patients undergoing RRP with concurrent PLND.[41] In this study of over 2400 men treated over 15 years, the authors reported 5-, 10-, and 15-year recurrence-free survival rates of 84, 74, and 66%, metastasis-free rates of 96, 90, and 82%, and cancer-specific survival rates of 99, 96, and 90%. Such data, while compelling, further begs whether PLND or favorable patient selection (or a combination of these factors) directly contributes to these excellent survival outcomes. Population-based studies suggest that PLND may have a benefit particularly when harvesting a threshold number of lymph nodes. Joslyn and Konety reviewed the Surveillance, Epidemiology, and End Results (SEER) database of 13,000 patients undergoing prostatectomy with or without lymphadenectomy.[42] In this cohort, LND was performed in 71% of patients with approximately 9% having positive lymph nodes at surgery. The authors found that a lymph node yield of $\geq$4 (positive or negative) was associated with a reduction of overall and cancer-specific mortality (OR = 0.77, $p = 0.007$), and harvesting $\geq$10 negative lymph nodes was also associated with lower cancer-specific death (OR = 0.89, $p = 0.08$).

The concept of number of lymph node yield and lymph node density has increasingly emerged as a potential stratification means in determining the therapeutic efficacy of PLND. The previously discussed study from Johns Hopkins found that among men with lymph node positive disease involving less than 15% of extracted nodes, the 5-year PSA progression-free rate for extended lymph node dissection was 43% versus 10% for the more limited lymph node dissection ($p = 0.01$).[39] Similar work by Daneshmand et al. demonstrated that when stratified by lymph node density, patients with a lymph node density of 20% or greater were at higher risk for clinical recurrence when compared with those whose density was less than 20% (relative risk: 2.32, $p < 0.0001$).[43] Briganti et al. also looked at the relationship between lymph node metastases and the number of lymph nodes removed at time of prostatectomy. Reviewing 858 patients with a median PSA of 5.8, Stage T1c or T2 with Gleason sum score of $\leq$6 (62.2%) or 7 (25.1%), they found that it was possible to predict metastases in up to 90% of patients who had at least 28 nodes removed, while PLND was not reliable with lymph node yield $\leq$10.[44]

Studies highlighting the natural history of lymph node positive prostate cancer also implicate a potential therapeutic role. Bader et al. studied 367 patients who underwent RP with PLND for organ-confined prostate cancer and demonstrated that although aggressive LND revealed a high rate of metastases (25%), some patients remained free from PSA relapse for >10 years without adjuvant treatment.[34] The authors suggested that PLND may have a positive impact on disease progression and long-term disease-free survival in this high-risk cohort.

The number of positive lymph nodes has been suggested as a potential predictor of survival outcomes. Cheng et al. demonstrated that the hazard ratio for death from prostate cancer among patients with a single lymph node metastasis compared with those without lymph node involvement was 1.5 (95% confidence interval (CI), 0.5–5.0, $p = 0.478$).[45] Furthermore, the hazard risk of death rose to 6.1 (95% CI, 1.9–19.6, $p = 0.002$) for those with two positive lymph nodes and 4.3 (95% CI, 1.4–13.0, $p = 0.009$) for patients with three or more positive lymph nodes. Boorjian et al. studied 507 men with lymph node positive disease treated with radical prostatectomy and showed on multivariate analysis that Gleason score 8–10 ($p = 0.004$), positive surgical margins ($p = 0.016$), tumor ploidy ($p = 0.023$), and 2 or greater positive nodes ($p = 0.001$) were all adverse predictors of cancer-specific survival.[46] Several other studies have further highlighted that two positive LN at PLND may represent a key threshold for biochemical and cancer-specific survival outcomes. A multicenter study of 703 node positive patients undergoing prostatectomy with extended PLND found that patients with two or fewer positive nodes had a significantly better cancer-specific survival at 15-year follow-up compared to those with more than two positive nodes (84% vs. 62%; $p<0.001$).[47] A similarly designed, albeit smaller, single institution series also reported that patients with $\leq 2$ positive LN had relatively better survival outcomes when compared to those with $\geq 3$ despite use of extended PLND and incorporation of ADT in almost 70% of cases.[48]

Despite some of the earlier mentioned studies, the improved survival advantage associated with lymphadenectomy has not universally been supported. For example, the study by Cheng et al. discussed previously failed to demonstrate a difference in progression-free survival rate among patients with or without lymph node metastasis in multivariate analysis after controlling for all relevant variables, including treatments (HR, 1.0; 95% CI, 0.7–1.3; $p = 0.90$).[45] In 2006, Masterson et al. from the Memorial Sloan Kettering Cancer Center analyzed over 4600 patients undergoing prostatectomy with LND (median of 9 LN removed).[49] Positive LN were found in 175 patients (3.8%), although the overall number of LN removed did not predict freedom from biochemical recurrence (HR per additional 10 nodes removed = 1.02, 95% CI 0.92 to 1.13, $p = 0.7$). Similar findings were reported by Bhatta-Dhar et al. and Weight et al. in a low-risk cohort in which the 6- and 10-year biochemical recurrence rates were similar between patients undergoing PLND versus no-PLND cohort.[16,17] Finally, a study from the Mayo Clinic encompassing 7034 patients failed to demonstrate an association between the number of lymph nodes removed and survival.[50]

## CAN PREOPERATIVE IMAGING AID THE DECISION PROCESS TO PERFORM LND?

While noninvasive imaging plays a significant role in the diagnostic algorithm for certain malignancies (including urologic cancers such as bladder and kidney), the informatics yield is considerably less for prostate cancer.[51] Although MRI has shown value in predicting pathological outcomes and organ confinement of disease,[52,53] the majority of our current imaging modalities have limited ability to predict lymph node involvement and may add unnecessary cost and radiation exposure for low-risk patients if performed after the diagnosis of prostate cancer and before prostatectomy.[2,4,54] Therefore, much reported use highlights the role of surgical planning as opposed to staging.[53,55] Not surprisingly, the 2013 AUA PSA Best Practice Statement suggests that "Routine radiographic staging, such as with bone scan, computed tomography (CT), or magnetic resonance imaging (MRI), or surgical staging with pelvic lymph node dissection is not necessary in all cases of newly diagnosed prostate cancer."[29] A likelihood of lymph node metastases greater than 32–45% has been suggested as criteria to consider axial imaging for prostate cancer staging.[56] Patients categorized as high risk (PSA >20.0 ng/mL, Gleason score ≥8, or locally advanced) have been shown to have a 50% likelihood of nodal invasion.[29,56]

Newer imaging technologies may play a more significant role to accurately identify LN metastasis for prostate cancer.[57] For example, advances in technology using lymphotropic superparamagnetic nanoparticles in conjunction with high-resolution MRI has shown promise in the detection of small and otherwise undetectable lymph-node metastases by conventional techniques.[58] Additionally, the use of ferumoxtran-10, a magnetic resonance lymphangiography (MRL) contrast agent, has been investigated to better identify tumor burden in normal-sized lymph nodes. In a prospective, multicenter cohort study, Heesakkers et al. found a significantly higher sensitivity (82%) and negative predictive value (96%) for identifying LN metastasis when comparing MRL to multidetector CT in a cohort of intermediate- or high-risk patients.[59] The authors therefore concluded that the posttest probability of having lymph-node metastases using MRL is low enough to omit a PLND.

Similarly, Choline-PET/CT has also shown promise as a diagnostic tool for the detection of lymph node metastases as well as for restaging patients with rising PSA levels after radical prostatectomy. Schiavina et al. studied[11]Choline PET/CT in comparison to available staging nomograms in 57 patients with intermediate- and high-risk prostate cancers.[60] In this study, the authors found a sensitivity, specificity, NPV, and number of correctly recognized cases at PET/CT to be 60.0, 97.6, 87.2, and 87.7%, respectively, thereby outperforming current nomograms in specificity and accuracy. Nonetheless, there has been difficulty in detecting all tumor deposits with PET/CT scan due to limited spatial resolution, differences in metabolic states, and the inability to detect lesions smaller than 5 mm.[61]

While such modalities are still in development and validation stages, it is apparent that they will likely serve a more robust role than simple axial imaging in the future for identification of lymph node metastasis.

## CONCLUSIONS

The merits of pelvic lymphadenectomy at time of radical prostatectomy continue to be debated in the contemporary literature. Low-risk patients (either determined by nomograms or prediction tables) have a negligible risk of pelvic LN metastasis therefore implicating overtreatment when lymphadenectomy is performed in this cohort. Intermediate- and high-risk patients clearly have a greater likelihood of LN invasion. In such patients,

PLND (particularly extended PLND) is valuable in identifying the presence and extent of metastatic disease, which will clearly impact the utilization of adjuvant therapies (i.e., radiation and/or androgen deprivation). While PLND in appropriate patients is diagnostic, the actuarial therapeutic benefit remains unclear although numerous studies have suggested the potential for survival benefit when microscopic node positive disease can be eradicated. Refinements in imaging will likely aid in better identifying patients at greater risk of LN involvement who may benefit from PLND.

## References


1. Makarov DV, Trock BJ, Humphreys EB, et al. Updated nomogram to predict pathologic stage of prostate cancer given prostate-specific antigen level, clinical stage, and biopsy Gleason score (Partin Tables) based on cases from 2000 to 2005. *Urology* 2007;**69**(6): 1095–101.
2. Palapattu GS, Singer EA, Messing EM. Controversies surrounding lymph node dissection for prostate cancer. *Urol Clin North Am* 2010;**37**(1):57–65.
3. Touijer KA, Ahallal Y, Guillonneau BD. Indications for and anatomical extent of pelvic lymph node dissection for prostate cancer: practice patterns of uro-oncologists in North America. *Urol Oncol* 2013;**31**(8):1517–21 e1–e2.
4. Briganti A, Blute ML, Eastham JH, et al. Pelvic lymph node dissection in prostate cancer. *Eur Urol* 2009;**55**(6):1251–65.
5. La Rochelle JC, Amling CL. Role of lymphadenectomy for prostate cancer: indications and controversies. *Urol Clin North Am* 2011;**38**(4): 387–95.
6. Mohler JL. The 2010 NCCN clinical practice guidelines in oncology on prostate cancer. *J Natl Compr Canc Netw* 2010;**8**(2):145.
7. Briganti A, Chun FK, Salonia A, et al. Validation of a nomogram predicting the probability of lymph node invasion based on the extent of pelvic lymphadenectomy in patients with clinically localized prostate cancer. *BJU Int* 2006;**98**(4):788–93.
8. Suardi N, Larcher A, Haese A, et al. Indication for and extension of pelvic lymph node dissection during robot-assisted radical prostatectomy: an analysis of five European institutions. *Eur Urol* 2014;**66**(4):635–43.
9. Wagner M, Sokoloff M, Daneshmand S. The role of pelvic lymphadenectomy for prostate cancer – therapeutic? *J Urol* 2008;**179**(2): 408–13.
10. Klein EA, Kattan M, Stephenson A, et al. How many lymphadenectomies does it take to cure one patient? *Eur Urol* 2008;**53**(1): 13–5 discussion 18–20.
11. Narayan P, Fournier G, Gajendran V, et al. Utility of preoperative serum prostate-specific antigen concentration and biopsy Gleason score in predicting risk of pelvic lymph node metastases in prostate cancer. *Urology* 1994;**44**(4):519–24.
12. Kamel MH, Moore PC, Bissada NK, et al. Potential years of life lost due to urogenital cancer in the United States: trends from 1972 to 2006 based on data from the SEER database. *J Urol* 2012;**187**(3): 868–71.
13. Kawakami J, Meng MV, Sadetsky N, et al. Changing patterns of pelvic lymphadenectomy for prostate cancer: results from CaPSURE. *J Urol* 2006;**176**(4 Pt 1):1382–6.
14. Meng MV, Carroll PR. When is pelvic lymph node dissection necessary before radical prostatectomy? A decision analysis. *J Urol* 2000;**164**(4):1235–40.
15. Bishoff JT, Reyes A, Thompson IM, et al. Pelvic lymphadenectomy can be omitted in selected patients with carcinoma of the prostate: development of a system of patient selection. *Urology* 1995;**45**(2):270–4.
16. Bhatta-Dhar N, Reuther AM, Zippe C, et al. No difference in six-year biochemical failure rates with or without pelvic lymph node dissection during radical prostatectomy in low-risk patients with localized prostate cancer. *Urology* 2004;**63**(3):528–31.
17. Weight CJ, Reuther AM, Gunn PW, et al. Limited pelvic lymph node dissection does not improve biochemical relapse-free survival at 10 years after radical prostatectomy in patients with low-risk prostate cancer. *Urology* 2008;**71**(1):141–5.
18. Partin AW, Yoo J, Carter HB, et al. The use of prostate specific antigen, clinical stage and Gleason score to predict pathological stage in men with localized prostate cancer. *J Urol* 1993;**150**(1):110–4.
19. Partin AW, Kattan MW, Subong EN, et al. Combination of prostate-specific antigen, clinical stage, and Gleason score to predict pathological stage of localized prostate cancer. A multi-institutional update. *JAMA* 1997;**277**(18):1445–51.
20. Kattan MW, Stapleton AM, Wheeler TM, et al. Evaluation of a nomogram used to predict the pathologic stage of clinically localized prostate carcinoma. *Cancer* 1997;**79**(3):528–37.
21. Blute ML, Bergstralh EJ, Partin AW, et al. Validation of Partin Tables for predicting pathological stage of clinically localized prostate cancer. *J Urol* 2000;**164**(5):1591–5.
22. Cagiannos I, Karakiewicz P, Eastham JA, et al. A preoperative nomogram identifying decreased risk of positive pelvic lymph nodes in patients with prostate cancer. *J Urol* 2003;**170**(5):1798–803.
23. Briganti A, Larcher A, Abdollah F, et al. Updated nomogram predicting lymph node invasion in patients with prostate cancer undergoing extended pelvic lymph node dissection: the essential importance of percentage of positive cores. *Eur Urol* 2012;**61**(3):480–7.
24. Conrad S, Graefen M, Pichlmeier U, et al. Systematic sextant biopsies improve preoperative prediction of pelvic lymph node metastases in patients with clinically localized prostatic carcinoma. *J Urol* 1998;**159**(6):2023–9.
25. Wang L, Hricak H, Kattan MW, et al. Combined endorectal and phased-array MRI in the prediction of pelvic lymph node metastasis in prostate cancer. *AJR Am J Roentgenol* 2006;**186**(3):743–8.
26. Karam JA, Svatek RS, Karakiewicz PI, et al. Use of preoperative plasma endoglin for prediction of lymph node metastasis in patients with clinically localized prostate cancer. *Clin Cancer Res* 2008;**14**(5): 1418–22.
27. Bhojani N, Salomon L, Capitanio U, et al. External validation of the updated Partin Tables in a cohort of French and Italian men. *Int J Radiat Oncol Biol Phys* 2009;**73**(2):347–52.
28. Thompson IM, Thrasher JB, Aus G, et al. Guideline for the management of clinically localized prostate cancer: 2007 update. *J Urol* 2007;**177**(6):2106–31.
29. Carroll P, Albertsen, PC, Greene, K, et al. PSA testing for the pretreatment staging and post-treatment management of prostate cancer: 2013 revision of 2009 best practice statement; 2013.
30. Heidenreich A, Bellmunt J, Bolla M, et al. EAU guidelines on prostate cancer. Part 1: screening, diagnosis, and treatment of clinically localised disease. *Eur Urol* 2011;**59**(1):61–71.
31. Kawachi MH, Bahnson RR, Barry M, et al. NCCN clinical practice guidelines in oncology: prostate cancer early detection. *J Natl Compr Canc Netw* 2010;**8**(2):240–62.
32. Brossner C, Ringhofer H, Hernady T, et al. Lymphatic drainage of prostatic transition and peripheral zones visualized on a three-dimensional workstation. *Urology* 2001;**57**(2):389–93.
33. Mattei A, Fuechsel FG, Bhatta Dhar N, et al. The template of the primary lymphatic landing sites of the prostate should be revisited: results of a multimodality mapping study. *Eur Urol* 2008;**53**(1):118–25.
34. Bader P, Burkhard FC, Markwalder R, et al. Is a limited lymph node dissection an adequate staging procedure for prostate cancer? *J Urol* 2002;**168**(2):514–8 discussion 518.

35. Stone NN, Stock RG, Unger P. Laparoscopic pelvic lymph node dissection for prostate cancer: comparison of the extended and modified techniques. *J Urol* 1997;**158**(5):1891–4.
36. Burkhard FC, Schumacher M, Studer UE. The role of lymphadenectomy in prostate cancer. *Nat Clin Prac Urol* 2005;**2**(7):336–42.
37. Yee DS, Katz DJ, Godoy G, et al. Extended pelvic lymph node dissection in robotic-assisted radical prostatectomy: surgical technique and initial experience. *Urology* 2010;**75**(5):1199–204.
38. Heidenreich A, Varga Z, Von Knobloch R. Extended pelvic lymphadenectomy in patients undergoing radical prostatectomy: high incidence of lymph node metastasis. *J Urol* 2002;**167**(4):1681–6.
39. Allaf ME, Palapattu GS, Trock BJ, et al. Anatomical extent of lymph node dissection: impact on men with clinically localized prostate cancer. *J Urol* 2004;**172**(5 Pt 1):1840–4.
40. Clark T, Parekh DJ, Cookson MS, et al. Randomized prospective evaluation of extended versus limited lymph node dissection in patients with clinically localized prostate cancer. *J Urol* 2003;**169**(1):145–7 discussion 147–148.
41. Han M, Partin AW, Pound CR, et al. Long-term biochemical disease-free and cancer-specific survival following anatomic radical retropubic prostatectomy. The 15-year Johns Hopkins experience. *Urol Clin North Am* 2001;**28**(3):555–65.
42. Joslyn SA, Konety BR. Impact of extent of lymphadenectomy on survival after radical prostatectomy for prostate cancer. *Urology* 2006;**68**(1):121–5.
43. Daneshmand S, Quek ML, Stein JP, et al. Prognosis of patients with lymph node positive prostate cancer following radical prostatectomy: long-term results. *J Urol* 2004;**172**(6 Pt 1):2252–5.
44. Briganti A, Chun FK, Salonia A, et al. Critical assessment of ideal nodal yield at pelvic lymphadenectomy to accurately diagnose prostate cancer nodal metastasis in patients undergoing radical retropubic prostatectomy. *Urology* 2007;**69**(1):147–51.
45. Cheng L, Zincke H, Blute ML, et al. Risk of prostate carcinoma death in patients with lymph node metastasis. *Cancer* 2001;**91**(1):66–73.
46. Boorjian SA, Thompson RH, Siddiqui S, et al. Long-term outcome after radical prostatectomy for patients with lymph node positive prostate cancer in the prostate specific antigen era. *J Urol* 2007;**178**(3 Pt 1):864–70 discussion 870–1.
47. Briganti A, Karnes JR, Da Pozzo LF, et al. Two positive nodes represent a significant cut-off value for cancer specific survival in patients with node positive prostate cancer. A new proposal based on a two-institution experience on 703 consecutive N+ patients treated with radical prostatectomy, extended pelvic lymph node dissection and adjuvant therapy. *Eur Urol* 2009;**55**(2):261–70.
48. Schumacher MC, Burkhard FC, Thalmann GN, et al. Good outcome for patients with few lymph node metastases after radical retropubic prostatectomy. *Eur Urol* 2008;**54**(2):344–52.
49. Masterson TA, Bianco Jr FJ, Vickers AJ, et al. The association between total and positive lymph node counts, and disease progression in clinically localized prostate cancer. *J Urol* 2006;**175**(4):1320–4 discussion 1324–5.
50. DiMarco DS, Zincke H, Sebo TJ, et al. The extent of lymphadenectomy for pTXNO prostate cancer does not affect prostate cancer outcome in the prostate specific antigen era. *J Urol* 2005;**173**(4):1121–5.
51. Aparici CM, Carlson D, Nguyen N, et al. Combined SPECT and multidetector CT for prostate cancer evaluations. *Am J Nucl Med Mol Imaging* 2012;**2**(1):48–54.
52. Tempany CM, McNeil BJ. Advances in biomedical imaging. *JAMA* 2001;**285**(5):562–7.
53. Jeong IG, Lim JH, You D, et al. Incremental value of magnetic resonance imaging for clinically high risk prostate cancer in 922 radical prostatectomies. *J Urol* 2013;**190**(6):2054–60.
54. Lavery HJ, Brajtbord JS, Levinson AW, et al. Unnecessary imaging for the staging of low-risk prostate cancer is common. *Urology* 2011;**77**(2):274–8.
55. Katz S, Rosen M. MR imaging and MR spectroscopy in prostate cancer management. *Radiol Clin North Am* 2006;**44**(5):723–34 viii.
56. Wolf Jr JS, Cher M, Dall'era M, et al. The use and accuracy of cross-sectional imaging and fine needle aspiration cytology for detection of pelvic lymph node metastases before radical prostatectomy. *J Urol* 1995;**153**(3 Pt 2):993–9.
57. Hovels AM, Heesakkers RA, Adang EM, et al. The diagnostic accuracy of CT and MRI in the staging of pelvic lymph nodes in patients with prostate cancer: a meta-analysis. *Clin Radiol* 2008;**63**(4):387–95.
58. Harisinghani MG, Barentsz J, Hahn PF, et al. Noninvasive detection of clinically occult lymph-node metastases in prostate cancer. *N Engl J Med* 2003;**348**(25):2491–9.
59. Heesakkers RA, Hovels AM, Jager GJ, et al. MRI with a lymph-node-specific contrast agent as an alternative to CT scan and lymph-node dissection in patients with prostate cancer: a prospective multicohort study. *Lancet Oncol* 2008;**9**(9):850–6.
60. Schiavina R, Scattoni V, Castellucci P, et al. 11C-choline positron emission tomography/computerized tomography for preoperative lymph-node staging in intermediate-risk and high-risk prostate cancer: comparison with clinical staging nomograms. *Eur Urol* 2008;**54**(2):392–401.
61. Kitajima K, Murphy RC, Nathan MA. Choline PET/CT for imaging prostate cancer: an update. *Ann Nucl Med* 2013;**27**(7):581–91.

CHAPTER

# 29

# The Surgical Anatomy of the Prostate

*Fairleigh Reeves, MB, BS*, Wouter Everaerts, MD, PhD*,**,†, Declan G. Murphy, MB, FRCS Urol*,**,†, Anthony Costello, MD, FRACS, FRCSI*,***

*Departments of Urology and Surgery, Royal Melbourne Hospital and University of Melbourne, Melbourne, Australia

**Epworth Prostate Centre, Epworth Healthcare, Melbourne, Australia

†Division of Cancer Surgery, Peter MacCallum Cancer Centre, University of Melbourne, East Melbourne, Victoria, Australia

Anatomical principles underpin optimal oncological and functional outcomes in the surgical management of prostate cancer. Therefore, an in-depth understanding of pelvic anatomy is prerequisite for all prostate cancer surgeons.

Recent technology development, such as the advent of 10-times magnification, high-definition video, and 3D video systems in robotic and laparoscopic surgery, enables the surgeon to identify subtle anatomical structures like never before, ever increasing the precision of surgery.

Unfortunately, understanding the complex anatomy of the pelvis is made particularly challenging by the existence of differing terminology. The EAU Section of Uro-Technology (ESUT) expert group on laparoscopy highlighted the importance of uniform terminology during radical prostatectomy in their 2013 editorial.[1] We refer readers to this paper for a concise summary of the various anatomical terms used in this area.

In this chapter, we review the current literature to provide an overview of the functional anatomy of the male pelvis relevant to radical prostatectomy surgery.

## THE PROSTATE

The prostate gland comprises acini of varying shapes and sizes embedded in a fibromuscular stroma. Broader than it is long, the prostate measures approximately 4 × 3 × 2 cm. Its base is fused with the bladder neck and perforated by the urethra, which traverses the length of the prostate to emerge at its apex. The prostate's inferolateral surfaces are cradled by levator ani. Posteriorly the prostate is pierced by the ejaculatory ducts, which pass through the gland to open into the prostatic urethra[2] (Figure 29.1).

Histopathological analysis of the prostate reveals four distinct regions or zones, which center on different areas of ejaculatory ducts[4] (Figure 29.2).

### Peripheral Zone

The peripheral zone (PZ) forms the bulk of the glandular prostate (70%) and arises from duct buds that develop laterally into the mesenchyme posterior to the distal urethral segment.[4] This region is the most common site of origin for prostate carcinoma (70% cancers).[6] Peripheral zone tumors of 0.2 mL or larger may be detected by digital rectal exam (DRE). Although the sensitivity and specificity of DRE is limited, an abnormal DRE is associated with an increased risk of a higher Gleason score and should therefore be considered an indication for prostate biopsy even if PSA is normal.[7]

### Central Zone

The smaller central zone (25% of the normal glandular prostate) arises from ducts clustered in a small area on the convexity of the verumontanum, immediately surrounding the ejaculatory duct orifices. These ducts branch proximally and laterally toward the prostatic base. This zone is relatively immune to cancer,[4] but can be susceptible to tumor infiltration from adjacent zones. In autopsy and surgical specimen studies, central zone cancers account for up to 10% of prostate cancer.[8]

Prostate Cancer. http://dx.doi.org/10.1016/B978-0-12-800077-9.00029-3

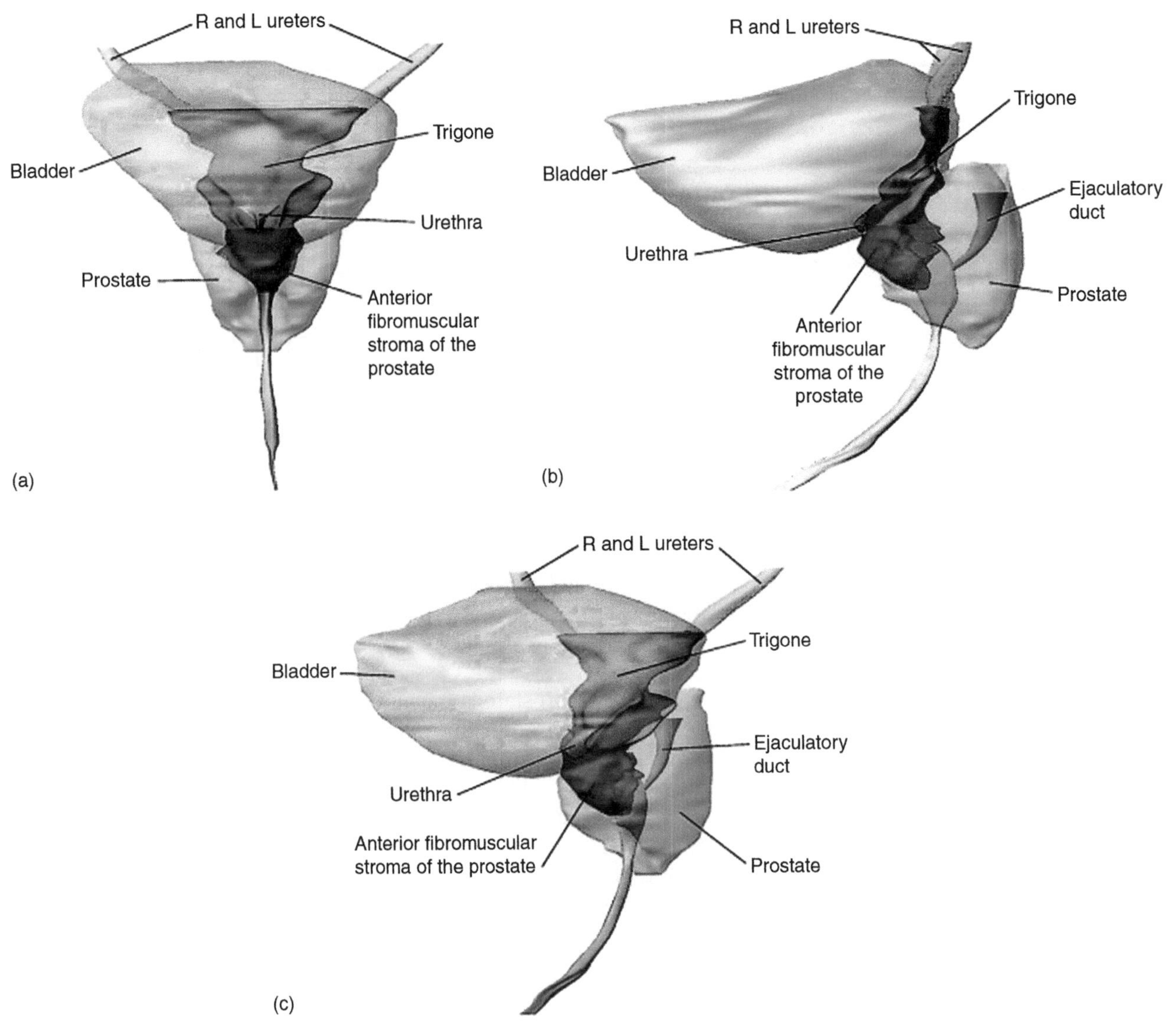


**FIGURE 29.1** (a) Anterior view reveals that trigone narrows below ureteral orifices and widens posterior to bladder neck. Trigone is continuous with anterior fibromuscular stroma of prostate. R, right; L, left. (b) Lateral view of trigonal anterior fibromuscular stroma of prostate shows that these structures are in continuity. Trigone thickens posterior to bladder neck and joins fibromuscular stroma anterior to prostate. (c) Oblique view of trigonal anterior fibromuscular stroma complex. *Reprinted with permission from Brooks et al.*[3]

Although uncommon, cancers that do arise in this zone may be more aggressive.[9]

## Transition Zone

The transition zone (TZ) arises from ducts in the proximal urethra. It accounts for less than 5% of the normal prostate gland, but can enlarge considerably in BPH.[4] The TZ is easily recognizable on ultrasound examination, as it is diffusely hypoechoic relative to the PZ.[8] TZ cancer is the site of origin of up to 25% of prostate cancer.[6] Cancers from this zone show much less capsular penetration and seminal vesicle invasion than PZ cancers of comparable volume, as the TZ boundary provides a barrier to cancer spread through the PZ.[8] When performing TRUS biopsies, separate transition zone biopsies give a very low detection rate and therefore may be omitted from the standard biopsy template at initial biopsy.[10]

## Anterior Fibromuscular Stroma

The anterior fibromuscular stroma produces the characteristic convexity of the anterior prostatic surface. It is a thick, nonglandular region that accounts for one-third of the bulk of the prostate. It is continuous with detrusor muscle proximally and the external urethral sphincter distally.[4]

# SEMINAL VESICLES

The seminal vesicles (SV) are paired thin-walled lobulated sacs that are applied to the posterior aspect of the bladder base, lateral to the ampulla of the

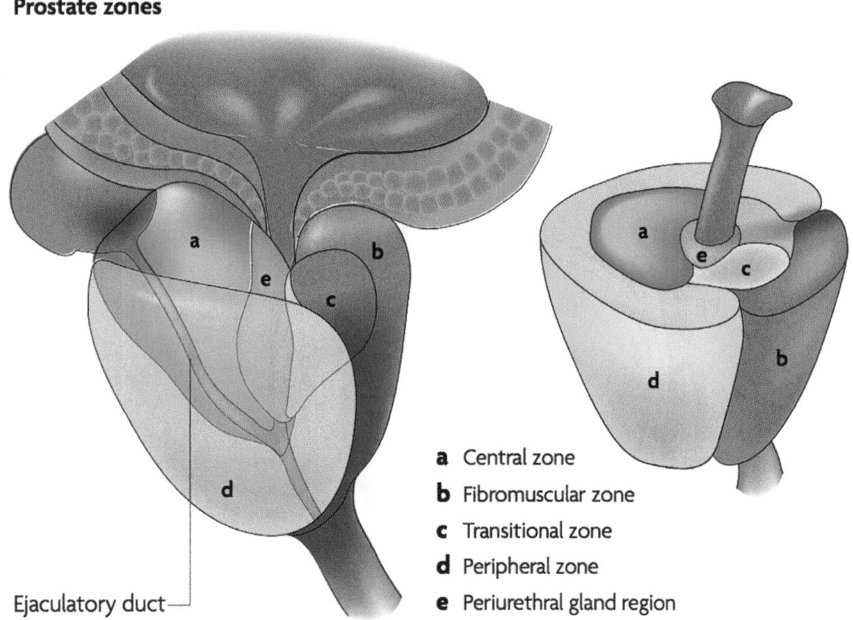


FIGURE 29.2 **Zonal anatomy of the prostate.** *Reprinted with permission from De Marzo et al.*[5]

ductus deferens. At their lower end the duct of the seminal vesicle joins the ductus deferens to form the ejaculatory duct, which opens into the prostatic urethra at the verumontanum.[2] The tip of the SV has a close relationship to the arterial supply to the bladder base and the proximal neurovascular bundle.[11] Frequency and extent of SV invasion correlates strongly with cancer volume.[12] Due to the association of SV invasion and poor prognosis, radical prostatectomy routinely involves en bloc resection of the seminal vesicles.

## FASCIAE

A thorough understanding of the fasciae investing the prostate, rectum, and muscles of the pelvis is imperative to the performance of nerve-sparing surgery. Histological studies and the advent of intraoperative magnified vision have enabled an improved appreciation of the complexity of the multilayered fascia around the prostate (Figure 29.3). However, the exact anatomy of these fascial layers remains controversial.[13] In addition, the use of different terminology contributes to confusion in this area.

### The Prostatic Capsule

The "prostatic capsule" is not a true capsule by definition, but an incomplete outer fibromuscular band that is an inseparable part of the prostatic stroma.[14] It represents the outermost aspect of the prostate gland and provides an important landmark in the pathological classification of the extent of prostate cancer involvement. Anteriorly, the capsule is not clearly evident, and is difficult to distinguish from the anterior fibromuscular stroma of the prostate.[15] The capsule is also deficient at the prostatic base and apex where the prostatic stroma merges with the detrusor muscle and urethral sphincter fibers, respectively.[14]

### Endopelvic Fascia

(*Pelvic fascia, Lateral pelvic fascia, superior pelvic fascia, parietal pelvic fascia, levator fascia, outer layer of periprostatic fascia, parapelvic fascia*[1,16])

The pelvic fascia comprises parietal and visceral fascia that line the walls and floor of the pelvis. This is often described collectively in prostate surgery as the endopelvic fascia. The parietal endopelvic fascia is continuous with transversalis fascia and provides cover to the pelvic muscles lateral to the fascial tendinous arch. The visceral endopelvic fascia sweeps medially from the tendinous arch, to cover the bladder and anterior prostate investing the pelvic organs and perineural and vascular sheaths.[16]

### Prostatic Fascia

(*Inner layer of periprostatic fascia, inner layer of lateral pelvic fascia*[16])

The prostatic fascia immediately surrounds the prostatic capsule anteriorly and laterally, and forms the medial boundary of the NVB.[17]

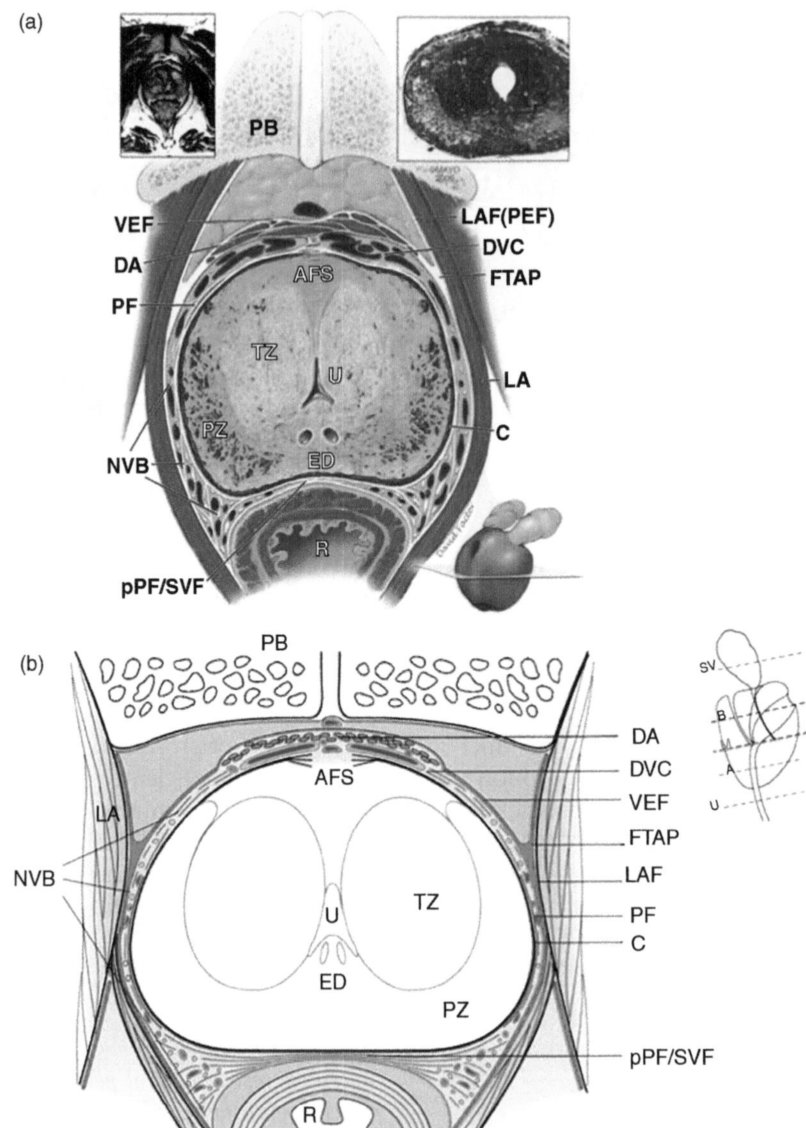


FIGURE 29.3 **Axial section of prostate and periprostatic fascias at midprostate.** (a) Anatomic (*reproduced with permission from the Mayo Clinic*). (b) Schematic. AFS, anterior fibromuscular stroma; C, capsule of prostate; DA, detrusor apron; DVC, dorsal vascular complex; ED, ejaculatory ducts; FTAP, fascial tendinous arch of pelvis; LA, levator ani muscle; LAF, levator ani fascia; NVB, neurovascular bundle; PB, pubic bone; PEF, parietal endopelvic fascia; PF, prostatic fascia; pPF/SVF, posterior prostatic fascia/seminal vesicles fascia (Denonvilliers' fascia); PZ, peripheral zone; R, rectum; TZ, transition zone; U, urethra; VEF, visceral endopelvic fascia. *Reprinted with permission from Walz et al.*[13]

## Denonvilliers' Fascia

(*Rectovesical septum, prostatorectal fascia, rectogenital fascia, posterior prostate visceral fascia, prostatoseminal vesicular fascia, posterior prostatic, and seminal vesical fascia*)[1,16]

Denonvilliers' fascia lies posteriorly, between the prostate and rectum. It has previously been described as having two layers; however, this is inaccurate as the posterior layer simply corresponds to the anterior fascia propria of the rectum. Denonvilliers' is most dense at the base of the prostate and thins caudally to its termination at the level of the external sphincter. Superiorly, it provides cover to the posterior aspect of the seminal vesicles. At its most lateral margin Denonvilliers merges with extensions of the endopelvic fascia at the NVB. Medial to this an intervening plane of adipose tissue lies between the prostate and Denonvilliers. In the midline Denonvilliers fuses anteriorly with the fibromuscular stroma of the prostate.[16]

## Detrusor Apron and Pubovesical/Puboprostatic Ligaments

The detrusor apron (DA) lies anterior to the prostate, obscuring the gland during radical prostatectomy

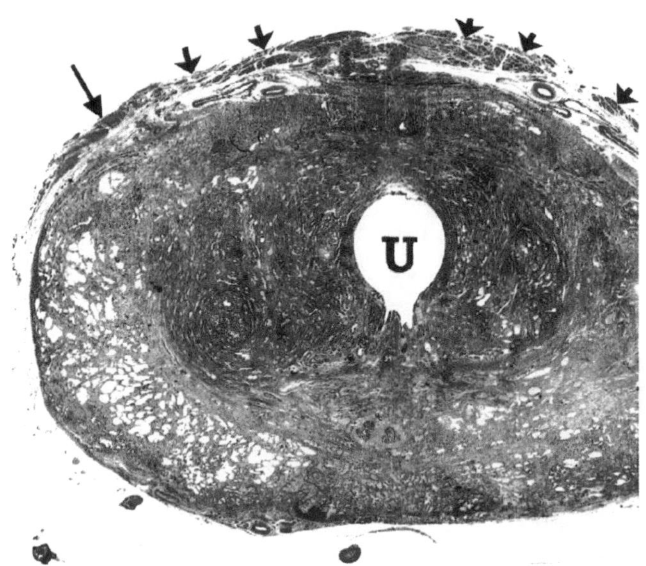


FIGURE 29.4 **Partial axial section of prostate through urethra (U) in Masson trichrome stain, showing detrusor apron.** *Reprinted with permission from Myers.*[18]

(Figure 29.4). It is an extension of the anterior bladder wall, beyond the point at which the bladder neck is formed. It comprises mostly longitudinal smooth muscle that is continuous with the bladder neck. In Myers' thorough review of the DA, he states that the DA must be considered a major component of McNeal's anterior fibromuscular stroma. The DA is most prominent at the midline where the smooth muscle fuses with the underlying prostate. Laterally, it fuses with the fascial tendinous arch of the pelvis. Anteriorly, the DA ends where it forms ligamentous attachments to the pubis.[18] These ligaments are more commonly known as the pubovesical/puboprostatic ligaments.

The puboprostatic ligaments are thought to provide important anterior stabilization of the urinary continence mechanism through anchoring the prostate, urethra, and bladder neck to the pubic bone. Preservation[19] or restoration of this support structure using an anterior suspension suture has been promoted to improve early urinary continence outcomes.[20,21]

## Vesicoprostatic Muscle

Two distinct tissue layers are encountered in the plane between the posterior bladder neck and the seminal vesicles and vas deferens. Directly beneath the bladder mucosa is a layer of longitudinal smooth muscle fibers that are in continuation with the outer detrusor muscle and insert into the base of the prostate.[22] This layer corresponds to the vesicoprostatic muscle[23] and may be considered analogous to the ventral detrusor apron.[13] Immediately posterior to this is a layer of fibroadipose tissue in continuity with the bladder adventitia.[22]

# PELVIC MUSCULATURE

The pelvic floor is often thought of simply as a muscular diaphragm, but in reality is a complex collection of muscles, ligaments, and fascia that work together to support the pelvic viscera and facilitate micturition and defection. The pelvic floor is traversed by the urethra and anal sphincters in the male, and comprises four principal layers: the endopelvic fascia, muscular pelvic diaphragm, urogenital diaphragm (perineal membrane), and the superficial transverse perinei muscles.[24]

## The Muscular Pelvic Diaphragm

The muscular pelvic floor is formed by levator ani (comprising iliococcygeus and pubococcygeus) and the vestigial coccygeus.[24] Levator ani shows a predominance of type 1 (slow-twitch) fibers.[25] These muscle fibers are able to maintain continuous activity, as demonstrated by EMG studies demonstrating continuous recordings from the striated muscles of the pelvic floor for up to 2 h, and even after falling asleep.[26]

Despite complex innervation patterns to individual pairs of muscles, the pelvic floor contracts as a single functional unit, to close the excretory tracts and support the pelvic viscera.[26]

## Puboperineales

(*Levator urethrae, levator prostatae, pubourethral muscle*[13])

The distinct *puboperineales* portion of levator ani is a specialized thickening of the levator ani that flanks the membranous urethra at the urogenital hiatus. It is anchored to the pubic bone anteriorly and inserts into the perineal body posteriorly (anterior to the anorectal junction).[27] Containing a population of fast twitch fibers, the puboperineales may assist in quick-stop urinary control by drawing the urethra anteriorly against the pubis.[3,25,27]

## Urinary Sphincter Complex

The urinary sphincter complex is comprised of an internal smooth muscle sphincter and a striated external sphincter commonly referred to as the rhabdosphincter, which extend from the bladder neck to the perineal membrane.[28,29] These two sphincters are responsible for passive continence and continence under conditions of stress, respectively.[29]

The smooth and striated components of the urethral sphincter muscles are indistinctly separated,[30,31] and are supported by the paraurethral musculature and connective tissue of the pelvic floor. In his 2008 review article, Koraitim provides a detailed historical and contemporary review of the urinary sphincter complex.[29]

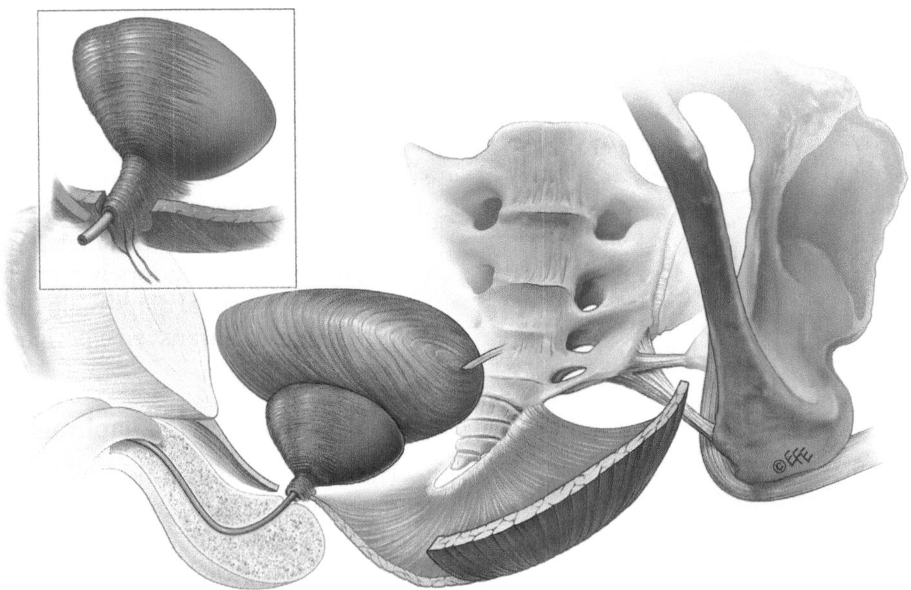

FIGURE 29.5 **External urethral sphincter and pelvic floor.**

## External Urethral Sphincter

*(Rhabdosphincter[1])*

The rhabdosphincter is an omega-shaped[32–37] or horseshoe-shaped[38] muscle that is most prominent at the anterior and lateral aspects of the membranous urethra. Its striated fibers are continuous distally to the perineal membrane, and extend proximally over the anterior prostate to the base of the bladder where they thin out as they intermingle with the smooth muscle fibers of the bladder[29,39] (Figure 29.5). Various anatomical descriptions exist for the dorsal insertion of this muscle. These include the perineal body, posterior musculofascial plate or rectourethralis muscle or fibrous median raphe.[31,33,34,36,40] Reconstruction of this posterior musculofascial plate, using the Rocco stitch, has been reported to improve early return of continence.[41]

The rhabdosphincter is an independent morphological unit, which is separated from the surrounding pelvic floor muscles by connective tissue.[42] Due to the small size of the individual muscle fibers of the rhabdosphincter (25–30% smaller than fibers of the associated muscles) and the fact that these are embedded in connective tissue, visualization of the rhabdosphincter at radical prostatectomy is obscured.[39] Its fascia is inseparable from the prostatic capsule,[39] and in fact some authors consider the sphincter to be a component of the prostatic capsule.[43]

The rhabdosphincter is comprised of slow and fast twitch muscle fibers, with a greater proportion of fast twitch fibers seen in the main caudal part.[43] The rhabdosphincter is activated proportionally with increased intra-abdominal pressure[44] to maintain continence in stress conditions. Its essential role in maintenance of continence has been illustrated in urodynamic investigation of postprostatectomy incontinence. These studies revealed sphincter weakness in the overwhelming majority, with bladder dysfunction rarely the isolated cause of incontinence in this patient group.[45–47] Endoluminal ultrasound studies demonstrate that contraction of the rhabdosphincter causes retraction of the urethra toward the perineal body[48,49] acting in opposition to the forward movement that occurs with contraction of the puboperinealis muscle of the pelvic floor.

## Internal Urethral Sphincter

*(Lissosphincter[1])*

The internal urethral sphincter (IUS) is a smooth muscle sphincter that completely encircles the urethra. It consists of an inner longitudinal layer and an outer circular layer.[31] In contrast to the rhabdosphincter, the IUS is most prominent proximally at the bladder neck and thins out distally.[3]

In addition to its role in passive continence, contraction of the IUS prevents retrograde ejaculation.

# NEUROANATOMY

The complex visceral activity of the pelvis requires the coordination of dual autonomic and somatic nervous systems[26] (Figure 29.6).

## Pelvic Autonomic Innervation

Thoracolumbar sympathetic outflow (T10-L2) via the hypogastric nerves, and parasympathetic supply via

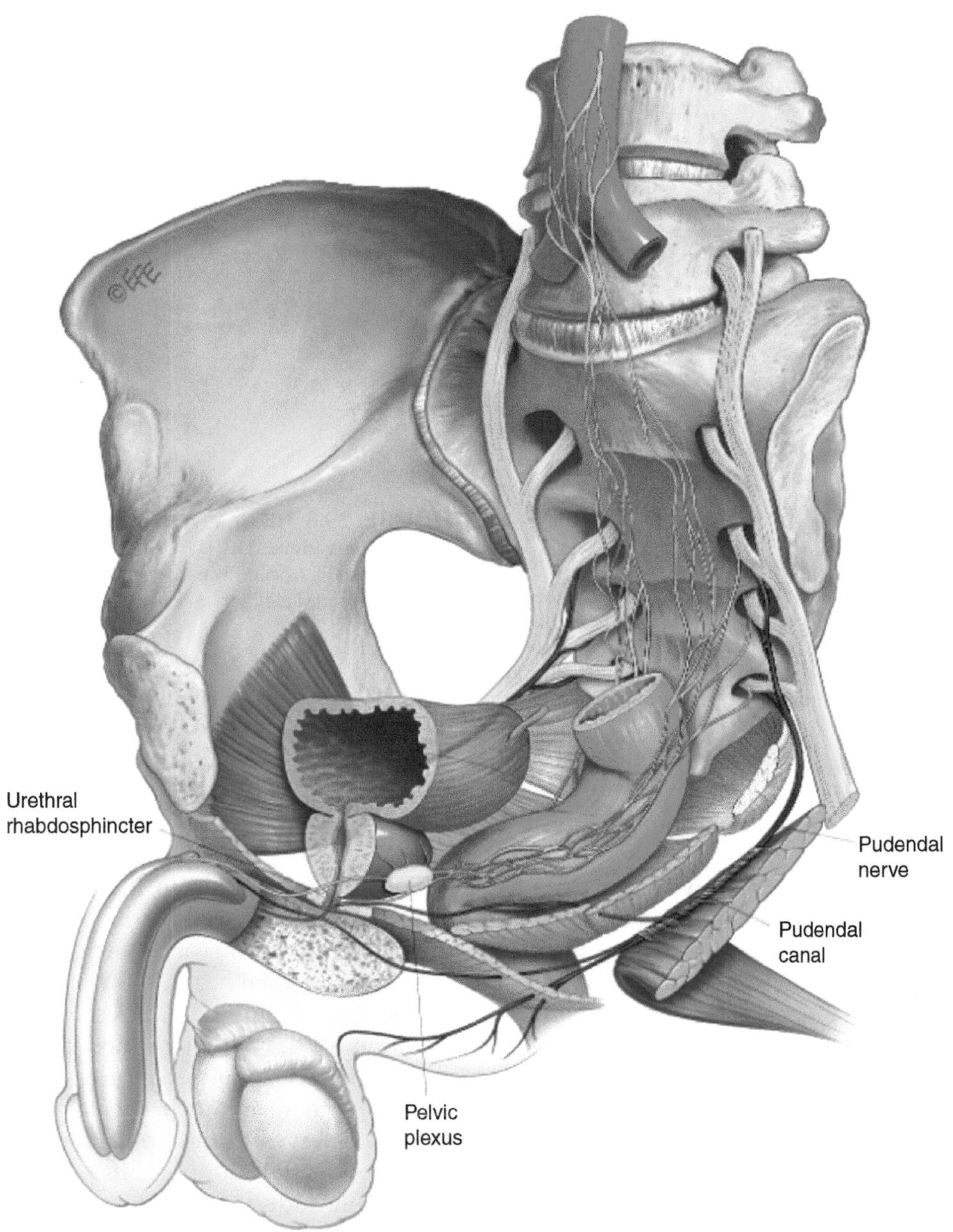


FIGURE 29.6 The pelvic plexus.

the pelvic splanchnic nerves (S2-4), converge at the pelvic plexus (*inferior hypogastric plexus*). This retroperitoneal, flat fenestrated network of nerves lies in a sagittal plane lateral to the rectum. It ranges from 3 cm to 5.5 cm long by 2.5–5 cm high, extending as far as 1.5 cm posterior to the dorsal edge of the rectum and 1 cm superior to the rectovesical pouch. The pelvic plexus has three major projections. Its anterior nerves extend across the lateral surface of the seminal vesicle and the inferolateral surface of the bladder. Anteroinferiorly, nerves from the pelvic plexus pass to the prostatovesical junction and obliquely along the lateral surface of the prostate. Inferiorly, nerves course between the rectum and the posterolateral surface of the prostate. These most caudal fibers of the pelvic plexus join branches of the inferior vesicle artery to form the NVB.[50]

## The Neurovascular Bundle

The NVB is a complex and variable structure that transmits the cavernosal nerves among other neurovascular structures. Given the complexity of the neurovascular bundle constituents it is not appropriate to interchange the terms "cavernosal nerves" and "NVB."[51] Damage to the NVB at surgery can occur by direct injury from cutting, diathermy or sutures, or

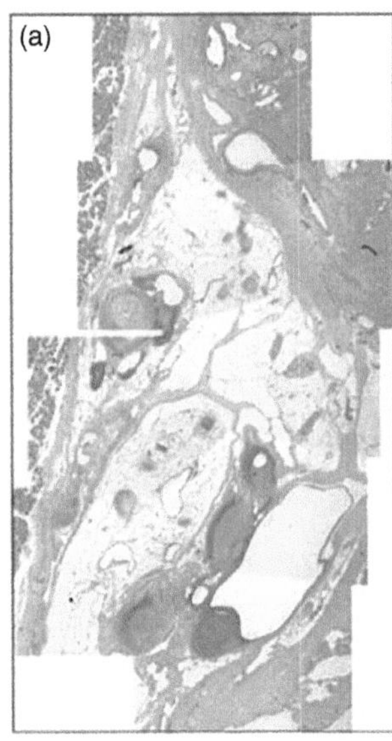


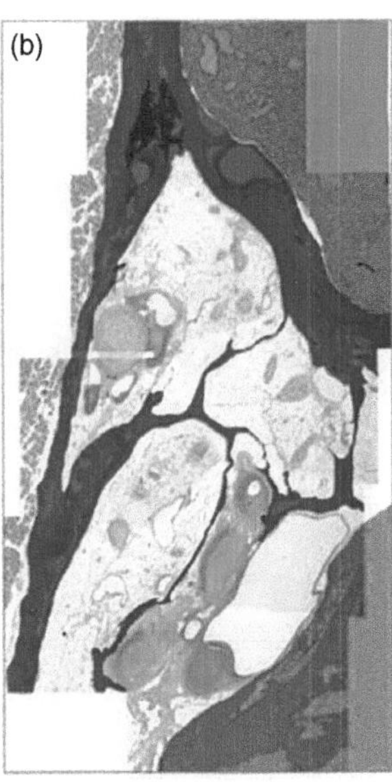


FIGURE 29.7 **Compartmental architecture of the neurovascular bundle (midprostate).** (a) Hematoxylin and eosin slide showing compartmental neurovascular bundle architecture. (b, overlay) Prostate (green), fascial bands (blue), nerves (yellow), para-rectal tissue (gray), levator ani musculature (spotted pink). *Reprinted with permission from Costello et al.*[51]

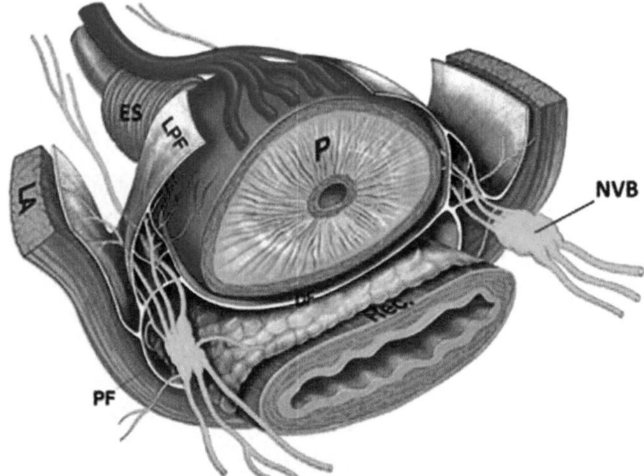


FIGURE 29.8 **Illustrative representation of periprostatic autonomic innervations.** DF, Denonviller's Fascia; LA, levator ani; LPF, lateral prostatic fascia; PF, para-rectal fat; P, rectum and prostate; Rec, rectum. *Reprinted with permission from Costello et al.*[51]

indirectly as a result of neuropraxia from stretching of the NVB.

The NVB commences at the base of the prostate between 3 and 9 o'clock. It then courses along the dorsolateral aspect of the prostate, converging toward the midprostatic level to form a more condensed NVB, only to diverge again toward the apex.[50] Here, the cavernosal nerves traverse the urogenital hiatus posterolateral to the prostatic apex,[50] to travel with the deep artery and vein of the penis in order to provide parasympathetic supply to the crura.[52]

The NVB is functionally compartmentalized into three distinct facial spaces (Figure 29.7). The posterior and posterolateral sections, within the leaves of Denonvilliers and pararectal fascia, supply the rectum. Laterally, the lateral pelvic fascia surrounds the supply to levator ani. Most anteriorly lie the cavernosal nerves and prostatic neurovascular supply (Figure 29.8). This functional organization is not absolute and is less distinct proximally.[50]

In some cases periprostatic nerves may be more widely spread, without a definite bundle formation.[15] For this reason, some authors promote a "high anterior release" or "veil of Aphrodite" nerve sparing technique to preserve anterior parasympathetic fibers.[53,54] However, subsequent histological studies have demonstrated that only a tiny minority of ventrolateral nerves are functionally significant parasympathetic fibers. Therefore, a high anterior release is unlikely to result in improved preservation of cavernous nerve fibers, but may contribute to improved potency outcomes due to decreased traction on the NVB.[51,55]

## Innervation of the Pelvic Floor and Urinary Sphincter Muscles

Somatic innervation to the pelvic floor is primarily from the levator ani nerve on the superior side of the muscle (from S3 and/or S4), with a minor contribution from the pudendal nerve.[24]

The smooth muscle IUS receives autonomic supply from the NVB.[56] Although most authors agree that the striated rhabdosphincter receives purely somatic supply,[35,38,57,58] there is much debate about the origin and course of nerves to this sphincter. The simple description of pudendal supply via its perineal branch may not adequately address the finding in some studies that preserving the NVB during radical prostatectomy leads to improved urinary continence outcomes.[59,60]

The pudendal nerve arises most commonly from the S2-4 ventral rami of the sacral plexus, but occasionally receives contributions from S5.[38] After exiting the pelvis via the greater sciatic foramen, the pudendal nerve passes along the lateral wall of the ischiorectal fossa within Alcock's canal. Here it gives off its inferior rectal branch. The pudendal nerve then provides supply to the rhabdosphincter via one of its other terminal branches: the perineal branch[58,61] or dorsal nerve of the penis.[62]

Potential neurological mechanisms that may account for better early continence outcomes with NVB sparing surgery seen in some studies include the presence of intrapelvic somatic branches supplying the rhabdosphincter or close proximity of pudendal branches to the prostatic apex.

In a histological study of male fetal urethral specimens, myelinated nerve fibers running with the unmyelinated fibers from the bladder neck to the rhabdosphincter were identified. Authors were unable to confirm the origin of these fibers, but their findings do support that the rhabdosphincter may also receive intrapelvic (supralevator) somatic supply.[63] Another study could not identify any pudendal branches to the rhabdosphincter in a significant proportion of specimens (4 out of 10), which led to the hypothesis that in these cases collateral somatic branches travelled with the pelvic plexus to innervate the rhabdosphincter.[38] In a cadaveric dissection study of 12 hemipelves, a consistent intrapelvic branch from the pudendal nerve was found to arise from Alcock's canal.[61] Similarly, one study reported "a nerve component supplying the area of the rhabdosphincter seemed to accompany the cavernous nerve."[64] Currently, the frequency and precise course and origin of intrapelvic somatic nerves to the rhabdosphincter are not well described.

Pudendal nerve branches to the sphincter can be found as close as 3 mm from the apex of the prostate.[58] In one study, the mean distance from the lowest point of the endopelvic fascia to the sphincteric branch of the pudendal nerve was only 5.5 mm.[65] Careful attention must be paid during dissection of the prostatic apex and when using diathermy or sutures in this area, in order to avoid inadvertent damage to these branches. In addition, preservation of levator ani fascia has been reported to assist preservation of the levator ani muscle, rhabdosphincter, and pudendal nerve branches to the rhabdosphincter.[65]

The pelvis and perineum are regions of complex neuroanatomy. Cases of autonomic–somatic communications have been described between the pudendal nerve, the nerve to levator ani, and the pelvic splanchnic nerves.[38,62] However, other studies have been unable to reproduce these findings.[34] More recently, histological methods have been employed to investigate pelvic anatomy due to the challenges associated with gross dissection of the fine nerve branches in this region. In a study of the immunohistochemistry of serial sections, with three-dimensional reconstruction, supralevator, intralevator, and infralevator autonomic–somatic communications were identified.[38] Similar autonomic–somatic communications are more often reported in the penile region.[52,66]

Despite significant advances in anatomical knowledge in recent years, even today, pelvic neuroanatomy remains incompletely understood.

## VASCULAR ANATOMY

### Arterial Supply of the Prostate

Arterial supply to the prostate is variable. Although it usually arises from two arteries, it may only arise from one. In the case of dual arterial supply, these are cranially the vesiculo-prostatic artery (anterior–lateral prostatic pedicle) and caudally the prostatic artery (posterior–lateral prostatic pedicle). The former most frequently arises from the superior vesical artery. It provides supply to the bladder base and inner and cranial prostate gland via the inferior vesical and prostatic arteries. This artery is frequently enlarged in BPH. The caudal prostatic artery supplies the peripheral and caudal prostate and may have a close relationship with rectal or anal branches.[67,68]

### Arterial Supply of the Seminal Vesicles

The vesiculo-deferential artery supplies the seminal vesicles. This artery most commonly originates from the internal iliac artery at the site of origin of the umbilical artery branch, but may also come from the superior vesical artery, umbilical artery, or from the more proximal anterior division of the internal iliac artery. It passes medially behind the superolateral border of the bladder, in front of the ureter, to the lateral end of the vesicle where it divides to give branches to the anterior surface of the SV, vas deferens, and trigone. Small branches from the prostatovesical artery supply the posterior aspect of the SV.[67]

### Accessory Pudendal Arteries

Accessory pudendal arteries (APA) usually occur as a result of congenital anatomical variation, but may correlate with the presence of internal pudendal atherosclerotic disease in some cases.[69] They are highly variable and can originate from the obturator, inferior vesicle, or external pudendal artery.[70] Reported prevalence ranges from 4% to 85%.[69,71] APAs can be classified based on their location as lateral or apical. The former pass over the prostatic surface or the pelvic sidewall and may run above or below the endopelvic fascia. Apical APAs characteristically emerge through the levator ani muscle to approach the prostatic apex inferolateral to the puboprostatic ligaments.[70]

Injury to APAs may be responsible for vasculogenic impotence after nerve sparing radical prostatectomy,[72] particularly as they may provide sole arterial supply to the corpora in some cases.[73] Given that preservation of APAs is feasible without increasing risk of positive surgical margin, these arteries should be preserved where possible during radical prostatectomy.[70,71]

### The Dorsal Vascular Complex

The dorsal venous complex, initially described by Santorini in 1724, is more accurately considered as a dorsal vascular complex (DVC) due to the presence of small arteries as well as veins.[74] It lies just posterior to the pubic arch on the ventral prostatic surface, covering it and the urethral sphincter complex.[18]

Ligation of the DVC during RP reduces blood loss during apical dissection of the prostate.[74] Notably, venous drainage of the erectile tissue occurs via supralevator and infralevator pathways, emptying into the lateral vesicoprostatic veins and the internal pudendal veins, respectively.[75] Due to the presence of this collateral infralevator drainage, ligation of the supralevator DVC does not impair erectile function.

More recently, some surgeons have reported successful division of the DCV without initial ligation, due to the increased intra-abdominal pressure from $CO_2$ insufflation in laparoscopic and robotic-assisted surgery. This technique is promoted to increase the likelihood of achieving a negative apical margin and maximize urethral length.[74]

## References


1. Rassweiler J, Laguna P, Chlosta P, et al. ESUT expert group on laparoscopy proposes uniform terminology during radical prostatectomy: we need to speak the same language. *Eur Urol* 2013;**64**(1):97–100.
2. McMinn R. *Lasts anatomy regional and applied*. 9th ed. Australia: Elsevier; 1994.
3. Brooks JD, Chao WM, Kerr J. Male pelvic anatomy reconstructed from the visible human data set. *J Urol* 1998;**159**(3):868–72.
4. McNeal JE. The zonal anatomy of the prostate. *Prostate* 1981; **2**(1):35–49.
5. De Marzo AM, Platz EA, Sutcliffe S, et al. Inflammation in prostate carcinogenesis. *Nat Rev Cancer* 2007;**7**(4):256–69.
6. McNeal JE, Redwine EA, Freiha FS, et al. Zonal distribution of prostatic adenocarcinoma. Correlation with histologic pattern and direction of spread. *Am J Surg Pathol* 1988;**12**(12):897–906.
7. Carvalhal GF, Smith DS, Mager DE. Digital rectal examination for detecting prostate cancer at prostate specific antigen levels of 4 ng/mL or less. *J Urol* 1999;**161**:835–9.
8. McNeal JE. Cancer volume and site of origin of adenocarcinoma in the prostate: relationship to local and distant spread. *Hum Pathol* 1992;**23**(3):258–66.
9. Cohen RJ, Shannon BA, Phillips M, et al. Central zone carcinoma of the prostate gland: a distinct tumor type with poor prognostic features. *J Urol* 2008;**179**(5):1762–7 discussion 1767.
10. Pelzer AE, Bektic J, Berger AP, et al. Are transition zone biopsies still necessary to improve prostate cancer detection? *Eur Urol* 2005;**48**:916–21.
11. Zlotta AR, Roumeguère T, Ravery V, et al. Is seminal vesicle ablation mandatory for all patients undergoing radical prostatectomy? *Eur Urol* 2004;**46**:42–9.
12. Villers AA, McNeal JE, Redwine EA, et al. Pathogenesis and biological significance of seminal vesicle invasion in prostatic adenocarcinoma. *J Urol* 1990;**143**:1183–7.
13. Walz J, Burnett AL, Costello AJ, et al. A critical analysis of the current knowledge of surgical anatomy related to optimization of cancer control and preservation of continence and erection in candidates for radical prostatectomy. *Eur Urol* 2010;**57**(2): 179–92.
14. Ayala AG, Ro JY, Babaian R, et al. The prostatic capsule: does it exist? Its importance in the staging and treatment of prostatic carcinoma. *Am J Surg Pathol* 1989;**13**(1):21–7.
15. Kiyoshima K. Anatomical features of periprostatic tissue and its surroundings: a histological analysis of 79 radical retropubic prostatectomy specimens. *Jpn J Clin Oncol* 2004;**34**:463–8.
16. Raychaudhuri B, Cahill D. Pelvic fasciae in urology. *Ann R Coll Surg Engl* 2008;**90**(8):633–7.
17. Tewari A, Peabody JO, Fischer M, et al. An operative and anatomic study to help in nerve sparing during laparoscopic and robotic radical prostatectomy. *Eur Urol* 2003;**43**:444–54.
18. Myers RP. Detrusor apron, associated vascular plexus, and avascular plane: relevance to radical retropubic prostatectomy – anatomic and surgical commentary. *Urology* 2002;**59**(4):472–9.
19. Steiner MS. Continence-preserving anatomic radical retropubic prostatectomy. *Urology* 2000;**55**(3):427–35.
20. Tewari AK, Bigelow K, Rao S, et al. Anatomic restoration technique of continence mechanism and preservation of puboprostatic collar: a novel modification to achieve early urinary continence in men undergoing robotic prostatectomy. *Urology* 2007;**69**:726–31.
21. Patel VR, Coelho RF, Palmer KJ, et al. Periurethral suspension stitch during robot-assisted laparoscopic radical prostatectomy: description of the technique and continence outcomes. *Eur Urol* 2009;**56**:472–8.
22. Secin FP, Karanikolas N, Gopalan A, et al. The anterior layer of Denonvilliers' fascia: a common misconception in the laparoscopic prostatectomy literature. *J Urol* 2007;**177**(2):521–5.
23. Dorschner W, Stolzenburg JU, Dieterich F. A new theory of micturition and urinary continence based on histomorphological studies. 2. The musculus sphincter vesicae: continence or sexual function? *Urol Int* 1994;**52**(3):154–8.
24. Stoker J. Anorectal and pelvic floor anatomy. *J Best Pract Res Clin Gastroenterol*. Elsevier Ltd; 2009 p. 463–475.
25. Gosling JA, Dixon JS, Critchley HO, et al. A comparative study of the human external sphincter and periurethral levator ani muscles. *Br J Urol* 1981;**53**:35–41.
26. Enck P, Vodušek DB. Electromyography of pelvic floor muscles. *J Electromyogr Kinesiol* 2006;**16**(6):568–77.
27. Myers RP, Cahill DR, Kay PA, et al. Puboperineales: muscular boundaries of the male urogenital hiatus in 3d from magnetic resonance imaging. *J Urol* 2000;**164**(4):1412–5.
28. Petersén I, Franksson C. Electromyographic study of the striated muscles of the male urethra. *Br J Urol* 1955;**27**:148–53.
29. Koraitim MM. The male urethral sphincter complex revisited: an anatomical concept and its physiological correlate. *J Urol* 2008;**179**:1683–9.
30. Yucel S, Baskin LS. An anatomical description of the male and female urethral sphincter complex. *J Urol* 2004;**171**:1890–7.
31. Burnett AL, Mostwin JL. *In situ* anatomical study of the male urethral sphincteric complex: relevance to continence preservation following major pelvic surgery. *J Urol* 1998;**160**(4):1301–6.
32. Ludwikowski B, Oesch Hayward I, Brenner E, et al. The development of the external urethral sphincter in humans. *BJU Int* 2001;**87**:565–8.
33. Sebe P, Schwentner C, Oswald J, et al. Fetal development of striated and smooth muscle sphincters of the male urethra from a common primordium and modifications due to the development of the prostate: an anatomic and histologic study. *Prostate* 2005;**62**(4):388–93.
34. Strasser H, Klima G, Poisel S, et al. Anatomy and innervation of the rhabdosphincter of the male urethra. *Prostate* 1996;**28**:24–31.
35. Dalpiaz O, Mitterberger M, Kerschbaumer A, et al. Anatomical approach for surgery of the male posterior urethra. *BJU Int* 2008;**102**(10):1448–51.

36. Helwege G, Strasser H, Knappe R, et al. Transurethral/sonomorphologic evaluation of male external sphincter of urethra. Eur Radiol 1994;4(6):525–528.
37. Strasser H, Bartsch G. Anatomy and innervation of the rhabdosphincter of the male urethra. *Semin Urol Oncol* 2000;**18**:2–8.
38. Akita K, Sakamoto H, Sato T. Origins and courses of the nervous branches to the male urethral sphincter. *Surg Radiol Anat* 2003;**25**(5–6):387–92.
39. Oelrich TM. The urethral sphincter muscle in the male. *Am J Anat* 1980;**158**(2):229–46.
40. Soga H, Takenaka A, Murakami G, et al. Topographical relationship between urethral rhabdosphincter and rectourethralis muscle: a better understanding of the apical dissection and the posterior stitches in radical prostatectomy. *Int J Urol* 2008;**15**(8):729–32.
41. Rocco B, Cozzi G, Spinelli MG, et al. Posterior musculofascial reconstruction after radical prostatectomy: a systematic review of the literature. *Eur Urol* 2012;**62**(5):779–90.
42. Dorschner W, Biesold M, Schmidt F, et al. The dispute about the external sphincter and the urogenital diaphragm. *J Urol* 1999;**162**(6):1942–5.
43. Elbadawi A, Mathews R, Light JK, et al. Immunohistochemical and ultrastructural study of rhabdosphincter component of the prostatic capsule. *J Urol* 1997;**158**(5):1819–28.
44. Stafford RE, Ashton-Miller JA, Constantinou CE, et al. Novel insight into the dynamics of male pelvic floor contractions through transperineal ultrasound imaging. *J Urol* 2012;**188**(4):1224–30.
45. Chao R, Mayo ME. Incontinence after radical prostatectomy. *J Urol* 1995;**154**:16–8.
46. Ficazzola MA, Nitti VW. The etiology of post-radical prostatectomy incontinence and correlation of symptoms with urodynamic findings. *J Urol* 1998;**160**(4):1317–20.
47. Groutz A, Blaivas JG, Chaikin DC, et al. The pathophysiology of post-radical prostatectomy incontinence: a clinical and video urodynamic study. *J Urol* 2000;**163**:1767–70.
48. Strasser H, Ninkovic M, Hess M, et al. Anatomic and functional studies of the male and female urethral sphincter. *World J Urol* 2000;**18**:324–9.
49. Strasser H, Pinggera GM, Gozzi C, et al. Three-dimensional transrectal ultrasound of the male urethral rhabdosphincter. *World J Urol* 2004;**22**:335–8.
50. Costello AJ, Brooks M, Cole OJ. Anatomical studies of the neurovascular bundle and cavernosal nerves. *BJU Int* 2004;**94**(7):1071–6.
51. Costello AJ, Dowdle BW, Namdarian B, et al. Immunohistochemical study of the cavernous nerves in the periprostatic region. *BJU Int* 2011;**107**(8):1210–5.
52. Paick JS, Donatucci CF, Lue T. Anatomy of cavernous nerves distal to prostate: microdissection study in adult male cadavers. *Urology* 1993;**42**(2):145–9.
53. Menon M, Shrivastava A, Kaul S, et al. Vattikuti institute prostatectomy: contemporary technique and analysis of results. *Eur Urol* 2007;**51**(3):648–58.
54. Budaus L, Isbarn H, Schlomm T, et al. Current technique of open intrafascial nerve-sparing retropubic prostatectomy. *Eur Urol* 2009;**56**(2):317–24.
55. Ganzer R, Stolzenburg J-U, Wieland WF, et al. Anatomic study of periprostatic nerve distribution: immunohistochemical differentiation of parasympathetic and sympathetic nerve fibres. *Eur Urol* 2012;**62**:1150–6.
56. Gosling JA, Dixon JS. The structure and innervation of smooth muscle in the wall of the bladder neck and proximal urethra. *Br J Urol* 1975;**47**:549–58.
57. Juenemann KP, Lue TF, Schmidt RA, et al. Clinical significance of sacral and pudendal nerve anatomy. *J Urol* 1988;**139**:74.
58. Narayan P, Konety B, Aslam K, et al. Neuroanatomy of the external urethral sphincter: implications for urinary continence preservation during radical prostate surgery. *J Urol* 1995;**153**(2):337–41.
59. Nandipati KC, Raina R, Agarwal A, et al. Nerve-sparing surgery significantly affects long-term continence after radical prostatectomy. *Urology* 2007;**70**(6):1127–30.
60. Choi WW, Freire MP, Soukup JR, et al. Nerve-sparing technique and urinary control after robot-assisted laparoscopic prostatectomy. *World J Urol* 2011;**29**(1):21–7.
61. Hollabaugh J, Robert S, Dmochowski RR, et al. Neuroanatomy of the male rhabdosphincter. *Urology* 1997;**49**:426–34.
62. Song LJ, Lu HK, Wang JP, et al. Cadaveric study of nerves supplying the membranous urethra. *Neurourol Urodyn* 2010;**29**:592–5.
63. Karam I, Droupy S, Abd-Alsamad I, et al. The precise location and nature of the nerves to the male human urethra: histological and immunohistochemical studies with three-dimensional reconstruction. *Eur Urol* 2005;**48**(5):858–64.
64. Takenaka A, Murakami G, Matsubara A, et al. Variation in course of cavernous nerve with special reference to details of topographic relationships near prostatic apex: histologic study using male cadavers. *Urology* 2005;**65**(1):136–42.
65. Takenaka A, Hara R, Soga H, et al. A novel technique for approaching the endopelvic fascia in retropubic radical prostatectomy, based on an anatomical study of fixed and fresh cadavers. *BJU Int* 2005;**95**(6):766–71.
66. Colleselli K, Strasser H, Morrigl B, et al. Anatomical approach in surgery on the membranous urethra. *World J Urol* 1990;**7**(4):189–91.
67. Clegg EJ. The arterial supply of the human prostate and seminal vesicles. *J Anat* 1955;**89**(2):209–16.
68. Bilhim T, Tinto HR, Fernandes L, et al. Radiological anatomy of prostatic arteries. *Tech Vasc Interv Radiol* 2012;**15**(4):276–85.
69. Droupy S, Benoit G, Giuliano F, et al. Penile arteries in humans origin-distribution-variations. *Surg Radiol Anat* 1997;**19**(3):161–7.
70. Secin FP, Touijer K, Mulhall J, et al. Anatomy and preservation of accessory pudendal arteries in laparoscopic radical prostatectomy. *Eur Urol* 2007;**51**(5):1229–35.
71. Rogers CG, Trock BP, Walsh PC. Preservation of accessory pudendal arteries during radical retropubic prostatectomy: surgical technique and results. *Urology* 2004;**64**(1):148–51.
72. Droupy S, Hessel A, Benoit G, et al. Assessment of the functional role of accessory pudendal arteries in erection by transrectal color Doppler ultrasound. *J Urol* 1999;**162**(6):1987–91.
73. Breza J, Aboseif S, Lue T, et al. Detailed anatomy of penile neurovascular structures: surgical significance. *J Urol* 1989;**141**(2):437–43.
74. Power NE, Silberstein JL, Kulkarni GS, et al. The dorsal venous complex (DVC): dorsal venous or dorsal vasculature complex? Santorini's plexus revisited. *BJU Int* 2011;**108**(6):930–2.
75. Benoit G, Droupy S, Quillard J, et al. Supra and infralevator neurovascular pathways to the penile corpora cavernosa. *J Anat* 1999;**195**(Pt 4):605–15.

CHAPTER

# 30

# Radical Retropubic Prostatectomy

*Leonard G. Gomella, MD, Chandan Kundavaram, MD*

Department of Urology, Kimmel Cancer Center, Thomas Jefferson University, Philadelphia, PA, USA

## INTRODUCTION

The concept of removal of the entire prostate to treat prostate cancer was first introduced by Irish urologist Dr Terence Millin in the late 1940s.[1] The procedure did not develop widespread popularity until Dr Patrick Walsh performed detailed studies of the anatomy relating to the procedure and popularized the nerve-sparing open radical retropubic prostatectomy in the early 1980s. This dramatically improved outcomes with decreased blood loss and improved continence and erectile function.[2] Traditionally considered the "gold standard," the open radical prostatectomy (ORP) has been replaced by the robotic-assisted laparoscopic radical prostatectomy (RALP) as the most common surgical approach to treat localized prostate cancer based on 2008 data that began to specifically identify robotic assistance.[3] Pure laparoscopic radical prostatectomy is still performed at a very limited number of centers. Open radical prostatectomy using the perineal approach is described in Chapter 37 and is less commonly used today.

RALP presents the major alternative approach to radical prostatectomy (RP) today and is discussed in Chapter 43: "The technique of robotic nerve sparing prostatectomy." In experienced hands the outcomes appear to be very similar between ORP and RALP with the exception of increased blood loss with the open technique and higher overall cost using the robot. Clinical pathways and improved perioperative management strategies guiding patient and spousal expectations have limited the previous significant differences in length of hospitalization, pain management, and recovery thus minimizing the initial marketing advantages of the RALP technique.[4–6] The evidence surrounding RALP generally supports shorter hospitalization but lacks conclusive evidence that the robotic approach results in earlier return to physical activity or improved disease-specific outcomes.[7] Schroeck et al. also found that patients who underwent RALP had higher expectations concerning erectile function return than did their radical retropubic prostatectomy counterparts. Furthermore, patients may decide on the experience of the surgeon over the technology in their decision to undergo open prostatectomy.[7] For men 65 years of age and older, RALP and ORP had similar rates of complications based on a SEER data analysis. RALP and ORP have comparable rates of complications and additional cancer therapies, even in the RALP postdissemination era.[8] Again in this study, RALP was associated with lower risk of blood transfusions and a slightly shorter length of stay, but the benefits did not translate to a decrease in cost.

There is ongoing conflict concerning outcome measures, such as potency and urinary continence, when comparing the RALP with ORP. Oncologic follow-up for robotically assisted laparoscopic radical prostatectomy and open radical prostatectomy appears to be similar.[4] Variations in measurement and reporting of postoperative outcomes continue to make direct comparisons difficult.[9] Clinical trials are unlikely to answer this question; a prospective randomized trial comparing RALP versus ORP at the Mayo Clinic was closed due to slow enrollment.[10]

There is a wide spectrum of treatments for localized prostate cancer. The management decision takes into account tumor characteristics, patient characteristics and preferences of the patient and surgeon. It is essential that the patient be actively engaged in discussions of the risks and benefits of the various approaches that may include active treatment such as surgical removal of the prostate or some form of radiation therapy or less aggressive approaches such as active surveillance. At the Kimmel Cancer Center of Thomas Jefferson University we established a multidisciplinary clinic in 1996.[11] This provides the opportunity for men and their families to be evaluated by all GU specialists including urology, radiation oncology, and medical oncology in one session, allowing a thorough discussion of all treatment options including active treatment with surgery or radiation,

Prostate Cancer. http://dx.doi.org/10.1016/B978-0-12-800077-9.00030-X

active surveillance, or investigational protocols. This clinic design is becoming commonplace in other centers.[12]

## INDICATIONS AND CONTRAINDICATIONS

The traditional indications for RP have included men with organ-confined disease and a greater than 10 year life expectancy.[13,14] There has been a shift toward lower-stage tumors and men are generally younger at the time of diagnosis and more interested in preserving sexual function. The European Association of Urology (EAU) has stated that radical prostatectomy is contraindicated with a high risk of extracapsular disease.[13] The 2014 National Comprehensive Cancer Network (NCCN) guidelines state that RP is appropriate for any cancer that is clinically localized and can be completely excised, with a life expectancy of >10 years and no serious comorbid conditions that would contraindicate an elective surgical procedure.[14]

However, the concept of cytoreductive radical prostatectomy in the setting of locally advanced or metastatic disease has recently gained interest.[15] The majority of the data are retrospective in nature suggesting that patients with metastatic prostate cancer who had prior radical prostatectomy had a better survival and improved response to systemic therapy.[15,16] The ECOG trial headed by Messing suggests excellent long-term (>10 years) survival in men undergoing ORP in the setting of positive nodes and immediate androgen deprivation.[17] Cytoreductive RP is an evolving concept that is not yet supported by any guideline. Salvage RP is an option for highly selected men following local recurrence after radiation or cryotherapy; however, the morbidity can be significant.[18]

There are some patients who are not ideal for RALP who may be better served with an open technique. Some relative contraindications and limitations of RALP described in the literature are noted here.[19,20] While these may also present challenges to open surgery, these are generally considered more easily managed using open techniques.

- Obesity (BMI >30 kg/m$^2$): may distort anatomy and may present difficulty tolerating the steep Trendelenburg position used in RALP.
- Prior abdominal surgery with extensive adhesions or history or ruptured viscera/peritonitis.
- Previous radiation or hormonal therapy.
- History of transurethral or suprapubic prostatectomy.
- Large-volume prostate (>60 g), large median or lateral lobes.
- Narrow pelvis.

## OPERATIVE TECHNIQUE

### Preoperative Management

Most guidelines agree that a man with low-risk prostate cancer (PSA <10 ng/mL and a Gleason score of 6 or less and clinical stage T1c or T2a) does not need formal staging by CT/MRI and bone scan.[14,21] Concerning more advanced risk features, there are wide variations in staging recommendations in the literature.[22,23]

Surgery should be delayed at least 8 weeks following prostate biopsy to allow resolution of any biopsy-induced inflammation. The patient can be offered the option of autologous blood donation; however, there has been a decline in the blood loss associated with ORP due to improved techniques. Some have advocated the use of erythrocyte stimulating medications combined with acute intraoperative hemodilution to minimize reductions in hemoglobin.[24] It is strongly recommended that patients undergo a complete health assessment with a focus on cardiac health before undergoing radical prostatectomy as a way to reduce perioperative morbidity and mortality.[25]

In 2014 the American Urologic Association (AUA) updated guidelines on the use of prophylactic antibiotics for procedures that enter the urinary tract such as radical prostatectomy.[26] A first- or second-generation cephalosporin is primarily recommended with other alternatives available based on patient's allergy history. Cefazolin 1 g IV is our drug of choice for ORP. The AUA has also issued guidelines for Anticoagulation and Antiplatelet Therapy in Urologic Practice.[27] Perioperative continuation of aspirin for stroke prevention or cardiac stent may be associated with a minor risk of increased bleeding, but the transfusion rate is not increased and the consequences of bleeding are minor. The use of perioperative heparin or low molecular weight heparins are thought to increase the risk of lymphocele; however, recent data suggest that this may not be true.[28] It is our practice not to use any systemic anticoagulation therapy to prevent deep venous thrombosis beyond compression stockings and early ambulation.

Patients have a mild bowel preparation by assuming a clear liquid diet for 24 h before surgery, as well as one bottle of magnesium citrate the day before, and an enema the morning of surgery before reporting to the hospital. General anesthesia is used at our center with epidural anesthesia an acceptable alternative at some centers.

### Patient Position

The patient is situated in the supine position with the table flexed allowing for optimal exposure of the pelvis. Care is taken to pad all pressure points. The legs may be placed into a frog leg position or into dorsal lithotomy position to gain access to the rectum for more difficult cases if needed. For complex cases such as in the setting of salvage prostatectomy following radiation therapy, a large 30F rectal tube is placed to facilitate rectal injury identification. The patient is prepped and draped including the lower abdomen and external genitalia. If not done previously, a flexible cystoscopy is performed to evaluate for any anatomic variations. A 22-French Foley catheter with 10 mL balloon is placed sterilely, with the balloon overinflated to 30 mL and left to drain.

## Incision and Exposure

A lower midline incision is made from below the umbilicus ending at the pubic bone. The length can vary based on body habitus and is typically 4–5 inches (10–13 cm). This incision is then carried down through the subcutaneous layers to reveal the fascia. The midline is palpated and the fascia is divided along the length of the incision to reveal the rectus muscle. The muscle is then bluntly divided in the midline and caudally the pyramidalis muscle is usually encountered and divided with electrocautery to reveal the pubic bone. Next the transversalis fascia is divided over the bladder. The space of Retzius is then developed bluntly usually with a sponge stick freeing the bladder from the pelvic sidewall and exposing the iliac vessels bilaterally. We use an Omni retractor system (Omni-Tract® Surgical, St Paul, MN) to maintain exposure for the case; alternatively a standard Balfour can be used as shown in the accompanying illustrations. The abdominal walls are retracted laterally through the remainder of the case making sure that there is no compression on the external iliac vessels. The ORP is carried out in the extravesical space that theoretically limits injury to major vascular structures and the bowels. Also, any urinary leakage or bleeding postoperatively will usually be contained in this limited space without entering the peritoneal cavity. Some have noted that the single infraumbilical incision is not that cosmetically different from the multiple incisions above the umbilicus for RALP.[29]

## Pelvic Lymph Node Dissection

Based on patient characteristics and surgeon preference a pelvic lymph node dissection can be performed. Chapter 34, "Indications For Pelvic Lymphadenectomy," reviews the role of pelvic lymphadenectomy in the management of prostate cancer. Guidelines suggest that pelvic lymphadenectomy can be avoided where there is a low risk of pelvic nodal involvement.[13,14,21] At the present time, there is some disagreement over the precise template and anatomic limits for pelvic lymphadenectomy for prostate cancer including the utility of an extended pelvic lymphadenectomy.[30]

A traditional limited bilateral obturator pelvic lymphadenectomy is usually performed. The table is tilted toward the side of interest. The bladder is retracted toward the contralateral side with a malleable blade retractor. Template borders for this obturator pelvic lymphadenectomy dissection are the external iliac vein, obturator nerve, and node of Cloquet. A combination of sharp and blunt dissection is used to free the packet. Care is taken to preserve tissues lateral to the external iliac vein. Small intervening vascular and lymphatic channels are controlled using a metal clip applier. Extreme care is taken to avoid injury to the obturator nerve. When possible, the obturator vessels are preserved. Frozen sections are obtained on the node packets; however, a positive lymph node frozen section usually does not preclude completion of the procedure based on current practice patterns.[15,17] It is our practice to provide this immediate feedback to the patient and family concerning the presence or absence of pelvic nodal metastatic disease.

## Prostatectomy

Exposure is adjusted after the lymphadenectomy by retracting the bladder cephalad using a malleable retractor to trap the overinflated Foley balloon and to maintain exposure of the prostate. Superficial fat overlying the prostate is removed using electrocautery. The superficial dorsal vein, if present, is controlled using a 5 mm jaw length LigaSure device (Covidien Mansfield, MA). Attention is then turned to remove excess fat overlying the endopelvic fascia bilaterally. Using scissors the endopelvic fascia is pierced laterally and incised along the groove between the levator ani muscles and the side of the prostate (Figure 30.1). Muscle fibers that extend from the levator ani muscles (referred to as the "pillars of the prostate") are either bluntly swept away or incised from the apex of the gland at the level of the urogenital diaphragm. Care is taken to avoid aggressive

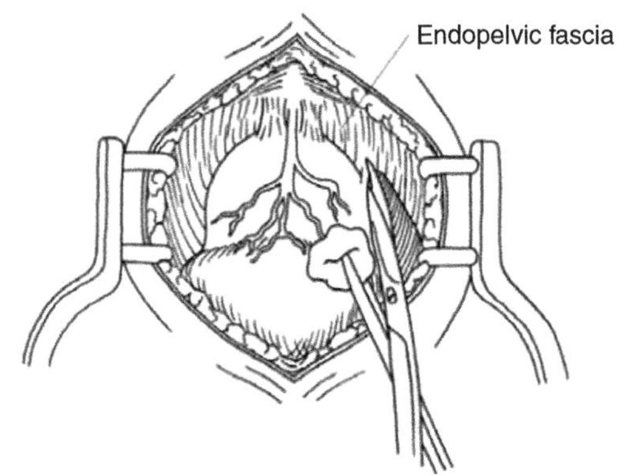


FIGURE 30.1 **The endopelvic fascia is incised in the groove between the levator ani muscles and the lateral border of the prostate.** *Reproduced with permission from Han and Catalona.*[29]

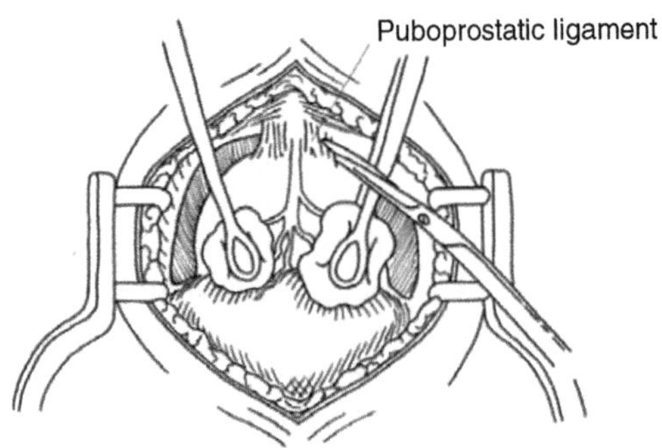


**FIGURE 30.2** **The puboprostatic ligaments are placed on stretch using sponge sticks on the prostate.** The ligaments are sharply incised. *Reproduced with permission from Han and Catalona.*[29]

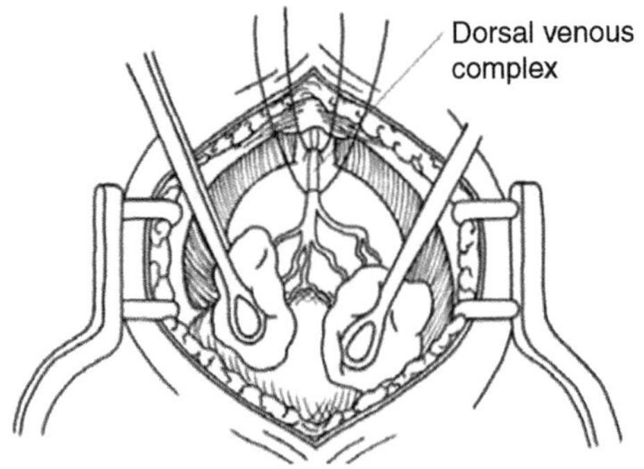


**FIGURE 30.4** **In order to reduce back bleeding a suture is placed in the prostate between the apex and bladder neck.** *Reproduced with permission from Han and Catalona.*[29]

dissection near the dorsal venous complex (DVC). Puboprostatic ligaments can be easily palpated and are divided sharply or using the LigaSure close to the pubis (Figure 30.2). Sometimes there are tributaries lateral to the puboprostatic ligaments that require ligation or cautery. The urethra with the catheter in place can then be palpated.

The plane between the urethra and DVC is developed with blunt dissection. Next, a right-angle clamp or more specialized McDougal prostatectomy clamp (Medline Mundelein, Illinois) is passed and spread to further develop the plane. Ties (0-Vicryl) or suture ligatures (0-Vicryl on a CT-1 needle) are then placed proximally and distally (Figure 30.3). A widely based "back-bleeding" stitch is placed proximally into the prostate using reabsorbable O-suture on a CT-1 needle (i.e., Vicryl) (Figure 30.4).

The DVC is then transected sharply with a knife with right-angle clamp jaws spread behind the complex but above the urethra (Figure 30.5). Alternatively the 5 mm LigaSure device can be used to control and cut the DVC. If bleeding persists from the DVC, suture ligature using reabsorbable O-suture on a CT-1 needle (i.e., Vicryl) can be used to control the distal DVC. A running transverse suture oversewing the distal DVC using 3-0 Vicryl or

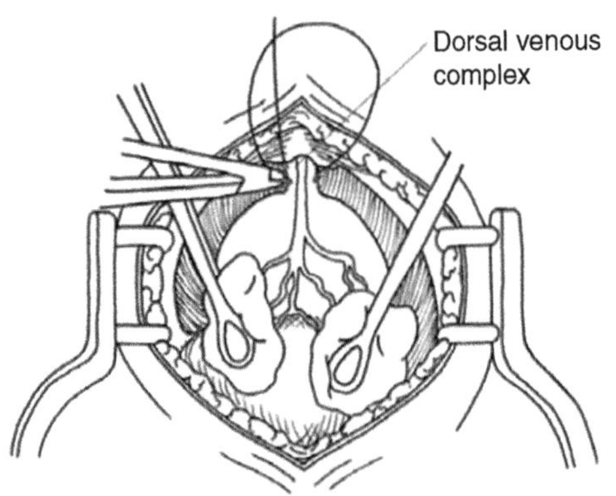


**FIGURE 30.3** **A suture ligature or 0-Vicryl Tie is placed distally on the DVC.** *Reproduced with permission from Han and Catalona.*[29]

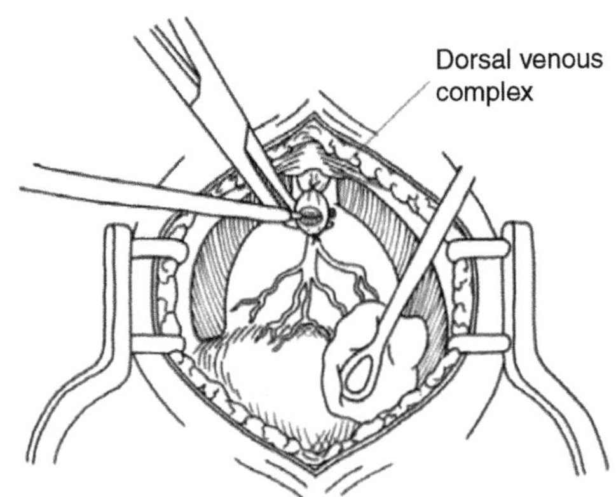


**FIGURE 30.5** **The DVC is sharply incised using a knife with the jaws of a right-angle spread behind protecting the urethra.** *Reproduced with permission from Han and Catalona.*[29]

chromic has also been described. It is essential to gain control of bleeding to allow for adequate visualization of the urethra and prostatic apex prior transection of the urethra.

With the DVC controlled attention can then be focused on the urethra. Apical prostate tissue is gently dissected off the urethra taking care not to injure the neurovascular bundles that run lateral to the urethra. The tissue posterior to the urethra is spread and the urethra with any encircling circumurethral external urethral sphincter fibers are elevated using a right-angle clamp. Using a fresh #15 blade on a long handle the anterior urethra is transected close to the prostate (Figure 30.6). Maintaining an adequate urethral "stump" above the urogenital diaphragm is considered critical in minimizing the risk of incontinence. The Foley catheter becomes visible and is pulled into the abdomen and clamped. An assistant transects the catheter below the inflation port and the entire catheter is pulled into the operative field. With a clamp across the Foley catheter it can be used as a handle allowing for gentle elevation and cephalad retraction of the prostate. Sharp transection of the posterior urethra can then be completed.

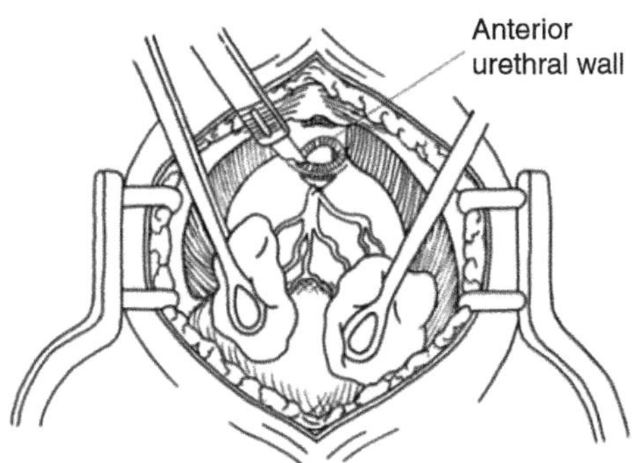


FIGURE 30.6 **The anterior urethra is sharply incised with a #15 scalpel blade exposing the Foley catheter.** *Reproduced with permission from Han and Catalona.*[29]

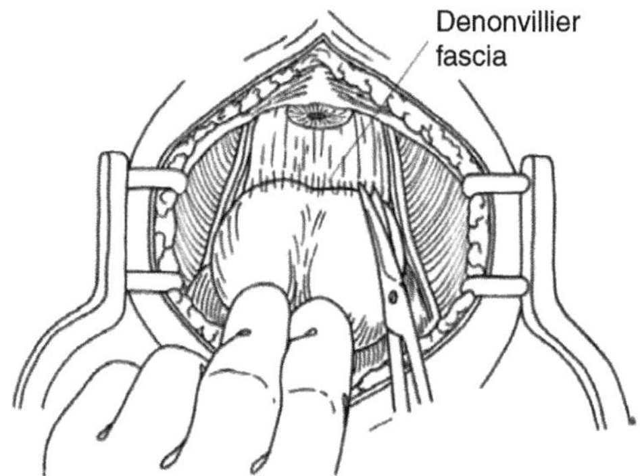


FIGURE 30.8 **Dissection of the neurovascular bundles is carried out in a cephalad direction until the portion of Denonvilliers fascia covering the ampullary portions of the vas and seminal vesicles is seen.** *Reproduced with permission from Han and Catalona.*[29]

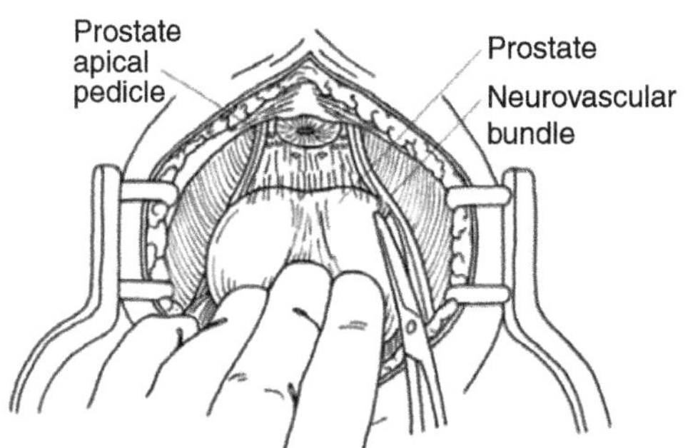


FIGURE 30.7 **Sharp dissection is used to perform a nerve-sparing radical prostatectomy starting at the apex.** *Reproduced with permission from Han and Catalona.*[29]

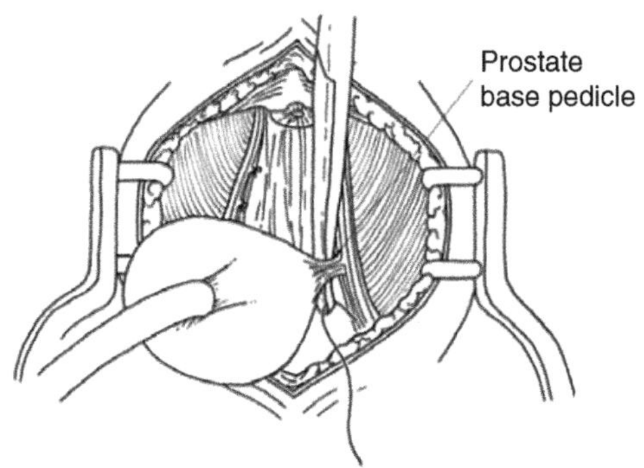


FIGURE 30.9 **The pedicle of the prostate located at the base of the gland is usually tied with 0 Vicryl suture.** Alternatively it can be clipped. *Reproduced with permission from Han and Catalona.*[29]

The rectourethralis muscle is sharply incised exposing the prerectal fat when in the proper plane of dissection. Neurovascular bundles are then identified bilaterally (Figure 30.7). Due to the low-volume disease seen today, most men are candidates for bilateral-nerve-sparing anatomic prostatectomy as described by Walsh.[31] However, the surgeon should use judgment if it is felt that the cancer control is compromised by performing nerve sparing in the setting of more high-grade disease or if there is a suggestion of extra prostatic extension of the cancer in the region of the neurovascular bundle. Using a right-angle clamp, neurovascular bundles are dissected free on each side and released off the prostate laterally with Metzenbaum scissors, avoiding the use of cautery. This is done by careful incision of the lateral pelvic fascia from the apex to the base of the gland. For nonnerve sparing procedure, the incision in the lateral pelvic fascia is performed more laterally. Excessive cephalad retraction of the prostate is avoided to help preserve the neurovascular bundles. Dissection of the neurovascular bundles is carried out in a cephalad direction until the portion of Denonvilliers fascia covering the ampullary portions of the vas and seminal vesicles is seen (Figure 30.8). The prostate is carefully dissected off the anterior surface of the rectum using blunt dissection via a right-angle clamp or with careful finger dissection until the lateral pedicles are reached. The lateral pedicles are carefully dissected anteriorly and posteriorly with the right-angle clamp. With the pedicles dissected, they are controlled with metal clips or by suture ligature (Figure 30.9). The ligation should be as close to the prostate as possible to minimize injury to the neurovascular bundles. Next, scissors are used to cut between the clips releasing the pedicles on both sides.

With cephalad retraction on the catheter in the prostate, Denonvilliers fascia is identified and incised superficially with a #15 blade. A right-angle clamp or tonsil clamp can be used to gently dissect Denonvilliers fascia to expose the ampullary portions of the vas in the midline and seminal vesicles more laterally. A right-angle clamp is then used to elevate each vas, which are then subsequently clipped with metal clips and transected. The space between the seminal vesicles and bladder is developed using a Willauer clamp on each side. We have found it useful to pass a Lowsley retractor into this space as it greatly facilitates the identification of the posterior bladder neck during transection of the prostate from the posterior bladder neck. A small sponge is placed at the

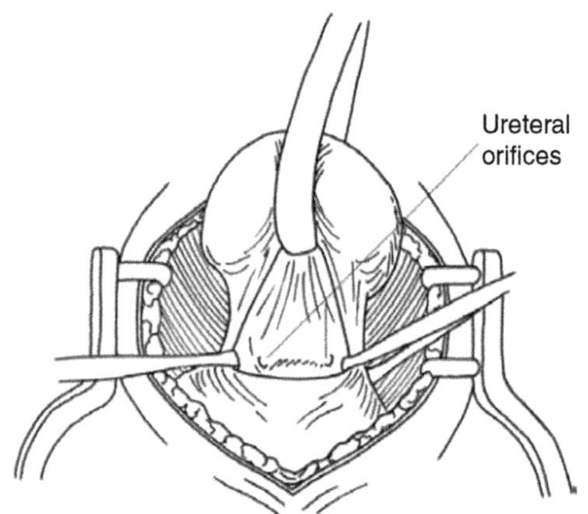


FIGURE 30.10 **After the anterior bladder neck has been opened, the ureteral orifices on the posterior bladder neck are carefully identified.** Allis clamps can provide traction on the bladder wall to facilitate visualization. *Reproduced with permission from Han and Catalona.*[29]

tip of the Lowsley to minimize the risk of vascular injury to the lateral pelvic wall. Also, 1 amp of indigo carmine is administered intravenously.

The prostate is returned to its anatomic position and the anterior bladder neck is grasped with a long Allis clamp that ideally incorporates the Foley catheter. The anterior bladder neck is transected with electrocautery until the Foley catheter is visualized. The Foley balloon is deflated (releasing the clamp on the previously transected catheter) and the catheter is looped around the prostate. Using electrocautery, the lateral attachments of the bladder and prostate are transected toward the posterior bladder neck bilaterally. Care is taken to identify the trigone and ureteral orifices that should be effluxing the blue–indigo carmine. This can be facilitated by using an open empty sponge stick. Allis clamps are used to grasp the bladder neck to maintain visualization (Figure 30.10). Attention should be paid to ensure the ureteral orifices are an adequate distance from the bladder neck. Transection of the posterior bladder neck is completed by pushing down on the prostate creating a natural "break" in the prostatovesical junction posteriorly over the Lowsley retractor and easily identifying the demarcation distal to the trigone. Throughout the dissection of the posterior bladder neck the position of the ureteral orifices must be monitored. In the event that indo carmine is not available or not reliably effluxing, the ureteral orifice position can be carefully monitored using 5-French feeding tubes inserted 10–20 cm into the ureter.

After the posterior bladder neck is incised, the Lowsley retractor is removed and the seminal vesicles are dissected. There are often numerous small vessels adjacent to and feeding the seminal vesicle that must be ligated with the LigaSure or small clips. Seminal vesicles can be completely removed or the tips preserved with little impact on outcomes particularly with Gleason 6 prostate cancer.[32] The neurovascular bundles and ureteral orifices lie close to the tips of the seminal vesicle and should be treated with care. Clips are placed near the tips and the seminal vesicles are transected. The prostate specimen is palpated and visually inspected for any breaks in the capsule or other areas where there may be inadequate tissue margins. If there is any suspicion, additional tissue including resection of the bladder neck or resection of the neurovascular bundles should be carefully considered. The prostate along with the seminal vesicles are then handed off en bloc.

Next, attention is turned to reconstruction of the bladder neck. The bladder neck is reconstructed in a tennis racket fashion using a 2-0 absorbable suture (Monocryl or Vicryl) on an SH needle. If ureteral orifices are close, they may be intubated with 5-French feeding tubes to ensure that they are protected. Attention is made to take a large bite of detrusor and small bite of bladder mucosa and to not injure ureteral orifices bilaterally. The closure is continued until the bladder neck opening is approximately 20-French. A good surrogate for this measurement is the end of a Singley tissue forceps. Once the bladder neck is reconstructed, the bladder mucosal edges are everted using 4-0 absorbable suture (Monocryl or Vicryl) on an RB1 needle. This is done circumferentially with individual sutures or in a running fashion ensuring good mucosal eversion of the newly reconstructed bladder neck on all edges (Figure 30.11). Ultimately, mucosal to mucosal apposition to the urethral stump helps minimize bladder neck contracture.

The pelvis is then irrigated with warm irrigation fluid and is assessed for bleeding. Hemostasis must be achieved prior to completing the vesicourethral anastomosis. Persistent neurovascular bundle bleeding should only be dealt with using 4-0 plain catgut if necessary.

Adequate visualization of the urethral stump is critical for a successful anastomosis. We have found the 24- or 28-French Roth Urethral Suture Guide (Greenwald Surgical, Gary, Indiana) useful in the setting

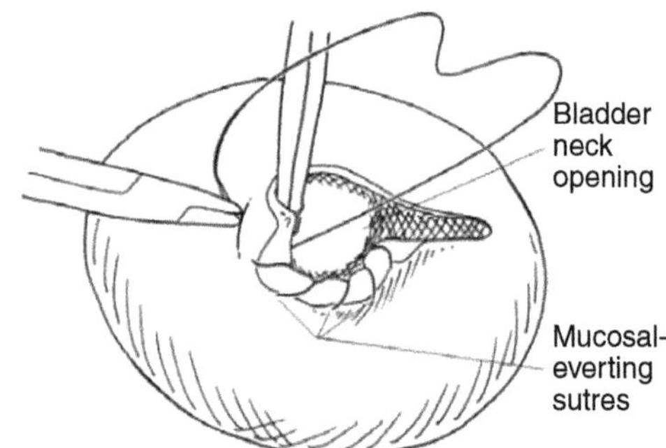


FIGURE 30.11 **The bladder neck is closed in a "tennis racquet" fashion.** The bladder neck opening is everted to expose the bladder mucosa in a running or interrupted fashion based on surgeon preference. *Reproduced with permission from Han and Catalona.*[29]

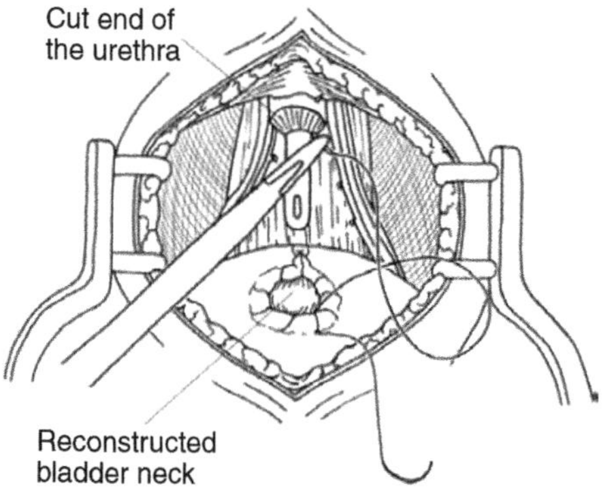


**FIGURE 30.12 The urethrovesical anastomosis is completed using 2-Monocryl on a UR6 needle with five sutures placed as noted in the text over a 20-French Foley catheter or catheter guide.** Based on surgeon preference, double-armed suture can also be used as shown. *Reproduced with permission from Han and Catalona.*[29]

of an adequate urethral stump. Alternatively when there is poor visualization or a short distal urethra, the Greenwald Roth "Grip Tip" Suture Guide is used (Greenwald Surgical, Gary, Indiana), which incorporates three retractable stainless steel rods that grip and fixate the urethra to the guide to improve exposure of the transected urethra. Other techniques, such as applying perineal pressure with sponge forceps to better expose the cut end of the urethral stump, are unnecessary when using these suture guides. Anastomotic sutures are placed using 2-0 Monocryl sutures on a UR6 needle (Figure 30.12). These sutures are placed at 6 o'clock, 4 o'clock, 8 o'clock, 2 o'clock, and 10 o'clock positions in an outside to inside fashion. The posterior lateral sutures are tagged with curved hemostats. The anterolateral sutures are tagged with straight hemostats and the 6 o'clock suture is tagged with a straight Hartmann mosquito clamp to keep all sutures appropriately aligned and identified. When placing sutures in the urethral stump it is essential that the urogenital diaphragm is not included as this can result in scarring of the diaphragm and compromise continence.

Next, the 6 o'clock suture from the urethra is placed to correspond with the 6 o'clock position at the reconstructed bladder neck inside to out. A fresh 20-French Foley catheter is inserted into the urethra and advanced into the bladder. At this point, 5 mL of sterile water is used to partially inflate the catheter balloon. We then place the remainder of the anastomotic sutures on the bladder neck in the corresponding positions.

The anterior bladder wall is then grasped with a sponge stick. The retractors are removed and the table is slightly flexed. The bladder is moved closer to the urethral stump with the sponge stick in an attempt to minimize tension on the anastomosis. Gentle traction with the Foley catheter can also be used. An assistant can use a small handheld malleable retractor to apply gentle lateral traction on the bladder to facilitate the suture tying. The anastomotic sutures are tied on each side. An additional 5 mL is added to the Foley balloon and the catheter is gently irrigated with a bulb syringe to assess for leakage or any bleeding or clots.

A medium Jackson–Pratt drain is then placed in the left lower quadrant and is secured to the skin with a 2-0 nylon stitch. The rectus fascia is closed with #1 PDS in a bidirectional running fashion. We have found that local infiltration of the wound with 20 mL long-lasting anesthetic liposomal bupivacaine (Exparel, Pacira Pharmaceuticals, San Diego, CA) provides excellent postoperative pain control. The subcutaneous tissues are reapproximated with interrupted 2-0 Vicryl sutures. The skin is closed with undyed 4-0 Monocryl suture in a running subcuticular fashion. The patient is cleansed and tincture of benzoin and adhesive wound strips are applied to the incision site. The incision site and the drain are then sterilely dressed. The Foley catheter is placed to gravity drainage and the Jackson–Pratt drain is placed on bulb suction.

## Postoperative Care

All prostatectomy patients are placed on a care pathway postoperatively. They have labs drawn (CBC and basic metabolic panel), receive a 1 L fluid bolus, and are monitored in the postanesthesia care unit. Most ORP patients are sent to a general surgical floor, unless medical comorbidities require additional monitoring. Each patient is placed on an analgesia regimen comprised of oral acetaminophen, celecoxib, pregabalin, and IV ketorolac, which is identical to our institutional RALP regimen. IV morphine by patient-controlled anesthesia (PCA) is occasionally needed for breakthrough pain. All patients are started on a clear liquid diet in the immediate postoperative period. Early ambulation on postoperative day zero is strongly encouraged. Lower extremity compression devices are placed in the operating room through the duration of the case, and are continued in the postoperative period until discharge. Labs are rechecked on postoperative day one including a creatinine determination from the drain to assess for urine leak. Patients are advanced to a regular diet and oral analgesics (Oxycodone). If patients are tolerating a diet, ambulating, and laboratory values are within acceptable range, the drain is removed and the patient is usually discharged home on postoperative day one or two. They are discharged home with Oxycodone, stool softener, and Sulfamethoxazole/Trimethoprim to be started one day prior to their scheduled follow-up visit for a catheter removal and voiding trial in approximately 10 days. Routine cystograms are not used to determine catheter removal in the absence of complicated urethral anastomosis.

## RADICAL PROSTATECTOMY ADDITIONAL CONSIDERATIONS

Some surgeons use a modification of the original radical prostatectomy technique described by Campbell known as the retropubic antegrade radical prostatectomy. The technique begins with the dissection of the base to the apex.[33] One advantage reported is reduced blood loss as the transection of the DVC is one of the last steps. The "mini-lap" radical retropubic prostatectomy reduces the incisional morbidity associated without compromising oncologic outcomes.[34] It relies on the use of a smaller 8 cm incision and a fixed retractor system to complete the ORP in standard fashion.

## POSTOPERATIVE COMPLICATIONS

Immediate postoperative complications from a series of over 7000 men have been summarized.[35] In this series there were no intraoperative deaths and one 30 day mortality due to myocardial infarction. Common complications (all generally less than 2–3%) include bladder neck contracture, groin hernia, lymphocele, infections including wound complications and PE/DVT. Very uncommon complications include urinoma, pelvic hematoma, rectal and ureteral injuries, and neurological positioning injuries.

Concerning transfusions, in a series of 16,144 men who underwent RP, the overall transfusion rate was 3.1%. The highest transfusion rate occurred on the day of surgery for patients undergoing ORP, and first postoperative day for RALP patients.[36]

## OUTCOMES

On average, a man treated for prostate cancer has a life expectancy of approximately 14 years; therefore, considerations of short-term and long-term complications and health-related quality of life (HRQOL).[37] As noted previously, the wide variation in reporting metrics seriously impacts the ability to compare treatment options objectively.[9,37]

Due to differing definitions, continence rates at 3–24 months range between 51% and 78%.[37] Continence tends to be worse immediately postoperatively and improves through 2 years. Younger age and nerve sparing improve continence rates.

Erectile dysfunction is a great concern for most men undergoing treatment for prostate cancer. However, it should be noted that 45–64% of men undergoing radical prostatectomy have pre-existing erectile dysfunction.[37] Recovery of erectile function at 3, 6, 12, 18, and 24–36 months ranges between 19% and 68%, 24% and 86%, 39%, and 90%, 44% and 86%, and, 35% and 94%.

Some men do experience transient decline in bowel function following radical prostatectomy. These are likely due to perioperative factors such as self-motivated dehydration to reduce incontinence, pain medications, and dietary alterations. However, in a study of 5 year bowel outcomes the results were similar in men undergoing radical prostatectomy and active surveillance.[38]

Another potential HRQOL benefit of radical prostatectomy relates to lower urinary tract symptoms (LUTS). LUTS includes storage and voiding urinary symptoms, most commonly due to benign prostatic hyperplasia in men. The natural history of LUTS in older men tends to be progressive and is commonly associated with prostatic enlargement. In one long-term study of LUTS, the symptoms remained constant between 2 years and 10 years following RP in men with and without baseline clinically significant LUTS.[39] This suggests that RP interferes with the natural history of BPH by removing the prostate, the primary contributor to LUTS progression in men, and improves this HRQOL domain.

Cancer control is of paramount concern in the management of prostate cancer. Strong 15 year prostate-cancer-specific mortality (PCSM) data are available. Eggner et al. have gathered data on over 11,500 men undergoing radical prostatectomy.[40] The overall PCSM was 7%. In men with Gleason score 6 tumors, the 15–20 year PCSM was negligible at 1.2% or less. With Gleason scores of 8–10 the PCSM risk rose to 31% or greater. Based on this and other data sets, relatively few patients including those with adverse features, are likely to die from prostate cancer, documenting the effectiveness of this technique.[35]

## CONCLUSIONS

The open radical prostatectomy has set the gold standard for the management of localized prostate cancer by providing robust long-term disease control and ever improving side-effect profiles with the introduction of the anatomic nerve-sparing radical prostatectomy.[41] All of the recent data on minimally invasive laparoscopic and robotic radical prostatectomy seeks to reproduce the oncologic and functional outcomes developed with the open technique and attempts to build on this advance in prostate cancer care. The pendulum has now swung with the challenge for open surgeons to match the reported postoperative benefits of RALP. Using critical care pathways and similar perioperative strategies the major differences beyond cost, favoring ORP, and blood loss, favoring RALP, appear to be minimizing.[9]

There is a concern that the number of surgeons who are skilled in open radical prostatectomy may be limited in the future with the extensive adoption of the RALP. It is hoped that centers of excellence will continue to provide this option and be able to train future urologic surgeons, as there will be patients who may not be candidates for RALP and who may derive more benefit from the open radical prostatectomy approach.

## References


1. Bouchier-Hayes DM. Terence Millin: pioneer of the retropubic space. *BJU Int* 2005;**96**(6):768–71.
2. Walsh PC. How surgical innovation reduced death and suffering from prostate cancer. *J Craniofac Surg* 2013;**24**(1):49–50.
3. Trinh QD, Sammon J, Sun M, et al. Perioperative outcomes of robot-assisted radical prostatectomy compared with open radical prostatectomy: results from the nationwide inpatient sample. *Eur Urol* 2012;**61**(4):679–85.
4. Diaz M, Peabody JO, Kapoor V, et al. Oncologic outcomes at 10 years following robotic radical prostatectomy. *Eur Urol* 2015;**67**(6):1168–76.
5. Huang KH, Carter SC, Hu JC. Does robotic prostatectomy meet its promise in the management of prostate cancer? *Curr Urol Rep* 2013;**14**(3):184–91.
6. Alkhateeb S, Lawrentschuk N. Consumerism and its impact on robotic-assisted radical prostatectomy. *BJU Int* 2011;**108**(11): 1874–8.
7. Schroeck FR, Krupski TL, Stewart SB, et al. Pretreatment expectations of patients undergoing robotic assisted laparoscopic or open retropubic radical prostatectomy. *J Urol* 2012;**187**(3):894–8.
8. Gandaglia G, Sammon JD, Chang SL, et al. Comparative effectiveness of robot-assisted and open radical prostatectomy in the post-dissemination era. *J Clin Oncol* 2014;**32**(14):1419–26.
9. Healy KA, Gomella LG. Retropubic, laparoscopic, or robotic radical prostatectomy: is there any real difference? *Semin Oncol* 2013;**40**(3):286–96.
10. https://clinicaltrials.gov/ct2/show/NCT01365143?term = radical + prostatectomy&rank = 7[accessed 29.11.2014].
11. Gomella LG, Lin J, Hoffman-Censits J, et al. Enhancing prostate cancer care through the multidisciplinary clinic approach: a 15-year experience. *J Oncol Pract* 2010;**6**(6):e5–e10.
12. Gomella LG. Prostate cancer: the benefits of multidisciplinary prostate cancer care. *Nat Rev Urol* 2012;**9**(7):360–2.
13. Heidenreich A, Bastian PJ, Bellmunt J, et al. European Association of Urology. EAU guidelines on prostate cancer. Part 1: screening, diagnosis, and local treatment with curative intent-update 2013. *Eur Urol* 2014;**65**(1):124–37.
14. Mohler JL, Kantoff PW, Armstrong AJ, et al. Prostate cancer, version 2. *J Natl Compr Canc Netw* 2014;**12**(5):686–718.
15. Faiena I, Singer EA, Pumill C, et al. Cytoreductive prostatectomy: evidence in support of a new surgical paradigm (review). *Int J Oncol* 2014;**45**(6):2193–8.
16. Rusthoven CG, Carlson JA, Waxweiler TV, et al. The impact of definitive local therapy for lymph node-positive prostate cancer: a population-based study. *Int J Radiat Oncol Biol Phys* 2014;**88**(5):1064–73.
17. Messing EM, Manola J, Yao J, et al. Immediate versus deferred androgen deprivation treatment in patients with node-positive prostate cancer after radical prostatectomy and pelvic lymphadenectomy. *Lancet Oncol* 2006;**7**(6):472–9.
18. Martinez PF, Billordo Peres N, Cristallo C, et al. Salvage radical prostatectomy after radiotherapy. *Arch Esp Urol* 2014;**67**(4): 313–22.
19. Menon M, Hemal AK. VIP Team. Vattikuti Institute prostatectomy: a technique of robotic radical prostatectomy: experience in more than 1000 cases. *J Endourol* 2004;**18**(7):611–9 discussion 619.
20. Huang AC, Kowalczyk KJ, Hevelone ND, et al. The impact of prostate size, median lobe, and prior benign prostatic hyperplasia intervention on robot-assisted laparoscopic prostatectomy: technique and outcomes. *Eur Urol* 2011;**59**(4):595–603.
21. Thompson I, Thrasher JB, Aus G, et al. Guideline for the management of clinically localized prostate cancer: 2007 update. *J Urol* 2007;**177**(6):2106–31.
22. Crawford ED, Stone NN, Yu EY, et al. Challenges and recommendations for early identification of metastatic disease in prostate cancer. *Urology* 2014;**83**(3):664–9.
23. Gupta M, McCauley J, Farkas A, et al. Clinical practice guidelines on prostate cancer: a critical appraisal. *J Urol* 2014;**193**(4):1153–8.
24. Lee BW, Park MG, Cho DY, et al. Preoperative erythropoietin administration in patients with prostate cancer undergoing radical prostatectomy without transfusion. *Korean J Urol* 2014;**55**(2): 102–5.
25. Auerbach A, Goldman L. Assessing and reducing the cardiac risk of noncardiac surgery. *Circulation* 2006 Mar;**113**(10):1361–76.
26. *Urologic Surgery Antimicrobial Prophylaxis*. https://www.auanet.org/education/guidelines/antimicrobial-prophylaxis.cfm [accessed 29.11.2014].
27. *Anticoagulation and Antiplatelet Therapy in Urologic Practice*. http://www.auanet.org/common/pdf/education/clinical-guidance/Anticoagulation-Antiplatelet-Therapy.pdf [accessed 29.11.2014].
28. Naselli A, Andreatta R, Introini C, et al. Predictors of symptomatic lymphocele after lymph node excision and radical prostatectomy. *Urology* 2010;**75**(3):630–5.
29. Han M, Catalona WJ. Open radical prostatectomy. In: Graham SD, Keane TE, editors. *Glen's urologic surgery*. 7th ed. Philadelphia: Wolters Kluwer; 2010 [Chapter 28].
30. Bivalacqua TJ, Pierorazio PM, Gorin MA, et al. Anatomic extent of pelvic lymph node dissection: impact on long-term cancer-specific outcomes in men with positive lymph nodes at time of radical prostatectomy. *Urology* 2013;**82**(3):653–8.
31. Walsh PC. Radical retropubic prostatectomy with reduced morbidity: an anatomic approach. *NCI Monogr* 1988;**7**:133–7.
32. Gofrit ON, Zorn KC, Shikanov SA, et al. Is seminal vesiculectomy necessary in all patients with biopsy Gleason score 6? *J Endourol* 2009;**23**(4):709–13.
33. Carini M, Masieri L, Minervini A, et al. Oncological and functional results of antegrade radical retropubic prostatectomy for the treatment of clinically localised prostate cancer. *Eur Urol* 2008;**53**(3):554–61.
34. Slabaugh Jr TK, Marshall FF. A comparison of minimally invasive open and laparoscopic radical retropubic prostatectomy. *J Urol* 2004;**172**(6 Pt 2):2545–8.
35. Poon SA, Scardino PT. Open radical retropubic prostatectomy. In: Eastham JA, Schaeffer EM, editors. *Radical prostatectomy – surgical perspectives*. New York: Springer; 2014 [Chapter 7].
36. Korets R, Weinberg AC, Alberts BD, et al. Utilization and timing of blood transfusions following open and robot assisted radical prostatectomy. *J Endourol* 2014;**28**(12):1418–23.
37. Prabhu V, Lee T, McClintock TR, et al. Short-, intermediate-, and long-term quality of life outcomes following radical prostatectomy for clinically localized prostate cancer. *Rev Urol* 2013;**15**(4):161–77.
38. Bergman J, Litwin MS. Quality of life in men undergoing active surveillance for localized prostate cancer. *J Natl Cancer Inst Monogr* 2012;**45**:242–9.
39. Prabhu V, Taksler GB, Sivarajan G, et al. Radical prostatectomy improves and prevents age-dependent progression of lower urinary tract symptoms. *J Urol* 2014;**191**(2):412–7.
40. Eggener SE, Scardino PT, Walsh PC, et al. Predicting 15-year prostate cancer specific mortality after radical prostatectomy. *J Urol* 2011;**185**(3):869–75.
41. Finkelstein J, Eckersberger E, Sadri H, et al. Open versus laparoscopic versus robot-assisted laparoscopic prostatectomy: the European and US experience. *Rev Urol* 2010;**12**(1):35–43.

CHAPTER

# 31

# Radiation-Resistant Prostate Cancer and Salvage Prostatectomy

*Yousef Al-Shraideh, MD*, Samir V. Sejpal, MD, MPH**, Joshua J. Meeks, MD, PhD**

*Department of Urology, Northwestern University, Feinberg School of Medicine, Chicago, IL, USA

**Department of Radiation Oncology, Northwestern University, Feinberg School of Medicine, Chicago, IL, USA

## RADIOTHERAPY FOR PROSTATE CANCER

Radiation therapy (RT) for the treatment of prostate cancer (PCa) has been utilized in the United States and Canada since 1915.[1] The first radiotherapy devices used radium applicators that were positioned adjacent to the prostate gland. Unfortunately, this technique resulted in significant morbidity including unwanted dose to adjacent critical structures. Low-penetrating electron beam X-rays were then applied in the palliative setting as high doses could result in skin cancers due to superficial dose-penetration.[2] After World War II, radiologists were able to use high-energy megavoltage machines with improved penetration to reach the prostate.[2] In the 1950s RT delivery evolved through the development of Cobalt 60 units, linear accelerators (LINACS), and betatrons. During the 1980s, radiation technology was further refined in the treatment of PCa as improved LINAC technology increased radiation delivery time and allowed more conformal planning of the targeting tissue. The number of PCa patients treated with external radiation therapy (EBRT) doubled in the United States in the 1980s. Currently, modern technologies, such as intensity modulated radiation therapy (IMRT), proton therapy (PT), and stereotactic ablative radiation therapy (SBRT), are used in the treatment of PCa. Intensity-guided radiation therapy (IGRT) allows for frequent imaging of the target volume (via two-dimensional X-rays, or three-dimensional imaging) during radiation treatment and may be utilized with all of the three techniques mentioned earlier to better localize target volume and thus treat the prostate gland to high doses of radiation. Brachytherapy (the use of radioactive sources placed in the prostate gland) can be accomplished via low-dose rate (LDR) brachytherapy or high-dose rate (HDR) brachytherapy and can be used as mono-therapy (LDR or HDR alone in early-stage prostate cancer), or as a boost following EBRT (LDR or HDR boost) in the treatment of more advanced disease.[3]

## DETECTION OF A RECURRENCE AFTER RADIOTHERAPY

### Routine Prostate Biopsy

There are currently no specific markers that can adequately diagnose partially treated versus untreated PCa after RT. While biopsy after prostate ablation is universally accepted to demonstrate successful treatment,[4] the American Society for Radiation Oncology (ASTRO) consensus panel recommended against systematic prostate rebiopsy within 30 months after RT in the absence of a PSA recurrence, as negative prostate biopsy rates ranged from 62% to 80% for patients with stage T1 and T2 tumors.[5] Conversion of gradable PCa to benign acting "treated" cancer can occur years after RT. Of men that underwent biopsy within 2 years from RT, almost one-third of biopsies that initially identified cancer were converted to a negative biopsy at 30 months.[6] Given the low probability of finding cancer, serum PSA has been used as a surrogate marker for PCa viability to determine cure after RT.

Prostate Cancer. http://dx.doi.org/10.1016/B978-0-12-800077-9.00031-1

## Prostate-Specific Antigen

Prostate-specific antigen (PSA), expressed by benign and malignant prostate glands, is the most sensitive biomarker to determine recurrence after RT. Serial PSA measurements are taken every 3–6 months. Following RT, PSA slowly declines and reaches nadir 18 months or longer.[7] Biochemical recurrence after radiotherapy is almost always asymptomatic and precedes clinical recurrence with an average time to relapse of 5 years.[8] In 1996, the American Society for Radiation Oncology (ASTRO) consensus panel established a recurrence after RT as three consecutive increases in PSA after reaching nadir.[9] This ASTRO definition of biochemical failure was revised by a second Consensus Conference sponsored in Phoenix, Arizona, on January 21, 2005.[10] The panel recommended a rise by 2 ng/mL or more above the nadir PSA following RT as the standard definition for biochemical failure (known as the "Phoenix definition" of recurrence). While the Phoenix definition is highly specific to identify a recurrence, multiple studies have demonstrated earlier salvage treatment at lower PSA is associated with improved biochemical free survival (BFS) compared to delayed treatment and occult progression. The significant time and challenge required to identify a recurrence after RT may contribute to delayed identification of biochemical failure. The BFS after salvage prostatectomy for men with a PSA less than 4 ng/mL was 86%, 55% for a PSA of 4–10 ng/mL, and only 28% in those with a PSA greater than 10 ng/mL.[11]

Even though the definition of failure after RT by the Phoenix criteria is defined as nadir +2.0 ng/mL, a prostate gland treated with RT may still have viable disease and not show a PSA that would prompt intervention. In a study of 78 patients with PCa who were treated with RT and then underwent a radical cystectomy at a later time for a bladder cancer, 45% of patients showed persistent gradable PCa when the prostate was removed. Sixty-nine percent showed Gleason grade ≥7 and 17% of specimens were at least pT3 disease.[12] At the time of cystectomy, only 21% had a known PCa recurrence and there was no difference in mean PSA at the time of surgery between those with and without known recurrence (2.0 vs. 1.5 ng/mgL, $p$ = 0.7). In those men without a known BCR, PCa was found in 37%. These data suggest that many PCas may not meet PSA criteria for intervention (PSA <2.0 ng/mL), but still remain viable after RT. Future studies should focus on identification of PCa biomarkers to identify early PCa recurrence.

## PSA Nadir

The lowest PSA achieved after RT (the nadir) is prognostic of long-term PCa response to RT.[13] In a retrospective study of 743 patients with localized PCa treated at the Memorial Sloan-Kettering Cancer Center with conformal RT, preoperative variables that predict PSA relapse were a PSA level higher than 10 ng/mL, a Gleason score of at least 7, and clinical stage T3 disease.[14] Yet, the PSA nadir after RT was the strongest predictor, overshadowing the significance of the pretreatment PSA level. While no study has demonstrated a clear cut-off that absolutely predicts cure or failure of treatment, a trend toward lower PSA associated with treatment success has been demonstrated. Of 364 patients at a mean of 46 months after external beam RT, 93% of men with the lowest PSA nadirs (0–0.99 ng/mL) had no biochemical evidence of recurrence versus 49% with PSA nadirs of 1.0–1.99 ng/mL and 16% of those with a PSA nadir greater than 2.0 ng/mL ($p$ = 0.0001).[15]

## PSA Bounce

Identifying PCa recurrence can be challenging due to noncancer-related variability of PSA due to inflammation and apoptosis of PCa cells.[16,17] A "PSA bounce" is a common post-RT occurrence defined as a transient rise in PSA. PSA bounce may be detected in about one-third of men receiving RT (EBRT or brachytherapy) and can be a major source of anxiety for patients and families.[18] While a bounce may occur any time after any modality of RT (Brachytherapy or EBRT), a bounce often occurs between 1.5 years and 2.6 years after treatment and occurs more commonly with brachytherapy.[19]

## PSA Doubling Time

PSA doubling time (PSADT) following RT may predict the aggressiveness of PCa recurrence and help determine which patients may benefit from local salvage therapy versus systemic treatment. PSADT is significantly shorter in patients who developed metastases.[14] A PSADT less than 3 months may be considered a surrogate predictor for PCa-specific mortality with a 20-fold increased risk of dying of PCa compared to those with a PSA-DT of ≥3 months.[20]

## Digital Rectal Exam

Digital rectal exam (DRE) is a specific but insensitive tool to detect a recurrence of PCa after RT. DRE findings often correlate with PSA values with marginal additional efficacy for detecting an early recurrence after RT. In a nonrandomized study of 235 men treated with RT, a total of 1544 digital rectal examinations were performed during 1627 visits. Overall, only eight nodules were identified in men, all with increasing PSAs.[21] Hence, PSA remains the cornerstone for early detection of recurrence following RT.

## POSTRADIOTHERAPY PROSTATE BIOPSY WHEN RECURRENCE IS STRONGLY SUSPECTED

A postradiotherapy prostate biopsy (PRB) demonstrating persistent gradable PCa is strongly recommended prior to considering local salvage therapy. The PRB is technically similar to a transrectal ultrasound-guided biopsy, but the prostate is often smaller after RT. We advocate performing at least a 12-core biopsy with six cores on either side and four designated cores of the prostate apex as the apex and urethra often harbor residual cancer as these areas are undertreated by RT.[22] The seminal vesicles (SVs) are not always targeted with the RT; therefore, we recommend biopsy of SVs with the PRB as multiple patients have been identified with an "SV-only" recurrence.[23]

The morphology of treated PCa is influenced greatly by cellular changes to the glands of the prostate that may be for the uropathologist to interpret.[24] It is imperative for the pathologist to appreciate these morphologic effects of radiation injury on the prostate to discern between residual gradable adenocarcinoma, prostate cancer with treatment effect, and radiation-induced atypia of nonneoplastic glands.[25] Histologic features that are helpful include infiltrative growth, perineural invasion, intraluminal crystalloids, blue mucin secretions, the absence of corpora amylacea, and the presence of coexistent high-grade prostatic intraepithelial neoplasia.[26] If a tumor may be given a Gleason grade, it should be considered persistent after radiotherapy and local treatment are considered.

The ability of the PRB to accurately characterize the extent and location of residual cancer is critical to treatment planning. If the PRB does not identify cancer within the prostate, the PSA elevation may be spurious or the recurrence is metastatic in nature. If the PRB identifies a small focus of PCa on one side of the prostate, some men will consider focal ablative techniques, whereas young patients with multifocal or high-grade disease may be best treated with whole-gland ablation or salvage radical prostatectomy (SRP).[5] We studied the accuracy of the PRB in 198 patients who ultimately had undergone salvage RP. We found that PRB accuracy varied by region from 62% to 76%. Biopsy missed up to 20% of tumors, more than half of the cancers were upgraded at salvage RP, and many that were unilateral on PRB were bilateral at SRP.[27]

## IMAGING STUDIES TO DETECT A RECURRENCE AFTER RADIOTHERAPY

### MRI

At the time of PSA recurrence after RT, a staging evaluation should be performed including bone scan and imaging of the abdomen and pelvis. Abdominal and pelvic computed tomography (CT) imaging has limited sensitivity to identify local recurrence within the prostate but may detect lymph node metastasis.[28] Lymph nodes of 2 cm or more may be considered abnormal and a high volume of pelvic nodes or higher retroperitoneal nodes may limit the success of local therapy. Multiparametric MRI (mpMRI) appears to be helpful in evaluating men with biochemical recurrence after RT. In one study comparing mpMRI (with dynamic contrast-enhanced MRIm and diffusion-weighted imaging) with T2-weighted imaging, mpMRI had increased sensitivity (77%) but lower specificity (92%) with a positive predictive value of 68% and negative predictive value 95%.[29] The sensitivity of mpMRI to detect tumor recurrence was substantially lower in the transitional zone (60%) than in the peripheral zone (82%), although these differences were not statistically significant ($p = 0.548$). Placement of an endorectal coil during mpMR imaging may improve local staging of PCa recurrence, which is often hindered by tissue changes related to RT and allow accurate detection of extracapsular extension and seminal vesicle invasion.[30]

### Nuclear Medicine Bone Scan

Radionuclide bone scintigraphy is a sensitive, but nonspecific tool to detect bone metastasis in PCa. The bone scan tends to detect metastases when the patient is symptomatic or serum PSA level is markedly elevated.[31] In a multivariate analysis study the single most useful parameter in predicting the bone scintigram result was tPSA ($p = 0.01$).[32] The probability of a positive bone scintigram remains 5% or lower until serum PSA reaches at least 30–40 ng/mL.[32]

### Positron Emission Tomography

Until recently, positron emission tomography has not added value to determine recurrence after RT. The most commonly used tracer, $^{18}$F-fluorodeoxyglucose, has not had a role in imaging of PCa recurrence. Recently, however, other positron emission tomography tracers, such as $^{11}$C-acetate and $^{11}$C- or (18)F-choline, have shown promising results in early detection and localization of PCa recurrence.[33,34] The detection rate of [(11)C]Choline-PET/CT shows a positive relationship with serum PSA levels in patients with biochemical recurrence after RT. The detection rate was 36% for a PSA value <1 ng/mL, 43% for a PSA value 1–2 ng/mL, 62% for a PSA value 2–3 ng/mL, and 73% for a PSA value ≥3 ng/mL.[35]

## SALVAGE RADICAL PROSTATECTOMY

The primary goal of salvage radical prostatectomy (SRP) after RT is biochemical-free survival (BFS). SRP can be considered for men who have (1) organ-confined recurrence with PRB identifying gradable PCa, (2) no evidence of distant or unresectable metastatic cancer by abdomino-pelvic imaging, and (3) at least 10-year life expectancy often with few comorbidities and good performance status.[36,37] Thus, a diligent preoperative evaluation should include a careful history and physical examination to evaluate for evidence of distant disease.[38] Imaging including abdominal and pelvic CT or mpMRI, bone scan should be performed. After imaging, a PRB that identifies gradable PCa is a requisite.

The SRP is performed in a similar fashion to a standard prostatectomy. The surgical planes of dissection are often obliterated with fibrosis with limited anatomical definition after radiotherapy.[39] Blood loss during SRP is often minimal due to changes in pelvic microvasculature.[40] One of the most technically difficult parts of the operation is establishing a plane between the rectum and prostate, increasing the risk of rectal injury. Nerve-sparing is very difficult to perform and may be of little functional significance as many men will have poor erections from RT.[40] Special attention should be placed on achieving a tension-free urethrovesical anastomosis with care to allow for a wider bladder neck to prevent bladder neck contraction.[36] Regardless of nomogram risk of metastatic PCa, a pelvic lymph node dissection should be performed. If the pelvic lymph nodes received radiation or a staging node dissection was performed before RT, the node volume may be sparse. After surgery, we strongly recommend performing a cystogram or CT cystogram prior to removal to ensure a healed urethrovesical anastomosis.[39] Mean operative time for open SRP was 3.2 h, hospital stay of 3 days, and transfusion rate of 29%.[36]

In a large retrospective single-institutional study, 199 patients were treated from 1967 to 2000 for radioresistant prostate cancer with SRP (138) or cystoprostatectomy (61).[41] At the time of surgery, PSA levels were <10 ng/mL in 78% of these patients, and pathologic T2 disease was identified in 39%. After a median follow-up of 92 months, the 5-year estimate of BFS was 63% (when PSA failure was defined as a PSA level >0.4 ng/mL) with a cancer-specific survival of 65% at 10 years. Those patients requiring cystoprostatectomy had a significantly lower median CSS of 4.4 years versus 8.7 years for men undergoing SRP ($p < 0.001$).

A more recent retrospective, international, multi-institutional cohort described the outcomes of 404 men who underwent SRP between 1985 and 2009.[38] Surgical margin status was positive in 25%, extraprostatic extension in 45%, seminal vesicle involvement in 30%, and lymph node metastasis in 16%. Overall, only 37% of men had not had a biochemical recurrence at 10 years after SRP but CSS was 83% with only 23% of patients developing metastasis. Based on multivariate analysis, patients with lower pre-SRP PSA levels (4 ng/mL or less) and PRB of Gleason score ≥7 had a BFS of 64% at 5 years and 51% at 10 years after SRP. While the main goal of SRP is BFS, other benefits include delayed or avoidance of hormone therapy and pelvic control of an aggressive neoplasm.[42] Even in men with advanced radiation refractory PCa, such as SVI, the rate of local pelvic recurrence was low at 11%.[23]

### Minimally Invasive Salvage Radical Prostatectomy

As radical prostatectomy is now performed more commonly with robotic assistance, an increasing number of salvage prostatectomies are performed robotically with less reported morbidity than open SRP. Potential benefits of salvage robotic-assisted laparoscopic prostatectomy (sRALP) include wide mobilization of the bladder to decrease tension on the vesicourethral anastomosis, decreased bleeding and improved visualization of the rectal/prostatic plane. In one of the largest series from Vanderbilt University, 34 patients underwent sRALP with no open conversions, a median time of surgery of 176 min and a 94% next day (POD1) discharge rate.[43] Surgical margins were similar to open surgery at 26%. At a median of 16 months of follow-up, BFS was 82% with three patients that did not achieve a PSA nadir of "<0.1 ng/mL" after surgery. Importantly, men appeared to have potential functional improvement compared to standard SRP with a lower rate of bladder neck contractures (9%), and higher rates of continence (39%) and erectile function (29%). Further studies with greater follow-up are needed to determine the long-term oncologic effectiveness.

## COMPLICATIONS AND QUALITY OF LIFE

For the patient, choosing prostatectomy after RT is daunting. All patients considering salvage surgery have previously had the option for surgery and chosen radiation, usually due to concerns about side effects of treatment. Thus, presented with the complications and functional recovery associated with SRP and sRALP, many will not choose treatment. In the perioperative period, the most concerning complication of SRP is rectal injury with a frequency as high as 19%.[44] For many tumors, adherence of cancer and fibrosis to the rectum make resection of part of the rectal wall essential to achieve a negative margin. A rectal injury of any significance during an sRALP should manage bowel diversion as successful healing is impaired by radiation.[40] Ward et al. described a rectal injury rate of 5% for SRP while the rate was as high as 10% for patients that required a cystoprostatectomy.[41] The rate of rectal injury appears to be lower after sRALP at only 1 in 34 patients.[43] After catheter removal, the most concerning complication after SRP is anastomotic stricture (or bladder neck contracture, BNC). The BNC rate for open surgery ranged from 22% to 41% depending on definition.[40] Often BNCs after SRP require multiple dilations until incontinence is reached with a scarred open bladder neck necessitating artificial urinary sphincter placement.[39] After sRALP only 9% (3/34) of patients developed a BNC in their early series (16 months) with management requiring only gentle office.[43] This may be due to greater mobilization of the bladder or a continuous anastomosis. While many validated studies are lacking, rates of incontinence range from 21% to 90% after SRP with 39% reporting complete recovery after sRALP.[40] Nerve-sparing surgery is very difficult to perform during SRP and most men have moderate to severe erectile dysfunction (ED) prior to SRP with rates of preop ED ranging with reported estimates ranging from 51% to 90%.[40] After surgery, ED approaches nearly 100% with open surgery, 29% of men who had erections before sRALP were able to penetrate after surgery.

## CONCLUSIONS

Recurrence of PCa after RT is common and likely underdiagnosed due to the lack of markers to identify cancer persistence or recurrence. Recurrence is established by either consecutive rises in PSA values or a rise to greater than 2.0 ng/mL over PSA nadir. As the dose of radiation delivered has increased, fewer men will likely have a recurrence of PCa after RT. These recurrences may be diagnosed earlier with MRI imaging and MRI–US fusion biopsies. In the future, development of improved biomarkers may allow for earlier detection of PCa recurrence and therefore improve outcomes of salvage treatments.

## References


1. Murphy GP, Kuss R, Khoury S, et al. Radiology and nuclear medicine. *International Symposium on Prostate Cancer*. 16–18 June 1986, Paris, FR; 1987.
2. Schiller J. *Prostate Cancer: Using Research from the Internet*. CreateSpace; 2010.
3. Halperin EC, Brady LW, Perez CA, et al. *Perez & Brady's principles and practice of radiation oncology*. 6th ed. Philadelphia: Lippincott Williams and Wilkins; 2013.
4. Singh PB, Anele C, Dalton E, et al. Prostate cancer tumour features on template prostate-mapping biopsies: implications for focal therapy. *Eur Urol* 2014;**66**(1):12–9.
5. Cox JD, Gallagher MJ, Hammond EH, et al. Consensus statements on radiation therapy of prostate cancer: guidelines for prostate re-biopsy after radiation and for radiation therapy with rising prostate-specific antigen levels after radical prostatectomy. *American Society for Therapeutic Radiology and Oncology Consensus Panel. J Clin Oncol* 1999;**17**(4):1155.
6. Crook JM, Perry GA, Robertson S, et al. Routine prostate biopsies following radiotherapy for prostate cancer: results for 226 patients. *Urology* 1995;**45**(4):624–32.
7. Crook JM, Choan E, Perry GA, et al. Serum prostate-specific antigen profile following radiotherapy for prostate cancer: implications for patterns of failure and definition of cure. *Urology* 1998;**51**(4):566–72.
8. Pisters LL, Leibovici D, Blute M, et al. Locally recurrent prostate cancer after initial radiation therapy: a comparison of salvage radical prostatectomy versus cryotherapy. *J Urol* 2009;**182**(2):517–27.
9. American Society for Therapeutic Radiology and Oncology. Consensus statement: guidelines for PSA following radiation therapy. American Society for Therapeutic Radiology and Oncology Consensus Panel. *Int J Radiat Oncol Biol Phys* 1997;**37**(5):1035–41.
10. Roach III M, Hanks G, Thames Jr H, et al. Defining biochemical failure following radiotherapy with or without hormonal therapy in men with clinically localized prostate cancer: recommendations of the RTOG-ASTRO Phoenix Consensus Conference. *Int J Radiat Oncol Biol Phys* 2006;**65**(4):965–74.
11. Bianco Jr FJ, Scardino PT, Stephenson AJ, et al. Long-term oncologic results of salvage radical prostatectomy for locally recurrent prostate cancer after radiotherapy. *Int J Radiat Oncol Biol Phys* 2005;**62**(2):448–53.
12. Meeks JJ, Kern SQ, Dalbagni G, et al. The prevalence of persistent prostate cancer after radiotherapy detected at radical cystoprostatectomy for bladder cancer. *J Urol* 2014;**191**(6):1760–3.
13. Critz FA, Levinson AK, Williams WH, et al. The PSA nadir that indicates potential cure after radiotherapy for prostate cancer. *Urology* 1997;**49**(3):322–6.
14. Zelefsky MJ, Ben-Porat L, Scher HI, et al. Outcome predictors for the increasing PSA state after definitive external-beam radiotherapy for prostate cancer. *J Clin Oncol* 2005;**23**(4):826–31.
15. PSA nadir levels after radiotherapy for prostate cancer: a powerful prognostic variable. Oncology 1997; 11(8): 1236, 1239.
16. Caloglu M, Ciezki J. Prostate-specific antigen bounce after prostate brachytherapy: review of a confusing phenomenon. *Urology* 2009;**74**(6):1183–90.
17. Pickles T. British Columbia Cancer Agency Prostate Cohort Outcomes I. Prostate-specific antigen (PSA) bounce and other fluctuations: which biochemical relapse definition is least prone to PSA false calls? An analysis of 2030 men treated for prostate cancer with external beam or brachytherapy with or without adjuvant androgen deprivation therapy. *Int J Radiat Oncol Biol Phys* 2006;**64**(5):1355–9.
18. McGrath SD, Antonucci JV, Fitch DL, et al. PSA bounce after prostate brachytherapy with or without neoadjuvant androgen deprivation. *Brachytherapy* 2010;**9**(2):137–44.
19. Hanlon AL, Pinover WH, Horwitz EM, et al. Patterns and fate of PSA bouncing following 3D-CRT. *Int J Radiat Oncol Biol Phys* 2001;**50**(4):845–9.
20. D'Amico AV, Moul JW, Carroll PR, et al. Surrogate end point for prostate cancer-specific mortality after radical prostatectomy or radiation therapy. *J Natl Cancer Inst* 2003;**95**(18):1376–83.
21. Johnstone PA, McFarland JT, Riffenburgh RH, et al. Efficacy of digital rectal examination after radiotherapy for prostate cancer. *J Urol* 2001;**166**(5):1684–7.
22. Huang WC, Kuroiwa K, Serio AM, et al. The anatomical and pathological characteristics of irradiated prostate cancers may influence the oncological efficacy of salvage ablative therapies. *J Urol* 2007;**177**(4):1324–9 quiz 1591.

23. Meeks JJ, Walker M, Bernstein M, et al. Seminal vesicle involvement at salvage radical prostatectomy. *BJU Int* 2013;**111**(8):E342–7.
24. Bostwick DG, Egbert BM, Fajardo LF. Radiation injury of the normal and neoplastic prostate. *Am J Surg Pathol* 1982;**6**(6):541–51.
25. Gaudin PB, Zelefsky MJ, Leibel SA, et al. Histopathologic effects of three-dimensional conformal external beam radiation therapy on benign and malignant prostate tissues. *Am J Surg Pathol* 1999;**23**(9):1021–31.
26. Cheng L, Cheville JC, Bostwick DG. Diagnosis of prostate cancer in needle biopsies after radiation therapy. *Am J Surg Pathol* 1999;**23**(10):1173–83.
27. Meeks JJ, Walker M, Bernstein M, et al. Accuracy of post-radiotherapy biopsy before salvage radical prostatectomy. *BJU Int* 2013;**112**(3):308–12.
28. Kramer S, Gorich J, Gottfried HW, et al. Sensitivity of computed tomography in detecting local recurrence of prostatic carcinoma following radical prostatectomy. *Br J Radiol* 1997;**70**(838):995–9.
29. Tamada T, Sone T, Jo Y, et al. Locally recurrent prostate cancer after high-dose-rate brachytherapy: the value of diffusion-weighted imaging, dynamic contrast-enhanced MRI, and T2-weighted imaging in localizing tumors. *AJR Am J Roentgenol* 2011;**197**(2):408–14.
30. Sala E, Eberhardt SC, Akin O, et al. Endorectal MR imaging before salvage prostatectomy: tumor localization and staging. *Radiology* 2006;**238**(1):176–83.
31. Terris MK, Klonecke AS, McDougall IR, et al. Utilization of bone scans in conjunction with prostate-specific antigen levels in the surveillance for recurrence of adenocarcinoma after radical prostatectomy. *J Nucl Med* 1991;**32**(9):1713–7.
32. Cher ML, Bianco Jr FJ, Lam JS, et al. Limited role of radionuclide bone scintigraphy in patients with prostate specific antigen elevations after radical prostatectomy. *J Urol* 1998;**160**(4):1387–91.
33. Kitajima K, Murphy RC, Nathan MA, et al. Update on positron emission tomography for imaging of prostate cancer. *Int J Urol* 2014;**21**(1):12–23.
34. Nunez R, Macapinlac HA, Yeung HW, et al. Combined 18F-FDG and 11C-methionine PET scans in patients with newly progressive metastatic prostate cancer. *J Nucl Med* 2002;**43**(1):46–55.
35. Krause BJ, Souvatzoglou M, Tuncel M, et al. The detection rate of [11C]choline-PET/CT depends on the serum PSA-value in patients with biochemical recurrence of prostate cancer. *Eur J Nucl Med Mol Imaging* 2008;**35**(1):18–23.
36. Stephenson AJ. Role of salvage radical prostatectomy for recurrent prostate cancer after radiation therapy. *J Clin Oncol* 2005;**23**(32):8198–203.
37. Rogers E, Ohori M, Kassabian VS, et al. Salvage radical prostatectomy: outcome measured by serum prostate specific antigen levels. *J Urol* 1995;**153**(1):104–10.
38. Chade DC, Shariat SF, Cronin AM, et al. Salvage radical prostatectomy for radiation-recurrent prostate cancer: a multi-institutional collaboration. *Eur Urol* 2011;**60**(2):205–10.
39. Gotto GT, Yunis LH, Vora K, et al. Impact of prior prostate radiation on complications after radical prostatectomy. *J Urol* 2011;**184**(1):136–42.
40. Chade DC, Eastham J, Graefen M, et al. Cancer control and functional outcomes of salvage radical prostatectomy for radiation-recurrent prostate cancer: a systematic review of the literature. *Eur Urol* 2012;**61**(5):961–71.
41. Ward JF, Sebo TJ, Blute ML, et al. Salvage surgery for radiorecurrent prostate cancer: contemporary outcomes. *J Urol* 2005;**173**(4):1156–60.
42. Paparel P, Cronin AM, Savage C, et al. Oncologic outcome and patterns of recurrence after salvage radical prostatectomy. *Eur Urol* 2009;**55**(2):404–10.
43. Kaffenberger SD, Keegan KA, Bansal NK, et al. Salvage robotic assisted laparoscopic radical prostatectomy: a single institution, 5-year experience. *J Urol* 2013;**189**(2):507–13.
44. Neerhut GJ, Wheeler T, Cantini M, et al. Salvage radical prostatectomy for radiorecurrent adenocarcinoma of the prostate. *J Urol* 1988;**140**(3):544–9.

CHAPTER

# 32

# Postradical Prostatectomy Incontinence

*Lisa Parrillo, MD, Alan Wein, MD, PhD(Hon)*

**Department of Urology, Perelman Center for Advanced Medicine, Philadelphia, PA, USA**

## INTRODUCTION

There was an estimated 233,000 new cases of prostate cancer in the United States in 2014 and many of these men would have been treated with radical prostatectomy.[1] Incontinence is a well-documented complication of this procedure and, regardless of a robotic or open approach, incontinence rates seem to be similar, ranging from 1% to 40% of all patients at one year.[2] Various definitions of postradical prostatectomy incontinence (PRPI) have been used in the literature, and when the definition is expanded to include occasional dribbling, up to 74% of men will report PRPI.[3] This is often not a temporary issue; urinary incontinence after prostatectomy continues to significantly affect health-related quality-of-life for men many years after surgery.[4]

After patients undergo an appropriate evaluation, they can be offered a variety of therapies that range from conservative management to surgical intervention. Conservative treatments include behavioral modifications and lifestyle changes, penile clamps, condom catheters, biofeedback, pharmacotherapy, electrical stimulation, and extracorporeal magnetic stimulation. If these are unsuccessful, patients can undergo more invasive therapies that include bulking agents, periurethral constrictors, balloon compression of the urethra, bulbar urethral slings, artificial urinary sphincters, and possibly, in the future, stem cell therapy.

## EPIDEMIOLOGY AND PATHOPHYSIOLOGY OF URINARY INCONTINENCE AFTER RADICAL PROSTATECTOMY

Radical prostatectomy can be accomplished by four different approaches: perineal, open retropubic, laparoscopic, or robotic-assisted laparoscopic. The most common procedures used currently are the open retropubic and robotic-assisted laparoscopic prostatectomies. Perineal prostatectomy fell out of favor in the 1980s with the advent of the nerve sparing retropubic prostatectomy and in 2007 less than 2% of the radical prostatectomies performed were by a perineal approach.[5] Pure laparoscopic prostatectomy has been shown to have similar operative and postoperative results as open and robotic prostatectomy, but is considered to be a technically more challenging procedure with a longer learning curve, and is not widely used.[6]

Postradical prostatectomy urinary incontinence can be broadly classified by etiology and contributing factors (Table 32.1). Continence in the male requires

1. Accommodation of increasing volumes of urine at a low intravesical pressure (normal compliance) and with appropriate sensation.
2. A bladder outlet that is closed at rest and remains so during increases in intra-abdominal pressure.
3. Absence of involuntary bladder contractions (detrusor overactivity).[7]

Outlet terminology differs among authors and requires some explanation. We prefer to use the terms smooth and striated sphincter area.[7] The smooth sphincter refers to the smooth muscle of the bladder neck and proximal urethra. This is a physiologic but not anatomic sphincter and is not under voluntary control. Others refer to this area as the internal sphincter, the proximal sphincter, and the bladder neck sphincter. There is virtually no physiologic or pharmacologic response that affects the smooth muscle of the most proximal urethra without affecting the smooth muscle of the bladder neck. Normally, the smooth sphincter is competent and resistance increases in this area during bladder filling and urine storage and decreases during an emptying contraction. In the proximal urethra there is an inner longitudinal smooth muscle layer and a thinner outer circular layer. Many believe the longitudinal layer is continuous with the musculature of the bladder base. Teleologically

Prostate Cancer. http://dx.doi.org/10.1016/B978-0-12-800077-9.00032-3

TABLE 32.1 Classification of Postprostatectomy Urinary Incontinence

| | |
|---|---|
| Functional | 1. Because of physical disability<br>2. Because of lack of awareness of concern |
| Bladder abnormalities | 1. Detrusor overactivity (involuntary contractions)<br>2. Decreased compliance<br>3. Hypersensitivity with incontinence<br>4. Overflow incontinence |
| Outlet abnormalities | 1. Sphincter insufficiency<br>2. Urethral instability<br>3. Postvoid dribbling |
| Combination | |

this is consistent with a tonic role of the circular muscle in helping to maintain closure during filling and storage and a phasic role for the longitudinal layer in contributing to opening of this area during voiding. The function of this entire proximal sphincter mechanism is essentially negated by radical prostatectomy.

The classic view of the "external urethral sphincter" is that of a sheet of striated muscle within the leaves of a "urogenital diaphragm" that extends horizontally across the pelvis. This is under voluntary control and is responsible for stopping the urinary stream in response to the command "stop voiding." The striated sphincter concept expands this definition to include intramural and extramural portions (Figure 32.1). The extramural portion, actually periurethral striated muscle, corresponds roughly to the "classic" external urethral sphincter. The intramural portion denotes skeletal muscle that is intimately associated with the urethra from the maximal condensation of the extramural striated muscle to at least the apex of the prostate. This intramural portion is called the rhabdosphincter and includes the intrinsic smooth muscle in that area as well.[8] The prevailing view is that the striated component of the rhabdosphincter

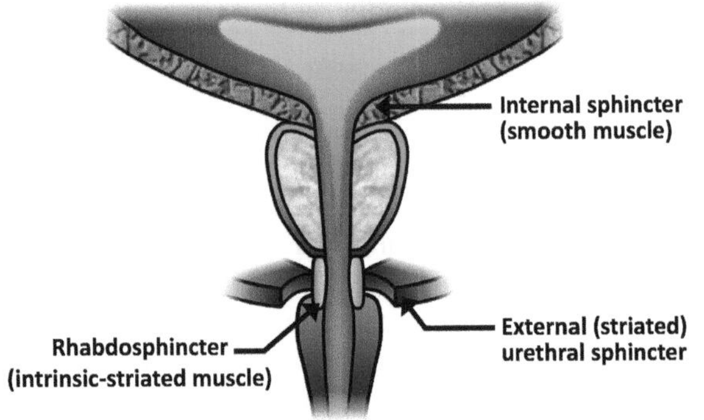


**FIGURE 32.1 Internal urethral sphincter, external urethral sphincter, and rhabdosphincter.**

contains predominately slow twitch (Type 1) striated muscle, which can maintain resting tone and preserve continence. The fibers of the periurethral (extrinsic to the urethra) striated muscle are fast twitch (Type 2) and capable of rapid on-demand contraction.[8,9] The striated muscle of the rhabdosphincter is considerably thicker ventrally and invested in a fascial framework that is supported below by a musculofascial plate that fuses with the midline raphe, a point of origin for the rectourethralis muscle.[10] It is this complex that is largely responsible for preservation of the sphincteric function of the outlet after radical prostatectomy. Damage to this area or its innervations, rendering it poorly or nonfunctional, is a major cause of PRPI.

Urinary incontinence after prostatectomy is predominantly sphincter related and the amount can range from a few drops with exercise or changing positions to total or near total leakage. Detrusor dysfunction is often found in such patients but is seldom the sole cause. There are many series that have characterized urodynamic findings in these patients. Two are illustrative. Ficazzola and Nitti found that 27 of 60 men (45%) with PRPI had some element of detrusor overactivity but this significantly contributed to the incontinence in only 16 (27%). Intrinsic sphincter deficiency (ISD) was found in 54 (90%). The PRPI was attributed to ISD alone in 67%, ISD and detrusor overactivity and/or decreased compliance in 23%, and bladder dysfunction alone in 3%.[11] Groutz et al. found that detrusor overactivity was present in 28 of 83 patients (33.7%), but was the only urodynamic finding in 3 patients (3.6%) and ISD was the most common finding and the dominant cause in 73 (88%). They also mention "low urethral compliance" presumably from scarring and fibrosis as a significant cause of ISD in 25 men (30.1%).[12]

Overflow incontinence is one of those entities that is easy to recognize but hard to urodynamically define. It is an uncommon cause of PRPI, usually due to postoperative bladder neck contracture, but one that must be ruled out.

There is a long list of factors that have been associated by some clinicians with the occurrence of PRPI, but unanimous agreement on all is lacking. These are summarized compositely by Koelbl et al.[8], Herschorn et al.[13], and Xie and Sandhu[14] and include advanced age, obesity, decreased overall well-being, prior irradiation, prior prostatectomy or ablation for benign disease, anastomotic structure, anastomotic extravasation, and prostate size and configuration. Some factors that may be protective against PRPI include nerve sparing prostatectomy, seminal vesical preservation, bladder neck preservation, puboprostatic ligament preservation, bladder neck intussusception or other alteration, posterior reconstruction of Denonvillier's musculofascial plate, anterior urethral suspension or support, urethral preservation, and having an experienced surgeon. In the final analysis, meticulous surgical technique of the apex is probably the most important single factor that we as surgeons can influence.

The use of MRI has highlighted the apparent adverse effect of periurethral or perisphincter fibrosis and a "short" urethral length (from verumontanum to proximal bulb). Tuygun et al. have cited a much higher incidence of perisphincter fibrosis in incontinent than continent men at 6 months or more after prostatectomy (22/22 vs. 4/14).[15] The Penn group reported that preoperative increased urethral length was related to decreased time to continence as a dichotomous (<20 mm vs. >20 mm) and continuous variable.[16] Dubbelman et al. reviewed a number of assessment techniques for urethral function as predictors of postradical prostatectomy continence status.[17] Their findings include the following:

1. There is no role in this respect for sphincter electromyography.
2. Postoperative abdominal leak point pressures correlated with the presence of ISD but not severity of continence.
3. Preoperative urethral profilometry "seems to be important" in prediction of PRPI. The maximal urethral closure pressure seems to be the most important parameter.
4. Preoperative urethral length (apex of prostate to bulb of penis, as defined by Myers et al.[18]) is associated with PRPI (shorter length, higher chance), although no cut-off values are proposed. Coakley et al. found that of those with a preoperative length of >12 mm 89% were continent at 1 year versus 77% with <12 mm.[19]
5. MRI evidence of perisphincter fibrosis "might" have a negative effect on sphincter function

Opinions differ regarding the effect on continence of postoperative radiation delivered to the prostatic fossa of postradical prostatectomy patients. Van Cangh et al. and Formenti et al. reported no effect.[20,21] Suardi et al. found a detrimental effect on continence as did Hofmann et al.[22,23]

## EVALUATION OF MEN WITH POSTPROSTATECTOMY INCONTINENCE

Virtually all men are incontinent immediately following prostatectomy, and it is recommended that they undergo at least 6–12 months of recovery prior to any invasive intervention. The evaluation for PRPI begins with a detailed history and physical examination. It is critical to characterize the type of incontinence (stress or urgency) and the severity. A voiding diary for 2–7 days has been validated as a tool to assess urge and incontinence episodes. Additionally, a 24-h pad test can provide more objective data on degree of incontinence.[24] There are currently efforts in place to validate a questionnaire that accurately assesses quality of life based on pad weight, which may further help delineate degrees of PPI and direct treatments.[25] Physical exam should assess for any previous surgical incision sites and the presence of an inguinal hernia that could affect placement of an artificial urinary sphincter. In addition, all patients should have a urine culture sent as any UTI must be treated prior to intervention. Finally, a postvoid residual, prostate-specific antigen and creatinine level should be sent to evaluate for voiding dysfunction, cancer status, and renal function.

Further evaluation with cystoscopy and multichannel urodynamics is recommended for all men considering surgical intervention. If an anastomotic stricture, bladder neck contracture, or urethral stricture is present it can complicate a surgical intervention and these need to be treated prior to any surgery for incontinence. Furthermore, during cystoscopy the integrity of the external sphincter can be evaluated, which may influence the choice of surgical treatment. Urodynamic testing is used to ascertain bladder capacity, bladder compliance, contractility, and the presence of detrusor overactivity, and can differentiate between stress, urgency, and mixed incontinence. It is important to identify patients with concomitant detrusor overactivity and urgency incontinence, as they may have less successful outcomes following surgery. Relevant Oxford Levels of Evidence (LOE) and Grades of Recommendation (GOR) from the International Consultation on Incontinence with respect to evaluation are[13]:

1. Basic history, physical exam, urinalysis, and postvoid residual (LOE1-2, GOR A);
2. Voiding diary (LOE 1-2, GOR B);
3. Pad tests (LOE 1-2, GOR B);
4. Cystoscopy, appropriate imaging (LOE 2-3, GOR B);
5. Multichannel urodynamics may be useful prior to invasive treatment (LOE 3, GOR C).

## CONSERVATIVE THERAPY

In the first 6–12 months after prostatectomy, it is generally recommended that patients attempt a variety of conservative, nonsurgical measures that may help improve PRPI. Common sense suggests that all patients should adopt simple lifestyle modifications that include limiting total fluid intake, timed voiding, and avoiding foods and drinks such as alcohol or caffeinated beverages that can cause irritation to the bladder .[26] The ICI Committee on adult conservative management cautions, however, that to date, no trials have addressed the topic of lifestyle interventions alone for men with any type of urinary incontinence.[27] Pelvic floor muscle training may likewise be initiated early in the postoperative period. For those men who are frustrated by pad use but are not interested in pursuing any additional interventions, the use of condom catheters or penile clamps may be considered. These patients need to be carefully monitored for penile erosion. Additional conservative therapies that have been reported to improve symptoms and quality of life include biofeedback, electrical stimulation, external magnetic stimulation, and pharmacotherapy.

Overall the effects of conservative treatments for urinary incontinence in men, both bladder and sphincter related, have received much less attention than in women. As a consequence, ICI recommendations for men are less numerous and less robust. Enthusiastic individual reports are often tempered by lesser ICI Committee recommendations, formulated after consideration of global data.

Pelvic floor muscle training (PFMT), or Kegel exercises, is defined as "any program of repeated voluntary pelvic floor muscle contractions taught by a health care professional."[28] The goal is to increase the strength of pelvic muscles and improve voluntary control of the pelvic floor via skeletal muscle contraction and support of urethral coaptation. While there is significant variation in treatment regimens, pelvic floor muscle training has been prospectively evaluated in the postprostatectomy population and been found to lessen the time to regain continence, but not change the final percent of those achieving "continence."[29] Given that there is no harm to pelvic floor muscle training, it is advisable that all men begin an exercise regimen promptly after prostatectomy. Furthermore, pelvic floor muscle training can improve incontinence even in men who have persistent PPI after a year. In a prospective study of over 200 men with incontinence more than a year after prostatectomy, pelvic floor muscle training decreased incontinence episodes by 51%.[30]

Biofeedback (BF) may be used in combination with pelvic floor muscle training to help coach patients how to properly use these muscles. It involves monitoring muscle contractions with EMG so patients can learn how to appropriately contract them. Randomized trials have compared biofeedback enhanced pelvic floor muscle therapy to no training or usual care and have suggested that significantly more men were continent at one to two months after prostatectomy with biofeedback, but this significance was no longer present at 3–4 months. When comparing patients who were treated with biofeedback-enhanced pelvic floor muscle therapy to those who received written or verbal instructions on pelvic floor muscle therapy, three prior randomized studies have shown no significant difference between treatments.[31] However, a recent randomized trial showed good effect. Men undergoing radical prostatectomy who were randomized to biofeedback training and instruction on a structured program of pelvic floor muscle training the day prior to surgery with monthly biofeedback sessions at home had significantly better continence at three and six months after surgery than men who simply received written instructions.[32]

The Committee on Conservative Management of the ICI offered the following conclusions[27]:

1. The evidence that therapist delivered PFMT with or without BF before or after radical prostatectomy improves continence recovery remains inconsistent.
2. Some preoperative or immediate postoperative instruction in PFMT may be helpful (GORB B) but whether this is in the form of hands-on therapy or verbal instruction and support remains unclear.
3. There is some evidence that those who undergo some sort of conservative management, including PFMT, will achieve continence in a shorter time frame but the difference is not significant at 12 months postoperatively (LOE 2).
4. It is not clear whether PFMT taught by digital rectal exam offers any benefit over verbal or written instruction (GOR B).
5. The use of BF to assist PFMT is currently a therapist/patient decision based on economics and preference.
6. For men with PRPI there does not appear to be a benefit to adding electrical stimulation (ES) to a PFMT program (GOR B) although ES may help achieve continence earlier.
7. In the absence of sufficient data from rigorous and well-reported trials it is not known whether ES, as a stand-alone treatment for male urgency or stress urinary incontinence, is better than no treatment, placebo, or control treatments.
8. It is not known if pre- or postoperative ES or magnetic stimulation (MS) has a role in reducing PRPI.

The Cochrane Collaboration reviewed conservative management for postprostatectomy incontinence up to November 2009 and concluded that the benefits were uncertain.[33] They further stated that long-term incontinence could be managed by external penile clamp but there are safety concerns.

The European Association of Urology Guidelines are in general agreement and state explicitly that the evidence is inconsistent as to whether ES alone can improve urinary incontinence (LOE 2) and that MS does not cure or improve incontinence (LOE 2).[34]

Pharmacologic therapy is useful management of the overactive bladder symptom syndrome and DO. This subject is exhaustively reviewed in the ICI Consultation and in the ICI Oxford Assessments[35] and the International Consultation on Male LUTS.[36] Since PRPI is mostly an outlet related phenomenon, the reader is referred to these excellent publications for a description of the therapy of OAB-related incontinence and mixed incontinence in the male. It should be noted that the populations studied in the trials considered for the ICI Oxford Assessments consisted primarily of women.

Similarly, the literature relevant to the pharmacotherapy of sphincteric incontinence relates primarily to stress urinary incontinence in women and the treatment of men with stress urinary incontinence after radical prostatectomy has received little attention. Tsakiris et al. searched for articles on drug treatment of male SUI published between 1966 and June 2007 and did a generalized database search in addition.[37] Only nine trials were identified using alpha adrenergic agonists, beta-2 antagonists or SNRSs. Only one of these included a comparison arm, 40 mg twice daily duloxetine plus pelvic floor exercise versus pelvic floor exercise with placebo.[38] Duloxetine is a serotonin–norepinephrine reuptake inhibitor which, in the cat acetic acid model of irritated bladder function, has been shown to significantly increase striated sphincteric activity only during the filling-storage phase of the micturation cycle.[39,40] The results suggested a positive effect of the drug but were not straightforward. Of those patients completing the 4 month trial, (91/112) 78% of the drug patients versus 52% of the placebo group were "dry." However, one month after the trial ended, 46% versus 73% were dry, respectively, a shift still observed 2 months later. The authors of the article suggested larger, well-designed studies on duloxetine for this potential usage.

Cornu et al. reported on a series of men with stress or mixed (stress predominant) urinary incontinence post-radical prostatectomy who were randomized to duloxetine or placebo after a 2-week placebo run in. Dosage was 20 mg BID for 7 days, 40 mg BID for 67 days, and 20 mg for 14 days. Subjects were at least a year out from surgery. Outcome measures included percent decrease in incontinence episode frequency (IEF), 1 h pad test, and various QOL measures. Statistical significance for IEF percent decrease occurred only at weeks 8 and 12, but there was clearly a trend at 4 weeks as well. There was no statistical difference in 1 h pad test weights but there was in various quality of life scores. A 50–100% decrease in IEF was seen at 12 weeks in over half the patients. Adverse events for drug and placebo included fatigue (50% vs. 13%), insomnia (25% vs. 7%), libido loss (19% vs. 7%), constipation (13% vs. 7%), nausea (13% vs. 7%), diarrhea (13% vs. 7%), dry mouth (6% vs. 0%), anorexia (6% vs. 0%), and sweating (25% vs. 20%).[41] Limitations are the small number of patients (the original proposed sample size was 90) and the lack of any placebo effect on IEF and QOL. One would logically not expect improvement to continue after drug withdrawal unless a permanent change occurred in behavior, anatomy or neuromuscular function. However, in an uncontrolled usage study on men with post-RP SUI, Collado Serra et al. reported that the benefit similarly remained in 85% of the patients after the drug was stopped. In that series, 25% of the patients withdrew because of AEs and 33% because of lack of effect.[42]

Regarding duloxetine the EAU guidelines state that[34]

1. Duloxetine does not cure urinary incontinence (LOE 1a).
2. Duloxetine causes significant gastrointestinal and CNS side effects leading to a high rate of treatment discontinuation (LOE 1a).
3. Duloxetine 80 mg daily can improve SUI in men (LOE 1b).
4. Duloxetine can be offered to women or men who are seeking temporary improvement in incontinence symptoms (GOR A).
5. Duloxetine should be initiated using drug titration because of high adverse event rates (GOR A).

Usage of duloxetine for PRPI in the male is off label. Larger controlled and better-designed studies are necessary to provide conclusive positive or negative data on this subject.

## SURGICAL PLANNING

If a male patient has undergone an evaluation for postprostatectomy incontinence and wishes to pursue surgical intervention, there are a variety of options available. The most commonly performed procedures are perineal slings and an artificial urinary sphincter (AUS), with the AUS being considered the gold standard. Additional surgical options include urethral bulking agents, periurethral constrictors, balloon compression, and stem cell therapies. In addition to the workup previously discussed and patient desire, when planning for surgery procedure choice should be tailored to each patient based on his BMI, surgical and radiation history, and capacity for manipulation of a sphincter pump. If the patient has a need for future transurethral surgical procedures, as is the case of men with bladder cancer or urethral stricture disease, some recommend they forgo AUS or sling placement as they have a greater risk of complications.

## URETHRAL BULKING AGENTS

Urethral bulking agents are more widely used in female patients with stress urinary incontinence but have been administered in men with PPI. Historically collagen was used as a bulking agent but is no longer available. Macroplastique is currently the only agent available on the market. Additionally, there are several additional agents that are not FDA approved for PPI such as ethylene vinyl alcohol copolymer (Tegress®) and carbon-coated zirconium beads.

Sanchez-Ortiz et al. evaluated 31 men who received retrograde collagen injection after radical retropubic prostatectomy and showed a 35% improvement rate but achieved complete continence in less than 10% of patients.[43] Smith et al. injected 54 men with PPI with collagen through a transurethral approach and showed that 35.2% achieved social continence, defined as at most one pad used daily.[44] Westney et al. reviewed a series of 322 men with intrinsic sphincter deficiency who received collagen injections after a variety of treatments for prostate cancer. In a subanalysis of men who had undergone radical prostatectomy there was a 49% mean pad reduction with a mean duration of response of 7 months.[45]

Macroplastique, a bulking agent composed of silicone macroparticles, has also been shown to have limited efficacy. Kylmälä et al. injected 50 men, 46 of whom underwent radical prostatectomies, with mild to moderate stress urinary incontinence and 12% achieved continence with one injection. With a second injection an additional 20% achieved continence, an additional 18% with three injections, and an additional 10% with four injections. This population was only tracked for three months, so prolonged efficacy could not be assessed.[46] With each of the additional injectable bulking agents (Tegress and carbon-coated zirconium beads) there has been no evidence of any significant benefit in the postprostatectomy population.[13]

Urethral bulking agents are the minimally invasive option for patients pursuing surgical intervention for their PPI. However, this modality only achieves success in a small percentage of patients and the effect decreases over time. It is not recommended as a first line therapy in this population unless the patient elects to pursue it.

Regarding bulking on agents in PRPI the EUA guidelines state that[34]

1. There is no evidence that bulking agents are PPI (LOE 2a).
2. There is weak evidence that they can offer temporary short-term improvement on QOL in men with PPI (LOE 3).
3. There is no evidence that one bulking agent is superior to another (LOE 3).
4. Only offer these to men with PRPI who desire temporary relief (GOR C).
5. Do not offer these to men with severe PRPI (GOR C).

## PERIURETHRAL CONSTRICTORS

One of the more recent devices to be introduced for the management of PPI is a periurethral constrictor (Silimed, Brazil), a device composed of an adjustable silicone cuff that is fitted around the proximal urethra and partially occludes it thereby increasing outlet pressure. Patients will need to increase their intra-abdominal pressure to void. In a series of 66 patients who at baseline used more than three protective pads 79% achieved total continence at a median of 26 months. Four patients experienced complications requiring ex-plant, two with infection, and two with erosion.[47] However, a retrospective review of the long-term outcomes of another group of 56 men showed only 39% achieved continence and 41% required device removal for complications, the most common of which was urethral erosion in 27%.[48] Currently, this device is not available in the United States.

## CONTINENCE BALLOON DEVICE

Another new device is ProACT (Uromedica, Plymouth, MN), an adjustable continence balloon device. It consists of two silicone balloons that are implanted at the bladder neck and increase urethral resistance via injections into a percutaneous port. Several studies have shown good results. Lebret et al. placed this device in 66 men and at 6 months 71% used 0 or 1 pads from a mean of 4.6 pads daily prior.[49] Gregori et al. implanted this device in 62 men and followed them for a mean of 25 months. Forty-one patients were considered dry, indicated by having a 24 h pad weight of less than 8 g. Five patients failed treatment, all of whom had been previously irradiated. Of these five, three had unilateral balloon migrations and two experienced urethral erosions.[50] From the Netherlands, Kjaer et al. reported on 114 men who had the ProACT device inserted. There was an overall dry rate of 50% (less than 8 g of daily leakage). There were 23 patients with complications, all of whom were successfully treated with removal and replacement of the device. Fourteen patients eventually transitioned to a urethral sling or an artificial sphincter.[51] This device is also not currently commercially available in the United States.

The EAU conclusions regarding the ProACT device specifically are as follows[34]:

1. Very limited short-term evidence suggests that the noncircumferential compression device ProACT is effective for PRPI (LOE 3).
2. The noncircumferential compression device ProACT is associated with a high failure and complication rate leading to frequent explantation (LOE 3).

3. Do not offer the ProACT to men who have had pelvic radiotherapy (GOR C).

The ICI states that it is reasonable to use the ProACT in the short and medium term in men with mild to moderate leakage, who have not had radiation therapy previously. This device cannot be compared at this point to male slings or the AUS. No recommendation is made given the variability in the rate of complications.[13]

## PERINEAL SLINGS

The male sling was developed based on the concept of urethral compression by the penile crura from Kaufman in the 1970s. He transitioned this idea into a silicone compressive implant that was placed under the bulbar urethra and had a 63% continence rate.[52] Contemporary slings include the InVance (AMS, Minnetonka, MN), the Argus adjustable male sling (Promedon SA, Cordoba, Argentina), Male Remeex System (Neomedic International, Barcelona, Spain), ATOMS (AMI, Vienna, Austria), the I-Stop TOMS (CL Medical, Winchester, MA), the Virtue quadratic sling (Coloplast, Humlebaek, Denmark), and the Advance XP slings (AMS, Minnetonka, MN). While the artificial urinary sphincter is considered the gold standard in men with PPI, a perineal sling may be a better choice in men who are concerned with the complexity of the procedure of AUS implantation, the potential need for revisions, and the cost. Additionally, slings are a better option in patients with any issue with manual dexterity. The most common postoperative complications in these patients are urethral erosion and infection. All patients need to undergo a cystoscopy during the procedure to ensure the bladder and urethra have not been perforated.

The InVance male sling is composed of a polypropylene mesh that is attached and tensioned by titanium screws placed in the pubic rami so it compresses the bulbospongiosus muscle. Guimarães et al. followed men implanted with Invance slings after prostatectomy and 65% were cured at a mean of 28 months. Two patients (3%) required explantation secondary to infection.[53] Similarly, in 45 men with moderate to severe PPI, 76% of men were dry or had improved to one or two pads a day and 72% were satisfied with the procedure. Ten patients had perineal numbness in the immediate postoperative period but this resolved within 3 months. Infection occurred in one patient.[54] The Invance system has been removed from the market with the development of other slings that have a lower rate of infections and no risk of bone osteitis.

The Argus device (Figure 32.2) is an adjustable sling with a silicone cushion that tensions the bulbar urethra via two silicone rings that rest on the rectus fascia. Once implanted, the sling can be adjusted by reopening the suprapubic incision and pulling the columns that connect the cushion to the rings further through the rings. Bochove-Overgaauw and Schrier reported on a series of 100 men who were followed for a mean of 27 months. Forty percent were cured and an additional 32% improved to using only one to two pads every day. Twenty-four patients underwent a single revision and seven required an additional revision. There were, however, significant complications: 11 patients suffered infections requiring sling removal, six suffered bladder

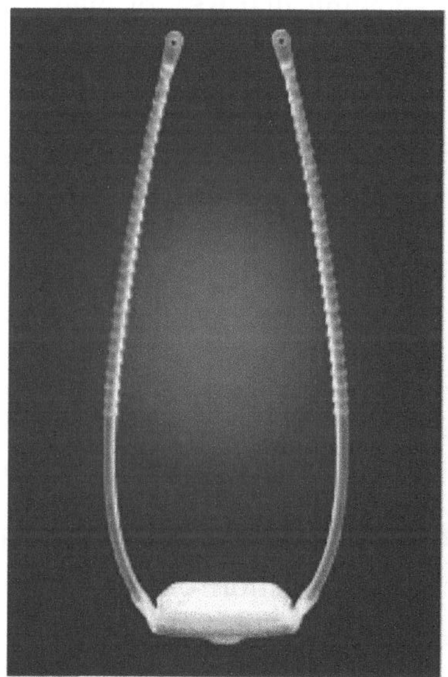
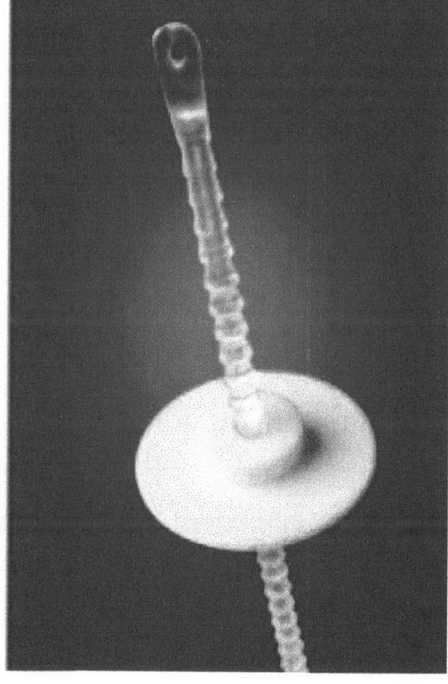

FIGURE 32.2 **Argus sling.** *(Promedon SA; Cordoba, Argentina. http://www.promedon.com/us/urology-urogynecology/male-urinary-incontinence.)*

FIGURE 32.3 **Reemex sling.** *(Neomedic International, Barcelona, Spain. http://www.neomedic.com/en/.)*

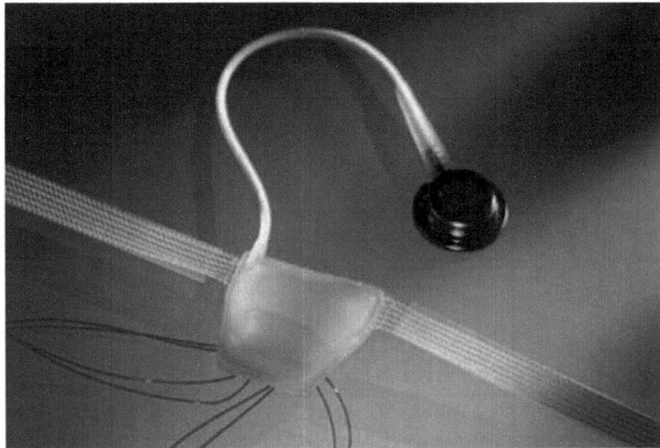

FIGURE 32.4 **ATOMS.** *(AMI, Vienna, Austria. http://www.ami.at/en/produkt/a-m-i-atoms-system-2/?portfolioID=1712.)*

or urethral erosion, and one had sling rupture. These complications were more likely in men who had previously had radiation or a known urethral stricture.[55] An additional 101 men were examined over a mean of 2 years in a series by Hübner et al. and they found that 79% of patients were dry with 39% requiring adjustment and a 16% removal rate secondary to urethral erosion or infection.[56]

There are two additional adjustable slings. The Remeex (Figure 32.3) is a suburethral mesh attached to a suprapubic mechanical regulator by two monofilament traction threads. The regulator is adjusted post-op day one via an external manipulator that is then removed. In a multicentric European study, 51 men with PPI had the Remeex placed and 65% were either completely dry or used one security pad daily that was typically dry. Another 20% showed improvement. Complications included one urethral erosion and two infections at the regulator site.[57] Another adjustable sling is the transobturator ATOMS device, which is a mesh implant with a silicone cushion within it connected to a titanium port for adjusting the volume of the cushion (Figure 32.4). Unlike the Argus, this device does not require additional surgery for adjustment. Hoda et al. placed this device in 99 men and found 31% were dry after initial placement and that 63% were dry after adjustment at a mean of 17 months. Four patients developed wound infections at the port site, all of whom had success for explantation and reimplantation three months later.[58]

The EAU guidelines for adjustable slings state that[34]

1. There is limited evidence that adjustable male slings can cure or improve SUI in men (LOE 3).
2. There is limited evidence that early explantation rates are high (LOE 3).
3. There is no evidence that adjustability of the male sling offers additional benefit over other types of slings.

The most widely used sling in the United States is the advance retrourethral transobturator sling, which is placed at the membranous urethra instead of the bulbar urethra. It is thought to work by repositioning the sphincteric portion of the urethra back into the pelvis in the anatomically appropriate location instead of directly compressing the urethral lumen.[59] Several studies with the Advance sling showed good results with Cornu et al. showing that 63% of men treated with a sling who at baseline wore five pads or less daily were pad-free at one year.[60] At a median of 27 months, Bauer et al. showed a 51.6% cure rate and a 23.8% improved rate, which was defined as one to two pads daily or a reduction of pad use greater than 50%.[61] In contrast, Cornel et al. placed slings in 36 men and defined cure as no pad use and less than 2 g urine loss daily. Only 9% of men achieved this cure at one year, although 46% were improved by at least 50% on pad test.[62]

The most recently developed sling is the Virtue male sling (Figure 32.5) composed of a large polypropylene mesh with four extensions, two superior and prepubic, and two inferior and transobturator. This design is thought to allow for urethral compression and proximal relocation of the membranous urethra. Comiter et al. implanted the sling in 22 patients and showed its ability to increase retrograde leak point pressure to a mean of 68.8 cm water from a mean baseline of 33.4 cm water.[63] In a follow-up to this study, they performed

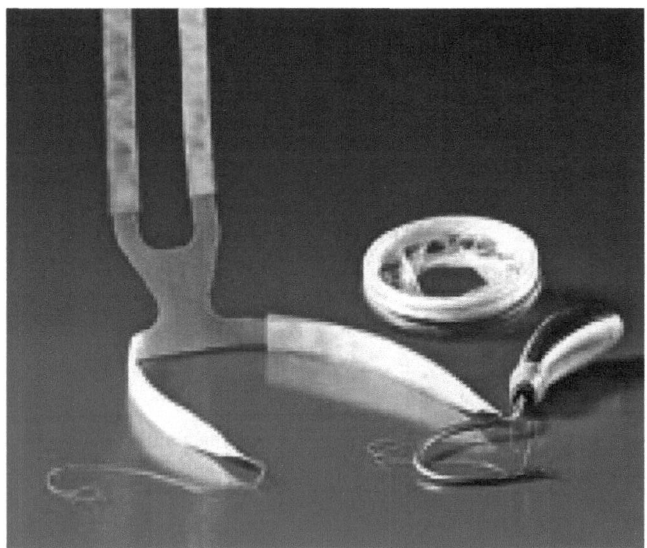

FIGURE 32.5 **Virtue sling.** *(Coloplast, Humlebaek, Denmark. http://www.us.coloplast.com).*

two prospective trials. In the first, 98 men had the Virtue sling placed. Of these 42% had a greater than 50% improvement in pad weight at 12 months and 42% subjective success defined as a very much/much improved score on the Patient Global Impression of Improvement. They then enrolled a second cohort of 31 men who underwent Virtue placement with fixation sutures placed to the periosteum at the junction of the pubis and rami. While they did not recruit the number of patients needed to power this portion of the study, 79% of study subjects had a greater than 50% improvement in pad weight at 12 months and 71% achieved subjective success. Neither group experienced infections or erosions.[64]

The EAU guidelines regard fixed slings are[34]:

1. There is limited evidence that fixed male slings cure or improve postprostatetcomy incontinence in patients with mild to moderate incontinence (LOE 3).
2. Men with severe incontinence, previous radiotherapy, or urethral stricture surgery have poor outcomes from fixed slings (LOE 3).
3. There is no evidence that one type of male sling is better than another.
4. Offer fixed slings to men with mild to moderate PRPI (GOR B).
5. Warn men that severe incontinence, prior pelvic radiotherapy, or urethral stricture surgery may worsen the outcome of fixed male sling surgery (GOR C).

The ICI has concluded that slings are best in men with low to moderate incontinence who have not had radiation or an AUS placed previously. There seems to be a lower risk of erosion or atrophy with slings than with the AUS. Long-term outcomes for transobturator, pubourethral, and quadratic slings are not yet known.[13]

## ARTIFICIAL URINARY SPHINCTER

The artificial urinary sphincter remains the gold standard for surgical treatment for men with postprostatectomy incontinence. It is the most effective and long-lasting solution for these men. As previously mentioned, it is expensive, and requires patient manipulation several times daily. Kumar et al. showed that when given the choice between AUS and placement of a sling, men prefer the sling in order to avoid using a mechanical device.[65] It is also the most invasive procedure and can have significant complications. However, it can also be successfully implanted after men have had a previous intervention with a sling.[66,67] Given these factors, patients need to be carefully selected based on the objective and subjective measurements of incontinence, their dexterity, and history of previous surgical procedures.

The AUS is composed of an inflatable cuff, a pressure-regulating balloon, and a control pump. The cuff ranges in size from 3.5 cm to 11 cm and the balloon comes in five different preset pressures. Both of these are chosen by the surgeon for each individual patient in the operating room. The most common cuff size is 4.0 or 4.5 cm. The control pump is positioned in the scrotum, typically on the side of handedness. When a patient needs to void he will squeeze the control pump several times, thereby transferring the fluid from the cuff into the reservoir and releasing the coaptation of the urethra.

Several studies have been performed validating the AUS as a surgical treatment of PPI. With success defined as wearing 0–1 pad per day (PPD), rates have varied between 60% and 90% in modern series (Table 32.2). Complications include urinary retention, infection, urethral erosion, and urethral atrophy. With infection, the device needs to be removed immediately, but can be reimplanted after appropriate antiseptic irrigation. If a patient presents with urethral erosion, the device must be removed and the injury can be managed with a urethral catheter or suprapubic tube. When a patient has urethral atrophy, the AUS can be revised and a small cuff or a tandem cuff may be placed.[68]

Approximately 10–15% of men continue to have incontinence after placement of an AUS. In these men a tandem or double cuff system may be placed to provide acceptable continence. Kabalin placed a second cuff in five men, four of whom required no or only one safety pad daily.[80] O'Connor et al. followed 22 patients who had double cuffs placed for a mean of 4.8 years. Eleven percent required no pads daily. An additional 61% wore either a safety pad or one pad daily. There were a total of 12 complications in this population: two had a cuff

TABLE 32.2 Summary of Trials Testing Artificial Urinary Sphincters

| Authors, year published | Number of patients | Follow up | Results | Complications |
|---|---|---|---|---|
| Litwiller et al. (1996)[69] | 65 | 23.4 months | 20% 0 PPD<br>55% a few drops 22% leakage less than 1 teaspoon | 6% Infections<br>9% Atrophy<br>18% Revision or Removal |
| Haab et al. (1997)[70] | 68 | 7.2 years | 84% 0 or 1 PPD | 7.4% Erosion<br>14.8% Atrophy<br>25% Revision<br>5.8% Removal |
| Klijn et al. (1998)[71] | 27 | 20 months | 81% 0–1 PPD | 30% Revision |
| Mottet et al. (1998)[72] | 96 | 1 year | 62% no leakage<br>28% 0 PPD | 12%Infection or Erosion<br>21% Revision |
| Clemens et al. (2001)[73] | 60 | 26.6 months | 34% 0 PPD<br>46% 1–2 PPD | 12% Infection<br>13% Erosion<br>36% Revisions |
| Gousse et al. (2001)[74] | 71 | 7.7 years | 27% 0 PPD<br>32% 1 PPD | 1.4% Infection<br>4% Erosion<br>29% Revision |
| Montague et al. (2001)[75] | 209 | 73 months | 64% 0–1 PPD | 12% Revision |
| Lai et al. (2006)[76] | 218 | 36.5 months | 35% 0 PPD<br>34% 1 PPD | 5.5% Infection<br>6.0% Erosion<br>9.6% Atrophy<br>27.1% Revision<br>6.0% Removal |
| O'Connor et al. (2006)[77] | 33 | 5.0 years | 83% 0–1 PPD | 3% Infection<br>9% Erosion<br>14% Revision<br>14% Removal |
| Trigo Rocha et al. (2007)[78] | 40 | 53.4 months | 90% 0–1 PPD | 7.5% Infection<br>20% Revision |
| Kim et al. (2008)[79] | 124 | 6.8 years | 27.1% 0 PPD<br>52% 1 PPD | 4.3% Infection<br>4.3% Erosion<br>29% Revision |

erosion, two had an infection, three had urethral atrophy, one developed a urethral stricture, and one developed a recto-urethral fistula. The remaining three had a device malfunction.[81]

Finally, an AUS can also be placed in a transcorporal fashion in a patient with a history of urethral injury or erosion. There is evidence that there are decreased rates of erosion and fewer urethral injuries with this technique.[82]

EAU recommendations regarding compressive devices as[34]

1. There is limited evidence that primary AUS implantation is effective for cure of SUI in men (LOE 2b).
2. Long-term failure rate for AUS is high although device replacement can be performed (LOE 3).
3. Previous pelvic radiotherapy does not appear to affect the outcome of AUS implantation.
4. Men who develop cognitive impairment or loss manual dexterity will have difficulty operating an AUS (LOE 3).
5. Tandem cuff placement is not superior to single cuff placement (LOE 3).
6. The penoscrotal and perineal approaches generate equivalent outcomes (LOE 3).
7. Offer AUS to men with moderate to severe PRPI (GOR B).
8. Implantation of compressive devices should only be offered in high-volume centers (GOR C).
9. Warn men receiving compressive devices that even in high-volume centers, there is a high risk of complications, mechanical failure, or a need for explantation (GOR C).

The ICI states that the AUS is the gold standard for treatment of PRPI in patients with severe incontinence, those who have had radiation therapy, or those previously treated with an AUS. Given the long-term success rates and high patient satisfaction, the risk of revision is acceptable.[13]

## STEM CELL THERAPY

There has recently been interest in the use of autologous stem cells or muscle-derived cells as an injectable approach to restoring a deficient sphincter. Multiple animal models have shown promise with this technology.[83] Mitterberger et al. injected 63 men with PPI with autologous fibroblasts and myoblasts. At 12 months after injection, 41 of these men were continent and 17 showed improvement. There was also a statistically significant increase in maximum urethral closure pressure on urodynamics.[84] There has, however, been considerable concern regarding the legitimacy of this group's previously published data in women with stress urinary incontinence.[13] Current research continues to focus on cellular engineering.[83]

## CONCLUSIONS

Postprostatectomy incontinence is a widespread and common issue for the many men today being treated with prostate cancer. It can significantly impact the quality of life for these individuals. However, there are many options available to combat this. With the appropriate evaluation and a treatment tailored to each patient's physical condition and desire, continence can be achieved in the majority.

## References


1. SEER Cancer Statistics Factsheets: Prostate Cancer. National Cancer Institute. Bethesda, MD, http://seer.cancer.gov/statfacts/html/prost.html.
2. Chapple CR, Milsom I. Urinary incontinence and pelvic prolapse: epidemiology and pathophysiology. In: Wein AJ, Kavoussi LR, Novick A, et al., editors. *Campbell-Walsh Urology: Tenth Edition Review*. Philadelphia, USA: Saunders; 2011. p. 343–7 [chapter 63].
3. Holm HV, Fosså SD, Hedlund H, et al. How should continence and incontinence after radical prostatectomy be evaluated? A prospective study of patient-ratings and changes over time. *J Urol* 2014;**192**(4):3312–6.
4. Miller DC, Sanda MG, Dunn RL, et al. Long-term outcomes among localized prostate cancer survivors: health-related quality-of-life changes after radical prostatectomy, external radiation, and brachytherapy. *J Clin Oncol* 2005;**23**(12):2772–80.
5. Prasad SM, Gu X, Lavelle R, et al. Comparative effectiveness of perineal versus retropubic and minimally invasive radical prostatectomy. *J Urol* 2011;**185**(1):111–5.
6. Rozet F, Jaffe J, Braud G, et al. A direct comparison of robotic assisted versus pure laparoscopic radical prostatectomy: a single institution experience. *J Urol* 2007;**178**(2):478–82.
7. Wein AJ. Pathophysiology and classification of lower urinary tract dysfunction: overview. In: Wein AJ, Kavoussi LR, Novick A, et al., editors. *Campbell-Walsh Urology: Tenth Edition Review*. Philadelphia, USA: Saunders; 2011. p. 335–9 [chapter 61].
8. Koelbl H, Igawa T, Laterza RM, et al. Committee 4: pathophysiology. In: Abrams P, Cardozo C, Khoury K, Wein A, editors. *Incontinence*. Paris: IUCD-EAU; 2013. p. 262–334.
9. Gosling JA, Dixon JS, Critchley HO, et al. A comparative study of the human external sphincter and periurethral levator ani muscles. *Br J Urol* 1981;**53**(1):35–41.
10. Burnett AL, Mostwin JL. *In situ* anatomical study of the male urethral sphincteric complex: relevance to continence preservation following major pelvic surgery. *J Urol* 1998;**160**(4):1301–6.
11. Ficazzola MA, Nitti VW. The etiology of post-radical prostatectomy incontinence and correlation of symptoms with urodynamic findings. *J Urol* 1998;**160**(4):1317–20.
12. Groutz A, Blavias JG, Chaikin DC, et al. The pathophysiology of post-radical prostatectomy incontinence: a clinical and video urodynamic study. *J Urol* 2000;**163**(6):1767–70.
13. Herschorn S, Bruschini H, Comiter C, et al. Committee 13: surgical treatment of urinary incontinence in men. In: Abrams P, Cardozo C, Khoury K, Wein A, editors. *Incontinence*. Paris: IUCD-EAU; 2013. p. 1230–84.
14. Xie D, Sandhu J. Post prostatectomy incontinence. *Curr Bladder Dysfunct Rep* 2013;**8**(4):268–76.
15. Tuygun C, Imamoglu A, Keyik B, et al. Significance of fibrosis around and/or at external urinary sphincter on pelvic magnetic resonance imaging in patients with postprostatectomy incontinence. *Urology* 2006;**68**(6):1308–12.
16. Mendoza PJ, Stern JM, Li AY, et al. Pelvic anatomy on preoperative magnetic resonance imaging can predict early continence after robot-assisted radical prostatectomy. *J Endourol* 2011;**25**(1):51–5.
17. Dubbelman YD, Groen J, Wildhagen MF, et al. Urodynamic quantification of decrease in sphincter function after radical prostatectomy: relation to postoperative continence status and the effect of intensive pelvic floor muscle exercises. *Neurourol Urodyn* 2012;**31**(5):646–51.
18. Myers RP, Cahill DR, Devine RM, et al. Anatomy of radical prostatectomy as defined by magnetic resonance imaging. *J Urol* 1998;**159**(6):2148–58.
19. Coakley FV, Eberhardt S, Kattan MW, et al. Urinary continence after radical retropubic prostatectomy: relationship with membranous urethral length on preoperative endorectal magnetic resonance imaging. *J Urol* 2002;**168**(3):1032–5.
20. Van Cangh PJ, Richard F, Lorge F, et al. Adjuvant radiation therapy does not cause urinary incontinence after radical prostatectomy: results of a prospective randomized study. *J Urol* 1998;**159**(1):164–6.
21. Formenti SC, Lieskovsky G, Skinner D, et al. Update on impact of moderate dose of adjuvant radiation on urinary continence and sexual potency in prostate cancer patients treated with nerve-sparing prostatectomy. *Urology* 2000;**56**(3):453–8.
22. Suardi N, Gallina A, Lista G, et al. Impact of adjuvant radiation therapy on urinary continence recovery after radical prostatectomy. *Eur Urol* 2014;**65**(3):546–51.
23. Hofmann T, Gaensheimer S, Buchner A, et al. An unrandomized prospective comparison of urinary continence, bowel symptoms and the need for further procedures in patients with and with no adjuvant radiation after radical prostatectomy. *BJU Int* 2003;**92**(4):360–4.
24. Groutz A, Blaivas JG, Chaikin DC, et al. Noninvasive outcome measures of urinary incontinence and lower urinary tract symptoms: a multicenter study of micturition diary and pad tests. *J Urol* 2000;**164**(3):698–701.
25. Nitti VW, Mourtzinos A, Brucker BM. Correlation of patient perception of pad use with objective degree of incontinence measured

by pad test in men with post prostatectomy incontinence: the SUFU Pad Test Study. *J Urol* 2014;**192**:836–42.

26. Wyman JF, Burgio KL, Newman DK. Practical aspects of lifestyle modifications and behavioural interventions in the treatment of overactive bladder and urgency urinary incontinence. *Int J Clin Pract* 2009;**63**(8):1177–91.
27. Moore K, Bradley C, Burgio K, et al. Committee 12: adult conservative management. In: Abrams P, Cardozo C, Khoury K, Wein A, editors. *Incontinence*. Paris: IUCD-EAU; 2013. p. 1103–214.
28. Wilson PD, Berghmans B, Hagen S, et al. Adult conservative management. *Incontinence* 2005;**2**:855–964.
29. Filocamo MT, Li Marzi V, Popolo GD, et al. Effectiveness of early pelvic floor rehabilitation treatment for post-prostatectomy incontinence. *Eur Urol* 2005;**48**(5):734–8.
30. Goode PS, Burgio KL, Johnson TM, et al. Behavioral therapy with or without biofeedback and pelvic floor electrical stimulation for persistent postprostatectomy incontinence: a randomized controlled trial. *JAMA* 2011;**305**(2):151–9.
31. MacDonald R, Fink HA, Huckabay C, et al. Pelvic floor muscle training to improve urinary incontinence after radical prostatectomy: a systematic review of effectiveness. *BJU Int* 2007;**100**(1): 76–81.
32. Tienforti D, Sacco E, Marangi F, et al. Efficacy of an assisted low-intensity programme of perioperative pelvic floor muscle training in improving the recovery of continence after radical prostatectomy: a randomized controlled trial. *BJU Int* 2012;**110**(7):1004–10.
33. Campbell SE, Glazener CMA, Hunter KF. Conservative management for postprostatectomy urinary incontinence. *Cochrane Database Syst Rev* 2012;(1) Article CD 001843.
34. Lucas MG, Bedretdinova D, Bosch JLHR, et al. *Guidelines on urinary incontinence*. European Association of Urology: The Netherlands; 2014. Available from: http://www.uroweb.org/wp-content/uploads/20-Urinary-Incontinence_LR.pdf [accessed April 2014].
35. Andersson KE, Chapple CR, Cardozo L, et al. Committee 8: pharmacological treatment of urinary incontinence. In: Abrams P, Cardozo C, Khoury K, Wein A, editors. *Incontinence*. Paris: IUCD-EAU; 2013. p. 625–703.
36. Roerborn C, Andersson KE, Wei J, et al. Committee 8: medical management and new therapeutic targets. In: Chapple C, McVary K, Roehrborn C, editors. *Lower urinary tract symptoms*. Fukuoka: SIU-IUCD; 2013. p. 377–517.
37. Tsakiris P, de la Rosette JJ, Michel M, et al. Pharmacologic treatment of male stress urinary incontinence: systemic review of the literature and levels of evidence. *Eur Urol* 2008;**53**:53–9.
38. Filocamo MT, LiMarzi V, Del Popoilo G, et al. Pharmacologic treatment in postprostatectomy stress urinary incontinence. *Eur Urol* 2007;**51**:1559.
39. Thor KB, Katofiasc MA. Effects of duloxetine, a combined serotonin and norepinephrine reuptake inhibitor, on central neural control of lower urinary tract function in the chloralose-anesthetized female cat. *J Pharmacol Exp Ther* 1995;**242**(2):1014–24.
40. Katofiasc MA, Nissen J, Audia JE, et al. Comparison of the effects of serotonin selective, norepinephrine selective, and dual serotonin and norepinephrine reuptake inhibitors on lower urinary tract function in cats. *Life Sci* 2002;**71**(11):1227–36.
41. Cornu JN, Merlet B, Ciofu C, et al. Duloxetine for mild to moderate postprostatectomy incontinence: preliminary results of a randomised, placebo-controlled trial. *Eur Urol* 2011;**59**(1):148–54.
42. Collado Serra A, Rubio-Briones J, Payás MP, et al. Postprostatectomy established stress urinary incontinence treated with duloxetine. *Urology* 2011;**78**(2):261–6.
43. Sanchez-Ortiz RF, Broderick GA, Chaikin DC, et al. Collagen injection therapy for post-radical retropubic prostatectomy incontinence: role of Valsalva leak point pressure. *J Urol* 1997;**158**(6):2132–6.
44. Smith DN, Appell RA, Rackley RR, et al. Collagen injection therapy for post-prostatectomy incontinence. *J Urol* 1998;**160**(2):364–7.
45. Westney OL, Bevan-Thomas R, Palmer J, et al. Transurethral collagen injections for male intrinsic sphincter deficiency: the University of Texas-Houston experience. *J Urol* 2005;**174**(3):994–7.
46. Kylmälä T, Tainio H, Raitanen M, et al. Treatment of postoperative male urinary incontinence using transurethral macroplastique injections. *J Endourol* 2003;**17**(2):113–5.
47. Introini C, Naselli A, Zaninetta G, et al. Safety and efficacy of periurethral constrictor implantation for the treatment of post-radical prostatectomy incontinence. *Urology* 2012;**79**(5):1175–9.
48. Lima RS, Barros EG, Souza CA, et al. Periurethral constrictor: late results of the treatment of post prostatectomy urinary incontinence. *Int Braz J Urol* 2011;**37**(4):483–7.
49. Lebret T, Cour F, Benchetrit J, et al. Treatment of postprostatectomy stress urinary incontinence using a minimally invasive adjustable continence balloon device, ProACT: results of a preliminary, multicenter, pilot study. *Urology* 2008;**71**(2):256–60.
50. Gregori A, Romanò AL, Scieri F, et al. Transrectal ultrasound-guided implantation of adjustable continence therapy (ProACT): surgical technique and clinical results after a mean follow-up of 2 years. *Eur Urol* 2010;**57**(3):430–6.
51. Kjær L, Fode M, Nørgaard N, et al. Adjustable continence balloons: clinical results of a new minimally invasive treatment for male urinary incontinence. *Scand J Urol Nephrol* 2012;**46**(3):196–200.
52. Kaufman JJ. Surgical treatment of post-prostatectomy incontinence: use of the penile crura to compress the bulbous urethra. *J Urol* 1972;**107**(2):293–7.
53. Guimarães M, Oliveira R, Pinto R, et al. Intermediate-term results, up to 4 years, of a bone-anchored male perineal sling for treating male stress urinary incontinence after prostate surgery. *BJU Int* 2009;**103**(4):500–4.
54. Carmel M, et al. Long-term efficacy of the bone-anchored male sling for moderate and severe stress urinary incontinence. *BJU Int* 2010;**106**(7):1012–6.
55. Bochove-Overgaauw DM, Schrier BP. An adjustable sling for the treatment of all degrees of male stress urinary incontinence: retrospective evaluation of efficacy and complications after a minimal followup of 14 months. *J Urol* 2011;**185**(4):1363–8.
56. Hübner WA, Gallistl H, Rutkowski M, et al. Adjustable bulbourethral male sling: experience after 101 cases of moderate-to-severe male stress urinary incontinence. *BJU Int* 2011;**107**(5):777–82.
57. Sousa-Escandon A, Cabrera J, Mantovani F, et al. Adjustable suburethral sling (male remeex system) in the treatment of male stress urinary incontinence: a multicentric European study. *Eur Urol* 2007;**52**(5):1473–80.
58. Hoda MR, Primus G, Fischereder K, et al. Early results of a European multicentre experience with a new self-anchoring adjustable transobturator system for treatment of stress urinary incontinence in men. *BJU Int* 2013;**111**(2):296–303.
59. Davies TO, Bepple JL, McCammon KA. Urodynamic changes and initial results of the AdVance male sling. *Urology* 2009;**74**(2):354–7.
60. Cornu J-N, Sèbe P, Ciofu C, et al. The AdVance transobturator male sling for postprostatectomy incontinence: clinical results of a prospective evaluation after a minimum follow-up of 6 months. *Eur Urol* 2009;**56**(6):923–7.
61. Bauer RM, Soljanik I, Füllhase C, et al. Mid-term results for the retroluminar transobturator sling suspension for stress urinary incontinence after prostatectomy. *BJU Int* 2011;**108**(1):94–8.
62. Cornel EB, Elzevier HW, Putter H. Can advance transobturator sling suspension cure male urinary postoperative stress incontinence? *J Urol* 2010;**183**(4):1459–63.
63. Comiter CV, Nitti V, Elliot C, et al. A new quadratic sling for male stress incontinence: retrograde leak point pressure as a measure of urethral resistance. *J Urol* 2012;**187**(2):563–8.

64. Comiter CV, Rhee EY, Tu LM, et al. The virtue sling – a new quadratic sling for postprostatectomy incontinence – results of a multinational clinical trial. *Urology* 2014;**84**(2):438–9.
65. Kumar A, Litt ER, Ballert KN, et al. Artificial urinary sphincter versus male sling for post-prostatectomy incontinence – what do patients choose? *J Urol* 2009;**181**(3):1231–5.
66. Fisher MB, Aggarwal N, Vuruskan H, et al. Efficacy of artificial urinary sphincter implantation after failed bone-anchored male sling for postprostatectomy incontinence. *Urology* 2007;**70**(5): 942–4.
67. Abdou A, Cornu JN, Sèbe P, et al. Salvage therapy with artificial urinary sphincter after Advance™ male sling failure for post-prostatectomy incontinence: a first clinical experience. *Prog Urol* 2012;**22**(11):650–6.
68. Wessells H, Peterson AC. Surgical procedures for sphincteric incontinence in the male: the artificial genitourinary sphincter and perineal sling procedures. In: Wein AJ, Kavoussi LR, Novick A, et al., editors. *Campbell-Walsh Urology: Tenth Edition Review*. Philadelphia, USA: Saunders; 2011. p. 427–9 [chapter 79].
69. Litwiller SE, Kim KB, Fone PD, et al. Post-prostatectomy incontinence and the artificial urinary sphincter: a long-term study of patient satisfaction and criteria for success. *J Urol* 1996;**156**(6): 1975–80.
70. Haab F, Trockman BA, Zimmern PE, et al. Quality of life and continence assessment of the artificial urinary sphincter in men with minimum 3.5 years of followup. *J Urol* 1997;**158**(2):435–9.
71. Klijn AJ, Hop WC, Mickisch G, et al. The artificial urinary sphincter in men incontinent after radical prostatectomy: 5 year actuarial adequate function rates. *Br J Urol* 1998;**82**(4):530–3.
72. Mottet N, Boyer C, Chartier-Kastler E, et al. Artificial urinary sphincter AMS 800 for urinary incontinence after radical prostatectomy: the French experience. *Urol Int* 1998;**60**(Suppl. 2):25–9.
73. Clemens JQ, Schuster TG, Konnak JW, et al. Revision rate after artificial urinary sphincter implantation for incontinence after radical prostatectomy: actuarial analysis. *J Urol* 2001;**166**(4):1372–5.
74. Gousse AE, Madjar S, Lambert MM, et al. Artificial urinary sphincter for post-radical prostatectomy urinary incontinence: long-term subjective results. *J Urol* 2001;**166**(5):1755–8.
75. Montague D, Angermeier KW, Paolone DR. Long-term continence and patient satisfaction after artificial sphincter implantation for urinary incontinence after prostatectomy. *J Urol* 2001;**166**(2):547–9.
76. Lai HH, Hsu EI, Teh BS, et al. 13 years of experience with artificial urinary sphincter implantation at Baylor College of Medicine. *J Urol* 2001;**177**(3):1021–5.
77. O'Connor RC, Nanigian DK, Patel BN, et al. Artificial urinary sphincter placement in elderly men. *Urology* 2007;**69**(1):126–8.
78. Trigo Rocha F, Gomes CM, Mitre AI, et al. A prospective study evaluating the efficacy of the artificial sphincter AMS 800 for the treatment of postradical prostatectomy urinary incontinence and the correlation between preoperative urodynamic and surgical outcomes. *Urology* 2008;**71**(1):85–9.
79. Kim SP, Sarmast Z, Daignault S, et al. Long-term durability and functional outcomes among patients with artificial urinary sphincters: a 10-year retrospective review from the University of Michigan. *J Urol* 2008;**179**(5):1912–6.
80. Kabalin JN. Addition of a second urethral cuff to enhance performance of the artificial urinary sphincter. *J Urol* 1996;**156**(4):1302–4.
81. O'Connor RC, Lyon MB, Guralnick ML, et al. Long-term follow-up of single versus double cuff artificial urinary sphincter insertion for the treatment of severe postprostatectomy stress urinary incontinence. *Urology* 2008;**71**(1):90–3.
82. Guralnick ML, Miller E, Toh KL, et al. Transcorporal artificial urinary sphincter cuff placement in cases requiring revision for erosion and urethral atrophy. *J Urol* 2002;**167**:2075–8.
83. Stangel-Wojcikiewicz K, Małgorzata S, Nikolavsky D, et al. Cellular therapy for treatment of stress urinary incontinence. *Curr Stem Cell Res Ther* 2010;**5**(1):57–62.
84. Mitterberger M, Marksteiner R, Margreiter E, et al. Myoblast and fibroblast therapy for post-prostatectomy urinary incontinence: 1-year followup of 63 patients. *J Urol* 2008;**179**(1):226–31.

CHAPTER

# 33

# Prognostic Significance of Positive Surgical Margins and Other Implications of Pathology Report

*George R. Schade, MD*, Jonathan L. Wright, MD, MS*,**, Daniel W. Lin, MD*,***

*Department of Urology, University of Washington School of Medicine, Seattle, WA, USA

**Division of Public Health Sciences, Fred Hutchinson Cancer Research Center, Seattle, WA, USA

## INTRODUCTION

With the adoption of widespread prostate-specific antigen (PSA) screening, prostate cancer (PCa) is increasingly diagnosed at a younger age and at an earlier stage of the disease.[1–3] For patients managed with radical prostatectomy (RP), the opportunity to perform pathologic evaluation of the specimen offers valuable prognostic information and directs informed decision-making regarding surveillance and secondary therapies. For patients with localized disease, RP has demonstrated excellent long-term oncologic outcomes.[4,5] While preoperative PSA, Gleason score, and pathologic stage may be the dominant predictors of post-RP outcomes,[4–6] other findings observed at the time of pathologic assessment have been associated with cancer-specific outcomes as well. The goal of this chapter is to review the clinical significance of these additional pathologic findings, including positive surgical margins, tertiary Gleason pattern, lymphovascular invasion, seminal vesicle invasion, and lymph node involvement.

## POSITIVE SURGICAL MARGINS

Positive surgical margins (PSMs), defined by cancerous cells touching the inked surface of the RP specimen (Figure 33.1),[7] are considered an adverse outcome associated with failure to achieve surgical cure of PCa following RP. Data comparing open, laparoscopic, and robotic-assisted laparoscopic approaches have demonstrated equivalence between approaches with respect to margin status.[8,9] In an extensive contemporary literature review, Novara et al. demonstrated varying PSMs rates depending on tumor stage.[10] Stage-specific rates of PSMs were 4–23% (mean 9%) for pT2, 29–50% (mean 37%) for pT3, and 40–75% (mean 50%) for pT4 disease, supporting the notion that more extensive disease is associated with higher PSMs rates. Further PSM rates have been significantly associated with tumor Gleason score and tumor volume.[11]

The role of nontumor-related clinical factors on the risk of PSMs has been evaluated by many groups with inconclusive results. Intuitively, factors that contribute to making surgery more difficult could contribute to increased incidence of PSMs and is supported by data from Patel et al. in which they observed high body mass index and large prostate volume to be independent predictors of PSMs overall and in those with organ-confined disease in a large multi-institutional study.[12] However, other studies have not found an association between PSM and BMI[13] or prostate volume,[14] as well as other factors including prior abdominal surgery[15] and previous benign prostatic hyperplasia (BPH) debulking procedures.[16] An additional factor that may relate to PSMs is surgeon experience. Data suggest higher-volume surgeons have fewer positive margins with learning curves being described for open RP,[17] laparoscopic RP,[18] and robotic-assisted RP[19] of 250, 200–250, and 1000–1500 cases, respectively. These data suggest that the rate of PSMs is a factor of surgeon judgment and experience, for example in determining the surgical approach (i.e., nerve-sparing vs. wide-field resection).

Prostate Cancer. http://dx.doi.org/10.1016/B978-0-12-800077-9.00033-5

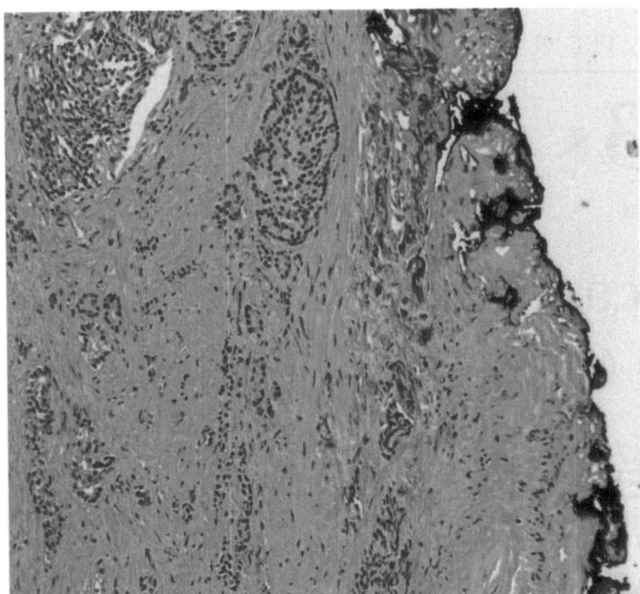

**FIGURE 33.1 Histological appearance of a positive surgical margin with prostate cancer extending to inked margin or radical prostatectomy specimen.**

## Association of PSMs and Clinical Outcomes

The impact of PSMs on patient clinical outcomes has been studied extensively. PSMs have consistently been associated with elevated risk of PSA recurrence,[7,10,20–27] imparting an approximately 1.5–5-fold increased risk of biochemical recurrence (BCR) (see Table 33.1). Additionally, the extent of PSMs affects the risk of recurrence, whereby those with extensive PSMs have a significantly increased risk of BCR in patients compared to those with only focal PSMs.[5,28–37] Using a 3 mm cut point, Babaian et al., observed that patients with ≤3 mm of margin positivity had a recurrence rate of 14% compared to 53% in those with >3 mm of margin involvement.[38] Similarly, Lake et al. demonstrated significantly worse 10-year recurrence-free survival when stratified by margin status: 84, 64, 38%, for negative margins, focal PSMs, and extensive PSMs, respectively.[37] Finally, compared to solitary PSMs, Stephenson et al. identified multifocal PSMs as an independent predictor of post-RP disease progression (HR 1.4, 95% CI 1.1–1.8) with significantly worse 7-year PSA progression-free probability of 62% versus 49%.[5] Such observations helped inform the 2009 International Society of Urologic Pathologists (ISUP) consensus statement calling for the amount of tumor at the inked margin to be measured and recorded in millimeters.[7]

An additional PSM characteristic to be documented in the pathology report according to the 2009 ISUP consensus statement is the location of PSMs.[7] Likely due to technical aspects of RP, such as attempts to preserve membranous urethral length and nerve-sparing, the location of PSMs varies anatomically, with observed rates of 8–58% at the apex, 9–40% at the posterolateral prostate, 2–19% at the base, and 4–20% of cases at the bladder neck among cases with PSMs in a review by Fleshner et al.[39] With respect to risk of BCR, all PSM locations do not appear equal. In most studies, apical margins and anterior margins have not been independently associated with increased risk of PSA recurrence, with recurrence rates similar to patients with negative margins.[40–42] Conversely, posterolateral margins, possibly resulting from attempts at nerve sparing, may be the most significant. In a retrospective study by Eastham et al. examining the outcome of 2442 RPs, isolated posterolateral and posterior PSMs were an independent risk factor (HR 2.8, 95% CI 1.8–4.4 and HR 2.0 (95% CI 1.1–3.6, respectively)) for BCR compared to negative margins whereas other margin sites were not independently associated with BCR in their cohort.[43] Finally, bladder neck margins may be the site with the most controversy. While some studies have reported higher risk of biochemical recurrence[43–45] others have not.[46,47] In one recent study, isolated bladder neck PSMs were not associated with BCR when controlling for patient age, PSA, Gleason sore, and pathologic stage, with 12-year biochemical recurrence-free survival (BRFS) similar to men with pT3a and pT3b disease.[48] Similarly, in another study by Buschemeyer et al., isolated bladder neck PSMs imparted a similar risk of recurrence compared to extraprostatic extension (EPE).[49]

Finally, while PSM defines when cancer cells are present at the inked surface, recent studies[50–52] have evaluated the impact of a close surgical margin challenging the notion that the distance from the tumor to the inked specimen margin does not impact the risk of recurrence.[53–55] When defined as ≤0.1 mm from the inked margin in >5300 men from the combined series, a close margin was independently associated with a 1.5–2.2-fold significantly increased risk of BCR.[50–52] Furthermore, the difference in the risk of BCR between close margins and PSMs was insignificant in the study by Izard et al.[51] The proposed mechanism for close margins is that with additional sampling levels, a close margin may actually be positive.[56] Based on these data, although not included in ISUP consensus statements, several institutions now report a close surgical margin in RP specimens as a separate category in addition to positive and negative margins.

While characteristics of PSMs, including extent and location, may influence the risk of BCR, the overall risk of PSMs on more clinically meaningful post-RP outcomes, such as metastatic progression and survival, remains less certain. Largely due to limited follow-up, and the long clinical course of most men with PCa, most studies addressing the impact of PSMs on treatment efficacy have only been able to use BCR. Within

**TABLE 33.1** Risk of Biochemical Recurrence, Metastatic Progression, and Prostate-Cancer-Specific Mortality Following Radical Prostatectomy Among Men With Positive Versus Negative Surgical Margins

| Study | *N* | % PSMs | Median follow-up (years) | HR for BCR (95% CI), *p* value | HR for MP (95% CI), *p* value | HR for PCSM (95% CI), *p* value |
|---|---|---|---|---|---|---|
| Mauermann et al.[23] | 1,712 | 16.4* | 6.2 | 1.7 (1.2–3.3), 0.001 | 1.07 (0.3–3.4), 0.9 | 1.4(0.36–5.4), 0.63 |
| | | 18.1* | 6.2 | 2 (1.5–2.7), <0.0001 | 0.9 (0.3–3.1). 0.98 | 1.15 (0.29–4.47), 0.84 |
| Mitchell et al.[57,**] | 843 | 56 | 14.3 | NR | NR | 2.14 (1.23–3.72), 0.007 |
| Chalfin et al.[22] | 4,461 | 10.4 | 10 | 5 (3.7–6.7), <0.001 | NR | 1.4 (1.0–1.9), 0.036 |
| Wright et al.[25] | 65, 633 | 21.2 | 4.2 | NR | NR | 1.7 (1.3–2.2), NR*** |
| Boorjian et al.[21] | 11,729 | 31.1 | 8.2 | 1.6 (1.5–1.8), <0.0001 | NS, 0.95 | NS, 0.15 |
| Pfitzenmaier et al.[24] | 406 | 17 | 5.2 | 3.2 (2.1–4.8), <0.001 | 6.6 (1.9–23), 0.003 | NR |

PSM, positive surgical margin; HR, hazard ratio; NR, not reported; NS, not significant; BCR, biochemical recurrence; MP, metastatic progression; PCSM, prostate-cancer-specific mortality; CI, confidence interval.
* *Solitary PSM, 16.4%; multiple PSMs, 18.1%.*
** *Only patients with cT3 disease.*
*** *p-Values were not reported; statistical significance was reached only among men with high-grade tumors or extraprostatic extension.*

these limitations, several recent studies have attempted to address the relationship between PSMs and metastases and death. Specifically, Wright et al. observed a significant 1.7-fold (95% CI 1.3–2.2) increased risk of PCa-specific mortality (PCSM) among men with PSMs versus negative margins among >65,000 men in the Surveillance Epidemiology and End Results cancer registry (Table 33.1).[25] However, on subanalysis, adjusting for adverse pathologic features, these observations remained significant only for men with pT3 and high-grade disease. Similarly, among men with cT3 disease, Mitchell et al. found PSMs at the time of RP to be an independent predictor of PCSM (HR 2.14, 95% CI 1.23–3.72).[57] Finally, while Chalfin et al. confirmed the significant impact of PSMs on PCSM, the impact was small when compared to other factors such as final Gleason score and pathologic stage.[22] Conversely, Boorjian et al. failed to observe an association between PSMs and PCSM or MP,[21] while two studies encompassing >2100 patients combined with >6 and >5 years median follow-up, respectively, demonstrated that PSMs were an independent predictor of metastatic progression (MP) but not PCSM.[23,24] Thus, while PSMs clearly impact the risk BCR, the overall impact of PSM on the development of clinical metastases and subsequent PCSM remains less clear and is likely in part due to the fact that the natural history of BCR is highly variable, limiting its proxy for disease progression and PCSM. Furthermore, the associated risk and time to event vary widely, dependent to a large extent on the presence or absence of other risk factors.[58] Additionally, attempts to preserve the neurovascular bundles or bladder neck can result in residual benign prostatic tissue as another potential source of measurable PSA irrespective of margin status,[59,60] confounding the clinical meaning of a detectable PSA following RP, and is in part responsible for the varying definitions of BCR used in the literature. Finally, the protracted natural history of PCa has undoubtedly contributed to the murky waters overlying the relationship between PSMs and metastatic progression and PCSM.

## Association Between PSMs and Adjuvant Treatment Outcomes

Existing level 1 data indicate that radiation therapy (RT) delivered to the prostatic fossa immediately following RP may alter the course of high-risk patients with pathologically advanced PCa. In this patient population, the SWOG 8794, European Organization for Research and Treatment of Cancer (EORTC) 20911, and German Cancer Society (ARO96–02) studies demonstrated a 50–60% reduction in the risk of PSA progression among men who received immediate postoperative RT.[61–63] With longer follow-up (median >10 years), updated results from SWOG 8794, but not EORTC 20911, translated this benefit into a lower risk of metastatic progression (HR 0.7) and improved overall survival in both.[64,65] While several concerns regarding routine administration of adjuvant radiotherapy following RP persist and reside outside the scope of this chapter (see Chapter 53 for detailed analysis), a subanalysis of EORTC 20911 demonstrated that patients with PSMs on central pathologic review, including patients with poorly differentiated cancers and seminal vesicle invasion, appeared to derive the greatest benefit from adjuvant RT.[66] These data support the notion that, at a minimum, adjuvant RT should be discussed with all men diagnosed with PSMs, in particular perhaps those who have extensive or multifocal PSMs.

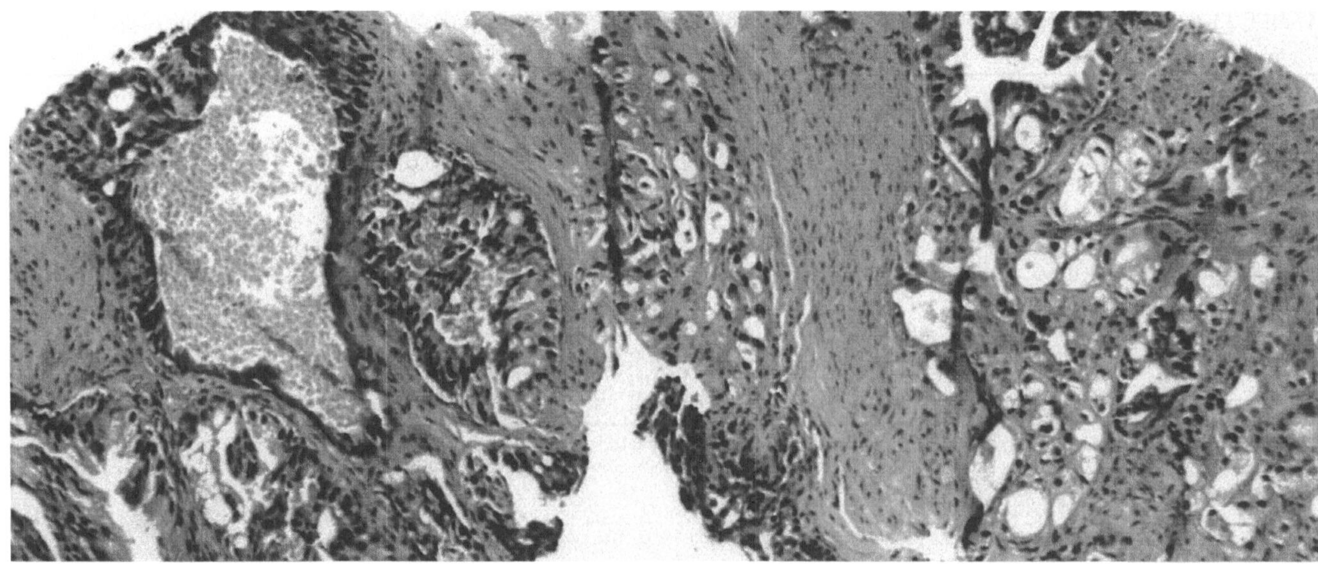

**FIGURE 33.2 Histologic appearance of a prostate containing Gleason 3 + 4 + 5 prostate cancer.** Gleason pattern 3 is located on the right-hand side of the panel, Gleason 4 in the lower left, with Gleason pattern 5 (comedo necrosis) located in the upper left.

**TABLE 33.2** Association Between the Presence of a Tertiary Gleason Component and Biochemical-Free Survival

| Study | Total *N* | Gleason sum | TGC *N* (%) | Median follow-up (years) | 5-year BCR free survival TGC versus no TGC, *p* value | HR for BCR (95% CI), *p* value |
|---|---|---|---|---|---|---|
| van Oort et al.[69] | 223 | NR | 106 (48) | NR | 63% versus 87%, 0.0002 | 1.78 (0.91–3.46), NR |
| Hattab et al.[70] | 228 | 7 | 50 (22) | NR | 19% versus 70%, <0.001 | 2.10 (1.24–3.55), 0.006 |
| Sim et al.[71] | 509 | 7 | 66 (13) | 2.2 | NR, <0.001 | 1.78 (1.00–3.17), NR |
| Whittemore et al.[72] | 214 | 7 | 36 (17) | NR | 40% versus 70%, <0.001 | 1.90 (0.99–3.64, 0.053 |
| Trock et al.[73] | 3230 | 6–10 | 333 (10) | 2 | NR | 1.45 (1.04–2.02), 0.029 |

TGC, tertiary Gleason component; BCR, biochemical recurrence; HR, hazard ratio; CI, confidence interval; NR, not reported.

## TERTIARY GLEASON PATTERN

For more than four decades, PCa grading has used the Gleason system, describing the architecture of malignant cells and categorizing them into the most (primary) and second most (secondary) prevalent pattern (grade) and summing them to produce the Gleason sum (GS).[67,68] Despite the many changes in PCa diagnosis and care since its inception, the Gleason system remains a highly predictive prognostic tool for pathologic and clinical outcomes following RP. However, many tumors are composed of more than two different grade patterns (Figure 33.2), raising the question of how these "tertiary Gleason" components (TGC) contribute to disease outcomes and whether they should be reported (Table 33.2).

Pan et al. first reported outcomes of 114 patients with a TGC undergoing RP compared to >2000 cases without TGC and found that among men with low-grade tumors (Gleason sum 5/6), those with a TGC pattern 4 or 5, had significantly more advanced stage compared to those without a TGC but less advanced stage than Gleason sum 7.[74] Patients with Gleason 7 and TGC 5 had significantly worse pathologic stage than typical Gleason 7 tumors and similar stage to Gleason 8 tumors, with the volume of TGC 5 correlating with pathologic stage. Furthermore, the presence of TGC 4 or 5 in men with low-grade tumors and TGC 5 in men with intermediate grade (Gleason 7) tumors was associated with significantly worse disease progression following RP. These observations lead to formal recommendations to include the TGC in the pathology report for RP specimens and to changes for reporting of Gleason score in prostate needle biopsies in the 2005 ISUP consensus statement on Gleason Grading.[75]

Since the 2005 ISUP consensus statement on Gleason grading, many groups have reported on TGC, demonstrating a prevalence of 10–48% in RP specimens[69,71,73,76,77,70] owing largely to differences in the definition of TGC between institutions. The majority of these studies have identified TGC as a harbinger of disease burden, in particular advanced pathologic stage[69,71,73,76,72] including extra-prostatic extension and seminal vesicle invasion[69,71,73,76,72] and significantly higher rates of nonorgan confined disease (47% vs. 23%).[73] On multivariable analysis, among men with Gleason 7 disease, TGC 5 has been identified as an independent predictor of advanced pathologic stage (HR 2.55, 95% CI 1.40–4.65)[71] and EPE (HR 3.04, 95% CI 1.36–6.77) among men with TGC 5 ≥ 5% of the tumor volume.[76] In addition to pathologic stage, TGC has also been associated with increased preoperative PSA,[73] the presence of lymphovascular invasion,[72] increased tumor volume[69] and node-positive disease.[76] Finally, the presence of TGC has been significantly associated with PSMs following RP,[69,73,76] with PSMs more than twice as high among men with TGC than those without (18% vs. 8%) in one series,[73] and another identifying ≥5% TGC pattern 5 as an independent predictor (HR 2.16, 95% CI 1.23–3.78) of PSMs on multivariable analysis.[76]

### Association of TGC and Clinical Outcomes

Due to the associations between the presence of TGC and adverse pathologic criteria following RP, it may be expected that the presence of TGC could impact post-RP outcomes. Indeed, among patients with Gleason 7 disease, the presence of TGC 5 leads to significantly worse BRFS in men with TGC5 (mean 54 months vs. 121 months),[72] 5-year BRFS (19% vs. 70%),[70] and clinical failure[77] compared to those without TGC 5. Furthermore, the presence of TGC 5 among men with Gleason 7 disease has been identified as an independent predictor of BCR (HR 1.78, 95% CI 1.00–3.17)[71] and clinical failure (HR 4.03, 95% CI 1.72–9.46).[77] Interestingly, in one series the presence of TGC 5 (HR 2.1, 95% CI 1.24–3.55), but not primary Gleason pattern (3 or 4) among men with GS 7, was an independent predictor of biochemical recurrence when controlling for pathologic stage, PSMs, and primary Gleason pattern (Gleason 3 or 4).[70] While these data suggest that TGC 5 among men with Gleason 7 is a predictor of PCa recurrence, not all studies have identified TGC as an independent predictor.[69]

Among men with non-GS 7 disease the role of TGC in post-RP outcomes is less well described. However, in one large study, the presence of TGC demonstrated a similar relationship with PCa recurrence. Specifically, among 333 men with TGC and 2897 men without TGC, Trock et al. observed significantly more recurrences (PSA >0.2 ng/mL) (17.1% vs. 5.7%) in men with TGC across all Gleason sums compared to those without TGC. Additionally, patients with TGC had significantly worse BRFS when stratified across all Gleason sums (6, 3 + 4, 4 + 3, and 8). On multivariable analysis controlling for PSA, pathologic stage, margin status, and Gleason sum, TGC was independently associated with BCR (HR 1.45, 1.04–2.02). However, the c-index did not change substantially (0.792 vs. 0.793) when including TGC in a nomogram predicting disease recurrence. Based on these findings, the authors proposed a modified Gleason score (6, 6.5, 7, 7.5, etc.) capturing TGC in the sum.[73]

## LYMPHOVASCULAR INVASION

The presence of lymphovascular invasion (LVI) is an accepted prognostic factor in many malignancies (i.e., early stage breast cancer).[78] LVI is defined as the presence of tumor cells within an endothelium lined space without underlying muscular walls and includes lymphatic invasion, vascular invasion, or lymph-vascular invasion.[79] By AJCC/UICC convention, LVI does not affect the T category indicating local extent of tumor unless specifically included in the definition of a T category.[80] Like many cancers, PCa is known to spread via lymphatic channels and as a result, reporting the presence of LVI is recommended by the College of American Pathologists[80] and ISUP.[69]

The presence of LVI has been described in 5–21% of RP specimens and has been associated with multiple adverse clinic pathologic criteria in both biopsy and RP specimens.[79,81,87] Among clinical data available in the preoperative setting from prostate biopsies, LVI has been significantly associated with increased preoperative PSA, PSA density, the presence of cT2 disease compared to cT1c, higher biopsy Gleason sum, and greater percent of positive biopsy cores. In RP specimens, LVI has been significantly associated with higher Gleason sum and the presence of EPE and SVI.[79,82–85] Additionally, it has been associated with significantly increased tumor volume,[83,85] increased PSM rates,[79,82–85] and the presence of nodal metastases.[79,82–84]

### Association of LVI and Clinical Outcomes

Although LVI seems to be clearly associated with higher-risk pathologic features, the existing clinical data regarding its impact on disease recurrence following RP is mixed. In a retrospective review of 504 patients, Cheng et al. found that LVI was an independent predictor of BCR (HR 1.6, 95% CI 1.12–2.38) when controlling for pathologic stage, Gleason sum, and PSMs.[87] Similarly, May et al. found that LVI on RP was associated with significantly worse 5-year BRFS (38.3% vs. 87.3%, $p < 0.001$) in a retrospective study of node-negative patients.[81] Furthermore, LVI was an independent predictor of BCR (PSA >0.2) (HR 4.39, 95% CI 2.47–7.90) when controlling for preoperative PSA, RP Gleason sum, SVI, PSA density, and percentage of positive biopsy cores. Finally, Yee et al. examined LVI in 1298 patients undergoing RP, identifying LVI as an independently associated risk factor (HR 1.77, 95% CI 1.11–2.82) for BCR on multivariate analysis controlling for preop PSA, Gleason sum, and the presence of EPE, LVI, nodal metastases and PSMs with an associated significantly worse 2-year BRFS survival in patients with LVI (62% vs. 95%, $p < 0.001$).[84] Conversely, in a review of 1393 patients with longer follow-up, while LVI was associated with significantly worse 10-year BRFS (PSA >0.2) (55% vs. 86%, $p < 0.001$), on multivariable analysis controlling for PSMs, pathologic tumor and nodal stage, the presence of LVI was not significantly associated with post-RP recurrence.[83] Similarly, in a more recent retrospective study looking at BCR following 407 robotic-assisted RP, the presence of LVI was not associated with BCR on univariate or multivariate analysis.[85]

Collectively, these data are summarized in a systematic review of 19 studies evaluating the relationship between LVI and BCR following RP by Ng et al. in which 11/19 studies identified LVI as independent risk factor for BCR.[86] However, owing to heterogeneity in data reporting and definitions of disease recurrence a meta-analysis was not possible. Within these limitations, the majority of large studies (>400 patients) demonstrated a significant independent association between the presence of LVI and BCR (Table 33.3). However, most had only a modest HR of <2.0, and the largest study did not demonstrate an association between LVI and BCR, suggesting that LVI may provide only a weak association with BCR and be of limited clinical utility with respect to BCR. This notion is further supported by the observation that the addition of LVI to a predictive model containing standard clinicopathologic features only marginally increased the concordance index of the model (0.884 vs. 0.880) despite LVI being an independent predictor of BCR in this series.[84]

While the majority of the existing literature examining the prognostic impact of LVI has focused on disease recurrence, a handful of studies have evaluated the relationship between LVI and other oncologic outcomes. Shariat et al. reported significantly worse 7-year metastases-free survival (47% vs. 94%, $p < 0.001$) and 7-year overall survival (70% vs. 98%, $p < 0.001$) in men with LVI versus those without in a cohort of 630 men undergoing RP on Kaplan–Meier analysis.[79] Similarly, Cheng et al., reported worse 5-year PCSM in men with LVI (90% vs. 98%, $p < 0.001$) following RP, and LVI was found to be independently associated with PCSM (HR 2.75, 95% CI 1.04–3.28).[82] Finally, LVI has been reported to be associated with significantly shorter PSA doubling time (3.3 vs. 6.9 months, $p = 0.012$) leading up to salvage radiotherapy with corresponding significantly worse 3-year postsalvage radiation biochemical progression rates (100% vs. 45%, $p = 0.017$).[79] Similarly, Brooks et al. reported LVI on RP as an independent predictor of post-salvage radiation recurrence (HR 5.5, 95% CI 2.5–12.2) with significantly shorter median time to recurrence (2.6 vs. 7.8 years, $p < 0.001$).[91]

## SEMINAL VESICLE INVASION

The presence of SVI has long been considered an important pathologic finding reflected in its inclusion in pathologic staging since the first edition of the AJCC staging guide as T3 disease.[92] However, historically there has been considerable variability within the pathologic community as to what truly constitutes SVI, which was addressed by the ISUP 2009 consensus conference. Specifically, it was determined that only muscular invasion of the extraprostatic seminal vesicle should be considered as SVI (Figure 33.3). Further specimen processing

**TABLE 33.3** Association Between Lymphovascular Invasion and Biochemical Recurrence (Only Studies That Report Risk Included)

| Study | Total *N* | LVI positive: *N* (%) | Median follow-up (years) | Independent predictor | HR for BCR (95% CI), *p* value |
|---|---|---|---|---|---|
| Cheng et al.[82] | 504 | 106 (21) | 3.6 | Yes | 1.6 (1.12–2.38), 0.002 |
| Huang et al.[87] | 131 | 17 (15) | 2.3 | No | 3.5 (0.79–15.65), NS |
| Ito et al.[88] | 82 | 38 (46) | 1.8 | Yes | 4.39 (1.40–13.70), 0.019 |
| Lee et al.[89] | 557 | 90 (16) | 7.5 | Yes | 1.92 (1.26–2.92), <0.01 |
| Loeb et al.[83] | 1709 | 118 (7) | 6.2 | No | 1.5 (0.9–2.4), 0.1 |
| May et al.[81] | 412 | 42 (10) | 4.4 | Yes | 4.39 (2.47–7.80), <0.001 |
| Quinn et al.[90] | 732 | 38 (5) | 3.3 | No | 1.37 (0.82–2.30), 0.23 |
| Shariat et al.[79] | 630 | 32 (5) | 1.8 | No | 1.67 (0.93–2.99), 0.083 |
| Whittemore et al.[72] | 214 | 12 (6) | NR | Yes | 2.49 (1.09–5.65), 0.03 |

LVI, lymphovascular invasion; BCR, biochemical recurrence; HR, hazard ratio; CI, confidence interval; NS, not significant.

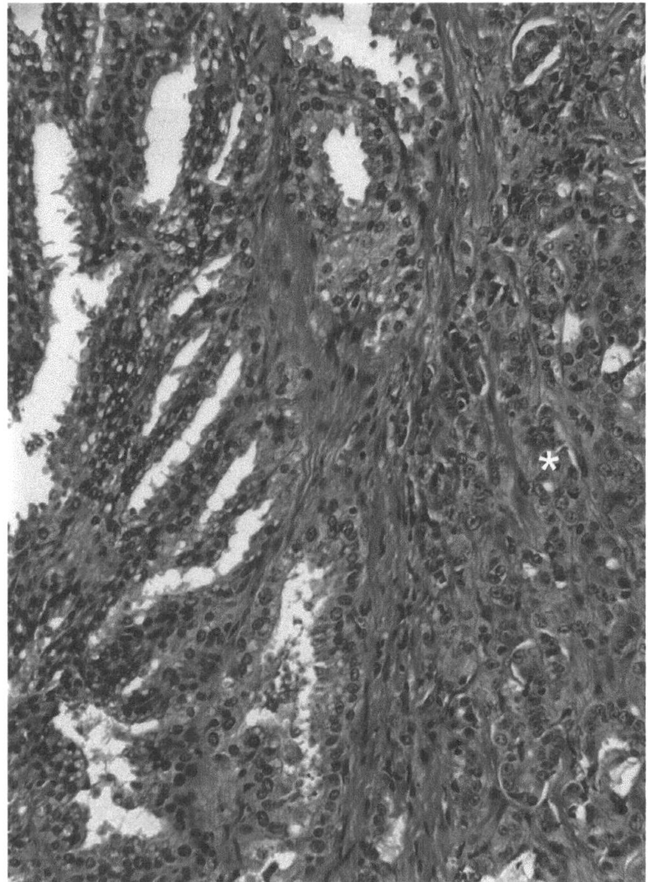

FIGURE 33.3 Histologic appearance of seminal vesicle invasion with prostate cancer (*) invading the muscular wall of the seminal vesicle.

has been standardized to include mandatory sampling of the junction between the base of the seminal vesicles and prostate.[93] These guidelines should help standardize reporting of SVI, improving the comparability of results between institutions in the future. Despite these limitations, it is clear that with widespread adoption of PSA screening and the associated stage migration, the incidence of SVI has decreased from 7.7–12.7% in the pre-PSA era to 3.9–6.0% in contemporary series reporting outcomes before and after PSA screening.[94,95]

## Association of SVI and Clinical Outcomes

With respect to oncologic outcomes, SVI is well recognized as a high-risk pathologic feature. SVI has been associated with significantly worse progression-free rates[96] and 5-year BFS (52% vs. 81%, $p < 0.001$) following RP, with a nearly twofold increased risk of BCR[27] among a cohort of patients with pT2 or pT3 disease. Additionally, post-RP SVI is an independent predictor of BCR[97] and 10-year PCSM.[57,98] Pierorazio et al. recently reported on RP outcomes among 894 patients with SVI spanning the pre-PSA versus early PSA versus the contemporary era. BFS at 10 years (25.8% vs. 28.6% vs. 19.7%, respectively) and 10-year freedom from PCSM (79.7% vs. 79.6% vs. 83.9%) did not differ significantly, despite advances in surgical techniques and significantly lower incidence of PSMs and nodal metastases in more recently treated patients.[95] These observations underscore the inherently aggressive nature of the disease for most men with SVI.

# LYMPH NODE POSITIVE DISEASE

Like SVI, the incidence of lymph node positive disease varies greatly across the literature. Historically, in the pre-PSA era 21–40% of patients undergoing RP were found to have nodal metastases.[99–101] In contemporary series, the incidence of nodal metastases at the time of RP has decreased substantially ranging from 1.8% to 12% reflecting the stage migration associated with widespread PSA screening.[102–106] An important factor in evaluating the incidence of nodal metastases is the extent of pelvic lymph node dissection performed. Studies comparing a limited pelvic lymph node dissection, generally involving the external iliac and obturator nodal packets, compared to an extended node dissection, including external iliac, internal iliac, and obturator nodal packets plus or minus the common iliac lymph nodes, have demonstrated significantly more nodal metastases with an extended dissection[107–109] with 58–63% of positive nodes found outside the limited template.[110–112] Furthermore, the only site of nodal metastases may lie outside of the limited template in 19–22% of patients.[110,111]

## Association of Nodal Metastases and Clinical Outcomes

For patients with nodal involvement, the presence of nodal metastases at the time of RP has long been recognized as a poor prognostic finding. In the preadjuvant androgen deprivation therapy (ADT) era, 10-year PSA recurrence-free, clinical recurrence-free, and overall survival were 40, 69, and 74%, respectively, among 163 patients with N+ disease who received no adjuvant or neoadjuvant therapy.[102] Following the publication of a randomized trial demonstrating significantly improved progression-free survival (HR 3.42, 95% CI 1.96–5.98), prostate-cancer-specific survival (HR 4.09, 95% CI 1.76–9.49), and overall survival (HR 1.84, 95% CI 1.01–3.35) with receipt of immediate adjuvant ADT in men with clinically localized disease found to have nodal metastases at the time of lymph node dissection,[113] many patients are now likely to receive adjuvant ADT. However, nodal metastases remain a poor prognostic factor with 10-year event-free survival for patients with positive

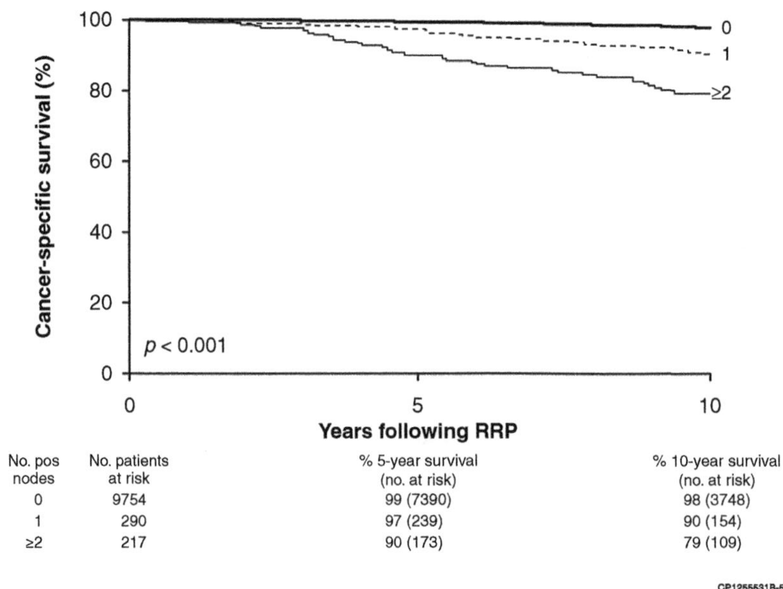


**FIGURE 33.4** **Prostate-cancer-specific mortality following radical prostatectomy stratified by number of positive lymph nodes.** *Reprinted with permission from Boorjian et al.*[105]

lymph nodes of 56, 89, 80, and 86% for BCR, local recurrence, systemic progression, and cancer death, respectively, in a cohort of 507 men of which 90% received immediate adjuvant ADT.[105]

Due to the association of nodal metastases and disease outcomes, many investigators have evaluated the influence of the number of positive nodes on patient outcomes with positive correlation.[104,105,114–116] Boorjian et al. observed that a single nodal metastasis had an approximately threefold significant increase in disease progression, and that ≥2 nodes imparted an additional twofold risk in a cohort of 507 node-positive patients of which the vast majority received adjuvant ADT.[105] Furthermore, a single nodal metastasis was associated with a significant fourfold increased risk of PCSM and two or more positive nodes an additional twofold risk (Figure 33.4). Subsequently, Schumacher et al. observed significantly better 5-year BRFS (24.7, 11.8, and 4.9%, $p < 0.001$) for 1, 2, and ≥3 nodes, respectively, in a cohort of 122 patients with clinically localized disease, found to have nodal metastases at the time RP with extended PLND, who did not receive immediate adjuvant ADT; additionally, a nearly 1.4-fold increased risk of PCSM with each additional positive lymph node when controlling for pathologic factors. However, on subanalysis the risk of death was similar between one and two positive nodes, but ≥3 positive lymph nodes was associated with a 5.6-fold increased risk of PCSM compared to one node.[115] Similarly, Briganti et al. observed that >2 nodal metastases was an independent predictor of PCSM (HR 1.9) compared to ≤2 positive nodes.[116] Such data, suggest that while a single nodal metastases defines a high-risk group for disease progression and PCSM, those patients with >2 nodal metastases may compose a "highest" risk group.

## CONCLUSIONS

Pathologic evaluation of the RP specimen can provide considerable insight into patients' prognosis and offer clinicians key data to tailor surveillance strategies and adjuvant treatment recommendations to each individual patient. While factors, such as preoperative PSA, Gleason score, and pathologic stage, may continue to be the dominant driver of patient outcomes following RP, other pathologic findings may contribute to individual risk. Specifically, the identification of PSMs at RP places patients at higher risk of BCR, metastatic progression, and PCSM, while also identifying the subset of patients most likely to benefit from adjuvant radiotherapy. Patients with SVI and/or nodal metastases are likely the subset of patients with greatest risk of BCR, disease progression, and PCSM. Finally, the presence of TGC 5 and/or LVI likely place patients at higher risk of BCR and are often associated with adverse pathologic features. The potential of next-generation sequencing and molecular interrogation of pathologic samples will bring a new platform of risk parameters on which to base disease prognostication and patient consultation. Continued efforts to use standardized pathologic template reporting to include all critical items is necessary for effective communication between pathologists, urologists, and patients. This uniform reporting and examination will further facilitate the combined analyses from surgical series and allow us to accurately define the independent risk of adverse outcome for the different criteria included in the pathologic report.

## Acknowledgment

The authors would like to acknowledge Lawrence True, MD for providing histologic slides for representative figures.

## References

1. Dong F, Reuther AM, Magi-Galluzzi C, et al. Pathologic stage migration has slowed in the late PSA era. *Urology* 2007;**70**(5): 839–42.
2. Berger AP, Spranger R, Kofler K, et al. Early detection of prostate cancer with low PSA cut-off values leads to significant stage migration in radical prostatectomy specimens. *Prostate* 2003;**57**(2):93–8.
3. Neppl-Huber C, Zappa M, Coebergh JW, et al. Changes in incidence, survival and mortality of prostate cancer in Europe and the United States in the PSA era: additional diagnoses and avoided deaths. *Ann Oncol* 2012;**23**(5):1325–34.
4. Stephenson AJ, Scardino PT, Eastham JA, et al. Preoperative nomogram predicting the 10-year probability of prostate cancer recurrence after radical prostatectomy. *J Natl Cancer Inst* 2006;**98**(10):715–7.
5. Stephenson AJ, Wood DP, Kattan MW, et al. Location, extent and number of positive surgical margins do not improve accuracy of predicting prostate cancer recurrence after radical prostatectomy. *J Urol* 2009;**182**(4):1357–63.
6. Kattan MW, Eastham JA, Stapleton AM, et al. A preoperative nomogram for disease recurrence following radical prostatectomy for prostate cancer. *J Natl Cancer Inst* 1998;**90**(10):766–71.
7. Tan PH, Cheng L, Srigley JR, et al. International Society of Urological Pathology (ISUP) Consensus Conference on handling and staging of radical prostatectomy specimens. Working group 5: surgical margins. *Mod Pathol* 2011;**24**(1):48–57.
8. Touijer K, Eastham JA, Secin FP, et al. Comprehensive prospective comparative analysis of outcomes between open and laparoscopic radical prostatectomy conducted in 2003 to 2005. *J Urol* 2008;**179**(5):1811–7 discussion 1817.
9. Tewari A, Sooriakumaran P, Bloch DA, et al. Positive surgical margin and perioperative complication rates of primary surgical treatments for prostate cancer: a systematic review and meta-analysis comparing retropubic, laparoscopic, and robotic prostatectomy. *Eur Urol* 2012;**62**(1):1–15.
10. Novara G, Ficarra V, Mocellin S, et al. Systematic review and meta-analysis of studies reporting oncologic outcome after robot-assisted radical prostatectomy. *Eur Urol* 2012;**62**(3):382–404.
11. Ploussard G, Agamy MA, Alenda O, et al. Impact of positive surgical margins on prostate-specific antigen failure after radical prostatectomy in adjuvant treatment-naïve patients. *BJU Int* 2011;**107**(11):1748–54.
12. Patel VR, Coelho RF, Rocco B, et al. Positive surgical margins after robotic assisted radical prostatectomy: a multi-institutional study. *J Urol* 2011;**186**(2):511–6.
13. Wiltz AL, Shikanov S, Eggener SE, et al. Robotic radical prostatectomy in overweight and obese patients: oncological and validated-functional outcomes. *Urology* 2009;**73**(2):316–22.
14. Milhoua PM, Koi PT, Lowe D, et al. Issue of prostate gland size, laparoscopic radical prostatectomy, and continence revisited. *Urology* 2008;**71**(3):417–20.
15. Ginzburg E, Klimas N, Parvus C, et al. Long-term safety of testosterone and growth hormone supplementation: a retrospective study of metabolic, cardiovascular, and oncologic outcomes. *J Clin Med Res* 2010;**2**(4):159–66.
16. Huang AC, Kowalczyk KJ, Hevelone ND, et al. The impact of prostate size, median lobe, and prior benign prostatic hyperplasia intervention on robot-assisted laparoscopic prostatectomy: technique and outcomes. *Eur Urol* 2011;**59**(4):595–603.
17. Vickers A, Bianco F, Cronin A, et al. The learning curve for surgical margins after open radical prostatectomy: implications for margin status as an oncological end point. *J Urol* 2010;**183**(4):1360–5.
18. Secin FP, Savage C, Abbou C, et al. The learning curve for laparoscopic radical prostatectomy: an international multicenter study. *J Urol* 2010;**184**(6):2291–6.
19. Sooriakumaran P, John M, Wiklund P, et al. Learning curve for robotic assisted laparoscopic prostatectomy: a multi-institutional study of 3794 patients. *Minerva Urol Nefrol* 2011;**63**(3):191–8.
20. Yossepowitch O, Bjartell A, Eastham JA, et al. Positive surgical margins in radical prostatectomy: outlining the problem and its long-term consequences. *Eur Urol* 2009;**55**(1):87–99.
21. Boorjian SA, Karnes RJ, Crispen PL, et al. The impact of positive surgical margins on mortality following radical prostatectomy during the prostate specific antigen era. *J Urol* 2010;**183**(3):1003–9.
22. Chalfin HJ, Dinizo M, Trock BJ, et al. Impact of surgical margin status on prostate-cancer-specific mortality. *BJU Int* 2012;**110**(11):1684–9.
23. Mauermann J, Fradet V, Lacombe L, et al. The impact of solitary and multiple positive surgical margins on hard clinical end points in 1712 adjuvant treatment-naive pT2-4 N0 radical prostatectomy patients. *Eur Urol* 2013;**64**(1):19–25.
24. Pfitzenmaier J, Pahernik S, Tremmel T, et al. Positive surgical margins after radical prostatectomy: do they have an impact on biochemical or clinical progression? *BJU Int* 2008;**102**(10):1413–8.
25. Wright JL, Dalkin BL, True LD, et al. Positive surgical margins at radical prostatectomy predict prostate cancer specific mortality. *J Urol* 2010;**183**(6):2213–8.
26. Alkhateeb S, Alibhai S, Fleshner N, et al. Impact of positive surgical margins after radical prostatectomy differs by disease risk group. *J Urol* 2010;**183**(1):145–50.
27. Blute ML, Bergstralh EJ, Iocca A, et al. Use of Gleason score, prostate specific antigen, seminal vesicle and margin status to predict biochemical failure after radical prostatectomy. *J Urol* 2001;**165**(1):119–25.
28. Fontenot PA, Mansour AM. Reporting positive surgical margins after radical prostatectomy: time for standardization. *BJU Int* 2013;**111**(8):E290–9.
29. Brimo F, Partin AW, Epstein JI. Tumor grade at margins of resection in radical prostatectomy specimens is an independent predictor of prognosis. *Urology* 2010;**76**(5):1206–9.
30. Cao D, Kibel AS, Gao F, et al. The Gleason score of tumor at the margin in radical prostatectomy is predictive of biochemical recurrence. *Am J Surg Pathol* 2010;**34**(7):994–1001.
31. Huang JG, Pedersen J, Hong MK, et al. Presence or absence of a positive pathological margin outperforms any other margin-associated variable in predicting clinically relevant biochemical recurrence in Gleason 7 prostate cancer. *BJU Int* 2013;**111**(6):921–7.
32. Chuang AY, Epstein JI. Positive surgical margins in areas of capsular incision in otherwise organ-confined disease at radical prostatectomy: histologic features and pitfalls. *Am J Surg Pathol* 2008;**32**(8):1201–6.
33. van Oort IM, Bruins HM, Kiemeney LA, et al. The length of positive surgical margins correlates with biochemical recurrence after radical prostatectomy. *Histopathology* 2010;**56**(4):464–71.
34. Emerson RE, Koch MO, Jones TD, et al. The influence of extent of surgical margin positivity on prostate specific antigen recurrence. *J Clin Pathol* 2005;**58**(10):1028–32.

35. Shikanov S, Song J, Royce C, et al. Length of positive surgical margin after radical prostatectomy as a predictor of biochemical recurrence. *J Urol* 2009;**182**(1):139–44.
36. Cao D, Humphrey PA, Gao F, et al. Ability of linear length of positive margin in radical prostatectomy specimens to predict biochemical recurrence. *Urology* 2011;**77**(6):1409–14.
37. Lake AM, He C, Wood DP. Focal positive surgical margins decrease disease-free survival after radical prostatectomy even in organ-confined disease. *Urology* 2010;**76**(5):1212–6.
38. Babaian RJ, Troncoso P, Bhadkamkar VA, et al. Analysis of clinicopathologic factors predicting outcome after radical prostatectomy. *Cancer* 2001;**91**(8):1414–22.
39. Fleshner NE, Evans A, Chadwick K, et al. Clinical significance of the positive surgical margin based upon location, grade, and stage. *Urol Oncol* 2010;**28**(2):197–204.
40. Kordan Y, Chang SS, Salem S, et al. Pathological stage T2 subgroups to predict biochemical recurrence after prostatectomy. *J Urol* 2009;**182**(5):2291–5.
41. Karakiewicz PI, Eastham JA, Graefen M, et al. Prognostic impact of positive surgical margins in surgically treated prostate cancer: multi-institutional assessment of 5831 patients. *Urology* 2005;**66**(6):1245–50.
42. Fesseha T, Sakr W, Grignon D, et al. Prognostic implications of a positive apical margin in radical prostatectomy specimens. *J Urol* 1997;**158**(6):2176–9.
43. Eastham JA, Kuroiwa K, Ohori M, et al. Prognostic significance of location of positive margins in radical prostatectomy specimens. *Urology* 2007;**70**(5):965–9.
44. Aydin H, Tsuzuki T, Hernandez D, et al. Positive proximal (bladder neck) margin at radical prostatectomy confers greater risk of biochemical progression. *Urology* 2004;**64**(3):551–5.
45. Poulos CK, Koch MO, Eble JN, et al. Bladder neck invasion is an independent predictor of prostate-specific antigen recurrence. *Cancer* 2004;**101**(7):1563–8.
46. Bianco FJ, Grignon DJ, Sakr WA, et al. Radical prostatectomy with bladder neck preservation: impact of a positive margin. *Eur Urol* 2003;**43**(5):461–6.
47. Yossepowitch O, Sircar K, Scardino PT, et al. Bladder neck involvement in pathological stage pT4 radical prostatectomy specimens is not an independent prognostic factor. *J Urol* 2002;**168**(5):2011–5.
48. Pierorazio PM, Epstein JI, Humphreys E, et al. The significance of a positive bladder neck margin after radical prostatectomy: the American Joint Committee on Cancer Pathological Stage T4 designation is not warranted. *J Urol* 2010;**183**(1):151–7.
49. Buschemeyer WC, Hamilton RJ, Aronson WJ, et al. Is a positive bladder neck margin truly a T4 lesion in the prostate specific antigen era? Results from the SEARCH Database. *J Urol* 2008;**179**(1):124–9 discussion 129.
50. Lu J, Wirth GJ, Wu S, et al. A close surgical margin after radical prostatectomy is an independent predictor of recurrence. *J Urol* 2012;**188**(1):91–7.
51. Izard JP, True LD, May P, et al. Prostate cancer that is within 0.1 mm of the surgical margin of a radical prostatectomy predicts greater likelihood of recurrence. *Am J Surg Pathol* 2014;**38**(3):333–8.
52. Shikanov S, Marchetti P, Desai V, et al. Short (≤1 mm) positive surgical margin and risk of biochemical recurrence after radical prostatectomy. *BJU Int* 2013;**111**(4):559–63.
53. Bong GW, Ritenour CW, Osunkoya AO, et al. Evaluation of modern pathological criteria for positive margins in radical prostatectomy specimens and their use for predicting biochemical recurrence. *BJU Int* 2009;**103**(3):327–31.
54. Emerson RE, Koch MO, Daggy JK, et al. Closest distance between tumor and resection margin in radical prostatectomy specimens: lack of prognostic significance. *Am J Surg Pathol* 2005;**29**(2):225–9.
55. Epstein JI, Sauvageot J. Do close but negative margins in radical prostatectomy specimens increase the risk of postoperative progression? *J Urol* 1997;**157**(1):241–3.
56. Humphrey PA. Complete histologic serial sectioning of a prostate gland with adenocarcinoma. *Am J Surg Pathol* 1993;**17**(5):468–72.
57. Mitchell CR, Boorjian SA, Umbreit EC, et al. 20-Year survival after radical prostatectomy as initial treatment for cT3 prostate cancer. *BJU Int* 2012;**110**(11):1709–13.
58. Corcoran NM, Hovens CM, Metcalfe C, et al. Positive surgical margins are a risk factor for significant biochemical recurrence only in intermediate-risk disease. *BJU Int* 2012;**110**(6):821–7.
59. Djavan B, Milani S, Fong YK. Benign positive margins after radical prostatectomy means a poor prognosis – pro. *Urology* 2005;**65**(2):218–20.
60. Godoy G, Tareen BU, Lepor H. Does benign prostatic tissue contribute to measurable PSA levels after radical prostatectomy? *Urology* 2009;**74**(1):167–70.
61. Thompson IM, Tangen CM, Paradelo J, et al. Adjuvant radiotherapy for pathologically advanced prostate cancer: a randomized clinical trial. *JAMA* 2006;**296**(19):2329–35.
62. Bolla M, van Poppel H, Collette L, et al. Postoperative radiotherapy after radical prostatectomy: a randomised controlled trial (EORTC trial 22911). *Lancet* 2005;**366**(9485):572–8.
63. Wiegel T, Bottke D, Steiner U, et al. Phase III postoperative adjuvant radiotherapy after radical prostatectomy compared with radical prostatectomy alone in pT3 prostate cancer with postoperative undetectable prostate-specific antigen: ARO 96-02/AUO AP 09/95. *J Clin Oncol* 2009;**27**(18):2924–30.
64. Thompson IM, Tangen CM, Paradelo J, et al. Adjuvant radiotherapy for pathological T3N0M0 prostate cancer significantly reduces risk of metastases and improves survival: long-term followup of a randomized clinical trial. *J Urol* 2009;**181**(3):956–62.
65. Bolla M, van Poppel H, Tombal B, et al. Postoperative radiotherapy after radical prostatectomy for high-risk prostate cancer: long-term results of a randomised controlled trial (EORTC trial 22911). *Lancet* 2012;**380**(9858):2018–27.
66. Van der Kwast TH, Bolla M, Van Poppel H, et al. Identification of patients with prostate cancer who benefit from immediate postoperative radiotherapy: EORTC 22911. *J Clin Oncol* 2007;**25**(27):4178–86.
67. Mellinger GT, Gleason D, Bailar J. The histology and prognosis of prostatic cancer. *J Urol* 1967;**97**(2):331–7.
68. Gleason DF, Mellinger GT. Prediction of prognosis for prostatic adenocarcinoma by combined histological grading and clinical staging. *J Urol* 1974;**111**(1):58–64.
69. van Oort IM, Schout BM, Kiemeney LA, et al. Does the tertiary Gleason pattern influence the PSA progression-free interval after retropubic radical prostatectomy for organ-confined prostate cancer? *Eur Urol* 2005;**48**(4):572–6.
70. Hattab EM, Koch MO, Eble JN, et al. Tertiary Gleason pattern 5 is a powerful predictor of biochemical relapse in patients with Gleason score 7 prostatic adenocarcinoma. *J Urol* 2006;**175**(5):1695–9 discussion 1699.
71. Sim HG, Telesca D, Culp SH, et al. Tertiary Gleason pattern 5 in Gleason 7 prostate cancer predicts pathological stage and biochemical recurrence. *J Urol* 2008;**179**(5):1775–9.
72. Whittemore DE, Hick EJ, Carter MR, et al. Significance of tertiary Gleason pattern 5 in Gleason score 7 radical prostatectomy specimens. *J Urol* 2008;**179**(2):516–22 discussion 522.
73. Trock BJ, Guo CC, Gonzalgo ML, et al. Tertiary Gleason patterns and biochemical recurrence after prostatectomy: proposal for a modified Gleason scoring system. *J Urol* 2009;**182**(4):1364–70.
74. Pan CC, Potter SR, Partin AW, et al. The prognostic significance of tertiary Gleason patterns of higher grade in radical prostatectomy specimens: a proposal to modify the Gleason grading system. *Am J Surg Pathol* 2000;**24**(4):563–9.

75. Epstein JI, Allsbrook WC, Amin MB, et al. The 2005 International Society of Urological Pathology (ISUP) Consensus Conference on Gleason Grading of Prostatic Carcinoma. *Am J Surg Pathol* 2005;**29**(9):1228–42.
76. Isbarn H, Ahyai SA, Chun FK, et al. Prevalence of a tertiary Gleason grade and its impact on adverse histopathologic parameters in a contemporary radical prostatectomy series. *Eur Urol* 2009;**55**(2):394–401.
77. Servoll E, Saeter T, Vlatkovic L, et al. Impact of a tertiary Gleason pattern 4 or 5 on clinical failure and mortality after radical prostatectomy for clinically localised prostate cancer. *BJU Int* 2012;**109**(10):1489–94.
78. Soerjomataram I, Louwman MW, Ribot JG, et al. An overview of prognostic factors for long-term survivors of breast cancer. *Breast Cancer Res Treat* 2008;**107**(3):309–30.
79. Shariat SF, Khoddami SM, Saboorian H, et al. Lymphovascular invasion is a pathological feature of biologically aggressive disease in patients treated with radical prostatectomy. *J Urol* 2004;**171**(3):1122–7.
80. Edge SB, Compton CC. The American Joint Committee on Cancer: the 7th edition of the AJCC cancer staging manual and the future of TNM. *Ann Surg Oncol* 2010;**17**(6):1471–4.
81. May M, Kaufmann O, Hammermann F, et al. Prognostic impact of lymphovascular invasion in radical prostatectomy specimens. *BJU Int* 2007;**99**(3):539–44.
82. Cheng L, Jones TD, Lin H, et al. Lymphovascular invasion is an independent prognostic factor in prostatic adenocarcinoma. *J Urol* 2005;**174**(6):2181–5.
83. Loeb S, Roehl KA, Yu X, et al. Lymphovascular invasion in radical prostatectomy specimens: prediction of adverse pathologic features and biochemical progression. *Urology* 2006;**68**(1):99–103.
84. Yee DS, Shariat SF, Lowrance WT, et al. Prognostic significance of lymphovascular invasion in radical prostatectomy specimens. *BJU Int* 2011;**108**(4):502–7.
85. Jung JH, Lee JW, Arkoncel FR, et al. Significance of perineural invasion, lymphovascular invasion, and high-grade prostatic intraepithelial neoplasia in robot-assisted laparoscopic radical prostatectomy. *Ann Surg Oncol* 2011;**18**(13):3828–32.
86. Ng J, Mahmud A, Bass B, et al. Prognostic significance of lymphovascular invasion in radical prostatectomy specimens. *BJU Int* 2012;**110**(10):1507–14.
87. Huang SP, Huang CY, Wang JS, et al. Prognostic significance of p53 and X-ray repair cross-complementing group 1 polymorphisms on prostate-specific antigen recurrence in prostate cancer post radical prostatectomy. *Clin Cancer Res* 2007;**13**:6632–8.
88. Ito K, Nakashima J, Mukai M, et al. Prognostic implication of microvascular invasion in biochemical failure in patients treated with radical prostatectomy. *Urol Int* 2003;**70**:297–302.
89. Lee KL, Marotte JB, Ferrari MK, McNeal JE, Brooks JD, Presti Jr JC. Positive family history of prostate cancer not associated with worse outcomes after radical prostatectomy. *Urology* 2005;**65**:311–5.
90. Quinn DI, Henshall SM, Haynes AM, et al. Prognostic significance of pathologic features in localized prostate cancer with radical prostectomy: implications for staging systems and predictive models. *J Clin Oncol* 2001;**19**:3692–705.
91. Brooks JP, Albert PS, O'Connell J, et al. Lymphovascular invasion in prostate cancer: prognostic significance in patients treated with radiotherapy after radical prostatectomy. *Cancer* 2006;**106**(7):1521–6.
92. American Committee for Cancer Staging and End-Results Reporting. *The manual for staging of cancer*. 1st ed. Philadelphia: Lippincott-Raven; 1977.
93. Berney DM, Wheeler TM, Grignon DJ, et al. International Society of Urological Pathology (ISUP) Consensus Conference on Handling and Staging of Radical Prostatectomy Specimens. Working group 4: seminal vesicles and lymph nodes. *Mod Pathol* 2011;**24**(1):39–47.
94. Eggener SE, Roehl KA, Smith ND, et al. Contemporary survival results and the role of radiation therapy in patients with node negative seminal vesicle invasion following radical prostatectomy. *J Urol* 2005;**173**(4):1150–5.
95. Pierorazio PM, Ross AE, Schaeffer EM, et al. A contemporary analysis of outcomes of adenocarcinoma of the prostate with seminal vesicle invasion (pT3b) after radical prostatectomy. *J Urol* 2011;**185**(5):1691–7.
96. Catalona WJ, Smith DS. 5-Year tumor recurrence rates after anatomical radical retropubic prostatectomy for prostate cancer. *J Urol* 1994;**152**(5 Pt 2):1837–42.
97. Inman BA, Frank I, Boorjian SA, et al. Dynamic prediction of metastases after radical prostatectomy for prostate cancer. *BJU Int* 2011;**108**(11):1762–8.
98. Hull GW, Rabbani F, Abbas F, et al. Cancer control with radical prostatectomy alone in 1,000 consecutive patients. *J Urol* 2002;**167**(2 Pt 1):528–34.
99. Fowler JE, Torgerson L, McLeod DG, et al. Radical prostatectomy with pelvic lymphadenectomy: observations on the accuracy of staging with lymph node frozen sections. *J Urol* 1981;**126**(5): 618–9.
100. Gervasi LA, Mata J, Easley JD, et al. Prognostic significance of lymph nodal metastases in prostate cancer. *J Urol* 1989;**142**(2 Pt 1): 332–6.
101. Cheng L, Zincke H, Blute ML, et al. Risk of prostate carcinoma death in patients with lymph node metastasis. *Cancer* 2001;**91**(1):66–73.
102. Daneshmand S, Quek ML, Stein JP, et al. Prognosis of patients with lymph node positive prostate cancer following radical prostatectomy: long-term results. *J Urol* 2004;**172**(6 Pt 1):2252–5.
103. Naya Y, Babaian RJ. The predictors of pelvic lymph node metastasis at radical retropubic prostatectomy. *J Urol* 2003;**170**(6 Pt 1): 2306–10.
104. Bader P, Burkhard FC, Markwalder R, et al. Disease progression and survival of patients with positive lymph nodes after radical prostatectomy. Is there a chance of cure? *J Urol* 2003;**169**(3): 849–54.
105. Boorjian SA, Thompson RH, Siddiqui S, et al. Long-term outcome after radical prostatectomy for patients with lymph node positive prostate cancer in the prostate specific antigen era. *J Urol* 2007;**178**(3 Pt 1):864–70 discussion 870–861.
106. Dell'Oglio P, Abdollah F, Suardi N, et al. External validation of the European association of urology recommendations for pelvic lymph node dissection in patients treated with robot-assisted radical prostatectomy. *J Endourol* 2014;**28**(4):416–23.
107. Heidenreich A, Varga Z, Von Knobloch R. Extended pelvic lymphadenectomy in patients undergoing radical prostatectomy: high incidence of lymph node metastasis. *J Urol* 2002;**167**(4):1681–6.
108. Allaf ME, Palapattu GS, Trock BJ, et al. Anatomical extent of lymph node dissection: impact on men with clinically localized prostate cancer. *J Urol* 2004;**172**(5 Pt 1):1840–4.
109. Stone NN, Stock RG, Unger P. Laparoscopic pelvic lymph node dissection for prostate cancer: comparison of the extended and modified techniques. *J Urol* 1997;**158**(5):1891–4.
110. Heck MM, Retz M, Bandur M, et al. Topography of lymph node metastases in prostate cancer patients undergoing radical prostatectomy and extended lymphadenectomy: results of a combined molecular and histopathologic mapping study. *Eur Urol* 2013;**66**(2):222–9.
111. Bader P, Burkhard FC, Markwalder R, et al. Is a limited lymph node dissection an adequate staging procedure for prostate cancer? *J Urol* 2002;**168**(2):514–8 discussion 518.
112. Weckermann D, Dorn R, Trefz M, et al. Sentinel lymph node dissection for prostate cancer: experience with more than 1,000 patients. *J Urol* 2007;**177**(3):916–20.

113. Messing EM, Manola J, Yao J, et al. Immediate versus deferred androgen deprivation treatment in patients with node-positive prostate cancer after radical prostatectomy and pelvic lymphadenectomy. *Lancet Oncol* 2006;**7**(6):472–9.
114. Kroepfl D, Loewen H, Roggenbuck U, et al. Disease progression and survival in patients with prostate carcinoma and positive lymph nodes after radical retropubic prostatectomy. *BJU Int* 2006;**97**(5):985–91.
115. Schumacher MC, Burkhard FC, Thalmann GN, et al. Good outcome for patients with few lymph node metastases after radical retropubic prostatectomy. *Eur Urol* 2008;**54**(2):344–52.
116. Briganti A, Blute ML, Eastham JH, et al. Pelvic lymph node dissection in prostate cancer. *Eur Urol* 2009;**55**(6):1251–65.

C H A P T E R

# 34

# Open Versus Robotic Prostatectomy

*Cesar E. Ercole, MD**,*
*Andrew J. Stephenson, MD, FACS, FRCS(C)**,**
*Center for Urologic Oncology, Glickman Urological & Kidney Institute, Cleveland Clinic, Cleveland, OH, USA
**Cleveland Clinic Lerner College of Medicine, Case Western Reserve University, Cleveland, OH, USA

## INTRODUCTION

Prostate cancer is the most common solid malignancy among men in the United States and the second most common cause of death, with an estimated 233,000 new cases and 29,480 deaths in 2014.[1] The "gold standard" for surgical management for prostate cancer has been a radical prostatectomy, dating back to the first perineal prostatectomy described by Hugh Hampton Young in 1905.[2] Several advances have been made with the procedure to improve oncological outcomes, including adjusting to the retropubic approach. To optimize and address issues with the retropubic radical prostatectomy (RRP), including short- and long-term complications, as well as oncological, continence, and potency outcomes, there was, and continues to be, a demand for a better understanding of the local anatomy and physiology of the prostate and periprostatic area. Pioneers in this work include Reiner and Walsh with their description of the dorsal venous complex,[3] Oelrich's work defining the importance of the striated sphincter to improve continence,[4] and Walsh and Donker, who better defined the neurovascular bundles to assist with improving postoperative potency.[5]

To reduce the invasiveness of the procedure, technological advances have been introduced to incorporate laparoscopic and robotic-assisted laparoscopic techniques. Schuessler et al. first described their experience with performing a laparoscopic prostatectomy (LP) in 1997.[6] Several other authors have contributed to standardizing the LP technique.[7,8] The laparoscopic procedure proved to have a steep learning curve with reduction in motion and two-dimensional vision, leading to limited acceptance as an improvement on the widely practiced use of the RRP. The introduction of robotic instrumentation with the daVinci Surgical System (Intuitive Surgical Inc., Sunnyvale, CA) enabled surgeons to overcome the challenging technical learning curve to LP.[9] In the United States, Menon et al. standardized the technique for robotic-assisted laparoscopic radical prostatectomy (RALRP), which subsequently led to widespread acceptance in Europe and the United States.[10] The technology involved with the system is the driving force in acceptance as it provides surgeons with improved optics (three-dimensional) and instrument controls that have less-restrictive movements, thus mimicking surgeons' hand movements with increased accuracy. Concurrently, it has reduced the learning curve associated with the minimally invasive approach to radical prostatectomy and many surgeons perceive an improvement in functional results without sacrificing early oncologic outcomes.

As time has allowed for data sets to mature there have been more systematic reviews to compare the perioperative, functional, and oncological outcomes between the minimally invasive approaches (LRP and RALRP) to the traditional approach (RRP). RP has become the preferred technique for managing prostate cancer, and with the introduction of LRP and RALRP there has been wide acceptance around the world. Review of the literature provides support for advantages and disadvantages associated with each approach. The discussion in this chapter will endeavor to highlight the differences between RRP and LRP/RALRP and provide information for the surgeon and patient to hold a fruitful discussion to determine the best approach for that patient.

## SURGICAL TECHNIQUE

From a technical perspective, the traditional open approach (RRP) is done through a small midline infraumbilical incision down to the symphysis pubis allowing

Prostate Cancer. http://dx.doi.org/10.1016/B978-0-12-800077-9.00034-7

for exposure of the pelvis, access to the prostate and pelvic lymph nodes. The procedure is done under direct visualization of the tissue with the surgeon being able to appreciate the tissue by touch. The laparoscopic and robotic approach use smaller incisions for the camera and working ports, and the surgeon relies on perceived haptic feedback to appreciate the tissues. Another advantage to the minimally invasive techniques is the visualization of the tissues and the positive pressure provided by the pneumoperitoneum used for insufflation as it limits blood loss. Some of the limitations of the LRP were overcome with the development of the RALRP, especially with the degrees of freedom with the instruments, improved ergonomics, and visualization. A disadvantage to the RALRP is that an experienced bedside assistant is paramount and cost.

## Determining Approach

The most appropriate surgical approach should be dictated by surgeon experience/expertise, patient anatomical factors (e.g., body habitus, pelvic anatomy, prostate size), treatment-related factors (e.g., prior abdominal/pelvic surgery, radiation therapy), comorbid illness, and patient preference. Essentially the indication for any of the available procedures is the same: localized disease without clinical or radiographic metastatic disease. With respect to comorbid illness, severe cardiopulmonary compromise, uncorrectable bleeding diatheses, and severe liver dysfunction are relative contraindications to both procedures. For minimally invasive approaches, the need for pneumoperitoneum and steep Trendelenberg positioning may be associated with increased difficulty in adequately ventilating patients (causing hypercarbia and acidosis) compared to RRP. Patients with complex pelvic surgery, large prostate size (e.g., >100 g), morbid obesity, prior pelvic radiation will increase the technical challenge to the open and minimally invasive approaches.[11,12] Extreme obesity may present challenges for the open and minimally invasive approaches. For RRP, adequate exposure of the surgical field may be challenging in the severely obese patient. For LRP/RALRP, extreme obesity presents a lesser challenge for the technical aspects of the surgery but may cause anesthetic issues related ventilation and steep Trendelenberg positioning. Ahlering et al. found that obese patients (BMI >30) had worse baseline urinary and sexual function, and slower return to baseline when compared to nonobese patients. At 6 months, 91.4% of nonobese patients were pad-free, while 47% of obese were able to achieve pad-free urinary continence ($p \leq 0.001$).[13] Prior abdominal surgery is not a contraindication for LRP/RALRP but severe adhesions caused by extensive abdominal surgery, prior peritonitis, or prior trauma may hamper efforts to perform the operation and significantly prolong the duration of the surgery if extensive enterolysis is required, which may also contribute to postoperative ileus and increase the risk of bowel injury. By virtue of its extra-peritoneal approach, RRP avoids these potential problems. Likewise, RRP may afford some advantages in patients with prior colorectal reconstruction, such as an ileal J-pouch. Extraperitoneal RALRP has also been described and mastered by some surgeons and this may also circumvent these problems. Access to the pelvis may also be limited by previous pelvic surgery, including hernia repairs, which may obliterate the space of Retzius, making RRP more challenging. Some surgeons may prefer to approach the prostate via a transperitoneal approach using RALRP for this reason. The extraperitoneal approach has the advantage of leaving the peritoneum as a natural barrier keeping the bowel away from the operative field and urine out of the peritoneal cavity should there be an anastomotic leak. For extraperitoneal RALRP, a limiting factor is the reduced working space, as well as tension on the vesicourethral anastomosis as the bladder is not mobilized to the same extent as a transperitoneal approach. There are also some reports of high $CO_2$ absorption with extraperitoneal versus transperitoneal approach requiring higher minute volume to compensate for the hypercarbia and associated acidosis.[14,15] Most series do not show a difference, or very little difference in outcomes when comparing the extraperitoneal to the transperitoneal.

## Management of Dorsal Venous Complex (DVC)

The dorsal venous complex (DVC) can be the source of greatest blood loss during the procedure. Blood in the field will also make the meticulous and precise dissection of the neurovascular bundles (NVB) much more challenging. Variation on approach to the DVC such as with an antegrade progression where it is suture ligated but not divided until later in the procedure when the bladder neck dissection has been done. At the time that the bladder neck is divided, the prostatic pedicles are then identified and ligated. With the pedicles ligated, the blood supply to the prostate is controlled. The NVB are dissected from between the prostate and the levator fascia, and any perforating vessels are controlled for satisfactory hemostasis as the dissection is carried toward the prostatic apex. With this approach the NVB is almost completely dissected prior to reaching the apex of the prostate and the DVC is divided, always mindful to limit any thermal injury to the NVB. In 1998, Walsh described a retrograde approach to help preserve the NVB with more accuracy.[16] An initial step in the procedure is to suture ligate the DVC and then divide it and the urethra. The prostate is released from apex to base as the attachments to the levator fascia are taken down and thus releasing the NVB as the dissection is completed. The prostatic pedicle is controlled by ligation and the bladder neck can be divided. Last, the prostatic pedicles are secured with large hemoclips and the bladder neck divided. This approach is favored by some due to early control of the blood supply to the prostate and thus limit excessive blood loss during the procedure.[17] Yet others prefer early ligation and division of the DVC to allow for a meticulous apical dissection, in efforts to limit the chance for positive margins and the preservation of the striated sphincter. The approach is up to the surgeon's discretion; an absolute is maintaining a bloodless operative field and preserving the NVB as best as possible for optimal urinary and sexual function without compromising oncological outcomes. The use of electrocautery should be judicious with some authors preferring the use of bipolar electrocautery in RALRP to control any penetrating prostatic vessels.[18]

## Managing the Neurovascular Bundle (NVB)

Most men will place significant value in their postoperative potency; therefore, preservation of the cavernosal blood vessels and nerves is essential to the maintenance – and return – of erectile function. In efforts to improve postsurgical sexual and urinary function Walsh described a high release of the levator fascia or preservation of the prostatic fascia at the anterior apex to lessen the traction on the cavernosal nerves and avoiding the ligation of small, anteriorly traveling arteries to the striated sphincter.[19] As we appreciate the periprostatic anatomy better and better, several studies have highlighted the benefits of a high-anterior release of the NVB and avoiding electrocautery.[20–23] Bartsch and coworkers describe a "curtain" or concave distribution of the cavernosal nerves over the anterior rim of the prostate.[24] Kaul et al. describe their technique of reduced traction on the NVB during RALRP to aid in the dissection of the "veil of Aphrodite" where the robotic technology allows for closer dissection and preservation of the nerves without violating the prostatic capsule.[18] These series need to mature further to determine if one is more effective than the other for preservation of the NVB. Historical series with RRP have demonstrated that in men in their 40s that undergo a nerve-sparing procedure, 95% of them will have return of function adequate for intercourse. This diminishes with age, as older patients in their 70s only 50% of them will regain the same function, even with attempts to aggressively preserve the bundle while maintaining low positive surgical margins (single-digits).[25] With the robotic experience, Menon and coworkers showed that with a nerve-sparing RALRP 96% of men were able to achieve erections rigid enough for intercourse, and from an oncologic outcome of a positive margin rate of 4.6%.[26] To maximize potency results, injury to the NVB needs to be minimized, especially when using electrocautery as thermal damage to the NVB can be potentially irreversible. Caution should always be taken when using electrocautery close the NVB to avoid thermal injury.[27] Ong et al. described that there may be a delay in return in potency and only reach 50–68% at 24 months or greater in RALRP where bipolar electrocautery is used, indicating that a neuropraxia process occurs rather than a complete and irreversible nerve injury.[28,29] As a result of these findings, most LRP and RALRP techniques have evolved toward avoidance or minimization of electrocautery or other thermal devices during dissection around the NVBs, thus replicating what has long been described in open surgery.

## Reconstruction and Vesicourethral Anastomosis

Technical challenges are present with each approach when dealing with the bladder neck and the vesicourethral anastomosis. Continence rates with contemporary series list results at 90–98% for RALRP[30,31] at 1 year and 72–98% with RALRP at 3–12 months.[32,33] With either approach ultimately few men suffer urinary incontinence and most men have excellent urinary control whether they undergo a RRP, LRP, or RALP.

The traditional RRP describes a mucosa-to-mucosa approximation of the urethra and the bladder neck. Walsh et al. in 2005 described a modification involving placement of an anterior figure-of-eight at the bladder neck resulting in intussusceptions of the bladder neck.[34] With this modification, the bladder neck is less likely to pull open with bladder filling, thus reducing incontinence, and an early return of urinary control was noted in 98% of men defined as being pad-free at one year postoperatively. Generally, the open RRP anastomosis is done with interrupted sutures. There have been studies evaluating the value of reconstructing Denovillier's fascia to the posterior rhabdosphincter in efforts to provide support to the vesicourethral anastomosis. Rocco et al. were able to show a statistically significant improvement in return to continence,[35] yet the mechanism remains unclear and surgeon-dependent.

Other differences between the open and laparoscopic approach with the vesicourethral anastomosis are due to the challenges inherent with the LRP especially with intracorporeal knot tying; some of these challenges are lessened with the RALRP, particularly with improved instrument control and ability to be more precise with suturing. Therefore, the bladder neck is generally sutured with a double-armed suture in two directions from the posterior aspect to the anterior completed with a single knot.[36] Another advantage to the continuous suture is that it distributes the tension of the anastomosis evenly along the suture line decreasing the chances of tearing. If there is redundancy of the bladder neck, then an anterior reconstruction may be performed. The anastomosis can be tested to see if it is water-tight after watching it come together under direct visualization.

# FACTORS INFLUENCING OUTCOMES

## Learning Curves

With the robotic approach widely accepted, the learning curve for a radical prostatectomy needs to be addressed individually for each approach. Proficiency with any of these techniques is measured in oncological, continence, and potency outcomes (i.e., "The Trifecta").[37] When evaluating the experience of the center with one of the world's largest robotics experience, the Vattikuti Institute was able to show that within the first three years they were able to match or improve on operative times, estimated blood loss (EBL), and complication rates comparing robotic versus historical open RRP series.[38] Others have attempted to determine the number of cases to become proficient or achieve expert status with the procedure. For the open RRP series Memorial Sloan Kettering noted that the surgeon reaches this status at 250 cases, with a statistically significant association between surgical experience and the probability of patients remaining biochemical recurrence-free after RP.[39] Vickers et al. also noted that although surgical margins are not a good surrogate for cancer control, they do highlight technical expertise, with surgeons having only 10 prior cases have a 40% probability of a positive surgical margin versus 25% for surgeons with 250 prior cases.[40] Defining the learning curve and how it influences clinical endpoints, including functional outcomes (potency, continence) and perioperative/postoperative outcomes (complications, surgical margins, cancer control), is surgeon experience dependent.[41] Fellowship training was considered by Bianco et al. to evaluate the learning curves for surgical margins and cancer control. Fellowship trained physicians were noted to have a faster learning curve ($p = 0.006$), which resulted in an overall superior cancer control ($p = 0.001$; difference: 4.7%; 95% CI: 2.6%, 7.4%). For RALRP, Smith et al. showed that the same number of cases were needed to achieve the same oncological outcomes as open surgeons.[42]

## Oncologic Outcomes

Concern with the learning curve, for the open and minimally invasive procedures, speaks to the role it has with positive surgical margins. Smith et al. compared their RALRP to their RRP series.[42] They noted a lower rate of positive surgical margins with the RALRP compared to open RRP (15% vs. 35%, $p < 0.001$) and in both groups the apex was the most common site of positive surgical margins (52% RALRP vs. 37% RRP). It should be noted that there was some selection bias with these groups as the RALRP had lower PSA values and lower clinical stage and Gleason score.

Recent series have shown that there is no difference in positive surgical margins between the open and LRP/RALRP experience at about 11% for both.[43] These values increase with higher stage and are center/surgeon volume dependent. Positive surgical margin reported by Patel et al. was excellent for pT2 with margin rates as low as 2.5%.[44]

Few large LRP and RALRP series have evaluated biochemical recurrence-free survival. Guillonneau et al. evaluated their outcomes of 1000 consecutive LRPs with median follow-up of 12 months, and noted an overall actuarial biochemical progression-free survival rate of 90.5% at 3 months.[45] When broken down by stage, 92% for pT2a, 88% for pT2b, 77% for pT3a, and 44% for pT3b. Badani et al. reported similar outcomes with a median follow-up of 22 months with a 5-year actuarial biochemical-free survival of 84%.[46] Analysis of US Surveillance, Epidemiology, and End Results (SEER) Medicare link data noted that physicians with higher volumes were less likely to use secondary therapy independent of tumor characteristics, suggesting that more experienced surgeons may opt to observe patients with positive margins and extracapsular extension than the use of adjuvant therapy.[47] As for most minimally invasive series, longer-term follow-up is necessary to assess the progression of disease and prostate-cancer-specific mortality.

# FUNCTIONAL OUTCOMES

## Potency and Continence

For RRP the historical rates of potency and continence were noted to range from 50% to 90% and 90% to 98%, respectively, at one year. For LRP the equivalent reported rates at one year are 12–67% and 70–92%.[30,48–50] Reported potency and continence rates after RALRP vary from 20% to 90% and 72% to 98%, respectively, over one year and are summarized in the review by Ficarra et al.[9] Much of the variation on these outcomes is related to surgeon experience and patient populations. Several factors may influence how these outcomes are reported depending on the method used to assess patient outcomes (e.g., validated questionnaire, patient reported), and comparison of sexual function prior to surgery and after surgery. Sonn et al. reviewed patient self-assessments and surgeon assessments on fatigue, pain, and sexual, urinary, and bowel dysfunction using SF-36™ and the UCLA-PCI. Surgeons were more likely to focus on sexual and urinary function, whereas patients reported more often on fatigue and pain.[51] The wide range of potency and continence rates for minimally invasive RP is representative of initial and mature patient series from a variety of surgeons and heterogeneous patient populations. Resnick et al. used data from the Prostate Cancer Outcomes Study (PCOS) to compare long-term urinary, bowel, and sexual function after prostatectomy or external-beam radiation therapy. Patients were noted to have differences with surgery in the immediate time period postoperatively, yet at 15 years no significant difference with incontinence, and at 2 years no difference with erectile dysfunction. In addition, the definition, methodology, and criteria used to define successful preservation of cavernous nerves and sexual function confound any comparison between series and techniques. In a population-based observation cohort study using SEER data, Hu et al. noted statistical difference in shorter length of stay, lower rates of blood transfusions, postoperative respiratory complications, miscellaneous surgical complications, and anastomotic strictures in patients undergoing a minimally invasive RP compared to open RP. Those who underwent an open procedure were less likely to have genitourinary complications and diagnoses of incontinence and erectile dysfunction, with no difference in need for additional therapy postprocedure.[52] These outcomes are ultimately dependent on preoperative patient characteristics, surgeon experience, and utmost should not compromise oncologic outcomes.

### Pain Management and Convalescence

Patient recovery is of greatest importance and with the minimally invasive approaches the verdict on reduced pain and earlier convalescence is still being defined. With regard to pain, Webster et al. noted similar low postoperative pain scores and use of narcotics between patients who underwent RRP and those that underwent a robotic approach.[53] When considering the postoperative period there are no significant differences in how the patient is managed whether he has an open or laparoscopic procedure. Therefore, most will have a similar length of hospital stay and postoperative narcotics use. Bhayani et al. noted that in their series patients who underwent an RRP took a week longer to return to baseline physical function compared to those who had an RALRP procedure. This was based on the patient-reported changes in physical function and measured with the SF-12 Physical and Mental Health Survey Acute Form.[54] Others have noted that there may be no difference in early convalescence, and more likely a difference lies in narcotic requirements in the first 24 h and length of hospital stay.[55] The verdict is still pending on the impact of the open versus robotic procedure on functional outcomes and return to baseline for daily activities. Miller et al. evaluated the short-term impact and recovery of health-related quality of life in patients undergoing open and robotic-assisted laparoscopic RP by having patients fill out the SF-12, version 2 Physical and Mental Health Survey Acute Form.[56] The surveys were filled out preoperatively and then weekly for 6 weeks postoperative. They noted faster return to baseline with the Physical Component Scores in the robotic group – one week earlier (week 5 and 6) compared to the open group (week 6 and 7). With regard to the Mental Component Scores, they were not different with each group. Wood et al. also evaluated short-term health outcomes in robotic and open prostatectomy patients. Although the robotic group was noted to have a shorter mean length of hospital stay and less mean narcotics use within the first 24 h postoperative, there was no difference in median time to normal activity, 100% activity, or time to driving.[57]

## COST CONSIDERATION

As the healthcare landscape adjusts to demands for cost awareness, the use of the robot has been questioned. The appeal of the robotic procedure from a surgical perspective is from a technical standpoint but it also translates to an increase in cost to the procedure. In United States, it was estimated that in 2009 the majority (69–85%) of RPs were done with the robotic platform.[58] Currently in the United States, hospital stay regardless of RP type is about 1–2 days. Therefore, length of hospital stay may not be as a relevant factor in cost. Several studies have looked at comparing costs with most recently Tomaszewski et al. performing a retrospective review of 473 consecutive patients who underwent an RALRP ($n = 115$) or RRP ($n = 358$) during a period of 15 months.[59] They noted that ancillary, cardiology, imaging, administrative, laboratory, and pharmacy costs were not significantly different between the approaches, only robotic equipment and supplies increased the cost of the procedure, noting an average profit of $1325 for an RRP and each RALRP lost $4013.

Some studies evaluated the cost of the procedure based on the experience of the surgeon, such that more seasoned surgeons use less operating room (OR) time. Thus, cost comparison should be made with mature series, as the cost of the learning curve was estimated at $217,034 per surgeon based on an average learning curve of 77 RALRP.[60] Keeping in mind that as the robot has been increasingly adopted by practicing surgeons, it has also been increasingly used by residents in training; therefore, this figure will likely decrease with time. Lotan et al. demonstrated that LRP has proved to be almost as cost-competitive as open RRP, whereas RALRP is the most expensive to the hospital due to the initial purchase of the robot, then the maintenance and operative instruments cost.[61] With an upfront cost to purchase the robot and then annual maintenance – at $1.5 million for a single console and $150,000 maintenance.[62] Direct cost comparisons have proven difficult to perform secondary to differences in reported data and reimbursement scales.

A cost that is more difficult to capture is based on patient expectations. Those patients that are scheduled for a RALRP tend to have higher expectations with regards to their continence and especially with erectile function recovery.[63] Schroeck et al. surveyed patients who underwent RRP and RALRP to assess satisfaction and regret.[64] They noted that of 400 respondents, 84% were satisfied and 19% regretted their treatment choice. Domains associated with satisfaction were lower income, shorter follow-up, having undergone RRP versus RALRP, urinary domain scores, and hormonal domain scores. Patients who underwent a RALRP were more likely to have regret compared to those who had an RP (OR, 3.02; 95% CI 1.50–6.07). Regret was also associated with lower urinary domain scores, hormonal domain scores, and years from surgery. African-American race and lower bowel domain scores were also independently associated with regret ($p \leq 0.028$). This highlights the need to have a good discussion with patients to establish realistic expectations regardless of approach.

As experience with the LRP and RALRP increases, the training of new generations of surgeons will involve robotics from an earlier phase, and as costs of robotic technologies decrease, cost equivalence between open, laparoscopic, and robotic surgeries may be achieved.

## CONCLUSIONS

The patient who presents with localized prostate cancer has several options for surgical management. The decision to have an open or laparoscopic/robotic procedure depends on the surgeon and the patient. Contemporary series from centers of excellence demonstrate that oncologic, continence, and potency outcomes are consistent and similar with each technique. More time is necessary to evaluate long-term oncological outcomes with the minimally invasive techniques. Ultimately, the best predictor of success is based on the experience and clinical judgment of the surgeon, and as we endeavor to build on the current practices to improve oncological and functional outcomes for patients undergoing radical prostatectomy.

## References

1. Siegel R, Ma J, Zou Z, et al. Cancer statistics. *CA Cancer J Clin* 2014;**64**(1):9–29.
2. Young HH. VIII. Conservative perineal prostatectomy: the results of two years' experience and report of seventy-five cases. *Ann Surg* 1905;**41**(4):549–57.
3. Reiner WG, Walsh PC. An anatomical approach to the surgical management of the dorsal vein and Santorini's plexus during radical retropubic surgery. *J Urol* 1979;**121**(2):198–200.
4. Oelrich TM. The urethral sphincter muscle in the male. *Am J Anat* 1980;**158**(2):229–46.
5. Walsh PC, Donker PJ. Impotence following radical prostatectomy: insight into etiology and prevention. *J Urol* 1982;**128**(3):492–7.
6. Schuessler WW, Schulam PG, Clayman RV, et al. Laparoscopic radical prostatectomy: initial short-term experience. *Urology* 1997;**50**(6):854–7.
7. Guillonneau B, Vallancien G. Laparoscopic radical prostatectomy: the Montsouris technique. *J Urol* 2000;**163**(6):1643–9.
8. Abbou CC, Salomon L, Hoznek A, et al. Laparoscopic radical prostatectomy: preliminary results. *Urology* 2000;**55**(5):630–4.
9. Ficarra V, Cavalleri S, Novara G, et al. Evidence from robot-assisted laparoscopic radical prostatectomy: a systematic review. *Eur Urol* 2007;**51**(1):45–55 discussion 56.
10. Menon M, Shrivastava A, Tewari A, et al. Laparoscopic and robot assisted radical prostatectomy: establishment of a structured program and preliminary analysis of outcomes. *J Urol* 2002;**168**(3):945–9.
11. Singh A, Fagin R, Shah G, et al. Impact of prostate size and body mass index on perioperative morbidity after laparoscopic radical prostatectomy. *J Urol* 2005;**173**(2):552–4.
12. Brown JA, Rodin DM, Lee B, et al. Laparoscopic radical prostatectomy and body mass index: an assessment of 151 sequential cases. *J Urol* 2005;**173**(2):442–5.
13. Ahlering TE, Eichel L, Edwards R, et al. Impact of obesity on clinical outcomes in robotic prostatectomy. *Urology* 2005;**65**(4):740–4.
14. Meininger D, Byhahn C, Wolfram M, et al. Prolonged intraperitoneal versus extraperitoneal insufflation of carbon dioxide in patients undergoing totally endoscopic robot-assisted radical prostatectomy. *Surg Endosc* 2004;**18**(5):829–33.
15. Bivalacqua TJ, Schaeffer EM, Alphs H, et al. Intraperitoneal effects of extraperitoneal laparoscopic radical prostatectomy. *Urology* 2008;**72**(2):273–7.
16. Walsh PC. Anatomic radical prostatectomy: evolution of the surgical technique. *J Urol* 1998;**160**(6 Pt 2):2418–24.
17. Rassweiler J, Sentker L, Seemann O, et al. Laparoscopic radical prostatectomy with the Heilbronn technique: an analysis of the first 180 cases. *J Urol* 2001;**166**(6):2101–8.
18. Kaul S, Bhandari A, Hemal A, et al. Robotic radical prostatectomy with preservation of the prostatic fascia: a feasibility study. *Urology* 2005;**66**(6):1261–5.
19. Wein AJ, Kavoussi LR, Campbell MF. *Campbell-Walsh urology*. 10th ed. Philadelphia, PA: Elsevier Saunders; 2012.
20. Costello AJ, Brooks M, Cole OJ. Anatomical studies of the neurovascular bundle and cavernosal nerves. *BJU Int* 2004;**94**(7):1071–6.
21. Takenaka A, Murakami G, Soga H, et al. Anatomical analysis of the neurovascular bundle supplying penile cavernous tissue to ensure a reliable nerve graft after radical prostatectomy. *J Urol* 2004;**172**(3):1032–5.
22. Takenaka A, Hara R, Soga H, et al. A novel technique for approaching the endopelvic fascia in retropubic radical prostatectomy, based on an anatomical study of fixed and fresh cadavers. *BJU Int* 2005;**95**(6):766–71.
23. Lunacek A, Schwentner C, Fritsch H, et al. Anatomical radical retropubic prostatectomy: 'curtain dissection' of the neurovascular bundle. *BJU Int* 2005;**95**(9):1226–31.
24. Horninger W, Strasser H, Bartsch G. Radical retropubic prostatectomy: apical preparation and curtain dissection of the neurovascular bundle. *BJU Int* 2005;**95**(6):911–23.
25. Han M, Partin AW, Chan DY, et al. An evaluation of the decreasing incidence of positive surgical margins in a large retropubic prostatectomy series. *J Urol* 2004;**171**(1):23–6.
26. Kaul S, Savera A, Badani K, et al. Functional outcomes and oncological efficacy of Vattikuti Institute prostatectomy with Veil of Aphrodite nerve-sparing: an analysis of 154 consecutive patients. *BJU Int* 2006;**97**(3):467–72.
27. Ong AM, Su L-M, Varkarakis I, et al. Nerve sparing radical prostatectomy: effects of hemostatic energy sources on the recovery of cavernous nerve function in a canine model. *J Urol* 2004;**172**(4 Pt 1): 1318–22.
28. Ong AM, Su LM, Varkarakis I, et al. Nerve sparing radical prostatectomy: effects of hemostatic energy sources on the recovery of cavernous nerve function in a canine model. *J Urol* 2004;**172**(4 Pt 1): 1318–22.

29. Ahlering TE, Eichel L, Skarecky D. Evaluation of long-term thermal injury using cautery during nerve sparing robotic prostatectomy. *Urology* 2008;**72**(6):1371–4.
30. Stanford JL, Feng Z, Hamilton AS, et al. Urinary and sexual function after radical prostatectomy for clinically localized prostate cancer: the Prostate Cancer Outcomes Study. *JAMA* 2000;**283**(3): 354–60.
31. Walsh PC, Marschke P, Ricker D, et al. Patient-reported urinary continence and sexual function after anatomic radical prostatectomy. *Urology* 2000;**55**(1):58–61.
32. Joseph JV, Rosenbaum R, Madeb R, et al. Robotic extraperitoneal radical prostatectomy: an alternative approach. *J Urol* 2006;**175**(3 Pt 1): 945–50.
33. Patel VR, Tully AS, Holmes R, et al. Robotic radical prostatectomy in the community setting – the learning curve and beyond: initial 200 cases. *J Urol* 2005;**174**(1):269–72.
34. Walsh PC, Marschke PL. Intussusception of the reconstructed bladder neck leads to earlier continence after radical prostatectomy. *Urology* 2002;**59**(6):934–8.
35. Rocco B, Gregori A, Stener S, et al. Posterior reconstruction of the rhabdosphincter allows a rapid recovery of continence after transperitoneal videolaparoscopic radical prostatectomy. *Eur Urol* 2007;**51**(4):996–1003.
36. Van Velthoven RF, Ahlering TE, Peltier A, et al. Technique for laparoscopic running urethrovesical anastomosis: the single knot method. *Urology* 2003;**61**(4):699–702.
37. Bianco Jr FJ, Scardino PT, Eastham JA. Radical prostatectomy: long-term cancer control and recovery of sexual and urinary function ("trifecta"). *Urology* 2005;**66**(5 Suppl.):83–94.
38. Menon M, Shrivastava A, Tewari A. Laparoscopic radical prostatectomy: conventional and robotic. *Urology* 2005;**66**(5 Suppl.):101–4.
39. Vickers AJ, Bianco FJ, Serio AM, et al. The surgical learning curve for prostate cancer control after radical prostatectomy. *J Natl Cancer Inst* 2007;**99**(15):1171–7.
40. Vickers A, Bianco F, Cronin A, et al. The learning curve for surgical margins after open radical prostatectomy: implications for margin status as an oncological end point. *J Urol* 2010;**183**(4):1360–5.
41. Herrell SD, Smith Jr JA. Robotic-assisted laparoscopic prostatectomy: what is the learning curve? *Urology* 2005;**66**(5 Suppl.):105–7.
42. Smith JA, Chan RC, Chang SS, et al. A comparison of the incidence and location of positive surgical margins in robotic assisted laparoscopic radical prostatectomy and open retropubic radical prostatectomy. *J Urol* 2007;**178**(6):2385–9.
43. Touijer K, Eastham JA, Secin FP, et al. Comprehensive prospective comparative analysis of outcomes between open and laparoscopic radical prostatectomy conducted in 2003–2005. *J Urol* 2008;**179**(5):1811–7.
44. Patel VR, Shah K, Palmer KJ, et al. Robotic-assisted laparoscopic radical prostatectomy: a report of the current state. *Expert Rev Anticancer Ther* 2007;**7**(9):1269–78.
45. Guillonneau B, el-Fettouh H, Baumert H, et al. Laparoscopic radical prostatectomy: oncological evaluation after 1,000 cases a Montsouris Institute. *J Urol* 2003;**169**(4):1261–6.
46. Badani KK, Kaul S, Menon M. Evolution of robotic radical prostatectomy: assessment after 2766 procedures. *Cancer* 2007;**110**(9):1951–8.
47. Williams SB, Gu X, Lipsitz SR, et al. Utilization and expense of adjuvant cancer therapies following radical prostatectomy. *Cancer* 2011;**117**(21):4846–54.
48. Anastasiadis AG, Salomon L, Katz R, et al. Radical retropubic versus laparoscopic prostatectomy: a prospective comparison of functional outcome. *Urology* 2003;**62**(2):292–7.
49. Stolzenburg JU, Rabenalt R, Do M, et al. Endoscopic extraperitoneal radical prostatectomy: oncological and functional results after 700 procedures. *J Urol* 2005;**174**(4 Pt 1):1271–5 discussion 1275.
50. Rassweiler J, Stolzenburg J, Sulser T, et al. Laparoscopic radical prostatectomy – the experience of the German Laparoscopic Working Group. *Eur Urol* 2006;**49**(1):113–9.
51. Sonn GA, Sadetsky N, Presti JC, et al. Differing perceptions of quality of life in patients with prostate cancer and their doctors. *J Urol* 2013;**189**(1 Suppl.):S59–65 discussion S65.
52. Hu JC, Gu X, Lipsitz SR, et al. Comparative effectiveness of minimally invasive vs open radical prostatectomy. *JAMA* 2009;**302**(14):1557–64.
53. Webster TM, Herrell SD, Chang SS, et al. Robotic assisted laparoscopic radical prostatectomy versus retropubic radical prostatectomy: a prospective assessment of postoperative pain. *J Urol* 2005;**174**(3):912–4.
54. Bhayani SB, Pavlovich CP, Hsu TS, et al. Prospective comparison of short-term convalescence: laparoscopic radical prostatectomy versus open radical retropubic prostatectomy. *Urology* 2003;**61**(3):612–6.
55. Wood DP, Schulte R, Dunn RL, et al. Short-term health outcome differences between robotic and conventional radical prostatectomy. *Urology* 2007;**70**(5):945–9.
56. Miller J, Smith A, Kouba E, et al. Prospective evaluation of short-term impact and recovery of health related quality of life in men undergoing robotic assisted laparoscopic radical prostatectomy versus open radical prostatectomy. *J Urol* 2007;**178**(3 Pt 1):854–8 discussion 859.
57. Wood DP, Schulte R, Dunn RL, et al. Short-term health outcome differences between robotic and conventional radical prostatectomy. *Urology* 2007;**70**(5):945–9.
58. Lowrance WT, Eastham JA, Savage C, et al. Contemporary open and robotic radical prostatectomy practice patterns among urologists in the United States. *J Urol* 2012;**187**(6):2087–92.
59. Tomaszewski JJ, Matchett JC, Davies BJ, et al. Comparative hospital cost-analysis of open and robotic-assisted radical prostatectomy. *Urology* 2012;**80**(1):126–9.
60. Steinberg PL, Merguerian PA, Bihrle III W, et al. The cost of learning robotic-assisted prostatectomy. *Urology* 2008;**72**(5):1068–72.
61. Lotan Y, Cadeddu JA, Gettman MT. The new economics of radical prostatectomy: cost comparison of open, laparoscopic and robot assisted techniques. *J Urol* 2004;**172**(4 Pt 1):1431–5.
62. Bolenz C, Gupta A, Hotze T, et al. Cost comparison of robotic, laparoscopic, and open radical prostatectomy for prostate cancer. *Eur Urol* 2010;**57**(3):453–8.
63. Schroeck FR, Krupski TL, Stewart SB, et al. Pretreatment expectations of patients undergoing robotic assisted laparoscopic or open retropubic radical prostatectomy. *J Urol* 2012;**187**(3):894–8.
64. Schroeck FR, Krupski TL, Sun L, et al. Satisfaction and regret after open retropubic or robot-assisted laparoscopic radical prostatectomy. *Eur Urol* 2008;**54**(4):785–93.

CHAPTER

# 35

# The Technique of Robotic Nerve-Sparing Prostatectomy

*Firas Abdollah, MD, Mani Menon, MD*

Vattikuti Urology Institute & VUI Center for Outcomes Research Analytics and Evaluation, Henry Ford Hospital, Detroit, MI, USA

## INTRODUCTION

The following chapter will describe the evolution of radical prostatectomy with a special concentration on the technique of robotic-assisted radical prostatectomy (RARP) that was developed and established at the Vattikuti Urology Institute at Henry Ford Hospital in Detroit, Michigan.

## HISTORICAL PERSPECTIVE

Radical prostatectomy was considered a morbid procedure, marred by blood loss, poor visualization, as well as high rates of postoperative urinary incontinence and erectile dysfunction. This changed in 1982, when Walsh and Donker published their technique for a novel anatomic approach to radical prostatectomy.[1] This technique took in consideration periprostatic anatomy and blood supply in order to minimize blood loss and overall morbidity. Moreover, they identified the pelvic plexus as the autonomic innervation of the corpora cavernosum, which is responsible for the neurophysiologic control of erectile function. They described a neurovascular bundle (NVB) that courses in a dorsolateral fashion to the prostate, between the rectum and urethra, and proposed that injury to this structure can result in erectile dysfunction. Walsh proposed that the lateral pelvic fascia should be incised anterior to the NVB, and that lateral pedicles should be divided close to the prostate in order to avoid damaging the NVB.[2] This was the first description of open "nerve-sparing" radical prostatectomy (ORP), which later on became the gold standard technique for treating prostate cancer patients who seek surgery.

## EVOLUTION OF NERVE-SPARING WITH ROBOTIC SURGERY

After the description of nerve-sparing radical prostatectomy by Walsh and Donker, there were many efforts to enhance the understanding of the periprostatic neuroanatomy, with the objective of improving the surgical technique of "nerve-sparing" radical prostatectomy, and consequently improve postoperative functional outcomes. In this context, the introduction of the da Vinci Surgical System (Intuitive Surgical, Sunnyvale, CA) represented a breakthrough, because it allowed for a 12-fold magnification, three-dimensional view, and articulated robotic arm, providing wristed, seven-degree motion. In 2000–2001, Menon et al. established the first robotic urologic program in the world, describing for the first time the Vattikuti Institute Prostatectomy (VIP).[3–5] This procedure underwent several technical improvements over the years (Table 35.1). In the following sections, we will be focusing on the description of the most contemporary VIP technique.

## VATTIKUTI INSTITUTE PROSTATECTOMY

### Indications and Contraindications

Indications for RARP are generally similar to those for ORP. Specifically, patients with low-/intermediate-risk PCa with 10 or more years life expectancy, as well as those with high-risk PCa, are considered good candidates for surgery. It is noteworthy that in the case of high-risk tumors RARP and ORP are considered as the first step in a multimodal treatment approach. However, recent

Prostate Cancer. http://dx.doi.org/10.1016/B978-0-12-800077-9.00035-9

TABLE 35.1 Technical Changes in Vattikuti Institute Prostatectomy Over the Years

| Year | Technique | Benefit |
|---|---|---|
| 2001 | Initial approach to bladder neck | Decrease operating time |
| 2002 | Running anastomosis | Decrease leak, and stricture |
| 2002 | Avoid monopolar cautery after seminal vesical transaction | Not evident |
| 2003 | "Veil" nerve-sparing | Improve postoperative erectile function |
| 2004 | Anterior traction on the bladder to identify bladder neck | Easier transference of skill |
| 2004 | Delayed ligation of the dorsal venous complex | More precise urethral transaction, decrease positive apical margins |
| 2004 | Not opening the endopelvic fascia | Earlier recovery of urinary continence |
| 2005 | Fully athermal nerve-sparing technique | Earlier recovery of urinary continence |
| 2007 | Double-layer anastomosis | Unchanged continence, decrease urinary leak |
| 2008 | Use of percutaneous suprapubic tube instead of urinary catheter | Decrease patient discomfort, and earlier recovery of urinary continence |
| 2008 | Primary hypogastric node dissection for low- to intermediate-risk disease | Increased node positivity |
| 2010 | Barbed anastomotic suture | Decrease anastomotic time |
| 2011 | Modified organ retrieval for examination (MORE) using GelPoint access | Palpation-oriented frozen section biopsies to assess surgical margins |

reports[6,7] showed that up to about 50% of patients can benefit from complete cancer control with surgery alone.

During RARP, patients are exposed to pneumoperitoneum and steep Trendelenberg position, which might create difficulties in ventilation. For this reason, obstructive pulmonary disease and cardiac output abnormalities are considered as relative contraindications. Previous abdominal surgery might increase the complexity of the case, but should not be considered as contraindication. Factors such as morbid obesity, large prostate, large median lobe, and surgery in the salvage setting may present difficulties and are better reserved for surgeons with high expertise.[8]

## Preoperative Workup

These might vary according to local guidelines and specific hospital policy. In general, anticoagulant and antiplatelet drugs are ceased before surgery, while prophylactic antibiotics and thromboprophylaxis are started. A laxative is given the night before surgery. In case of patients with aggressive tumor characteristics, a mechanical bowel preparation should be used. This would allow a clean environment, in case of rectal opening due to rectal involvement.

## Patient Positioning and Preparation

The patient is placed in the lithotomy position with arms secured to the sides and all pressure points protected using foam pads. The patient is fastened to the table with tape and straps. The thumbnails should face the ceiling and the arms held lax to avoid peripheral nerve compression. The legs are separated in flexion and abduction to allow the patient sidecart sufficient access. Sequential compression devices are placed on the calves. The abdomen down to the upper thighs is prepped with antiseptic and then draped. The bladder is drained with an 18F Foley catheter that is secured to the side so that the assistant can manipulate it during the operation. Intravenous fluids are limited to <1 L during surgery to reduce excessive urine production, which can obscure the view and necessitate copious suctioning.

## Instrumentation

1. Nondisposable
   a. Monopolar hook
   b. Fenestrated bipolar forceps
   c. Round tip scissors
   d. Large needle driver
   e. 8-mm instrument cannula
   f. GORE_ Suture Passer (WL Gore & Associates, Flagstaff, AZ)
2. Disposable
   a. 12 × 100 mm and 5 × 100 mm laparoscopic ports (Kii Fios First Entry port, Applied Medical, Rancho Santa Margarita, CA)
   b. Endopath® Veress needle (Ethicon Endo-Surgery, Blue Ash, OH)
   c. StrykeFlow 2 suction irrigator (Stryker, Kalamazoo, MI)
   d. Hem-o-Lok clips (Weck, Teleflex Medical, NC)

e. 3-0 barbed polyglyconate suture (V-Loc®, Covidien, Mansfield, MA)
f. No. 15 scalpel blade,
g. 10 mm Endopouch® bag (Ethicon Endo-Surgery, Blue Ash, OH)
h. Rutner 14F percutaneous suprapubic catheter (Cook Medical, Bloomington, IN)
i. Polypropylene CT-1 needle (Ethicon, Somerville, NJ)
j. 1 PDS™ CT-1 needle (Ethicon, Somerville, NJ)
k. 4-0 Monocryl™ suture (Ethicon, Somerville, NJ)
l. Polypropylene button (Ethicon Endo-Surgery, Blue Ash, OH)
m. Mastisol® adhesive (Eloquest, Ferndale, MI)
n. Steri-Strip™ (3M, St Paul, MN)

3. Optional
   a. Monopolar curved scissors
   b. Maryland bipolar forceps
   c. Harmonic™ACE curved shears
   d. 12 × 130 mm balloon port for camera (Kii Balloon Blunt Tip System, Applied Medical, Rancho Santa Margarita, CA)
   e. GelPoint Advanced Access Platform (Applied Medical, Rancho Santa Margarita, CA)

## Surgical Steps

### *Port Placement*

The procedure is performed with two bedside assistants (Figure 35.1). The table is set to a steep Trendelenberg

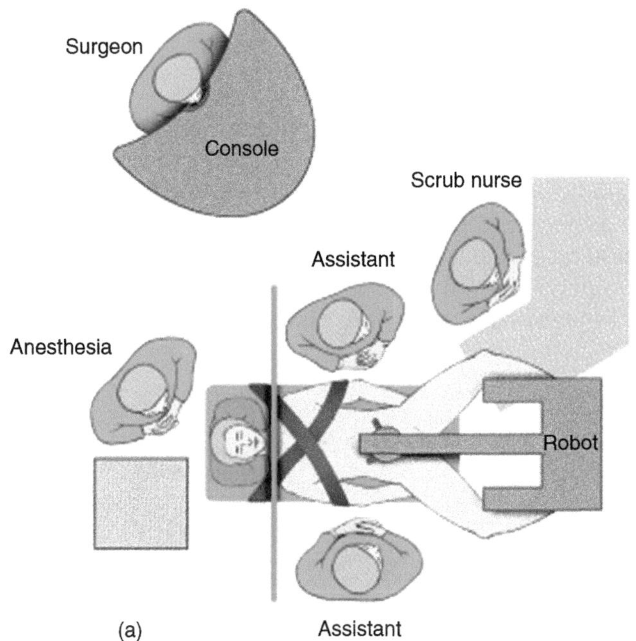

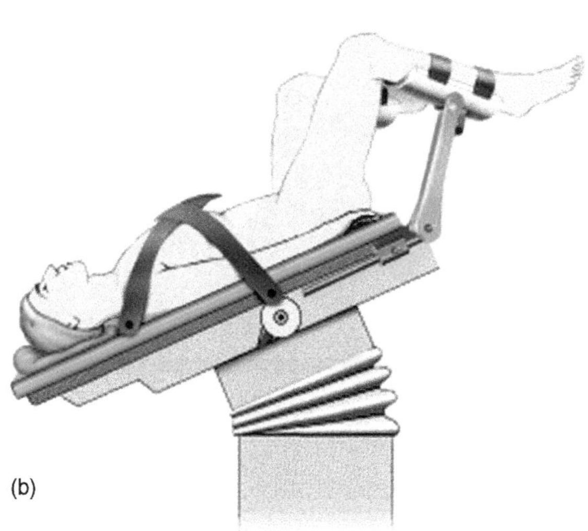


FIGURE 35.1 **Operating room setup for Vattikuti Institute prostatectomy.** *Reprinted by permission of John Wiley & Sons, Inc.; adapted from Ref. [9].*

position, and a Veress needle is used to establish pneumoperitoneum. The Veress needle is then replaced with a 12-mm port, which is used to pass a 30-degree robotic laparoscope that is placed periumbilically. This access is also used at the end of the procedure for specimen retrieval. Initially, the intraperiotoneum pressure is set to 20 mmHg for port placement, and then it is decreased to 15 mmHg for the rest of the operation. The 30-degree robotic laparoscope is angled upward to visualize the anterior abdominal wall, and all other ports are inserted under vision. This minimizes the risk of visceral injury. In addition to the camera port, two robotic 8-mm ports are placed on each side approximately 3–5 cm below the umbilicus and lateral to the rectus. A right-side 12-mm assistant port is inserted in the midaxillary line, 2–3 cm cranial to the anterior superior iliac spine. A distance of about 8 cm should be kept between these ports. Next, a 5-mm port for the sucker is inserted in the space between the right robotic arm, and camera port, but cranially by 3–5 cm. Another 5-mm port is placed on the left side in a position similar to that of the right-side 12-mm port. In total, six trocars are placed, three for robotic arms, and three for assistants (Figure 35.2).

To adjust for patient height, the camera port should be placed infraumbilically in patients ≥6 ft., and supraumbilically in those shorter than 6 ft. The assistant 12-mm port allows for needles and clips to be passed easily. It also allows the removal of small tissue specimens, such as lymph nodes. In obese patients, a longer balloon port for the camera can prevent the port coming out during the procedure. Likewise, bariatric robotic ports should be considered, when operating on morbidly obese patients. Moreover, it is important to avoid inserting trocars through adhesions, where bowel can be injured inadvertently. If necessary, adhesions can be managed laparoscopically before trocar placement. If the procedure is to be performed with only one bedside assistant, then the left-side 5-mm assistant port should be replaced with a robotic 8-mm port for the fourth robotic arm. To avoid instrument clashing, there should be sufficient space between these ports. Likewise, when using the Harmonic ACE curved shears, the right-side robotic port should be placed 2–3 cm more caudal than normal to compensate for the shorter reach of this instrument.

### *Release of Bowel*

With the patient in deep Trendelenburg position, the small bowel falls away from the pelvis. However, the sigmoid colon on the left, and the cecum on the right can be adherent to the posterior and lateral peritoneum. In these cases, the large bowel has to be mobilized to facilitate pelvic lymph node dissection. In the case of dense adhesions, we prefer sharp dissection without energy.

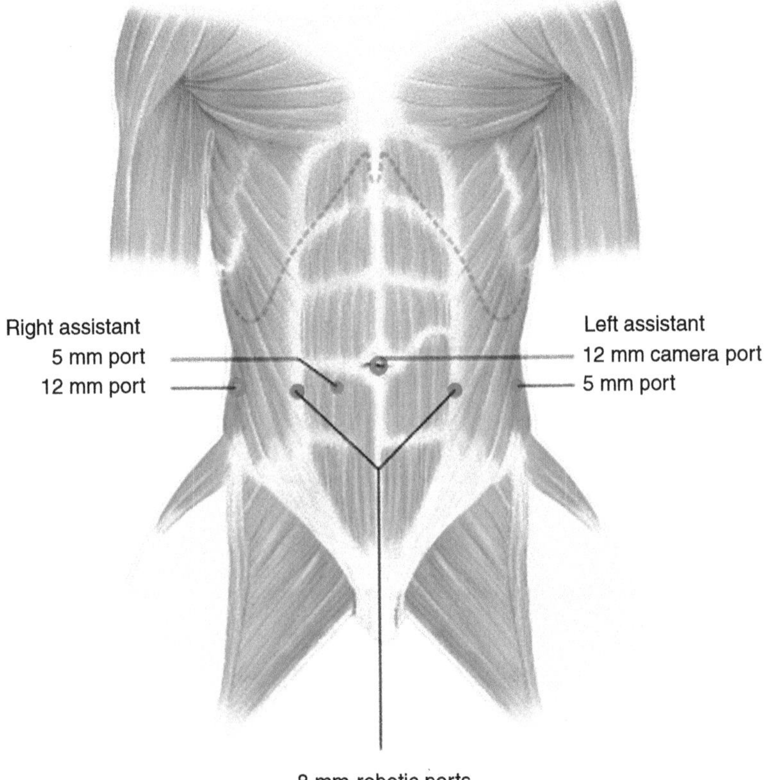


**FIGURE 35.2 Trocar placement for Vattikuti Institute prostatectomy.** *Reprinted by permission of John Wiley & Sons, Inc.; adapted from Ref. [9].*

### *Development of Extraperitoneal Space (Bladder Mobilization)*

After adequate mobilization of the large bowel, the extraperitoneal space is entered by making an inverted U-shape incision on the anterior peritoneum (Figure 35.3). Briefly, while the assistant retracts the medial umbilical ligament toward the ipsilateral shoulder, an incision is made lateral to this ligament, and medial to the internal inguinal ring and extended down to the vasa. This maneuver should be performed carefully, without injuring the epigastric vessel that passes nearby. The space of Retzius is entered, and dissection proceeds in the loose areolar tissue. The vertical limb of the incision is extended inferiorly until the pubic bone is seen. This maneuver must be performed bilaterally. The two vertical incisions are then joined by a transverse incision between the two umbilical ligaments, high enough to avoid the bladder. The resulted inverted U-shape incision releases the urachus and drops the bladder off the anterior abdominal wall. Now, the whitish fibers of the transversalis fascia come into view. The bladder is mobilized caudally by following the contour of the abdominal wall inferiorly until the landmark of the pubic bone is visualized. Dissection is continued on either side to expose the endopelvic fascia.

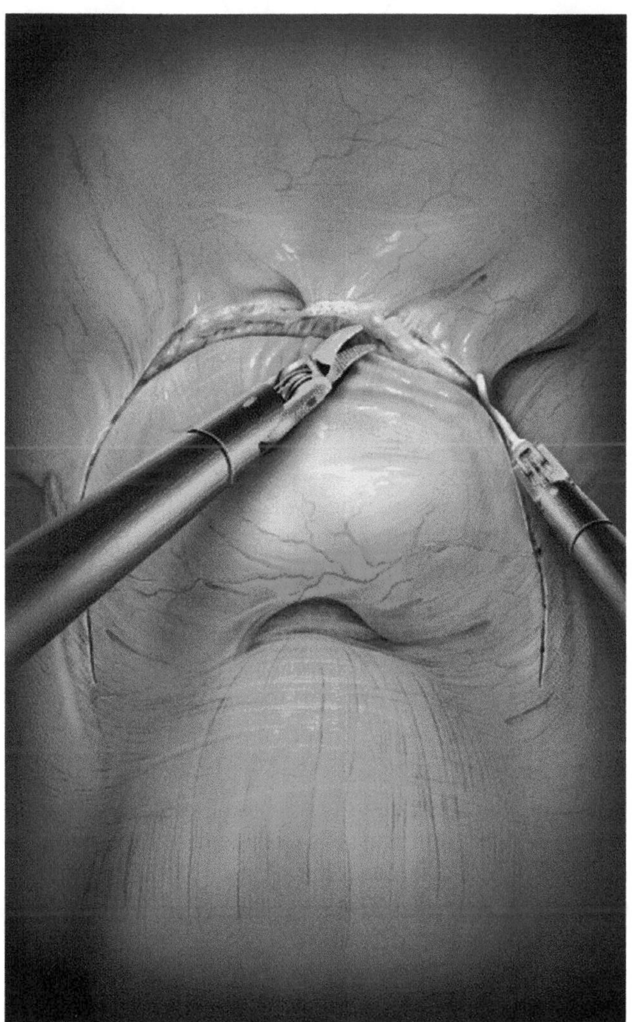

**FIGURE 35.3** **Access to the space of Retzius.** *Reprinted by permission of John Wiley & Sons, Inc.; adapted from Ref. [9].*

During the previously described procedure, a 0° lens can be used. However, we prefer a 30-degree lens for this part, and switch to 30° looking down, once the bladder is down. This offers two advantages. First, the 30° down lens provides a better perception of depth, which facilitate the dissection around the vessels. Second, the risk of smudging the lens is lower.

### *Division of the Bladder Neck*

The pubic bone is cleared of all fat and connective tissue. Here, dissection is performed lateral to medial to avoid avulsion of the superficial dorsal vein. The latter is coagulated using bipolar cautery, after which the fat over the prostate is peeled off. Occasionally, an accessory pudendal artery may be present, and one should be careful not to damage it. We do not divide the puboprostatic ligaments.

In the next step, the bladder neck is approached without opening the endopelvic fascia or ligating the dorsal venous complex (DVC). Later on, this allows for precise suturing of the DVC, after the division of the urethra. At this point, the Foley balloon is deflated, while the catheter is kept inside the bladder. The anterior wall of the bladder is grasped at the midline and lifted upward by the right-side assistant. This maneuver pulls the bladder away from the prostate and helps in identifying the bladder neck. Alternatively, the balloon can be kept inflated, and the division between the bladder and prostate can be identified by moving the catheter forward and backward. The anterior bladder neck is then incised at 12-o'clock position for an approximate length of 1 cm (Figure 35.4). The incision is deepened to reveal the catheter, which is grasped by the tip, taken out of the bladder, and given to the left-side assistant, who provides firm traction by placing it on the anterior abdominal wall. The remaining lateral fibers attaching the bladder to the prostate are now divided, and the electrocautery is used to dissect the posterior lip of the bladder neck (Figure 35.5). During this process, the nearby ureteral orifices should be identified and avoided. Excessive bleeding can be a sign of being too close to the capsule of the prostate. In this case, the plane of incision should be altered.

### *Posterior Dissection of Vas Deferens, Seminal Vesicles, and Denonvilliers' Fascia*

Dissection is now continued underneath the prostate until the anterior layer of Denonvilliers' fascia is identified (Figure 35.6). This fascia is then incised, which exposes the vasa and seminal vesicles. Care is taken

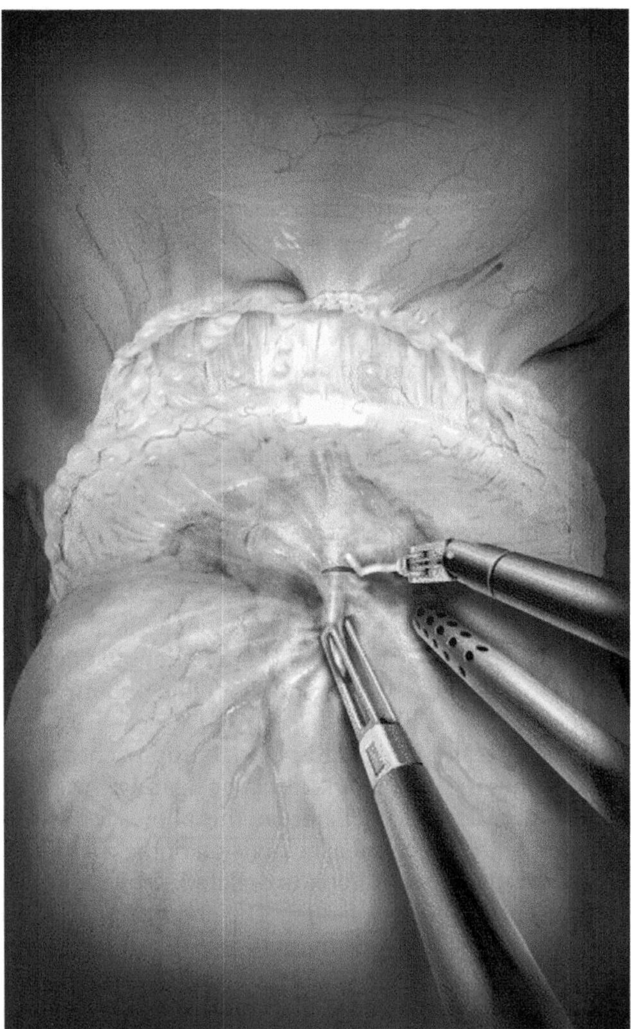

FIGURE 35.4 **Anterior bladder neck dissection.** *Reprinted by permission of John Wiley & Sons, Inc.; adapted from Ref. [9].*

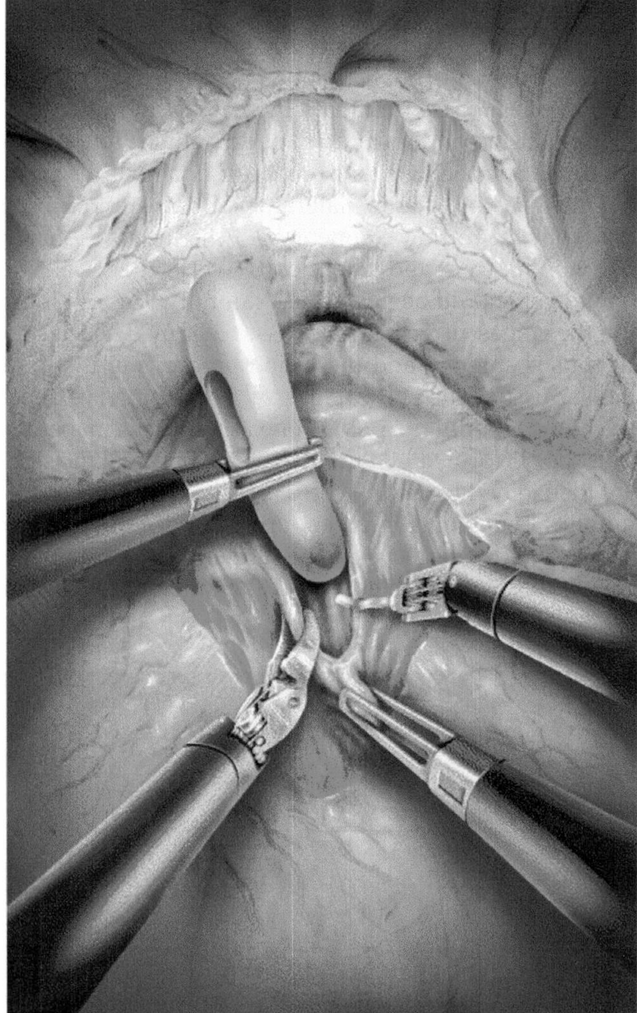

FIGURE 35.5 **Posterior bladder neck dissection.** *Reprinted by permission of John Wiley & Sons, Inc.; adapted from Ref. [9].*

to carefully dissect the vasa, and the artery to the vas deferens is controlled. The vasa are transected and the cranial stump held by the right-side assistant, while the proximal end is lifted upward by the left-side assistant (Figure 35.7). This helps with dissection of the seminal vesicles. The seminal vesicle artery is controlled with fine bipolar coagulation or clips. Limited use of cautery is preferred in this zone, as the NVB are near this location. After the seminal vesicles are released, they are held upward by each assistant, so that the posterior prostate is retracted, which reveals the longitudinal fibers of the posterior Denonvilliers' fascia (Figure 35.8). A transverse incision is made in this fascia into the avascular plane between the perirectal fat and posterior prostate. At this point, a blunt dissection is used to push the rectum away from the prostate. Dissection is continued distally toward the apex and laterally. Recently, we have replaced the monopolar hook with the Harmonic ACE curved shears for dissection of the seminal vesicles and performing nerve sparing (Veil of Aphrodite). This accounts for a subjective improvement in visibility and better hemostasis, with less energy spread to surrounding tissues. In patients with low-risk disease where potency is a priority, the tips of the seminal vesicles might be intentionally preserved provided that the intraoperative frozen section of the margins are negative for tumor.

### *Nerve Sparing*

While the contralateral assistant keeps the seminal vesicle in traction, prostatic pedicles are identified, and divided between Hem-o-lok clips. Bipolar coagulation can be used to control individual vessels (Figure 35.9). Nerve sparing may then be performed using one of the two following methods:

1. *Standard nerve sparing*: In this technique, the plan of dissection is in between the prostatic fascia and the levator fascia. It begins by incising the lateral

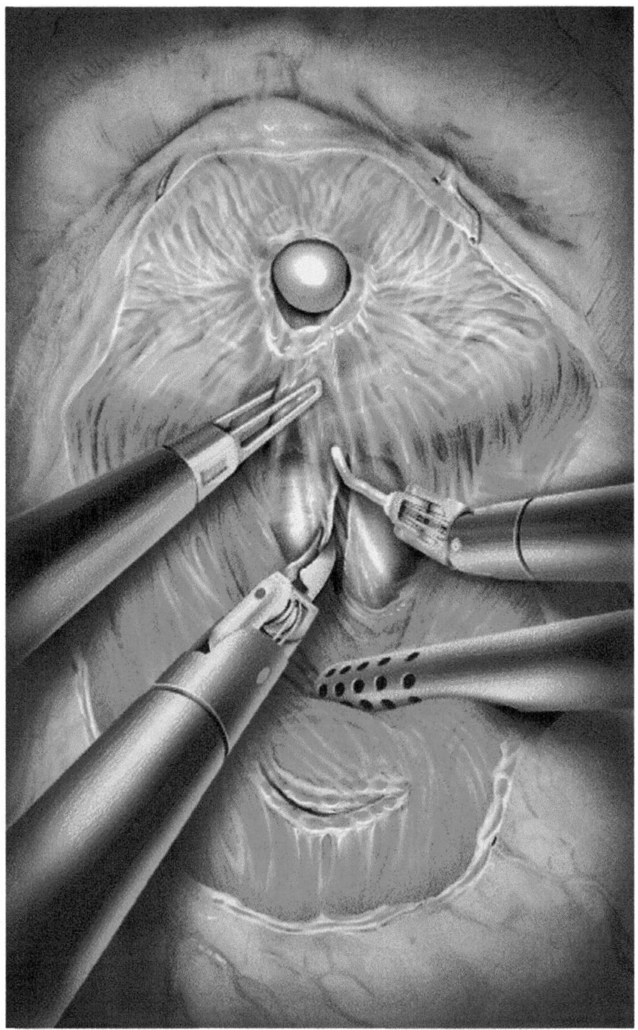

FIGURE 35.6 **Anterior Denonvilliers fascia incision.** *Reprinted by permission of John Wiley & Sons, Inc.; adapted from Ref. [9].*

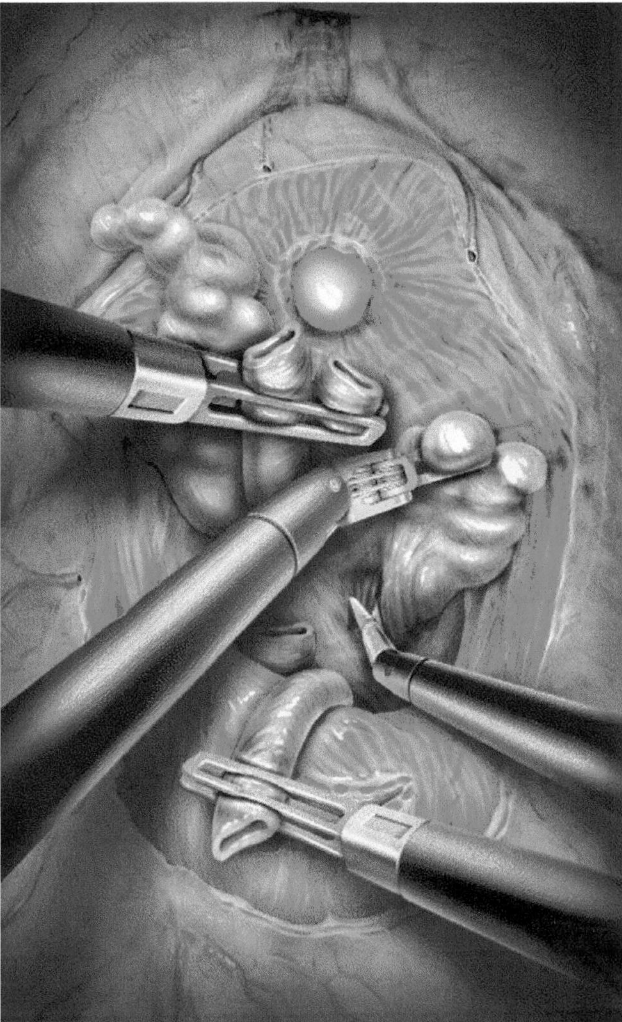

FIGURE 35.7 **Vas deferens, seminal vesicles dissection.** *Reprinted by permission of John Wiley & Sons, Inc.; adapted from Ref. [9].*

prostatic fascia anteriorly around the midprostate, and then proceeds by using cold scissors and bipolar coagulation as necessary. The posterolateral bundle opens up, and dissection is carried to the apex while staying close to the prostate capsule to prevent damage to the bundle. Smaller penetrating vessels can be controlled with fine bipolar coagulation or clips.

2. *"Veil of Aphrodite" or high anterior release*: This begins by developing a space between the prostatic capsule and fascia, starting at the base of the seminal vesicles. The curved shears of the Harmonic ACE can be particularly helpful for this procedure. The plane can be developed by slowly pushing the tissues apart, deep to the venous sinuses. On the right side, interfascial dissection is performed between 1 o'clock and 5 o'clock. Symmetrically, it is performed between 6 o'clock and 11 o'clock on the left. Dissection continues until the entire prostatic fascia up to the pubourethral ligament is mobilized (Figure 35.10). If in the correct plane, the dissection is mostly bloodless except where the fascia fuses with the puboprostatic ligament. At the end, curtains of periprostatic tissue should hang from the pubourethral ligament; this structure is named "the Veil of Aphrodite".

Generally, the Harmonic ACE is used when performing a veil procedure. This allows an easier access to the interfascial plane. Furthermore, the DVC can be ligated without suturing. One drawback with the Harmonic, however, is that it lacks wristed articulation. Regardless of which instrument is used, it is important to avoid excessive traction on the neurovascular bundles during nerve sparing.

### *Apical Dissection, Urethral Transection, and Control of the DVC*

A 0° lens can be used at this point. The right assistant retracts the anterior prostate so that the DVC and

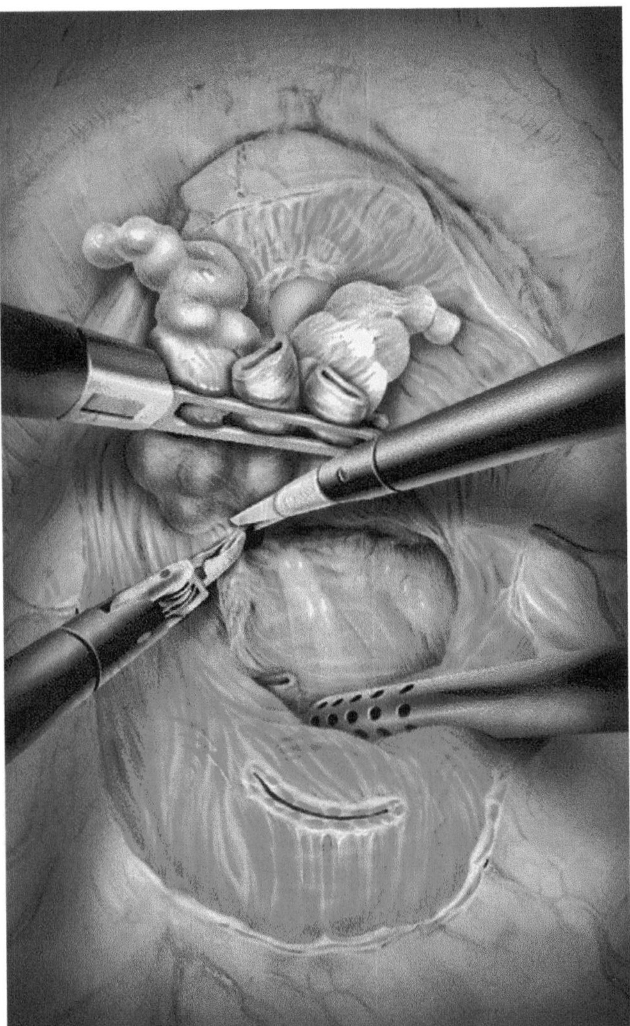

**FIGURE 35.8 Posterior Denonvilliers fascia incision.** *Reprinted by permission of John Wiley & Sons, Inc.; adapted from Ref. [9].*

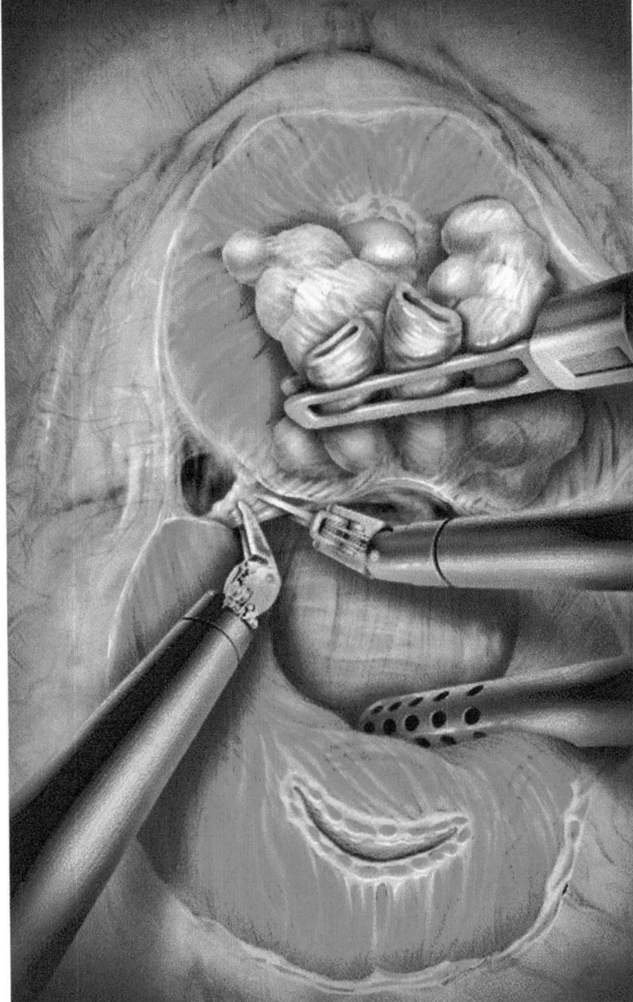

**FIGURE 35.9 Prostate pedicles dissection.** *Reprinted by permission of John Wiley & Sons, Inc.; adapted from Ref. [9].*

urethra are taut. The next step is to divide the puboprostatic ligament and DVC using cautery or the harmonic instrument (Figure 35.11). After the apex is freed circumferentially, the urethra is dissected into the prostatic notch and transected anteriorly a few millimeters distal to this (Figure 35.12). This will expose the urethral catheter, which should be withdrawn so that the posterior urethral wall is visible. The latter is then cut to complete the urethral transection. It is noteworthy that at this point, the prostate may still be attached to the perirectal tissue by fascia. In such a case, it should be divided carefully with the left robotic instrument protecting the rectum. The freed prostate is placed in a 10-mm Endopouch bag.

Given the close intimacy of the prostatic apex with the rectum, apical dissection might be considered the part of RARP with the highest risk of rectal injury. In this context, it might be useful to attempt the anterior apical dissection, only after a complete posterior dissection all the way to the apex. Most of the time, the high pressure created by the pneumoperitoneum does not allow a significant bleeding from the DVC during apical dissection. For this reason, the ligation of DVC with running suture (2-0 V-Loc, RB-1 needle) is usually delayed till the prostate is totally free. At this point, perineal pressure can be applied to help identifying all bleeding sinuses. However, if for any reason bleeding during apical dissection is profuse, the DVC should be controlled early so that oncologic principles are not compromised.

### *Pelvic Lymphadenectomy*

In 2008, we changed our lymph node dissection scheme in patients with low-/intermediate-risk tumors. Specifically, instead of a limited external iliac and obturator node dissection, we shifted to an internal iliac node dissection, which seems to significantly increase the retrieval of positive lymph nodes.[5] On the other hand, an anatomically extended pelvic lymph node dissection is always performed in patients with high-risk tumors. The

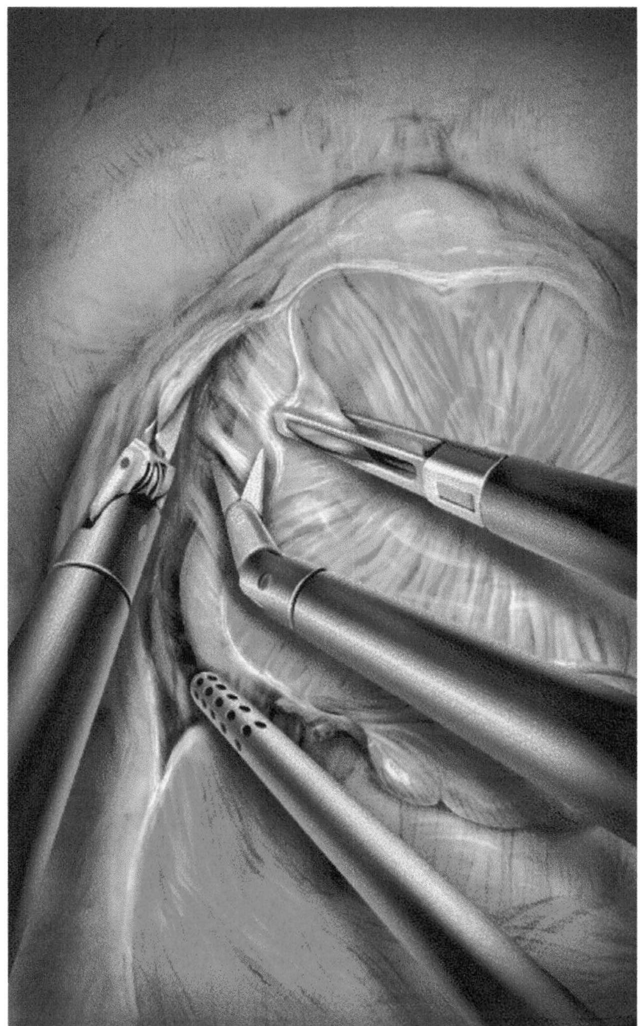

FIGURE 35.10 **Preserving the lateral prostate fascia (Veil of Aphrodite).** *Reprinted by permission of John Wiley & Sons, Inc.; adapted from Ref. [9].*

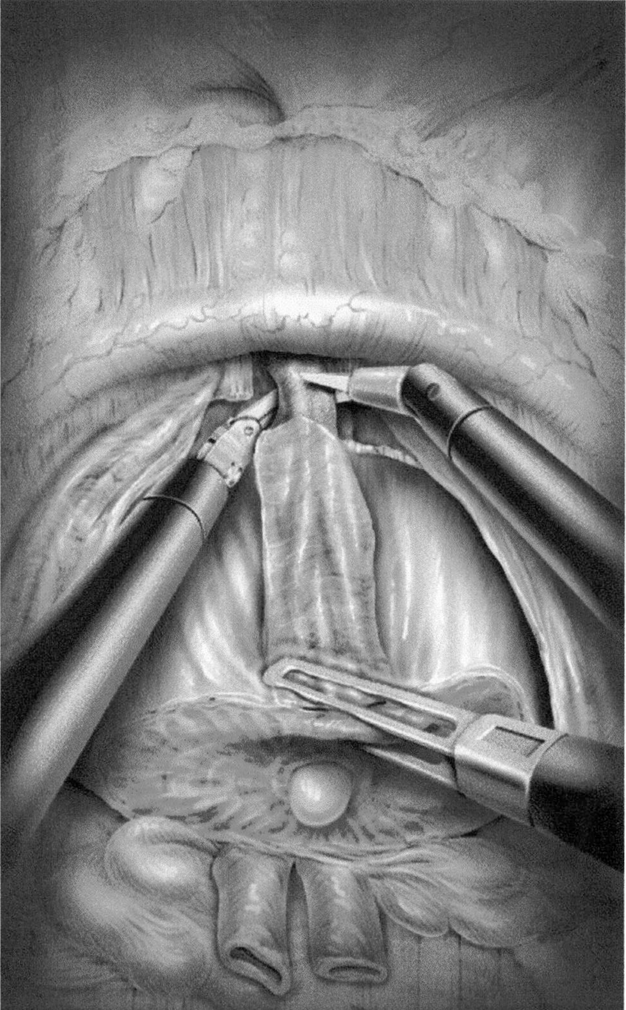

FIGURE 35.11 **Apical dissection and control of the DVC.** *Reprinted by permission of John Wiley & Sons, Inc.; adapted from Ref. [9].*

latter include external iliac, obturator, and internal iliac nodes. The vasa are divided bilaterally. On each side, the ureter is identified on its entrance to the pelvis, where it crosses the bifurcation of the iliac vessels, and then continues its course distally under the superior vesical artery. The lymph node dissection should be distal to this artery and medial to the external iliac vein, so that the ureter is protected. Tissue around the obturator nerve and along the lateral pelvic wall is dissected to clear the obturator fossa. This procedure should be performed meticulously to avoid damage to the obturator nerve. Finally, all tissue from the obturator fossa to the lateral aspect of the bladder should also be cleared.

### *Posterior Reconstruction*

Before proceeding with the urethrovescial anastomosis, a posterior reconstruction of Denonvilliers' fascia and the rhabdosphincter should be undertaken. This facilitates the anastomosis and reduces any tension on it. First, a barbed V-Loc suture (3-0, RB-1 needle, 6 inches) is anchored by passing it through the posterior layer of Denonvilliers' fascia and the far lateral aspect of the posterior bladder neck. Then the suture is passed to the posterior rhabdosphincter and back to the posterior bladder neck, incorporating Denonvilliers' fascia with each throw. A total of four passes are performed from the right side to the left side. After each throw, the suture should be cinched. A correctly performed posterior construction should put the bladder neck and urethral stump in a close proximity.

### *Urethrovesical Anastomosis (UVA)*

We were the first to report about the safety and feasibility of using a barbed polyglyconate suture for the urethrovesical anastomosis (UVA) during RARP.[10] Two 3-0 V-Locs (6-inches, RB-1 needle) are made into a double-armed stitch with a total of 11″ of barbs, or 14″ in larger bladder

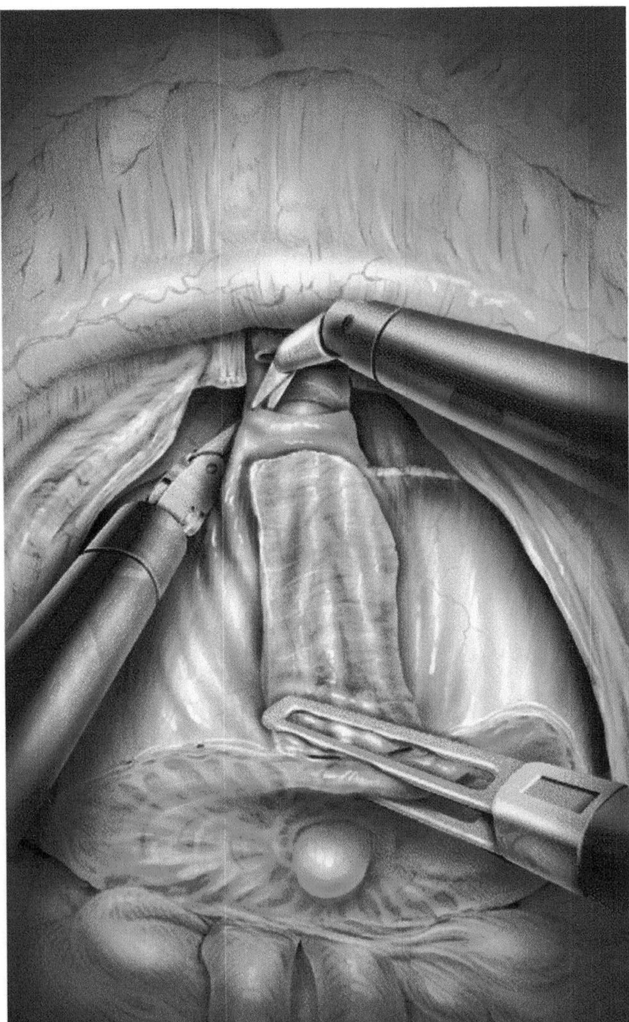

**FIGURE 35.12** **Urethral dissection.** *Reprinted by permission of John Wiley & Sons, Inc.; adapted from Ref. [9].*

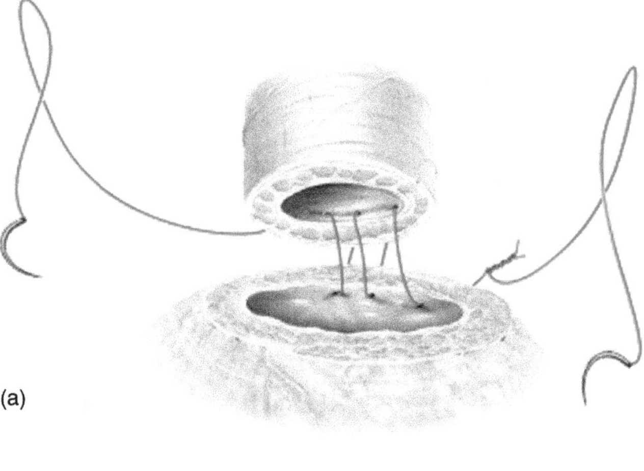


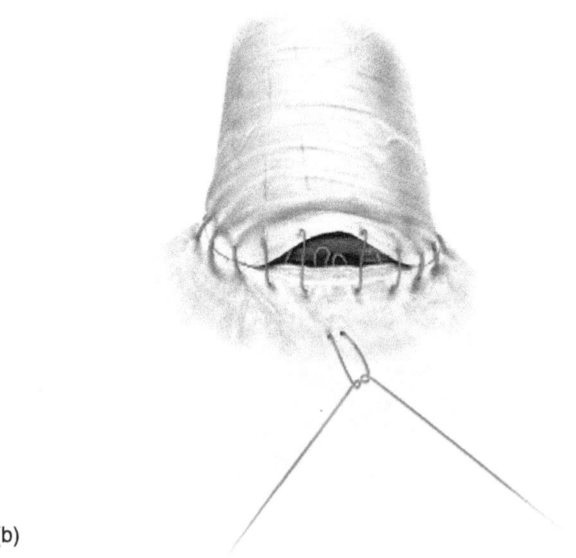


**FIGURE 35.13** **Urethrovesical anastomosis.** *Reprinted by permission of John Wiley & Sons, Inc.; adapted from Ref. [9].*

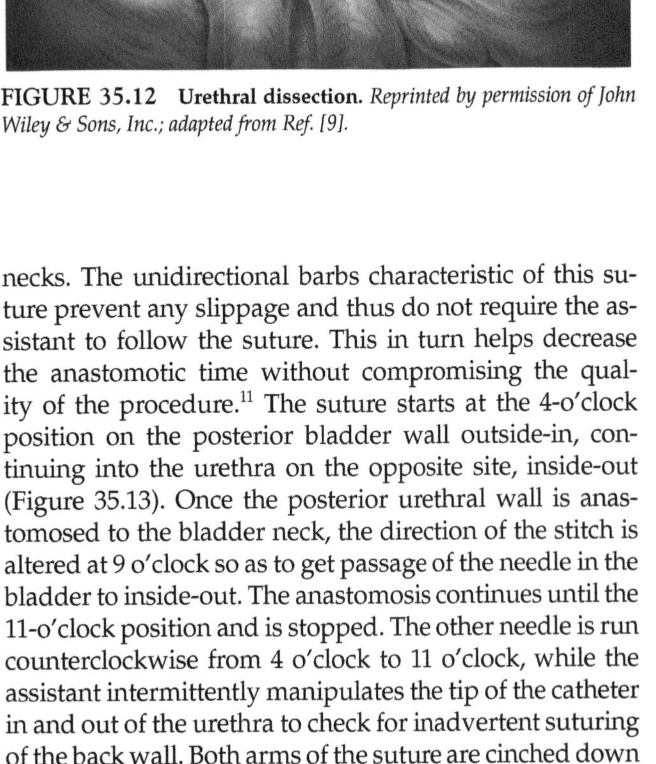

necks. The unidirectional barbs characteristic of this suture prevent any slippage and thus do not require the assistant to follow the suture. This in turn helps decrease the anastomotic time without compromising the quality of the procedure.[11] The suture starts at the 4-o'clock position on the posterior bladder wall outside-in, continuing into the urethra on the opposite site, inside-out (Figure 35.13). Once the posterior urethral wall is anastomosed to the bladder neck, the direction of the stitch is altered at 9 o'clock so as to get passage of the needle in the bladder to inside-out. The anastomosis continues until the 11-o'clock position and is stopped. The other needle is run counterclockwise from 4 o'clock to 11 o'clock, while the assistant intermittently manipulates the tip of the catheter in and out of the urethra to check for inadvertent suturing of the back wall. Both arms of the suture are cinched down at 11 o'clock to complete the anastomosis; the needles are cut without tying knots. To test for leaking, the bladder is filled with 250 mL of saline via the urethral catheter.

### *Percutaneous Suprapubic Tube (PST) Bladder Drainage*

Currently, our standard of care is to place a percutaneous suprapubic tube (PST) for bladder drainage, rather than a urethral catheter. In our experience, this appears to be associated with greater patient comfort and continence recovery, without an increase in the urethral stricture rate.[12] First, the bladder should be distended, and a horizontal mattress suture (1-0 prolene, CT-1 needle) is placed through the full thickness of the anterior bladder wall. This serves to hold the bladder to the abdominal wall and then brought outside to the abdominal skin using a GORE suture passer. Under visualization, a 14F Rutner suprapubic catheter is placed percutaneously in the midline approximately one-third of the distance between the pubic symphysis to the umbilicus. Upward traction on the bladder sutures with robotic needle drivers tents the bladder wall, allowing the assistant to insert the needle obturator through the bladder between the sutures. Immediate drainage of fluid implies a correct positioning of the catheter. The PST is secured by inflating its balloon with 4 mL of sterile water.

### *Specimen Retrieval and Closure*

The bagged specimen is retrieved by extending the camera port incision. Thereafter, the pneumoperitoneum is evacuated, the robot is dedocked, and the patient is moved to the supine position. At this point, the external sutures of the PST are pulled and fixed to the skin over a sterile plastic button. This allows bringing the cystotomy toward the anterior abdominal wall. The tails of the suture are wrapped around the tube with adhesive (Mastisol) and Steri-Strip. Fascia closure is performed with two interrupted 1 PDS sutures in a figure-of-eight. Skin closure is performed with subcuticular stitches.

### *Modified Organ Retrieval for Examination (MORE)*

Lack of tactile sensation during RARP preclude the possibility of palpating the prostate during the procedure, and this might increase surgical margin rates.[13,14] To overcome this drawback, in early 2011, we incorporated the GelPoint access port with the VIP technique.

This has allowed us to modify retrieval of the whole prostate at an earlier stage of the operation for pathologic examination (modified organ retrieval for examination). The prostate can be removed without dedocking the robot or losing pneumoperitoneum. Once the prostate is bagged, it is withdrawn out of the GelPoint. Thereafter, the prostate is palpated bimanually, and frozen section biopsy samples from suspicious areas are taken. Lymphadenectomy is performed while the biopsy specimens are assessed; if positive, more tissue may be removed to achieve negative margins.

The GelPoint device is placed periumbilical with a preplaced 12-mm camera and 10-mm additional port for prostate extraction on the wound retractor ring. All other ports are placed in the same manner as previously described.

## Postoperative Care

In the recovery room, a saline bolus is administered. Intravenous ketorolac and oral acetaminophen with codeine can be used for the management of pain. Morphine might slow bowel motion considerably, and lead to ileus. For this reason, it is better to avoid the use of morphine for pain control in these patients. Diet can be started immediately with clear liquid, and then advanced to soft diet the day after. Early mobilization within 6 h is encouraged to encourage bowl motion and minimize the risk of thromboembolic complications. Blood tests are not performed routinely. Patients are routinely discharged within 24 h. Patients with a PST are instructed to clamp the catheter on postoperative day 5 and record postvoid residuals. If <50 mL per void, the catheter can be removed on postoperative day 7.

## Management of Intraoperative Complications

The main intraoperative complications consist of bleeding and injury to adjacent organs. Small-bowel injury may occur during adhesiolysis and/or as a result of instrument passage. This can be closed robotically in two layers. Most rectal injuries occur posterolaterally, close to the apex. The risk of these injuries is higher in patients with locally advanced tumors, and/or in salvage prostatectomy, after radiation therapy or cryotherapy. If dissection is difficult, sharp dissection is preferred. Small rectal tears can be close in two lawyers using a barbed suture. If the rectal suture and the urethrovesical anastomosis are watertight, a colostomy is not necessary. After instituting broad-spectrum intravenous antibiotics and a clear liquid diet, the majority of patients can be discharged within 72 h.[15] Ureteral injury is a rare complication and may occur during extended PLND or during posterior dissection. It can be repaired primarily or with a ureteral reimplantation if very low; both can be performed robotically.

## Management of Postoperative Complications

Generally, these are similar to ORP postoperative complication, but are less frequent.[16–18] Deep venous thrombosis should be managed with anticoagulation. In case of symptomatic lymphocele, which is more frequent after extended PLND,[19] the patient may present within the first few weeks with fever, pain, and/or leg swelling. The lymphocele should be drained by placing a percutaneous tube. Bladder neck contractures are rare and can usually be managed with office dilation. Incisional hernias are also a rare occurrence. The use of dilating trocars may decrease the incidence of port site hernias.

# CONCLUSIONS

In this chapter, we described the evolvement, as well as the most contemporary technique to perform a "nerve-sparing" RARP in our center. In our hands, this technique decreased dramatically blood loss and transfusion in comparison to ORP. This advantage, in addition to the magnified three-dimensional view, and the articulated robot arm movement allow surgeons to perform an elegant and meticulous procedure. This might result in more favorable postoperative functional outcomes, without compromising cancer control outcomes. Moreover, the minimally invasive nature of RARP implies less morbidity to the patients, which frequently translate into less peri- and postoperative complications.

## References


1. Walsh PC, Donker PJ. Impotence following radical prostatectomy: insight into etiology and prevention. *J Urol* 1982;**128**:492–7.
2. Walsh PC, Mostwin JL. Radical prostatectomy and cystoprostatectomy with preservation of potency. Results using a new nerve-sparing technique. *Br J Urol* 1984;**56**:694–7.
3. Menon M, Tewari A, Peabody J. Team VIP. Vattikuti Institute prostatectomy: technique. *J Urol* 2003;**169**:2289–92.
4. Menon M, Tewari A, Baize B, et al. Prospective comparison of radical retropubic prostatectomy and robot-assisted anatomic prostatectomy: the Vattikuti Urology Institute experience. *Urology* 2002;**60**:864–8.
5. Menon M, Shrivastava A, Bhandari M, et al. Vattikuti Institute prostatectomy: technical modifications in 2009. *Eur Urol* 2009;**56**:89–96.
6. Briganti A, Joniau S, Gontero P, et al. Identifying the best candidate for radical prostatectomy among patients with high-risk prostate cancer. *Eur Urol* 2012;**61**:584–92.
7. Yuh B, Artibani W, Heidenreich A, et al. The role of robot-assisted radical prostatectomy and pelvic lymph node dissection in the management of high-risk prostate cancer: a systematic review. *Eur Urol* 2014;**65**:918–27.
8. Montorsi F, Wilson TG, Rosen RC, et al. Best practices in robot-assisted radical prostatectomy: recommendations of the Pasadena Consensus Panel. *Eur Urol* 2012;**62**:368–81.
9. Shrivastava A, Baliga M, Menon M. The Vattikuti Institute prostatectomy. *BJU Int* 2007;**99**(5):1173–89.
10. Kaul S, Sammon J, Bhandari A, et al. A novel method of urethrovesical anastomosis during robot-assisted radical prostatectomy using a unidirectional barbed wound closure device: feasibility study and early outcomes in 51 patients. *J Endourol* 2010;**24**:1789–93.
11. Sammon J, Kim TK, Trinh QD, et al. Anastomosis during robot-assisted radical prostatectomy: randomized controlled trial comparing barbed and standard monofilament suture. *Urology* 2011;**78**:572–9.
12. Sammon JD, Trinh QD, Sukumar S, et al. Long-term follow-up of patients undergoing percutaneous suprapubic tube drainage after robot-assisted radical prostatectomy (RARP). *BJU Int* 2012;**110**:580–5.
13. Lepor H. A review of surgical techniques for radical prostatectomy. *Rev Urol* 2005;**7**(Suppl. 2):S11–7.
14. Rocco B, Djavan B. Robotic prostatectomy: facts or fiction? *Lancet* 2007;**369**:723–4.
15. Kheterpal E, Bhandari A, Siddiqui S, et al. Management of rectal injury during robotic radical prostatectomy. *Urology* 2011;**77**:976–9.
16. Hu JC, Wang Q, Pashos CL, et al. Utilization and outcomes of minimally invasive radical prostatectomy. *J Clin Oncol* 2008;**26**:2278–84.
17. Kowalczyk KJ, Levy JM, Caplan CF, et al. Temporal national trends of minimally invasive and retropubic radical prostatectomy outcomes from 2003 to 2007: results from the 100% Medicare sample. *Eur Urol* 2012;**61**:803–9.
18. Trinh QD, Sammon J, Sun M, et al. Perioperative outcomes of robot-assisted radical prostatectomy compared with open radical prostatectomy: results from the nationwide inpatient sample. *Eur Urol* 2012;**61**:679–85.
19. Abdollah F, Suardi N, Gallina A, et al. Extended pelvic lymph node dissection in prostate cancer: a 20-year audit in a single center. *Ann Oncol* 2013;**24**:1459–66.

CHAPTER

# 36

# Anterior Approach to Robotic Radical Prostatectomy

*Andrew C. Harbin, MD, Daniel D. Eun, MD*

**Department of Urology, Temple University Hospital, Philadelphia, PA, USA**

## INTRODUCTION

Robot-assisted radical prostatectomy (RARP) has become a common approach to surgical management of prostate cancer (PCa). By some reports, this approach has become more common than open radical prostatectomy (RP) over the last several years.[1] As a result of the wide and rapid adoption of this technology, multiple different approaches have emerged. These approaches include the anterior transperitoneal prostatectomy (ARARP), the posterior RARP (PRARP), and the extraperitoneal RARP.[2–4] The primary difference between the ARARP and the PRARP is the timing of the dissection of the seminal vesicles (SVs) and vasa deferentia.[5] However, as will be demonstrated, the variety in approaches to ARARP is extensive, and many surgeons have described innovations in their approach.

This chapter is dedicated to describing the ARARP technique. PRARP and extraperitoneal RARP are described in detail elsewhere in this text. Note that the present chapter was written with the DaVinci Si model in mind. However, the great majority of the concepts and techniques described herein are applicable to the DaVinci Standard, S and Xi platforms.

## PATIENT SELECTION

Generally, patients who are good candidates for a RARP overlap with good candidates for open RP. Patients who benefit the most have organ-confined disease and a good life expectancy.[6] The age cutoff for RARP and RP is not well defined, but life expectancy is considered a better selection criterion than absolute age.[7] Contraindications to laparoscopic surgery include uncorrected bleeding diathesis, hemoperitoneum, and hemodynamic instability. However, factors that were traditionally considered contraindications for a minimally invasive approach, such as peritonitis, pelvic radiation, prior abdominal surgery, and so on, have rapidly become less significant.[8,9] There are several anatomic and physiologic considerations when selecting patients for RARP.

### Anatomic Factors

Obese patients, especially men, are more likely to have significant intra-abdominal and pelvic fat. This fat can create significant problems during positioning, since RARP requires a steep Trendelenburg position. This may have an effect on ventilation during the procedure.[10]

Presence of abdominal aortic or iliac aneurysm is considered by many to be a relative contraindication. Many surgeons will opt to use the Hasson approach[11] in this setting. However, in the setting of a known abdominal aneurysm, the Veress needle can safely be placed through a lateral puncture.[6]

Other important anatomic considerations are presence of bladder diverticulae, horseshoe or ectopic kidneys, and presence of hernia. Inguinal hernias are frequently found at the time of RARP, and can be repaired concomitantly with or without mesh.[12]

Prior abdominal surgery and prior pelvic radiation are two factors that have precluded laparoscopic surgery in the past. However, as experience has progressed, many surgeons are comfortable attempting RARP with lysis of adhesions in a patient even with significant prior surgery.[13,14] Some prior procedures in particular may be associated with a higher risk of complication, including prior rectal or colonic anastomosis in the pelvis, history of colostomy, abdominal mesh, transurethral resection of the prostate, and prior pelvic radiation. These operations in particular may contribute to difficulty due to the specific location of the resulting scar tissue.[14]

**Prostate Cancer. http://dx.doi.org/10.1016/B978-0-12-800077-9.00036-0**

## Physiologic Factors

Most medical comorbidities that would preclude open surgery would also be important in RARP. A patient's preoperative cardiac and pulmonary history should be thoroughly assessed prior to laparoscopic surgery. Congestive heart failure or diminished ejection fraction can lead to difficulty maintaining cardiac output, due to the increased afterload associated with abdominal insufflation.[15] A patient's respiratory status is important to assess preoperatively, as insufflation can exacerbate the hypercarbia associated with chronic obstructive pulmonary disease or pulmonary fibrosis.[16] Surgeons should have a low threshold for involving their cardiac or pulmonary colleagues in the preoperative evaluation and postoperative management.

Anticoagulation and antiplatelet therapy usages are ubiquitous, and these factors must be considered preoperatively. Clearance from the prescribing physician, usually a cardiologist, should be sought, and a plan should be outlined for holding or bridging therapy. In general, antiplatelet agents and warfarin should be held for 5–7 days, and a coagulation profile should be drawn the day of surgery.[6]

One final preoperative consideration is the patient's preference for blood transfusion. Certain religious affiliations, such as Jehovah's Witnesses, are associated with refusal of blood transfusion, and this should be discussed in detail prior to proceeding with surgery. Although many surgeons report reduced blood loss with RARP,[17] significant bleeding is not impossible. Patients should be counseled extensively on the potential for bleeding, and the refusal of blood transfusion should be thoroughly documented prior to RARP.

## Preoperative Preparation

Preoperative testing should include complete blood count, basic metabolic panel, coagulation profile, electrocardiogram, chest X-ray, and urinalysis with or without urine culture. As stated earlier, cardiac and pulmonary clearance should be sought when clinically indicated.

A full review of the patient's medication list should be performed. Any anticoagulation should be stopped as described earlier; dietary supplements should also be held prior to surgery to avoid any untoward anticoagulant effects. Anti-HMG-CoA inhibitors (also known as "statins") and beta-blocker medications should routinely be continued through the day of surgery.

Preoperative administration of unfractionated or low molecular weight heparin is controversial.[18] Because RARP is considered a higher-risk operation for venous thromboembolism (VTE),[19] many surgeons use heparin-based products for perioperative VTE-prophylaxis. Concern over bleeding complications has precluded widespread adoption. Sequential compression devices or stockings have no significant downsides and should be used in all patients.

Bowel preparation has typically been done routinely in all patients undergoing RP. The traditional approach is a self-administered enema the night prior to surgery, although this is highly variable among robotic surgeons. Many surgeons will recommend a clear liquid diet the day prior in lieu of a full mechanical bowel preparation.

Preoperative abdominopelvic imaging should be based on oncologic risk of metastatic disease. Generally, patients with D'Amico high-risk disease should undergo preoperative bone scan[20] and those with a high likelihood of lymph node involvement should undergo cross-sectional imaging with computed tomography or magnetic resonance imaging.[21] A growing number of surgeons are now using multiparametric magnetic resonance imaging for assistance in preoperative staging, prediction of positive margins, and to help decide on proper candidates for nerve sparing.[22] This has not been widely adopted, and the negative and positive predictive value in one series was only 61 and 50%, respectively.[23]

# PATIENT SETUP

Prior to the procedure, it is important to review all pertinent patient data, including age, BMI, PSA, digital rectal exam, biopsy information, prostate size, as well as preoperative urinary and sexual function scores.

General endotracheal anesthesia is required. Preoperative antibiotics should be given, generally in the form of first- or second-generation cephalosporin, or an aminoglycoside with metronidazole or clindamycin.[24] Once preoperative VTE-prophylaxis is administered and an orogastric tube is in place, the abdomen is shaved above the pubic symphysis.

After positioning, the anesthesia team should double check all IV lines and blood pressure cuffs, as they may lose access to these when the patient is positioned. The patient is then placed in a dorsal lithotomy or split leg position. If the position is lithotomy, stirrups should be used; we prefer Yellowfin type (Allen Medical, Acton, MA). Attention should be paid to the exact position of the hips, ankles, feet, and overall position of the pelvis, before and after placing the patient into Trendelenburg. Femoral nerve stretch or common peroneal nerve compression is preventable at this stage.[16]

Arms are tucked at the sides in anatomic position with plenty of foam padding, with the thumbs facing upward toward the ceiling. Alternatively, arm boards, sleds, or even beanbags have been described, and may be helpful if the patient is morbidly obese. The patient should be secured to the table in some fashion to prevent excessive cranial migration after being placed in steep

Trendelenburg. We prefer using foam padding and elastic foam tape in a crisscross fashion over the torso, ensuring proper support for the top of the shoulders. The table is then placed in steep Trendelenburg position. The patient should be checked for any excessive migration and shoulder pressure. If the patient slides excessively, we recommend leveling the patient and redoing the positioning and padding to avoid positioning-related injuries. At this point, peak inspiratory pressures and endotracheal tube position should be monitored by the anesthesia team. If any modifications need to be made in the ventilator settings it is best to do so prior to docking.

At this point a proper digital rectal exam can be performed, and a belladonna and opiate suppository may be placed to minimize postoperative bladder spasm. A digital rectal exam just prior to surgery can help with surgical planning, especially in the setting of bulky disease.

Skin preparation can be performed with any agent, though we prefer a chlorhexidine–alcohol solution. The prepped area should include the abdomen from ziphoid superiorly, laterally to the tucked arms, and inferiorly to include the perineum and upper thighs. The abdomen and penis should be draped with towels, excluding the scrotum; the towels can be stapled to the skin to prevent contamination. After draping, an 18-French Foley catheter is placed, so that it can be accessed from the sterile field.

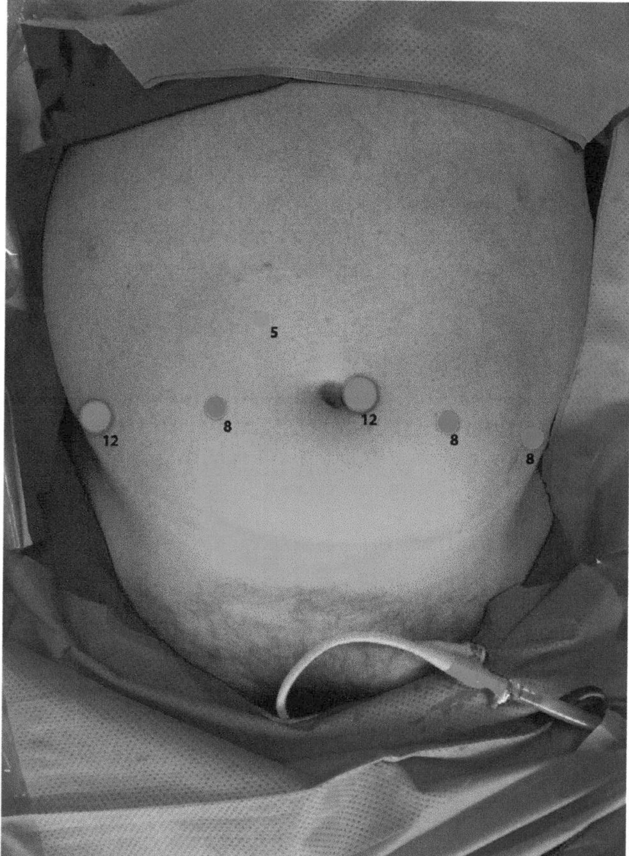


FIGURE 36.1 **Illustration depicting the port placement for a standard RARP. All port sizes are displayed in millimeters.**

## ACCESS

Veress needle access to the peritoneal cavity is achieved through a 12 mm midline incision. Since the left side of the abdomen receives two ports versus three on the right, this midline camera port incision can be placed just above or just to the left of the umbilicus. Alternatively, if there is concern for bowel adhesions from a prior midline incision, the Veress can be placed through a lateral rectus puncture. Some surgeons prefer a Hasson approach, though this can sometimes result in leakage of the pneumoperitoneum. The Hasson approach can also be a helpful alternative in cases of difficult Veress needle access.[11] However, in our experience, the Veress needle has been safe and effective in the vast majority of cases.

We prefer to insufflate the abdomen to 20 mmHg during port placement and reduce this to 15 mmHg for the remainder of the case. The Veress needle is then removed and the midline 12 mm port is placed. This can be done blindly or under direct vision using a 10 mm 0° laparoscopic camera through a clear plastic semiblunt trocar (for example, Ethicon XCEL Bladeless 12 mm Trocar; Ethicon, Somerville, NJ). Global inspection of the peritoneal cavity should be performed at this point to ensure that there are no inadvertent bowel or vascular injuries. At this point, the remainder of the ports can be placed under direct vision.

Our preferred port placement is depicted in Figure 36.1. We prefer placing the lateral-most ports as wide as possible before placing the more medial ports to maximize the available working space. The medial ports are then distributed with a minimum of one hand-breadth (approximately 8 cm) between each robotic port. Alternatively, some surgeons prefer to place the assistant port on the left.[25]

After port placement, the pneumoperitoneum is reduced to 12–15 mmHg. The robotic patient side cart is then driven in by the circulating nurse and docked between the patient's legs.

## SURGICAL STEPS

### Bowel Mobilization

Upon docking the robot, the remainder of the procedure is performed robotically. Instrumentation varies widely depending on surgeon preference. We use the Maryland bipolar grasper on the left hand, monopolar hook on the right, and a Cobra grasper (Intuitive, Sunnyvale, CA) on the fourth arm. The entire procedure can be performed with these three instruments, plus round-tip cold scissors and large needle drivers. Most surgeons prefer to use the monopolar curved scissors on the right arm instead of cautery hook.

We perform the initial portions of the operation with a 30° up camera lens. The first step in the procedure is to mobilize any adhered bowel or colon out of the deep pelvis. Many surgeons may find it easier to mobilize adhesions laparoscopically prior to docking the robot. However, in our experience these adhesions are lysed most efficiently by the robotic surgeon, even if it requires delaying placement of a port until robotic lysis of adhesions can be performed.

Once the bowel is free of any attachments, the small bowel and redundant colon is gently pulled out of the pelvis and allowed to fall cephalad with gravity.

## Bladder Mobilization

Next, the space of Retzius is entered by incising the anterior parietal peritoneum. It is important to find an avascular plane anterior to the preperitoneal fat but posterior to the transversalis fascia. This plane is developed with a combination of blunt and sharp dissection until the anterior surface of the prostate is encountered. We perform a thorough dissection of the periprostatic fat, as well as the fat covering the endopelvic fascia and the bladder neck. This anterior fat pad dissection helps to better expose the bladder neck–prostate base junction for later steps, and it may also be important for staging.[26]

## Pelvic Lymph Node Dissection

At this point in the procedure, some authors prefer to perform a pelvic lymph node dissection (PLND). Alternatively, this can be performed later on in the operation, either before or after vesicourethral anastomosis. A detailed description of PLND is outside the scope of this chapter, and can be found elsewhere in this text.

## Opening Endopelvic Fascia

Next, the endopelvic fascia may be opened bilaterally. This step is optional, and some surgeons exclude this portion. If the endopelvic fascia is opened, a plane is created between the prostate and the levator ani muscles all the way to the apex (Figure 36.2). In our institution, we change the camera lens to a 30° down position at this point.

## Addressing Puboprostatic Ligament and DVC

This is a point of crucial distinction between RARP and open RP. In the open approach to the prostate, control of bleeding from the dorsal venous complex (DVC) is of the utmost importance and cannot be ignored. However, in robotic procedures, due to the luxury of pneumoperitoneum, early control of the DVC has become optional. Many surgeons prefer to place a stitch at this point, or potentially use a stapler to control the complex. In our institution, we do not routinely ligate the DVC at this point in the operation.

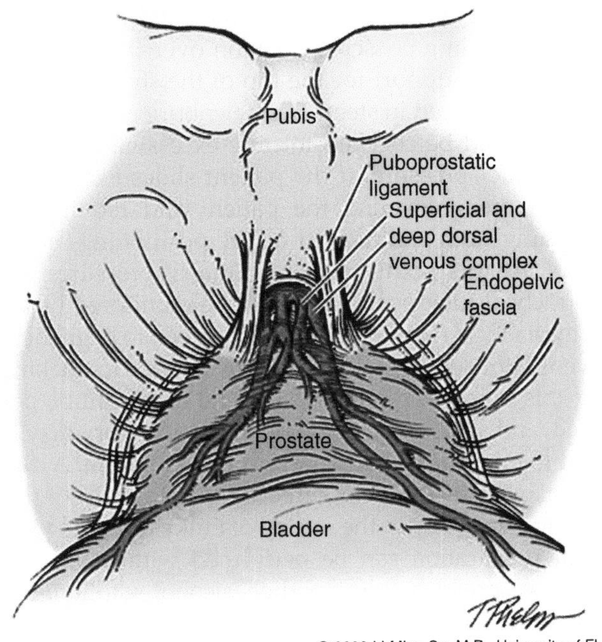


**FIGURE 36.2 Anterior surface of the prostate after development of the space of Rhetzius and removal of anterior fat pad.** *Reprinted with permission from Li Ming Su.*

## Bladder Neck Dissection

Next, the anterior bladder neck is opened, and a plane is created between the prostate base and the bladder neck. This is optimally performed if the fourth arm or the assistant holds traction on the bladder, to better define the plane. To minimize distortion of the bladder neck, we prefer to deflate the balloon of the Foley catheter prior to this step. Once the Foley catheter is encountered, it is passed through the prostatic urethra and retracted anteriorly with the fourth arm (Figure 36.3). At this point, the bedside assistant applies countertraction to the external portion of the catheter. This maneuver helps to better define the plane between the prostate and bladder neck posteriorly.

Dissection of the posterior bladder neck requires careful attention to the boundaries of the prostatic base and bladder neck, and violation of the prostate capsule must be avoided. Inadvertent injury to the ureteral orifice is possible especially the setting of a wide bladder neck margin or a median lobe.

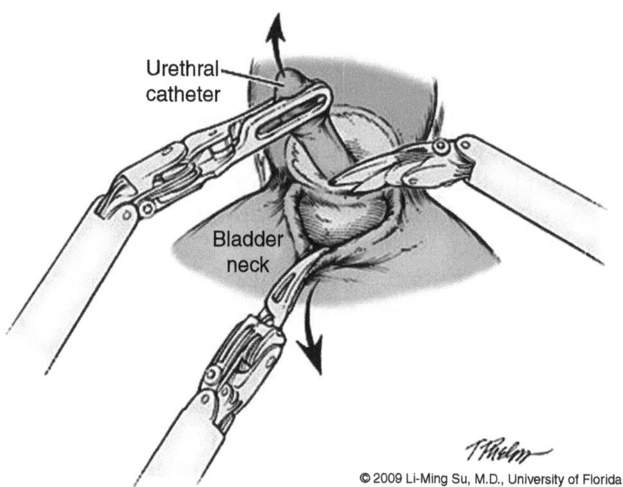


FIGURE 36.3 **After entry into the bladder neck, the catheter is grasped and the prostate is retracted anteriorly.** *Reprinted with permission from Li Ming Su.*

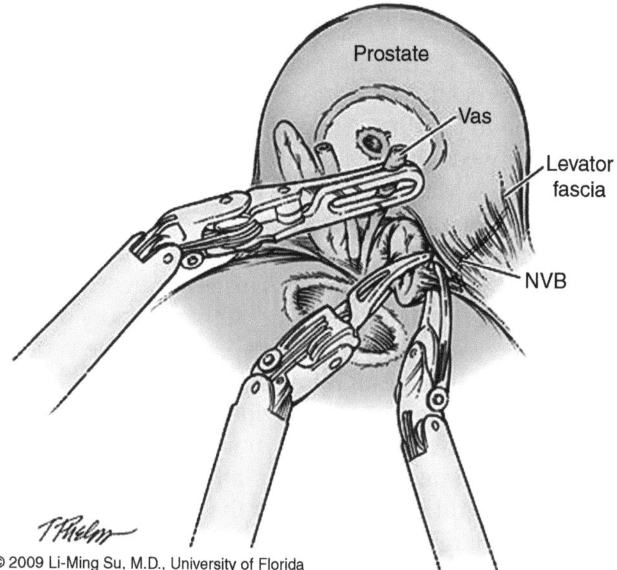


FIGURE 36.5 **After dissection and exposure of the SVs and vasa deferentia, these structures are grasped and retracted anteriorly in order to better expose the posterior plane.** *Reprinted with permission from Li Ming Su.*

Once this dissection is complete, the bladder neck size can be assessed for the necessity for any reconstruction. The presence of a large median lobe can make this dissection difficult and can result in a large bladder neck. Multiple techniques have been described, and the key common step is grasping and upward retraction of the median lobe away from the bladder neck.[27–29] Coelho et al. describe a technique of lateral-to-medial dissection, which should then progress posteriorly and forward once the midline is reached.[30]

## Dissection of Posterior Space

Once the vas deferens is encountered posteriorly (Figure 36.4) and divided, the seminal vesicles (SVs) are also freed from attachments. These structures are then grasped and used for better retraction of the prostate anteriorly (Figure 36.5). It is generally accepted that the neurovascular bundles travel very close to the tips of the SVs, therefore minimal cautery is employed in this area. Some authors utilize clips for dissection of the SV tips and others spare the SV tips altogether for better preservation of the bundles.

While retracting the prostate and SVs anteriorly, the posterior fascia, known as Denonvillier's fascia, is divided. There are two leaflets of this fascia, and the ideal plane is between these two layers. This plane is carried distally toward the apex of the prostate and the neurovascular bundle (NVB) can typically be gently swept laterally from this posterior view.

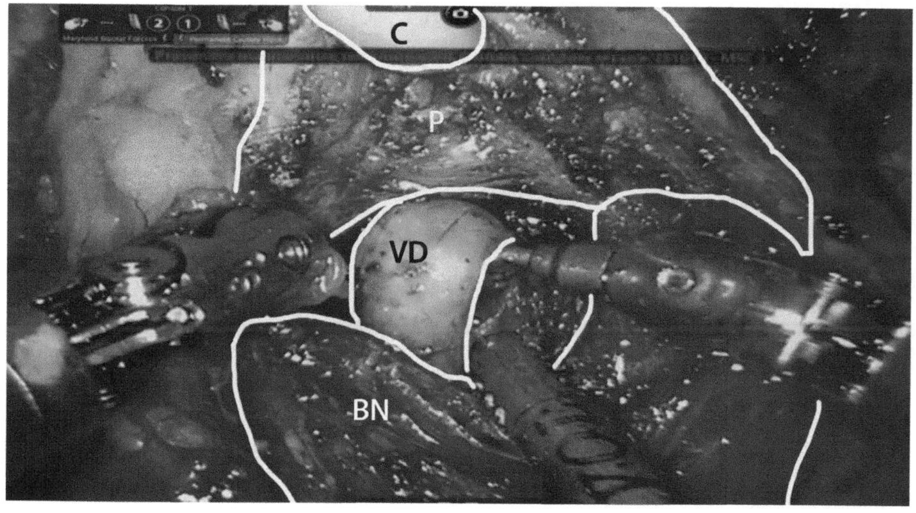


FIGURE 36.4 **Intraoperative photograph showing retraction of the right vas deferens after division of the posterior bladder neck.** C, catheter; VD, vas deferens; BN, bladder neck; P, prostate.

## Division of Pedicle and Nerve Sparing

The nerve-sparing portion of the operation is one of the most crucial, and it has been linked to both erectile function and continence recovery.[6] The NVB of the prostate lie between the levator fascia and the prostatic fascia (Figure 36.6). Many techniques to preserve the NVB have been described, but the critical step in most of them is identification and reflection of the levator fascia without entry into the prostatic fascia. This is frequently referred to as an interfascial dissection.[31] Other authors describe a plane between the prostatic capsule and the prostatic fascia, and advocate an intrafascial dissection.[32]

In 2005, Menon et al. described the "Veil of Aphrodite" as a network of nerve branches spreading from posterior to anterior along the lateral prostatic fascia.[33] The veil should be dissected off the prostatic fascia, and there are multiple descriptions of how best to do so. The antegrade approach involves an incision near the base with dissection toward the apex,[34] while the retrograde approach starts with an apical incision and dissection posteriorly toward the pedicle.[35] The "superveil" technique involves sparing all of the fascia lateral to 11 and 1 o'clock position and preservation of the anterior tissues along the puboprostatic ligament.[36] Yet another approach describes a posterior release of the veil, which is then coupled with early release of the pedicles.[37]

In addition to the varying approaches to nerve sparing, multiple modifications have been made to help preserve nerve function postoperatively. Most surgeons agree that erectile function is improved if the prostate pedicles and seminal vesicles are dissected athermally, since both are closely intertwined with the NVB.[38] This generally entails the use of clips and sharp division, as even harmonic scalpel has been implicated in thermal injury to the NVB.[39] Also important during this dissection is limiting the amount of traction or displacement of the NVB, as this has been shown to have a significant effect on potency.[40] Other techniques described to boost nerve sparing include nerve mapping[41] and endorectal cooling with ice slush.[42]

In our institution, we perform an antegrade nerve sparing technique by finding the plane between the prostatic fascia and the levator fascia near the base of the prostate, and gently peeling the fascial layer off bilaterally. The prostate pedicles, branches of the inferior vesical artery, are ligated using absorbable Lapro-Clips™ (Covidien, Dublin, Ireland). The use of absorbable clips may decrease the incidence of postoperative clip erosion into the urinary tract.

## Apical Dissection

We prefer to divide the DVC at this point via our technique. The apex of the prostate is gently peeled from the neurovascular bundles posterolaterally. The DVC is then divided prior to apical dissection and division of the urethra. Most surgeons prefer to place a DVC stitch or use a stapler to control bleeding from the DVC. However, since significant bleeding can still occur while dissecting in this area, some have advocated increasing pneumoperitoneum up to 20 mmHg during DVC ligation. Overdissection or overuse of cautery during DVC dissection can inadvertently traumatize the striated sphincter, so great care must be taken during this portion of the procedure.

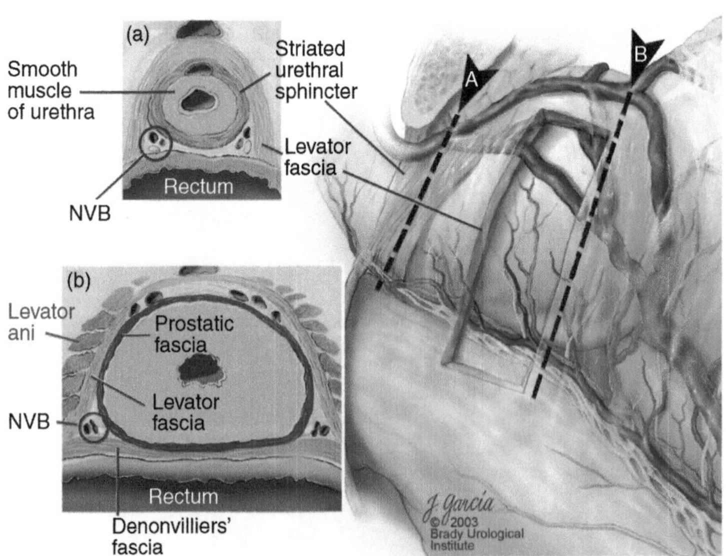


FIGURE 36.6 **Cross-section through the prostate.** (a) The level of the apex of the prostate, showing the relationship between the NVB, the urethra, and the levator fascia. (b) The level of the base of the prostate, showing the location of the NVB between the levator fascia and the prostatic fascia. *Reprinted with permission from the Brady Urologic Institute.*

We prefer to grasp the DVC with a Cobra grasper and develop a plane between the DVC and the anterior prostate. Once the DVC is divided and the prostate apex is visualized, the periurethral fibers are bluntly teased off of the prostate apex, in order to maximize the length of the urethral stump.[43] The urethra should then be divided sharply to minimize cautery effect and maximize continence potential.

With the prostate completely free, the DVC can now be inspected for obvious bleeding. At this point, the Cobra grasper is released from the DVC and visualization of any bleeding from the DVC can be facilitated by applying perineal pressure and using a 0° lens. DVC can be closed using a running 3-0 V-loc™ (Covidien, Dublin, Ireland) suture while carefully avoiding periurethral tissue.

A PLND can also be performed at this point in the procedure. As mentioned previously, a full description of this procedure can be found elsewhere in this text.

## Anastomosis

There are many descriptions of the vesicourethral anastomosis, including variations using braided, monofilament, and even barbed suture. The most important principle is the creation of a tension-free, watertight anastomosis through mucosa-to-mucosa apposition. The original description by Van Velthoven et al. uses two polyglycolic acid sutures tied together at the tails, run in opposite directions in a running fashion.[44] This is the most common technique, though there is significant variation in the type of suture used as well as the method of attaching the tails. We use two 6-in. 3-0 poliglecaprone (Monocryl®, Ethicon, Somerville, NJ) sutures with an RB needle to create a double-armed suture, and perform the anastomosis very similar to Van Velhoven's technique. Proper mucosa-to-mucosa apposition will lead to improved continence, low leak rates, and reliable prevention of bladder neck contracture.[45]

We typically leave a new 20-French two-way Foley catheter with 20 $cm^3$ in the balloon as our final catheter. The anastomosis is then tested by filling the bladder with 300 $cm^3$ of saline and observing for any possible leaks. Krane et al. described the use of a suprapubic tube for postoperative drainage and early removal of the urethral catheter. They reported less pain compared to traditional urethral catheterization.[46]

## BN Reconstruction and Anterior/Posterior Repair

In some cases, especially when the prostate is larger in size (>50 $cm^3$) or when there is a significant median lobe, the bladder neck diameter can be much larger than that of the urethra. In these cases, most surgeons prefer to utilize some form of bladder neck reconstruction. The traditional anterior tennis racquet has been described extensively in open, and also in RARP series. It involves a longitudinal closure of the midline redundant bladder after the posterior and lateral anastomosis is complete.[47] Other techniques include the posterior tennis racquet,[48] and the transverse plication.[49]

Other techniques to aid continence recovery include anterior urethropexy and the Rocco posterior musculofascial reconstruction.[50] We use the DVC closure stitch, which is a 3-0 VLOC suture, to affix the puboprostatic remnant to the anterior bladder neck, thereby creating an anterior reconstruction. The remaining length of the VLOC suture is then used to plicate the anterior bladder neck after the anastomosis is complete, thus creating a more tapered bladder outlet. This technique has been described in the open[51] and robotic approach.[52]

## Case Completion

The anesthetist is instructed to give a bolus of IV fluids to better facilitate production of dilute urine and prevention of intravesical blood clots. The prostate and any lymph nodes are placed in an endoscopic specimen pouch and the robot is undocked and removed. Many authors will place a closed suction drain, in order to better detect urine leak and/or bleeding in the postoperative period. At our institution, we do not routinely place drains unless an extensive lymphadenectomy has been performed. The specimen is removed through the midline port and the fascia of the extraction incision should be closed. Fascial stitches to close 8 mm robotic ports are typically not necessary, and all ports can be closed at the skin with absorbable suture in a subcuticular fashion. We prefer local anesthetic injection at all port sites.

# POSTOP

The patient is typically observed overnight as an inpatient. Intravenous fluids should be given to support hydration, but the patient can be placed on a clear liquid diet after surgery. Antibiotics should continue for 24 h, and subcutaneous heparin should be continued until discharge.

Incentive spirometry is encouraged and all patients ambulate with assistance on the evening of surgery. Most patients receive ketorolac 15–30 mg IV every 6 h, unless they have baseline renal dysfunction. Oral and intravenous narcotic pain medication is given as needed, as well as oxybutynin and belladonna and opiate suppositories.

Diet is advanced to regular on the morning of postoperative day 1. Perioperative labs are not routinely drawn at our institution, unless there are any concerns for bleeding, vital sign abnormality, decreased urine output, or electrolyte disturbance. Patients may shower on postoperative day 2, and are instructed to avoid heavy lifting for 4–6 weeks. The catheter typically remains in place for 5–7 days, and is removed in the office. If a bladder neck reconstruction has been performed or there is concern regarding the quality of the anastomosis, the catheter can be left in for 10–14 days and a cystogram can be performed prior to catheter removal. Oral antibiotics are typically given prior to catheter removal. Although most patients can restart their anticoagulation or antiplatelet medications approximately one week postoperatively, this must be carefully individualized in accordance with their medical specialists.

## CONCLUSIONS

The anterior approach to RARP has been a popular approach since it was first described in 2003.[2] This was somewhat of a diversion from the Montsouris technique,[3] which was the traditional posterior approach that had been carried over from laparoscopic RP. To our knowledge there has never been a head-to-head comparison of the two techniques, and they both provide potential advantages.

Proponents of the posterior approach argue that the seminal vesicles are easier to find regardless of the size of the prostate. They also report that the seminal vesicle and posterior prostatic dissection is easier due to the broader surgical field and the avoidance of working in a hole.[53] Supporters of the anterior approach argue that the posterior approach may be difficult in patients with excessive adipose tissue. In addition, some argue that the ability to put traction on the SVs is unique to the anterior approach, and that makes it easier to dissect the neurovascular bundle off the prostate.[5]

While it is clear that both sides have advantages, the difference likely comes down to surgeon preference. The immense variation employed, in both instrumentation and surgical technique, leads one to the conclusion that it is the quality of the surgeon that affects outcomes and not the specific technique employed.

## References


1. Pilecki MA, McGuire BB, Jain U, et al. National multi-institutional comparison of 30-day postoperative complication and readmission rates between open retropubic radical prostatectomy and robot-assisted laparoscopic prostatectomy using NSQIP. *J Endourol* 2014;**28**(4):430–6.
2. Menon M, Tewari A, Peabody J, et al. Vattikuti Institute prostatectomy: technique. *J Urol* 2003;**169**(6):2289–92.
3. Guillonneau B, Vallancien G. Laparoscopic radical prostatectomy: the Montsouris technique. *J Urol* 2000;**163**(6):1643–9.
4. Bollens R, Vanden Bossche M, Roumeguere T, et al. Extraperitoneal laparoscopic radical prostatectomy. Results after 50 cases. *Eur Urol* 2001;**40**(1):65–9.
5. Maddox M, Elsamra S, Kaplon D, et al. The posterior surgical approach to robot-assisted radical prostatectomy facilitates dissection of large glands. *J Endourol* 2013;**27**(6):740–2.
6. Wein AJ, Kavoussi LR, Campbell MF, et al. *Campbell-Walsh urology*. 10th ed. Philadelphia, PA: Elsevier Saunders; 2012.
7. Thompson I, Thrasher JB, Aus G, et al. Guideline for the management of clinically localized prostate cancer: 2007 update. *J Urol* 2007;**177**(6):2106–31.
8. Ball MW, Reese AC, Mettee L, et al. Safety of minimally invasive radical prostatectomy in patients with prior abdominopelvic or inguinal surgery. *J Endourol* 2014;**29**:192–7.
9. Kaffenberger SD, Keegan KA, Bansal NK, et al. Salvage robotic assisted laparoscopic radical prostatectomy: a single institution, 5-year experience. *J Urol* 2013;**189**(2):507–13.
10. Kalmar AF, Foubert L, Hendrickx JF, et al. Influence of steep Trendelenburg position and CO(2) pneumoperitoneum on cardiovascular, cerebrovascular, and respiratory homeostasis during robotic prostatectomy. *Br J Anaesth* 2010;**104**(4):433–9.
11. Hasson HM. Open laparoscopy. *Biomed Bull* 1984;**5**(1):1–6.
12. Lee DK, Montgomery DP, Porter JR. Concurrent transperitoneal repair for incidentally detected inguinal hernias during robotically assisted radical prostatectomy. *Urology* 2013;**82**(6):1320–2.
13. Ginzburg S, Hu F, Staff I, et al. Does prior abdominal surgery influence outcomes or complications of robotic-assisted laparoscopic radical prostatectomy? *Urology* 2010;**76**(5):1125–9.
14. Siddiqui SA, Krane LS, Bhandari A, et al. The impact of previous inguinal or abdominal surgery on outcomes after robotic radical prostatectomy. *Urology* 2010;**75**(5):1079–82.
15. Kadono Y, Yaegashi H, Machioka K, et al. Cardiovascular and respiratory effects of the degree of head-down angle during robot-assisted laparoscopic radical prostatectomy. *Int J Med Robot* 2013;**9**(1):17–22.
16. Hsu RL, Kaye AD, Urman RD. Anesthetic challenges in robotic-assisted urologic surgery. *Rev Urol* 2013;**15**(4):178–84.
17. Trinh QD, Sammon J, Sun M, et al. Perioperative outcomes of robot-assisted radical prostatectomy compared with open radical prostatectomy: results from the nationwide inpatient sample. *Eur Urol* 2012;**61**(4):679–85.
18. Patel T, Kirby W, Hruby G, et al. Heparin prophylaxis and the risk of venous thromboembolism after robotic-assisted laparoscopic prostatectomy. *BJU Int* 2011;**108**(5):729–32.
19. Geerts WH, Bergqvist D, Pineo GF, et al. Prevention of venous thromboembolism: American College of Chest Physicians Evidence-Based Clinical Practice Guidelines (8th Edition). *Chest* 2008;**133**(6 Suppl.):381S–453S.
20. D'Amico AV, Whittington R, Malkowicz SB, et al. Biochemical outcome after radical prostatectomy, external beam radiation therapy, or interstitial radiation therapy for clinically localized prostate cancer. *JAMA* 1998;**280**(11):969–74.
21. Mohler JL. The 2010 NCCN clinical practice guidelines in oncology on prostate cancer. *J Natl Compr Canc Netw* 2010;**8**(2):145.
22. Lawrence EM, Gallagher FA, Barrett T, et al. Preoperative 3-T diffusion-weighted MRI for the qualitative and quantitative assessment of extracapsular extension in patients with intermediate- or high-risk prostate cancer. *AJR Am J Roentgenol* 2014;**203**(3):W280–6.
23. Heidenreich A. Consensus criteria for the use of magnetic resonance imaging in the diagnosis and staging of prostate cancer: not ready for routine use. *Eur Urol* 2011;**59**(4):495–7.
24. Wolf Jr JS, Bennett CJ, Dmochowski RR, et al. Best practice policy statement on urologic surgery antimicrobial prophylaxis. *J Urol* 2008;**179**(4):1379–90.
25. Stern JL, Lee DI. *Atlas of robotic urologic surgery*: New York, NY Humana Press; 2011.
26. Yuh B, Wu H, Ruel N, et al. Analysis of regional lymph nodes in periprostatic fat following robot-assisted radical prostatectomy. *BJU Int* 2012;**109**(4):603–7.

27. Sarle R, Tewari A, Hemal AK, et al. Robotic-assisted anatomic radical prostatectomy: technical difficulties due to a large median lobe. *Urol Int* 2005;**74**(1):92–4.
28. Jenkins LC, Nogueira M, Wilding GE, et al. Median lobe in robot-assisted radical prostatectomy: evaluation and management. *Urology* 2008;**71**(5):810–3.
29. Patel SR, Kaplon DM, Jarrard D. A technique for the management of a large median lobe in robot-assisted laparoscopic radical prostatectomy. *J Endourol* 2010;**24**(12):1899–901.
30. Coelho RF, Chauhan S, Guglielmetti GB, et al. Does the presence of median lobe affect outcomes of robot-assisted laparoscopic radical prostatectomy? *J Endourol* 2012;**26**(3):264–70.
31. Walsh PC, Mostwin JL. Radical prostatectomy and cystoprostatectomy with preservation of potency. Results using a new nerve-sparing technique. *Br J Urol* 1984;**56**(6):694–7.
32. Galfano A, Ascione A, Grimaldi S, et al. A new anatomic approach for robot-assisted laparoscopic prostatectomy: a feasibility study for completely intrafascial surgery. *Eur Urol* 2010;**58**(3):457–61.
33. Menon M, Shrivastava A, Kaul S, et al. Vattikuti Institute prostatectomy: contemporary technique and analysis of results. *Eur Urol* 2007;**51**(3):648–57; discussion 657–8.
34. Rassweiler J, Wagner AA, Moazin M, et al. Anatomic nerve-sparing laparoscopic radical prostatectomy: comparison of retrograde and antegrade techniques. *Urology* 2006;**68**(3):587–91; discussion 591–2.
35. Ko YH, Coelho RF, Sivaraman A, et al. Retrograde versus antegrade nerve sparing during robot-assisted radical prostatectomy: which is better for achieving early functional recovery? *Eur Urol* 2013;**63**(1):169–77.
36. Menon M, Shrivastava A, Bhandari M, et al. Vattikuti Institute prostatectomy: technical modifications in 2009. *Eur Urol* 2009;**56**(1):89–96.
37. Ischia J, Sengupta S, Webb D. Early release of pedicles and posterior development of the "Veil of Aphrodite" in robotic-assisted laparoscopic prostatectomy (RALP). *BJU Int* 2010;**106**(11):1856–61.
38. Ficarra V, Novara G, Ahlering TE, et al. Systematic review and meta-analysis of studies reporting potency rates after robot-assisted radical prostatectomy. *Eur Urol* 2012;**62**(3):418–30.
39. Chen C, Kallakuri S, Vedpathak A, et al. The effects of ultrasonic and electrosurgery devices on nerve physiology. *Br J Neurosurg* 2012;**26**(6):856–63.
40. Kowalczyk KJ, Huang AC, Hevelone ND, et al. Stepwise approach for nerve sparing without countertraction during robot-assisted radical prostatectomy: technique and outcomes. *Eur Urol* 2011;**60**(3):536–47.
41. Ponnusamy K, Sorger JM, Mohr C. Nerve mapping for prostatectomies: novel technologies under development. *J Endourol* 2012;**26**(7):769–77.
42. Finley DS, Chang A, Morales B, et al. Impact of regional hypothermia on urinary continence and potency after robot-assisted radical prostatectomy. *J Endourol* 2010;**24**(7):1111–6.
43. Mendoza PJ, Stern JM, Li AY, et al. Pelvic anatomy on preoperative magnetic resonance imaging can predict early continence after robot-assisted radical prostatectomy. *J Endourol* 2011;**25**(1): 51–5.
44. Van Velthoven RF, Ahlering TE, Peltier A, et al. Technique for laparoscopic running urethrovesical anastomosis: the single knot method. *Urology* 2003;**61**(4):699–702.
45. Msezane LP, Reynolds WS, Gofrit ON, et al. Bladder neck contracture after robot-assisted laparoscopic radical prostatectomy: evaluation of incidence and risk factors and impact on urinary function. *J Endourol* 2008;**22**(1):97–104.
46. Krane LS, Bhandari M, Peabody JO, et al. Impact of percutaneous suprapubic tube drainage on patient discomfort after radical prostatectomy. *Eur Urol* 2009;**56**(2):325–30.
47. Patel VR, Coelho RF, Palmer KJ, et al. Periurethral suspension stitch during robot-assisted laparoscopic radical prostatectomy: description of the technique and continence outcomes. *Eur Urol* 2009;**56**(3):472–8.
48. Samadi DB, Muntner P, Nabizada-Pace F, et al. Improvements in robot-assisted prostatectomy: the effect of surgeon experience and technical changes on oncologic and functional outcomes. *J Endourol* 2010;**24**(7):1105–10.
49. Lin VC, Coughlin G, Savamedi S, et al. Modified transverse plication for bladder neck reconstruction during robotic-assisted laparoscopic prostatectomy. *BJU Int* 2009;**104**(6):878–81.
50. Rocco F, Gadda F, Acquati P, et al. Personal research: reconstruction of the urethral striated sphincter. *Arch Ital Urol Androl* 2001;**73**(3):127–37.
51. Walsh PC, Marschke PL. Intussusception of the reconstructed bladder neck leads to earlier continence after radical prostatectomy. *Urology* 2002;**59**(6):934–48.
52. Lee DI, Wedmid A, Mendoza P, et al. Bladder neck plication stitch: a novel technique during robot-assisted radical prostatectomy to improve recovery of urinary continence. *J Endourol* 2011;**25**(12):1873–7.
53. Hemal AK, Bhandari A, Tewari A, et al. The window sign: an aid in laparoscopic and robotic radical prostatectomy. *Int Urol Nephrol* 2005;**37**(1):73–7.

C H A P T E R

# 37

# Posterior Approach to Robotic-Assisted Laparoscopic Radical Prostatectomy

*Mark Mann, MD, Costas D. Lallas, MD, FACS, Edouard J. Trabulsi, MD, FACS*

**Department of Urology, Sidney Kimmel Cancer Center, Thomas Jefferson University, Philadelphia, PA, USA**

## INTRODUCTION

The introduction of the prostate-specific antigen (PSA) test led to an unprecedented increase in the number of men diagnosed with prostate cancer. With that rise in rate of diagnosis came a rise in men found to have localized disease.[1] As the concurrent adoption of minimally invasive surgical approaches progressed, laparoscopic techniques began to be applied to radical prostatectomy performed for prostate cancer with initial marginal results and cause for question.[2] Pioneered in Montsouris, France by Dr Guillonneau and Dr Vallancien in an attempt to have shorter catheter durations and to get men out of the hospital quicker, the Montsouris technique for laparoscopic prostatectomy described a technique that was quicker and easier, allowing dissemination and acceptance of minimally invasive radical prostatectomy (MIRP) as a viable surgical strategy.[3] The Montsouris technique utilized a dissection that began posteriorly with the midline identification and liberation of the vas deferens (VD) followed by the seminal vesicles (SVs) bilaterally. This allowed direct visualization of the tips of the SVs and thereby avoidance of damage to the erectile nerves that course in the area. Furthermore, it removed the need for blind dissection of those structures after the bladder was dropped and the bladder neck was incised (thereby sparing the ureters and rectum from potential blind damage). This approach, dissecting in the pouch of Douglas behind the bladder and anterior to the rectum, was not an area that most urologists had experience in, let alone were comfortable with, and it is a testament to the visionary approach by Dr Guillonneau and Dr Vallancien, as well as Dr Schuessler, Dr Clayman, and Dr Kavoussi before them.[4]

After the introduction of the Zeus and DaVinci robotic surgical systems, urologists rapidly applied this new technology to prostatectomy.[5] Laparoscopic prostatectomy had already been proven to have advantages in visualization of pelvic anatomy, decreased blood loss, and a more rapid convalescence. However, laparoscopic surgery in the pelvis had significant limitations. There was a very difficult learning curve due to the challenge of manipulating distant tissues with very long, nonarticulating instruments using two-dimensional images. The introduction of articulating robotic arms along with binocular vision allowed the surgeon to keep the improved outcomes from laparoscopic MIRP as well as vastly improved on these limitations and significantly shortened the learning curve.[6–14] Shortly thereafter, a proliferation of surgical approaches to prostatectomy occurred with changes in many areas of dissection.[15,16] Early adopters published included the Frankfurt Technique, which applied the Montsouris approach of a posterior dissection as done with laparoscopic prostatectomy, however, with the aid of the DaVinci robotic platform.[13] As well as the Vattikuti Urology Institute where among other changes, the approach was to initially develop the space of Retzius and separate the bladder from the anterior abdominal wall rather than by dissection of the VDs and the SVs.[17] As the DaVinci robot proliferated in Western medicine, the majority of MIRP transitioned from laparoscopic to robotic approaches.[18,19] Today, there are two

**Prostate Cancer. http://dx.doi.org/10.1016/B978-0-12-800077-9.00037-2**

main approaches to robotic-assisted laparoscopic radical prostatectomy, and they essentially consist of beginning the dissection anteriorly or posteriorly. Here we will be describing the Montsouris-based posterior approach.[20,21]

## RELEVANT ANATOMY

In this section, anatomy will be described from the perspective of a laparoscopic camera placed in the midline in the region of the umbilicus. Inferior to the bladder rests the prostate, within which, the urethra transits. Along the posteriolateral surface of the prostate lie the erectile neurovascular bundles as well as the prostatic vascular pedicles. The VDs travel medially bilaterally to meet the SVs at the prostate in the midline. Slightly lateral and anterior to this junction of VDs and prostate, but in between VD and the prostate sit the SVs. From the posterior perspective looking inside the pelvis from a superior perspective, one will observe at the deepest (most inferior) part of the peritoneal pelvis there are two ridges anteriorly below the bladder. At the more inferior ridge is usually found the VDs.

## POSTERIOR APPROACH PROCEDURE

After injection of heparin or enoxaparin and application of sequential compression devices, general anesthesia is induced, and an appropriate antibiotic is administered. The patient is positioned in low dorsal lithotomy and secured taking care to pad all pressure points. The bed is then placed in steep Trendelenberg and the patient is prepped and draped in a sterile fashion. After a Foley catheter is advanced into the bladder, insufflation of the abdomen is then performed using either a Veress needle or via a Hasson approach to a pneumoperitoneum of 15 mmHg. After introduction of the laparoscope via the midline camera port, and checking for any intra-abdominal damage as a result of insufflation and port placement, the robotic ports are all placed under direct vision transilluminating the skin to help identify and avoid blood vessels with placement. Ports are arranged in a standard fan array triangulating 8–10 cm from the superior aspect of the umbilicus with 15 cm from the midline of the superior edge of the pubis for the medial robotic ports. The lateral ports are measured as three fingerbreadths superior from the anterior–superior iliac spine and two fingerbreadths medial. The right lateral port is a 10 mm assistant port, and the left lateral port is the fourth robotic arm. In between and superior to the camera and the right robotic port is a 5 mm assistant port. The robotic instruments are then advanced into the abdomen under direct vision, with a monopolar electrocautery scissors in the right hand, a Maryland bipolar in the left hand, and a Prograsp in the fourth robotic arm on the left. After inspection of the abdomen and pelvis, adhesions are first taken down where necessary in order to be able to retract the sigmoid colon posteriorly and superiorly. After optimal vision of the junction of the anterior peritoneum and the rectum is obtained, the two anterior ridges are viewed (Figure 37.1). In some patients one can even see the VDs running in a medial direction along the anterior peritoneum toward the inferior ridge. It is often useful to have the bedside assistant pull gently on the Foley back and forth to ensure you are distal to the bladder neck and thus a safe distance from the ureters. A "U" shaped electrocautery incision is made with the scissors approximately 5 mm superior to the inferior-most ridge (Figure 37.2). The superior aspect of this ridge is then

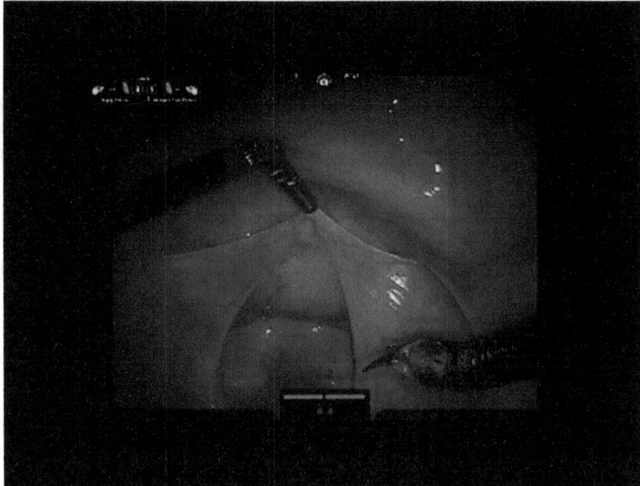

**FIGURE 37.1** The anterior peritoneum is grabbed with the Prograsp and pulled anteriorly to allow easier access to the recto-vesicle junction.

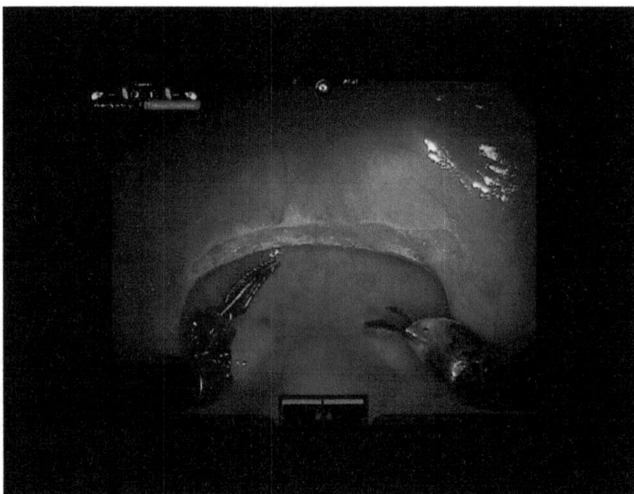

**FIGURE 37.2** A transverse incision is made in the anterior peritoneum approximately 5 mm anterior to the posterior-most ridge by the recto-vesicle junction.

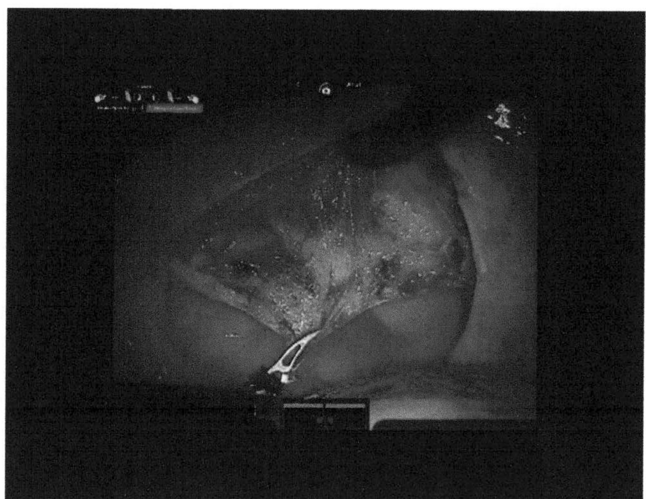

FIGURE 37.3 **After the incision in the peritoneum, the fat is pushed anteriorly and inferiorly off of the peritoneal edge.**

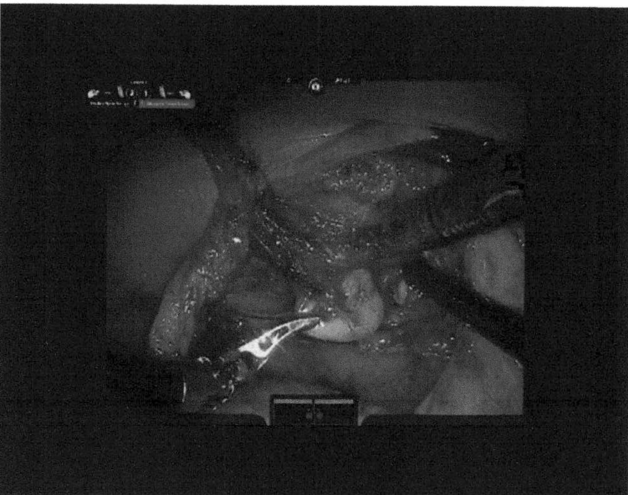

FIGURE 37.5 **The seminal vesicle is then dissected toward the prostate while the VD is retracted anteriorly by the Prograsp.**

held up with the third arm while an assistant retracts back the bowel if necessary. The fat behind the peritoneum is then bluntly pushed anteriorly until the VDs are revealed bilaterally running in a wide "V" pattern with the apex at the center of the dissection (Figure 37.3). The VDs are then dissected out and transected using either bipolar or monopolar energy with the distal stumps then grabbed with the third arm and retracted directly anteriorly (Figure 37.4). The retraction and transection of the VDs reveals the SVs slightly lateral to midline in between the prostate and the natural course of the VDs. Due to the course of the erectile neurovascular bundles lateral to SVs, only bipolar energy is used here as necessary for hemostasis. The SVs are consistently found deep and lateral to the retracted VDs. The SVs are then dissected down toward their base in the prostate and are left attached. This process of identifying the VDs, severing it, using it to retract anteriorly and dissection of the SV is then repeated on the contralateral side (Figure 37.5).

The third arm then grips both VDs directly anteriorly together. Posterior to the SVs at this point, Denonvillier's fascia is encountered and incised (Figure 37.6). Using the assistant's suction tool to help retract Denonvillier's fascia posteriorly and superiorly, the robotic arms are then free to progress the dissection farther inferiorly and laterally with the suction regripping the fascia deeper as allowed. This dissection proceeds inferiorly/deep until the fascia begins curving upward and visualization of the perirectal fat occurs. At this point further dissection is no longer possible as the rectum changes angle here from parallel to perpendicular to the plane of dissection (Figure 37.7).

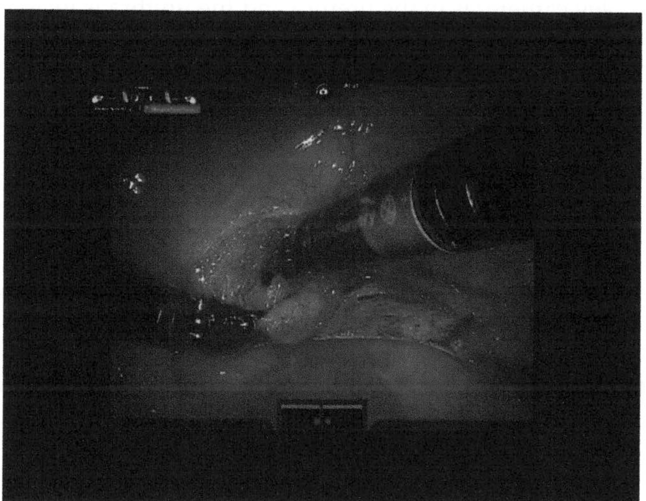

FIGURE 37.4 **The VD is identified and dissected and then cut distally, revealing the seminal vesicle behind it.**

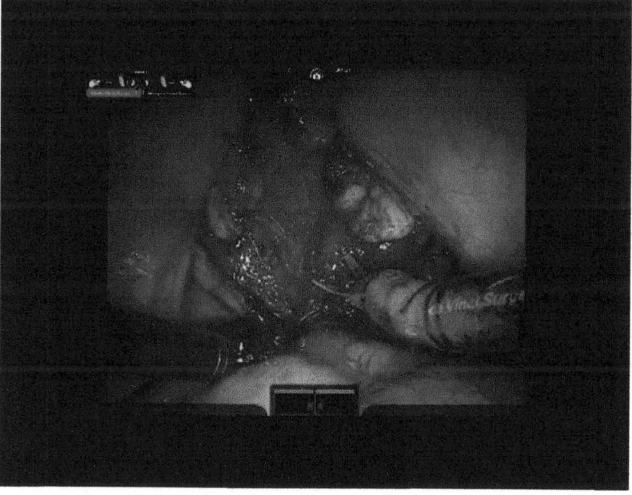

FIGURE 37.6 **Both vasa are retracted anteriorly by the Prograsp while Denonvillier's is incised and developed.**

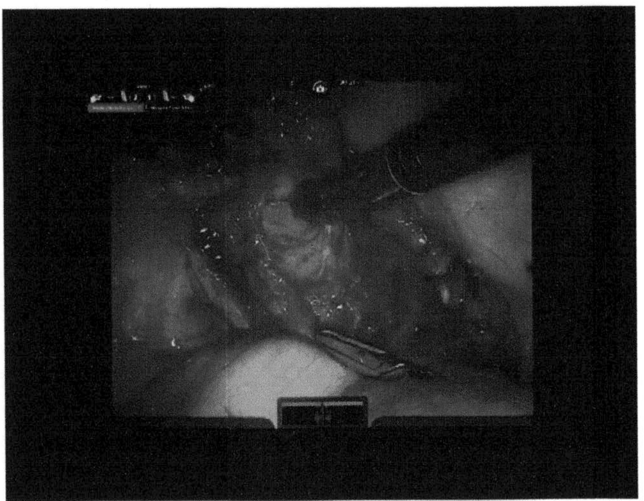

FIGURE 37.7 **Denonvillier's fascia is developed until yellow perirectal fat is encountered.**

At this point in time attention is directed anteriorly to developing the space of Retzius. First, the bilateral medial umbilical ligaments are identified and retracted posteriorly and medially. When this is under with the correct amount of tension, it will reveal a potential space lateral to the bladder that can be safely incised with monopolar electrocautery. The incised plane is developed using a combination of blunt and electrocautery inferiorly aiming toward the area that the pubic bone is expected to be encountered. This is then repeated on the contralateral side. Once the pubis has been identified, it can be used as a marker for the level for which it is safe to incise the peritoneum medially. When dissecting the bladder medially, it is important to cauterize the medial umbilical ligaments carefully as the obliteration of the vessels contained within them that usually occurs as a neonate may not always be complete. Care should then be taken to mobilize the bladder medially away from the iliac vessels to help with the later lymph node dissection. After the bladder has been mobilized and the space of Retzius is fully developed, the dissection proceeds inferiorly toward the endopelvic fascia. At this point, the fat covering the endopelvic fascia and the prostate is swept superiorly and off using electrocautery as needed taking care to avoid injuring superficial prostatic vessels. Anteriorly, fat is carefully swept off of the prostate until the superficial dorsal vein is identified cauterized using bipolar energy and incised.

The prostate is retracted medially and a plane should be exposed on the endopelvic fascia lateral to the prostate, which is incised bluntly allowing the levator ani to be pushed off of the prostate laterally and inferiorly. The separation of the levator ani from the prostate usually requires incision of the puboprostatic ligaments in order to adequately expose the urethra and the dorsal venous complex (DVC).

The bedside assistant slides the Foley in and out moving the Foley balloon against the bladder neck to help identify the junction of the bladder neck and the prostate. The Prograsp grips the edge of the urachus on the bladder, retracts it slightly posteriorly and superiorly to create slight tension on the bladder neck. The anterior bladder neck is incised using electrocautery and once the urethra and bladder interior is visualized the balloon on the catheter is deflated and it is pulled anteriorly through the incision. The assistant grasps the Foley and using a clamp external to the body as well, the assistant is then able to retract the prostate anteriorly. The bladder neck and bladder is then inspected to ensure the ureteral orifices are a safe distance from the posterior aspect of the bladder neck incision and there are no appreciable bladder tumors.

The posterior bladder neck is then incised using electrocautery as well and progressed until the previously dissected VDs and SVs are visualized. It is important to account for median lobes during the posterior bladder neck incision as well as to take care to allow for a sufficiently thick posterior bladder wall. At this time, the assistant releases the catheter and the Prograsp is then used to retract the SVs and VDs anteriorly. This retraction should allow the surgeon to see if further development of the dissection of Denonvillier's is necessary or if one can begin performing dissection of the lateral prostatic pedicles and nerve sparing. There are a variety of approaches to nerve sparing. The approach we most commonly take involves dissecting the pedicle into vascular packets and after applying Weck clips, incising with sharp dissection. This is continued until the last attachments of the prostate are the urethra and the DVC.

The DVC is then either incised using electrocautery with the pressure briefly raised to 20 mmHg and then over sewn, or suture ligated and then incised depending on surgeon preference. Finally, the assistant ensures the Foley catheter is in the urethra and the urethra is incised sharply and the prostate is then placed into an endocatch bag, which is set aside for later removal.

At this point in time, the bilateral pelvic lymphadenectomy is then performed. The first step is identifying the pulsation of the external iliac arteries with the knowledge that the vein will be directly inferior to them. The lymphofatty tissue medial and inferior to the external iliac vein is dissected down to the pelvic sidewall, then anteriorly to the node of Cloquet, inferiorly to the pubic bone, posteriorly to the obturator nerve, then finally proximally to the bifurcation of the common iliac. This template can be extended proximately and posteriorly to include the common and internal iliac nodes as indicated by the patient's risk level. The nodes are then removed by the lateral assistant port or placed in endocatch bags for removal with the final specimen taking care to mark laterality.

Attention is then turned to the bladder neck, which is examined to determine if a plastic reconstruction is necessary to make it 20-French in size. If necessary, reconstruction is performed using a 2-0 braided absorbable figure of eight suture in a "tennis racquet" formation bringing the posterior aspects together.

After the bladder neck is adjusted if necessary, urethrovesical anastomosis is then performed using 3-0 unidirectionally barbed absorbable monofilament with a needle attached to each end. It is not this institution's routine to perform a Rocco stitch, however, this may be done if desired at this time. Urethrovesicle anastomosis is then performed in a running fashion beginning in the bladder at a position corresponding to the 5 o'clock position on the urethra and moving clockwise with at least three posterior throws and continued until approximately 11 o'clock with one needle. Care should be taken not tighten the anastomotic stitch tightly posteriorly until at least two or three throws have been made, as the urethra is at risk for tearing. At this point the first needle is reversed upon itself such that it is coming from inside the bladder out. The second running suture is begun at 5 o'clock and continued counter-clockwise until 11 o'clock where the second needle's direction is reversed through the bladder as well. Over the course of this suturing, the urethra is repeatedly identified and propped open by the assistant sliding the Foley catheter's tip in and out of the urethral aperture and the surgeon quickly following the catheter's removal with the needle. After suturing is complete, the catheter is swapped for a fresh catheter and the sutures are then tied down. The Foley balloon is then inflated and the catheter is then irrigated and the anastomosis is checked for leak.

The Prograsp is then removed from the body and if desired a JP drain is then advanced and placed in the pelvis through the robotic port.

The robot is then undocked and the endocatch bag(s) is pulled out via the midline camera port site. After the fascia of the specimen removal site is then closed, the abdomen is then reinsufflated to ensure no intra-abdominal contents were injured by fascial closure and the rest of the ports are removed under direct vision. It is important to pay attention to the 5 mm assistant port removal as it is removed through the rectus and thus has an anecdotal slightly higher risk of bleeding postoperatively. It is not routine to close the right lateral 10 mm port site's fascia. All skin is closed with a 4-0 monofilament subcuticular stitch.

## DISCUSSION

Comparative studies evaluating outcomes and advantages of the specific surgical approach used for MIRP are lacking, and the surgical approach used for MIRP is likely less important than the experience and expertise of the surgeon performing the procedure. One small retrospective study of anterior and posterior approach to LRP did not show any significant differences in operative parameters or outcomes between the two approaches, in a small sample size (59 patients).[22] However, there are several putative advantages of using the initial posterior approach to MIRP. Posterior dissection allows direct visualization of the tips of the SVs and thereby avoidance of damage to the erectile nerves that course nearby. By dissecting the SV tip medially away from the neurovascular bundle, and the judicious and conservative use of electrocautery, there is a theoretical advantage to minimizing electrical and stretch injury to the nerves. When approaching the prostate anteriorly, successful dissection of the entire SV out to its tip can require dissection further laterally, closer to the vicinity of the neurovascular bundles. Additionally, when using an anterior approach, there may be temptation to amputate the SV instead of complete resection laterally to avoid injuring the nerves, raising oncologic concerns. Perhaps most importantly, initial posterior dissection removes the need for blind dissection after the posterior bladder neck is incised, thereby sparing the ureters and rectum from potential blind damage.

Anecdotally, at our center, we noted that the posterior approach was preferred in special circumstances at either end of the spectrum, but not truly different for the average procedure. In patients with either very small or very large prostates, those that are markedly obese and those having large prostate median lobes, the posterior approach offers cleaner dissection planes and more efficient completion of the procedure. Accordingly, over the past 5 years, the vast majority of our MIRP are performed with initial posterior dissection with excellent outcomes.

## CONCLUSIONS

Clearly there are many ways to skin a cat, and initial posterior approach to MIRP is a useful technique to learn and perform. While it is unlikely that randomized trials evaluating different approaches to MIRP will be or need to be completed, there is utility in re-engagement with this technique, which was the first approach used widely in laparoscopic radical prostatectomy. Experience and knowledge of this technique and tissue planes involved can also be helpful for other procedures in urology, including robotic radical cystectomy, seminal vesiculectomy, diverticulectomy, distal ureterectomy with psoas hitch, Boari flap, and ureteral reimplantation.

## References

1. Ryan CJ, Elkin EP, Small EJ, et al. Reduced incidence of bony metastasis at initial prostate cancer diagnosis: data from CaPSURE. *Urol Oncol* 2006;**24**(5):396–402.

2. Schuessler WW, Schulam PG, Clayman RV, et al. Laparoscopic radical prostatectomy: initial short-term experience. *Urology* 1997;**50**(6):854–7.
3. Guillonneau B, Vallancien G. Laparoscopic radical prostatectomy: the Montsouris technique. *J Urol* 2000;**163**(6):1643–9.
4. Kavoussi LR, Schuessler WW, Vancaillie TG, et al. Laparoscopic approach to the seminal vesicles. *J Urol* 1993;**150**(2 Pt 1):417–9.
5. Abbou CC, Hoznek A, Salomon L, et al. Laparoscopic radical prostatectomy with a remote controlled robot. *J Urol* 2001;**165**(6 Pt 1):1964–6.
6. Coelho RF, Chauhan S, Palmer KJ, et al. Robotic-assisted radical prostatectomy: a review of current outcomes. *BJU Int* 2009;**104**(10):1428–35.
7. Colombo Jr JR, Santos B, Hafron J, et al. Robotic assisted radical prostatectomy: surgical techniques and outcomes. *Int Braz J Urol* 2007;**33**(6):803–9.
8. Dasgupta P, Kirby RS. Outcomes of robotic assisted radical prostatectomy. *Int J Urol* 2009;**16**(3):244–8.
9. Diaz M, Peabody JO, Kapoor V, et al. Oncologic outcomes at 10 years following robotic radical prostatectomy. *Eur Urol* 2014;**67**(6):1168–76.
10. Finkelstein J, et al. Open versus laparoscopic versus robot-assisted laparoscopic prostatectomy: the European and US experience. *Rev Urol* 2010;**12**(1):35–43.
11. Hakimi AA, Blitstein J, Feder M, et al. Direct comparison of surgical and functional outcomes of robotic-assisted versus pure laparoscopic radical prostatectomy: single-surgeon experience. *Urology* 2009;**73**(1):119–23.
12. Pow-Sang J. Pure and robotic-assisted laparoscopic radical prostatectomy: technology and techniques merge to improve outcomes. *Exp Rev Anticancer Ther* 2008;**8**(1):15–9.
13. Wolfram M, Bräutigam R, Engl T, et al. Robotic-assisted laparoscopic radical prostatectomy: the Frankfurt technique. *World J Urol* 2003;**21**(3):128–32.
14. Korets R, Weinberg AC, Alberts BD, et al. Utilization and timing of blood transfusions following open and robot-assisted radical prostatectomy. *J Endourol* 2014;**28**(12):1418–23.
15. Guillonneau B, Cathelineau X, Barret E, et al. Laparoscopic radical prostatectomy: technical and early oncological assessment of 40 operations. *Eur Urol* 1999;**36**(1):14–20.
16. Guillonneau B, Vallancien G. Laparoscopic radical prostatectomy: initial experience and preliminary assessment after 65 operations. *Prostate* 1999;**39**(1):71–5.
17. Tewari A, Peabody J, Sarle R, et al. Technique of Da Vinci robot-assisted anatomic radical prostatectomy. *Urology* 2002;**60**(4):569–72.
18. Trabulsi EJ, Zola JC, Gomella LG, et al. Transition from pure laparoscopic to robotic-assisted radical prostatectomy: a single surgeon institutional evolution. *Urol Oncol* 2010;**28**(1):81–5.
19. Trabulsi EJ, Zola JC, Colon-Herdman A, et al. Minimally invasive radical prostatectomy: transition from pure laparoscopic to robotic-assisted radical prostatectomy. *Arch Esp Urol* 2011;**64**(8):823–9.
20. Trabulsi EJ, Guillonneau B. Laparoscopic radical prostatectomy. *J Urol* 2005;**173**(4):1072–9.
21. Trabulsi EJ, Hassen WA, Touijer AK, et al. Laparoscopic radical prostatectomy: a review of techniques and results worldwide. *Minerva Urol Nefrol* 2003;**55**(4):239–50.
22. Li B, Suzuki K, Tsuru N, et al. Retrospective comparative study of 59 cases of laparoscopic radical prostatectomy: transperitoneal anterior versus transperitoneal posterior approach. *Int J Urol* 2007;**14**(11):1005–8.

CHAPTER

# 38

# The Technique of Robotic Nerve Sparing Prostatectomy: Extraperitoneal Approach

*Christian P. Pavlovich, MD*

James Buchanan Brady Urological Institute, Johns Hopkins University School of Medicine, MD, USA

## INTRODUCTION

Prior to the advent of robotics, a consensus panel of expert urologic surgeons judged laparoscopic radical prostatectomy to be "extremely difficult," and the highest level of urologic surgical complexity in their ranking system.[1] The robotic approach has certainly facilitated the performance and dissemination of laparoscopic radical prostatectomy throughout the United States. Once the transperitoneal approach to the prostate was perfected laparoscopically, some of the initial pioneers developed a totally extraperitoneal approach to minimally invasive radical prostatectomy.[2,3] It was just a matter of time until robot-assisted radical prostatectomy was performed extraperitoneally,[4,5] a development facilitated by the smaller, second-generation robotic platform, the Da Vinci S™. The more concise arms of robotic devices marketed allow less clashing in relatively more confining space of Retzius, while sitting at the remote console decreases the physical demands of the laparoscopic approach on the surgeon.

## THE TECHNIQUE

The concept is to mimic open radical retropubic prostatectomy as far as possible, and to use the advantages of robotic instrumentation and pneumoperitoneum to minimize blood loss and facilitate careful dissection. Compared to open radical retropubic prostatectomy, extraperitoneal robotic radical prostatectomy utilizes essentially the same steps, albeit in a slightly different order. After all, both procedures are performed in the retropubic or extraperitoneal space of Retzius and with the same goal: complete prostate extirpation, nerve sparing as indicated, and in many cases, bilateral pelvic lymphadenectomy.

### Patient Positioning and Port Placement

The patient is placed supine on a butterfly beanbag specifically made for robotic pelvic surgery that is secured to the operating table. Legs are placed in Allen stirrups and spread apart and then lowered slightly with flexion at the hip joints. Arms are carefully tucked and secured at the sides and held in place by the beanbag, as are the shoulders. After prepping and draping, a Foley catheter is placed sterilely in order to decompress the bladder.

Rather than using Veress needle insufflations, the retropubic space is best entered through a 12 mm infraumbilical skin incision, in similar fashion as for total extraperitoneal (TEP) inguinal hernia repair. Through this incision the anterior rectus fascia is identified and incised. Our preference is to incise directly with a Visiport angled 60–80° to the vertical, aiming slightly caudally, though others prefer the open cut down Hasson technique (note that the Visiport does not accommodate the standard 12 mm robotic camera, but only a laparoscopic 10 mm or robotic 8 mm camera). At this point the Visiport is gently advanced between the rectus muscle bellies, generally to the left of the linea alba (Figure 38.1) and is then slid over the posterior sheath toward the arcuate line at a more acute angle (30° to the skin) aiming caudally and into the space of Retzius. The arcuate line is passed over as the horizontal fibers of the posterior sheath give way to transversalis fascia, and then the Visiport is pushed left and right in a fan-shaped manner to gently open up the retropubic potential space for subsequent balloon dilatation. The Visiport is removed

Prostate Cancer. http://dx.doi.org/10.1016/B978-0-12-800077-9.00038-4

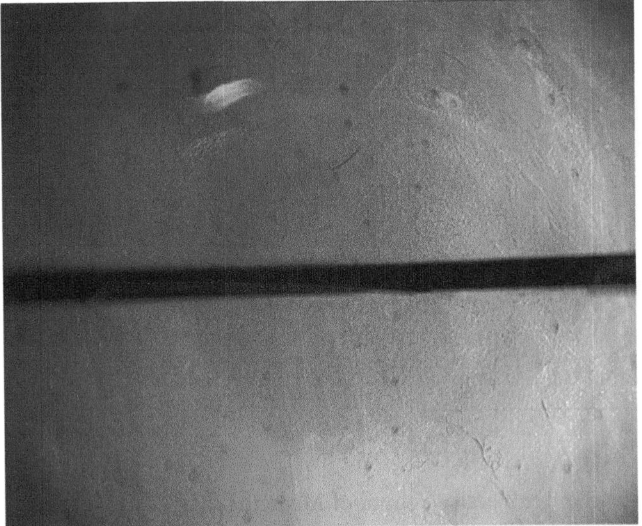

**FIGURE 38.1 View through the visiport left of the linea alba and after the anterior rectus fascia has been horizontally incised.** The left rectus m. belly is noted at the left and the posterior rectus sheath is visible at the right (yellow/areolar tissue).

and a kidney-shaped laparoscopic balloon (PDBS2 AutoSuture™, Covidien, Mansfield, MA) is blindly inserted into the infraumbilical incision just past the incised anterior rectus fascia and angled southward into the space of Retzius in front of the bladder and behind the pubis. At this point the balloon is manually insufflated and the expanding balloon is visualized through the port with the laparoscopic zero degree lens (Figure 38.2). The inferior epigastric artery and veins are easily identified superolaterally through the clear developing balloon as it pushes the parietal peritoneum away superiorly. Finally, the balloon is deflated and removed, and the Visiport trocar is reinserted into the space and insufflation is started. The whole case can be performed with pressures ranging from 12 mmHg to 20 mmHg; it is always safest to use the lowest possible insufflation pressure to maintain the space. If the space is intact, it should appear concave, football shaped, and have flimsy areolar tissue within it overlying the structures of the deep pelvis.

At this point the remaining ports are placed, all with blunt trocars, and essentially in a fan-shaped position around the pubic midline. Reusable metal robotic ports (3 × 8 mm) are used, as well as a 12 mm disposable assistant port (Ethicon Xcel for example). The configuration is as in Figure 38.3, with ports angled caudally about 45–70° from the vertical and generally placed along the line from the anterior superior iliac spine (ASIS) to the umbilical incision, with a minimum of 8 cm between robotic ports. The first port is 8–9 cm left lateral to the umbilical port along the line to the left ASIS; then that is used to insert a blunt alligator grasper, which can facilitate dissection of the peritoneum cranially and posterior for right lateralmost port placement 2 cm superior and 2 cm medial to the right ASIS (Figure 38.4). This port gets tunneled subcutaneously until it reaches the line between the right ASIS and the umbilical port. It then gets pushed in deeper and into the true pelvis gliding over the spermatic cord and lateral to the peritoneal reflection, which is often noted here (Figure 38.5). Finally, the right medial 8 mm robotic trocar is placed 8–9 cm to the right of the umbilical port along the line toward the right ASIS, and then the left-most 8 mm robotic port for the fourth robotic arm is placed near the left ASIS but again some 2 cm superior and medial to it, then tunneled subcutaneously toward the left ASIS–umbilical port line, before passing into the pelvis and over the spermatic cord.

**FIGURE 38.2 View of the pelvis from within the just-deployed balloon that is creating the potential space in which the operation will take place.**

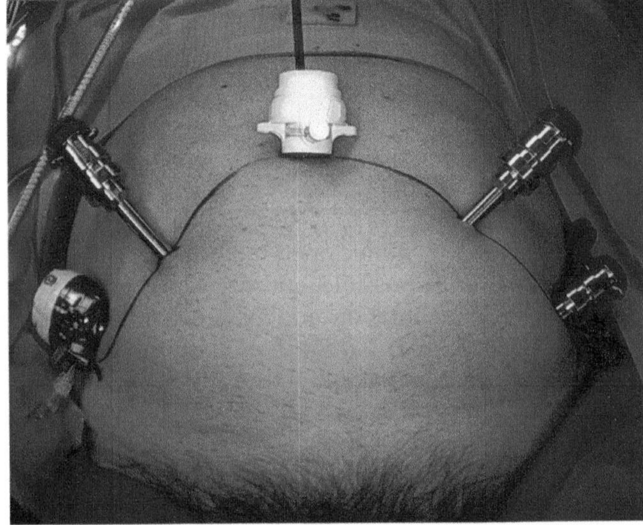

**FIGURE 38.3 Port configuration for extraperitoneal robotic-assisted radical prostatectomy.** The four robotic ports (the metal ports plus the midline camera port) must each be at least 8 cm from each other.

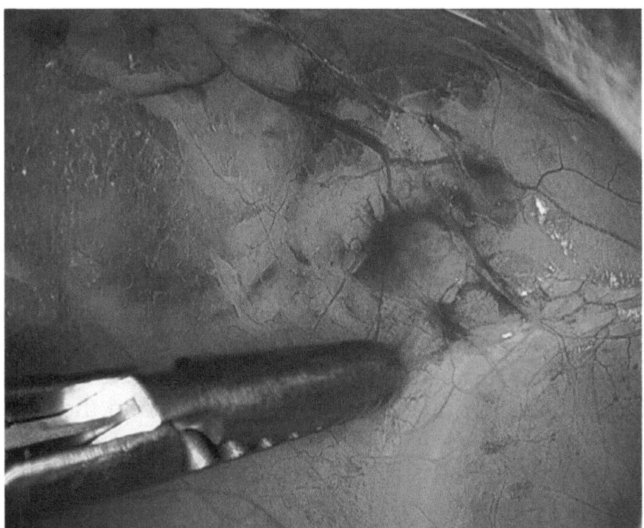

FIGURE 38.4 **The peritoneal reflection is noted in the right lower aspect of the figure, appearing as a pale line going across the image from lower left to upper right.** An alligator grasper is being used to push the peritoneum more inferiorly and superiorly such that the sidewall can be identified over the spermatic cord.

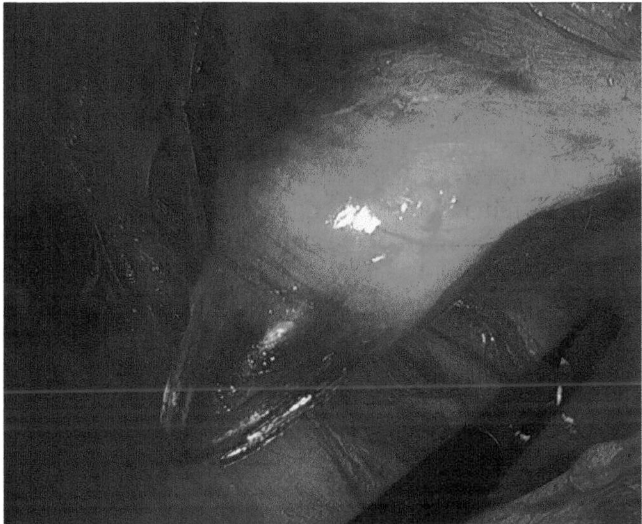

FIGURE 38.5 **The spermatic cord is being held down as the assistant port is being placed caudal and superior to the peritoneal lining in the right lower quadrant.**

It is crucial to angle all of the ports approximately 45–70° caudally on insertion (e.g., toward the legs) in order to avoid the peritoneal space, which is often visible at the time of port placement, and to watch the trocars go in under direct vision.

The patient is then placed in modest Trendelenburg (full Trendelenburg is rarely needed) and the space is analyzed. A peritoneotomy would be noticed at this point as the space would appear small and its superoposterior wall (peritoneum) would be billowing into the field. This relatively rare occurrence (about 1/20 cases in experienced hands) need not change one's operative plan. A small/microscopic peritoneotomy, which lets gas into the peritoneal cavity, can be managed by inserting a Veress needle into the abdomen for continuous venting of this space during the operation. Alternately, since the bladder has been "taken down" already, a small inadvertent peritoneotomy can purposely be enlarged on one or both sides to render the extra- and transperitoneal spaces contiguous, but still without losing some of the advantages of extraperitoneal surgery, such as the fact that the bladder is already taken down, and the veil of tissue composed of the urachus and medial umbilical ligaments maintains separation between the patient's bowels and the pelvic space.

The Da Vinci robot is then moved into place between the legs. If the legs need to be spread a bit more or lowered to avoid clashing with the robot, this is easily accomplished by adjusting the Allen stirrups. Docking is done carefully and remembering to keep the ports angled even while docking to avoid inadvertent peritoneotomy. Robotic arm number three comes in left lateral-most and must be docked such that it does not clash with the stirrup, left knee, left foot, or pelvis. Then, under direct vision, the robotic instruments should be advanced into the retropubic space. Our preference is for a ProGrasp forcep in the port at far left (arm #3), a bipolar Maryland grasper in the more medial left robotic port (arm #2), a monopolar scissor in the right robotic port (arm #1), and a zero degree lens in the camera port. The 12 mm assistant port is used for a right-sided assistant and allows for retraction, suction, clip placement, and also specimen (lymph node packet(s)) extraction.

## The Extraperitoneal Space

The extraperitoneal space at this point only needs to be expanded a bit in order to appear identical to the standard open retropubic space to which surgeons are accustomed. It is worthwhile to spend several minutes dissecting through areolar tissue until the entire pubis and midline symphysis are clearly identified. It then helps to de-fat the area as well until the endopelvic fasciae are visible (Figures 38.6 and 38.7). The superficial dorsal vein is usually apparent as fat gets dissected off of the anterior prostate and it should be taken with electrocautery. The fat overlying it generally ramifies along with its proximal branching centrally and to the right and left, over the prostate and toward the bladder; with bipolar diathermy these branches can easily be taken and the fat and superficial dorsal venous complex removed to better expose the prostate. In recent years, we have been sending this fatty tissue for pathological analysis because a minority of patients (5–15%) will have periprostatic lymph node(s) contained within.

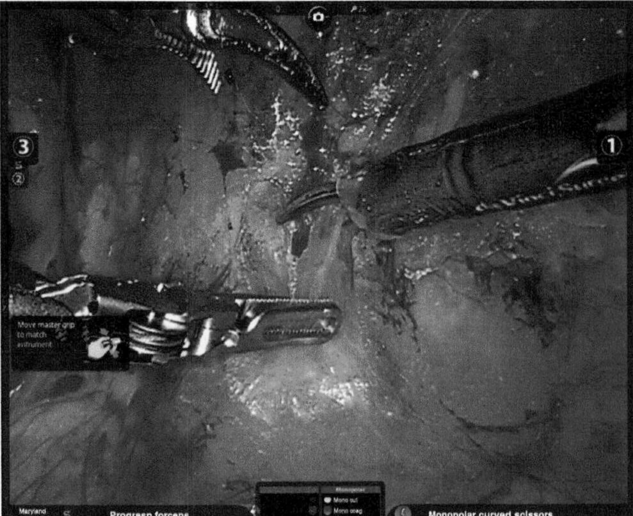


FIGURE 38.6 The space of retzius as it appears after removal of the insufflation balloon, port placement, and insertion of robotic instruments.

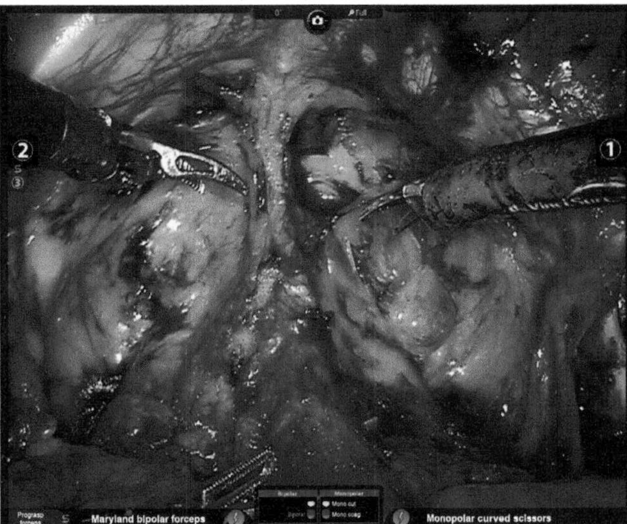


FIGURE 38.7 The same space after areolar tissue has been cleared away, but prior to defatting.

Flimsy tissues overlying the external iliac vessels and spermatic cords should then be incised to facilitate easy access to the pelvis for the assistant and fourth robotic arm. Any inguinal hernias or herniated fat that are in the way should be reduced. Finally, identification of important landmarks ensues: the puboprostatic ligaments and endopelvic fasciae for the prostatectomy, and the external iliac veins (EIV) and obturator nerves for men who will be receiving pelvic lymph node dissection (PLND). It is particularly important to make sure to assess the position of the right EIV in comparison to the right assistant port, and ideally to either advance that port past the vein or keep it quite superolateral to the vein such that instruments that are placed through the port do not risk traumatizing the vein.

## Pelvic Lymph Node Dissection

We prefer to perform PLND prior to prostatectomy, though it certainly can be accomplished subsequent to the mobilization of the prostate. In either case, preparation for PLND involves identifying the EIV and obturator nerves. The former can usually be found, even if there is much fat overlying them, by looking for venous waveforms just above the pelvic brim. Once found, the fascia overlying the EIVs must be incised and the sidewall accessed under the veins. The assistant's suction device can be placed in this space while the operator dissects the nodes off the EIV and looks for the obturator nerve. If not obviously under the EIV it can often be found by looking for the obturator foramen, which often has a fat plug in it. The nerve courses through the foramen and out of the pelvis by hugging the foramen superolaterally; once it is found the nodal packet under the EIV and above the obturator nerve should be clipped distally and handled with the ProGrasp for dissection as proximally as indicated. A standard PLND can certainly be performed in this space, taking all level I and II nodes, as the bifurcation of the iliac vessels is reachable (as in open radical prostatectomy, which as mentioned, is performed in the same space and with the same limitations). With the ProGrasp holding the nodes anteriorly, two arms remain for dissection, cautery, and retraction. In addition, the wrist of the arm on the side of the PLND can be used to hold back preperitoneal fat if it gets in the way, while still allowing the instrument to be used as a grasper/dissector. Extended PLND depending on the definition may or may not be able to be performed, as presacral nodes are not reachable without mobilizing the bladder more than can be done extraperitoneally; a 30° lens may be used to facilitate deeper and more medial dissections at the internal iliac artery(ies) and vein(s). In any case, we favor using Hem-0-Lok™ clips (Weck, Teleflex, Limerick, and PA) to secure lymphatic pedicles, rather than bipolar cautery or suction whenever possible. A harmonic scalpel is also a great tool for PLND and for sealing small lymphatics, but is certainly not necessary. The nodes are typically extracted after the dissection, through the 12 mm assistant port, though they can also be left in their location for later extraction with the prostate specimen. A drain should be placed at the conclusion of the case whenever PLND is performed, particularly when confined to the extraperitoneal space.

## Radical Prostatectomy

The subsequent steps can be performed in the same manner as a radical prostatectomy would be done in open retropubic fashion, or in the anterior approach to transperitoneal robotic radical prostatectomy. The order of steps and the tools used may differ, but fundamentally a retropubic potential space, once expanded, allows a comparable retropubic approach to the cancerous prostate. An advantage for remaining extraperitoneal, or preperitoneal, as in open radical retropubic radical prostatectomy, is that steep Trendelenburg is not necessary prior to docking the robot, although minor Trendelenburg is helpful. The reason is that the bladder and peritoneum form a natural superior and posterior floor to the operation until Denonvillier's fascia is reached, preventing small and large bowel, and any prior intra-abdominal adhesions from complicating access to the prostate. The critical aspect of the case that differs from most other approaches and that must be mastered in order to perform successful extraperitoneal radical prostatectomy is identification, dissection, and taking of the posterior genital structures (the vasa deferentia and seminal vesicles) anteriorly and without posterior dissection through the cul-de-sac. These structures are taken after bladder neck transection and not accessed transperitoneally (posteriorly) nor retrograde as in open retropubic radical prostatectomy. Nerve sparing on the other hand, may be accomplished antegrade, retrograde, or in a combined manner regardless of the approach to the prostate, and is performed per the preferences of the operating surgeon.

### Step 1: Incising the Endopelvic Fasciae and Puboprostatic Ligaments

The surgeon should search for a relatively avascular plane between the prostate and its lateral attachments. The optimal place to start is at the proximal shoulder of the prostate on either side, just distal to the vasculature of the bladder at the lateral prostate–bladder interface. The endopelvic fascia at its proximal origin is easily identifiable and can be swept toward the apex until at one point it needs to be incised and will no longer sweep away. This incision line, made by scissor or with monopolar cautery, should be continued to the puboprostatic ligaments, which are then carefully taken at their midpoint in order to allow for better ligation of the deep dorsal vein beneath them. The more distal and inferior aspects of the puboprostatics should not be taken as they lend support to the urethra as it exits the prostate apically. Ideally, after incision of the endopelvic fascia, the lateral prostatic fascia is kept on the prostate, the levator ani fascia is left on the levator ani muscle, and a plane between them develops with venous channels (often quite prominent) notable on the prostate side and just covered by lateral prostatic fascia. For intrafascial dissections, if desired, these veins can be left lateral to the plane of dissection and contiguous with the levator fascia, allowing the surgeon to get right onto the prostate capsule. This dissection is however more difficult, associated with more bleeding, and should ideally only be undertaken in low- to intermediate-risk patients with minimal or no apparent disease on that side.

### Step 2: The Dorsal Vein

Once the endopelvic fasciae and puboprostatic ligaments have been incised, the prostatic apex becomes readily apparent. Most of what is seen is deep dorsal vein and supporting connective tissue over the prostate, and the urethra and its Foley catheter are surprisingly deep. With meticulous dissection, deep and lateral to the puboprostatics, the sides of the deep dorsal vein of the penis become apparent and its posterior extent is usually visible. Once the surgeon is content with both sides of the apical dissection, a prudent move is to simulate where a dorsal vein stitch would be thrown from one side to the other, and place an instrument there such as a Maryland or PK dissector and gently spread its jaws to prepare the area and move the vein away from it. Then the assistant switches to needle drivers and a stitch is thrown. The ligation is performed as distally as possible to ensure that the prostatic apex is not inadvertently entered during subsequent division of the DVC. We prefer to use a 2–0 braided suture (e.g., Polysorb) on a large needle (GS-21) for most dorsal venous complexes and throw the needle right-to-left and horizontally under the dorsal vein but above the urethra, then again throw it horizontally in a figure-of-eight and locking fashion just anterior to the dorsal vein and through the remaining anterior portion of the incised puboprostatic ligaments (Figure 38.8). The placement of these throws and the locking of the suture promote upward traction on the

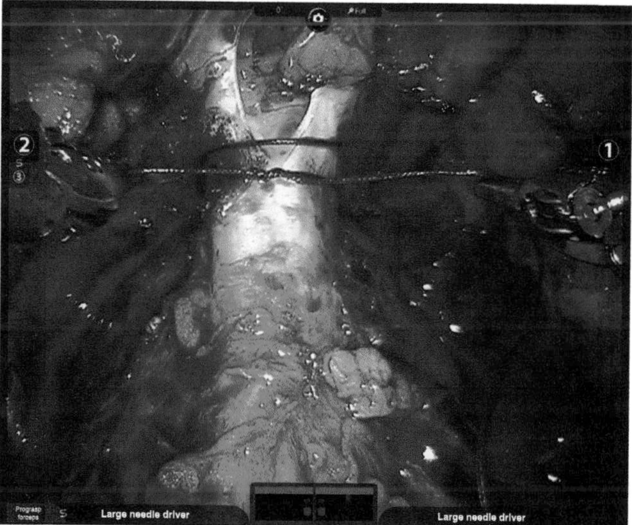


FIGURE 38.8 **The endopelvic fasciae have been incised as have some of the puboprostatic ligaments.** The dorsal vein stitch (2–0 Polysorb) has been thrown from right to left under the deep dorsal vein and then again superior to the vein and through the more anteriorly placed remnant puboprostatic ligaments.

final knot, and the braided nature of the suture as well as an initial surgeon's knot facilitates the knot holding fast. Most surgeons prefer not to actually divide the dorsal venous complex until later in the operation, at the time of the apical prostatic dissection. However, some surgeons are now using a stapler to take the dorsal vein rather than a stitch, which may save time but increases disposable costs significantly as such a device would not otherwise be opened. In addition, one must take care not to inadvertently staple into or too close to the urethra so as not to leave any metallic foreign bodies near the forthcoming anastomosis.

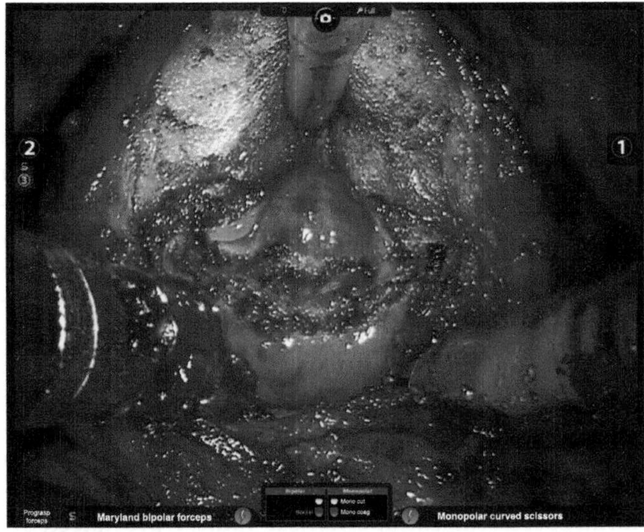


FIGURE 38.10 **The bladder neck has been transected anteriorly and the posterior bladder mucosa has been scored for subsequent transection just under a small median lobe.** The instruments are in the bladder, holding down the anterior bladder neck for purpose of photo.

### *Step 3: The Bladder Neck Transection*

Attention is then turned to the bladder neck. Several different maneuvers can aid in identification of the interface between the prostate and bladder neck. First, with the assistant gently pulling on the urethral catheter, the surgeon can usually identify the junction between these two structures. This junction can often be further delineated as the point where the perivesical fat gives way to the prostatic contour. Once this junction is identified, the anterior bladder neck is divided using electrocautery (Figure 38.9). The bladder is retracted cranially and posteriorly during this maneuver to facilitate dissection.

Once the lumen of the bladder is entered, the urethral catheter is identified and the balloon is deflated. The catheter is grasped with the fourth robotic arm and tented anteriorly over the pubic symphysis to suspend the prostate. At this point, as the assistant retracts the anterior neck inferiorly, the surgeon should attempt to check for the presence of a median lobe, identify the trigone, and judge the position of or actually see the ureteral orifices. The mucosa of the posterior bladder neck is then divided in an upward smile shape under and proximal to the prostatic urethra (Figure 38.10). The detrusor muscle of the posterior bladder neck is then divided, taking care not to enter the prostate but also ensuring that this layer is left sufficiently thick to later hold anastomotic stitches.

### *Step 4: Anterior Approach to the Posterior Genital Structures*

After division of the posterior bladder neck, the vasa deferentia are identified deep and near the midline (Figure 38.11). Once reasonably well exposed, these are grasped with the ProGrasp as it is no longer needed in

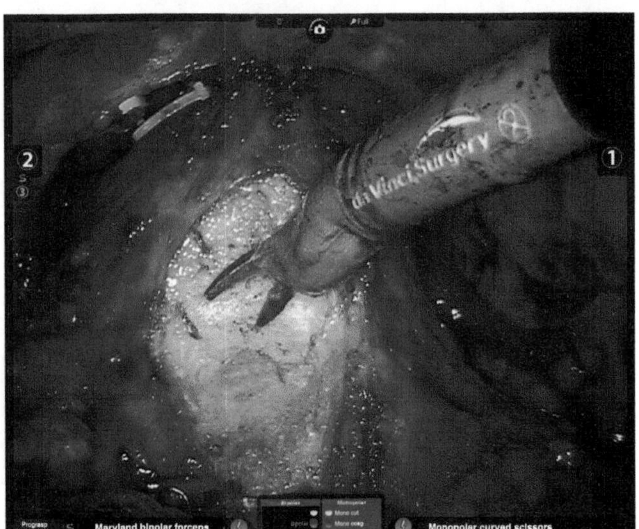


FIGURE 38.9 **Starting to separate the bladder neck from the prostate.**

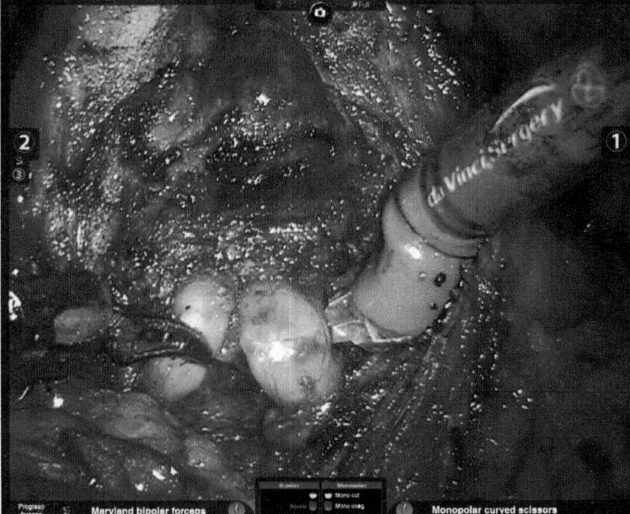


FIGURE 38.11 **The bladder has been transected from the prostate, a layer of fascia anterior to the posterior genital structures has been incised, and the midline vasa deferentia have come into view.**

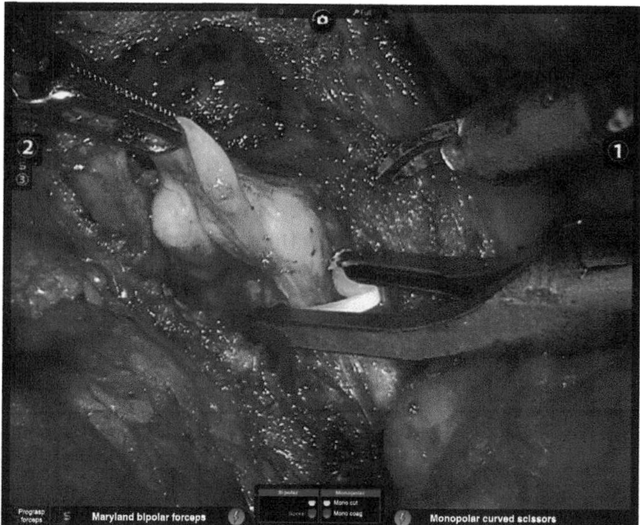


FIGURE 38.12 **The right seminal vesicle pedicle is now being taken as the ProGrasp™ holds it up and to the left.**

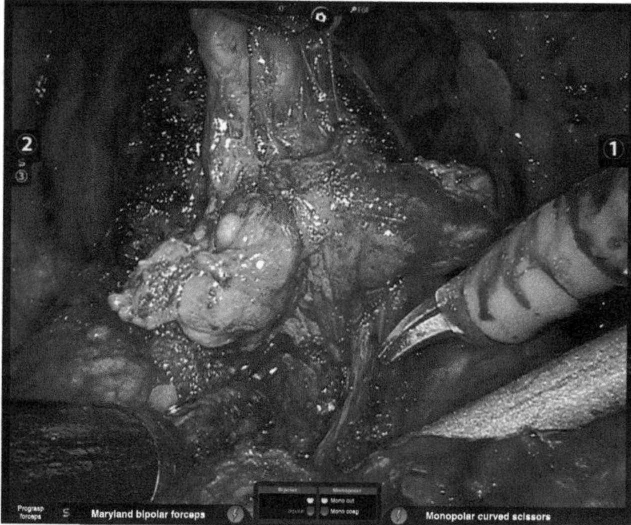


FIGURE 38.13 **The vasa and both seminal vesicles have now been dissected free and are being elevated in order to get under the prostate and continue toward the apex.**

order to suspend the prostate anteriorly by the urethral catheter; the vasa are retracted anteriorly one at a time and carefully dissected free from surrounding structures as they head laterally toward the deep inguinal rings. The vasa are then clipped and divided. This exposes the seminal vesicles, and the more proximally the vasa are taken the closer to the tips of the seminal vesicles the surgeon will find themselves. One at a time, each seminal vesicle is then grasped and dissected free from surrounding tissues. It is easier to identify the medial border first and sweep Denonvillier's fascia posteriorly off of the seminal vesicle, and then work toward the proximal tip and lateral border. A neurovascular pedicle is typically present at the tip of each seminal vesicle; this structure is clipped and divided without cautery to prevent damage to the neurovascular bundle (Figure 38.12). Once both seminal vesicles are dissected free, they are retracted anteriorly, exposing more of Denonvillier's fascia and the lateral pedicles to the prostate (Figure 38.13). Denonvillier's is then divided horizontally in the midline, a few millimeters from the junction of the posterior prostate and seminal vesicle insertion, in order to leave a healthy posterior margin on the prostate. Division of Denonvillier's fascia exposes the underlying prerectal fat. The plane between the prostate and rectum can then be gently developed in the midline using a combination of sharp and blunt dissection, proceeding toward the apex of the prostate. Care is taken not to proceed too far laterally with this dissection, as this can result in troublesome bleeding from the vascular pedicles, which have not yet been ligated.

### *Step 5: The Pedicles and Nerve-Sparing*

At this point, the tip of a seminal vesicle is grasped with the fourth arm ProGrasp and retracted cranially and medially. This exposes the ipsilateral neurovascular pedicle. Depending on preoperative disease characteristics, the surgeon can then choose to perform nerve-sparing or take a wider soft-tissue margin. This has been described extensively in prior chapters of this textbook. Working from here toward the apex is challenging in that one must stay aware of the changing contour of the ellipsoid prostate gland as it courses apically, and use clips and cutting between the neurovascular bundles and the prostate in order to sever their branching neurovascular and fascial attachments (Figure 38.14).

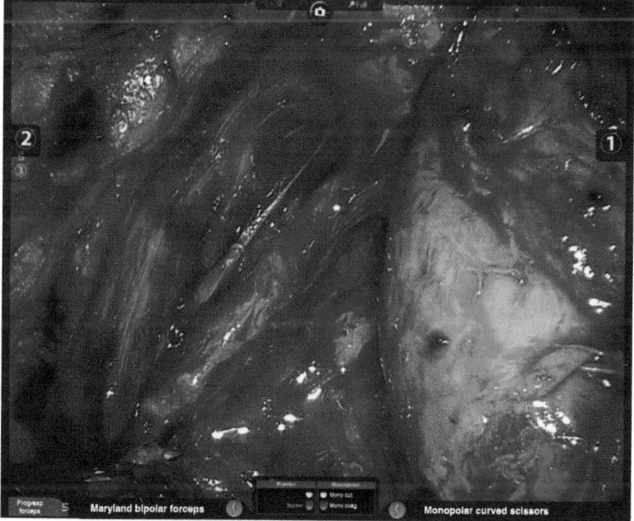


FIGURE 38.14 **An intrafascial dissection from prostate base to apex has been performed.** The healthy veil of tissue overlying the left neurovascular bundle that was spared is being shown by the Maryland bipolar at lower left. Note the lack of clips over the bundle this distally and the smooth left prostatic margin.

The surgeon must remain vigilant in all cases about leaving any permanent foreign bodies (e.g., Hem-o-Lok clips) near the bladder neck and distal remaining urethral stump to prevent subsequent erosion into the neo-bladder neck, clips, and similar nonabsorbables should never be left within approximately 2 cm of this area. Options to control bleeding more distally are to use gentle bipolar cautery to specific arterial bleeders, and to use fine absorbable suture ligation, bipolar cautery, and/or topical hemostatic agents for any open venous channels that may remain atop the neurovascular bundles after they have been separated from the prostate. Later, it is important also to reassess the bladder neck as it is brought down for reanastomosis, to remove any clips from it that would find themselves in proximity to the new anastomosis.

### *Step 6: The Apical Dissection*

The deep dorsal venous complex, which was previously ligated, is now divided, leaving a 1/2 cm cuff of tissue between the cut and the stitch. If troublesome bleeding occurs at this point, it is best managed initially by simply finishing the transection of the complex such that the cut distal end retracts up into the suture and under the pubis. If bleeding continues, particularly if arterial (rare) or venous (through large open sinuses that are not compressed by the stitch) the placement of an additional hemostatic suture may be necessary. One easy way of dealing with this is to religate the complex, this time with a 2–0 absorbable braided suture on a GU needle passed horizontally from one side to the other in a figure-of-eight fashion.

Following division of the DVC, the urethra is visible and is divided in a location that maximizes urethral length but leaves sufficient tissue coverage on the prostate to avoid a positive margin. Division of the urethra frees up most, if not all of the specimen, which is then grasped with the fourth arm at its right and then left shoulders (or seminal vesicles) and retracted medially such that the posterior urethral plate or striated sphincter can be accessed. Any remaining fibers here between the transected urethra anteriorly and the previously freed up posterior overlying the rectum are carefully divided, thus freeing up the specimen completely. The specimen is then retracted out of the operative field by grasping it with the fourth arm and retracting it cranially and just lateral to the camera on the left. Placing the prostate specimen in an EndoCatch bag at this point actually hinders construction of the anastomosis; it is far better to just retract and hold onto the specimen with the fourth arm's ProGrasp until the anastomosis is complete and the Endo Catch bag can be deployed for specimen extraction.

### *Step 7: Assessment of the Specimen and Prostatectomy Bed*

The field is then inspected for bleeding, which can be facilitated with irrigation of sterile water by the assistant, and the lowering of insufflation pressure to 10–12 mmHg. Arterial bleeders should be clipped or managed with pinpoint electrocautery. Venous bleeders can be oversewn, although this is rarely necessary. Clips or bipolar cautery can be judiciously used, but if the venous oozers are distal and large caliber it is often best to proceed with the anastomosis and then undermine it with a hemostatic agent once it is half complete. Reinspection at this point makes sense but rarely will venous bleeding not stop by now.

The specimen itself is inspected prior to reanastomosis of bladder and urethra and any areas suspicious for capsular incision should elicit consideration of resecting a portion of matching tissue from the prostatectomy bed, for frozen or permanent section. Similarly, areas in the bed suspicious for prostate or cancer tissue should be resected and sent for analysis. At this point, a barbed suture can optionally be run in order to approximate the cut edges of Denonvillier's fascia near the urethral stump and proximally near the bladder neck. The aim is to reduce tension on the upcoming anastomosis, and reinforce the posterior plate.

### *Step 8: The Anastomosis*

The vesicourethral anastomosis is then performed. We prefer to perform the anastomosis with two different colored 9″ V-Loc-90™ (Covidien) barbed sutures on small tapered needles tied together at 7 in., such that there are barbs throughout the anastomosis and no remaining areas of unbarbed suture. The anastomosis is performed in running fashion as per van Velthoven,[6] beginning at the bladder neck at the 6 o'clock position. One suture is run clockwise and the other counterclockwise, meeting at the 12 o'clock position, where the sutures are tied together. It is our preference to tie across the anastomosis rather than on one side of it, so the suture on one side is thrown in an outside-in, inside-out fashion bridging the 12 o'clock position on the bladder neck, exiting at either 11 or 1 o'clock and tied to the matching suture from the contralateral side, which is exiting the urethra at pretty much the same position. An 18-French silicone urethral catheter is placed just prior to tying the anastomosis. The bladder is then irrigated with 120 $cm^3$ of normal saline to check for an anastomotic leak, and then the catheter balloon is inflated to 15 $cm^3$.

Often the caliber of the bladder neck opening will be significantly larger than that of the urethra. In this setting, we prefer to "parachute" the bladder down, spacing neighboring sutures further apart on the bladder neck than on the urethra. In some cases it is helpful to narrow the bladder neck at each side with a barbed suture until it more closely approximates the caliber of the urethral stump. If a cystotomy persists at the completion of the anastomosis, an anterior reconstruction can be performed to close the defect in the bladder onto itself, often with a simple figure-of-eight suture at 12 o'clock. The anastomosis is checked for leakage by instilling up to 180 $cm^3$ through the Foley, then the Foley balloon is inflated to 15 $cm^3$ and gently pulled to make sure it settles in the appropriate place at the bladder neck.

### *Step 9: Closure*

After completing the anastomosis, the specimen is placed into a laparoscopic entrapment sack and the robot is undocked. A closed suction drain is placed through the left-lateral robotic port and sewn to the skin. The accessory ports are removed after passing the specimen bag stitch through the infraumbilical camera port. Finally, the specimen is extracted through the infraumbilical incision, which is extended at the skin level to accommodate the specimen, and at the level of the anterior rectus fascia in the midline superiorly and inferiorly as much as necessary. The anterior rectus fascia is closed with interrupted #1 Maxon suture with two or three figure-of-eight throws, making sure to capture at least 1 cm of fascia on either side with each bite. Skin incisions are closed with absorbable suture and/or skin glue and all specimen(s) are sent for histopathologic analysis.

## SPECIAL CONSIDERATIONS

The extraperitoneal approach is ideal for men with factors that make them more challenging surgical patients:

- *Obesity*: Obese patients pose an additional challenge, even more so when they have central obesity. The distances and angles traveled in order to reach the prostate are greater, therefore the use of bariatric length ports is recommended. The extraperitoneal approach is helpful in these men, however no intra-abdominal fat is encountered, and the bowels are kept out of the way by the peritoneum, which acts as a retractor. The safety of the extraperitoneal approach in obese men has been demonstrated.[7] It helps to use more Trendelenburg in obese men than one would otherwise use in order to keep the trocar/instrument angles more standard and not excessively angled due to a protuberant abdomen.
- *Prior abdominal surgery*: Men with prior surgery, particularly low abdominal operations including appendectomy for perforated appendix, low anterior resection for colon cancer or diverticulitis, or umbilical/ventral hernia repair, are at an increased risk for enterotomy during port placement and lysis of adhesions. Angling the ports into the space of Retzius and performing robotic radical prostatectomy extraperitoneally is simply ideal for such patients, and has resulted in zero enterotomies in over 800 laparoscopic and robot-assisted extraperitoneal cases at our institution, even without a Hasson-type access approach.[8] By gliding infraumbilically with the camera port, and then carefully placing subsequent ports under direct vision and at least 2 cm away from any prior incisions, access for extraperitoneal robot-assisted radical prostatectomy can be achieved in almost every patient. Small peritoneotomies are more commonly encountered in patients with prior surgery, but are usually of little consequence if they are either enlarged somewhat or simply vented transperitoneally with a Veress needle or angiocath for the duration of the case.
- *Men with inguinal hernia(s)*: Men with prior bilateral mesh inguinal hernia repairs, especially if performed laparoscopically and even more so if performed totally extraperitoneally, should only be offered the extraperitoneal approach cautiously. Not only the pieces of mesh occasionally meet in the midline, they also result in significant reaction and adhesion in the exact space in which radical prostatectomy is supposed to be performed. In such men, especially those who have had bilateral repairs performed extraperitoneally, a transperitoneal robot-assisted approach is far safer and easier. On the other hand, men with prior unilateral hernia repairs with mesh, or with unilateral or bilateral reducible and untreated inguinal hernias pose only minor inconvenience, and can actually have their hernia defects repaired during the prostatectomy prior to conclusion of the case. We have performed many mesh inguinal hernia repairs in the same setting as the extraperitoneal robot-assisted radical prostatectomy (often with a general surgery colleague's assistance). No recurrences have been noted to date, and the incidence of incisional (port-site hernia) is also very low (0.3%) despite the fact that we only formally close fascia at the infraumbilical specimen extraction site.[12] The angled extraperitoneal nature of the ports, and the blunt trocars used, are likely explanations for why we see virtually no port-related hernias with the extraperitoneal approach.

## References


1. Guillonneau B, Abbou CC, Doublet JD, et al. Proposal for a "European Scoring System for Laparoscopic Operations in Urology". *Eur Urol* 2001;**40**(1):2–6.
2. Stolzenburg JU, Do M, Pfeiffer H, et al. The endoscopic extraperitoneal radical prostatectomy (EERPE): technique and initial experience. *World J Urol* 2002;**20**(1):48–55.
3. Gettman MT, Hoznek A, Abbou CC, et al. Laparoscopic radical prostatectomy: description of the extraperitoneal approach using the da Vinci robotic system. *J Urol* 2003;**170**:416–9.

4. Esposito M, Dakwar G, Ahmed M, et al. Extraperitoneal robotic prostatectomy: comparison of technique and results at one institution. *J Endourol* 2004;**18**:691–706.
5. Joseph JV, Rosenbaum R, Madeb R, et al. Robotic extraperitoneal radical prostatectomy: an alternative approach. *J Urol* 2006;**175**:945–50.
6. Van Velthoven RF, Ahlering TE, Peltier A, et al. Technique for laparoscopic running urethrovesical anastomosis: the single knot method. *Urology* 2003;**61**(4):699–702.
7. Boczko J, Madeb R, Golijanin D, et al. Robot-assisted radical prostatectomy in obese patients. *Can J Urol* 2006;**13**(4):33169–73.
8. Ball MW, Reese AC, Mettee LZ, et al. Safety of minimally invasive radical prostatectomy in patients with prior abdominopelvic or inguinal surgery. *J Endourol* 2014;**29**(2):192–7.
9. Lin BM, Hyndman ME, Steele KE, et al. Incidence and risk factors for inguinal and incisional hernia after laparoscopic radical prostatectomy. *Urology* 2011;**77**(4):957–62.

CHAPTER

# 39

# Clinical and Pathologic Staging of Prostate Cancer

*Adam C. Reese, MD*

Department of Urology, Temple University School of Medicine, Philadelphia, PA, USA

## INTRODUCTION

Cancer staging criteria are designed to characterize the extent of a tumor and the burden of disease. Staging criteria serve several purposes, including defining patient prognosis, helping to determine appropriate treatment, facilitating the exchange of information among treatment centers, and serving as a basis for cancer research.[1]

The Whitmore–Jewett classification was the original system used to stage prostate cancer. This system has since been supplanted by the TNM staging system, which has been used to stage prostate cancer since 1975. The TNM system stages cancers by the extent of the primary tumor (T), involvement of regional lymph nodes (N), and the presence or absence of metastatic disease (M).[1]

It is important to differentiate between the clinical and pathological stage of prostate cancers. Clinical staging assigns a stage using data available at the time of diagnosis from physical examination and imaging studies, before definitive treatment is initiated. This differs from the pathological staging, which assigns stage based on pathological examination of the radical prostatectomy specimen and any lymph nodes removed at the time of surgery. Thus, pathological stage is the more accurate of the two staging systems, as limitations in physical exam and imaging modalities can result in a clinical stage assignment that poorly estimates the true extent of disease.

The clinical and pathological staging criteria for prostate cancer, as defined by the American Joint Committee on Cancer (AJCC) Cancer Staging Manual,[1] are shown in Table 39.1.

## CLINICAL STAGING OF PROSTATE CANCER

### Criteria Used to Assign Clinical Stage and Staging Errors

The AJCC clinical staging criteria for prostate cancer are shown in Table 39.1. Tumors that are neither palpable nor visible on imaging are assigned clinical stage T1. Those that are either palpable on digital rectal examination (DRE) or visible on either transrectal ultrasound (TRUS) or MRI are deemed to be clinical stage T2. Extraprostatic tumors are assigned to clinical stage T3, and those that invade adjacent organs are classified as clinical stage T4.

Prior research has suggested that errors in assigning clinical stage are common.[2–5] Many of these staging errors are due to the inappropriate consideration of prostate biopsy results when assigning clinical stage. Per the AJCC, DRE and imaging findings at diagnosis are the criteria that should be considered when assigning clinical stage. Prostate biopsy results are not meant to factor into clinical stage assignment; however, many clinicians consider laterality of positive prostate biopsies when determining clinical stage.[4] For example, a patient with a unilateral nodule on DRE and a bilaterally positive biopsy, should be assigned to clinical stage T2a or T2b. However, many of these patients are incorrectly assigned to clinical stage T2c due to inappropriate consideration of the bilaterally positive biopsy result.

Another common cause of clinical staging errors is the disregard for imaging findings when assigning clinical stage. In fact, positive imaging findings may be ignored in nearly two-thirds of patients when assigning clinical

Prostate Cancer. http://dx.doi.org/10.1016/B978-0-12-800077-9.00039-6

**TABLE 39.1** Clinical and Pathologic Staging Criteria for Prostate Cancer

| Primary tumor (T) | | Regional lymph nodes (N) | | Distant metastasis (M) | |
|---|---|---|---|---|---|
| *CLINICAL STAGE* | | | | | |
| cTx | Primary tumor cannot be assessed | Nx | Regional lymph nodes not assessed | M0 | No distant metastasis |
| cT0 | No evidence of primary tumor | N0 | No regional lymph node metastases | M1 | Distant metastasis |
| cT1 | Clinically unapparent tumor neither palpable nor visible by imaging | N1 | Metastasis in regional lymph nodes | M1a | Nonregional lymph node(s) |
| cT1a | Tumor incidental histologic finding in 5% or less of tissue resected | | | M1b | Bone(s) |
| cT1b | Tumor incidental histologic finding in more than 5% of tissue resected | | | M1c | Other site(s) with or without bone disease |
| cT1c | Tumor identified by needle biopsy (e.g., because of elevated PSA | | | | |
| cT2 | Tumor confined within the prostate | | | | |
| cT2a | Tumor involves one-half of one lobe or less | | | | |
| cT2b | Tumor involves more than one-half of one lobe but not both lobes | | | | |
| cT2c | Tumor involves both lobes | | | | |
| cT3 | Tumor extends through the prostate capsule | | | | |
| cT3a | Extracapsular extension (unilateral or bilateral) | | | | |
| cT3b | Tumor invades seminal vesicle(s) | | | | |
| cT4 | Tumor is fixed or invades adjacent structures other than seminal vesicles such as rectum, bladder, levator muscles, and/or pelvic wall | | | | |
| **Primary tumor (T)** | | **Regional lymph nodes (N)** | | **Distant metastasis (M)** | |
| *PATHOLOGIC STAGE* | | | | | |
| pT2 | Organ confined | pNx | Regional nodes not sampled | M0 | No distant metastasis |
| pT2a | Unilateral, one-half of one side or less | pN0 | No positive regional nodes | M1 | Distant metastasis |
| pT2b | Unilateral, involving more than one-half of one side but not both sides | pN1 | Metastases in regional node(s) | M1a | Nonregional lymph node(s) |
| pT2c | Bilateral disease | | | M1b | Bone(s) |
| pT3 | Extraprostatic extension | | | M1c | Other site(s) with or without bone disease |
| pT3a | Extracapsular extension or microscopic invasion of bladder neck | | | | |
| pT3b | Seminal vesicle invasion | | | | |
| pT4 | Invasion of rectum, levator muscles, and/or pelvic wall | | | | |

stage.[4] These errors often occur when lesions on TRUS or MRI are disregarded, and patients with these abnormal findings are incorrectly assigned to clinical stage T1c.

Recent data suggest that MRI is a promising imaging modality for the diagnosis and staging of prostate cancer.[6,7] This is likely to result in increased use of MRI and MRI/ultrasound fusion biopsies in the near future. The improved ability of MRI, compared to ultrasound, to visualize clinically localized prostate tumors should result in a higher prevalence of clinical T2 tumors, with a corresponding decrease in T1 lesions. However, it is unclear whether clinicians will tend to ignore MRI results, as is common with TRUS findings, when assigning clinical stage. A failure to incorporate MRI findings in clinical stage assignment will only increase the prevalence of staging errors. Future staging guidelines will need to reach a consensus regarding how imaging findings should be incorporated into clinical staging parameters.

Staging errors resulting from the misinterpretation of clinical staging criteria clearly lead to heterogeneous patient populations within the same clinical stage group. This heterogeneity can bias clinical research, generate miscommunication among practitioners, and potentially lead to the administration of inadequate or overly aggressive treatment to men with prostate cancer.

## Time Trends in the Distribution of Prostate Cancer Clinical Stage

In the 1980s, the widespread adoption of prostate-specific antigen (PSA) screening for prostate cancer resulted in a significant stage migration toward early stage disease.[8,9] The majority of contemporary prostate cancer patients are diagnosed with clinical stage T1c disease, and the incidence of metastatic and locally advanced disease has decreased over time. This stage migration, along with the increased incidence of prostate cancer, has resulted in a marked increase in the use of modalities used to treat clinically localized prostate cancer. Recent data suggest that this often results in the overtreatment of clinically localized tumors,[10] subjecting patients to potentially unnecessary morbidity.

The United States Preventative Task Force (USPTF) recently issued a grade D recommendation for prostate cancer screening, recommending against PSA screening.[11] The USPTF guidelines strongly impact the screening practices of primary care physicians who do the majority of the prostate cancer screening in the United States.[12] These guidelines are likely to decrease the prevalence of PSA screening in the general population. There is significant concern that a decrease in PSA-based screening could result in a second stage migration toward a higher prevalence of advanced stage or metastatic prostate cancer, with a subsequent increase in prostate cancer mortality.

## Prognostic Implications of Clinical Stage

According to the AJCC Cancer Staging Manual, one of the primary goals of clinical staging is to help determine an individual patient's prognosis. For patients with prostate cancer, it is clear that the presence of regional lymphadenopathy (N1) or metastatic disease (M1) is an adverse prognostic factor and is associated with increased mortality.[13] Furthermore, most evidence would suggest that in the absence of metastatic disease, extraprostatic tumor extension (cT3-T4) carries an adverse prognosis as well. However, due to the PSA-induced stage migration toward early stage disease, the majority of contemporary prostate cancers are diagnosed with clinically organ- confined disease (cT1-T2).[8,9] It remains controversial whether substratification of patients with clinically organ-confined disease carries any prognostic significance.

Reese et al. performed an analysis of the CaPSURE database, a national disease registry of men with prostate cancer, to identify clinical variables associated with biochemical recurrence after radical prostatectomy.[14] In a multivariable model including serum PSA, biopsy Gleason score, and percent of positive biopsy cores, clinical stage offered no independent prognostic information in predicting biochemical recurrence. Several additional studies have similarly reported a lack of prognostic significance for clinical stage among patients with clinically localized disease.[15,16]

A more recent series from the Mayo Clinic presented findings contrary to Reese et al.[17] This study argued that clinical stage is in fact an adverse prognostic factor for clinically localized prostate cancer. In an analysis of nearly 15,000 patients treated with radical prostatectomy, the authors found advanced clinical stage to be associated with systemic progression and death from prostate cancer.

These data suggest that there is some association between clinical stage and the risk of disease recurrence after radical prostatectomy. However, the strength of this association is likely moderate, especially in the context of multivariable prediction models including stronger predictors of disease outcome such as PSA and biopsy Gleason score. It is likely that these variables, and not clinical stage, should be emphasized when attempting to predict patient prognosis prior to definitive treatment.

## Clinical Staging in Multivariable Prediction Models

It is challenging to accurately assess the clinical significance of prostate cancer at the time of diagnosis and to predict the risk of disease recurrence after definitive treatment. Multiple individual variables have been shown to be useful in this setting. Incorporation of several of these risk factors into a multivariable model is a more accurate method of assessing disease risk than using any of these variables alone.

Numerous multivariable models have been developed for prostate cancer risk assessment.[18] These prediction tools include risk groupings,[19–21] multivariable nomograms,[22] probability tables,[23] and artificial neural networks.[24,25] Clinical stage criteria are incorporated into several of these multivariable risk assessment tools.

The commonly used D'Amico prostate cancer risk criteria use only three variables (clinical stage, PSA, and biopsy Gleason score) to assign disease risk.[20] In the D'Amico system, a patient must have ≤cT2a disease to be considered low risk; cT2b disease is considered intermediate risk and cT2c tumors are assigned to the high-risk category regardless of PSA and biopsy results. The NCCN risk groupings use a modification of the D'Amico risk criteria, where cT2c tumors are considered intermediate, as opposed to high, risk.[19] A recent analysis of the Johns Hopkins prostate cancer database suggests that patients assigned to the intermediate- or high-risk groups based on clinical staging criteria alone have superior BCR-survival rates compared to those with an advanced Gleason score or elevated PSA.[26] These data suggest that clinical stage data are perhaps weighted too strongly in the D'Amico risk criteria.

The UCSF-CAPRA score uses clinical stage as well as patient age, serum PSA, biopsy Gleason score, and percent of positive biopsy cores to assess disease risk.[21] Using this system, patients with clinical T1 versus T2 tumors are considered to be at comparable risk of metastasis and mortality. Only tumors with evidence of extraprostatic extension (clinical stage T3 or greater) are deemed to be at increased risk compared to men with organ-confined disease.

The Kattan nomogram for predicting disease recurrence after radical prostatectomy also incorporates clinical stage data.[22] Interestingly, this nomogram shows patients with cT1c tumors to be at a greater risk of recurrence than those with cT2a lesions. Furthermore, clinical T2b tumors are at a markedly increased risk of recurrence compared to cT2c lesions. The lack of correlation between increasing clinical stage and recurrence risk in the Kattan nomogram argues that clinical stage is, at best, a weak predictor of disease recurrence after definitive treatment.

In conclusion, when using multivariable models to assess the risk of prostate cancer recurrence after definitive treatment, there is a questionable association between advanced clinical stage and recurrence risk. Certainly, other variables present at the time of diagnosis, including serum PSA and biopsy Gleason score, appear to be more powerful predictors of patient outcomes. These data suggest that clinical stage should be de-emphasized in risk assessment and question the utility of clinical staging criteria in predicting prostate cancer risk at the time of diagnosis.

## Clinical Stage in Determining Eligibility for Active Surveillance

In recent years, there has been increasing concern regarding the overtreatment of screen-detected prostate cancer.[11] Definitive treatment of clinically indolent tumors cancer often results in significant, and unnecessary, patient morbidity. Active surveillance for low-risk prostate cancer is a potential solution to the problem of overtreatment. Active surveillance is meant to identify men with low-risk prostate cancers and avoid or delay overtreatment in these men.

Accurate identification of men with low-risk tumors is crucial for a successful active surveillance program. Failure to identify higher-risk disease could result in disease progression while on surveillance, and inaccurate classification of indolent tumors as high risk will only perpetuate the problem of overtreatment.

Various active surveillance eligibility criteria have been developed to attempt to accurately differentiate low-risk men who are potential candidates for surveillance from higher-risk men who are more likely to benefit from definitive treatment.[27] With rare exceptions,[28] most eligibility criteria from high-volume AS programs restrict AS eligibility to men with ≤cT2a tumors. As previously discussed, the prognostic significance of an advanced clinical stage is questionable,[14–16] and clinician errors in assigning clinical stage are common.[3–5] These limitations of clinical staging could potentially preclude many low-risk men from enrolling in active surveillance programs. Furthermore, as MRI-ultrasound fusion prostate biopsies increase in popularity, more prostate tumors will be visible on preoperative imaging. This will result in a higher prevalence of clinical T2 tumors that would have been staged cT1c in the absence of MRI imaging, resulting in even fewer men who qualify for active surveillance. Recent data suggest that AS eligibility criteria could safely be expanded to include all men with clinical T2 tumors, potentially decreasing overtreatment rates.[29] Further research is crucial in order to determine the optimal clinical staging criteria to determine active surveillance eligibility.

## Evaluation for Extraprostatic or Metastatic Disease

As previously discussed, the majority of prostate cancers diagnosed in the modern era are early-stage, organ-confined tumors (≤ clinical T2). However, in patients proceeding to radical prostatectomy, a significant percentage of men thought to have organ-confined tumors will in fact have extraprostatic disease on final pathology.[29–31]

The failure to accurately diagnose extraprostatic disease prior to surgery is largely due to the limitations of DRE and TRUS as staging modalities. Focal extraprostatic extension is nearly impossible to diagnose on DRE. Capsular distortion can occasionally be seen on TRUS in patients with significant extraprostatic tumor extension, although this finding is subtle and can easily be missed by inexperienced ultrasonographers. Thus, it is difficult to accurately differentiate localized from extraprostatic disease at the time of diagnosis.

The literature suggests that prostate MRI may be more accurate in identifying extraprostatic disease extension, thus improving the correlation between clinical and pathological stage. Two recent series of men thought to have clinically localized prostate tumors found that multiparametric pelvic MRI improved the ability to accurately identify pathologic T3 disease and resulted in improved preoperative risk stratification.[32,33] The ability to more accurately diagnose nonorgan-confined disease at the time of diagnosis would not only result in improved risk stratification at diagnosis, but may also alter surgical decisions regarding nerve sparing or radiation protocols.

The presence of lymph node involvement or distant metastases at the time of diagnosis is clearly associated with an adverse prognosis.[13] Furthermore, accurate identification of these patients is imperative as treatment options for men with metastatic disease differ from those with earlier-stage tumors. Several nomograms and multivariable prediction models have been developed to identify men with lymph node involvement.[23,34,35] These models are useful in identifying men in need of cross-sectional imaging prior to definitive treatment and determining the need for lymphadenectomy at the time of prostatectomy.

The NCCN guidelines include recommendations for which men should undergo imaging to assess for lymph node involvement or distant metastases prior to definitive therapy.[36] The guidelines suggest that bone scan should be performed in men who meet any of the following criteria: clinical T1 disease and PSA >20, clinical T2 disease and PSA >10, Gleason score ≥ 8, clinical T3 or T4 disease, or symptoms suggestive of metastases. It is recommended that cross-sectional imaging of the pelvis (CT or MRI) be performed on men with clinical T3 or T4 disease, or a nomogram-indicated probability of lymph node involvement >10%.

The literature suggests that these guidelines are largely ignored when determining the need for imaging prior to definitive treatment. Several analyses have shown widespread misuse of these imaging tests, where unindicated imaging is often ordered for men at low risk for metastatic disease.[37–39] Furthermore, the percentage of men for whom unnecessary imaging studies are ordered appears to be increasing over time.[37] The overuse of imaging studies results in significant costs to the health care system. In fact, a study of Medicare beneficiaries concluded that prostate cancer imaging costs increased by 5.1% from 1999 to 2006, far outpacing the 1.8% increase in total costs relating to prostate cancer management.[40] Furthermore, radiation exposure from unnecessary imaging studies can compromise patient safety, as even small doses of radiation can increase the risk of subsequent radiation-induced malignancies.[41]

## PATHOLOGIC STAGING OF PROSTATE CANCER

The AJCC pathologic staging criteria for prostate cancer are shown in Table 39.1. Pathologic stage is determined by pathologic examination of the prostate specimen after radical prostatectomy. Men with organ-confined disease are assigned to pathologic stage T2, with subdivisions based on extent of tumor involvement. Tumors with extraprostatic extension or seminal vesicle invasion are considered pathologic stage T3 and tumors that invade adjacent organs are assigned to pathologic stage T4. Pathologic examination of lymph node specimens in men who undergo lymphadenectomy is necessary to determine N-stage. As is the case with clinical staging criteria, pathologic M-stage is based on the presence or absence of metastatic disease.

### Prognostic Implications of Pathologic Stage

The association between advanced pathologic stage and the risk of prostate cancer recurrence has been apparent for several decades.[42,43] More recently, numerous multivariable predictions tools have been developed to predict the risk of disease recurrence after radical prostatectomy.[44–50] Although many of these models include advanced pathologic stage as an adverse prognostic factor,[35,44,45,47,48,50] several did not identify an association between pathologic stage and recurrence risk in this setting.[46,49] Furthermore, in the majority of models where pathologic stage was associated with recurrence, the risk of recurrence is only increased for extra-prostatic (pathologic stage T3) relative to organ-confined (pathologic stage T2 tumors). There is little evidence to suggest that substratification of pathologic stage T2 based on tumor extent is of any utility in predicting the risk of disease recurrence. For example, in an analysis of 142 patients with pathologic T2 tumors, Billis et al. found no difference in biochemical recurrence rates when comparing pathologic T2c to T2a tumors.[51]

## Management Implications of Advanced Pathologic Stage

Patients with extraprostatic tumor extension on final pathology after radical prostatectomy are at an increased risk of disease recurrence. Contemporary series suggest that 50–75% of men with pathologic T3 disease will ultimately experience biochemical recurrence.[52,53] In this setting, several studies have advocated the use of adjuvant radiation therapy to decrease recurrence rates.[54,55] EORTC trial 22911 randomized over 1000 men with pathologic T3 disease or positive surgical margins to adjuvant radiation therapy versus observation.[55] The authors found biochemical failure rates were significantly decreased in patients who received adjuvant radiation therapy, although there was no difference in overall or prostate-cancer-specific survival. The Southwest Oncology Group (SWOG) trial 8794 similarly randomized men with pT3 disease or positive surgical margins to adjuvant radiation versus observation.[54] This trial not only found a decreased incidence of biochemical recurrence in men treated with adjuvant radiation, but also reported improved metastasis-free and overall survival in the adjuvant radiotherapy group.[56]

These findings support the routine use of adjuvant radiation therapy for men with pT3 disease or positive margins after radical prostatectomy. Importantly, however, these trials were conducted prior to the widespread use of ultrasensitive PSA assays that are capable of detecting recurrence earlier than traditional PSA assays. It is unclear whether adjuvant radiation (administered when ultrasensitive PSA is undetectable) is superior to early salvage radiation (administered when ultrasensitive PSA first becomes detectable). Furthermore, adverse effects, such as rectal complications and urethral strictures, are more common in patients receiving adjuvant radiation, tempering the enthusiasm for this approach.[54]

Equally contentious is the management of patients with lymph node metastases discovered after radical prostatectomy and pelvic lymph node dissection. Numerous management strategies have been advocated for these patients including observation, immediate androgen-deprivation therapy, or adjuvant radiotherapy with or without androgen deprivation.[57–61] Proponents of observation have cited surprisingly favorable long-term disease-specific survival for lymph node positive patients in the absence of androgen deprivation or other adjuvant therapies.[60,61] Outcomes appear to be particularly favorable in men with Gleason score <8 and a low metastatic nodal burden. Proponents of androgen deprivation cite level one data showing improvement in survival for men with lymph node metastases treated with immediate androgen deprivation therapy after radical prostatectomy.[59] However, this randomized trial has been largely criticized due to a small sample size and failure to initiate therapy in the control group until clinical progression of disease. More recent studies have advocated a multimodal approach for patients with node positive disease.[57,58] This approach includes radical prostatectomy with extended bilateral pelvic lymph node dissection followed by adjuvant radiation and androgen deprivation therapy. It remains unclear whether the improved survival using this trimodal approach is due to the improved local control offered by pelvic radiation or a more systemic effect at sites of metastases.

## References


1. Edge SB. *American joint committee on cancer. AJCC cancer staging manual*. 7th ed. New York, London: Springer; 2010.
2. Campbell T, Blasko J, Crawford ED, et al. Clinical staging of prostate cancer: reproducibility and clarification of issues. *Int J Cancer* 2001;**96**(3):198–209.
3. Gosselaar C, Kranse R, Roobol MJ, et al. The interobserver variability of digital rectal examination in a large randomized trial for the screening of prostate cancer. *Prostate* 2008;**68**(9):985–93.
4. Reese AC, Sadetsky N, Carroll PR, et al. Inaccuracies in assignment of clinical stage for localized prostate cancer. *Cancer* 2011;**117**(2):283–9.
5. Sexton T, Rodrigues G, Brecevic E, et al. Controversies in prostate cancer staging implementation at a tertiary cancer center. *Can J Urol* 2006;**13**(6):3327–34.
6. Salami SS, Vira MA, Turkbey B, et al. Multiparametric magnetic resonance imaging outperforms the Prostate Cancer Prevention Trial risk calculator in predicting clinically significant prostate cancer. *Cancer* 2014;**120**(18):2876–82.
7. Walton Diaz A, Hoang AN, Turkbey B, et al. Can magnetic resonance-ultrasound fusion biopsy improve cancer detection in enlarged prostates? *J Urol* 2013;**190**(6):2020–5.
8. Cooperberg MR, Lubeck DP, Mehta SS, et al. CaPsure. Time trends in clinical risk stratification for prostate cancer: implications for outcomes (data from CaPSURE). *J Urol* 2003;**170**(6 Pt 2):S21–5; discussion S26–7.
9. Shao YH, Demissie K, Shih W, et al. Contemporary risk profile of prostate cancer in the United States. *J Natl Cancer Inst* 2009;**101**(18):1280–3.
10. Cooperberg MR, Broering JM, Carroll PR. Time trends and local variation in primary treatment of localized prostate cancer. *J Clin Oncol* 2010;**28**(7):1117–23.
11. Moyer VA. U.S. Preventive Services Task Force. Screening for prostate cancer: U.S. Preventive Services Task Force recommendation statement. *Ann Inter Med* 2012;**157**(2):120–34.
12. Tasian GE, Cooperberg MR, Cowan JE, et al. Prostate specific antigen screening for prostate cancer: knowledge of, attitudes towards, and utilization among primary care physicians. *Urol Oncol* 2012;**30**(2):155–60.
13. Gandaglia G, Karakiewicz PI, Briganti A, et al. Impact of the site of metastases on survival in patients with metastatic prostate cancer. *Eur Urol* 2015;**68**(2):325–34.
14. Reese AC, Cooperberg MR, Carroll PR. Minimal impact of clinical stage on prostate cancer prognosis among contemporary patients with clinically localized disease. *J Urol* 2010;**184**(1):114–9.
15. Armatys SA, Koch MO, Bihrle R, et al. Is it necessary to separate clinical stage T1c from T2 prostate adenocarcinoma? *BJU Int* 2005;**96**(6):777–80.
16. Billis A, Magna LA, Watanabe IC, et al. Are prostate carcinoma clinical stages T1c and T2 similar? *Int Braz J Urol* 2006;**32**(2): 165–71.

17. Tollefson MK, Karnes RJ, Rangel LJ, et al. The impact of clinical stage on prostate cancer survival following radical prostatectomy. *J Urol* 2013;**189**(5):1707–12.
18. Shariat SF, Kattan MW, Vickers AJ, et al. Critical review of prostate cancer predictive tools. *Future Oncol* 2009;**5**(10):1555–84.
19. The NCCN clinical practice guidelines in oncology for prostate cancer V.4.2011.
20. D'Amico AV, Whittington R, Malkowicz SB, et al. Biochemical outcome after radical prostatectomy, external beam radiation therapy, or interstitial radiation therapy for clinically localized prostate cancer. *JAMA* 1998;**280**(11):969–74.
21. Cooperberg MR, Pasta DJ, Elkin EP, et al. The University of California, San Francisco Cancer of the Prostate Risk Assessment score: a straightforward and reliable preoperative predictor of disease recurrence after radical prostatectomy. *J Urol* 2005;**173**(6): 1938–42.
22. Stephenson AJ, Scardino PT, Eastham JA, et al. Preoperative nomogram predicting the 10-year probability of prostate cancer recurrence after radical prostatectomy. *J Natl Cancer Inst* 2006;**98**(10):715–7.
23. Eifler JB, Feng Z, Lin BM, et al. An updated prostate cancer staging nomogram (Partin Tables) based on cases from 2006 to 2011. *BJU Int* 2013;**111**(1):22–9.
24. Finne P, Finne R, Auvinen A, et al. Predicting the outcome of prostate biopsy in screen-positive men by a multilayer perceptron network. *Urology* 2000;**56**(3):418–22.
25. Snow PB, Smith DS, Catalona WJ. Artificial neural networks in the diagnosis and prognosis of prostate cancer: a pilot study. *J Urol* 1994;**152**(5 Pt 2):1923–6.
26. Reese AC, Pierorazio PM, Han M, et al. Contemporary evaluation of the National Comprehensive Cancer Network prostate cancer risk classification system. *Urology* 2012;**80**(5):1075–9.
27. Dall'Era MA, Albertsen PC, Bangma C, et al. Active surveillance for prostate cancer: a systematic review of the literature. *Eur Urol* 2012;**62**(6):976–83.
28. Klotz L, Zhang L, Lam A, et al. Clinical results of long-term follow-up of a large, active surveillance cohort with localized prostate cancer. *J Clin Oncol* 2010;**28**(1):126–31.
29. Reese AC, Landis P, Han M, et al. Expanded criteria to identify men eligible for active surveillance of low risk prostate cancer at Johns Hopkins: a preliminary analysis. *J Urol* 2013;**190**(6):2033–8.
30. Smaldone MC, Cowan JE, Carroll PR, et al. Eligibility for active surveillance and pathological outcomes for men undergoing radical prostatectomy in a large, community based cohort. *J Urol* 2010;**183**(1):138–43.
31. Sundi D, Ross AE, Humphreys EB, et al. African American men with very low-risk prostate cancer exhibit adverse oncologic outcomes after radical prostatectomy: should active surveillance still be an option for them? *J Clin Oncol* 2013;**31**(24):2991–7.
32. Marcus DM, Rossi PJ, Nour SG, et al. The impact of multiparametric pelvic magnetic resonance imaging on risk stratification in patients with localized prostate cancer. *Urology* 2014;**84**(1):132–7.
33. Pugh TJ, Frank SJ, Achim M, et al. Endorectal magnetic resonance imaging for predicting pathologic T3 disease in Gleason score 7 prostate cancer: implications for prostate brachytherapy. *Brachytherapy* 2013;**12**(3):204–9.
34. Briganti A, Larcher A, Abdollah F, et al. Updated nomogram predicting lymph node invasion in patients with prostate cancer undergoing extended pelvic lymph node dissection: the essential importance of percentage of positive cores. *Eur Urol* 2012;**61**(3):480–7.
35. Cagiannos I, Karakiewicz P, Eastham JA, et al. A preoperative nomogram identifying decreased risk of positive pelvic lymph nodes in patients with prostate cancer. *J Urol* 2003;**170**(5):1798–803.
36. National Comprehensive Cancer Network. *NCCN clinical practice guidelines in oncology, prostate cancer, Version 2.2014*. Available from: www.nccn.org; 2014 [accessed 20.08.2014].
37. Porten SP, Smith A, Odisho AY, et al. Updated trends in imaging use in men diagnosed with prostate cancer. *Prostate Cancer Prostatic Dis* 2014;**17**(3):246–51.
38. Choi WW, Williams SB, Gu X, et al. Overuse of imaging for staging low risk prostate cancer. *J Urol* 2011;**185**(5):1645–9.
39. Lavery HJ, Brajtbord JS, Levinson AW, et al. Unnecessary imaging for the staging of low-risk prostate cancer is common. *Urology* 2011;**77**(2):274–8.
40. Dinan MA, Curtis LH, Hammill BG, et al. Changes in the use and costs of diagnostic imaging among Medicare beneficiaries with cancer, 1999–2006. *JAMA* 2010;**303**(16):1625–31.
41. Smith-Bindman R, Lipson J, Marcus R, et al. Radiation dose associated with common computed tomography examinations and the associated lifetime attributable risk of cancer. *Arch Intern Med* 2009;**169**(22):2078–86.
42. Humphrey PA, Frazier HA, Vollmer RT, et al. Stratification of pathologic features in radical prostatectomy specimens that are predictive of elevated initial postoperative serum prostate-specific antigen levels. *Cancer* 1993;**71**(5):1821–7.
43. Smith Jr JA, Hernandez AD, Wittwer CJ, et al. Long-term follow-up after radical prostatectomy. Identification of prognostic variables. *Urol Clin North Am* 1991;**18**(3):473–6.
44. Bauer JJ, Connelly RR, Seterhenn IA, et al. Biostatistical modeling using traditional preoperative and pathological prognostic variables in the selection of men at high risk for disease recurrence after radical prostatectomy for prostate cancer. *J Urol* 1998;**159**(3):929–33.
45. Cooperberg MR, Hilton JF, Carroll PR. The CAPRA-S score: a straightforward tool for improved prediction of outcomes after radical prostatectomy. *Cancer* 2011;**117**(22):5039–46.
46. Graefen M, Noldus J, Pichlmeier U, et al. Early prostate-specific antigen relapse after radical retropubic prostatectomy: prediction on the basis of preoperative and postoperative tumor characteristics. *Eur Urol* 1999;**36**(1):21–30.
47. McAleer SJ, Schultz D, Whittington R, et al. PSA outcome following radical prostatectomy for patients with localized prostate cancer stratified by prostatectomy findings and the preoperative PSA level. *Urol Oncol* 2005;**23**(5):311–7.
48. Potter SR, Miller MC, Mangold LA, et al. Genetically engineered neural networks for predicting prostate cancer progression after radical prostatectomy. *Urology* 1999;**54**(5):791–5.
49. Stamey TA, Yemoto CM, McNeal JE, et al. Prostate cancer is highly predictable: a prognostic equation based on all morphological variables in radical prostatectomy specimens. *J Urol* 2000;**163**(4): 1155–60.
50. Stephenson AJ, Scardino PT, Eastham JA, et al. Postoperative nomogram predicting the 10-year probability of prostate cancer recurrence after radical prostatectomy. *J Clin Oncol* 2005;**23**(28): 7005–12.
51. Billis A, Meirelles LL, Freitas LL, et al. Should pathologists continue to use the current pT2 substaging system for reporting of radical prostatectomy specimens? *Int Urol Nephrol* 2011;**43**(3):707–14.
52. Eggener SE, Roehl KA, Smith ND, et al. Contemporary survival results and the role of radiation therapy in patients with node negative seminal vesicle invasion following radical prostatectomy. *J Urol* 2005;**173**(4):1150–5.
53. Salomon L, Anastasiadis AG, Johnson CW, et al. Seminal vesicle involvement after radical prostatectomy: predicting risk factors for progression. *Urology* 2003;**62**(2):304–9.
54. Thompson Jr IM, Tangen CM, Paradelo J, et al. Adjuvant radiotherapy for pathologically advanced prostate cancer: a randomized clinical trial. *JAMA* 2006;**296**(19):2329–35.
55. Bolla M, van Poppel H, Tombal B, et al. Postoperative radiotherapy after radical prostatectomy for high-risk prostate cancer: long-term results of a randomised controlled trial (EORTC trial 22911). *Lancet* 2012;**380**(9858):2018–27.

56. Thompson IM, Tangen CM, Paradelo J, et al. Adjuvant radiotherapy for pathological T3N0M0 prostate cancer significantly reduces risk of metastases and improves survival: long-term followup of a randomized clinical trial. *J Urol* 2009;**181**(3):956–62.
57. Briganti A, Karnes RJ, Da Pozzo LF, et al. Combination of adjuvant hormonal and radiation therapy significantly prolongs survival of patients with pT2-4 pN+ prostate cancer: results of a matched analysis. *Eur Urol* 2011;**59**(5):832–40.
58. Da Pozzo LF, Cozzarini C, Briganti A, et al. Long-term follow-up of patients with prostate cancer and nodal metastases treated by pelvic lymphadenectomy and radical prostatectomy: the positive impact of adjuvant radiotherapy. *Eur Urol* 2009;**55**(5):1003–11.
59. Messing EM, Manola J, Yao J, et al. Immediate versus deferred androgen deprivation treatment in patients with node-positive prostate cancer after radical prostatectomy and pelvic lymphadenectomy. *Lancet Oncol* 2006;**7**(6):472–9.
60. Pierorazio PM, Gorin MA, Ross AE, et al. Pathological and oncologic outcomes for men with positive lymph nodes at radical prostatectomy: the Johns Hopkins Hospital 30-year experience. *Prostate* 2013;**73**(15):1673–80.
61. Touijer KA, Mazzola CR, Sjoberg DD, et al. Long-term outcomes of patients with lymph node metastasis treated with radical prostatectomy without adjuvant androgen-deprivation therapy. *Eur Urol* 2014;**65**(1):20–5.

CHAPTER

# 40

# Management of Bladder Neck Contracture in the Prostate Cancer Survivor


*Daniel C. Parker, MD*,†, Timothy J. Tausch, MD**, Jay Simhan, MD**

*Department of Urology, Temple University Hospital, Philadelphia, PA, USA

**Department of Urology, UT Southwestern Medical Center, Dallas, TX, USA

†Department of Urology, Fox Chase Cancer Center, Einstein Urologic Institute, Philadelphia, PA, USA


## INTRODUCTION

Bladder neck contracture (BNC) is an infrequent but commonly recognized complication encountered by the urologist in patients following prostate surgery.[1] While the number of patients seeking surgery or radiation therapy for prostate cancer has increased in the prostate-specific antigen (PSA) era,[2] the number of bladder neck contractures has decreased due to advancements in surgical methods and the introduction of robotic surgery.[3,4] Nonetheless, there exists a small population of patients who will seek treatment for symptomatic bladder neck stricture. A large number of these will be managed nonoperatively with success, while others may require minimally invasive or endoscopic efforts.[5–10] Even still, a minority will develop complex, refractory disease, and it is this population for whom a defined treatment strategy remains elusive. Here, the natural history of bladder neck contracture in the prostate cancer survivor is presented along with established treatment modalities for uncomplicated patients. In addition, a recently published, novel experience with a hybrid balloon dilation and transurethral incision approach for the standard treatment of BNC is reviewed. Finally, options for refractory cases are discussed including novel therapies for the management of this complex condition.

## EPIDEMIOLOGY AND PATHOPHYSIOLOGY

Treatment for prostate cancer is ultimately the most common etiology of BNC.[2] However, BNC is not a uniform phenomenon and can exist subsequent to a multitude of pathophysiologic mechanisms depending on the primary treatment modality that preceded it. For example, patients with a history of pelvic radiation therapy for prostate cancer may develop BNC as a result of tissue necrosis and fibrosis secondary to obliteration of the microvasculature supplying the bladder neck.[11,12] This process usually occurs for years before the establishment of symptomatic stricture.[13] On the other hand, patients treated with surgery for prostate cancer may develop BNC a mere months after intervention due to increased blood loss[14] or technical issues preventing the operator from achieving the desired tension-free, watertight vesicourethral anastomosis.[15] Foreign bodies, such as hemostatic clips placed at the time of surgery and/or undissolved suture material, may act as a nidus for fibrosis at the anastomosis as well.[14] The era of robotic prostate surgery has alleviated many of these procedural confounders, offering the surgeon a more unobstructed view of the working anastomosis, broader range of motion, and a drier operative field. In addition, the adoption of a running vesicourethral anastomosis, often


Prostate Cancer. http://dx.doi.org/10.1016/B978-0-12-800077-9.00040-2

employed in robotic surgery and rarely employed in the open approach, has been credited with a more reliable mucosal approximation necessary for a water-tight seal.[1,3] As a result, the incidence of BNC has decreased from rates as high as 32% in some open prostatectomy series[15–18] toward zero in many reports utilizing a minimally invasive approach.[1,3,19]

However, BNC still affects upward of 30% of patients receiving salvage prostatectomy following radiation therapy, and perhaps the most devastated patients are those who undergo multimodal treatment with brachytherapy followed by either salvage radiation or surgery.[13,20–22] BNC is a documented complication following surgery for benign disease states as well, such as benign prostatic hyperplasia (BPH), and the urologist must suspect this condition in active surveillance prostate cancer patients undergoing bladder outlet reduction surgery. Conventional transurethral resection of the prostate (TURP) has been shown to offer a higher risk of developing postoperative BNC (1–12.3%)[23,24] compared with newer modalities such as KTP-laser prostatectomy (3–5%).[25–27]

Patient demographic factors and behaviors also contribute to risk for developing BNC. Borboroglu et al. conducted a multivariable analysis on 52 patients who underwent retropubic prostatectomy. Diabetes mellitus, coronary artery disease, and smoking were all positive predictors of BNC.[15] The experience reported by Ramirez et al. substantiated the link between BNC and smoking in patients with recurrent bladder neck stricture who smoked greater than 10 pack-years and underwent deep lateral transurethral incision of the bladder neck.[28] Age and BMI were also associated with BNC occurrence requiring interventions when the Cancer of the Prostate Strategic Urologic Research Endeavor database was queried in a recent analysis.[13]

## MANAGEMENT OF BNC

### Urethral Dilation

Nonoperative approaches may be considered for properly selected prostate cancer patients with BNC. Although commonly referred to as "conservative management," outpatient serial dilation of the urethra requires highly motivated and compliant cohorts and can carry significant physical consequences leading to detriments in quality of life. The character and caliber of the urethra is important to assess prior to enrolling patients in this therapy. Favorable urethras will have soft strictures with some amount of patency remaining. Dilation programs usually are initiated in the clinic setting with flexible cystoscopy followed by office-based dilation. Gradually, patients are instructed to self-catheterize periodically to prevent disease recurrence. There is no defined regimen with this approach as it is usually highly individualized to patient and treating urologist. In one such series, Park et al. demonstrated that BNC was successfully treated with urethral dilation up to 18Fr followed by a three-month period of intermittent catheterization in 32 post-prostatectomy patients. Demonstrating the notion that self-dilation therapy requires a long-term commitment, 93% of this cohort required only one or two dilations with one-year median follow-up.[29] Similarly, a large ($n$ = 510) British experience reported positive outcomes in 48 patients who developed BNC after prostatectomy and further proceeded to undergo urethral dilation or an intermittent catheterization/dilation program.[30] Due to complications compromising quality of life such as acute urinary retention, gross hematuria, recurrent infection, urethral trauma, and concomitant stricture development, many self-dilation patients report high rates of dissatisfaction and fail to tolerate this therapy.[31]

### UroLume Urethral Stent

Introduced in 1988, the UroLume (American Medical Systems, Minnetonka, Minnesota) was conceived as a uroprosthesis intended to tent open the bladder outlet at the level of stricture. Initial reports by Milroy et al. demonstrated a 100% success rate in a small series of eight patients.[32] Subsequent reports, however, revealed many of the UroLume's problems that ultimately led to it falling out of favor for the management of BNC in prostate cancer survivors.

Hussain et al. noted a 50% complication rate in 30 UroLume patients, a rate that was confirmed by De Vocht et al. and Magera et al. who also experienced reoperation in 24–27% of patients due to obstruction of the stent.[33–35] The proportion of patients with BNC due to prostate cancer therapy who failed UroLume stenting was even higher in a study by Erickson et al. – 57% of men ($n$ = 38) required a revision.[36] Besides its tendency to become obstructed, the UroLume was also afflicted with a penchant for migration. This complication was encountered in the early and late postoperative periods, occurring in patients from 6 weeks to 3 years following placement.[37,38] Issues with the UroLume becoming encased in stone or tissue in-growth and patients reporting gross hematuria were also noted.[39–42] For these reasons, as well as the relative success of urethral dilation and incision, the UroLume is no longer considered a minimally invasive option for the management of BNC and, in fact, is no longer available in the United States.

### Bladder Neck Incision

Incision of the bladder neck has been utilized via many different modalities including cold or hot knives, laser systems, and traditional resection loops.[6,7,43–48] This procedure is typically performed in the outpatient setting and is normally well tolerated making it a popular recommendation among general urologists and specialists alike. Data comparing success rates between modalities is lacking due to small cohort sizes and short-term median follow-up. What is clear, however, is that many prostate cancer patients can expect to undergo multiple procedures when opting for bladder neck incision.

In the experience of Borboroglu et al., 53.5% ($n$ = 28) required more than one incision for success, leading to the investigators' conclusion that treatment success and number of maneuvers are likely inversely related.[15] To combat this, Morey and coworkers devised a novel, hybrid procedure that combined bladder neck incision with urethral dilation in the same operation for prostate cancer patients with reffractory BNC.[28,49] In summary, bladder neck incision is made down to perivesical fat with a Collings knife at the 3 and 9 o'clock positions after dilation of the urethra is undertaken with a 4×24 cm UroMax Ultra™ balloon dilator. Patients are then discharged with a Foley catheter in place for 2 days. Uroflowmetry and cystoscopy are performed at 2 months postoperatively to delineate success, which the authors define as successful passage of a 16Fr flexible cystoscope in the office in the 2–3 month postoperative period. Using this approach, Morey and coworkers reported a one-time procedural success rate of 72% and an increase in success to 86% with a second balloon dilation procedure (median follow-up 16 months). It should be noted that this protocol is applicable to refractory BNC as almost 80% of the series had undergone a prior transurethral incision of bladder neck contracture (TUIBNC).[49]

Despite the relative success of minimally invasive dilation and incisional procedures on bladder neck stricture compared to urethral stenting, these approaches are fraught with recurrence and patient dissatisfaction in the long term. As such, there still exists a role for open surgery in the management of the severe or refractory BNC.

### Open Surgery of Refractory BNC

Patients deemed candidates for open bladder neck reconstruction are those with obliterated or completely disrupted urethras as well as those with bladder neck stricture refractory or not amenable to endoscopic maneuvers. Consideration of continence is of utmost importance, as unmasking of stress urinary incontinence (SUI) following reconstruction is common with any of the open approaches. Therefore, consultation with a genitourinary reconstructive specialist is paramount, and all of the experience describing open surgical management of BNC comes from high volume reconstructive institutions.[50–52] As such, the literature regarding this topic suffers from a relative lack of power, follow-up, and generalizability. Three approaches have been described in detail: transpubic, pure perineal, and abdominoperineal.

The transpubic approach has been most recently detailed. Morey and coworkers previously reported the results of a staged procedure whereby the bladder neck is mobilized with the contracture scar excised followed by artificial urinary sphincter (AUS) placement for preservation of continence 3 months later.[49] However, the exact timing of AUS placement varies among various reconstructive urologists in the reported literature, all with acceptable results.[50,53,54] In the largest published series of 15 patients, satisfaction rates were over 90% with transpubic bladder neck reconstruction.[55] From a pure perineal perspective, Simonato et al. has shown success performing a posterior urethroplasty (functional ureteroureterostomy) followed by AUS placement in six patients.[53] Schlossberg et al. performed a bladder neck reconstruction employing an abdominoperineal method in two patients who previously underwent prostatectomy. In order to gain maximal bladder neck mobility, a partial pubectomy may be performed with preservation of continence through sparing of the external urinary sphincter.[51]

### Emerging Techniques

Techniques combining endoscopic ablation of bladder neck strictures followed by injection of the area with cytotoxic agents have also been described recently. Validation of these methods is still needed before consideration is given to them as standard in the armamentarium against BNC. In a series of 24 patients, Eltahawy et al. performed incision of the bladder neck with a holmium laser immediately followed by administration of triamcinolone at the stricture location with 83% success.[56] Similarly, Vanni et al. inject the stricture with Mitomycin C (MMC) following cold knife transurethral incision in radial fashion.[57] However, serious concerns regarding the effects of MMC have been put forth regarding adverse outcomes involving extravasation,[58,59] anaphylaxis,[60] and impaired urothelial healing.[61] A recent multi-institutional series retrospectively examined 55 patients undergoing TUIBNC with MMC over a 5-year period with a mean follow-up of 9 months and concluded MMC to add little benefit.[62] Overall success rate was 75% (41/55) but 23 had a recurrence of BNC and 15 needed a repeat TUIBNC/MMC. Importantly, four patients (7%) had a serious adverse event from MMC with the development of bladder neck necrosis with three patients opting for eventual cystectomy.[62]

## STRESS URINARY INCONTINENCE AFTER BNC THERAPY

It has been observed that some patients who undergo management, either endoscopic or open, of BNC following radical prostatectomy will exhibit urinary incontinence following surgery. This "unmasking" of stress urinary incontinence following bladder neck surgery has been reported at variable rates, with some institutions experiencing it negligibly[7,10,15,45,46] while others observing it more commonly.[29,44,47] Nevertheless, the largest centers suggest a rate of 2–7%.[49,63–65] One major risk factor appears to be salvage prostatectomy, where rates of SUI following bladder neck reconstruction are much higher (25–45%).[66,67] Management of this condition often necessitates placement of an AUS and this has been most closely studied in patients following UroLume implantation for BNC. Elliott and Boone showed failure in a single patient with stricture recurrence out of nine after UroLume and subsequent AUS with 17.5 months follow-up.[39] A larger series with more extensive follow-up (37 months) demonstrated a nearly 50% failure rate due to either obstruction of the stent or erosion of the AUS. In the same series, 18 patients required the AUS be decoupled to permit endoscopic management of BNC recurrence. These patients were more likely to experience subsequent AUS erosion.[68] In the experience of Morey and coworkers, *de novo* SUI after reconstruction of the bladder neck is a much less common scenario than patients presenting with both conditions concomitantly.[28] As with other authors,[69] they advocate for a waiting-period of 2–3 months after instrumentation of the bladder neck in order to adequately assess the full impact of the intervention. At the 2-month interval, if the bladder neck appears stable on cystoscopy and the patient complains of SUI, he is offered an AUS; however, the optimal wait time before AUS placement is up for debate.[44,57,69] In the event of a contracture recurrence after AUS, holmium laser incision via flexible cystoscope or semirigid ureteroscope has been shown to not compromise the urinary sphincter.[70] In severe cases, the sphincter may need to be uncoupled or removed to permit more extensive incision of dilation.[28,49]

## CONCLUSIONS

Despite a decrease in incidence since the advent of robotic surgery, bladder neck contracture in prostate cancer patients is still a condition encountered by the urologist. Factors such as age, diabetes, coronary artery disease, and smoking are known risk factors. BNC is not a uniform disease, however, and manifestations can vary in both pathophysiology and severity. In our experience, the uncomplicated BNC is treated with excellent results after urethral balloon dilation followed be transurethral incision of the bladder neck. Complex or refractory cases should be referred to a reconstructive specialist familiar with advanced endoscopic and open strategies. Stress urinary incontinence after treatment for BNC is successfully managed with artificial urinary sphincter placement, although the optimal timing of placement has not been defined.

## References


1. Breyer BN, Davis CB, Cowan JE, et al. Incidence of bladder neck contracture after robot-assisted laparoscopic and open radical prostatectomy. *BJU Int* 2010;**106**(11):1734–8.
2. Cooperberg MR, Moul JW, Carroll PR. The changing face of prostate cancer. *J Clin Oncol* 2005;**23**(32):8146–51.
3. Msezane LP, Reynolds WS, Gofrit ON, et al. Bladder neck contracture after robot-assisted laparoscopic radical prostatectomy: evaluation of incidence and risk factors and impact on urinary function. *J Endourol* 2008;**22**(1):97–104.
4. Carlsson S, Nilsson AE, Schumacher MC, et al. Surgery-related complications in 1253 robot-assisted and 485 open retropubic radical prostatectomies at the Karolinska University Hospital, Sweden. *Urology* 2010;**75**(5):1092–7.
5. Brodak M, Kosina J, Pacovsky J, et al. Bipolar transurethral resection of anastomotic strictures after radical prostatectomy. *J Endourol* 2010;**24**(9):1477–81.
6. Carr LK, Webster GD. Endoscopic management of the obliterated anastomosis following radical prostatectomy. *J Urol* 1996;**156**(1):70–2.
7. Dalkin BL. Endoscopic evaluation and treatment of anastomotic strictures after radical retropubic prostatectomy. *J Urol* 1996;**155**(1):206–8.
8. Giannarini G, Manassero F, Mogorovich A, et al. Cold-knife incision of anastomotic strictures after radical retropubic prostatectomy with bladder neck preservation: efficacy and impact on urinary continence status. *Eur Urol* 2008;**54**(3):647–56.
9. Kostakopoulos A, Argiropoulos V, Protogerou V, et al. Vesicourethral anastomotic strictures after radical retropubic prostatectomy: the experience of a single institution. *Urol Int* 2004;**72**(1):17–20.
10. Ramchandani P, Banner MP, Berlin JW, et al. Vesicourethral anastomotic strictures after radical prostatectomy: efficacy of transurethral balloon dilation. *Radiology* 1994;**193**(2):345–9.
11. Hall EJ, Astor M, Bedford J, et al. Basic radiobiology. *Am J Clin Oncol* 1988;**11**(3):220–52.
12. Turina M, Mulhall AM, Mahid SS, et al. Frequency and surgical management of chronic complications related to pelvic radiation. *Arch Surg* 2008;**143**(1):46–52; discussion 52.
13. Elliott SP, Meng MV, Elkin EP, et al. Incidence of urethral stricture after primary treatment for prostate cancer: data From CaPSURE. *J Urol* 2007;**178**(2):529–34; discussion 534.
14. Parihar JS, Ha YS, Kim IY. Bladder neck contracture-incidence and management following contemporary robot assisted radical prostatectomy technique. *Prostate Int* 2014;**2**(1):12–8.
15. Borboroglu PG, Sands JP, Roberts JL, et al. Risk factors for vesicourethral anastomotic stricture after radical prostatectomy. *Urology* 2000;**56**(1):96–100.
16. Wang R, Wood Jr DP, Hollenbeck BK, et al. Risk factors and quality of life for post-prostatectomy vesicourethral anastomotic stenoses. *Urology* 2012;**79**(2):449–57.
17. Erickson BA, Meeks JJ, Roehl KA, et al. Bladder neck contracture after retropubic radical prostatectomy: incidence and risk factors from a large single-surgeon experience. *BJU Int* 2009;**104**(11):1615–9.
18. Davidson PJ, van den Ouden D, Schroeder FH. Radical prostatectomy: prospective assessment of mortality and morbidity. *Eur Urol* 1996;**29**(2):168–73.
19. Gonzalgo ML, Pavlovich CP, Trock BJ, et al. Classification and trends of perioperative morbidities following laparoscopic radical prostatectomy. *J Urol* 2005;**174**(1):135–9; discussion 139.

20. Moreira Jr SG, Seigne JD, Ordorica RC, et al. Devastating complications after brachytherapy in the treatment of prostate adenocarcinoma. *BJU Int* 2004;**93**(1):31–5.
21. Zelefsky MJ, Whitmore Jr WF. Long-term results of retropubic permanent 125iodine implantation of the prostate for clinically localized prostatic cancer. *J Urol* 1997;**158**(1):23–9; discussion 29–30.
22. Nguyen PL, D'Amico AV, Lee AK, et al. Patient selection, cancer control, and complications after salvage local therapy for postradiation prostate-specific antigen failure: a systematic review of the literature. *Cancer* 2007;**110**(7):1417–28.
23. Lee YH, Chiu AW, Huang JK. Comprehensive study of bladder neck contracture after transurethral resection of prostate. *Urology* 2005;**65**(3):498–503; discussion 503.
24. Puppo P, Bertolotto F, Introini C, et al. Bipolar transurethral resection in saline (TURis): outcome and complication rates after the first 1000 cases. *J Endourol* 2009;**23**(7):1145–9.
25. Kim HS, Cho MC, Ku JH, et al. The efficacy and safety of photoselective vaporization of the prostate with a potassium-titanyl-phosphate laser for symptomatic benign prostatic hyperplasia according to prostate size: 2-year surgical outcomes. *Korean J Urol* 2010;**51**(5):330–6.
26. Malde S, Rajagopalan A, Patel N, et al. Potassium-titanyl-phosphate laser photoselective vaporization for benign prostatic hyperplasia: 5-year follow-up from a district general hospital. *J Endourol* 2012;**26**(7):878–83.
27. Sandhu JS, Ng C, Vanderbrink BA, et al. High-power potassium-titanyl-phosphate photoselective laser vaporization of prostate for treatment of benign prostatic hyperplasia in men with large prostates. *Urology* 2004;**64**(6):1155–9.
28. Ramirez D, Zhao LC, Bagrodia A, et al. Deep lateral transurethral incisions for recurrent bladder neck contracture: promising 5-year experience using a standardized approach. *Urology* 2013;**82**(6):1430–5.
29. Park R, Martin S, Goldberg JD, et al. Anastomotic strictures following radical prostatectomy: insights into incidence, effectiveness of intervention, effect on continence, and factors predisposing to occurrence. *Urology* 2001;**57**(4):742–6.
30. Besarani D, Amoroso P, Kirby R. Bladder neck contracture after radical retropubic prostatectomy. *BJU Int* 2004;**94**(9):1245–7.
31. Lubahn JD, Zhao LC, Scott JF, et al. Poor quality of life in patients with urethral stricture treated with intermittent self-dilation. *J Urol* 2013;**191**(1):143–7.
32. Milroy EJ, Chapple CR, Cooper JE, et al. A new treatment for urethral strictures. *Lancet* 1988;**1**(8600):1424–7.
33. Hussain M, Greenwell TJ, Shah J, et al. Long-term results of a self-expanding wallstent in the treatment of urethral stricture. *BJU Int* 2004;**94**(7):1037–9.
34. De Vocht TF, van Venrooij GE, Boon TA. Self-expanding stent insertion for urethral strictures: a 10-year follow-up. *BJU Int* 2003;**91**(7):627–30.
35. Magera Jr JS, Inman BA, Elliott DS. Outcome analysis of urethral wall stent insertion with artificial urinary sphincter placement for severe recurrent bladder neck contracture following radical prostatectomy. *J Urol* 2009;**181**(3):1236–41.
36. Erickson BA, McAninch JW, Eisenberg ML, et al. Management for prostate cancer treatment related posterior urethral and bladder neck stenosis with stents. *J Urol* 2011;**185**(1):198–203.
37. Chancellor MB, Gajewski J, Ackman CF, et al. Long-term followup of the North American multicenter UroLume trial for the treatment of external detrusor-sphincter dyssynergia. *J Urol* 1999;**161**(5):1545–50.
38. Badlani GH, Press SM, Defalco A, et al. Urolume endourethral prosthesis for the treatment of urethral stricture disease: long-term results of the North American Multicenter UroLume Trial. *Urology* 1995;**45**(5):846–56.
39. Elliott DS, Boone TB. Combined stent and artificial urinary sphincter for management of severe recurrent bladder neck contracture and stress incontinence after prostatectomy: a long-term evaluation. *J Urol* 2001;**165**(2):413–5.
40. Corujo M, Badlani GH. Epithelialization of permanent stents. *J Endourol* 1997;**11**(6):477–80.
41. Beier-Holgersen R, Brasso K, Nordling J, et al. The "Wallstent": a new stent for the treatment of urethral strictures. *Scand J Urol Nephrol* 1993;**27**(2):247–50.
42. Morgia G, Saita A, Morana F, et al. Endoprosthesis implantation in the treatment of recurrent urethral stricture: a multicenter study. Sicilian-Calabrian Urology Society. *J Endourol* 1999;**13**(8):587–90.
43. Gillitzer R, Thomas C, Wiesner C, et al. Single center comparison of anastomotic strictures after radical perineal and radical retropubic prostatectomy. *Urology* 2010;**76**(2):417–22.
44. Anger JT, Raj GV, Delvecchio FC, et al. Anastomotic contracture and incontinence after radical prostatectomy: a graded approach to management. *J Urol* 2005;**173**(4):1143–6.
45. Yurkanin JP, Dalkin BL, Cui H. Evaluation of cold knife urethrotomy for the treatment of anastomotic stricture after radical retropubic prostatectomy. *J Urol* 2001;**165**(5):1545–8.
46. Pansadoro V, Emiliozzi P. Iatrogenic prostatic urethral strictures: classification and endoscopic treatment. *Urology* 1999;**53**(4):784–9.
47. Surya BV, Provet J, Johanson KE, et al. Anastomotic strictures following radical prostatectomy: risk factors and management. *J Urol* 1990;**143**(4):755–8.
48. Brede C, Angermeier K, Wood H. Continence outcomes after treatment of recalcitrant postprostatectomy bladder neck contracture and review of the literature. *Urology* 2014;**83**(3):648–52.
49. Ramirez D, Simhan J, Hudak SJ, et al. Standardized approach for the treatment of refractory bladder neck contractures. *Urol Clin North Am* 2013;**40**(3):371–80.
50. Theodoros C, Katsifotis C, Stournaras P, et al. Abdomino-perineal repair of recurrent and complex bladder neck-prostatic urethra contractures. *Eur Urol* 2000;**38**(6):734–40; discussion 740–1.
51. Schlossberg S, Jordan G, Schellhammer P. Repair of obliterative vesicourethral stricture after radical prostatectomy: a technique for preservation of continence. *Urology* 1995;**45**(3):510–3.
52. Wessells H, Morey AF, McAninch JW. Obliterative vesicourethral strictures following radical prostatectomy for prostate cancer: reconstructive armamentarium. *J Urol* 1998;**160**(4):1373–5.
53. Simonato A, Gregori A, Lissiani A, et al. Two-stage transperineal management of posterior urethral strictures or bladder neck contractures associated with urinary incontinence after prostate surgery and endoscopic treatment failures. *Eur Urol* 2007;**52**(5):1499–504.
54. Mundy AR, Andrich DE. Posterior urethral complications of the treatment of prostate cancer. *BJU Int* 2012;**110**(3):304–25.
55. Reiss P, Pfalzgraf D, Kluth L, et al. Perineal-reanastomosis for the treatment of recurrent anastomotic strictures: outcome and patient satisfaction (abstract). *J Urol* 2011;**185**(4):e84.
56. Eltahawy E, Gur U, Virasoro R, et al. Management of recurrent anastomotic stenosis following radical prostatectomy using holmium laser and steroid injection. *BJU Int* 2008;**102**(7):796–8.
57. Vanni AJ, Zinman LN, Buckley JC. Radial urethrotomy and intralesional mitomycin C for the management of recurrent bladder neck contractures. *J Urol* 2011;**186**(1):156–60.
58. Doherty AP, Trendell-Smith N, Stirling R, et al. Perivesical fat necrosis after adjuvant intravesical chemotherapy. *BJU Int* 1999;**83**(4):420–3.
59. Oddens JR, van der Meijden AP, Sylvester R. One immediate postoperative instillation of chemotherapy in low risk Ta, T1 bladder cancer patients. Is it always safe? *Eur Urol* 2004;**46**(3):336–8.
60. Moran DE, Moynagh MR, Alzanki M, et al. Anaphylaxis at image-guided epidural pain block secondary to corticosteroid compound. *Skeletal Radiol* 2012;**41**(10):1317–8.
61. Hou JC, Landas S, Wang CY, et al. Instillation of mitomycin C after transurethral resection of bladder cancer impairs wound healing: an animal model. *Anticancer Res* 2011;**31**(3):929–32.

62. Redshaw JD, Broghammer JA, Smith III TG, et al. Intralesional injection of mitomycin C at the time of transurethral incision of bladder neck contracture may offer limited benefit: from the TURNS Study Group. *J Urol* 2014;**193**(2):587–92.
63. Kundu SD, Roehl KA, Eggener SE, et al. Potency, continence and complications in 3,477 consecutive radical retropubic prostatectomies. *J Urol* 2004;**172**(6 Pt 1):2227–31.
64. Lepor H, Kaci L. The impact of open radical retropubic prostatectomy on continence and lower urinary tract symptoms: a prospective assessment using validated self-administered outcome instruments. *J Urol* 2004;**171**(3):1216–9.
65. Sacco E, Prayer-Galetti T, Pinto F, et al. Urinary incontinence after radical prostatectomy: incidence by definition, risk factors and temporal trend in a large series with a long-term follow-up. *BJU Int* 2006;**97**(6):1234–41.
66. Sanderson KM, Penson DF, Cai J, et al. Salvage radical prostatectomy: quality of life outcomes and long-term oncological control of radiorecurrent prostate cancer. *J Urol* 2006;**176**(5):2025–31; discussion 2031–2.
67. Stephenson AJ, Scardino PT, Bianco Jr FJ, et al. Morbidity and functional outcomes of salvage radical prostatectomy for locally recurrent prostate cancer after radiation therapy. *J Urol* 2004;**172**(6 Pt 1):2239–43.
68. Borawski K, Webster G. Long term consequences in the management of the devastated, obstructed outlet using combined UroLume stent with subsequent artificial urinary sphincter placement (abstract). *J Urol* 2010;**183**(4):e427.
69. Gousse AE, Tunuguntla HS, Leboeuf L. Two-stage management of severe postprostatectomy bladder neck contracture associated with stress incontinence. *Urology* 2005;**65**(2):316–9.
70. Weissbart SJ, Chughtai B, Elterman D, et al. Management of anastomotic stricture after artificial urinary sphincter placement in patients who underwent salvage prostatectomy. *Urology* 2013;**82**(2):476–9.

CHAPTER

# 41

# Reimbursement for Prostate Cancer Treatment

*Ahmed Haddad, MBChB, PhD, Yair Lotan, MD*

Department of Urology, University of Texas Southwestern Medical Center, Dallas, TX, USA

## INTRODUCTION

Prostate cancer (PCa) is the most common solid organ malignancy in men with over 200,000 new cases diagnosed each year.[1] The long life expectancy of most men diagnosed with PCa has led to a burgeoning prevalence of the disease. An estimated 2.7 million men currently live with PCa in the United States, and the prevalence is only likely to increase in the future with the growth of the aging population. The healthcare resource implications associated with the diagnosis and treatment of PCa are vast. PCa accounts for approximately 10% ($12 billion) of the total annual budget for cancer care in the United States.[2] Most cases are diagnosed by prostate-specific antigen (PSA) screening, which results in significant overdetection of indolent, low-risk PCa. The overdetection and overtreatment of PCa are estimated to account for around 40% of the total national cost of PCa.[3]

PCa patients are confronted with multiple possible treatment options with comparable outcomes but with widely varying cost. The American Urologic Association guidelines recommend either surgery or radiation therapy for localized disease; however, there is currently no level I evidence demonstrating superiority of one modality over another.[4] The guidelines recommend active surveillance as an option for patients with low-risk disease, yet studies have shown that it is rarely utilized with even the lowest-risk patients most often treated with definitive surgery or radiation therapy. The last decade has also witnessed the introduction of ever more expensive technologies, such as robotic surgery and intensity-modulated radiation therapy (IMRT) as first-line therapies in PCa. These novel technologies have disseminated widely and with little consideration to cost-effectiveness over traditional treatments such as open radical prostatectomy and conformal radiation therapy.

Prostate Cancer. http://dx.doi.org/10.1016/B978-0-12-800077-9.00041-4


Several factors can influence the treatment decision in PCa patients. A significant referral bias exists with urologists more likely to recommend surgery, and radiation oncologists more likely to recommend radiation therapy.[5] In the United States, Medicare reimbursement for medical services can also influence decision-making and the choice of therapy. An AUA survey reported that 16% of urologists had discontinued tests or procedures due to a reduction in reimbursement.[6] In another study, Escarce et al. demonstrated that a reduction in Medicare reimbursement was associated with a decrease in volume and complexity of the services provided.[7] Elliott et al. demonstrated the effect of reduced reimbursement for androgen deprivation therapy (ADT) following the 2003 Medicare Modernization act.[8] They found that there was a significant reduction in nonindicated ADT use (defined as ADT given without surgery or radiation in low-risk disease) one year after implementation of the reimbursement changes.[8]

Medicare expenditure for treating PCa continues to rise. Between 1992 and 2003, expenditure increased 20% per capita, mostly due to higher usage of physician services (rather than rising cost of services).[9] This increase in cost to the health system was prior to the introduction of expensive technologies such as robotic-assisted laparoscopic prostatectomy (RALP) and IMRT, which have led to further rises in healthcare expenditure for PCa. With the unsustainable growth of national healthcare expenditure, a comprehensive analysis of the cost effectiveness of various treatment options for prostate cancer is imperative.

## ECONOMICS OF OPEN VERSUS ROBOTIC RADICAL PROSTATECTOMY

There has been an exponential rise in the uptake of robotic technology by the urologic community over the last decade. In 2004 only 8% of prostatectomies were performed with robotic assistance, compared to 67% in 2010.[10] The introduction of the robot may have even led to an increase in the overall proportion of men receiving surgery compared to nonsurgical treatments for PCa. This is evidenced by a 60% increase in the number of hospital discharges for radical prostatectomy between 2005 and 2008 based on the National Inpatient Sample.[11] During the same period, prostate cancer incidence had remained stable. Several factors have driven this dramatic increase in RALP over the last decade. Aggressive (and often inaccurate) direct-to-consumer marketing by the manufacturer has been reported.[12] Patient preferences for the most innovative and minimally invasive procedure further fuel the drive for robotic surgery. From the surgeon's perspective, the robotic approach offers many benefits such as stereoscopic visualization, wristed instruments, tremor filtration, and improved ergonomics. Surgeons transitioning to RALP do not experience as steep a learning curve as that associated with pure laparoscopic prostatectomy; however, experience has still been shown to impact results.[13] Medical facilities have embraced the technology due to patient and physician demand and possible financial incentives. However, the marketing drive to robotic surgery can lead to unrealistic patient expectations, with one study demonstrating increased patient dissatisfaction and regret following RALP as compared to patients undergoing open radical prostatectomy.[14]

Despite the perceived advantages, there are no prospective randomized trials that demonstrate improved oncologic outcomes for RALP compared to open prostatectomy. Retrospective studies have not demonstrated an improvement in biochemical recurrence or survival for RALP compared to open or laparoscopic prostatectomy. The only consistently demonstrated benefit of RALP over open retropubic prostatectomy has been reduced blood loss and shorter hospital stay.[15] The effect of RALP on positive surgical margin rates is unclear. A recent study demonstrated lower positive surgical margin rates for RALP compared to open prostatectomy (HR 0.75, $p < 0.001$),[16] which goes against the findings of the meta-analysis by Ficarra et al. that demonstrated no difference in PSM between open radical prostatectomy and minimally invasive approaches.[15] Recent results also suggest improved 12-month potency and continence rates following RALP compared to open surgery, although the use of nerve-sparing approaches, surgeon experience, and cancer stage influence these endpoints.[17,18] In a recent systematic review comparing RALP to standard laparoscopic prostatectomy, RALP was associated with significantly reduced risk of organ injury and improved positive surgical margins.[19] However, a study of Medicare claims of 5923 men comparing RALP and standard laparoscopic prostatectomy did not find a difference in bowel complications, or other general medical and surgical complications.[20] The lack of standardized reporting is a major limitation of these comparative retrospective studies and partly explains the inconsistencies in outcomes between studies. Furthermore, the reports include surgeons with different levels of experience and at different points in the learning curve of RALP.

The widespread adoption of RALP as the primary surgical treatment modality has occurred with little consideration for the cost implications.[21] The main costs to consider when comparing different modalities for prostatectomy are equipment costs, operative time, and length of hospital stay.[22] Several studies have highlighted the considerable increase in cost of surgery associated with RALP compared to open prostatectomy.[23–27] Bolenz et al. reviewed a number of studies reporting direct costs for RP, and demonstrated that the costs varied from center to center but on average was higher for RALP (range $5058–$11,806) than open prostatectomy (range $4075–$6296).[28] Lotan et al. demonstrated that RALP was $1155 more costly than open prostatectomy, largely due to the high cost of the instruments.[21] This analysis did not take into account the fixed fee for purchasing the robot, amounting to 1–2 million dollars. In addition to the large initial outlay of purchasing the robot, there is a yearly service maintenance fee of $100,000–$200,000. In some centers where the robot is received as a gift, a maintenance cost of $150,000 would still add $1000 per case if 150 cases were performed per year.[29] The current lack of competition in the robotic market, with only one manufacturer producing the surgical robot means that equipment costs are likely to remain high.

Operative time and length of stay are important factors to consider in the cost of surgical procedures. With the inclusion of robotics into many urology residency programs, the issue of the learning curve is becoming less of a factor for urologists starting independent practice. Studies vary widely in their estimates of number of RALP cases required to master the procedure.[30] Guillonneau and Vallancien reported that it took 100 cases for an expert laparoscopist to reduce their RALP operating time from 4 h to 3 h; however, others have reported that it can take up to 200 cases to master the procedure.[31] The cost of the learning curve based on differences in operating time has been estimated to be approximately $215,000.[30] Davis et al. examined a nationwide database of over 71,000 prostatectomies and demonstrated longer operative time for RALP compared to open (mean 4.4 h vs. 3.4 h, $p < 0.0001$), but also a shorter inpatient stay with RALP (mean 2.2 days vs. 3.2 days, $p < 0.0001$). Surgery time, hospital stay, and complication rates improved significantly with surgeon experience.[32]

Medicare reimbursement for radical prostatectomy is the same regardless of surgical approach. Thus, the added cost of robotic prostatectomies falls on the provider. Lotan et al. demonstrated that each robotic case resulted in a loss of over $4000 to the hospital.[33] However, in a recent analysis of hospital reimbursement for RALP from mixed payers, reimbursement was higher for RALP compared with open prostatectomy (approximately $2000 per case).[34] RALP is also potentially advantageous to the surgeon due both to increased volume and additional reimbursement for the S2900 code (use of robot). However, reimbursement for the S2900 code is inconsistent since Medicare does not pay for the S2900 code and payment is variable between insurance companies.[33] In addition, the performance of laparoscopic pelvic lymph node dissection is bundled with open radical prostatectomy but not with RALP so further physician payment can be obtain for performing a laparoscopic node dissection at time of RALP.

The additional cost of RALP could be justified if the procedure provided improved outcomes or quality of life for patients with PCa. To date, given the minimal benefit associated with RALP and the additional cost of the robot, most studies have not demonstrated the cost-effectiveness of RALP. Cost-effectiveness analyses employ the incremental cost-effectiveness ratio (ICER) as a measure of cost-effectiveness of one health intervention over another.[35] Although not an absolute cut-off, an ICER of <$50,000 per quality adjusted life year (QALY) is regarded as a cut-off indicative of cost-effective intervention.[36] Markov models and decision tree analyses are often employed in cost-effectiveness studies to model patient outcomes over a specified time horizon. In a study comparing the various surgical and radiation treatment modalities, Cooperberg et al. compared cost-effectiveness based on literature estimates of outcome probabilities over a lifetime horizon.[23] They determined that QALY was not significantly different between different surgical modalities. Costs accounted for included direct costs of medical resource utilization (procedures, office visits, imaging). The mean lifetime costs were comparable for robotic and open prostatectomy ($19,901 vs. $20,245). Since cost estimates were derived from Medicare Fee schedules, which reimburse similarly for RALP and open prostatectomy, the study did not demonstrate a significant difference in cost-effectiveness between the two surgical modalities. This study took the perspective of the payer (Medicare) and did not account for the additional costs of RALP, which Medicare does not cover.

A cost-effectiveness analysis of RALP versus laparoscopic prostatectomy from the European perspective estimated the incremental cost per QALY gained from robotic compared to laparoscopic prostatectomy at £18,329, well below the threshold of £30,000 for cost-effectiveness set by the National Institute for Health and Care Excellence (NICE) in the United Kingdom.[37] This analysis was conducted with the assumption that >150 prostatectomies per year would be performed, and the sensitivity analysis demonstrated that when only 50 procedures per year were performed, the ICER rises considerably to £106,839/QALY. RALP was deemed cost-effective despite higher costs due to the model assumption that RALP offered an advantage in terms of fewer organ injuries and lower rates of positive surgical margins, which over a lifetime horizon result in higher QALY with RALP. This study, like the Cooperberg study, did not take into account the capital and maintenance costs of the robotic procedure. Despite geographic variations, these studies highlight the potential cost-effectiveness of RALP, and emphasize the dependence of cost on surgical case volume.

## ECONOMICS OF RADIATION THERAPY FOR PROSTATE CANCER

### Intensity-Modulated Radiation Therapy (IMRT)

Radiation oncology techniques in prostate cancer have evolved dramatically over the past decade.[38] Conventional radiation has been superseded by 3D-conformal radiation therapy (3DCRT) and more recently by intensity-modulated radiation therapy (IMRT) as the standard of care. Similar to the dissemination of the surgical robot, the newer, more expensive radiation technologies have been widely adopted without much consideration for cost and without strong evidence of improved outcomes compared to earlier technologies. In 2000, only 0.15% of patients undergoing primary radiation therapy for prostate cancer received IMRT, compared to 95.9% in 2008.[39] IMRT involves more complicated planning, longer treatment time, and higher equipment costs. As such, IMRT is around $10,000 more expensive than 3DCRT, and the excess cost to Medicare due to the adoption of IMRT is over $250 million.[40]

The rationale for 3DCRT and IMRT are that these technologies provide more precise contouring of the radiation field and improved prostate targeting compared to conventional techniques. This allows for the utilization of higher doses of radiation with the goal of enhancing cancer control while maintaining an acceptable toxicity profile.[38] Zietman et al. demonstrated improved oncologic outcomes in patients treated with 3DCRT compared to conventional radiation therapy, although this trial was a dose escalation study rather than a comparison of the two technologies.[41] 3DCRT at 78 Gy was associated with improved biochemical recurrence-free survival but higher rectal morbidity (grade 2) compared to conventional radiation therapy at 70 Gy.

Despite the almost complete dissemination of IMRT as the standard external beam radiation treatment modality in the United States, there is no level I evidence demonstrating superiority of IMRT over 3DCRT. In the largest institutional cohort, Zelefsky et al. reported lower late GI morbidity with IMRT compared to 3DCRT (5% vs. 13%) even though IMRT patients received a higher radiation dose in their series.[42] The same group reported on IMRT dose escalated to 86.4 Gy in 1002 patients with biochemical recurrence-free survival rates of 98.8, 85.6, and 67.9% for low-, intermediate-, and high-risk patients, respectively.[43] Late grade 2 GU complications were common in this series (21%) but late GI toxicity was low (4.4%). In a subanalysis of a Dutch randomized trial, patients receiving 78 Gy IMRT were shown to have significantly less rectal morbidity compared to patients receiving the same dose of 3DCRT (20% vs. 61%, $P = 0.001$).[44] A systematic review comparing IMRT to 3DCRT concluded that acute and late GI and GU toxicity was at minimum no different, and in some cases superior to 3DCRT.[45]

Few cost-effectiveness studies have been performed comparing IMRT and 3DCRT. Konski et al. reported that the expected mean cost for IMRT at their institution was $47,931 compared to $21,865 for 3DCRT.[46] They determined health utilities and quality of life from studies conducted at their institution using validated questionnaires and estimated QALY for IMRT was 6.27 compared to 5.62 for 3DCRT. Employing a Markov model, they demonstrated that IMRT was at the upper limit of being cost effective at $40,101 per QALY compared to 3DCRT. Hummel et al. conducted a cost analysis of IMRT based on United Kingdom costs.[47] They reported that the additional cost of IMRT compared with 3DCRT was only £1120. Their analysis assumed equivalent cancer outcomes and varying assumptions of GI toxicity rates. When GI toxicity rates were comparable for IMRT and 3DCRT, the ICER was £104,000/QALY. With greater improvements in GI toxicity (15% improvement), the ICER moved into the range where it was borderline cost effective (£35,000/QALY). Yong et al. examined the cost effectiveness of IMRT from the Canadian perspective.[48] They reported that the cost of IMRT in Canada was only $1019 more than the cost for 3DCRT. In this study, IMRT was cost-effective compared to 3DCRT with an ICER of $26,768/QALY. These findings have not been replicated in studies modeled on the United States health system where IMRT costs are substantially higher than for 3DCRT. IMRT has also been shown to be less cost effective than low dose rate and high dose rate brachytherapy for patients with low- to intermediate-risk prostate cancer.[49,50] Likewise, compared to RALP, IMRT is less cost effective.[23,51] Cooperberg et al. demonstrated IMRT to be $20,000 more costly than surgery and associated with worse QALY.[23]

Many urology groups have now integrated IMRT facilities into their practices, likely driven by the high reimbursement for IMRT in the United States.[52,53] In a study comparing use of IMRT before and after ownership in urology practices with IMRT facilities, Mitchell found that IMRT use by self-referring urologists increased from 13.1% to 32.3% ($p < 0.002$) after ownership.[53] During the same period among nonreferring urologists, the rate of IMRT use only increased from 14.3% to 15.6%. This study highlights the concern of potential overutilization of IMRT as a treatment modality for PCa patients, a phenomenon potentially fuelling the escalating costs associated with this expensive technology. Others have suggested that urologist integration of IMRT services leads to appropriate utilization of an effective noninvasive treatment for PCa by urologists who would otherwise be inclined to recommend surgical therapy.[54]

## Stereotactic Body Radiation Therapy (SBRT)

Stereotactic body RT (SBRT) is another innovative radiation technology for PCa that has been robustly marketed with minimal comparative efficacy data.[55] SBRT provides even greater conformity of the target field than IMRT and steep dose gradients. SBRT has been employed in hypofractionation protocols where large doses of radiation are delivered in as few as five fractions.[56] Hypofractionation takes advantage of the specific radiobiologic properties of prostate tissue (specifically a high $\alpha/\beta$ ratio) compared to surrounding tissues.[57] To date, studies have demonstrated comparable biochemical recurrence-free survival with SBRT for localized PCa compared to IMRT; however, the data are preliminary with insufficient follow-up and small sample size.[58–60] The literature is conflicting regarding late GU and GI toxicity of SBRT. A recent study using Medicare claims demonstrated higher GU toxicity at 24 months with SBRT compared to IMRT (43.9% vs. 36.3%, $p = 0.001$).[56] A review of trials comparing SBRT to 3DCRT did not demonstrate a difference in late GU or GI toxicity (grade≥2).[61] However, these studies have to be interpreted with caution since most late radiation toxicity occurs after 10 years and is not captured in studies with short follow-up.[38]

SBRT is a more technologically intensive modality than IMRT and incurs higher fixed equipment costs especially with robotic delivery of SBRT. Despite the higher initial costs, SBRT may be more cost-effective than IMRT given the fewer treatment sessions required (5–6 vs. 39). Based on Medicare reimbursement costs, Yu et al. reported the mean treatment cost of SBRT as $13,645 versus $21,023 for IMRT.[61] Sher et al. compared the cost-effectiveness of SBRT and IMRT using a Markov model assuming equivalent local recurrence but higher rectal and urinary toxicity with SBRT.[60] Despite the higher assumed toxicity, SBRT (even using the robotic device) was considerably more cost-effective than IMRT. The ICER of IMRT ranged from $100,000/QALY to $591,000/QALY depending on the sensitivity analysis, and was therefore not cost-effective under all model assumptions. Hodges et al. demonstrated that SBRT was over $12,000 cheaper than IMRT, and with equal QALY assumed for both modalities, SBRT was clearly more cost-effective.[58] Thus, SBRT is a promising technology and mature comparative efficacy data are awaited to determine whether this modality is a viable cost-effective alternative to IMRT.

### Proton Beam Therapy (PBT)

The most recent radiation modality that has been increasingly implemented in the treatment of localized PCa is proton beam therapy (PBT). Similar to the previously described innovative technologies, PBT has gained traction with aggressive marketing and minimal rigorous comparative efficacy studies.[62] The premise for the use of PBT is related to the radiobiologic properties of protons, which are heavy particles that release most of their energy within the target tissue, known as the Bragg peak.[63] Consequently, PBT is associated with less radiation exposure to normal surrounding tissues. Early studies of the dose distribution of PBT demonstrated that compared to IMRT, PBT reduced rectal and bladder radiation dose by 59% and 39%, respectively.[64] The largest institutional series of PBT for localized PCa is from Loma Linda where 1255 patients with stage T1–T3 PCa have been treated with PBT (731 received 3DCRT in addition to PBT, and 524 PBT alone).[65] Eight-year biochemical-free survival rates were 84, 65, and 48% for patients with pretreatment PSA of 4.1–10, 10.1–20, and >20, respectively.[65] Rates of grade ≥3 late GU and GI toxicity were low (1 and 0.2%, respectively. Zietman et al. conducted a randomized controlled trial of low- versus high-dose radiation therapy with all patients receiving 50.4 Gy to the pelvis and either 19.8 Gy or 28.8 Gy equivalent proton treatment to the prostate.[66] Patients in the high-dose arm had superior oncologic outcomes and late GI and GU grade ≥3 toxicity was 2 and 1% for both study arms.

Proton therapy was initially used for the treatment of pediatric CNS tumors where a sharp cut-off of radiation dose was critical. However, the use of PBT has shifted dramatically toward PCa with over 70% of all PBT treatments in the United States being performed for PCa.[67] The vast capital investment required to establish a proton facility has fueled heated debate regarding the appropriateness of dissemination of this technology when there are no rigorous comparative trials demonstrating a benefit over current therapies. The cost of establishing a proton facility is currently over $100 million, which puts considerable financial pressure on these facilities particularly for-profit centers.[68] The reasons for its increased use in PCa are likely driven by the need to maintain financial viability in the face of the high facility costs and debt payments. Treatment times for PCa are short compared to more complex tumors, such as CNS tumors, which allows for greater patient turnover, and together with the wide potential market and high reimbursement rate ($32,428 vs. $18,575 by Medicare compared to IMRT),[61] PCa is seen as a viable market for use of PBT.[69]

Cost-effectiveness analyses for PBT are limited by the paucity of data on the efficacy, toxicity, and quality of life of patients receiving this treatment. Additional costs associated with proton therapy include travel and accommodation costs due to the restricted availability of proton centers nationwide. Ollendorf et al. performed a cost analysis of PBT compared to other radiation modalities and showed that over a life-time horizon, PBT was associated with the highest costs of any treatment modality for PCa ($53,828 for PBT and $37,861 for IMRT).[70] Konski et al. conducted a cost-effectiveness analysis using Medicare physician reimbursement rates and demonstrated that proton therapy was close to being cost effective compared to IMRT given a time horizon of 15 years (ICER $55,726/QALY).[71] This study however did not take into account the initial cost of establishing a proton facility.

The longer life of cyclotrons used in PBT compared to IMRT technology and the development of newer cheaper single galley cyclotrons ($25 million) will contribute to lowering long-term costs of PBT. Hypofractionation protocols with PBT are a further avenue for improving the cost-effectiveness of PBT. However, until more rigorous prospective data on the efficacy of PBT in the management of PCa emerge, the technology will continue to attract intense scrutiny from payers and health revenue agencies increasingly focused on maximizing cost-effectiveness of afforded therapies.

## CONCLUSIONS

The continued rise in healthcare expenditure has placed increased emphasis on rigorous cost-effectiveness evaluation of novel technologies used in healthcare. PCa therapies are under specific scrutiny due to the high prevalence of the disease and costly treatments making PCa a major contributor to the overall healthcare budget. Current guideline-recommended therapies for PCa include radical prostatectomy and radiation therapy (external beam or brachytherapy). Active surveillance is also an option for men with low-risk PCa according to current guidelines, and has demonstrated cost-effectiveness.[4,72,73] However, most men with low-risk PCa in the United States receive either radical surgery or radiation therapy despite over 97% cancer-specific survival demonstrated in mature clinical series of active surveillance.[74,75] The low uptake of active surveillance may be due to the perception of low reimbursement rates. Contrary to this idea, studies have shown that the cumulative costs of active surveillance actually result in higher reimbursement for the urologist.[70,72] The estimated potential cost-savings with the use of active surveillance over other treatments is $1.9 billion.[72]

IMRT and RALP are examples of costly innovative treatment modalities for PCa that have widely disseminated despite limited evidence to suggest superiority to prior standards of care (3DCRT and open prostatectomy). Almost all external beam radiation therapy in the United States is performed with IMRT, which has been shown to be the least cost-effective among all treatment modalities (surgery or radiation) currently employed with the exception of PBT. Medicare reimbursement for IMRT is high, and controversially, self-referral among urologic groups with IMRT facilities has increased. Surgical therapy for PCa has similarly incorporated new technology with the majority of radical prostatectomy currently being performed robotically in the United States. The dissemination of RALP has occurred despite equivalent Medicare reimbursement for robotic and open prostatectomy. Reasons for the adoption of RALP despite the high cost incurred to healthcare facilities include pressure by patients and physicians who are the primary drivers behind the uptake of new technologies. Since patients have similar out-of-pocket costs whether their surgery is performed open or robotic, their decision is motivated by nonfinancial factors.[76] Hospitals may also have incentive to increase utilization of RALP to cover the cost of purchasing the robot. Furthermore, although Medicare reimbursement may increase over time, future reimbursement is calculated based on current charges.[11]

Although patients and physicians are the principal drivers behind the uptake of new technologies, Medicare and private payer reimbursement can also influence physician recommendations and treatment choice. Future changes in health policy, such as the Accountable Care Act, may significantly change the reimbursement landscape. Although future legislation is unlikely to have a major impact on already disseminated technologies, such RALP and IMRT, the effect of reimbursement policies on developing technologies, such as proton therapy for PCa, is unclear. Clearly, randomized comparative studies are required to fully evaluate new technologies. This comparative data will form the basis of comprehensive cost-effectiveness analysis, which will be increasingly employed to affect healthcare decisions and the diffusion of new technologies. The National Institute of Care Excellence in the United Kingdom employs such a model for guiding resource allocation in the National Health Service. Evolving legislation in the United States will increasingly require rigorous assessment of new technologies. This will direct health resource allocation in the future, and PCa is set to be at the forefront of this period of high scrutiny of healthcare decision-making.

## References


1. Siegel R, Naishadham D, Jemal A. Cancer statistics, 2013. *CA Cancer J Clin* 2013;**63**(1):11–30.
2. Mariotto AB, Yabroff KR, Shao Y, et al. Projections of the cost of cancer care in the United States: 2010–2020. *J Natl Cancer Inst* 2011;**103**(2):117–28.
3. Heijnsdijk EA, der Kinderen A, Wever EM, et al. Overdetection, overtreatment and costs in prostate-specific antigen screening for prostate cancer. *Br J Cancer* 2009;**101**(11):1833–8.
4. Thompson I, Thrasher JB, Aus G, et al. Guideline for the management of clinically localized prostate cancer: 2007 update. *J Urol* 2007;**177**(6):2106–31.
5. Kim SP, Gross CP, Nguyen PY, et al. Specialty bias in treatment recommendations and quality of life among radiation oncologists and urologists for localized prostate cancer. *Prostate Cancer Prostatic Dis* 2014;**17**(2):163–9.
6. O'Leary MP, Baum NH, Blizzard R, et al. 2001 American Urological Association Gallup Survey: changes in physician practice patterns, satisfaction with urology, and treatment of prostate cancer and erectile dysfunction. *J Urol* 2002;**168**(2):649–52.
7. Escarce JJ. Effects of lower surgical fees on the use of physician services under Medicare. *JAMA* 1993;**269**(19):2513–8.
8. Elliott SP, Jarosek SL, Wilt TJ, et al. Reduction in physician reimbursement and use of hormone therapy in prostate cancer. *J Natl Cancer Inst* 2010;**102**(24):1826–34.
9. Zhang Y, Skolarus TA, Miller DC, et al. Understanding prostate cancer spending growth among Medicare beneficiaries. *Urology* 2011;**77**(2):326–31.
10. Lowrance WT, Eastham JA, Savage C, et al. Contemporary open and robotic radical prostatectomy practice patterns among urologists in the United States. *J Urol* 2012;**187**(6):2087–92.
11. Barbash GI, Glied SA. New technology and health care costs – the case of robot-assisted surgery. *N Engl J Med* 2010;**363**(8):701–4.
12. Mirkin JN, Lowrance WT, Feifer AH, et al. Direct-to-consumer internet promotion of robotic prostatectomy exhibits varying quality of information. *Health Aff* 2012;**31**(4):760–9.
13. Abboudi H, Khan MS, Guru KA, et al. Learning curves for urological procedures: a systematic review. *BJU Int* 2013;**114**(4):617–29.
14. Schroeck FR, Krupski TL, Sun L, et al. Satisfaction and regret after open retropubic or robot-assisted laparoscopic radical prostatectomy. *Eur Urol* 2008;**54**(4):785–93.
15. Ficarra V, Novara G, Artibani W, et al. Retropubic, laparoscopic, and robot-assisted radical prostatectomy: a systematic review and cumulative analysis of comparative studies. *Eur Urol* 2009;**55**(5):1037–63.
16. Sooriakumaran P, Srivastava A, Shariat SF, et al. A multinational, multi-institutional study comparing positive surgical margin rates among 22393 open, laparoscopic, and robot-assisted radical prostatectomy patients. *Eur Urol* 2013;**66**(3):450–6.
17. Ficarra V, Novara G, Ahlering TE, et al. Systematic review and meta-analysis of studies reporting potency rates after robot-assisted radical prostatectomy. *Eur Urol* 2012;**62**(3):418–30.
18. Ficarra V, Novara G, Rosen RC, et al. Systematic review and meta-analysis of studies reporting urinary continence recovery after robot-assisted radical prostatectomy. *Eur Urol* 2012;**62**(3):405–17.
19. Robertson C, Close A, Fraser C, et al. Relative effectiveness of robot-assisted and standard laparoscopic prostatectomy as alternatives to open radical prostatectomy for treatment of localised prostate cancer: a systematic review and mixed treatment comparison meta-analysis. *BJU Int* 2013;**112**(6):798–812.
20. Lowrance WT, Elkin EB, Jacks LM, et al. Comparative effectiveness of prostate cancer surgical treatments: a population based analysis of postoperative outcomes. *J Urol* 2010;**183**(4):1366–72.
21. Lotan Y, Cadeddu JA, Gettman MT. The new economics of radical prostatectomy: cost comparison of open, laparoscopic and robot assisted techniques. *J Urol* 2004;**172**(4 Pt 1):1431–5.
22. Sleeper J, Lotan Y. Cost-effectiveness of robotic-assisted laparoscopic procedures in urologic surgery in the USA. *Exp Rev Med Devices* 2011;**8**(1):97–103.

23. Cooperberg MR, Ramakrishna NR, Duff SB, et al. Primary treatments for clinically localised prostate cancer: a comprehensive lifetime cost-utility analysis. *BJU Int* 2013;**111**(3):437–50.
24. Makarov DV, Yu JB, Desai RA, et al. The association between diffusion of the surgical robot and radical prostatectomy rates. *Med Care* 2011;**49**(4):333–9.
25. Ramsay C, Pickard R, Robertson C, et al. Systematic review and economic modelling of the relative clinical benefit and cost-effectiveness of laparoscopic surgery and robotic surgery for removal of the prostate in men with localised prostate cancer. *Health Technol Assess* 2012;**16**(41):1–313.
26. Scales Jr CD, Jones PJ, Eisenstein EL, et al. Local cost structures and the economics of robot assisted radical prostatectomy. *J Urol* 2005;**174**(6):2323–9.
27. Bolenz C, Freedland SJ, Hollenbeck BK, et al. Costs of radical prostatectomy for prostate cancer: a systematic review. *Eur Urol* 2012;**65**(2):316–24.
28. Bolenz C, Freedland SJ, Hollenbeck BK, et al. Costs of radical prostatectomy for prostate cancer: a systematic review. *Eur Urol* 2014;**65**(2):316–24.
29. Lotan Y. Economics of robotics in urology. *Curr Opin Urol* 2010;**20**(1):92–7.
30. Steinberg PL, Merguerian PA, Bihrle III W, et al. The cost of learning robotic-assisted prostatectomy. *Urology* 2008;**72**(5):1068–72.
31. Guillonneau B, Vallancien G. Laparoscopic radical prostatectomy: the Montsouris technique. *J Urol* 2000;**163**(6):1643–9.
32. Davis JW, Kreaden US, Gabbert J, et al. Learning curve assessment of robot-assisted radical prostatectomy compared with open-surgery controls from the premier perspective database. *J Endourol* 2014;**28**(5):560–6.
33. Lotan Y, Bolenz C, Gupta A, et al. The effect of the approach to radical prostatectomy on the profitability of hospitals and surgeons. *BJU Int* 2010;**105**(11):1531–5.
34. Kim SP, Shah ND, Karnes RJ, et al. Hospitalization costs for radical prostatectomy attributable to robotic surgery. *Eur Urol* 2013;**64**(1):11–6.
35. Weinstein MC, Siegel JE, Gold MR, et al. Recommendations of the panel on cost-effectiveness in health and medicine. *JAMA* 1996;**276**(15):1253–8.
36. Diamond GA, Kaul S. Cost, effectiveness, and cost-effectiveness. *Circ Cardiovasc Qual Outcomes* 2009;**2**(1):49–54.
37. Close A, Robertson C, Rushton S, et al. Comparative cost-effectiveness of robot-assisted and standard laparoscopic prostatectomy as alternatives to open radical prostatectomy for treatment of men with localised prostate cancer: a health technology assessment from the perspective of the UK National Health Service. *Eur Urol* 2013;**64**(3):361–9.
38. Zaorsky NG, Harrison AS, Trabulsi EJ, et al. Evolution of advanced technologies in prostate cancer radiotherapy. *Nat Rev Urol* 2013;**10**(10):565–79.
39. Sheets NC, Goldin GH, Meyer AM, et al. Intensity-modulated radiation therapy, proton therapy, or conformal radiation therapy and morbidity and disease control in localized prostate cancer. *JAMA* 2012;**307**(15):1611–20.
40. Nguyen PL, Gu X, Lipsitz SR, et al. Cost implications of the rapid adoption of newer technologies for treating prostate cancer. *J Clin Oncol* 2011;**29**(12):1517–24.
41. Zietman AL, DeSilvio ML, Slater JD, et al. Comparison of conventional-dose vs high-dose conformal radiation therapy in clinically localized adenocarcinoma of the prostate: a randomized controlled trial. *JAMA* 2005;**294**(10):1233–9.
42. Zelefsky MJ, Levin EJ, Hunt M, et al. Incidence of late rectal and urinary toxicities after three-dimensional conformal radiotherapy and intensity-modulated radiotherapy for localized prostate cancer. *Int J Radiat Oncol Biol Phys* 2008;**70**(4):1124–9.
43. Spratt DE, Pei X, Yamada J, et al. Long-term survival and toxicity in patients treated with high-dose intensity modulated radiation therapy for localized prostate cancer. *Int J Radiat Oncol Biol Phys* 2013;**85**(3):686–92.
44. Al-Mamgani A, van Putten WL, van der Wielen GJ, et al. Dose escalation and quality of life in patients with localized prostate cancer treated with radiotherapy: long-term results of the Dutch randomized dose-escalation trial (CKTO 96-10 trial). *Int J Radiat Oncol Biol Phys* 2011;**79**(4):1004–12.
45. Bauman G, Rumble RB, Chen J, Members of the IIEP. et al. Intensity-modulated radiotherapy in the treatment of prostate cancer. *Clin Oncol* 2012;**24**(7):461–73.
46. Konski A, Watkins-Bruner D, Feigenberg S, et al. Using decision analysis to determine the cost-effectiveness of intensity-modulated radiation therapy in the treatment of intermediate risk prostate cancer. *Int J Radiat Oncol Biol Phys* 2006;**66**(2):408–15.
47. Hummel SR, Stevenson MD, Simpson EL, et al. A model of the cost-effectiveness of intensity-modulated radiotherapy in comparison with three-dimensional conformal radiotherapy for the treatment of localised prostate cancer. *Clin Oncol* 2012;**24**(10):e159–67.
48. Yong JH, Beca J, McGowan T, et al. Cost-effectiveness of intensity-modulated radiotherapy in prostate cancer. *Clin Oncol* 2012;**24**(7):521–31.
49. Shah C, Lanni Jr TB, Ghilezan MI, et al. Brachytherapy provides comparable outcomes and improved cost-effectiveness in the treatment of low/intermediate prostate cancer. *Brachytherapy* 2012;**11**(6):441–5.
50. Wilson LS, Tesoro R, Elkin EP, et al. Cumulative cost pattern comparison of prostate cancer treatments. *Cancer* 2007;**109**(3):518–27.
51. Hayes JH, Ollendorf DA, Pearson SD, et al. Observation versus initial treatment for men with localized, low-risk prostate cancer: a cost-effectiveness analysis. *Ann Intern Med* 2013;**158**(12):853–60.
52. Falit BP, Gross CP, Roberts KB. Integrated prostate cancer centers and over-utilization of IMRT: a close look at fee-for-service medicine in radiation oncology. *Int J Radiat Oncol Biol Phys* 2010;**76**(5):1285–8.
53. Mitchell JM. Urologists' use of intensity-modulated radiation therapy for prostate cancer. *N Engl J Med* 2013;**369**(17):1629–37.
54. Rickles DJ. Integrated prostate cancer centers and over-utilization of IMRT: in regard to Falit et al. (*Int J Radiat Oncol Biol Phys* 2010;76:1285–1288). *Int J Radiat Oncol Biol Phys* 2011;**79**(4):1280; author reply-1.
55. Bentzen SM, Wasserman TH. Balancing on a knife's edge: evidence-based medicine and the marketing of health technology. *Int J Radiat Oncol Biol Phys* 2008;**72**(1):12–4; discussion 4–8.
56. Yu JB, Cramer LD, Herrin J, et al. Stereotactic body radiation therapy versus intensity-modulated radiation therapy for prostate cancer: comparison of toxicity. *J Clin Oncol* 2014;**32**(12):1195–201.
57. Mangoni M, Desideri I, Detti B, et al. Hypofractionation in prostate cancer: radiobiological basis and clinical appliance. *BioMed Res Int* 2014;**2014**:781340.
58. Hodges JC, Lotan Y, Boike TP, et al. Cost-effectiveness analysis of stereotactic body radiation therapy versus intensity-modulated radiation therapy: an emerging initial radiation treatment option for organ-confined prostate cancer. *J Oncol Pract* 2012;**8**(3 Suppl.):e31s–e37s.
59. Parthan A, Pruttivarasin N, Davies D, et al. Comparative cost-effectiveness of stereotactic body radiation therapy versus intensity-modulated and proton radiation therapy for localized prostate cancer. *Front Oncol* 2012;**2**:81.
60. Sher DJ, Parikh R, Mays-Jackson S, et al. Cost-effectiveness analysis of SBRT versus IMRT for low-risk prostate cancer. *Am J Clin Oncol* 2012;**37**(3):215–21.
61. Zaorsky NG, Studenski MT, Dicker AP, et al. Stereotactic body radiation therapy for prostate cancer: is the technology ready to be the standard of care? *Cancer Treat Rev* 2013;**39**(3):212–8.
62. Wisenbaugh ES, Andrews PE, Ferrigni RG, et al. Proton beam therapy for localized prostate cancer 101: basics, controversies, and facts. *Rev Urol* 2014;**16**(2):67–75.

63. Efstathiou JA, Gray PJ, Zietman AL. Proton beam therapy and localised prostate cancer: current status and controversies. *Br J Cancer* 2013;**108**(6):1225–30.
64. Vargas C, Fryer A, Mahajan C, et al. Dose-volume comparison of proton therapy and intensity-modulated radiotherapy for prostate cancer. *Int J Radiat Oncol Biol Phys* 2008;**70**(3):744–51.
65. Slater JD, Rossi Jr CJ, Yonemoto LT, et al. Proton therapy for prostate cancer: the initial Loma Linda University experience. *Int J Radiat Oncol Biol Phys* 2004;**59**(2):348–52.
66. Zietman AL, Bae K, Slater JD, et al. Randomized trial comparing conventional-dose with high-dose conformal radiation therapy in early-stage adenocarcinoma of the prostate: long-term results from proton radiation oncology group/American College of Radiology 95-09. *J Clin Oncol* 2010;**28**(7):1106–11.
67. Mitin T, Zietman AL. Promise and pitfalls of heavy-particle therapy. *J Clin Oncol* 2014;**32**(26):2855–63.
68. Yu JB, Soulos PR, Herrin J, et al. Proton versus intensity-modulated radiotherapy for prostate cancer: patterns of care and early toxicity. *J Natl Cancer Inst* 2013;**105**(1):25–32.
69. Elnahal SM, Kerstiens J, Helsper RS, et al. Proton beam therapy and accountable care: the challenges ahead. *Int J Radiat Oncol Biol Phys* 2013;**85**(4):e165–72.
70. Ollendorf DA, Hayes JH, McMahon P. Management options for low-risk prostate cancer: a report on comparative effectiveness and value: Institute for Clinical and Economic Review, 2009.
71. Konski A, Speier W, Hanlon A, et al. Is proton beam therapy cost effective in the treatment of adenocarcinoma of the prostate? *J Clin Oncol* 2007;**25**(24):3603–8.
72. Keegan KA, Dall'Era MA, Durbin-Johnson B, et al. Active surveillance for prostate cancer compared with immediate treatment: an economic analysis. *Cancer* 2012;**118**(14):3512–8.
73. Hayes JH, Ollendorf DA, Pearson SD, et al. Active surveillance compared with initial treatment for men with low-risk prostate cancer: a decision analysis. *JAMA* 2010;**304**(21):2373–80.
74. Klotz L. Active surveillance: current and future directions. *Curr Opin Urol* 2013;**23**(3):237–8.
75. Shao YH, Albertsen PC, Roberts CB, et al. Risk profiles and treatment patterns among men diagnosed as having prostate cancer and a prostate-specific antigen level below 4.0 ng/mL. *Arch Intern Med* 2010;**170**(14):1256–61.
76. Chang SL, Kibel AS, Brooks JD, et al. The impact of robotic surgery on the surgical management of prostate cancer in the USA. *BJU Int* 2014;**115**(6):929–36.

P A R T VII

# RADIATION THERAPY

CHAPTER

# 42

# Fundamentals of Radiation Treatment for Prostate Carcinoma – Techniques, Radiation Biology, and Evidence Base

*Mohan P. Achary, PhD, Curtis T. Miyamoto, MD*

**Department of Radiation Oncology and Radiology, Temple University School of Medicine, Philadelphia, PA, USA**

## INTRODUCTION

*Radiation therapy* was used to fight cancer within a few years of the discovery of X-rays in 1895. By 1902, Emil Grubbe had published an article outlining the conclusions from his own experience and that of dozens of writers for the treatment of cancer with radiation therapy. His conclusions were that the X-ray is a most remarkable therapeutic agent, in properly selected cases of so-called incurable conditions. The relief from pain is one of the most prominent features of the treatment. The X-ray has a pronounced effect upon internal cancers, the greatest value of the X-ray is obtained in treating postoperative cases to prevent recurrences, and it has a selective influence upon cells of the body; abnormal cells being effected more readily than the normal.[1] Over a hundred years later, these principles still hold true in the treatment of prostate cancer. Radiation therapy remains ideally suited for curative treatment of localized disease as well as for palliation.[2,3] Treatment techniques have evolved and the field has benefited from rapid advances in technology.[4,5] At the same time surgical and other techniques have also improved making the selection of the optimal management of localized prostate carcinoma difficult for patients and clinicians. There are no conclusive prospective randomized studies demonstrating significant differences in overall survival and biochemical-free relapse between surgery and radiation therapy.[6–13] In this chapter, we will provide fundamentals of radiation therapy, an overview of radiation biology and different treatment options in standard care, which will include both definitive and palliative radiation therapy in prostate carcinoma.

## FUNDAMENTALS OF RADIOTHERAPY

There are multiple forms of *external beam radiation therapy*. The most common are photon, electron, and proton irradiation. Photons and electrons can be produced by linear accelerators (artificially produced) or emitted by isotopes (i.e., Co-60, Cs-137, Ir-192, I-131, Sr-89, Sm-153, etc.). Protons are produced by cyclotrons. These three forms of radiation are low LET (linear energy transfer) forms of radiation. These have similar biological effects on normal and malignant tissue. Neutron irradiation, on the other hand, is a high LET form of radiation that has a significantly greater biologic effect. This type of radiation is found only in a very limited number of locations and is not conventionally used to treat prostate cancer. The biological effects of these different forms of radiation will be explained in detail later in this chapter.

The ability to better localize the target volumes and normal structures for radiation treatment planning has resulted in improved outcomes and decreased secondary effects. External beam techniques, such as *intensity modulated radiation therapy, volumetric modulated arc therapy, and proton therapy* coupled with real-time image guidance, have improved our ability to concentrate the dose over the target volumes while minimizing risk for nor-

**Prostate Cancer. http://dx.doi.org/10.1016/B978-0-12-800077-9.00042-6**

mal tissue damage. Additionally, they have resulted in shortening the daily and overall treatment times. These *hypofractionated techniques* deliver higher doses per fraction to a lower overall total dose.[14]

These techniques recently have been employed more and more in everyday practice but have not yet become the standard of care. *Brachytherapy*, or the placement of radioactive sources in the prostate and seminal vesicles, has also undergone significant improvement in recent years with the increasing use of high dose rate temporary implants. This has also resulted in shorter treatment courses for favorable prostate cancer.[15] Combined modality treatment with *hormonal therapy* has become standard for locally advanced definitive treatment.

Patients are first seen by urologists and may or may not be referred to other specialists for consideration of nonsurgical treatment opinions. In a study published in 2010, a review of the SEER Medicare linked database identified 85,088 men with clinically localized prostate cancer diagnosed at an age ≥65 years between 1994 and 2002. Of the 85,088 patients, 42,309 (49.73%) were seen exclusively by urologists, 37,540 (44.12%) were seen by both urologists and radiation oncologists, 2329 (2.74%) by both urologists and medical oncologists while the remaining 2910 (3.42%) were seen by all 3 specialists.[16]

In 2010, the CapSURE study was published reviewing treatment patterns in 36 community-based urologic practices across the United States collected since 1995.[17] There were 11,892 men included in this registry exam of the practice patterns. The most common treatment was surgical with 6.8% of the patients choosing active surveillance, 49.9% choosing *radical prostatectomy*, 11.6% choosing definitive treatment with external beam irradiation, 13.3% choosing brachytherapy (radioactive implant), 14.4% choosing primary *androgen deprivation monotherapy*, and 4% choosing *cryotherapy*. There was a bias toward treating younger, healthier, privately insured, lower-risk and higher socioeconomic status patients for prostatectomy. Brachytherapy and cryotherapy patients were somewhat older, had lower socioeconomic status and higher comorbidity. External beam irradiation patients also had similar characteristics. External beam radiation and cryoablation were used more commonly in patients with higher-risk level. The personal preferences of each center, medical comorbidities, and socioeconomic status of the patients also influenced treatment decisions. In a subsequent CapSURE paper, 8982 men with localized disease (clinical stage ≤T3aN0) treated as mentioned earlier were reviewed.[18] Of these, 7538 patients were analyzable. In total 226 men died of prostate cancer. This study concluded that *prostatectomy* for localized prostate cancer had a significantly improved mortality rate relative to radiation or androgen deprivation monotherapy. The hazard ratio for cancer-specific mortality compared to prostatectomy adjusted for age and risk, for radiation was 2.21 (1.50–3.24) and for androgen deprivation, 3.22 (2.16–4.81). However, when the results from these two publications were combined, it is evident that there are too many variables to conclude that surgery is superior.

A SEER report indicated that in recent years more men were being diagnosed with favorable risk disease.[19] Their treatment varied with 42% of patients being treated with radiation therapy, 21% by radical prostatectomy, and 17% by androgen deprivation therapy. The remaining 20% underwent active surveillance. There appeared to be a correlation between the type of specialist first seen by the patient and the treatment delivered. In the case of patients who were seen only by a urologist, 70% of the patients between the ages of 65 and 69 underwent radical prostatectomy as compared to patients 70–74 years of age where only 45% of patients underwent radical prostatectomy. When patients are seen by both a radiation oncologist and the urologist at presentation, 83% of patients underwent radiation therapy as the initial primary treatment. When patients were seen by their primary care physician, they were more likely to undergo active surveillance. An earlier survey study in 1998 of 504 urologists and 559 radiation oncologists, the American Medical Association Registry of Physicians showed that 92% of urologists chose radical prostatectomy as the preferred option whereas 72% of radiation oncologists believed external beam radiation therapy and surgery were equivalent.[20]

Petrelli et al. published a systematic review and meta-analysis of high-risk prostate cancer patients treated by radical prostatectomy or radiotherapy.[21] This study analyzed 17 studies including one randomized trial. They found that radical prostatectomy was associated with improved overall survival, a lower prostate-cancer-specific mortality, and nonprostate cancer specific mortality with $P$ values of <0.00001, 0.007, and 0.002, respectively. The fact that the nonprostate-cancer-specific mortality was lower implies that the prostatectomy patients were healthier and likely younger than the patients undergoing radiation therapy. This potentially biased the data and resulted in improved outcomes for the healthier surgical patients. As already mentioned earlier, this difference is a common practice in many institutions. Therefore, as stated earlier, there is still a controversy as to what is the best primary treatment for intermediate- and high-risk prostate cancer.

## OVERVIEW OF RADIATION BIOLOGY

*Radiation biology* is the study of the effects of ionizing radiation on living organisms. Experimental and theoretical studies in radiation biology are critical to the development of radiation therapy in processes such as identification of the mechanisms of irradiation in a given tissue with the surrounding microenvironment, development of novel treatment strategies, and designing protocols based on appropriate models. These theoretical and experimental models, however, will eventually rely on the outcome of the clinical trials in an appropriate clinical setting.[22]

Here we briefly discuss different aspects of radiation biology to help the reader grasp the intricate physical and chemical effects of irradiation in radiation therapy.

*Ionizing radiation* may be defined as a radiation with sufficient energy which during an interaction with an atom can remove tightly bound electrons from the orbit of an atom of the matter/cell, causing the atom to become ionized.[23] Examples of ionizing radiation include alpha particles, beta particles, γ-rays, X-rays, as well as shortwave UV light.

For understanding biological effects of ionizing radiation, it may be divided into *directly* and *indirectly ionizing* types. Those individually charged particles, such as electrons, protons, neutrons, and heavy-charged ions possessing sufficient kinetic energy, can directly disrupt the atomic structure of the traversed material causing intense damage to the molecules in living tissue.[24] Indirectly ionizing electromagnetic radiations on the other hand include X- and γ-rays. These rays are identical in their nature and properties but differ in the designations of how they are produced. While X-rays are produced extra-nuclearly, γ-rays are produced intra-nuclearly. The X-rays are generated in electrical devices that accelerate electrons to a high energy and then abruptly stop them in a target (tungsten or gold). A part of the kinetic energy produced by these electrons is converted into X-rays. The γ-rays on the other hand are emitted by radioactive isotopes. Generally, isotopes are forms of an element that possess the same number of protons and electrons but a different number of neutrons in the nucleus of an atom. These unstable isotopes undergo a transition to a more stable state with the release of radioactivity (i.e., Hydrogen [H] has three stable isotopes, namely, protium with one proton and one electron [$_1^1H$]; deuterium with one proton, one neutron, and one electron [$_1^2H$]; and tritium with one proton, two neutrons, and one electron [$_1^3H$]).

Electrons are the lightest stable subatomic particles with a negative electric charge. These particles can be accelerated to a high energy by using electrical devices, such as a linear accelerator, and are widely used in cancer treatment. Protons are positively charged particles with a mass of about 2000 times greater than that of an electron. Because of their higher mass they are used in cancer therapy only in a few specialized facilities. The α-particles are nuclei of helium atoms and consist of two protons and two neutrons. They are emitted during the decay of uranium and radium. These α-particles are a major source of natural background radiation and a major cause of lung cancer mostly in smokers. Neutrons are particles similar in size to protons, but unlike protons and electrons they do not carry any electrical charge. However, they can penetrate tissue much deeper than charged particles such as X-rays or γ-rays possessing the same mass energy. Neutrons are absorbed by tissue either by elastic or inelastic scattering phenomenon. While X-ray photons interact with orbital electrons of atoms of the tissue by Compton or photoelectric processes, neutrons interact with the nuclei of the atoms of the tissue and set in motion of fast recoil protons, α-particles and heavier nuclear fragments or ions. On interaction with the nuclei of carbon and oxygen atoms, the neutron may generate three and four α-particles, respectively. These densely ionizing α-particles are known as "spallation products." Because of their densely ionizing nature they may increase the interaction between the particle track and target molecules.[25] Ions are positively charged particles/nuclei produced by elements such as carbon, neon, silicon, and iron, which are stripped of some or all of their orbital electrons. Acceleration of this kind of particles requires an extremely high-energy source and therefore is not generally applicable in radiation therapy. However, high-energy charged ions are useful in cancer therapy because of the energy distribution along their passage with a high peak at their ends (*the Bragg peak*). Thus, these ions can deposit high energy densities at deeper tissues. However, production of such high energies require special accelerator facilities, which are limited due to high cost and advanced technical requirements.[24,25]

The energy produced by a photon is given here by Planck's equation where "$E$" is the energy of the photon, "$h$" the Planck's constant, "$v$" frequency, "$c$" velocity of light, and "$\lambda$" the wavelength.[24]

$$E = hv = hc / \lambda$$

In photoelectric and Compton scattering processes much of the energy of the absorbed photons is converted into fast electrons.[25]

The biological effects of radiation include damage to the critical target in the cells – the DNA. When absorbed into cells, X-rays and γ-rays may interact directly with DNA. Thus, these rays may ionize or excite atoms of the target, leading to a biological effect. However, while traversing through matter/tissue the ionizing radiation loses energy along the length of its path. Consequently, the rate of loss of energy depends on the density of a given absorber (matter/tissue). Thus, the density of the energy that is actually deposited in a given tissue by ionizing radiation is known as LET. It can be defined as the average energy that is deposited per unit length of the track of radiation in a tissue and is expressed in KeV/μm. The LET indicates the quality of different types of radiation and the comparison of different biological effects produced by different types of radiations is expressed as relative biological effectiveness (RBE). The radiations are categorized into low and high LETs. The particulate radiations (electrons, protons, alpha particles, neutrons, and negative $\pi$ mesons and heavy charged ions) are considered as high and electromagnetic radiations (X-rays and $\gamma$-rays) as low. In general, the RBE of a given radiation increases with its LET up to a value of about 100 KeV/$\mu$m, above which it starts to decline due to energy deposition in tissue causing a biological effect.[24,25]

Radiations, such as X-rays and $\gamma$-rays, on the other hand, may interact with other molecules in the cell, water in particular and produce free radicals, which may lead to a similar biological effect by eventually damaging the target DNA molecules. These *free radicals* are free atoms or molecules carrying unpaired orbital electrons in the outer shell, which have high reactivity with cellular molecules found in processes such as metabolism, oxidation, reduction, as well as certain disease conditions including induction of cancer. Since water occupies a major volume in a given cell, a major amount of radiation energy will be absorbed by cellular water content. The chemical change that occurs in water exposed to radiation is known as *water radiolysis*. In this process, when a photon or electron of X-rays or gamma rays interacts with water, it loses an electron and ionizes the water molecule to an electrically charged $H_2O^+$ ion followed by the generation of hydroxyl radicals as shown in the equation

$$\mathrm{IR} + \mathrm{H_2O} \rightarrow \mathrm{H_2O^+} + e^-$$

Here, $H_2O^+$ is an electrically charged ion because it has lost an electron. Since $H_2O^+$ has an unpaired electron, it can also be a free radical. These free radicals react with another water molecule to form a highly reactive hydroxyl radical (OH.)

$$\mathrm{H_2O^+} + \mathrm{H_2O} \rightarrow \mathrm{H_3O^+} + \mathrm{OH}$$

The free-radical-based indirect radiations can be modified by using *radio protectors* (Cysteine, cysteamine, pyridoxamine, amifostine, etc.) or *radio sensitizers* (camptothecin, Vitamin K, hydroxyurea, methotrexate, temozolomide, cetuximab, celecoxib, etc.). While dimethyl sulfoxide can scavenge primary radicals of water radiolysis, amifostine, on the other hand, can scavenge hydroxyl radicals.[24]

*Dmax* is the maximum dose of radiation (electrons/photons) that can be achieved at a certain depth of the tissue. At such a depth the Dmax is considered to be 100 and as the depth increases further the radiation dose decreases due to absorption of energy by tissue along the path of the radiation. The region between the radiation incident surface and the Dmax is called as the *buildup region* and is responsible for "skin-sparing" effects. Thus, Dmax as well as size of the buildup region are radiation energy dependent. Examples of typical depths of Dmax for various photon beam energies and a field size of 5×5 cm$^2$ are: Co-60 = 0.5 cm; 4 MV = 1 cm; 6 MV = 1.5 cm; 10 MV = 2.5 cm; 18 MV = 3.5 cm; 25 MV = 5 cm.[24–27]

*Radiation quantities* are measured as *exposure dose* and *absorbed dose*. The unit for exposure dose is Coulombs/kg in air (or Roentgen R in old units; 1R= 2.58 × $10^{-4}$ C/kg air); therefore, it cannot be used to describe the same in a tissue. The amount of energy (Joules) absorbed per unit mass of tissue has the units of gray (1 Gy = 1 J/kg). Previously, the *radiation absorbed dose* (rad) was used as the dose unit (100 rad = 1 Gy or 1 rad = 1 cGy). Equivalent dose is used to compare biological effectiveness of different types of radiation on tissues. The dose equivalent ($H_T$) in Sievert ($S_V$) is the product of the absorbed dose ($D_T$) in the tissue multiplied by a radiation weighting factor ($W_R$) called the quality factor. Thus, equivalent dose is expressed as the summation of irradiation effect of tissues by more than one type of radiation as shown in the next equation.[24] The quality factor for low LET radiations is 1 (1 $S_V$ = 1 Gy).

$$H_T = \sum W_R \times D_T$$

*Effective dose* is the sum of equivalent doses in each organ ($H_T$) and the tissue weighting factor ($W_T$), expressed in Sievert ($S_V$) units.

$$E = \sum W_T \times H_T$$

*Survival curves* are used to measure radiosensitivity of a given cell population exposed to different doses of radiation. However, for practical purposes, generally, survival fraction at 2 Gy (SF2) is measured due to common usage of 2 Gy in radiotherapy. Radiosensitivity of a given cell population is measured by detecting its ability to undergo a few cell divisions (about 5–6) after being exposed to radiation. The irradiated cells are allowed to form colonies of cells, which are counted by staining with dyes such as crystal violet. The percentage of these cell colonies is calculated by correcting for the plating efficiency of controls (unirradiated cells). While colony formation is the most accepted method to determine radiosensitivity of a cell population, other simpler and faster methods, such as MTT assay and Annexin V, are often used with the assumption that there is a relationship between cell growth or cell death and cell survival over a wide range of doses of radiation. Survival curves are generally shown as semilog plots of survival against a range of radiation doses (1–10 Gy). As shown below, the common model being used is the *linear-quadratic model* (LQM) in which $S$ is the fraction of cells surviving a dose $D$ and fitted with the constants $\alpha$ and $\beta$ representing the linear and quadratic components of cell killing. Equal cell killing of the linear and quadratic contributions could be achieved when $D = \alpha/\beta$. The LQM, however, does not work with high LET irradiation studies because its quadratic contribution is small or nonexistent.[24,25]

$$S = e^{-\alpha D - \beta D^2}$$

*DNA damage* occurs in cells exposed to radiation. They may include lesions in DNA such as single-strand breaks in the phosphodiester bonds, double-stranded breaks, base damage, protein–DNA crosslinks, as well as protein–protein crosslinks (histones and nonhistones). However, several DNA repair mechanisms, such as nonhomologous end joining (NHEJ), homologous recombination (HR), base excision repair (BER), mismatch repair (MR), and nucleotide excision repair (NER), are in place to respond to such radiation-based damages leading to cell survival. High doses of radiation, however, produce damages in the cells at the chromosomal level, such as dicentrics, rings, and translocations, which may not be repairable and the cells possessing such aberrations may not survive past the following mitotic division.

Radiation damage in mammalian cells can be divided into three categories: (1) *lethal* (irreversible and irreparable); (2) *sublethal* (repairable provided no additional sublethal damages are added), and (3) *potentially lethal damage* (cells will die unless modified by postirradiation environmental conditions). Sublethal damage repair is the operational term for the increase in cell survival that is observed if a given radiation dose is split into two fractions separated by a time interval. As discussed in the Chinese hamster cell experiments most of the surviving cells from a first dose are in S-phase of the cell cycle.[28] If about 6 h are allowed to elapse before a second dose of radiation is given, this cell population progresses to a sensitive G2/M phase, which may lead to the fall in the survival fraction of the cells. The pattern of repair thus is a combination of four processes taking place simultaneously. The first process includes *repair of sublethal radiation damage*. The second one, the progression of cells to the following phases of cell cycle during the interval period between the split doses is known as *reassortment* (or regeneration). The third is termed *repopulation* (or redistribution) in which the survival fraction of cells increases if the duration between the split doses is 10–12 h (hamster cell cycle time is 9–10 h). The fourth process is called as *reoxygenation*. Van Putten and Kallman showed that reoxygenation makes hypoxic cells to oxygenated cells after receiving a dose of radiation.[29] Thus, the process of reoxygenation has considerable implications in radiotherapy. If patient tumors are oxygenated immediately after irradiation, then multiple fractions of radiation therapy, theoretically, may effectively deal with hypoxic cells in the tumors. In addition to these four Rs, some investigators like Steel et al. even suggested that *intrinsic radiosensitivity* should be considered as the fifth R because of its obvious role in fractionated radiotherapy.[30,31] Survival curves for the mammalian cells exposed to X-rays showed that cells are much more sensitive in the presence of oxygen than its absence.[32] Because oxygen can interact with free radicals formed by radiation it will lead to considerable damage to DNA. The ratio of radiation doses required to achieve the same biological effect in the absence or the presence of oxygen is called *oxygen enhancement ratio* (*OER*). The OER for X-rays and γ-rays at high doses (greater than 3 Gy) is in the range of 2.5–3.3, for a wide range of cell lines and tissues. For doses less than 3 Gy the OER is reduced in a dose-dependent manner. Such a reduction of the OER at lower doses, however, is clinically important because the daily dose of fractionated radiotherapy is usually about 2 Gy.[24,25]

## TREATMENT OPTIONS IN RADIOTHERAPY

For radiation therapy to be effective, all evidence of malignancy needs to be included in the target volume. The normal tissue tolerances need to be respected and the comorbidities taken into account. In the past several decades our ability to identify the target volumes using anatomic and functional imaging has significantly improved. This has been covered in Chapter 13 (see chapter "Prostate-Specific Antigen Screening Guidelines"). These images are used to determine the stage and therefore treatment options with radiation therapy. They help determine whether the patient will be treated palliatively or definitively. Our knowledge of the natural history of prostate cancer and patterns of spread have also improved, contributing to a better definition of the volumes at risk. This has resulted in a much more effective use of radiation therapy as definitive treatment, as adjuvant treatment, and for palliation. Radiation can be utilized in multiple forms including external beam radiation therapy, brachytherapy, and systemic therapy.

In the case of external beam radiation therapy the anatomic and functional images are fused with a treatment planning CT or MRI scan. The CT scan has the advantage of being spatially more accurate whereas the MRI scan shows greater detail, which can be used to determine the extent more accurately.[33–35] New technology allows these fusions to take into account variations between the scans due to patient positioning, organ motion, and distortion. This is termed *deformable registration*. There are a number of software products available for such analyses. Once the image fusion is completed, contouring takes place outlining the target volume and critical normal structures including the bladder, rectum, penile bulb, small bowel, and hips. Target volumes are defined as the *gross tumor volume* or GTV, *clinical target volume* or CTV, and *planning*

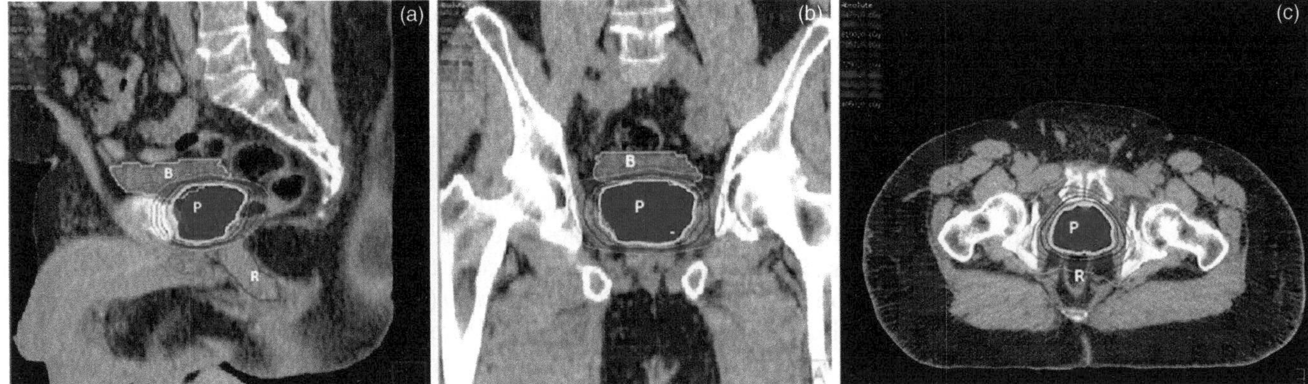


FIGURE 42.1 **Volumetric modulated arc therapy for treatment of the prostate only.** (a) Sagittal view, (b) coronal view, (c) axial view; P, prostate; B, bladder; R, rectum.

*target volume* or PTV.[36] The GTV consists of the prostate in intermediate-risk tumors, and the prostate and seminal vesicles in more advanced intermediate-risk tumors or early high-risk tumors.[37,38] There is still a controversy over whether grossly involved *lymph nodes* should be included in the gross tumor volume.[39] Intensity-modulated radiation therapy (IMRT) has become the standard of care in radiation oncology. Between the years 2000 and 2008, over 95% of the external beam radiation treated cases over the age of 65 were treated using IMRT.[4,5] This technique allows for the intensity of the radiation to be varied across the field allowing better conformality in a three-dimensional volume (Figures 42.1 and 42.2). This can be performed using multiple static fields or dynamically using a continuously modulating ARC technique. For low-risk patients, 180 cGy per fraction to a total dose of 7920 cGy is appropriate. For intermediate- or high-risk patients doses of up to 8100 cGy provide improved PSA assessed disease control (Table 42.1).[40–46] Dose volume histograms and dose statistics are calculated for the target volumes and critical normal structures (Figures 42.3 and 42.4). These allow radiation oncologists, dosimetrists, and physicists to plan treatments with the best coverage of the prostate, seminal vesicles, and lymph nodes as indicated with the lowest possible doses to the bladder, rectum, penile bulb, femoral heads, and small bowel. The Radiation Therapy Oncology Group (RTOG 0126) has published a set of dose constraints.[47] A meta-analysis published in 2009 evaluated the dose–response relationship found in seven randomized controlled trials.[48] There were a total of 2812 patients that met the study criteria. The results showed a significant reduction in the incidence of biochemical failure in patients treated with the higher doses ($p < 0.0001$). The improvement was also seen in low-, intermediate-, and high-risk groups. There was however no difference in the mortality ($p = 0.38$) and specific prostate cancer mortality ($p = 0.45$) rates. Recently hypofractionation has become more prevalent. This consists of higher doses per fraction (240–400 cGy per fraction) delivered two lower total doses over 4–6 weeks. These must be delivered utilizing intensity-modulated radiation therapy with image guidance or equivalent. There are even more intense techniques using 650 cGy per fraction or greater.[49] These must be carried out with

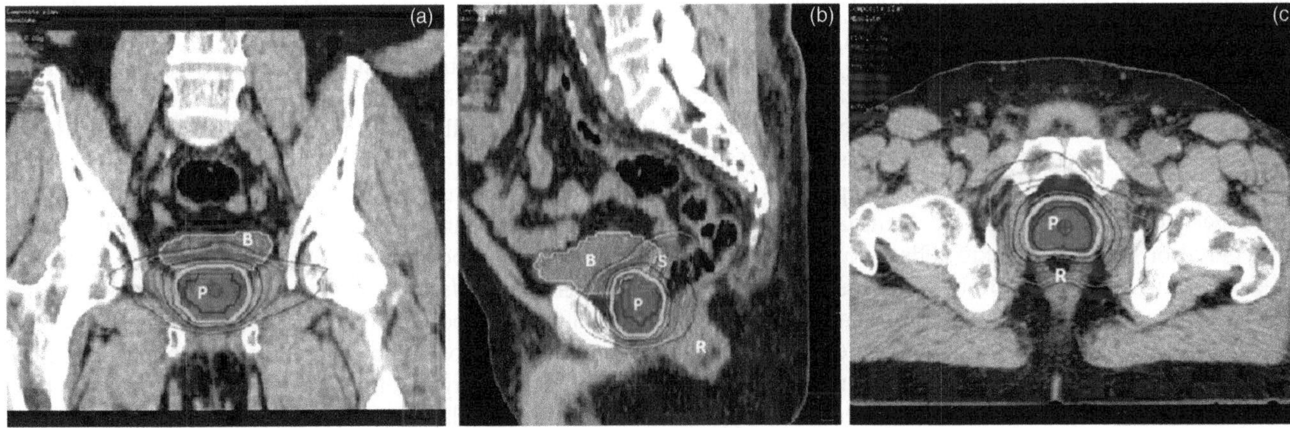


FIGURE 42.2 **Volumetric modulated arc therapy for treatment of the prostate and seminal vesicles.** (a) Coronal view, (b) sagittal view, (c) axial view; P, prostate; S, seminal vesicles; B, bladder; R, rectum.

**TABLE 42.1** Dose Escalation Studies and Outcome in Prostate Cancer[40–46]

| Trial | No. | Risk groups | Dose (Gy) | Outcome |
|---|---|---|---|---|
| MGH[40] | 202 | High versus very high | 67.2 Versus 75.6 | 5-Year tumor-free survival, 59% versus 73% (proton boost) |
| MDACC[41] | 301 | Low versus high | 70 Versus 78 | 8-Year freedom from failure, 59% versus 78% |
| PROG 95-09[42] | 393 | Low versus high | 70.2 Versus 79.2 | 10-Year biochemical control rate, 68% versus 83% |
| GETUG[43] | 306 | Intermediate | 70 Versus 80 | 5-Year biochemical control rate, 61% versus 72% |
| DUTCH CKVO96-10[44] | 664 | Low versus high | 68 Versus 78 | 5-Year freedom from failure, 54% versus 64% |
| MRC RT01[45] | 843 | Low versus high | 64 Versus 74 | 5-Year biochemical profession-free survival, 60% versus 71% |

*Modified from Martin and D'Amico (2014).*[46]

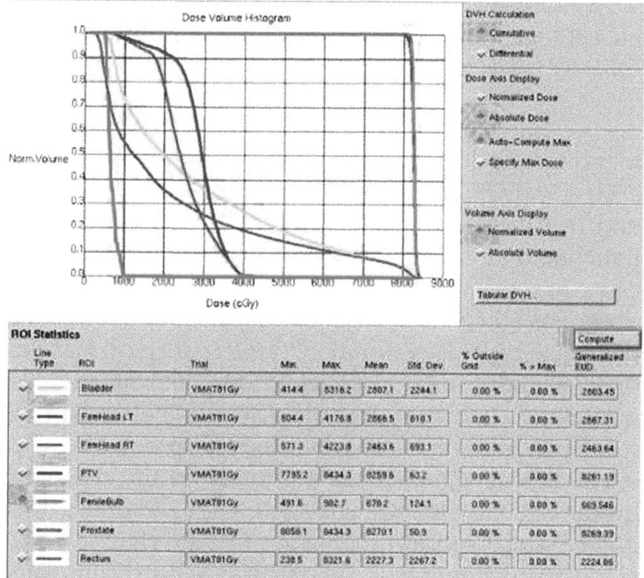


Region Of Interest Dose Statistics:

| ROI | Volume (cm^3) | Min Value | Max Value | Mean Value | Std. Dev. | Units |
|---|---|---|---|---|---|---|
| Prostate | 40.8205 | 8056.1 | 8434.3 | 8270.1 | 50.9 | cGy |
| PTV | 87.286 | 7795.2 | 8434.3 | 8259.6 | 63.2 | cGy |
| PenileBulb | 4.85728 | 491.6 | 982.7 | 670.2 | 124.1 | cGy |
| Bladder | 116.964 | 414.4 | 8316.2 | 2807.1 | 2244.1 | cGy |
| Small Bowel | 803.756 | 8.5 | 759.0 | 106.0 | 126.0 | cGy |
| Sigmoid | 41.7654 | 30.3 | 722.4 | 321.5 | 199.4 | cGy |
| Rectum | 113.78 | 238.5 | 8321.6 | 2227.3 | 2267.2 | cGy |
| Large Bowel | 19.0625 | 10.2 | 45.0 | 20.7 | 7.6 | cGy |
| FemHead RT | 66.7791 | 571.3 | 4223.8 | 2463.6 | 693.1 | cGy |
| FemHead LT | 67.4842 | 604.4 | 4176.8 | 2866.5 | 610.1 | cGy |
| Rectum_COPY_1 | 106.248 | 238.5 | 8081.0 | 1821.8 | 1736.1 | cGy |
| Bladder_COPY_1 | 106.981 | 414.4 | 7786.3 | 2353.9 | 1748.8 | cGy |
| sp | 32.2382 | 6005.7 | 8394.9 | 7851.8 | 314.8 | cGy |
| r1 | 95.1028 | 1838.9 | 8115.8 | 6047.4 | 1246.2 | cGy |
| r2 | 132.907 | 850.5 | 6969.7 | 3772.7 | 1616.0 | cGy |
| r3 | 176.81 | 623.8 | 5932.2 | 2620.8 | 1606.2 | cGy |
| r4 | 226.809 | 482.6 | 5068.3 | 2023.4 | 1446.1 | cGy |
| r5 | 282.857 | 377.1 | 4342.0 | 1629.1 | 1304.0 | cGy |
| Rectum_COPY_2 | 60.0965 | 552.3 | 8081.0 | 2784.9 | 1779.4 | cGy |
| Rectum_COPY_3 | 18.2264 | 1266.1 | 8081.0 | 3887.7 | 1821.3 | cGy |
| ROI_1 | 18.9879 | 7483.8 | 8406.6 | 8223.4 | 95.7 | cGy |
| Bladder_COPY_2 | 44.6684 | 1019.8 | 7786.3 | 3918.5 | 1579.4 | cGy |

**FIGURE 42.3 Dose volume histogram and statistics for treatment of the prostate only.**

special expertise. For patients with high-risk cancers, neoadjuvant, concomitant, and adjuvant androgen deprivation therapy should be delivered for 2–3 years.[50] Treatments should be ideally performed utilizing some form of image guidance. These include use of implanted fiducials, electromagnetic targeting/tracking, cone beam CT imaging, and ultrasound imaging. Immobilization is also used consisting mainly of endorectal balloons. These are placed daily prior to treatment. Patients may be instructed to move their bowels prior to treatment. This often requires the use of laxatives. Additionally, patients are instructed to drink fluid prior to treatment to fill the bladder to minimize dose to the bladder and small bowel.

Brachytherapy is the placement of permanent or temporary radioactive sources in the prostate and when indicated the seminal vesicles. Permanent implants most commonly consist of placement of low dose rate iodine-125 or palladium-103 seeds (Figure 42.5). These are placed freehand or in strands. They can be preplanned or placed real-time. When they are placed real-time, an ultrasound is employed to determine the location of the target volume and critical normal structures. In all cases, a postimplantation CT scan is obtained to identify the location of the seeds and final dose to the target. Brachytherapy can be used as monotherapy for low-risk cancers. Intermediate-risk cancer patients are often treated with a combination of 4–50 Gy of external beam irradiation and 4–6 months of androgen deprivation therapy. For patients with high-risk cancers androgen deprivation treatment is increased to 2–3 years. The prescribed total dose for monotherapy is 145 Gy for iodine-125 implants and 125 Gy for palladium-103. As a boost after 40–50 Gy of external beam irradiation the iodine-125 dose is 110 Gy and for palladium-103 is 90–100 Gy. High-dose brachytherapy is also being employed. This consists of placement of catheters in the prostate and seminal vesicles followed by introduction of high-intensity radioactive sources for a short period of time. As with low dose rate brachytherapy this can be used as monotherapy and as a boost before or after external beam radiation. When used as monotherapy a common regimen is 13.5 Gy fractions × 2. Typical boost regimens for this include 9.5–11.5 Gy fractions × 2,

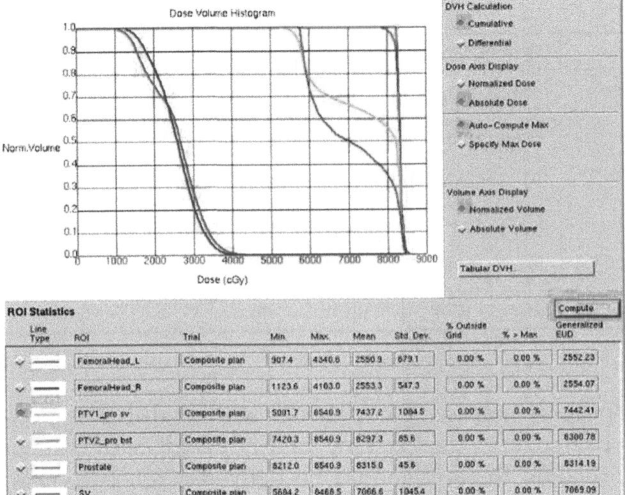


Region Of Interest Dose Statistics:

| ROI | Volume (cm^3) | Min Value | Max Value | Mean Value | Std. Dev. | Units |
|---|---|---|---|---|---|---|
| Bladder | 121.106 | 433.6 | 8490.6 | 3058.2 | 1987.8 | cGy |
| FemoralHead_R | 74.6945 | 1123.6 | 4103.0 | 2553.3 | 547.3 | cGy |
| FemoralHead_L | 73.2218 | 907.4 | 4340.6 | 2550.9 | 679.1 | cGy |
| Penile Bulb | 5.79006 | 513.9 | 5753.4 | 1909.5 | 1626.0 | cGy |
| Bowel | 3057.38 | 4.3 | 5849.5 | 133.5 | 281.0 | cGy |
| Prostate | 16.1372 | 8212.0 | 8540.9 | 8315.0 | 45.6 | cGy |
| SV | 10.6977 | 5684.2 | 8468.5 | 7066.6 | 1045.4 | cGy |
| PTV1_pro sv | 79.5599 | 5001.7 | 8540.9 | 7437.2 | 1084.5 | cGy |
| PTV2_pro bst | 42.5914 | 7420.3 | 8540.9 | 8297.3 | 85.6 | cGy |
| 10 mm r | 100.794 | 1864.9 | 8349.1 | 6382.4 | 1299.1 | cGy |
| 20 mm r | 175.692 | 575.5 | 7313.5 | 3760.5 | 1588.2 | cGy |
| 30mm r | 277.695 | 418.5 | 5988.5 | 2615.9 | 1475.2 | cGy |
| external | 22552.8 | 0.0 | 8540.9 | 524.2 | 972.3 | cGy |
| ntt | 21883.9 | 0.0 | 5597.7 | 419.3 | 668.4 | cGy |
| ant rectum | 6.9384 | 4824.6 | 8101.9 | 6433.9 | 660.2 | cGy |
| rec_sig | 138.919 | 100.4 | 8453.1 | 1523.7 | 1929.5 | cGy |
| post rectum | 114.016 | 100.4 | 5465.9 | 779.8 | 931.1 | cGy |
| max 3000 | 2.75602 | 2507.7 | 3304.9 | 2900.7 | 153.9 | cGy |

FIGURE 42.4 **Dose volume histogram and dose statistics for treatment of the prostate and seminal vesicles.**

5.5 to 7.5 Gy fractions × 3, and 4–6 Gy fractions × 4. The details of these regimens are found in Chapter 44, of this book (see chapter "Brachytherapy for Prostate Cancer: An Overview").

External beam radiation therapy has been used as adjuvant therapy after prostatectomy with pT3 tumors, poorly differentiated Gleason score 8–10 adenocarcinomas, positive margins of resection, and/or seminal vesicle involvement.[51] This is delivered within a year after radical prostatectomy and after patients have had maximal recovery from surgery. This ideally consists of full recovery of urinary continence. Patients with positive margins may also benefit from androgen deprivation therapy. The target volume typically consists of the prostate and seminal vesicle beds. As with definitive treatment, IMRT and image guidance are recommended.

It has been estimated that possibly one-third of patients undergoing radical prostatectomy will have positive margins.[51] Such cases are treated to a total dose of 6400–7000 cGy. EORTC (European Organisation for Research and Treatment of Cancer) 22911[52] randomly assigned 503 patients to be actively observed and 502 patients to receive immediate postoperative radiotherapy. The total dose was 60 Gy delivered using 200 cGy/fraction. Patients had capsule perforation, positive surgical margins, or invasion of seminal vesicles. They had no evidence of nodal or distant metastatic disease. The median follow-up was 5 years. The biochemical progression-free survival was significantly improved in the irradiated group (74.0%, 98% CI 68.7–79.3 vs. 52.6%, 46.6–58.5; $p < 0.0001$). Clinical progression-free survival was also significantly improved ($p = 0.0009$) as was the cumulated rate of locoregional failure ($p < 0.0001$). Grade 2 or 3 late effects were higher in the group receiving adjuvant radiation therapy. Grade 2 late effects consisted of moderate diarrhea, moderate colic, greater than five bowel movements daily, excessive rectal mucous, and intermittent bleeding. Grade 3 GI effects consisted of obstruction or bleeding requiring surgery. Grade 2 urinary track effects included moderate urinary frequency, generalized telangectasia, and intermittent microscopic hematuria. Grade 3 urinary-tract toxicity included severe urinary frequency and dysuria with generalized telangiectasia and frequent hematuria. There was a reduction in bladder capacity to less than 150 mL. Higher than grade 3 toxicities were rare in patients treated with radiation therapy. In a study by Thompson et al. 431 men with pT3N0M0 adenocarcinoma prostate were randomized to receive 60–64 Gy of adjuvant radiation therapy or undergo surveillance.[53] There were 214 patients who received treatment and 211 were observed. The metastasis-free survival (HR 0.71; 95% CI 0.54, 0.94; $p = 0.016$) and survival (HR 0.72; 95% CI 0.55, 0.96; $p = 0.023$) were better for the treated patients.

Radiation therapy is also often-times indicated for palliation. In the case of bone metastases 800 cGy in a single

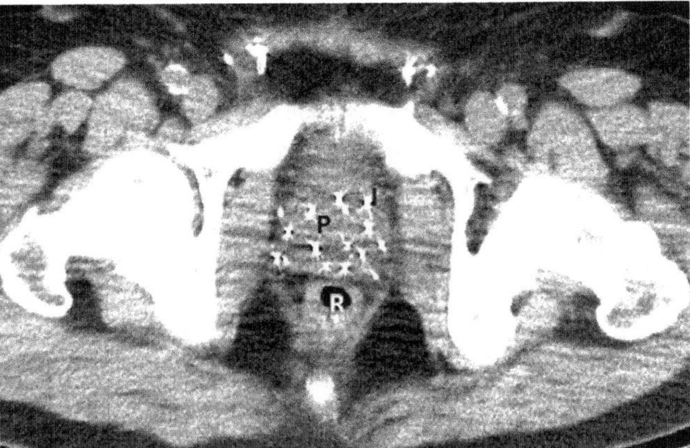


FIGURE 42.5 **Brachytherapy-postimplant CT after permanent I-125 seed implants.** P, prostate; I, radioactive implants; R, rectum.

fraction is most often recommended. The RTOG trial 97–14 revealed no difference between radiation delivered for painful bone metastasis at a dose of 8 Gy in a single fraction versus 30 Gy in 10 fractions from pain relief or narcotic use 3 months after randomization.[54] This was also observed in multiple other studies.[55–57] A possible exception to this is vertebral metastases, which continues to be treated with 300 cGy per fraction to a total dose of 3000 cGy in many programs. However, in the same RTOG study, there was no difference in outcome for treatment of the spine using a single fraction.[58] Another alternative for palliation of bone metastasis is the use of radioactive isotopes, administered intravenously. Initially, the strontium-89 isotope was used. Today, however, samarium-153 and radium-223 are preferred due to the increased risk for thrombocytopenia from strontium-89. Administration of radioactive isotopes can be combined with focal external beam radiation therapy for improved outcome.

## Acknowledgments

The authors thankfully acknowledge Pierre Charpentier, for providing figures taken in our Department of Radiation Oncology, Temple University Hospital and Suryalekshmy Rajasimhan for labeling and formatting them.

## References

1. Grubbe EH. X-rays in the treatment of cancer and other malignant diseases. *Med Rec (W. Wood)* 1902;**62**:692–5.
2. Widmark A, Klepp O, Solberg A, et al. Endocrine treatment, with or without radiotherapy, in locally advanced prostate cancer (SPCG-7/SFUO-3): an open randomised phase III trial. *Lancet* 2009;**373**:301–8.
3. Warde P, Mason M, Ding K, et al. Combined androgen deprivation therapy and radiation therapy for locally advanced prostate cancer: a randomised, phase 3 trial. *Lancet* 2011;**378**:2104–11.
4. Nguyen PL, Gu X, Lipsitz SR, et al. Cost implications of the rapid adoption of newer technologies for treating prostate cancer. *J Clin Oncol* 2011;**29**:1517–24.
5. Sheets NC, Goldin GH, Meyer AM, et al. Intensity-modulated radiation therapy, proton therapy, or conformal radiation therapy and morbidity and disease control in localized prostate cancer. *JAMA* 2012;**307**:1611–20.
6. D'Amico AV, Whittington R, Kaplan I, et al. Equivalent biochemical failure-free survival after external beam radiation therapy or radical prostatectomy in patients with a pretreatment prostate specific antigen of >4–20 ng/mL. *Int J Radiat Oncol Biol Phys* 1997;**37**:1053–8.
7. Kupelian P, Katcher J, Levin H, et al. External beam radiotherapy versus radical prostatectomy for clinical stage T1-2 prostate cancer: therapeutic implications of stratification by pretreatment PSA levels and biopsy Gleason scores. *Cancer J Sci Am* 1997;**3**:78–87.
8. Zelefsky MJ, Wallner KE, Ling CC, et al. Comparison of the 5-year outcome and morbidity of three-dimensional conformal radiotherapy versus transperineal permanent iodine-125 implantation for early-stage prostatic cancer. *J Clin Oncol* 1999;**17**:517–22.
9. Cooperberg MR, Vickers AJ, Broering JM. Risk-adjusted mortality outcomes after primary surgery, androgen-deprivation therapy for localized prostate cancer. *Cancer* 2010;**116**:5226–34.
10. Kibel AS, Ciezki JP, Klein EA, et al. Survival among men with clinically localized prostate cancer treated with radical prostatectomy or radiation therapy in the prostate specific antigen era. *J Urol* 2012;**187**:1259–65.
11. Sooriakumaran P, Nyberg T, Akre O, et al. Comparative effectiveness of radical prostatectomy and radiotherapy in prostate cancer: observational study of mortality outcomes [serial online]. *BMJ* 2014;**348**:1502.
12. Zelefsky MJ, Eastham JA, Cronin AM, et al. Metastasis after radical prostatectomy or external beam radiotherapy for patients with clinically localized prostate cancer: a comparison of clinical cohorts adjusted for case mix. *J Clin Oncol* 2010;**28**:1508–13.
13. Zelefsky MJ, Yamada Y, Pei X, et al. Comparison of tumor control and toxicity outcomes of high-dose intensity-modulated radiotherapy and brachytherapy for patients with favorable risk prostate cancer. *Urology* 2011;**77**:986–90.
14. Ritter M, Forman J, Kupelian K, et al. Hypofractionation for prostate cancer. *Cancer J* 2009;**15**:1–6.
15. Martinez AA, Demanes J, Vargas C, et al. High-dose-rate prostate brachytherapy: an excellent accelerated-hypofractionated treatment for favorable prostate cancer. *Am J Clin Oncol* 2010;**33**:481–8.
16. Jang TL, Bekelman JE, Liu Y, et al. Physician visits prior to treatment for clinically localized prostate cancer. *Arch Intern Med* 2010;**170**:440–50.
17. Cooperberg MR, Broering JM, Carroll PR. Time trends and local variation in primary treatment of localized prostate cancer. *J Clin Oncol* 2010;**28**:1117–23.
18. Cooperberg MR, Vickers AJ, Broering JM, et al. Comparative risk-adjusted mortality outcomes after primary surgery, radiotherapy, or androgen-deprivation therapy for localized prostate cancer. *Cancer* 2010;**116**:5226–34.
19. Jang TL, Bekelman JE, Liu Y, et al. Physician visits prior to treatment for clinically localized prostate cancer. *Arch Intern Med* 2010;**170**:440–50.
20. Fowler Jr FJ, McNaughton CM, Albertsen PC, et al. Comparison of recommendations by urologists and radiation oncologists for treatment of clinically localized prostate cancer. *JAMA* 2000;**283**:3217–22.
21. Petrelli F, Vavassori I, Coinu A, et al. Radical prostatectomy or radiotherapy in high-risk prostate cancer: a systematic review and metaanalysis. *Clin Genitourin Cancer* 2014;**12**:215–24.
22. Steel, GG. In: *Basic Clinical Radiobiology*. 3rd ed. Arnold Publishers, London; Oxford University Press, New York; 2002.
23. http://www.who.int/ionizing_radiation/about/what_is_ir/en/.
24. Radiation biology: A hand book for teachers and students, Training Course Series # 42, IAEA Publishers, Vienna, 2010 ISSN 1018–5518.
25. Hall EJ. *Radiobiology for the radiologist*. 5th ed. Philadelphia: Lippincott Williams & Wilkins; 2000.
26. Hall EJ, Giaccia AJ. *Radiobiology for the radiologist*. 6th ed. Philadelphia: Lippincott Williams & Wilkins; 2006.
27. Podgorsak EB. External photon beams: physical aspects. In: *Review of radiation oncology physics: a handbook for teachers and students*. IAEA Publications, Vienna; 2005 pp: 161–216; ISBN 92-0-107304-6.
28. Elkind MM, Sutton-Gilbert H, Moses WB, et al. Radiation response of mammalian cells grown in culture: V. Temperature dependence of the repair of X-ray damage in surviving cells (aerobic and hypoxic). *Radiat Res* 1965;**25**:359–76.
29. Van Putten LM, Kallman RF. Oxygenation status of a transplantable tumor during fractionated radiation therapy. *J Natl Cancer Inst* 1968;**40**:441–51.
30. Pajonk F, Vlashi E, Mc Bride WH. Radiation resistance of cancer stem cells: the 4 R's of radiobiology revisited. *Stem Cells* 2010;**28**:639–48.
31. Steel GG, McMillan TJ, Peacock JH. The 5Rs of radiobiology. *Int J Radiat Biol* 1989;**56**:1045–8.
32. Palcic B, Skarsgard LD. Reduced oxygen enhancement ratio at low doses of ionizing radiation. *Radiat Res* 1984;**100**:328–39.

33. Hegde JV, Chen MH, Myulkern RV, et al. Multiparametric endorectal magnetic resonance imaging findings and the odds of upgrading and upstaging at radical prostatectomy in men with clinically localized prostate cancer [serial online]. *Int J Radiat Oncol Biol Phys* 2013;**85**:101–7.
34. Hambrock T, Somford DM, Huisman HJ, et al. Relationship between apparent diffusion coefficients at 3.0-T MR imaging and Gleason grade in peripheral zone prostate cancer. *Radiology* 2011;**259**:453–61.
35. Hentschel B, Oehler W, Strauss D, et al. Definition of the CTV prostate in CT and MRI by using CT-MRI image fusion in IMRT planning for prostate cancer. *Strahlenther Onkol* 2011;**187**: 183–90.
36. Berthelsen KI, Dobbs J, Kjellén E, et al. What's new in target volume definition for radiologists in ICRU Report 71? How can the ICRU volume definitions be integrated in clinical practice? *Cancer Imaging* 2007;**7**:104–16.
37. Michalski JM, Lawton C, Naqa IE, et al. Development of RTOG consensus guidelines for the definition of the clinical target volume for postoperative conformal radiation therapy for prostate cancer. *Int J Radiat Oncol Biol Phys* 2010;**76**:361–8.
38. Boehmer D, Maingon PH, Poortmans PH, et al. Guidelines for primary radiotherapy of patients with prostate cancer. *Radiother Oncol* 2006;**79**:259–69.
39. Hinton BK, Fiveash JB, Wu X, et al. Optimal planning target volume margins for elective pelvic lymphatic radiotherapy in high-risk prostate cancer patients. *ISRN Oncol* 2013;**2013**:1–5.
40. Shipley WU, Verhey LJ, Munzenrider JE, et al. Advanced prostate cancer: the results of a randomized comparative trial of high dose irradiation boosting with conformal protons compared with conventional dose irradiation using photons alone. *Int J Radiat Oncol Biol Phys* 1995;**32**:3–12.
41. Pollack A, Zagars GK, Starkschall G, et al. Prostate cancer radiation dose response: results of the M. D. Anderson phase III randomized trial. *Int J Radiat Oncol Biol Phys* 2002;**53**:1097–105.
42. Zietman AL, Bae K, Slater JD, et al. Randomized trial comparing conventional-dose with high-dose conformal radiation therapy in early-stage adenocarcinoma of the prostate: long-term results from Proton Radiation Oncology Group/American College of Radiology 95-09. *J Clin Oncol* 2010;**28**:1106–11.
43. Beckendorf V, Guerif S, Prisé EL, et al. 70 Gy versus 80 Gy in localized prostate cancer: 5-year results of GETUG 06 randomized trial. *J Radiat Oncol Biol Phys* 2011;**80**:1056–63.
44. Al-Mamgani A, van Putten WLJ, Heemsbergen WD. Update of Dutch multicenter dose-escalation trial of radiotherapy for localized prostate cancer. *Int J Radiat Oncol Biol Phys* 2008;**72**:980–8.
45. Dearnaley DP, Sydes MR, Graham JD, et al. Escalated-dose versus standard-dose conformal radiotherapy in prostate cancer: first results from the MRC RT01 randomised controlled trial. *Lancet* 2007;**8**:475–87.
46. Martin NE, D'Amico AV. Progress and controversies: radiation therapy for prostate cancer. *CA Cancer J Clin* 2014;**64**:389–407.
47. Michalski JM, Yan Y, Tucker S, et al. Dose volume analysis of Grade 2+ late GI toxicity on RTOG 0126 after high-dose 3DCRT or IMRT. *Int J Radiat Oncol Biol Phys* 2012;**84**:S14–5.
48. Viani GA, Stefano EJ, Sergio Afonso SL. Higher-than-conventional radiation doses in localized prostate cancer treatment: a meta-analysis of randomized, controlled trials. *Int J Radiat Oncol Biol Phys* 2009;**74**:1405–18.
49. Launders J, Inamdar R, Tipton K, et al. Stereotactic Body Radiation Therapy: Scope of the Literature. *Ann Intern Med* 2011;**154**:737–45.
50. Zelefsky M, Lee W, Zietman A, et al. Evaluation of adherence to quality measures for prostate cancer radiotherapy in the United States from the Quality Research in Radiation Oncology (QRRO) survey. *Prac Radiat Oncol* 2012;**3**(1):2–8.
51. Cooperberg MR, Broering JM, Litwin MS, et al. The contemporary management of prostate cancer in the United States: lessons from the Cancer of the Prostate Strategic Urologic Research Endeavor (CaPSURE), a national disease registry. *J Urol* 2004;**171**:1393.
52. Bolla M, Poppel HV, Collette L, et al. Postoperative radiotherapy after radical prostatectomy: a randomised controlled trial (EORTC trial 22911). *Lancet* 2005;**366**(9485):572–8.
53. Thompson IM, Tangen CM, Paradelo J, et al. Adjuvant radiotherapy for pathological T3N0M0 prostate cancer significantly reduces risk of metastases and improves survival: long-term followup of a randomized clinical trial. *J Urol* 2009;**81**:956–62.
54. Hartsell WF, Scott C, Bruner DW, et al. Phase III randomized trial of 8 Gy in 1 fraction vs. 30 Gy in 10 fractions for palliation of painful bone metastases: preliminary results of RTOG 97−14. *Int J Radiat Oncol Biol Phys* 2003;**57**:124.
55. Yarnold JR. 8 Gy single fraction radiotherapy for the treatment of metastatic skeletal pain: randomised comparison with a multifraction schedule over 12 months of patient follow-up. Bone Pain Trial Working Party. *Radiother Oncol* 1999;**52**:111–21.
56. Sze WM, Shelley M, Held I, et al. Palliation of metastatic bone pain: single fraction versus multifraction radiotherapy – a systematic review of the randomised trials. *Cochrane Database Syst Rev* 2002; **1**: Art number CD004721.
57. Wu JS, Wong R, Johnston M, et al. Cancer Care Ontario Practice Guidelines Initiative Supportive Care Group. Meta-analysis of dose-fractionation radiotherapy trials for the palliation of painful bone metastases. *Int J Radiat Oncol Biol Phys* 2003;**55**:594–605.
58. Howell DD, James JL, Hartsell WF, et al. Single-fraction radiotherapy versus multifraction radiotherapy for palliation of painful vertebral bone metastases-equivalent efficacy, less toxicity, more convenient: a subset analysis of Radiation Therapy Oncology Group trial 97–14. *Cancer* 2013;**119**:888–96.

CHAPTER

# 43

# Radiation with Hormonal Therapy

*Brock R. Baker, BS, Jahan J. Mohiuddin, BS, Ronald C. Chen, MD, MPH*

Department of Radiation Oncology, University of North Carolina, Chapel Hill, NC, USA

## INTRODUCTION

External beam radiation therapy (EBRT) is an effective primary treatment for patients with nonmetastatic prostate cancer, and is often used together with androgen deprivation therapy (ADT) in patients with aggressive (intermediate- or high-risk) disease. A large number of randomized trials have been conducted over the past 40 years, which guide clinical practice today. This chapter will summarize results of these trials and other key literature related to the use of EBRT and ADT, as well as potential side effects of treatment and their management.

## TYPES OF ANDROGEN DEPRIVATION THERAPY USED WITH RADIATION THERAPY

There are three main classes of drugs used in conjunction with external beam radiation therapy (EBRT): (1) gonadotropin releasing hormone (GnRH) agonists, (2) GnRH antagonists, and (3) antiandrogens.

### Gonadotropin Releasing Hormone Analogs

GnRH agonists and antagonists are both synthetic analogs of the GnRH peptide hormone, and achieve castrate testosterone levels by shutting down the GnRH-mediated release of luteinizing hormone (LH) and follicle-stimulating hormone (FSH) from the anterior pituitary. GnRH antagonists achieve this effect through direct blockade of the GnRH receptor. In contrast, GnRH agonists function through their interruption of the normal pulsatile signaling of physiologic GnRH. Persistent elevation of GnRH agonist activity leads to downregulation of the GnRH receptor, thereby causing decreased levels of LH, FSH, and testosterone.[1,2] GnRH agonists often cause an initial surge in LH, FSH, and testosterone before this downregulation occurs. In contrast, GnRH antagonists do not cause this androgen surge. Available GnRH agonists include leuprolide, goserelin, and triptorelin. Available GnRH antagonists include abarelix and degarelix.

### GnRH Agonist Versus GnRH Antagonist

Most trials that have combined radiation therapy with androgen deprivation therapy have used a GnRH agonist. The recent development of GnRH antagonists may provide an alternative option. In the CS21 trial, 610 patients with prostate cancer (including those with metastatic and nonmetastatic disease) in whom ADT was indicated were randomized to 12 months of degarelix versus 12 months of leuprolide (with bicalutamide allowed at the discretion of the treating physicians to prevent androgen flares).[3] The primary endpoint was suppression of testosterone below 0.5 ng/mL during the 12 months, and degarelix showed noninferiority to leuprolide in this regard. Moreover, degarelix demonstrated faster time to castrate levels of testosterone, and faster suppression of prostate-specific antigen (PSA). Over 95% of patients receiving degarelix achieved castrate levels of testosterone at 3 days after starting treatment, compared to none of the patients in the leuprolide arm. Similarly, PSA levels in the degarelix arm declined by a greater percentage than in the leuprolide arm at both 14 days (65% degarelix vs. 18% leuprolide) and 28 days (83% degarelix vs. 68% leuprolide). Eighty-one percent of patients receiving leuprolide had a testosterone surge upon initiating therapy, while no patients receiving degarelix had a testosterone surge. Investigators also reported similar rates of adverse effects between both groups.

Prostate Cancer. http://dx.doi.org/10.1016/B978-0-12-800077-9.00043-8

This trial demonstrates that a GnRH antagonist is able to achieve faster testosterone suppression than a GnRH agonist. However, it is unknown whether this translates to an improvement in oncologic outcomes (cancer control, survival) when used with radiation therapy.

## Antiandrogens

The other main class of drugs used with EBRT is the antiandrogens. These synthetic compounds bind the testosterone receptor in target tissues, and competitively inhibit the binding of testosterone and dihydrotestosterone. Antiandrogens are often used together with GnRH agonists to block the androgen surge and to achieve maximal androgen blockade, and it is uncommon to use antiandrogen alone with radiation therapy. Available formulations include flutamide, bicalutamide, and nilutamide.

# TREATMENT OF HIGH-RISK PROSTATE CANCER WITH RADIATION THERAPY

At least three randomized trials have compared the use of ADT plus EBRT versus ADT alone in the treatment of high-risk prostate cancer (Table 43.1). These trials support the use of radiation therapy for patients with aggressive prostate cancer.

The SPCG-7/SFUO-3 Scandinavian trial randomized 875 patients with locally advanced prostate cancer (PSA <70, N0, M0) to ADT plus EBRT versus ADT alone.[4] ADT consisted of 3 months of maximal androgen blockade with leuprolide plus flutamide, followed by indefinite flutamide therapy. After a median follow-up of 7.6 years, patients in the ADT plus EBRT arm had significantly lower prostate-cancer-specific mortality (10-year incidence 23.9% ADT alone vs. 11.9% EBRT/ADT, $p < 0.001$), overall mortality (39.4% ADT alone vs. 29.6% EBRT/ADT, $p < 0.001$), and biochemical recurrence (74.7% ADT alone vs. 25.9% ADT/EBRT, $p < 0.001$).

The Intergroup T94-0110 trial randomized 1205 patients with T2–T4, N0, M0 prostate cancer to lifelong ADT (GnRH agonist or bilateral orchiectomy) with or without EBRT.[5] As in the Scandinavian trial, a clear benefit was seen with combined ADT plus EBRT versus ADT alone in terms of 8-year prostate-cancer-specific survival (77.7% ADT alone vs. 89.3% EBRT/ADT, $p < 0.0001$) and overall survival (56.8% ADT alone vs. 66% EBRT/ADT, $p = 0.0003$).

In a much smaller French trial, 264 patients with T3–T4, N0 M0 disease were randomized to 3 years ADT (leuprolide) plus EBRT to 68–70 Gy versus 3 years ADT alone.[6] After a median follow-up of 5.8 years, patients who received EBRT had significantly lower biochemical recurrence (91.5% ADT alone vs. 39.1 EBRT/ADT, $p < 0.0001$), locoregional recurrence (29.2% ADT alone vs. 9.8% EBRT/ADT, $p < 0.0001$), and progression to metastatic disease (10.8% ADT alone vs. 3.0% EBRT/ADT, $p < 0.018$). No statistically significant difference in overall or disease-specific survival was seen with this relatively short follow-up.

**TABLE 43.1** Randomized Trials Comparing Androgen Deprivation Therapy to Androgen Deprivation With Radiation

| | Patients included | Median follow-up (Years) | Treatment arms | ADT used | Overall mortality (%) | Cancer-specific mortality (%) | Distant failure (%) | Biochemical recurrence (%) |
|---|---|---|---|---|---|---|---|---|
| SPCG-7/SFUO-3[4] $n = 875$ | PSA < 70, N0, M0; 73% T3 | 7.6 | Indefinite ADT versus EBRT + indefinite ADT | 3 Months Leuprolide + indefinite flutamide | (10 years) 39.4 29.6 | (10 years) 23.9 11.9 | NA | (10 years) 74.7 25.9 |
| Intergroup T94-0110[5] $n = 1205$ | T2–T4, N0, M0 | 8.0 | Indefinite ADT versus EBRT + indefinite ADT | GnRH agonist or bilateral orchiectomy | (8 years) 43.2 34.0 | (8 years) 22.3 10.7 | NA | HR 0.46 with EBRT+ADT |
| Mottet et al.[6] $n = 264$ | T3–4, N0, M0 | 5.8 | 3 Year ADT versus EBRT + 3 year ADT | Leuprolide | (5 years) 28.5 28.6 (ns) | (5 years) 13.8 6.8 (ns) | (5 years) 10.8 3.0 | (5 years) 91.5 39.1 |

ADT, androgen deprivation therapy; EBRT, external beam radiation therapy; GnRH, gonadotropin-releasing hormone; NA, not available; NS, not statistically significant; PSA, prostate specific antigen. All results significant unless otherwise noted.

These trials have clearly established radiation therapy as a standard treatment for patients with aggressive prostate cancer. Mature results from the SPCG and Intergroup trials demonstrated an approximately 10% absolute improvement in overall survival from radiation therapy compared to conservative management with ADT.

## THE BENEFIT OF ANDROGEN DEPRIVATION THERAPY IN COMBINATION WITH RADIATION THERAPY

Multiple randomized trials have consistently demonstrated that treatment outcomes are improved when ADT is added to EBRT in intermediate-risk, high-risk, and locally advanced prostate cancer (Table 43.2).

### ADT and EBRT in Intermediate-Risk Prostate Cancer

In a randomized trial conducted at Harvard by D'Amico et al., 206 men with intermediate- (76%) and high-risk (24%) localized prostate cancer were randomized to EBRT versus EBRT plus 6 months of ADT.[7,8] After a median follow-up of 7.6 years, there was higher overall 8-year mortality in patients who received EBRT alone compared to EBRT plus ADT (39% EBRT alone vs. 26% EBRT/ADT, $p = 0.01$). In subgroup analysis, the benefit from adding ADT was apparent in intermediate- and high-risk patient groups.[9]

This survival benefit from the addition of ADT was also demonstrated in the RTOG 94-08 trial that randomized 1979 patients to EBRT alone versus EBRT plus 4 months of ADT.[10] This trial included patients with low- (36%), intermediate- (53%), and high-risk (11%) prostate cancers. The 10-year overall survival was increased

**TABLE 43.2** Randomized Trials Comparing Radiation Therapy to Radiation Plus Androgen Deprivation Therapy

| | Patients included | Median follow-up (years) | Treatment arms | ADT used | Overall mortality (%) | Cancer-specific mortality (%) | Distant failure (%) | BR/LF (%) |
|---|---|---|---|---|---|---|---|---|
| Harvard[8] $n = 206$ | Intermediate- (76%) and high-risk (24%) | 7.6 | EBRT versus EBRT + 6 months ADT | GnRH agonist + flutamide | (8 years) 39 26 | (8 years) HR=4.1 | NA | NA |
| RTOG 94-08[10] $n = 1979$ | Low-(36%), intermediate- (53%), and high-risk (11%) | 9.1 | EBRT versus EBRT + 4 months ADT | GnRH agonist + flutamide | (10 years) 43 38 | (10 years) 8 4 | (10 years) 8 6 | (10 years) 41 26 |
| TROG 96.01[14] $n = 818$ | Intermediate-(16%), high-risk (44%), and locally advanced (40%) | 10.6 | EBRT versus EBRT + 3 months ADT EBRT + 6 months ADT | Goserelin + flutamide | (10 years) 42.5 36.7 (ns) 29.2 | (10 years) 22.0 18.9 (ns) 11.4 | (10 years) 20.6 18.3 (ns) 10.9 | (10 years) 73.8 60.4 52.8 |
| RTOG 86-10[16] $n = 456$ | T2–T4, bulky (5 cm × 5 cm) primary tumor | 13.2* 11.9* | EBRT versus EBRT + 4 months ADT | Goserelin + flutamide | (10 years) 64 57 (ns) | (10 years) 36 23 | (10 years) 47 35 | (10 years) 80 65 |
| RTOG 85-31[19] $n = 977$ | cT3 or any N1, M0 | 7.6 | EBRT versus EBRT + indefinite ADT | Goserelin | (10 years) 61 51 | (10 years) 22 16 | (10 years) 39 24 | (10 years) 38 23 |
| EORTC 22863[22] $n = 412$ | T3–T4 | 5.5 | EBRT versus EBRT + 3 year ADT | Goserelin | (5 years) 38 22 | (5 years) 21 6 | (5 years) 29 10 | (5 years) 16.4 1.7 |

ADT, androgen deprivation therapy; BR, biochemical recurrence; EBRT, external beam radiation therapy; GnRH, gonadotropin-releasing hormone; LF, local failure; NA, not available; NS, not statistically significant; PSA, prostate specific antigen. All results significant unless otherwise noted.
* *Median follow-up for living patients.*

in the combination EBRT/ADT arm (57% EBRT alone vs. 62% EBRT/ADT, $p = 0.03$). In subgroup analysis, a statistically significant survival benefit was seen in the intermediate-risk group (61% combination vs. 54% EBRT alone, $p = 0.03$). The survival outcomes seen in the low-risk (67% combination vs. 64% EBRT alone) and high-risk (53% combination vs. 51% EBRT alone) groups did not reach statistical significance.

The Harvard and RTOG 94-08 trials both showed that adding short-term ADT to radiation therapy improves overall survival in intermediate-risk prostate cancer. In order to answer the question of the optimal duration of ADT in these patients, D'Amico et al. conducted a meta-analysis[11] of data from three randomized trials of EBRT with either 3, 4, or 6 months ADT.[8,12,13] For patients with Gleason score 7 cancer, a significant reduction in prostate-cancer-specific mortality was found with 6 months versus 3 or 4 months ADT (adjusted hazard ratio 0.47, $p = 0.01$). The authors concluded that Gleason 7 patients receiving EBRT should receive 6 months ADT.

## ADT and EBRT in High-Risk and Locally Advanced Prostate Cancer

In addition to the Harvard and RTOG 94-08 trials mentioned earlier, a number of other randomized trials have examined the disease control and survival outcomes of adding ADT to EBRT in high-risk and locally advanced prostate cancer.

The TROG 96.01 trial randomized 818 men with intermediate-risk (16%), high-risk (44%), and locally advanced prostate cancer (40%) to either EBRT alone, EBRT plus 3 months ADT, or EBRT plus 6 months EBRT.[12,14] After median follow-up of 10.6 years, both the 3 and 6 months ADT plus EBRT arms showed significantly improved rates of local and biochemical failure versus EBRT alone; the 6 months ADT arm versus EBRT alone also showed a decrease in distant failures (20.6% EBRT alone vs. 10.9% EBRT/ADT, HR 0.49, $p = 0.001$), prostate-cancer-specific mortality (22.0% EBRT alone vs. 11.4% EBRT/ADT, HR 0.49, $p = 0.0008$), and all-cause mortality (42.5% EBRT alone vs. 29.2% EBRT/ADT, HR 0.63, $p = 0.0008$).

The RTOG 86-10 trial examined the addition of short-term ADT to EBRT for locally advanced prostate cancer.[15,16] A total of 456 patients with bulky (5 cm × 5 cm) primary tumors (T2–T4) were randomized to EBRT with and without 4 months ADT with goserelin and flutamide. The addition of short-term ADT resulted in significantly improved rates of distant metastasis (47% EBRT alone vs. 35% EBRT/ADT, $p = 0.006$), biochemical failure (80% EBRT alone vs. 65% EBRT/ADT, $p < 0.0001$), prostate-cancer-free survival (3% EBRT alone vs. 11% EBRT/ADT, $p < 0.0001$), and prostate-cancer-specific mortality (36% EBRT alone vs. 23% EBRT/ADT, $p = 0.01$). The difference in overall survival favored the ADT arm (34% EBRT alone vs. 43% EBRT/ADT), but did not reach statistical significance ($p = 0.12$).

RTOG 85-31 randomized 977 patients with locally advanced (cT3) or node-positive (27.8%) prostate cancer to EBRT plus indefinite ADT with goserelin or EBRT only.[17–19] This study demonstrated a clear advantage with the use of long-term ADT in terms of local control (10-year local failure rate 38% EBRT alone vs. 23% EBRT/ADT, $p < 0.0001$), distant metastases (39% EBRT alone vs. 24% EBRT/ADT, $p < 0.001$), prostate-cancer-specific mortality (22% EBRT alone vs. 16% EBRT/ADT, $p = 0.0052$), and overall survival (39% EBRT alone vs. 49% EBRT/ADT, $p = 0.002$). A significant percentage of patients randomized to EBRT plus indefinite ADT ceased ADT therapy voluntarily (i.e., for reasons other than death, disease progression, or initiation of a different hormone therapy): 35.4% stopped ADT prior to 1 year, and 67.7% prior to 5 years.[20]

EORTC 22863 confirmed the benefit of long-term ADT when added to EBRT in the treatment of locally advanced prostate cancer. In this trial, 412 patients with T3–T4 disease were randomized to EBRT versus EBRT plus 3 years ADT using goserelin.[21,22] After a median follow-up of 5.5 years, the ADT arm showed superior 5-year disease-free survival (40% EBRT alone vs. 74% EBRT/ADT, $p = 0.0001$) and overall survival (62% EBRT alone vs. 78% EBRT/ADT, $p = 0.0002$).

These trials demonstrated the effectiveness of adding ADT of some duration to EBRT in the treatment of high-risk and locally advanced prostate cancer, but none compared short- to long-term ADT directly. EORTC 22961 examined this question. In this trial, 970 patients with high-risk and locally advanced prostate cancer (19% T2c, 77% T3–T4, 3% node-positive) were randomized to EBRT plus either 6 months ADT (triptorelin plus antiandrogen) or 3 years ADT (triptorelin plus antiandrogen for 6 months followed by triptorelin for 30 months).[23] This study was designed as a noninferiority trial. After a median follow-up of 6.4 years, short-term ADT had a higher rate of 5-year overall mortality relative to long-term ADT (19.0% short-term vs. 15.2% long-term), and the null hypothesis of the noninferiority of short-term ADT was not rejected. Similarly, the RTOG 92-02 trial compared EBRT plus either short-term or long-term ADT. A total of 1554 patients with high-risk (45%) or locally advanced (T2c–T4,N0, 55%) prostate cancer were randomized to EBRT with either 4 or 28 months ADT.[24,25] Relative to short-term (4 months) ADT, long-term (28 months) therapy reduced local progression (22.2% short-term vs. 12.3% long-term, $p < 0.0001$), biochemical failure (68.1% short-term vs. 51.9% long-term, $p \leq 0.0001$), distant metastasis (22.8% short-term vs. 14.8% long-term, $p < 0.0001$), and improved prostate-cancer-specific survival (83.9% short-term vs. 88.7% long-term, $p = 0.0042$).

TABLE 43.3 Randomized Trials Comparing Long-Term to Short-Term Androgen Deprivation Therapy in High-Risk and Locally Advanced Prostate Cancer

| | Patients included | Median follow-up (years) | Treatment arms | ADT used | Overall mortality (%) | Cancer-specific mortality (%) | Distant failure (%) | Biochemical recurrence (%) |
|---|---|---|---|---|---|---|---|---|
| EORTC 22961[23] n = 970 | 91% T2c-T4 N0 | 6.4 | EBRT + 6 months ADT versus EBRT+ 3 year ADT | 6 Months triptorelin + antiandrogen, then 30 months triptorelin | (5 years) 19.0 15.2* | (5 years) 4.7 3.2 | (5 years) 14 6 | (5 years) 38 15 |
| RTOG 92-02[25] n = 1554 | High-risk (45%), locally advanced (55%) | 11.3 | EBRT + 4 months ADT versus EBRT + 28 months ADT | 4 Months goserelin + flutamide, then 24 months goserelin | (10 years) 51.6 53.9 (ns) | (10 years) 16.1 11.3 | (10 years) 22.8 14.8 | (10 years) 68.1 51.9 |

ADT, androgen deprivation therapy; EBRT, external beam radiation therapy; NS, not statistically significant; PSA, prostate-specific antigen. All results significant unless otherwise noted.

* *This study was a noninferiority trial. Noninferiority of short-term ADT was not rejected.*

A subgroup analysis demonstrated an overall survival benefit of long-term ADT in the Gleason score ≥8 group (10-year survival 31.9% short-term vs. 45.1% long-term, $p$ = 0.006). These results of RTOG 92-02 and EORTC 22961 provide strong support for the use of long-term (2–3 years) ADT with EBRT in patients with high-risk and locally advanced prostate cancer (Table 43.3).

## DOSE-ESCALATED RADIOTHERAPY AND ANDROGEN DEPRIVATION THERAPY

All of the previously mentioned trials , which demonstrated a benefit of adding ADT to EBRT, were conducted using dose ranges of 64–70 Gy. In the modern era of 3D conformal and intensity-modulated radiation therapy (IMRT), doses of 75–80 Gy or higher can be safely delivered. Five randomized trials of conventional (lower) dose versus modern (dose-escalated) EBRT for prostate cancer have demonstrated improved local control and biochemical progression-free survival with dose escalation (Table 43.4).[26–34] These results have changed the modern standard of care to delivering at least 75.6 Gy of radiation.[35]

It is notable that delivering a higher dose of radiation, while demonstrating improved disease control, has not yet been shown to improve overall survival in patients; on the other hand, multiple trials have shown improved survival from adding ADT to EBRT. However, the potential benefit of ADT in a setting of modern dose-escalated EBRT for prostate cancer is unknown, and is a relevant question especially for intermediate-risk patients. Two ongoing phase III trials, EORTC 22991 (NCT 00021450) and RTOG 08-15 (NCT 00936390), will help to answer this question. EORTC 22991 will randomize 800 intermediate- or high-risk patients to dose-escalated radiation therapy with and without 6 months ADT, and will examine biochemical and clinical disease-free survival as the primary outcomes. RTOG 08-15 will randomize 1520 intermediate-risk patients to dose-escalated radiation therapy with and without 6 months ADT and will measure overall survival as the primary outcome.

Several nonrandomized studies have examined the use of ADT with dose-escalated EBRT. The Grupo de Investigacion Clinica en Oncologia Radioterapica (GICOR) conducted a prospective, nonrandomized trial of patients with intermediate- and high-risk prostate cancer treated with EBRT and ADT.[36] Intermediate-risk patients received EBRT plus 4–6 months of ADT, and high-risk patients received EBRT plus 2 years of ADT. The radiation dose was not prescribed by the protocol, and so some patients received conventional dose EBRT (<70 Gy), while others received dose-escalated EBRT (≥72 Gy) per physician discretion. The authors compared those who received conventional versus dose-escalated EBRT, and found that dose-escalated EBRT significantly improved 5-year biochemical disease-free survival in high-risk patients (63% conventional dose vs. 84% dose-escalated EBRT, $p$ = 0.003), thus demonstrating that in the setting of ADT, dose-escalated EBRT provides an incremental benefit relative to conventional dose EBRT.

TABLE 43.4 Randomized Trials Comparing Conventional-Dose to Dose-Escalated External Beam Radiation Therapy in Prostate Cancer

| | Patients included | Median follow-up (years) | Treatment arms (Gy) | Overall mortality (%) | Cancer-specific mortality (%) | Distant failure (%) | Biochemical/clinical recurrence (%) |
|---|---|---|---|---|---|---|---|
| Dutch Trial[28] $n = 664$ | T1b–T4N0M0; 21.5% received ADT | 9.2 | 68<br>78 | 30.5<br>31.2 (ns) | 13.3<br>13.2 (ns) | 11.4<br>14.8 (ns) | (10 years)<br>57<br>39 |
| M.D. Anderson Trial[30] $n = 301$ | T1b–T3N0M0 | 8.7 | 70<br>78 | (8 years)<br>22<br>21 (ns) | (8 years)<br>5<br>1 (ns) | (8 years)<br>5<br>1 (ns) | (10 years)<br>50<br>27 |
| PROG/ACR 95-09[32] $n = 393$ | T1b–T2b, PSA ≤ 15 | 8.9 | 70.2<br>79.2 | (10 years)<br>21.6<br>16.6 (ns) | NA | NA | (10 years)<br>32.4<br>16.7 |
| GETUG 06[33] $n = 306$ | T1b–T3aN0M0 | 5.1 | 70<br>80 | (5 years)<br>8.4<br>8.4 (ns) | (5 years)<br>2.6<br>3.9 (ns) | (5 years)<br>1.9<br>5.2 (ns) | (5 years)<br>32.1<br>23.5 |
| MRC RT01[34] $n = 843$ | T1b–T3aN0M0; 100% received ADT | 10.0 | 64<br>74 | (10 years)<br>29<br>29 (ns) | (10 years)<br>11<br>11 (ns) | (10 years)<br>12<br>12 (ns) | (10 years)<br>57<br>45 |

Gy, Gray; NA, not available; NS, not statistically significant; PSA, prostate-specific antigen. All results significant unless otherwise noted.

Two retrospective studies support the addition of ADT to dose-escalated EBRT. Zumsteg et al. examined a single-institution cohort of 710 patients with intermediate-risk prostate cancer who received EBRT at doses of ≥81 Gy, including 357 who also received short-term ADT.[37] At a median follow-up of 7.9 years, patients who received ADT had significantly lower 10-year rates of biochemical recurrence (80.0% EBRT alone vs. 67.5% EBRT/ADT, $p = 0.003$), distant metastasis (12.3% EBRT alone vs. 6.5% EBRT/ADT, $p = 0.011$), and prostate-cancer-specific mortality (5.0% EBRT alone vs. 2.4% EBRT/ADT, $p = 0.032$). These results mirror those obtained in another retrospective study by Castle et al. that included 544 patients with intermediate-risk disease who received dose-escalated EBRT with or without ADT.[38] In this study, the authors separately analyzed patients with "favorable" intermediate-risk (Gleason score 3 + 4, T2a-b) and "unfavorable" intermediate-risk (Gleason score 4 + 3, T2c) disease. The addition of ADT to dose-escalated EBRT decreased biochemical recurrence in the unfavorable group (5-year rate 26% EBRT alone vs. 6% EBRT/ADT, $p = 0.005$), but no benefit was seen in the favorable group (5% EBRT alone vs. 6% EBRT/ADT, $p = 0.85$). Taken together, these data from nonrandomized trials support the addition of ADT to dose-escalated EBRT, although the Castle study suggests that for a select group of patients with favorable intermediate-risk disease, ADT could potentially be avoided. Results from ongoing randomized trials EORTC 22991 and RTOG 08-15 will more definitively answer this question.

## ANDROGEN DEPRIVATION THERAPY AND RADIOTHERAPY AFTER PROSTATECTOMY

For patients who received prostatectomy and subsequently develop a biochemical recurrence, EBRT is a standard salvage treatment that offers the patient a second chance for cure.[39] Whether adding ADT to salvage EBRT improves patient outcomes has not been definitively answered, and a recent population-based patterns of care study showed that 43% of patients who receive postprostatectomy radiation also receive ADT.[40]

RTOG 96-01 was a phase III trial that randomized 771 postprostatectomy patients with biochemical recurrence to either salvage EBRT alone or salvage EBRT plus 24 months of bicalutamide (150 mg/day). Results from this trial have been presented in abstract form but not yet published.[41] The addition of bicalutamide was associated with an increased 7-year rate of freedom from PSA progression (40% EBRT alone vs. 57% EBRT/ADT, $p < 0.0001$) and a decreased 7-year cumulative incidence of distant metastasis (12.6% EBRT alone vs. 7.4% EBRT/ADT, $p = 0.04$). These results suggest a benefit from adding ADT to salvage radiation therapy, but bicalutamide alone with EBRT is not commonly used in modern practice, nor is the dose used in that trial (150 mg). A currently ongoing trial, Radiotherapy and Androgen Deprivation in Combination after Local Surgery (RADICALS, NCT00541047) will further help elucidate the potential benefit of adding ADT to salvage EBRT.[42] This study uses a two-step randomization of postprostatectomy patients. First, patients will be randomized to either immediate postoperative adjuvant radiation therapy or salvage radiation therapy upon biochemical recurrence. Then, at the time of receiving radiation therapy, patients will be randomized to EBRT alone, EBRT plus 6 months ADT (GnRH agonist +/− antiandrogen), or EBRT plus 2 years ADT.

## SIDE EFFECTS OF ANDROGEN DEPRIVATION THERAPY

While the disease-control and survival benefits from the addition of ADT to radiotherapy have been repeatedly demonstrated in randomized trials, these benefits should be weighed against the side effects of ADT.

### ADT and Cardiovascular Risk

In recent years, a large body of literature has been published examining the metabolic changes that occur during and after ADT and the possible increase in risk of mortality from cardiovascular events as a result. It is well-established that ADT is associated with an increased risk of insulin resistance,[43,44] diabetes,[45–47] dyslipidemia;[48] decreased lean body mass; and increased obesity and fat mass.[48–51] Whether these changes in metabolism and cardiovascular risk factors translate into an increase in mortality from cardiovascular events is unclear.

Multiple retrospective studies of large patient registries have found an association between ADT and a modest increase in cardiovascular events and mortality. This link has been demonstrated in studies using data from Surveillance, Epidemiology, and End Results-Medicare (SEER-Medicare),[45,52] Veteran's Administration,[46] and the Cancer of the Prostate Strategic Urologic Research Endeavor (CaPSURE) cohort (Table 43.5).[53] Keating et al. examined a cohort of 73,196 prostate cancer patients in SEER-Medicare.[45] Patients who received GnRH agonist therapy were at an increased risk of diabetes (adjusted hazard ratio [AHR] 1.44, $p < 0.001$), coronary heart disease (AHR 1.16, $p < 0.001$), myocardial infarction (AHR 1.11, $p = 0.03$), and sudden cardiac death (AHR 1.16, $p = 0.004$). A separate analysis of 22,816 prostate cancer patients in SEER-Medicare by Saigal et al. demonstrated a 20% relative increase in cardiovascular morbidity with ADT as measured by an index abstracted from Medicare claims data.[52] These findings were replicated in a study of 37,443 prostate cancer patients treated by the Veterans Administration, which showed that treatment with a GnRH agonist was associated with increased risk of diabetes (AHR 1.28, 95% confidence interval [CI] 1.19–1.38), coronary heart disease (AHR 1.19, 95% CI 1.10–1.28), myocardial infarction (AHR 1.28, 95% CI 1.08–1.52), sudden cardiac death (AHR 1.35, 95% CI 1.18–1.54), and stroke (AHR 1.21, 95% CI 1.05–1.40).[46] Similar results were obtained in a retrospective study of 4892 patients in the CaPSURE registry who underwent either radical prostatectomy (66.7%) or EBRT (33.3%).[53] Among these patients, receipt of ADT was associated with an increased 5-year incidence of death from cardiovascular causes (2.0% vs. 5.5% for patients who received ADT, $p = 0.002$).

Not all studies, however, have found this association. Secondary analyses of the RTOG 86-10, RTOG 85-31, and TROG 96.01 randomized trials of EBRT versus EBRT plus ADT showed no significant association between ADT and cardiovascular disease. In the RTOG 86-10 trial, the 10-year incidence of fatal cardiac events was 9.1% in the EBRT arm versus 12.5% in the ADT plus EBRT arm ($p = 0.32$).[16] Similar results were reported in TROG 96.01 (10-year, 7.5% EBRT alone vs. 6.4% EBRT/ADT, $p = 0.65$) and RTOG 85-31 (9-year, 11.4% EBRT alone vs. 8.4% EBRT/ADT, $p = 0.17$).[41,54] A meta-analysis of 4141 patients in eight randomized trials of EBRT or radical prostatectomy with versus without ADT in nonmetastatic prostate cancer found no difference in cardiovascular events between those patients receiving ADT versus not (11.0% ADT vs. 11.2% no-ADT, $p = 0.41$).[55] In addition, two studies have shown no increase in cardiovascular mortality in prostate cancer patients who received EBRT with short- versus long-term ADT based on secondary analyses of the EORTC 22961 (5-year fatal cardiac event 4.0% short-term vs. 3% long-term) and RTOG 92-02 (5-year cardiovascular mortality 4.8% short-term vs. 5.9% long-term, $p = 0.16$) trials.[23,56]

It is possible that the association between ADT and cardiovascular risk occurs in certain patient subgroups only. This hypothesis is supported by a secondary analysis of data from the Harvard randomized trial (EBRT vs. EBRT plus 6 months ADT), which showed that the apparent survival benefit from ADT was present only in the subgroup of patients with little or no comorbidity at baseline.[8] Similarly, a retrospective study by Nanda et al. of 5077 patients treated with brachytherapy with and without ADT found an increase in cardiovascular mortality in patients treated with ADT only in the subgroup of patients with preexisting cardiovascular disease.[57] These two studies generate interesting hypotheses about potential patient subgroups that may not benefit from the addition of ADT to radiation, but further prospective studies are needed to more definitively address this issue.

Given the available literature on the benefits and potential cardiovascular effects of ADT, the American Heart Association, American Cancer Society, and American Urological Association published a joint statement, recommending "the treating physician weigh the potential risks and benefits of ADT in each patient's specific clinical scenario. For patients with aggressive prostate cancer in whom the addition of ADT is necessary, no further evaluation by an internist, cardiologist, or endocrinologist is

TABLE 43.5 Studies Which Examined the Association Between Androgen Deprivation Therapy and Cardiovascular Risk

| | Data source | Patients | Results |
|---|---|---|---|
| Studies which showed an association between ADT and cardiovascular risk | | | |
| Keating et al.[45] $n = 73{,}196$ | SEER-Medicare | Nonmetastatic prostate cancer; 36.3% received GnRH agonist | GnRH agonist associated with increased risk of: DM (AHR 1.44, $p < 0.001$), CHD (AHR 1.16, $p < 0.001$), MI (AHR 1.11, $p = 0.03$), and sudden cardiac death (AHR 1.16, $p = 0.004$) |
| Saigal et al.[52] $n = 22{,}816$ | SEER-Medicare | All prostate cancer; 21% received ADT | ADT associated with 20% increase in cardiovascular morbidity |
| Keating et al.[46] $n = 37{,}443$ | Veterans administration | Nonmetastatic prostate cancer; 39% received GnRH agonist | GnRH agonist associated with increased risk of: DM (AHR 1.28, 95% CI 1.19–1.38), CHD (AHR 1.19, 95% CI 1.10–1.28), MI (AHR 1.28, 95% CI 1.08–1.52), sudden cardiac death (AHR 1.35, 95% CI 1.18–1.54), and stroke (AHR1.21, 95% CI 1.05–1.40) |
| Tsai et al.[53] $n = 4892$ | CaPSURE | Nonmetastatic prostate cancer who underwent RP (66.7%) or EBRT (33.3%); 20.7% received ADT | ADT associated with increased 5-year risk of death from cardiovascular causes, 2.0% versus 5.0% ADT ($p = 0.002$) |
| Studies which showed no association between ADT and cardiovascular risk | | | |
| RTOG 86-10[16] $n = 456$ | Randomized trial (secondary data analysis) | T2-T4, bulky (5 cm × 5 cm) primary tumor | No significant association between ADT and cardiovascular mortality: 10-year incidence of fatal cardiac events 9.1% EBRT alone versus 12.5% EBRT/ADT ($p = 0.32$) |
| TROG 96.10[14,41] $n = 818$ | Randomized trial (secondary data analysis) | Intermediate- (16%), high-risk (44%), and locally advanced (40%) | No significant association between ADT and cardiovascular mortality: 10-year incidence of fatal cardiac events 7.5% EBRT alone versus 6.4% EBRT/ADT ($p = 0.65$) |
| RTOG 85-31[19,54] $n = 977$ | Randomized trial (secondary data analysis) | cT3 or any N1, M0 | No significant association between ADT and cardiovascular mortality: 9-year incidence of fatal cardiac events 11.4% EBRT alone versus 8.4% EBRT/ADT ($p = 0.17$) |
| Nguyen et al.[55] $n = 4141$ | Meta-analysis of 8 randomized trials of ERBT or RP +/− ADT | Nonmetastatic prostate cancer | No difference in cardiovascular events between those receiving ADT versus not: 11% ADT versus 11.2% no-ADT ($p = 0.41$) |
| EORTC 22961[23] $n = 970$ | Randomized trial (secondary data analysis) | 91% T2c-T4 N0 | No difference in cardiovascular mortality between short- versus long-term ADT (5 years, 4% short-term versus 3% long-term) |
| RTOG-9202[25,56] $n = 1554$ | Randomized trial (secondary data analysis) | High-risk (45%), locally-advanced (55%) | No difference in cardiovascular mortality between short- versus long-term ADT (5 years, 4.8% short-term versus 5.9% long-term, $p = 0.16$) |

ADT, androgen deprivation therapy; AHR, adjusted hazard ratio; CaPSURE, Cancer of the Prostate Strategic Urologic Research Endeavor; CHD, coronary heart disease; CI, confidence interval; DM, diabetes mellitus; EBRT, external beam radiation therapy; GnRH, gonadotropin-releasing hormone; MI, myocardial infarction; RP, radical prostatectomy; SEER, Surveillance, Epidemiology, and End Results.

recommended. Patients with cardiac disease should receive appropriate secondary preventive measures (such as statins, aspirin, and antihypertensive medications) and be monitored by their primary care physicians."[58]

## Management of ADT-Associated Side Effects

Other potential side effects of ADT include decreased bone mineral density, hot flashes, fatigue, and sexual dysfunction – these are briefly summarized next.

The long-term use of GnRH agonists is associated with accelerated bone density loss and increased risk of fracture.[59–62] A study of 23 patients with nonmetastatic prostate cancer undergoing GnRH agonist therapy found that femoral neck bone mineral density decreased by an average of 4% during the first 2 years of therapy, and by 2% in each year thereafter, which was significantly greater than the average 0.5% decrease per year in a group of healthy age-matched controls.[59] Two retrospective analyses of SEER-Medicare data also support a link between ADT and an increased fracture risk. In one study, Shahinian et al. examined 50,613 men diagnosed with prostate cancer.[60] Taking only the patients who survived for at least 5 years beyond diagnosis, the authors reported a significantly increased rate of fracture in those who received ADT (19.4% vs. 12.6% for no-ADT, $p < 0.001$). A similar analysis looking only at patients with nonmetastatic prostate cancer also found an increased fracture risk with ADT (7.88 fractures per 100 person-years with ADT vs. 6.51 fractures per 100 person-years no-ADT, $p < 0.001$).[62] Agents that have been studied to combat this decrease in bone mineral density include bisphosphonates; toremifene (a selective estrogen receptor modulator [SERM]); and denosumab, a monoclonal antibody against RANK ligand, a promoter of osteoclast activity. Multiple randomized studies have demonstrated that bisphosphonates reduce or prevent bone mineral density loss with long-term ADT in localized, nonmetastatic prostate cancer.[63–65] Specific agents studied include alendronate (+1.7% change in lumbar spine bone mineral density vs. −1.9% change placebo at 1 year, $p < 0.0001$),[65] zoledronic acid (+6% change in lumbar spine vs. −5% change placebo at 3 years, $p < 0.0001$),[64] and risendronate (−0.9% change in lumbar spine vs. −13.6% change placebo at 2 years, $p = 0.06$).[63] A meta-analysis of 15 studies of bisphosphonates in patients with localized and metastatic prostate cancer demonstrated a decrease in fracture risk (RR 0.80, $p = 0.005$).[66] Denosumab and toremifene are relatively newer agents, and each has a randomized trial showing an increase in bone mineral density with their use in nonmetastatic prostate cancer patients undergoing long-term ADT.[67,68] In a randomized trial of 734 patients, denosumab was found to increase bone mineral density at 2 years by an average of 5.6% in the lumbar spine, compared to an average decrease of 1% in the placebo group ($p < 0.001$).[67] A similar trial examining the effects of toremifene found that in addition to decreasing bone mineral loss relative to placebo, it also decreased the 2-year incidence of new vertebral fractures (2.5% toremifene vs. 4.9% placebo, $p = 0.05$).[68]

Hot flashes are a well-recognized side effect of androgen deprivation, and affect up to 80% of men receiving ADT.[69] Selective serotonin reuptake inhibitors (SSRIs) and serotonin–norepinephrine reuptake inhibitors (SNRIs) have been studied in the treatment of hot flashes in postmenopausal women,[70–72] and several small, nonrandomized pilot studies of these agents in prostate cancer patients receiving ADT showed promise in reducing hot flash intensity and frequency.[73–75] However, these results did not hold up in a randomized trial of prostate cancer patients receiving ADT, as venlafaxine was found to be no more effective than placebo in managing hot flash symptoms as measured by a symptom score consisting of hot flash frequency times severity.[76]

Gabapentin (300 mg three times daily)[77] and pregabalin (75 or 150 mg, twice daily)[78] have also been studied in randomized trials for ADT-associated hot flashes. After four weeks of therapy, gabapentin was found to decrease the mean frequency and severity of hot flashes by 45%, compared to a 22% (frequency, $p = 0.02$) and 27% (severity, $p = 0.10$) reduction with placebo. In another trial, patients receiving pregabalin reported a mean 58.5% (75 mg twice daily) or 61.1% (150 mg twice daily) reduction in hot flash frequency after six weeks of therapy, compared to a 36.3% reduction in the placebo arm ($p = 0.007$ for both doses). A corresponding decrease in hot flash severity was seen as well, with a mean reduction of 64.9% (75 mg twice daily) or 71.0% (150 mg twice daily) compared to a 50.1% reduction in the placebo arm ($p < 0.009$ for both doses). At the conclusion of the study, 74% of patients in the still-blinded 150 mg twice daily arm indicated that they were satisfied with hot flash control, compared to only 33% in the placebo arm ($p = 0.001$).

Fatigue is another common side effect of ADT. In a longitudinal study of 144 men with nonmetastatic prostate cancer, patients' quality of life and fatigue were assessed using a questionnaire. Those who were receiving ADT reported significantly worse fatigue ($p = 0.001$) and quality of life ($p = 0.001$) compared to patients receiving local treatment alone (EBRT or radical prostatectomy) or no treatment.[79] Two different randomized trials have shown that exercise can significantly reduce fatigue associated with ADT. Galvão et al. randomized 57 patients undergoing ADT for prostate cancer to either a 12 week program of aerobic and resistance training or usual care.[80] Patients' quality of life and fatigue were assessed using a validated questionnaire. At the conclusion of 12 weeks, patients in the exercise arm reported a significantly greater reduction in their fatigue score relative to the control group. A similar trial by Segal et al. randomized patients to 12 weeks of resistance training or usual care and tracked changes in patient-reported fatigue and quality of life.[81] Relative to the control arm, patients in the exercise arm had a greater reduction in fatigue and a greater increase in quality of life.

Men receiving ADT also report a high rate of sexual dysfunction, in terms of decreased sexual interest and erectile dysfunction.[82,83] The use of phosphodiesterase type 5 (PDE5) inhibitors can effectively treat erectile dysfunction in men receiving ADT. RTOG 0215 was a clinical trial in which 115 men receiving ADT/EBRT for prostate cancer were randomized to 12 weeks of either sildenafil (a PDE5 inhibitor) or placebo, followed by 1 week of no treatment, and then 12 more weeks of the alternate arm.[84] Sildenafil use was associated with significantly greater patient-reported scores of erection frequency ($p = 0.009$) as assessed by the International Index of Erectile Function scale.[85]

## CONCLUSIONS

Radiation therapy and ADT is a proven combination in the treatment of patients with intermediate-risk, high-risk, and locally advanced prostate cancer. For patients with intermediate-risk disease, randomized trials support the use of EBRT with short-term (6 months) ADT. For patients with high-risk and locally advanced prostate cancer, randomized data support EBRT with long-term (2–3 years) ADT. Ongoing and future trials will provide further insight on the potential benefit of adding ADT in the setting of dose-escalated EBRT, the benefit of ADT for salvage EBRT, and whether in certain patient subgroups (based on disease status or comorbidities) ADT can be avoided without compromising patient outcomes.

## References

1. Conn PM, Crowley WF. Gonadotropin-releasing hormone and its analogues. *N Engl J Med* 1991;**324**(2):93–103.
2. Limonta P, Montagnani Marelli M, Moretti RM. LHRH analogues as anticancer agents: pituitary and extrapituitary sites of action. *Exp Opin Investig Drugs* 2001;**10**(4):709–20.
3. Klotz L, Boccon-Gibod L, Shore ND, et al. The efficacy and safety of degarelix: a 12-month, comparative, randomized, open-label, parallel-group phase III study in patients with prostate cancer. *BJU Int* 2008;**102**(11):1531–8.
4. Widmark A, Klepp O, Solberg A, et al. Endocrine treatment, with or without radiotherapy, in locally advanced prostate cancer (SPCG-7/SFUO-3): an open randomised phase III trial. *Lancet* 2009;**373**(9660):301–8.
5. Gospodarowicz MK, Mason M, Parulekar W, et al. Final analysis of intergroup randomized phase III study of androgen deprivation therapy (ADT) ± radiation therapy (RT) in locally advanced prostate cancer (NCIC-CTG, SWOG, MRC-UK, INT: T94-0110; NCT00002633). *Int J Radiat Oncol Biol Phys* 2012;**84**(3):S4.
6. Mottet N, Peneau M, Mazeron JJ, et al. Addition of radiotherapy to long-term androgen deprivation in locally advanced prostate cancer: an open randomised phase 3 trial. *Eur Urol* 2012;**62**(2):213–9.
7. D'Amico AV, Manola J, Loffredo M, et al. 6-month androgen suppression plus radiation therapy vs radiation therapy alone for patients with clinically localized prostate cancer: a randomized controlled trial. *JAMA* 2004;**292**(7):821–7.
8. D'Amico AV, Chen MH, Renshaw AA, et al. Androgen suppression and radiation vs radiation alone for prostate cancer: a randomized trial. *JAMA* 2008;**299**(3):289–95.
9. Nguyen PL, Chen MH, Beard CJ, et al. Radiation with or without 6 months of androgen suppression therapy in intermediate- and high-risk clinically localized prostate cancer: a postrandomization analysis by risk group. *Int J Radiat Oncol Biol Phys* 2010;**77**(4):1046–52.
10. Jones CU, Hunt D, McGowan DG, et al. Radiotherapy and short-term androgen deprivation for localized prostate cancer. *N Engl J Med* 2011;**365**(2):107–18.
11. D'Amico AV, Chen MH, Crook J, et al. Duration of short-course androgen suppression therapy and the risk of death as a result of prostate cancer. *J Clin Oncol* 2011;**29**(35):4682–7.
12. Denham JW, Steigler A, Lamb DS, et al. Short-term androgen deprivation and radiotherapy for locally advanced prostate cancer: results from the Trans-Tasman Radiation Oncology Group 96.01 randomised controlled trial. *Lancet Oncol* 2005;**6**(11):841–50.
13. Armstrong JG, Gillham CM, Dunne MT, et al. A randomized trial (Irish clinical oncology research group 97-01) comparing short versus protracted neoadjuvant hormonal therapy before radiotherapy for localized prostate cancer. *Int J Radiat Oncol Biol Phys* 2011;**81**(1):35–45.
14. Denham JW, Steigler A, Lamb DS, et al. Short-term neoadjuvant androgen deprivation and radiotherapy for locally advanced prostate cancer: 10-year data from the TROG 96.01 randomised trial. *Lancet Oncol* 2011;**12**(5):451–9.
15. Pilepich MV, Winter K, John MJ, et al. Phase III radiation therapy oncology group (RTOG) trial 86-10 of androgen deprivation adjuvant to definitive radiotherapy in locally advanced carcinoma of the prostate. *Int J Radiat Oncol Biol Phys* 2001;**50**(5):1243–52.
16. Roach M, Bae K, Speight J, et al. Short-term neoadjuvant androgen deprivation therapy and external-beam radiotherapy for locally advanced prostate cancer: long-term results of RTOG 8610. *J Clin Oncol* 2008;**26**(4):585–91.
17. Pilepich MV, Caplan R, Byhardt RW, et al. Phase III trial of androgen suppression using goserelin in unfavorable-prognosis carcinoma of the prostate treated with definitive radiotherapy: report of Radiation Therapy Oncology Group Protocol 85-31. *J Clin Oncol* 1997;**15**(3):1013–21.
18. Lawton CA, Winter K, Murray K, et al. Updated results of the phase III Radiation Therapy Oncology Group (RTOG) trial 85-31 evaluating the potential benefit of androgen suppression following standard radiation therapy for unfavorable prognosis carcinoma of the prostate. *Int J Radiat Oncol Biol Phys* 2001;**49**(4):937–46.
19. Pilepich MV, Winter K, Lawton CA, et al. Androgen suppression adjuvant to definitive radiotherapy in prostate carcinoma – long-term results of phase III RTOG 85-31. *Int J Radiat Oncol Biol Phys* 2005;**61**(5):1285–90.
20. Souhami L, Bae K, Pilepich M, et al. Impact of the duration of adjuvant hormonal therapy in patients with locally advanced prostate cancer treated with radiotherapy: a secondary analysis of RTOG 85-31. *J Clin Oncol* 2009;**27**(13):2137–43.
21. Bolla M, Gonzalez D, Warde P, et al. Improved survival in patients with locally advanced prostate cancer treated with radiotherapy and goserelin. *N Engl J Med* 1997;**337**(5):295–300.
22. Bolla M, Collette L, Blank L, et al. Long-term results with immediate androgen suppression and external irradiation in patients with locally advanced prostate cancer (an EORTC study): a phase III randomised trial. *Lancet* 2002;**360**(9327):103–6.
23. Bolla M, de Reijke TM, Van Tienhoven G, et al. Duration of androgen suppression in the treatment of prostate cancer. *N Engl J Med* 2009;**360**(24):2516–27.
24. Hanks GE, Pajak TF, Porter A, et al. Phase III trial of long-term adjuvant androgen deprivation after neoadjuvant hormonal cytoreduction and radiotherapy in locally advanced carcinoma of the prostate: the Radiation Therapy Oncology Group Protocol 92-02. *J Clin Oncol* 2003;**21**(21):3972–8.
25. Horwitz EM, Bae K, Hanks GE, et al. Ten-year follow-up of radiation therapy oncology group protocol 92-02: a phase III trial of the duration of elective androgen deprivation in locally advanced prostate cancer. *J Clin Oncol* 2008;**26**(15):2497–504.
26. Peeters ST, Heemsbergen WD, Koper PC, et al. Dose–response in radiotherapy for localized prostate cancer: results of the Dutch multicenter randomized phase III trial comparing 68 Gy of radiotherapy with 78 Gy. *J Clin Oncol* 2006;**24**(13):1990–6.
27. Al-Mamgani A, van Putten WL, Heemsbergen WD, et al. Update of Dutch multicenter dose-escalation trial of radiotherapy for localized prostate cancer. *Int J Radiat Oncol Biol Phys* 2008;**72**(4):980–8.
28. Heemsbergen WD, Al-Mamgani A, Slot A, et al. Long-term results of the Dutch randomized prostate cancer trial: impact of dose-escalation on local, biochemical, clinical failure, and survival. *Radiother Oncol* 2014;**110**(1):104–9.

29. Pollack A, Zagars GK, Smith LG, et al. Preliminary results of a randomized radiotherapy dose-escalation study comparing 70 Gy with 78 Gy for prostate cancer. *J Clin Oncol* 2000;**18**(23):3904–11.
30. Kuban DA, Tucker SL, Dong L, et al. Long-term results of the M. D. Anderson randomized dose-escalation trial for prostate cancer. *Int J Radiat Oncol Biol Phys* 2008;**70**(1):67–74.
31. Zietman AL, DeSilvio ML, Slater JD, et al. Comparison of conventional-dose vs high-dose conformal radiation therapy in clinically localized adenocarcinoma of the prostate: a randomized controlled trial. *JAMA* 2005;**294**(10):1233–9.
32. Zietman AL, Bae K, Slater JD, et al. Randomized trial comparing conventional-dose with high-dose conformal radiation therapy in early-stage adenocarcinoma of the prostate: long-term results from Proton Radiation Oncology Group/American College Of Radiology 95-09. *J Clin Oncol* 2010;**28**(7):1106–11.
33. Beckendorf V, Guerif S, Le Prisé E, et al. 70 Gy versus 80 Gy in localized prostate cancer: 5-year results of GETUG 06 randomized trial. *Int J Radiat Oncol Biol Phys* 2011;**80**(4):1056–63.
34. Dearnaley DP, Jovic G, Syndikus I, et al. Escalated-dose versus control-dose conformal radiotherapy for prostate cancer: long-term results from the MRC RT01 randomised controlled trial. *Lancet Oncol* 2014;**15**(4):464–73.
35. Mohler JL, Kantoff PW, Armstrong AJ, et al. Prostate cancer, version 2.2014. *J Natl Compr Canc Netw* 2014;**12**(5):686–718.
36. Zapatero A, Valcárcel F, Calvo FA, et al. Risk-adapted androgen deprivation and escalated three-dimensional conformal radiotherapy for prostate cancer: Does radiation dose influence outcome of patients treated with adjuvant androgen deprivation? A GICOR study. *J Clin Oncol* 2005;**23**(27):6561–8.
37. Zumsteg ZS, Spratt DE, Pei X, et al. Short-term androgen-deprivation therapy improves prostate cancer-specific mortality in intermediate-risk prostate cancer patients undergoing dose-escalated external beam radiation therapy. *Int J Radiat Oncol Biol Phys* 2013;**85**(4):1012–7.
38. Castle KO, Hoffman KE, Levy LB, et al. Is androgen deprivation therapy necessary in all intermediate-risk prostate cancer patients treated in the dose escalation era? *Int J Radiat Oncol Biol Phys* 2013;**85**(3):693–9.
39. Thompson IM, Valicenti RK, Albertsen P, et al. Adjuvant and salvage radiotherapy after prostatectomy: AUA/ASTRO guideline. *J Urol* 2013;**190**(2):441–9.
40. Sheets NC, Hendrix LH, Allen IM, et al. Trends in the use of post-prostatectomy therapies for patients with prostate cancer: a surveillance, epidemiology, and end results Medicare analysis. *Cancer* 2013;**119**(18):3295–301.
41. Wilcox C, Kautto A, Steigler A, et al. Androgen deprivation therapy for prostate cancer does not increase cardiovascular mortality in the long term. *Oncology* 2012;**82**(1):56–8.
42. Parker C, Clarke N, Logue J, et al. RADICALS (Radiotherapy and Androgen Deprivation in Combination after Local Surgery). *Clin Oncol (R Coll Radiol)* 2007;**19**(3):167–71.
43. Dockery F, Bulpitt CJ, Agarwal S, et al. Testosterone suppression in men with prostate cancer leads to an increase in arterial stiffness and hyperinsulinaemia. *Clin Sci (Lond)* 2003;**104**(2):195–201.
44. Smith MR, Lee H, Nathan DM. Insulin sensitivity during combined androgen blockade for prostate cancer. *J Clin Endocrinol Metab* 2006;**91**(4):1305–8.
45. Keating NL, O'Malley AJ, Smith MR. Diabetes and cardiovascular disease during androgen deprivation therapy for prostate cancer. *J Clin Oncol* 2006;**24**(27):4448–56.
46. Keating NL, O'Malley AJ, Freedland SJ, et al. Diabetes and cardiovascular disease during androgen deprivation therapy: observational study of veterans with prostate cancer. *J Natl Cancer Inst* 2010;**102**(1):39–46.
47. Alibhai SM, Duong-Hua M, Sutradhar R, et al. Impact of androgen deprivation therapy on cardiovascular disease and diabetes. *J Clin Oncol* 2009;**27**(21):3452–8.
48. Smith MR, Finkelstein JS, McGovern FJ, et al. Changes in body composition during androgen deprivation therapy for prostate cancer. *J Clin Endocrinol Metab* 2002;**87**(2):599–603.
49. Smith MR. Changes in fat and lean body mass during androgen-deprivation therapy for prostate cancer. *Urology* 2004;**63**(4): 742–5.
50. Smith MR, Lee H, McGovern F, et al. Metabolic changes during gonadotropin-releasing hormone agonist therapy for prostate cancer: differences from the classic metabolic syndrome. *Cancer* 2008;**112**(10):2188–94.
51. Smith MR, Saad F, Egerdie B, et al. Sarcopenia during androgen-deprivation therapy for prostate cancer. *J Clin Oncol* 2012;**30**(26):3271–6.
52. Saigal CS, Gore JL, Krupski TL, et al. Androgen deprivation therapy increases cardiovascular morbidity in men with prostate cancer. *Cancer* 2007;**110**(7):1493–500.
53. Tsai HK, D'Amico AV, Sadetsky N, et al. Androgen deprivation therapy for localized prostate cancer and the risk of cardiovascular mortality. *J Natl Cancer Inst* 2007;**99**(20):1516–24.
54. Efstathiou JA, Bae K, Shipley WU, et al. Cardiovascular mortality after androgen deprivation therapy for locally advanced prostate cancer: RTOG 85-31. *J Clin Oncol* 2009;**27**(1):92–9.
55. Nguyen PL, Je Y, Schutz FA, et al. Association of androgen deprivation therapy with cardiovascular death in patients with prostate cancer: a meta-analysis of randomized trials. *JAMA* 2011;**306**(21):2359–66.
56. Efstathiou JA, Bae K, Shipley WU, et al. Cardiovascular mortality and duration of androgen deprivation for locally advanced prostate cancer: analysis of RTOG 92-02. *Eur Urol* 2008;**54**(4):816–23.
57. Nanda A, Chen MH, Braccioforte MH, et al. Hormonal therapy use for prostate cancer and mortality in men with coronary artery disease-induced congestive heart failure or myocardial infarction. *JAMA* 2009;**302**(8):866–73.
58. Levine GN, D'Amico AV, Berger P, et al. Androgen-deprivation therapy in prostate cancer and cardiovascular risk: a science advisory from the American Heart Association, American Cancer Society, and American Urological Association: endorsed by the American Society for Radiation Oncology. *CA Cancer J Clin* 2010;**60**(3):194–201.
59. Daniell HW, Dunn SR, Ferguson DW, et al. Progressive osteoporosis during androgen deprivation therapy for prostate cancer. *J Urol* 2000;**163**(1):181–6.
60. Shahinian VB, Kuo YF, Freeman JL, et al. Risk of fracture after androgen deprivation for prostate cancer. *N Engl J Med* 2005;**352**(2):154–64.
61. Maillefert JF, Sibilia J, Michel F, et al. Bone mineral density in men treated with synthetic gonadotropin-releasing hormone agonists for prostatic carcinoma. *J Urol* 1999;**161**(4):1219–22.
62. Smith MR, Lee WC, Brandman J, et al. Gonadotropin-releasing hormone agonists and fracture risk: a claims-based cohort study of men with nonmetastatic prostate cancer. *J Clin Oncol* 2005;**23**(31):7897–903.
63. Choo R, Lukka H, Cheung P, et al. Randomized, double-blinded, placebo-controlled, trial of risedronate for the prevention of bone mineral density loss in nonmetastatic prostate cancer patients receiving radiation therapy plus androgen deprivation therapy. *Int J Radiat Oncol Biol Phys* 2013;**85**(5):1239–45.
64. Kachnic LA, Pugh SL, Tai P, et al. RTOG 0518: randomized phase III trial to evaluate zoledronic acid for prevention of osteoporosis and associated fractures in prostate cancer patients. *Prostate Cancer Prostatic Dis* 2013;**16**(4):382–6.
65. Klotz LH, McNeill IY, Kebabdjian M, et al. A phase 3, double-blind, randomised, parallel-group, placebo-controlled study of oral weekly alendronate for the prevention of androgen deprivation bone loss in nonmetastatic prostate cancer: the Cancer and

Osteoporosis Research with Alendronate and Leuprolide (CORAL) study. *Eur Urol* 2013;**63**(5):927–35.
66. Serpa Neto A, Tobias-Machado M, Esteves MA, et al. Bisphosphonate therapy in patients under androgen deprivation therapy for prostate cancer: a systematic review and meta-analysis. *Prostate Cancer Prostatic Dis* 2012;**15**(1):36–44.
67. Smith MR, Egerdie B, Hernández Toriz N, et al. Denosumab in men receiving androgen-deprivation therapy for prostate cancer. *N Engl J Med* 2009;**361**(8):745–55.
68. Smith MR, Morton RA, Barnette KG, et al. Toremifene to reduce fracture risk in men receiving androgen deprivation therapy for prostate cancer. *J Urol* 2013;**189**(1 Suppl.):S45–50.
69. Schow DA, Renfer LG, Rozanski TA, et al. Prevalence of hot flushes during and after neoadjuvant hormonal therapy for localized prostate cancer. *South Med J* 1998;**91**(9):855–7.
70. Stearns V, Beebe KL, Iyengar M, et al. Paroxetine controlled release in the treatment of menopausal hot flashes: a randomized controlled trial. *JAMA* 2003;**289**(21):2827–34.
71. Simon JA, Portman DJ, Kaunitz AM, et al. Low-dose paroxetine 7.5 mg for menopausal vasomotor symptoms: two randomized controlled trials. *Menopause* 2013;**20**(10):1027–35.
72. Pinkerton JV, Constantine G, Hwang E, et al. Desvenlafaxine compared with placebo for treatment of menopausal vasomotor symptoms: a 12-week, multicenter, parallel-group, randomized, double-blind, placebo-controlled efficacy trial. *Menopause* 2013;**20**(1):28–37.
73. Quella SK, Loprinzi CL, Sloan J, et al. Pilot evaluation of venlafaxine for the treatment of hot flashes in men undergoing androgen ablation therapy for prostate cancer. *J Urol* 1999;**162**(1):98–102.
74. Loprinzi CL, Barton DL, Carpenter LA, et al. Pilot evaluation of paroxetine for treating hot flashes in men. *Mayo Clin Proc* 2004;**79**(10):1247–51.
75. Naoe M, Ogawa Y, Shichijo T, et al. Pilot evaluation of selective serotonin reuptake inhibitor antidepressants in hot flash patients under androgen-deprivation therapy for prostate cancer. *Prostate Cancer Prostatic Dis* 2006;**9**(3):275–8.
76. Vitolins MZ, Griffin L, Tomlinson WV, et al. Randomized trial to assess the impact of venlafaxine and soy protein on hot flashes and quality of life in men with prostate cancer. *J Clin Oncol* 2013;**31**(32):4092–8.
77. Loprinzi CL, Dueck AC, Khoyratty BS, et al. A phase III randomized, double-blind, placebo-controlled trial of gabapentin in the management of hot flashes in men (N00CB). *Ann Oncol* 2009;**20**(3):542–9.
78. Loprinzi CL, Qin R, Balcueva EP, et al. Phase III, randomized, double-blind, placebo-controlled evaluation of pregabalin for alleviating hot flashes, N07C1. *J Clin Oncol* 2010;**28**(4):641–7.
79. Herr HW, O'Sullivan M. Quality of life of asymptomatic men with nonmetastatic prostate cancer on androgen deprivation therapy. *J Urol* 2000;**163**(6):1743–6.
80. Galvão DA, Taaffe DR, Spry N, et al. Combined resistance and aerobic exercise program reverses muscle loss in men undergoing androgen suppression therapy for prostate cancer without bone metastases: a randomized controlled trial. *J Clin Oncol* 2010;**28**(2):340–7.
81. Segal RJ, Reid RD, Courneya KS, et al. Resistance exercise in men receiving androgen deprivation therapy for prostate cancer. *J Clin Oncol* 2003;**21**(9):1653–9.
82. Potosky AL, Knopf K, Clegg LX, et al. Quality-of-life outcomes after primary androgen deprivation therapy: results from the Prostate Cancer Outcomes Study. *J Clin Oncol* 2001;**19**(17):3750–7.
83. Lubeck DP, Grossfeld GD, Carroll PR. The effect of androgen deprivation therapy on health-related quality of life in men with prostate cancer. *Urology* 2001;**58**(2 Suppl. 1):94–100.
84. Watkins Bruner D, James JL, Bryan CJ, et al. Randomized, double-blinded, placebo-controlled crossover trial of treating erectile dysfunction with sildenafil after radiotherapy and short-term androgen deprivation therapy: results of RTOG 0215. *J Sex Med* 2011;**8**(4):1228–38.
85. Rosen RC, Riley A, Wagner G, et al. The international index of erectile function (IIEF): a multidimensional scale for assessment of erectile dysfunction. *Urology* 1997;**49**(6):822–30.

CHAPTER

# 44

# Brachytherapy for Prostate Cancer: An Overview

*Nicholas G. Zaorsky, MD, Eric M. Horwitz, MD*

**Department of Radiation Oncology, Fox Chase Cancer Center, Philadelphia, PA, USA**

## INTRODUCTION

Patients with localized (i.e., T1–T2) prostate cancer (PCa) have better outcomes if there is local tumor control, even in the presence of high-risk features (e.g., prostate-specific antigen, PSA >20 ng/mL; Gleason score, GS 8–10).[1] Treatment options for localized PCa typically include radical prostatectomy (RP)[2] and radiation therapy, which is delivered either as external beam radiation therapy (EBRT; typically, dose-escalated conventionally fractionated RT (CFRT)[3–7]) or brachytherapy (BT). In this chapter, we discuss the use of BT for PCa.

Briefly, low dose rate (LDR; defined as ≤2 Gy/h, and typically much less in practice) BT for PCa consists of the permanent deposition of sealed sources (i.e., "seeds") in the prostate including the tumor.[8] High dose rate (HDR; defined as ≥12 Gy/h) BT consists of temporary robotic source insertion and computer guidance to optimize dose distribution.[9] EBRT (to a lower dose than dose-escalated CFRT; or as hypofractionated RT (HFRT)) combined with either type of BT (i.e., "LDR-BT boost," "HDR-BT boost," respectively) is hypothesized to further improve local control and patient outcomes among certain intermediate- and high-risk patients. For reference, the typical fractionation schedules of EBRT (both CFRT and HFRT), LDR-BT, HDR-BT, and BT boost are illustrated in Figure 44.1.

In this chapter, we discuss the history of BT, the technical aspects and sequencing of BT (with respect to EBRT), the clinical outcomes of BT, and the follow-up after BT. Within each section, we address LDR-BT and HDR-BT separately. Although there are many standard treatment options for PCa, randomized clinical trials to define the optimal therapy for patients with localized disease are limited. BT and BT boost may be used as first-line therapies in the management of PCa patients of all National Comprehensive Cancer Network (NCCN)-defined risk groups, as listed in Table 44.1. Before treatment, all patients require a biopsy Gleason score, pretherapy serum PSA, and clinical tumor classification,[12] as these prognostic factors determine low-, intermediate-, or high-risk classification. The contraindications to BT are listed in Table 44.2.[8,9] Androgen deprivation therapy (ADT) may be used with either form of BT, in certain intermediate- and high-risk patients (also listed in Table 44.1).[16,17]

## BRIEF HISTORICAL BACKGROUND OF PROSTATE BRACHYTHERAPY

### LDR-BT History

In 1917, the first report of brachytherapy for PCa was published.[18] In the 1950s, LDR-BT was performed with $^{198}$Au.[19] By the 1970s, $^{125}$I seeds were used for prostate implants.[20] However, the dose distribution of these isotopes and clinical outcomes were not ideal.[21] The application of a transrectal ultrasound (TRUS) to guide LDR-BT was successful in Denmark[22] and the United States[23] in the 1980s.[24] TRUS-guided LDR-BT became a standard technique in the United States in the 1990s because it had improved outcomes and was less invasive when compared to laparotomy-based approaches.[25,10] LDR-BT has been endorsed for low-risk PCa by organizations including the National Cancer Institute (NCI),[26] the American Cancer Society (ACS), the NCCN,[27] and the American Urologic Association (AUA.)[28]

The disadvantages of LDR-BT include possible seed migration (for loose seeds), high dependence on the operator skill for proper seed placement, permanent deposition of radioactive material in the body, inability to adjust seeds once they are deposited, inability to deliver

**Prostate Cancer. http://dx.doi.org/10.1016/B978-0-12-800077-9.00044-X**

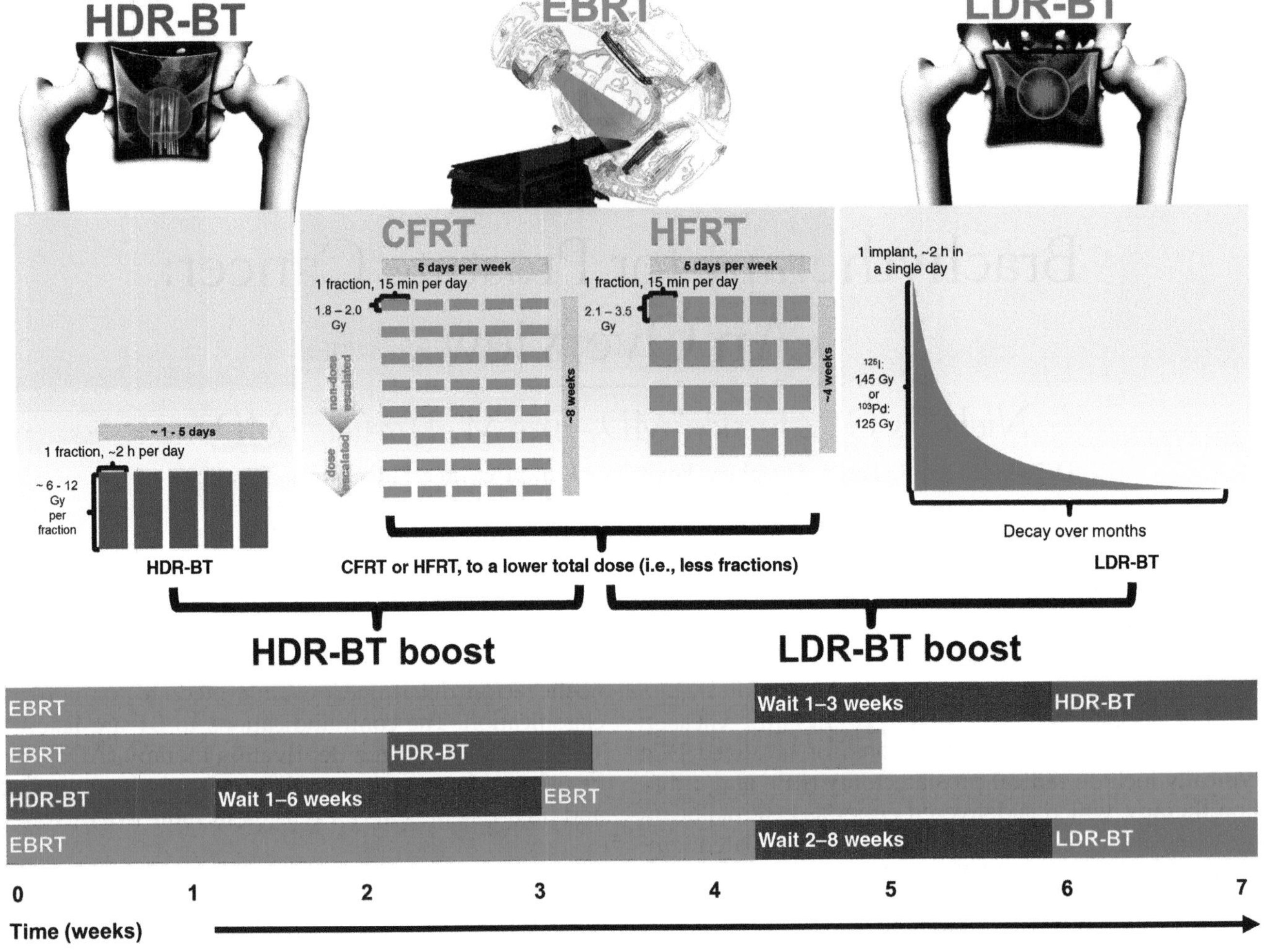


FIGURE 44.1 **An Illustration of HDR-BT schedules, two EBRT schedules, and an LDR-BT schedule; and the BT boost counterparts.** (Top panel) BT is a form of radiotherapy where a radiation source is placed inside or next to the area requiring treatment. For prostate cancer, BT is typically given as either HDR-BT delivered using a RALS) or LDR-BT (delivered using permanently implanted radioactive seeds). HDR-BT monotherapy is typically delivered in 1–5 fractions. LDR-BT monotherapy consists of one implant session. EBRT is a form of radiation therapy where the patient sits or lies on a couch and an external source of radiation is pointed at the cancer. It includes CFRT and HFRT. (Bottom panel) BT boost schedules have various temporal arrangements. When BT boost is used, the total dose of EBRT is typically lower than if it were to be used as a monotherapy. *Reprinted with permission from Zaorsky et al.*[10,11]

high doses over a short period of time, variability of dosimetry among implants, and exposure of the staff to radiation.[24] These disadvantages led physicians to explore other BT systems (i.e., HDR-BT) (Table 44.3.).

## HDR-BT History

A TRUS-guided remote afterloading system (RALS) was first introduced in 1980 to deliver a high radiation dose to the prostate and helped address some limitations of LDR-BT.[10] HDR-BT was first used as a boost with EBRT in Sweden, Germany, and the United States in the 1980s and 1990s.[10] HDR-BT was shown to be safe and effective in Phase I/II trials. In Osaka, Japan, a systematic treatment series of HDR-BT boost was started in 1994. After a year of using HDR-BT boost, a trial of HDR-BT monotherapy was initiated. Subsequently, the Osaka group published the first report on HDR-BT monotherapy in 2000.[29] In the United States, the group led by Alvaro Martinez at William Beaumont Hospital reported the results from a series of prospective trials of first HDR-BT boost and then HDR-BT monotherapy over a period of more than 20 years.[30,31]

# TECHNICAL ASPECTS AND SEQUENCING OF BRACHYTHERAPY FOR PROSTATE CANCER

## LDR-BT Technical Aspects

As of 2014, LDR-BT is typically performed with $^{125}$I or $^{103}$Pd; few centers use $^{131}$Cs. The details of these radionuclides and comparison to $^{192}$Ir (used in HDR-BT) are listed in Table 44.4. The standard procedure for

**TABLE 44.1** BT as a Treatment Option for Men with NCCN Risk Group-Stratified Prostate Cancer, in Relation to Other Treatment Options

| Options and subtypes | | NCCN risk group: Low Gleason score ≤6, and PSA <10 ng/mL, and clinical tumor classification, T1, T2a | NCCN risk group: Intermediate Gleason score 7, or, PSA > 10 ng/mL < 20 ng/mL, or clinical tumor classification of T2b, T2c | NCCN risk group: High Gleason score 8–10, or, PSA >20 ng/mL, or clinical tumor classification of T3a |
|---|---|---|---|---|
| RP[2] | Open, laparoscopic, robotic approaches | Monotherapy | Monotherapy | Monotherapy |
| BT | LDR | Monotherapy or boost[14,15] | Monotherapy or boost | Boost > monotherapy |
| | HDR | Monotherapy | Boost<br>*monotherapy | Boost<br>*Monotherapy (infrequently) |
| EBRT [3–7,10,13] | CFRT | Monotherapy > boost | Monotherapy or boost | Monotherapy or boost |
| | HFRT | *Monotherapy > *boost | *Monotherapy > *boost | *Monotherapy> *boost |
| | SBRT | *Monotherapy | *Monotherapy (infrequently) | *Monotherapy (infrequently) |
| ± ADT | | No | Sometimes, for 4–6 m[16,17] | Almost always, for 24–36 m[16,17] |

ADT, androgen deprivation therapy; BT, brachytherapy; EBRT, external beam radiation therapy; CFRT, conventionally fractionated radiation therapy; HFRT, hypofractionated radiation therapy; HDR-BT, high dose rate brachytherapy; LDR-BT, low dose rate brachytherapy; PSA, prostate-specific antigen; SBRT, stereotactic body radiation therapy.
* *Denotes treatment options that are largely investigational.*
*Reprinted with permission from Davis et al.[8] and Zaorsky et al.[11]*

**TABLE 44.2** Contraindications to Prostate BT[8,9]

| | Contraindications |
|---|---|
| Absolute | Limited life expectancy (e.g., <10 years) |
| | Unacceptable operative risks, or medically unsuited for anesthesia |
| | Distant metastases |
| | Absence of rectum such that TRUS guidance is precluded |
| | Large TURP defects which preclude seed placement and acceptable radiation dosimetry |
| | Ataxia telangiectasia |
| | Pre-existing rectal fistula |
| Relative | High IPSS score (typically defined as >20) |
| | History of prior pelvic radiotherapy |
| | TURP defects |
| | Large median lobes |
| | Gland size >60 $cm^3$ at time of implantation |
| | Inflammatory bowel disease |
| | Patient peak flow rate <10 $cm^3$/s and post void residual volume prior to BT >100 $cm^3$ |
| | Pubic arch interference (e.g., prior pelvic fracture, irregular pelvic anatomy, or a penile prosthesis) |

BT, brachytherapy; IPSS, International Prostate Symptom Score; TRUS, transrectal ultrasound; TURP, transurethral resection of prostate.

**TABLE 44.3** Properties of Radionuclides Used in BT for Prostate Cancer

| BT type | Radionuclide | Half-life (days) | Avg. energy (keV) | Years introduced | Typical monotherapy seed strength (mCi) | (mCi) |
|---|---|---|---|---|---|---|
| LDR-BT | Iodine-125 | 59.4 | 28.4 | 1965 | 0.3–0.6 | 0.4–0.8 |
| | Palladium-103 | 17.0 | 20.7 | 1986 | 1.1–2.2 | 1.4–2.8 |
| | Cesium-131 | 9.7 | 30.4 | 2004 | 2.5–3.9 | 1.6–2.5 |
| HDR-BT | Iridium-192 | 73.8 | 380 | 1980 | – | – |

BT, brachytherapy; HDR, high dose rate; LDR, low dose rate.

**TABLE 44.4** HDR-BT as Monotherapy for Prostate Cancer: Outcomes and Toxicity

| Study | References | Type | Arms | *n* | Total BT dose | Fractions | Gy/ fraction | Med FU | Actuarial FU (y) | FFBF (%) | | | RTOG late Grade 3–4 toxicity% | | EP |
|---|---|---|---|---|---|---|---|---|---|---|---|---|---|---|---|
| | | | | | | | | | | L | I | H | GU | GI | (%) |
| Demanes (2011) | [30] | Prospective | HDR-BT | 157 | 42 | 6 | 7 | 5.2 | 8 | 97 | 97 | N/A | NR | NR | NR |
| | | | HDR-BT | 141 | 38 | 4 | 9.5 | | | 97 | 97 | N/A | | | |
| Barkati (2012) | [32] | Prospective | HDR-BT | 19 | 30 | 3 | 10 | 3.3 | 5 | 97 | 87.5 | N/A | 0 | 0 | 30 |
| | | | HDR-BT | 19 | 31.5 | 3 | 10.5 | | | | | | 26 | 16 | |
| | | | HDR-BT | 19 | 33 | 3 | 11 | | | | | | 10 | 0 | |
| | | | HDR-BT | 22 | 34.5 | 3 | 11.5 | | | | | | 4.7 | 0 | |
| Mark (2007, 2011) | [33,34] | Retrospective | HDR-BT | 301 | 45 | 6 | 7.5 | 8 | 8 | 89 | 89 | N/A | 0 | 0.5 | NR |
| Zamboglou (2012) | [35] | Retrospective | HDR-BT | 141 | 38 | 4 | 9.5 | 4.4 | 5 | 95 | 93 | 93 | 3.7 | 1.6 | 81 |
| | | | HDR-BT | 351 | 38 | 4 | 9.5 | 4.4 | 5 | | | | | | |
| | | | HDR-BT | 226 | 34.5 | 3 | 11.5 | 4.4 | 5 | | | | | | |
| Ghadjar (2009) | [36] | Retrospective | HDR-BT | 36 | 38 | 4 | 9.5 | 3 | 3 | 100 | 100 | N/A | 11 | 0 | 75 |
| Grills (2004) | [37] | Retrospective | HDR-BT | 100 | 38 | 4 | 9.5 | 2.9 | 3.2 | 98 | 98 | N/A | 10 | 1.1 | NR |
| | | | LDR-BT | 100 | 120 | N/A | N/A | | | 97 | 97 | N/A | 12 | 0 | NR |
| Rogers (2012) | [38] | Retrospective | HDR-BT | 284 | 39 | 6 | 6.5 | 2.9 | 5 | N/A | 94 | N/A | 1 | 0 | 43 |
| Yoshioka (2006, 2011) | [39,40] | Retrospective | HDR-BT | 112 | 48–54 | 8–9 | 6 | 2.2 | 5 | 85 | 93 | 79 | 0 | 1 | NR |
| Sullivan (2007) | [41] | Retrospective | HDR-BT + EBRT | 425 | 12–15 | 3 | 4–5 | 3.6 | 5 | NR | 69 | NR | 8 | NR | NR |
| | | | HDR-BT | 47 | 30–33 | 3 | 10–11 | 1.8 | 5 | NR | | NR | 0 | NR | NR |
| Ghilezan (2012) | [42] | Retrospective | HDR-BT | 94 | 24–27 | 2 | 12–13.5 | 1.6 | 1.6 | NR | NR | NR | 1.1 | 2.2 | NR |
| Prada (2012) | [43] | Retrospective | HDR-BT | 40 | 38 | 4 | 9.5 | 1.6 | 2.7 | 100 | 88 | N/A | 0 | 0 | 89 |
| Hayes (2006) | [44] | Retrospective | HDR-BT | 326 | 39 | 6 | 6.5 | 1.1 | 3 | 99 | 98 | N/A | 0 | 0 | 75 |

BT, brachytherapy; CSS, cancer specific survival; DM, distant metastasis; EP, erectile preservation; FFBF, freedom from biochemical failure; FU, follow-up; GI, gastrointestinal; GU, genitourinary; H, high-risk; I, intermediate risk; L, low risk; N/A, not applicable; NR, not reported; OS, overall survival; US, ultrasound.
Studies sorted by level of evidence; then by median FU time.
*Reprinted with permission from Zaorsky et al.*[24]

seed implantation is to use a transperineal approach under TRUS and template guidance. Patient position and the TRUS-probe angle should coincide with the preimplant planning study. A high-resolution biplanar ultrasound system (at 5–12 MHz) with dedicated prostate BT software is mandatory. Fluoroscopy is frequently used to monitor seed deposition, as a complimentary imaging modality to TRUS,[45] and is used in some centers for intraoperative dose calculation using image fusion.[46] However, this is not considered mandatory for successful LDR-BT.[8]

The American Brachytherapy Society (ABS)[8] recommends that CT-based postoperative dosimetry be performed within 60 days of the implant. A planning system generates dose volume histograms, dose volume statistics, and 2D and 3D isodose curves superimposed on CT and other images. Careful postimplant assessment provides providers with objective measures of implant quality allowing for continual technical improvement. Postimplant dosimetry is performed on the day of LDR-BT and a few weeks after the implant (once edema has resolved); satisfactory day 0 dosimetry does not obviate dosimetric calculation on a subsequent date.[47] The time necessary to minimize edema is radionuclide-specific: 16 ± 4 days for $^{103}$Pd and 30 ± 7 days for $^{125}$I. Improvement of reproducibility of postimplant dosimetry, such as MR-CT image fusion, is also possible.[48–50]

Regarding dosimetric analysis, the ABS[8] recommends that the following post-BT dosimetric parameters be determined: (1) prostate – D90 (i.e., dose to 90% of the prostate volume) in Gy, V100 (volume receiving 100% of the dose) in percentage, and V150; (2) urethra – V150 in percentage, V5 in percentage, V30 in percentage; and (3) rectum – V100. For the prostate, it is recommended that the D90 is greater than 100% of the prescription dose; the V100 >90–95% of the dose; and the V150 be <60–65% of the dose. Many critical organ dose parameters have been reported.[51,52] For the urethra, the V150 should be less than the prescription dose; V5 <150%; and V30 <125%. It is recognized that meeting these constraints is not always possible, especially in smaller prostates (<20 cm$^3$). For the rectum, the V100 should be <1 cm$^3$ on day 0 dosimetry and <1.3 cm$^3$ on day 30. Critical structures for postimplant erectile dysfunction have not been agreed upon, although the internal pudendal artery, penile bulb, and neurovascular bundles have been studied.[53–55]

## LDR-BT Fractionation and Sequencing

Per the ABS guidelines,[8] the recommended dose of LDR-BT monotherapy using $^{125}$I is 145 Gy; for LDR-BT boost, the EBRT dose is 41.4–50.4 Gy (at 1.8 Gy fractions per day); and the LDR-BT dose is 109–110 Gy. The recommended dose of LDR-BT monotherapy using $^{103}$Pd is 125 Gy; for LDR-BT boost, the EBRT dose is 41.4–50.4 Gy (at 1.8 Gy fractions per day); and the LDR-BT dose is 90–100 Gy.

EBRT is generally performed before LDR-BT, with a 2–8-week interval between the two therapies, though other approaches are acceptable.[8] No studies have been published investigating either the sequencing of LDR-BT and EBRT, or the time interval between the modalities. Delivering LDR-BT before EBRT exposes tissues to radiation simultaneously from both treatments and may theoretically increase normal tissue toxicity; on the contrary, performing LDR-BT first also allows physician assessment of the seeds such that the EBRT dose may be adjusted if necessary.[8] The sequencing of LDR-BT boost, compared to various HDR-BT boost schedules, is illustrated in Figure 44.1 (bottom panel).

## HDR-BT Technical Aspects

During the HDR-BT procedure, a RALS automatically deploys and retracts a single small radioactive source of $^{192}$Ir along the implant needle at specific positions delivering ≥12 Gy/h, compared to 0.4–2.0 Gy/h of LDR-BT. The RALS allows a physician to control the position where the HDR source stops (the dwell position) for a predetermined time period (the dwell time). Moreover, US-based planning of HDR-BT allows the procedure to be completed in about 1–5 sessions, minimizing catheter displacement.[24,56,11]

HDR-BT has a number of benefits (compared to EBRT and LDR-BT), some of which are theoretical: (1) there is a potential to increase prostate cancer cell death and minimize radiation-related toxicity with higher doses per fraction (based on a higher biologically equivalent dose (BED));[24,57] (2) there is improved radiophysics and dosimetry, with better dose distribution than EBRT and LDR-BT;[58,59] (3) there is convenience for the patient similar to that of LDR-BT (i.e., the treatment is completed in a few fractions over 2–4 days, which is more convenient than a protracted course of conventional EBRT);[60] and (4) there is improved resource allocation and less cost of treatment (compared to EBRT).[19,61–66]

Regarding dosimetric analysis, the dose plan is typically prepared by a dosimetrist or physicist. It is then reviewed and approved by the treating physician. The ABS[9] states that the V100 should be at least 90% in most cases, with an expected V100 >95%. A range of isodose distributions of 50, 100, 110, 120, and 150% of the prescription dose relative to the PTV should be used for treatment plan evaluation. The ABS does not provide normal tissue constraints given the heterogeneity in dose fractionation.

## HDR-BT Fractionation and Sequencing

The standard EBRT dose varies among institutions.[9] A single dose fractionation schedule has not been recommended by the ABS. In HDR-BT monotherapy,[19] the dose is typically delivered in two to six fractions, each up to 9–12 Gy, to a total dose of 30–50 Gy.[30,32–44] For HDR-BT boost,[56,11] generally, EBRT is delivered to a total dose of 36–54 Gy in 1.8–2.0 Gy fractions to the gross tumor volume (i.e., the prostate, with or without the seminal vesicles); and HDR-BT is typically delivered to a total dose of 12–30 Gy in one to four fractions, also delivered to the gross tumor volume, which is typically the prostate alone.

Two HDR-BT boost fractionation schedules have evolved.[56,11] A single insertion followed by two to four fractions delivered over 1–2 days is favored in North America. On the other hand, a separate procedure for catheter insertion for each fraction is usually used in European countries. The separate insertion schedule results in a greater workload of resource-intensive procedures and a greater anesthesia time, while the single-insertion procedure may involve overnight hospital admission and has risks of interfractional catheter displacement.[67]

There are three temporal approaches of combining EBRT and HDR-BT (Figure 44.1, bottom panel).[11,56] If EBRT is delivered first, HDR-BT is typically delivered 1–6 weeks after. A possible benefit of this method is to use the HDR-BT to account for suboptimal dosimetry of EBRT. Alternatively, EBRT may be interdigitated with HDR-BT. Finally, HDR-BT may be delivered first, and EBRT delivered 1–3 weeks after. With the latter method, EBRT may be used to account for suboptimal implant dosimetry of HDR-BT, and there is a minimization of preimplant radiation-induced edema and GU symptoms that typically follow EBRT.

## LDR-BT: CLINICAL OUTCOMES

### LDR-BT, Monotherapy for Low-Risk Patients

Patients with low-risk features deemed suitable candidates for LDR-BT may be appropriately treated with LDR-BT alone, also known as LDR-BT monotherapy. Published experience demonstrates that excellent long-term outcomes can be expected when optimal dosimetric parameters are achieved.[68–71] Based on a review published by the PCa Results Study Group, among LDR-BT monotherapy studies, the 10-year rates of freedom from biochemical failure (FFBF) for low-risk patients have been estimated at >86%.[72] Rates of prostate cancer distant metastasis (DM), specific mortality (PCSM), and overall survival (OS) are estimated to be <10, <5, and >85% at 10 years for these patients.

### LDR-BT Monotherapy and Boost for Intermediate-Risk Patients

For intermediate-risk patients, the appropriateness of LDR-BT monotherapy depends on many factors including the required treatment margin. In pathologic series of whole-mount prostatectomy specimens of organ-confined disease,[73–76] the radial extraprostatic extension (EPE) rarely extends beyond 5 mm, with the posterolateral region at highest risk of EPE. Many intermediate-risk tumors have equivalent or even lower risk of adverse pathologic features such as significant EPE, seminal vesicle invasion (SVI), or lymph node involvement (LNI). Thus, the recommended margin of 5 mm around the prostate to form the planning target volume (PTV) in all directions except posteriorly usually encompasses all EPE in intermediate-risk disease.[8,73]

Based on a review published by the PCa Results Study Group,[72] among LDR-BT monotherapy studies, the 10-year rates of FFBF for intermediate-risk patients is estimated to be >~65%. In a large LDR-BT monotherapy multi-institutional analysis of nearly 3000 PCa patients, which included 960 intermediate-risk patients, the 8 year FFBF rate was 70%.[71] These outcomes are encouraging, as the majority of these patients were treated before 1999 and fewer than 25% had formal postimplant quality assurance. In examining present-day practice patterns, a patterns-of-care study by Frank et al.[77] surveyed 18 BT practitioners with cumulative experience of over 10,000 LDR-BT cases. The authors report that experienced practitioners examine intermediate-risk patients on a case-by-case basis and employ monotherapy cautiously.

### LDR-BT Boost for High-Risk Patients

Patients with high-risk features being considered for primary EBRT are known to benefit from treatment combined with ADT from multiple randomized prospective trials.[78,79] Additionally, outcomes and toxicities of LDR-BT boost have been reported from RTOG 0019[14] and CALGB 99809.[15] In comparison to EBRT trials combined with ADT, however, the data are less robust in demonstrating that ADT provides improvement in clinical endpoints for high-risk PCa treated with BT. In a series of high-risk patients by Merrick et al.,[80] 10-year FFBF was improved with ADT use, but OS and PCSM were not. In addition, in a multi-institutional series reported by Stone et al.,[81] patients with GS 8–10 had improved OS and freedom from DMs with a higher biologically equivalent dose of LDR-BT. LDR-BT boost is now an acceptable first-line treatment option in certain intermediate-risk and most high-risk cancer patients, based on prospective evidence.

## HDR-BT: CLINICAL OUTCOMES

### HDR-BT Monotherapy for Low- and Intermediate-Risk Patients

HDR-BT monotherapy may be used in select low- and intermediate-risk patients.[24] Per the ABS, monotherapy for high-risk patients should be recommended only on IRB-approved protocol, or in case-by-case circumstances.[9] Results of efficacy and toxicity from studies using HDR-BT as monotherapy are listed in Table 44.5.[30,32–44] FFBF rates for low-, intermediate-, and high-risk patients have generally been ≥85% at up to 5 years, with the exception of the Yoshioka et al. reports, where the 5-year FFBF rate for high-risk patients was 79%.[39,40] There has only been one study with a median follow-up time >5 years; the authors report 8-year actuarial FFBF for low- and intermediate-risk patients (each, 97%).[30] Rates of OS, PCSM, local recurrence, and DM are typically >95, <4, <4, and <4%, respectively, among all studies. The highest rate of DM is reported in the Yoshioka et al. study,[39,40] which likely reflects their inclusion of high-risk patients compared to other studies. Less than 5% of patients have Grade 3–4 GI or GU toxicity.[24] In general, studies of HDR-BT monotherapy have encouraging results in terms of biochemical control, patient survival, treatment toxicity, and erectile preservation.[24]

### HDR-BT Boost for Intermediate- and High-Risk Patients

HDR-BT boost is theorized to combine the benefits of (1) CFRT[3–7] or hypofractionated HFRT monotherapy (i.e., delivery of radiation to the prostate and surrounding tissues (e.g., pelvic LNs),[13] which may potentially harbor micrometastatic disease) and (2) with HDR-BT monotherapy, which is theorized to have a better dose distribution and ability to kill cancer cells than LDR-BT.[11,82] Moreover, the BED achieved with HDR-BT boost (listed in Table 44.5) is typically much higher than what can be achieved with EBRT alone (at an alpha/beta ratio of 1.5, the BEDs are 200–300 vs. ~180).[11]

Results of from prospective studies using HDR-BT boost are listed in Table 44.5.[83–99] Five-year rates of PCSM, OS, LR, and DM have been 99–100%, 85–100%, 0–8%, and 0–12%, respectively. These outcomes are comparable to studies of LDR-BT monotherapy, EBRT monotherapy, and LDR-BT boost.[11,56] Few studies have reported outcomes outside of these ranges; poor outcomes are usually secondary to inclusion of high-risk and locally advanced patients, inclusion of a "low dose" HDR-BT boost arm, or exclusion of patients who received ADT.[87,90–93]

Urethral stricture is one of the more common toxicities following HDR-BT boost, and it occurs in up to 8% of patients; [67] it is less common for patients treated with EBRT alone.[3–7,100] Overall, less than 8% of patients have Grade 3–4 GU toxicities, though most of the toxicities reported among individual studies were due to stricture.[11,56] To compare, LDR-BT boost RTOG Grade 3–4 GU toxicities from two Phase II studies were seen in 13[14] and 3%[15] of patients; and GI toxicities were seen in 3[14] and 0%[15] of patients. Phase III studies of EBRT have shown RTOG late GU toxicity Grade ≥2 in 8–40% of patients and GI toxicity Grade ≥2 in 9–26% of patients.[3–7] In summary, HDR-BT boost is now a relatively well-established treatment modality for certain intermediate-risk and high-risk prostate cancer patients. Compared to studies of LDR-BT and EBRT, HDR-BT monotherapy studies have a relatively shorter follow-up time and fewer patients.

### HDR-BT as Salvage for Local Recurrence After EBRT

HDR monotherapy as a salvage treatment for local recurrence after EBRT or LDR-BT has been recently reported.[101,102] Salvage HDR-BT appears to be a promising treatment option, particularly for patients who are not deemed fit for salvage RP. For salvage HDR-BT, referral to a specialty center with salvage HDR experience is recommended.

## FOLLOW-UP AFTER PROSTATE BRACHYTHERAPY

After BT, close follow-up with digital rectal examinations (DRE) and PSA at regular intervals is recommended. The optimal surveillance frequency following BT has not been established, although an interval of every 6–12 months is considered appropriate.[8] For patients with higher risk features, more frequent surveillance is appropriate. A routine biopsy is typically not recommended unless the documentation of a local recurrence is necessary. If a rising PSA occurs and prostate biopsy is performed, it is recommended that the biopsy be done at least 30 months following LDR-BT, or else it may not be interpretable, and a false positive may occur when actually a benign PSA bounce is likely.[103]

### PSA Kinetics and Defining Biochemical Failure

Regarding how FFBF is defined, the ABS favors the use of the Phoenix definition (i.e., nadir + 2 ng/mL) following treatment.[104] After LDR-BT, PSA has been noted to decrease to <0.3 ng/mL in most men with localized disease.[105] There is substantial overlap between features of benign PSA bounces and BF, though benign bounce is uncommon after 36 months.[106] McGrath et al. report that PSA bounces >1.0 ng/mL are rare after LDR-BT with or without neoadjuvant ADT, occurring in less than 10% of patients. They note that as PSA bounce amplitude increases, the BF rate increases; and that BF definitions based on absolute nadir are less sensitive to prostate bounce after LDR-BT.[107] On the other hand, Naghavi et al. have noted that patients with Gleason 6 disease are more likely to experience a PSA bounce and have

**TABLE 44.5** HDR-BT Boost for Prostate Cancer: Outcomes and Toxicity of Prospective Studies

| Study | References | Type | Arms | $n$ | Total BT dose (Gy) | Gy/ Fraction | Total EBRT dose (Gy) | Total BED at α/β of 1.5 (Gy) | Median FU (y) | Actuarial FU (y) | FFBF% | | | RTOG late Grade 3-4 toxicity% | |
|---|---|---|---|---|---|---|---|---|---|---|---|---|---|---|---|
| | | | | | | | | | | | L | I | H | GU (stricture) | GI |
| Demanes (2005) | [83] | Prospective | HDR-BTb | 209 | 23 | 6 | 36 | 190 | 7.3 | 7.3 | 90 | 87 | 69 | 7.7 (6.7) | 0 |
| Hoskin (2012) | [84] | Phase III | EBRT | 108 | N/A | N/A | 55 | 156 | 7.1 | 10 | 60 | 62 | 70 | 4 (2) | 2 |
| | | | HDR-BTb | 110 | 17 | 8 | 36 | 215 | 7.1 | 10 | 100 | 89 | 80 | 11 (8) | 0 |
| Duchesne (2007) | [85] | Phase I/II | HDR-BTb | 108 | 16–20 | 4–5 | 46 | 166 | 6.5 | 5 | NR | NR | NR | 4.5 (NR) | 2.8 |
| Demanes (2009) | [86] | Prospective | HDR-BTb | 200 | 23 | 6 | 36 | 190 | 6.4 | 10 | 92 | 87 | 63 | 0 (0) | 0 |
| | | | HDR-BTb +ADT | 211 | 23 | 6 | 36 | 190 | 6.4 | 10 | 92 | 87 | 63 | 0 (0) | 0 |
| Galalae (2006) | [87] | Phase II | HDR-BTb | 122 | 18 | 6 | 46–50 | 198 | 4.5 | 5 | N/A | 90 | N/A | NR | NR |
| | | | | 25 | 18 | 6 | 46–50 | 198 | 5.5 | 5 | N/A | N/A | 68 | NR | NR |
| | | | | 95 | 21 | 10 | 46–50 | 276 | 5.5 | 5 | N/A | N/A | 85 | NR | NR |
| | | | | 23 | 18 | 6 | 46–50 | 198 | 6.5 | 5 | N/A | N/A | N/A | NR | NR |
| | | | | 57 | 21 | 10 | 46–50 | 276 | 6.5 | 5 | N/A | N/A | N/A | NR | NR |
| Kalkner (2007) | [88] | Phase I/II | HDR-BTb | 154 | 20 | 10 | 50 | 270 | 6.1 | 5 | 97 | 83 | 83 | 5 (NR) | 1 |
| Tang (2006) | [89] | Phase I/II | HDR-BTb | 47 | 16 | 4 | 46 | 179 | 5 | 5 | 76 | 68 | 33 | NR | NR |
| | | | HDR-BTb | 41 | 20 | 5 | 46 | 194 | 5 | 5 | | | | NR | NR |
| | | | EBRT | 104 | N/A | N/A | 66 | 154 | 4.7 | 5 | NR | 71 | NR | NR | NR |
| Galalae (2004) | [90] | Phase II | HDR-BTb | 46 | 16–30 | 6–12 | 45.6–50 | 297 | 5 | 5 | 96 | N/A | N/A | NR | NR |
| | | | | 188 | 16–30 | 6–12 | 45.6–50 | 297 | 5 | 5 | N/A | 88 | N/A | NR | NR |
| | | | | 359 | 16–30 | 6–12 | 45.6–50 | 297 | 5 | 5 | N/A | N/A | 69 | NR | NR |
| Vargas (2006) | [91] | Phase I/II | HDR-BTb | 67 | 16–20 | 6–8 | 46 | 211 | 4.9 | 5 | N/A | 69 | | 7.5 (1.5) | 0 |
| | | | | 130 | 18–23 | 9–12 | 46 | 275 | 4.9 | 5 | N/A | 86 | | 3 (1.5) | 3 |

| | | | | | | | | | | | | | | | |
|---|---|---|---|---|---|---|---|---|---|---|---|---|---|---|---|
| Martinez (2003, 2010) | [92,93] | Phase II | HDR-BTb | 58 | 18 | 6 | 46 | 197 | 4.8 | 5 | NR | 85 | 75 | 12 (12) | 1 |
| | | | | 149 | 19 | 8–12 | 46 | 227 | 4.8 | 5 | NR | 85 | 75 | 2 (2) | 1 |
| Myers (2012) | [94] | Phase I/II | HDR-BTb | 26 | 15 | 15 | 38 | 276 | 4.4 | 5 | 100 | 100 | N/A | 3.8 (3.8) | 0 |
| Martinez-Monge (2012) | [95] | Phase II | HDR-BTb | 200 | 19 | 5 | 54 | 198 | 3.7 | 9 | N/A | N/A | 75.5 | 5 (NR) | 1.5 |
| Kestin (2000) | [96] | Prospective, matched pair | HDR-BTb | 161 | 16–21 | 6–11 | 46 | 215 | 2.8 | 5 | 67 | N/A | N/A | NR | NR |
| | | | EBRT | 161 | N/A | N/A | 66 | 154 | 2.8 | 5 | 44 | N/A | N/A | NR | NR |
| Hsu (2010) | [97] | Phase II | HDR-BTb | 129 | 19 | 10 | 45 | 238 | 2.3 | 2.5 | NR | NR | NR | 2.6 (0.7) | 2.6 |
| Morton (2011) | [98] | 2 Phase II trials | HDR-BTb | 123 | 15 | 15 | 37.5 | 265 | 2 | 5 | N/A | 95 | N/A | 1 (0) | 0 |
| | | | | 60 | 20 | 10 | 45 | 252 | 2 | 5 | N/A | 97 | N/A | 0 (0) | 0 |
| Borghede (1997) | [99] | Prospective | HDR-BTb | 50 | 10 | 5 | 50 | 111 | 1.5 | 1.5 | 97 | 97 | 92 | 2 (0) | 0 |

ADT, androgen deprivation therapy; EBRT, external beam radiation therapy; FFBF, freedom from biochemical failure; FU, follow-up; GI, gastrointestinal; GU, genitourinary; H, high-risk; HDR-BTb, high dose rate brachytherapy boost; I, intermediate risk; L, low risk; N/A, not applicable; NR, not reported; RTOG, Radiation Therapy Oncology Group.

Gy rounded to whole numbers. Studies sorted by longest median FU time.

*Reprinted with permission from Zaorsky et al.*[56]

improved FFBF.[108] The prostate D90 appears to be predictive of a bounce.[109]

After HDR-BT, there is typically a very low PSA nadir (e.g., <0.05 ng/mL).[110] The rate of PSA bounce appears to be more frequent in LDR-BT, compared to HDR-BT/EBRT (42% vs. 23/20%).[111] In a series of 114 men treated with HDR-BT boost with HFRT, PSA bounce occurred in 39% of men, after a median of 16 months. The median magnitude of bounce was 0.45 ng/mL; BF occurred in 11% of patients with a bounce.[112] In a separate series of 67 patients, Mehta et al. reported a bounce in 43% of patients, varying from <1 ng/mL in 28% of cases to ≥1 ng/mL in 15%, with most bounces associated with a lower Gleason score and younger age (<55 years).[113]

PSA bounces occur after both LDR-BT and HDR-BT. It is important to understand that they are common and do not automatically represent cancer recurrence. PSA bounces are associated with patient, cancer, and dosimetric factors. Before any treatment is recommended or initiated, cancer recurrence needs to be documented by biopsy or imaging.

## CONCLUSIONS

Although there are many standard treatment options for PCa, randomized clinical trials to define the optimal therapy for patients with localized disease are limited. BT and BT boost may be used as first-line therapies in the management of PCa patients of all NCCN-defined risk groups. LDR-BT is readily acknowledged as a standard option in low-risk PCa by many health organizations. LDR-BT boost is recommended in certain intermediate-risk and most high-risk cancer patients. HDR-BT monotherapy may be used as a first-line treatment option in certain low- and intermediate-risk PCa patients. However, compared to studies of LDR-BT and external radiation therapy, HDR-BT monotherapy studies have a shorter follow-up time and smaller patient numbers. HDR-BT boost is a relatively well-established treatment modality for certain intermediate-risk and high-risk PCa. ADT is used in addition to either form of BT in certain intermediate- and high-risk patients.

## References

1. Cahlon O, Zelefsky MJ, Shippy A, et al. Ultra-high dose (86.4 Gy) IMRT for localized prostate cancer: toxicity and biochemical outcomes. *Int J Radiat Oncol Biol Phys* 2008;**71**(2):330–7.
2. Novara G, Ficarra V, Mocellin S, et al. Systematic review and meta-analysis of studies reporting oncologic outcome after robot-assisted radical prostatectomy. *Eur Urol* 2012;**62**(3):382–404.
3. Kuban DA, Tucker SL, Dong L, et al. Long-term results of the M. D. Anderson randomized dose-escalation trial for prostate cancer. *Int J Radiat Oncol Biol Phys* 2008;**70**(1):67–74.
4. Al-Mamgani A, van Putten WL, Heemsbergen WD, et al. Update of Dutch multicenter dose-escalation trial of radiotherapy for localized prostate cancer. *Int J Radiat Oncol Biol Phys* 2008;**72**(4):980–8.
5. Zietman AL, DeSilvio ML, Slater JD, et al. Comparison of conventional-dose vs high-dose conformal radiation therapy in clinically localized adenocarcinoma of the prostate: a randomized controlled trial. *JAMA* 2005;**294**(10):1233–9.
6. Zietman AL, Bae K, Slater JD, et al. Randomized trial comparing conventional-dose with high-dose conformal radiation therapy in early-stage adenocarcinoma of the prostate: long-term results from Proton Radiation Oncology Group/American College of Radiology 95-09. *J Clin Oncol* 2010;**28**(7):1106–11.
7. Dearnaley DP, Sydes MR, Graham JD, et al. Escalated-dose versus standard-dose conformal radiotherapy in prostate cancer: first results from the MRC RT01 randomised controlled trial. *Lancet Oncol* 2007;**8**(6):475–87.
8. Davis BJ, Horwitz EM, Lee WR, et al. American Brachytherapy Society consensus guidelines for transrectal ultrasound-guided permanent prostate brachytherapy. *Brachytherapy* 2012;**11**(1):6–19.
9. Yamada Y, Rogers L, Demanes DJ, et al. American Brachytherapy Society consensus guidelines for high-dose-rate prostate brachytherapy. *Brachytherapy* 2012;**11**(1):20–32.
10. Zaorsky NG, Harrison AS, Trabulsi EJ, et al. Evolution of advanced technologies in prostate cancer radiotherapy. *Nat Rev Urol* 2013;**10**(10):565–79.
11. Zaorsky NG, Doyle LA, Yamoah K, et al. High dose rate brachytherapy boost for prostate cancer: a systematic review. *Cancer Treat Rev* 2014;**40**(3):414–25.
12. Edge SB, Byrd DR, Compton CC, Fritz AG, Greene FL, Trotti A, editors. *AJCC cancer staging manual*. 7th ed. New York: Springer-Verlag; 2010.
13. Zaorsky NG, Ohri N, Showalter TN, et al. Systematic review of hypofractionated radiation therapy for prostate cancer. *Cancer Treat Rev* 2013;**39**(7):728–36.
14. Lee WR, Bae K, Lawton C, et al. Late toxicity and biochemical recurrence after external-beam radiotherapy combined with permanent-source prostate brachytherapy: analysis of Radiation Therapy Oncology Group study 0019. *Cancer* 2007;**109**(8):1506–12.
15. Hurwitz MD, Halabi S, Archer L, et al. Combination external beam radiation and brachytherapy boost with androgen deprivation for treatment of intermediate-risk prostate cancer: long-term results of CALGB 99809. *Cancer* 2011;**117**(24):5579–88.
16. Zaorsky NG, Trabulsi EJ, Lin J, et al. Multimodality therapy for patients with high-risk prostate cancer: current status and future directions. *Semin Oncol* 2013;**40**(3):308–21.
17. Nomiya T, Tsuji H, Toyama S, et al. Management of high-risk prostate cancer: radiation therapy and hormonal therapy. *Cancer Treat Rev* 2013;**39**(8):872–8.
18. Young HH. The use of radium and the punch operation in desperate cases of enlarged prostate. *Ann Surg* 1917;**65**(5):633–41.
19. Flocks RH, Kerr HD, Elkins HB, et al. The treatment of carcinoma of the prostate by interstitial radiation with radioactive gold (Au198); a follow-up report. *J Urol* 1954;**71**(5):628–33.
20. Whitmore Jr WF, Hilaris B, Grabstald H. Retropubic implantation to iodine 125 in the treatment of prostatic cancer. *J Urol* 1972;**108**(6):918–20.
21. Zelefsky MJ, Whitmore Jr WF. Long-term results of retropubic permanent 125iodine implantation of the prostate for clinically localized prostatic cancer. *J Urol* 1997;**158**(1):23–9 discussion 29–30.
22. Holm HH, Juul N, Pedersen JF, et al. Transperineal 125iodine seed implantation in prostatic cancer guided by transrectal ultrasonography. 1983. *J Urol* 2002;**167**(2 Pt 2):985–8 discussion 988–9.
23. Ragde H, Blasko JC, Grimm PD, et al. Interstitial iodine-125 radiation without adjuvant therapy in the treatment of clinically localized prostate carcinoma. *Cancer* 1997;**80**(3):442–53.

24. Zaorsky NG, Doyle LA, Hurwitz MD, et al. Do theoretical potential and advanced technology justify the use of high-dose rate brachytherapy as monotherapy for prostate cancer? *Exp Rev Anticancer Ther* 2014;**14**(1):39–50.
25. Heysek RV. Modern brachytherapy for treatment of prostate cancer. *Cancer Control* 2007;**14**(3):238–43.
26. NCI. Prostate Cancer Treatment – Treatment Option Overview. National Cancer Institute Website; 2010.
27. Mohler JL, Kantoff PW, Armstrong AJ, et al. Prostate cancer. *J Natl Compr Canc Netw* 2013;**11**(12):1471–9 version 1.2014.
28. Thompson I, Thrasher JB, Aus G, et al. Guideline for the management of clinically localized prostate cancer: 2007 update. *J Urol* 2007;**177**(6):2106–31.
29. Yoshioka Y, Nose T, Yoshida K, et al. High-dose-rate interstitial brachytherapy as a monotherapy for localized prostate cancer: treatment description and preliminary results of a phase I/II clinical trial. *Int J Radiat Oncol Biol Phys* 2000;**48**:675–81.
30. Demanes DJ, Martinez AA, Ghilezan M, et al. High-dose-rate monotherapy: safe and effective brachytherapy for patients with localized prostate cancer. *Int J Radiat Oncol Biol Phys* 2011;**81**(5):1286–92.
31. Martinez AA, Pataki I, Edmundson G, et al. Phase II prospective study of the use of conformal high-dose-rate brachytherapy as monotherapy for the treatment of favorable stage prostate cancer: a feasibility report. *Int J Radiat Oncol Biol Phys* 2001;**49**(1):61–9.
32. Barkati M, Williams SG, Foroudi F, et al. High-dose-rate brachytherapy as a monotherapy for favorable-risk prostate cancer: a Phase II trial. *Int J Radiat Oncol Biol Phys* 2012;**82**(5):1889–96.
33. Mark R, Akins RS, Anderson PJ, et al. Interstitial high dose rate (HDR) brachytherapy as monotherapy for early stage prostate cancer: a report of 206 cases. *Int J Rad Biol Phys* 2007;**69**(3):S329.
34. Mark R, Anderson PJ, Akins RS, et al. High-dose-rate brachytherapy under local anesthesia for early stage prostate cancer: a report of 546 cases. *Brachytherapy* 2011;**10**:S93.
35. Zamboglou N, Tselis N, Baltas D, et al. High-dose-rate interstitial brachytherapy as monotherapy for clinically localized prostate cancer: treatment evolution and mature results. *Int J Radiat Oncol Biol Phys* 2012;**85**(3):672–8.
36. Ghadjar P, Keller T, Rentsch CA, et al. Toxicity and early treatment outcomes in low- and intermediate-risk prostate cancer managed by high-dose-rate brachytherapy as a monotherapy. *Brachytherapy* 2009;**8**(1):45–51.
37. Grills IS, Martinez AA, Hollander M, et al. High dose rate brachytherapy as prostate cancer monotherapy reduces toxicity compared to low dose rate palladium seeds. *J Urol* 2004;**171**(3):1098–104.
38. Rogers CL, Alder SC, Rogers RL, et al. High dose brachytherapy as monotherapy for intermediate risk prostate cancer. *J Urol* 2012;**187**(1):109–16.
39. Yoshioka Y, Konishi K, Oh RJ, et al. High-dose-rate brachytherapy without external beam irradiation for locally advanced prostate cancer. *Radiother Oncol* 2006;**80**(1):62–8.
40. Yoshioka Y, Konishi K, Sumida I, et al. Monotherapeutic high-dose-rate brachytherapy for prostate cancer: five-year results of an extreme hypofractionation regimen with 54 Gy in nine fractions. *Int J Radiat Oncol Biol Phys* 2011;**80**(2):469–75.
41. Sullivan L, Williams SG, Tai KH, et al. Urethral stricture following high dose rate brachytherapy for prostate cancer. *Radiother Oncol* 2009;**91**(2):232–6.
42. Ghilezan M, Martinez A, Gustason G, et al. High-dose-rate brachytherapy as monotherapy delivered in two fractions within one day for favorable/intermediate-risk prostate cancer: preliminary toxicity data. *Int J Radiat Oncol Biol Phys* 2012;**83**(3):927–32.
43. Prada PJ, Jimenez I, Gonzalez-Suarez H, et al. High-dose-rate interstitial brachytherapy as monotherapy in one fraction and transperineal hyaluronic acid injection into the perirectal fat for the treatment of favorable stage prostate cancer: treatment description and preliminary results. *Brachytherapy* 2012;**11**(2):105–10.
44. Hayes J, Hansen R, Rogers L, Post-treatment PSA. et al. Kinetics of three prostate cancer treatment regimens involving brachythe of three prostate cancer treatment regimens involving brachytherapy. *Brachytherapy* 2006;**5**(113):P106.
45. Prestidge BR, Prete JJ, Buchholz TA, et al. A survey of current clinical practice of permanent prostate brachytherapy in the United States. *Int J Radiat Oncol Biol Phys* 1998;**40**(2):461–5.
46. Orio III PF, Tutar IB, Narayanan S, et al. Intraoperative ultrasound-fluoroscopy fusion can enhance prostate brachytherapy quality. *Int J Radiat Oncol Biol Phys* 2007;**69**(1):302–7.
47. Shaikh T, Zaorsky NG, Ruth K, et al. Is it necessary to perform week three dosimetric analysis in low-dose rate brachytherapy for prostate cancer when day 0 dosimetry is done? A quality assurance assessment. *Brachytherapy* 2014;**14**(3):316–21.
48. Crook J, McLean M, Yeung I, et al. MRI-CT fusion to assess post-brachytherapy prostate volume and the effects of prolonged edema on dosimetry following transperineal interstitial permanent prostate brachytherapy. *Brachytherapy* 2004;**3**(2):55–60.
49. Polo A, Cattani F, Vavassori A, et al. MR and CT image fusion for postimplant analysis in permanent prostate seed implants. *Int J Radiat Oncol Biol Phys* 2004;**60**(5):1572–9.
50. Tanaka O, Hayashi S, Matsuo M, et al. Comparison of MRI-based and CT/MRI fusion-based postimplant dosimetric analysis of prostate brachytherapy. *Int J Radiat Oncol Biol Phys* 2006;**66**(2):597–602.
51. Crook JM, Potters L, Stock RG, et al. Critical organ dosimetry in permanent seed prostate brachytherapy: defining the organs at risk. *Brachytherapy* 2005;**4**(3):186–94.
52. Nath R, Bice WS, Butler WM, et al. AAPM recommendations on dose prescription and reporting methods for permanent interstitial brachytherapy for prostate cancer: report of Task Group 137. *Med Phys* 2009;**36**(11):5310–22.
53. Gillan C, Kirilova A, Landon A, et al. Radiation dose to the internal pudendal arteries from permanent-seed prostate brachytherapy as determined by time-of-flight MR angiography. *Int J Radiat Oncol Biol Phys* 2006;**65**(3):688–93.
54. Merrick GS, Butler WM, Wallner KE, et al. The importance of radiation doses to the penile bulb vs. crura in the development of postbrachytherapy erectile dysfunction. *Int J Radiat Oncol Biol Phys* 2002;**54**(4):1055–62.
55. Buyyounouski MK, Horwitz EM, Uzzo RG, et al. The radiation doses to erectile tissues defined with magnetic resonance imaging after intensity-modulated radiation therapy or iodine-125 brachytherapy. *Int J Radiat Oncol Biol Phys* 2004;**59**(5):1383–91.
56. Zaorsky NG, Den RB, Doyle LA, et al. Combining theoretical potential and advanced technology in high-dose rate brachytherapy boost therapy for prostate cancer. *Exp Rev Med Devices* 2013;**10**(6):751–63.
57. Fowler J, Chappell R, Ritter M. Is alpha/beta for prostate tumors really low? *Int J Radiat Oncol Biol Phys* 2001;**50**(4):1021–31.
58. Yoshioka Y, Nishimura T, Kamata M, et al. Evaluation of anatomy-based dwell position and inverse optimization in high-dose-rate brachytherapy of prostate cancer: a dosimetric comparison to a conventional cylindrical dwell position, geometric optimization, and dose-point optimization. *Radiother Oncol* 2005;**75**(3):311–7.
59. Sumida I, Shiomi H, Yoshioka Y, et al. Optimization of dose distribution for HDR brachytherapy of the prostate using Attraction-Repulsion Model. *Int J Radiat Oncol Biol Phys* 2006;**64**(2):643–9.
60. Holmboe ES, Concato J. Treatment decisions for localized prostate cancer: asking men what's important. *J Gen Intern Med* 2000;**15**(10):694–701.
61. Van de Werf E, Lievens Y, Verstraete J, et al. Time and motion study of radiotherapy delivery: economic burden of increased quality assurance and IMRT. *Radiother Oncol* 2009;**93**(1):137–40.

62. Shah C, Lanni Jr TB, Ghilezan MI, et al. Brachytherapy provides comparable outcomes and improved cost-effectiveness in the treatment of low/intermediate prostate cancer. *Brachytherapy* 2012;**11**(6):441–5.
63. Parthan A, Pruttivarasin N, Taylor D, et al. CyberKnife for prostate cancer: Is it cost-effective? ASCO: Genitourinary Cancers Symposium *J Clin Oncol*; 2011. p. Suppl. 7:abstr 87.
64. Perez CA, Kobeissi B, Smith BD, et al. Cost accounting in radiation oncology: a computer-based model for reimbursement. *Int J Radiat Oncol Biol Phys* 1993;**25**(5):895–906.
65. Lievens Y, van den Bogaert W, Kesteloot K. Activity-based costing: a practical model for cost calculation in radiotherapy. *Int J Radiat Oncol Biol Phys* 2003;**57**(2):522–35.
66. Norlund A. Costs of radiotherapy. *Acta Oncol* 2003;**42**(5–6):411–5.
67. Morton GC. The emerging role of high-dose-rate brachytherapy for prostate cancer. *Clin Oncol* 2005;**17**(4):219–27.
68. Grimm PD, Blasko JC, Sylvester JE, et al. 10-year biochemical (prostate-specific antigen) control of prostate cancer with (125)I brachytherapy. *Int J Radiat Oncol Biol Phys* 2001;**51**(1):31–40.
69. Potters L, Morgenstern C, Calugaru E, et al. 12-year outcomes following permanent prostate brachytherapy in patients with clinically localized prostate cancer. *J Urol* 2005;**173**(5):1562–6 (Reprinted in J Urol. 2008 May;179(5 Suppl.):S20–S24; PMID: 18405743.).
70. Stone NN, Potters L, Davis BJ, et al. Customized dose prescription for permanent prostate brachytherapy: insights from a multicenter analysis of dosimetry outcomes. *Int J Radiat Oncol Biol Phys* 2007;**69**(5):1472–7.
71. Zelefsky MJ, Kuban DA, Levy LB, et al. Multi-institutional analysis of long-term outcome for stages T1-T2 prostate cancer treated with permanent seed implantation. *Int J Radiat Oncol Biol Phys* 2007;**67**(2):327–33.
72. Grimm P, Billiet I, Bostwick D, et al. Comparative analysis of prostate-specific antigen free survival outcomes for patients with low, intermediate and high risk prostate cancer treatment by radical therapy. Results from the Prostate Cancer Results Study Group. *BJU Int* 2012;**109**(Suppl. 1):22–9.
73. Davis BJ, Pisansky TM, Wilson TM, et al. The radial distance of extraprostatic extension of prostate carcinoma: implications for prostate brachytherapy. *Cancer* 1999;**85**(12):2630–7.
74. Sohayda C, Kupelian PA, Levin HS, et al. Extent of extracapsular extension in localized prostate cancer. *Urology* 2000;**55**(3):382–6.
75. Teh BS, Bastasch MD, Mai W-Y, et al. Predictors of extracapsular extension and its radial distance in prostate cancer: implications for prostate IMRT, brachytherapy, and surgery. *Cancer J* 2003;**9**(6):454–60.
76. Chao KK, Goldstein NS, Yan D, et al. Clinicopathologic analysis of extracapsular extension in prostate cancer: should the clinical target volume be expanded posterolaterally to account for microscopic extension? *Int J Radiat Oncol Biol Phys* 2006;**65**(4):999–1007.
77. Frank SJ, Grimm PD, Sylvester JE, et al. Interstitial implant alone or in combination with external beam radiation therapy for intermediate-risk prostate cancer: a survey of practice patterns in the United States. *Brachytherapy* 2007;**6**(1):2–8.
78. Horwitz EM, Bae K, Hanks GE, et al. Ten-year follow-up of radiation therapy oncology group protocol 92-02: a phase III trial of the duration of elective androgen deprivation in locally advanced prostate cancer. *J Clin Oncol* 2008;**26**(15):2497–504.
79. Bolla M, Van Tienhoven G, Warde P, et al. External irradiation with or without long-term androgen suppression for prostate cancer with high metastatic risk: 10-year results of an EORTC randomised study. *Lancet Oncol* 2010;**11**(11):1066–73.
80. Merrick GS, Butler WM, Wallner KE, et al. Androgen deprivation therapy does not impact cause-specific or overall survival in high-risk prostate cancer managed with brachytherapy and supplemental external beam. *Int J Radiat Oncol Biol Phys* 2007;**68**(1): 34–40.
81. Stone NN, Potters L, Davis BJ, et al. Multicenter analysis of effect of high biologic effective dose on biochemical failure and survival outcomes in patients with Gleason score 7–10 prostate cancer treated with permanent prostate brachytherapy. *Int J Radiat Oncol Biol Phys* 2009;**73**(2):341–6.
82. Yoshioka Y. Current status and perspectives of brachytherapy for prostate cancer. *Int J Clin Oncol* 2009;**14**(1):31–6.
83. Demanes DJ, Rodriguez RR, Schour L, et al. High-dose-rate intensity-modulated brachytherapy with external beam radiotherapy for prostate cancer: California endocurietherapy's 10-year results. *Int J Radiat Oncol Biol Phys* 2005;**61**(5):1306–16.
84. Hoskin PJ, Rojas AM, Bownes PJ, et al. Randomised trial of external beam radiotherapy alone or combined with high-dose-rate brachytherapy boost for localised prostate cancer. *Radiother Oncol* 2012;**103**(2):217–22.
85. Duchesne GM, Williams SG, Das R, et al. Patterns of toxicity following high-dose-rate brachytherapy boost for prostate cancer: mature prospective phase I/II study results. *Radiother Oncol* 2007;**84**(2):128–34.
86. Demanes DJ, Brandt D, Schour L, et al. Excellent results from high dose rate brachytherapy and external beam for prostate cancer are not improved by androgen deprivation. *Am J Clin Oncol* 2009;**32**(4):342–7.
87. Galalae RM, Martinez A, Nuernberg N, et al. Hypofractionated conformal HDR brachytherapy in hormone naive men with localized prostate cancer. Is escalation to very high biologically equivalent dose beneficial in all prognostic risk groups? *Strahlenther Onkol* 2006;**182**(3):135–41.
88. Kalkner KM, Wahlgren T, Ryberg M, et al. Clinical outcome in patients with prostate cancer treated with external beam radiotherapy and high dose-rate iridium 192 brachytherapy boost: a 6-year follow-up. *Acta Oncol* 2007;**46**(7):909–17.
89. Tang JI, Williams SG, Tai KH, et al. A prospective dose escalation trial of high-dose-rate brachytherapy boost for prostate cancer: evidence of hypofractionation efficacy? *Brachytherapy* 2006;**5**(4):256–61.
90. Galalae RM, Loch T, Riemer B, et al. Health-related quality of life measurement in long-term survivors and outcome following radical radiotherapy for localized prostate cancer. *Strahlenther Onkol* 2004;**180**(9):582–9.
91. Vargas CE, Martinez AA, Boike TP, et al. High-dose irradiation for prostate cancer via a high-dose-rate brachytherapy boost: results of a phase I to II study. *Int J Radiat Oncol Biol Phys* 2006;**66**(2): 416–23.
92. Martinez A, Gonzalez J, Spencer W, et al. Conformal high dose rate brachytherapy improves biochemical control and cause specific survival in patients with prostate cancer and poor prognostic factors. *J Urol* 2003;**169**(3):974–9 discussion 979–80.
93. Martinez AA, Demanes J, Vargas C, et al. High-dose-rate prostate brachytherapy: an excellent accelerated-hypofractionated treatment for favorable prostate cancer. *Am J Clin Oncol* 2010;**33**(5):481–8.
94. Myers MA, Hagan MP, Todor D, et al. Phase I/II trial of single-fraction high-dose-rate brachytherapy-boosted hypofractionated intensity-modulated radiation therapy for localized adenocarcinoma of the prostate. *Brachytherapy* 2012;**11**(4):292–8.
95. Martinez-Monge R, Moreno M, Ciervide R, et al. External-beam radiation therapy and high-dose rate brachytherapy combined with long-term androgen deprivation therapy in high and very high prostate cancer: preliminary data on clinical outcome. *Int J Radiat Oncol Biol Phys* 2012;**82**(3):e469–76.
96. Kestin LL, Martinez AA, Stromberg JS, et al. Matched-pair analysis of conformal high-dose-rate brachytherapy boost versus external-beam radiation therapy alone for locally advanced prostate cancer. *J Clin Oncol* 2000;**18**(15):2869–80.
97. Hsu IC, Bae K, Shinohara K, et al. Phase II trial of combined high-dose-rate brachytherapy and external beam radiotherapy

for adenocarcinoma of the prostate: preliminary results of RTOG 0321. *Int J Radiat Oncol Biol Phys* 2010;**78**(3):751–8.

98. Morton G, Loblaw A, Cheung P, et al. Is single fraction 15 Gy the preferred high dose-rate brachytherapy boost dose for prostate cancer? *Radiother Oncol* 2011;**100**(3):463–7.
99. Borghede G, Hedelin H, Holmang S, et al. Combined treatment with temporary short-term high dose rate iridium-192 brachytherapy and external beam radiotherapy for irradiation of localized prostatic carcinoma. *Radiother Oncol* 1997;**44**(3):237–44.
100. Sanda MG, Dunn RL, Michalski J, et al. Quality of life and satisfaction with outcome among prostate-cancer survivors. *N Engl J Med* 2008;**358**(12):1250–61.
101. Tharp M, Hardacre M, Bennett R, et al. Prostate high-dose-rate brachytherapy as salvage treatment of local failure after previous external or permanent seed irradiation for prostate cancer. *Brachytherapy* 2008;**7**(3):231–6.
102. Lee B, Shinohara K, Weinberg V, et al. Feasibility of high-dose-rate brachytherapy salvage for local prostate cancer recurrence after radiotherapy: the University of California-San Francisco experience. *Int J Radiat Oncol Biol Phys* 2007;**67**(4):1106–12.
103. Reed D, Wallner K, Merrick G, et al. Clinical correlates to PSA spikes and positive repeat biopsies after prostate brachytherapy. *Urology* 2003;**62**(4):683–8.
104. Roach III M, Hanks G, Thames Jr H, et al. Defining biochemical failure following radiotherapy with or without hormonal therapy in men with clinically localized prostate cancer: recommendations of the RTOG-ASTRO Phoenix Consensus Conference. *Int J Radiat Oncol Biol Phys* 2006;**65**(4):965–74.
105. Reis LO, Sanches BC, Zani EL, et al. PSA-nadir at 1 year as a sound contemporary prognostic factor for low-dose-rate iodine-125 seeds brachytherapy. *World J Urol* 2014;**32**(3):753–9.
106. Hackett C, Ghosh S, Sloboda R, et al. Distinguishing prostate-specific antigen bounces from biochemical failure after low-dose-rate prostate brachytherapy. *J Contemp Brachytherapy* 2014;**6**(3):247–53.
107. McGrath SD, Antonucci JV, Fitch DL, et al. PSA bounce after prostate brachytherapy with or without neoadjuvant androgen deprivation. *Brachytherapy* 2010;**9**(2):137–44.
108. Naghavi AO, Strom TJ, Nethers K, et al. Clinical implications of a prostate-specific antigen bounce after radiation therapy for prostate cancer. *Int J Clin Oncol* 2014;**20**:598–604.
109. Tanaka N, Asakawa I, Fujimoto K, et al. Minimal percentage of dose received by 90% of the urethra (%UD90) is the most significant predictor of PSA bounce in patients who underwent low-dose-rate brachytherapy (LDR-brachytherapy) for prostate cancer. *BMC Urol* 2012;**12**:28.
110. Fuller DB, Naitoh J, Mardirossian G. Virtual HDR CyberKnife SBRT for localized prostatic carcinoma: 5-year disease-free survival and toxicity observations. *Front Oncol* 2014;**4**:321.
111. Pinkawa M, Piroth MD, Holy R, et al. Prostate-specific antigen kinetics following external-beam radiotherapy and temporary (Ir-192) or permanent (I-125) brachytherapy for prostate cancer. *Radiother Oncol* 2010;**96**(1):25–9.
112. Patel N, Souhami L, Mansure JJ, et al. Prostate-specific antigen bounce after high-dose-rate prostate brachytherapy and hypofractionated external beam radiotherapy. *Brachytherapy* 2014;**13**(5):450–5.
113. Mehta NH, Kamrava M, Wang PC, et al. Prostate-specific antigen bounce after high-dose-rate monotherapy for prostate cancer. *Int J Radiat Oncol Biol Phys* 2013;**86**(4):729–33.

C H A P T E R

# 45

# Intensity Modulated Radiotherapy and Image Guidance

*Melissa R. Young, MD, PhD, James B. Yu, MD*

Department of Therapeutic Radiology, Yale School of Medicine, New Haven, CT, USA

## INTRODUCTION

External beam radiotherapy has been accepted as a nonsurgical means of curative treatment for localized prostate cancer since the 1950s and 1960s, coinciding with the advent of cobalt and megavoltage X-ray units that allowed for sufficient energy deposition in the prostate without unacceptable skin toxicity. Since, there have been advances in radiation delivery and treatment planning that has significantly improved the control of cancer as well as reducing the toxicity of treatment. The experience of patients undergoing external beam radiotherapy in the modern era of image-guided, intensity-modulated radiation therapy (IMRT) is one of minimal toxicity with a high likelihood of cure. Nonradiation oncology physicians or those who have trained in the era prior to modern prostate radiotherapy may not be aware of the substantial improvements that have been made to patient quality of life and cancer control. This chapter outlines the historical development and practical aspects of modern prostate radiotherapy.

## CONFORMAL RADIATION THERAPY: FROM CONVENTIONAL 2D TO 3D-CRT TO IMRT

The general principle of radiotherapy is to provide adequate dose to a target of interest (tumor) to attain adequate tumor control, while minimizing dose to adjacent normal tissues in order to limit acute and late toxicity to adjacent normal tissue and preserving quality of life. Advances in technology have allowed for the development of conformal radiation delivery whereby radiation beams can be "shaped" to mimic the contour of the targeted tissue. The mechanism by which conformal radiation fields have been designed and delivered has evolved as improved technologies have become available. The most notable advances have been possible due to the advent of computed tomography (CT), computer-based radiation treatment planning systems, and in the more modern era, development of multileaf collimators (MLCs) that have allowed for intensity modulation of radiation beams. The exponential increase in computer processing capabilities have also allowed for the development of inverse planning software that have the capacity to predict radiation dose distributions based on complex mathematical modeling of radiation interaction with matter in three dimensions. The evolution of external beam radiation techniques for prostate cancer is summarized here.

### Conventional 2D Radiation Therapy

At its introduction in prostate cancer therapy, radiation was planned based in two dimensions (2D), using plain X-ray films and fluoroscopy-based planning techniques. Four coplanar beams (anterior–posterior (AP), posterior–anterior (PA), right lateral (RL), and left lateral (LL)) were designed to encompass a large pelvic field, using bony anatomic landmarks to predict prostate location,[1] with radiation dose being calculated by hand with the aid of depth-dose tables and approximations of patient contours and anatomy. For AP and PA radiation treatment fields, the superior border was placed at the midsacroiliac joints, the inferior border at the lower edge of the ischial tuberosities, and lateral borders 1.5–2.0 cm lateral to the pelvic brim. On the lateral fields, the anterior border was often the pubic symphysis and posteriorly the S2/S3 interspace with the superior and inferior borders designed to match the AP and PA fields. After initial treatment to this larger pelvic field, a set of smaller

Prostate Cancer. http://dx.doi.org/10.1016/B978-0-12-800077-9.00045-1

radiation fields were then designed by shrinking the field by lowering the anterior border to the top of the acetabulum and laterally including two-thirds of the obturator foramen. This second set of radiation fields (known in radiation oncology parlance as a "cone down") were typically delivered once the course of treatment to the larger fields were completed. Dose was defined and prescribed at a single point at the geometric center of the beams and calculated by hand based on single plane dose distributions for each beam. Though 2D treatment planning required significant skill as well as mathematical and physical knowledge, the technology of the time still required large volumes of normal tissue to be treated; the source of significant treatment-related toxicity. Furthermore, modern virtual CT-based reconstruction of these standard 2D fields has shown that such fields are more likely than modern treatment to have incomplete inclusion of bulky tumors, tumors that invade seminal vesicles, or locally advanced disease.[2]

While 2D planning is no longer used routinely, the outcomes served as historical controls as 3D planning was introduced and investigated. For T1/T2 disease, 2D treatment planning led to 5 and 10 year PSA relapse-free rates that ranged from 60% to 80% and approximately 40%, respectively, as opposed to 25–32% and approximately 10%, respectively, for T3/T4 disease.[3–5] When stratified by pretreatment PSA, PSA relapse-free rates ranged from 44% to 65% for PSA 4–10 ng/mL, 27% to 72% for PSA 10–20 ng/mL, and 11% to 28% for PSA >20 ng/mL.[4–8] Additionally, the results of RTOG studies 7506 and 7706 helped establish 70 Gy to the prostate delivered by 2D planning as the maximum tolerated dose owing to significant GI toxicity.[9,10] Despite this, there was evidence that higher prostate doses improved biochemical control, thus much interest was directed at how to deliver conformal therapy to allow for dose escalation while minimizing rectal dose.

## 3D-CRT

With the advent of CT imaging and its increased availability in the 1980s and 1990s, 3D conformal radiation therapy (3D-CRT) techniques were developed and rapidly adopted. Though 3D-CRT treatment planning and patient positioning can vary between institutions, patients are typically placed in a reproducible supine position using an immobilization device (typically customized to the individual patient), after which a CT scan is acquired and target tissues such as the prostate, and normal tissues such as rectum, bladder, and femoral heads (organs at risk – OARs) are contoured on the axial images. Radiation target volumes are defined such that gross tumor volume (GTV) is considered the area of all known disease, radiographic or otherwise, whereas clinical target volume (CTV) is the GTV plus any areas considered to potentially harbor microscopic disease. In early prostate 3D-CRT, classically the CTV was the prostate plus any at risk volume such as seminal vesicles and/or lymph nodes if advanced disease was suspected. A final expansion of these volumes to a planning target volume (PTV) is then applied to account for systematic errors, which include daily patient set-up error and internal organ motion. This CTV to PTV expansion varied by institution, but was typically between 1.0 cm and 1.5 cm. Computer-based planning then allows generation of virtual 3D volumes representing the CTV and OARs in axial image sets and in the beams eye view of each of the radiation treatment beams. Furthermore, the planning software permits virtual arrangement of a desired number of beams (often four to six) around the patient, after which blocks (either by MLC or cerrobend cutouts) can be designed for each beam using the beam's eye view function in order to "shape" the radiation field to the contour of the PTV while blocking exposure to OARs. Dose volume histograms (DVHs) are generated and show a graphical representation of the radiation dose delivered to any defined volume. Plans can be optimized by adjusting number and angles, as well as shape of blocking, with the best plan providing highest prostate coverage and simultaneously minimizing dose to OARs.

3D-CRT was quickly adopted for prostate cancer therapy over 2D treatment as it allowed higher prostate dose delivery while limiting dose to surrounding normal tissue. Although early dose calculation studies showed that reduction of dose to surrounding tissues, such as the bladder or rectum, was on the order of 14%, this translated to a clinically significant reduction in patient toxicity.[11] This was similar to the results of a randomized study of 266 patients comparing toxicity with 3D-CRT or conventional treatment to 66 Gy, where grade 2 GI toxicity was reduced to 19% from 32% using 3D-CRT.[12] Furthermore, studies showed that 3D-CRT not only led to lower rates of normal tissue toxicity, but it also allowed for safe delivery of higher doses of radiotherapy with at least equal, and in most cases improved, PSA relapse-free rates compared to conventional 2D therapy historical controls.[13–15] The data regarding dose escalation therapy are reviewed later in this chapter.

## IMRT

Although 3D-CRT is a significant improvement over conventional 2D planning, it has limitations. 3D-CRT beams have uniform radiation intensity across the field, thus cannot conform to concave structures as may be encountered at the prostate/rectum or prostate/bladder interfaces. Concave conformality requires the ability to modulate radiation beam intensity across a field, which was not readily possible or practical with earlier technology. In the 1990s, rapid improvements in computing power combined with improvements in the design and capabilities of linear accelerators and multileaf collimation systems, resulted in the emergence of IMRT.[16]

IMRT improved upon 3D-CRT delivery by allowing nonuniform intensity across a radiation field. Originally, this was done by breaking up each beam into multiple segments with MLCs at different positions, treating each segment separately, with the beam turned off between each segment treatment (step and shoot). While step and shoot methods are still used, newer technology has emerged such that MLCs can be rapidly moved across the field while the radiation beam is on (dynamic multileaf collimation), similarly allowing for modulation of the beam intensity across the field, but resulting in faster delivery of an IMRT plan. Further advances include the ability to rotate the gantry of the linear accelerator (arc-based therapy) simultaneously while using dynamic MLCs to modulate beam intensity, thus effectively delivering radiation from an infinite number of gantry angles further increasing target conformality.[17] While all modalities if planned properly provide comparable and sufficient target coverage, arc-based therapies have higher treatment efficiency than the other methods described. Figure 45.1 demonstrates a comparison of the dose distribution of 3D-CRT versus IMRT versus arc-based therapy in a localized prostate cancer case.

IMRT planning required further technical advances in computer-based planning. Instead of forward planning as occurs with 3D-CRT whereby the physician and dosimetrist choose beam angles, beam energies, and beam weighting, and designs field blocks upfront, followed by calculation of the dose distribution to the target and OARs, IMRT makes use of inverse planning. Through inverse planning, the physician and dosimetrist still chooses the beam angles (or arcs, if using arc-based therapy), and beam energies up front. They must then define the criteria that a plan must achieve, such as percentage of the PTV that must reach a specified dose, maximum and/or minimum point doses, and dose-volume limits to OARs and assign the priority of each of these input variables. The planning algorithm then uses these criteria to run through various iterations of an IMRT plan adjusting MLC positions and beam intensity until a plan is generated that meets the input criteria. Due to the complexity of the calculations and modeling assumptions used to generate IMRT plans, IMRT also requires an extra level of quality assurance (QA). QA measures require that each plan undergo a check with a medical physicist, and be tested on a phantom where the dose delivered is measured and verified before it can ever be used for the treatment of a patient.

The use of IMRT was quickly introduced in the treatment of prostate cancer in the 1990s at a number of centers. Multiple groups assessed and confirmed the improved conformality of IMRT compared to 3D-CRT methods.[18–20] Furthermore, reduced toxicity has been shown to be achieved when using IMRT over 3D-CRT. In one of the largest series evaluating toxicity with IMRT, Zelefsky et al showed rates of grade 2 rectal toxicity lowered to 2% with IMRT when delivering a dose of 81 Gy

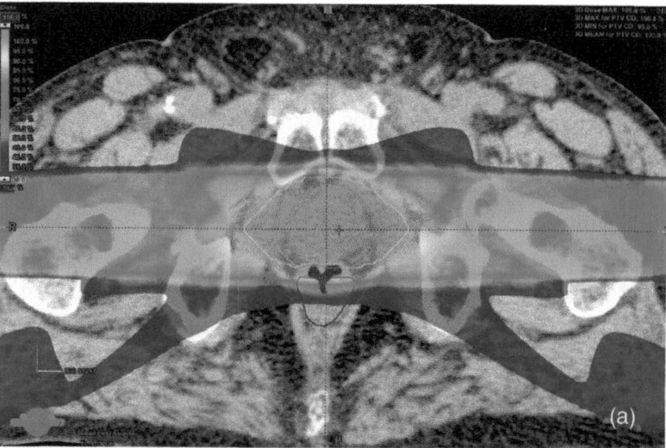


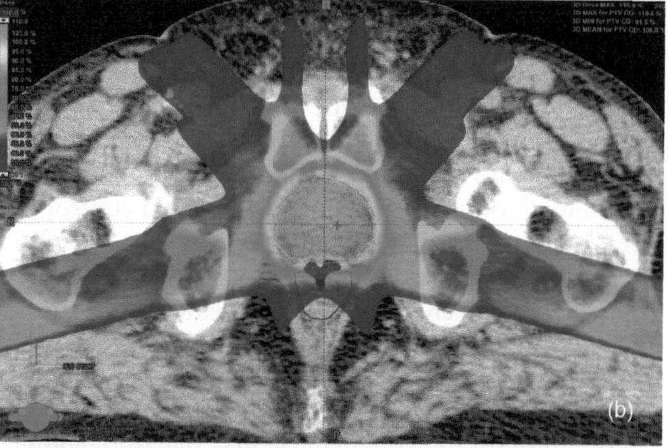


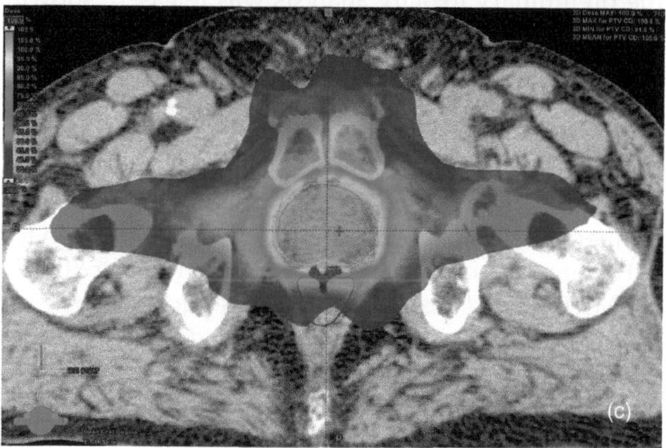


**FIGURE 45.1 Comparison of Dose Distribution for 3D-CRT, IMRT, and VMAT.** The images represent the dose distribution at the same axial slice of a patient receiving prostate only therapy utilizing (a) 3D-CRT, (b) static beam IMRT, and (c) volumetric modulated arc therapy (VMAT). Doses are represented by color wash with the lowest dose shown at 30% isodose in blue, and maximum dose in red. The white line contoured in each image represents the 100% isodose line. The red lines are the prostate PTV and the blue lines the rectal contour. The conformality index (volume receiving 100% dose/volume of PTV) was calculated for each; 3D-CRT – 1.63, IMRT – 0.97, VMAT – 1.06.

to the prostate compared to 14% for 3D-CRT.[21] This reduction in toxicity was not at the cost of unacceptable tumor control. With longer follow up, actuarial PSA relapse-free survival rates for favorable, intermediate, and unfavorable risk groups were 85%, 76%, and 72%, respectively. Similarly, Vora et al. reported on long-term experience using image-guided IMRT in 302 patients treated to a median dose of 75.6 Gy (range 70.2–77.4 Gy), and noted that with a median follow up of 91 months, IMRT resulted in durable biochemical control rates of 77.4%, 69.6%, and 53.3% for low-, intermediate-, and high-risk patients, respectively.[22] Reported chronic grade 3 GU and GI toxicity was 0% and 0.7%, respectively.

As much as 3D-CRT became standard over conventional therapy, even in the absence of randomized control trials, IMRT became accepted as a treatment standard for localized prostate cancer, due in large part to the preponderance of prospective and retrospective data showing its efficacy and safety. In 2001, the Radiation Therapy Oncology Group (RTOG), in a randomized control trial, RTOG 0126, was initiated and is using a 2 × 2 design to compare clinical outcomes and toxicity of 3D-CRT versus IMRT to 70.2 Gy or 79.2 Gy. While hormonal therapy was not allowed in this study, thus likely confounding the applicability of the results in modern era treatment, the biochemical outcomes are eagerly awaited. Recently, the preliminary toxicity analysis was released, showing a reduction in the volumes of bladder and rectum receiving 65, 70, and 75 Gy with IMRT, which correlated with reduced grade 2 acute GI/GU toxicity compared to 3D-CRT (Table 45.1).[23]

IMRT is also promising for locally advanced disease where risk of lymph node disease necessitates pelvic radiotherapy. Ashman et al. compared dosimetric and toxicity outcomes of whole pelvic treatment using 3D-CRT or IMRT.[24] IMRT resulted in a 60% reduction in the volume of bowel receiving 45 Gy compared to 3D-CRT and a decrease in incidence of grade 2 GI toxicity from 58% for 3D-CRT to 8% with IMRT. Similar outcomes have been reported by others showing decreased dose to OARs with associated reductions in acute toxicity.[25,26] To date, phase I and II data suggest pelvic IMRT is feasible and tolerated, but results regarding the effect on biochemical disease-free survival are still waiting to mature.

IMRT has also been introduced into the adjuvant and salvage settings following radical prostatectomy. Dosimetric studies have demonstrated IMRT can significantly reduce dose delivered to the bladder and rectum compared to whole pelvic and 3D-CRT techniques.[27] Extrapolating from the IMRT experience in the intact prostate, this is expected to translate into reduction of severity and overall rates of acute toxicity.[28,29] Goenka et al. demonstrated that in a group of 285 patients treated with 3D-CRT or IMRT (72% receiving 70.2 Gy), IMRT resulted in a reduction of grade 2 GI toxicity from 10.2% to 1.9%.[30]

**TABLE 45.1** Toxicity Results of 79.2 Gy Arm of RTOG 0126

| Reported toxicity | 3D-CRT (%) | IMRT (%) | $p$ |
|---|---|---|---|
| Acute effects | | | |
| Combined Grade 2+ GI/GU | 15.1 | 9.7 | 0.042 |
| Grade 2+ erectile impotence | 5.4 | 4.8 | – |
| Grade 2+ urinary frequency/ urgency | 10.0 | 7.6 | – |
| Late effects | | | |
| Grade 2+ GI | 22.0 | 15.1 | 0.039 |
| Grade 3+ GI | 5.1 | 2.6 | 0.09 |
| Grade 2+ erectile impotence | 67.6 | 58.6 | – |
| Grade 2+ urinary frequency/ urgency | 24.8 | 24.2 | – |
| Grade 2+ urinary incontinence | 5.8 | 6.6 | – |

Michalski et al. reported preliminary acute and cumulative incidence at 3 years of late GI/GU toxicity.[23] Results are summarized in this table. Of note, 75% of 3D-CRT patients underwent a cone down to prostate alone (from prostate plus entire seminal vesicles) after 55.8 Gy, whereas patients treated with IMRT were treated to prostate and the proximal 1.0 cm seminal vesicles throughout the entire treatment. A dash (–) represents data with no $p$ value given, but reported as nonsignificant.

Given the relatively new adoption of IMRT for prostate bed radiotherapy, biochemical outcomes still need time to mature. At present, while there is no evidence to indicate definitively that IMRT should be used over 3D-CRT, there is likewise no evidence to refute its use.[31]

## DOSE ESCALATION

Given the phase I and II evidence presented earlier that conformal therapy can provide lower morbidity and equal or improved biochemical control in part due to the ability to deliver higher doses to the prostate, a series of dose escalation trials were initiated with the advent of conformal therapy. The results of a select set of randomized and prospective trials are compiled in Table 45.2.[14,15,32–41] Each of these studies showed improved biochemical control with dose escalation, although in some studies, this was a significant improvement in only intermediate and high risk disease. Confounding some of the dose escalation data is the widespread recognition that short- or long-term androgen deprivation therapy may be synergistic with radiotherapy, such that the contribution of dose escalation or addition of hormones is less certain. In all, however, these series of studies were paramount for establishing the safety and efficacy of dose-escalated treatment, and improvements in IMRT techniques combined with improved image guidance

TABLE 45.2 Summary of Select Dose Escalation Studies

| Institution/study group | Stage | n | Dose (Gy/fractions) | Biochemical PFS | Follow-up | Late toxicity |
|---|---|---|---|---|---|---|
| *RANDOMIZED PHASE III TRIALS* | | | | | | |
| MRC RT01 (1998–2001)[32,33] | T1b–T3a | 843 | 64/32 versus 74/37 (3D-CRT) | 43% versus 55% ($p = 0.003$) | 10 years | Grade 3 GI 6% versus 10%; Grade 3 GU 2% versus 4% × 5 years |
| GETUG (1999–2002)[14,34] | T2–T3, or T1 + G7+ or PSA 10–50 | 306 | 70/35 versus 80/40 (3D-CRT) | 61% versus 72% ($p = 0.036$) | 5 years | Grade 3 GI 2% versus 6%; Grade 2+ GU 10% versus 18% |
| Dutch CKVO96-10 (1997–2003)[35,36] | T1b–4 | 664 | 68/34 versus 78/39 (3D-CRT) | 47% versus 54% ($p = 0.04$) | 70 months | Grade 3+ GI 4% versus 6%; Grade 3+ GU 12% versus 13% |
| PROG 95-09 (1996–1999)[37] | T1b–T2b, PSA < 15 | 393 | 70.2/39 versus 79.2/44 (proton/photon, conformal) | 68% versus 83% ($p < 0.001$) | 10 years | Grade 3+ GI 2% versus 2%; Grade 3+ GU 1% versus 1% |
| MDACC (1993–1998)[38] | T1b–T3 | 301 | 70/35 versus 78/39 (3D-CRT) | 59% versus 78% ($p = 0.004$) | 8 years | Grade 3+ GI 1% versus 7%; Grade 2+ GU 8% versus 13% |
| *PROSPECTIVE TRIALS* | | | | | | |
| MSKCC (1996–1998)[39] | T1c–T4, clinically localized | 170 | 81/45 (5 field IMRT) | 81% (low-risk), 78% (intermediate-risk), 62% (high-risk) | 10 years | Grade 2+ GI 2%; Grade 2+ GU 17% |
| MSKCC (1997–2008)[40] | Localized prostate cancer | 1002 | 86.4/48 (5–7 field IMRT) | 99% (low-risk), 86% (intermediate-risk), 68% (high-risk) | 7 years | Grade 2+ GI 4.4%; Grade 2+ GU 21.1% |
| RTOG 9406 (1994–2000)[15,41] | Localized prostate cancer | 1051 | (I) 68.4/38, (II) 73.8/41, (III) 79.2/44, (IV) 74/37, (V) 78/39 (3D-CRT) | (I) 21–26% (II) 23–37% (III) 16–48% (IV) 13–45% (IV) 45–61% | 9.2–11.7 years | Grade 3+ GI/GU (I) 3–6% (II) 2–4% (III) 6% (IV) 7–9% (IV) 9–12% |

A select set of studies that reported outcomes for dose-escalated prostate radiation therapy are summarized here. All used 3D-CRT with the exception of the MSKCC 81 Gy and 86.4 Gy reports that used IMRT. The years indicated in parenthesis in the institution/study group column represent the years patients received radiation. GI, gastrointestinal; GU, genitourinary; Gy, Gray; *n*, number of patients.

holds promise for more precise delivery with fewer side effects while not sacrificing disease control.

Presently, dose-escalated external beam radiotherapy, defined as ≥75.6 Gy, has been widely adopted. In 2006, 70.7% of patients in the United States receiving definitive radiotherapy for prostate cancer received dose-escalated therapy; this increased to 90% of cases in 2011.[42] The most recent National Comprehensive Cancer Network (NCCN) clinical guidelines for prostate cancer recommend doses of 75.6–79.2 Gy for low-risk prostate cancer, whereas doses up to 81 Gy are appropriate for improved biochemical control in intermediate- and high-risk disease.[43] Given that dose escalation is now considered standard, several groups have proposed dose constraints based on clinical and radiobiologic data as a guideline for minimizing toxicity. Tables 45.3 and 45.4 summarize a select set of commonly referenced parameters.[23,44–50]

## TARGET DELINEATION

### Immobilization, Simulation, and CT Scanning

Due to the relatively steep dose gradient that can be achieved using IMRT, accurate dose delivery relies on consistent patient setup daily. Patient instructions may vary by institution, but many suggest simulation with a full bladder and empty rectum as this can be relatively consistent throughout a course of radiotherapy. Custom immobilization devices are also used, which can include

TABLE 45.3 Dose Constraints for Pelvic Organs at Risk Based on QUANTEC

| Organ | Dose constraint | Toxicity rate |
|---|---|---|
| Rectum[45] (whole organ) | V50 < 50%; V60 < 35%; V65 < 25%; V70 < 20%; V75 < 15% | <15% (≥Grade 2 late toxicity); <10% (≥Grade 3 late toxicity) |
| Bladder[46,*] (whole organ) | V65 ≤ 50%; V70 ≤ 35%; V75 ≤ 25%; V80 ≤ 15% | |
| Penile bulb[47] (whole organ) | Mean dose to 95% of organ <50 Gy | <35% rate of severe erectile dysfunction |
| | Minimum dose to hottest 90% of organ <50 Gy | |
| | Dose to 60–70% penile bulb <70 Gy | <55% rate of severe erectile dysfunction |
| Small bowel[48] | V15 <120 cm$^3$ (bowel loops contoured) | <10% (≥Grade 3 acute toxicity) |
| | V45 <195 cm$^3$ (whole potential space-peritoneal bowel "bag") | |

Constraints summarized in Quantitative Analysis of Normal Tissue Effects in the Clinic (QUANTEC).[44] The basis of the dose constraint guidelines were established by evaluation of toxicity data from 3D-CRT published treatment outcomes.
* *Based on RTOG 0415 recommendations, based on ≥ Grade 3 late toxicity.*
Gy, gray; VX Gy <Y refers to volume receiving X Gy less than Y% or volume.

TABLE 45.4 Dose Constraints of Select Dose Escalated IMRT Studies

| Volume | RTOG 0126 – 79.2 Gy[23] | MSKCC – 81 Gy[49] | MSKCC – 86.4 Gy[50] |
|---|---|---|---|
| GTV (prostate) | $D_{min}$ – 79.2 Gy | $D_{min}$ – 81 Gy | $D_{min}$ – 86.4 Gy |
| PTV* | No more than 2% PTV can receive <79.2 Gy; $D_{max}$ – 84.7 Gy | No more than 10% of PTV can receive ≤77 Gy**; $D_{max}$ – 89 Gy | $D_{max}$ – 95.5 Gy |
| Rectum† | V75 Gy <15%; V70 Gy <25%; V65 Gy <35%; V60 Gy <50% | V47 Gy <53%; V75.6 Gy <30%; $D_{max}$ – 82 Gy | V47 Gy <53%; V75.6 Gy <30%; $D_{max}$ – 85.5 Gy |
| Bladder | V80 Gy <15%; V75 Gy <25%; V70 Gy <35%; V65 Gy <50% | V47 Gy <53% | V47 Gy <53%; V75.6 Gy <30% |
| Bowel | None given | | $D_{max}$ (large bowel) – 60 Gy<br>$D_{max}$ (small bowel) – 53 Gy |
| Penile bulb | Mean dose less ≤52.5 Gy | | None given |

Dose constraints described in each study. Values in the RTOG 0126 study represent constraints for the 79.2 Gy arm.
$D_{min}$, minimum dose; $D_{max}$, maximum point dose; V × Gy < Y refers to volume receiving × Gray less than Y%; GTV, gross tumor volume; PTV, planning target volume.
* *For RTOG 0126, an initial PTV included at least 0.5 cm expansion from a CTV containing the prostate and proximal 1.0 cm seminal vesicles, the PTV for MSKCC 81–86.4 Gy included prostate and seminal vesicles with a 1.0 cm expansion, except 0.6 cm posteriorly at rectum.*
** *The <81 Gy dose restricted to PTV/rectal volume interface, otherwise,* $D_{min}$ *– 81 Gy.*
† *Rectum contoured from anus (level of ischial tuberosities) for 15 cm length or rectosigmoid flexure for RTOG 0126, but only rectum 0.5 cm superior and inferior of the PTV was contoured for the MSKCC 81–86.4 Gy studies.*

alpha cradles, vac-loc or similar devices, which allows for consistent thigh and pelvic alignment. While patients can be simulated in either the supine or prone (using a thermoplastic shell) position, most centers have transitioned to supine simulation and treatment as prostate motion is less in the supine.[51] Some institutions also require emptying the rectum via a bowel regimen that varies intensity from laxatives to enema and rectal balloon. After the patient is positioned, the patient is scanned through an area centered on the prostate (or prostate bed if treating in the postoperative setting). The planning CT is then reviewed, an isocenter is placed near the center of the prostate or prostate bed, and the triangulation points for the isocenter are tattooed at the point of localizing laser intersection on the patient.

Target delineation is then performed using the CT dataset acquired at simulation. For localized prostate cancer, the CTV is defined as the prostate ± seminal vesicles (typically the proximal 1 cm). The CTV to PTV expansion may range from 6 mm to 1 cm depending on the estimation of daily setup or organ motion error and whether or not image guidance will be used (discussed further later). Normal tissues are also identified and contoured an each CT slice, including, but not limited to the walls of the rectum, bladder, femoral heads, and at risk bowel.

## Prostate Delineation

Accurate delineation of the prostate on noncontrasted CT datasets has been a topic of extensive study, and some controversy. While MRI provides the best resolution for accurate delineation of the prostate and seminal vesicles, it is not routinely performed as part of simulation or available as a separate study that can be "merged" with the CT simulation dataset. Rather, guidelines have been established to aid practitioners in CT-based contouring of the prostate. Comparison of MRI and CT images of the prostate have shown that CT estimation of the prostate volume is 32–40% larger when compared to MRI definition, with most of the variation in volume at the seminal vesicles, apex, and neurovascular bundle.[52–54] In response to these documented variabilities, McLaughlin et al. developed a set of methods to improve prostate contouring accuracy.[55] On review of CT- and MRI-generated contours of 300 patients, it was found that most contouring errors can be detected by verifying the contour on the lateral view projection to confirm an expected globular form and identification and recognition of the genitourinary diaphragm structures to delineate the prostate apex (Figure 45.2). Additionally, the RTOG has published a consensus panel contouring atlas regarding normal pelvic tissue anatomy.[56]

## Pelvic Nodal Delineation

In some circumstances, especially in clinical situations where risk of lymph node involvement may be estimated to be greater than 15%, elective nodal irradiation may be warranted.[57–59] While there is some debate as to the pattern of lymph node drainage in prostate cancer, the targeted nodes for elective nodal radiation should include the distal common iliac, presacral lymph nodes (S1–S3), external, internal, and obturator iliac lymph nodes with a 7 mm margin around the associated vessels, excluding bowel, bladder, muscle, bone or other anatomic boundaries where microscopic disease is unlikely to reside. A contouring atlas based on expert opinion has been made available by the RTOG is available for reference.[60] Of note, volumes are recommended to begin at the L5/S1 interspace and end at the superior aspect of the pubic bone. RTOG 75-06 indicated no benefit to elective periaortic lymph node irradiation.[10]

# PROSTATE MOTION

As discussed earlier, IMRT allows for rapid dose falloff around targets of interest, but care must be taken to account for prostate motion. Effective and accurate treatment in prostate cancer requires a thorough understanding of how the prostate location may vary between radiation treatments (interfraction) and during a single radiation treatment (intrafraction). Daily variations in rectal fullness, bladder filling variation, and pelvic muscle relaxation that may occur during the course of

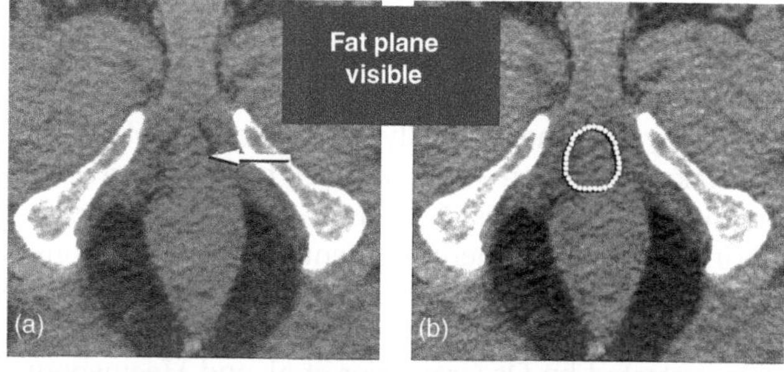


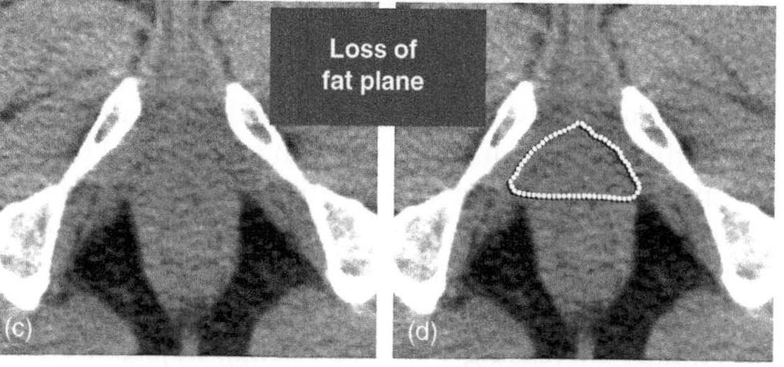


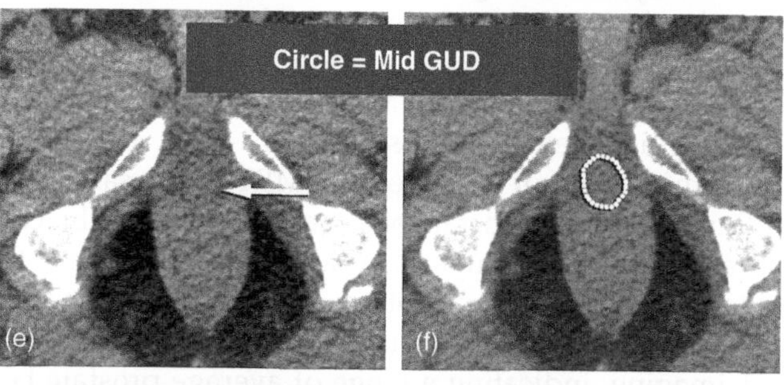


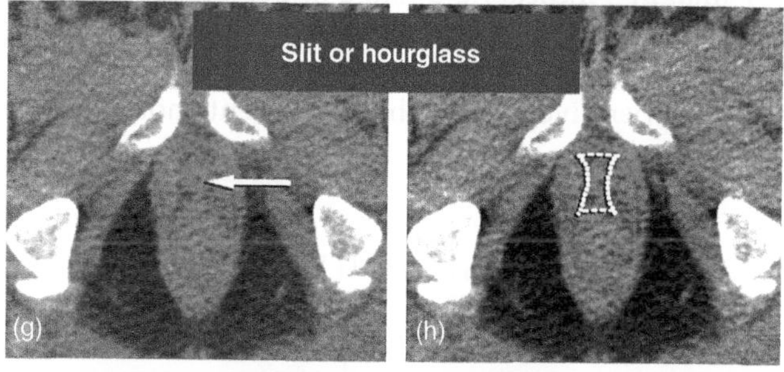


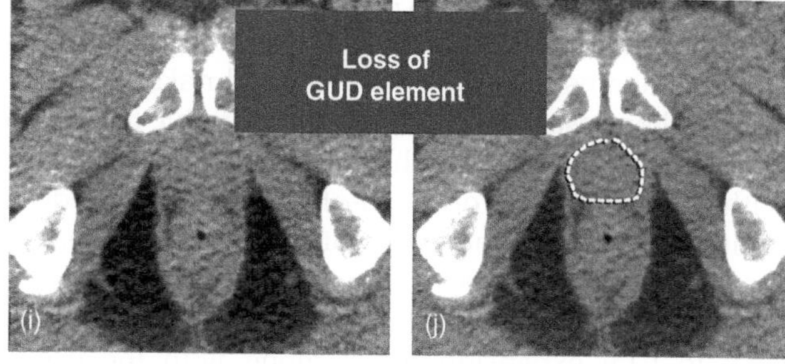


**FIGURE 45.2 CT Anatomic Features for Delineation of Prostate Apex.** McLaughlin et al.[55] outlined features of the genitourinary diaphragm (GUD) on axial CT imaging to assist with delineation of the prostate apex. Axial images are described superiorly to inferiorly (Figure 45.2a–i). (a and b) Penile bulb with loss of fat plane of central bulb and lateral pelvis. (c and d) Triangle with loss of fat plane; (e and f) highlight of the circular central GUD formed by the external sphincter. (g and h) An hourglass slit forms the superior GUD just below the apex. (i and j) Level of apex defined by loss of central prostate and lateral muscle separation. *Reprinted with permission from McLaughlin et al.*[55]

fractionated radiotherapy contribute to interfraction and potentially deformation of the prostate, while peristaltic activity within the rectum during treatment (intrafraction) may become clinically significant in IMRT plans that take more than a few minutes to deliver given the potential for the prostate to move outside of the field during critical treatment segments. These variations in target location can be significant, and if not accounted for, may also result in poorer patient outcomes. This was demonstrated by De Crevoisier et al. and Heemsbergen et al. where 5-year biochemical and local control were negatively affected by the presence of excessive rectal gas on planning CT, presumably leading to underdosing of the prostate due to correction of the rectal fullness during the actual course of treatment delivery without adjustment of the radiotherapy plan.[61,62] Thus, the practitioner must decide whether to account for this motion by increasing the size of the CTV to PTV expansion, thus potentially increasing the dose normal adjacent OARs, or using image guidance to locate the prostate or prostate bed during the course of radiotherapy allowing for a smaller PTV expansion. Most centers elect for use of image guidance where possible.

In the setting of prostate bed IMRT in the salvage or adjuvant setting, prostate bed motion must also be considered. Several groups have attempted to define the extent of prostate bed motion, whether using surgical clips as surrogates for fiducial markers or cone beam CT imaging, indicating a range of average prostate bed motion from 1.0 mm to 2.7 mm.[63–65] Given the extent of prostate bed motion, image guidance when using IMRT for postoperative prostate bed treatment should also be considered.

## IMAGE GUIDANCE

Given the indications of improved results with dose escalation in prostate cancer radiotherapy, the benefit of improved biochemical disease-free survival must be weighed against the potential increased risks of toxicity with higher dose therapy. While IMRT is able to provide excellent control over the shape of the high dose regions with relatively rapid dose fall-off to surrounding dose-limiting structures, the extent of normal tissue sparing can be hampered if there is significant uncertainty regarding target volume definitions necessitating a large CTV to PTV expansion. Thus, along with the use of IMRT for prostate cancer therapy, there has been a need to identify ways to accurately localize the prostate for each treatment, allowing for a reduction in the CTV to PTV expansion, and therefore preservation of smaller treatment fields and potential for reduced dose to surrounding OARs without compromise of tumor control probability. A variety of methods aimed at providing image guidance to daily radiotherapy treatments have thus emerged.

The clinical benefit of image guided-IMRT (IG-IMRT) has been demonstrated in select series. Zelefsky et al. retrospectively reviewed outcomes of 376 patients treated to 86.4 Gy with or without image-guided radiotherapy and demonstrated a reduction in 3-year likelihood of grade 2 or higher GU toxicity from 20.0% to 10.4% comparing non-IG-IMRT to IG-IMRT.[66] Furthermore, for high-risk patients, there was an improvement in 3-year biochemical disease-free outcomes. Similarly, image guidance techniques have been shown to negate the risk of biochemical and local failure previously observed in non-IGRT studies due to large rectal volume at simulation,[61,62] whether through daily ultrasound (US)-based positioning[67] or CT-based off-line adaptive radiotherapy with replanning after one week based on observed patient-specific interfractional prostate motion.[68] The decrease in acute toxicity with the addition of image guidance for prostate only, prostate/pelvic, or prostate bed radiotherapy has been reported by many, in large part due to the ability to reduce treatment margins and more accurately deliver dose as planned.[63,69–71] The following sections aim to summarize the modalities available for image guidance and review the benefits and disadvantages of each of the systems in use (Table 45.5).

### Bony Landmarks

Use of implanted markers or CT scans have been used extensively to demonstrate a poor correlation of prostate position with bony anatomy.[72–74] Balter et al. showed a maximum displacement of 7.5 mm in a series of 10 patients, although typical movement was less than 5 mm. The clinical significance of this discrepancy was investigated by Paluska et al., who compared IMRT treatment setting up to bony landmarks using a 1 cm PTV compared to treatment using implanted fiducial markers and a 7 mm PTV then used weekly cone-beam CT imaging to reconstruct the dose distribution.[75] Prostate coverage using bony landmarks for daily alignment with a 1 cm PTV expansion was significantly worse than the prostate coverage achieved with fiducial marker alignment and a 0.7 cm PTV expansion. Thus, use of bony landmarks alone for image guidance is not recommended.

### Fiducial Markers

Fiducial markers (FM) have been available long before the era of IMRT, and were developed as a means for positional verification in tissues not easily visualized by portal kV/MV imaging, such as the prostate. Over the last two decades, their use as a means for prostate localization in IGRT has increased. Fiducial markers are typically cylindrical seeds made of gold, and are implanted

**TABLE 45.5** Comparison of Image-Guided Techniques for Prostate IMRT

| | Bony landmarks | Fiducial markers | Cone-beam CT | Calypso 4D | Ultrasound |
|---|---|---|---|---|---|
| Patient eligibility | All | Contraindicated if some patients receiving anticoagulation | All | Contraindicated if large abdominal girth, pacemaker, or other electromagnetic device | All |
| Correlation with prostate position | Poor, up to 1 cm discrepancy | Excellent | Excellent | Excellent | Fair |
| Interuser variability | Yes | Yes | Yes | Minimal | Significant |
| Intrafraction motion tracking capability | No | No | No | Yes | No |
| Excess ionizing radiation exposure | Yes, minimal | Yes, minimal | Yes, likely clinically insignificant | No | No |
| Invasive procedure | No | Yes | No | Yes | Minimally |

The advantages and disadvantages of the most widely used image guidance techniques are summarized.

into the prostate by transrectal ultrasound, with a minimum of three required for all dimensions of correction. The surface of the seeds are modified to reduce risk for seed migration after insertion and while seed migration can occur, stability of FM placement has been well documented.[76–79] Given that the prostate/rectal interface is the most critical area to align, placing the markers in the posterior part of the prostate is preferred. After placement, typically at least one week is given before moving forward with CT simulation. The locations of the FMs are delineated on the CT image dataset, so that daily kV or MV images of the FMs can be compared to DRRs allowing for appropriate daily shifts to be made prior to treatment delivery of the IMRT plan.

Analysis of 453 patients undergoing FM-based IG-IMRT indicated that prostate position can be detected with an accuracy of 0.6 mm.[80] Furthermore, interuser variability is minimal.[81,82] Recently, fiducial-based IG-IMRT was also shown to provide excellent means of localization resulting in accurate delivery of dose to pelvic lymph node PTVs in patients with high-risk prostate cancer >99% of the time, strengthening support for fiducial-based IG-IMRT even if lymph nodes are targeted.[83]

In the postoperative setting, implanted fiducial markers have also been explored. Schiffner et al. evaluated the utility of TRUS-implanted gold seed fiducials in patients undergoing salvage or adjuvant radiotherapy.[84] No significant seed migration was identified, and average prostate bed motion was 0.2, 1.2, and 0.3 mm in the L-R, S-I, and A-P dimensions, respectively, although maximum range of motion from setup position was 22.3, 12.7, and 9.9 mm, respectively. In all, the percentage of treatments that required shifts >5 mm from setup position was 14.1%, 38.7%, and 28.2%, respectively. A recent report also noted that implanted fiducials were more reliable as a basis for image guidance as opposed to surgical clips as there was less interobserver variability in matching.[85]

While in many respects, FMs have been considered a gold standard for prostate localization, the disadvantages must be considered. FM-based IG-IMRT include the need for an invasive procedure for placement of the FMs, thus potential ineligibility of certain patients with coagulopathies, and small, but real risk of seed migration or seed embolization. Additionally, this system requires daily kV or MV imaging, resulting in excess radiation dose throughout the course of an entire treatment regimen. However, the excess radiation dose has been estimated by several groups, and is likely in the range of 2–10 mGy for pretreatment daily imaging, less than a single cone-beam CT scan (22 mGy), or up to 185–406 mGy if performing intrafractional imaging.[86,87] In all, compared to the prescribed dose delivered, the additional dose from imaging is likely insignificant.

## Cone-Beam

In-room cone-beam CT (CBCT) systems have been used as part of IGRT since 1997. In comparison to traditional CT, CBCT uses a cone-shaped X-ray beam that allows for acquisition of complete volume information in a single rotation, allowing for faster image acquisition than in helical imaging. CBCT is mounted to the treatment unit (onboard imaging), and thus can rotate at the same isocenter as the treatment unit. CBCT performed prior to treatment delivery can be compared to CT simulation soft tissue anatomy and planned isocenter. If anatomical variations are noted, and a shift is needed, this can be performed prior to radiation delivery. The advantage of CBCT is thus that patients can be precisely positioned based on cross-sectional imaging on the treatment unit itself.

Given that soft tissue anatomy can vary, and become deformed, appropriate delineation of where to "match" fields should be determined. Langen et al. reported on interuser variability of registration methods utilizing only the planning contours or the entire grayscale anatomy obtained with CBCT.[88] Misalignments were less frequent when users were asked to localize based off the entire grayscale anatomy rather than contours alone, occurring less than 5% of the time at a 5 mm threshold. CBCT has also been shown to provide excellent ability to ensure the prostate and pelvic lymph nodes receive acceptable coverage >97% of the time when soft tissue anatomy is used for alignment.[89] Anderson et al. showed rectal volume and rectal diameter normalized during a course of IMRT for prostate cancer.[90] One of the most significant advantages of daily CBCT is the ability to identify changes in rectal volume, which then allows for pretreatment correction and more accurate delivery. In the setting of prostate bed image-guided radiotherapy, CBCT also shows promise.[65]

There are few contraindications to the use of CBCT, making it nearly a universal option for image guidance in prostate IMRT. Even in the presence of hip prosthesis, which might distort kV CBCT, this artifact can be nearly eliminated if MV CBCT is available and still allow CBCT image guidance.[91] The disadvantages of CBCT for image guidance include the inability to detect intrafraction motion and the excess radiation exposure to patients, thus potential increased risk for secondary malignancy. Several groups have evaluated the estimated excess dose for daily CBCT. Kan et al. reviewed CBCT doses from a single CBCT scan of the pelvis, and estimated an effective dose of 22.7 mGy per scan at standard mode delivery, thus 1 Gy over a treatment course if 44 treatments assumed.[92] This was assumed to correlate to a 2–4% increased risk for secondary cancer. However, if a low-dose CBCT mode is utilized, the excess dose can be reduced by 80%. Deng et al. showed that half-fan kV-CBCT delivers an estimated clinically insignificant additional dose of 3.4 cGy to the prostate (only 1.7% of the dose from the IMRT treatment), although testicular dose was 2.9 cGy (approximately 300% of the dose that would be given by the IMRT plan). However, reducing the CBCT field span from 30 cm to 10 cm resulted in a reduction of testicular dose to 0.1–0.2 cGy.[93]

### Implanted Radiofrequency ransponders – Calypso Beacon 4D Localization System

Within the last decade, a new transponder system was developed to allow for target localization and real-time tracking for radiotherapy. In the Calpyso 4D system (Varian Medical Systems, Palo Alto, CA), three nonionizing electromagnetic beacons measuring approximately 8 mm in length are permanently placed under transrectal ultrasound guidance within the prostate. During daily treatment, these transponders communicate via 10 Hz radio waves to an antenna array that is placed within close proximity to the patient. Positional information can then be obtained, allowing for objective, user-independent tracking of the position of the beacons (thus the prostate) in real time. If the beacons move beyond a certain threshold during treatment (intrafraction motion), treatment software and therapists will be alerted and radiotherapy can be interrupted to allow for patient repositioning. Such real-time tracking can allow for theoretically more accurate delivery of IMRT.

The use of these implantable radiofrequency transponders has become widely adopted for IG-IMRT in the prostate. Initial studies showed comparable accuracy to FM-based systems as well as reproducibility across multiple institutions.[94,95] Real-time tracking using this system has shown that intrafraction prostate motion is nontrivial. In a study of 17 patients, the average time the prostate was displaced >3 mm and >5 mm during treatment was 13.6% and 3.3%, respectively.[96] Displacement of the prostate was also noted to increase with time, such that 10 min after initial tracking, 25% of the time displacement of >3 mm was identified. As expected, real time tracking is shown to improve target coverage and aid with rectal sparing during radiotherapy.[97] Sandler et al. reported an experience of treating 64 patients with reduced treatment margins of 3 mm, resulting in reduced patient-reported morbidity in patients receiving 81 Gy.[98] No PSA relapse rates have been reported in this cohort, thus longer-term follow up and further studies would be necessary to know if reduced margins may result in increased PSA failure rates.

Similar to gold fiducials, implanted radiofrequency transponders can be used to help identify prostate bed motion in the adjuvant or salvage treatment settings. Real-time tracking indicates that both interfraction and intrafraction motion in the prostate bed is nontrivial. In a study of 20 patients who underwent US-guided placement of three transponders within the prostate bed, prostate bed motion of more than 5 mm lasting 30 s was identified in 11% of treatments, with motion twice as likely to occur posteriorly than anteriorly.[99] Nearly 70% of patients were repositioned during their course of radiotherapy, with 15% of treatments interrupted for repositioning.

Significant advantages of implanted radiofrequency transponders are the ability to perform real-time patient/target motion allowing for one to address any significant intrafraction motion that may occur and thus allow consideration of reduced treatment margin, the lack of additional ionizing radiation as exists with fiducial- or CBCT-based image guidance, and the lack of interuser variability as transponder localization is reported objectively by the systems software. Unfortunately, not all patients are candidates for transponder placement. Ineligible patients include persons with large body habitus that would lead to transponder to array distance >27 cm[100] and those with pacemakers or other electromagnetic devices. Some have indicated that the presence of hip prostheses or other large metal implants may preclude use of this system, although a recent report suggests that tracking error may be negligible in this setting.[101] Disadvantages of this system also include the need for an invasive procedure for the transponders to be placed and the lack of information regarding potential prostate gland deformity. Furthermore, the prostate is not well visualized on MRI imaging after fiducial placement due to imaging artifact from the fiducials themselves.

### Ultrasound

Transabdominal ultrasound can be an efficient tool for daily prostate localization. This was first adopted in the 1990s with conformal prostate radiation therapy, and requires performance of daily US to locate the prostate with matching to CT simulation images. Advantages to this system included the lack of additional radiation exposure for image acquisition and absence of the need for an invasive procedure, but unfortunately, a wide range of disadvantages exist.

Interobserver and interuser variability is a significant concern. Image quality is dependent on the pressure applied to the abdomen, or angle of the probe on the skin, leading to inaccuracies in patient setup.[102,103] These errors in image acquisition may also play a role in the documented discrepancies of US image localization compared to FM- or CT-based methods. When compared to FMs, a disagreement of an average vector distance of 8.8 mm has been reported,[104] although other studies have suggested average discrepancies of up to 2.7 mm (±4.5 mm) in a single axis.[105] Compared to CT scan, average disagreements of 4.6 mm in a single direction with a maximum difference of 9.0 mm have been shown.[106] In all, the concern over accuracy of localization limits the use of this modality for IG-IMRT.

## CONCLUSIONS

In conclusion, there has been significant improvement in the way external beam radiation is planned and delivered for men with prostate cancer. As the field has moved from 2D to 3D to IMRT planning, and treatment has evolved from minimal patient immobilization to custom immobilization and image guidance, the amount of normal tissue treated with radiotherapy has decreased, and the doses able to be delivered to the prostate have increased. Therefore, cure rates have increased even as complication rates have fallen. Future innovations in radiation therapy, including the maturation of stereotactic body radiotherapy and the optimal integration of radiotherapy with androgen deprivation therapy, chemotherapy, and immune therapy for high-risk disease, will hopefully improve outcomes even further.

## References


1. Horwitz EM, Hanks GE. External beam radiation therapy for prostate cancer. *CA Cancer J Clin* 2000;**50**(6):349–75; quiz 76–9.
2. Kantzou I, Platoni K, Sandilos P, et al. Conventional versus virtual simulation for radiation treatment planning of prostate cancer: final results. *J BUON* 2011;**16**(2):309–15.
3. Zietman AL, Coen JJ, Dallow KC, et al. The treatment of prostate cancer by conventional radiation therapy: an analysis of long-term outcome. *Int J Radiat Oncol Biol Phys* 1995;**32**(2):287–92.
4. Horwitz EM, Vicini FA, Ziaja EL, et al. Assessing the variability of outcome for patients treated with localized prostate irradiation using different definitions of biochemical control. *Int J Radiat Oncol Biol Phys* 1996;**36**(3):565–71.
5. Zagars GK, Pollack A, von Eschenbach AC. Prognostic factors for clinically localized prostate carcinoma: analysis of 938 patients irradiated in the prostate specific antigen era. *Cancer* 1997;**79**(7):1370–80.
6. Keyser D, Kupelian PA, Zippe CD, et al. Stage T1-2 prostate cancer with pretreatment prostate-specific antigen level < or = 10 ng/ml: radiation therapy or surgery? *Int J Radiat Oncol Biol Phys* 1997;**38**(4):723–9.
7. Pisansky TM, Cha SS, Earle JD, et al. Prostate-specific antigen as a pretherapy prognostic factor in patients treated with radiation therapy for clinically localized prostate cancer. *J Clin Oncol* 1993;**11**(11):2158–66.
8. Kaplan ID, Cox RS, Bagshaw MA. Prostate specific antigen after external beam radiotherapy for prostatic cancer: followup. *J Urol* 1993;**149**(3):519–22.
9. Pilepich MV, Asbell SO, Krall JM, et al. Correlation of radiotherapeutic parameters and treatment related morbidity – analysis of RTOG Study 77-06. *Int J Radiat Oncol Biol Phys* 1987;**13**(7):1007–12.
10. Pilepich MV, Krall JM, Sause WT, et al. Correlation of radiotherapeutic parameters and treatment related morbidity in carcinoma of the prostate – analysis of RTOG study 75-06. *Int J Radiat Oncol Biol Phys* 1987;**13**(3):351–7.
11. Soffen EM, Hanks GE, Hunt MA, et al. Conformal static field radiation therapy treatment of early prostate cancer versus nonconformal techniques: a reduction in acute morbidity. *Int J Radiat Oncol Biol Phys* 1992;**24**(3):485–8.
12. Koper PC, Stroom JC, van Putten WL, et al. Acute morbidity reduction using 3DCRT for prostate carcinoma: a randomized study. *Int J Radiat Oncol Biol Phys* 1999;**43**(4):727–34.
13. Hanks GE, Hanlon AL, Epstein B, et al. Dose response in prostate cancer with 8–12 years' follow-up. *Int J Radiat Oncol Biol Phys* 2002;**54**(2):427–35.
14. Beckendorf V, Guerif S, Le Prise E, et al. The GETUG 70 Gy vs. 80 Gy randomized trial for localized prostate cancer: feasibility and acute toxicity. *Int J Radiat Oncol Biol Phys* 2004;**60**(4):1056–65.
15. Michalski J, Winter K, Roach M, et al. Clinical outcome of patients treated with 3D conformal radiation therapy (3D-CRT) for prostate cancer on RTOG 9406. *Int J Radiat Oncol Biol Phys* 2012;**83**(3):-e363–70.
16. Bortfeld T. IMRT: a review and preview. *Phys Med Biol* 2006;**51**(13):R363–79.
17. Tsai CL, Wu JK, Chao HL, et al. Treatment and dosimetric advantages between VMAT, IMRT, and helical tomotherapy in prostate cancer. *Med Dosim* 2011;**36**(3):264–71.

18. Ling CC, Burman C, Chui CS, et al. Conformal radiation treatment of prostate cancer using inversely-planned intensity-modulated photon beams produced with dynamic multileaf collimation. *Int J Radiat Oncol Biol Phys* 1996;**35**(4):721–30.
19. Zelefsky MJ, Fuks Z, Happersett L, et al. Clinical experience with intensity modulated radiation therapy (IMRT) in prostate cancer. *Radiother Oncol* 2000;**55**(3):241–9.
20. Wu VW, Kwong DL, Sham JS. Target dose conformity in 3-dimensional conformal radiotherapy and intensity modulated radiotherapy. *Radioth Oncol* 2004;**71**(2):201–6.
21. Zelefsky MJ, Fuks Z, Hunt M, et al. High-dose intensity modulated radiation therapy for prostate cancer: early toxicity and biochemical outcome in 772 patients. *Int J Radiat Oncol Biol Phys* 2002;**53**(5):1111–6.
22. Vora SA, Wong WW, Schild SE, et al. Outcome and toxicity for patients treated with intensity modulated radiation therapy for localized prostate cancer. *J Urol* 2013;**190**(2):521–6.
23. Michalski JM, Yan Y, Watkins-Bruner D, et al. Preliminary toxicity analysis of 3-dimensional conformal radiation therapy versus intensity modulated radiation therapy on the high-dose arm of the Radiation Therapy Oncology Group 0126 prostate cancer trial. *Int J Radiat Oncol Biol Phys* 2013;**87**(5):932–8.
24. Ashman JB, Zelefsky MJ, Hunt MS, et al. Whole pelvic radiotherapy for prostate cancer using 3D conformal and intensity-modulated radiotherapy. *Int J Radiat Oncol Biol Phys* 2005;**63**(3):765–71.
25. Muren LP, Wasbo E, Helle SI, et al. Intensity-modulated radiotherapy of pelvic lymph nodes in locally advanced prostate cancer: planning procedures and early experiences. *Int J Radiat Oncol Biol Phys* 2008;**71**(4):1034–41.
26. Wang-Chesebro A, Xia P, Coleman J, et al. Intensity-modulated radiotherapy improves lymph node coverage and dose to critical structures compared with three-dimensional conformal radiation therapy in clinically localized prostate cancer. *Int J Radiat Oncol Biol Phys* 2006;**66**(3):654–62.
27. Studenski MT, Shen X, Yu Y, et al. Intensity-modulated radiation therapy and volumetric-modulated arc therapy for adult craniospinal irradiation – a comparison with traditional techniques. *Med Dosim* 2013;**38**(1):48–54.
28. Corbin KS, Kunnavakkam R, Eggener SE, et al. Intensity modulated radiation therapy after radical prostatectomy: Early results show no decline in urinary continence, gastrointestinal, or sexual quality of life. *Pract Radiat Oncol* 2013;**3**(2):138–44.
29. Riou O, Laliberte B, Azria D, et al. Implementing intensity modulated radiotherapy to the prostate bed: dosimetric study and early clinical results. *Med Dosim* 2013;**38**(2):117–21.
30. Goenka A, Magsanoc JM, Pei X, et al. Improved toxicity profile following high-dose postprostatectomy salvage radiation therapy with intensity-modulated radiation therapy. *Eur Urol* 2011;**60**(6):1142–8.
31. Bauman G, Rumble RB, Chen J, et al. Intensity-modulated radiotherapy in the treatment of prostate cancer. *Clin Oncol (R Coll Radiol)* 2012;**24**(7):461–73.
32. Dearnaley DP, Jovic G, Syndikus I, et al. Escalated-dose versus control-dose conformal radiotherapy for prostate cancer: long-term results from the MRC RT01 randomised controlled trial. *Lancet Oncol* 2014;**15**(4):464–73.
33. Dearnaley DP, Sydes MR, Graham JD, et al. Escalated-dose versus standard-dose conformal radiotherapy in prostate cancer: first results from the MRC RT01 randomised controlled trial. *Lancet Oncol* 2007;**8**(6):475–87.
34. Beckendorf V, Guerif S, Le Prise E, et al. 70 Gy versus 80 Gy in localized prostate cancer: 5-year results of GETUG 06 randomized trial. *Int J Radiat Oncol Biol Phys* 2011;**80**(4):1056–63.
35. Peeters ST, Heemsbergen WD, Koper PC, et al. Dose-response in radiotherapy for localized prostate cancer: results of the Dutch multicenter randomized phase III trial comparing 68 Gy of radiotherapy with 78 Gy. *J Clin Oncol* 2006;**24**(13):1990–6.
36. Al-Mamgani A, van Putten WL, Heemsbergen WD, et al. Update of Dutch multicenter dose-escalation trial of radiotherapy for localized prostate cancer. *Int J Radiat Oncol Biol Phys* 2008;**72**(4):980–8.
37. Zietman AL, Bae K, Slater JD, et al. Randomized trial comparing conventional-dose with high-dose conformal radiation therapy in early-stage adenocarcinoma of the prostate: long-term results from Proton Radiation Oncology Group/American College Of Radiology 95-09. *J Clin Oncol* 2010;**28**(7):1106–11.
38. Kuban DA, Tucker SL, Dong L, et al. Long-term results of the M. D. Anderson randomized dose-escalation trial for prostate cancer. *Int J Radiat Oncol Biol Phys* 2008;**70**(1):67–74.
39. Alicikus ZA, Yamada Y, Zhang Z, et al. Ten-year outcomes of high-dose, intensity-modulated radiotherapy for localized prostate cancer. *Cancer* 2011;**117**(7):1429–37.
40. Spratt DE, Pei X, Yamada J, et al. Long-term survival and toxicity in patients treated with high-dose intensity modulated radiation therapy for localized prostate cancer. *Int J Radiat Oncol Biol Phys* 2013;**85**(3):686–92.
41. Michalski JM, Bae K, Roach M, et al. Long-term toxicity following 3D conformal radiation therapy for prostate cancer from the RTOG 9406 phase I/II dose escalation study. *Int J Radiat Oncol Biol Phys* 2010;**76**(1):14–22.
42. Swisher-McClure S, Mitra N, Woo K, et al. Increasing use of dose-escalated external beam radiation therapy for men with nonmetastatic prostate cancer. *Int J Radiat Oncol Biol Phys* 2014;**89**(1):103–12.
43. Mohler JL, Kantoff PW, Armstrong AJ, et al. Prostate Cancer, Version 2.2014. *J Natl Compr Canc Netw* 2014;**12**(5):686–718.
44. Marks LB, Yorke ED, Jackson A, et al. Use of normal tissue complication probability models in the clinic. *Int J Radiat Oncol Biol Phys* 2010;**76**(3 Suppl):S10–9.
45. Michalski JM, Gay H, Jackson A, et al. Radiation dose-volume effects in radiation-induced rectal injury. *Int J Radiat Oncol Biol Phys* 2010;**76**(3 Suppl):S123–9.
46. Viswanathan AN, Yorke ED, Marks LB, et al. Radiation dose-volume effects of the urinary bladder. *Int J Radiat Oncol Biol Phys* 2010;**76**(3 Suppl):S116–22.
47. Roach III M, Nam J, Gagliardi G, et al. Radiation dose-volume effects and the penile bulb. *Int J Radiat Oncol Biol Phys* 2010;**76** (3 Suppl):S130–4.
48. Kavanagh BD, Pan CC, Dawson LA, et al. Radiation dose-volume effects in the stomach and small bowel. *Int J Radiat Oncol Biol Phys* 2010;**76**(3 Suppl):S101–7.
49. Zelefsky MJ, Chan H, Hunt M, et al. Long-term outcome of high dose intensity modulated radiation therapy for patients with clinically localized prostate cancer. *J Urol* 2006;**176**(4 Pt. 1):1415–9.
50. Cahlon O, Zelefsky MJ, Shippy A, et al. Ultra-high dose (86.4 Gy) IMRT for localized prostate cancer: toxicity and biochemical outcomes. *Int J Radiat Oncol Biol Phys* 2008;**71**(2):330–7.
51. Bayley AJ, Catton CN, Haycocks T, et al. A randomized trial of supine vs. prone positioning in patients undergoing escalated dose conformal radiotherapy for prostate cancer. *Radiother Oncol* 2004;**70**(1):37–44.
52. Roach III M, Faillace-Akazawa P, Malfatti C, et al. Prostate volumes defined by magnetic resonance imaging and computerized tomographic scans for three-dimensional conformal radiotherapy. *Int J Radiat Oncol Biol Phys* 1996;**35**(5):1011–8.
53. Milosevic M, Voruganti S, Blend R, et al. Magnetic resonance imaging (MRI) for localization of the prostatic apex: comparison to computed tomography (CT) and urethrography. *Radiother Oncol* 1998;**47**(3):277–84.
54. Rasch C, Barillot I, Remeijer P, et al. Definition of the prostate in CT and MRI: a multi-observer study. *Int J Radiat Oncol Biol Phys* 1999;**43**(1):57–66.
55. McLaughlin PW, Evans C, Feng M, et al. Radiographic and anatomic basis for prostate contouring errors and methods to

improve prostate contouring accuracy. *Int J Radiat Oncol Biol Phys* 2010;**76**(2):369–78.

56. Gay HA, Barthold HJ, O'Meara E, et al. Pelvic normal tissue contouring guidelines for radiation therapy: a Radiation Therapy Oncology Group consensus panel atlas. *Int J Radiat Oncol Biol Phys* 2012;**83**(3):e353–62.
57. Yu JB, Makarov DV, Gross C. A new formula for prostate cancer lymph node risk. *Int J Radiat Oncol Biol Phys* 2011;**80**(1):69–75.
58. Nguyen PL, Chen MH, Hoffman KE, et al. Predicting the risk of pelvic node involvement among men with prostate cancer in the contemporary era. *Int J Radiat Oncol Biol Phys* 2009;**74**(1):104–9.
59. Roach III M, DeSilvio M, Lawton C, et al. Phase III trial comparing whole-pelvic versus prostate-only radiotherapy and neoadjuvant versus adjuvant combined androgen suppression: Radiation Therapy Oncology Group 9413. *J Clin Oncol* 2003;**21**(10):1904–11.
60. Lawton CA, Michalski J, El-Naqa I, et al. RTOG GU Radiation oncology specialists reach consensus on pelvic lymph node volumes for high-risk prostate cancer. *Int J Radiat Oncol Biol Phys* 2009;**74**(2):383–7.
61. de Crevoisier R, Tucker SL, Dong L, et al. Increased risk of biochemical and local failure in patients with distended rectum on the planning CT for prostate cancer radiotherapy. *Int J Radiat Oncol Biol Phys* 2005;**62**(4):965–73.
62. Heemsbergen WD, Hoogeman MS, Witte MG, et al. Increased risk of biochemical and clinical failure for prostate patients with a large rectum at radiotherapy planning: results from the Dutch trial of 68 Gy versus 78 Gy. *Int J Radiat Oncol Biol Phys* 2007;**67**(5):1418–24.
63. Sandhu A, Sethi R, Rice R, et al. Prostate bed localization with image-guided approach using on-board imaging: reporting acute toxicity and implications for radiation therapy planning following prostatectomy. *Radiother Oncol* 2008;**88**(1):20–5.
64. Huang K, Palma DA, Scott D, et al. Inter- and intrafraction uncertainty in prostate bed image-guided radiotherapy. *Int J Radiat Oncol Biol Phys* 2012;**84**(2):402–7.
65. Simpson DR, Einck JP, Nath SK, et al. Comparison of daily cone-beam computed tomography and kilovoltage planar imaging for target localization in prostate cancer patients following radical prostatectomy. *Pract Radiat Oncol* 2011;**1**(3):156–62.
66. Zelefsky MJ, Kollmeier M, Cox B, et al. Improved clinical outcomes with high-dose image guided radiotherapy compared with non-IGRT for the treatment of clinically localized prostate cancer. *Int J Radiat Oncol Biol Phys* 2012;**84**(1):125–9.
67. Kupelian PA, Willoughby TR, Reddy CA, et al. Impact of image guidance on outcomes after external beam radiotherapy for localized prostate cancer. *Int J Radiat Oncol Biol Phys* 2008;**70**(4):1146–50.
68. Park SS, Yan D, McGrath S, et al. Adaptive image-guided radiotherapy (IGRT) eliminates the risk of biochemical failure caused by the bias of rectal distension in prostate cancer treatment planning: clinical evidence. *Int J Radiat Oncol Biol Phys* 2012;**83**(3):947–52.
69. Chung HT, Xia P, Chan LW, et al. Does image-guided radiotherapy improve toxicity profile in whole pelvic-treated high-risk prostate cancer? Comparison between IG-IMRT and IMRT. *Int J Radiat Oncol Biol Phys* 2009;**73**(1):53–60.
70. Sveistrup J, af Rosenschold PM, Deasy JO, et al. Improvement in toxicity in high risk prostate cancer patients treated with image-guided intensity-modulated radiotherapy compared to 3D conformal radiotherapy without daily image guidance. *Radiat Oncol* 2014;**9**:44.
71. Nath SK, Sandhu AP, Rose BS, et al. Toxicity analysis of postoperative image-guided intensity-modulated radiotherapy for prostate cancer. *Int J Radiat Oncol Biol Phys* 2010;**78**(2):435–41.
72. Crook JM, Raymond Y, Salhani D, et al. Prostate motion during standard radiotherapy as assessed by fiducial markers. *Radiother Oncol* 1995;**37**(1):35–42.
73. Balter JM, Sandler HM, Lam K, et al. Measurement of prostate movement over the course of routine radiotherapy using implanted markers. *Int J Radiat Oncol Biol Phys* 1995;**31**(1):113–8.
74. Schallenkamp JM, Herman MG, Kruse JJ, et al. Prostate position relative to pelvic bony anatomy based on intraprostatic gold markers and electronic portal imaging. *Int J Radiat Oncol Biol Phys* 2005;**63**(3):800–11.
75. Paluska P, Hanus J, Sefrova J, et al. Utilization of cone beam CT for reconstruction of dose distribution delivered in image-guided radiotherapy of prostate carcinoma – bony landmark setup compared to fiducial markers setup. *J Appl Clin Med Phys* 2013;**14**(3):4203.
76. Dehnad H, Nederveen AJ, van der Heide UA, et al. Clinical feasibility study for the use of implanted gold seeds in the prostate as reliable positioning markers during megavoltage irradiation. *Radiother Oncol* 2003;**67**(3):295–302.
77. Kupelian PA, Willoughby TR, Meeks SL, et al. Intraprostatic fiducials for localization of the prostate gland: monitoring intermarker distances during radiation therapy to test for marker stability. *Int J Radiat Oncol Biol Phys* 2005;**62**(5):1291–6.
78. Poggi MM, Gant DA, Sewchand W, et al. Marker seed migration in prostate localization. *Int J Radiat Oncol Biol Phys* 2003;**56**(5):1248–51.
79. Pouliot J, Aubin M, Langen KM, et al. (Non)-migration of radiopaque markers used for on-line localization of the prostate with an electronic portal imaging device. *Int J Radiat Oncol Biol Phys* 2003;**56**(3):862–6.
80. van der Heide UA, Kotte AN, Dehnad H, et al. Analysis of fiducial marker-based position verification in the external beam radiotherapy of patients with prostate cancer. *Radiother Oncol* 2007;**82**(1):38–45.
81. Ullman KL, Ning H, Susil RC, et al. Intra- and inter-radiation therapist reproducibility of daily isocenter verification using prostatic fiducial markers. *Radiat Oncol* 2006;**1**:2.
82. Deegan T, Owen R, Holt T, et al. Interobserver variability of radiation therapists aligning to fiducial markers for prostate radiation therapy. *J Med Imaging Radiat Oncol* 2013;**57**(4):519–23 quiz 24–5.
83. Eminowicz G, Dean C, Shoffren O, et al. Intensity-modulated radiotherapy (IMRT) to prostate and pelvic nodes-is pelvic lymph node coverage adequate with fiducial-based image-guided radiotherapy? *Br J Radiol* 2014;**87**(1037):20130696.
84. Schiffner DC, Gottschalk AR, Lometti M, et al. Daily electronic portal imaging of implanted gold seed fiducials in patients undergoing radiotherapy after radical prostatectomy. *Int J Radiat Oncol Biol Phys* 2007;**67**(2):610–9.
85. Fortin I, Carrier JF, Beauchemin MC, et al. Using fiducial markers in the prostate bed in postprostatectomy external beam radiation therapy improves accuracy over surgical clips. *Strahlenther Onkol* 2014;**190**(5):467–71.
86. Ng JA, Booth J, Poulsen P, et al. Estimation of effective imaging dose for kilovoltage intratreatment monitoring of the prostate position during cancer radiotherapy. *Phys Med Biol* 2013;**58**(17):5983–96.
87. Crocker JK, Ng JA, Keall PJ, et al. Measurement of patient imaging dose for real-time kilovoltage X-ray intrafraction tumour position monitoring in prostate patients. *Phys Med Biol* 2012;**57**(10):2969–80.
88. Langen KM, Zhang Y, Andrews RD, et al. Initial experience with megavoltage (MV) CT guidance for daily prostate alignments. *Int J Radiat Oncol Biol Phys* 2005;**62**(5):1517–24.
89. Ferjani S, Huang G, Shang Q, et al. Alignment focus of daily image guidance for concurrent treatment of prostate and pelvic lymph nodes. *Int J Radiat Oncol Biol Phys* 2013;**87**(2):383–9.
90. Anderson NS, Yu JB, Peschel RE, et al. A significant decrease in rectal volume and diameter during prostate IMRT. *Radiother Oncol* 2011;**98**(2):187–91.
91. Aubin M, Morin O, Chen J, et al. The use of megavoltage cone-beam CT to complement CT for target definition in pelvic radiotherapy in the presence of hip replacement. *Br J Radiol* 2006;**79**(947):918–21.
92. Kan MW, Leung LH, Wong W, et al. Radiation dose from cone beam computed tomography for image-guided radiation therapy. *Int J Radiat Oncol Biol Phys* 2008;**70**(1):272–9.

93. Deng J, Chen Z, Yu JB, et al. Testicular doses in image-guided radiotherapy of prostate cancer. *Int J Radiat Oncol Biol Phys* 2012;**82**(1):e39–e47.
94. Willoughby TR, Kupelian PA, Pouliot J, et al. Target localization and real-time tracking using the Calypso 4D localization system in patients with localized prostate cancer. *Int J Radiat Oncol Biol Phys* 2006;**65**(2):528–34.
95. Kupelian P, Willoughby T, Mahadevan A, et al. Multi-institutional clinical experience with the Calypso System in localization and continuous, real-time monitoring of the prostate gland during external radiotherapy. *Int J Radiat Oncol Biol Phys* 2007;**67**(4):1088–98.
96. Langen KM, Willoughby TR, Meeks SL, et al. Observations on real-time prostate gland motion using electromagnetic tracking. *Int J Radiat Oncol Biol Phys* 2008;**71**(4):1084–90.
97. Rajendran RR, Plastaras JP, Mick R, et al. Daily isocenter correction with electromagnetic-based localization improves target coverage and rectal sparing during prostate radiotherapy. *Int J Radiat Oncol Biol Phys* 2010;**76**(4):1092–9.
98. Sandler HM, Liu PY, Dunn RL, et al. Reduction in patient-reported acute morbidity in prostate cancer patients treated with 81-Gy intensity-modulated radiotherapy using reduced planning target volume margins and electromagnetic tracking: assessing the impact of margin reduction study. *Urology* 2010;**75**(5):1004–8.
99. Klayton T, Price R, Buyyounouski MK, et al. Prostate bed motion during intensity-modulated radiotherapy treatment. *Int J Radiat Oncol Biol Phys* 2012;**84**(1):130–6.
100. Balter JM, Wright JN, Newell LJ, et al. Accuracy of a wireless localization system for radiotherapy. *Int J Radiat Oncol Biol Phys* 2005;**61**(3):933–7.
101. Bittner N, Butler WM, Kurko BS, Merrick GS. Effect of metal hip prosthesis on the accuracy of electromagnetic localization tracking. *Pract Radiat Oncol* 2014;**5**(1):43–8.
102. Serago CF, Chungbin SJ, Buskirk SJ, et al. Initial experience with ultrasound localization for positioning prostate cancer patients for external beam radiotherapy. *Int J Radiat Oncol Biol Phys* 2002;**53**(5):1130–8.
103. Foster RD, Solberg TD, Li HS, et al. Comparison of transabdominal ultrasound and electromagnetic transponders for prostate localization. *J Appl Clin Med Phys* 2010;**11**(1):2924.
104. Scarbrough TJ, Golden NM, Ting JY, et al. Comparison of ultrasound and implanted seed marker prostate localization methods: implications for image-guided radiotherapy. *Int J Radiat Oncol Biol Phys* 2006;**65**(2):378–87.
105. Langen KM, Pouliot J, Anezinos C, et al. Evaluation of ultrasound-based prostate localization for image-guided radiotherapy. *Int J Radiat Oncol Biol Phys* 2003;**57**(3):635–44.
106. Lattanzi J, McNeeley S, Pinover W, et al. A comparison of daily CT localization to a daily ultrasound-based system in prostate cancer. *Int J Radiat Oncol Biol Phys* 1999;**43**(4):719–25.

C H A P T E R

# 46

# Proton Beam Therapy

*Curtiland Deville, MD*

**Johns Hopkins University, The Sidney Kimmel Comprehensive Cancer Center, Sibley Memorial Hospital, Washington, DC, USA**

## INTRODUCTION

Particle beam therapy exploits subatomic particles with mass rather than X-rays or gamma rays to deliver radiation dose. It has gained increasing interest, development, and application due to its physical and radiobiologic properties. The physical property allows for precise dose localization and superior depth dose distribution with heavy charged particles such as protons. The advantageous radiobiologic property of heavier charged particles, such as carbon and helium, is that high linear energy transfer (LET) radiation deposits more dose along its path than conventional X-rays, which are low LET. High-LET radiation is more damaging to hypoxic cells, less cell cycle dependent, and there is less repair of induced damage. Protons are 1800 times heavier than electrons, so accelerating and delivering them to the patient requires higher energies and heavier magnets than photon linear accelerators, which accelerate electrons. Heavy ion particle accelerators are significantly more costly to build and maintain. Currently, only proton therapy facilities exist in the United States. Proton therapy, in addition to heavier particles, such as carbon ion therapy facilities, can be found in Europe and Asia. An experimental heavy ion facility is being planned in the United States.

## HISTORICAL PERSPECTIVE

Particle therapy use in clinical oncology was initially proposed in the 1940s[1] emanating from post-World War II study of the potential applications of nuclear technology and has been used in cancer management for over 60 years. In the 1950s, charged particle therapy was initiated at the Lawrence Berkeley Laboratory, in Uppsala, Sweden,[2] and at the Harvard Cyclotron.[3] In 1961, the Harvard Cyclotron Laboratory began a 40-year collaboration with the Massachusetts General Hospital (MGH) treating over 9000 patients through 2002 when treatment transferred to the MGH campus. These initial facilities typically offered relatively low-energy protons delivered through a fixed beam line of generally limited energies insufficient for penetration to deep-seated tumors, so clinical applications were limited to more sites like base of skull and ocular tumors. In fact, initial studies in prostate cancer, used a 160-MeV proton beam from the Harvard Cyclotron, applied as a conformal perineal boost after initial MV photon therapy. Prior to the MGH, Loma Linda University Medical Center established the first hospital-based proton therapy facility and began treating patients in the early 1990s.[4] The center offered higher energy protons, allowing the treatment of deep-seated tumors, such as the prostate, and marking the transition of proton therapy to more mainstream practice. Proton therapy is becoming increasingly available at a number of hospital-based facilities that are often built and supported by commercial vendors. As of September 2014, there were 14 operational proton centers in the United States, at least 39 abroad, and numerous others in development. As such, there is a call for high-level evidence with healthcare practitioners, policy makers, and the public alike, seeking clear and definitive data to support its use in the setting of increased cost.

## PHYSICS RATIONALE

Protons are not generally considered high-LET particles despite having a slightly higher LET than X-rays. Their advantage lies in their *physical dose distribution* compared to X-rays. When passing through tissue, a heavy, charged particle, such as a proton, deposits increasing dose slowly with depth, until reaching a sharp increase at its maximum depth of penetration, referred to as the Bragg peak (Figure 46.1). As they have mass, they do

Prostate Cancer. http://dx.doi.org/10.1016/B978-0-12-800077-9.00046-3

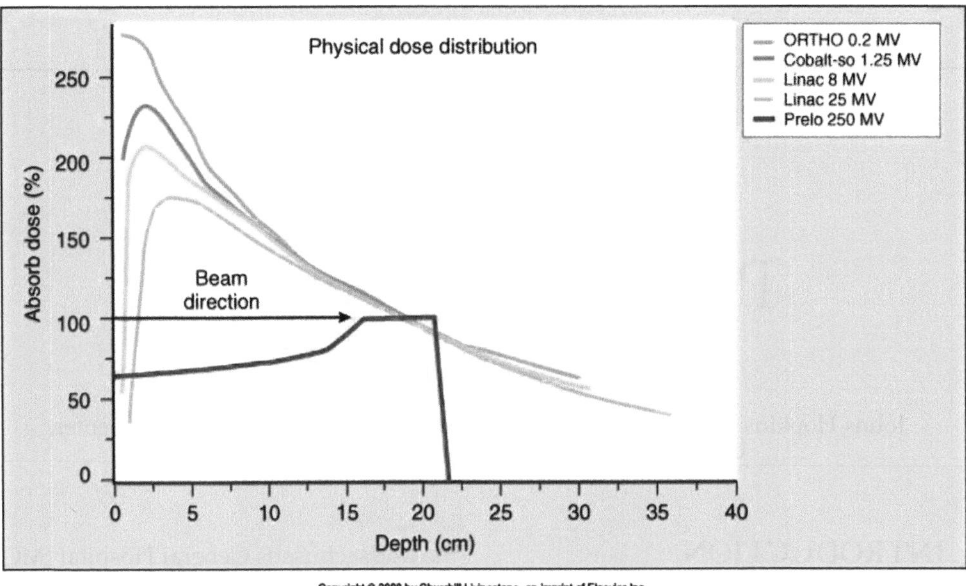



FIGURE 46.1 **Depth-dose distributions for a proton beam compared with other photon beams.** The dose for the proton beam is limited for the entrance tissues, reaches a Bragg peak at the desired depth, and then displays an extremely sharp fall-off. *Reprinted with permission from Abeloff et al.*[9]

not travel an infinite distance. Varying the proton beam energy or altering compensating material placed in the beam path adjusts the maximum depth of penetration. Since the width of the Bragg peak is only 4–7 mm, clinically, the Bragg peak is often spread out (SOBP) using specialized filters to cover the full thickness of a particular target with a uniform dose and achieve the desired dose deposition pattern desired, all-the-while maintaining the sharp dose fall-off at the deep edge of the beam. A conformal 3D dose distribution is ultimately designed using multiple beams or varying compensators [5] (Figure 46.2a). Thus, a typical proton beam disperses a low constant dose of radiation along the entrance path of the beam, a high uniform dose throughout the range of the SOBP, and no exit dose, eliminating much of the integral dose inherent in with X-rays. In contrast to photons, the majority of radiation energy from a proton beam is actually deposited within the target. The potential dosimetric advantages to normal tissues in three-dimensional conformal therapy proton therapy for prostate cancer have been reported in several studies[6,7] generally noting significant differences in the volumes of bladder and rectum receiving medium- and low-dose radiation relative to X-ray therapy using intensity-modulated radiation therapy (IMRT) (Figure 46.2b). Due to the proximity of the anterior rectal wall and bladder base to the prostate, the volumes of these organs receiving high radiation doses are generally similar for the IMRT and proton therapy. The reduction in normal tissue integral dose may also be important in preventing radiation-induced second malignancies, particularly in younger patients with prostate cancer undergoing radiation therapy. A modeling study from the MD Anderson evaluated the risk of secondary malignancies with IMRT compared to proton therapy in patients with early stage prostate cancer and found that proton therapy reduced the excessive relative risk by 26–39% for all models.[8]

## CLINICAL EVIDENCE

Harvard conducted the first phase III proton therapy study for prostate cancer randomly assigning over 200 patients with tumor stage 3 and 4 prostate cancer to either 67.2 Gy photon therapy or 75.6 CGE, consisting of 50.4 Gy photon therapy and a 25.2 CGE proton boost.[10] With a median follow up of 5 years, in the dose-escalated arm there was an increase in local control for poorly differentiated prostate cancer (94% vs. 64%; $P = 0.0014$), a trend toward improved local control for the overall cohort (5-year LC, 92% vs. 80%; $P = .089$), and no significant differences in overall survival or disease-specific survival. The collaborative Proton Radiation Oncology Group trial 95-09 also examined dose-escalation in a combined photon–proton setting.[11] From 1996 to 1999, the study randomized 393 men with T1b–2b prostate cancer and prostate-specific antigen (PSA) <15 to either a low-dose (70.2 Gy/CGE) or high-dose (79.2 Gy/CGE) proton "boost" with either 19.8 CGE or 28.8 CGE delivering using opposed lateral 250 MV beams at Loma Linda University Medical Center or a single perineal 160 MV proton beam at the MGH, followed by 50.4 Gy with 3DCRT. Importantly, dose-escalation reduced biochemical failure with still overall low rates of grade >3 genitourinary (2%) and gastrointestinal (1%) toxicity. Regarding proton therapy as solo treatment for prostate

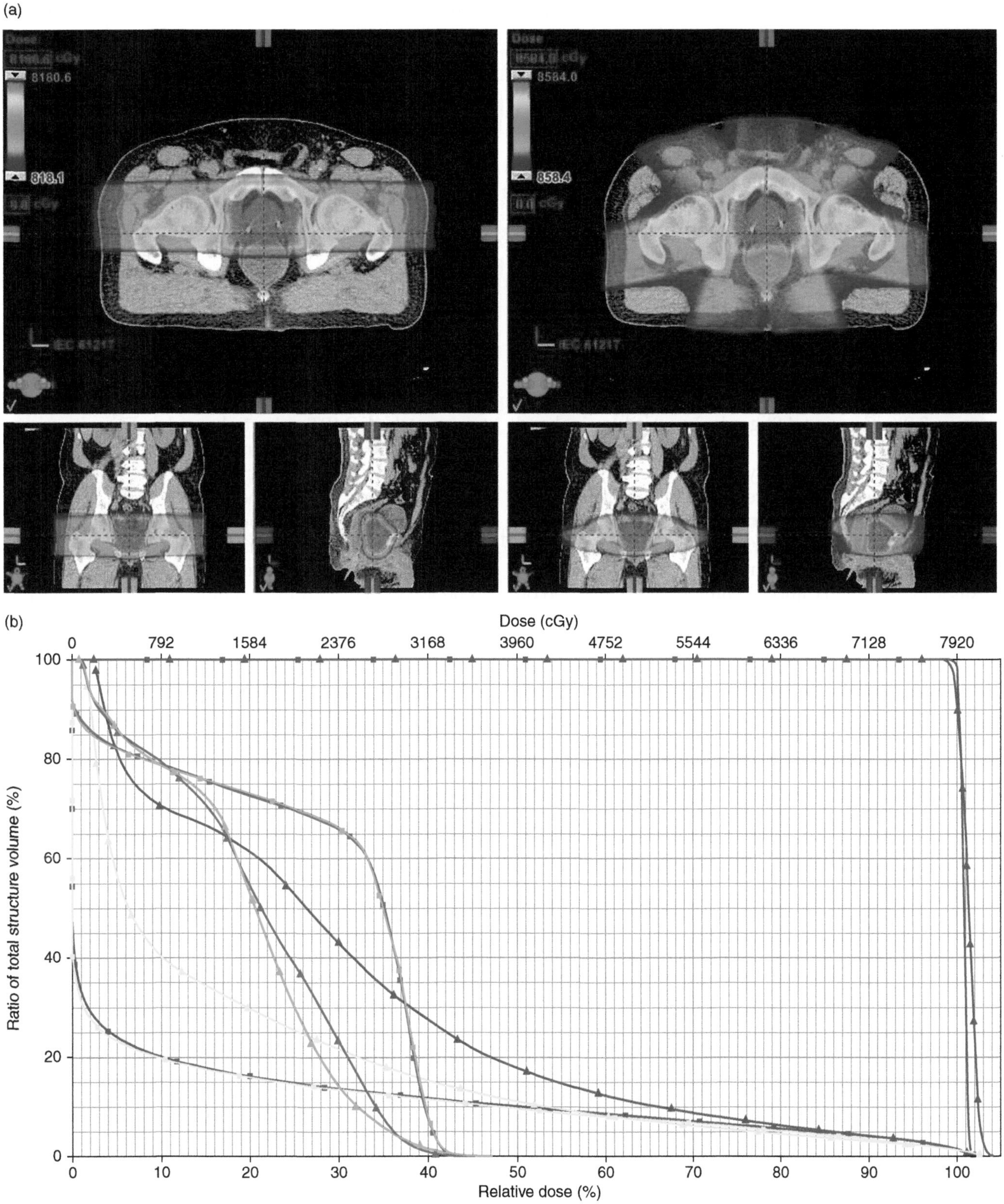

(a)

(b)

Dose (cGy)

0 792 1584 2376 3168 3960 4752 5544 6336 7128 7920

Ratio of total structure volume (%)

100 80 60 40 20 0

0 10 20 30 40 50 60 70 80 90 100

Relative dose (%)

FIGURE 46.2 (a) Color wash dose distributions intensity modulated radiation therapy (IMRT) with photons (right) and pencil beam scanning proton therapy (left) for a prostate cancer patient treated with definitive intent to prescription dose of 79.2 Gy (RBE). (b) Dose–volume histogram comparing IMRT (triangles) and pencil beam scanning proton therapy (squares). Note the comparable target coverage (cyan lines) and reduction in integral dose and low–intermediate dose regions to the adjacent organs at risk – rectum (brown line), bladder (yellow line), and femoral heads (orange and green lines) – with the proton plan, which is lost in the highest dose region.

cancer, Loma Linda University Medical Center reported on 1255 prostate cancer patients treated to total doses of 74–75 CGE utilizing protons alone ($n = 524$) or applied as a boost ($n = 731$) following photon therapy.[12] With a median follow-up of 63 months, 5-year biochemical failure-free survival was 75% and late grade 3 or greater genitourinary or gastrointestinal toxicities was <1%.

Over the last decade, more proton centers have been built in the United States and abroad with increased clinical experience in proton therapy for prostate cancer confirming the overall low toxicity rates.[13] The University of Florida Proton Therapy Institute reported the 5-year clinical outcomes of 211 prostate cancer patients enrolled on three prospective trials of image-guided proton therapy for prostate cancer (89 low-risk, 82 intermediate-risk, and 40 high-risk). Actuarial 5-year rates of late Common Terminology Criteria for Adverse Events (CTCAE), version 3.0 (or version 4.0) grade 3 gastrointestinal and genitourinary toxicity were overall low at 1.0% (0.5%) and 5.4% (1.0%), respectively. There were no significant changes between median pretreatment summary scores and Expanded Prostate Cancer Index Composite (EPIC) scores at greater than 4 years for bowel and urinary irritation, obstruction, and continence.[14] The University of Texas MD Anderson Cancer Center, recently reported on 226 men receiving passively scattered (PS) and 65 receiving, the more modern, spot scanning (SS) proton therapy with minimum 2 year follow up on a prospective nonrandomized quality of life (QOL) protocol, finding that both resulted in statistically significant changes in sexual, urinary, and bowel EPIC summary scores.[15] Only bowel summary, function, and bother resulted in clinically meaningful decrements beyond treatment completion and through 24-month follow-up; cumulative grade 2 or greater genitourinary and gastrointestinal toxicity at 24 months were 13.4% and 9.6%, respectively, with only one grade 3 gastrointestinal toxicity (PS group) and no other grade 3 or greater gastrointestinal or genitourinary toxicity. Argon plasma coagulation application was infrequent and not significantly different between groups (PS 4.4% vs. SS 1.5%). No statistically significant differences were appreciated between PS and SS regarding toxicity or QOL. Finally, the University of Pennsylvania compared contemporaneously treated, nonrandomized proton therapy and IMRT cohorts of 394 total prostate cancer patients using a case-matched analysis and found that although bladder and rectum dosimetric variables were significantly lower for proton therapy versus IMRT ($P \leq 0.01$), on multivariable analysis including direct adjustment for confounders and independent predictors, there were no statistically significant differences in the risk of acute or late grade ≥2 GI or GU toxicities.[16]

Still, these historical and modern clinical studies have shown only limited improvement, if any, in toxicity profiles and there are no completed randomized trials comparing proton therapy to photon therapy in men with clinically localized prostate cancer, although one is currently underway.[17] Historic and current series have not established whether proton beam therapy (either alone or in combination with photon therapy) is less toxic than photon therapy alone or brachytherapy. Attempts have been made to utilize comparative effectiveness research and population-based data to compare outcomes in proton and photon therapy. A Surveillance, Epidemiology and End Results (SEER)-Medicare analysis of 684 men treated with proton therapy from 2002 to 2007 compared to a matched IMRT cohort found that IMRT was associated with less gastrointestinal "morbidity" and no significant differences in other toxicities and no difference in additional cancer therapy, suggest comparable oncologic outcome.[18] Contrastingly, a Medicare analysis of 421 men treated with proton therapy from 2008 to 2009 compared to a matched IMRT cohort showed less genitourinary "toxicity" at 6 months for protons, which disappeared by 1 year, and no other significant differences.[19] Limitations of these reports include their retrospective nature, use of surrogates for toxicity such as diagnosis or procedure codes, and lack of information of treatment details such as delivered dose and the use of image guidance known to be associated with treatment outcome.[20] Finally, a recent report retrospectively examined patient reported outcomes after proton therapy ($n = 95$), 3D conformal photon therapy (3DCRT) ($n = 123$), and IMRT ($n = 153$) at a single institution.[21] Regarding gastrointestinal outcomes at 2–3 months, patients who received 3DCRT and IMRT reported a clinically meaningful decrement in bowel-related QOL, but not the proton cohort; however by 12–24 months, all three cohorts reported clinically meaningful decrements in bowel QOL. Regarding urinary outcomes, at 2–3 months, IMRT only reported clinically meaningful decrements in urinary irritation/obstruction and incontinence. At 12 months, the proton therapy cohort only reported clinically meaningful decrement in urinary irritation/obstruction; however by 24 months, none of the three cohorts reported clinically meaningful changes in urinary QOL. They concluded that patients who received 3DCRT, IMRT, or proton therapy reported distinct patterns of treatment-related QOL; although the timing of toxicity varied between the cohorts, patients reported similar modest QOL decrements in the bowel domain and minimal QOL decrements in the urinary domains at 24 months.

## EVOLVING APPLICATIONS AND CONSIDERATIONS

There is increasing interest in the application of proton therapy for other prostate cancer-related indications to take advantage of its physical properties and dose distribution. Sites of study include postprostatectomy radiation [22] and pelvic nodal irradiation.[23] Not without controversy[24] due to higher cost of delivery relative to other radiotherapy techniques, the availability of proton therapy as a standard treatment modality is anticipated to become more widespread with increasing numbers of centers worldwide and emerging technologies may allow for less costly delivery.[25] In the interim, the concept of *reference pricing with evidence development* has been introduced to allow the opportunity of ongoing treatment with deliberate study to determine whether there are clinically significant benefits.[26] The practice establishes a common level of payment, set by payers, for different therapies with similar outcomes, which for proton therapy for prostate cancer would be set at the rate currently paid for IMRT.

## References


1. Wilson RR. Radiological use of fast protons. *Radiology* 1946;**47**(5):487–91.
2. Falkmer S, Fors B, Larsson B, et al. Pilot study on proton irradiation of human carcinoma. *Acta Radiol* 1962;**58**:33–51.
3. Kjellberg RN, Shintani A, Frantz AG, et al. Proton-beam therapy in acromegaly. *N Engl J Med* 1968;**278**(13):689–95.
4. Slater JD. Development and operation of the Loma Linda University Medical Center proton facility. *Technol Cancer Res Treat* 2007;**6**(4 Suppl.):67–72.
5. Suit H. The Gray Lecture 2001: coming technical advances in radiation oncology. *Int J Radiat Oncol Biol Phys* 2002;**53**:798–809.
6. Trofimov A, Nguyen PL, Coen JJ, et al. Radiotherapy treatment of early-stage prostate cancer with IMRT and protons: a treatment planning comparison. *Int J Radiat Oncol Biol Phys* 2007;**69**(2):444–53.
7. Weber DC, Zilli T, Vallee JP, et al. Intensity modulated proton and photon therapy for early prostate cancer with or without transperineal injection of a polyethylene glycol spacer: a treatment planning comparison study. *Int J Radiat Oncol Biol Phys* 2012;**84**(3):e311–8.
8. Fontenot JD, Lee AK, Newhauser WD. Risk of secondary malignant neoplasms from proton therapy and intensity-modulated X-ray therapy for early-stage prostate cancer. *Int J Radiat Oncol Biol Phys* 2009;**74**(2):616–22.
9. Abeloff MD, Armitage JO, Niederhuber JE, editors. *Abeloff's Clinical Oncology*. 4th edn Philadelphia, PA: Churchill Livingstone; 2008. p. 443.
10. Shipley WU, Verhey LJ, Munzenrider JE, et al. Advanced prostate cancer: the results of a randomized comparative trial of high dose irradiation boosting with conformal protons compared with conventional dose irradiation using photons alone. *Int J Radiat Oncol Biol Phys* 1995;**32**(1):3–12.
11. Zietman AL, Bae K, Slater JD, et al. Randomized trial comparing conventional-dose with high-dose conformal radiation therapy in early-stage adenocarcinoma of the prostate: long-term results from Proton Radiation Oncology Group/American College Of Radiology 95-09. *J Clin Oncol* 2010;**28**(7):1106–11.
12. Slater JD, Rossi Jr CJ, Yonemoto LT, et al. Proton therapy for prostate cancer: the initial Loma Linda University experience. *Int J Radiat Oncol Biol Phys* 2004;**59**(2):348–52.
13. Hoppe B, Henderson R, Mendenhall WM, et al. Proton therapy for prostate cancer. *Oncology (Williston Park)* 2011;**25**(7):644–50 652.
14. Mendenhall NP, Hoppe BS, Nichols RC, et al. Five-year outcomes from 3 prospective trials of image-guided proton therapy for prostate cancer. *Int J Radiat Oncol Biol Phys* 2014;**88**(3):596–602.
15. Pugh TJ, Munsell MF, Choi S, et al. Quality of life and toxicity from passively scattered and spot-scanning proton beam therapy for localized prostate cancer. *Int J Radiat Oncol Biol Phys* 2013;**87**(5):946–53.
16. Fang P, Mick R, Deville C, et al. A case-matched study of toxicity outcomes after proton therapy and intensity-modulated radiation therapy for prostate cancer. *Cancer* 2014;**121**(7):1118–27.
17. *Proton therapy vs. IMRT for low or low-intermediate risk prostate cancer*. http://clinicaltrials.gov/ct2/show/NCT01617161. (Accessed 13.05.2013).
18. Sheets NC, Goldin GH, Meyer AM, et al. Intensity-modulated radiation therapy, proton therapy, or conformal radiation therapy and morbidity and disease control in localized prostate cancer. *JAMA* 2012;**307**(15):1611–20.
19. Yu JB, Soulos PR, Herrin J, et al. Proton versus intensity-modulated radiotherapy for prostate cancer: patterns of care and early toxicity. *J Natl Cancer Inst* 2013;**105**(1):25–32.
20. Deville C, Ben-Josef E, Vapiwala N. Radiation therapy modalities for prostate cancer. *JAMA* 2012;**308**(5):451 author reply 451–452.
21. Gray PJ, Paly JJ, Yeap BY, et al. Patient-reported outcomes after 3-dimensional conformal, intensity-modulated, or proton beam radiotherapy for localized prostate cancer. *Cancer* 2013;**119**(9):1729–35.
22. Doyle LA, Studenski M, Harvey A, et al. Dosimetric comparison of VMAT, IMRT and proton therapy for post-prostatectomy radiation therapy for prostate cancer. *Bodine J* 2010;**3**(1) Article 28.
23. Chera BS, Vargas C, Morris CG, et al. Dosimetric study of pelvic proton radiotherapy for high-risk prostate cancer. *Int J Radiat Oncol Biol Phys* 2009;**75**(4):994–1002.
24. Suit H, Kooy H, Trofimov A, et al. Should positive phase III clinical trial data be required before proton beam therapy is more widely adopted? No. *Radiother Oncol* 2008;**86**(2):148–53.
25. Hede K. Research groups promoting proton therapy "lite". *J Natl Cancer Inst* 2006;**98**(23):1682–4.
26. Bekelman JE, Hahn SM. Reference pricing with evidence development: a way forward for proton therapy. *J Clin Oncol* 2014;**32**(15):1540–2.

CHAPTER

# 47

# Radiotherapy After Radical Prostatectomy: Adjuvant Versus Salvage Approach

*Baruch Mayer Grob, MD**, *Taryn G. Torre, MD***, *Albert Petrossian, MD*[†]

*Urology Section, Hunter Holmes McGuire VA Medical Center, and Division of Urology, Virginia Commonwealth University Health System, Richmond, VA, USA

**Division of Urology, Radiation Oncology Associates, a Division of Virginia Urology, Richmond, VA, USA

[†]Division of Urology, Virginia Commonwealth University Health System, Richmond, VA, USA

## INTRODUCTION

Over 200,000 men are expected to have been diagnosed with prostate cancer in 2014.[1] Approximately one-third will undergo radical prostatectomy[2] and one-third of these men will have a positive margin.[3,4] Another 10% are likely to have seminal vesicle invasion.[5] Other variables that increase the patient's risk for recurrence include high Gleason score, extracapsular extension, positive lymph nodes, and preoperative prostate-specific antigen (PSA).[6] Some patients may also recur without those features. The potential benefit of adjuvant irradiation has been studied prospectively in men with extracapsular extension, positive surgical margins, and positive seminal vesicles. Despite level one evidence suggesting that adjuvant radiotherapy be offered to all those with high-risk features, its use has still not gained widespread acceptance among urologists. This may be due in part to concern over design flaws in the randomized trials as well as the belief that early salvage irradiation is comparable. Salvage radiotherapy has also been shown to be beneficial albeit without a randomized trial.

Whether used in the adjuvant or salvage setting, the potential benefit of irradiation is based on the premise that there is residual microscopic disease in the prostatic fossa and that this disease can be eradicated with local radiation. In the true adjuvant setting, there is no evidence of residual disease, but the patient is at significant risk of recurrence based on the high-risk features delineated earlier. In the salvage setting, the disease is known to be recurrent, usually only by biochemical relapse although new imaging modalities may be capable of demonstrating clinically apparent disease. Often the challenge is determining which men also have micrometastatic disease.

While the prospective randomized controlled trials (RCTs) comparing adjuvant versus salvage radiotherapy are pending, nomograms may be useful in assisting urologists and radiation oncologists help their patients make appropriate choices. New imaging techniques are evolving, such as dynamic contrast enhanced MRI, choline PET CT and fluoride bone scans, that may provide additional useful clinical information in these difficult cases.

## ADJUVANT RADIOTHERAPY

There are three RCTs comparing the use of adjuvant radiotherapy after radical prostatectomy with observation.[7–10] To be eligible for these trials, patients had at least one of three concerning pathologic features: extracapsular extension, positive margins, or positive seminal vesicles.

In 2006, Thompson et al. published the initial review of SWOG 8794,[7] where 425 patients with pathologically advanced prostate cancer were randomized to receive either adjuvant radiotherapy or usual care. In most cases, the men with biochemical recurrences were observed at

Prostate Cancer. http://dx.doi.org/10.1016/B978-0-12-800077-9.00047-5

least initially, but some men were eventually treated with salvage irradiation. Undetectable PSA was not required. Approximately one-third of these patients might be considered salvage patients by contemporary standards. The men received between 60 Gy and 64 Gy of external beam radiotherapy to the prostatic fossa in 30–32 fractions. At 10.6 years of follow-up, there was a statistically significant improvement in PSA relapse-free survival (RFS) in the radiation arm compared to the observation arm, 10.3 years versus 3.1. Adjuvant radiotherapy reduced the risk of initiation of hormonal therapy in half. There was a strong trend toward improved metastasis-free survival in the radiation arm in which 35.5% were diagnosed with metastatic disease compared to 43.1% in the observation arm. There was no difference in overall survival between the two groups at the initial review. Biochemical failure was defined as PSA of 0.4 ng/mL.

In 2009, the SWOG authors published a longer-term follow-up of SWOG 8794 with about 12.6 years of follow-up.[8] In this analysis, there was a statistically significant improvement in metastasis-free survival between the two arms, 93/214 in the radiation arm versus 114/211 in the observation arm. Overall survival was also improved in the treatment arm with 88 deaths compared to 110 deaths in the observation arm. The benefit was achieved despite relatively low doses of irradiation (60–64 Gy) that were in common use when the study was designed.

A companion quality of life study was performed on 217 of the original 425 men.[11] Side effects were generally minor and self-limited. Rectal urgency peaked at 6 weeks in the radiation arm, but by 2 years, there was no significant difference compared to the observation group. Rates of erectile function were also very similar between the two arms. The one urinary side effect that did persist was urinary urgency. Despite the persistent urinary symptoms, global assessment of quality of life, which was initially worse in the radiation group, became similar by 2 years and after 3 years, was superior in the radiotherapy arm. Increased rates of biochemical relapse, hormonal therapy, and subsequent salvage irradiation likely explain the overall improvement in quality of life.

A common criticism of this paper (and all three of these RCTs), is that the observation group did not receive radiation at the time of biochemical relapse as is commonly recommended in current practice. Critics suggest that with earlier salvage therapy, many of the men might have avoided clinical failure. Additionally, critics argue that a benefit in biochemical failure alone would not be justification to treat every at risk patient if salvage irradiation could be successful in the majority of cases. In fact, 70/211 men of the observation arm patients in SWOG 8794 ultimately did receive radiotherapy, five of whom started in the adjuvant setting while about half initiated therapy at the time of biochemical relapse. The median PSA at the time of radiotherapy was 1.0. A subset analysis of these salvage patients revealed that they had inferior metastasis-free survival to those who received radiotherapy when PSA was still undetectable.[8] By contemporary standards, even the patients who received salvage irradiation started treatment at higher levels of PSA. This delay may have contributed to the benefit in metastasis-free survival. Despite that limitation and in the absence of a randomized trial comparing adjuvant radiotherapy with salvage radiotherapy, this is one more piece of evidence favoring the adjuvant approach.

Another limitation of SWOG 8794 is the lack of central pathological review. Only 73% of the SWOG trial patients had central review. But there was 95% concordance between the central and local review. The SWOG authors also acknowledged that the extent of disease in patients during the time period of the study was probably of greater volume than contemporary patients, potentially increasing the differences in the outcomes between the two arms.[8]

The largest randomized trial comparing adjuvant irradiation to observation is EORTC 22911. This multi-institutional trial randomized 1005 patients to either immediate postoperative irradiation versus a wait-and-see policy. This study also employed a dose of 60 Gy. Similar to the SWOG trial, at least 10% of these patients also had detectable PSA. Also like SWOG 8749, this study showed a benefit in biochemical recurrence. With a median follow-up of 10.6 years, among the patients treated with adjuvant radiation, 39.4% experienced biochemical relapse versus 61.8% in the observation arm.[9] The initial analysis of the study[12] showed a benefit in clinical progression-free survival as well as biochemical RFS but that benefit was not maintained at the 10.6 year follow-up analysis. There was no proven effect on distant metastases or overall survival in either report. Biochemical failure was defined as an increase of 0.2 ng/mL over the lowest postoperative value measured on three occasions.

Similar to SWOG 8749, many patients in the observation arm of EORTC 22911 were offered salvage radiotherapy. Of the 265 patients in the control arm who experienced either biochemical or clinical failure, 155 were treated with irradiation. The EORTC investigators utilized a salvage dose of 70 Gy. The median PSA value that triggered a salvage approach was 1.7.[9] Clearly many patients in this study would have received salvage treatment at a lower PSA if treated today.

No grade 4 toxicity was reported in EORTC 22911. The 10-year cumulative incidence of grade 3 late toxicity was higher for the immediate irradiation group and genitourinary toxicity was higher as well (21.3% vs. 13.5%), but gastrointestinal toxicity was similar between the groups.[9]

The EORTC investigators theorized several possibilities that might explain the difference between their initial analysis and the 10-year follow-up study. In the long-term analysis, there was a greater contribution of nonprostate-cancer mortality compared to their initial review. There was also an excess of death as the first event in men older than 70 years. The authors conclude that limiting adjuvant irradiation to men younger than age 70 and to those with positive margins might maximize the potential benefit.[9] In comparing their study to SWOG 8749, they suggest that salvage treatment in the SWOG trial might have been less intensive because there were similar rates of salvage treatment in both trials despite a higher relapse rate in the SWOG study.

Lack of central review was even more of an issue in the EORTC trial. Only about half of the patients had central review and in the centrally reviewed patients, only those with positive margins benefited from the adjuvant approach. The extracapsular extension versus organ-confined status changed in approximately half of the patients reviewed.[10] This discrepancy could have potentially skewed the results, if some patients with organ-confined disease were miscategorized as having extracapsular extension. Accurate pathologic analysis is clearly important in the treatment of high-risk patients.

There is one final randomized trial comparing adjuvant radiation to an observation period after radical prostatectomy for patients with pT3 disease, ARO 96-02.[10] A dose of 60 Gy was also employed in this study. There are minor differences between this trial based in Germany and the other two RCTs, notably the exclusion of patients with detectable PSA following surgery in ARO 96-02. Similar to the other trials, there is a proven benefit in reduction in biochemical progression, (74% vs. 53%) but these investigators defined biochemical relapse as PSA of 0.1 ng/mL or greater. The investigators did not choose to examine overall survival as the primary end-point. In multivariate analysis, preoperative PSA level greater than 10 and tumor stage were independent predictors of biochemical outcome (Table 47.1).

**TABLE 47.1** Review of Prospective Randomized Trials Comparing Adjuvant Radiation to Observation

| Study | Type | Number of patients | Primary endpoint | Inclusion criteria | ART used | Definitions | Limitations | Results | Conclusion |
|---|---|---|---|---|---|---|---|---|---|
| EORTC 22911; Bolla et al.[9] | RCT – Obs versus ART | 1005 | Biochemical progression-free survival | Positive surgical margin or pT3N0M0 CaP | 60 Gy in 30 fractions | BPFS: Increase of more than 0.2 μg/L over the lowest postop value on three occasions | Conventional irradiation, low dose of 60 Gy, and variable postop nadir | BPFS significantly improved in irradiated group (98% CI 68.7–79.3 $p < 0.0001$) | ART improves BPFS and local control |
| ARO 96-02 AUO AP 09/95; Wiegel et al.[10] | RCT – Obs versus ART | 385 | Biochemical progression-free survival | pT3N0 and undetectable PSA | 60 Gy in 30 fractions | BPFS: Two consecutive increasing PSA values | Lower number of patients, low dose of 60 Gy, and low concordance between pathologists | BPFS significantly improved in irradiated group (95% CI 37–79 $p = 0.0015$) | ART improves BPFS in those with pT3 CaP and undetectable PSAs |
| SWOG 8794; Thompson et al.[8] | RCT – Obs versus ART | 431 | Metastasis-free survival | pT3N0M0 CaP | 60–64 Gy in 30–32 fractions | MFS: Time from randomization to first evidence of metastasis or death due to any cause. | Incomplete central pathologic review, long duration of study | MFS was significantly greater with ART (95% CI 0.54–0.94 $p = 0.016$) | ART reduces risk of metastasis and increases survival in men with pT3N0M0 CaP |

Notes: RCT, randomized controlled trial; CaP, prostate cancer; MFS, metastasis free survival; Obs, observation; Gy, Gray; ART, adjuvant radiotherapy; BPFS, biochemical progression free survival.

A meta-analysis of all three RCTs was published in 2008.[13] The three studies representing 1743 patients were included in this review. Combining the two RCTs that reported on overall survival, there was no statistically significant improvement with adjuvant radiotherapy. All three trials reported on biochemical RFS and the meta-analysis did demonstrate an advantage to adjuvant radiotherapy of a 53% decrease in biochemical relapse. The authors recognize that longer follow-up is necessary to accurately assess the impact of adjuvant irradiation on overall survival.

Two large retrospective reviews have attempted to assess the implications of the three RCTs. In a cohort of over 1000 patients treated with radical prostatectomy alone or in combination with adjuvant irradiation, Abdollah et al. performed a multivariate analysis examining the role of Gleason score ≥8, pT3b/T4, and presence of positive lymph nodes. A novel risk score was developed in which men who shared at least two of the three pathologic features were found to have significantly improved cancer-specific survival when treated with adjuvant radiotherapy.[14] This risk score system was validated when Gandaglia et al. used the SEER database and came to the same conclusion.[15]

In 2013, the AUA and ASTRO published a joint guideline statement on the use of adjuvant and salvage radiotherapy after prostatectomy. The authors concluded that based on their review of 294 relevant articles, physicians should offer adjuvant radiotherapy to patients with adverse pathologic features, specifically seminal vesicle invasion, positive surgical margins, and extraprostatic extension. The decision to use adjuvant approach should be made in context of the possible side-effects as well as the patient's history, values, and quality of life (Table 47.2).[16]

TABLE 47.2 AUA/ASTRO Guidelines for Adjuvant and Salvage Radiotherapy

| | |
|---|---|
| 1 | Prostatectomy candidates should be advised of potential benefit from postoperative radiation therapy for pathologic high risk features |
| 2 | Adjuvant radiotherapy reduces risk of bPSA recurrence, local recurrence and clinical progression in patients with + SVI, + margins, or + EPE |
| 3 | Physicians should offer adjuvant radiotherapy to patients with theses adverse pathologic findings |
| 4 | Inform patients that PSA recurrence is associated with increased risk of metastatic PCa or death from disease. PSA should be regularly monitored after prostatectomy to enable early salvage therapy |
| 5 | PSA recurrence is defined as two or more levels >0.2 ng/mL |
| 6 | Consider a restaging evaluation in the patients with PSA recurrence |
| 7 | Salvage radiotherapy should be offered for PSA or local recurrence after prostatectomy if no evidence of distant disease is present |
| 8 | Potential acute and chronic side effects from postoperative radiation as well as potential benefits from decreased recurrence should be discussed with patients |

SVI, Seminal vesical invasion; EPE, extra prostatic extension; PCa, prostate cancer.

The controversy regarding the use of adjuvant radiotherapy for high-risk patients after radical prostatectomy versus an initial observation period with salvage radiotherapy reserved for those men who experience biochemical failure cannot be resolved without a prospective randomized trial. Three such trials are underway: (1) the RADICALS trial in the United Kingdom, (2) the RAVES trial led by the Trans-Tasman Radiation Group, and (3) the GETUG-17 trial led by the French Genitourinary Oncology Group. Thompson et al. examined the study designs for a potential trial to test the benefit of adjuvant over salvage radiation and calculated that a sample size of 8300 men would be required.[8] In a letter to the editor of European Urology, the lead investigators acknowledge the difficulty in adequately powering individual trials and hope to develop an individual participant data meta-analysis.[17]

Despite the results of the RCTs and the consensus statements from AUA and ASTRO, resistance to the adjuvant approach remains among urologists. In a survey of Canadian urologists and radiation oncologists, radiation oncologists were 48% more likely to recommend adjuvant treatment. In the specific scenario of a patient with Gleason 6 cancer and a positive surgical margin, 70% of radiation oncologists and only 21% of urologists were likely to recommend adjuvant radiotherapy. Patient age above 75 and urinary incontinence were factors that decreased the probability of recommendation for adjuvant approach.[18] Perhaps the main appeal to the salvage approach is that it avoids overtreatment of patients who are not destined to recur. Coupled with the well-recognized limitations of the RCTs, this may contribute to the current disparity between specialists.

## SALVAGE RADIOTHERAPY

When biochemical relapse occurs after radical prostatectomy and adjuvant therapy was not offered at the time of surgery, salvage radiotherapy may be a reasonable option for some men. Within 6 weeks of surgery, PSA values should be in the undetectable range, which may vary among different laboratories. The AUA defines biochemical failure as >0.2 ng/mL on two readings. Many clinicians with access to ultrasensitive PSA consider any consistent value >0.1 ng/mL to be indicative of recurrence. If PSA never falls to undetectable and/or is rapidly rising, the likelihood exists of at least a component of systemic disease. If PSA rises slowly after an extended period of undetectable values, the implication is for local recurrence in the prostatic bed that may respond to salvage radiotherapy.[19]

Sixty-five percent of men with biochemical or local recurrence will eventually develop clinical metastases if left untreated and the majority of those patients will die of the disease.[20] Based on AUA/ASTRO guideline statement, these men should be offered salvage radiotherapy if they have a reasonable life-expectancy.[16] There are several large retrospective reviews of salvage radiotherapy that can inform the clinicians' decisions regarding its use. It should be noted that there is a paucity of data regarding the side effects of salvage radiotherapy in any of these studies.

In 2008, Trock et al. from Johns Hopkins published their cohort study of 635 men who experienced biochemical recurrence after radical prostatectomy and either received salvage radiotherapy, radiotherapy with hormonal ablation, or observation at the discretion of the treating physician. While this was not a randomized trial, it was the first such study to show a benefit in prostate-cancer-specific survival for salvage patients compared to an observation cohort. With a median follow-up of 6 years after recurrence, 22% of the men who received no salvage treatment died of prostate cancer compared to 11% of the men who were treated with salvage radiotherapy. The improvement was most striking in men with PSA doubling time of less than 6 months. This result seems counter-intuitive given that these patients are most likely to have subclinical metastatic disease. It may reflect that patients with longer PSA doubling times have biologically less aggressive disease.

The beneficial effect of salvage irradiation was limited to patients who were treated within 2 years of recurrence. Similarly, when salvage radiotherapy was initiated with PSA $\leq$2 ng/mL, there was an association with improved prostate-cancer-specific survival but this improvement was no longer significant when PSA was >2 ng/mL. The benefit was seen regardless of whether hormonal ablation was added to the course of radiation. Since the hormonal therapy was used only for high-risk patients and the survival was identical, there is indirect evidence that hormonal therapy is beneficial at least in the high-risk patients.[21]

In 2009, Boorjian et al. from the Mayo Clinic reported on 856 patients who were treated with salvage radiotherapy after biochemical relapse.[22] This cohort constituted 32% of the total number of patients who experienced biochemical failure during the observation period. The median time to salvage therapy was 0.7 years and the median PSA was 0.8 ng/mL. In multivariate analysis, this study found 90% decreased risk of local recurrence, 20% decreased risk of late androgen deprivation therapy, and 75% decreased risk of systemic progression in the patients who received salvage irradiation compared to those who were simply observed. In contrast to the study cited earlier by Trock et al., Boorjian et al. did not see a difference in overall mortality.

In a letter to the editor of the *Journal of Urology*, the Johns Hopkins authors noted that some of the patients from the Mayo Clinic may have received adjuvant deprivation therapy, which might have minimized the difference in survival between the two groups. The Hopkins paper compared salvage radiotherapy with observation alone, with neither group receiving hormonal ablation. The Mayo Clinic authors in their response, while acknowledging differences in the patient groups between the two institutions, noted that any differences in the use of androgen deprivation therapy should have been accounted for by the multivariate analysis.[23]

Another large retrospective study involves 285 consecutive patients treated with salvage radiotherapy from Memorial Sloan-Kettering Cancer Center. Two hundred and seventy patients were treated to a dose of $\geq$66 Gy of whom 205 patients received $\geq$70 Gy. Approximately one-third of the patients received androgen deprivation therapy. Seven-year actuarial biochemical RFS was 37% and metastasis-free survival was 77%. Doses over 70 Gy were not associated with improvement in biochemical control, suggesting that dose escalation beyond 66 Gy is unnecessary. The only independent predictor of metastatic disease was a PSA doubling time of <3 months.[24] This seems to contradict the data from Hopkins, which demonstrated that only the patient with doubling times <6 months benefited from salvage treatment.[21]

In 2011, King performed a comprehensive systematic review of salvage radiotherapy after radical prostatectomy encompassing 41 different studies.[25] In his analysis, PSA level before salvage treatment and the radiotherapy dose had a significant and independent association with RFS, with a dose of 70 Gy achieving 54% relapse-free survival compared to only 34% for 60 Gy. Most interesting perhaps, was the average of 2.6% loss of RFS survival for each incremental increase in 0.1 ng/mL with an RFS of 64% when salvage radiotherapy was initiated at PSA $\leq$0.2 ng/mL. Based on this analysis, it may be reasonable in select patients with detectable PSA <0.2 ng/mL to offer salvage radiotherapy prior to reaching the biochemical recurrence point as defined by the AUA.

In the 2013 joint guideline statement, the AUA and ASTRO recommend that salvage radiotherapy be offered to men with biochemical or local recurrence in whom there is no evidence of distant disease. The authors recognized that the strength of the evidence was only a level C. In the absence of a prospective randomized trial, only observational studies are available for analysis. There is overwhelming evidence from all of these studies that lower presalvage PSA levels are associated with improved biochemical relapse-free survival.[16]

One unique study demonstrated that the optimal timing of salvage therapy may also be influenced by the Gleason score. Karlin et al. from Virginia Commonwealth University and Duke Medical Center reported recently that patients with Gleason 8–10 had significant improvement in biochemical RFS when salvage radiotherapy was initiated at PSA of 0.33 or less. This association was not seen for Gleason 6 or 7 patients. The authors also demonstrated as many others have that Gleason 8–10 and negative margin status were independent predictors of biochemical relapse.[26]

One major unanswered question remains regarding the use of androgen deprivation with salvage radiotherapy. RTOG 96-01 is an attempt to answer the question. The initial results were presented at ASCO in 2011, but the paper has not been published in manuscript form to date. Patients were randomized to receive salvage radiation alone versus radiation in combination with 2 years of bicalutamide, 150 mg daily. While there was no difference in overall survival at 7 year median follow-up, the authors demonstrated improvement in biochemical RFS (57% compared to 40%) and metastasis-free survival. The cumulative incidence of metastases was 7% in the radiation plus anti-androgen arm versus 13% in the radiation alone arm.[27] This study has been criticized for use of nonconventional androgen deprivation and a dose of radiation, which would be considered inadequate today.[28] In the absence of a randomized trial, there are several retrospectives studies that have examined this issue, but the results are too inconsistent to draw any meaningful conclusions.

One tool to assist clinicians in making the decision to proceed with salvage radiotherapy is a nomogram developed by Stephenson et al. In their analysis of a multi-institutional cohort with over 1500 patients, they used a multivariable Cox regression analysis to construct a model to predict the probability of disease progression after salvage radiotherapy. The 6 year progression-free probability was 32% for the entire cohort. They found that significant variables in the model were PSA levels prior to radiotherapy, prostatectomy Gleason score, PSA doubling time, surgical margin status, and androgen deprivation therapy. Nearly 50% of the patients with PSA $\leq$0.5 ng/mL at time of salvage radiotherapy were disease-free at 6 years, emphasizing the need to initiate salvage therapy as early as possible.[29]

## FUTURE DIRECTIONS

Imaging of recurrent cancer is one rapidly growing area of research interest. One specific technique is the use of dynamic contrast enhanced MRI to detect recurrent prostate cancer in the prostatic fossa even at very low levels of PSA. One recent report is very interesting in that it only looked at patients who had a complete response to salvage radiotherapy, implying a local recurrence only situation. In 22/33 patients, nodules were detected on MRI. All postradiation images showed resolution of the initial suspicious lesions. The average preradiation PSA levels were significantly higher in the group with suspicious MRI-detected nodules, 0.74 ng/mL versus 0.24 ng/mL. These authors found that a preradiation PSA level of >0.54 ng/mL predicted for a positive finding on MRI. The nodules were detected without the benefit of an endorectal coil, which potentially reduces the ability of the radiation oncologist to fuse the images with CT for treatment planning purposes. Therefore, this technology may offer the ability to increase radiation dose to targeted areas but reduce overall toxicity.[30]

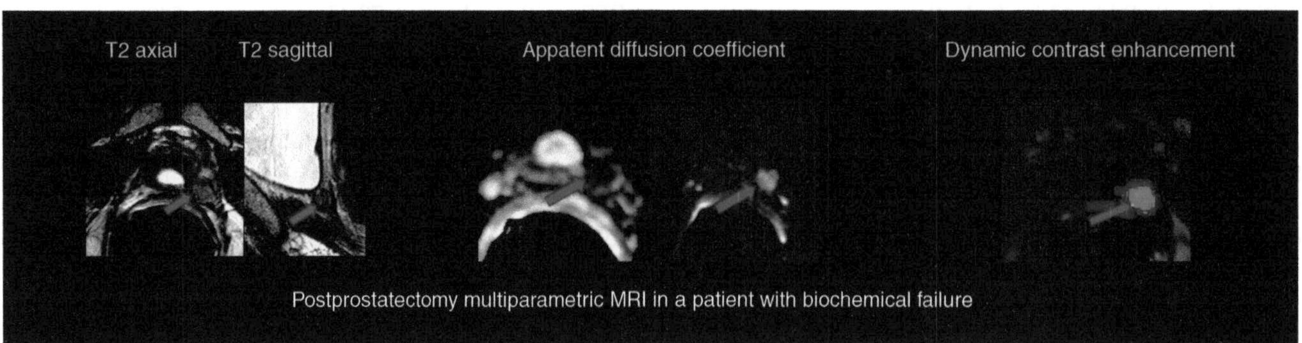


Another growing area of interest is the use of [11C] Choline-Positron Emission Tomography/Computed Tomography (PET/CT) for diagnosis of systemic disease after primary treatment for prostate cancer. Recognized leaders in the United States, the Mayo Clinic recently published a paper on salvage lymph node dissection for recurrence identified on [11C] Choline-PET/CT. Salvage node dissection in the setting of presumed oligo-metastatic disease is certainly a controversial topic; what is interesting however is the accuracy of the PET/CT as demonstrated by histology of the node dissections in which 92% of the specimens were correctly predicted by PET/CT.[31]

Another recent study examined the relationship of a positive PET/CT with failure after salvage radiotherapy.[32]

In a pilot study of 27 patients, 16 patients developed rising PSA after treatment. Of those 16, 5 showed evidence of either nodal disease or bone metastasis on choline PET/CT. In the future, PET/CT may help define those patients most suitable for salvage therapy, potentially including salvage node dissection with salvage radiotherapy of the prostatic fossa.

Another area of research interest is the use of biomarkers to predict recurrence. Two promising new markers are *PITX2* and Caveolin-1. *PITX2* is a transcription factor that is required for cell-type specific proliferation. *PITX2* DNA methylation has been found to be an outcome prediction marker in patients with breast cancer. Recently, a multi-institution study was performed by Bañez et al. seeking associations between *PITX2* methylation status and biochemical recurrence in patients postprostatectomy. The study demonstrated on multivariate analysis that men with high methylation status were at a significantly higher risk for biochemical recurrence than those with low methylation status (HR 3.0 95% CI 1.2–3.3, $p = 0.005$). Although the findings were compelling, it remains to be seen whether *PITX2* methylation status can predict risk of metastasis and prostate-cancer-specific death.[33]

Caveolin-1 is a protein that performs numerous functions in cells including antiapoptotic activity. Its over-expression has been associated with adverse pathologic characteristics and poor outcomes in other malignancies such as meningiomas, breast, renal, colon, and lung cancers. In prostate cancer its over-expression has been associated with higher Gleason scores, positive surgical margins, metastases to lymph nodes, and a higher probability of disease recurrence postprostatectomy. A paper by Karam et al. confirmed that over-expression in 232 consecutive patients postprostatectomy was associated with aggressive PSA recurrence in univariate and multivariate analysis ($p < 0.001$ and $P = 0.001$). As such, Caveolin-1 immunostaining postprostatectomy may help identify patients with aggressive disease that could benefit from early therapeutic intervention.[34]

## CONCLUSIONS

In the absence of randomized RCTs comparing adjuvant radiotherapy to salvage radiotherapy at the time of biochemical recurrence, the best available data suggest a benefit to the use of adjuvant radiotherapy. This treatment option should be reviewed with the patient after radical prostatectomy in which the pathology is consistent with high risk of recurrence, specifically extracapsular extension, positive surgical margins and/or positive seminal vesicles. Patients with at least two concerning features may be more likely to benefit. The potential risks should be discussed with the patient to allow for shared decision-making. If a patient has a biochemical recurrence and did not previously choose adjuvant therapy, salvage radiotherapy should be offered as a potential definitive management at the earliest possible time if there is no evidence of systemic disease. The patient's comorbidities and personal goals should be taken into account when planning this approach. In the near future, MRI, choline PET/CT, and other novel techniques will likely be used to better define the most appropriate patients for adjuvant and salvage radiotherapy.

## References


1. Siegal R, Ma J, Zou Z, et al. Cancer statistics. *CA Cancer J Clin* 2014;**64**:9–29.
2. Underwood W, Jackson J, Wei JT, et al. Racial Treatment trends in localized/regional prostate carcinoma: 1992-1999. *Cancer* 2005;**103**(3):538–45.
3. Yossepowitch O, Bjartell A, Eastham J, et al. Positive surgical margins in radical prostatectomy: outlining the problem and its long-term consequences. *Eur Urol* 2009;**55**:87–99.
4. Cooperberg MR, Broering JM, Litwin MS, et al. The contemporary management of prostate cancer in the united states: lessons from the Cancer of the Prostate Strategic Urologic Research Endeavor (CaPSURE), a national disease registry. *J Urol* 2004;**171**(4):1393–401.
5. Klein EA, Bianco FJ, Serio AM, et al. Surgeon experience is strongly associated with biochemical recurrence after radical prostatectomy for all preoperative risk categories. *J Urol* 2008;**179**:2212–7.
6. Swindle P, Eastham JA, Ohori M, et al. Do margins matter? The prognostic significance of positive surgical margins in radical prostatectomy specimens. *J Urol* 2005;**174**:903–7.
7. Thompson IM, Tangen CM, Paradelo J, et al. Adjuvant radiotherapy for pathologically advanced prostate cancer: a randomized clinical trial. *JAMA* 2006;**296**:2329–35.
8. Thompson IM, Tangen CM, Paradelo J, et al. Adjuvant radiotherapy for pathological T3N0M0 prostate cancer significantly reduces risk of metastases and improves survival: long-term followup of a randomized clinical trial. *J Urol* 2009;**181**:956–62.
9. Bolla M, van Poppel H, Tombal B, et al. Postoperative radiotherapy after radical prostatectomy for high-risk prostate cancer: long-term results of a randomized controlled trial (EORTC trial 22911). *Lancet* 2012;**380**:2018–27.
10. Weigel T, Bottke D, Steiner U, et al. Phase III postoperative adjuvant radiotherapy after radical prostatectomy compared with radical prostatectomy alone in pT3 prostate cancer with undetectable prostate-specific antigen: ARO 96-02/AUO AP 09/95. *J Clin Oncol* 2009;**27**:2924–30.
11. Moinpour CM, Hayden KA, Unger JM, et al. Health-related quality of life results in pathologic stage c prostate cancer from a southwest oncology group trial comparing radical prostatectomy alone with radical prostatectomy plus radiation therapy. *J Clin Oncol* 2008;**26**:112–20.
12. Bolla M, van Poppel H, Collette L, et al. Postoperative radiotherapy after radical prostatectomy: a randomized controlled trial (EORTC trial 22911). *Lancet* 2005;**366**:572–8.
13. Morgan SC, Wadron TS, Eapen L, et al. Adjuvant radiotherapy following radical prostatectomy for pathologic T3 or margin-positive prostate cancer: a systematic review and meta-analysis. *Radiother Oncol* 2008;**88**:1–9.
14. Abdollah F, Suardi N, Cozzarini C, et al. Selecting the optimal candidate for adjuvant radiotherapy after radical prostatectomy for prostate cancer: a long-term survival analysis. *Eur Urol* 2013;**63**:998–1008.

15. Gandaglia G, Karakiewicz PI, Briganti A, et al. Early radiotherapy after radical prostatectomy improves cancer-specific survival only in patients with highly aggressive prostate cancer: validation of recently released criteria. *Int J Urol* 2015;**22**(1):89–95.
16. Thompson IM, Valicenti RK, Albertsen P, et al. Adjuvant and salvage radiotherapy after prostatectomy: AUA/ASTRO guideline. *J Urol* 2013;**190**:441–9.
17. Sydes MR, Vale C, Kneebone A, et al. Letter to the editor of European Urology. *Eur Urol* 2012;**62**:e99.
18. Lavallee LT, Ferguson D, Mallick R, et al. Radiotherapy after radical prostatectomy: treatment recommendations differ between urologists and radiation oncologists. *PLoS One* 2013;**8**:1–10.
19. Wallace TJ, Torre T, Grob M, et al. Current approaches, challenges, and future directions for monitoring treatment response in prostate cancer. *J Cancer* 2014;**5**:3–24.
20. Pound CC, Partin AW, Eisenberger MA, et al. Natural history of progression after PSA elevation following radical prostatectomy. *JAMA* 1999;**281**:1591–7.
21. Trock BJ, Han M, Freedland SJ, et al. Prostate cancer-specific survival following salvage radiotherapy vs observation in men with biochemical recurrence after radical prostatectomy. *JAMA* 2008;**299**:2760–9.
22. Boorjian SA, Karnes RJ, Crispen PL, et al. Radiation therapy after radical prostatectomy: impact on metastasis and survival. *J Urol* 2009;**182**:2708–15.
23. Trock Bruce J, Walsh Patrick C. Letter to the editor. *J Urol* 2010;**183**:2466–7.
24. Goenka A, Magsanoc JM, Pei X, et al. Long-term outcomes after high-dose postprostatectomy salvage radiation treatment. *Int J Radiat Oncol Biol Phys* 2012;**84**:112–8.
25. King CR. The timing of salvage radiotherapy after radical prostatectomy: a systematic review. *Int J Radiat Oncol Biol Phys* 2012;**84**:104–11.
26. Karlin JD, Koontz BF, Freedland SJ, et al. Identifying appropriate patients for early salvage radiotherapy after prostatectomy. *J Urol* 2013;**190**:1–6.
27. Shipley WU, Hunt D, Lukka HR, et al. Initital Report of RTOG 9601, a phase III trial in prostate cancer: effect of anti-androgen therapy with bicalutamide during and after radiation therapy on freedom from progression and incidence of metastatic disease in patients following radical prostatectomy with pT2-3, N0 disease and elevated PSA levels. *J Clin Oncol* 2011;**29**(Suppl. 7).
28. Jang JW, Hwang W-T, Guzzo TJ, et al. Upfront androgen deprivation therapy with salvage radiation may improve biochemical outcomes in prostate cancer patients with post-prostatectomy rising PSA. *Int J Radiat Oncol Biol Phys* 2012;**83**:1493–9.
29. Stephenson AJ, Scardino PT, Kattan MW, et al. Predicting the outcome of salvage radiation therapy for recurrent prostate cancer after radical prostatectomy. *J Clin Oncol* 2007;**15**:2035–41.
30. Rischke HC, Schafer AO, Nestle U, et al. Detection of local recurrent prostate cancer after radical prostatectomy in terms of salvage radiotherapy using dynamic contrast enhanced-MRI without endorectal coil. *Radiat Oncol* 2012;**7**:1–8.
31. Karnes RJ, Murphy CR, Bergstralh EJ, et al. Salvage lymph node dissection for prostate cancer nodal recurrence detected by 11C-Choline positron emission tomography/computed tomography. *J Urol* 2015;**193**(1):111–6.
32. Reske SN, Moritz S, Kull T. [11C] Choline-PET/CT for outcome prediction of salvage radiotherapy of local relapsing prostate carcinoma. *Q J Nucl Med Mol Imaging* 2012;**5**:430–9.
33. Bañez L, Sun L, Leenders GJ, et al. Multicenter clinical validation of *PITX2* methylation as a prostate specific antigen recurrence predictor in patients with post-radical prostatectomy prostate cancer. *J Urol* 2010;**183**:149–56.
34. Karam J, Lotan Y, Roehrborn C, et al. Caveolin-1 overexpression is associated with aggressive prostate cancer recurrence. *Prostate* 2007;**67**:614–22.

C H A P T E R

# 48

# Emerging Modalities in Radiation Therapy for Prostate Cancer

*Curtis T. Miyamoto, MD, Mohan P. Achary, PhD*

**Department of Radiation Oncology and Radiology, Temple University School of Medicine, Philadelphia, PA, USA**

## INTRODUCTION

Radiation therapy techniques and their combination with other modalities have developed rapidly over the past two decades in radiation oncology. This continues today at an ever increasing pace offering considerable hope to the patients with both localized and metastatic disease. This chapter will attempt to highlight the following areas in the treatment of prostate cancer. The areas are better methods for the stratification of patients who would truly benefit from treatment, improved target definition using both anatomical and biological techniques, accelerated and more cost-effective treatments with greater accuracy, better combinatorial treatments involving systemic and surgical approaches, adaptive radiation therapy, and expansion of new types of external beam irradiation.

## IDENTIFICATION OF PATIENTS FOR RADIATION THERAPY

The role of radiation therapy in the treatment of prostate cancer is still evolving. One of the most important issues still is the identification of the patients that would most benefit from radiation therapy.[1–3] By definition radiation therapy is used for local control. Several new tests have been designed to determine whether patients have systemic disease. Curative intent would only be in patients who have no evidence of distant metastatic disease. Multiple predictive tools have been developed to, first, determine patients' risk for distant disease; second, reduce risks in cancer recurrence after prostatectomy and third, to increase the survival benefit. One of the predictive tools is the Cancer of the Prostate Risk Assessment Postsurgical Score (CAPRA-S). Recently, CAPRA-S was validated in a multi-institutional review to be able to predict disease recurrence and mortality after radical prostatectomy.[4]

There are several new strategies being employed with magnetic resonance imaging (MRI) scans. Increasing data are supporting MRI and MR spectroscopic imaging (MRSI) as being superior to the trans-rectal ultrasound (TRUS) procedure in locating prostate tumors. In one study the accuracy of dynamic contrast enhanced MRI with three-dimensional MRSI was compared with trans-rectal ultrasound in localizing prostate tumors.[5] In this study, 73.6% of the patients had non-palpable disease (T1c) and 68.64% of the patients had organ-confined disease (pT2). MRI was significantly better at detecting malignancies in the midgland (52% vs. 41.1%, $P = 0.0015$) and in the transitional zone (40.1% vs. 24.3%, $P < 0.0001$). MRI had a higher sensitivity in larger ($\geq 50$ g) than smaller prostates (50.3% vs. 42.2%, $P = 0.0017$). A genomic test using the 3T multiparametric MRI – Blue Laser™ – could be used in identifying prostate cancer at its early stages.[6] This genomic test works at the DNA level to identify Colton early stage prostate carcinoma. It identifies an epigenetic field or "halo" associated with malignancy at the DNA level in cells adjacent to the cancer foci. This test is still not commonly used but may help detect prostate cancer at the early stages when local treatment may be more effective.

Tests that measure multiple markers can help determine the risk for developing metastatic disease. One such test is the "Decipher" (developed by GenomeDx Biosciences) that measures 22 markers associated with aggressive prostate cancer.[7] In one study, addition of

Prostate Cancer. http://dx.doi.org/10.1016/B978-0-12-800077-9.00048-7

"Decipher" test to the CAPRA-S and Stephenson nomogram, improved the accuracy of predicting metastatic disease within 5 years after surgery in an independent cohort of men with adverse pathologic features after radical prostatectomy.[8] Decipher was independently validated as a genomic metastasis signature for predicting metastatic disease within 5 years after surgery in a cohort of high-risk men treated with radical prostatectomy without adjuvant therapy.[8]

Another area with great promise is the biomarker discovery, which has been extensively investigated in recent years. We can hope for the availability of genomic, proteomic, as well as metabolomic biomarkers in the near future, which may be helpful in the stratification of patients who will or will not be benefited from a given radiation treatment.[7–13]

## TARGET IDENTIFICATION

Target identification is one of the most challenging parts of treatment planning. MRI has been established as an important tool in improving target volume delineation and also for treatment planning.[14–16] For many years this depended on anatomic imaging including CT (computed tomography) and MRI scans.[14–17] These scanned images allow radiation oncologists to identify the prostate, seminal vesicles, and nodal areas as well as the normal structures. These however, are based on static images and generally could not specifically identify the area of malignancy. At times, nodules were visible but may not have represented the entire extent of the disease, and as a result the entire drainage of structures, such as prostate, seminal vesicles, and lymph node, are included when planning for radiation treatments. Morphological MRI has been shown to enable a more reproducible definition of the prostate especially with the prostate apex.[18,19] Recently, multiparametric MRI (mp-MRI) was introduced to further enhance the tumor staging in prostate cancer.[20] In addition, more recently, functional studies have begun to play a larger role in treatment planning. These functional studies include positron emission tomography – computed tomography (PET-CT) scans, functional MRI scans, and molecular imaging. These are beginning to allow us to explore the possibility of more defined partial volume treatments. For example, there are clinical trials delivering high doses of radiation to more defined targets based on these scans. Such trials will also allow us to better protect normal structures permitting higher doses of radiation per fraction. Present guidelines recommend extension of the target volume beyond the prostate to account for extracapsular region and seminal vesicle invasion.[21] Such target volume definitions are particularly important when techniques like IMRT (intensity-modulated radiation therapy) are used.[17,22]

## HYPOFRACTIONATION

Hypofractionation is the delivery of higher doses of radiation per fraction to a lower total dose. The ability of a cell to repair DNA damage is inversely related to the dose of radiation. In other words, higher doses per fraction of radiation or a large single-dose is more cytotoxic than doses separated by a period of time.[23] The time between fractions allows the repair of the radiation damage in the cells. Therefore, the biologic effectiveness of large doses per fraction to a certain total dose is equivalent to treatment with standard fractionation of 180–200 cGy per fraction to a significantly higher total dose. In theory, the cure rates as well as rates of morbidity should ideally be equivalent. The hypofractionation approach can apply to brachytherapy as well as external beam radiation therapy (EBRT). This in part depends on adequate target definition as well as maneuvers to protect normal structures.[24] An example is the use of saline injections to separate the rectal wall from the posterior portion of the prostate. This also depends on the immobilization of the prostate, which again would permit smaller margins of prostate and better protection of normal tissues. EBRT became a popular treatment of prostate cancer in recent years because of a growing body of evidence suggesting that prostate cancer has a low $\alpha/\beta$ ratio (1.4 Gy) compared to that of adjacent organs at risk, such as the rectum and bladder, therefore a potential therapeutic benefit.[25] A number of Radiation Therapy Oncology Group (RTOG) studies have been carried out or are in progress to help better define the role of hypofractionation (Table 48.1).[26–31] These shorter courses of radiation therapy, assuming biological equivalence, will likely be more economical and allow for better patient compliance. A smaller number of treatments will mean fewer trips for patients and therefore a greater likelihood they will be able to maintain the treatment regimen without unscheduled interruptions.

In one of the clinical trials, 168 patients were randomized to receive hypofractionation or conventional radiation therapy utilizing 3D conformal techniques.[29] In this study, all patients received 9 months of androgen deprivation. The hypofractionated group was treated with 310 cGy per fraction for 20 fractions and the standard fractionation group with 2 Gy per fraction to a total dose of 80 Gy. The 3-year freedom from biochemical failure rates were 87% for the hypofractionation group and 79% for the standard fractionation group. The *P* value was 0.035.

In another trial, men were randomized to receive IMRT with 180 cGy per fraction to a total dose of 7560 cGy or hypofractionation with 240 cGy per fraction to a total dose of 72 Gy.[32] All patients had a PSA of less than or equal to 20 ng/mL and a Gleason score of less than 10. In this study, 203 patients were

TABLE 48.1 Randomized Trials of Moderate Hypofractionation[26–31]

| Study | Number of participants | Median follow-up | Regimen | Outcome (%) |
|---|---|---|---|---|
| NCIC PR5[27] | 936 | 5.7 years | 52.5 Gy in 20 fractions versus 66 Gy in 33 fractions | 5 Year freedom from biochemical failure, 40% versus 43% |
| Royal Adelaide Hospital[28] | 217 | 7.5 | 55 Gy in 20 fractions versus 64 Gy in 32 fractions | 7.5 Year freedom from biochemical failure, 53% versus 34% |
| Regina Elena[29] | 168 | 5.8 | 62 Gy in 20 fractions versus 80 Gy in 40 fractions | 5 Year freedom from biochemical failure, 85% versus 79% |
| FCCC[30] | 303 | 5.5 | 70.2 Gy in 26 fractions versus 76 Gy in 36 fractions | 5 Year biochemical disease-free survival, 77% versus 79% |
| MDACC[31] | 204 | 4.7 | 72 Gy in 30 fractions versus 75.6 Gy in 42 fractions | 5 Year freedom from biochemical failure, 96% versus 92% |

*Reprinted with permission from Martin and D'Amico.[26]*

analyzed and classified into low-risk, intermediate-risk, and high-risk groups and the median follow-up was 6 years. There was an increase in the absolute frequency of the late gastrointestinal (GI) toxicity for men receiving hypofractionation but the difference was not statistically significant. The 5-year actuarial grade 2 or 3 late GI toxicity was 5.1% for patients with IMRT and 10% for patients treated with hypofractionation. The increase in toxicity was a result of moderate- and high-radiation dose to the rectum. There was no statistically significant difference in the absolute frequency of late GU toxicity. The 5-year actuarial grade 2 or 3 late GU toxicity was 16.5% for standard fractionation and 15.8% for hypofractionation. They concluded that "decreasing the length of treatment decreases the cost and is more convenient for patients."[32]

In a review on hypofractionated radiotherapy for prostate cancer, Hegemann et al. summarized detailed additional clinical trials covering, hypofractionated primary radiotherapy, hypofractionated adjuvant/salvage therapy, hypofractionated radiotherapy including pelvic nodes, and hypofractionated IMRT/IGRT (image guided radiation therapy) treatments.[25]

## FUTURE COMBINED MODALITY TREATMENT

The combination of radiation therapy with other treatment modalities has continued to evolve considerably in recent years. Radiation therapy is well known to improve local control. Today, it is often combined with androgen suppression for locally advanced disease. The ideal combination of agents, the timing with radiation therapy as well as the length of the androgen suppression is still under study. The indications are clear but still not fully developed. Radiation therapy in combination with chemotherapy with or without hormonal therapy is also under study, especially in more advanced cases. An RTOG study (1992–2002) showed that additional 2.5 years of hormonal therapy (HT) to EBRT group for locally advanced prostate cancer resulted in significant improvement over EBRT alone.[33] These improvements were shown for all the endpoints except for overall survival (OS) in comparison to short-term of HT (4 months) with EBRT. Similar results were demonstrated in a European Organization for Research and Treatment of Cancer (EORTC) trial (EORTC-22961).[34] In these studies the role of HT in conjunction with EBRT for high-risk prostate cancer patients however, was not clear. To address these and other ambiguities, a phase III, multicenter, randomized, controlled, trimodality therapy study using I-125, brachytherapy, EBRT, and short- or long-term hormonal therapy for high-risk localized prostate cancer is being conducted (2010–2022).[35] This ongoing study may hopefully provide additional insight with regard to the efficacy and limitations of additional 2 years of adjuvant HT to the trimodality therapy and whether long-term HT would benefit the prostate cancer patients with micrometastases.

The need for adjuvant radiation therapy after prostatectomy for high-risk patients has been demonstrated in a number of studies. These indications include positive margins, presence of postoperative PSA, and extracapsular extension. The exact timing of when this should occur, and the role of hypofractionation is still under study.[36,37]

Another emerging area of great potential is the combination of radiation therapy with biologic response modifiers, although it is in its early stages of development. Recently, a comprehensive review was published on the mechanisms and targets of radiosensitization in prostate cancer.[38] The targets of radiosensitization included are proteins/genes, Rad17-RFC complex, 9-1-1 complex, MRN complex, ATM (ataxia-telangiectasia mutated), ATR (ataxia- and Rad3-related), P53, MDM2, Chk2, P21, cyclin E-CDK2 complex, pRB, E2F, Cdc25A, CDK2, CDK4, cyclin B-CDK1 complex, CAK (CDK activating kinase), histone H2AX, 53BP1, Ki-67, Bcl-2, Bax, PTEN, Akt, PAR-4, caspase-1, Ras, and Cox-2 (Table 48.2).[38] A few studies show that prostate cancer cells after radiation therapy overexpress proteins associated with proliferation including Ki67 and apoptosis including Bcl-2 and Bax.[39–41] The over-expression of P53 and Bcl-2 in prostate cancer cells after failing radiation therapy suggests that these may have a role in radiation resistance.[42–46] It has been shown that adenoviral p53 gene therapy can induce prostate cancer cells to be more radiation sensitive.[47] There are many other biologicals that are in various stages of preclinical development that act in conjunction with radiation therapy to increase its effectiveness.[38,48] Most of the ongoing clinical studies combining radiation-sensitizing agents with radiation therapy are listed in Table 48.3.[38] Such studies will lead us to understand in-depth, the cellular pathways that are involved in the development of radiation resistance in tumor cells. While such clinical trials are limited in number, this area of radiosensitization in tumor cells, holds great promise in combined modality therapies.

**TABLE 48.2** Radiation Sensitization Targets

| Proteins/genes | Site of the cell | Cellular process |
|---|---|---|
| Rad17-RFC complex | Nucleus | DNA damage sensor |
| 9-1-1 complex | Nucleus | DNA damage sensor |
| MRN complex | Nucleus | DNA damage sensor |
| ATM (ataxia-telangiectasia mutated) | Nucleus | DNA repair |
| ATR (ataxia- and Rad3-related) | Nucleus | DNA repair |
| P53 | Nucleus | DNA repair |
| MDM2 | Nucleus | Negative regulator of p53 |
| Chk2 | Nucleus | Cell cycle |
| P21 | Nucleus | Cell cycle |
| Cyclin E-CDK2 complex | Nucleus | Cell cycle |
| pRB | Nucleus | Cell cycle |
| E2F | Nucleus | Cell cycle |
| Cdc25A | Nucleus | Cell cycle |
| CDK2 | Nucleus | Cell cycle |
| CDK4 | Nucleus | Cell cycle |
| Cyclin B-CDK1 complex | Nucleus | Cell cycle |
| CAK (CDK activating kinase) | Nucleus | Cell cycle |
| Histone H2AX | Nucleus | Nucleosome formation |
| 53BP1 | Nucleus | P53 phosphorylation |
| Ki-67 | Nucleus | Cellular proliferation |
| Bcl-2 | Mitochondria | Apoptosis |
| Bax | Mitochondria | Apoptosis |
| PTEN | Cytoplasm | Apoptosis and proliferation |
| Akt | Cytoplasm, nucleus | Apoptosis and proliferation |
| PAR-4 | Cytoplasm, nucleus | Apoptosis |
| Caspase-1 | Cytoplasm | Apoptosis |
| Ras | Cytoplasm, nucleus | Cell growth, differentiation, survival |
| Cox-2 | Cytoplasm, nucleus | Cell growth, differentiation, survival |

*Reprinted with permission from Palacios et al.*[38]

## ADAPTIVE RADIATION THERAPY

Adaptive radiation therapy provides a radiation oncologist the ability to modify a patient's treatment plan during a course of radiation therapy taking patient-specific treatment variations into account. These variations include inconsistencies in the daily setup of the patient's position during treatment as well as uncertainties in the position, shape, and size of adjacent internal organs. These changes could make a significant difference in the ability for radiation therapy to control the malignancy as well as to protect the normal structures. For example, in patients with large target volumes, these tumors may shrink during treatment allowing for a reduction in the volume being irradiated. The main issue with this type of procedure is our ability to replan treatment of patients quickly to avoid any delays in treatment. This could mean a difference in planning the treatment of a patient once at the beginning of treatment versus planning the patient multiple times during the treatment. Insurance carriers generally pay for a limited number of treatment plans. Therefore, the process could be aided by more automated treatment planning as well as possible development of multiple treatment plans prior to starting actual treatment.

In the mid-1990s, in order to reduce the target volume and treatment dose, thus minimizing the deleterious effects of positional inconsistencies, an off-line adaptive radiation therapy was developed.[49–51] A clinical study

TABLE 48.3 Clinical Trials Combining Radiation Sensitizing Agents with Radiation Therapy

| Trial | Sponsor | Status |
|---|---|---|
| Sunitinib with hormonal ablation in patients with localized prostate cancer | MD Anderson Cancer Center | Ongoing, not recruiting |
| SU5416 with hormonal ablation in patients with localized prostate cancer | University of Chicago | Unknown |
| Everolimus with hormonal ablation in patients with high-risk localized prostate cancer | University of Michigan | Not yet open |
| TAK-700 with hormonal ablation in patients with high-risk localized prostate cancer | Radiation Therapy Oncology Group | Recruiting |
| Everolimus with hormonal ablation in patients with high-risk locally advanced prostate cancer | Centre Val d'Aurelle – Paul Lamarque | Recruiting |
| Bevacizumab with hormonal ablation in patients with high-risk localized prostate cancer | Benaroya Research Institute | Completed |
| Everolimus for salvage treatment z of biochemical recurrence after prostatectomy | Abramson Cancer Center of the University of Pennsylvania | Recruiting |
| IL-12 gene therapy | Baylor College of Medicine | Completed |
| Isoflavones in patients with localized prostate cancer | Barbara Ann Karmanos Cancer Institute | Completed |
| Eflornithine and bicalutamide compared with eflornithine alone, bicalutamide alone, and no neoadjuvant therapy in treating patients with localized prostate cancer | University of Alabama | Completed |
| Selenomethionine in patients with localized prostate cancer | Roswell Park Cancer Center | Withdrawn |
| R-Flubiprofen in patients with high-risk localized prostate chance | Myrexis Inc. | Unknown |
| Panobinostat in patients with localized prostate cancer | Novartis Pharmaceuticals | Completed |

*Reprinted with permission from Palacios et al.*[38]

used weekly cone-beam CT scan images instead of conventional images for their off-line adaptive radiation therapy in 20 patients. This showed an average reduction of 29% in planning target volume (PTV) and the volume of rectal wall receiving >65 Gy was reduced by 19%.[51,52] So far about 2000 patients have been successfully treated using off-line adaptive radiation therapy.[51] Toward the late 1990s, on-line image-guided adaptive radiation therapy was developed and promoted by onboard volumetric imaging technology.[53] These adaptations enabled on-line corrections to adjust changes in the patients as well as in the target volumes in routine clinical operations.[54] These on-line volumetric image-guided adaptive techniques included on-line replanning through beam aperture modifications for conformal radiation therapy (CRT) or multileaf collimator (MLC) adjustment for IMRT, and on-line adaptive inverse planning.[55–61] Such advances in adaptive radiation therapy over the years, especially using on-line adaptive IMRT, could allow for an increase in target dose of 10% or more through additional target margin reduction and minimization of deleterious effects on the normal tissues.[51]

## NEW TREATMENT MODALITIES

The recent expansion of the use of proton radiation therapy has allowed for studying novel treatment modalities to establish their superiority or equivalence to IMRT and IGRT.[62,63] One of the major differences is the ability to do real-time IGRT without the need for placement of fiducials.[54] The fiducial markers generally are comprised of radio-opaque seeds placed by the urologist that are visible on orthogonal images and used for alignment prior to each treatment. Cone-beam CT imaging, which is common for standard photon IMRT and volumetric-modulated arc therapy (VMAT) irradiation, is being adapted for use along with proton irradiation, but currently this modality is available in investigational settings only.[51,64–66] However, this technology is expected eventually to become common place for improving proton radiation therapy targeting and obviating the need for fiducials. High-dose-rate brachytherapy for localized prostate cancer has been under study for a number of years. A recently published article on the high-dose-rate brachytherapy suggested that the accelerated-hypofractionated treatment is favorable to prostate cancer.[67] In that study, 248 patients were treated with high-dose-rate brachytherapy receiving 38 Gy in four fractions twice a day or 42 Gy in six fractions in two separate implants separated by 1 week. The biochemical control rate at 4.8 years was 91% for the 38 Gy group and 88% for the 42 Gy group. The majority of the complications observed were in grade 1. In another study, 351 consecutive patients with clinically localized prostate cancer were treated with trans-rectal ultrasound-guided high-dose-rate brachytherapy.[68] These patients received 38 Gy in four fractions. The median follow-up was 59.3 months. The 36- and 60-month biochemical control rates were 98% and 94%, respectively. Acute grade 3 GU toxicity with no acute grade 3 GI toxicity was found in 4.8% of patients. Late GU and GI toxicities occurred in 3.4% and 1.4% of patients, respectively. Although gaining popularity recently, this form of treatment has not become widespread but definitely it does offer some advantages; for example, the entire course of radiation therapy is condensed into a few days. Other radiation modalities, such as neutrons, are not likely to become part of standard treatment due to the risk to the normal structures. High-intensity focused ultrasound (HIFU) has been employed for prostate carcinoma for a number of years.[69,70] However, this has not yet become a standard of care approach but is gaining greater acceptance world-wide.

The area of biomarker discovery seems to play an important role in current modalities that are being used in the treatment of prostate cancer. Recently, Velonas et al. summarized several genetic and proteomic markers that could be used in the diagnosis and prognosis of prostate cancer.[71] The list of genetic markers included are: B7-H3 (CD276), Ki-67, EPCA, LAT1 (CD98), PCA3, PSCA, TMPRSS-ERG, BRCA1/BRCA2, PTEN, PI3K, CTAM, CXCR3, FCRL3, KIAA1143, KLF12, TMEM204, SAMSN1, MME, and PSGR. The list of proteomic markers included are: AMACR, Endoglin (CD105), Engrailed 2 (EN-2), PSMA, Caveolin-1, IL-6, CD 147, S-100 protein family, Annexin A3, TGF-β 1, KLK2, and MSMB.[71] In a retrospective study, the marker Periostin (POSTN) was found to overexpress in stroma, which was associated with increased extracapsular extension in prostatectomy specimens and shorter survival. They also observed that the underexpression of POSTN was associated with shorter PSA-free survival.[72,73] Watahiki et al. identified a panel of 10 miRNAs as potential biomarkers to differentiate localized prostate cancer from castration-resistant metastatic disease.[74] In another retrospective study, Som et al. showed that baseline levels of bone-specific alkaline phosphatase (BAP) and urine N-telopeptide (uNTx) but not PSA were prognostic in both androgen-dependent and castration-resistant disease. They also found that a reduction in PSA, BAP, and uNTx or BAP/uNTx on therapy was predictive of overall survival in both groups of prostate cancer patients.[75] Metabolomics research is currently being explored to discover cancer-specific diagnostic and prognostic biomarkers.[76,77] Recent metabolomic studies using cell lines and animal prostate tumors showed over-expression of a Coactivator SRC-2 in prostate cancer and amplification as well as over-expression in 37% of the metastatic tumors.[77] Their study suggests that SRC-2 could be used as a therapeutic target for prostate cancer. These genomic, proteomic, and metabolomic markers could be used in diagnosis, defining patients who may benefit from a given therapy, determining the extent of disease, monitoring response of the cancer to a given intervention and in prognosis. Such markers may eventually be helpful in customizing prostate cancer treatment in the near future.[7–13,71–77]

In conclusion, while the indications, techniques, and integration with other treatment modalities for radiation therapy are continuing to evolve, there are many ongoing clinical trials exploring these issues. One thing is for certain; radiation therapy will continue to be an essential part of the management of prostate carcinoma. With the advent of improved methods of controlling the radiation dose, improved image guidance, hypofractionation, and clear indications, radiation therapy is becoming more effective, more convenient, and also tend to show fewer secondary effects. Additional studies will elucidate a better understanding of its interaction and additive effects with other treatment modalities. These studies will need to have long-term follow-ups as many of the benefits from radiation therapy (survival, biochemical response and secondary effects) either in the primary or in the adjuvant settings will not often be statistically significant in the short-term studies. With the availability of advanced methods for the stratification of patients, sophisticated imaging modalities, biomarkers, better understanding of the molecular mechanisms of various radiation sensitization targets, and a wide-range of combinatorial treatment modalities, the future of radiation therapy treatment options for prostate cancer look promising.

## References


1. Bill-Axelson A, Holmberg L, Garmo H, et al. Radical prostatectomy or watchful waiting in early prostate cancer. *N Engl J Med* 2014;**370**:932–42.
2. Cooperberg MR, Broering JM, Carroll PR. Time trends and local variation in primary treatment of localized prostate cancer. *J Clin Oncol* 2010;**28**:1117–23.
3. Wilt TJ, Brawer MK, Jones KM, et al. Radical prostatectomy versus observation for localized prostate cancer. *N Engl J Med* 2012;**367**:203–13.
4. Punnen S, Freedland SJ, Presti Jr JC. Multi-institutional validation of the CAPRA-S score to predict disease recurrence and mortality after radical prostatectomy. *Eur Urol* 2014;**65**:1171–7.
5. Goris Gbenou MC, Peltier A, Addla SK, et al. Localising prostate cancer: comparison of endorectal magnetic resonance (MR)

imaging and 3D-MR spectroscopic imaging with transrectal ultrasound-guided biopsy. *Urol Int* 2012;**88**:12–7.

6. Kurhanewicz J, Vigneron D, Carroll P, et al. Multiparametric magnetic resonance imaging in prostate cancer: present and future. *Curr Opin Urol* 2008;**18**:71–7.
7. Ross AE, Feng FY, Ghadessi M, et al. A genomic classifier predicting metastatic disease progression in men with biochemical recurrence after prostatectomy. *Prostate Cancer Prostatic Dis* 2014;**17**:64–9.
8. Klein EA, Yousefi K, Haddad Z, et al. A genomic classifier improves prediction of metastatic disease within 5 years after surgery in node-negative high-risk prostate cancer patients managed by radical prostatectomy without adjuvant therapy. *Eur Urol* 2014;**67**(4):778–86.
9. Drake RR, Kislinger T. The proteomics of prostate cancer exosomes. *Expert Rev Proteomics* 2014;**11**:167–77.
10. Kachakova D, Mitkova A, Popov E, et al. Combinations of serum prostate-specific antigen and plasma expression levels of let-7c, miR-30c, miR-141, and miR-375 as potential better diagnostic biomarkers for prostate cancer. *DNA Cell Biol* 2015;**34**:189–200.
11. Kartha GK, Nyame Y, Klein EA. Evaluation of the Oncotype DX genomic prostate score for risk stratification in prostate cancer patients considered candidiates for active surveilance. *Clin Oncol* 2014;**4**:122.
12. Leung CM, Li SC, Chen TW, et al. Comprehensive microRNA profiling of prostate cancer cells after ionizing radiation treatment. *Oncol Rep* 2014;**31**:1067–78.
13. McDunn JE, Li Z, Adam KP, et al. Metabolomic signatures of aggressive prostate cancer. *Prostate* 2013;**73**:1547–60.
14. Korsholm ME, Waring LW, Edmund JM. A criterion for the reliable use of MRI-only radiotherapy. *Radiat Oncol* 2014;**9**:16.
15. Khoo EL, Schick K, Plank AW, et al. Prostate contouring variation: can it be fixed? *Int J Radiat Oncol Biol Phys* 2012;**82**:1923–9.
16. Rischke HC, Nestle U, Fechter T, et al. 3 Tesla multiparametric MRI for GTV-definition of dominant intraprostatic lesions in patients with prostate cancer – an interobserver variability study. *Radiat Oncol* 2013;**8**:183.
17. Panje C, Panje T, Putora PM, et al. Guidance of treatment decisions in risk-adapted primary radiotherapy for prostate cancer using multiparametric magnetic resonance imaging: a single center experience. *Radiat Oncol* 2015;**10**:47.
18. Debois M, Oyen R, Maes F, et al. The contribution of magnetic resonance imaging to the three-dimensional treatment planning of localized prostate cancer. *Int J Radiat Oncol Biol Phys* 1999;**45**:857–65.
19. Parker CC, Damyanovich A, Haycocks T, et al. Magnetic resonance imaging in the radiation treatment planning of localized prostate cancer using intra-prostatic fiducial markers for computed tomography co-registration. *Radiother Oncol* 2003;**66**:217–24.
20. Verma S, Turkbey B, Muradyan N, et al. Overview of dynamic contrast-enhanced MRI in prostate cancer diagnosis and management. *AJR Am J Roentgenol* 2012;**198**:1277–88.
21. Boehmer D, Maingon P, Poortmans P, et al. Guidelines for primary radiotherapy of patients with prostate cancer. *Radiother Oncol* 2006;**79**:259–69.
22. Teh BS, Bastasch MD, Wheeler TM, et al. IMRT for prostate cancer: defining target volume based on correlated pathologic volume of disease. *Int J Radiat Oncol Biol Phys* 2003;**56**:184–91.
23. Smith RP, McKenna WG. *The basics of radiation therapy*. In: Abeloff, MD, Armitage, JO, Niederhuber JE, Kastan MB, McKenna WG, editors. *Clinical Oncology*, 3rd ed. Elsevier, Inc., Philadelphia, 2004 pp. 537–578.
24. Launders J, Inamdar R, Tipton K, et al. Stereotactic body radiation therapy: scope of the literature. *Ann Intern Med* 2011;**154**:737–45.
25. Hegemann NS, Guckenberger M, Claus Belka C, et al. Hypofractionated radiotherapy for prostate cancer. *Radiat Oncol* 2014;**9**:275.
26. Martin NE, D'Amico AV. Progress and controversies: radiation therapy for prostate cancer. *CA Cancer J Clin* 2014;**64**:389–407.
27. Lukka H, Hayter C, Julian JA, et al. Randomized trial comparing two fractionation schedules for patients with localized prostate cancer. *J Clin Oncol* 2005;**23**:6132–8.
28. Yeoh EE, Botten RJ, Butters J, et al. Hypofractionated versus conventionally fractionated radiotherapy for prostate carcinoma: final results of phase III randomized trial. *Int J Radiat Oncol Biol Phys* 2011;**81**:1271–8.
29. Arcangeli G, Saracino B, Gomellini S, et al. A prospective phase III randomized trial ofconventional fractionation in patients with high-risk prostate cancer. *Int J Radiat Oncol Biol Phys* 2010;**78**:11–8.
30. Pollack A, Walker G, Horwitz EM, et al. Hypofractionated radiotherapy for prostate cancer. *J Clin Oncol* 2013;**31**:3860–8.
31. Kuban DA, Nogueras-Gonzalez GM, Hamblin L, et al. Preliminary report of a randomized dose escalation trial for prostate cacner using hypofractionation[abstract]. *Int J Radiat Oncol Biol Phys* 2010;**78**:S58–9.
32. Hoffman EK, Voong KR, Pugh TJ, et al. Risk of late toxicity in men receiving dose-escalated hypofractionated intensity modulated prostate radiation therapy: results from a randomized trial. *Int J Radiat Oncol Biol Phys* 2014;**88**:1074–84.
33. Horwitz EM, Bae K, Hanks GE, et al. Ten-year follow-up of radiation therapy oncology group protocol 92-02: a phase III trial of the duration of elective androgen deprivation in locally advanced prostate cancer. *J Clin Oncol* 2008;**26**:2497–504.
34. Bolla M, de Reijke TM, Van Tienhoven G, et al. Duration of androgen suppression in the treatment of prostate cancer. *N Engl J Med* 2009;**360**:2516–27.
35. Konaka H, Egawa S, Saito S, et al. Tri-Modality therapy with I-125 brachytherapy, external beam radiation therapy, and short- or long-term hormone therapy for high-risk localized prostate cancer (TRIP): study protocol for a phase III, multicenter, randomized, controlled trial. *BMC Cancer* 2012;**12**:110.
36. Thompson IM, Tangen CM, Paradelo J, et al. Adjuvant radiotherapy for pathological T3N0M0 prostate cancer significantly reduces risk of metastases and improves survival: long-term follow up of a randomized clinical trial. *J Urol* 2009;**81**:956–62.
37. Cooperberg MR, Broering JM, Litwin MS, et al. The contemporary management of prostate cancer in the United States: lessons from the Cancer of the Prostate Strategic Urologic Research Endeavor (CaPSURE), a national disease registry. *J Urol* 2004;**171**:1393.
38. Palacios DA, Miyake M, Rosser CJ. Radiosensitization in prostate cancer: mechanisms and targets. *BMC Urol* 2013;**13**:4.
39. Pollack A, Cowen D, Troncoso P, et al. Molecular markers of outcome after radiotherapy in patients with prostate carcinoma: Ki-67, Bcl-2, Bax, and Bcl-x. *Cancer* 2003;**97**:1630–8.
40. Rakozy C, Grignon DJ, Sarkar FH, et al. Expression of Bcl-2, p53, and p21 in benign and malignant prostatic tissue before and after radiation therapy. *Mod Pathol* 1998;**11**:892–9.
41. Rosser CJ, Reyes AO, Vakar-Lopez F, et al. Bcl-2 is significantly overexpressed in localized radio-recurrent prostate carcinoma, compared with localized radio-naive prostate carcinoma. *Int J Radiat Oncol Biol Phys* 2003;**56**:1–6.
42. Cheng L, Sebo TJ, Cheville JC, et al. p53 Protein overexpression is associated with increased cell proliferation in patients with locally recurrent prostate carcinoma after radiation therapy. *Cancer* 1999;**85**:1293–9.
43. Grossfeld GD, Olumi AF, Connolly JA, et al. Locally recurrent prostate tumors following either radiation therapy or radical prostatectomy have changes in Ki-67 labeling index, p53 and bcl-2 immunoreactivity. *J Urol* 1998;**159**:1437–43.
44. Huang A, Gandour-Edwards R, Rosenthal SA, et al. p53 And Bcl-2 immunohistochemical alterations in prostate cancer treated with radiation therapy. *Urology* 1998;**51**:346–51.

45. Prendergast NJ, Atkins MR, Schatte EC, et al. p53 Immunohistochemical and genetic alterations are associated at high incidence with post-irradiated locally persistent prostate carcinoma. *J Urol* 1996;**155**:1685–92.
46. Scherr DS, Vaughan Jr ED, Wei J, et al. Bcl-2 and p53 expression in clinically localized prostate cancer predicts response to external beam radiotherapy. *J Urol* 1999;**162**:503.
47. Sasaki R, Shirakawa T, Zhang ZJ, et al. Additional gene therapy with Ad5CMVp53 enhanced the efficacy of radiotherapy in human prostate cancer cells. *Int J Radiat Oncol Biol Phys* 2001;**51**:1336–45.
48. Alcorn S, Walker AJ, Gandhi N, et al. Molecularly targeted agents as radiosensitizers in cancer therapy—focus on prostate cancer. *Int J Mol Sci* 2013;**14**:14800–32.
49. Martinez A, Yan D, Brabbins D, et al. Improvement in dose escalation using the process of adaptive radiotherapy combined with 3D-conformal or intensity modulated beams for prostate cancer. *Int J Radiat Oncol Biol Phys* 2001;**50**:1226–34.
50. Hoogeman MS, van Herk M, de Bois J, et al. Strategies to reduce the systematic error due to tumor and rectum motion in radiotherapy of prostate cancer. *Radiother Oncol* 2005;**74**:177–85.
51. Ghilezan M, Yan D, Martinez A. Adaptive radiation therapy for prostate cancer. *Semin Radiat Oncol* 2010;**20**:130–7.
52. Nijkamp JN, Pos FJ, Nuver TT, et al. Adaptive radiotherapy for prostate cancer using kilovoltage cone-beam computed tomography: first clinical results. *Int J Radiat Oncol Biol Phys* 2008;**70**:75–82.
53. Jaffray DA, Drake DG, Moreau M, et al. A radiographic and tomographic imaging system integrated into a medical linear accelerator for localization of bone and soft-tissue targets. *Int J Radiat Oncol Biol Phys* 1999;**45**:773–89.
54 Tsai TS, Miyamoto C, Micaily B. Optimization and quality assurance of an image-guided radiation therapy system (IGRT) for IMRT radiotherapy. *Med Dosim* 2012;**37**:321–33.
55. Feng Y, Castro-Pareja C, Shekhar R, et al. Direct aperture deformation: an inter-fraction image guidance strategy. *Med Phys* 2006;**33**:4490–8.
56. Court LE, Dong L, Lee AK, et al. An automatic CT-guided adaptive radiation therapy technique by online modification of multileaf collimator leaf positions for prostate cancer. *Int J Radiat Oncol Biol Phys* 2005;**62**:154–63.
57. Wu C, Jeraj R, Olivera GH. Re-optimization in adaptive radiotherapy. *Phys Med Biol* 2002;**47**:3181–95.
58. Derek S, Liang J, Yan D, et al. Comparison of various online IGRT strategies: the benefits of online treatment plan re-optimization. *Radiol Oncol* 2009;**90**:367–76.
59. Wu Q, Liang J, Yan D. Application of dose compensation in image-guided radiotherapy of prostate cancer. *Phys Med Biol* 2006;**51**:1405–19.
60. Adamson J, Wu Q, Yan D. A hybrid strategy using discriminant analysis for prostate intra-fraction motion management (abstr). *Med Phys* 2009;**36**:2726.
61. Zhang P, Liang J, Yan D. Evaluation of online image guided adaptive inverse planning methodology. *Int J Radiat Oncol Biol Phys* 2007;**69**:S641.
62. Orth M, Lauber K, Niyazi M, et al. Current concepts in clinical radiation oncology. *Radiat Environ Biophys* 2014;**53**:1–29.
63. Nguyen NP, Jang S, Vock J, et al. Feasibility of intensity-modulated and image-guided radiotherapy for locally advanced esophageal cancer. *BMC Cancer* 2014;**14**:265.
64. Teoh M, Clark CH, Wood K, et al. Volumetric modulated arc therapy: a review of current literature and clinical use in practice. *Br J Radiol* 2011;**84**:967–96.
65. Quan EM, Li X, Li Y, et al. A comprehensive comparison of IMRT and VMAT plan quality for prostate cancer treatment. *Int J Radiat Oncol Biol Phys* 2012;**83**:1169–78.
66. Elith CA, Dempsey SE, Warren-Forward HM. A retrospective planning analysis comparing intensity modulated radiation therapy (IMRT) to volumetric modulated arc therapy (VMAT) using two optimization algorithms for the treatment of early-stage prostate cancer. *J Med Radiat Sci* 2013;**60**:84–92.
67. Martinez AA, Demanes J, Vargas C, et al. High-dose-rate prostate brachytherapy: an excellent accelerated-hypofractionated treatment for favorable prostate cancer. *Am J Clin Oncol* 2010;**33**:481–8.
68. Tselis N, Tunn UW, Chatzikonstantinou G, et al. High dose rate brachytherapy as monotherapy for localised prostate cancer: a hypofractionated two-implant approach in 351 consecutive patients. *Radiat Oncol* 2013;**8**:115.
69. Punwani S, Emberton M, Walkden M, et al. Prostatic cancer surveillance following whole-gland high-intensity focused ultrasound: comparison of MRI and prostate-specific antigen for detection of residual or recurrent disease. *Br J Radiol* 2012;**85**:720–8.
70. Yiallouras C, Ioannides K, Dadakova T, et al. Three-axis MR-conditional robot for high-intensity focused ultrasound for treating prostate diseases transrectally. *J Ther Ultrasound* 2015;**3**:2.
71. Velonas VM, Woo HH, Cristobal G, et al. Current status of biomarkers for prostate cancer. *Int J Mol Sci* 2013;**14**:11034–60.
72. Nuzzo PV, Rubagotti A, Zinoli L, et al. Prognostic value of stromal and epithelial periostin expression in human prostate cancer: correlation with clinical pathological features and the risk of biochemical relapse or death. *BMC Cancer* 2012;**12**:625.
73. Wallace TJ, Torre T, Grob M, et al. Current approaches, challenges and future directions for monitoring treatment response in prostate cancer. *J Cancer* 2014;**5**:3–24.
74. Watahiki A, Macfarlane RJ, Gleave ME, et al. Plasma miRNAs as biomarkers to identify patients with castration-resistant metastatic prostate cancer. *Int J Mol Sci* 2013;**14**:7757–70.
75. Som A, Tu S-M, Liu J, et al. Response in bone turnover markers during therapy predicts overall survival in patients with metastatic prostate cancer: analysis of three clinical trials. *Br J Cancer* 2012;**107**:1547–53.
76. Trock BJ. Application of metabolomics to prostate cancer. *Urol Oncol* 2011;**29**:572–81.
77. Dasgupta S, Putluri N, Long W, et al. Coactivator SRC-2–dependent metabolic reprogramming mediates prostate cancer survival and metastasis. *J Clin Invest* 2015;**125**:1174–88.

PART VIII

# CLINICAL DILEMMAS

CHAPTER

# 49

# Management of PSA Recurrences After Radical Prostatectomy

*John B. Eifler, MD*, Joseph A. Smith, Jr, MD***

*Department of Urologic Surgery, Vanderbilt University Medical Center, Nashville, TN, USA

**William L. Bray Professor of Urologic Surgery, Department of Urologic Surgery, Vanderbilt University Medical Center, Nashville, TN, USA

## INTRODUCTION

Following surgery for prostate cancer, approximately 30–35% of men will develop a rising prostate-specific antigen (PSA) level within 10 years.[1] Considering that an estimated 138,000 radical prostatectomies were performed in the United States in 2010,[2] approximately 41,400–48,300 men would be expected to have experienced a rising PSA after surgery, also known as biochemical recurrence (BCR). Furthermore, Pound et al. found the median time between BCR and metastasis was 8 years,[3] suggesting that 331,200–386,400 men in the United States would harbor BCR with no evidence of metastasis at any given time.

## DEFINITION AND NATURAL HISTORY

Though rising PSA after RP is relatively common, the precise definition of BCR remains controversial. The European Association of Urology (EAU) and the American Urological Association (AUA) guidelines define BCR as a single PSA value above 0.2 ng/mL followed by a subsequent confirmatory PSA value above 0.2 ng/mL.[4,5] The National Comprehensive Cancer Network (NCCN) defines BCR as either a failure of PSA to reach undetectable levels following surgery or an undetectable PSA after surgery with at least two subsequent increases in PSA level.[6]

One reason for the different criteria for BCR is that not all men who have a PSA elevation after surgery will develop clinical progression.[3,7] Pound et al. examined the natural history of disease for 304 men who experienced a PSA above 0.2 ng/mL during follow-up after RP.[3] None of the men received postoperative radiotherapy or androgen deprivation therapy. The authors found a 5-year metastasis-free survival rate of 63% with a median metastasis-free survival of 8 years after BCR. In the analysis, not all men who experienced BCR had the same likelihood of metastasis, with higher likelihood in men with Gleason score 8–10, an interval between surgery and recurrence less than 2 years, and those with Gleason score less than 8 and a PSA doubling time (PSADT) of less than 10 months. Furthermore, another work has demonstrated that some men with a PSA above 0.2 ng/mL will have a stable PSA without subsequent rises. Amling et al. found that only 49% of men following RP with an increased PSA level in the 0.20–0.29 range progressed to a higher level and recommended using 0.40 ng/mL to define BCR.[8] The authors suggested that benign prostatic glands may be left behind after surgery, or other tissue sites may produce PSA in some men.

As a result of the heterogeneity of outcomes among men with a PSA above 0.2 ng/mL, some authors have suggested changing the definition of BCR to better reflect likelihood of developing metastatic disease. Stephenson et al. evaluated 10 definitions of BCR to determine which had the highest association with metastatic disease in a multivariable model adjusting for postoperative Gleason score, preoperative PSA, surgical margin status, pathologic stage, and use of adjuvant therapies (radiotherapy or ADT).[9] The authors found the definition of BCR that provided the best combination of sensitivity and specificity (based on the goodness-of-fit statistic) to be PSA ≥0.4 and rising. Thus, men would only be considered to have BCR if the PSA reached 0.4 and a subsequent PSA was higher. One interesting note from the article is that use of the

Prostate Cancer. http://dx.doi.org/10.1016/B978-0-12-800077-9.00049-9

American Society of Therapeutic Radiation and Oncology (ASTRO) definition of BCR (three consecutive PSA rises, with BCR backdated to the midpoint between the date of the last undetectable PSA and the date of the first of the three PSA rises)[10] required a median time of 42 months, though as a result of back-dating, the time to recurrence would be considered 15 months. This delay in diagnosis may unnecessarily postpone salvage therapy, making the ASTRO definition an unattractive option following RP. Other authors found a similar delay using the revised ASTRO definition of nadir PSA +2 ng/mL.[11]

The importance in choosing an appropriate definition of BCR lies in its ability to determine timing of salvage therapy after RP. In a retrospective study assessing the impact of salvage therapy on prostate-cancer-specific survival after BCR, Trock et al. demonstrated that an increased time between BCR and initiation of salvage therapy was associated with prostate-cancer-specific mortality (PCSM).[12] The authors found the greatest inflection point to be at 2 years after BCR, with men treated greater than 2 years after BCR not benefiting from salvage radiotherapy (SRT). However, when comparing men treated within 1 year of BCR to those treated greater than 1 year after BCR, those treated earlier seemed to benefit more from salvage therapy (H.R. 0.15 vs. 0.22, respectively), though the difference was not statistically significant. Nonetheless, patients and clinicians would prefer to identify patients who will progress as early as possible, when SRT likely has the greatest chance to prevent PCSM. Ideally, one could use a definition of BCR that accurately identifies men likely to die of prostate cancer as early after RP as possible.

Due to the drawbacks of using PSA alone to determine who should undergo additional therapy, some authors recommend using variable definitions for BCR based on an individualized risk of cancer progression. Similar to Stephenson et al.,[9] Mir et al.[13] examined 14 definitions for BCR for their association with subsequent disease progression. However, the authors stratified the patients into risk groups based on clinicopathologic factors known after RP. The authors found that for each definition of BCR, the likelihood of disease progression varied greatly based on these clinicopathologic risk factors. For instance, the likelihood of disease progression for a detectable PSA ≤0.1 ng/mL was 18–25% for men with good pathologic features and 73–88% for men with pT3b, pN1, or Gleason 8–10 disease. The investigators recommend using a single PSA ≥0.05 for men with the highest risk pathologic features (pT3b or pN1, or Gleason 8–10) as the definition of BCR versus PSA ≥0.4 and rising for those with lowest risk (pT2, Gleason 6 or 3 + 4, negative surgical margins).[13] More generally, this work demonstrates that the use of any PSA rise after RP as the threshold for SRT in all patients is ill-advised. Instead, a PSA rise after RP must be considered in the context of disease risk when deciding whether SRT is likely to be advantageous. This notion is supported by the results reported by Karlin et al., who assessed the recurrence-free survival (RFS) rate for SRT after RP. The authors found that men with pathologic Gleason score 8–10 disease who underwent SRT at PSA level ≤0.33 ng/mL had a 77% RFS rate at 53 months compared to a 26% survival rate when SRT was initiated at a PSA between 0.34 ng/mL and 1.0 ng/mL.[14]

Even if patients with a rising PSA are destined to progress, they may die of nonprostate cancer causes prior to the onset of morbidity from metastatic prostate cancer. Xia et al. used a competing risks model to assess the likelihood that men who experienced BCR would progress to metastatic disease before dying of nonprostate cancer causes.[15] Since the median follow-up after BCR was 7.0 years, the authors had limited power to detect the incidence of metastases beyond 10 years. Instead, using US life tables adjusted for the good overall health of men undergoing RP, the authors calculated a lower bound for overdetection of BCR. Using this analysis, the authors estimate that at least 9.1% of men who experienced BCR within 5 years of RP and at least 15.6% of men with BCR between 5 years and 10 years after RP would never experience metastatic disease in their lifetime. For men who are older, sicker, and have lower stage or grade, SRT or hormonal therapy may be of no benefit.

For men with a rising PSA after RP, decisions regarding salvage therapies rely on careful consideration of the definition of BCR, how well BCR predicts oncologic outcome after RP, and the competing risk of death from nonprostate cancer causes before metastatic prostate cancer would arise. For men with low-risk pathology, a more conservative definition of BCR (with higher specificity) is appropriate, whereas those with aggressive features should choose low PSA thresholds before consideration of salvage therapy. Prior to initiation of salvage therapy, physicians and patients should consider the long natural history of recurrent prostate cancer, with median time between BCR and diagnosis of metastatic disease of 8 years. Only after considering these factors, as well as the patient's comfort with risk of disease progression and PCSM, should physicians and patients choose a management course.

## DIAGNOSTIC WORK-UP

The initial site of macroscopic prostate cancer metastasis is most commonly bone, followed by pelvic lymph node metastasis.[16] The tests of choice to evaluate for metastases are radionuclide bone scan and contrast enhanced CT of the abdomen and pelvis. Guidelines for

TABLE 49.1 Guideline Recommendation Regarding Imaging for BCR after RP

| | NCCN | AUA | EAU |
|---|---|---|---|
| Consider Initial Imaging | 1. If PSA never undetectable<br>2. If PSA undetectable, but subsequently detectable, PSA and 2 additional PSA rises | 1. If PSA never undetectable<br>2. Rising PSA value after surgery that is ≥0.2 ng/mL with second confirmatory level ≥0.2 ng/mL | 1. High baseline PSA (>10 ng/mL)<br>2. High PSA velocity (>0.5 ng/mL/month)<br>3. Symptomatic bone disease |

management of BCR show no consensus for when imaging should be initiated or how often they should be performed (see Table 49.1).[5,6,17]

In the absence of data from randomized, controlled trials, retrospective studies have been utilized to determine the ideal timing to initiate imaging studies after BCR. In the largest series, Loeb et al. retrospectively evaluated men who underwent RP at Johns Hopkins, where men routinely obtain bone scan annually after BCR.[18] A total of 193 men with positive bone scans were included in the study, none of whom had prior hormonal therapy. The authors found that transition to a positive bone scan occurred at a PSA level of <10 ng/mL in 25.9% of patients, at a PSA between 10 ng/mL and 100 ng/mL in 50.8%, and at a PSA >100 ng/mL in 23.3%. Of note, men with higher pathologic Gleason score and lower preoperative PSA tended to transition to a positive bone scan at a lower PSA level. The authors concluded that men should obtain regular bone scans after BCR since no PSA cut-off can be relied upon. Other authors suggest waiting until higher PSA levels to initiate imaging studies.[19,20] In a study from Memorial Sloan Kettering Cancer Center, Dotan et al. evaluated factors associated with positive bone scan in 239 men who underwent RP and experienced BCR.[20] During follow-up, 60 patients were found to have a positive bone scan. Of note, bone scans were performed at the discretion of the treating clinician and not at regular intervals. In this setting, only 4% of bone scans performed at a PSA level below 10 ng/mL were positive, compared to 36% of bone scans between 10 ng/mL and 20 ng/mL, 50% between 20 ng/mL and 50 ng/mL, and 79% above 50 ng/mL. In a multivariable analysis, total PSA, high PSA slope (the slope of a linear regression line fit to the PSA data), and low PSA velocity were associated with positive bone scans, and the resulting nomogram demonstrated excellent discrimination on internal validation (c-index 0.93). Thus, while a minority of men destined to develop metastatic disease could be found to have oligometastatic disease at a PSA level below 10 ng/mL, the wide majority of bone scans performed at PSA <10 ng/mL in this setting would be negative.

Deciding when to initiate imaging studies also depends upon the utility of diagnosing metastatic disease at an early time point. With the advent in recent years of novel immunotherapy options and the emerging role for early chemotherapy in this population, there may be an advantage to discovering oligometastatic disease. However, to date, the advantage of initiating systemic therapy at a low metastatic burden has not been clearly established. Interest has increased recently in local therapy for oligometastatic disease with surgical extirpation or stereotactic radiotherapy. The evidence supporting/refuting these approaches is discussed in a separate section. Should randomized trials prove local therapy to be advantageous, a more aggressive screening protocol with regular bone scans initiated at PSA thresholds below 10 ng/mL may be favored.

Though CT and bone scan are the studies of choice for prostate cancer staging after BCR, pelvic MRI has shown promise in improving detection of local recurrence. Dynamic contrast-enhanced MRI with endorectal coil has shown high sensitivity (84–88%) and specificity (89–100%) for detecting local recurrence after RP (with recurrence confirmed by TRUS-guided biopsy).[21–24] However, it should be noted that the average PSA was above 1.0 ng/mL in each study, and the median tumor size was >1.0 cm. The likelihood of detecting recurrence disease in the prostatic fossa increased with PSA, ranging from 8% in men with PSA ≤0.19 ng/mL to 38% for men with PSA above 0.72 ng/mL.[24] Whether MRI would improve patient selection in candidates for salvage therapy, who typically have PSA levels below 0.5 ng/mL, remains unclear. No published randomized controlled trials (RCT) investigating the use of MRI to select patients for SRT have been reported, though MRI is currently being used to individualize the radiotherapy dose and volume.[24]

While MRI may improve detection of local recurrence after RP, developments of novel radiotracers for use in positron emission tomography (PET) promise to improve detection of distant recurrences. The traditional radiotracer used in PET, $^{18}$-fluorodeoxyglucose (FDG), has limited uptake in prostate cancer. In contrast, choline uptake is increased in prostate cancer cells, and the novel radiotracers [$^{11}$C]choline and [$^{18}$F]choline seek to leverage this biologic property to improve PET sensitivity and specificity. As with bone scan, the accuracy of PET depends on clinicopathologic features of the study population, particularly the PSA at the time of the PET/CT. The optimal PSA level for lesion detection has been reported to be approximately 1.0–2.0 ng/mL.[25,26] For men with PSA <1 ng/mL after RP, the likelihood of having a positive PET/CT ranged from 5% to 26%, while PSADT, PSA velocity, and Gleason score are all significantly associated with detection rates.[27–30] When comparing PET/CT findings to histological analysis, Scattoni et al. have found sensitivity of 64%, specificity 90%, positive predictive value (PPV) 86%, and negative predictive value (NPV) 72%.[31] Other authors have found false positive rates to be far higher.[32,33] Of note, studies evaluating [$^{18}$F]choline PET/CT have demonstrated similar accuracy and a similar rate of false positives.[34] Also, an agent previously considered promising for detection of occult metastases, the specific PSMA targeting agent $^{111}$In-capromab (ProstaScint™), has not shown efficacy in predicting patients likely to respond to SRT,[35] and its use is not recommended.[5]

The use of PET/CT holds promise in improving patient care after BCR by identifying patients likely to progress to metastatic disease, and by improving localization of disease for salvage treatment planning. However, routine use of these imaging studies cannot be recommended in this setting currently. The sensitivity and specificity of PET/CT at PSA levels below 1 ng/mL are likely quite low due to the high likelihood of micrometastatic disease in this population, and some investigators note that the optimal PSA at which to initiate [$^{11}$C]choline PET/CT is 1.05 ng/mL.[25] As is discussed in the subsequent section, the success of SRT is more likely at lower PSA values, and most guidelines recommend initiating SRT at PSA levels below 0.5 ng/mL. Thus, using imaging studies to determine who should receive salvage therapies could lead to undertreatment, and some men may lose the window of curability as a result. Furthermore, some authors report false positive rates of 30% for PET/CT,[36] raising the possibility that imaging would increase overtreatment of men with BCR who are unlikely to progress. Furthermore, even when only a single lymph node is PET/CT positive, which represents an ideal candidate for salvage lymphadenectomy, the PPV of PET/CT is only 34.8% for the lymphatic landing zone where the isolated node was located, while 60% had tumor in other landing zones.[33] These data suggest that PET/CT cannot reliably determine the template of dissection, and an extensive dissection should be performed regardless of which lymph node is PET positive.[33] Currently, the NCCN guidelines recommend [$^{11}$C]choline PET as an option in patients with BCR after RP.[6]

## TREATMENT OPTIONS

### Salvage Radiotherapy

Management of men with BCR after RP and negative imaging studies remains a clinical dilemma. While the disease may be localized to the surgical site and curable with salvage therapy, the presence of occult metastases cannot be excluded. Surprisingly, SRT localized to the pelvis leads to a 68% decrease in PCSM compared to observation even in the absence of biopsy-proven local recurrence.[12] As a result, SRT remains the standard of care for men with a long life expectancy.[5,6]

If given early after BCR, SRT provides durable response in the majority of men (see Table 49.2). The likelihood of an undetectable PSA after SRT has been reported to be 60%, and of patients who achieve an undetectable PSA after SRT, 75% will have an undetectable PSA after 3.5 years.[37] However, SRT is not without risk, and determining the optimal PSA cut-off to initiate SRT remains controversial. What is clear is that oncologic outcomes are best when SRT is initiated at a low PSA level. Wiegel et al. report that 60% of patients who initiated SRT when PSA <0.5 ng/mL remain disease-free at 5 years, compared to 33% who initiated SRT at a PSA ≥0.5 ng/mL,[37] and more recent studies have demonstrated an RFS of ~75% in men with PSA ≤0.5 ng/mL at the time of SRT.[38,39] In a pooled analysis of retrospective series reporting RFS after SRT, the updated RFS at 5 years was 71.1% overall.[39] Lowering the PSA cut-off below 0.5 ng/mL may improve RFS even further. Siegmann et al.[40] report RFS of 83% at 2 years for those with PSA <0.2 ng/mL at the time of SRT, compared to 78% for those with PSA <0.28 ng/mL and 61% for those with PSA ≥0.28 ng/mL,

**TABLE 49.2** Outcomes after Salvage Radiotherapy for Men with Biochemical Recurrence

| Study | Year | Number of cases | Pre-SRT PSA | Dose | RFS | RFS timepoint |
|---|---|---|---|---|---|---|
| Wiegel et al. | 2009 | 162 | median 0.33 | 66.6 Gy | 54% | 3.5 years |
| Siegmann et al. | 2012 | 301 | median 0.29 | 66.6/70.2 Gy | 74% | 2 years |
| Stephenson et al. | 2007 | 1540 | <0.5 | variable | 48% | 6 years |
| | | | 0.51–1.0 | | 40% | 6 years |
| | | | 1.01–1.50 | | 28% | 6 years |
| | | | >1.50 | | 18% | 6 years |
| Briganti et al. | 2014 | 472 | <0.5 | 66.6/70.2 Gy | 73.40% | 5 years |

though of note some patients included in the study had a PSA >1.0 ng/mL at the time of SRT, which could have influenced the results. When reviewing these data, one must consider the possibility of the Will Rogers phenomenon,[41] since a rising PSA after RP will often plateau and never lead to clinical progression of disease.[8] By assessing patients under 0.2 ng/mL, the RFS may appear improved due to a higher proportion of patients harboring indolent disease that never would recur in the absence of treatment, not due to an improved treatment effect in this population. To address this concern, prospective randomized controlled trials are currently accruing.[42]

Factors associated with RFS include PSA prior to SRT, time between BCR and initiation of SRT, undetectable PSA nadir after SRT, PSADT, pT3b disease, positive nodes at prostatectomy, Gleason score at prostatectomy, and surgical margin status.[37,40,43,44] While guidelines recommend that at least 64–65 Gy[17] or 66 Gy[5] be used for SRT, whether dose escalation from 65 Gy to 70.2 Gy improves RFS remains controversial.[44,45] Whether addition of ADT to SRT improves oncologic outcomes also remains unclear, though in one large retrospective series of men suffering BCR, the addition of androgen deprivation therapy to SRT was not associated with PCSM.[12] A prospective RCT is needed to address the benefit of ADT with RT, and RTOG-9061, which randomized patients after RP to RT + placebo or RT + bicalutamide, is currently accruing patients for that purpose.

Algorithms and nomograms allow individualized risk stratification to select patients most likely to benefit from radiotherapy. Trock et al. recommended initiating SRT for PSADT <6 months, regardless of surgical margin status or Gleason score. For PSADT ≥6 months, the authors recommend SRT only for men with positive surgical margins or Gleason score 8–10.[12] Two prominent groups have developed nomograms to predict RFS following SRT.[38,44] Stephenson et al. retrospectively evaluated a multi-institutional cohort of 1540 patients who underwent SRT after RP.[44] In a multivariable model, PSA level at the time of SRT, pathologic Gleason score, PSADT, surgical margin status, use of ADT before or during SRT, and positive lymph nodes were associated with recurrence after SRT. The authors constructed a nomogram to predict RFS after SRT, finding a c-index of 0.69 on internal validation. External validation of the model shows adequate discrimination, with a c-index of 0.65 in 102 men in the SEARCH database,[46] and 0.70 in 143 men with pT3 or margin-positive disease from a single institution.[47] Citing evidence that risk factors, such as pathologic stage[14] and PSADT,[48] perform differently at low PSA, Briganti et al. restricted the development cohort to men with node-negative disease at RP who achieved undetectable PSA after surgery and underwent SRT with PSA ≤0.5 ng/mL.[38] On internal validation with 200 bootstrap resamples, the nomogram demonstrated good discrimination with a c-index of 0.74. In addition to improving predictive accuracy by tailoring the population studied, groups have incorporated novel genomic tests into nomogram models. Den et al. report that the Decipher® genomic classifier (GC) improves the discriminatory accuracy of the Stephenson nomogram from a c-index of 0.70–0.78 for RFS, and from 0.70 to 0.80 for development of distant metastases in a retrospective single institution study.[47]

Determining the appropriate trigger for SRT should balance the potential benefits to early SRT with the morbidity of therapy. Many large retrospective series assessing oncologic outcomes after SRT did not include long-term toxicity data. In a retrospective study of 742 patients who underwent adjuvant RT (ART) or SRT at a single institution, the incidence of acute toxicity ≥Grade 2 was 19% after ART and 17% after SRT,[49] whereas the incidence of ≥Grade 3 toxicity was 8% and 6%, respectively. Late toxicity was more common, with 23.9% and 23.7% experiencing ≥Grade 2 toxicity at 8 years after ART and SRT, respectively. Late Grade 3 toxicity was slightly higher after ART than SRT (12.2% and 10.0%), though the difference was nonsignificant ($p = 0.11$). When comparing specific complications of ART and SRT, the likelihood of urethral stricture (5% and 3%), Grade 3 bleeding (2% and 1%), severe incontinence (7% and 6%), and need for cystectomy (1 patient in each arm) were similar. In a separate study reporting the toxicity of SRT in which all patients received at least 70 Gy, urinary incontinence rates of 13.6% and Grade 3 erectile dysfunction occurred in 26% within 5 years of treatment.[50] With even higher doses (76 Gy), the GU and GI toxicity increased to 22% and 8% of patients, respectively.[51]

## Local Treatment for Oligometastatic Disease

Improvement in the diagnosis of oligometastatic disease with choline PET/CT has allowed investigators to consider resection of metastases in this setting. Previous studies have demonstrated that patients with three or fewer metastases have significantly lower chance of death at a median follow-up of 6 years than those with four or more metastatic sites.[52] Though no randomized trials have assessed the efficacy of treating oligometastatic disease with radiotherapy or surgery, investigators point to durable responses in patients with pN1 pathologic stage at RP treated with multimodality therapy.[53] Several small case series have been published that report safety of lymphadenectomy and adequate short- to intermediate-term follow-up.[54–57] In the largest study to date, Rigatti et al. reported that 56.9% of patients had a PSA ≤0.2 ng/mL after the procedure, though 5-year RFS was only 19%.[58] Karnes et al. reported 45.5% RFS at 3 years, likely in line with the Rigatti et al. data.[54]

SRT directed at oligometastatic disease has also been reported, though each study has a slightly different protocol for radiotherapy delivery.[59,60] Typically, radiotherapy is directed at the pelvis as well as at suspicious lymph nodes found on PET/CT, though in some series radiotherapy is targeted only at suspicious lymph nodes. The dose of radiation varied among studies, as did the fractionation pattern. Short-term RFS rates have been reported to be 35–59% after 2–3 years.[59,60]

Many of the patients included in these studies have ADT either at the time of treatment or afterward, making direct interpretation of the efficacy of localized therapy difficult. No study incorporated a control group. As Klein noted, "Metastasis-directed therapy as currently practiced is of no proven benefit ... and we must not confuse our *ability to treat* such disease safely with an *actual benefit* in doing so."[61] Currently, these treatments should be considered investigational and should be performed under the auspices of a clinical trial.

### Androgen Deprivation Therapy

Prior to the advent of SRT in the management of BCR after RP, patients were often treated with hormonal therapy. Then, as now, the appropriate PSA level at which to initiate ADT was controversial. Though no randomized controlled trials have been performed, perhaps the best evidence guiding this decision is the Center for Prostate Disease Research Multicenter National Database. In this observational study through the Department of Defense, 4967 men underwent RP between 1988 and 2002.[62] With a median follow-up after RP of 4.7 years (range 0.1–13 years), 1352 developed BCR and were treated either with early ADT (at the time of BCR) or no/late ADT (not initiated before the development of metastatic disease). The majority of men in the early ADT arm initiated therapy when the PSA was between 0.21 ng/mL and 2.5 ng/mL. The authors found that ADT administered at PSA ≤10 ng/mL demonstrated a significant delay in onset of clinical metastases in patients with pathological Gleason 7 or greater disease, or in patients with PSADT ≤12 months (Hazard ratio = 2.10, $p = 0.020$ in a multivariable model). However, in the overall cohort, early ADT did not affect clinical metastasis-free survival.[62] Survival data in this population has yet to be reported. In another retrospective study from a single institution, men with high-risk disease who underwent adjuvant ADT were found to have decreased PCSM compared to men who did not receive ADT after BCR.[63] However, no difference was observed in overall survival, and when other subgroups were analyzed (e.g., initiation of ADT at PSA ≥0.4 ng/mL), no difference was observed in PCSM. While these data suggest that ADT delays the onset of metastatic disease in men with high-risk features, whether early ADT improves survival remains controversial.

## CONCLUSIONS

Despite advances in the care of prostate cancer, a rising PSA remains a common dilemma for men who undergo RP. If caught when the PSA is low, preferably below 0.5 ng/mL, the majority of patients will experience long-term RFS with SRT. However, epidemiologic data demonstrate that not all men with a rising PSA after RP would progress to metastatic disease if left untreated, and population-based studies suggest overtreatment in this population. In the future, MRI and PET/CT may play a role in determining the appropriate candidate for SRT though they are currently investigational. While we await the results of clinical trials that will guide decision-making, clinicians and patients must weigh the individual's disease characteristics, comorbidities, and life expectancy when determining the appropriate trigger for salvage therapy.

### References

1. Han M, Partin AW, Pound CR, et al. Long-term biochemical disease-free and cancer-specific survival following anatomic radical retropubic prostatectomy. The 15-year Johns Hopkins experience. *Urol Clin North Am* 2001;**28**:555.
2. National Hospital Discharge Survey: 2010 table, Procedures by selected patient characteristics – number by procedure category and age; 2010.
3. Pound CR, Partin AW, Eisenberger MA, et al. Natural history of progression after PSA elevation following radical prostatectomy. *JAMA* 1999;**281**:1591.
4. Cookson MS, Aus G, Burnett AL, et al. Variation in the definition of biochemical recurrence in patients treated for localized prostate cancer: the American Urological Association Prostate Guidelines for Localized Prostate Cancer Update Panel report and recommendations for a standard in the reporting of surgical outcomes. *J Urol* 2007;**177**:540.
5. Mottet N, Bastian PJ, Bellmunt J, et al. Guidelines on Prostate Cancer; European Association of Urology, 2014, Available from: http//uroweb.org/wp-content/uploads/1607-Prostate-Cancer_LRV3.pdf.
6. Mohler JL, Kantoff PW, Armstrong AJ, et al. Prostate cancer, version 2.2014. *J Natl Compr Canc Netw* 2014;**12**:686.
7. Boorjian SA, Thompson RH, Tollefson MK, et al. Long-term risk of clinical progression after biochemical recurrence following radical prostatectomy: the impact of time from surgery to recurrence. *Eur Urol* 2011;**59**:893.
8. Amling CL, Bergstralh EJ, Blute ML, et al. Defining prostate specific antigen progression after radical prostatectomy: what is the most appropriate cut point? *J Urol* 2001;**165**:1146.
9. Stephenson AJ, Kattan MW, Eastham JA, et al. Defining biochemical recurrence of prostate cancer after radical prostatectomy: a proposal for a standardized definition. *J Clin Oncol* 2006;**24**:3973.
10. Consensus statement: guidelines for PSA following radiation therapy. American Society for Therapeutic Radiology and Oncology Consensus Panel. Int J Radiat Oncol Biol Phys 1997;37(5):1035–1041.
11. Nielsen ME, Makarov DV, Humphreys E, et al. Is it possible to compare PSA recurrence-free survival after surgery and radiotherapy using revised ASTRO criterion–"nadir+2"? *Urology* 2008;**72**:389.
12. Trock BJ, Han M, Freedland SJ, et al. Prostate cancer-specific survival following salvage radiotherapy vs observation in men with biochemical recurrence after radical prostatectomy. *JAMA* 2008;**299**:2760.

13. Mir MC, Li J, Klink JC, et al. Optimal definition of biochemical recurrence after radical prostatectomy depends on pathologic risk factors: identifying candidates for early salvage therapy. *Eur Urol* 2013;**66**:204.
14. Karlin JD, Koontz BF, Freedland SJ, et al. Identifying appropriate patients for early salvage radiotherapy after prostatectomy. *J Urol* 2013;**190**:1410.
15. Xia J, Trock BJ, Gulati R, et al. Overdetection of recurrence after radical prostatectomy: estimates based on patient and tumor characteristics. *Clin Cancer Res* 2014;**20**:5302.
16. Morris MJ, Scher HI. Clinical state of the rising psa value after definitive local therapy: a practical approach. In: Wein AJ, Kavoussi LR, Novick AC, et al. editors. *Campbell-Walsh Urology*. 10th ed., Elsevier Saunders, Philadelphia, 2011. p. 2921–2933.
17. Thompson IM, Valicenti RK, Albertsen P, et al. Adjuvant and salvage radiotherapy after prostatectomy: AUA/ASTRO Guideline. *J Urol* 2013;**190**:441.
18. Loeb S, Makarov DV, Schaeffer EM, et al. Prostate specific antigen at the initial diagnosis of metastasis to bone in patients after radical prostatectomy. *J Urol* 2010;**184**:157.
19. Cher ML, Bianco Jr FJ, Lam JS, et al. Limited role of radionuclide bone scintigraphy in patients with prostate specific antigen elevations after radical prostatectomy. *J Urol* 1998;**160**:1387.
20. Dotan ZA, Bianco Jr FJ, Rabbani F, et al. Pattern of prostate-specific antigen (PSA) failure dictates the probability of a positive bone scan in patients with an increasing PSA after radical prostatectomy. *J Clin Oncol* 2005;**23**:1962.
21. Boonsirikamchai P, Kaur H, Kuban DA, et al. Use of maximum slope images generated from dynamic contrast-enhanced MRI to detect locally recurrent prostate carcinoma after prostatectomy: a practical approach. *AJR Am J Roentgenol* 2012;**198**:W228.
22. Sella T, Schwartz LH, Swindle PW, et al. Suspected local recurrence after radical prostatectomy: endorectal coil MR imaging. *Radiology* 2004;**231**:379.
23. Cirillo S, Petracchini M, Scotti L, et al. Endorectal magnetic resonance imaging at 1.5 Tesla to assess local recurrence following radical prostatectomy using T2-weighted and contrast-enhanced imaging. *Eur Radiol* 2009;**19**:761.
24. Liauw SL, Pitroda SP, Eggener SE, et al. Evaluation of the prostate bed for local recurrence after radical prostatectomy using endorectal magnetic resonance imaging. *Int J Radiat Oncol Biol Phys* 2013;**85**:378.
25. Castellucci P, Ceci F, Graziani T, et al. Early biochemical relapse after radical prostatectomy: which prostate cancer patients may benefit from a restaging $^{11}$C-Choline PET/CT scan before salvage radiation therapy? *J Nucl Med* 2014;**55**:1424.
26. Mitchell CR, Lowe VJ, Rangel LJ, et al. Operational characteristics of (11)c-choline positron emission tomography/computerized tomography for prostate cancer with biochemical recurrence after initial treatment. *J Urol* 2013;**189**:1308.
27. Castellucci P, Fuccio C, Nanni C, et al. Influence of trigger PSA and PSA kinetics on 11C-Choline PET/CT detection rate in patients with biochemical relapse after radical prostatectomy. *J Nucl Med* 2009;**50**:1394.
28. Giovacchini G, Picchio M, Briganti A, et al. [$^{11}$C]choline positron emission tomography/computerized tomography to restage prostate cancer cases with biochemical failure after radical prostatectomy and no disease evidence on conventional imaging. *J Urol* 2010;**184**:938.
29. Giovacchini G, Picchio M, Scattoni V, et al. PSA doubling time for prediction of [(11)C]choline PET/CT findings in prostate cancer patients with biochemical failure after radical prostatectomy. *Eur J Nucl Med Mol Imaging* 2010;**37**:1106.
30. Giovacchini G, Picchio M, Coradeschi E, et al. Predictive factors of [(11)C]choline PET/CT in patients with biochemical failure after radical prostatectomy. *Eur J Nucl Med Mol Imaging* 2010;**37**:301.
31. Scattoni V, Picchio M, Suardi N, et al. Detection of lymph-node metastases with integrated [$^{11}$C]choline PET/CT in patients with PSA failure after radical retropubic prostatectomy: results confirmed by open pelvic-retroperitoneal lymphadenectomy. *Eur Urol* 2007;**52**:423.
32. Osmonov DK, Heimann D, Janssen I, et al. Sensitivity and specificity of PET/CT regarding the detection of lymph node metastases in prostate cancer recurrence. *Springerplus* 2014;**3**:340.
33. Passoni NM, Suardi N, Abdollah F, et al. Utility of [$^{11}$C]choline PET/CT in guiding lesion-targeted salvage therapies in patients with prostate cancer recurrence localized to a single lymph node at imaging: results from a pathologically validated series. *Urol Oncol* 2014;**32**:38.e9–38.e16.
34. Tilki D, Reich O, Graser A, et al. $^{18}$F-Fluoroethylcholine PET/CT identifies lymph node metastasis in patients with prostate-specific antigen failure after radical prostatectomy but underestimates its extent. *Eur Urol* 2013;**63**:792.
35. Koontz BF, Mouraviev V, Johnson JL, et al. Use of local (111)In-Capromab pendetide scan results to predict outcome after salvage radiotherapy for prostate cancer. *Int J Radiat Oncol Biol Phys* 2008;**71**:358.
36. Schilling D, Schlemmer HP, Wagner PH, et al. Histological verification of 11C-choline-positron emission/computed tomography-positive lymph nodes in patients with biochemical failure after treatment for localized prostate cancer. *BJU Int* 2008;**102**:446.
37. Wiegel T, Lohm G, Bottke D, et al. Achieving an undetectable PSA after radiotherapy for biochemical progression after radical prostatectomy is an independent predictor of biochemical outcome – results of a retrospective study. *Int J Radiat Oncol Biol Phys* 2009;**73**:1009.
38. Briganti A, Karnes RJ, Joniau S, et al. Prediction of outcome following early salvage radiotherapy among patients with biochemical recurrence after radical prostatectomy. *Eur Urol* 2014;**66**:479.
39. Pfister D, Bolla M, Briganti A, et al. Early salvage radiotherapy following radical prostatectomy. *Eur Urol* 2014;**65**:1034.
40. Siegmann A, Bottke D, Faehndrich J, et al. Salvage radiotherapy after prostatectomy – what is the best time to treat? *Radiother Oncol* 2012;**103**:239.
41. Feinstein AR, Sosin DM, Wells CK. The Will Rogers phenomenon. Stage migration and new diagnostic techniques as a source of misleading statistics for survival in cancer. *N Engl J Med* 1985;**312**:1604.
42. Pearse M, Fraser-Browne C, Davis ID, et al. A Phase III trial to investigate the timing of radiotherapy for prostate cancer with high-risk features: background and rationale of the Radiotherapy – Adjuvant Versus Early Salvage (RAVES) trial. *BJU Int* 2014;**113**(Suppl. 2):7.
43. Stephenson AJ, Shariat SF, Zelefsky MJ, et al. Salvage radiotherapy for recurrent prostate cancer after radical prostatectomy. *JAMA* 2004;**291**:1325.
44. Stephenson AJ, Scardino PT, Kattan MW, et al. Predicting the outcome of salvage radiation therapy for recurrent prostate cancer after radical prostatectomy. *J Clin Oncol* 2007;**25**:2035.
45. Goenka A, Magsanoc JM, Pei X, et al. Long-term outcomes after high-dose postprostatectomy salvage radiation treatment. *Int J Radiat Oncol Biol Phys* 2012;**84**:112.
46. Moreira DM, Jayachandran J, Presti Jr JC, et al. Validation of a nomogram to predict disease progression following salvage radiotherapy after radical prostatectomy: results from the SEARCH database. *BJU Int* 2009;**104**:1452.
47. Den RB, Feng FY, Showalter TN, et al. Genomic prostate cancer classifier predicts biochemical failure and metastases in patients after postoperative radiation therapy. *Int J Radiat Oncol Biol Phys* 2014;**89**:1038.

48. Cary KC, Johnson CS, Cheng L, et al. A critical assessment of post-prostatectomy prostate specific antigen doubling time acceleration – is it stable? *J Urol* 2012;**187**:1614.
49. Cozzarini C, Fiorino C, Da Pozzo LF, et al. Clinical factors predicting late severe urinary toxicity after postoperative radiotherapy for prostate carcinoma: a single-institute analysis of 742 patients. *Int J Radiat Oncol Biol Phys* 2012;**82**:191.
50. Goenka A, Magsanoc JM, Pei X, et al. Improved toxicity profile following high-dose postprostatectomy salvage radiation therapy with intensity-modulated radiation therapy. *Eur Urol* 2011;**60**:1142.
51. Ost P, Lumen N, Goessaert AS, et al. High-dose salvage intensity-modulated radiotherapy with or without androgen deprivation after radical prostatectomy for rising or persisting prostate-specific antigen: 5-year results. *Eur Urol* 2011;**60**:842.
52. Schweizer MT, Zhou XC, Wang H, et al. Metastasis-free survival is associated with overall survival in men with PSA-recurrent prostate cancer treated with deferred androgen deprivation therapy. *Ann Oncol* 2013;**24**:2881.
53. Abdollah F, Karnes RJ, Suardi N, et al. Predicting survival of patients with node-positive prostate cancer following multimodal treatment. *Eur Urol* 2014;**65**:554.
54. Karnes RJ, Murphy CR, Bergstralh EJ, et al. Salvage lymph node dissection for prostate cancer nodal recurrence detected by 11C-choline positron emission tomography/computerized tomography. *J Urol* 2015;**193**(1):111–6.
55. Jilg CA, Rischke HC, Reske SN, et al. Salvage lymph node dissection with adjuvant radiotherapy for nodal recurrence of prostate cancer. *J Urol* 2012;**188**:2190.
56. Suardi N, Gandaglia G, Gallina A, et al. Long-term outcomes of salvage lymph node dissection for clinically recurrent prostate cancer: results of a single-institution series with a minimum follow-up of 5 years. *Eur Urol* 2015;**67**(2):299–309.
57. Ost P, Bossi A, Decaestecker K, et al. Metastasis-directed therapy of regional and distant recurrences after curative treatment of prostate cancer: a systematic review of the literature. *Eur Urol* 2015;**67**(5):852–63.
58. Rigatti P, Suardi N, Briganti A, et al. Pelvic/retroperitoneal salvage lymph node dissection for patients treated with radical prostatectomy with biochemical recurrence and nodal recurrence detected by [11C]choline positron emission tomography/computed tomography. *Eur Urol* 2011;**60**:935.
59. Schick U, Jorcano S, Nouet P, et al. Androgen deprivation and high-dose radiotherapy for oligometastatic prostate cancer patients with less than five regional and/or distant metastases. *Acta Oncol* 2013;**52**:1622.
60. Decaestecker K, De Meerleer G, Lambert B, et al. Repeated stereotactic body radiotherapy for oligometastatic prostate cancer recurrence. *Radiat Oncol* 2014;**9**:135.
61. Klein EA. Seeing and not believing: oligometastases and the future of metastatic prostate cancer. *Eur Urol* 2015;**67**(5):864–5.
62. Moul JW, Wu H, Sun L, et al. Early versus delayed hormonal therapy for prostate specific antigen only recurrence of prostate cancer after radical prostatectomy. *J Urol* 2004;**171**:1141.
63. Siddiqui SA, Boorjian SA, Inman B, et al. Timing of androgen deprivation therapy and its impact on survival after radical prostatectomy: a matched cohort study. *J Urol* 2008;**179**:1830.

CHAPTER

# 50

# Salvage Therapy for Locally Recurrent Prostate Cancer After External Beam Radiation Therapy

*David M. Marcus, MD*, David B. Cahn, DO, MBS**, Daniel J. Canter, MD†*

*Department of Radiation Oncology, Winship Cancer Institute, Emory University, Atlanta, GA, USA

**Department of Urology, Einstein Healthcare Network, Philadelphia, PA, USA

†Department of Urology, Fox Chase Cancer Center and Einstein Healthcare Network, Philadelphia, PA, USA

## INTRODUCTION

External beam radiation therapy (EBRT) is a commonly applied treatment modality for patients with localized prostate cancer, with up to 35% of patients with prostate cancer undergoing EBRT as first-line therapy.[1] EBRT is associated with excellent outcomes for patients with localized disease, and outcomes associated with the use of EBRT in this patient population are generally thought to be equivalent to those achieved with radical prostatectomy (RP). Following definitive EBRT, patients typically undergo routine disease surveillance with serial digital rectal exams and prostate-specific antigen (PSA) screening. The majority of disease recurrences are detected by a rising serum PSA – biochemical recurrence (BCR) or failure. For patients who are treated with EBRT or brachytherapy, BCR is defined either as three consecutive serum PSA increases after achievement of a nadir (the American Society for Radiation Oncology [ASTRO] definition) or a PSA value of nadir plus 2 ng/mL (the Phoenix definition).[2–5] Regardless of which definition is used, several disease factors are known to predict for BCR, including clinical T stage, biopsy Gleason score, and the PSA doubling time.[6–9]

While most patients undergoing primary EBRT for prostate cancer achieve long-term clinical disease-free survival and biochemical relapse-free survival (BRFS), about 35–50% of patients with high-risk disease, 25–45% of patients with intermediate-risk disease, and 20–30% of patients with low-risk disease will experience a BCR within 10 years after treatment.[10] In the setting of post-EBRT prostate cancer recurrence, it is common practice to prescribe long-term systemic androgen deprivation therapy (ADT) in the form of luteinizing hormone releasing hormone (LHRH) agonists and/or androgen receptor blockers. However, as distant metastasis only develops in a small subset of patients with BCR, many patients may be eligible for local salvage therapy, either with or without concomitant ADT. This chapter will review the evaluation of post-EBRT patients with BCR, the indications and options for local salvage therapy, and published outcomes and toxicities of each local salvage therapy modality.

Although BCR is known to predict for inferior survival,[11] the time interval between biochemical failure and the development of distant metastasis is generally fairly lengthy. In a study from Johns Hopkins University describing the outcomes of patients with biochemical failure after radical prostatectomy who received no salvage therapy, only 34% of patients developed distant metastatic disease within the study period (median follow up of 5.3 years). In patients who did develop distant metastases, the median time to development of metastatic disease was 8 years.[12] Similarly, in a recently published randomized trial by the Radiation Therapy Oncology Group (RTOG), the 10-year rate of BCR was 26% in patients receiving EBRT with concurrent ADT, with a corresponding 10-year rate of distant metastasis of only 6%.[10] Finally, in a study of 151 men who had undergone definitive EBRT for prostate cancer

Prostate Cancer. http://dx.doi.org/10.1016/B978-0-12-800077-9.00050-5

and were found to have a rising PSA with no salvage therapy, there was a median interval of 3 years between biochemical progression and clinical recurrence.[13] The results of these studies suggest that in patients with BCR of prostate cancer, there may be a window of opportunity in which local salvage therapy may eliminate the focus of recurrent disease before distant metastatic disease has a chance to develop. Therefore, aggressive local salvage therapy may be indicated in a potentially large proportion of these patients. Nonetheless, while the inference that a significant portion of patients with recurrent prostate cancer after EBRT may be candidates for local salvage therapy, the adoption of local salvage therapy in this setting has not been widely adopted. A recent analysis of the British Columbia Tumor Registry found that only 2% of patients who were eligible for post-EBRT local therapy between 1999 and 2000 received local salvage treatment, suggesting that local salvage therapy is underutilized for localized prostate cancers after EBRT.[14] As more data become available describing outcomes and toxicities associated with local salvage therapy, however, it is conceivable that this approach may become more widely utilized.

For patients with BCR after primary EBRT, there is no standard of care for local salvage treatment. Although several modalities have been utilized in this setting, the published literature for local salvage therapy is somewhat sparse, and the majority of reports are single institution retrospective series. For any patient who is being considered for local salvage therapy after prostate EBRT, the choice of a salvage modality should be made after a careful consideration of several patient and disease factors that may influence the oncologic and functional outcomes as well as the toxicities associated with salvage treatment.

## EVALUATION

In patients being considered for local salvage therapy after EBRT, it is critical to exclude the presence of distant or regional metastatic disease prior to committing to definitive local therapy. Therefore, all patients should undergo imaging of the skeletal system with a bone scan (although this test is thought to be fairly insensitive if the PSA level is less than 20 ng/mL[15]) and evaluation of the pelvic lymph nodes using either computed tomography (CT) or magnetic resonance imaging (MRI). As MRI is more sensitive than CT for imaging of the prostate gland and seminal vesicles (with up to 97% sensitivity for prostate cancer localization within the peripheral zone of the prostate),[16] it may be considered the preferred modality for imaging the pelvis in this setting. Finally, all patients being considered for local salvage therapy should have a prostate biopsy performed. The purpose of the biopsy is not only to confirm the presence of locally recurrent disease within the prostate but also to garner important pathologic details, such as Gleason score, that will provide important oncologic prognostic information, helping to guide treatment decision-making for definitive local salvage therapy.

## LOCAL SALVAGE TREATMENT OPTIONS

In patients with biopsy-proven localized recurrence who qualify for local salvage therapy, the treatment options include RP, cryotherapy, or brachytherapy. The choice of salvage modality should incorporate each individual patient's risk of disease progression, the risks and toxicities of each modality, and the patient's individual preferences. Furthermore, in patients with a short life expectancy, long PSA doubling time, or significant medical comorbidities, it may be most appropriate to opt for a course of disease surveillance.

### Salvage Prostatectomy

With careful patient selection, salvage RP may provide durable local control with acceptable toxicity for patients with post-EBRT BCR. However, the literature describing oncologic and functional outcomes and toxicity for patients undergoing RP in this setting is limited, with only retrospective series. Furthermore, there is significant heterogeneity across studies, with varying definitions of biochemical failure, variable follow-up periods, and inconsistent reporting of the use of ADT. Furthermore, surgical methods and experience are inconsistent across studies.

The largest series of post-EBRT salvage RP is a multi-institutional international study published by Chade et al. that includes 404 men who underwent salvage RP for post-EBRT biochemical failure. At 10 years post-RP, the rates of BRFS, metastasis-free survival, and cancer-specific survival estimates were 37%, 77%, and 83%, respectively. On multivariate analysis, predictors of subsequent BCR and distant metastasis after salvage RP included the presalvage PSA, presalvage Gleason score, and the Gleason score in the RP specimen. In the more favorable group of patients with a presalvage PSA ≤4 and a presalvage Gleason score ≤7 (comprising about 30% of the cohort), the 10-year probability of BRFS was 51%.[17] The second-largest study from the Mayo Clinic included 199 patients with biochemically recurrent prostate cancer after EBRT who underwent surgical salvage (including 138 patients undergoing RP and 61 patients undergoing cystoprostatectomy). In this study, which was reported after a median follow-up of 92 months, the 5-year rate of BRFS was 63%. In patients undergoing salvage RP, specifically (i.e., excluding patients who had undergone cystoprostatectomy), median progression-free survival was 8.7 years.[18] Also, in a single institutional study from the Memorial Sloan Kettering Cancer Center that described outcomes for 146 patients undergoing salvage surgery for biochemically recurrent prostate cancer, the 5-year recurrence-free survival rate was 54%, and only one patient had a clinical local recurrence.[19] A separate study by Paparel et al. of 100 patients treated with RP for post-EBRT prostate cancer recurrence at Memorial Sloan Kettering Cancer Center and Baylor College of Medicine demonstrated a 5-year progression-free survival rate of 55%, with a median progression-free interval of 6.4 years.[20] From a comparative standpoint, the total number of patients in these studies is small; however, the oncologic results seem relatively consistent in that close to half of all patients will ultimately experience a BCR after salvage RP. Nevertheless, when reported, the incidence of metastases and prostate-cancer-related mortality is infrequent.

While the literature describing oncologic outcomes for salvage RP after EBRT is limited, there are even less data describing treatment-related toxicity in this setting. In general, however, the toxicities of RP in the post-EBRT setting are thought to be more frequent and severe than the toxicities of RP in patients who are radiation-naïve. The majority of data on the morbidity of salvage RP is derived from a few select studies. A study by Gotto et al. reviewed medical and surgical complications for 3458 consecutive patients undergoing primary open RP and 98 patients who underwent salvage RP from 1999 to 2007. In this cohort, patients undergoing salvage RP had higher rates of medical complications (26% vs. 9%) and surgical complications (53% vs. 19%) compared to patients undergoing *de novo* RP. Patients undergoing salvage RP were more likely to develop urinary tract infection, bladder neck contracture, urinary retention, urinary fistula, abscess, and rectal injury. Salvage RP was associated with postoperative worsening of erectile function, and 3-year actuarial recovery of continence in preoperatively continent patients was only 30%.[21] In a separate analysis of patients who had undergone salvage RP after EBRT or interstitial brachytherapy at Memorial Sloan Kettering Cancer Center, Stephenson et al. reported that only 39% of patients were free from urinary incontinence at 5 years, although only 68% required one pad per day or less. The overall 5-year erectile potency rate was 28% following nerve sparing RP but was 45% in previously potent patients.[22] Similarly, Masterson et al. reported a 5-year actuarial rate of recovery of erectile function of 16% after salvage RP, with five of six patients who recovered erectile function having been treated with a bilateral nerve sparing technique.[23]

Finally, although several studies have evaluated the impact of robotically assisted laparoscopic RP as initial therapy, there is very little literature describing the use of this technique for patients undergoing salvage RP after primary EBRT. Both robotically assisted and open techniques are considered to be technically feasible, however, and the ultimate choice between a robotic or an open approach should be made on an individualized level and should take into account patient and disease characteristics, in addition to the comfort level and experience of the surgeon.

## Salvage Cryotherapy

Cryotherapy, which involves the ablation of tissue by local induction of extremely cold temperatures, is a recognized therapeutic option for salvage treatment of prostate cancer after primary EBRT. Although no clear patient selection guidelines exist, the American Urological Association Best Practice Consensus statement suggests that ideal patients for salvage cryotherapy should have a life expectancy of greater than 10 years, pretreatment PSA $\leq 4$ ng/mL, absence of seminal vesicle invasion, and a long PSA doubling time.[24] Other studies have suggested that patients with a Gleason score $\geq 9$ in the previously irradiated prostate gland are unlikely to be successfully salvaged with cryotherapy.[25] Due to the less invasive nature of cryotherapy compared to RP, many of the medical and surgical complications associated with salvage RP are not seen in patients undergoing salvage cryotherapy.[17,26] However, in contrast to patients undergoing salvage RP, patients undergoing cryotherapy cannot be assessed pathologically and oncologic success is therefore assessed by ongoing PSA surveillance. Finally, in a recent SEER-Medicare analysis comparing outcomes and costs associated with salvage cryotherapy and salvage RP, salvage cryotherapy was found to be associated with lower overall mortality and lower Medicare expenditures compared to salvage RP.[27]

The majority of published reports describing outcomes and toxicities associated with salvage cryotherapy are retrospective, single-institution series with small, heterogeneous patient cohorts, and limited follow-up. Furthermore, as previously mentioned, reported outcome measures among these studies are highly variable and subjective. The single prospective study to evaluate outcomes associated with salvage cryotherapy after primary EBRT is a series from the United Kingdom that reported on 100 patients treated between 2000 and 2005. In this cohort, the overall BRFS rate (using the ASTRO definition of biochemical failure) was 83% at 12 months, 72% at 24 months, and 59% at 36 months. In multivariate analysis, PSA nadir of $>0.1$ ng/mL and a higher pre-EBRT Gleason score were negative prognostic factors for biochemical failure after salvage cryotherapy.

Several retrospective studies also describe outcomes for salvage cryotherapy after EBRT. The largest of these is an analysis of 797 patients treated at six tertiary care centers between 1990 and 2005. With a median follow-up time of 3.4 years, the BRFS was a modest 34% in this collection of patients. Based on their results, the authors developed a nomogram to predict for biochemical failure, with predictive variables including serum PSA at diagnosis, biopsy Gleason grade, and pre-EBRT clinical T stage.[28] A similar study by Spiess et al. aimed to identify predictors of postcryotherapy biochemical failure (using the Phoenix definition of biochemical failure) and found that the postsalvage PSA nadir and the presalvage biopsy Gleason score were the most important predictors of outcome.[29] Likewise, a recent analysis of 283 patients from Ahmad et al. found that the postcryotherapy PSA nadir was strongly associated with the disease-free survival interval, with 12- and 36-month disease-free survival rates of 84% and 67% for patients with PSA nadir ≤1 ng/mL, compared to 56% and 14% for patients with a PSA nadir of >1 ng/mL.[30]

Although less invasive, salvage cryotherapy, like salvage RP, is associated with several acute treatment-related toxicities, including recto-urethral fistula, urethral stricture formation, urinary tract infection, urinary retention, hematuria, lower urinary tract symptoms, chronic perineal pain, and proctitis. However, in recent years, newer cryotherapy units have been developed with more precise targeting technology as well as urethral warmers, which have at least partially mitigated the urethral complications associated with this procedure.[31] In the long term, salvage cryotherapy may cause urinary incontinence and/or bothersome lower urinary tract symptoms. In a quality of life study after salvage cryotherapy, patient-reported urinary incontinence rate was 72% with a median follow-up time of 17 months.[32] In a study by Ahmad et al., the length of the prostate gland and the length of the iceball used for cryoablation were significant predictors of urinary toxicity after salvage cryotherapy.[30] Finally, with regard to erectile function, salvage cryotherapy may have an even more deleterious effect than salvage RP, as the ablative ice ball generally extends beyond the prostatic capsule to the neurovascular bundles that are responsible for erectile function.[31]

## Salvage Brachytherapy

The same characteristics that make brachytherapy an attractive therapeutic option for the definitive treatment of localized prostate cancer also contribute to its significant potential as an effective salvage treatment after EBRT. Using either low-dose-rate (LDR) or high-dose-rate (HDR) techniques, brachytherapy facilitates a steep gradient of decreasing radiation dose beyond the target volume, such that adjacent structures, such as the rectum and bladder, receive a minimal radiation dose. HDR and LDR brachytherapy have been studied in a prospective fashion for post-EBRT salvage treatment, and the data for this modality are arguably the most robust of all local salvage modalities in this setting.

In a prospective phase II study by Nguyen et al., 25 men with biochemically recurrent prostate cancer after EBRT underwent MRI-guided salvage LDR brachytherapy. With a median follow-up of 47 months, biochemical control (using the Phoenix definition) was 70% at 4 years. The 4-year estimate of grade 3 or 4 gastrointestinal or genitourinary toxicities where patients required a colostomy and/or a urostomy to repair a fistula as a result of salvage brachytherapy were 30% and 13%, respectively. Toxicity was noted to be statistically significantly worse in patients who had an interval between primary EBRT and salvage brachytherapy of less than 4.5 years.[33] In a separate study of the same patient cohort focusing on quality of life outcomes (using a validated questionnaire completed at various time points after treatment), the use of salvage brachytherapy was found to be associated with worsening sexual function over time. And, although bowel and urinary symptoms were found to be worse than baseline at 3 or 15 months after treatment, they improved back to near-baseline levels by 27 months. Finally, an interval of less than 4.5 years between primary EBRT and salvage brachytherapy was significantly associated with a higher decline in bowel-related quality of life.[34]

In the largest retrospective series to date of salvage LDR brachytherapy after EBRT, Grado et al. reported biochemical outcomes and toxicity for 49 patients treated with either iodine-125 or palladium-103 LDR implants for locally recurrent prostate cancers after primary EBRT. In this cohort, the 3-year and 5-year rates of BRFS were 48% and 34%, respectively, and post-treatment PSA nadir <0.5 ng/mL was found to be the only significant predictor of BRFS. In patients who achieved a post-treatment PSA nadir of <0.5 ng/mL, the 3-year and 5-year rates of BRFS were 77% and 56%, respectively. Within this cohort, the frequency of late complications was generally low. Overall, 14% of patients required transurethral resection of the prostate for urinary symptoms/urinary retention; 4% of patients had a late rectal toxicity.[35]

Recently, the RTOG completed accrual on a multi-institutional, prospective phase II study evaluating the use of LDR brachytherapy for salvage treatment of locally recurrent prostate cancer after EBRT. The primary objective of this study is to evaluate the late treatment-related gastrointestinal and genitourinary adverse events of brachytherapy in patients with local tumor recurrence after EBRT. The results of this study will provide the most robust data to date for any local salvage treatment modality after EBRT.

Based on the potential radio-biologic and dosimetric advantages of HDR over LDR brachytherapy techniques, the use of HDR brachytherapy is also emerging as a viable local salvage option in post-EBRT patients.

A recent prospective phase II study from the Memorial Sloan Kettering Cancer Center analyzed the toxicity and biochemical control of HDR Iridium-192 brachytherapy. With 42 patients enrolled and a median follow-up of 36 months, the authors reported a 5-year BRFS rate of 68.5%. Although late grade 1 and 2 genitourinary toxicities (38% and 48%, respectively) and late grade 1 and 2 gastrointestinal toxicities (17% and 8%, respectively) were fairly common, late grade 3 toxicities were only seen in two patients.[36] The most robust retrospective data of salvage HDR brachytherapy in this setting is from the University of California at San Francisco study, which reported on 52 consecutive patients treated to a dose of 36 Gray in six fractions. With a median follow-up time of 59.6 months, the 5-year BRFS rate was 51%, with late grade 3 toxicity reported in only two patients.[37]

Overall, salvage brachytherapy is the most intensively studied local salvage modality for post-EBRT local recurrence of prostate cancer. As previously noted, the RTOG trial has completed accrual, and several additional prospective studies evaluating LDR and HDR brachytherapy in this setting are currently enrolling patients. The results of these and other studies are eagerly anticipated, as they will undoubtedly provide guidance for patient selection and important information regarding the efficacy and toxicity of this modality.

## INVESTIGATIONAL MODALITIES

In addition to RP, cryotherapy, and brachytherapy, additional modalities under investigation for local salvage treatment of locally recurrent prostate cancer after EBRT include high-intensity focused ultrasound (HIFU) and ferromagnetic thermal ablation. Neither of these treatments is presently approved by the Food and Drug Administration, and they are considered experimental at present. There is very limited data describing outcomes and toxicity for both modalities, although studies are ongoing.

The largest series of salvage HIFU after EBRT was from Murat et al. and described outcomes and toxicities for 167 patients treated between 1995 and 2006. With a mean follow-up of 18.1 months, the 3-year progression-free survival rates were 53%, 42%, and 25% for low-, intermediate-, and high-risk patients, respectively. Among the entire cohort, 83% of patients had urinary incontinence after the procedure, and 18% of patients required artificial urinary sphincter implantation. Furthermore, 33% of patients developed bladder outlet obstruction.[38] Similarly, Uddin Ahmed et al. reported outcomes and toxicities for 84 men who were treated with salvage HIFU for biopsy-proven locally recurrent prostate cancer after EBRT. In this cohort, the 1- and 2-year progression-free survival rates were 49% and 43%, respectively. Within this group of patients, 62% of men were pad-free and leak-free after HIFU; however, mean International Index of Erectile Function scores were significantly negatively impacted by the HIFU procedure.[39]

Ferromagnetic thermal ablation is a technique that involves the use of oscillating magnetic fields to produce heat in a series of cobalt-palladium alloy rods, which may be permanently implanted into the prostate gland to ablate recurrent foci of prostate cancer. Master et al. reported the results of a pilot study of 14 men treated with ferromagnetic ablation for post-EBRT prostate cancer failures. The authors reported that 11 of 14 patients had a PSA decrease from pretreatment levels within 6 months after treatment, while three patients were nonresponders. Toxicities included necrosis of the prostatic urethra in nine patients – including severe necrosis in four patients – along with erectile dysfunction and incontinence in most patients.[40]

At the present time, the data/literature on HIFU and ferromagnetic thermal ablation for salvage treatment of recurrent prostate cancer after EBRT is sparse, and both modalities appear to be associated with significant risks of long-term toxicities with modest or nonexistent efficacy. Further studies are needed before either modality can be considered within the standard of care for local salvage therapy after EBRT. For the time being, HIFU and ferromagnetic thermal ablation remain investigational therapies in this setting.

## SYSTEMIC THERAPY

The standard systemic therapy for patients with recurrent prostate cancer is androgen deprivation therapy (ADT), which may consist of an LHRH agonist and/or an androgen receptor blocker. In the setting of locally recurrent disease, there is no accepted standard of care regarding the use of ADT, and the decision of whether to administer ADT in this setting should be made on an individual basis, depending on disease-specific and patient-specific characteristics such as the PSA level, the PSA doubling time, and patient comorbidities. Furthermore, for all patients, the potential benefits of ADT should be carefully balanced against the known toxicities of treatment, which may include weight gain, fatigue, hot flashes, mood changes, and an increased long-term risk of cardiovascular disease and diabetes mellitus.[41] Finally, for patients who are deemed to be appropriate candidates for ADT, the optimal timing and duration of ADT in this setting are not firmly established.

Beyond ADT, there are several recently developed systemic agents that have been shown to be effective in the management of advanced prostate cancer, including sipuleucel-T, enzalutamide, abiraterone, cabazitaxel, and radium-223 chloride. However, the use of these agents is currently restricted to patients with metastatic disease, and there is currently no role for any of these therapies in the setting of local or biochemical only recurrence. Further study of each of these agents is ongoing, and it is conceivable that one or more of them might prove to be useful in the management of locally recurrent prostate cancer after EBRT in the future.

## CONCLUSIONS

At present, the primary options for salvage therapy for recurrent prostate cancer after EBRT include RP, cryotherapy, brachytherapy, and ADT. As this chapter highlights, the data describing outcomes and toxicities for each of these modalities remain limited, with most studies consisting of single-institution retrospective reports. As there are no randomized trials comparing therapeutic modalities in this setting, the salvage treatment choice for a patient with locally recurrent prostate cancer after EBRT should be made after careful consideration of the potential risks and benefits of each modality, taking into account each individual patient's clinical characteristics and personal preferences. Furthermore, all patients being considered for salvage therapy should be evaluated in a multidisciplinary fashion, ideally including discussion at a tumor board that includes representation from urology, medical oncology, and radiation oncology. Finally, when applicable, patients should be considered for and made aware of any clinical trials that may be available for their participation.

### References


1. Kapoor DA, Zimberg SH, Ohrin LM, et al. Utilization trends in prostate cancer therapy. *J Urol* 2011;**186**(3):860–4.
2. Consensus statement: guidelines for PSA following radiation therapy. American Society for Therapeutic Radiology and Oncology Consensus Panel. *Int J Radiat Oncol Biol Phys* 1997;37(5):1035–1041.
3. Kuban DA, Levy LB, Potters L, et al. Comparison of biochemical failure definitions for permanent prostate brachytherapy. *Int J Radiat Oncol Biol Phys* 2006;**65**(5):1487–93.
4. Kuban DA, Thames HD, Shipley WU. Defining recurrence after radiation for prostate cancer. *J Urol* 2005;**173**(6):1871–8.
5. Roach III M, Hanks G, Thames Jr H, et al. Defining biochemical failure following radiotherapy with or without hormonal therapy in men with clinically localized prostate cancer: recommendations of the RTOG-ASTRO Phoenix Consensus Conference. *Int J Radiat Oncol Biol Phys* 2006;**65**(4):965–74.
6. D'Amico AV, Chen MH, de Castro M, et al. Surrogate endpoints for prostate cancer-specific mortality after radiotherapy and androgen suppression therapy in men with localised or locally advanced prostate cancer: an analysis of two randomised trials. *Lancet Oncol* 2012;**13**(2):189–95.
7. D'Amico AV, Cote K, Loffredo M, et al. Determinants of prostate cancer-specific survival after radiation therapy for patients with clinically localized prostate cancer. *J Clin Oncol* 2002;**20**(23):4567–73.
8. D'Amico AV, Moul JW, Carroll PR, et al. Surrogate end point for prostate cancer-specific mortality after radical prostatectomy or radiation therapy. *J Natl Cancer Inst* 2003;**95**(18):1376–83.
9. Zelefsky MJ, Ben-Porat L, Scher HI, et al. Outcome predictors for the increasing PSA state after definitive external-beam radiotherapy for prostate cancer. *J Clin Oncol* 2005;**23**(4):826–31.
10. Jones CU, Hunt D, McGowan DG, et al. Radiotherapy and short-term androgen deprivation for localized prostate cancer. *N Engl J Med* 2011;**365**(2):107–18.
11. Albertsen PC, Hanley JA, Penson DF, et al. Validation of increasing prostate specific antigen as a predictor of prostate cancer death after treatment of localized prostate cancer with surgery or radiation. *J Urol* 2004;**171**(6 Pt. 1):2221–5.
12. Pound CR, Partin AW, Eisenberger MA, et al. Natural history of progression after PSA elevation following radical prostatectomy. *JAMA* 1999;**281**(17):1591–7.
13. Lee WR, Hanks GE, Hanlon A. Increasing prostate-specific antigen profile following definitive radiation therapy for localized prostate cancer: clinical observations. *J Clin Oncol* 1997;**15**(1):230–8.
14. Tran H, Kwok J, Pickles T, et al. Underutilization of local salvage therapy after radiation therapy for prostate cancer. *Urol Oncol* 2014;**32**(5):701–6.
15. Kundra V, Silverman PM, Matin SF, et al. Imaging in oncology from the University of Texas M. D. Anderson Cancer Center: diagnosis, staging, and surveillance of prostate cancer. *AJR Am J Roentgenol* 2007;**189**(4):830–44.
16. Kelloff GJ, Choyke P, Coffey DS. Challenges in clinical prostate cancer: role of imaging. *AJR Am J Roentgenol* 2009;**192**(6):1455–70.
17. Chade DC, Shariat SF, Cronin AM, et al. Salvage radical prostatectomy for radiation-recurrent prostate cancer: a multi-institutional collaboration. *Eur Urol* 2011;**60**(2):205–10.
18. Ward JF, Sebo TJ, Blute ML, et al. Salvage surgery for radiorecurrent prostate cancer: contemporary outcomes. *J Urol* 2005;**173**(4):1156–60.
19. Paparel P, Cronin AM, Savage C, et al. Oncologic outcome and patterns of recurrence after salvage radical prostatectomy. *Eur Urol* 2009;**55**(2):404–10.
20. Bianco Jr FJ, Scardino PT, Stephenson AJ, et al. Long-term oncologic results of salvage radical prostatectomy for locally recurrent prostate cancer after radiotherapy. *Int J Radiat Oncol Biol Phys* 2005;**62**(2):448–53.
21. Gotto GT, Yunis LH, Vora K, et al. Impact of prior prostate radiation on complications after radical prostatectomy. *J Urol* 2010;**184**(1):136–42.
22. Stephenson AJ, Scardino PT, Bianco Jr FJ, et al. Morbidity and functional outcomes of salvage radical prostatectomy for locally recurrent prostate cancer after radiation therapy. *J Urol* 2004;**172**(6 Pt. 1):2239–43.
23. Masterson TA, Stephenson AJ, Scardino PT, et al. Recovery of erectile function after salvage radical prostatectomy for locally recurrent prostate cancer after radiotherapy. *Urology* 2005;**66**(3):623–6.
24. Babaian RJ, Donnelly B, Bahn D, et al. Best practice statement on cryosurgery for the treatment of localized prostate cancer. *J Urol* 2008;**180**(5):1993–2004.
25. Pisters LL, Perrotte P, Scott SM, et al. Patient selection for salvage cryotherapy for locally recurrent prostate cancer after radiation therapy. *J Clin Oncol* 1999;**17**(8):2514–20.
26. Finley DS, Belldegrun AS. Salvage cryotherapy for radiation-recurrent prostate cancer: outcomes and complications. *Curr Urol Rep* 2011;**12**(3):209–15.
27. Friedlander DF, Gu X, Prasad SM, et al. Population-based comparative effectiveness of salvage radical prostatectomy vs cryotherapy. *Urology* 2014;**83**(3):653–7.
28. Spiess PE, Katz AE, Chin JL, et al. A pretreatment nomogram predicting biochemical failure after salvage cryotherapy for locally recurrent prostate cancer. *BJU Int* 2010;**106**(2):194–8.

29. Spiess PE, Levy DA, Mouraviev V, et al. Predictors of biochemical failure in patients undergoing prostate whole-gland salvage cryotherapy: a novel risk stratification model. *BJU Int* 2013;**112**(4): E256–61.
30. Ahmad I, Kalna G, Ismail M, et al. Prostate gland lengths and iceball dimensions predict micturition functional outcome following salvage prostate cryotherapy in men with radiation recurrent prostate cancer. *PLoS One* 2013;**8**(8):e69243.
31. Caso JR, Tsivian M, Mouraviev V, et al. Complications and postoperative events after cryosurgery for prostate cancer. *BJU Int* 2012;**109**(6): 840–5.
32. Perrotte P, Litwin MS, McGuire EJ, et al. Quality of life after salvage cryotherapy: the impact of treatment parameters. *J Urol* 1999;**162**(2): 398–402.
33. Nguyen PL, Chen MH, D'Amico AV, et al. Magnetic resonance image-guided salvage brachytherapy after radiation in select men who initially presented with favorable-risk prostate cancer: a prospective phase 2 study. *Cancer* 2007;**110**(7):1485–92.
34. Nguyen PL, Chen RC, Clark JA, et al. Patient-reported quality of life after salvage brachytherapy for radio-recurrent prostate cancer: a prospective Phase II study. *Brachytherapy* 2009;**8**(4):345–52.
35. Grado GL, Collins JM, Kriegshauser JS, et al. Salvage brachytherapy for localized prostate cancer after radiotherapy failure. *Urology* 1999;**53**(1):2–10.
36. Yamada Y, Kollmeier MA, Pei X, et al. A Phase II study of salvage high-dose-rate brachytherapy for the treatment of locally recurrent prostate cancer after definitive external beam radiotherapy. *Brachytherapy* 2014;**13**(2):111–6.
37. Chen CP, Weinberg V, Shinohara K, et al. Salvage HDR brachytherapy for recurrent prostate cancer after previous definitive radiation therapy: 5-year outcomes. *Int J Radiat Oncol Biol Phys* 2013;**86**(2):324–9.
38. Murat FJ, Poissonnier L, Rabilloud M, et al. Mid-term results demonstrate salvage high-intensity focused ultrasound (HIFU) as an effective and acceptably morbid salvage treatment option for locally radiorecurrent prostate cancer. *Eur Urol* 2009;**55**(3): 640–7.
39. Uddin Ahmed H, Cathcart P, Chalasani V, et al. Whole-gland salvage high-intensity focused ultrasound therapy for localized prostate cancer recurrence after external beam radiation therapy. *Cancer* 2012;**118**(12):3071–8.
40. Master VA, Shinohara K, Carroll PR. Ferromagnetic thermal ablation of locally recurrent prostate cancer: prostate specific antigen results and immediate/intermediate morbidities. *J Urol* 2004; **172**(6 Pt. 1):2197–202.
41. Keating NL, O'Malley AJ, Smith MR. Diabetes and cardiovascular disease during androgen deprivation therapy for prostate cancer. *J Clin Oncol* 2006;**24**(27):4448–56.

CHAPTER

# 51

# Management of Locally Advanced (Nonmetastatic) Prostate Cancer

*Vincent Tang, MBBS, MSc, DIC, FRCS Urol*,***, *Declan G. Murphy, MB, FRCS Urol*,***, *Daniel Moon, MBBS(Hon), FRACS**,†*

*Department of Urology, Royal Melbourne Hospital, Melbourne, Australia
**Division of Cancer Surgery, University of Melbourne, Peter MacCallum Cancer Centre, Melbourne, Australia
†Robotic Surgery, Epworth Healthcare, Melbourne, Australia

## INTRODUCTION

Locally advanced prostate cancer represents a heterogeneous group of malignancies. Currently, there is no universally agreed definition of what is known as "locally advanced" disease within the urology community, as this disease group of patients is getting smaller due to widespread use of prostate-specific antigen (PSA) testing. However, it remains a common clinical challenge and management is controversial. In this chapter we have summarized some of the current evidence of this controversial topic.

### Clinical Definition of Locally Advanced Prostate Cancer

#### *Definitions*

Currently, there is no universally agreed definition of what is known as "locally advanced" prostate cancer. In 1999, Oh and Kantoff[1] defined locally advanced prostate cancer as a clinical diagnosis whereby the tumor has extended beyond the capsule with no evidence of nodal or distant metastatic spread. Since then there have been a number of variations in the definition for locally advanced prostate cancer by a number of different oncological and urological societies. The following listed are some of the commonly used definitions of locally advanced prostate cancer and this highlights the discrepancies:

1. NCCN guidelines:[2] Those with clinical stage T3b to T4 (locally advanced);
2. EAU guidelines:[3] All T3–4, Nx–N0, M0 disease;
3. AJCC guideline:[4] Stage III disease – T3a-b, N0, M0 with any PSA and any Gleason score.

It is clear that the aggressiveness of the disease is not just based upon the clinical staging; this has been demonstrated by a number of well validated scoring or stratification systems, for example, the UCSF-CAPRA score (Tables 51.1 and 51.2)[5] and D'Amico risk stratification (Table 51.3). These confirm that the Gleason grade and the PSA are also important indicators, can aid in predicting the natural history of the disease, and can be used to guide management decision-making. Therefore, these scoring systems may offer a more suitable alternative in the risk stratification of locally advanced prostate cancer.

Other than the definition variations causing confusion, the standard of clinical staging methods are also unreliable. Digital rectal examination often underestimates tumor extension, with only a positive correlation of less than 50% as demonstrated by Spigelman et al. in 1986.[6] Using transrectal ultrasound also appears to be unreliable as it only reveals 60% of pT3 lesions as shown in the Enlund et al. study in 1990.[7] Therefore, these diagnosing methods can wrongly stage the prostate cancer and influence treatment recommendations in a negative fashion.

### Summary

1. Locally advanced disease generally refers to the presence of extraprostatic extension; however, currently there is no universally agreed exact definition within the urology community.

Prostate Cancer. http://dx.doi.org/10.1016/B978-0-12-800077-9.00051-7

TABLE 51.1 UCSF-CAPRA Score

| Variable | Level | Points |
|---|---|---|
| PSA (ng/mL) | 2.1–6 | 0 |
| | 6.1–10 | 1 |
| | 10.1–20 | 2 |
| | 20.1–30 | 3 |
| | >30 | 4 |
| Gleason | 1–3/1–3 | 0 |
| | 1–3/4–5 | 1 |
| | 4–5/1–5 | 3 |
| T stage | T1/T2 | 0 |
| | T3a | 1 |
| Percent positive biopsies | <34% | 0 |
| | ≥34% | 1 |
| Age | <50 | 0 |
| | ≥50 | 1 |

2. Clinical diagnostic methods are often inaccurate and inadequate in assessing the local extent of the malignancy.
3. For better prognostic and management stratification, validated tools, such as the D'Amico classification or CAPRA score, should be employed.

## Imaging to Define Locally Advanced Prostate Cancer

Locally advanced prostate cancer falls into the high-risk category of the D'Amico classification and the clinical diagnostic methods can be unreliable. Therefore, more detailed imaging is required to identifying the site and extent of extracapsular extension as well as accurately staging the patient for the presence of nodal or distant metastases.

TABLE 51.2 UCSF-CAPRA Score with 5-Year Recurrence Rate

| CAPRA score | Crude 5-year recurrence rate (%) |
|---|---|
| 0 | 5.6 |
| 1 | 7.1 |
| 2 | 9.5 |
| 3 | 18.2 |
| 4 | 17.4 |
| 5 | 26.2 |
| 6 | 37.2 |
| 7 | 76.2 |
| 8 | 75.0 |
| 9 | 100.0 |
| 10 | n/a |

TABLE 51.3 D'Amico Risk Stratification

| Risk | Clinical stage | PSA (ng/mL) | Gleason grade |
|---|---|---|---|
| Low | T1–T2a | ≤10 | ≤6 |
| Intermediate | T2b | 10–20 | 7 |
| High | ≥T2c | >20 | ≥8 |

### *MRI for Extracapsular Staging*

MRI is frequently used for radiological staging of prostate cancer. Conventional anatomical MRI technique using T1 and T2 weighted sequences has a sensitivity of 13–95% and specificity of 49–97% for extracapsular extension prostate cancer, whereas, a sensitivity of 23–80% and 81–99% specificity for the seminal vesicles invasion disease.[8–13] The result of MRI for staging appears to be variable within the literature. It appears to be dependent on a number of factors such as the technique used, the experience of the radiologist, the strength of the MRI magnet, use of endorectal coil, and whether the extracapsular disease is focal or extensive.[14–19] The recent development of multiparametric prostatic MRI employs a combination of the T2 weighted sequence, diffusion weighted images, as well as dynamic contrast-enhanced imaging, and shows promise in improving local staging accuracy and eliminating a degree of the interobserver variability with the development of a standardized reporting system: Prostate Imaging – Reporting and Data System (PI-RADS) (Figure 51.1).[20]

### *MRI/CT for Nodal Staging*

The conventional radiological threshold of diagnosing lymph node metastasis is using size criteria >1 cm. A meta-analysis by Hövels et al. suggested that its sensitivity is around 30%, this is because 70% of the lymph node metastasis are too small (<8 mm) to be evaluated by MRI.[21] A more recent development is that of lymphotropic nanoparticle-enhanced MRI, which utilizes a contrast agent containing ultrasmall superparamagnetic particles of iron oxide. This is transported by macrophages into healthy lymph nodes but not within areas of malignancy, which contain few macrophages. Particularly when combined with diffusion-weighted imaging this allows differentiation of normal-sized but malignant lymph nodes with a sensitivity of at least 65–75% and specificity over 90%.[22,23]

### *Bone Scan*

Tc 99m bone scan is widely used for the assessment of bony metastatic disease. The rate of positivity depends upon multiple factors, including PSA level, clinical staging, and Gleason grades.[24,25] This was well docu-

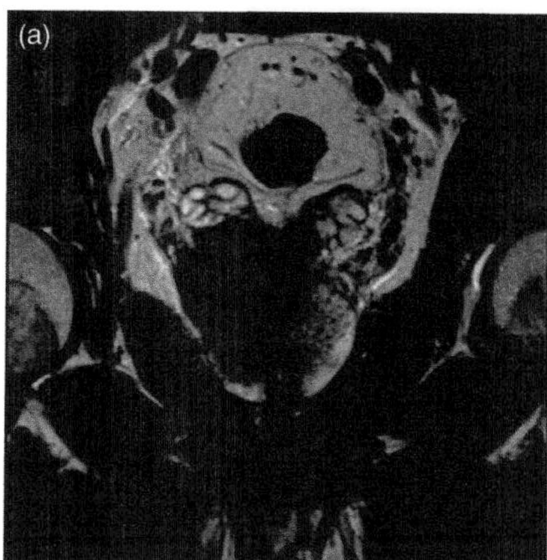


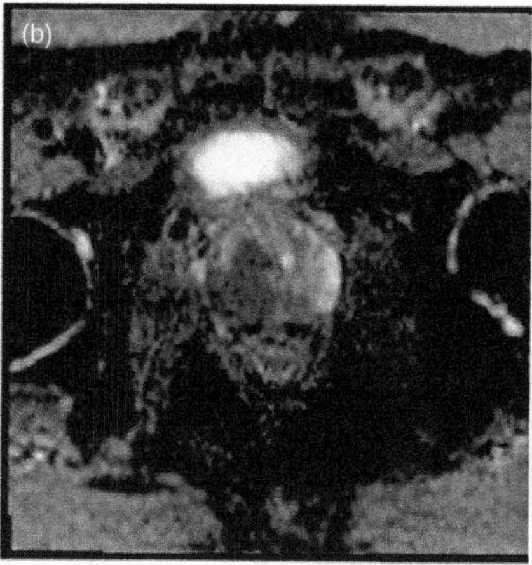


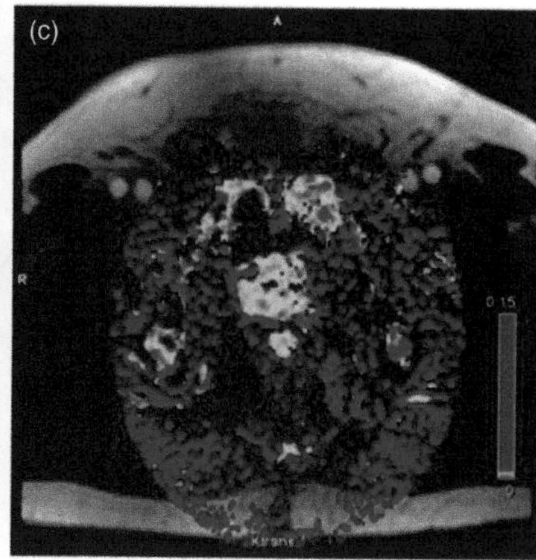


**FIGURE 51.1** Multiparametric prostate MRI employing (a) T2-weighted sequence, (b) diffusion-weighted image, (c) dynamic contrast enhancement, demonstrating a locally advanced, predominantly right-sided malignancy.

mented in a meta-analysis performed by Abuzallouf et al.[26] in which the bone scan positive rate was 2.3%, 5.3%, 16.2%, 39.2%, and 73.4% for PSA 0–9.9, 10–19.9, 20–49.9, 50–99.9, and >100 ng/mL, respectively. It is also related to clinical staging; a number of studies suggest that the bone scan positivity rate ranges from 1.3% to 13.6% in localized disease versus 19% to 90.7% in patients with locally advanced disease. The Gleason score also influences the bone scan positivity rate; for Gleason score ≤7 the bone scan positivity rate is between 4.9% and 5.8%, whereas with Gleason score ≥8 the bone scan positivity rate is between 23.5% and 29.6%.[27–30]

### PET Scan

Traditionally, FDG PET scanning has not formed a component of the prostate cancer staging algorithm due to its lack of sensitivity; however, there is increasing experience in the use of non-FDG PET scanning for staging of prostate cancer. A meta-analysis by Mohsen et al. in 2013 reviewed the use of $^{11}$C-acetate PET imaging. This demonstrated a pooled sensitivity of 73% and specificity of 79% for lymph node staging[31] and there is a suggestion by Haseebuddin et al. that $^{11}$C-acetate PET/CT may detect lymph node metastases that are not picked up by conventional staging methods.[32] More recently, prostate-specific membrane antigen (68 Ga-PSMA) has been investigated as a ligand for PET/CT. PSMA appears to be overexpressed in prostate cancer and is also directly related to metastasis and disease progression. Therefore, this newer ligand could be a superior substrate for PET/CT scanning compared to other PET scanning techniques for prostate cancer (Figure 51.2).[33]

## Summary

1. Staging scans form an essential part of assessment of locally advanced prostate cancer.
2. Multiparametric prostate MRI is emerging as the most accurate radiological method for local staging and predicting the presence/site of extraprostatic disease.
3. Nodal staging using MRI/CT is limited by inability to detect small volume lymph node metastases.
4. Bone scan is useful to rule out bony metastasis in a selective group of patients as its positive rate is dependent on PSA, Gleason grade, and T staging.
5. The utility of PET/CT for prostate cancer staging is still being studied; however, the use of newer, non-FDG substrates, particularly PSMA, appears more sensitive than current staging methods for detecting nodal and distant disease.

## Natural History

While historically, locally advanced prostate cancer accounted for 40% of prostate cancer diagnoses, the proportion of patients presenting with locally advanced prostate cancer has decreased to around 15–20% in the past 20 years largely as a result of widespread use of PSA testing. Although the frequency of locally advanced disease has reduced, this remains to be a challenging group of patients to treat, and these men continue to suffer a high risk of disease progression and mortality from prostate cancer.[34]

These patients constitute a heterogeneous group, with varying patterns and speed of progression. Patients may suffer local complications of ureteric or urethral obstruction, hematuria, and pelvic pain, or conversely develop rapidly progressive metastatic disease. The symptoms particularly of a locally invasive, bulky pelvic malignancy can be difficult to palliate. It has been estimated that T3 and T4 disease without treatment has a progression-free survival of 46.6% and disease-specific survival of 56.5% at 15 years. In the study of the natural history of

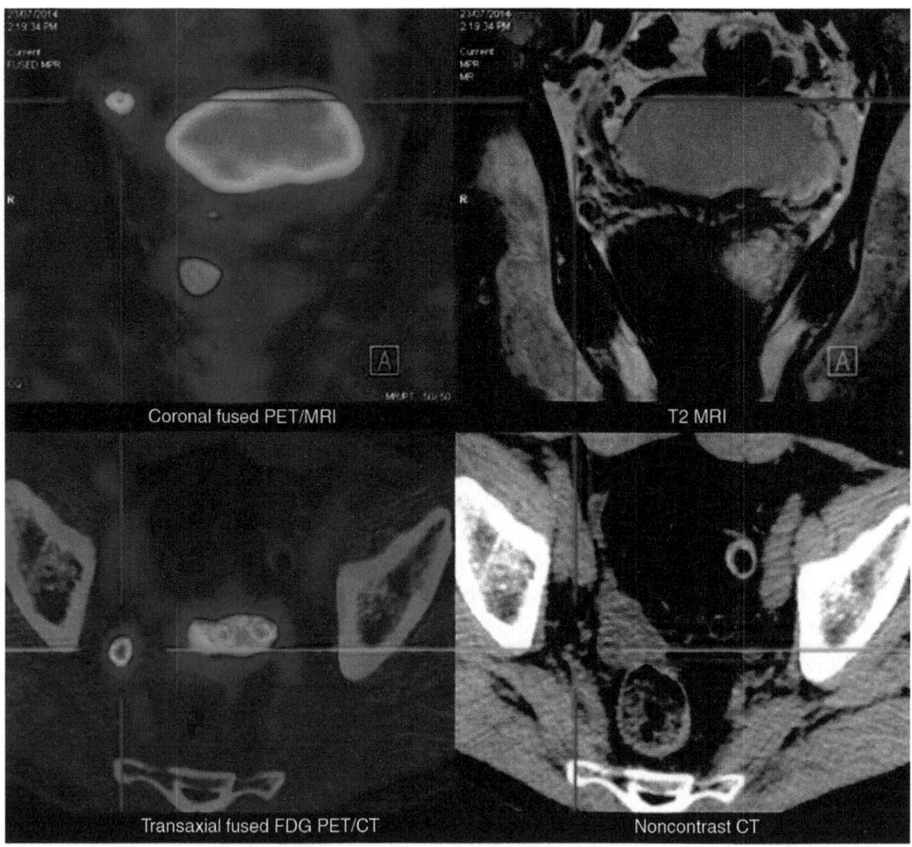


FIGURE 51.2 **PSMA PET Scanning images fused with MRI and CT, demonstrating a bulky prostatic malignancy and solitary pelvic lymph node metastasis.**

prostate cancer by Albertsen et al. 2005, the death rate from high-grade (Gleason 8–10) prostate cancer was 66% at 20 years.[35]

## Summary

1. New diagnosis of locally advanced disease has decreased due to PSA testing.
2. Generally, the outcomes are poor without any treatments, and patients may suffer from local or systemic progression of disease.

# TREATMENT OPTIONS

## Deferred Treatment and Androgen Deprivation Therapy

### *Active Surveillance*

It is largely accepted that there is a limited role for active surveillance in locally advanced prostate cancer. This group of patients carry a high risk of progression without treatment as outlined earlier. Therefore, if the patient is fit enough to undertake radical treatment this should not be deferred.[36]

### *Watchful Waiting versus Early Androgen Deprivation Therapy*

"Watchful waiting" implies a period of surveillance with a view to commencement of androgen deprivation therapy for evidence of disease progression. This may be appropriate for patients who wish to avoid the side effects of immediate androgen deprivation, or those with multiple additional comorbidities and short life expectancy. The main issues in this setting relate to the safety of watchful waiting and the most appropriate timing for commencement of androgen deprivation therapy. Adolfsson et al. looked at deferred treatment for T3 M0 N0 disease patients and followed the cohort for 169 months.[37] They found that patients with well and moderately differentiated disease had a 5-year cancer-specific survival (CSS) of 90% and 10 year cancer-specific survival of 74%. This equated to the likelihood of patients not requiring treatment at 5 and 10 years being 40 and 30%, respectively. On the other hand the EORTC 30981 prospectively randomized phase III trial, reported by Studer et al. in 2008 favored immediate treatment due to a slight improvement of overall survival.[38] This study compared patients with T0-4 N0-2 M0 disease using immediate versus delayed androgen deprivation and reported fewer deaths of nonprostate-cancer-related causes in the men treated with early androgen deprivation therapy. However, there were no differences in time from the start of the study to symptomatic progression of hormone refractory disease and no difference in cancer-specific survival. This study also found that the median time to commencement of deferred androgen deprivation therapy was 7 years with 25% of the patients dying of nonprostate-cancer-related causes without ever needing treatment. Through subgroup analysis the men most likely to benefit from early treatment were those with PSA >50 ng/mL or PSA between 8 ng/mL and 50 ng/mL with a PSA doubling time <12 months.

In the patient who is undergoing watchful waiting, therefore, it is suggested that the triggers for intervention be based upon radiological or symptomatic progression and/or PSA rise above 50 ng/mL.

## Summary

1. Deferring treatment and androgen deprivation therapies alone are not suitable for young, fit patients who are better offered treatment with curative intent.
2. Patients unsuitable for radical local treatment, with markers of more aggressive disease based on PSA kinetics, may benefit from early androgen deprivation therapy; otherwise, watchful waiting with a view to institution of treatment for evidence of significant disease is appropriate.

## Radiotherapy

Radiotherapy has previously been the more commonly utilized treatment for locally advanced disease given concerns relating to morbidity of surgery for bulky tumors, and the high risk of disease recurrence despite local treatment. Where radiation is used as primary therapy, it is clear however that optimal treatment requires a form of dose escalation, and concomitant androgen deprivation therapy.

### *External Beam Radiation Dose*

External beam radiation therapy has been the most commonly used form of radiation, with variable results depending on the nature of the disease. It has been demonstrated by Shipley and coworkers that in locally advanced malignancies, some patients may have early disease progression and this is highly dependent on the patients' risk groupings.[39] The independent prognostic factors for prostate-cancer-related death in patients treated with external beam radiotherapy are Gleason score, clinical stage, and lymph node status as shown by Roach et al.[40] It has also been demonstrated that conventional external beam radiotherapy alone is inadequate for high-risk prostate cancer treatment. Increased radiation dose is associated with an increase in cancer cell death but at the same time increases toxicity to normal surrounding tissue. With advances in techniques, such as 3D conformal radiotherapy and intensity modulated radiation therapy (IMRT) to allow shaping of the radiation beam, the precision of radiation to the intended target has improved. This was demonstrated by Dearnaley et al. who showed that 3D conforming radiotherapy produced the same efficacy as unshaped external beam radiotherapy but significantly reduced rectal toxicity.[41] Multiple studies have since confirmed that dose escalation improves biochemical survival although a benefit in overall survival is yet to be shown.[42–46]

### *Radiotherapy and Androgen Deprivation Versus Radiotherapy Alone*

Recently, radiation therapy alone, as a treatment of high-risk disease, has been re-examined through a number of randomized controlled clinical trials. It is clear that using concurrent adjuvant hormone therapy with radiation therapy improves progression-free survival. EORTC 22863 reported by Bolla et al. has popularized the use of neoadjuvant/adjuvant hormone treatment in conjunction with external beam radiation therapy. In this analysis, the use of 3 years of hormone therapy was associated with a significant improvement in 10-year clinical progression-free survival (47.7% vs. 22.2%), overall survival (58.1% vs. 39.8%), along with reduced prostate cancer mortality (11.1% vs. 31%) with no significant increase in cardiovascular events (11.1% vs. 8.2%) compared to use of external beam radiotherapy alone.[47]

To minimize side effects of androgen deprivation therapy, it is important to determine the optimal duration required during radiation treatment. RTOG 92-02 reported by Hanks et al. (short-term hormone and radiotherapy vs. long-term hormone (24 months) and radiotherapy)[48] and EORTC 22961 (short term vs. long term (3 years)) by Bolla et al.,[49] both demonstrated that longer duration hormonal therapy combined with radiotherapy was associated with better outcomes. In particular EORTC 22961, where 6 months of androgen deprivation was compared to a duration of 3 years, found that longer-term hormonal therapy conferred improved 5-year outcomes of prostate-cancer-specific mortality (3.2% long term vs. 4.7% short term) and overall mortality rate (15.2% long term vs. 19% short term). This study also failed to demonstrate a difference in fatal cardiac events or quality of life between the short- and long-term hormone therapy groups, although the side effects of the longer duration treatment as expected were significantly worse, mainly relating to insomnia, hot flushes, sexual interest, and sexual activity.

### *Radiotherapy and Androgen Deprivation Versus Androgen Deprivation Alone*

Endocrine treatment with or without radiation therapy for locally advanced prostate cancer has been examined by two large phase III clinical trials. The NCIC

CTG PR3/MRC UK PR07 study reported by Warde et al. and SPCG-7/SFUO-3 study reported by Widmark et al. confirmed that the combination of hormonal therapy and radiation is superior to the use of androgen deprivation therapy alone in terms of disease-specific survival as well as overall survival.[50,51] It is well documented, however, that the side effects of the combined treatment is greater than hormonal therapy alone including diarrhea and rectal bleeding as well as dyspnea and fatigue.

#### *High Dose Rate Brachytherapy with External Beam Radiation Therapy*

High dose rate brachytherapy is another means of achieving dose escalation for treatment of high-risk or locally advanced prostate cancer. This generally involves a combination of the use of high dose rate iridium-192 brachytherapy and external beam radiation therapy. For high-risk patients this treatment confers a 10-year biochemical-free survival (BFS) of between 62% and 74% as reported by Demanes et al.,[52] Ghilezan et al.,[53] and Hasan et al.,[54] although the overall survival benefit compared to surgery or external beam radiation alone is not well established. The side effects from high dose rate brachytherapy with external beam radiation therapy are largely related to genital urinary toxicity, gastrointestinal toxicity, and erectile dysfunction. Delayed urethral toxicity in the form of complex, recurrent strictures can be particularly debilitating with reported rates up to 32%.[55] Erectile dysfunction affects 10–51% as reported by Borghede et al.[56] and Martinez et al.[57]

### Summary

- Higher dose rate external beam radiation therapy in the form of 3D conforming radiation therapy or IMRT, provides better disease control than lower dose treatment.
- In the setting of high-risk disease (including locally advanced disease, it is better to use a combination of external beam radiation therapy and prolonged hormonal therapy (2–3 years), which improves cancer-specific survival as well as overall survival compared to hormonal treatment alone or radiation therapy alone.
- High dose rate brachytherapy in combination with external beam radiation may be useful in high-risk disease but potentially at the cost of greater urethral toxicity and with overall survival benefit not confirmed.

### Radical Prostatectomy and Pelvic Lymphadenectomy for Locally Advanced Disease

Radical prostatectomy for locally advanced prostate cancer has been traditionally less frequently employed due to concerns of oncological efficacy and surgical morbidity. There is a greater chance of positive surgical margins compared to surgery for low-risk disease, reported between 33.5% and 66%, and lymph node metastases between 7.9% and 49%. It has been estimated that the chance of local and distant relapse, which eventually requires secondary treatments, is high ranging from 56% to 78%.[58–61]

However, with the refinement in surgical techniques and reduction in perioperative morbidity of radical prostatectomy particularly utilizing minimally invasive approaches, it is now being appreciated that surgery can be employed as primary treatment with a low risk of complications and a number of distinct advantages over primary radiation therapy.

#### *Overstaging of Disease*

As previously highlighted, the clinical and radiological staging methods have limitations. It has been estimated that overstaging of cT3 prostate cancer can occur in up to 27% of the patients, who are found at radical prostatectomy to have organ-confined disease and may be potentially cured by surgery alone.[59]

#### *Outcomes Following Radical Surgery*

It is understood that the margin positive rate for locally advanced disease is higher than localized prostate cancer, ranging between 33.5% and 66%. Despite this, these patients often demonstrate remarkably good outcomes after radical prostatectomy in terms of BFS, CSS, and overall survival (OS).[58–61] One of the largest series reported by Ward et al. described 842 patients with cT3 disease undergoing radical prostatectomy, and reported BFS of 43%, CSS of 90%, and OS of 76% at 10 years.[59] More recently, Van Poppel et al. reported their 139 patients series with a BFS of 51.8%, CSS of 94.6%, and OS of 85.9% at 10 years.[61] Johnstone et al. and Joniau et al. reported small cohorts of T3b and T4 disease who underwent radical prostatectomy with 5-year BFS, CSS, and OS of 52%, 88–92%, and 73–88%, respectively.[62,63]

Inevitably, many men will need additional treatment; on average over 50% of the patients with locally advanced disease will require multimodality treatment. Importantly, if surgery is the primary treatment, subsequent radiation therapy either in the adjuvant setting or as salvage treatment, can be easily administered for true, local multimodal treatment. This carries far less toxicity than when salvage surgery is performed for local recurrence after primary radiation.

#### *The Role of Pelvic Lymphadenectomy*

In terms of lymphadenectomy, most urological guidelines recommend extended lymph node dissection be performed during radical prostatectomy for high-risk prostate cancer. Limited node dissection should not be performed as this can miss up to half of the positive lymph nodes. The presence of positive nodes at surgery is not an indication to abandon radical prostatectomy as it appears removal of the primary tumor confers a survival advantage, when compared to the outcomes of similar patients who have had the prostate left *in situ*.

There is no universal acceptance of what constitutes an extended lymph node dissection. Some authors use external iliac, obturator fossa, and internal iliac as their dissection boundary and some use a more extensive template, which includes common iliac and presacral areas. It is clear that the more extensive the lymph node dissection, the greater the positive nodal detection rate, which in turn assists clinical decision-making regarding adjuvant treatment and guided prognosis. However, this also carries an increased rate of complications such as lymphocele, lymphoedema, deep vein thrombosis, and pulmonary embolism, ranging from 8.1% to 35.9% for extended lymphadenectomy compared to 2% to 9% for limited lymph node dissection. In terms of therapeutic benefit, the evidence is so far limited. There are early suggestions that the number of lymph nodes removed correlates with time to disease progression.[64–77]

### *Morbidity of Surgery*

Increasing series reporting outcomes of surgery for high-risk patients have confirmed that a greater risk of perioperative and long-term complications should not be expected in experienced centers. Ward et al. reported a subgroup of 842 patients with T3 cancer undergoing radical prostatectomy demonstrating the morbidity to be similar to surgery for organ-confined.[59] Gontero et al. performed a single expert surgeon direct comparison study for radical prostatectomy outcomes between locally advanced disease and localized disease and showed no significant difference in surgical morbidity except for blood transfusion, operative time, and lymphoceles between the two groups.[78] These results are further supported by a more recent study by van Poppel and coworkers who suggested that perioperative complications in this group of patients is uncommon.[61] In the van Poppel study their data are compared to the major contemporary radical prostatectomy series with organ-confined diseases including Dillioglugil et al.,[79] Hisasue et al.,[80] Gaylis et al.,[81] Lepor et al.,[82] and Lerner et al.[83] series. They suggested that without any mortality, a perioperative complication rate of 1.4%, and postoperative complication rate of 12.9%, it is comparable to T1 and T2 disease that underwent radical prostatectomy. Furthermore, in terms of functional outcomes, the 12 months continence rate including using a single pad is 94.2% and anastomotic stricture rate is low at 2.9%. Understandably, erectile function rate is likely to be lower as non-nerve sparing or unilateral nerve-sparing techniques are often used to ensure not compromising the cancer surgery.

### *Surgery Versus Radiotherapy*

Although there is no randomized controlled trial comparing surgery with primary radiation therapy, there is mounting evidence that radical prostatectomy may offer patients with locally advanced disease superior local control and long-term survival.

- Surgery compared to radiotherapy alone: There are a number of studies showing that the outcomes following surgery are superior to radiotherapy alone for men with high-risk prostate cancer. Zelefsky et al. found that patients with high-risk prostate cancer who underwent radical prostatectomy treatment are less likely to develop metastatic progression and also have a lower prostate-cancer-specific death compared to EBRT alone.[84] Tewari et al. also demonstrated that patients treated with radical prostatectomy have a better median overall survival than EBRT alone.[85] Cooperberg et al., using the Cancer of the Prostate Strategic Urology Research Endeavor (CaPSURE) registry, demonstrated that in high-risk disease, cancer-specific mortality is lower in patients receiving radical prostatectomy as opposed to EBRT alone.[86] Most recently, Sooriakumaran et al. used the Swedish prostate cancer registry data to further confirm patients with high-risk disease; especially fit and younger patients <65 years of age, achieve superior overall and disease-specific survival through radical prostatectomy compared with radiation therapy.[87]
- Surgery compared to combination radiotherapy and androgen deprivation therapy: Boorjian et al. compared outcomes of patients undergoing radical prostatectomy to patients receiving EBRT (with or without adjuvant androgen deprivation therapy) for high-risk prostate cancer. In a subgroup analysis they found that radical prostatectomy conferred an equivalent risk of systemic progression and prostate cancer mortality compared to EBRT with adjuvant androgen deprivation therapy. However, the risk of all-cause mortality was higher after EBRT with adjuvant androgen deprivation therapy than following radical prostatectomy.[88]

In addition to these potential oncological advantages, patients undergoing radical prostatectomy can avoid the need for prolonged neoadjuvant and adjuvant androgen deprivation therapy required for radiation therapy. Moreover, postoperative complications and functional deficits following surgery are generally more correctable when compared to the potential radiation therapy complications such as radiation-related strictures, proctitis, and hemorrhagic cystitis.

## Summary

1. Radical prostatectomy for locally advanced disease may offer superior oncological outcomes to primary radiation therapy. It avoids the need for neoadjuvant/adjuvant hormonal therapy, allows postoperative radiation therapy to be administered for dual local treatment, and can be performed with low morbidity comparable to surgery for localized disease.
2. The use of extended lymph node dissection offers prognostic information, is important for accurate staging, and guides further decision-making regarding adjuvant treatments.
3. Radical prostatectomy in many occasions can be used as monotherapy for locally advanced disease and complications are generally more amenable to treatment than those caused by radiation toxicity.

## Multimodal Therapy

Patients with locally advanced prostate cancer have a higher risk of failing primary treatment whether this be radiotherapy or surgery. It is extremely important to ensure the patient is well informed regarding this and understands that multimodality therapies are often required to optimally control the disease. Currently, there are a number of multimodality therapies being investigated. It is likely that in the future the newer prostate cancer treatment agents will be incorporated into the multimodality therapies.

### *Use of Androgen Deprivation Therapy Before Surgery*

Currently, there is little evidence to suggest that the use of neoadjuvant hormonal therapy improves survival outcomes. A meta-analysis reported by Shelley et al. in 2009 shows that neoadjuvant hormone therapy improves the pathological surrogate markers, such as negative margin rate, organ confinement, and lymph node involvement, but no improvement of overall and disease-free survival have been demonstrated.[89]

### *Radiation Therapy Postsurgery (Early vs. Salvage)*

Debate continues as to the optimal timing and indications for radiation therapy following radical prostatectomy. There are clearly men who benefit from early postoperative irradiation of the prostatic bed; however, it remains to be proven whether adjuvant treatment offers a survival advantage compared to early salvage treatment. With the development of ultrasensitive PSA assays, local recurrence can be detected early, and offered to the men who will benefit, sparing those who remain disease-free after surgery alone, and those who develop metastatic disease progression warranting early systemic therapy rather than further local treatment. Three large randomized trials, SWOG 8794,[90] EORTC 22911,[91] and ARO 96-02,[92] have looked at this question. These trials demonstrated an improved biochemical progression-free survival for those undergoing early radiation therapy; SWOG trial 53% versus 30% at 10 years, EORTC trial 60.6% versus 41% at 10 years, and ARO trial 72% versus 54% at 5 years, respectively. However, in terms of early radiotherapy, evidence for improvements in the overall survival rate is scarce. The SWOG trial suggested early radiotherapy improves overall survival; the EORTC suggested otherwise. Toxicity in the irradiated group included a stricture rate of up to 17.9% at 12.7 years in the SWOG trial. Other series have reported a demonstrable reduction in continence rates in men undergoing adjuvant radiation therapy; therefore, it is clear that selective treatment, if not compromising oncological outcomes, is most desirable. Further guidance will hopefully be provided as the RADICAL and RAVES trials mature.

### *Management of Lymph Node Positive Disease After Surgery*

Messing et al. reported a prospective randomized trial suggesting that in patients undergoing radical prostatectomy who are found to have lymph node positive disease, immediate adjuvant androgen deprivation therapy improves cancer-specific survival as well as overall survival.[93] However, in this modern era, high-volume nodal disease is rare. It is more likely to find microscopic nodal disease through extended lymph node dissection. Schumacher et al. found that if there is ≤2 microscopic lymph node involvement, the prognosis was best with 79% cancer-specific survival at 10 years and 41% of the patients not requiring hormonal therapy, whereas, if there was ≥3 lymph node involvement, the cancer-specific survival dropped to 33% at 10 years.[94] The implication therefore is that early androgen deprivation therapy may be avoided in patients with small-volume lymph node metastases.

### *Salvage Surgery After Failed Radiotherapy*

Salvage surgery after failed radiotherapy treatment is not commonly performed due to the high complication rate. However, with improved radiotherapy and surgical techniques, there are small case series to suggest it can be an effective multimodality treatment option. Heidenreich et al. published a well selected group of patients using the following criteria: a life expectancy >10 years, clinically organ-confined disease, the absence of locoregional and systemic metastases, and PSA ≤20 ng/mL. This group was able to achieve better outcomes compared to previous series, including lower perioperative complications (9% vs. 13%), transfusion rate (4.5% vs. 29%), as well as an improved urinary continence rate (80% vs. 68%).[95]

## Summary

1. A patient undergoing treatment for locally advanced prostate cancer needs to be counseled as to the likelihood of requiring multimodality therapy.
2. More evidence is required to identify the best multimodality treatment and timing/role for adjuvant treatment following surgery is as yet undefined.

## CONCLUSIONS

Locally advanced prostate cancer is decreasing because of stage migration with PSA testing. There is, however, a need for carefully designed, evidence-based multimodality treatment strategies to treat these men, many of whom are curable, and who will otherwise suffer considerable morbidity from local and systemic disease progression. Radical prostatectomy as primary treatment offers a number of advantages, with the option of additional local treatment through adjuvant or salvage radiation therapy, along with systemic therapies in the event of metastatic disease progression. For patients unsuitable for surgery or preferring a non-operative approach, radiation therapy incorporating dose escalation and combined androgen deprivation therapy offers an alternative treatment strategy. The risks and benefits of treatments need to be discussed carefully with the patients, ideally using a shared decision-making approach, and by a multidisciplinary uro-oncology team.

## Acknowledgments

We thank Associate Professor Richard O'Sullivan, Director of MRI services at Epworth Hospital, Melbourne Australia for providing multiparametric MRI images (Figure 51.1) and Professor Rodney Hicks, Director of cancer imaging at Peter MacCallum Cancer Centre, Melbourne Australia for providing PSMA PET images (Figure 51.2).

## References

1. Oh WK, Kantoff PW. Treatment of locally advanced prostate cancer: is chemotherapy the next step? *J Clin Oncol* 1999;**17**(11): 3664–75.
2. NCCN. Clinical practice guidelines in oncology: prostate cancer. National comprehensive cancer network 2012; Version 3, Available from: NCCN.org.
3. EAU. EAU guideline: prostate cancer. European Association of Urology 2014, Available from: uroweb.org.
4. In: Edge SB, Byrd DR, Compton CC, editors. *AJCC cancer staging manual*. 7th ed. New York, NY: Springer; 2010. pp. 457–468.
5. Cooperberg MR, Pasta DJ, Elkin EP, et al. The University of California, San Francisco Cancer of the Prostate Risk Assessment score: a straightforward and reliable preoperative predictor of disease recurrence after radical prostatectomy. *J Urol* 2005;**173**(6): 1938–42.
6. Spigelman SS, McNeal JE, Freiha FS, et al. Rectal examination in volume determination of carcinoma of the prostate: clinical and anatomical correlations. *J Urol* 1986;**136**(6):1228–30.
7. Enlund A, Pedersen K, Boeryd B, et al. Transrectal ultrasonography compared to histopathological assessment for local staging of prostatic carcinoma. *Acta Radiol* 1990;**31**(6):597–600.
8. Sala E, Akin O, Moskowitz CS, et al. Endorectal MR imaging in the evaluation of seminal vesicle invasion: diagnostic accuracy and multivariate feature analysis. *Radiology* 2006;**238**:929–37.
9. Bartolozzi C, Menchi I, Lencioni R, et al. Local staging of prostate carcinoma with endorectal coil MRI: correlation with whole-mount radical prostatectomy specimens. *Eur Radiol* 1996;**6**: 339–45.
10. Cornud F, Flam T, Chauveinc L, et al. Extraprostatic spread of clinically localized prostate cancer: factors predictive of pT3 tumor and of positive endorectal MR imaging examination results. *Radiology* 2002;**224**:203–10.
11. Ikonen S, Karkkainen P, Kivisaari L, et al. Magnetic resonance imaging of clinically localized prostatic cancer. *J Urol* 1998;**159**:915–9.
12. Ikonen S, Karkkainen P, Kivisaari L, et al. Endorectal magnetic resonance imaging of prostatic cancer: comparison between fat suppressed T2-weighted fast spin echo and three-dimensional dual-echo, steady-state sequences. *Eur Radiol* 2001;**11**:236–41.
13. May F, Treumann T, Dettmar P, et al. Limited value of endorectal magnetic resonance imaging and transrectal ultrasonography in the staging of clinically localized prostate cancer. *BJU Int* 2001;**87**:66–9.
14. Perrotti M, Kaufman Jr RP, Jennings TA, et al. Endo-rectal coil magnetic resonance imaging in clinically localized prostate cancer: is it accurate? *J Urol* 1996;**156**:106–9.
15. Presti JC, Hricak H, Narayan PA, et al. Local staging of prostatic carcinoma: comparison of transrectal sonography and endorectal MR imaging. *AJR Am J Roentgenol* 1996;**166**:103–8.
16. Rorvik J, Halvorsen OJ, Albrektsen G, et al. MRI with an endorectal coil for staging of clinically localized prostate cancer prior to radical prostatectomy. *Eur Radiol* 1999;**9**:29–34.
17. Futterer JJ, Engelbrecht MR, Huisman HJ, et al. Staging prostate cancer with dynamic contrast-enhanced endorectal MR imaging prior to radical prostatectomy: experienced versus less experienced readers. *Radiology* 2005;**237**:541–9.
18. Beyersdorff D, Taymoorian K, Knosel T, et al. MRI of prostate cancer at 1.5 T and 3. 0 T: comparison of image quality in tumor detection and staging. *AJR Am J Roentgenol* 2005;**185**:1214–20.
19. Heijmink SW, Futterer JJ, Hambrock T, et al. Prostate cancer: body-array versus endorectal coil MR imaging at 3T-comparison of image quality localization, and staging performance. *Radiology* 2007;**244**:184–95.
20. Bloch BN, Furman-Haran E, Helbich TH, et al. Prostate cancer: accurate determination of extracapsular extension with high-spatial resolution dynamic contrast-enhanced and T2-weighted MR imaging – initial results. *Radiology* 2007;**245**:176–85.
21. Hövels AM, Heesakkers RA, Adang EM, et al. The diagnostic accuracy of CT and MRI in the staging of pelvic lymph nodes in patients with prostate cancer: a meta-analysis. *Clin Radiol* 2008;**63**(4):387–95.
22. Fortuin AS, Smeenk RJ, Meijer HJ, et al. Lymphotropic nanoparticle-enhanced MRI in prostate cancer: value and therapeutic potential. *Curr Urol Rep* 2014;**15**(3):389.
23. Birkhäuser FD, Studer UE, Froehlich JM, et al. Combined ultrasmall superparamagnetic particles of iron oxide-enhanced and diffusion-weighted magnetic resonance imaging facilitates detection of metastases in normal-sized pelvic lymph nodes of patients with bladder and prostate cancer. *Eur Urol* 2013;**64**(6):953–60.
24. Miller PD, Eardley I, Kirby RS. Prostate specific antigen and bone scan correlation in the staging and monitoring of patients with prostatic cancer. *Br J Urol* 1992;**70**(3):295–8.

25. Jacobson AF. Association of prostate-specific antigen levels and patterns of benign and malignant uptake detected on bone scintigraphy in patients with newly diagnosed prostate carcinoma. *Nucl Med Commun* 2000;**21**(7):617–22.
26. Abuzallouf S, Dayes I, Lukka H. Baseline staging of newly diagnosed prostate cancer: a summary of the literature. *J Urol* 2004;**171**(6 Pt. 1):2122–7.
27. Gleave ME, Coupland D, Drachenberg D, et al. Ability of serum prostate-specific antigen levels to predict normal bone scans in patients with newly diagnosed prostate cancer. *Urology* 1996;**47**(5):708–12.
28. Ataus S, Citci A, Alici B, et al. The value of serum prostate specific antigen and other parameters in detecting bone metastases in prostate cancer. *Int Urol Nephrol* 1999;**31**(4):481–9.
29. Lin K, Szabo Z, Chin BB, et al. The value of a baseline bone scan in patients with newly diagnosed prostate cancer. *Clin Nucl Med* 1999;**24**(8):579–82.
30. Lee N, Fawaaz R, Olsson CA, et al. Which patients with newly diagnosed prostate cancer need a radionuclide bone scan? An analysis based on 631 patients. *Int J Radiat Oncol Biol Phys* 2000;**48**(5):1443–6.
31. Mohsen B, Giorgio T, Rasoul ZS, et al. Application of C-11-acetate positron-emission tomography (PET) imaging in prostate cancer: systematic review and meta-analysis of the literature. *BJU Int* 2013;**112**(8):1062–72.
32. Haseebuddin M, Dehdashti F, Siegel BA, et al. $^{11}$C-acetate PET/CT before radical prostatectomy: nodal staging and treatment failure prediction. *J Nucl Med* 2013;**54**(5):699–706.
33. Afshar-Oromieh A, Zechmann CM, Malcher A, et al. Comparison of PET imaging with a (68)Ga-labelled PSMA ligand and (18)F-choline-based PET/CT for the diagnosis of recurrent prostate cancer. *Eur J Nucl Med Mol Imaging* 2014;**41**(1):11–20.
34. van den Ouden D, Schröder FH. Management of locally advanced prostate cancer. 1. Staging, natural history, and results of radical surgery. *World J Urol* 2000;**18**(3):194–203.
35. Albertsen PC, Hanley JA, Fine J. 20-year outcomes following conservative management of clinically localized prostate cancer. *JAMA* 2005;**293**(17):2095–101.
36. Johansson JE, Holmberg L, Johansson S, et al. Fifteen-year survival in prostate cancer: a prospective, population-based study in Sweden. *JAMA* 1997;**277**:467–71.
37. Adolfsson J, Steineck G, Hedlund PO. Deferred treatment of locally advanced nonmetastatic prostate cancer: a long-term follow up. *J Urol* 1999;**161**(2):505–8.
38. Studer UE, Collette L, Whelan P, et al. Using PSA to guide timing of androgen deprivation in patients with T0-4 N0-2 M0 prostate cancer not suitable for local curative treatment (EORTC 30891). *Eur Urol* 2008;**53**(5):941–9.
39. Zietman AL1, Westgeest JC, Shipley WU. Radiation-based approaches to the management of T3 prostate cancer. *Semin Urol Oncol* 1997;**15**(4):230–8.
40. Roach M1, Lu J, Pilepich MV, et al. Four prognostic groups predict long-term survival from prostate cancer following radiotherapy alone on Radiation Therapy Oncology Group clinical trials. *Int J Radiat Oncol Biol Phys* 2000;**47**(3):609–15.
41. Dearnaley DP, Khoo VS, Norman AR, et al. Comparison of radiation side-effects of conformal and conventional radiotherapy in prostate cancer: a randomised trial. *Lancet* 1999;**353**(9149): 267–72.
42. Kuban DA, Levy LB, Cheung MR, et al. Long-term failure patterns and survival in a randomized dose-escalation trial for prostate cancer. Who dies of disease? *Int J Radiat Oncol Biol Phys* 2011;**79**(5):1310–7.
43. Zietman AL, Bae K, Slater JD, et al. Randomized trial comparing conventional-dose with high-dose conformal radiation therapy in early-stage adenocarcinoma of the prostate: long-term results from proton radiation oncology group/american college of radiology 95-09. *J Clin Oncol* 2010;**28**(7):1106–11.
44. Peeters ST, Heemsbergen WD, Koper PCM, et al. Dose-response in radiotherapy for localized prostate cancer: results of the Dutch multicenter randomized phase III trial comparing 68 Gy of radiotherapy with 78 Gy. *J Clin Oncol* 2006;**24**(13):1990–6.
45. Beckendorf V, Guerif S, Le Prisé E, et al. 70 Gy versus 80 Gy in localized prostate cancer: 5-year results of GETUG 06 randomized trial. *Int J Radiat Oncol Biol Phys* 2011;**80**(4):1056–63.
46. Dearnaley DP, Sydes MR, Graham JD, et al. Escalated-dose versus standard-dose conformal radiotherapy in prostate cancer: first results from the MRC RT01 randomized controlled trial. *Lancet Oncol* 2007;**8**(6):475–87.
47. Bolla M, Collette L, Blank L, et al. Long-term results with immediate androgen suppression and external irradiation in patients with locally advanced prostate cancer (an EORTC study): a phase III randomised trial. *Lancet* 2002;**360**(9327):103–6.
48. Hanks GE, Pajak TF, Porter A, et al. Phase III trial of long-term adjuvant androgen deprivation after neoadjuvant hormonal cytoreduction and radiotherapy in locally advanced carcinoma of the prostate: the Radiation Therapy Oncology Group Protocol 92-02. *J Clin Oncol* 2003;**21**(21):3972–8.
49. Bolla M, de Reijke TM, Van Tienhoven G, et al. Duration of androgen suppression in the treatment of prostate cancer. *N Engl J Med* 2009;**360**(24):2516–27.
50. Warde P, Mason M, Ding K, et al. Combined androgen deprivation therapy and radiation therapy for locally advanced prostate cancer: a randomised, phase 3 trial. *Lancet* 2012;**378**(9809): 2104–11.
51. Widmark A, Klepp O, Solberg A, et al. For the Scandinavian Prostate Cancer Group Study, the Swedish Association for Urological Oncology. Endocrine treatment, with or without radiotherapy, in locally advanced prostate cancer (SPCG-7/SFUO-3): an open randomized phase III trial. *Lancet* 2008;**373**(9660):301–8.
52. Demanes DJ, Rodriguez RR, Schour L, et al. High-dose-rate intensity-modulated brachytherapy with external beam radiotherapy for prostate cancer: California endocurietherapy's 10-year results. *Int J Radiat Oncol Biol Phys* 2005;**61**(5):1306–16.
53. Ghilezan M, Galalae R, Demanes J, et al. 10-Year results in 1577 intermediate/high risk prostate cancer patients treated with external beam RT (EBRT) and hypofractionated high dose rate (HDR) brachytherapy boost. *Int J Radiat Oncol Biol Phys* 2007;**69**:S83–4.
54. Hasan Y, Mitchell C, Wilson G, et al. Long-term outcome for high-dose-rate brachytherapy boost treatment of prostate cancer. *Brachytherapy* 2007;**6**:85.
55. Hindson BR, Millar JL, Matheson B. Urethral strictures following high-dose-rate brachytherapy for prostate cancer: analysis of risk factors. *Brachytherapy* 2013;**12**:50.
56. Borghede G, Hedelin H, Holmang S, et al. Combined treatment with temporary short-term high dose rate iridium-192 brachytherapy and external beam radiotherapy for irradiation of localized prostatic carcinoma. *Radiother Oncol* 1997;**44**:237–44.
57. Martinez A, Gonzalez J, Spencer W, et al. Conformal high dose rate brachytherapy improves biochemical control and causes specific survival in patients with prostate cancer and poor prognostic factors. *J Urol* 2003;**169**:974–9 discussion 979–980.
58. Joniau S, Hsu CY, Lerut E, et al. A pretreatment table for the prediction of final histopathology after radical prostatectomy in clinical unilateral T3a prostate cancer. *Eur Urol* 2007;**51**(2):388–96.
59. Ward JF, Slezak JM, Blute ML, et al. Radical prostatectomy for clinically advanced (cT3) prostate cancer since the advent of prostate-specific antigen testing: 15-year outcome. *BJU Int* 2005;**95**(6): 751–6.
60. Hsu CY, Joniau S, Oyen R, et al. Outcome of surgery for clinical unilateral T3a prostate cancer: a single-institution experience. *Eur Urol* 2007;**51**(1):121–8.

61. Joniau SG, Van Baelen AA, Hsu CY, Van Poppel HP. Complications and functional results of surgery for locally advanced prostate cancer. *Adv Urol* 2012;**2012**:1–8.
62. Joniau S, Hsu CY, Gontero P, et al. Radical prostatectomy in very high-risk localized prostate cancer: long-term outcomes and outcome predictors. *Scand J Urol Nephrol* 2012;**46**:164–71.
63. Johnstone PA, Ward KC, Goodman M, et al. Radical prostatectomy for clinical T4 prostate cancer. *Cancer* 2006;**106**:2603–9.
64. Bader P, Burkhard FC, Markwalder R, et al. Is a limited lymph node dissection an adequate staging procedure for prostate cancer? *J Urol* 2002;**168**:514–8.
65. Heidenreich A, Varga Z, Von Knobloch R. Extended pelvic lymphadenectomy in patients undergoing radical prostatectomy: high incidence of lymph node metastasis. *J Urol* 2002;**167**(4):1681–6.
66. Jeschke S, Nambirajan T, Leeb K, et al. Detection of early lymph node metastases in prostate cancer by laparoscopic radioisotope guided sentinel lymph node dissection. *J Urol* 2005;**173**(6):1943–6.
67. Ghavamian R, Bergstralh EJ, Blute ML, et al. Radical retropubic prostatectomy plus orchiectomy versus orchiectomy alone for pTxN+ prostate cancer: a matched comparison. *J Urol* 1999;**161**:1223.
68. Engel J, Bastian PJ, Baur H, et al. Survival benefit of radical prostatectomy in lymph node-positive patients with prostate cancer. *Eur Urol* 2010;**57**(5):754–61.
69. Mattei A1, Fuechsel FG, Bhatta Dhar N, et al. The template of the primary lymphatic landing sites of the prostate should be revisited: results of a multimodality mapping study. *Eur Urol* 2008;**53**(1):118–25.
70. Briganti A, Chun FK, Salonia A, et al. Critical assessment of ideal nodal yield at pelvic lymphadenectomy to accurately diagnose prostate cancer nodal metastasis in patients undergoing radical retropubic prostatectomy. *Urology* 2007;**69**(1):147–51.
71. Briganti A, Chun FK, Salonia A, et al. Complications and other surgical outcomes associated with extended pelvic lymphadenectomy in men with localized prostate cancer. *Eur Urol* 2006;**50**(5):1006–13.
72. Heidenreich A, Von Knobloch R, Varga Z, et al. Extended pelvic lymphadenectomy in men undergoing radical retropubic prostatectomy (RRP)-an update on >300 cases. *J Urol* 2004;**171**:312.
73. Burkhard FC, Schumacher M, Studer UE. The role of lymphadenectomy in prostate cancer. *Nat Clin Pract Urol* 2005;**2**(7):336–42.
74. Stone NN, Stock RG, Unger P. Laparoscopic pelvic lymph node dissection for prostate cancer: comparison of the extended and modified techniques. *J Urol* 1997;**158**(5):1891–4.
75. Bader P, Burkhard FC, Markwalder R, et al. Is a limited lymph node dissection an adequate staging procedure for prostate cancer? *J Urol* 2002;**168**(2):514–8.
76. Joslyn SA, Konety BR. Impact of extent of lymphadenectomy on survival after radical prostatectomy for prostate cancer. *Urology* 2006;**68**(1):121–5.
77. Ji J, Yuan H, Wang L, et al. Is the impact of the extent of lymphadenectomy in radical prostatectomy related to the disease risk? A single center prospective study. *J Surg Res* 2012;**178**(2):779–84.
78. Gontero P, Marchioro G, Pisani R, et al. Is radical prostatectomy feasible in all cases of locally advanced non-bone metastatic prostate cancer? Results of a single-institution study. *Eur Urol* 2007;**51**(4):922–9.
79. Dillioglugil O, Leibman BD, Leibman NS, et al. Risk factors for complications and morbidity after radical retropubic prostatectomy. *J Urol* 1997;**157**(5):1760–7.
80. Hisasue SI, Takahashi A, Kato R, et al. Early and late complications of radical retropubic prostatectomy: experience in a single institution. *Jpn J Clin Oncol* 2004;**34**(5):274–9.
81. Gaylis FD, Friedel WE, Armas OA. Radical retropubic prostatectomy outcomes at a community hospital. *J Urol* 1998;**159**(1):167–71.
82. Lepor H, Nieder AM, Ferrandino M, et al. Intraoperative and postoperative complications of radical retropubic prostatectomy in a consecutive series of 1,000 cases. *J Urol* 2001;**166**(5):1729–33.
83. Lerner SE, Blute ML, Zincke H. Extended experience with radical prostatectomy for clinical stage T3 prostate cancer: outcome and contemporary morbidity. *J Urol* 1995;**154**(4):1447–52.
84. Zelefsky MJ, Eastham JA, Cronin AM, et al. Metastasis after radical prostatectomy or external beam radiotherapy for patients with clinically localized prostate cancer: a comparison of clinical cohorts adjusted for case mix. *J Clin Oncol* 2010;**28**(9):1508–13.
85. Tewari A, Divine G, Chang P, et al. Long-term survival in men with high grade prostate cancer: a comparison between conservative treatment, radiation therapy and radical prostatectomy – a propensity scoring approach. *J Urol* 2007;**177**(3):911–5.
86. Cooperberg MR, Vickers AJ, Broering JM, et al. Comparative risk-adjusted mortality outcomes after primary surgery, radiotherapy, or androgen-deprivation therapy for localized prostate cancer. *Cancer* 2010;**116**(22):5226–34.
87. Sooriakumaran P, Nyberg T, Akre O, et al. Comparative effectiveness of radical prostatectomy and radiotherapy in prostate cancer: observational study of mortality outcomes. *BMJ* 2014;**348**:g 1502.
88. Boorjian SA, Karnes RJ, Viterbo R, et al. Long-term survival after radical prostatectomy versus external-beam radiotherapy for patients with high-risk prostate cancer. *Cancer* 2011;**117**(13):2883–91.
89. Shelley MD, Kumar S, Wilt T, et al. A systematic review and meta-analysis of randomised trials of neo-adjuvant hormone therapy for localised and locally advanced prostate carcinoma. *Cancer Treat Rev* 2009;**35**(1):9–17.
90. Bolla M, van Poppel H, Tombal B, et al. European Organisation for Research and Treatment of Cancer, Radiation Oncology and Genito-Urinary Groups. Postoperative radiotherapy after radical prostatectomy for high-risk prostate cancer: long-term results of a randomised controlled trial (EORTC trial 22911). *Lancet* 2012;**380**(9858):2018–27.
91. Wiegel T, Bottke D, Steiner U, et al. Phase III postoperative adjuvant radiotherapy after radical prostatectomy compared with radical prostatectomy alone in pT3 prostate cancer with postoperative undetectable prostate-specific antigen: ARO 96-02/AUO AP 09/95. *J Clin Oncol* 2009;**27**(18):2924–30.
92. Thompson IM, Tangen CM, Paradelo J, et al. Adjuvant radiotherapy for pathological T3N0M0 prostate cancer significantly reduces risk of metastases and improves survival: long-term followup of a randomized clinical trial. *J Urol* 2009;**181**(3):956–62.
93. Messing EM, Manola J, Yao J, et al. Immediate versus deferred androgen deprivation treatment in patients with node-positive prostate cancer after radical prostatectomy and pelvic lymphadenectomy. *Lancet Oncol* 2006;**7**(6):472–9.
94. Schumacher MC, Burkhard FC, Thalmann GN, et al. Good outcome for patients with few lymph node metastases after radical retropubic prostatectomy. *Eur Urol* 2008;**54**(2):344–52.
95. Heidenreich A, Richter S, Thüer D, et al. Prognostic parameters, complications, and oncologic and functional outcome of salvage radical prostatectomy for locally recurrent prostate cancer after 21st-century radiotherapy. *Eur Urol* 2010;**57**(3):437–43.

P A R T IX

# ADVANCED PROSTATE CANCER

CHAPTER

# 52

# Androgen Deprivation Therapy: Appropriate Patients, Timing to Initiate ADT, and Complications

*Amirali H. Salmasi, MD, Neal Patel, MD, Isaac Yi Kim, MD, PhD*

**Section of Urologic Oncology, Rutgers Cancer Institute of New Jersey and Division of Urology, Department of Surgery, Rutgers Robert Wood Johnson Medical School, New Brunswick, NJ, USA**

## INTRODUCTION

Prostate cancer is the second most frequently diagnosed cancer of men and the sixth leading cause of death in men with close to 258,000 deaths reported worldwide.[1,2] In the United States alone, prostate cancer is the second leading cause of cancer mortality.[3] Metastatic castration-resistant prostate cancer (CRPC) is almost always associated with death in patients with prostate cancer. The seminal observation that resulted in the development of ADT occurred in the 1940s when Dr Huggins and Hodges discovered the role of androgen in prostate cancer at the University of Chicago. They showed that prostate cancer cell growth and spread were dependent on hormonal signaling. Huggins and Hodges were awarded the Nobel Prize in 1966 for their discovery of the androgen-dependent nature of prostate cancer. This vulnerability of prostate cancer was exploited for over half a century until new therapeutic agents were discovered and utilized at the turn of the century. Nevertheless, androgen deprivation therapy (ADT) still remains the foundation of advanced prostate cancer care.[4]

## ANDROGEN DEPRIVATION THERAPY

The use of ADT leveled after peaking in the 1990s.[5,6] Indeed, Shahinian et al. reported a substantial decline in the usage of ADT as a primary treatment modality for localized, low- to moderate-grade prostate cancer from 38.7% in 2003 to 30.6% and 25.7% in 2004 and 2005, respectively.[5] Nevertheless, Cetin et al. estimated that on December 31, 2008, 188, 916 men aged ≥45 with nonmetastatic prostate cancer in the United States were receiving continuous ADT therapy for ≥6 months.[7] In addition, Kuykendal et al. analyzed the Surveillance Epidemiology and End Results-Medicare database and reported that 12.4% of men with nonmetastatic prostate cancer aged 66–80 years received ADT discordant with published NCCN guidelines.[8]

Although primary ADT has no survival benefit in men with clinically localized prostate cancer, ADT is beneficial in some clinical settings.[9] Specifically, the combination of ADT with radiation provides a superior survival in men with intermediate- or high-risk prostate cancer[10–16] (Table 52.1). Similarly, adjuvant ADT improves survival in patients with positive lymph node after radical prostatectomy[21,22] (Table 52.2). Notwithstanding, ADT is most frequently used in a palliative role in metastatic prostate cancer.

## COMBINATION THERAPY

### Neoadjuvant ADT With Radical Prostatectomy

The concept of neoadjuvant hormone therapy before prostate surgery was evaluated in multiple clinical trials[27–30] (Table 52.2). Shelley et al. published a systematic review and meta-analysis of randomized trials to evaluate the efficacy of neoadjuvant hormone therapy before prostatectomy for localized and locally advanced prostate cancer.[23] This study showed that 3 months of

Prostate Cancer. http://dx.doi.org/10.1016/B978-0-12-800077-9.00052-9

**TABLE 52.1** ADT with Radiation

| Trials | Groups | Sample size | 10-year overall survival* | 10-year cancer-specific mortality* |
|---|---|---|---|---|
| Pilepich et al.[10] (RTOG 85-31) | RT + life time adjuvant ADT versus RT + ADT as relapse | 477 Versus 468 | 49% Versus 39% | 16% Versus 22% |
| Bolla et al.[11,12] (EORTC 22863) | RT versus RT + immediate ADT for 3 years | 208 Versus 207 | 39.8% Versus 58.1% | 30.4% Versus 10.3% |
| Widmark et al.[13] | 3 Months flutamide versus 3 months flutamide + RT | 439 Versus 436 | 60.6% Versus 70.4% | 23.9% Versus 11.9% |
| Warde et al.[14] | Lifelong ADT versus Lifelong ADT + RT | 662 Versus 603 | 66% Versus 74%** | 19% Versus 9%** |
| Horwitz et al.[15] (RTOG 92-02) | 4 Months ADT + RT versus 24 months ADT + RT | 763 Versus 758 | 51.6% Versus 53.9%† | 16.1% Versus 11.3% |
| Bolla et al.[16] (EORTC 22961) | 6 Months ADT + RT versus 3 years ADT + RT | 483 Versus 487 | 81% Versus 84.8%‡ | 4.7% Versus 3.2%‡ |
| Roach et al.[17,18] (RTOG 86-10) | 2 Months neoadjuvant and 2 months concurrent ADT + RT versus RT | 224 Versus 232 | 43% Versus 34%§ | 23% Versus 36% |
| Jones et al.[19] (RTOG 94-08) | 2 Months neoadjuvant and 2 months concurrent ADT + RT versus RT | 987 Versus 992 | 62% Versus 57% | 4% Versus 8% |
| Denham et al.[20] (TROG 96.01) | 3 Months versus 6 months neoadjuvant ADT + RT versus only RT | 265 Versus 267 versus 270 | 63.3% Versus 70.8% versus 57.5%¶ | 18.9% Versus 11.4% versus 22%¶ |

**All differences are significant, unless otherwise stated.*
***Seven-year overall survival and cancer specific mortality.*
*†Difference not significant; 24 months ADT had survival benefit in a subgroup of patients with Gleason score of 8 to 10.*
*‡Five-year overall survival and cancer specific mortality.*
*§Difference not significant.*
*¶No significant difference between 3 months neoadjuvant ADT and only RT groups.*

neoadjuvant hormone therapy prior to prostatectomy is associated with significant reduction in positive margin rates (RR 0.49, 95% CI (confidence interval) 0.42–0.56, $p < 0.00001$), increase in organ-confined disease (RR 1.63, 95% CI 1.37–1.95, $p < 0.0001$), and reduction in lymph node invasion (RR 0.49, 95% CI 0.42–0.56, $p < 0.02$). However, it did not improve disease-specific, disease-free, and overall survival.[23] Recently, Yee et al. confirmed these results in long-term follow-up of a randomized trial of radical prostatectomy with or without a 3-month course of neoadjuvant hormone (goserelin acetate and flutamide) for clinically localized prostate cancer.[24] At median follow-up of 8 years, there was no significant difference in biochemical recurrence rates between the

**TABLE 52.2** Radical Prostatectomy and ADT

| Study | Clinical stage | Design | Follow-up (years) | Overall survival | Progression-free survival |
|---|---|---|---|---|---|
| Shelley et al.[23] | Localized ± LN metastasis | Meta-analysis of trials with neoadjuvant ADT | – | No change | No change |
| Yee et al.[24] | Localized | 3 Months of neoadjuvant ADT + RP versus RP | 8 | No change | No change |
| Wirth et al.[25] | Localized | RP versus RP + adjuvant flutamide | 6.1 | No change | Better in flutamide group |
| Siddiqui et al.[26] | Localized | Effect of timing of adjuvant ADT after RP | 10 | No change | Better if it starts before recurrence |
| Messing et al.[21] | LN metastasis | RP + Immediate ADT versus ADT after recurrence | 11.9 | Better with immediate ADT | Better with immediate ADT |
| Dorff et al.[22] | LN metastasis | RP + 2 years of adjuvant ADT alone versus ADT + mitoxantrone | 4.4 | Better than expected in ADT only group* | Better than expected in ADT only group* |

**This trial was closed early after three cases of acute myelogenous leukemia were reported in the mitoxantrone treatment arm.*

two groups. It also showed no significant relationship between neoadjuvant hormone therapy and biochemical recurrence.[24] The lack of a clear survival advantage, cost, and side effects of neoadjuvant ADT before radical prostatectomy strongly suggest that neoadjuvant ADT is not indicated in men undergoing surgery.

## ADT With Radiation

ADT is now recommended for radiation therapy in patients with intermediate-risk disease with adverse features and high-risk prostate cancer (Table 52.1).[9] Pilepich et al. (RTOG 85-31) randomized patients with clinical stage T3 or lymph node stage N1 to either radiation and lifetime adjuvant goserelin or radiation alone and delayed goserelin at relapse.[10] At 10 years, the overall survival rate was superior in the lifetime adjuvant group compared to only radiation group (49% vs. 39%, $p = 0.002$).[10] Bolla et al. (EORTC 22863) also evaluated the efficacy of long-term ADT (3 years) with radiation therapy in patients with clinical stage T1–T2 World Health Organization grade 3 or T3–T4 N0–N1 prostate cancer.[11,12] Again, 10-year clinical disease-free survival and overall survival was significantly higher in the combined treatment group versus the radiotherapy-alone group.[12]

Two additional studies investigated the role of radiation therapy in combination with ADT.[13,14] Widmark et al. randomized men with locally advanced prostate cancer (T3, N0, M0) to ADT alone or to the same endocrine treatment combined with radiation therapy. The cumulative incidence of prostate-cancer-specific mortality and overall mortality at 10 years was significantly lower in the combination group.[13] In the second trial, Warde et al. compared the overall survival in men with locally advanced prostate cancer managed with ADT alone or ADT with radiation, and found that addition of radiation to ADT improved overall survival at 7 years with minimal side effects.[14]

Largely based on these studies, long-term (2–3 years) ADT in combination with radiation therapy is indicated for patients with high to very high risk of progression while short-term ADT (4–6 months) in combination with radiation is a viable option in men with intermediate-risk prostate cancer.[9] Horwitz et al. (RTOG 92-02) compared the efficacy of 4 months of goserelin and flutamide before and during radiation therapy versus 24 months of ADT in 1554 patients with T2c–T4, N0, M0 prostate cancer.[15] At 10 years, long-term ADT improved disease-free survival, disease-specific survival, local progression, distant metastasis, and biochemical failure. It had no effect on overall survival. However, long-term ADT had survival benefit in a subgroup of patients with Gleason score of 8 to 10.[15] Furthermore, Bolla et al. (EORTC 22961) demonstrated that short-term (6 months) androgen suppression with radiation therapy provides inferior survival in locally advanced disease compared with radiotherapy plus 3 years of androgen suppression at intermediate-term follow up.[16]

More recently, short-term (4–6 months) neoadjuvant hormone therapy has been shown to have survival benefit over radiation therapy alone.[17–20,31] Roach et al. (RTOG 86-10) reported that overall survival is higher in patients with locally advanced prostate cancer who received ADT 2 months before and concurrent with radiation (total 4 months) versus radiation alone.[17] However, this difference did not reach statistical significance.[18] In contrast, Jones et al. (RTOG 94-08) evaluated the efficacy of the same neoadjuvant ADT in a cohort of patients with low- and intermediate-risk prostate cancer and showed at a median follow-up of 9.1 years, short-term (4 months) ADT has a significant survival benefit compared with radiation therapy alone.[19] Denham et al. (TROG 96.01) compared the 3-month and 6-month neoadjuvant ADT combined with radiation for locally advanced prostate cancer.[20] In this study 3 months ADT had no effect on prostate-cancer-specific mortality or all-cause mortality compared with radiotherapy alone. However, 6-month ADT had a survival benefit after a median follow-up of 10.6 years compared with radiotherapy alone.[20] Finally, the outcomes of short-term (4 months) and long-term (8 months) neoadjuvant ADT before radiotherapy was assessed in patients with localized prostate cancer and it was found that at median of 102 months, survival, biochemical failure-free survival, and prostate-cancer-specific survival did not differ significantly between the two groups.[31]

## Combined Androgen Blockade

The concept of adding an antiandrogen to surgical or medical castration is based on the observation that androgens secreted from the adrenal gland after castration can cause prostate cancer progression.[32] In the 1990s, the efficacy of combined androgen blockage (CAB) using nilutamide and flutamide was investigated in multiple clinical trials.[33] The collaborative meta-analysis of 27 randomized trials on the combination of an antiandrogen (nilutamide, flutamide, or cyproterone acetate) with either surgical castration or a luteinizing hormone-releasing hormone (LHRH) agonist showed no significant 5-year survival benefit in the CAB group versus the androgen suppression group.[33] In this study, most patients (88%) had metastatic disease while the remainder had locally advanced disease. In a separate systematic review and meta-analysis, Samson et al. compared monotherapy and CAB in men with advanced prostate cancer.[34] Although, they found no statistically significant difference in survival at 2 years, there was a statistically significant difference in survival at 5 years in the CAB group versus monotherapy.[34] Klotz et al. reanalyzed the clinical trials of CAB and found that CAB with 50 mg bicalutamide reduced the risk of death by 20% compared with castration alone.[35] In a phase 3 trial, Akaza et al. compared CAB with LHRH agonist plus bicalutamide 80 mg versus LHRH agonist monotherapy in patients with advanced prostate cancer.[36] At a median follow-up of 5.2 years, a significant overall survival advantage was observed in the CAB group over LHRH agonist monotherapy.[36] Taken together, CAB provides a modest survival benefit compared with monotherapy in the management of advanced prostate cancer.

## TIMING

### Clinically Localized Disease

Although ADT is widely used, there is no survival benefit from primary ADT for most men with localized prostate cancer.[37–39] Potoskey et al. reviewed a retrospective cohort of men with prostate cancer diagnosed between 1995 and 2008 using three integrated health plans. These men were not treated with any additional therapies with a curative intent. At a median follow-up of 61 months, they found no significant reduction in all-cause mortality or prostate-cancer-specific mortality from primary ADT compared with no primary ADT for men with clinically localized prostate cancer who did not receive any therapies with curative intent.[37] However, primary ADT was associated with small but statistically significant overall mortality benefit in the subgroup of men with high risk of cancer progression.[37] In a population-based study of 19,271 men with clinical stage T1–T2 prostate cancer, Lu-Yao et al. found primary ADT was associated with no improvement in 10-year overall survival among most elderly men with localized prostate cancer compared with conservative management.[38] Moreover, Iversen et al. evaluated the efficacy and tolerability of an antiandrogen monotherapy, bicalutamide, in patients with localized or locally advanced prostate cancer in three randomized, double-blind, placebo-controlled trials, and found no improvement in progression-free survival or overall survival.[39]

The role of ADT after radical prostatectomy of clinically localized disease is also controversial. Wirth et al. assessed the role of adjuvant flutamide after radical prostatectomy in 309 patients with stage pT3–4, N0, M0 prostate cancer between 1989 and 1996. At a median follow-up 6.1 year, the flutamide group had considerable toxicity. Although recurrence-free survival was better in the flutamide group ($p$ = 0.0041), there was no significant improvement in overall survival ($p$ = 0.92).[25] In a separate study, Siddiqui et al. investigated the effect of timing of adjuvant ADT after radical prostatectomy in a cohort of patients with lymph-node-negative prostate cancer between 1990 and 1999.[26] At a median follow-up of 10 years, they found that adjuvant androgen deprivation therapy after radical prostatectomy was associated with improved 10-year systemic progression-free survival (95% vs. 90%, $p < 0.001$) and 10-year cancer-specific survival (98% vs. 95%, $p$ = 0.009), if it was delivered before the time of biochemical recurrence or systemic progression. Nevertheless, adjuvant ADT postsurgery was not associated with an increased overall survival.[26] These results are summarized in Table 52.2.

### Lymph Node Metastasis

Using the Surveillance, Epidemiology and End Results-Medicare data of men who underwent radical prostatectomy between 1991 and 1999 and had positive pelvic lymph nodes, Wong et al. reported that adjuvant ADT (within 120 days of radical prostatectomy) had no overall survival benefit in the ADT group versus non-ADT group (ADT initiated >120 days from surgery or no ADT).[40] The authors suggested that deferring immediate ADT in men with positive lymph nodes after radical prostatectomy may not significantly compromise survival.[40] In contrast, the Eastern Cooperative Oncology Group (ECOG 3886) evaluated the effect of immediate ADT (goserelin or bilateral orchiectomy) for node-positive prostate cancer in patients following radical prostatectomy.[21] Between 1988 and 1993, 98 men with positive lymph node after prostatectomy were randomly assigned to receive immediate ADT (47 men) or to be treated with ADT after detection of distant metastases or symptomatic recurrences. After a median follow-up of 11.9 years, men in the immediate ADT group had a significant improvement in overall survival (hazard ratio 1.84 [95% CI 1.01–3.35], $p$ = 0.04), prostate-cancer-specific survival (4.09 [1.76–9.49], $p$ = 0.0004), and progression-free survival (3.42 [1.96–5.98], $p < 0.0001$). Similarly, in the SWOG (Southwest Oncology Group) S9921 trial, 983 men with adverse features after radical prostatectomy (Gleason score of 8 or higher; preoperative PSA of 15 ng/mL or greater; stage T3b, T4, or N1 disease; or Gleason score of 7 with either preoperative PSA greater than 10 ng/mL or a positive margin) were randomly assigned to receive 2 years of adjuvant ADT alone (goserelin and bicalutamide) or in combination with mitoxantrone chemotherapy.[22] This trial was closed early after three cases of acute myelogenous leukemia were reported in the mitoxantrone treatment arm. Among the 481 eligible men assigned to receive ADT alone, 61% of patients had stage T3 disease or higher, and 16% was node-positive. The estimated 5-year biochemical failure-free survival was 92.5% (95% CI, 90–95), and 5-year overall survival was 95.9% (95% CI, 93.9–97.9). This study demonstrated that in the intermediate-term follow-up, adjuvant ADT following surgery may provide better-than-expected survival in men with adverse features.[22] Based on these two randomized clinical trials, it is reasonable to offer patients with lymph-node-positive or high-risk prostate cancer adjuvant ADT after radical prostatectomy. Again, Table 52.2 summarizes these results.

### Locally Advanced, Asymptomatic Metastatic Disease or Disease not Suitable for Local Treatment

The Medical Research Council (MRC) trial in 1997 randomized patients with advanced or asymptomatic metastatic prostate cancer either to immediate androgen suppression therapy or to delayed ADT until clinically significant progression occurs.[41] The results demonstrated that there were less overall mortality and cancer-specific mortality in the immediate group. At the time of the last follow up in 2003, there were still significant improvement in disease-specific survival in the immediate group.[42] Overall survival was also higher in immediate therapy; however, the difference was not significant. In a similar study, Schroder et al. (EORTC 30846) evaluated the effect of early androgen deprivation versus delayed androgen suppression in pN1-3 prostate cancer. After 13 years of follow-up, prostate-cancer-specific survival between immediate and delayed groups were similar. Notwithstanding, this trial was underpowered to reach its goal of showing noninferiority.[43] In a separate trial, Keating et al. demonstrated that the receipt of ADT in their cohort of men aged 66 years or older with metastatic prostate cancer was associated with similar magnitude of improved survival in early and delayed ADT.[44] Finally, immediate versus delayed ADT (orchiectomy) was investigated in patients with T0-4 N0-2 M0 prostate cancer who were not candidates for curative local treatment (EORTC 30891).[45] The authors reported that at median follow-up of 7.8 years, immediate ADT resulted in a significant, though small improvement in overall survival but there was no difference in prostate-cancer-specific mortality or overall symptom-free survival between the groups.[45]

### Intermittent Therapy

The efficacy of intermittent ADT (IADT) versus continuous ADT (CADT) in patients with metastatic prostate cancer and biochemical recurrence was investigated in two large trials. Both were designed as noninferiority studies.[46,47] Crook et al. enrolled patients with a PSA level greater than 3 ng/mL more than 1 year after primary or salvage radiotherapy for localized prostate cancer.[46] Of the 1386 enrolled patients, 690 were randomly assigned to IADT and 696 to CADT. At median follow-up of 6.9 years, median overall survival was 8.8 years in the intermittent-therapy group versus 9.1 years in the continuous-therapy group (hazard ratio for death, 1.02; 95% CI, 0.86 to 1.21). IADT provided small benefits on quality of life that included physical function, fatigue, urinary problems, hot flashes, libido, and erectile function. The investigators concluded that IADT was noninferior to CADT with respect to overall survival. In the second trial, Hussain et al. compared IADT versus CADT in men with newly diagnosed, metastatic, hormone-sensitive prostate cancer, the ECOG performance status of 0–2, and the serum PSA level of 5 ng/mL or higher after receiving LHRH agonist and an antiandrogen agent for 7 months.[47] They randomized patients in whom the PSA level fell to 4 or lower to CADT or IADT. At median follow-up of 9.8 years, median survival was 5.8 years in the CADT group and 5.1 years in the IADT group (hazard ratio for death with IADT, 1.10; 90% CI, 0.99–1.23). This study also demonstrated that in patients with metastatic hormone-sensitive prostate cancer, there was a 20% relative increase in the risk of death with IADT compared with CADT; however, too few events occurred to unequivocally rule out significant inferiority of intermittent therapy. In this trial, IADT provided small temporary improvements in quality of life (erectile function and mental health) early in the course of treatment. Taken together, these reports reveal that IADT provides noninferior survival compared with CADT in patients with biochemical recurrence or hormone sensitive metastatic disease while offering better quality of life and lower cost of treatment.

## GENERAL COMPLICATIONS OF ADT

Despite the clinical benefits of ADT, the treatment is associated with complications that often compromise the quality of the life of patients. These adverse effects of ADT include diabetes and metabolic syndrome as well cardiovascular disease. The following section summarizes some of the well know side effects of ADT. Table 52.3 lists these complications and potential treatments.

### Osteoporosis

When treating men with ADT it is important to stress the need for close follow-up in regard to bone health. ADT is associated with increased risk of clinical fractures. This increased fracture risk is directly related to ADT's effect on decreasing bone mineral density (BMD) while also increasing bone turnover rates. It has been reported that the use of ADT may increase the incidence of clinical fractures by up to 45%.[50] Zolendronic acid has been shown to significantly improve BMD and prevent bone loss at 1 year in men on ADT. Another promising therapy, denosumab, a monoclonal antibody that binds to receptor activator of nuclear factor kappa-B ligand (RANKL),

TABLE 52.3 Adverse Effects of ADT and Potential Treatment and Prevention Strategies

| Adverse effect | Treatment/prevention strategy |
|---|---|
| Anemia/fatigue* | Recombinant human erythropoietin[48] |
| Cardiovascular | Lifestyle modifications, diet, exercise, lipid lowering agents[49] |
| Metabolic syndrome/diabetes* | Resistance exercise, glycemic control, life style modifications |
| Changes in body habitus/ Gynecomastia (breast pain) | Resistance exercise, pre/initial treatment radiation, aromatase inhibitors/tamoxifen for breast pain[48] |
| Cognitive/emotional changes* | Resistance exercise |
| Osteoporosis | Exercise, calcium supplementation, decreased alcohol intake, smoking cessation, bisphosphonates, denosumab |
| Sexual dysfunction (ED/loss of libido)* | PDE-5 inhibitors, intracavernosal injections, vacuum constriction devices or penile prosthesis[48] |
| Hot flashes* | 25-mg Dose of venlafaxine, Alt: gabapentin, low-dose megestrol acetate, cytoproterone[48] |

*Most side effects are reversible after cessation of ADT.

a protein that regulates osteoclast function and survival, has been approved by the FDA for treatment of men undergoing androgen ablation of nonmetastatic prostate cancer. An international randomized, double-blind placebo-controlled trial showed that giving 60 mg of denosumab subcutaneously every 6 months for 2 years versus placebo resulted in a significantly reduced incidence at 36 months of new vertebral fractures (1.5% vs. 3.9%). This study also showed a 5.6% increase in BMD in patients receiving denosumab versus placebo. However, appropriate counseling of possible side effects, which include osteonecrosis of the jaw, along with coordination with the primary care physician is important in monitoring bone health for patients undergoing ADT.

## Sexual Dysfunction (Erectile Dysfunction and Loss of Libido)

Erectile dysfunction (ED) usually occurs in patients undergoing long-term ADT. As expected, ED is especially worse with combination therapy versus monotherapy. However, traditional phosphodiesterase-5 inhibitor therapies have been shown to improve ADT-induced impotence. Loss of libido is also a common adverse effect of ADT. It is important to have a baseline assessment of sexual function prior to initiating ADT especially in older men who frequently have preexisting sexual dysfunction.[51] Aside from PDE-5 inhibitors, intracavernosal injections, vacuum constriction devices and even penile prosthesis may be utilized to treat resultant ED.[48]

## Hot Flashes

Up to 80% of patients receiving ADT may experience hot flashes. It has been reported that hot flashes may increase in frequency starting at 3 months after the start of therapy and persist long term. The severity of hot flashes may deter some patients from continuing therapy. Megestrol acetate may reduce hot flashes by up to 85%. However, other medications, such as gabapentin and venlafaxine, have also shown to be efficacious against hot flashes.[51]

## Cognitive, Mental, and Emotional Effects

In multiple small studies, it was found that there was evidence of decline in verbal, spatial, and executive functioning in patients undergoing ADT.[51] However, a recent study examining quality of life (QoL) assessment from the CaPSURE registry data for the emotional and mental status of the patient undergoing ADT for prostate cancer pre- and posttherapy found that after 24 months of exposure to ADT there were no statistically significant clinical declines in mental and emotional well-being.[52] However, it has been noted that lifestyle modifications and exercise may improve cognitive and emotional effects of ADT. Therefore, additional studies are necessary to determine the impact of ADT on cognitive function.

## Changes in Body Habitus

As the regulation of lean and fat body mass involves testosterone, patients on ADT exhibit a decrease in lean body mass of as much as 3.8%, and a concomitant increase in primarily subcutaneous fat body mass of as much as 11.0%.[50,53] Simultaneously, there is a significant decline in muscle mass, which directly correlates with decreased muscle strength and physical performance.[51] Gynecomastia is also known to develop during androgen ablation and usually remains permanent if the treatment lasts longer than 1 year due to fibrosis. Radiation of the breast is one option in the prevention/treatment of gynecomastia in men undergoing ADT.[48]

### Diabetes and Metabolic Syndrome

Fasting plasma insulin levels are elevated in men undergoing ADT for prostate cancer while significant reduction in insulin sensitivity was found in nondiabetic men undergoing ADT.[50,54] It has also been demonstrated that ADT increases the incidence of diabetes (adjusted HR 1.42).[50,55] Similarly, orchiectomy has also been shown to have a significant association with diabetes.[51] More commonly, metabolic syndrome has been reported in men receiving ADT for prostate cancer. Men receiving ADT were likely to have higher triglycerides and fasting plasma glucose levels and increased abdominal girth.[48,56]

### Cardiovascular Morbidity and Mortality

ADT also increases the serum lipid levels. Smith et al. reported a rise in low-density lipoprotein as high as 9%, high-density lipoprotein 7%, cholesterol 11%, and triglycerides 28%.[50,53,54] Keating et al.[55] found that ADT conferred not only a significant increase in the risk of diabetes but also the incidence of coronary artery disease (HR of 1.16) and frequency of myocardial infarction (HR 1.11). Nevertheless, the data regarding cardiovascular morbidity and ADT remain controversial. Multiple studies have shown that ADT appears to be associated with increased risk of cardiovascular-related morbidity and mortality. Keating et al. in 2010 reported the results of a large observational population-based study of 37,443 men. The study showed an increased risk of diabetes and cardiovascular disease related to GnRH agonists. Specifically, they reported an adjusted hazard ratio of 1.19 for coronary heart disease, 1.28 for myocardial infarction, 1.29 sudden cardiac death, and 1.22 for stroke.[57] Similarly, Tsai et al. also showed that ADT was associated with cardiac-specific mortality (CSM) in men. Specifically, men 65 years older and undergoing radical prostatectomy for localized prostate cancer who were treated with ADT had 5% incidence of CSM versus 2% for men who did not receive ADT.[58] Saigal et al. also reported that ADT may be associated with increased CSM in men with prostate cancer and may in fact lower overall survival in men with low-risk disease.[49] A recent 2014 study on CSM performed by Ziehr et al. examined 5077 men with cT1c–T3N0M0 prostate cancer who underwent treatment with brachytherapy +/− neoadjuvant ADT. In this analysis, it was found that in men who had CHF or a prior MI, ADT resulted in a 5% excess risk of cardiac-specific mortality at 5 years.[59]

In contrast, Nguyen performed a systematic review and meta-analysis in 2011 of randomized trials involving ADT to determine if it is associated with cardiovascular mortality in men with unfavorable-risk, nonmetastatic prostate cancer. In this study, 4141 patients were included from eight separate randomized trials and compared to control groups; men treated with ADT were not at a significantly higher risk for cardiovascular mortality. Subgroup analyses examining long (3+ years) and short (<6 months) duration were not associated with excess cardiovascular death. Furthermore in 11 of the trials, compared to placebo, ADT was associated with a significant decrease in prostate-cancer-specific mortality and all-cause mortality.[60]

### Anemia and Fatigue

Anemia and resultant fatigue are commonly seen in patients undergoing ADT, specifically normocytic normochromic anemia. Deficiency of testosterone along with other hormones is known to decrease erythropoiesis and lead to anemia with hematocrits reaching as low as 25 g/L after 6 months of ADT.[51] This usually resolves after the cessation of therapy; however, recombinant human erythropoietin may be utilized to treat symptomatic anemia resulting from ADT.[48]

## CONCLUSIONS

More than 70 years after the discovery of prostate cancer's androgen dependence, ADT remains the cornerstone of many prostate cancer treatments. Notwithstanding, multiple studies and investigations have improved the effectiveness of ADT. In men with metastatic and high-risk disease, early ADT appears to have survival benefit. In addition, ADT is now the standard of care in patients with intermediate- and high-risk prostate cancer who elect radiation. Yet, ADT is not free. Any length of ADT has been associated with significant complications that may compromise the quality of life. Thus, in counseling patients considering ADT, it is of paramount importance that the provider not only discuss the potential benefit of ADT but also the multiple side effects.

### References

1. Ferlay J, Shin HR, Bray F, et al. Estimates of worldwide burden of cancer in 2008: GLOBOCAN 2008. *Int J Cancer* 2010;**127**(12):2893–917.
2. Center MM, Jemal A, Lortet-Tieulent J, et al. International variation in prostate cancer incidence and mortality rates. *Eur Urol* 2012;**61**(6):1079–92.
3. Siegel R, Naishadham D, Jemal A. Cancer statistics, 2013. *CA Cancer J Clin* 2013;**63**(1):11–30.
4. Singer EA, Srinivasan R. Intravenous therapies for castration-resistant prostate cancer: toxicities and adverse events. *Urol Oncol* 2012;**30**(4 Supplement):S15–9.
5. Shahinian VB, Kuo YF, Gilbert SM. Reimbursement policy and androgen-deprivation therapy for prostate cancer. *N Engl J Med* 2010;**363**(19):1822–32.
6. Cooperberg MR, Grossfeld GD, Lubeck DP, et al. National practice patterns and time trends in androgen ablation for localized prostate cancer. *J Natl Cancer Inst* 2003;**95**(13):981–9.

7. Cetin K, Li S, Blaes AH, et al. Prevalence of patients with nonmetastatic prostate cancer on androgen deprivation therapy in the United States. *Urology* 2013;**81**(6):1184–9.
8. Kuykendal AR, Hendrix LH, Salloum RG, et al. Guideline-discordant androgen deprivation therapy in localized prostate cancer: patterns of use in the medicare population and cost implications. *Ann Oncol* 2013;**24**(5):1338–43.
9. Mohler JL, Panel NPC. Joint statement by members of the NCCN Prostate Cancer Guidelines Panel. *J Natl Compr Canc Netw* 2013;**11**(11):1310–2.
10. Pilepich MV, Winter K, Lawton CA, et al. Androgen suppression adjuvant to definitive radiotherapy in prostate carcinoma – long-term results of phase III RTOG 85-31. *Int J Radiat Oncol Biol Phys* 2005;**61**(5):1285–90.
11. Bolla M, Collette L, Blank L, et al. Long-term results with immediate androgen suppression and external irradiation in patients with locally advanced prostate cancer (an EORTC study): a phase III randomised trial. *Lancet* 2002;**360**(9327):103–6.
12. Bolla M, Van Tienhoven G, Warde P, et al. External irradiation with or without long-term androgen suppression for prostate cancer with high metastatic risk: 10-year results of an EORTC randomised study. *Lancet Oncol* 2010;**11**(11):1066–73.
13. Widmark A, Klepp O, Solberg A, et al. Endocrine treatment, with or without radiotherapy, in locally advanced prostate cancer (SPCG-7/SFUO-3): an open randomised phase III trial. *Lancet* 2009;**373**(9660):301–8.
14. Warde P, Mason M, Ding K, et al. Combined androgen deprivation therapy and radiation therapy for locally advanced prostate cancer: a randomised, phase 3 trial. *Lancet* 2011;**378**(9809):2104–11.
15. Horwitz EM, Bae K, Hanks GE, et al. Ten-year follow-up of radiation therapy oncology group protocol 92-02: a phase III trial of the duration of elective androgen deprivation in locally advanced prostate cancer. *J Clin Oncol* 2008;**26**(15):2497–504.
16. Bolla M, de Reijke TM, Van Tienhoven G. Duration of androgen suppression in the treatment of prostate cancer. *N Engl J Med* 2009;**360**(24):2516–27.
17. Roach III M, Bae K, Speight J, et al. Short-term neoadjuvant androgen deprivation therapy and external-beam radiotherapy for locally advanced prostate cancer: long-term results of RTOG 8610. *J Clin Oncol* 2008;**26**(4):585–91.
18. Roach III M. Current trends for the use of androgen deprivation therapy in conjunction with radiotherapy for patients with unfavorable intermediate-risk, high-risk, localized, and locally advanced prostate cancer. *Cancer* 2014;**120**(11):1620–9.
19. Jones CU, Hunt D, McGowan DG, et al. Radiotherapy and short-term androgen deprivation for localized prostate cancer. *N Engl J Med* 2011;**365**(2):107–18.
20. Denham JW, Steigler A, Lamb DS, et al. Short-term neoadjuvant androgen deprivation and radiotherapy for locally advanced prostate cancer: 10-year data from the TROG 96.01 randomised trial. *Lancet Oncol* 2011;**12**(5):451–9.
21. Messing EM, Manola J, Yao J, et al. Immediate versus deferred androgen deprivation treatment in patients with node-positive prostate cancer after radical prostatectomy and pelvic lymphadenectomy. *Lancet Oncol* 2006;**7**(6):472–9.
22. Dorff TB, Flaig TW, Tangen CM, et al. Adjuvant androgen deprivation for high-risk prostate cancer after radical prostatectomy: SWOG S9921 study. *J Clin Oncol* 2011;**29**(15):2040–5.
23. Shelley MD, Kumar S, Wilt T, et al. A systematic review and meta-analysis of randomised trials of neo-adjuvant hormone therapy for localised and locally advanced prostate carcinoma. *Cancer Treat Rev* 2009;**35**(1):9–17.
24. Yee DS, Lowrance WT, Eastham JA, et al. Long-term follow-up of 3-month neoadjuvant hormone therapy before radical prostatectomy in a randomized trial. *BJU Int* 2010;**105**(2):185–90.
25. Wirth MP, Weissbach L, Marx FJ, et al. Prospective randomized trial comparing flutamide as adjuvant treatment versus observation after radical prostatectomy for locally advanced, lymph node-negative prostate cancer. *Eur Urol* 2004;**45**(3):267–70 discussion 70.
26. Siddiqui SA, Boorjian SA, Inman B, et al. Timing of androgen deprivation therapy and its impact on survival after radical prostatectomy: a matched cohort study. *J Urol* 2008;**179**(5):1830–7 discussion 7.
27. Lee F, Siders DB, McHug TA, et al. Long-term follow-up of stages T2-T3 prostate cancer pretreated with androgen ablation therapy prior to radical prostatectomy. *Anticancer Res* 1997;**17**(3A):1507–10.
28. Witjes WP, Schulman CC, Debruyne FM. Preliminary results of a prospective randomized study comparing radical prostatectomy versus radical prostatectomy associated with neoadjuvant hormonal combination therapy in T2-3 N0 M0 prostatic carcinoma. The European Study Group on Neoadjuvant Treatment of Prostate Cancer. *Urology* 1997;**49**(3A Suppl):65–9.
29. Soloway MS, Pareek K, Sharifi R, et al. Neoadjuvant androgen ablation before radical prostatectomy in cT2bNxMo prostate cancer: 5-year results. *J Urol* 2002;**167**(1):112–6.
30. Klotz LH, Goldenberg SL, Jewett MA, et al. Long-term followup of a randomized trial of 0 versus 3 months of neoadjuvant androgen ablation before radical prostatectomy. *J Urol* 2003;**170**(3):791–4.
31. Armstrong JG, Gillham CM, Dunne MT, et al. A randomized trial (Irish clinical oncology research group 97-01) comparing short versus protracted neoadjuvant hormonal therapy before radiotherapy for localized prostate cancer. *Int J Radiat Oncol Biol Phys* 2011;**81**(1):35–45.
32. Labrie F, Dupont A, Giguere M, et al. Benefits of combination therapy with flutamide in patients relapsing after castration. *Br J Urol* 1988;**61**(4):341–6.
33. Prostate Cancer Trialists' Collaborative Group. Maximum androgen blockade in advanced prostate cancer: an overview of the randomised trials. *Lancet* 2000;**355**(9214):1491–8.
34. Samson DJ, Seidenfeld J, Schmitt B, et al. Systematic review and meta-analysis of monotherapy compared with combined androgen blockade for patients with advanced prostate carcinoma. *Cancer* 2002;**95**(2):361–76.
35. Klotz L, Schellhammer P, Carroll K. A re-assessment of the role of combined androgen blockade for advanced prostate cancer. *BJU Int* 2004;**93**(9):1177–82.
36. Akaza H, Hinotsu S, Usami M, et al. Combined androgen blockade with bicalutamide for advanced prostate cancer: long-term follow-up of a phase 3, double-blind, randomized study for survival. *Cancer* 2009;**115**(15):3437–45.
37. Potosky AL, Haque R, Cassidy-Bushrow AE, et al. Effectiveness of primary androgen-deprivation therapy for clinically localized prostate cancer. *J Clin Oncol* 2014;**32**(13):1324–30.
38. Lu-Yao GL, Albertsen PC, Moore DF, et al. Survival following primary androgen deprivation therapy among men with localized prostate cancer. *JAMA* 2008;**300**(2):173–81.
39. Iversen P, McLeod DG, See WA, et al. Antiandrogen monotherapy in patients with localized or locally advanced prostate cancer: final results from the bicalutamide Early Prostate Cancer programme at a median follow-up of 9.7 years. *BJU Int* 2010;**105**(8):1074–81.
40. Wong YN, Freedland S, Egleston B, et al. Role of androgen deprivation therapy for node-positive prostate cancer. *J Clin Oncol* 2009;**27**(1):100–5.
41. Immediate versus deferred treatment for advanced prostatic cancer: initial results of the Medical Research Council Trial. The Medical Research Council Prostate Cancer Working Party Investigators Group. *Br J Urol* 1997;79(2):235–246.
42. Kirk D. Timing and choice of androgen ablation. *Prostate Cancer Prostatic Dis* 2004;**7**(3):217–22.

43. Schroder FH, Kurth KH, Fossa SD, et al. Early versus delayed endocrine treatment of T2–T3 pN1-3 M0 prostate cancer without local treatment of the primary tumour: final results of European Organisation for the Research and Treatment of Cancer protocol 30846 after 13 years of follow-up (a randomised controlled trial). *Eur Urol* 2009;**55**(1):14–22.
44. Keating NL, O'Malley AJ, McNaughton-Collins M, et al. Use of androgen deprivation therapy for metastatic prostate cancer in older men. *BJU Int* 2008;**101**(9):1077–83.
45. Studer UE, Whelan P, Albrecht W, et al. Immediate or deferred androgen deprivation for patients with prostate cancer not suitable for local treatment with curative intent: European Organisation for Research and Treatment of Cancer (EORTC) Trial 30891. *J Clin Oncol* 2006;**24**(12):1868–76.
46. Crook JM, O'Callaghan CJ, Duncan G, et al. Intermittent androgen suppression for rising PSA level after radiotherapy. *N Engl J Med* 2012;**367**(10):895–903.
47. Hussain M, Tangen CM, Berry DL, et al. Intermittent versus continuous androgen deprivation in prostate cancer. *N Engl J Med* 2013;**368**(14):1314–25.
48. Kumar RJ, Barqawi A, Crawford EFD. Adverse events associated with hormonal therapy for prostate cancer. *Rev Urol* 2005;**7**(Suppl. 5):S37–43.
49. Saigal CS, Gore JL, Krupski TL, et al. Androgen deprivation therapy increases cardiovascular morbidity in men with prostate cancer. *Cancer* 2007;**110**(7):1493–500.
50. Isbarn H, Boccon-Gibod L, Carroll PR, et al. Androgen deprivation therapy for the treatment of prostate cancer: consider both benefits and risks. *Eur Urol* 2009;**55**(1):62–75.
51. Mohile SG, Mustian K, Bylow K, et al. Management of complications of androgen deprivation therapy in the older man. *Crit Rev Oncol Hematol* 2009;**70**(3):235–55.
52. Cary KC, Singla N, Cowan JE, et al. Impact of androgen deprivation therapy on mental and emotional well-being in men with prostate cancer: analysis from the CaPSURE registry. *J Urol* 2014;**191**(4):964–70.
53. Smith MR, Finkelstein JS, McGovern FJ, et al. Changes in body composition during androgen deprivation therapy for prostate cancer. *J Clin Endocrinol Metab* 2002;**87**(2):599–603.
54. Smith MR, Lee H, Nathan DM. Insulin sensitivity during combined androgen blockade for prostate cancer. *J Clin Endocrinol Metab* 2006;**91**(4):1305–8.
55. Keating NL, O'Malley AJ, Smith MR. Diabetes and cardiovascular disease during androgen deprivation therapy for prostate cancer. *J Clin Oncol* 2006;**24**(27):4448–56.
56. Moorjani S, Dupont A, Labrie F, et al. Changes in plasma lipoproteins during various androgen suppression therapies in men with prostatic carcinoma: effects of orchiectomy, estrogen, and combination treatment with luteinizing hormone-releasing hormone agonist and flutamide. *J Clin Endocrinol Metab* 1988;**66**(2):314–22.
57. Keating NL, O'Malley A, Freedland SJ, et al. Diabetes and cardiovascular disease during androgen deprivation therapy: observational study of veterans with prostate cancer. *J Natl Cancer Inst* 2012;**104**(19):1518–23.
58. Tsai HK, D'Amico AV, Sadetsky N, et al. Androgen deprivation therapy for localized prostate cancer and the risk of cardiovascular mortality. *J Natl Cancer Inst* 2007;**99**(20):1516–24.
59. Ziehr DR, Chen MH, Zhang D, et al. Association of androgen deprivation therapy with excess cardiac-specific mortality in men with prostate cancer. *BJU Int* 2015;**116**(3):358–65.
60. Nguyen PL, Je Y, Schutz FA, et al. Association of androgen deprivation therapy with cardiovascular death in patients with prostate cancer: a meta-analysis of randomized trials. *JAMA* 2011;**306**(21):2359–66.

CHAPTER

# 53

# Bone Health in Prostate Cancer

*Eugene Pietzak, MD*, Phillip Mucksavage, MD***

*University of Pennsylvania Health Care System, Pennsylvania, USA

**Perelman School of Medicine, University of Pennsylvania, Pennsylvania, USA

## INTRODUCTION

The maintenance of bone health is important throughout the entire prostate cancer disease spectrum. The majority of prostate cancer patients are already at increased risk for fragility fractures secondary to age-related bone loss.[1] This is further compounded by the threat of treatment-associated osteoporosis and fractures. In more advanced stages, additional risk arises from disease-associated skeleton-related events (SREs) secondary to bone metastasis.[2] Further complicating the prevention of skeletal morbidity is the fact that a substantial portion of prostate cancer patients have abnormal baseline levels of bone mineral density (BMD) prior to the initiation of therapies.[3]

Most studies looking at bone health use the term "skeletal-related events" as a clinical endpoint, which include pathologic fractures, spinal cord compression, need for surgery, or need for radiation therapy. SREs are associated with a significant decline in quality of life (QOL).[4] Once a single SRE occurs, the risk of subsequent SREs increases.[5] SRE also imposes a substantial economic burden, with a mean annualized SRE cost of $12,469.[6] Without interventions to prevent them, the rate of SREs is estimated to be up to 44% at 15 months in patients with castrate-resistant prostate cancer (CRPC).[7]

## BONE PHYSIOLOGY AND METABOLISM

The integrity of the skeletal system is a complex and dynamic coupled process that results in bone remodeling (Figure 53.1). Bone remodeling is a complex interaction of various regulators within a rich milieu of growth factors. This process is initiated by bone resorption from osteoclasts, which secrete acid and collagenase. Osteoclasts are large multinucleated cells that are induced from macrophages after exposure to certain cytokines. The bone resorption effects of osteoclasts are balanced by bone formation mediated by osteoblasts. Osteoblasts are single nucleated cells, which are the product of the differentiation and maturation of mesenchymal stem cells. Osteoblasts organize themselves into a functional group of connected cells that produce the extracellular bone matrix. The bone matrix includes an organic component of dense cross-linked collagen with other proteins. The organic component is then mineralized by deposition of a calcium-phosphate-hydroxide salt called hydroxyapatite. Modeling occurs in localized area of cortical and trabecular bone known as bone metabolic units.

Imbalances within the interactions between osteoclast and osteoblast could lead to a reduction in BMD. If BMD loss is substantial, it increases the risk of bone fractures in situations that would be unexpected in healthy individuals – so-called fragility fractures. Increased bone turnover can be observed with certain laboratory evaluations, such as bone-specific alkaline phosphatase for osteoblast activity and urinary N-telopeptide for osteoclasts.[2]

## ADT-ASSOCIATED OSTEOPOROSIS AND FRACTURES

Androgen deprivation therapy (ADT) is common in many stages of prostate cancer therapy.[9,10] Unfortunately, ADT is associated with a substantial decrease in BMD, which increases with the duration of therapy.[11] It is not uncommon for patients on ADT to have BMD losses of 3–5% per year, up to even 14% in some studies.[1,12–14] This represents a five- to sixfold increased rate compared to healthy men.[1,3,12–14] BMD loss occurs throughout the entire skeletal system in multiple sites.[13] BMD loss is most profound in the first year of initiating ADT but continues annually.[12]

Prostate Cancer. http://dx.doi.org/10.1016/B978-0-12-800077-9.00053-0

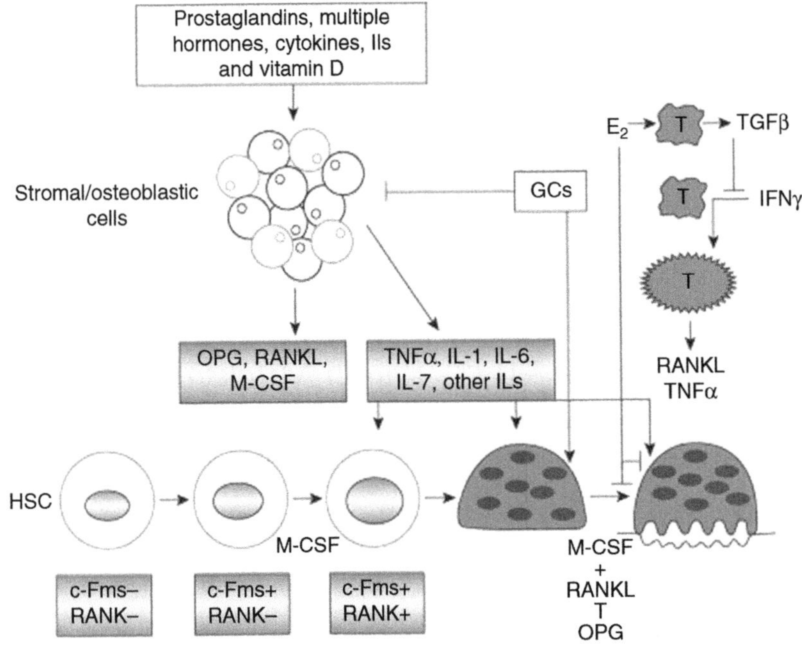


FIGURE 53.1 **Role of cytokines, hormones, and prostaglandins in the osteoclast formation and activation.** Hemotopoietic stem cells (HSC) express c-Fms (receptor for M-CSF) and RANK (receptor for RANKL) and differentiate to osteoclast. *Reprinted with permission from Orwoll et al.*[8]

Unfortunately, many men starting on ADT have baseline osteoporosis. The prevalence of osteoporosis in hormone-naive prostate cancer patients is estimated to be approximately 35%. Risk factors for an abnormal baseline BMD include age over 70 years, a low body mass index, and a higher PSA level.[15] The prevalence of osteoporosis increases with ADT duration to nearly 50% within 4 years of therapy and over 80% after 10 years of ADT.[3] Short-term ADT does not appear to increase fracture risk, which suggests that bone loss from hypogonadism may be reversible.[1]

Associated with the BMD loss is the significantly increased risk of fragility fractures on ADT. One of the first studies to clearly demonstrate that fracture risk was increased with ADT used Surveillance, Epidemiology, and End Results (SEER)-Medicare linked data to show that 19.4% of patients receiving ADT sustained a fracture, as compared to 12.6% not receiving ADT ($p < 0.001$).[16] Furthermore, individuals on ADT are more likely to experience a fracture requiring hospitalization (5.2% vs. 2.4% $p < 0.001$).[16] There is over a twofold increase in the risk of femoral neck and intertrochanter fractures after orchiectomy compared to healthy controls. This increased risk can be seen as early as 6 months after androgen deprivation, and remains elevated for at least 15 years while in a hypogonadal state.[17] Caucasian males are at a greater risk of fractures on ADT.[18]

In addition to the decreased quality of life and economic burden from fractures, approximately 40% of patients will be unable to walk independently 1 year after a hip fracture. After a hip fracture, between 60% and 80% of patients will be unable to perform independent activities of daily living.[19] Most importantly, ADT-related fractures are also associated with a reduced survival by over 3 years (121 vs. 160 months $p = 0.04$).[12]

The mechanism of bone loss from ADT is thought to be from decreases in serum testosterone, and therefore decreased aromatized estrogens, resulting in increased osteoclast activity.[12] In population studies of elderly men, low estradiol is more strongly associated with low BMD and greater fracture risk than even low testosterone.[1] Even with the addition of antiandrogens, bone loss will continue to occur.[12]

As cancer progresses to the castrate-resistant state, the current treatment paradigm is for these patients to remain on continuous ADT as they receive additional disease-modifying agents. As more novel and effective therapies emerge, patients with CRPC are living longer, which results in a lengthening of the duration of ADT and the associated risk of ADT-related fractures increases.

## DETECTION AND PREVENTION OF ADT-ASSOCIATED OSTEOPOROSIS AND FRACTURES

Early detection of osteoporosis may result in the potential avoidance of the morbidity of fractures by prompt therapeutic interventions and lifestyle changes. Bone mineral testing can be performed by several noninvasive modalities including ultrasound and quantitative

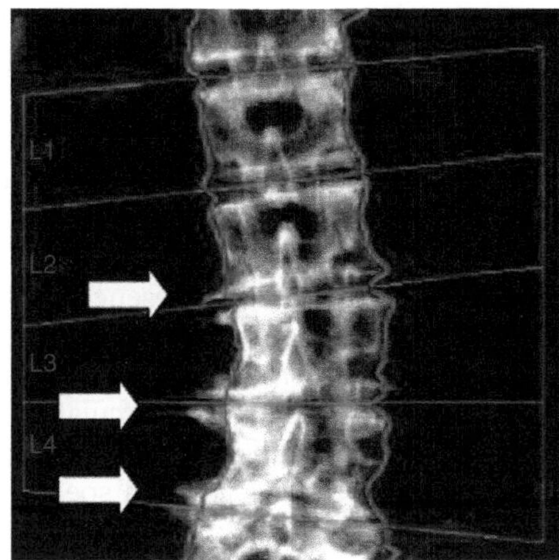

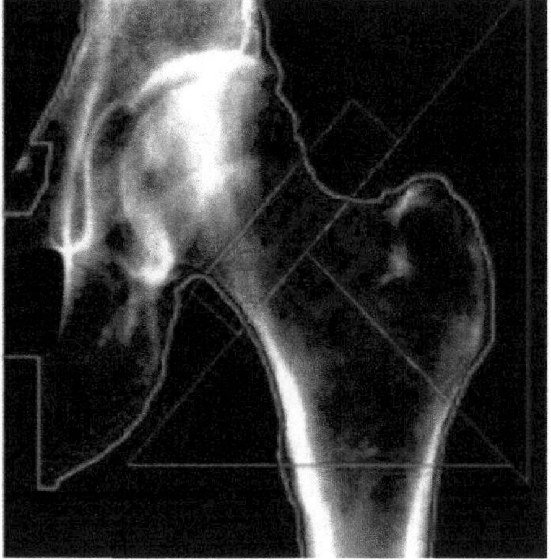

FIGURE 53.2 **Spinal degenerative changes elevating BMD as seen on DXA scan.** Spinal degenerative changes are noted (osteophytes and sclerosis seen at the arrows). This additional bone contributes to a "normal" BMD at the L1-4 spine, whereas the femur neck BMD yields a T-score of −2.5 and therefore osteoporosis. *Reprinted with permission from Orwoll et al.*[8]

computed tomography (CT); however, dual-energy X-ray absorptiometry (DEXA) is the favored gold standard as it provides an assessment of BMD at multiskeletal sites (spine, hip, proximal femur, and total body) with minimal radiation while still being inexpensive, accessible, precise, and validated in clinical trials[1] (Figure 53.2).

BMD can be expressed in absolute terms, such as grams per centimeter squared, or more commonly as a relative score.[1] The *T*-score is the difference in standard deviation that an individual's BMD is from a "young normal" adult of the same gender. A BMD within one standard deviation is considered a normal *T*-score. A *T*-score of −1.0 to −2.5 is considered osteopenia. Osteoporosis is defined as the occurrence of either an osteoporotic fracture or a *T*-score ≤−2.5. DEXA also provides what is known as a *Z* score, which is the difference in number of standard deviations between the individual and the mean value for a population of the same age and gender. *T*-score and *Z*-score have been well established in perimenopausal women, but their validity in men has been questioned by some.[1]

An additional assessment tool to predict fracture risk is the World Health Organization's FRAX (fracture risk algorithm) score, which considers age, body mass index, and other clinical risk factors to calculate a 10-year probability of hip or major osteoporosis-related fracture.

Patients on ADT at particularly high risk for the development of osteoporosis include those who have been on ADT for over 6 months, previous fractures, a family history of osteoporosis, low body weight, tobacco and alcohol abuse, corticosteroid use, medical comorbidities, and vitamin D deficiency.[12,20] Although low BMD is a risk factor for bone fractures, it is important to recognize that the majority of men on ADT who experience fractures have BMD that do not meet osteoporosis criteria.[21]

The National Comprehensive Cancer Network (NCCN) guidelines do recommend that all patients receiving ADT receive a baseline assessment of BMD. Routine follow-up BMD scans every 6–12 months while on ADT should also be considered, especially in high-risk patients.[12,20] Unfortunately, despite these recommendations, few patients on ADT actually receive DEXA screening.[22]

Several medications have been shown to increase BMD in nonmetastatic patients receiving ADT. These include the bisphosphonates pamidronate (60 mg every 12 weeks), alendronate (70 mg once per week), and the more potent zoledronic acid 5 mg once per year, in addition to the receptor activator of nuclear factor-kappa B ligand (RANKL) monoclonal antibody denosumab (60 mg every 6 months). However, to date, only denosumab has been shown to reduce the actual risk of fracture for patients on ADT.

The recommendations by the National Osteoporosis Foundation and NCCN are for consideration of pharmacologic therapy with a *T*-score between −1.5 and −2.0. The guidelines recommend strong consideration for a *T*-score <−2.0, FRAX fracture risk probability ≥3% for their hip, or FRAX fracture risk probability ≥20% at other sites.[1]

## BIOLOGY OF BONE METASTASIS IN PROSTATE CANCER

Bone is the most common site of metastasis in prostate cancer, with osseous metastasis occurring in nearly 80% of patients with metastatic prostate cancer. The most common site of osseous metastasis corresponds with the amount of bone marrow present, such as the vertebral column, pelvis, ribs, and long bones. In contrast to other solid malignancies, bone metastases from prostate cancer usually appear as osteoblastic lesions. Despite the radiographic appearance of osteoblastic lesions, the complex interaction between the metastasis with bone stroma result in abnormal bone remodeling leading to biomechanical instability and pathological fractures. Expansion of tumor within bone and progressive destruction of bony matrix may lead to pain, compression, or pathologic fractures.

Replacement of bone marrow with metastasis also impairs hematologic production, which is further complicated by myelosuppression from long-term ADT, radiation, and cytotoxic therapies. Hypercalcemia is uncommon from prostate metastasis, if present; it is more suggestive of parathyroid hormone-related protein expression from differentiation.[23,24]

The exact mechanism for the development of bony metastasis has not been conclusively defined but there exists robust support for the "seed" and "soil" hypothesis, originated by Stephen Paget in 1889.[25] This hypothesis postulates that metastasis development results from a complex series of interactions between metastatic cells (the "seed") and the bone microenvironment (the "soil"). Primary tumor sites are a heterogeneous group of cells with various mutations. For metastasis to occur a tumor cell must be able to separate itself from the extracellular matrix. Detachment of cells from the primary tumor is believed to be through an epithelial–mesenchymal transition with alteration in integrin-mediated attachments. The detached cancer cells must then invade through the tissue stroma and into the vasculature. Circulating tumor cells must then survive immune surveillance and attach to cortical or medullary bone spaces via specific receptors, including the integrin family. The establishment of metastatic sites results from the reciprocal interactions between cancer cells and bone microenvironment. Central to this complex interaction is the rich milieu of growth factors within the bone microenvironment. Many of these interactions are mediated by androgen responsive elements and other bone-derived proteins (Figure 53.3).

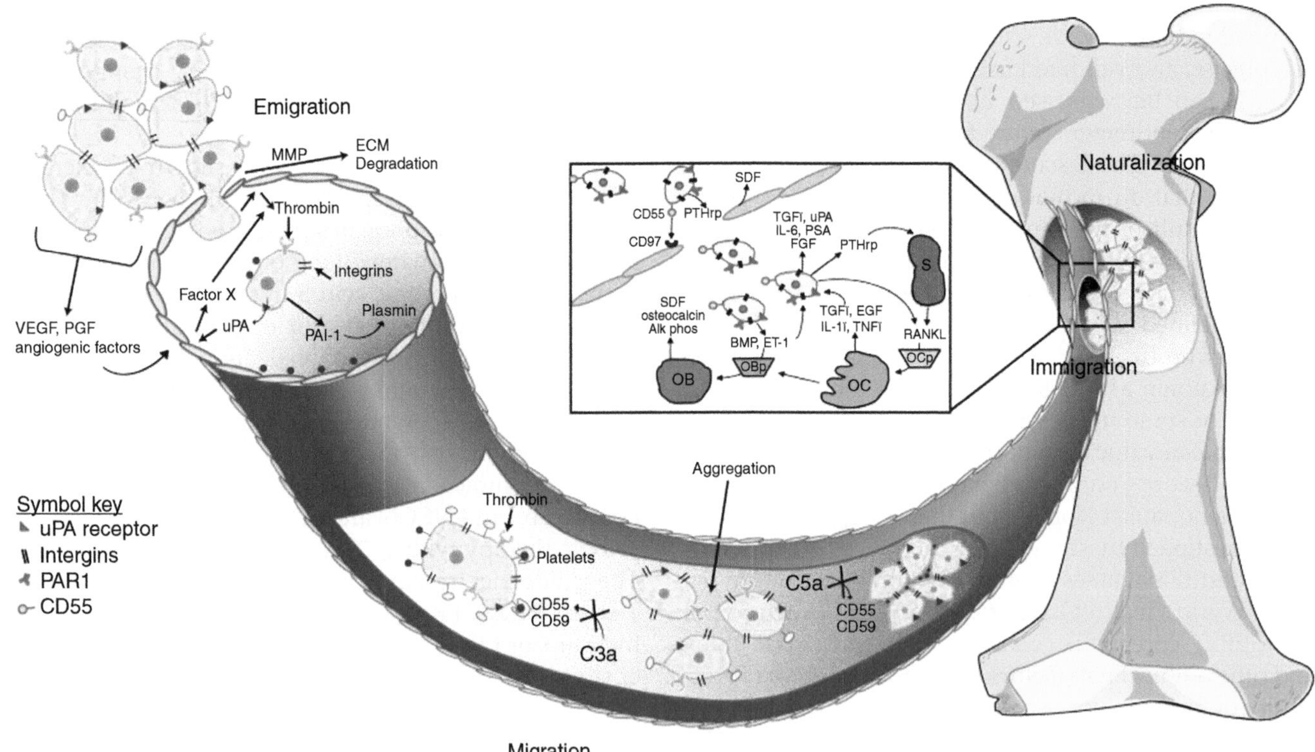


**FIGURE 53.3** **Mechanisms of cancer metastasis.** A number of angiogenic and a growth factors (VEGF, PDGF) are released by the primary localized tumor promoting neoangiogenesis. Matrix metalloproteinases (MMP) are secreted breaking down the extracellular matrix (ECM), thus allowing tumor cells to enter the circulation. Several factors are released in response to tumor cell intravasation, including urokinase plasminogen activator(uPA), plasinogen activator inhibitor-1 (PAI-1), and thrombin which promote tumor cell survival and metastasis. The aggregation of tumor cells and platelets during transit (migration) promote survival and ultimately extravasation at the secondary tumor site. The mechanisms of tumor cell extravasation (immigration) involve docking to the endothelium with subsequent transit into the bone microenvironment. Successful proliferation (naturalization) requires a coordinated symptomatic relationship between the invading tumor cells and all of the host cells of the bone to create a tumor friendly environment. OBp, osteoblastic progenitor cells; OCp, osteoclastic progenitor cells; OB, osteoblast; OC, osteoclast; S, stromal cell; TFAg, Thomas Friedrich antigen; gal3, galactin 3 receptor; PTHrp, parathyroid hormone related protein; SDF, stromal derived factor; CD55, decay accelerating factor; CD59, Protectin. *Reprinted with permission from Heymann.*[26]

## DEVELOPMENT OF BONE METASTASIS

The time to the development of bony metastasis is extremely variable for castrate-sensitive and castrate-resistant prostate cancer. This is because they are both very heterogeneous groups of patients. In CRPC, the median time from a rising PSA despite ADT to the development of a bony metastasis is approximately 2 years, but ranges as widely as 9 months to 30 months.[27–29] It is estimated that without the addition of bone-targeted therapy, the rate of SREs for patients with CRPC would be about 44% at 15 months.[7]

Pretreatment PSA, PSA nadir on ADT, and PSA velocity has been associated with the time to development of bony metastasis.[27–29] After development of metastasis despite radical localized therapy, the median time till death is approximately 5 years.[30] The extent of bony metastasis does have prognostic significance, and elevated serum alkaline phosphatase levels are independently associated with shorter survival.[31,32]

## IMAGING FOR BONE METASTASIS

The most commonly used imaging modality for the detection of bony metastases is a radionucleotide bone scintigraphy using Technetium 99m-diphosphonate, commonly referred to as a bone scan. While it is estimated that nearly 30–50% of bone marrow needs to be replaced by tumor cells to be detected on plain film X-ray, only 10% of the marrow needs to be replaced for the area to appear abnormal on bone scan.[1] Bone scintigraphy is limited in early stage lesion as the osteoblastic response may not be substantial enough for detection on bone scan. Furthermore, healing fractures and degenerative changes can be difficult to differentiate from metastasis. MRI can also be used and has a high sensitivity (82–100%) and specificity (73–100%) in detecting tumor within marrow, even if an osteoblastic response is not yet present.

Bone scintigraphy scans are an established and important component of most clinical trials in advanced prostate cancer.[33] As the disease progresses after ADT initiation, many patients may be relatively asymptomatic despite the presence of occult bony metastasis.[28] The prevalence of a positive bone scan correlates with serum PSA. Depending on the patient population characteristics the rate of bone scan positivity is estimated to be <1% if serum PSA ≤10, approximately 10% if PSA is between 10 and 50, and over 50% for a PSA ≥50. These rates vary depending on the study design and population; for example, others have estimated the rate of a positive scan for a patient with CRPC and PSA ≤10 to be between 12% and 26%.[28]

PSA kinetics also correlates with bone scintigraphy scan positivity as a shorter PSA doubling time (PSADT) is also associated with an increased risk of bony metastasis. In nonmetastatic CRPC, a PSADT of ≤10, ≤6, and ≤4 months is associated with a median time to first bone metastasis of 26, 22.1, and 18.5 months, respectively.[34]

Guideline recommendations for obtaining bone scans from the NCCN are for patients who are either symptomatic, have clinical T3/T4 disease, clinical T2 with PSA >10, clinical T1 with PSA >20, or any Gleason Score ≥8 as long as life expectancy is over 5 years. The AUA guidelines recommend a bone scan for patients with a PSA >10 or have poorly differentiated/high-grade tumors. The EUA guidelines suggest that a bone scan is usually not indicated for asymptomatic patients with a PSA <20 ng/mL "in the presence of well-differentiated or moderately differentiated tumors."

Despite these recommendations approximately 45% of low-risk patients receive unnecessary imaging (39% bone scans), while only 66% of high-risk patients receive appropriate imaging.[35] The overutilization of unnecessary routine bone scans in low-risk prostate cancer patients has been identified by the AUA and ASCO in their "choosing wisely campaign" as a practice that needs to be stopped. A Swedish effort to disseminate utilization data and guideline recommendations to urologists led to a decrease in inappropriate imaging of low risk (from 45% to 3% $p < 0.001$), but also had the unintended consequence of decreased appropriate imaging of high risk (from 63% to 47% $p < 0.001$) over a 10-year time period.[36] Although it is unclear if a national level initiate in the United States would be successful, regional collaboration that provides feedback on imaging utilization and guideline recommendation have been shown to reduce inappropriate imaging and variation among a group of urologists in Michigan, Ohio, and Indiana.[37]

For patients who receive localized treatment, the NCCN guidelines suggest imaging should be based on individual risk, age, PSADT, Gleason score, and overall health, but they acknowledge that bone scans are rarely positive in asymptomatic men with PSA <10 ng/mL (Figure 53.4).

## NOVEL BIOMARKERS FOR BONE METASTASIS

### Positron Emission Tomography/Computed Tomography (PET/CT)

Positron emission tomography (PET) combined with CT (PET/CT) has shown promising results in the detection of bony metastasis. $^{18}$F-Fluorodeoxyglucose (FDG) is a glucose analog that is taken up by metabolically active tumor cells, and allows PET/CT to detect active disease within the marrow before a secondary bone reaction occurs. $^{18}$F-Fluoride is another PET tracer, which

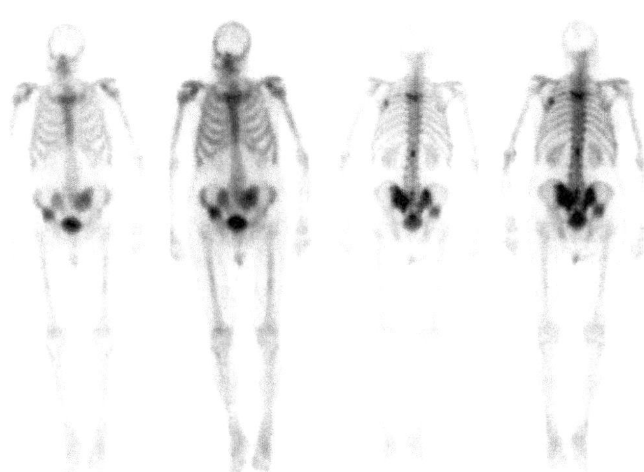

FIGURE 53.4 **A staging bone scintigraphy on a 77-year-old male patient with prostate cancer.** Metastasis are seen in the left 6th rib posteriorly, the right 5th and 6th rib laterally, T6, spinous process of L2, sacrum and both ilia adjacent to sacroiliac joints in the right superior acetabulum. *Reprinted with permission from Heymann.*[26]

is taken up by bone surface cells with the most active turnover. The combination of $^{18}$F-fluoride with FDG is under investigation, along with several other PET tracers.[38] The sensitivity for PET/CT ranges from 62% to 100% with a specificity from 96% to 100%.[39] However, data on PET/CT remains limited in metastatic prostate cancer. At this time, bone scintigraphy is still considered the preferred first-line imaging modality for prostate cancer.[39] Currently, $^{18}$F-FDG-PET/CT is usually reserved for equivocal cases after bone scan to help differentiate active metastases from healing bones (EUA).

## Bone Biomarkers

Bone matrix is comprised of nearly 90% Type 1 collagen. Serum concentration of the precursor molecule, procollagen, is elevated in the setting of early bone formation and osteoblast activity. Similarly, the degradation products of Type 1 collagen can be used as markers of bone resorption. Elevated levels of these and other bone biomarkers have been shown to correlate with tumor burden within bone, which may provide some prognostic and predictive value.[40] For example, a reduction in bone biomarkers of osteoclast activity, such as urinary N-terminus procollagen, may predict the impact of denosumab in reducing subsequent SREs,[41] although other results have been mixed and prospective validation is needed.[40]

Alkaline phosphatase is another biomarker of bone metabolism with a role in the diagnosis and monitoring of patients with bony metastasis and has been used extensively for some time. Alkaline phosphatase is found in osteoblasts, but can be a rather nonspecific serum marker of metastasis as it is also found in liver and the intestine. The bone isoform accounts for 40–50% and is more specific than total alkaline phosphatase but can still have 15–20% cross-reactivity with liver alkaline phosphatase.[40] Alkaline phosphatase is recommended by the NCCN and EUA for the detection of bony metastasis.[39]

## Bone Scan Index

Although bone scintigraphy scans can demonstrate the presence of osseous metastases, bone scans are limited in quantifying disease burden and measuring the degree of treatment response. Bone Scan Index (BSI) is an attempt to improve upon this. BSI measures the tumor burden in bone as a percent of the total skeletal mass based on a reference male skeleton.[42] BSI can be automated to reduce time and interobserver differences.[43] In a preliminary analysis, BSI was associated with overall survival in patients receiving cytotoxic chemotherapy. These findings will need to be validated in the expanding array of systemic therapies for metastatic prostate cancer.

# PREVENTION OF BONY METASTASIS

Historically, patients dying from prostate cancer usually presented with bony metastasis and died of their disease after rapidly progressing to CRPC. This presentation is far less common in the PSA screening era. More so, PSA monitoring has led to the creation of "the clinical state of a rising PSA" despite radical localized therapy. Although some argue this is an artificial disease context, these patients are frequently started on ADT and might progress to CRPC without any evidence of metastatic disease. Patients with nonmetastatic CRPC are a very heterogeneous group with a variable length of time till they develop metastatic disease; however, when metastasis does develop, they will most likely occur in the axial skeleton.

Currently, no bone-targeting agent is approved for the prevention of osseous metastasis. Several negative trials have been conducted using bisphosphonates and endothelin antagonists. Only the RANKL inhibitor denosumab in the Denosumab 147 metastasis prevention trial demonstrated a 4.2 month improvement in bone metastasis-free survival in nonmetastatic CRPC patients with either a PSA ≥8 or PSADT ≤10 months. However, no benefit from denosumab was seen in OS in this setting and there are several criticisms in the trial design. To date, denosumab is not approved for the prevention of bony metastasis.[44]

Two on-going bone metastasis prevention trials with zoledronic acid include the RADAR (NCT00193856) and STAMPEDE (NCT00268476). RADAR randomized 1071 "high-risk" patients (clinical stage T2b–4 or clinical stage T2a with Gleason ≥7 and PSA ≥10) without metastasis to definitive radiation with ADT (short term or long term), with or without zoledronic acid. The STAMPEDE trial is a complex seven-arm phase II/III trial, which plans to enroll 4000 men with high-risk localized, metastatic, or relapsed prostate cancer starting on ADT. STAMPEDE examines several combinations of ADT, zoledronic acid, docetaxel, abiraterone, and celecoxib. The primary endpoint for STAMPEDE is OS.[45]

## LOCALIZED THERAPY WITH EBRT

External beam radiation therapy (EBRT) is the treatment for patients with bony metastasis resulting in localized pain at one or just a few sites. Therapy is only palliative, but can provide pain relief for 60–85% of patients, with 15–58% experiencing complete pain response. EBRT can also provide some preservation of function and maintenance of skeletal integrity.[46]

A single fraction of radiation using a dose of 8 Gy has been shown in several randomized trials to be as effective as multiple fractionated doses in providing relief of bone pain. It has been compared to 30 Gy over 10 fractions, 24 Gy over six fractions, and 20 Gy over five fractions. A single 8 Gy fraction is more convenient and cost-effective, but associated with a slightly higher retreat rate at 20% compared to 8% after multiple fractionated doses. If pathologic fractures are present in weight-bearing bones, surgical fixation should be considered prior to EBRT. Prophylactic surgical fixation should also be considered if the risk of future pathologic fractures at that site is elevated.[46]

## MANAGEMENT OF BONE PAIN

Bone pain induced from cancer is multifactorial involving a complex interaction of peripheral, central, and supraspinal mechanisms.[47] Mild to moderate bone pain can initially be controlled with nonsteroidal anti-inflammatory drugs (NSAIDs). Only if the pain persists or worsens should the patient be advanced up the "analgesic ladder" to narcotic medications.[48] In addition to analgesic medications, treatment should be considered with disease-modifying therapies, such as EBRT for localized pain and radiopharmaceuticals for systemic bone pain. Cancer pain management is improved with the involvement of pain medicine physicians or palliative care services (Table 53.1).[49]

## TREATMENT AND MANAGEMENT OF ACUTE SPINAL CORD COMPRESSION AND PATHOLOGIC FRACTURES

Epidural metastasis can be the most devastating complication of metastatic prostate cancer. Early diagnosis and treatment is critically important to preserving baseline function. Back pain will occur in nearly 100% of patients with epidural metastasis. Other common symptoms include weakness below the level of the lesion, alternations in autonomic and sensory function, as well as bowel and bladder dysfunction. Symptomatic spinal cord metastases occur in 5–10% of prostate cancer patients, although asymptomatic lesions are likely more common.[51]

A thorough neurologic and general examination should be performed in all patients suspected of spinal cord metastasis. An urgent MRI is the gold standard for diagnosing an epidural metastasis but only focuses at the area of interest therefore additional asymptomatic lesions may be missed. A spinal X-ray can be obtained to identify asymptomatic lesions elsewhere in the spinal column.

Treatment includes high dose dexamethasone, radiation therapy, and possible need for surgical decompression/stabilization. There is no known optimal dosing of dexamethasone, but IV loading doses of up to 10 mg followed with 4–10 mg every 6 h have been reported. The use of high dose steroids to treat acute epidural metastasis results in higher rate of ambulation on long-term follow-up.[52] Unlike the single dose of 8 Gy typically used for painful bone metastasis, in the setting of spinal cord compression, EBRT is usually provided in multifractions over a longer schedule, such as 30 Gy over 10 fractions. These regimes are well tolerated with pain relief reported in approximately 90%.[53] Surgical fixation through kyphoplasty/vertebroplasty, then followed by radiation appears to be superior to radiation alone.[54] Surgery should be considered for any patient with progressive signs and symptoms during or after radiation therapy, or if they present with unstable pathologic fractures. Spinal cord compression with need for surgery represents the single most expensive SRE, averaging approximately $82,868.[55]

## PREVENTION OF SKELETAL-RELATED EVENTS IN PROSTATE CANCER

### Conservative Measures

#### *Education*

Many patients on ADT are not unaware of their risk of osteoporosis, nor are they aware of many of the recommended activities necessary to improve bone health.[22] With appropriate education patients may be more likely to actively engage in optimizing their own bone health.

#### *Lifestyle Modifications*

Tobacco and alcohol are associated with reduced BMD; therefore, it is recommended that patients limit

TABLE 53.1 Common Pain Syndromes in Metastatic Castration-Resistant Prostate Cancer

| Pain syndrome | Initial management | Other therapeutic alternatives |
|---|---|---|
| Localized bone pain | Pharmacologic pain management<br>Localized radiotherapy (special attention to weight-bearing areas, lytic metastasis, and extremities) | Surgical stabilization of pathologic fractures or extensive bone erosions<br>Epidural metastasis and cord compression should be evaluated in patients with focal back pain.<br>Radiopharmaceuticals should be considered if local radiation therapy fails. |
| Diffuse bone pain | Pharmacologic pain management "multispot" or widefield radiotherapy<br>Radiopharmaceuticals | Corticosteroids<br>Bisphosphonates<br>Calcitonin<br>Chemotherapy |
| Epidural metastasis and cord compression | High-dose corticosteroids<br>Radiation therapy<br>Surgical decompression and stabilization should be indicated in high-grade epidural blocks, extensive bone involvement, or recurrence after irradiation. | Pharmacologic pain management |
| Plexopathies caused by direct tumor extension or prior therapy (rare) | Pharmacologic pain management<br>Radiation therapy (if not previously employed)<br>Neurolytic procedures (nerve blocks) | Tricyclic antidepressants (amitriptyline)<br>Anticonvulsants |
| Miscellaneous neurogenic causes: postherpetic neuralgia, peripheral neuropathies | Careful neurologic evaluation<br>Pharmacologic pain management<br>Discontinuation of neurotoxic drugs: paclitaxel, docetaxel, *Vinca* alkaloids, platinum compounds | Tricyclic antidepressants (amitriptyline)<br>Anticonvulsants |
| Other uncommon pain syndromes: extensive skull metastasis with cranial nerve involvement, extensive painful liver metastasis or pelvic masses | Radiation therapy<br>Pharmacologic pain management<br>Corticosteroids (cranial nerve involvement) | Chemotherapy<br>Intrathecal chemotherapy may ameliorate symptoms of meningeal involvement; regional infusions may be considered. |

*Reprinted with permission from McDougal et al.*[50]

their intake of alcoholic beverages and cease smoking.[14] Furthermore, in general, 30–40 min of weight-bearing activities are recommended at least three to four times a week.[1,14] Routine exercise is safe for patients on ADT. Resistance-exercise training through supervised exercise programs appear to provide better outcomes with improved adherence than an exercise plan away from the clinic.[56,57] However, a recent systematic review of exercise interventions revealed only a single study looking at the effects of exercise on bone health. After 20 weeks of resistance training, a small cohort of 10 men had no change in their BMD, which the authors suggested demonstrated an attenuation of ADT-induced bone loss.[58,59] The optimal duration of supported exercise interventions is unknown at this time.

### Calcium and Vitamin D

It is generally recommended that all patients on ADT consume 1000–1500 mg of calcium and 400–500 IU of Vitamin D per day. Dietary calcium intake is directly correlated with BMD with a low intake as an independent risk factor for osteoporosis in men with prostate cancer.[60] Ideally calcium and vitamin D would come from dietary sources; however, intake of these nutrients in prostate cancer patients is frequently inadequate so supplements may be needed.[14,60] Although supplemental calcium and vitamin D have never been shown to prevent bone loss in patients on ADT, calcium is a large constituent of bone mass. Additionally, vitamin D increases intestinal absorption of calcium.[12] There is also some very early animal data that suggest that a deficiency of vitamin D can promote bone metastasis growth.[61] Vitamin D analogs may also have a future role in the treatment of bony metastasis.[62]

### Intermittent ADT

A complete discussion of intermittent ADT versus continuous ADT is beyond the scope of this chapter, but is well covered in other chapters. It has been shown that during a median ADT off period of 8 months (range 3–12.9 months) a stabilization of BMD can occur. Two recent landmark randomized trials comparing intermittent to continuous ADT (one in patient with metastatic disease and the other in nonmetastatic rising PSA after radiation therapy only disease) did not present data on SREs.[20,63,64] Based on these trials, intermittent ADT is not inferior in the nonmetastatic rising PSA state and could be considered to limit the adverse effects of ADT including BMD loss. However, intermittent ADT may reduce overall survival in patients with metastatic disease and should be implemented only cautiously.

## BONE-TARGETED AGENTS

### Bisphosphates

Bisphosphates are pyrophosphate analogs that bind to bone surfaces at active remodeling sites and are internalized by osteoclasts leading to apoptosis of the osteoclast, thus blocking osteolysis. Bisphosphates were first approved for treatment of reduced bone mineral density in postmenopausal women.

Zoledronic acid is a very potent bisphosphate. The nitrogen-containing component of the molecule makes it approximately 850× as potent as older generations of bisphosphates.[13] Zoledronic acid also has a much shorter infusion time of 15 min compared to older generations of bisphosphonates, which tend to require 2–4 h of transfusion.

A study of 106 men with nonmetastatic castrate-sensitive disease randomized to either 4 mg of zoledronic acid or placebo every 4 months for 1 year found an improved lumbar spine BMD of 5.6% for zoledronic acid compared to a 2.2% loss in those given placebo ($p < 0.001$). Improvements in BMD from zoledronic acid and losses in BMD from placebo were also seen at other skeletal sites.[65] Unfortunately, if bisphosphates are discontinued but the patient remains on ADT, BMD will continue to be lost.[66] Zoledronic acid is approved to reduce the risk of bone loss in nonmetastatic patients on ADT, at a dose of 5 mg IV annually. However, no bisphosphate, including zoledronic acid, has ever been proven to reduce the risk of actual fractures.

The ability of zoledronic acid to reduce the rate of SRE in metastatic CRPC was assessed in a phase III RCT versus placebo trial. Three arms were initially compared: (1) 4 mg zoledronic acid, (2) 8 mg zoledronic acid, and (3) placebo every 3 weeks for 15 months. However, the 221 patients assigned to 8 mg had to have their dose reduced to 4 mg due to renal toxicity early in the trial. Compared to placebo, 4 mg of zoledronic acid was found to significantly reduce the rate of SRE (38% vs. 49% $p = 0.028$) and delayed median time to first SRE by nearly 6 months (16 vs. 10.5 months $p = 0.009$);[5,67] Based on the strength of this trial, zoledronic acid was approved in 2002 to reduce the risk of SREs in progressive metastatic CRPC. The approved dose for this indication is 4 mg intravenous every 3–4 weeks.[66] In the multiarm TRAPEZE trial in metastatic CRPC, the addition of zoledronic acid to docetaxel improved the SRE-free interval from 13.1 months to 18.1 months ($p = 0.008$), with most of the difference occurring after progression on chemotherapy, leading the investigators to suggest a role for zoledronic acid as postchemotherapy maintenance therapy.[68] Zoledronic acid had no impact on prostate-cancer-specific survival or overall survival in their randomized trials.

A phase III RCT, CALGB 90202, evaluated the role of zoledronic acid in SRE reduction in patients with metastatic disease who were still in the castrate-sensitive state. CALGB 90202 aimed to randomize 680 metastatic castrate-sensitive patients to zoledronic acid or placebo with open-label switch to zoledronic acid upon development of CRPC. Unfortunately, the trial was closed prematurely when the "corporate supporter withdrew study drug supply" after only 299 of the 470 intended SRE had occurred. No statistical difference in time to first SRE, PFS, or OS was seen between zoledronic acid and placebo in this study. However, a subgroup analysis suggested that those presenting with a SRE while still in the castrate-sensitive state might benefit from zoledronic acid before progression to CRPC.[69]

The impact of zoledronic acid in the prevention of bone metastasis was assessed in two clinical trials. The first trial had a lower than expected event rate leading to early trial cessation and the results of that trial were never published in their entirety, but did inform the development of a subsequent trial, the Zoledronic acid European Study (ZEUS).[66] The ZEUS trial randomly assigned 1433 patients with high-risk nonmetastatic prostate cancer (PSA ≥20, Gleason score ≥8, or positive lymph nodes) to either zoledronic acid every 3 months for 4 years, or to observation. After median follow-up of 4.8 years, there was no difference in the primary end point of bone metastases incidence (48 months incidence of 14.7% vs. 13.2% $p = 0.65$).[66,70]

Although zoledronic acid is the most established, several other bisphosphates have been studied in prostate cancer. Pamidronate, a less potent, second-generation bisphosphate, was also found to prevent bone loss compared to placebo for nonmetastatic castrate-sensitive patients on ADT.[71] However, pamidronate was no better than placebo at decreasing SREs in a 2 year phase III RCT of 422 patients with metastatic CRPC.[72] Interestingly, a first-generation bisphosphonate, Clodronate, was initially reported as being no better than placebo in improving bone progression-free survival; however, upon longer-term follow-up, clodronate was associated with a 23% reduction in mortality for patients with metastatic castrate sensitive disease.[73] Despite these findings, clodronate is not approved for this indication in the United States.

Side effects of bisphosphates include mild flu-like symptoms (fatigue, arthralgia/myalgias, fever), anemia, nausea, abdominal pain, and esophagitis. Concomitant vitamin D and calcium is recommended to avoid the risk of hypocalcemia while on bisphosphates, particularly in vitamin D deficient individuals. A well-documented side effect of bisphosphates is osteonecrosis of the jaw, especially in those undergoing dental work. A dental evaluation should be performed prior to initiating therapy, to ensure that dental work will not be needed while on bisphosphates. Patients experiencing osteonecrosis of the jaw should be immediately referred to an oral surgeon. Serum creatinine should be monitored while patients are on zoledronic acid, as adjustments in dosing may need to be made.[74]

## RECEPTOR ACTIVATOR OF NUCLEAR FACTOR-KAPPA B (RANK) SIGNALING PATHWAY INHIBITORS

The receptor activator of nuclear factor-kappa B (RANK) pathway is involved in the activation, differentiation, proliferation, and apoptosis of osteoclasts. RANK is expressed on the cell surface of mature and immature osteoclasts, while RANKL is expressed on bone stroma and osteoblasts. The interaction between RANK and RANKL leads to osteoclastogenesis and subsequent increased bone turnover. RANK signaling pathway is regulated in part by osteoprotegerin (OPG), which functions as a decoy receptor for RANKL and directly competes with RANK.

Denosumab (Xgeva) is a fully humanized monoclonal antibody that binds RANKL mimicking the effects of endogenous osteoprotegerin leading to the suppression of osteoclast activity. It has several benefits over zoledronic acid including subcutaneous administration (as oppose to intravenous), and it does not require adjustment or monitoring in renal insufficiency.

The Hormone Ablation Bone Loss Trial (HALT 138) prostate cancer trial, a large phase III trial randomizing 1468 patients with nonmetastatic castrate-sensitive disease to 60 mg of denosumab every 6 months versus placebo, found a significant increase in BMD at 24 and 36 months, and even more importantly a decreased incidence in new vertebral fractures compared to placebo (1.5% vs. 3.9% $p = 0.006$). However, there was no statistical difference in the time to first fracture or the incidence of any fracture. Preplanned subgroup analysis revealed denosumab was most beneficial for patients at high risk for fractures (70 years or older, a low baseline BMD, or prior fracture).[75] Off the strength of this trial, denosumab is approved at a dose of 60 mg subcutaneous every 6 months to increase BMD in patients with nonmetastatic prostate cancer on ADT at high risk for fracture. Only denosumab, and not zoledronic acid, has been shown to reduce risk of actual bone fractures.

A landmark trial comparing denosumab to zoledronic acid for the prevention of SRE in patients with metastatic CRPC was published in 2011.[76] In this study, 1904 bisphosphate-naïve patients were randomized to either 120 mg of subcutaneous denosumab with intravenous placebo every 4 weeks or to 4 mg of intravenous zoledronic acid with subcutaneous placebo. Denosumab showed improved time to first skeletal events (20.7 m vs. 17.1 m $p = 0.008$, HR = 0.82) over zoledronic acid. No difference was seen in OS or progression-free survival in either arm of this trial. Given the delayed time to SRE compared to standard therapy in this trial and successful trials in metastatic breast and other solid malignancies, denosumab was approved at a dose of 120 mg subcutaneously every 4 weeks for the prevention of SRE from bone metastasis in prostate cancer.[77] Although for prostate cancer this benefit has only been demonstrated in the castrate-resistant state, this is not specifically stated in its FDA approval; however, data in castrate-sensitive disease are lacking.[66]

The Denosumab 147 metastasis prevention study randomized 1432 patients with nonmetastatic CRPC at high risk for bone metastasis (defined as PSA ≥8 or PSADT ≤10 months) to either denosumab 120 mg every 4 weeks or placebo. The primary endpoint was bone metastasis-free survival, which was improved by 4.2 months in favor of denosumab (29.5 vs. 25.2 months $p = 0.028$). This composite endpoint included symptomatic and asymptomatic lesions found on bone scan scheduled every 4 months and yearly skeletal surveys. The benefit of denosumab appeared to be greatest for those with shorter PSADT.[34] Although the time to symptomatic bone metastasis was also delayed, no difference was seen in overall survival. It is not clear what effect the early institution of denosumab will have on the longer-term risk for SRE since the study design required patients to discontinue the investigational agent upon development of bone metastasis, yet denosumab has now become a standard therapy for SRE prevention in patients with bone metastasis. To date, denosumab is not approved for the prevention of bony metastasis.[44]

Side effects of denosumab are similar to zoledronic acid and include fatigue, nausea, hypophosphatemia, GI distress, and osteonecrosis of the jaw. The rate of hypocalcemia is higher with denosumab than zoledronic acid (13% vs. 6%), so vitamin D and supplemental calcium are strongly encouraged, in addition to monitoring serum calcium levels.[74] The risk of osteonecrosis does appear to be cumulative as the incidence increases with time of therapy, 1% at 1 year, 3% at 2 years, and 4% at 3 years.[44]

Additional inhibitors of the RANK pathway other than denosumab have been under investigation, such as recombinant osteoprotegerin. However, to date, only denosumab has clinical evidence supporting its use and is available (Figure 53.5).[78]

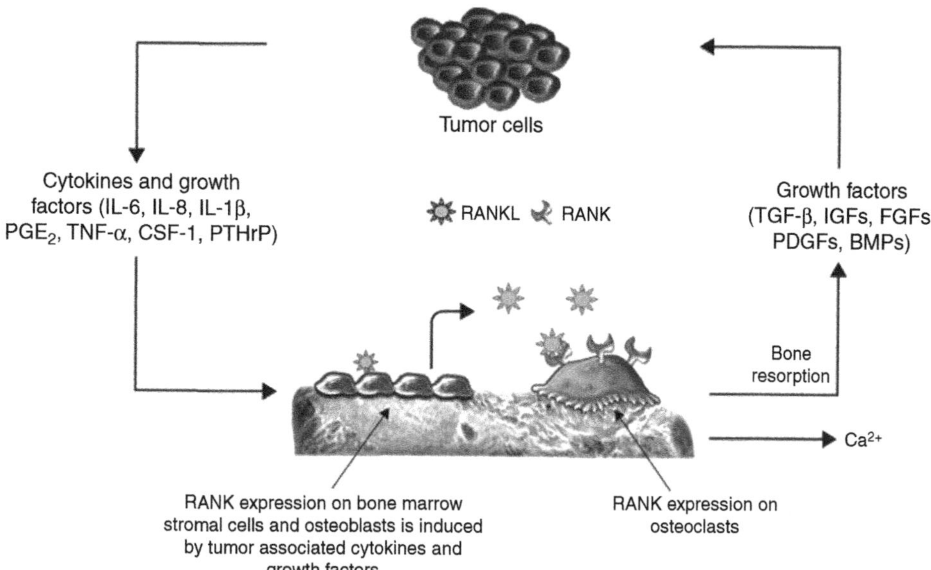


FIGURE 53.5 **RANK ligand is an essential mediator in the "vicious cycle" hypothesis of bone destruction and metastatic cancer.** RANKL is necessary for osteoclastogenesis and is generally increased as a result of tumor/bone interactions. The release of growth factors and calcium through the resorptive action of osteoclasts may then allow the establishment and progression of the tumor in the bone. *Reprinted with permission from Heymann.*[26]

## RADIOPHARMACEUTICALS (SYSTEMIC "BONE-SEEKING" AGENTS)

EBRT is very successful for the treatment of focal bone pain from individual metastasis. However, for patients with diffuse bone pain from widespread metastatic disease, EBRT is not practical and would be too toxic. This unmet need resulted in the development of radiopharmaceuticals that function as calcium mimetics that localize selectively to the bone.

Until recently, only two radiopharmaceuticals were approved in metastatic prostate cancer: strontium-89 and samarium-153. Both agents emit low-energy beta-particles that travel several millimeters and result in single strand breaks. They are primarily used for pain palliation in those with symptomatic diffuse bone metastasis.[79] Strontium and samarium are limited by their myelotoxicity. Strontium is further limited by a widely variable half-life, which depends on the extent of bony disease that could impact the duration and degree of myelotoxicity.[80] Samarium-153 has less severe and more transient hematologic toxicity due to its shorter half-life, and is usually selected over strontium-89.[74]

Strontium and samarium have only palliative effects and do not provide a survival benefit. However, clinical trials have looked at the impact of combining radiopharmaceuticals with chemotherapy in chemosensitive CRPC patients and have shown some benefit. One phase II trial looking at doxorubicin with strontium-89 in those responding to induction chemotherapy found the median survival to be 27.7 versus 16.8 months favoring the group receiving doxorubicin with strontium instead of doxorubicin alone ($p = 0.001$) (Tu 2001 Bone-targeted therapy for advanced).[81] A subsequent phase II trial on docetaxel with samarium-153 had a median overall survival of 29 months and significant improvement in pain symptoms.[82] Within the factorial TRAPEZE trial, the addition of strontium-89 to six cycles of docetaxel, improved clinical progression-free survival (HR = 0.845; 95% CI 0.72–0.99, $p = 0.036$) but had no impact on overall survival.[68]

Radium-223 (Xofigo) is another radiopharmaceutical that functions as a calcium mimetic, but unlike the beta-emitting strontium and samarium, radium is a larger alpha particle that has a shorter effect range (<100 μm) resulting in less bone marrow toxicity. Radium also results in double strand breaks. The landmark ALSYMPCA trial took 922 patients with symptomatic CRPC without visceral metastasis and at least two bone metastases, and randomized them 2:1 to either 50 kBq/kg of radium or placebo every 4 weeks for six total doses. Patients could be enrolled regardless of whether they received prior chemotherapy (58% had received prior chemotherapy). Unlike antiresorptive agents and other radioisotopes, radium-223 was found to have an antitumor effect. There was a significantly improved overall survival advantage of 3.6 months (14.9 vs. 11.3 months $P < 0.001$) with radium. This was even larger for chemotherapy-naïve patient who had a 4.1 month survival advantage. The median time to first symptomatic SRE was 15.6 versus 9.8 month, again favoring radium (<0.001). Preplanned subgroup analysis suggested that patients with an elevated bone alkaline phosphatase and those also receiving zoledronic acid may benefit even more from radium. Nausea, vomiting, and diarrhea are the most common side effects. Myelosuppression was only slightly higher than placebo, as grades 3 and 4 neutopenia occurred in 2.2% versus 0.7% for placebo. Based off the strength of this study, radium-223 was approved by the FDA in 2013 for patients with CRPC with symptomatic bone metastasis in the pre- and postchemotherapy setting.[83] Currently, there are ongoing trials combining radium-223 with other agents, such as Sipuleucel-T and docetaxel.

## DISEASE MODIFYING AGENTS

### Taxane Chemotherapy

Docetaxel inhibits microtubule assembly leading to apoptosis. The two landmark trials on docetaxel, TAX 327 and SWOG 99-16, demonstrated a survival benefit in metastatic CRPC but neither trial had a reduction in SREs as a trial endpoint.[84,85] However, pain reduction occurred in approximately 35% compared to 22% in placebo. Interestingly, Cabazitaxel, another taxane-based chemotherapeutic, approved for postdocetaxel progressive CRPC, did improve OS compared to mitoxantrone, but no difference was seen in pain response.[86]

It is also important to note that the current CRPC paradigm in 2014, at the time of writing this chapter, divides patients into "predocetaxel" and "postdocetaxel" categories. These labels have less to do with cancer biology than they do with FDA regulations. The current status of advanced prostate cancer will likely dramatically change based on the results of the ChemoHormonal Therapy Versus Androgen Ablation Randomized Trial for Extensive Disease in Prostate Cancer (CHAARTED) trial. CHAARTED was presented as an abstract at the 2014 annual meeting of the American Society of Clinical Oncology (ASCO), but has not yet appeared in a peer-reviewed journal. CHAARTED randomized 790 patients who had metastatic hormone-sensitive prostate cancer to receive standard ADT or ADT plus six cycles of docetaxel. The addition of early docetaxel in the hormone-sensitive state improved the median survival by 14 months, and up to 17 months in those with "high risk" metastatic disease (defined as visceral metastasis or more than four bone lesions). The results of CHAARTED will most likely dramatically change the way metastatic prostate cancer is treated in the near future. The effect this will have on optimal bone health is not yet clear.

### Abiraterone

This CYP17 androgen biosynthesis inhibitor has been evaluated in patients with progressive CRPC after docetaxel (COU-AA-301) and was found to have a significant benefit in overall survival compared to prednisone alone (15.8 months vs. 11.2 months; $p < 0.0001$). On exploratory secondary endpoints, abiraterone was found to improve pain palliation (45% vs. 28.8%; $p = 0.0005$) and improve the time to first SRE (25 months vs. 20.3 months; $p = 0.0001$).[87] When abiraterone was evaluated in progressive CRPC prior to receiving docetaxel (COU-AA-302), radiographic progression-free survival was improved (16.5 months vs. 8.3 months; $p < 0.001$), although prespecified efficacy boundary was not reached for OS. Of 1088 patients within the COU-AA-302 trial, 373 patients were on bone-targeted therapy. Patients on bone-target therapy appeared to have improved survival, symptomatic progression, and declines in performance status.[66]

Enzalutamide is a very potent androgen receptor antagonist that demonstrated a survival benefit compared to placebo in patients with progressive CRPC after docetaxel (AFFIRM trial) (18.4 months vs. 13.6 months; $p < 0.001$). The time to first SRE was a secondary endpoint, which was also improved on enzalutamide (16.7 months vs. 13.3 months; $p < 0.001$). Enzalutamide was evaluated in progressive CRPC prior to receiving docetaxel (PREVAIL trial) and showed a significant benefit over placebo in the coprimary endpoints radiographic progression-free survival and overall survival. The time until the first SRE in the PREVAIL trial was also a secondary endpoint, which was improved with enzalutamide (HR = 0.72 (95% CI = 0.61, 0.84); $p < 0.001$).[88]

## NOVEL AND EMERGING THERAPIES

### Endothelin-1

This novel pathway is involved in the paracrine signaling between osteoblasts and prostate cancer metastasis cells. Although preclinical studies appeared promising, the addition of the endothelin receptor antagonist Atrasentan to docetaxel failed to demonstrate an improvement over docetaxel alone in CRPC.[89–91] Another endothelin receptor antagonist, zibotentan (ZD4054), was also shown to add no benefit over docetaxel alone.[92]

### Cabozantinib

A tyrosine kinase inhibitor that targets the hepatocyte growth factor (MET) and VEGFR-2 pathways has shown some favorable preliminary results with improving bone scan in patients with metastatic CRPC. There are two ongoing prospective phase III trials in heavily pretreated metastatic CRPC patients who have progressed on docetaxel and one postdocetaxel-approved agent (either Abiraterone or Enzalutamide). COMET 1 trial will evaluate cabozantinib for those without pain compared to prednisone alone and the COMET2 trial will evaluate the role for cabozantinib in patients with pain compared to mitoxantrone.[7]

### Dasatinib

A nonreceptor tyrosine kinase inhibitor targeting the Src pathway. Src pathways leads to complex signal transduction involved in the regulation of osteoclast activity and bone metastasis pathogenesis. Unfortunately, no benefit was seen in the phase III randomized READY trial comparing 100 mg of dasatinib per day combined with docetaxel compared to docetaxel alone in chemotherapy-naïve CRPC patients.[7,93]

### Selective Estrogen Modulators

At one time estrogenic agents were largely abandoned in advanced prostate cancer, due to an increased risk of venothrombotic events. However, there has been a renewed interested in selective estrogen modulators (SERMs) for their potential benefits in improving skeletal health while on ADT. In a phase III RCT of 1284 patients on ADT comparing Toremifene 80 mg to placebo, the 2-year incidence of new vertebral fractures was 4.9% compared to 2.5%, for a 50% relative risk reduction with Toremifene ($p = 0.05$). BMD was significantly increased at all measured sites. The incidence of venothrombotic events was 2.6% versus 1.1% with Toremifene. The number needed to treat to prevent a vertebral facture was 41, while the number needed to harm was 66. To date, Toremifene is not FDA approved and requires further study.[94]

### Other Possible Targets

The interplay between bone microenvironment and prostate cancer cells is very complex. Several possible therapeutics targets have been considered including transforming growth factor beta, bone morphogenic proteins, insulin-like growth factor, and serine proteases. In addition to these targets, there are many other pathways and growth factors under active investigation (Figures 53.6 and 53.7).[95]

## AMERICAN UROLOGIC ASSOCIATION GUIDELINE RECOMMENDATIONS FOR OPTIMAL BONE HEALTH IN PROSTATE CANCER

The American Urologic Association (AUA) guidelines on CRPC initially reviewed English-language articles published between 1996 and May 2013 to provide guidance on six "index patients" that demonstrate their proposed management algorithm depending on their disease extent, symptoms, chemotherapy exposure, and performance status. There is a section on bone health

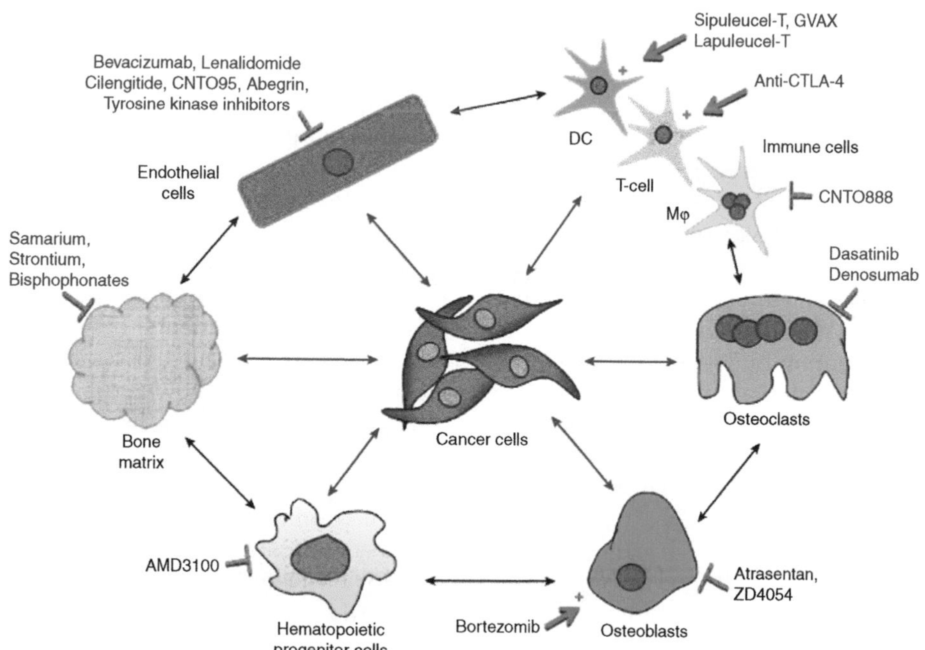


FIGURE 53.6 **Tumor cell-bone micro-environment interactions.** Tumor cells interact with the bone extracellular matrixstromal cells, osteoblasts, osteoclasts, hematopoietic progenitor cells, endothelial cells, and the cells of the immune system to coordinate a sophisticated series of interactions to promote tumor cell survival and proliferation. DC, dendritic cells; Mφ, macrophage. *Reprinted with permission from Heymann.*[26]

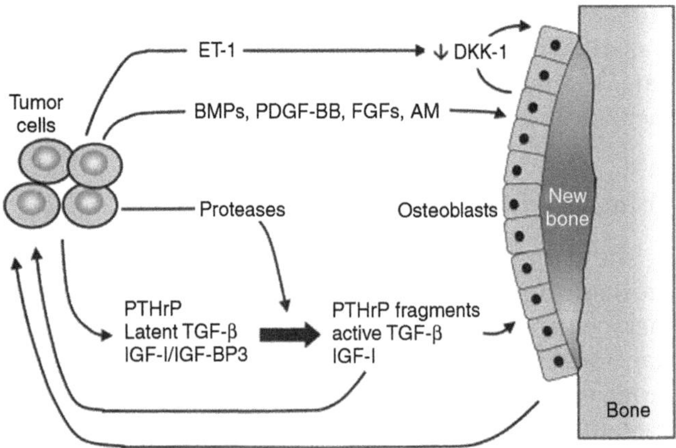


**FIGURE 53.7 Model of interactions at sites of osteoblastic metastasis.** Multiple factor such as ET-1, BMPs or PDGF, are secreted by cancer cells in the bone and microenvironment to induce osteoblastic proliferation and new bone formation. ET-1 produced by cancer cells stimulates osteoblast by repressing the expression of DKK-1, and inhibitor of the Wnt/β-Catenin pathway. In turn, proteins produced by the osteoblasts support cancer cell development in bone. Other factor secreted by cancer cells, including PTHrP, IGF and TGF-β, require processing by proteases such as PSA to acquire their osteoblastic/active state. Mature IGF-I promotes bone formation and cancer cell growth. BMP, bone morphogenetic protein; DKK-1, dickkopf homolog-1; ET-1, endothelin-1; IGF, insulin-like growth factor; PDGF, platelet-derived growth factor; PTHrP, parathyroid hormone related protein; TGF-β, transforming growth factor-β. *Reprinted with permission from Heymann.*[26]

that is not specific to any one patient but could be considered for all index patients.[74]

The AUA guidelines state "clinicians should offer preventative treatment (e.g., supplemental calcium and vitamin D) for fractures and skeletal-related events to CRPC patients. (Recommendation; Evidence Level Grade C)." The AUA guidelines cite a large meta-analysis of 9000 patients over 60 years suggesting a 26% reduction in the relative risk of hip fracture and 23% reduction in nonvertebral fractures from 700 IU/day to 800 IU/day of vitamin D. This meta-analysis was not in prostate cancer patients and the majority of patients in this meta-analysis were actually perimenopausal women. The AUA guidelines recommend supplemental calcium, particularly to patients on antiresorptive therapy, but hedge their recommendation by health concerns possibly associated with supplemental calcium.

The AUA guidelines also state "clinicians may choose either denosumab or zoledronic acid when selecting a preventative treatment for skeletal-related events for mCRPC patients with bony metastases. (Option; Evidence Level Grade C)". The guidelines state that "both denosumab and zoledronic acid can be considered options, with denosumab providing slightly superior efficacy results in a head-to-head comparison, and, therefore, is listed as the first option."

Data from the ALSYMPCA trial on radium-223 were not available at the time of the initial guideline preparation, but in a subsequent update of the CRPC guidelines with new evidence up until April 2014, radium-223 was recommended for symptomatic metastatic CRPC patients without visceral metastasis, regardless of prior chemotherapy exposure. Since radium-223 was primarily only tested in patients with good performance status, "there is no data indicating an advantage over standard radiopharmaceuticals" for poor performance status patients according to the AUA guidelines.

## PRACTICAL MANAGEMENT FOR OPTIMAL BONE HEALTH

All patients started on long-term ADT should have a baseline DEXA scan to measure their initial BMD. If BMD is within normal limits, then follow-up DEXA scans should be obtained every 1–2 years to monitor for bone loss, recognizing that osteoporosis risk increases with the duration of ADT. If a patient has a history of fragility fractures, low BMD (*T*-scan $<-1.5$ on DEXA), or an increased fracture risk (FRAX hip fracture risk ≥3% or FRAX any fracture risk ≥20%), then treatment should be considered with either zoledronic acid or denosumab. Recognizing that only denosumab has been shown to reduce the risk of actual fractures. Patient with a low baseline BMD warrant more frequent monitoring with DEXA scans every 6–12 months.[12]

Patients starting on ADT should have their serum 25(OH) vitamin D levels checked to ensure there is no deficiency. It is advisable to start all patients on ADT on supplemental vitamin D and calcium. Patients on zoledronic acid or denosumab should certainly be receiving supplemental vitamin D and calcium to reduce the risk of hypocalcemia. In addition, patients being considered for antiresorptive therapy should have a dental consultation. While on antiresorptive therapy, serum levels of creatinine, calcium, magnesium, and phosphate should be monitored.

To date, no bone-targeted therapy is approved for the prevention of bone metastasis in the castrate-sensitive or castrate-resistant patients. For patients in the castrate-sensitive state who have osseous metastasis, zoledronic acid has not been shown to be beneficial in reducing SREs, while the role for denosumab for this indication is unknown. However, since these patients will usually be treated with ADT, the same indications apply to initiate zoledronic acid or denosumab to prevent BMD loss as those without metastasis.[96] However, the dosages are less for BMD loss prevention than for reduction of SRE. Currently, only in the castrate-resistant state are higher doses of zoledronic acid and denosumab approved for the reduction of SREs.

For patients with metastatic CRPC, guideline recommendations from the AUA, EAU, and NCCN recommend patients receive either zoledronic acid or denosumab for the prevention of SREs, recognizing the slight superiority of denosumab. Patients with focal bone pain from a metastasis can be palliated with a single fraction of 8 Gy EBRT. The use of disease-modifying agents, such as radium-223, needs to be individualized.

## References


1. Gralow JR, Biermann JS, Farooki A, et al. NCCN Task Force Report: bone health in cancer care. *J Natl Compr Canc Netw* 2013;**11**(Suppl. 3): S1–S50 quiz S1.
2. Saylor PJ, Lee RJ, Smith MR. Emerging therapies to prevent skeletal morbidity in men with prostate cancer. *J Clin Oncol* 2011;**29**(27): 3705–14.
3. Morote J, Morin JP, Orsola A, et al. Prevalence of osteoporosis during long-term androgen deprivation therapy in patients with prostate cancer. *Urology* 2007;**69**(3):500–4.
4. Weinfurt KP, Li Y, Castel LD, et al. The significance of skeletal-related events for the health-related quality of life of patients with metastatic prostate cancer. *Ann Oncol* 2005;**16**(4):579–84.
5. Saad F, Gleason DM, Murray R, et al. Long-term efficacy of zoledronic acid for the prevention of skeletal complications in patients with metastatic hormone-refractory prostate cancer. *J Natl Cancer Inst* 2004;**96**(11):879–82.
6. Lage MJ, Barber BL, Harrison DJ, et al. The cost of treating skeletal-related events in patients with prostate cancer. *Am J Manag Care* 2008;**14**(5):317–22.
7. Saylor PJ, Armstrong AJ, Fizazi K, et al. New and emerging therapies for bone metastases in genitourinary cancers. *Eur Urol* 2013;**63**(2):309–20.
8. Orwoll ES, Bilezikian JP, Vanderschueren D, editors. *Osteoporosis in Men: The Effects of Gender on Skeletal Health*. 2nd edn London: Academic Press; 1999.
9. Gilbert SM, Kuo YF, Shahinian VB. Prevalent and incident use of androgen deprivation therapy among men with prostate cancer in the United States. *Urol Oncol* 2011;**29**(6):647–53.
10. Meng MV, Grossfeld GD, Sadetsky N, et al. Contemporary patterns of androgen deprivation therapy use for newly diagnosed prostate cancer. *Urology* 2002;**60**(3 Suppl. 1):7–11 discussion -2.
11. Preston DM, Torrens JI, Harding P, et al. Androgen deprivation in men with prostate cancer is associated with an increased rate of bone loss. *Prostate Cancer Prostatic Dis* 2002;**5**(4):304–10.
12. Israeli RS, Ryan CW, Jung LL. Managing bone loss in men with locally advanced prostate cancer receiving androgen deprivation therapy. *J Urol* 2008;**179**(2):414–23.
13. Eastham JA. Bone health in men receiving androgen deprivation therapy for prostate cancer. *J Urol* 2007;**177**(1):17–24.
14. Watts NB, Adler RA, Bilezikian JP, et al. Osteoporosis in men: an Endocrine Society clinical practice guideline. *J Clin Endocrinol Metab* 2012;**97**(6):1802–22.
15. Conde FA, Sarna L, Oka RK, et al. Age, body mass index, and serum prostate-specific antigen correlate with bone loss in men with prostate cancer not receiving androgen deprivation therapy. *Urology* 2004;**64**(2):335–40.
16. Shahinian VB, Kuo YF, Freeman JL, et al. Risk of fracture after androgen deprivation for prostate cancer. *N Engl J Med* 2005;**352**(2): 154–64.
17. Dickman PW, Adolfsson J, Astrom K, et al. Hip fractures in men with prostate cancer treated with orchiectomy. *J Urol* 2004;**172**(6 Pt. 1): 2208–12.
18. Saylor PJ, Morton RA, Hancock ML, et al. Factors associated with vertebral fractures in men treated with androgen deprivation therapy for prostate cancer. *J Urol* 2011;**186**(2):482–6.
19. Cooper C. The crippling consequences of fractures and their impact on quality of life. *Am J Med* 1997;**103**(2A):12S–7S discussion 7S–9S.
20. Higano CS. Understanding treatments for bone loss and bone metastases in patients with prostate cancer: a practical review and guide for the clinician. *Urol Clin North Am* 2004;**31**(2):331–52.
21. Morgans AK, Smith MR. Bone-targeted agents: preventing skeletal complications in prostate cancer. *Urol Clin North Am* 2012;**39**(4):533–46.
22. Nadler M, Alibhai S, Catton P, et al. Osteoporosis knowledge, health beliefs, and healthy bone behaviours in patients on androgen-deprivation therapy (ADT) for prostate cancer. *BJU Int* 2013;**111**(8):1301–9.
23. Nelson EC, Cambio AJ, Yang JC, et al. Clinical implications of neuroendocrine differentiation in prostate cancer. *Prostate Cancer Prostatic Dis* 2007;**10**(1):6–14.
24. di Sant'Agnese PA, Cockett AT. Neuroendocrine differentiation in prostatic malignancy. *Cancer* 1996;**78**(2):357–61.
25. Paget S. The distribution of secondary growths in cancer of the breast. *Cancer Metastasis Rev* 1989;**8**(2):98–101.
26. Heymann D, editor. *Bone Cancer: Progression and Therapeutic Approaches*. London: Academic Press; 2010.
27. Smith MR, Kabbinavar F, Saad F, et al. Natural history of rising serum prostate-specific antigen in men with castrate nonmetastatic prostate cancer. *J Clin Oncol* 2005;**23**(13):2918–25.
28. Smith MR, Cook R, Lee KA, et al. Disease and host characteristics as predictors of time to first bone metastasis and death in men with progressive castration-resistant nonmetastatic prostate cancer. *Cancer* 2011;**117**(10):2077–85.
29. Dotan ZA, Bianco Jr FJ, Rabbani F, et al. Pattern of prostate-specific antigen (PSA) failure dictates the probability of a positive bone scan in patients with an increasing PSA after radical prostatectomy. *J Clin Oncol* 2005;**23**(9):1962–8.
30. Pound CR, Partin AW, Eisenberger MA, et al. Natural history of progression after PSA elevation following radical prostatectomy. *JAMA* 1999;**281**(17):1591–7.
31. Sabbatini P, Larson SM, Kremer A, et al. Prognostic significance of extent of disease in bone in patients with androgen-independent prostate cancer. *J Clin Oncol* 1999;**17**(3):948–57.
32. Cook RJ, Coleman R, Brown J, et al. Markers of bone metabolism and survival in men with hormone-refractory metastatic prostate cancer. *Clin Cancer Res* 2006;**12**(11 Pt. 1):3361–7.
33. Scher HI, Halabi S, Tannock I, et al. Design and end points of clinical trials for patients with progressive prostate cancer and castrate levels of testosterone: recommendations of the Prostate Cancer Clinical Trials Working Group. *J Clin Oncol* 2008;**26**(7):1148–59.
34. Smith MR, Saad F, Oudard S, et al. Denosumab and bone metastasis-free survival in men with nonmetastatic castration-resistant prostate cancer: exploratory analyses by baseline prostate-specific antigen doubling time. *J Clin Oncol* 2013;**31**(30):3800–6.
35. Makarov DV, Desai RA, Yu JB, et al. The population level prevalence and correlates of appropriate and inappropriate imaging to stage incident prostate cancer in the Medicare population. *J Urol* 2012;**187**(1):97–102.
36. Makarov DV, Loeb S, Ulmert D, et al. Prostate cancer imaging trends after a nationwide effort to discourage inappropriate prostate cancer imaging. *J Natl Cancer Inst* 2013;**105**(17):1306–13.
37. Miller DC, Murtagh DS, Suh RS, et al. Regional collaboration to improve radiographic staging practices among men with early stage prostate cancer. *J Urol* 2011;**186**(3):844–9.
38. Messiou C, Cook G, deSouza NM. Imaging metastatic bone disease from carcinoma of the prostate. *Br J Cancer* 2009;**101**(8):1225–32.
39. Rajarubendra N, Bolton D, Lawrentschuk N. Diagnosis of bone metastases in urological malignancies – an update. *Urology* 2010;**76**(4):782–90.
40. Brown JE, Sim S. Evolving role of bone biomarkers in castration-resistant prostate cancer. *Neoplasia* 2010;**12**(9):685–96.

41. Fizazi K, Lipton A, Mariette X, et al. Randomized phase II trial of denosumab in patients with bone metastases from prostate cancer, breast cancer, or other neoplasms after intravenous bisphosphonates. *J Clin Oncol* 2009;**27**(10):1564–71.
42. Dennis ER, Jia X, Mezheritskiy IS, et al. Bone scan index: a quantitative treatment response biomarker for castration-resistant metastatic prostate cancer. *J Clin Oncol* 2012;**30**(5):519–24.
43. Ulmert D, Kaboteh R, Fox JJ, et al. A novel automated platform for quantifying the extent of skeletal tumour involvement in prostate cancer patients using the Bone Scan Index. *Eur Urol* 2012;**62**(1): 78–84.
44. Smith MR, Saad F, Coleman R, et al. Denosumab and bone-metastasis-free survival in men with castration-resistant prostate cancer: results of a phase 3, randomised, placebo-controlled trial. *Lancet* 2012;**379**(9810):39–46.
45. Tombal B. Zometa European Study (ZEUS): another failed crusade for the holy grail of prostate cancer bone metastases prevention? *Eur Urol* 2015;**67**(3):492–4.
46. Lutz S, Berk L, Chang E, et al. Palliative radiotherapy for bone metastases: an ASTRO evidence-based guideline. *Int J Radiat Oncol Biol Phys* 2011;**79**(4):965–76.
47. Falk S, Dickenson AH. Pain and nociception: mechanisms of cancer-induced bone pain. *J Clin Oncol* 2014;**32**(16):1647–54.
48. Pandit-Taskar N, Larson SM, Carrasquillo JA. Bone-seeking radiopharmaceuticals for treatment of osseous metastases, Part 1: alpha therapy with 223Ra-dichloride. *J Nucl Med* 2014;**55**(2): 268–74.
49. Portenoy RK. Treatment of cancer pain. *Lancet* 2011;**377**(9784): 2236–47.
50. McDougal S, Wein AJ, Kavoussi LI, et al. *Campbell-Walsh Urology*. 10th edn Elsevier; 2012.
51. Sorensen S, Borgesen SE, Rohde K, et al. Metastatic epidural spinal cord compression. Results of treatment and survival. *Cancer* 1990;**65**(7):1502–8.
52. Sorensen S, Helweg-Larsen S, Mouridsen H, et al. Effect of high-dose dexamethasone in carcinomatous metastatic spinal cord compression treated with radiotherapy: a randomised trial. *Eur J Cancer* 1994;**30A**(1):22–7.
53. Grant R, Papadopoulos SM, Greenberg HS. Metastatic epidural spinal cord compression. *Neurol Clin* 1991;**9**(4):825–41.
54. Patchell RA, Tibbs PA, Regine WF, et al. Direct decompressive surgical resection in the treatment of spinal cord compression caused by metastatic cancer: a randomised trial. *Lancet* 2005;**366**(9486):643–8.
55. Jayasekera J, Onukwugha E, Bikov K, et al. The economic burden of skeletal-related events among elderly men with metastatic prostate cancer. *Pharmacoeconomics* 2014;**32**(2):173–91.
56. Galvao DA, Spry N, Denham J, et al. A multicentre year-long randomised controlled trial of exercise training targeting physical functioning in men with prostate cancer previously treated with androgen suppression and radiation from TROG 03.04 RADAR. *Eur Urol* 2014;**65**(5):856–64.
57. Bourke L, Gilbert S, Hooper R, et al. Lifestyle changes for improving disease-specific quality of life in sedentary men on long-term androgen-deprivation therapy for advanced prostate cancer: a randomised controlled trial. *Eur Urol* 2014;**65**(5):865–72.
58. Galvao DA, Nosaka K, Taaffe DR, et al. Resistance training and reduction of treatment side effects in prostate cancer patients. *Med Sci Sports Exerc* 2006;**38**(12):2045–52.
59. Gardner JR, Livingston PM, Fraser SF. Effects of exercise on treatment-related adverse effects for patients with prostate cancer receiving androgen-deprivation therapy: a systematic review. *J Clin Oncol* 2014;**32**(4):335–46.
60. Planas J, Morote J, Orsola A, et al. The relationship between daily calcium intake and bone mineral density in men with prostate cancer. *BJU Int* 2007;**99**(4):812–5 discussion 5–6.
61. Zheng Y, Zhou H, Ooi LL, et al. Vitamin D deficiency promotes prostate cancer growth in bone. *Prostate* 2011;**71**(9):1012–21.
62. Peleg S. Responsiveness of osteoblastic and osteolytic bone metastases to vitamin D analogs. *Crit Rev Eukaryot Gene Expr* 2007;**17**(2):149–58.
63. Crook JM, O'Callaghan CJ, Duncan G, et al. Intermittent androgen suppression for rising PSA level after radiotherapy. *N Engl J Med* 2012;**367**(10):895–903.
64. Hussain M, Tangen CM, Berry DL, et al. Intermittent versus continuous androgen deprivation in prostate cancer. *N Engl J Med* 2013;**368**(14):1314–25.
65. Smith MR, Eastham J, Gleason DM, et al. Randomized controlled trial of zoledronic acid to prevent bone loss in men receiving androgen deprivation therapy for nonmetastatic prostate cancer. *J Urol* 2003;**169**(6):2008–12.
66. Gartrell BA, Saad F. Managing bone metastases and reducing skeletal related events in prostate cancer. *Nat Rev Clin Oncol* 2014;**11**(6):335–45.
67. Saad F, Gleason DM, Murray R, et al. A randomized, placebo-controlled trial of zoledronic acid in patients with hormone-refractory metastatic prostate carcinoma. *J Natl Cancer Inst* 2002;**94**(19):1458–68.
68. James NPS, Barton D, Brown J, et al. (CRPC) metastatic to bone randomized in the factorial TRAPEZE trial to docetaxel (D) with strontium-89 (Sr89), zoledronic acid (ZA), neither, or both (ISRCTN 12808747). *J Clin Oncol* 2013;**31**(15 Suppl.) (May 20 Supplement), LBA5000.
69. Smith MR, Halabi S, Ryan CJ, et al. Randomized controlled trial of early zoledronic acid in men with castration-sensitive prostate cancer and bone metastases: results of CALGB 90202 (alliance). *J Clin Oncol* 2014;**32**(11):1143–50.
70. Wirth M, Tammela T, Cicalese V, et al. Prevention of bone metastases in patients with high-risk nonmetastatic prostate cancer treated with zoledronic acid: efficacy and safety results of the Zometa European Study (ZEUS). *Eur Urol* 2014.
71. Smith MR, McGovern FJ, Zietman AL, et al. Pamidronate to prevent bone loss during androgen-deprivation therapy for prostate cancer. *N Engl J Med* 2001;**345**(13):948–55.
72. Small EJ, Smith MR, Seaman JJ, et al. Combined analysis of two multicenter, randomized, placebo-controlled studies of pamidronate disodium for the palliation of bone pain in men with metastatic prostate cancer. *J Clin Oncol* 2003;**21**(23):4277–84.
73. Dearnaley DP, Mason MD, Parmar MK, et al. Adjuvant therapy with oral sodium clodronate in locally advanced and metastatic prostate cancer: long-term overall survival results from the MRC PR04 and PR05 randomised controlled trials. *Lancet Oncol* 2009;**10**(9):872–6.
74. Cookson MS, Roth BJ, Dahm P, et al. Castration-resistant prostate cancer: AUA Guideline. *J Urol* 2013;**190**(2):429–38.
75. Smith MR, Egerdie B, Hernandez Toriz N, et al. Denosumab in men receiving androgen-deprivation therapy for prostate cancer. *N Engl J Med* 2009;**361**(8):745–55.
76. Fizazi K, Carducci M, Smith M, et al. Denosumab versus zoledronic acid for treatment of bone metastases in men with castration-resistant prostate cancer: a randomised, double-blind study. *Lancet* 2011;**377**(9768):813–22.
77. Lipton A, Fizazi K, Stopeck AT, et al. Superiority of denosumab to zoledronic acid for prevention of skeletal-related events: a combined analysis of 3 pivotal, randomised, phase 3 trials. *Eur J Cancer* 2012;**48**(16):3082–92.
78. Zinonos I, Luo KW, Labrinidis A, et al. Pharmacologic inhibition of bone resorption prevents cancer-induced osteolysis but enhances soft tissue metastasis in a mouse model of osteolytic breast cancer. *Int J Oncol* 2014;**45**(2):532–40.
79. Jager PL, Kooistra A, Piers DA. Treatment with radioactive (89) strontium for patients with bone metastases from prostate cancer. *BJU Int* 2000;**86**(8):929–34.

80. Bauman G, Charette M, Reid R, et al. Radiopharmaceuticals for the palliation of painful bone metastasis-a systemic review. *Radiother Oncol* 2005;**75**(3):258–70.
81. Tu SM, Millikan RE, Mengistu B, Delpassand ES, Amato RJ, Pagliaro LC, Daliani D, Papandreou CN, Smith TL, Kim J, Podoloff DA, Logothetis CJ. Bone-targeted therapy for advanced androgen-independent carcinoma of the prostate: a randomised phase II trial. *Lancet* 2001;**357**(9253):336–41.
82. Fizazi K, Beuzeboc P, Lumbroso J, et al. Phase II trial of consolidation docetaxel and samarium-153 in patients with bone metastases from castration-resistant prostate cancer. *J Clin Oncol* 2009;**27**(15):2429–35.
83. Parker C, Nilsson S, Heinrich D, et al. Alpha emitter radium-223 and survival in metastatic prostate cancer. *N Engl J Med* 2013;**369**(3):213–23.
84. Tannock IF, de Wit R, Berry WR, et al. Docetaxel plus prednisone or mitoxantrone plus prednisone for advanced prostate cancer. *N Engl J Med* 2004;**351**(15):1502–12.
85. Petrylak DP, Tangen CM, Hussain MH, et al. Docetaxel and estramustine compared with mitoxantrone and prednisone for advanced refractory prostate cancer. *N Engl J Med* 2004;**351**(15):1513–20.
86. de Bono JS, Oudard S, Ozguroglu M, et al. Prednisone plus cabazitaxel or mitoxantrone for metastatic castration-resistant prostate cancer progressing after docetaxel treatment: a randomised open-label trial. *Lancet* 2010;**376**(9747):1147–54.
87. Logothetis CJ, Basch E, Molina A, et al. Effect of abiraterone acetate and prednisone compared with placebo and prednisone on pain control and skeletal-related events in patients with metastatic castration-resistant prostate cancer: exploratory analysis of data from the COU-AA-301 randomised trial. *Lancet Oncol* 2012;**13**(12):1210–7.
88. Beer TM, Armstrong AJ, Rathkopf DE, et al. Enzalutamide in metastatic prostate cancer before chemotherapy. *N Engl J Med* 2014;**371**(5):424–33.
89. Quinn DI, Tangen CM, Hussain M, et al. Docetaxel and atrasentan versus docetaxel and placebo for men with advanced castration-resistant prostate cancer (SWOG S0421): a randomised phase 3 trial. *Lancet Oncol* 2013;**14**(9):893–900.
90. Carducci MA, Padley RJ, Breul J, et al. Effect of endothelin-A receptor blockade with atrasentan on tumor progression in men with hormone-refractory prostate cancer: a randomized, phase II, placebo-controlled trial. *J Clin Oncol* 2003;**21**(4):679–89.
91. Nelson JB, Love W, Chin JL, et al. Phase 3, randomized, controlled trial of atrasentan in patients with nonmetastatic, hormone-refractory prostate cancer. *Cancer* 2008;**113**(9):2478–87.
92. Fizazi KS, Higano CS, Nelson JB, et al. Phase III, randomized, placebo-controlled study of docetaxel in combination with zibotentan in patients with metastatic castration-resistant prostate cancer. *J Clin Oncol* 2013;**31**(14):1740–7.
93. Araujo JC, Trudel GC, Saad F, et al. Docetaxel and dasatinib or placebo in men with metastatic castration-resistant prostate cancer (READY): a randomised, double-blind phase 3 trial. *Lancet Oncol* 2013;**14**(13):1307–16.
94. Smith MR, Morton RA, Barnette KG, et al. Toremifene to reduce fracture risk in men receiving androgen deprivation therapy for prostate cancer. *J Urol* 2013;**189**(1 Suppl):S45–50.
95. Weilbaecher KN, Guise TA, McCauley LK. Cancer to bone: a fatal attraction. *Nat Rev Cancer* 2011;**11**(6):411–25.
96. Higano CS. To treat or not to treat, that is the question: the role of bone-targeted therapy in metastatic prostate cancer. *J Clin Oncol* 2014;**32**(11):1107–11.

CHAPTER

# 54

# Castration Resistant Prostate Cancer: Role of Chemotherapy

*Marijo Bilusic, MD, PhD*

**Department of Medical Oncology, Fox Chase Cancer Center/Temple University, Philadelphia, PA, USA**

## INTRODUCTION

Prostate cancer is the second most common cancer (after skin cancer) and one of the major causes of cancer mortality in the United States with an expected 29,720 deaths in 2013.[1,2] About 40% of prostate cancer patients will fail surgical or radiation therapy and will develop metastatic disease.[3,4] Testosterone suppression is an effective frontline treatment in newly diagnosed metastatic disease; however, cancer will ultimately progress to castration-resistant disease. Androgen deprivation therapy (ADT) is the backbone of treatment in castration-resistant disease; although, additional agents, such as antiandrogens, immunotherapy, radiopharmaceuticals, and chemotherapy, are necessary in order to palliate symptoms and prolong survival. Historically, chemotherapy was considered ineffective for metastatic castration resistant prostate cancer (mCRPC) with a response rate of less than 20% and without any significant impact on overall survival (OS). Surprisingly in the last decade a few chemotherapy agents have demonstrated improvement in OS, establishing chemotherapy as a standard treatment option in advanced disease.

## FDA-APPROVED CHEMOTHERAPY REGIMENS FOR mCRPC

### Mitoxantrone

Mitoxantrone is a DNA-intercalating derivative of an anthracenedione antibiotic, which binds to topoisomerase II, thus causing inhibition of DNA and RNA replication. Based on the encouraging early phase trials, a phase III trial enrolled 161 patients with symptomatic mCRPC to mitoxantrone 12 mg/m$^2$ IV every-3-weeks plus prednisone arm and to prednisone alone arm. Mitoxantrone with prednisone significantly improved pain control compared to prednisone alone (29% vs. 12%) with the median duration of response of 7.6 versus 2.1 months ($p = 0.0009$). No significant improvement in OS was reported.[5] Another phase III trial (CALGB 9182) randomized 244 mCRPC patients to mitoxantrone plus hydrocortisone arm and to hydrocortisone alone. This study also showed improvement in pain control with no significant survival advantage (12.3 vs. 12.6 months, $p = 0.3298$).[6] Mitoxantrone was approved by the FDA in 1996 for the palliative treatment of mCRPC. Mitoxantrone is generally well-tolerated. Cardiotoxicity was observed in 5.5% of patients (defined as any decrease in left ventricle ejection fraction below the normal range, congestive heart failure, or myocardial ischemia), while secondary leukemia was reported in 1% of the treated patients. Today, mitoxantrone is used with the goal of improving quality of life and pain control as second- or third-line chemotherapy.

### Docetaxel

Docetaxel is a taxane derivate, which prevents androgen receptor (AR) nuclear translocation by binding to microtubules[7,8] and causing apoptosis through Bcl-2 phosphorylation.[9] Single-agent docetaxel or in combination with estramustine demonstrated objective response rates (ORR) in up to 38% patients, PSA decline >50% in up to 69% patients with a median OS between 20 months and 23 months.[10–12] These exciting findings prompted two phase III trials (TAK 327 and SWOG 99-16) that independently confirmed that docetaxel chemotherapy improved OS in mCRPC.

The TAX 327 study randomized 1006 mCRPC patients to three arms (all with daily oral prednisone):

Prostate Cancer. http://dx.doi.org/10.1016/B978-0-12-800077-9.00054-2

(1) mitoxantrone 12 mg/m$^2$ IV every 3 weeks, (2) docetaxel 30 mg/m$^2$ IV weekly, and (3) docetaxel 75 mg/m$^2$ IV every 3 weeks. Eight milligrams of dexamethasone was prescribed 12, 3, and 1 h prior treatment in order to minimize toxicity. Only the docetaxel every-3-weeks arm showed significant median OS advantage compared with weekly docetaxel and mitoxantrone (19.2 vs. 17.8 vs. 16.3 months, respectively), with additional improvement in pain and quality of life.[13,14] Subset analyses confirmed survival advantage in all subgroups. Docetaxel toxicity was acceptable. The most common toxicity in the docetaxel every-3-weeks group was grade 3/4 neutropenia compared with weekly docetaxel and mitoxantrone group (32% vs. 2% vs. 22%). Neutropenic infections were rare (3% vs. 0% vs. 2%, respectively). Treatment discontinuation due to side effects was relatively uncommon in all arms (11% vs. 16% vs. 10%, respectively). This trial has established docetaxel (75 mg/m$^2$ every-3-weeks) plus daily prednisone (5 mg twice a day) as the standard treatment option for mCRPC and was approved by the FDA in 2004.

The SWOG 99-16 phase III trial enrolled 770 mCRPC patients to two arms: (1) docetaxel (60 mg/m$^2$ IV every-3-weeks) plus estramustine (280 mg PO TID days 1–5) and (2) mitoxantrone (12 mg/m$^2$ IV every-3-weeks) plus prednisone (5 mg BID). The docetaxel plus estramustine arm demonstrated improvement in median OS (17.5 vs. 15.6 months, $p = 0.02$).[15] Pain improvement was similar in both groups. Grade 3/4 toxicity (gastrointestinal, cardiovascular/thromboembolic events, infections, pain, and neuropathy) was significantly higher and was reported in 54% patients (compared to 34% on mitoxantrone and prednisone arm); however, study discontinuation due to toxicities and treatment-related deaths (2% vs. 1%) were similar in both groups. Despite demonstrated survival advantage, the combination of docetaxel plus estramustine is rarely used due to significant toxicity of the regimen.

## Docetaxel Combinations

It is a widely accepted rule in oncology that chemotherapy drugs are most effective when given in combination since using agents that work by different mechanisms can improve the treatment effect while decreasing the risk of resistance development. Several large phase III studies have combined docetaxel plus prednisone with different agents such as bevacizumab,[16] aflibercept,[17] lenalidomide,[18] dasatinib,[19] atrasentan,[20] zibotentan,[21] and calcitriol.[22] Surprisingly, none of the combinations were superior. The majority of combination studies demonstrated significant toxicities requiring docetaxel dose reduction and delays as well as potentially compromising survival. Several ongoing phase II and III studies are testing the combination of docetaxel with other agents; however, none of them have established a role yet (Table 54.1).

## Cabazitaxel

Cabazitaxel is a derivative of docetaxel, which is cytotoxic for docetaxel-resistant cell lines due to gp-1 overexpression and has a better blood–brain barrier penetration.[23] Based on the results of a phase I study that enrolled 25 patients (only eight patients with mCRPC),[24] a decision was made to test this drug in mCRPC. TROPIC was a phase III study that enrolled 755 docetaxel pretreated mCRPC patients to two groups: (1) mitoxantrone 12 mg/m$^2$ IV every-3-weeks plus prednisone 10 mg and (2) cabazitaxel 25 mg/m$^2$ IV every-3-weeks plus prednisone 10 mg daily. The vast majority of patients had advanced mCRPC: 25% of the patients had visceral disease and 92% of the patients previously received a >225 mg/m$^2$ cumulative dose of docetaxel. At the first interim analysis, the cabazitaxel arm demonstrated significant improvement in median OS (15.1 vs. 12.7 months), with a 30% decrease in relative risk of dying (HR = 0.7, $p < 0.0001$). Interestingly, subset analysis demonstrated

**TABLE 54.1** Ongoing Phase II and III Studies Combining Docetaxel with Other Agents

| Disease stage | Drug | Phase | Number of pts | End point | Clincaltrials.gov |
|---|---|---|---|---|---|
| mCRPC; chemo naive | Custirsen | III | 1023 | OS | NCT01188187 (SYNERGY) |
| mCRPC; chemo naive | DCVAC/PCa | III | 1170 | OS | NCT02111577 (VIABLE) |
| mCRPC; chemo naive | Dendritic cell vaccine | II | 40 | Immune response | NCT01446731 |
| mCRPC; chemo naive | Imatinib | II | 17 | Disease progression | NCT00251225 |
| mCRPC; chemo naive | Cediranib | II | 84 | PFS | NCT00527124 |
| mCRPC; chemo naive | Reolysin | II | 80 | Disease progression | NCT01619813 |
| mCRPC; chemo naive | Bevacizumab, thalidomide | II | 73 | PSA response | NCT00089609 |
| mCRPC; prior docetaxel | Tasquinimod | II | 140 | PFS radiographic | NCT01732549 |

Pts, Patients; mCRPC, metastatic castration resistant prostate cancer; OS, overall survival; PSA, prostate specific antigen; PFS, progression free survival.

the greatest survival advantage in heavily pretreated patients.[25] Recently published updated analyses showed that survival benefit was sustained with >2 years survival in 27% of patients treated with cabazitaxel versus 16% in the control arm.[26]

Cabazitaxel treatment was quite toxic: 94% of patients had neutropenia, 97% anemia, and 47% thrombocytopenia. Neutropenic fever occurred in 8% of the patients and five patients died. Other common toxicities were diarrhea (47%), nausea (34%), and vomiting (22%). Cabazitaxel was FDA approved in 2010 for patients who progressed after docetaxel treatment. Due to significant toxicities, prophylaxis with colony-stimulating factors is recommended for high-risk groups and patients older than 65 years of age. An ongoing randomized phase III study will evaluate the safety and efficacy of a lower dose of cabazitaxel (20 mg/m$^2$) (NCT01308567) and the role of cabazitaxel as the frontline therapy (NCT01308580).

## OTHER CHEMOTHERAPY REGIMENS FOR mCRPC

There are no standard chemotherapy regimens for patients with mCRPC who progressed or did not respond to docetaxel and cabazitaxel. The majority of those patients will eventually be treated with other novel agents such as abiraterone, enzalutamide, or radium-223; however, some may be candidates for additional chemotherapy. Several phase II trials tested different cytotoxic chemotherapies in prostate cancer patients and a few of them are presented in the subsequent section.

### Estramustine

Estramustine is a conjugate of estradiol and alkylating agent, which binds to microtubule-associated proteins and tubulin, thereby inhibiting activity of microtubules and leading to anaphase arrest. It also has additional anticancer benefit given antiandrogen effects due to the estradiol component. Although it was one of the first effective chemotherapeutics that demonstrated PSA response, this drug has rarely been used due to arterial and venous thromboembolic events.[1] Several prophylactic measures have been tried, such as prophylaxis with daily aspirin or low-dose warfarin, however, without any significant impact.[27–29]

### Cisplatin

Cisplatin is an alkylating agent that binds with DNA and forms intrastrand cross-links affecting DNA replication. Other cytotoxic mechanisms include decreased ATPase activity, mitochondrial damage, and altered cellular transport mechanisms. Several phase II studies tested single-agent cisplatin in mCRPC and showed a response rate of up to 19%.[30,31] Cisplatin is used today mostly in combination with etoposide in patients with small cell neuroendocrine prostate carcinoma.

### Carboplatin

Carboplatin activity is very similar to cisplatin as it binds with DNA and affects replication; however, it has a more favorable adverse effect profile. Single-agent carboplatin demonstrated stable disease in 50% of patients with mCRPC.[32] The combination of carboplatin, paclitaxel, and estramustine demonstrated antitumor activity in up to 45% mCRPC patients.[33] Carboplatin (AUC 4) and docetaxel 60 mg/m$^2$ were tested in 34 patients and demonstrated 50% PSA declines in six (18%) and a partial response in three (14%) patients.[34] The combination of carboplatin plus paclitaxel was also tested in 38 patients (24 were pretreated with ≥2 chemotherapy regimens). The median OS was 10 months with clinical and/or PSA response observed in 26% of patients.[35] Another phase II study evaluated carboplatin and docetaxel combination in 34 men who had progressed after the completion or during docetaxel chemotherapy. PSA responses were observed in 18% patients, the median PFS was 3 months with the median OS of 12 months.[36]

### Oxaliplatin

Two studies have evaluated oxaliplatin after docetaxel-based chemotherapy. The first study enrolled 47 mCRPC patients treated with oxaliplatin and pemetrexed and demonstrated PSA response in 64%, objective response in 25% of the patients with the median PFS of 6 months and median OS of 12 month.[37] Another phase II study of oxaliplatin plus capecitabine in 14 patients showed PSA response in 57% of the patients with median OS of 24 weeks.[38]

### Ixabepilone

The epothilones are tubulin-polymerizing agents with broad spectrum activity.[39] Initial study of ixabepilone in prostate cancer showed promising antitumor activities,[40] which led to phase II studies. A SWOG phase II study enrolled 48 taxane-naïve mCRPC patients. Fourteen patients had PSA responses (33%) and two patients had an undetectable PSA. The median PFS was 6 months and the median OS was 18 months. Hematologic toxicity, infection, flu-like symptoms, and neuropathy were among the most commonly reported grade 3/4 adverse events.[41] Another trial randomized taxane-refractory mCRPC patients to ixabepilone and mitoxantrone. Partial responses were observed in one ixabepilone patient (out of 24) and in two mitoxantrone treated patients (out of 21) with

PSA declines >50% demonstrated in 17% of ixabepilone treated patients.[42]

### Vinorelbine

Vinorelbine is a semisynthetic vinca alkaloid that affects microtubule activity. Initial studies with vinorelbine alone showed ≥50% PSA decline in up to 17% of patients.[43,44] A phase III study randomized 414 chemo naïve patients to vinorelbine 30 mg/m$^2$ IV on days 1 and 8 every-3-weeks plus hydrocortisone versus hydrocortisone alone. The PSA response (defined as ≥50% decline of PSA lasting for at least 6 weeks) was 30.1% in the vinorelbine arm and 19.2% in the control arm. The median PFS was 3.7 versus 2.8 months and the treatment was well-tolerated.[45]

## WHEN TO START CHEMOTHERAPY?

Since predictive and prognostic biomarkers are still in their infancy, the decision to start cytotoxic chemotherapy in mCRPC is based on several factors, including the presence of disease symptoms, performance status and other comorbidities, available treatment options, and patients' preference. There is limited evidence that patients who developed castration-resistant disease in <16 months will respond better to chemotherapy.[46] Chemotherapy is frequently offered earlier to patients with aggressive and symptomatic disease (such as younger patients with visceral metastases and a higher Gleason score), although trials supporting this practice are still lacking.

In newly diagnosed metastatic disease, the situation is now clearer. The Eastern Cooperative Oncology Group (ECOG) phase III study randomized 790 newly diagnosed patients with metastatic disease to six cycles of docetaxel 75 mg/m$^2$ IV every-3-weeks plus ADT and to ADT alone. This trial (CHAARTED) demonstrated significant improvement in median OS: 57.6 months versus 44.0 months (HR = 0.61, $p$ = 0.0003). The median OS in the high-risk group ($N$ = 514), defined as presence of visceral metastases, ≥4 bone metastases including ≥1 bone lesion beyond the vertebral column or pelvis improved from 32.2 months to 49.2 months (HR = 0.60, $p$ = 0.006).[47] Based on this study, six cycles of docetaxel with ADT are treatment of choice for metastatic prostate cancer patients with a high volume of metastases commencing ADT who are suitable for docetaxel therapy.

## DURATION OF CHEMOTHERAPY

The optimal duration of chemotherapy in metastatic CRPC is also unclear. Studies have suggested that >10 cycles of docetaxel did not significantly improve OS; although, some patients still had some benefit after 1 year of therapy.[48] The previously mentioned TAX 327 study allowed chemotherapy beyond 10 cycles, which did not demonstrate survival improvement; however, it seems that continuation of treatment, despite pain or PSA progression, was associated with longer postprogression survival. Alternative therapies should be used if the patient exhibits signs of disease progression such as pain, PSA, or radiographic progression.[48]

The hypothesis of intermittent docetaxel treatment was evaluated in the ASCENT-1 study. Patients with mCRPC who achieved a PSA level of ≤4 ng/mL following induction treatment with a docetaxel-based regimen had their treatment suspended until the PSA level increased by ≥50% and >2 ng/mL, or on evidence of progressive disease. The first treatment holiday duration was 18 weeks (range 4–70 weeks) and once the treatment was again restarted, 45.5% of patients had again ≥50% PSA decline for more than 12 weeks. This study suggested that chemotherapy holidays may be considered for some patients; however, this approach needs to be validated.[49]

## THE FUTURE OF CHEMOTHERAPY IN mCRPC

Several ongoing randomized trials should define which agent should be used as frontline chemotherapy; however, the optimal sequences of other therapies, such as abiraterone, enzalutamide and radium-223, are not clear yet.

A few phase III trials are now evaluating cabazitaxel and its role in mCRPC. FIRSTANA (NCT01308567) will randomize 1170 chemotherapy naïve mCRPC to docetaxel 75 mg/m$^2$ IV every-3-weeks, cabazitaxel 25 mg/m$^2$ IV every-3-weeks, or cabazitaxel 20 mg/m$^2$ IV every-3-weeks as firstline chemotherapy to evaluate whether a similar efficacy can be achieved with less drug. PROSELICA (NCT01308580) will randomize 1200 previously treated mCRPC patients to cabazitaxel 20 mg/m$^2$ IV every-3-weeks and cabazitaxel 25 mg/m$^2$ IV every-3-weeks to evaluate the noninferiority in median OS. TAXYNERGY (NCT01718353) is a randomized phase II study, which evaluates an early change of chemotherapy from firstline docetaxel to cabazitaxel, or vice versa, if patients do not achieve >30% PSA decline after four cycles. This trial is also evaluating mechanisms of taxane resistance and biomarkers. AFFINITY (NCT01578655) is a phase III study, which is evaluating the role of custirsen (OGX-011 – an antisense oligonucleotide that blocks the expression of clusterin, a prosurvival protein associated with drug resistance) in combination with cabazitaxel plus prednisone as second-line chemotherapy.

Eribulin mesylate (nontaxane halichondrin-B analogue microtubule inhibitor) demonstrated ≥50%

TABLE 54.2 Ongoing Phase II and III Studies Combining Chemotherapy with Biological Agents

| Disease | Chemotherapy | Biological agent | Number of pts | End point | Clincatrials.gov |
|---|---|---|---|---|---|
| mCRPC; mitoxantrone naive | Mitoxantrone | IMC-3G3 | 123 | PFS | NCT01204710 |
| mCRPC; postdocetaxel | Cabazitaxel | Custirsen | 630 | OS | NCT01578655 (AFFINITY) |
| mCRPC; mitoxantrone and cabozantinib naive | Mitoxantrone | Cabozantinib | 246 | Pain response at 12 weeks | NCT01522443 |

Pts, Patients; mCRPC, metastatic castration resistant prostate cancer; OS, overall survival; PFS, progression free survival.

PSA decrease in 22.4% of taxane-naïve and in 8.5% of taxane-treated mCRPC patients with an excellent toxicity profile.[50] Other combinations of chemotherapy with biological agents are currently being tested in mCRPC (Table 54.2).

## CONCLUSIONS

Chemotherapy has an important role in castration-sensitive and castration-resistant prostate cancer. Due to the lack of validated predictive biomarkers, optimal sequence, selection, and ideal timing of chemotherapy is still unclear. With the recently approved drugs such as abiraterone, enzalutamide, radium 223, and sipuleucel-T, the perfect timing of chemotherapy is even more complicated than before.

Although both docetaxel and cabazitaxel demonstrated significant improvements in median OS, their responses are relatively short. Second-line chemotherapeutic options are still limited. In the future, the combinations of agents with proven survival advantage may be the most beneficial, such as the combination of enzalutamide plus abiraterone, chemotherapy plus androgen inhibitors (abiraterone or enzalutamide), or radium-223 and androgen inhibitors.

Future novel drug combination therapies with the use of synergistic chemotherapy, better understanding of the drug resistance mechanisms, and targeting new pathways will hopefully change the treatment paradigm. Predictive biomarkers are urgently needed to guide the most feasible sequence of therapeutic agents and their combination for personalized treatment that will ultimately provide better survival for prostate cancer patients.

## References


1. Ravery V, Fizazi K, Oudard S, et al. The use of estramustine phosphate in the modern management of advanced prostate cancer. *BJU Int* 2011;**108**(11):1782–6.
2. Siegel R, Naishadham D, Jemal A. Cancer statistics. *CA Cancer J Clin* 2013;**63**(1):11–30.
3. Freedland SJ, Humphreys EB, Mangold LA, et al. Risk of prostate cancer-specific mortality following biochemical recurrence after radical prostatectomy. *JAMA* 2005;**294**(4):433–9.
4. Freedland SJ, Humphreys EB, Mangold LA, et al. Death in patients with recurrent prostate cancer after radical prostatectomy: prostate-specific antigen doubling time subgroups and their associated contributions to all-cause mortality. *J Clin Oncol* 2007;**25**(13): 1765–71.
5. Tannock IF, Osoba D, Stockler MR, et al. Chemotherapy with mitoxantrone plus prednisone or prednisone alone for symptomatic hormone-resistant prostate cancer: a Canadian randomized trial with palliative end points. *J Clin Oncol* 1996;**14**(6):1756–64.
6. Kantoff PW, Halabi S, Conaway M, et al. Hydrocortisone with or without mitoxantrone in men with hormone-refractory prostate cancer: results of the cancer and leukemia group B 9182 study. *J Clin Oncol* 1999;**17**(8):2506–13.
7. Darshan MS, Loftus MS, Thadani-Mulero M, et al. Taxane-induced blockade to nuclear accumulation of the androgen receptor predicts clinical responses in metastatic prostate cancer. *Cancer Res* 2011;**71**(18):6019–29.
8. Jordan MA, Wilson L. Microtubules as a target for anticancer drugs. *Nat Rev Cancer* 2004;**4**(4):253–65.
9. Haldar S, Basu A, Croce CM. Bcl2 is the guardian of microtubule integrity. *Cancer Res* 1997;**57**(2):229–33.
10. Kreis W, Budman DR, Fetten J, et al. Phase I trial of the combination of daily estramustine phosphate and intermittent docetaxel in patients with metastatic hormone refractory prostate carcinoma. *Ann Oncol* 1999;**10**(1):33–8.
11. Petrylak DP, Macarthur R, O'Connor J, et al. Phase I/II studies of docetaxel (Taxotere) combined with estramustine in men with hormone-refractory prostate cancer. *Sem Oncol* 1999;**26**(5 Suppl. 17):28–33.
12. Petrylak DP, Macarthur RB, O'Connor J, et al. Phase I trial of docetaxel with estramustine in androgen-independent prostate cancer. *J Clin Oncol* 1999;**17**(3):958–67.
13. Berthold DR, Pond GR, Soban F, et al. Docetaxel plus prednisone or mitoxantrone plus prednisone for advanced prostate cancer: updated survival in the TAX 327 study. *J Clin Oncol* 2008;**26**(2):242–5.
14. Tannock IF, de Wit R, Berry WR, et al. Docetaxel plus prednisone or mitoxantrone plus prednisone for advanced prostate cancer. *N Engl J Med* 2004;**351**(15):1502–12.
15. Petrylak DP, Tangen CM, Hussain MH, et al. Docetaxel and estramustine compared with mitoxantrone and prednisone for advanced refractory prostate cancer. *N Engl J Med* 2004;**351**(15): 1513–20.
16. Kelly WK, Halabi S, Carducci M, et al. Randomized, double-blind, placebo-controlled phase III trial comparing docetaxel and prednisone with or without bevacizumab in men with metastatic castration-resistant prostate cancer: CALGB 90401. *J Clin Oncol* 2012;**30**(13):1534–40.
17. Tannock IF, Fizazi K, Ivanov S, et al. Aflibercept versus placebo in combination with docetaxel and prednisone for treatment of men with metastatic castration-resistant prostate cancer (VENICE): a phase 3, double-blind randomised trial. *Lancet Oncol* 2013;**14**(8): 760–8.
18. Petrylak DP Fizazi K, Sternberg CN, et al. A Phase 3 Study to Evaluate the Efficacy and Safety of Docetaxel and Prednisone (DP)

With or Without Lenalidomide (LEN) in Patients with Castrate-Resistant Prostate Cancer (CRPC): The MAINSAIL Trial ESMO; 2012, LBA24.

19. Araujo JC, Trudel GC, Saad F, et al. Docetaxel and dasatinib or placebo in men with metastatic castration-resistant prostate cancer (READY): a randomised, double-blind phase 3 trial. *Lancet Oncol* 2013;**14**(13):1307–16.
20. Quinn DI, Tangen CM, Hussain M, et al. Docetaxel and atrasentan versus docetaxel and placebo for men with advanced castration-resistant prostate cancer (SWOG S0421): a randomised phase 3 trial. *Lancet Oncol* 2013;**14**(9):893–900.
21. Fizazi KS, Higano CS, Nelson JB, et al. Phase III, randomized, placebo-controlled study of docetaxel in combination with zibotentan in patients with metastatic castration-resistant prostate cancer. *J Clin Oncol* 2013;**31**(14):1740–7.
22. Scher HI, Jia X, Chi K, et al. Randomized, open-label phase III trial of docetaxel plus high-dose calcitriol versus docetaxel plus prednisone for patients with castration-resistant prostate cancer. *J Clin Oncol* 2011;**29**(16):2191–8.
23. Vrignaud P, Semiond D, Lejeune P, et al. Preclinical antitumor activity of cabazitaxel, a semisynthetic taxane active in taxane-resistant tumors. *Clin Cancer Res* 2013;**19**(11):2973–83.
24. Mita AC, Figlin R, Mita MM. Cabazitaxel: more than a new taxane for metastatic castrate-resistant prostate cancer? *Clin Cancer Res* 2012;**18**(24):6574–9.
25. de Bono JS, Oudard S, Ozguroglu M, et al. Prednisone plus cabazitaxel or mitoxantrone for metastatic castration-resistant prostate cancer progressing after docetaxel treatment: a randomised open-label trial. *Lancet* 2010;**376**(9747):1147–54.
26. Bahl A, Oudard S, Tombal B, et al. Impact of cabazitaxel on 2-year survival and palliation of tumour-related pain in men with metastatic castration-resistant prostate cancer treated in the TROPIC trial. *Ann Oncol* 2013;**24**(9):2402–8.
27. Chittoor S, Berry W, Loesch D, et al. Phase II study of low-dose docetaxel/estramustine in elderly patients or patients aged 18–74 years with hormone-refractory prostate cancer. *Clin Genitourin Cancer* 2006;**5**(3):212–8.
28. Eymard JC, Priou F, Zannetti A, et al. Randomized phase II study of docetaxel plus estramustine and single-agent docetaxel in patients with metastatic hormone-refractory prostate cancer. *Ann Oncol* 2007;**18**(6):1064–70.
29. Oudard S, Banu E, Beuzeboc P, et al. Multicenter randomized phase II study of two schedules of docetaxel, estramustine, and prednisone versus mitoxantrone plus prednisone in patients with metastatic hormone-refractory prostate cancer. *J Clin Oncol* 2005;**23**(15):3343–51.
30. Moore MR, Troner MB, DeSimone P, et al. Phase II evaluation of weekly cisplatin in metastatic hormone-resistant prostate cancer: a Southeastern Cancer Study Group Trial. *Cancer Treat Rep* 1986;**70**(4):541–2.
31. Qazi R, Khandekar J. Phase II study of cisplatin for metastatic prostatic carcinoma. An Eastern Cooperative Oncology Group study. *Am J Clin Oncol* 1983;**6**(2):203–5.
32. Canobbio L, Guarneri D, Miglietta L, et al. Carboplatin in advanced hormone refractory prostatic cancer patients. *Eur J Cancer* 1993;**29a**(15):2094–6.
33. Kelly WK, Curley T, Slovin S, et al. Paclitaxel, estramustine phosphate, and carboplatin in patients with advanced prostate cancer. *J Clin Oncol* 2001;**19**(1):44–53.
34. Oh WK, Manola J, Ross RW, et al. A phase II trial of docetaxel plus carboplatin in hormone refractory prostate cancer (HRPC) patients who have progressed after prior docetaxel chemotherapy. Prostate Cancer Symposium. *J Clin Oncol* 2007;**24**.
35. Kentepozidis N, Soultati A, Giassas S, et al. Paclitaxel in combination with carboplatin as salvage treatment in patients with castration-resistant prostate cancer: a Hellenic oncology research group multicenter phase II study. *Cancer Chemother Pharmacol* 2012;**70**(1):161–8.
36. Ross RW, Beer TM, Jacobus S, et al. A phase 2 study of carboplatin plus docetaxel in men with metastatic hormone-refractory prostate cancer who are refractory to docetaxel. *Cancer* 2008;**112**(3):521–6.
37. Dorff TB, Tsao-Wei DD, Groshen S, et al. Efficacy of oxaliplatin plus pemetrexed in chemotherapy pretreated metastatic castration-resistant prostate cancer. *Clinical Genitourin Cancer* 2013;**11**(4):416–22.
38. Gasent Blesa JM, Giner Marco V, Giner-Bosch V, et al. Phase II trial of oxaliplatin and capecitabine after progression to first-line chemotherapy in androgen-independent prostate cancer patients. *Am J Clin Oncol* 2011;**34**(2):155–9.
39. Jordan MA, Toso RJ, Thrower D, et al. Mechanism of mitotic block and inhibition of cell proliferation by taxol at low concentrations. *Proc Natl Acad Sci U S A* 1993;**90**(20):9552–6.
40. Goodin S, Kane MP, Rubin EH. Epothilones: mechanism of action and biologic activity. *J Clin Oncol* 2004;**22**(10):2015–25.
41. Hussain M, Tangen CM, Lara Jr PN, et al. Ixabepilone (epothilone B analogue BMS-247550) is active in chemotherapy-naive patients with hormone-refractory prostate cancer: a Southwest Oncology Group trial S0111. *J Clin Oncol* 2005;**23**(34):8724–9.
42. Rosenberg JE, Weinberg VK, Kelly WK, et al. Activity of second-line chemotherapy in docetaxel-refractory hormone-refractory prostate cancer patients: randomized phase 2 study of ixabepilone or mitoxantrone and prednisone. *Cancer* 2007;**110**(3):556–63.
43. Fields-Jones S, Koletsky A, Wilding G, et al. Improvements in clinical benefit with vinorelbine in the treatment of hormone-refractory prostate cancer: a phase II trial. *Ann Oncol* 1999;**10**(11):1307–10.
44. Morant R, Hsu Schmitz SF, Bernhard J, et al. Vinorelbine in androgen-independent metastatic prostatic carcinoma – a phase II study. *Eur J Cancer* 2002;**38**(12):1626–32.
45. Abratt RP, Brune D, Dimopoulos MA, et al. Randomised phase III study of intravenous vinorelbine plus hormone therapy versus hormone therapy alone in hormone-refractory prostate cancer. *Ann Oncol* 2004;**15**(11):1613–21.
46. Loriot Y, Massard C, Albiges L. Personalizing treatment in patients with castrate-resistant prostate cancer: A study of predictive factors for secondary endocrine therapies activity; 2012 *J Clin Onc* 30, 2012 (Suppl. 5, abstr 213).
47. Sweeney CJ, Chen Y-H, Carducci MA, et al. Impact on overall survival (OS) with chemohormonal therapy versus hormonal therapy for hormone-sensitive newly metastatic prostate cancer (mPrCa): An ECOG-led phase III randomized trial. ASCO, Chicago. *J Clin Oncol* 2014;**32**(5s).
48. Droz JP, Balducci L, Bolla M, et al. Management of prostate cancer in older men: recommendations of a working group of the International Society of Geriatric Oncology. *BJU Int* 2010;**106**(4):462–9.
49. Beer TM, Ryan CW, Venner PM, et al. Intermittent chemotherapy in patients with metastatic androgen-independent prostate cancer: results from ASCENT, a double-blinded, randomized comparison of high-dose calcitriol plus docetaxel with placebo plus docetaxel. *Cancer* 2008;**112**(2):326–30.
50. de Bono JS, Molife LR, Sonpavde G, et al. Phase II study of eribulin mesylate (E7389) in patients with metastatic castration-resistant prostate cancer stratified by prior taxane therapy. *Ann Oncol* 2012;**23**(5):1241–9.

CHAPTER

# 55

# Antiandrogen Monotherapy in the Treatment of Prostate Cancer

*Zachary Piotrowski, MD**, *Richard E. Greenberg, MD, FACS***

*Department of Urology, Fox Chase Cancer Center, and Temple University School of Medicine, Philadelphia, PA, USA

**Division of Urologic Oncology, Fox Chase Cancer Center, and Temple University School of Medicine, Philadelphia, PA, USA

## INTRODUCTION

The development and subsequent normal physiologic function of the adult prostate gland is heavily dependent upon the presence of androgens. Although between 70% and 90% of testosterone is produced by the testicles, under certain circumstances the contribution by the adrenal glands can also be of importance. For the last 80 years, it has been known that removal of male hormones by surgical or medical castration, or inhibition of their activities by the administration of female hormones, can drastically influence the development of benign and malignant prostate tissue.[1]

Androgen activity is governed by the presence and expression of the androgen receptor and, regardless of the sort of androgens it is exposed to, blockage of this receptor will materially alter the activity of the prostate cell. Because all malignant prostate cells will ultimately become androgen independent, the efficacy of antiandrogen treatment is ultimately of limited palliative use. Debate still continues over whether a percentage of androgen-independent cells are present when cancer is still very small or whether that independence is achieved by changes occurring in the androgen receptor or its expression during the course of hormonal therapy. The corollary of this statement is that all prostate cancer cells are at one time androgen dependent. This is critical when evaluating patients for primary treatment of systemic prostate cancer.

Androgen receptor antagonists, antiandrogens, were developed in the late 1970s and are structurally of two different classes, steroidal and nonsteroidal. Because they can inhibit androgen activity regardless of a testicular or adrenal source, as monotherapy the receptor antagonists have always been of interest. In addition, steroidal androgen receptor antagonists will inhibit production of luteinizing hormone from the pituitary and, therefore, induce a degree of medical castration as well as interrupting the androgen pathway at a cellular level.

In the late 1980s, owing largely to the work of Labrie et al.,[2] the concept of removing the testicular source of androgens by surgical castration or the use of the recently developed luteinizing hormone-releasing hormone (LHRH) agonists, combined with the administration of an antiandrogen, the so-called maximal androgen blockade (MAB), became appealing. Many studies since the 1980s have uncovered differences not only between the steroidal and nonsteroidal antiandrogen but also among the nonsteroidal antiandrogens themselves, showing that they probably have different modes and sites of action. The question remained as to whether the different compounds could play different roles in the treatment of advanced prostate cancer. It was not until around the year 2000 that studies began to call into question the equivalence of medical castration, MAB or the use of antiandrogen monotherapy, with systematic reviews demonstrating that in fact, antiandrogen monotherapy is likely to be inferior to medical or surgical castration with regard to progression-free (PFS) and overall survival (OS).[3,4] As such, antiandrogen monotherapy is now more judicially practiced, usually instituted by practitioners on a case-by-case basis, and represents an important milestone in the evolving treatment of prostate cancer. Often the driving force for the use of androgen monotherapy is the desire to control the cancer while limiting toxicities usually associated with low testosterone.

Prostate Cancer. http://dx.doi.org/10.1016/B978-0-12-800077-9.00055-4

## THE ANDROGEN RECEPTOR

The androgen receptor is a member of the family of nuclear receptor transcription factors.[5] The protein has a DNA-binding domain, a ligand-binding domain, and a number of transactivation domains. The androgen receptor gene is situated on the X Chromosome. Normally the androgen receptor is situated in the cytoplasm but in the presence of androgens migrates into the nucleus. Testosterone is converted by 5-alpha reductase in the prostate cell and as 5-alpha-dihydrotestosterone attaches to the androgen receptor, a dimer of these molecules then makes its way to the nucleus and attaches via the DNA-binding segment to the nuclear DNA. The androgen receptor's antagonists compete with 5-dihydrotestosterone for the ligand-binding domain of the receptor.

When the activated receptor dimer arrives on the DNA of the prostate cell, it sets into motion a series of transcription events, which stimulate genes in the cell cycle to give rise to cell proliferation, suppresses apoptosis, and also causes the production of a number of proteins of different enzymatic functions, notably prostate-specific antigen (PSA), the most important marker of prostatic cellular activity. In the proliferation pathway the sequence of events involves the production of proteins, which in turn stimulate genes to produce a variety of growth factors, specifically epidermal growth factor (EGF), insulin-like growth factor, and fibroblast growth factor (FGF), which diffuse out of the cell and, by an autocrine or paracrine mechanism via their own specific receptors, stimulate further cellular proliferation. Some of the androgen target genes are the cyclin-dependent kinases CDK-2, CDK-4, and also BCL-2, the most important androgen-regulated gene in apoptosis.[6,7]

The development of androgen-independent prostate cancer may involve growth factor pathways that completely bypass the androgen receptor and act independently, such as EGP or insulin growth factor pathways, or the development of changes in the receptor itself, leading to overexpression or mutation. Such a mutation has been clearly defined in the LINCaP prostate cancer cell line, where a threonine to alanine substitution at position 877 has led to a dramatic effect on ligand specificity, enabling the cell to proliferate in response, not only to testosterone and dihydrotestosterone, but also to antiandrogens.[8]

That such a mutation occurs clinically has been shown in the clinical manifestation of the antiandrogen withdrawal phenomenon, where patients who have been treated with MAB and who have progressed can respond to withdrawal of the antiandrogen. This phenomenon has been shown with all the antiandrogens, although initially reported with flutamide.[8]

## ANDROGEN-DEPENDENT AND ANDROGEN-INDEPENDENT PROSTATE GROWTH

The development of androgen-independent proliferation in prostate cancer is a gradual process and there are a number of locations along the pathway where the careful use of different modalities of endocrine therapy can cause repeated responses in the malignant cells.

Table 55.1 shows the development of the completely endocrine-independent cell occurring in four stages. The hormone-naïve patient, who has been diagnosed as suffering from prostate cancer that is no longer to be

**TABLE 55.1** Methodological Classification of Prostate Cancer Based or Hormone Sensitivity

| Category | | Tumor factors | Host factors |
|---|---|---|---|
| 1. Androgen dependent | Endocrine naïve: 0<br>No prior hormone therapy | Antitumor effect:<br>1. Androgen withdrawal<br>2. Antiandrogens are administered | Physiologic level of androgens in the blood |
| 2. Androgen dependent | Endocrine sensitive:<br>1. Relapse after neoadjuvant therapy<br>2. Intermittent therapy-planned discontinuation of hormones<br>3. Relapse on antiandrogens alone | Decrease in proliferation if:<br>1. Androgens are withdrawn<br>2. Antiandrogens are administered (except group 3) | Noncastrate levels of androgens in the blood |
| 3. Androgen independent | Endocrine sensitive | Decrease in proliferation in response to:<br>1. Adrenal androgen blockade<br>2. Corticosteroids<br>3. Withdrawal of agents that bind steroid hormone receptors<br>4. Other hormone manipulators | Castrate levels of testosterone |
| 4. Hormone independent | Androgen independent and endocrine insensitive | Insensitive to all hormonal manipulation(s) | Castrate levels of testosterone |

cured by radical prostatectomy or radiotherapy, will respond to the use of monotherapy with androgen receptor antagonists, either steroidal or nonsteroidal, or medical or surgical castration. Within a period of 1.5–3 years the cells will show resistance to this monotherapeutic option and require hormone treatment. In patients who have received monotherapy with an antiandrogen, there will still be normal levels of testosterone in the blood, and in some instance the levels may be even slightly elevated. Withdrawal of the patient's own androgens by means of medical or surgical castration will give rise to further response in approximately half of the patients. In those patients who have been medically or surgically castrated, there will be a response in approximately 30% to the administration of an androgen receptor antagonist, which will block the androgens, or androgen precursors, produced by the adrenal glands. At this point, the patient's own testosterone will be at castrate level. Although the tumor is probably now androgen resistant, it will still respond to other endocrine manipulations, such as the administration of corticosteroids or, in the case of MAB, to the withdrawal of an antiandrogen. In addition, at this stage, many patients will respond to estrogens, probably as a result of their cytotoxic effects. Only after further proliferation of the cancer cells is there talk of a truly endocrine-independent tumor.[9]

## STEROIDAL ANTIANDROGENS

### Cyproterone Acetate

In the late 1970s the steroidal androgen receptor antagonist and antiprogestational agent, cyproterone acetate (CPA) became available. Jacobi et al. and later Tunn et al. showed cyproterone acetate in a dose of 300 mg weekly by intramuscular injection is as effective as diethylstilbestrol in a dose of 3 mg/day.[10,11]

In 1976, the European Organization for Research and Treatment in Cancer (EORTC) started a randomized phase III study of cyproterone acetate, 250 mg/day versus diethylstilbestrol (DES) 3 mg/day, and medroxyprogesterone acetate in a low dose of 200 mg/weekly after 8 weeks at a starting dose of 500 mg intramuscularly ×3 weekly. A total of 217 patients were recruited for the study and the end-points were overall response, progression-free survival (PFS), and overall survival (OS). Responses in localized and metastatic disease were measured but their inaccuracy led to subsequent dropping of this metric of hormonal activity from future studies. Nevertheless, the response rates were similar between cyproterone acetate and diethylstilbestrol, both being superior to medroxyprogesterone acetate. Similarly, the time to progression did not appear to be statistically different. Medroxyprogesterone acetate remained the weakest of the three and this may well have been a reflection of the dose of the compound used. Being a steroid, CPA exhibited many of the same side effects as diethylstilbestrol but to a lesser extent.

Nevertheless, significant cardiovascular events occurred in around 8% of patients although there were no documented cardiovascular deaths. Gynecomastia was less of a problem compared with DES; however, loss of libido and potency occurred in approximately 70%.[12] Subsequently, the EORTC Genitourinary (GU) Group carried out a further trial of CPA monotherapy versus flutamide in patients presenting with asymptomatic metastatic prostate cancer. Final results demonstrated no difference in efficacy between the two therapies. What has been found to be a dramatic feature of the study has been the side effect profile. The side effect profile of flutamide dosed 250 mg three times per day, notably with diarrhea in 23%, nausea, loss of appetite and liver function disturbances, was considerably worse than that of cyproterone acetate. In a small number of patients who were sexually active at the beginning of the trial, the same percentage remained sexually active on both arms, showing that loss of potency and libido occurs in a slower fashion than originally anticipated. Unfortunately, sexual activity and erectile function were no better preserved with either therapy.[13]

As monotherapy, CPA has also found a role in the treatment of hot flushes caused by medical or surgical castration. In doses as low as 50 mg/day patients have reported relief of this distressing symptom in over 50% of cases.[14] CPA, along with all other antiandrogens, can also be used when starting an LHRH agonist to inhibit the testosterone surge, which can occur in the first 4 weeks of treatment.

## NONSTEROIDAL ANTIANDROGENS

### Flutamide

Flutamide was originally used in a number of phase II trials, notably by Sogani et al.,[15] in patients with advanced metastatic prostate cancer. It showed a subjective response in over 70% and an objective response of just fewer than 50%. Also noted were good subjective responses where the patient had been previously surgically castrated. Unlike the steroidal antiandrogens, the nonsteroidal antiandrogens appear to exert no cardiovascular side effects. Flutamide does, however, cause gastrointestinal side effects, most noticeably diarrhea in between 17% and 23% of patients plus nausea, anorexia, and depression in a further 15%. In addition, hepatotoxicity has been shown in between 3% and 8% of cases. Two different dosage schedules have been employed in phase II studies and one small phase III study: 250 mg

3× per day and 1500 mg/day. The side effects are noticeably worse in the higher dosages and there appeared to be no additional clinical benefit. However, studies have not been carried out with this compound to determine the maximally tolerated dose.[16]

Three important phase III studies have been carried out using flutamide 250 mg. Boccon-Gibod carried out a study of 104 patients, half of whom underwent surgical castration and the other half received flutamide monotherapy. There was no therapeutic difference in the two arms. Flutamide did, however, give rise to more side effects than were experienced in the castration arm.[17] In a large Italian study with over 420 patients, Pavone-Macaluso compared flutamide with MAB. Again, there appeared to be no difference in the time to progression and cancer-related survival. Once more, the side-effect profile of flutamide led to withdrawal of patients because of diarrhea, and there was some hepatic toxicity.[18] The last important trial was mentioned earlier and comprises that of the EORTC GU Group with 310 patients randomized between flutamide and cyproterone acetate. Final results demonstrated no difference in efficacy between the two therapies. Sexual activity and erectile function were no better preserved with either therapy yet overall toxicity was more pronounced with flutamide.[13]

## Nilutamide

This nonsteroidal antiandrogen has been investigated in one major phase II study[19] comprising only 26 patients. Further information on its possible role as monotherapy was also obtained via phase IV study, a postmarketing surveillance study, carried out by Schasfoort et al. wherein 80 patients for one reason or another used the compound as monotherapy. It has one great advantage over flutamide in having a long half-life, which allows once daily dosing.[20]

It has an entirely different side-effect profile compared to other nonsteroidal antiandrogens. In approximately a quarter of the patients there is mild alcohol intolerance. In 15%, in various combined trials, there appears to be visual disturbances, amounting to a failure of light–dark adaptation in approximately 5%, making it extremely dangerous for patients on this compound to drive at night. Interstitial pneumonitis with pulmonary fibrosis has also been described in a small percentage of patients subjected to nilutamide therapy. That the compound is an effective antiandrogen has been shown in a large phase III study, when the compound was used as part of MAB.[21] In this study with over 400 patients, the combined therapy arm appeared superior to orchiectomy alone. At the present time, there are no ongoing studies with nilutamide monotherapy.

As with all nonsteroidal antiandrogen monotherapy, there is risk of gynecomastia. This is caused by the normal levels of circulating testosterone being unable to attach to receptors in the normal way and being converted by aromatase to estrogens.

## Bicalutamide

Of all available nonsteroidal antiandrogens, bicalutamide has been shown experimentally to have the greatest affinity for the androgen receptor. It, therefore, would theoretically appear to be the most powerful of this group of compounds.[22] It is also the compound with the longest half-life, and therefore, can be given in a once-daily dosage. The side-effect profile is favorable compared to the other two nonsteroidal antiandrogens. Gynecomastia is the only serious side effect and that may be preventable by pretreatment radiotherapy to the breast tissue or the use of an antiestrogen or aromatase inhibitors. Apart from the pharmacological side effects anticipated from androgen blockade, there are no significant cardiovascular side effects reported. There are approximately 7% of patients who will experience a reversible rise in transaminases, which reverts with discontinuation of the medication.

In extensive dose-ranging studies, a series of phase II and phase III studies confirmed the dose of 150 mg/day as appropriate for monotherapy. Even in doses as high as 600 mg/day, there were no additional dose-limited side effects.[23]

In two large phase III studies, comparing 50 mg Casodex/day with castration, Casodex was found to be inferior in terms of the time of progression and survival. In further phase III studies, comparing Casodex 150 mg/day and castration, in patients with nonmetastatic disease ($n = 480$), Casodex was found to be equivalent to castration with fewer side effects and in patients with small-volume metastatic disease and PSA levels <400, there was also equal efficacy.[24]

Further development of monotherapy with bicalutamide came from a large worldwide study with over 8000 patients that was completed in 2001, where patients following radical prostatectomy, definitive radiotherapy, or those being managed by a policy of surveillance, without evidence of disease progression were randomized to receive placebo or bicalutamide. Interim analysis revealed that those treated with bicalutamide developed fewer signs of progression and, most noticeably, development of metastases than those who received no androgen antagonist. Although only a handful of events occurred, the difference between the two arms was so dramatic that it was felt necessary to publish the results and, therefore, give collaborators a chance to offer patients therapy earlier than was initially planned.[25]

The early results of a phase II study of bicalutamide in advanced prostate cancer patients who have failed conventional hormonal manipulation showed that while there was no definitive objective response, there was a subjective improvement in serious symptoms, such as pain and lethargy, with an improvement in the quality of life of patients within the first year of therapy.[26] The exact mechanism of this response has not been elucidated but it may relate to the observation that bicalutamide appears to maintain its antagonist activity longer than the other antiandrogens. In other words, if mutation occurs at the androgen receptor, it does not mean that bicalutamide will automatically act as an agonist. Currently, it has a role in prostate cancer patients with biochemical failure to delay further progression of disease.

### MDV3100/Enzalutamide

Enzalutamide, formerly MDV3100, is a second-generation antiandrogen. The compound targets multiple steps in the androgen receptor-signaling pathway, specifically with its ability to inhibit nuclear translocation of the androgen receptor, DNA binding, and coactivator recruitment. These actions, along with the drug's higher affinity for the androgen receptor, make it distinct from its first-generation counterparts. Enzalutamide was initially shown to induce tumor shrinkage in xenograft models, which led to phase I and II trials conducted on men with castrate-resistant prostate cancer. These studies showed a strong antitumor response in men, regardless of prior treatment. Although these pilot studies did not necessarily examine enzalutamide in the setting of primary monotherapy, the results were, nonetheless, exciting. In 2012, the *New England Journal of Medicine* published results of the AFFIRM study. The study evaluated approximately 1200 men with castrate-resistant prostate cancer who had previously undergone chemotherapy with docetaxel. Groups were randomized to receive 160 mg of enzalutamide daily versus matched placebo and results demonstrated a 37% reduction in the risk of death in the enzalutamide group with a median 4.8 month increase in survival. The side effect profile is acceptable, including symptoms like fatigue, diarrhea, and hot flashes with a small risk of increase of seizure threshold in patients with predisposing conditions.[27] Even more recent studies from 2014 show a potential benefit for enzalutamide monotherapy in metastatic prostate cancer prior to the induction of chemotherapy.[28]

At the time of this publication, we know of at least one ongoing European trial that is transitioning the use of enzalutamide to determine its possible efficacy as primary monotherapy. In a single-arm, phase II trial, 67 hormone-naive patients from European centers were analyzed for their response to enzalutamide. While some of these patients had received primary therapy in the form of radiation or surgery, all were hormone-naïve. The primary endpoint was PSA reduction, and by 25 weeks of treatment, the majority of patient's experienced an 80% decline in PSA.[29]

Despite the overwhelming trend away from antiandrogens as initial monotherapy, the results of the AFFIRM trial confirm the significance of the androgen receptor in the development and progression of prostate cancer and reinforce that even as the disease becomes castrate resistant or "androgen independent," as it was formerly known, the action of androgens will continue to play an important role in future treatments. The *Lancet* study begins to scratch the surface of antiandrogen therapies potential to provide affective disease suppression in men with earlier stages of disease and represents a necessary area for further research. With the potential benefits of enzalutamide, we need to address cost/benefit issues as this second-generation antiandrogen is much more expensive than any other antiandrogen used for prostate cancer.

## ANTIANDROGEN MONOTHERAPY

Of the four commercially available antiandrogens, three have been primarily investigated and historically used in monotherapy. Nilutamide has been used prospectively in one small phase II trial and some further information has accrued from a phase IV study, where 80 patients received effective monotherapy with compound but has mostly been abandoned. The remaining three, cyproterone, flutamide, and bicalutamide, have all been used as monotherapy in phase II and phase III trials. Bicalutamide has undergone exhaustive dose ranging studies. The others have been used with a degree of empiricism.[23] Table 55.2 compares the toxicities of the

**TABLE 55.2** Side Effects of Androgen-Receptor Antagonists

| Side effects | Compound |
|---|---|
| Cardiovascular | DES>CPA>MPA>>>NSAA |
| Gynecomastia | Bical=Flu=Nil>DES>MPA=CPA |
| Liver function disturbance | Flu>DES>CPA>Nil=Bical>MPA |
| Sexual function | DES>MPA=CPA=Flu>Nil=Bical |
| Hot flushes | DES>>MPA=CPA>Flu=Nil=Bical |
| Alcohol intolerance | DES=CPA>Flu>Bical |
| Visual problems | Nil>MPA>>the others |
| Pulmonary fibrosis | Nil>>the others |

Bical, Bicalutamide; CPA, cyproterone acetate; DES, diethylstilbestrol; Flu, flutamide; MPA, medroxyprogesterone acetate; Nil, nilutamide; NSAA, nonsteroidal antiandrogens.

androgen receptor antagonists and explains why the suggestion has been made that bicalutamide, being less toxic, is the most suitable for early therapy.

## COMMENTARY

As senior author, I want to take the opportunity to mention a contemporary "off-label" use of antiandrogen monotherapy I have been employing for some time. There are no level I studies or references available to support this truly anecdotal experience. However, in my hands, this has been an exceptional tool for managing especially elderly patients with biochemical-only hormone-sensitive prostate cancer. Specifically, patients with slowly rising PSA levels without demonstrable metastatic disease who have escaped an observation protocol or in the octogenarian population associated with comorbidities that preclude standard primary interventions, daily low dose (50 mg) bicalutamide provides excellent PSA control. In addition, in this elderly population, the psychosocial impact associated with the stigma of a rising PSA can be greatly ameliorated without significant toxicity and also importantly cost. Almost every 80 year old knows his PSA.

## CONCLUSIONS

Following the discovery that orchiectomy drastically improved prostate cancer progression, the use of androgen ablation therapies has made little headway in improving the long-term survival of prostate cancer patients. While adjuvant usage with radiotherapy or other local treatment have been shown to confer a survival benefit and implementation for symptomatic disease have led to improvement for patients, the data have not been as robust for monotherapy. In fact, an initial systematic review of the literature in 2000 was among the first to suggest that antiandrogen monotherapy may lead to lower survival rates.[3]

Since that time, combined randomized trials have failed to demonstrate that bicalutamide was even 75% as effective as surgical castration in prolonging time to progression or survival in clinical stage T3–T4 prostate cancer. In conjunction with the fact that hormonal or medical castration therapy with LHRH analogues and antagonists are equivalent to surgical castration, many societies, including the American Society for Clinical Oncology, have recommended against the usage of the apparently inferior antiandrogen monotherapy, specifically cyproterone acetate.[30]

Further evidence was released in a 2014 Cochrane Review of 11 studies examining over 3000 patients. Although the evidence from these combined studies was felt to have some risk of bias, it was deemed to be of moderate strength. Overall survival and increased clinical progression were significantly decreased in antiandrogen monotherapy as compared to castration.[4]

## References


1. Huggins C, Hodges CV. Studies on prostatic cancer: I. The effect of castration, of estrogen and of androgen injection on serum phosphatases in metastatic carcinoma of the prostate. 1941. *J Urol* 2002;**168**(1):9–12.
2. Labrie F, Dupont A, Belanger A. Complete androgen blockade for the treatment of prostate cancer. *Important Adv Oncol* 1985;193–217.
3. Seidenfeld J, Samson DJ, Hasselblad V, et al. Single-therapy androgen suppression in men with advanced prostate cancer: a systematic review and meta-analysis. *Ann Intern Med* 2000;**132**(7):566–77.
4. Kunath F, Grobe HR, Rücker G, et al. Non-steroidal antiandrogen monotherapy compared with luteinising hormone-releasing hormone agonists or surgical castration monotherapy for advanced prostate cancer. *Cochrane Database Syst Rev* 2014;**6**:pCD009266.
5. Beato M, Herrlich P, Schutz G. Steroid hormone receptors: many actors in search of a plot. *Cell* 1995;**83**(6):851–7.
6. Chen Y, Robles AI, Martinez LA, et al. Expression of G1 cyclins, cyclin-dependent kinases, and cyclin-dependent kinase inhibitors in androgen-induced prostate proliferation in castrated rats. *Cell Growth Differ* 1996;**7**(11):1571–8.
7. McDonnell TJ, Troncoso P, Brisbay SM, et al. Expression of the protooncogene bcl-2 in the prostate and its association with emergence of androgen-independent prostate cancer. *Cancer Res* 1992;**52**(24):6940–4.
8. Veldscholte J, Ris-Stalpers C, Kuiper GG, et al. A mutation in the ligand binding domain of the androgen receptor of human LNCaP cells affects steroid binding characteristics and response to antiandrogens. *Biochem Biophys Res Commun* 1990;**173**(2):534–40.
9. Kelly WK, Scher HI. Prostate specific antigen decline after antiandrogen withdrawal: the flutamide withdrawal syndrome. *J Urol* 1993;**149**(3):607–9.
10. Jacobi GH, Altwein JE, Kurth KH, et al. Treatment of advanced prostatic cancer with parenteral cyproterone acetate: a phase III randomised trial. *Br J Urol* 1980;**52**(3):208–15.
11. Tunn UW, Weiglein W , Saborowski J, et al. Clinical experience with cyproterone acetate in a randomised and in an open trial. *Prog Clin Biol Res* 1987;**243A**:365–8.
12. Pavone-Macaluso M, de Voogt HJ, Viggiano G, et al. Comparison of diethylstilbestrol, cyproterone acetate and medroxyprogesterone acetate in the treatment of advanced prostatic cancer: final analysis of a randomized phase III trial of the European Organization for Research on Treatment of Cancer Urological Group. *J Urol* 1986;**136**(3):624–31.
13. Schroder FH, Whelan P, de Reijke TM, et al. Metastatic prostate cancer treated by flutamide versus cyproterone acetate. Final analysis of the "European Organization for Research and Treatment of Cancer" (EORTC) Protocol 30892. *Eur Urol* 2004;**45**(4):457–64.
14. Radlmaier A, Bormacher K, Neumann F. Hot flushes: mechanism and prevention. *Prog Clin Biol Res* 1990;**359**:131–40 discussion 141–153.
15. Sogani PC, Vagaiwala MR, Whitmore Jr WF. Experience with flutamide in patients with advanced prostatic cancer without prior endocrine therapy. *Cancer* 1984;**54**(4):744–50.
16. Neri R, Florance K, Koziol P, et al. A biological profile of a nonsteroidal antiandrogen, SCH 13521 (4′-nitro-3′trifluoromethylisobutyranilide). *Endocrinology* 1972;**91**(2):427–37.
17. Boccon-Gibod L. Are non-steroidal anti-androgens appropriate as monotherapy in advanced prostate cancer? *Eur Urol* 1998;**33**(2): 159–64.

18. Prostate Cancer Trialists' Collaborative Group. Maximum androgen blockade in advanced prostate cancer: an overview of the randomised trials. *Lancet* 2000;**355**(9214):1491–8.
19. Dijkman GA, Janknegt RA, De Reijke TM, et al. Long-term efficacy and safety of nilutamide plus castration in advanced prostate cancer, and the significance of early prostate specific antigen normalization. International Anandron Study Group. *J Urol* 1997;**158**(1):160–3.
20. Decensi AU, Boccardo F, Guarneri D, et al. Monotherapy with nilutamide, a pure nonsteroidal antiandrogen, in untreated patients with metastatic carcinoma of the prostate. The Italian Prostatic Cancer Project. *J Urol* 1991;**146**(2):377–81.
21. Furr BJ, Tucker H. The preclinical development of bicalutamide: pharmacodynamics and mechanism of action. *Urology* 1996;**47**(1A Suppl.):13–25 discussion 29–32.
22. Schasfoort EM, Van De Beek C, Newling DW. Safety and efficacy of a non-steroidal anti-androgen, based on results of a post marketing surveillance of nilutamide. *Prostate Cancer Prostatic Dis* 2001;**4**(2):112–7.
23. Kolvenbag GJ, Blackledge GR. Worldwide activity and safety of bicalutamide: a summary review. *Urology* 1996;**47**(1A Suppl.):70–9 discussion 80–4.
24. Iversen P, Tyrrell CJ, Kaisary AV, et al. Casodex (bicalutamide) 150-mg monotherapy compared with castration in patients with previously untreated nonmetastatic prostate cancer: results from two multicenter randomized trials at a median follow-up of 4 years. *Urology* 1998;**51**(3):389–96.
25. Wirth M, Tyrrell C, Wallace M, et al. Bicalutamide (Casodex) 150 mg as immediate therapy in patients with localized or locally advanced prostate cancer significantly reduces the risk of disease progression. *Urology* 2001;**58**(2):146–51.
26. Kucuk O, Fisher E, Moinpour CM, et al. Phase II trial of bicalutamide in patients with advanced prostate cancer in whom conventional hormonal therapy failed: a Southwest Oncology Group study (SWOG 9235). *Urology* 2001;**58**(1):53–8.
27. Scher HI, Fizazi K, Saad F, et al. Increased survival with enzalutamide in prostate cancer after chemotherapy. *N Engl J Med* 2012;**367**(13):1187–97.
28. Beer TM, Tombal B. Enzalutamide in metastatic prostate cancer before chemotherapy. *N Engl J Med* 2014;**371**(18):1755–6.
29. Tombal B, Borre M, Rathenborg P, et al. Enzalutamide monotherapy in hormone-naive prostate cancer: primary analysis of an open-label, single-arm, phase 2 study. *Lancet Oncol* 2014;**15**(6):592–600.
30. Wirth MP, Hakenberg OW, Froehner M. Antiandrogens in the treatment of prostate cancer. *Eur Urol* 2007;**51**(2):306–13 discussion 314.

CHAPTER

# 56

# Sipuleucel-T – A Model for Immunotherapy Trial Development

*Susan F. Slovin, MD, PhD*

Genitourinary Oncology Service, Sidney Kimmel Center for Prostate and Urologic Cancers, Memorial Sloan-Kettering Cancer Center, New York, NY; Department of Medicine, Weill-Cornell Medical College, USA

## INTRODUCTION

Among five new therapies for patients with metastatic castration-resistant prostate cancer, sipuleucel-T (Provenge®) stands out as the first immunotherapy approved by the FDA for asymptomatic or minimally symptomatic castrate metastatic prostate cancer with the added benefit of improvement in overall survival.[1] Its approval was determined by the results of a placebo-controlled, randomized trial (the IMPACT trial),[1] conducted in 512 asymptomatic or minimally symptomatic men with metastatic castration-resistant prostate cancer. Although no difference in time to progression or PSA response rate was reported, a statistically meaningful 4.1-month improvement in median survival was achieved in the active arm with respect to the placebo arm (25.8 months vs. 21.7 months) (Figure 56.1). The survival benefit was comparable to that seen with standard chemotherapy, docetaxel, and now with the new AR-directed agents, enzalutamide (Xtandi®)[2] and abiraterone (Zytiga®),[3,4] second-line chemotherapy cabazitaxel (Jevtana®),[5] and even radium-223 (Xofigo®).[6] While its indication is for patients with asymptomatic or minimally symptomatic castrate metastatic disease, its major endorsement has been its potential broad applicability to all clinical states of the disease, that is, from biochemical relapse postprimary therapy, to castrate nonmetastatic to even postchemotherapy.[7] Overall, the treatment paradigm for treating castrate metastatic prostate cancer has been altered by introducing AR-directed therapies earlier into the treatment scheme and also introducing immune-based therapies not only earlier but later into the treatment arena, that is, with or after chemotherapy or in combination with biologics or other hormonal agents (Figure 56.2).

## ELUCIDATING THE MECHANISM OF ACTION OF SIPULEUCEL-T

Immunotherapy has previously been widely tested in many different diseases with a predominance of effort in the hematologic diseases where monoclonal antibodies against cell surface molecules are given either alone or in concert with radiopharmaceuticals. Antibody-radionuclide conjugates have been successfully developed for the treatment of non-Hodgkin's lymphoma, resulting in the approval of agents that are CD20-targeted: $^{131}$I-tositumomab (Bexxar) and $^{90}$Y-ibritumomab tiuxetan (Zevalin),[9–11] which can produce response rates of 50–85% in a variety of lymphomas. Sipuleucel-T represented an autologous cellular product vaccine, which mandated that patients be leukapheresed to obtain peripheral blood mononuclear cells followed by a 48-h turnover during which cells were processed, expanded, then incubated with prostatic acid phosphatase (PAP) combined with GM-CSF. The cells were divided into three reinfusions, one given every 2 weeks. Patients were then monitored per clinical practice with imaging and PSAs. Overall, the treatment was well-tolerated with expected transfusion-associated side effects such as fever and chills. Since the FDA approval of sipuleucel-T, additional studies have sought to expand its use and to identify in which patients the greatest clinical benefits may be derived. A retrospective analysis of the IMPACT trial found that patients in the lowest quartile of PSA values derived a greater benefit from sipuleucel-T vaccine with a 13-month improvement in OS (41.3 months with sipuleucel-T compared with 28.3 months with placebo; [HR 0.51; 95% confidence interval (CI) 0.35–0.85)]. However, for those patients in the highest baseline PSA quartile, the

Prostate Cancer. http://dx.doi.org/10.1016/B978-0-12-800077-9.00056-6

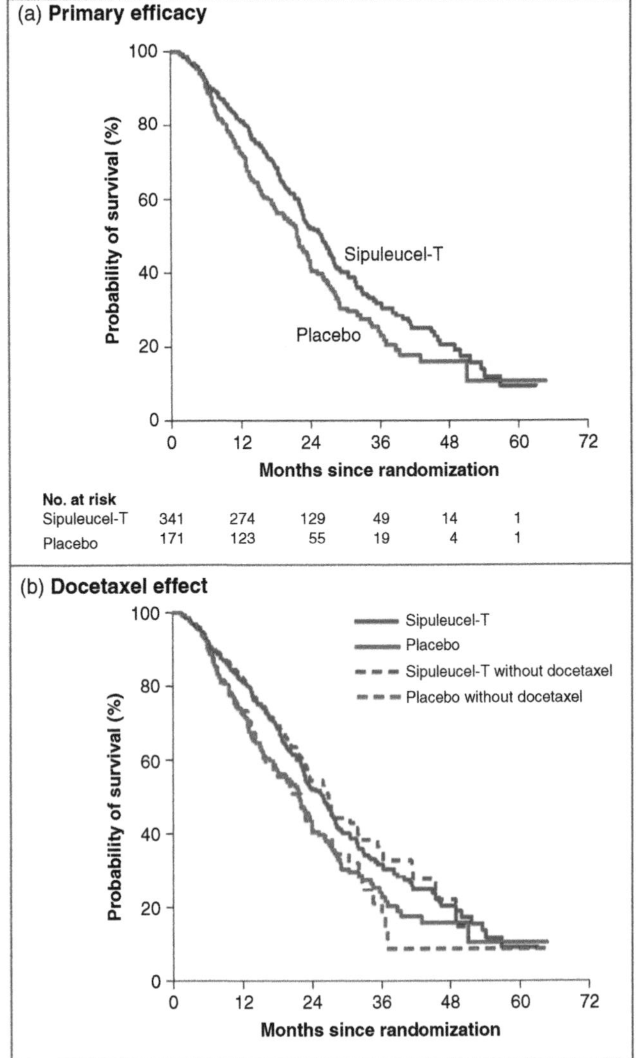


FIGURE 56.1 **Kaplan–Meier estimates of overall survival.** (a) The results of the primary efficacy analysis of treatment with sipuleucel-T as compared with placebo. (b) The results of the analysis with and without censoring at the time of the initiation of docetaxel therapy after study treatment. After censoring at the time of docetaxel initiation, a consistent treatment effect with sipuleucel-T was observed. *Reproduced with permission from Kantoff et al.*[1]

median OS was 18.4 compared with 15.6 months for placebo [HR 0.84; 95% CI 0.55–1.29], with an improvement of only 2.8 months.[12] Most of the studies with sipuleucel-T have been largely retrospective and its mechanism of action still controversial. A report by Drake et al.[13] postulated that antigen cascade may be responsible for its mechanism of action; also thought to be a key component of how the vaccine ProstVAC works.[14] Though antibodies to PAP were generated and a robust ki57 proliferative response was induced, suggesting involvement of the adaptive immune response. Additional evaluation of retrospective studies by Drake et al.[13] suggested that OS was enhanced in patients who received sipuleucel-T and had IgG antibody responses to greater than two secondary antigens compared with those patients who did not have antibodies induced. It should be noted that the word "response" should be used within the context of an association of antibody induction with a change in biology of the cancer, that is, clinical outcome, and not the generation of the antibody in response to an immunogen per se.

While sipuleucel-T was met with enthusiasm as the first immunotherapy approved for a solid tumor malignancy, the observation that overall survival was improved in the absence of significant clinical benefit solicited questions as to how the product worked and whether it in fact mediated an antitumor reaction of some sort. Many physicians in the field felt that knowing the mechanism of action was important in being able to build on the treatment, whereas others felt that knowing the mechanism of action made little difference in their use of the drug as long as the drug worked. As such, investigators have sought to identify a biomarker that may indicate that a target has been hit or that the immune system has been stimulated. Because of the mixed cellular nature of the autologous mononuclear cell product, it was unclear as to the nature of the effector population that may have been relevant in inducing a potential antitumor effect, and ultimately survival. Other than T cell proliferation assays, no one cellular marker was indicative of the drug inducing immunogenicity. The working premise has always been that the cellular product was enriched with antigen presenting cells (APC.) This was confirmed by the observation by flow cytometry that $CD54^+$ cells were responsible for antigen uptake and that $CD54^+$ cells harbored the PAP-specific antigen presentation activity as assayed using a PAP-specific HLA-DRβ1-restricted T cell hydridoma.[15,16] The marker CD54 or intracellular adhesion molecule-1 (ICAM-1) serves as a ligand for the CD11a/CD18 (LFA-1) leukocyte integrin complex and its interaction with other cell types is thought to be relevant in its role as a potential costimulatory receptor.[15,17,18] In the setting of sipuleucel-T, the fusion protein, PA2024, comprised of PAP fused to GM-CSF, was used as the immunogen, with the PAP portion of the molecule providing the necessary immunogenicity and the GM-CSF serving to activate the APC. Studies confirmed that the isolated $CD54^+$ cells took up antigen as well as presented and processed the antigen in an MHC-restricted manner.[15] Similar results were obtained with an HLA-DRβ1 restricted T cell hybridoma specific for a different PAP-derived peptide. These findings provided some insight to how the product might work *in vivo* but still required validation.

The opportunity to further validate the earlier role of $CD54^+$ cells was provided by the availability of cellular products from three phase III double-blind placebo-controlled trials in patients with metastatic CRPC including the IMPACT trial, which led to the product's FDA approval. Patients were randomized 2:1 in favor of

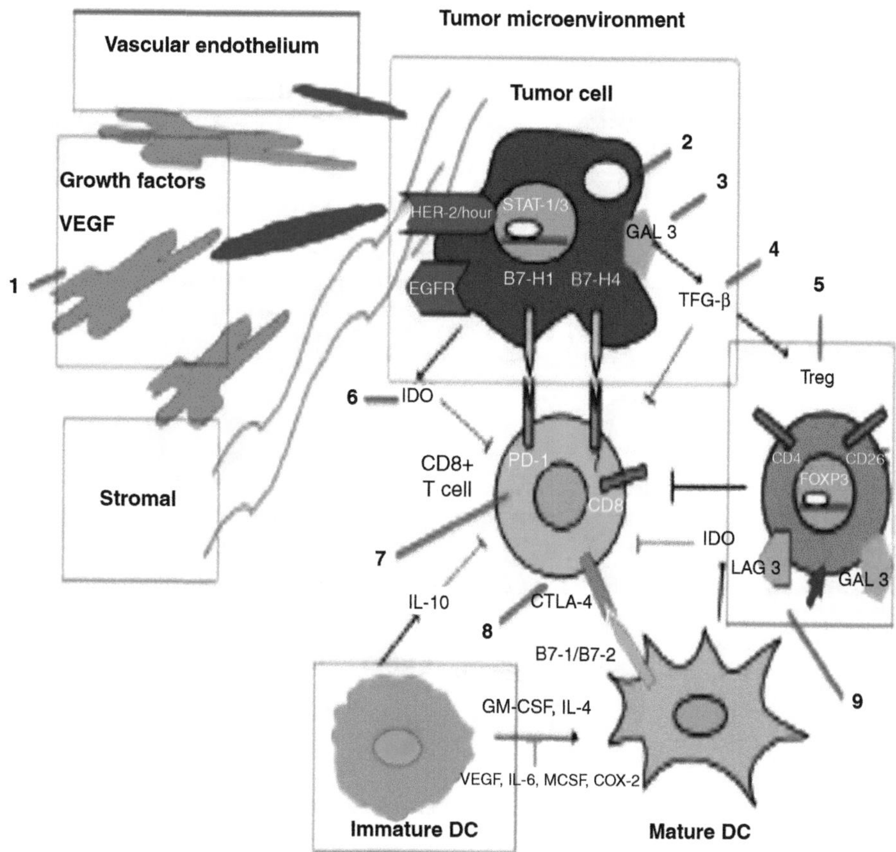


FIGURE 56.2 **Combinatorial approaches with immunotherapies and their impact on immune pathways.** *Reproduced with the permission from Le and Jaffee.*[8]

sipuleucel-T or to control. This included a minimum of at least one treatment with the cellular product as well as additional information provided from the product prepared at the primary manufacturing facility in Seattle, Washington. APC number, ACP activation, and total nucleated cells (TNC) were assessed in the control and investigational product. Also assessed were T cell proliferation and interferon-γ secretion by ELISPOT at treatment weeks 0, 2, and 4 in the IMPACT trial.[1,19,20] In the three trials, *ex vivo* APC activation was greater with sipuleucel-T relative to the control at weeks 0, 2, and 4 with the median APC activation increased approximately 6.2-fold. The median cumulative APC activation with sipuleucel-T alone across the three dose preparations was 26.7 [21.5, 33.6]. Elevated levels of T-cell activation-associated cytokines were noted during manufacture but not induced prior to and after exposure to GM-CSF alone. The treatment generated PA2024- and/or PAP-specific humoral responses in 68% (102/151) of patients compared with 3% (2/27) of control patients. The anti-PA2024 and anti-PAP antibody titers were greater in the sipuleucel-T arm compared with controls at all time points posttherapy, and a persistent response detectable after 26 weeks following initial posttreatment baseline.

It should be noted that overall product activation was confirmed by $TH_1$ cytokines (IFN-γ, TNF-α); TH2 cytokines (IL-5, IL-13) were also present implying that $TH_1$ and $TH_2$ cells were activated in an antigen-specific manner. IL-10 was less detectable relative to those cytokines that facilitate T cell expansion such as IL-2, IFN-γ, and TNF-α.[16] There also appeared to be a correlation between overall survival and T cell secretion of IFN-γ by ELISPOT and PA2024-specific antibody.[16]

## ENHANCING IMMUNOGENICITY

While induction of $CD54^+$ cells is a relatively straightforward mechanism by which the cellular product vaccine may work, it is not likely to be the other means by which enhancement of immune function may occur. The "immunologic orchestra" is far more complex with interactions among multiple cell types, inhibitory molecules, and cytokines, that may be either dependent or independent on MHC (Figure 56.2). Another mechanism of interest for the potential immunologic induction of sipuleucel-T is a concept known as "antigen/epitope spreading" or "antigen cascade" (Figure 56.3).[13,21,22] This

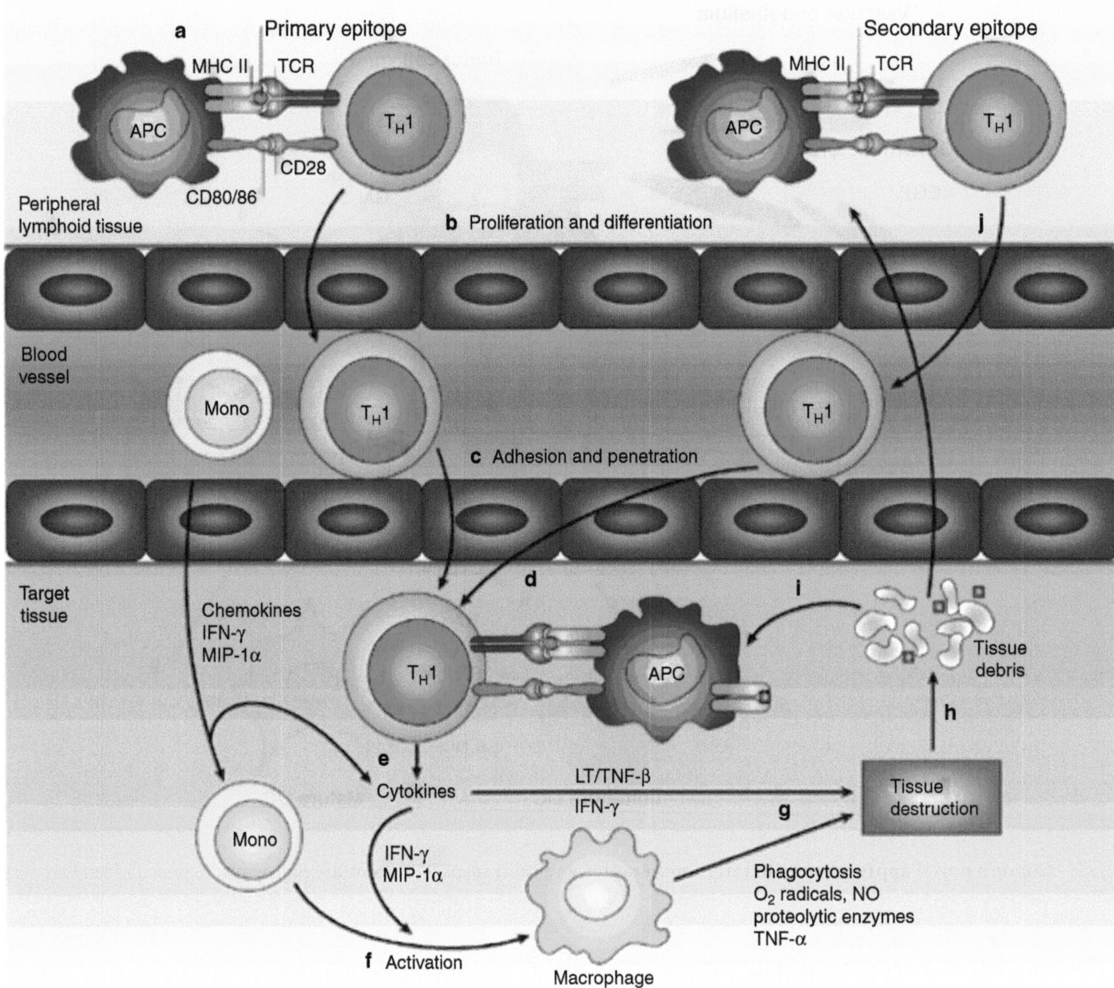


**FIGURE 56.3** **Cartoon describing antigen spreading.** *Reproduced with permission from Vanderlugt and Miller.*[25]

term implies that after the initial exposure of the immune system to a particular antigen or a series of antigens within a vaccine construct, the actual vaccine may lead to the priming of other T cells or B cells against tumor antigens (bystander effect) that are distinct form the original target(s) of the treatment. This observation may serve as a surrogate for clinical response and outcome. Thakurta et al.[21] used protein microarrays followed by validation studies using Luminex to assess human antigen spread and its association with overall survival (OS) after sipuleucel-T treatment from patients treated with the IMPACT trial. Higher IgG responses were seen on the ProtoArray against targets that were unusually expressed in prostate tumors. The Luminex xMAP confirmed that patients who received the autologous cellular product generated higher antibody titers against multiple tumor (cancer) antigens posttreatment (more than threefold), compared with control subjects (<5%) with an association of improved OS. This was further evaluated by Drake et al.[13] with an expanded database and found that IgG responses in this group of patients were against a diversity of secondary antigens including K-RAS oncogene and KLK2/hK2, serum human kallikrein-2 (hK2) and an hK2 gene (KLK2 variant),[23] a prostate-associated antigen. These responses were not seen in control patients. Based on further IMPACT patient analysis, the treated patients had longer OS survival with IgG responses to ≥2 secondary antigens than in those patients without response. The idea of antigen spreading may be applicable to other immunotherapies in the setting of delayed yet prolonged responses as seen in the PSA Tri-Com vaccine.[24]

## INTERROGATION OF THE NEOADJUVANT MILIEU

Among the mechanistic questions that remain unanswered about sipuleucel-T, exploration of an earlier clinical state may provide a more focused and controlled

**TABLE 56.1** While Neoadjuvant Studies can Evaluate Early Changes Within the Tumor Microenvironment, Interrogation of Tumors From Advanced Stage Disease may be Problematic

| Milieux for interrogation of immunologic integration | |
|---|---|
| **Neoadjuvant (Early)** | **Metastatic (Late)** |
| • Direct interrogation of tissue and stroma<br>• Impact on local disease to p0 or to MPR (<10% residual dz)<br>• Questionable impact on systemic progression unless validated in high risk population<br>• Hard to design trials due to long natural history | • High tumor burden<br>• Biology of disease different at met sites: bone versus LN versus visceral mets<br>• Not every site of known disease is active<br>• Immune cells can be crowded out by increase tumor cells in LN/bone |

Cancer: Th2-dominant disease either excess of IL-4, IL-5, IL-10, TGF-β production with a therapeutically drive shift back toward the Th1 profile.

evaluation as to how this drug impacts on the local tumor and tumor environment. The goals of testing drugs or therapies in the neoadjuvant window is to enable the study of the impact of a drug on the end organ, that is, prostate cancer, with the plan to perform a definitive intervention, that is, surgery, to explore the post- versus pretreatment effects of the drug(s). Fong et al.[26] studied 37 evaluable patients with previously untreated localized prostate cancer who were treated on a phase II open-label single arm trial of sipuleucel-T administered prior to radical prostatectomy. Endpoints included immune infiltrates in the prostatectomy specimens posttreatment and in paired pretreatment biopsies as determined by immunohistochemistry. Patients developed T cell proliferation and IFN-γ responses detectable in the blood posttreatment with more than threefold increase in the posttreatment prostate of infiltrating $CD3^+$, $CD4^+$, $FOXP3^+$, and $CD8^+$ T cells. The level of T cells infiltrate was seen at the tumor interface, that is, between the benign and malignant glands and was absent from a control group of untreated patients who went on to direct prostatectomy. Many of the infiltrating T cells were $Ki67^+$, a proliferation marker, as well as $PD\text{-}1^+$, a marker of activation. Interestingly, the level of magnitude of the immune response within the circulation did not appear to correlate with T cell infiltration within the prostate itself. While of significant interest as a measure of local immunogenicity, these observations may not account for the overall survival benefit seen in the metastatic prostate cancer trials, nor may not have any level of correlation with the serologic studies derived from those trials. What may be positively extrapolated from this trial is the fact that other solid tumor malignancies have shown a positive association between lymphocytic infiltrates within the primary or metastatic lesions and survival (Tables 56.1and 56.2).[27,28]

## USING THE SIPULEUCEL-T EXPERIENCE AS A PARADIGM FOR IMMUNOTHERAPY DEVELOPMENT

The unique experience with sipuleucel-T has fostered widespread investigation of not only how to improve its antitumor efficacy in the setting of survival benefit, but enhancing immunogenicity with biologics, cytokines, androgen-receptor (AR)-directed hormonal agents, and chemotherapy. Assessing the biologic impact of the

**TABLE 56.2** Updated Combination Trials with Sipuleucel-T

| | | | | | |
|---|---|---|---|---|---|
| II | Sipuleucel-T/ADT | NCT01431391 | Patients with nonmetastatic prostate cancer randomized to receive sipuleucel-T before or after ADT | Immune response | August 2014 |
| II | Sipuleucel-T/abiraterone | NCT01487863 | Patients with metastatic CRPC randomized to receive sipuleucel-T plus abiraterone and prednisone, administered either sequentially or concurrently | Immune response (including PAP-specific T-cell response); safety | July 2015 |
| II | Sipuleucel-T/enzalutamide | NCT01981122 | Patients with metastatic CRPC randomized to receive sipuleucel-T plus enzalutamide, administered either sequentially or concurrently | Immune response | September 2015 |
| Pilot | Sipuleucel-T/anti-PD1 antibody | NCT01420965 | Patients with metastatic CRPC randomized to sipuleucel-T with or without anti-PD1 and cyclophosphamide | Feasibility and immune response | December 2017 |

ADT, androgen-deprivation therapy; CRPC, castration-resistant prostate cancer; PAP, prostatic acid phosphatase; PSA, prostate-specific antigen; XRT, radiotherapy.

treatment has been variable and other than looking at CD54+ or other immune cells associated with humoral activation, there are no reliable immune biomarkers on which to base a treatment response. It should be remembered that no dramatic radiographic changes in scans were seen in the patient population. McNeel et al.[29] performed a retrospective analysis of patients from the three original phase III trials who were treated with sipuleucel-T and found an associated transient increase in serum eosinophil count at week 6 that resolved by week 14 in 105 of 377 patients (28%) studied. This eosinophil increase correlated with induced immune response, and longer prostate-cancer-specific survival [HR, 0.713; 9% CI 0.525–0.970; $P = 0.031$] and a trend toward overall survival [HR, 0.753; 95% CI, 0.563–1.008; $P = 0.057$]. Another association appeared to include transient globulin protein levels that also appeared to be associated with antigen-specific antibody responses. No correlation was seen with overall survival.

Investigating the optimal sequencing of sipuleucel-T and androgen-deprivation therapy (ADT), as well as the use of concurrent and sequential standard hormone therapy, the "STAND" trial is ongoing.[30,31] The population of this randomized phase II trial included patients with biochemically relapsed prostate cancer postprostatectomy with noncastrate levels of testosterone and no radiograph metastases. The primary endpoint was to determine whether androgen deprivation therapy (ADT) started before or after sipuleucel-T leads to an enhanced immune response. Secondary endpoints include humoral immune responses, cellular immune responses, cytokine responses, PSA responses, time to PSA progression, and metastasis-free survival (Figure 56.4). The arms included sipuleucel-T given as a lead-in 2 weeks before the start of ADT; the second arm starts ADT for 3 months followed by the initiation of sipuleucel-T. Sipuleucel-T induced robust responses to PA2024 and PAP. Antibody responses correlated with cumulative total nucleated cell count, PAP, and maximum eosinophil count. There were no overall differences between treatment arms. Cellular responses were robust and persisted for at least 12 months in both patient groups. Similarly, Drake et al.[32] provided further updates of these results of 65 men who have been followed by ≥9 months post-ADT (Figure 56.4). Sipuleucel-T continued to induce robust responses to PA2024 and PAP suggestive of T cell memory responses and ELISA antibody responses to PAP and PA2024 in both arms. Antigen-specific T cell proliferation increased after the second infusion and persisted to 26 weeks. These data were consistent with earlier studies but were suggestive that maximum eosinophil count after sipuleucel-T may be a marker associated with humoral responses.

Petrylak et al.[33] reported an update on the "STRIDE" trial (STRIDE-Randomized phase 2 open-label study of

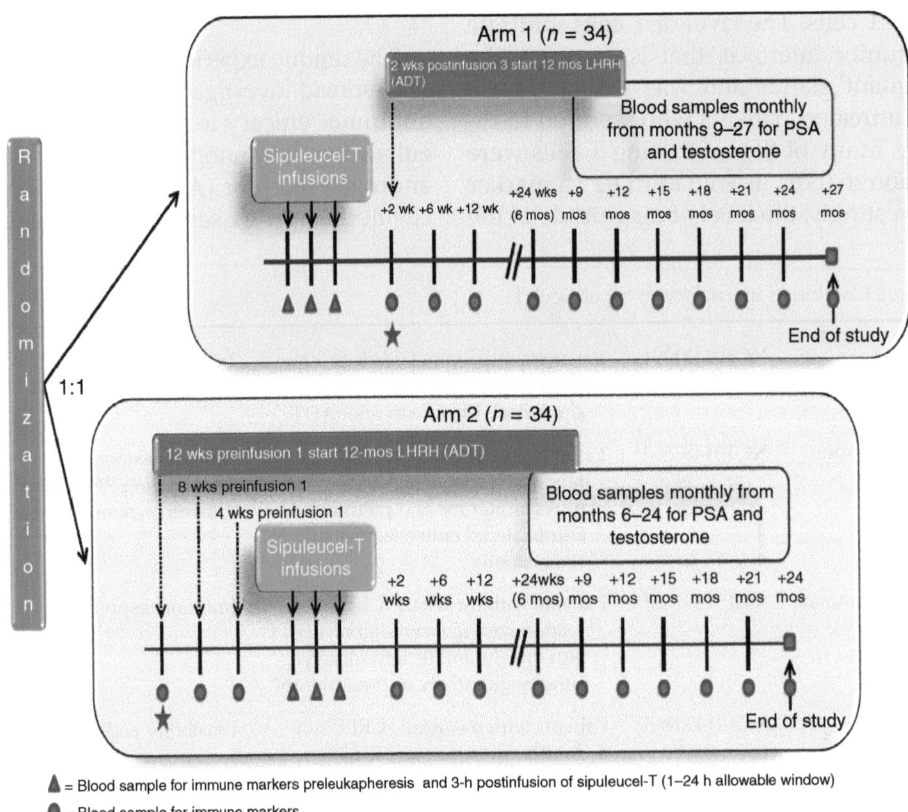


**FIGURE 56.4 Scheme of the androgen deprivation with sipuleucel-T trial.**[30]

sipuleucel-T with concurrent versus sequential enzalutamide in metastatic castration resistant prostate cancer) that randomized patients 1:1 to either sipuleucel-T and enzalutamide (160 mg orally daily × 52 weeks) starting 2 weeks before [concurrent arm, A, $N = 12$] versus enzalutamide 10 weeks after initiation of sipuleucel-T [sequential arm, B, $N = 18$]. The primary endpoint was peripheral T cell immune response to PA2024 with the secondary endpoint of time to recurrence of elevated PSA. No differences were seen in APC activation, APC count and TNC did not differ between arms. A prime boost of APCs was seen in both arms and enhanced after the second and third infusions. Overall, there was no difference between arms with regard to the endpoints.

Another ongoing trial involves evaluating the impact of concurrent abiraterone on product characteristics. The ongoing phase II trial will examine sipuleucel-T with concurrent or sequential abiraterone plus prednisone. Preliminary results have found no significant differences in median cumulative CD54 upregulation and $CD54^+$ count between the two arms. An increase in CD54 upregulation with the second and third sipuleucel-T vaccinations indicated a prime-boost effect in both arms. These data indicate that sipuleucel-T can be manufactured during treatment with abiraterone and prednisone while maintaining product integrity and effectiveness with an induced prime-boost effect similar to sipuleucel-T alone.[34]

The lack of earlier success with immunotherapy in prostate cancer lies in the nature of the therapies and in the difficulty of evaluating their effects, and may be stage-specific. For example, in patients with biochemical relapse only, the long natural history of this clinical state makes assessment of a survival benefit problematic in the clinical trial setting. Even in patients with radiographically measurable disease, the Response Evaluation Criteria in Solid Tumors (RECIST) cannot be presumed to apply given that treatment with immunotherapy may cause pseudoprogression of disease that in fact reflects immune cell infiltration, which precedes subsequent tumor shrinkage or remission. It is because of this that new criteria for assessing clinical endpoints based on the effects of the study agent have been developed. Furthermore, as is common across oncology, promising results gleaned in phase II trials have not necessarily translated into success in phase III.

## INCORPORATING SIPULEUCEL-T IN THE CLINICAL STATES PARADIGM

Controversy remains as to the length of time that a patient should be clinically monitored after receiving sipuleucel-T before going on to the next treatment (Figure 56.5). The prevailing data from the phase III trials did not monitor patients for greater than 6 months and while a survival benefit existed, nevertheless, patients often experienced anxiety awaiting any decline in PSA and/or in clinical response as assessed by imaging. The experience observed with melanoma, clearly a more immunogenic cancer, has been translated into a paradigm that is followed for other immunotherapies, that is, the observation that there may be a transient worsening of disease as manifested by infiltration of immune

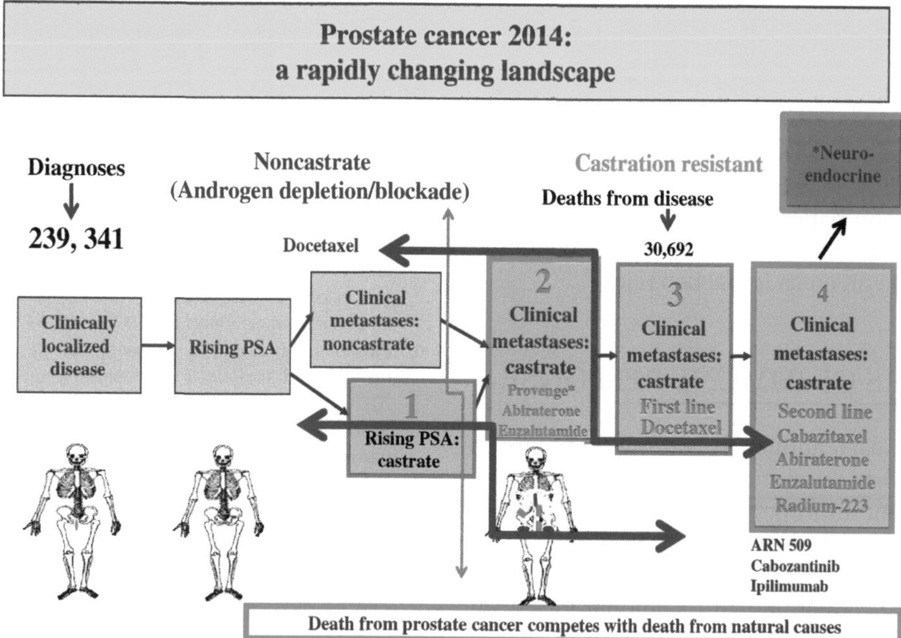


FIGURE 56.5 **Clinical states of prostate cancer evolution and potential therapies for each state.** *Modified from Scher and Heller.*[7]

cells (TILs) to the sites of disease followed by reduction in tumor burden leading to a treatment response that in many cases is durable, that is, >2 years. This observation that immune cell infiltrates migrate into the cancer and may contribute to the temporary worsening of disease has led to a resurgence in rebiopsying tumors throughout the clinical course at specific time intervals in order to document potential changes in the tumor's biology. Given the prognostic association of TILS and TAMS (tumor-associated macrophages) with specific cancers such as ovarian,[35,36] colorectal,[37] breast,[38] and melanoma,[39,40] TIL infiltration of a tumor may serve as a potential biomarker of disease response.

It has become clear that while some immunotherapies can be efficacious when given as single agents, recent data have demonstrated that synergy can exist with agents within the same class, for example, with monoclonal antibodies against the checkpoint inhibitors, CTLA-4 and PD-1. Prostate cancer[41,42] has not shown to be as responsive to these agents as have other solid tumors such as melanoma, renal cell, or lung cancers. While the overall response rates for anti-CTLA in melanoma have ranged between 15% and 20% in keeping with response rates seen with standard chemotherapy, the responses as well as their durability can be markedly enhanced when combining anti-CTLA-4 with anti-PD-1.[43] This has led to the consideration that other immune therapies, such as a vaccine or even a chimeric antigen receptor modified T cell (CAR), could be combined with other biologic agents including cytokines, that is, IL-2, or anti-CTLA-4, anti-PD-1,[8,44] anti-PDL1,[8,44] and even sipuleucel-T in addition to chemotherapy. The concept of a combination approach, while exciting, may not prove perfect for every solid tumor. Unlike other solid tumors, that is, melanoma, lung, and renal cell, prostate cancer is highly bone trophic and the direct impact of an immune modulatory drug on the bone marrow may be hard to assess in the setting of a marrow packed with prostate cancer cells. It may be more helpful to start evaluating patients in the neoadjuvant milieu as mentioned earlier as a proof of concept or to focus on patients with lymph node or visceral disease, albeit the latter may not be representative of the disease per se given its lethal behavior.[39]

## BIOMARKERS OF IMMUNE RESPONSE

As noted previously, antigen spreading[25] appears to be one means by which immune responses can be induced. Immune monitoring remains challenging as there are no validated biomarkers for immune therapies; however, there have been several immune parameters that have been identified in melanoma based on recent clinical trials. It is important to reinforce that an immune biomarker should be reflective of a biologic change in the cancer or should have some meaningful association with the cancer. Biomarkers can include the serum markers prostate-specific antigen (PSA), prostatic acid phosphatase (PAP) in addition to recent additions such as circulating tumor cells, quantification of absolute lymphocyte counts (ALC), quantitating soluble antigen or antibodies (humoral) in the bloodstream, T cell subpopulations such as CD8(Tregs), CD4(helper), T cell activation markers such as ICOS and FOXP3, and myeloid-derived suppressor cells(MDSC). Also reported in the literature are attempts to associate the following parameters with changes in prognosis: apoptosis of $CD8^+$ cells, B cell signature, suppressor cells in microenvironment[45] to include Tregs/MDSC in human tumors; the expression of $CD4^+$, $FOXP^+$, $CD24^{high}$ Treg is often associated with poor prognosis. Treg frequency among TIL cells correlated with tumor grade and reduced patient survival in melanoma, breast glioblastoma, and ovarian cancers.

## CONCLUSIONS

We are fortunate that we now have a variety of approaches that facilitate further exploration of how to use the immune system to fight cancer. Sipuleucel-T has laid the groundwork on which additional strategies can be implemented to facilitate the control of this disease.

### References

1. Kantoff PW, Higano CS, Shore ND, et al. Sipuleucel-T immunotherapy for castration-resistant prostate cancer. *N Engl J Med* 2010;**363**:411–22.
2. Scher HI, Fizazi K, Saad F, et al. Increased survival with enzalutamide in prostate cancer after chemotherapy. *N Engl J Med* 2012;**367**:1187–97.
3. Fizazi K, Scher HI, Molina A, et al. Abiraterone acetate for treatment of metastatic castration-resistant prostate cancer: final overall survival analysis of the COU-AA-301 randomised, double-blind, placebo-controlled phase 3 study. *Lancet Oncol* 2012;**13**:983–92.
4. Ryan CJ, Smith MR, de Bono JS, et al. Abiraterone in metastatic prostate cancer without previous chemotherapy. *N Engl J Med* 2013;**368**:138–48.
5. de Bono JS, Oudard S, Ozguroglu M, et al. Prednisone plus cabazitaxel or mitoxantrone for metastatic castration-resistant prostate cancer progressing after docetaxel treatment: a randomised open-label trial. *Lancet Oncol* 2010;**376**:1147–54.
6. Parker C, Nilsson S, Heinrich D, et al. Alpha emitter radium-223 and survival in metastatic prostate cancer. *N Engl J Med* 2013;**369**:213–23.
7. Scher HS, Heller G. Clinical states in prostate cancer: toward a dynamic model of disease progression. *Urology* 2000;**55**:323–7.
8. Topalian SL, Hodi FS, Brahmer JR, et al. Safety, activity, and immune correlates of anti–PD-1 antibody in cancer. *N Engl J Med* 2012;**366**:2443–54.
9. Teicher BA, Chari RVJ. Antibody conjugate therapeutics: challenges and potential. *Clin Cancer Res* 2011;**17**:6389–97.
10. Maloney D, Morschhauser F, Linden O, et al. Diversity in antibody-based approaches to non-Hodgkin lymphoma. *Leuk Lymphoma* 2010;**51**(Suppl. 1):20–7.

11. Slovin SF. Targeting castration-resistant prostate cancer with monoclonal antibodies and constructs. *Immunotherapy* 2013;**5**:1347–55.
12. Schellhammer PF, Chodak G, Whitmore JB, et al. Lower baseline prostate-specific antigen is associated with a greater overall survival benefit from sipuleucel-T in the Immunotherapy for Prostate Adenocarcinoma Treatment (IMPACT) trial. *Urology* 2013;**81**: 1297–302.
13. Drake CG, Fan L-Q, Thakurta DG, et al. Antigen spread and survival with sipuleucel-T in patients with advanced prostate cancer. *Proc Am Soc Clin Oncol* 2014;**32** (Suppl. 4; abstr88).
14. Kantoff PW, Schuetz TJ, Blumenstein BA, et al. Overall survival analysis of a phase II randomized controlled trial of a Poxviral-based PSA-targeted immunotherapy in metastatic castration-resistant prostate cancer. *J Clin Oncol* 2010;**28**:1099–105.
15. Sheikh NA, Jones LA. CD54 is a surrogate marker of antigen presenting cell activation. *Cancer Immunol Immunother* 2008;**57**:1381–90.
16. Sheikh NA, Petrylak D, Kantoff PW, et al. Sipuleucel T immune parameters correlate with survival: an analysis of the randomized phase 3 clinical trials in men with castration-resistant prostate cancer. *Cancer Immunol Immunother* 2013;**62**:137–47.
17. Loken MR, Brosnan JM, Bach BA, et al. Establishing optimal lymphocyte gates for immunophenotyping by flow cytometry. *Cytometry* 1990;**11**:453–9.
18. Gaglia JL, Greenfield EA, Mattoo A, et al. Intercelluar adhesion molecule I is critical for activation of CD28-deficient T cells. *J Immunol* 2000;**165**:6091–8.
19. Higano CS, Schellhammer PF, Small EJ, et al. Integrated data from 2 randomized, double-blind, placebo-controlled, phase 3 trials of active cellular immunotherapy with sipuleucel-T in advanced prostate cancer. *Cancer* 2009;**115**:3670–9.
20. Small EJ, Schellhammer PF, Higano CS, et al. Placebo-controlled phase III trial of immunologic therapy with sipuleucel-T (APC8015) in patients with metastatic, asymptomatic hormone refractory prostate cancer. *J Clin Oncol* 2006;**24**:3089–94.
21. Thakurta DG, Fan L-Q, Vu T, et al. Induction of antigen spread after sipuleucel-T treatment and its association with improved clinical outcome. Soc for Immunotherapy of Cancer, 28th Annual Meeting 2013;1(Suppl. 1):P101.
22. Singh BH, Gulley JL. Therapeutic vaccines as a promising treatment modality against prostate cancer: rationale and recent advances. *Ther Adv Vaccines* 2014;**2**:137–48.
23. Nam RK, Zhang WW, Klotz LH, et al. Variants of the hK2 protein gene (KLK2) are associated with serum hK2 levels and predict the presence of prostate cancer at biopsy. *Clin Cancer Res* 2006;**12**:6452–8.
24. Gulley JL. Therapeutic vaccines. The ultimate personalized therapy. *Human Vaccin Immunother* 2013;**9**:219–21.
25. Vanderlugt C, Miller SD. Epitope spreading in immune-mediated diseases: implications for immunotherapy. *Nat Rev Immunol* 2002;**2**:85–95.
26. Fong L, Carroll P, Weinberg V, et al. Activated lymphocyte recruitment into the tumor microenvironment following preoperative sipuleucel-T for localized prostate cancer. *J Natl Cancer Inst* 2014;**106**:1–9.
27. Pages F, Kirilovsky A, Mlecnik B, et al. *In situ* cytotoxic and memory T cells predict outcome in patients with early-stage colorectal cancer. *J Clin Oncol* 2009;**27**:5944–51.
28. Pages F, Berger A, Camus M, et al. Effector memory T cells, early metastasis and survival in colorectal cancer. *N Engl J Med* 2005;**353**:2654–66.
29. McNeel DG, Gardner TA, Higano CS, et al. A transient increase in eosinophils is associated with prolonged survival in men with metastatic castration-resistant prostate cancer who receive sipuleucel-T. *Cancer Immunol Res* 2014;**2**:988–99.
30. Antonarakis ES, et al. Sequencing of androgen-ablation and vaccination in men with biochemically recurrent prostate cancer. *J Clin Oncol* 2013;**31**(Suppl. 6) Abstract 34.
31. Antonarakis E, Kibel A, Adams G, et al. A randomized phase 2 study evaluating the optimal sequencing of sipuleucel-T and androgen deprivation therapy in biochemically-recurrent prostate cancer: Immune results with a focus on humoral responses. *Proc European Assoc Andrology Annual Congress, Stockholm* 2014;**April**: 11–5 poster 980.
32. Drake C, Adams G, Elfiky A, et al. A randomized phase 2 study evaluating optimal sequencing of sipuleucel;-T (sip-T) and androgen deprivation therapy (ADT) in biochemically recurrent prostate cancer (BRPC): variables that correlate with immune response. *Proc ESMO* 2014; abstr 775P.
33. Petrylak DP, Dreicer R, Shore N, et al. STRIDE, a randomized, phase 2, open-label study of sipuleucel T with concurrent vs sequential enzalutamide in metastatic castration-resistant prostate cancer (mCRPC). *Proc ESMO* 2014; Abstr 774P.
34. Small E, Lance R, Gardner T. A randomized phase II, open-label study of sipuleucel-T with concurrent or sequential abiraterone acetate (AA) in metastatic castrate-resistant prostate cancer (mCRPC). *Proc Am Soc Clin Oncol* 2013;**31**(Suppl. 6) abstract 114.
35. Mhawech-Fauceglia P, Wang D, Liaquat A, et al. Intraepithelial T cells and tumor-associated macrophages in ovarian cancer patients. *Cancer Immun* 2013;**13**:1–6.
36. Webb JR, Milne K, Watson P, et al. Tumor-infiltrating lymphocytes expressing the tissue resident memory marker CD103 are associated with increased survival in high-grade serous ovarian cancer. *Clin Cancer Res* 2014;**20**:434–44.
37. Katz SC, Bamboat ZM, Maker AV, et al. Regulatory T cell infiltration predicts outcome following resection of colorectal cancer liver metastases. *Ann Surg Oncol* 2013;**20**:946–55.
38. Gu-Trantien C, Loi S, Garaud S, et al. CD4+ follicular helper T cell infiltration predicts breast cancer survival. *J Clin Invest* 2013;**123**:2873–92.
39. Thomas NE, Busam KJ, From L, et al. Tumor-infiltrating lymphocyte grade in primary melanomas is independently associated with melanoma-specific survival in the population based genes, environment and melanoma study. *J Clin Oncol* 2013;**31**:4252–9.
40. Turcotte S, Gos A, Hogan K, et al. Phenotype and function of T cells infiltrating visceral metastases from gastrointestinal cancers and melanoma: implications for adoptive cell transfer therapy. *J Immunol* 2013;**191**:2217–25.
41. Slovin SF, Higano CS, Hamid O, et al. Ipilimumab alone or in combination with radiotherapy in metastatic castration-resistant prostate cancer: results from an open-label, multicenter phase I/II study. *Ann Oncol* 2013;**24**:1813–21.
42. Drake CG, Kwon ED, Fizazi K, et al. Results of subset analyses on overall survival (OS) from the study CA184-043: Ipilimumab versus placebo in post-docetaxel metastatic castration-resistant prostate cancer. *Proc Am Soc Clin Oncol* 2014; Abstr 2.
43. Wolchok JD, Kluger H, Callahan MK, et al. Nivolumab plus ipilimumab in advanced melanoma. *N Engl J Med* 2013;**369**:122–33.
44. Hamid O, Robert C, Daud A, et al. Safety and tumor responses with lambrolizumab (anti-PD-1) in melanoma. *N Engl J Med* 2012;**369**:134–44.
45. Le DT, Jaffee EM. Harnessing immune responses in the tumor microenvironment: all signals needed. *Clin Cancer Res* 2013;**19**: 6061–3.

CHAPTER

# 57

# Second-Line Hormonal for Castrate-Resistant Prostate Cancer

*Jaspreet Singh Parihar, MD, Isaac Yi Kim, MD, PhD*

Section of Urologic Oncology, Rutgers Cancer Institute of New Jersey and Division of Urology, Department of Surgery, Rutgers Robert Wood Johnson Medical School, New Brunswick, NJ, USA

## INTRODUCTION

Androgen deprivation therapy (ADT) using the luteinizing hormone-releasing hormone (LHRH) agonist or antagonist has been the mainstay treatment of systemic hormone-sensitive prostate cancer.[1–3] Following an initial decline in PSA levels and a clinically significant response rate of 80–90%, majority of patients on ADT develop progressive disease after an average of 18–24 months despite castrate levels of testosterone. Following further treatment with chemotherapy, median overall survival is consistently less than 2 years in patients with metastatic disease.[2,4] With the approval of new agents that block androgen receptors and subsequent signaling more effectively, second-line or secondary hormonal therapy is now the standard in patients who fail primary androgen therapy.[3] The resulting measurable clinical response suggests a retained degree of hormonal sensitivity in castrate-resistant prostate cancer (CRPC).

In 1989, Ferro et al. demonstrated a correlation between a decline in PSA levels and clinical response in prostate cancer.[5] Owing to this realization, PSA levels have been used as a surrogate marker in therapeutic evaluations.[6–8] During therapy, a serum PSA decline of at least 50% was associated with improved survival, leading to development of the 1999 Prostate-Specific Antigen Working Group eligibility criteria and response guidelines.[9] In this chapter, we will review the overall data on secondary hormonal manipulation in patients with prostate cancer who progress on the standard ADT with LHRH agonist or antagonist.

## FIRST-GENERATION ANTIANDROGENS

### Flutamide

When patients fail the first-line ADT with LHRH agonist/antagonist, antiandrogens are usually added for a complete androgen blockade. Among antiandrogens, flutamide is one of the first nonsteroidal pure antiandrogens as it inhibits androgen uptake and/or nuclear binding of androgen in target tissues. It has a relatively short half-life of 5.5 h. Following rapid absorption, the drug is metabolized to a biologically active alpha-hydroxylated derivative. Among the first-generation antiandrogens, flutamide has the lowest binding affinity for the androgen receptor.[10]

The benefit of adding flutamide to men who have failed the standard ADT was first reported by Labrie et al. in 1988. In this study, a total of 209 patients with disease progression were treated with flutamide following orchiectomy or DES or an LHRH agonist. Overall response rate was 34.5% (13 patients complete response, 20 partial response, 39 stable disease). Probability of survival at 2 years for partial, stable, and nonresponders was 87%, 67%, and 17%, respectively.[11]

Deferred use of antiandrogens following disease progression was further assessed by Fowler et al. Their study included 45 patients with localized cancer and 50 patients with metastatic disease. Following flutamide therapy, a PSA decrease of ≥50% was noted in 32 (80%) and 27 (54%) patients, respectively.

Currently, the approved indications for the use of flutamide include the following: in combination with LHRH agonists, adjunctive therapy to orchiectomy, and prior

Prostate Cancer. http://dx.doi.org/10.1016/B978-0-12-800077-9.00057-8

to and during definitive external beam radiotherapy for patients with bulky locally advanced prostate carcinoma. Adverse effects include gastrointestinal symptoms, gynecomastia, breast tenderness, and hepatitis. Postmarketing studies have noted occasional hospitalizations due to liver dysfunction as well as rare instances of deaths. Liver function tests are recommended prior to initiation, monthly for the first 4 months of therapy and periodically thereafter.[10]

## Bicalutamide

Bicalutamide is an oral nonsteroidal competitive inhibitor of androgen receptor. As with flutamide, bicalutamide is approved for combination therapy with LHRH analog for the treatment of metastatic prostate cancer. It has a relatively long half-life of 6 days and is extensively metabolized in the liver. Adverse effects are similar to flutamide and include hepatitis. Thus, monitoring liver function is mandatory. The concept of bicalutamide withdrawal effect has been analyzed, as discussed later.[12–14]

The use of high-dose bicalutamide in patients previously treated with long-term flutamide therapy has been described. In a pilot study of 31 patients with progressive disease treated with 150 mg bicalutamide daily, Joyce et al. reported a ≥50% PSA decline rate in 7/31 patients (22.5%), overall duration of response 4 months (3–13+ months). Response rate in patients on ADT + long-term flutamide was 6 out of 14 (43%).

## Nilutamide

Nilutamide is another oral nonsteroidal agent that binds and inhibits androgen receptor. Its rapid and complete absorption results in persistent plasma concentrations followed by extensive liver metabolism and eventual excretion in urine. The elimination half-life of nilutamide is approximately 41–49 h. Nilutamide is FDA approved to be used in combination with surgical castration for the treatment of metastatic prostate cancer.[15]

In 2001, Desai et al. reported their series on the use of nilutamide as a second-line hormonal agent. In men who failed prior antiandrogen therapy, 7 of 14 (50%) patients treated with nilutamide experienced a greater than 50% decline in their serum PSA levels.[16] In a separate study by Kassouf et al., a retrospective analysis of 28 men with castration-resistant prostate cancer subsequently treated with nilutamide revealed that 18 men (64%) experienced a reduction in PSA and 12 men had greater than 50% PSA decline (48%).[17]

Side effect profile of nilutamide is similar to the other nonsteroidal antiandrogens. A notable adverse effect is alcohol intolerance, presenting as facial flushing, malaise, and hypotension. Other unique side effects include a 13% to 57% incidence of delay in adaptation to dark, ranging from seconds to a few minutes.[15] Postmarketing data also indicate a 2% incidence of interstitial pneumonitis and pulmonary fibrosis, usually presenting within the first 3 months, leading to hospitalization and potential death. Monitoring parameters include liver function tests, chest X-ray ± pulmonary function tests at baseline, first 4 months of treatment then periodically thereafter.[18] Table 57.1 summarizes the clinical indications, mechanism of action, and adverse effects of common antiandrogens.

## Antiandrogen Withdrawal

In men who relapse while on the ADT that includes antiandrogens, cessation of the antiandrogen is often the first step. The antiandrogen withdrawal syndrome was first reported in 1993 by Kelly and Scher.[19] Their seminal observations noted three patients on complete androgen blockade with either gonadotropin-releasing hormone (GnRH) or orchiectomy plus the antiandrogen flutamide, who experienced sustained declines in serum PSA levels after discontinuation of the antiandrogen. The flutamide withdrawal syndrome was further characterized in their subsequent report on 36 men on hormonal treatment with progressive disease. Discontinuation of antiandrogen was associated with a significant decrease (≥50%) in PSA values in 10 of 35 patients (29%). Median duration of declines was over 5 months (range 2–10+).[20]

TABLE 57.1 First-Generation Antiandrogens

| Drug | Approved indications | Mechanism | Adverse effects |
|---|---|---|---|
| Flutamide | Combination use with LHRH agonist, adjunctive therapy to orchiectomy, prior to and during EBRT | Nonsteroidal antiandrogen | Gastrointestinal symptoms (nausea, vomiting, diarrhea, anorexia), hepatotoxicity, impotence, gynecomastia, breast tenderness |
| Bicalutamide | Combination use with LHRH agonist, adjunctive therapy to orchiectomy, prior to and during EBRT | Nonsteroidal antiandrogen | Similar to flutamide |
| Nilutamide | Combination use with LHRH agonist, adjunctive therapy to orchiectomy, prior to and during EBRT | Nonsteroidal antiandrogen | Similar as above, alcohol intolerance, delay in adaption to dark, interstitial pneumonitis, pulmonary fibrosis |

Note: EBRT, external beam radiotherapy.

The investigators concluded that a trial of flutamide withdrawal should be considered in patients progressing on total androgen blockade before the initiation of more toxic therapies.

In 1995, Small and Carroll provided further evidence toward this approach by examining a large group of unselected patients. Their group retrospectively analyzed 107 men with metastatic prostate cancer with disease progression on flutamide therapy. Of the 82 evaluable patients, three had a >80% fall in PSA value and another nine had a >50% decrease, for a response proportion of 14.6% (95% confidence interval (CI), 7.8–24.2%). The median response duration was 3.5 months (range 1–12+ months). In contrast to prior studies, the authors noted a more modest and a shorter median duration of response.[14] Figg et al. also noted a PSA response in 7/21 (31%) with response duration of 3–7 months.[21]

Data from these pivotal studies yield 29/138 patients (21%) who demonstrated ≥50% PSA decrease with overall response duration of 3–5 months, although responses >2 years have been reported.[22] Similarly, a 2008 Southwest Oncology Group Trial (SWOG 9426) reported on a prospective multi-institution clinical study evaluating 210 men who had prior treatment with an antiandrogen plus orchiectomy or LHRH agonist. After a median follow-up of 5 years, 21% of patients had confirmed PSA decreases of ≥50% (95% CI, 16–27%) after antiandrogen withdrawal. Median progression-free survival (PFS) was 3 months (95% CI, 2–4 months); however, 19% had 12-month or greater progression-free intervals. Median overall survival (OS) after antiandrogen withdrawal was 22 months. In multivariate analysis, longer antiandrogen use was found to be a significant predictor of PSA response.[23]

Similar clinical responses with discontinuation has been observed with bicalutamide,[12,14,24] nilutamide,[25,26] and chlormadinone.[27] Time to PSA response varies according to the drug half-life. Schellhammer noted that PSA decline in the flutamide group occurred within a few days after cessation in comparison to the bicalutamide group of 4–8 weeks.[24] Taken together, these studies demonstrate the potential therapeutic value of antiandrogen withdrawal.

## ESTROGENS

Secondary hormonal manipulation with diethylstilbestrol (DES) may be a viable option in select patients. Mechanism of action is directed at hypothalamic LHRH inhibition, thereby decreasing the pituitary LH levels and subsequent lowering in serum testosterone. Additionally, high-dose estrogens result in cytotoxic effect via cell cycle arrest and apoptosis.[28] In 1975, the Veterans' Administration Cooperative Urological Research Group reported a series of randomized clinical trials demonstrating the efficacy of diethylstilbestrol but also cautioned against the risk of cardiovascular deaths.[29]

In 2005, Manikandan et al. reported the efficacy of DES versus bicalutamide as a second-line agent in 58 men with evidence of biochemical or clinical progression on LHRH agonist therapy. In addition to primary hormonal treatment, Group A ($n$ = 26) received 1 mg of DES with 75 mg of aspirin per day while Group B ($n$ = 32) received bicalutamide at a dose of 50 mg/day. After a median follow-up of 24 months, low dose DES and 50 mg bicalutamide were equally effective with respect to PSA response rate and median response duration, 23% versus 31% and 9 months (range, 3–18 months) versus 12 months (range, 3–18 months) for group A and B, respectively. However, DES was noted to have more severe adverse effects, mainly cardiovascular toxicities including congestive heart failure, pulmonary embolism, and stroke.[30]

PC-SPEC (*PC* "prostate cancer" and *spes* for "hope") is a combination preparation consisting of seven unregulated Chinese herbal extracts.[31] *Glycyrrhiza* has been shown to increase estrogen levels and lower serum testosterone levels. The phytoestrogens in *Serenoa repens* have been shown to decrease endogenous estrogens.[32] Despite early enthusiasm, in 2002 the preparation was withdrawn from the market due to high variability of composition between different batches as well as drug contaminations.[33]

## CYPROTERONE

Cyproterone acetate is a steroidal antiandrogen. Currently, it is not approved for the treatment of prostate cancer in the United States. The final analysis of European Organization for Research and Treatment of Cancer (EORTC) Protocol 30892 was published in 2004 comparing the efficacy of flutamide 250 mg versus cyproterone acetate 100 mg in 310 men with treatment naïve metastatic prostate cancer and favorable prognostic factors. After a median follow-up of 8.6 years, the authors found no significant difference between the treatment arms with respect to overall survival, disease specific survival, or time to progression.[34] Side effect profile favored cyproterone, although liver toxicity was still a significant complication.

## MEGESTROL ACETATE

Megestrol acetate is a synthetic progestin that exerts its effects through various mechanisms. Primarily, it acts to inhibit intracellular androgen action. Secondary effects of megestrol acetate include direct cytotoxic effect at high concentration and inhibition of 5-alpha reductase, LH release, and adrenal axis.[35–37]

In a large multi-institutional randomized study, CALGB 9181, Dawson et al. compared standard 160 mg/day versus moderately high dose megestrol acetate 640 mg/day in 149 men with hormone refractory prostate cancer. The primary endpoint of the study was tumor response. No significant difference was found with respect to the median survival time, 11.2 versus 12.1 months. A greater than 50% decline in PSA occurred in 13.8% and 8.8% of patients in the low- and high-dose treatment arms, respectively.[38]

## KETOCONAZOLE

Ketoconazole is an antifungal imidazole that inhibits androgen biosynthesis at high doses. Heralded by its adverse effect of gynecomastia, it was later recognized to interfere with testosterone synthesis via cytochrome 3A4 inhibition.[39–41] *In vitro* studies on prostate cancer cell lines (PC-3 and DU-145) have demonstrated a direct cytotoxic effect of ketoconazole causing >90% suppression of tumor colony growth.[42]

In 1987, Vanuytsel et al. showed that 15 out of 17 patients (88%) experienced initial tumor response with ketoconazole therapy. However, the investigators noted the inability to maintain castrate levels as testosterone level rose slowly over a period of 1 month and to low normal range in 5 months.[43]

During the pre-PSA era, multiple studies of ketoconazole without glucocorticoids showed stable disease in 30% and objective responses in 16% of patients.[44] Small et al. reported a trial of 50 men with progressive disease following combined androgen blockade followed by antiandrogen withdrawal. These men were treated with ketoconazole and hydrocortisone therapy. The results demonstrated that 30 of the 48 evaluable patients (62.5%, 95% CI, 47.3–76.1%) had >50% decrease in PSA, while 23 (48%) experienced >80% decrease. The median duration of response was 3.5 months. Importantly, authors noted that the lack of prior response to antiandrogen withdrawal did not preclude response to subsequent therapy with ketoconazole.[45]

The investigators subsequently examined simultaneous suppression of adrenal androgens with flutamide withdrawal and ketoconazole with hydrocortisone. Eleven out of 20 men (55%) had a >50% decline in PSA (95% CI, 31.5–76.9%). The median PSA response duration was 8.5 months (95% CI, 7–17 months) and median survival was 19 months. The authors concluded that the combination therapy does appear to increase the PSA response as compared to antiandrogen withdrawal alone.[46]

In a similar more recent randomized study, antiandrogen withdrawal alone ($n$ = 132) was compared to antiandrogen withdrawal plus ketoconazole ($n$ = 128). PSA response and objective response was 11% versus 27% ($P$ = 0.0002) and 2% versus 20% ($P$ = 0.02) in patients undergoing antiandrogen withdrawal alone and antiandrogen withdrawal + ketoconazole, respectively. There was no difference noted in survival between the two cohorts.[47]

Common adverse reactions of ketoconazole include nausea, dizziness, weakness, skin reactions, pruritus, adrenal insufficiency, gynecomastia, hepatotoxicity as well as potential QT prolongation owing to drug interactions. Baseline and periodic liver function tests should be monitored.

## AMINOGLUTETHIMIDE

Aminoglutethimide inhibits the enzymatic conversion of cholesterol to pregnenolone, thereby decreasing adrenal glucocorticoids, mineralocorticoids, estrogens, and androgens. Typically administered with hydrocortisone, the initial starting dose of 250 mg three times daily for 3 weeks is increased to four times daily. Adverse reactions include peripheral edema, hypothyroidism, and abnormal LFT requiring frequent monitoring.

In the pre-PSA era, there were a number of studies suggesting a role for aminoglutethimide. In 2004, Kruit et al. reported on a prospective phase II study with aminoglutethimide and hydrocortisone using PSA measurements to determine antitumor activity. Thirty-five men with castration-resistant prostate cancer received the treatment with monthly PSA measurements. Twelve patients (37%) had >50% PSA decline with median progression time of 10.5 months compared to 4.5 months among all patients. Median duration of response and median survival among responders was 9 months and 23 months, respectively.[48]

## CORTICOSTEROIDS

Glucocorticoids exert a negative feedback effect on the hypothalamic-pituitary axis resulting in androgen synthesis suppression. The significance of suppression of adrenal source of androgens in hormone refractory settings has long been appreciated. First recognized for their palliative benefits in the early 1950s, much of the role of glucocorticoids in prostate cancer treatment paradigm is understood from control arms in subsequent trials.[49]

Important to recognize is the potential confounding role that corticosteroids can play in interpreting clinical trials. In a double-blind placebo-controlled randomized phase III trial evaluating suramin plus hydrocortisone to placebo plus hydrocortisone, 16% of patients who received hydrocortisone therapy experienced >50% decline in PSA.[50] A separate study by Sortor et al. evaluated the effects of prednisone on PSA in men with castration-resistant prostate cancer. Patients received 10 mg of prednisone orally two times a day. Ten out of 29 total patients (34%) had >50% PSA decline. Median PSA decline was 24%. Median time for progression-free survival and median survival was 2.0 (range 0–11) months and 12.8 months, respectively. A PSA decline of >50% was correlated with prolonged survival.[51]

Low-dose dexamethasone has also been evaluated as a secondary hormonal therapy in a metastatic castration resistance setting. In a retrospective review of 38 men with metastatic progression following orchiectomy, 24 patients (63%) experienced symptomatic improvement. Twenty-three (61%) were noted to have >50% PSA decrease, of which 8 (35%) had radiographic evidence of disease regression, 5 (22%) were stable, 7 (30%) had disease progression, and 3 (13%) did not have serial radiographic exams.[52] In men with symptomatic bone metastasis, low-dose prednisone (7.5–10 mg daily) is recognized to improve pain in association with suppression of adrenal androgens and to boost quality of life.[53]

## SECOND-GENERATION ANTIANDROGENS

### Enzalutamide

Enzalutamide (MDV3100) is the prototypical oral nonsteroidal second-generation antagonist.[54] In castrate-resistant states, enzalutamide has been demonstrated to remain a potent antagonist of androgen receptor.[55] In contrast to prior generation of antiandrogens, enzalutamide binds to the androgen receptor with greater affinity, inhibit nuclear translocation, and impair both DNA binding to androgen response elements with recruitment of coactivators.[56]

In late 2012, the drug received FDA approval based on its results of phase 3 clinical trial.[57] In the study, 1199 men with castrate-resistant prostate cancer were randomized 2:1 to receive either enzalutamide ($n$ = 800) or placebo ($n$ = 399). The primary end point was overall survival. The study was unblinded early after the interim analysis showed a median overall survival of 18.4 months (95% CI, 17.3 to not yet reached) in the enzalutamide group versus 13.6 months (95% CI, 11.3–15.8) in the placebo group (HR, 0.63; 95% CI, 0.53–0.75; $P < 0.001$). Enzalutamide was favored in all secondary endpoints, which included the proportion of patients with a reduction in PSA level by 50% or more (54% vs. 2%, $P < 0.001$), the soft-tissue response rate (29% vs. 4%, $P < 0.001$), the quality-of-life response rate (43% vs. 18%, $P < 0.001$), the time to PSA progression (8.3 vs. 3.0 months; HR, 0.25; $P < 0.001$), radiographic progression-free survival (8.3 vs. 2.9 months; hazard ratio, 0.40; $P < 0.001$), and the time to the first skeletal-related event (16.7 vs. 13.3 months; hazard ratio, 0.69; $P < 0.001$). Seizures were reported in five patients (0.6%) in the treatment arm.[57]

A subsequent randomized, double-blind, phase III trial (PREVAIL) enrolled 1717 men with chemotherapy-naive castration-resistant metastatic prostate cancer. The study was stopped early after a planned interim analysis again demonstrated an improved overall survival (hazard ratio, 0.71) and radiographic progression-free survival (hazard ratio, 0.19) as compared to the placebo group. Additionally, all secondary end points demonstrated treatment benefit – time until the initiation of cytotoxic chemotherapy (hazard ratio, 0.35), the time until the first skeletal-related event (hazard ratio, 0.72), a complete or partial soft-tissue response (59% vs. 5%), the time until prostate-specific antigen (PSA) progression (hazard ratio, 0.17), and a rate of decline of at least 50% in PSA (78% vs. 3%).[58]

### ARN-509

ARN-509 is a novel antiandrogen currently under investigation. As a competitive inhibitor of androgen receptor, it inhibits AR translocation and DNA without androgen receptor overexpression. Preclinical studies in murine models of CRPC suggest that ARN-509 has greater efficacy and a higher therapeutic-index than enzalutamide.[59] Phase I study indicated well tolerability of the drug with fatigue as the most commonly reported adverse effect.[60] Phase II preliminary analysis has reported on 46 patients with progressive mCRPC who received daily ARN-509 treatment dose. Primary endpoint was PSA response rate at 12 weeks among treatment-naïve as compared to postabiraterone patients. Results indicate the PSA response of 88% and 29%, respectively.[61] ARN-509 therapy is currently being evaluated in nonmetastatic treatment-naïve CRPC, mCRPC treatment-naïve, and mCRPC abiraterone acetate pretreated cohorts. Further investigations are ongoing. Table 57.2 summarizes the clinical indications, mechanism of action, and adverse effects.

## ABIRATERONE

While majority of testosterone is produced by the testicular Leydig cells, 10% of circulating androgen is synthesized in the adrenal gland.[62] Despite reaching castrate levels of testosterone (<50 ng/mL) with the LHRH agonist/antagonist-based ADT, evidence suggests that androgen receptor and androgen signaling pathway remain active and upregulated.[63,64] Additionally, an intracrine production, that is, intratumoral source of

TABLE 57.2 Second-Generation Antiandrogens

| Drug | Approved indications | Mechanism | Adverse effects |
|---|---|---|---|
| Enzalutamide | Treatment of patients with mCRPC | Androgen receptor antagonist | Asthenia, hypertension, fatigue, musculoskeletal pain, GI symptoms, upper respiratory tract infections, seizure 0.9% incidence |
| ARN-509 | Under investigation | Androgen receptor antagonist | Fatigue, GI symptoms, abdominal pain, musculoskeletal pain |

TABLE 57.3 Abiraterone Acetate

| Drug | Approved indications | Mechanism of action | Adverse effects |
|---|---|---|---|
| Abiraterone acetate | Treatment of patients with mCRPC | Inhibits 17 α-hydroxylase/ C17,20-lyase (CYP17) | Hypertension, hypokalemia, and fluid retention related to excess mineralocorticoids, fatigue, hepatotoxicity, adrenocorticoid insufficiency |

androgen may contribute to sustained androgen signaling in prostate cancer cells. For instance, Mostaghel et al. measured androgen levels and androgen-regulated gene expression in prostates of patients on androgen deprivation. Results indicate that although ADT reduces tissue androgens by 75% and reduces the expression of androgen-regulated genes (*NDRG1, FKBP5,* and *TMPRSS2*), there remains a considerable heterogeneity in protein expression of PSA and androgen receptor during castrated states.[65,66] Simultaneously, numerous mechanisms of ligand-independent receptor activation have been suggested.[67] One such mechanism include the overexpression of CYP17 [cytochrome P17 (17α-hydroxylase and 17,20-lyase)]; CYP17 expression levels are elevated in prostate cancer tissues and correlate with high stage, high Gleason score, and short relapse-free time.[68]

The development of abiraterone acetate – a CYP17 inhibitor – has renewed therapeutic prospects in addressing castrate-resistant prostate cancer.[69,70] Abiraterone has received FDA approval in pre- (2011) and post-chemotherapy (2012) settings based on two randomized placebo-controlled, multicenter phase III studies. In the predocetaxel setting (COU-AA-302), 1088 men with asymptomatic or minimally symptomatic metastatic CRPC (bone and lymph node) were randomized 1:1 to abiraterone/prednisone ($n$ = 546) or placebo/prednisone ($n$ = 542). Primary end points were radiographic progression-free survival and overall survival. After a median follow-up period of 22.2 months, median radiographic progression-free survival in the abiraterone and placebo group was 16.5 and 8.3 months, respectively (HR, 0.53; 95% CI, 0.45–0.62; $P < 0.001$). A trend for increased overall survival was also seen with abiraterone (median not reached vs. 27.2 months for placebo; HR, 0.75; 95% CI, 0.61–0.93; $P = 0.01$). As a result of these findings, the study was unblinded early to allow crossover of placebo patients.[71]

With approximately 10- to 30-fold greater potency than ketoconazole, abiraterone offers an improved efficacy and side effect profile.[69,72] Orally administered as a prodrug (abiraterone acetate) to increase bioavailability, the drug is recommended on an empty stomach. No food intake for at least 2 h before the dose and for at least 1 h following. Adverse effects including hypertension, hypokalemia, and fluid retention resulting from increased mineralocorticoid levels from CYP17 inhibition can be suppressed using concomitant prednisone (see Table 57.3). In addition, concomitant CYP3A4 inducers, such as rifampin, should be avoided as this leads to decreased exposure of abiraterone by 55%.[73]

## CONCLUSIONS

Since the pioneering discovery by Huggins and Hodges in 1941, which correlated the response of androgen ablation to symptomatic improvement in metastatic prostate cancer, our understanding of the pivotal role of the androgen receptor continues to expand.[74] With the recent discoveries in genetics, molecular, and subsequent therapeutic biology, the next challenge in clinical application becomes the appropriate patient selection and sequencing of therapies to improve the long-term survival in advanced prostate cancer.

### References

1. Caubet JF, Tosteson TD, Dong EW, et al. Maximum androgen blockade in advanced prostate cancer: a meta-analysis of published randomized controlled trials using nonsteroidal antiandrogens. *Urology* 1997;**49**(1):71–8.
2. Eisenberger MA, Blumenstein BA, Crawford ED, et al. Bilateral orchiectomy with or without flutamide for metastatic prostate cancer. *N Engl J Med* 1998;**339**(15):1036–42.
3. Lam JS, Leppert JT, Vemulapalli SN, et al. Secondary hormonal therapy for advanced prostate cancer. *J Urol* 2006;**175**(1):27–34.
4. Crawford ED, Eisenberger MA, McLeod DG, et al. A controlled trial of leuprolide with and without flutamide in prostatic carcinoma. *N Engl J Med* 1989;**321**(7):419–24.

5. Ferro MA, Gillatt D, Symes MO, et al. High-dose intravenous estrogen therapy in advanced prostatic carcinoma. Use of serum prostate-specific antigen to monitor response. *Urology* 1989;**34**(3):134–8.
6. Smith DC, Dunn RL, Strawderman MS, et al. Change in serum prostate-specific antigen as a marker of response to cytotoxic therapy for hormone-refractory prostate cancer. *J Clin Oncol* 1998;**16**(5):1835–43.
7. Kelly WK, Scher HI, Mazumdar M, et al. Prostate-specific antigen as a measure of disease outcome in metastatic hormone-refractory prostate cancer. *J Clin Oncol* 1993;**11**(4):607–15.
8. Balk SP, Ko YJ, Bubley GJ. Biology of prostate-specific antigen. *J Clin Oncol* 2003;**21**(2):383–91.
9. Bubley GJ, Carducci M, Dahut W, et al. Eligibility and response guidelines for phase II clinical trials in androgen-independent prostate cancer: recommendations from the Prostate-Specific Antigen Working Group. *J Clin Oncol* 1999;**17**(11):3461–7.
10. Euflex. *Non-Steroidal Antiandrogen. Product Monograph.* Quebec, Canada: Merck; 2012.
11. Labrie F, Dupont A, Giguere M, et al. Benefits of combination therapy with flutamide in patients relapsing after castration. *Br J Urol* 1988;**61**(4):341–6.
12. Nieh PT. Withdrawal phenomenon with the antiandrogen Casodex. *J Urol* 1995;**153**(3 Pt. 2):1070–2 discussion 1072–1073.
13. Scher HI, Liebertz C, Kelly WK, et al. Bicalutamide for advanced prostate cancer: the natural versus treated history of disease. *J Clin Oncol* 1997;**15**(8):2928–38.
14. Small EJ, Carroll PR. Prostate-specific antigen decline after Casodex withdrawal: evidence for an antiandrogen withdrawal syndrome. *Urology* 1994;**43**(3):408–10.
15. Nilandron. *Prescribing Information*. Bridgewater, NJ: Sanofi-Aventis; 2006.
16. Desai A, Stadler WM, Vogelzang NJ. Nilutamide: possible utility as a second-line hormonal agent. *Urology* 2001;**58**(6):1016–20.
17. Kassouf W, Tanguay S, Aprikian AG. Nilutamide as second line hormone therapy for prostate cancer after androgen ablation fails. *J Urol* 2003;**169**(5):1742–4.
18. Schasfoort EM, Van De Beek C, Newling DW. Safety and efficacy of a non-steroidal anti-androgen, based on results of a post marketing surveillance of nilutamide. *Prostate Cancer Prostatic Dis* 2001;**4**(2):112–7.
19. Kelly WK, Scher HI. Prostate specific antigen decline after antiandrogen withdrawal: the flutamide withdrawal syndrome. *J Urol* 1993;**149**(3):607–9.
20. Scher HI, Kelly WK. Flutamide withdrawal syndrome: its impact on clinical trials in hormone-refractory prostate cancer. *J Clin Oncol* 1993;**11**(8):1566–72.
21. Figg WD, Sartor O, Cooper MR, et al. Prostate specific antigen decline following the discontinuation of flutamide in patients with stage D2 prostate cancer. *Am J Med* 1995;**98**(4):412–4.
22. Small EJ, Vogelzang NJ. Second-line hormonal therapy for advanced prostate cancer: a shifting paradigm. *J Clin Oncol* 1997;**15**(1):382–8.
23. Sartor AO, Tangen CM, Hussain MH, et al. Antiandrogen withdrawal in castrate-refractory prostate cancer: a Southwest Oncology Group trial (SWOG 9426). *Cancer* 2008;**112**(11):2393–400.
24. Schellhammer PF, Venner P, Haas GP, et al. Prostate specific antigen decreases after withdrawal of antiandrogen therapy with bicalutamide or flutamide in patients receiving combined androgen blockade. *J Urol* 1997;**157**(5):1731–5.
25. Gomella LG, Ismail M, Nathan FE. Antiandrogen withdrawal syndrome with nilutamide. *J Urol* 1997;**157**(4):1366.
26. Huan SD, Gerridzen RG, Yau JC, et al. Antiandrogen withdrawal syndrome with nilutamide. *Urology* 1997;**49**(4):632–4.
27. Akakura K, Akimoto S, Furuya Y, et al. Incidence and characteristics of antiandrogen withdrawal syndrome in prostate cancer after treatment with chlormadinone acetate. *Eur Urol* 1998;**33**(6):567–71.
28. Robertson CN, Roberson KM, Padilla GM, et al. Induction of apoptosis by diethylstilbestrol in hormone-insensitive prostate cancer cells. *J Natl Cancer Inst* 1996;**88**(13):908–17.
29. Blackard CE. The Veterans' Administration Cooperative Urological Research Group studies of carcinoma of the prostate: a review. *Cancer Chemother Rep* 1975;**59**(1):225–7.
30. Manikandan R, Srirangam SJ, Pearson E, et al. Diethylstilboestrol versus bicalutamide in hormone refractory prostate carcinoma: a prospective randomized trial. *Urol Int* 2005;**75**(3):217–21.
31. DiPaola RS, Zhang H, Lambert GH, et al. Clinical and biologic activity of an estrogenic herbal combination (PC-SPES) in prostate cancer. *N Engl J Med* 1998;**339**(12):785–91.
32. Iehle C, Delos S, Guirou O, et al. Human prostatic steroid 5 alpha-reductase isoforms – a comparative study of selective inhibitors. *J Steroid Biochem Mol Biol* 1995;**54**(5–6):273–9.
33. Philippou Y, Hadjipavlou M, Khan S, et al. Complementary and alternative medicine (CAM) in prostate and bladder cancer. *BJU Int* 2013;**112**(8):1073–9.
34. Schroder FH, Whelan, P, de Reijke TM, et al. Metastatic prostate cancer treated by flutamide versus cyproterone acetate. Final analysis of the "European Organization for Research and Treatment of Cancer" (EORTC) Protocol 30892. *Eur Urol* 2004;**45**(4):457–64.
35. Geller J, Albert J, Geller S, et al. Effect of megestrol acetate (Megace) on steroid metabolism and steroid-protein binding in the human prostate. *J Clin Endocrinol Metab* 1976;**43**(5):1000–8.
36. Leinung MC, Liporace R, Miller CH. Induction of adrenal suppression by megestrol acetate in patients with AIDS. *Ann Intern Med* 1995;**122**(11):843–5.
37. Anderson DG. The possible mechanisms of action of progestins on endometrial adenocarcinoma. *Am J Obstet Gynecol* 1972;**113**(2):195–211.
38. Dawson NA, Conaway M, Halabi S, et al. A randomized study comparing standard versus moderately high dose megestrol acetate for patients with advanced prostate carcinoma: cancer and leukemia group B study 9181. *Cancer* 2000;**88**(4):825–34.
39. Graybill JR, Drutz DJ, Murphy AL. Ketoconazole: a major innovation for treatment of fungal disease. *Ann Intern Med* 1980;**93**(6):921–3.
40. Pont A, Williams PL, Azhar S, et al. Ketoconazole blocks testosterone synthesis. *Arch Intern Med* 1982;**142**(12):2137–40.
41. De Coster R, Caers I, Coene MC, et al. Effects of high dose ketoconazole therapy on the main plasma testicular and adrenal steroids in previously untreated prostatic cancer patients. *Clin Endocrinol* 1986;**24**(6):657–64.
42. Eichenberger T, Trachtenberg J, Toor P, et al. Ketoconazole: a possible direct cytotoxic effect on prostate carcinoma cells. *J Urol* 1989;**141**(1):190–1.
43. Vanuytsel L, Ang KK, Vantongelen K, et al. Ketoconazole therapy for advanced prostatic cancer: feasibility and treatment results. *J Urol* 1987;**137**(5):905–8.
44. Dawson NA. Treatment of progressive metastatic prostate cancer. *Oncology* 1993;**7**(5):17–24 27; discussion 27–19.
45. Small EJ, Baron AD, Fippin L, et al. Ketoconazole retains activity in advanced prostate cancer patients with progression despite flutamide withdrawal. *J Urol* 1997;**157**(4):1204–7.
46. Small EJ, Baron A, Bok R. Simultaneous antiandrogen withdrawal and treatment with ketoconazole and hydrocortisone in patients with advanced prostate carcinoma. *Cancer* 1997;**80**(9):1755–9.
47. Small EJ, Halabi S, Dawson NA, et al. Antiandrogen withdrawal alone or in combination with ketoconazole in androgen-independent prostate cancer patients: a phase III trial (CALGB 9583). *J Clin Oncol* 2004;**22**(6):1025–33.
48. Kruit WH, Stoter G, Klijn JG. Effect of combination therapy with aminoglutethimide and hydrocortisone on prostate-specific

antigen response in metastatic prostate cancer refractory to standard endocrine therapy. *Anticancer Drugs* 2004;**15**(9):843–7.
49. Miller GM, Hinman Jr F. Cortisone treatment in advanced carcinoma of the prostate. *J Urol* 1954;**72**(3):485–96.
50. Small EJ, Meyer M, Marshall ME, et al. Suramin therapy for patients with symptomatic hormone-refractory prostate cancer: results of a randomized phase III trial comparing suramin plus hydrocortisone to placebo plus hydrocortisone. *J Clin Oncol* 2000;**18**(7):1440–50.
51. Sartor O, Weinberger M, Moore A, et al. Effect of prednisone on prostate-specific antigen in patients with hormone-refractory prostate cancer. *Urology* 1998;**52**(2):252–6.
52. Storlie JA, Buckner JC, Wiseman GA, et al. Prostate specific antigen levels and clinical response to low dose dexamethasone for hormone-refractory metastatic prostate carcinoma. *Cancer* 1995;**76**(1):96–100.
53. Tannock I, Gospodarowicz M, Meakin W, et al. Treatment of metastatic prostatic cancer with low-dose prednisone: evaluation of pain and quality of life as pragmatic indices of response. *J Clin Oncol* 1989;**7**(5):590–7.
54. Chen Y, Clegg NJ, Scher HI. Anti-androgens and androgen-depleting therapies in prostate cancer: new agents for an established target. *Lancet Oncol* 2009;**10**(10):981–91.
55. Watson PA, Chen YF, Balbas MD, et al. Constitutively active androgen receptor splice variants expressed in castration-resistant prostate cancer require full-length androgen receptor. *Proc Natl Acad Sci USA* 2010;**107**(39):16759–65.
56. Tran C, Ouk S, Clegg NJ, et al. Development of a second-generation antiandrogen for treatment of advanced prostate cancer. *Science* 2009;**324**(5928):787–90.
57. Scher HI, Fizazi K, Saad F, et al. Increased survival with enzalutamide in prostate cancer after chemotherapy. *N Engl J Med* 2012;**367**(13):1187–97.
58. Beer TM, Armstrong AJ, Rathkopf DE, et al. Enzalutamide in metastatic prostate cancer before chemotherapy. *N Engl J Med* 2014;**371**(5):424–33.
59. Clegg NJ, Wongvipat J, Joseph JD, et al. ARN-509: a novel antiandrogen for prostate cancer treatment. *Cancer Res* 2012;**72**(6):1494–503.
60. Rathkopf DE, Morris MJ, Fox JJ, et al. Phase I study of ARN-509, a novel antiandrogen, in the treatment of castration-resistant prostate cancer. *J Clin Oncol* 2013;**31**(28):3525–30.
61. Dana E, Rathkopf ESA, Shore Neal D, et al. ARN-509 in men with metastatic castration-resistant prostate cancer (mCRPC). *J Clin Oncol* 2013;**31** (Suppl. 6; abstr 48).
62. Labrie F. Adrenal androgens and intracrinology. *Semin Reprod Med* 2004;**22**(4):299–309.
63. Chen CD, Welsbie DS, Tran C, et al. Molecular determinants of resistance to antiandrogen therapy. *Nat Med* 2004;**10**(1):33–9.
64. Debes JD, Tindall DJ. Mechanisms of androgen-refractory prostate cancer. *N Engl J Med* 2004;**351**(15):1488–90.
65. Page ST, Lin DW, Mostaghel EA, et al. Persistent intraprostatic androgen concentrations after medical castration in healthy men. *J Clin Endocrinol Metab* 2006;**91**(10):3850–6.
66. Mostaghel EA, Page ST, Lin DW, et al. Intraprostatic androgens and androgen-regulated gene expression persist after testosterone suppression: therapeutic implications for castration-resistant prostate cancer. *Cancer Res* 2007;**67**(10):5033–41.
67. Scher HI, Sawyers CL. Biology of progressive, castration-resistant prostate cancer: directed therapies targeting the androgen-receptor signaling axis. *J Clin Oncol* 2005;**23**(32):8253–61.
68. Stigliano A, Gandini O, Cerquetti L, et al. Increased metastatic lymph node 64 and CYP17 expression are associated with high stage prostate cancer. *J Endocrinol* 2007;**194**(1):55–61.
69. Handratta VD, Vasaitis TS, Njar VC, et al. Novel C-17-heteroaryl steroidal CYP17 inhibitors/antiandrogens: synthesis, *in vitro* biological activity, pharmacokinetics, and antitumor activity in the LAPC4 human prostate cancer xenograft model. *J Med Chem* 2005;**48**(8):2972–84.
70. Dong JT. Prevalent mutations in prostate cancer. *J Cell Biochem* 2006;**97**(3):433–47.
71. Ryan CJ, Smith MR, de Bono JS, et al. Abiraterone in metastatic prostate cancer without previous chemotherapy. *N Engl J Med* 2013;**368**(2):138–48.
72. Attard G, Reid AH, Yap TA, et al. Phase I clinical trial of a selective inhibitor of CYP17, abiraterone acetate, confirms that castration-resistant prostate cancer commonly remains hormone driven. *J Clin Oncol* 2008;**26**(28):4563–71.
73. Zytiga. *Full Product Information*. Horsham, PA: Janssen Biotech; 2012.
74. Huggins C, Hodges CV. Studies on prostatic cancer: I. The effect of castration, of estrogen and of androgen injection on serum phosphatases in metastatic carcinoma of the prostate. *Cancer Res* 1941;(1).

# PART X

# CRYOABLATION, HIFU AND FOCAL THERAPY

CHAPTER

# 58

# Salvage Cryoablation of the Prostate

*Juan Chipollini, MD, Sanoj Punnen, MD*

**Department of Urology, University of Miami Miller School of Medicine, Miami, FL, USA**

## INTRODUCTION

According to data from the Surveillance, Epidemiology, and End Results (SEER) Program, 233,000 estimated new cases of prostate cancer (PCa) will have been diagnosed in the United States in 2014, and approximately 29,480 men will die of this disease.[1] Common treatment options for localized PCa include radical prostatectomy (RP), radiation therapy (RT), or active surveillance. External beam radiation therapy (EBRT) and brachytherapy (BT) have become popular choices for the management of localized PCa with up to one third of patients selecting RT as their primary treatment.[2,3]

Among patients undergoing RT for localized PCa, some will experience a recurrence of their disease, which is most often detected by a rise in the serum prostate-specific antigen (PSA). One study has shown that among 4839 patients, 1582 (33%) had biochemical failure by PSA criteria (American Society for Therapeutic Radiology and Oncology – ASTRO), while 416 (9%) had local failure and 329 (7%) had distant failure after external beam radiation therapy (EBRT) for clinically localized diseases.[4] Another study using data from the Cancer of the Prostate Strategic Urologic Research Endeavor (CaPSURE) database, identified 587 out of 935 patients (63%) to have a rise in their serum PSA within 10 years after RT.[5] Traditionally, these patients receive androgen deprivation therapy (ADT) with only a few being offered salvage treatments.

Potential salvage options for radiorecurrent PCa include salvage RP, BT, high-intensity focused ultrasound, and cryotherapy. A major limitation to all of these therapies has been the morbidity of treatment in the salvage setting, which is compounded on the secondary adverse effects of primary RT. Although salvage RP has the longest history and best likelihood for local control, it is generally more technically challenging due to radiation-induced fibrosis and obliteration of tissue planes. As a result it is associated with worsening urinary incontinence, erectile dysfunction, anastomotic strictures, and rectal injury.[6,7]

Cryotherapy has emerged as an alternative salvage strategy with more contemporary technologies allowing an improvement in oncological outcomes and complication rates over the past few decades shifting this treatment modality from investigational status to an established therapeutic option. In 2008, the American Urological Association (AUA) Best Practice Consensus Statement recognized cryoablation of the prostate as an established treatment option for men with newly diagnosed or radiorecurrent organ-confined PCa.[8] However, controversy still exists over the selection of appropriate patients to allow maximum likelihood of cure with minimal morbidity. The objective of this chapter is to provide an update on salvage cryotherapy with a focus on patient selection and a discussion of oncological and adverse outcomes.

## DETECTION OF RADIORECURRENT PROSTATE CANCER

The two most widely used definitions for failure after primary radiotherapy are the ASTRO definition of three consecutive rises after the PSA nadir and the Phoenix definition of PSA nadir +2 ng/mL.[9] The ASTRO definition has been criticized for being poorly linked to clinical progression, as small consecutive rises in PSA can constitute meeting the definition without showing any significant concern for clinically progressive disease. As a response to this and other deficiencies of the ASTRO criteria, the Phoenix definition was created. The goal of the Phoenix definition is to select patients who are more likely to go on to experience clinical progression rather than just biochemical recurrence alone. Therefore, the threshold to reach this definition is much higher, which results in it having a lower sensitivity and higher specificity for detecting recurrence. Although both definitions

**Prostate Cancer. http://dx.doi.org/10.1016/B978-0-12-800077-9.00058-X**

are currently used in clinical practice, neither definition can reliably discern between patients with local recurrence, who may benefit from local salvage therapy versus those with systemic progression, who are unlikely to benefit from such treatments.

PSA kinetics continues to be one of the most helpful predictors of outcome after treatment. A pretreatment PSA velocity > 2.0 ng/mL per year in 18 months prior to diagnosis has been found to be associated with prostate-cancer-specific mortality and all-cause mortality after RT or RP.[10,11] Therefore, patients with a pretreatment PSA velocity > 2.0 ng/mL per year at initial presentation are not the optimal candidates for local salvage therapy at the time of PSA failure due to their high likelihood of having occult micrometastases.[12] Zagars and Pollack[13] observed that a PSA doubling time (PSA-DT) < 8 months had a 7-year actuarial metastatic rate of 54% while patients with a PSA-DT > 8 months had only a 7% metastatic rate. Spiess et al.[14] reported a statistically significant difference in PSA doubling time (12.3 months vs. 5.6 months) for men with lower (<10 ng/mL) versus higher (≥10 ng/mL) PSA level before salvage cryoablation. Izawa et al.[15] reported a 5-year disease-free survival (DFS) of 57% for precryotherapy PSA ≤ 10 ng/mL versus 23% for PSA > 10 ng/mL ($p$ = 0.004). Although all of these thresholds require further validation in prospective studies, it appears that the best candidates are those with a low serum PSA and a long doubling time prior to salvage therapy. Retrospective studies have suggested the optimal candidates should have a PSA < 10 ng/mL, and a PSA-DT > 8 months, preferably >12 months.[12] However, prospective trials are desperately needed in the salvage setting to help us understand who the best candidates are for local salvage after failed radiation.

The usual response to BCR is to restage the patient and the most commonly used imaging modalities for this purpose are computed tomography (CT) and bone scan. However, molecular imaging modalities, such as positron-emission tomography (PET) CT, are gaining popularity in restaging patients with BCR.[6] Choline is a compound used in phospholipid biosynthesis that shows increased uptake in tumor cells,[16] and it was recently approved by the US Food and Drug Administration for detection of recurrent PCa. Promising results have been obtained with newer PET tracers such as choline and fluoride for the detection of distant disease.[17] Generally, PET provides good sensitivity and specificity in detecting distant or regional PCa recurrence; however, these numbers are highly dependent on the serum PSA level, with better detection at higher PSA levels.[18]

MRI has also gained popularity in PCa detection, localization, and staging.[19] The interpretation of PCa on T2-weighted MR imaging can be affected by false-positive findings such as prostatitis, postbiopsy hemorrhage, and fibrosis.[20] To improve its diagnostic accuracy, functional MR imaging techniques have been developed, such as diffusion-weighted,[21] proton spectroscopic,[22] and dynamic contrast-enhanced MR imaging. The addition of these parameters has enhanced the ability to localize cancer within the prostate allowing improved detection of local recurrences as well as guide biopsy to areas of the gland that are suspicious for recurrence.[23]

Before considering any kind of local salvage therapy, a local recurrence of PCa needs to be confirmed with a prostate biopsy and reviewed by a pathologist with the knowledge of pathological radiation changes in the prostate after RT. The timing of the biopsy can be of significance since some studies have reported that approximately 30% of positive biopsies obtained 12 months after radiation will convert to negative by 24–30 months.[24,25] Due to the potential morbidity salvage therapy can bring, it becomes crucial to determine the patients who are more likely to acquire a local disease in order to spare unwarranted side effects in those who are unlikely to benefit due to systemic progression. Even with today's available diagnostic modalities, none of these tools or even the use of all of them can reliably discern between localized versus distant recurrence.

## HISTORY OF CRYOSURGERY

Cryosurgery uses freezing to induce cell death by a variety of mechanisms. As temperature falls below −20°C, water in the extracellular space crystallizes into ice resulting in withdrawal of water from the system and creating a hyperosmotic extracellular space. This leads to denaturation and direct rupture of cell membranes by the ice crystals.[26] Injury to endothelial cells leads to platelet aggregation and eventual ischemia due to thrombi formation, which leads to vascular stasis.[26] Delayed and indirect destructive effects of cryosurgery continue due to vascular disruption resulting in tissue hypoxia and vascular thrombosis.[27] The temperature and duration of freezing necessary to induce cell apoptosis are based on many *in vitro* and *in vivo* studies, but it is generally accepted that a minimum temperature of −40°C must be reached for at least 3 min for complete killing of tumor cells.[28,29] Larson et al.[30] found that uniform coagulative necrosis of prostatic tissue *in vivo* can be accomplished throughout a significantly larger zone with double freeze rather than a single freeze. The critical temperature to reach necrosis with a double freeze was approximately below −40°C. Gage and Baust established that a minimum of two freeze-thaw cycles were necessary for complete tumor eradication.[28]

The first generation of cryosurgical systems that came about in the 1960s utilized liquid nitrogen to create an ice ball and employed very little monitoring resulting in high complication rates. Cryosurgery became more prevalent with the introduction of transrectal ultrasound (TRUS) guidance described by Onik et al., which allowed for closer monitoring of the freezing cycle.[31] Sloughing of necrotic tissue per urethra was one of the major complications during the early stages of prostate cryosurgery, which led to its decreased popularity among urologists. With the introduction of urethral warming catheters, the risk of this urethral transmural necrosis and subsequent sloughing decreased significantly.[32,33] The use of TRUS guidance along with urethral warming catheters was the hallmark of second-generation cryosurgical units.[31,33] TRUS guidance allowed for more accurate placement of probes as well as real-time visualization of the ice ball in turn reducing the risk of injury to delicate structures such as the rectum and external urinary sphincter. The urethral warming catheters provided protection of the urethra during freezing, and led to a significant reduction of urethral sloughing as well.

Newer third-generation cryosurgical units transitioned from liquid nitrogen to argon/helium-based systems according to the Joule–Thompson principle. These gas-driven 17-gauge Cryoneedles™ (Oncura, Inc., Plymouth Meeting, PA) use pressurized gas to freeze (argon gas) and actively thaw (helium gas) prostatic tissue using the Joule–Thompson effect, in which all the gases undergo different and unique temperature changes when depressurized according to unique gas coefficients.[29,34,35] This allowed for smaller-diameter cryoneedles (1.5 mm) for transperineal placement using a BT-like template without the need for tract dilation and insertion kits.[29] Depending on surgeon preference, up to five thermocoupling sensors may be placed in the mid-gland, the external sphincter, neurovascular bundles (NVBs), and Denonvilliers' fascia to ensure the required temperature of −40°C is reached (midgland and NVBs). This minimizes the risk of incontinence and recto-urethral fistula (external sphincter and Denonvilliers' fascia).[29] These thermosensor probes combined with multifrequency biplanar TRUS, double free-thaw cycles, and urethral warming catheters have vastly improved the toxicity and complication profile of cryoablation.

Although cryosurgery had been used as primary treatment for PCa for decades, it was not until 1997 when the University of Texas M.D. Anderson Cancer Center published one of the earliest series evaluating cryosurgery as a salvage option for locally recurrent PCa in their cohort of 150 men using liquid nitrogen first-generation technology.[36] Acceptable local tumor control was reported based on end points such as posttreatment PSA and prostate biopsies. Those patients who underwent double freeze-thaw cycle had a higher negative biopsy rate (93% vs. 71%) and lower biochemical failure rate (44% vs. 65%) than those who underwent a single freeze-thaw cycle. However, there were significant long-term complications in patients including incontinence (73%), impotence (72%), obstructive symptoms (67%), and severe perineal pain (8%). These days, more contemporary generation units have helped decrease these rates even in the salvage setting.

## PATIENT SELECTION FOR SALVAGE CRYOABLATION

According to the 2008 AUA Best Practice Statement, optimal candidates should have biopsy-proven persistent organ-confined PCa, a PSA less than 10 ng/mL, no evidence of seminal vesicle invasion, a long PSA doubling time, and a negative metastatic evaluation.[8] One of studies supporting the panel's recommendation was based on the 7-year retrospective analysis by Bahn et al.[37] of 59 patients who received argon-based salvage cryoablation from 1993 to 2001. These men were rigorously followed with serial serum PSA testing and sextant biopsies at set intervals and anytime the PSA rose above 0.5 ng/mL. Using PSA threshold of 0.5 ng/mL as evidence of biochemical recurrence, 61, 62, and 50% of patients with a pretreatment PSA of <4, 4–10, and >10 ng/mL, respectively, remained free of biochemical relapse. Of the 38 patients who underwent biopsy, none showed recurrence of local cancer suggesting that failure was supposedly caused by overlooked micrometastatic disease. Given that at least 29% of their cohort presented with precryotherapy stage ≥T3, it is plausible that the PSA failures were due to preexisting systemic disease by the cohort prior to undergoing salvage local therapy. Spiess et al.[14] found PSA level >10 ng/mL ($p = 0.002$) at the time of diagnosis of local recurrence and PSA doubling time ≤16 months ($p = 0.06$) were predictors of biochemical failure after salvage cryotherapy in their series of 49 patients with a median follow up of 5.7 years. Although there is no supporting data, the panel also recommended biopsy of SVs in addition to a prostate biopsy based on the higher incidence of SV involvement in men who have failed primary RT. Pathological results from salvage RP series reveal that the rate of SV involvement can be as high as 28–49%,[38,39] and is associated with poor prognostic outcomes.

Pisters et al.[40] identified similar clinical pretreatment factors associated with early treatment failure after salvage cryosurgery. From their analysis of 145 patients, those failing initial RT with a PSA level >10 ng/mL, and having a recurrence of Gleason score ≥9 were unlikely to be successfully salvaged. Nguyen et al.[12] found in their systematic review that patients most likely to benefit from salvage local therapy were those with initial low-risk disease (PSA level <10 ng/mL, Gleason score <6, clinical stage T1c or T2a), pretreatment PSA velocity <2.0 ng/mL/year, interval to PSA failure >3 years, and PSA doubling time >12 months. Although there are no clearly defined guidelines for the proper selection of patients for salvage cryotherapy, the ultimate goal of these selection factors is to pick out those patients who are unlikely to harbor distant diseases since these are the men most likely to benefit from any salvage local therapy. The ideal candidate for salvage cryotherapy is similar in clinical characteristics to the ideal candidate for any local salvage modality. Along with these characteristics, these patients also share the need for more prospective studies in the salvage setting so we can better understand the true criteria for the ideal local salvage patient, allowing us to prevent unwarranted morbidity in men who are unlikely to benefit from such treatments.

Relative contraindications to cryotherapy are similar to those in BT.[27] They include patients with a previous transurethral resection of the prostate since it can result in a large tissue defect, patients with significant obstructive symptoms since they have an increased likelihood of postoperative urinary retention, large prostate gland size due to the difficulty of achieving complete cryoablation, and a history of abdominoperineal resection for rectal cancer, rectal stenosis, or other major rectal pathology.[8,27]

## ONCOLOGICAL EFFICACY OF SALVAGE CRYOABLATION

There is no cryoablation-specific definition for BCR and most series extrapolate from radiation definitions; thus, biochemical failure rates vary according to definition of failure, clinical characteristics, and length of follow up. In 2007, Ismail et al. reported the biochemical disease-free survival (bDFS) of 100 patients with a mean follow up of 33.5 months to be 83, 72, and 59% at 12, 24, and 36 months, respectively.[41] From their results, a PSA level >10 ng/mL before salvage cryoablation, clinical stage >T2b, and Gleason grade of >7 predicted unfavorable outcomes. In 2008, Pisters et al. used data from the Cryo On-Line Data (COLD) registry to report a 5-year bDFS of 58.9 and 54.5% using the ASTRO and Phoenix definitions of BCR, respectively, in 279 patients with a median follow up of 21 months.[42] In 2010, Levy et al.[43] used data from the COLD registry to assess the PSA-based bDFS of 455 patients using the Phoenix definition. Their analysis indicated that an initial PSA level of ≤0.6 ng/mL after salvage cryoablation portended a favorable bDFS of 67% at 36 months.

An undetectable PSA may not always be achievable due to the remaining periurethral tissue the cryoablation procedure tries to avoid.[44] As a result of this PSA-producing tissue, a posttreatment biopsy may be the most definitive determination of treatment failure after salvage cryoablation.[7] Many of the series that have included routine posttreatment biopsy as part of their protocol have shown a variable positive biopsy rate. Chin et al.[45] performed four-core posttreatment biopsies in 106 patients resulting in a 2.8% positive biopsy rate for malignancy and 73% of those positive cores were found within 1 year after cryoablation. In another series, Izawa et al.[46] reported a 21% positive biopsy rate in 107 patients after salvage cryotherapy; variables that remained associated with a positive biopsy on multivariate analysis were the number of probes used and postcryotherapy PSA nadir.

Spiess et al.[47] reported a pretreatment nomogram based on a large multi-institutional study, and identified serum PSA level at diagnosis, the initial biopsy Gleason score, and the clinical stage at diagnosis as important predictors of treatment outcome. Ng et al.[48] reported on the Western Ontario experience of 187 patients and showed a 5-year biochemical recurrence-free survival of 56% for patients with presalvage PSA level < 4 ng/mL and 14% for patients with presalvage PSA > 10 ng/mL. A PSA nadir > 0.1 ng/mL and Gleason score before initial radiotherapy were statistically significant factors for biochemical recurrence. Williams et al.[49] reported an update on 176 of these patients with mean follow up of nearly 8 years and showed a 64% 10-year DFS for patients with a presalvage PSA < 5 ng/mL versus 6.7% for patients with a presalvage PSA > 10 ng/mL. A PSA nadir > 1 ng/mL was highly predictive of early recurrence, which was comparable to other series[14,43] and reasserts that those patients with a PSA nadir > 1 ng/mL are likely to harbor metastatic disease.[49]

The most recent and extensive single institution series to date comes from Columbia where 328 patients were treated with second-and third-generation cryotherapy technologies.[50] At a median follow-up of 47 months, the 5- and 10-year recurrence-free survival rates were 63 and 35%, respectively. Factors associated with a better prognosis included an increasing time interval from original treatment to salvage therapy, a lower PSA at the time of surgery, and a lower postcryotherapy PSA nadir. Table 58.1 lists a breakdown of oncological outcomes in recent salvage cryoablation series.

## FOCAL CRYOTHERAPY

Although whole–gland salvage cryoablation is the standard treatment some authors have explored the role of focal or partial salvage cryoablation while showing encouraging oncological and functional outcomes. De Castro et al.[54] found a 5-year DFS by Phoenix criteria of 54% for salvage focal cryoablation ($n$ = 25) versus 86% for salvage total cryoablation ($n$ = 25) while showing less urinary morbidity in the focal treatment group. Eisenberg and Shinohara[55] found a DFS by ASTRO criteria of 89, 67, and 50 at 1, 2, and 3 years, respectively, in 19 patients, and only one had viable cancer on posttreatment biopsy at 1 year.

**TABLE 58.1** Oncological Results of Recent Salvage Cryoablation Series

| Series | Patients (*n*) | Medium follow-up (months) | Failure definition | Gleason before cryotherapy | Stage before cryotherapy | DFS |
|---|---|---|---|---|---|---|
| Williams et al.[49] | 187 | 90 | Phoenix | G 6–7:60.7%, G ≥ 8:39.3% | ≤T2:86.1%, ≥T3:13.9% | 64% (PSA < 5) and 6.7% (PSA > 10) at 10 years |
| Bahn et al.[37] | 59 | 82.3 | PSA ≥0.5 ng/mL or 1.0 ng/mL | G 6–7:71%, G ≥ 8:29% | ≤T2:69%, ≥T3:29% | 59% (PSA 0.5 ng/mL) and 69% (PSA 1.0 ng/mL) at 7 years |
| Pisters et al.[42] | 279 | 21.6 | ASTRO and Phoenix | G 6–7:51.2%, G ≥ 8:43.7% | NR | 58.9% (ASTRO) and 54.5% (Phoenix) at 5 years |
| Ismail et al.[41] | 100 | 33.5 | PSA <0.5 ng/mL and ASTRO | G 6–7:63%, G ≥ 8:37% | ≤T2:70%, ≥T3:30% | 83% 1 year, 72% 2 years, 59% 3 years (ASTRO) 73% low, 45% intermediate, 11% high risk at 5 years (PSA 0.5 ng/mL) |
| Wenske et al.[50]* | 328 | 47.8 | Phoenix | G 6–7:47% G ≥8:53% | ≤T2:75% ≥T3:25% | 63% 5 years, 35% 10 years |
| Philippou et al.[51] | 19 | 33.3 | Phoenix | Median G:7 | NR | 58% at 2 years |
| Ghafar et al.[52] | 38 | 20.7 | PSA >0.3 g/mL above nadir | Mean G:7 | NR | 86% 1 year, 74% 2 years |
| Cheetham et al.[53] | 51 | 121.2 | Phoenix | G 5–7:47% G ≥8:43% | ≥T3:9.8% Unknown:90.2% | 76% at 1 year |

ASTRO, American Society for Therapeutic Radiology and Oncology; DFS, disease-free survival; G, Gleason score; NR, not reported.
**17% of patients had focal ablation only.*

Although these results are intriguing, this treatment option remains investigational at this time and not considered standard of care; however, in highly selected patients with low volume, unilateral disease, it can become an attractive option in the future and requires further studies. The biggest limitation in the adoption of this approach is the inherent selection of patients who are likely to have a dominant tumor that can be targeted and kept free of significant disease elsewhere in the prostate.

## COMPLICATIONS

It is hypothesized that patients undergoing salvage procedures have higher complication rates than patients undergoing primary therapies due to the retreatment of previously irradiated and poorly vascularized tissue.[7] Salvage cryotherapy has shown incontinence rates of 36% and fistula rates of 2.6% across studies, in comparison with salvage prostatectomy series, which report incontinence rates of 41% and rectal injury rates of 2–15%.[12] Peters et al. found a combined toxicity rate of 30% in a multi-institutional study in the Netherlands among salvage RP, salvage cryotherapy and salvage BT patients; with Grade 3 GU and GI toxicity occurring in 32, 29, and 30% of salvage RP, BT, and cryoablation patients, respectively.[56]

With first-generation liquid nitrogen-based cryounits, complications such as urinary incontinence, recto-urethral fistula, and urethral sloughing were common. The high morbidity presented in early studies was due to many reasons. The lack of temperature probes and the unavailability of the urethral warming device, temporarily banned by the US Food and Drug Administration, allowed urethral sloughing to become prevalent leading to pain, urinary retention, and incontinence.[27] The lack of TRUS visualization also led to poor monitoring of the ice ball and inability to confirm correct placement of the cryoprobes. The ability to directly observe probe placement and ice ball formation was a major step forward to avoid damage to the urethra and adjacent tissues. When probe placement was combined with urethral preservation, the complication of sloughing decreased from 100% to <10%.[33] Cohen[57] compared the complications of salvage cryoablation between first-, second-, and third-generation cyrounits at a large single institution, and showed a significant improvement in the complication rate of urethral sloughing (16% vs. 2%), incontinence (8.6% vs. 0%), fistulas (2% vs. 0%), and prostatic stone formation (3% vs. 0%) between first- and second-generation versus third-generation equipment, respectively. Table 58.2 shows complications of salvage cryoablation associated with different device generations.

**TABLE 58.2** Complications of Salvage Cryoablation

| Series | Patients ($n$) | Mean age (year) | Device generation | Incontinence | Impotence (%) | Sloughing/ Retention (%) | Rectal-urethral fistula (%) |
|---|---|---|---|---|---|---|---|
| Pisters et al.[36] | 150 | 68 | First | 73% (any degree of dribbling) | 72** | 44 | 1 |
| Bahn et al.[37] | 59 | 67.5 | Second | 8% | NR | NR | 3.4 |
| Donnelly et al.[58] | 46 | 68.9 | Second/Third | 4.3% | 80* | 6.5 | 2.2 |
| Ng et al.[48] | 187 | 70.9 | Third | Mild-moderate:37%, severe:3% | NR | 14 | 2 |
| Ismail et al.[41] | 100 | 66.8 | Third | ≥1 pad in 24 h:13% | 86** | 2 | 1 |
| Han and Belldegrun[29] | 29 | NR | Third | Requiring pads:7% | 84* | 3 | 0 |

NR, not reported.
*Erection not sufficient for vaginal intercourse.*
***Assessed by survey or questionnaire.*

# CONCLUSIONS

Among men who experience a rise in serum PSA after EBRT or BT, many will have local only recurrence and thus benefit from a salvage local treatment. Available comparative studies are observational in nature and retrospective; thereby limiting our ability to reliably compare salvage therapies with respect to oncological outcomes and complication rates. Future prospective randomized trials in this setting are desperately needed to determine the comparative safety and efficacy of available options for local salvage therapy. Although savage RP is the most established salvage modality with long-term survival benefit, the procedure is complex and not without significant morbidity. Salvage cryoablation is becoming an increasingly popular alternative that is minimally invasive and demonstrates a comparable safety and efficacy profile. Newer technologies have allowed improvements in oncological outcomes while reducing the rates of serious complications. Although more prospective studies with long-term follow up are needed, salvage cyroablation should be considered a reasonable option for local salvage in select patients with confirmed localized recurrence after radiotherapy.

## References


1. DeSantis CE, Lin CC, Mariotto AB, et al. Cancer treatment and survivorship statistics. *CA Cancer J Clin* 2014;**64**(4):252–71.
2. Mettlin CJ, Murphy GP, Rosenthal DS, et al. The National Cancer Data Base report on prostate carcinoma after the peak in incidence rates in the US. The American College of Surgeons Commission on Cancer and the American Cancer Society. *Cancer* 1998;**83**(8): 1679–84.
3. Mettlin CJ, Murphy GP, McDonald CJ, et al. The National Cancer Data base Report on increased use of brachytherapy for the treatment of patients with prostate carcinoma in the US. *Cancer* 1999;**86**(9):1877–82.
4. Kuban DA, Thames HD, Levy LB, et al. Long-term multi-institutional analysis of stage T1–T2 prostate cancer treated with radiotherapy in the PSA era. *Int J Radiat Oncol Biol Phys* 2003;**57**(4): 915–28.
5. Agarwal PK, Sadetsky N, Konety BR, et al. Cancer of the prostate strategic urological research E. Treatment failure after primary and salvage therapy for prostate cancer: likelihood, patterns of care, and outcomes. *Cancer* 2008;**112**(2):307–14.
6. Punnen S, Cooperberg MR, D'Amico AV, et al. Management of biochemical recurrence after primary treatment of prostate cancer: a systematic review of the literature. *Eur Urol* 2013;**64**(6):905–15.
7. Finley DS, Belldegrun AS. Salvage cryotherapy for radiation-recurrent prostate cancer: outcomes and complications. *Curr Urol Rep* 2011;**12**(3):209–15.
8. Babaian RJ, Donnelly B, Bahn D, et al. Best practice statement on cryosurgery for the treatment of localized prostate cancer. *J Urol* 2008;**180**(5):1993–2004.
9. Nielsen ME, Partin AW. The impact of definitions of failure on the interpretation of biochemical recurrence following treatment of clinically localized prostate cancer. *Rev Urol* 2007;**9**(2):57–62.
10. D'Amico AV, Chen MH, Roehl KA, et al. Preoperative PSA velocity and the risk of death from prostate cancer after radical prostatectomy. *N Engl J Med* 2004;**351**(2):125–35.
11. D'Amico AV, Renshaw AA, Sussman B, et al. Pretreatment PSA velocity and risk of death from prostate cancer following external beam radiation therapy. *JAMA* 2005;**294**(4):440–7.
12. Nguyen PL, D'Amico AV, Lee AK, et al. Patient selection, cancer control, and complications after salvage local therapy for postradiation prostate-specific antigen failure: a systematic review of the literature. *Cancer* 2007;**110**(7):1417–28.
13. Zagars GK, Pollack A. Kinetics of serum prostate-specific antigen after external beam radiation for clinically localized prostate cancer. *Radiother Oncol* 1997;**44**(3):213–21.
14. Spiess PE, Lee AK, Leibovici D, et al. Presalvage prostate-specific antigen (PSA) and PSA doubling time as predictors of biochemical failure of salvage cryotherapy in patients with locally recurrent prostate cancer after radiotherapy. *Cancer* 2006;**107**(2):275–80.
15. Izawa JI, Madsen LT, Scott SM, et al. Salvage cryotherapy for recurrent prostate cancer after radiotherapy: variables affecting patient outcome. *J Clin Oncol* 2002;**20**(11):2664–71.
16. Hara T, Kosaka N, Kishi H. PET imaging of prostate cancer using carbon-11-choline. *J Nucl Med* 1998;**39**(6):990–5.
17. Picchio M, Giovannini E, Messa C. The role of PET/computed tomography scan in the management of prostate cancer. *Curr Opin Urol* 2011;**21**(3):230–6.

18. Martino P, Scattoni V, Galosi AB, et al. Role of imaging and biopsy to assess local recurrence after definitive treatment for prostate carcinoma (surgery, radiotherapy, cryotherapy, HIFU). *World J Urol* 2011;**29**(5):595–605.
19. Futterer JJ. MR imaging in local staging of prostate cancer. *Eur J Radiol* 2007;**63**(3):328–34.
20. Hambrock T, Futterer JJ, Huisman HJ, et al. Thirty-two-channel coil 3T magnetic resonance-guided biopsies of prostate tumor suspicious regions identified on multimodality 3T magnetic resonance imaging: technique and feasibility. *Invest Radiol* 2008;**43**(10):686–94.
21. Tan CH, Wei W, Johnson V, et al. Diffusion-weighted MRI in the detection of prostate cancer: meta-analysis. *AJR Am J Roentgenol* 2012;**199**(4):822–9.
22. Heijmink SW, Scheenen TW, Futterer JJ, et al. Prostate and lymph node proton magnetic resonance (MR) spectroscopic imaging with external array coils at 3 T to detect recurrent prostate cancer after radiation therapy. *Invest Radiol* 2007;**42**(6):420–7.
23. Li L, Wang L, Feng Z, et al. Prostate cancer magnetic resonance imaging (MRI): multidisciplinary standpoint. *Quant Imaging Med Surg* 2013;**3**(2):100–12.
24. Crook J, Malone S, Perry G, et al. Postradiotherapy prostate biopsies: what do they really mean? Results for 498 patients. *Int J Radiat Oncol Biol Phys* 2000;**48**(2):355–67.
25. Scardino PT. The prognostic significance of biopsies after radiotherapy for prostatic cancer. *Semin Urol* 1983;**1**(4):243–52.
26. Gage AA, Baust JM, Baust JG. Experimental cryosurgery investigations *in vivo*. *Cryobiology* 2009;**59**(3):229–43.
27. Lam JS, Belldegrun AS. Salvage cryosurgery of the prostate after radiation failure. *Rev Urol* 2004;(6 Suppl. 4):S27–36.
28. Gage AA, Baust J. Mechanisms of tissue injury in cryosurgery. *Cryobiology* 1998;**37**(3):171–86.
29. Han KR, Belldegrun AS. Third-generation cryosurgery for primary and recurrent prostate cancer. *BJU Int* 2004;**93**(1):14–8.
30. Larson TR, Rrobertson DW, Corica A, et al. *In vivo* interstitial temperature mapping of the human prostate during cryosurgery with correlation to histopathologic outcomes. *Urology* 2000;**55**(4):547–52.
31. Onik GM, Cohen JK, Reyes GD, et al. Transrectal ultrasound-guided percutaneous radical cryosurgical ablation of the prostate. *Cancer* 1993;**72**(4):1291–9.
32. Cohen JK, Miller RJ, Shuman BA. Urethral warming catheter for use during cryoablation of the prostate. *Urology* 1995;**45**(5):861–4.
33. Cohen JK, Miller RJ. Thermal protection of urethra during cryosurgery of prostate. *Cryobiology* 1994;**31**(3):313–6.
34. Chin JL, Downey DB, Mulligan M, et al. Three-dimensional transrectal ultrasound guided cryoablation for localized prostate cancer in nonsurgical candidates: a feasibility study and report of early results. *J Urol* 1998;**159**(3):910–4.
35. Wong WS, Chinn DO, Chinn M, et al. Cryosurgery as a treatment for prostate carcinoma: results and complications. *Cancer* 1997;**79**(5):963–74.
36. Pisters LL, von Eschenbach AC, Scott SM, et al. The efficacy and complications of salvage cryotherapy of the prostate. *J Urol* 1997;**157**(3):921–5.
37. Bahn DK, Lee F, Silverman P, et al. Salvage cryosurgery for recurrent prostate cancer after radiation therapy: a seven-year follow-up. *Clin Prostate Cancer* 2003;**2**(2):111–4.
38. Gheiler EL, Tefilli MV, Tiguert R, et al. Predictors for maximal outcome in patients undergoing salvage surgery for radio-recurrent prostate cancer. *Urology* 1998;**51**(5):789–95.
39. Rogers E, Ohori M, Kassabian VS, et al. Salvage radical prostatectomy: outcome measured by serum prostate specific antigen levels. *J Urol* 1995;**153**(1):104–10.
40. Pisters LL, Perrotte P, Scott SM, et al. Patient selection for salvage cryotherapy for locally recurrent prostate cancer after radiation therapy. *J Clin Oncol* 1999;**17**(8):2514–20.
41. Ismail M, Ahmed S, Kastner C, et al. Salvage cryotherapy for recurrent prostate cancer after radiation failure: a prospective case series of the first 100 patients. *BJU Int* 2007;**100**(4):760–4.
42. Pisters LL, Rewcastle JC, Donnelly BJ, et al. Salvage prostate cryoablation: initial results from the cryo on-line data registry. *J Urol* 2008;**180**(2):559–63.
43. Levy DA, Pisters LL, Jones JS. Prognostic value of initial prostate-specific antigen levels after salvage cryoablation for prostate cancer. *BJU Int* 2010;**106**(7):986–90.
44. Mouraviev V, Spiess PE, Jones JS. Salvage cryoablation for locally recurrent prostate cancer following primary radiotherapy. *Eur Urol* 2012;**61**(6):1204–11.
45. Chin JL, Touma N, Pautler SE, et al. Serial histopathology results of salvage cryoablation for prostate cancer after radiation failure. *J Urol* 2003;**170**(4 Pt 1):1199–202.
46. Izawa JI, Perrotte P, Greene GF, et al. Local tumor control with salvage cryotherapy for locally recurrent prostate cancer after external beam radiotherapy. *J Urol* 2001;**165**(3):867–70.
47. Spiess PE, Katz AE, Chin JL, et al. A pretreatment nomogram predicting biochemical failure after salvage cryotherapy for locally recurrent prostate cancer. *BJU Int* 2010;**106**(2):194–8.
48. Ng CK, Moussa M, Downey DB, et al. Salvage cryoablation of the prostate: followup and analysis of predictive factors for outcome. *J Urol* 2007;**178**(4 Pt 1):1253–7 discussion 7.
49. Williams AK, Martinez CH, Lu C, et al. Disease-free survival following salvage cryotherapy for biopsy-proven radio-recurrent prostate cancer. *Eur Urol* 2011;**60**(3):405–10.
50. Wenske S, Quarrier S, Katz AE. Salvage cryosurgery of the prostate for failure after primary radiotherapy or cryosurgery: long-term clinical, functional, and oncologic outcomes in a large cohort at a tertiary referral centre. *Eur Urol* 2013;**64**(1):1–7.
51. Philippou P, Yap T, Chinegwundoh F. Third-generation salvage cryotherapy for radiorecurrent prostate cancer: a centre's experience. *Urol Int* 2012;**88**(2):137–44.
52. Ghafar MA, Johnson CW, De La Taille A, et al. Salvage cryotherapy using an argon based system for locally recurrent prostate cancer after radiation therapy: the Columbia experience. *J Urol* 2001;**166**(4):1333–7.
53. Cheetham P, Truesdale M, Chaudhury S, et al. Long-term cancer-specific and overall survival for men followed more than 10 years after primary and salvage cryoablation of the prostate. *J Endourol* 2010;**24**(7):1123–9.
54. de Castro Abreu AL, Bahn D, Leslie S, et al. Salvage focal and salvage total cryoablation for locally recurrent prostate cancer after primary radiation therapy. *BJU Int* 2013;**112**(3):298–307.
55. Eisenberg ML, Shinohara K. Partial salvage cryoablation of the prostate for recurrent prostate cancer after radiotherapy failure. *Urology* 2008;**72**(6):1315–8.
56. Peters M, Moman MR, van der Poel HG, et al. Patterns of outcome and toxicity after salvage prostatectomy, salvage cryosurgery and salvage brachytherapy for prostate cancer recurrences after radiation therapy: a multi-center experience and literature review. *World J Urol* 2013;**31**(2):403–9.
57. Cohen JK. Cryosurgery of the prostate: techniques and indications. *Rev Urol* 2004;(6 Suppl. 4):S20–26.
58. Donnelly BJ, Saliken JC, Ernst DS, et al. Role of transrectal ultrasound guided salvage cryosurgery for recurrent prostate carcinoma after radiotherapy. *Prostate Cancer Prostatic Dis* 2005;**8**(3): 235–42.

CHAPTER

# 59

# High-Intensity Focused Ultrasound

*M. Francesca Monn, MD, MPH, Chandra K. Flack, MD, Michael O. Koch, MD*

**Department of Urology, Indiana University School of Medicine, Indianapolis, IN, USA**

## INTRODUCTION

Increases in prostate cancer (PCa) screening due to the widespread use of prostate-specific antigen (PSA) has resulted in the diagnosis of more focal and clinically insignificant prostate adenocarcinomas. Active surveillance is often offered as an option for managing carefully selected patients with low-volume disease; however, guidelines for selecting which men should qualify have not been widely accepted or implemented. Studies have suggested that there can be significant psychological distress associated with watchful waiting, even with low-risk tumors.[1] Alternatively, radical therapies for localized prostate cancer (PCa) such as radiotherapy and radical prostatectomy are known to impact patients' quality of life, specifically urinary continence and sexual function. High-Intensity Focused Ultrasound (HIFU) has been proposed as an alternative method of treating localized PCa with a potentially lower side effect profile than radiotherapy or radical prostatectomy.[1–4] HIFU is also being utilized in patients who require definitive therapy but may not be surgical candidates and as salvage therapy following recurrence after radical prostatectomy or radiotherapy. Advantages to HIFU therapy include minimal bleeding, reduced risk of infection, procedural simplicity, and repeatability.[5]

HIFU, like cryoablation, is one of the more common forms of thermal therapy that is currently being utilized in PCa management. Whereas cryoablation utilizes extreme cold to destroy tissue, HIFU uses intensely focused sound waves to burn it. HIFU was pioneered for the management of benign prostatic hypertrophy (BPH) in the 1990s.[6–9] However, due to its side effect profile and the availability of alternative minimally-invasive procedures and holmium laser enucleation or ablation of the prostate, the use of HIFU for BPH has not become widespread. Ohigashi et al. compared the efficacy of HIFU with transurethral needle ablation (TUNA) and transurethral microwave therapy (TUMT) for treating lower urinary tract symptoms (LUTS) and reported no differences in efficacy or long-term durability between the three therapies.[10] Despite the limited interest in using HIFU to treat BPH, the procedure is gaining popularity in other areas of urology as well as in other surgical fields.[11–14] HIFU has recently been used for the management of small renal tumors,[15,16] and it has carried FDA approval for the treatment of uterine fibroids since 2004.

## HOW HIFU WORKS

HIFU technology works by focusing high-energy (3–4 MHz) ultrasound waves emitted in a parabolic arrangement such that they converge onto a discrete region of tissue, the focal point (Figure 59.1). Ultrasound energy is absorbed at the focus, heating the tissue to temperatures in the range of 80–100°C.[17–19] The characteristics of the tissue being destroyed, such as its size, shape, and absorption coefficient, as well as the duration of exposure together determine the thermal dose. Thermal doses above the necessary threshold cause coagulative-tissue destruction, ultimately leading to local tissue necrosis. Intensity of the beam is adjusted to fall above the heat threshold for the tissue but below the point where cavitation occurs.[18] In cavitation, negative pressure formed by the ultrasound wave leads to air bubbles forming within the cells. These bubbles oscillate in size and shape with the changing pressures caused by incoming ultrasound waves until they eventually collapse, creating a shockwave that destroys the surrounding cells.[19] Through this process, sharply demarcated lesions are created whereby the border between HIFU ablated and non-HIFU ablated cells is only 5–7 cell layers thick, and no damage to intervening tissue was observed.[18]

**Prostate Cancer. http://dx.doi.org/10.1016/B978-0-12-800077-9.00059-1**

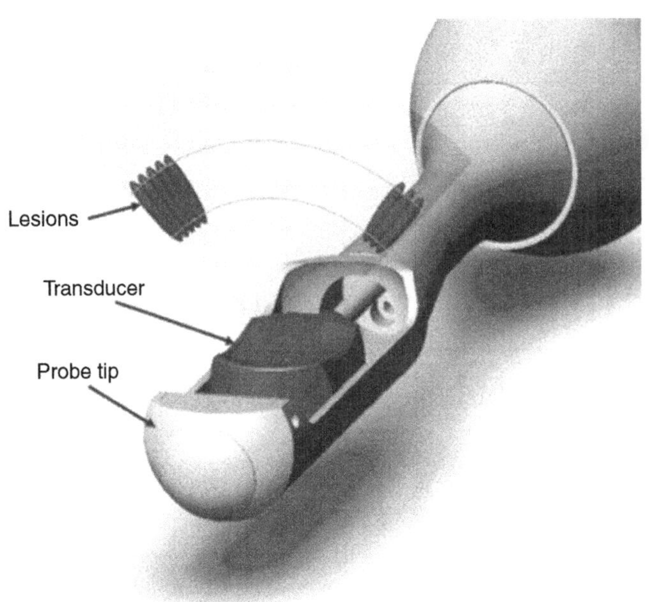


FIGURE 59.1 **Computer-generated rendering of transducer movement and lesion placement.** *Adapted from Koch et al.*[20]

There are currently two HIFU devices commercially available, Ablatherm Integrated Imaging® (Edap Technomed, Vaulx-en-Velin, France) and Sonablate 500® (Focus Surgery, Indianapolis, IN, USA).[21] These devices access the prostate gland via insertion of an ultrasound probe into the rectal vault. Ultrasound waves pass through rectal tissue with only minimal tissue change occurring outside the desired treatment area. Some rise in temperature outside the focal point does occur, and trials using early versions of the devices report complications of rectal burns and perforations. Newer versions mitigate this risk by using cooled, degassed fluid as a coupling medium to surround the probe and cool the rectal wall.[22] Real time rectal wall temperature, distance monitoring, ability to automatically detect patient movement, and pause treatment serve as safety features to protect surrounding tissue.[19,21]

To begin the procedure, the patient is placed in the right lateral decubitus position for the Ablatherm and in lithotomy or supine for Sonablate.[19] The device is lubricated and inserted into the rectum, providing access to the prostate gland for pretreatment imaging. Serial cross sectional images of the prostate are then obtained in order to define the shape of the prostate and surrounding landmarks. The resulting 3D computer reconstruction is used by the surgeon to program the device and define the treatment area.[18] The prostate is then sequentially destroyed using HIFU beams targeted to an area with lesions similar in size to a grain of rice.[23] By intentionally following a noncontiguous pattern of focus, the device limits buildup and spread of thermal energy and allows surrounding tissue to cool down. This allows highly precise and focused destruction of prostate tissue. Power level must be adjusted based on the characteristics of the tissue, accounting for patient history of external beam radiation or previous HIFU and the distance between the rectal wall and probe.[21] Real time imaging allows intraoperative adjustment that may be necessary to account for patient movement or whole-prostate swelling.[24] In the newest version of the Sonoblate device, acoustical imaging changes in the treatment zone and surrounding tissue are continuously monitored to detect energy absorption.[25,26] Significantly, focal length and ablation volumes vary between devices (Figure 59.2) and must be modified to account for prostate size.

## HISTOPATHOLOGIC CHANGES ASSOCIATED WITH HIFU

Biermann et al. examined the histopathologic findings of biopsies taken 6 months post-HIFU.[27] A majority of biopsies demonstrated coagulative necrosis, stromal fibrosis, and evidence of both acute and chronic inflammatory changes. Similarly, Madersbacher et al. found that HIFU ablated tissue exhibits classic coagulative necrosis within 7 days of treatment and by 10 weeks, the coagulative necrosis is resorbed and replaced by scar.[18] The resulting destruction to identifiable prostate architecture following treatment with HIFU raises concerns about potential difficulties in the identification of residual and recurrent PCa upon subsequent biopsy. As studied by Biermann et al., 44% of the HIFU ablated glands revealed residual prostate adenocarcinoma, and interestingly 82% of those biopsies that had adenocarcinoma did not show any evidence of HIFU treatment effect,[27] making accurate identification of residual disease of the utmost importance. Recently, Walter et al. reported that AMACR and MIB-1 stains most reliably differentiate carcinoma from necrosis in early HIFU biopsy samples, and they may be useful for differentiating between residual carcinoma and HIFU related architectural changes.[28] Additionally, Van Leenders et al. previously reported that HIFU lesions are marked by loss of cytokeratin 8 as compared to tissue that had not been ablated.[29] Although more challenging, accurate interpretation of post-HIFU biopsy samples is feasible.

## GUIDELINES FOR THE USE OF HIFU

Despite over a decade of experience with HIFU for the treatment of localized PCa, appropriate selection of patients to undergo HIFU therapy remains an unrealized goal. Definitions of patients deemed eligible for HIFU therapy vary between countries and institutions. For example, the University College London Trials Unit definition for patients eligible for HIFU includes a life expectancy of at least 5 years, PSA $\leq$ 15 ng/mL, imaging or transperineal biopsy demonstration of stage T1-2N0M0 disease, and biopsy Gleason $\leq$ 7.[1] Alternatively, French

| | Ablatherm integrated imaging [6,14,15] | Sonablate 500 [12–15] |
|---|---|---|
| Company | EDAP TMS SA, France | Focus surgery inc. USA |
| Since | 2005 | 2001 |
| Approval | Europe: CE mark for the treatment of localized prostate cancer: Australia; Canada; Russia; South Korea | Europe: CE mark for the treatment of pts with primary prostate cancer or recurrent prostate cancer following prior therapy |
| Table | Integrated table | Standard operating table |
| Anaesthesia | General, epidural, spinal | General, epidural, spinal |
| Patient positioning | Right lateral decubitus | Lithotomy or supine |
| Mode of administration | Transrectal | Transrectal |
| Software | Automated: Ablaview | Semiautomated |
| Treatment algorithm(s) | Three | One |
| Power | Predefined for each treatment algorithm: primary, salvage after EBRT failure, or repeat HIFU | Adjusted manually by user |
| Treatment probe/transducer | Single-treatment probe containing two ultrasound transducers | Single-treatment probe containing two ultrasound transducers of different focal lengths (either 30/40 mm or 45/50 mm for larger prostate glands) |
| Frequency | 7.5 MHz for imaging/treatment planning: 3 MHz for treatment | 4 MHz for imaging/treatment planning and treatment |
| Focal point | 45 mm from crystal | 30–50 mm from crystal |
| Treatment planning | Prostate is divided into a series of blocks | Prostate is divided into treatment regions: anterior, middle, posterior on both right/left side |
| Ablation volume | Adjustable ablation volume from 19 × 1.7 × 1.7 mm to 26 × 1.7 × 1.7 mm: total: 29–36 mm³ | 3 × 3 × 10–12 mm by each acoustic pulse; total: 21–47 mm³ |
| Ablation temperature | >85 °C | 80–98 °C |
| Imaging | Real time | Real time |
| Active cooling system | Yes | Yes: Sonachill |
| Real-time rectal wall distance monitoring | Yes | Yes |
| Real-time rectal wall temperature monitoring | Yes | Yes |
| Postoperative treatment | Temporary urinary catheter (urethral/suprapubic), prophylactic antibiotics, analgesics | |
| Costs | | |
| Device costs | €550 000 | €300 000 |
| Maintenance costs | €45 000/yr | €25 000/yr |
| Disposables | €550: Ablapack | €100 |

CE – Conformité Européenne (European conformity); EBRT – external-beam radiation therapy; HIFU – high-intensity focussed ultrasound; pts – patients.

**FIGURE 59.2 Comparison of the two currently available high-intensity focusing ultrasound devices.** *Adapted from Warmouth et al.*[21]

guidelines recommend HIFU as therapy in patients who are at least 70 years old with T1-T2N0M0 disease, PSA < 15 ng/mL, Gleason score < 7, and prostate volume < 40 mL.[30] Of note, in 2010 Ontario's Program in Evidence Based Care chose to not recommend HIFU as an alternative to accepted curative approaches.[31,32]

Relative contraindications to HIFU are less controversial and include prostate volume > 40 mL and stricture or stenosis that would prevent an instrument from being introduced into the rectum.[33] Prostate sizes > 40 mL increase the likelihood that portions of the periphery of a gland lie outside the device's treatment area. It has been suggested that this challenge can be mitigated with pre-HIFU TURP or androgen deprivation therapy (ADT). However, Uchida et al. reported that use of neoadjuvant hormonal therapy did not lead to a significant reduction in treatment failure.[34] When performing HIFU as a salvage therapy, shrinking of the prostate that occurs within the first 6 months following initial HIFU[35] or radiation therapy[33] brings the entirety of the gland within the device's treatment radius and diminishes the effect that prostate size has on the likelihood of complete tumor eradication.

## ONCOLOGICAL OUTCOMES

Emerging research examining the use of HIFU in the treatment of low volume PCa supports comparable PCa specific survival and metastasis free survival. Crouzet et al. reported long-term follow-up results to their prospective, single-institution cohort studies of 1002 clinical stage T1 and T2 nonsurgical candidates who underwent whole gland HIFU ablation with an Ablatherm device from 1997 to 2009.[36,37] In this study, 60% of patients received a single HIFU treatment, 38% underwent two

HIFU treatments, and 2% received three HIFU treatments.[36] 94% of the patients underwent pre-HIFU TURP and 39% of patients underwent pre-HIFU androgen deprivation therapy. Overall 10-year survival for the cohort was 80%, with 97% PCa specific survival, 94% metastasis free survival, and 60% progression free survival (PFS). Similar survival and progression outcomes were reported by Thuroff and Chaussy among a similar cohort of 704 patients[38] and by Blana et al. in smaller cohorts[39–41] (Table 59.1). Crouzet et al. report 5-year PFS rates of 86% for low risk, 78% for intermediate risk, and 68% for high-risk patients ($p < 0.001$).[36] Similarly, Ganzer et al. examined 538 patients, 229 low risk, 211 intermediate risk, and 91 high risk, and report 5-year PFS of 88, 83, and 48%, respectively. These numbers drop to 71, 63, and 32% for 10-year PFS.[42] Similar differences between risk groups have been demonstrated in smaller cohorts with shorter follow-up by multiple other researchers and institutions in Asia, Europe, and North America.[43–45] Of note, regardless of risk category, Crouzet et all report that combined 5-year PFS has improved over time, increasing from 66% among patients treated prior to 2000 to 83% in patients treated since 2005 ($p = 0.010$).[36]

Studies that examine PFS are limited by the fact that criteria for biochemical failure based on a validated definition specific to the post-HIFU setting do not exist. Phoenix criteria (PSA nadir plus 2 ng/mL), designed to be specific to postexternal beam radiation failures, is not recommended for application in the postablation or cryotherapy setting,[48] although studies frequently report HIFU failure using Phoenix criteria. In addition, because it takes time for the PSA to rise this much post treatment, when using the Phoenix criteria it is recommended that follow-up be reported as median follow-up minus 2 years. In 2009, Blana et al. proposed the Stuttgart definition for predicting post-HIFU biochemical failure, PSA nadir plus 1.2 ng/mL.[49] Although not universally accepted, multiple recent studies present outcomes using the Stuttgart definition.[43–45,46] Studies suggest that the majority of patients who achieve an initial PSA nadir of ≤0.5 ng/mL will achieve the best oncologic outcomes.[50] PSA nadir following HIFU is frequently reported as being independently associated with PFS,[36–38,43–45,46,50–52,47] although no standard for what should be considered an acceptable PSA nadir has been definitively established. Pfeiffer et al. report that a nadir PSA above 0.3 ng/mL is associated with seven times increased risk of recurrence.[44] Using different PSA nadir cutoffs, Ripert et al. report that a PSA nadir greater than one is associated with 7.5 times increased risk of biochemical failure and 18 times increased risk of oncologic failure (defined as a positive biopsy) while a PSA nadir of 0.21–1 is associated with four times increased risk of both biochemical and oncologic failure when compared with a nadir of ≤0.2 ng/mL.[46] Similarly, Ganzer et al. observed that patients who achieved a PSA nadir ≤0.2 ng/mL had 95%

**TABLE 59.1** Summary of HIFU Oncologic Outcomes

| | | | | 5-Year outcomes | | | | |
|---|---|---|---|---|---|---|---|---|
| Study | N. | Mean follow-up (months) | Median PSA nadir (ng/mL) | Mean time (months) | Nadir ≤ 0.2 (%) | OS (%) | DSS (%) | PFS (%) |
| Blana et al.[39] | 140 | 77 | 0.16 | 5 | 68† | 90 | 100 | 62 |
| Blana et al.[41] | 163 | 58 | NR | 6 | 64 | 97 | 100 | 86 |
| Ripert et al.[46] | 53 | 45* | 1 | 5 | 21 | NR | 100 | 22 |
| Komura et al.[43] | 171 | 43 | 0.03 | 2 | 70 | 99 | 100 | 62 |
| Pfeiffer et al.[44] | 191 | 53 | 0.09 | 2* | 74** | 86 | 98 | 63 |
| Blana et al.[40] | 356 | 34 | 0.11 | 3 | NR | NR | NR | 64 |
| Pinthus et al.[45] | 402 | 24* | 0.10 | 3* | 81* | NR | NR | 68 |
| Misrai et al.[47] | 119 | 47 | NR | NR | NR | NR | 100 | 30 |
| | | | | | | | 10-Year outcomes | |
| Crouzet et al.[36] | 1002 | 77* | 0.14** | 2 | 63‡ | 80 | 97 | 60 |
| Thuroff and Chaussy[38] | 704 | 64 | 0.1 | 2 | NR | NR | 99 | 60–68 |
| Ganzer et al.[42] | 538 | 97 | 0.4** | 5 | 71 | NR | NR | 61 |

PSA, prostate specific antigen; OS, overall survival; DSS, disease specific survival; PFS, progression free survival; NR, not reached.
*Median
**Mean
†≤0.5
‡≤0.3

PFS at 5 years compared to 55% PFS for PSA nadir 0.21–1.0 and 0% PFS for PSA nadir >1 ng/mL.[51] Additional studies have suggested that pre-HIFU PSA,[36–38,42,45,47] biopsy Gleason score,[36–38] and D'amico risk group[43] may be independently associated with increased risk of recurrence, although these findings are less well-supported in the literature. Clearly, a low PSA nadir posttreatment is optimal and most clearly approaches criteria utilized for surgical treatment, which enables the ideal comparisons to be drawn.

Guidelines for appropriate patient follow-up and surveillance after HIFU have yet to be established. As previously discussed, studies suggest that although post-HIFU changes make prostate biopsies more difficult to accurately interpret, use of immunohistochemical stains can improve discrimination between necrosis and carcinoma.[27–29] Alternately, post-HIFU recurrent cancer can be identified on conventional T2 MRI as a hypointense lesion, although sensitivity decreases significantly with smaller lesions.[53] Novel imaging techniques are emerging that may play a critical role in identifying both residual and recurrent disease following HIFU, with multiple reports suggesting that dynamic contrast enhanced (DCE) MRI is better capable of visualizing residual and recurrent disease than traditional imaging modalities.[53–58] Kim et al. reported that when compared with T2 MRI diffusion weighted imaging, DCE-MRI is more sensitive but less specific for the identification of recurrence.[57] The combination of a post-HIFU specific PSA recurrence definition with these improved imaging strategies will enable optimized patient surveillance guidelines to be generated.

Recently published studies have begun to examine the effect that pre-HIFU lesion location has on treatment efficacy. As previously discussed, lesions located anteriorly may lie outside the focal length of the device, particularly in prostates larger than 40 mL.[59] An interesting study by Boutier et al. examined 99 patients who received systematic post-HIFU sextant biopsies at 3–6 months of follow-up.[35] They found that the probability of a patient having biopsy-proven residual disease following primary HIFU treatment varied by pre-HIFU positive biopsy core location. Positive pre-HIFU biopsies from the apex had a 42% probability of leaving residual disease as compared to a 13% probability of residual disease for mid-gland biopsies and 9% for positive cores in the base.[35] Boutier et al. propose that this is likely due to the increased safety margin typically employed in apical ablation in an effort to avoid damage to the external urethral sphincter that would likely result in increased urinary incontinence. These findings underscore the importance of careful pretreatment mapping and patient selection. Accurate identification of the anatomy at the prostatic apex on ultrasound can be difficult and can lead to under treatment by less experienced surgeons.

## COMPLICATIONS OF HIFU AND QUALITY OF LIFE CONSIDERATIONS

Commonly reported complications following HIFU include urinary retention, bladder outlet obstruction, stricture or bladder neck contracture, and symptomatic urinary tract infection (UTI)[21,60] (Table 59.2). There is significant variability in reported complications, likely secondary to a lack of standardized definitions. Two prospective clinical trials (North America and Europe) have been performed that attempt to determine the safety profile of HIFU.[20,61] Koch et al. reported the findings of a very early clinical trial of 20 men, identifying UTI in 40%, prolonged urinary retention in 10%, and

**TABLE 59.2** Summary of Complications following Primary Whole Gland HIFU

| Complication | Proportion (%) | Studies |
|---|---|---|
| Rectourethral fistula (RUF) | 0–5 | Crouzetet al.;[36] Ganzer et al.;[42] Koch et al.;[20] Maestroni et al.,[65] Mearini et al.;[52] Pfeiffer et al.;[44] Thuroff et al.;[61] Uchida et al.[66] |
| Stricture or bladder neck contracture | 4–21 | Berge et al.;[67] Challacombe et al.;[63] Crouzet et al.;[36] Elterman et al.;[62] Ficarra et al.;[22] Komura et al.;[64] Mearini et al.;[52] Thuroff et al.;[61] Thuroff and Chaussy;[38] Uchida et al.[66] |
| Bladder outlet obstruction | 4–28 | Blana et al.;[41,68,69] Crouzet et al.;[36] Ficarra et al.;[22] Ganzer et al.;[42] Maestroni et al.;[65] Pfeiffer et al.;[44] Sung et al.[70] |
| Urinary retention beyond 30 days | 4–19 | Challacombe et al.;[63] Crouzet et al.;[36] Koch et al.;[20] Sung et al.;[70] Thuroff et al.;[61] Uchida et al.[66] |
| Symptomatic UTIs | 3–40 | Ahmed et al.;[71] Berge et al.;[67] Blana et al.;[41,68,69] Crouzet et al.;[36] Ficarra et al.;[22] Haddad et al.;[23] Koch et al;[60] Pfeiffer et al.;[44] Thuroff et al.;[61] Thuroff and Chaussy;[38] Sung et al.[70] |
| Urinary incontinence* | 0–7 | Ahmed Br J Ca et al.;[61] Blana et al.;[41,69] Crouzet et al.;[36] Ficarra et al.;[22] Ganzer et al.;[42] Thuroff et al;[61,] Thuroff and Chaussy[38] |
| Impotence at 1 year | 27–75 | Ahmed et al.;[71] Blana et al.;[41,68,69] Challacombe et al.;[63] Crouzet et al.;[37] Ganzer et al.;[42] Haddad et al.;[23] Sung et al.;[70] Thuroff and Chaussy[38] |

*SUI grades 2–3 or >1 ppd (pads per day).*

rectourethral fistula (RUF) in 5% in 2007.[20] No patients in the study developed urethral stricture or bladder neck contracture. The findings of this study led to FDA approval of a US based phase II/III clinical trial of HIFU in the setting of radiation failure, which is currently ongoing. In the first European multicenter Phase II/III clinical trial of 402 patients, Thuroff et al. reported that 14% of men that underwent HIFU required treatment for UTI, 9% had prolonged urinary retention, and 4% experienced stricture or bladder neck contracture.[61] RUF was reported in 1% of the cohort.[61] An additional prospective study by Ficarra et al. found that at 1 year of follow-up, 16% of patients had experienced symptomatic UTI, 10% had urethral stenosis, and 13% required endoscopic management of infravesical obstruction.[22] Among retrospective studies bladder neck contracture is a frequently reported complication, with up to 20% of patients requiring subsequent intervention for the contracture.[62–64] With the development of newer generation devices, studies have suggested a significant reduction in overall Clavien complications.[38,42]

Among HIFU complications, the most concerning is RUF, which has been reported in up to 5% of primary HIFU cohorts.[22,36,42,44,52,61,65,66] In studies that examine complications following repeat HIFU procedures, this rate is shown to increase. For example, Netsch et al. found that while only 1.2% of patients following primary HIFU developed RUF, 13.6% of patients requiring a second HIFU experienced a RUF.[72] The reason for RUF formation likely varies between patients; however, it has been proposed that placement of the focal point near the rectal wall can increase likelihood of perforation or fistula. This occurs more often in the setting of inappropriate positioning within the rectum.[20] In addition, the presence of an air bubble or debris at the rectal wall interface can cause increased sound reflection and energy absorption in this area. Recent generations of HIFU devices that incorporate rectal mucosal cooling devices appear to dramatically reduce the likelihood of rectal injury.[22] Current literature suggests that conservative management can work but has a high failure rate and that many patients will eventually require complicated surgical intervention.[72–76] Options for intervention include radical prostatectomy with omental or gracilis interposition flap or diversion of either the urinary tract or bowel to separate the systems and prevent infection.[72,77] Despite the trend toward surgical management, select reports have shown that successful conservative management is possible. Venkatesan et al. recently reported their experience in the management of 43 RUF.[77] Forty patients opted for surgical repair while the remaining three were successfully managed conservatively with suprapubic catheters and antibiotics.[77]

On discussing quality of life considerations, few studies have used validated questionnaires to assess the effects of whole-gland HIFU. Transient urinary retention for up to 2–4 weeks is common following whole-gland HIFU and should be included in discussions with patients when establishing expectations of the post-HIFU recovery period.[20,78–80] Grade 2–3 stress urinary incontinence (SUI) has been reported in up to 7% of men undergoing first line HIFU (Table 59.2). Within the Ganzer et al. cohort of 538 patients, 3% of men had grade 2 SUI at 6 months. At the last follow-up, this decreased to 2% of patients with grade 2 SUI, and 83% of the cohort was entirely pad-free.[42] Shoji et al. found that although there was a transient decrease in maximum flow rate and increase in residual urine volume at 6 months post-HIFU, these returned to baseline as 12 and 24 months of follow-up.[81] Reports of erectile dysfunction and impotence vary significantly, with studies reporting that between 27% and 75% of men who were potent prior to HIFU do not regain sexual function[23,36,38,41,42,63,68–71] (Table 59.2). However, other studies have suggested that sexual function continues to return beyond 1 year with Shoji et al. reporting 63% of men experience returned function at 12 months and 78% at 24 months based on the IIEF.[81] Although improved urinary continence and erectile function are considered potential benefits to selecting HIFU over radical therapy, currently available studies present inconsistent findings and comparative effectiveness trials have not been performed that directly compare HIFU to radical prostatectomy and radiotherapy. There does exist, however, a direct comparison between cryoablation and HIFU in which Li et al. report that cryoablation patients had a lower erectile function recovery rate compared with HIFU patients.[82]

## FOCAL HIFU

Although HIFU has primarily been used for whole gland ablation for localized PCa, focal gland ablation is emerging as an alternative to whole-gland HIFU therapy that has comparable rates for metastasis free survival but diminished morbidity compared to traditional whole-gland therapy. Hemiablation HIFU relies on the premise that PCa therapy need not be driven by a goal of completely eradicating all evidence of tumor, as has been the case in the past.[83] Despite the fact that concern has been raised about smaller tumors that may remain when utilizing a focal approach, emerging research suggests that approximately 80% of secondary tumors are low grade and can be safely managed with watchful waiting as it is the dominant tumor that dictates the course of the disease.[83–86] These studies have established feasibility for further experimentation with prostate-sparing treatment in the form of hemiablation, "hockey stick" ablation (destruction of an entire half of the prostate plus the anterior or posterior portion of the contralateral side), and true focal ablation of the prostate (Figure 59.3).

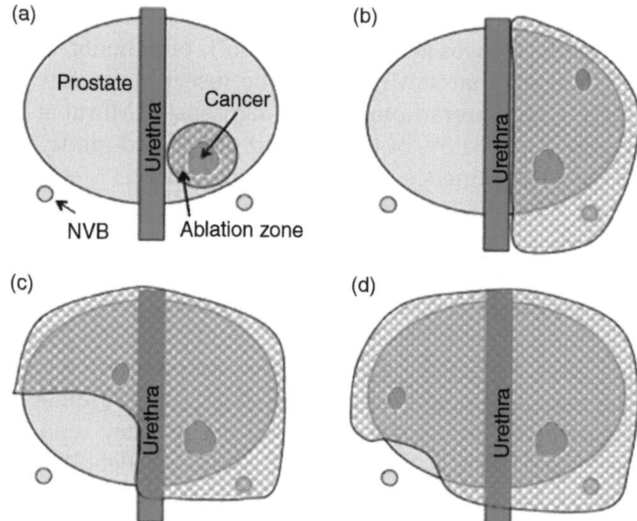


**FIGURE 59.3 Conceptual schematic of treatment options for focal therapy.** (a) True focal ablation. (b) Hemiablation: the ablation zone is extended to include the ipsilateral neurovascular bundle and margin (treatment of possible extracapsular extension of disease). (c) 75% ablation: note that 25% of the gland (untreated area) will be on an active surveillance protocol. (d) Near total ablation: some undetermined amount of parenchyma will not be treated to maintain the neurovascular bundle on that side. NVB, neurovascular bundle. *Adapted from de la Rosette et al.*[87]

Efficacy of cancer control as well as guidelines for determining which patients should be eligible for focal ablation with HIFU remains a topic that is much debated.[83,86,88] Novel imaging and targeted biopsy techniques have the potential to enable improved patient selection for focal HIFU therapy.[83,86,89] Multiple consensus meetings have met to form guidelines detailing which patients should be considered candidates for focal therapy (either HIFU or cryoablation) and what should be the recommended method of monitoring these patients for recurrence.[90–92] The most recent of these meetings expanded guidelines for eligible patients to include those who have nondominant Gleason 4; however, they continue to encourage further research into optimal patient selection (Figure 59.4).[83,90,91]

Feasibility studies have demonstrated that HIFU hemiablation is associated with minimal detriment to baseline urinary continence and erectile function.[93–96] Reported rates of urethral stricture and bladder neck contracture are below five percent[93,94,96,97] and urinary retention rates remain under 10%.[94–96] Oncologic outcomes are also available from two small studies. Muto et al. reported 2-year PFS stratified by D'amico risk category and found 83% PFS in low risk patients and 54% PFS in intermediate risk, neither of which differed from the PFS of a respective

| | |
|---|---|
| 1. | Candidates for focal therapy should ideally undergo transperineal template mapping biopsies, although a state-of-the-art multifunctional MRI with TRUS biopsy at expert centers may be acceptable. |
| 2. | Candidates for focal therapy should have a life expectancy of 10 or more years. |
| 3. | Patients with previous prostate surgery should be counseled with caution. |
| 4. | Patients with previous radiotherapy to the prostate or pelvis should not be treated until more data are available, although the panel accepted that focal salvage therapy may be a possibility in the future. |
| 5. | The effects of focal therapy on men with lower urinary tract symptoms are not well known.These men should be counseled with caution. |
| 6. | There will be specific attributes that are more related to the energy source than to focal therapy in general. Issues such as prostate size, presence of prostatic calcification, cysts, TUR cavity, access to rectum, and concurrent inflammation of rectal mucosa may need to be taken into consideration when selecting the optimal therapy. |
| 7. | Focal therapy should be limited to patients of low to moderate risk. |
| 8. | Focal therapy should be limited to men with clinical $T_{2a}$ or less $N_0M_0$ disease. |
| 9. | Focal therapy should be limited to men with radiologic $\leq T_{2b}$ $N_0M^0$ disease. |
| 10. | Defining the topography of the cancer is important. Disease that is predominantly apical or anterior in disposition may be technically difficult to manage with existing treatment modalities. |
| 11. | The long-term effects of focal therapy on potency/erectile functions are not known. Men should be counseled in this regard before therapy. |

MRI = magnetic resonance imaging; TRUS = transrectal ultrasonography; TUR = transurethral resection.

**FIGURE 59.4 Candidate selection for focal therapy.** *Adapted from Ward et al.*[83]

whole gland HIFU cohort.[97] El Fegoun et al. report oncologic outcomes for a cohort of 12 hemiablation patients. Overall 10-year survival was 83%, cancer-specific survival was 100%, and PFS was 38% at 10 years. Of note, PFS at 5 years was 90%.[95] Although these studies are small in number, they provide early, encouraging data in support of focal HIFU for carefully selected patients with PCa. Advances in multiparametric MRI offer promise for allowing treatment of tumor foci. A London-based trial looking at focal HIFU therapy as an alternative to surveillance to address treatment associated morbidity for low volume disease found that 92% of the treated patients had no evidence of clinically significant disease on 6 month biopsy.[94] Whether or not this actually alters the course of the disease is unclear.

## HIFU IN SPECIFIC SETTINGS

### HIFU Post Radiotherapy

Salvage radical prostatectomy for radiorecurrent PCa is associated with high morbidity.[98–101] Minimizing this morbidity while maintaining oncologic outcomes for patients with recurrent disease would greatly improve treatment options. A recent SEER study compared the use of salvage radical prostatectomy to salvage cryotherapy and reported improved overall and disease-specific survival for patients undergoing salvage cryotherapy.[102] Alternatively, Pisters et al. reported that salvage radical prostatectomy may offer superior oncologic control compared with cryotherapy.[103] Multiple studies have examined the use of HIFU as salvage therapy following EBRT failure.[33,104–110] In the most recent European Association of Urology guidelines, salvage HIFU was included as an alternative thermal ablation option for recurrent PCa.[111] Crouzet et al. presented a series of 290 biopsy confirmed recurrent PCa patients after initial radiation therapy who underwent HIFU.[105] Local cancer control in the form of negative post-HIFU prostate biopsies was achieved in 81% of patients. Cancer-specific and metastasis free survival was 80% for 7 years. Twenty percent of patients developed grade 2–3 urinary incontinence.[105] Similarly, Murat et al. reported 5-year overall survival of 84%.[33] These survival and progression outcomes are comparable to those observed in salvage radical prostatectomy series.[112,113] Not all postradiotherapy patients are eligible for treatment as many develop prostate calcifications after radiation and cannot be treated with HIFU because the calcifications reflect the sound waves.

Overall, studies suggest that salvage HIFU postradiotherapy can be a curative option with a reasonable morbidity profile in carefully selected patients. Chalasani et al. reviewed the role of salvage HIFU post radiotherapy and found that studies report incontinence in 10–50% of patients and rectal fistula in up to 16%. Overall treatment efficacy was achieved for 17–57% of patients. They recommend that salvage HIFU be reserved for low to intermediate preradiotherapy risk disease.[59] Murat et al. reported that 31.5% of their cohort developed grade 2–3 urinary incontinence, and of those patients, 11% experienced incontinence severe enough to necessitate artificial urinary sphincter placement.[33] This is much lower than what is observed in salvage radical prostatectomy cohorts. An interesting aspect of the Murat et al. study is a comparison of the incidence of complications before and after the development of postradiation parameters. The new parameters account for changes in vascularity occurring in previously irradiated tissue. They report a significant decrease in the rate of bladder outlet obstruction, recto-urethral fistula, and SUI severe enough to necessitate sphincter implantation.[33] Their findings underscore the value of the newer devices' ability to tailor treatment algorithms to specific tissue characteristics.

The role of focal salvage HIFU has been briefly explored and reviewed for premarket approval by the FDA of the Sonablate 450 for the treatment of PCa recurrence following EBRT is currently underway.[114] In a study that examined quality of life following focal salvage HIFU, Berge et al. reported that average urinary function scores from the UCLA-PCI dropped from 79.7 to 67.4 after 18 months of follow-up.[115] Their report also found a clinically and statistically significant decrease in UCLA-PCI sexual domain score at 18 months follow-up.[115]

### HIFU Post Radical Prostatectomy

Few studies have examined the feasibility of salvage HIFU after radical prostatectomy.[116,117] Asimakopoulos et al. reported on 19 patients with biopsy proven local recurrences after radical prostatectomy who were unwilling to undergo salvage radiotherapy. Each patient underwent a single HIFU session. Two patients did not achieve PSA nadir. At 48 months median follow-up time, eight of the remaining 17 patients had experienced recurrence.[116] Longer follow-up in larger patient cohorts is necessary before salvage HIFU can be considered an option for primary salvage therapy following radical prostatectomy. Additionally, proper imaging and localization of recurrences after surgery can be highly problematic, especially when using ultrasound. The role of DCE MRI in detecting recurrences in the setting of HIFU post radical prostatectomy requires more extensive evaluation but may offer improved sensitivity for detection.[118]

### HIFU Post HIFU

One of the significant benefits of HIFU is that patients can undergo repeat HIFU following biochemical or oncologic failure. Oncologic outcomes appear comparable to alternative salvage techniques. However, there are reports of increased patient morbidity following subsequent treatment.[21,42,67,68,72] Studies consistently report an increase in urinary side effects as a result of repeat HIFU. However, reported effects on erectile function following repeat-HIFU seem to be minimal, with reports suggesting that among men who are potent following initial HIFU, less than 5% will lose potency with repeat treatment.[68,67]

## MANAGEMENT OF HIFU FAILURES

Currently, patients with recurrent PCa post-HIFU have the option of management with repeat HIFU, salvage radiotherapy, or salvage radical prostatectomy. In a cohort of 32 patients who underwent salvage radiotherapy after local failure following HIFU, Pasticier et al. reported 5-year PFS for 64% of their cohort; however, they noted that this increased to 80% PFS in the setting of pre-salvage therapy positive biopsy findings compared with 44% PFS when there was isolated biochemical failure.[119] Similarly, Munoz et al. reported 3-year biochemical disease-free survival of 78%, which increased to 88% in the setting of patients achieving a PSA nadir of ≤0.35 ng/mL.[120] Riviere et al. found that while satisfactory oncologic control was achieved, there was significant urinary toxicity and erectile dysfunction among patients receiving salvage radiotherapy for HIFU failure.[121] Salvage radical prostatectomy is reported as being feasible in the post-HIFU setting, although most studies suggest increased morbidity when compared with primary surgery.[122–124] Limited patient size in each of these studies limits conclusions that can be made surrounding the efficacy and morbidity of salvage radiotherapy and radical prostatectomy for HIFU failures.

## HIFU: FUTURE RESEARCH

In order for HIFU to establish a firm presence as a viable therapeutic option for localized PCa, it is imperative that prospectively designed, adequately powered comparative effectiveness studies be performed. An actively enrolling clinical trial is currently underway that will include nearly 150 men who will undergo MRI and transperineal template prostate mapping biopsies along with targeted biopsies at 36 months in order to better evaluate the efficacy of hemiablative HIFU.[17] Ablatherm recently finished collecting data from a US-based phase II/III clinical trial (the ENLIGHT trial) comparing HIFU with crryoablation.[125] Results from this study are expected to emerge in the next year and may further solidify the role of HIFU in the PCa armamentarium. MR-guided HIFU for focal therapy is emerging as a means to develop more precise treatment margins and to focus therapy on specific tumor location rather than broad anatomic boundaries.[126]

## CONCLUSIONS

HIFU is a method of thermal ablation that has been gaining popularity as an acceptable alternative to radical therapy in patients with localized PCa. Through converging high-energy ultrasound waves, HIFU causes precise coagulative necrosis such that it avoids harm to outside structures and limits the morbidity (primarily urinary incontinence and erectile dysfunction) that occurs with radical prostatectomy and external beam radiation therapy. Despite being originally pioneered as a method for primary whole-gland ablation in men who may otherwise not be candidates for radical therapy, it is now being tested as a means of focal therapy (particularly when combined with novel MR-imaging techniques) or as salvage therapy following surgical or radiation failure. Its primary advantage remains that it is a repeatable therapy, making it a versatile option, either in isolation or as an adjunct to more traditional approaches.

### References


1. Lecornet E, Ahmed HU, Moore CM, et al. Conceptual basis for focal therapy in prostate cancer. *J Endourol* 2010;**24**:811.
2. Ahmed HU, Emberton M. Active surveillance and radical therapy in prostate cancer: can focal therapy offer the middle way? *World J Urol* 2008;**26**:457.
3. Lindner U, Lawrentschuk N, Schatloff O, et al. Evolution from active surveillance to focal therapy in the management of prostate cancer. *Future Oncol* 2011;**7**:775.
4. Lindner U, Trachtenberg J, Lawrentschuk N. Focal therapy in prostate cancer: modalities, findings and future considerations. *Nat Rev Urol* 2010;**7**:562.
5. Inoue Y, Goto K, Hayashi T, et al. Transrectal high-intensity focused ultrasound for treatment of localized prostate cancer. *Int J Urol* 2011;**18**:358.
6. Bihrle R, Foster RS, Sanghvi NT, et al. High-intensity focused ultrasound in the treatment of prostatic tissue. *Urology* 1994;**43**:21.
7. Foster RS, Bihrle R, Sanghvi N, et al. Production of prostatic lesions in canines using transrectally administered high-intensity focused ultrasound. *Eur Urol* 1993;**23**:330.
8. Foster RS, Bihrle R, Sanghvi NT, et al. High-intensity focused ultrasound in the treatment of prostatic disease. *Eur Urol* 1993;**23**(Suppl 1):29.
9. Bihrle R, Foster RS, Sanghvi NT, et al. High intensity focused ultrasound for the treatment of benign prostatic hyperplasia: early United States clinical experience. *J Urol* 1994;**151**:1271.
10. Ohigashi T, Nakamura K, Nakashima J, et al. Long-term results of three different minimally invasive therapies for lower urinary tract symptoms due to benign prostatic hyperplasia: comparison at a single institute. *Int J Urol* 2007;**14**:326.
11. Khokhlova TD, Hwang JH. HIFU for palliative treatment of pancreatic cancer. *J Gastrointest Oncol* 2011;**2**:175.
12. Mearini L. High intensity focused ultrasound, liver disease and bridging therapy. *World J Gastroenterol* 2013;**19**:7494.

13. Kovatcheva RD, Vlahov JD, Stoinov JI, et al. High-intensity focussed ultrasound (HIFU) treatment in uraemic secondary hyperparathyroidism. *Nephrol Dial Transplant* 2012;**27**:76.
14. Serrone J, Kocaeli H, Douglas Mast T, et al. The potential applications of high-intensity focused ultrasound (HIFU) in vascular neurosurgery. *J Clin Neurosci* 2012;**19**:214.
15. Margreiter M, Marberger M. Focal therapy and imaging in prostate and kidney cancer: high-intensity focused ultrasound ablation of small renal tumors. *J Endourol* 2010;**24**:745.
16. Hernandez Fernandez C, Lledo Garcia E, Subira Rios D, et al. Conservative treatment of renal cancer using HIFU. Procedure, indications, and results. *Actas Urol Esp* 2009;**33**:522.
17. Dickinson L, Ahmed HU, Kirkham AP, et al. A multi-centre prospective development study evaluating focal therapy using high intensity focused ultrasound for localised prostate cancer: the index study. *Contemp Clin Trials* 2013;**36**:68.
18. Madersbacher S, Kratzik C, Marberger M. Prostatic tissue ablation by transrectal high intensity focused ultrasound: histological impact and clinical application. *Ultrason Sonochem* 1997;**4**:175.
19. Tsakiris P, Thuroff S, de la Rosette J, et al. Transrectal high-intensity focused ultrasound devices: a critical appraisal of the available evidence. *J Endourol* 2008;**22**:221.
20. Koch MO, Gardner T, Cheng L, et al. Phase I/II trial of high intensity focused ultrasound for the treatment of previously untreated localized prostate cancer. *J Urol* 2007;**178**:2366.
21. Warmuth M, Johansson T, Mad P. Systematic review of the efficacy and safety of high-intensity focussed ultrasound for the primary and salvage treatment of prostate cancer. *Eur Urol* 2010;**58**:803.
22. Ficarra V, Antoniolli SZ, Novara G, et al. Short-term outcome after high-intensity focused ultrasound in the treatment of patients with high-risk prostate cancer. *BJU Int* 2006;**98**:1193.
23. Haddad RL, Hossack TA, Woo HH. Results of low threshold to biopsy following high-intensity focused ultrasound for localized prostate cancer. *Urol Ann* 2012;**4**:84.
24. Shoji S, Uchida T, Nakamoto M, et al. Prostate swelling and shift during high intensity focused ultrasound: implication for targeted focal therapy. *J Urol* 2013;**190**:1224.
25. Chen W, Sanghvi N, Carlson RF. Validation of tissue change monitoring (TCM) on the Sonablate® 500 during high intensity focused ultrasound (HIFU) treatment of prostate cancer with Real-time thermometry. The Eleventh Symposium on Therapeutic Ultrasound, 2011.
26. Chen W, Sanghvi N, Carlson RF, et al. Real-time tissue change monitoring on the Sonablate® 500 during high intensity focused ultrasound (HIFU) treatment of prostate cancer. The eleventh Symposium on Therapeutic Ultrasound, 2011.
27. Biermann K, Montironi R, Lopez-Beltran A, et al. Histopathological findings after treatment of prostate cancer using high-intensity focused ultrasound (HIFU). *Prostate* 2010;**70**:1196.
28. Walter B, Weiss T, Hofstadter F, et al. Utility of immunohistochemistry markers in the interpretation of post high-intensive focussed ultrasound prostate biopsy cores. *World J Urol* 2013;**31**:1129.
29. Van Leenders GJ, Beerlage HP, Ruijter ET, et al. Histopathological changes associated with high intensity focused ultrasound (HIFU) treatment for localised adenocarcinoma of the prostate. *J Clin Pathol* 2000;**53**:391.
30. Rebillard X, Soulie M, Chartier-Kastler E, et al. High-intensity focused ultrasound in prostate cancer; a systematic literature review of the French Association of Urology. *BJU Int* 2008;**101**:1205.
31. Lukka H, Waldron T, Chin J, et al. High-intensity focused ultrasound for prostate cancer: a practice guideline. *Can Urol Assoc J* 2010;**4**:232.
32. Lukka H, Waldron T, Chin J, et al. High-intensity focused ultrasound for prostate cancer: a systematic review. *Clin Oncol (R Coll Radiol)* 2011;**23**:117.
33. Murat FJ, Poissonnier L, Rabilloud M, et al. Mid-term results demonstrate salvage high-intensity focused ultrasound (HIFU) as an effective and acceptably morbid salvage treatment option for locally radiorecurrent prostate cancer. *Eur Urol* 2009;**55**:640.
34. Uchida T, Illing RO, Cathcart PJ, et al. The effect of neoadjuvant androgen suppression on prostate cancer-related outcomes after high-intensity focused ultrasound therapy. *BJU Int* 2006;**98**:770.
35. Boutier R, Girouin N, Cheikh AB, et al. Location of residual cancer after transrectal high-intensity focused ultrasound ablation for clinically localized prostate cancer. *BJU Int* 2011;**108**:1776.
36. Crouzet S, Chapelon JY, Rouviere O, et al. Whole-gland ablation of localized prostate cancer with high-intensity focused ultrasound: oncologic outcomes and morbidity in 1002 patients. *Eur Urol* 2014;**65**:907.
37. Crouzet S, Rebillard X, Chevallier D, et al. Multicentric oncologic outcomes of high-intensity focused ultrasound for localized prostate cancer in 803 patients. *Eur Urol* 2010;**58**:559.
38. Thuroff S, Chaussy C. Evolution and outcomes of 3 MHz high intensity focused ultrasound therapy for localized prostate cancer during 15 years. *J Urol* 2013;**190**:702.
39. Blana A, Murat FJ, Walter B, et al. First analysis of the long-term results with transrectal HIFU in patients with localised prostate cancer. *Eur Urol* 2008;**53**:1194.
40. Blana A, Robertson CN, Brown SC, et al. Complete high-intensity focused ultrasound in prostate cancer: outcome from the @-Registry. *Prostate Cancer Prostatic Dis* 2012;**15**:256.
41. Blana A, Rogenhofer S, Ganzer R, et al. Eight years' experience with high-intensity focused ultrasonography for treatment of localized prostate cancer. *Urology* 2008;**72**:1329.
42. Ganzer R, Fritsche HM, Brandtner A, et al. Fourteen-year oncological and functional outcomes of high-intensity focused ultrasound in localized prostate cancer. *BJU Int* 2013;**112**:322.
43. Komura K, Inamoto T, Takai T, et al. Single session of high-intensity focused ultrasound for localized prostate cancer: treatment outcomes and potential effect as a primary therapy. *World J Urol* 2013;5:1339–45.
44. Pfeiffer D, Berger J, Gross AJ. Single application of high-intensity focused ultrasound as a first-line therapy for clinically localized prostate cancer: 5-year outcomes. *BJU Int* 2012;**110**:1702.
45. Pinthus JH, Farrokhyar F, Hassouna MM, et al. Single-session primary high-intensity focused ultrasonography treatment for localized prostate cancer: biochemical outcomes using third generation-based technology. *BJU Int* 2012;**110**:1142.
46. Ripert T, Azemar MD, Menard J, et al. Six years' experience with high-intensity focused ultrasonography for prostate cancer: oncological outcomes using the new 'Stuttgart' definition for biochemical failure. *BJU Int* 2011;**107**:1899.
47. Misrai V, Roupret M, Chartier-Kastler E, et al. Oncologic control provided by HIFU therapy as single treatment in men with clinically localized prostate cancer. *World J Urol* 2008;**26**:481.
48. Roach 3rd M, Hanks G, Thames Jr H, et al. Defining biochemical failure following radiotherapy with or without hormonal therapy in men with clinically localized prostate cancer: recommendations of the RTOG-ASTRO Phoenix Consensus Conference. *Int J Radiat Oncol Biol Phys* 2006;**65**:965.
49. Blana A, Brown SC, Chaussy C, et al. High-intensity focused ultrasound for prostate cancer: comparative definitions of biochemical failure. *BJU Int* 2009;**104**:1058.
50. Ganzer R, Robertson CN, Ward JF, et al. Correlation of prostate-specific antigen nadir and biochemical failure after high-intensity focused ultrasound of localized prostate cancer based on the Stuttgart failure criteria – analysis from the @-Registry. *BJU Int* 2011;**108**:E196.
51. Ganzer R, Rogenhofer S, Walter B, et al. PSA nadir is a significant predictor of treatment failure after high-intensity focussed ultrasound (HIFU) treatment of localised prostate cancer. *Eur Urol* 2008;**53**:547.
52. Mearini L, D'Urso L, Collura D, et al. Visually directed transrectal high intensity focused ultrasound for the treatment of prostate

cancer: a preliminary report on the Italian experience. *J Urol* 2009;**181**:105.

53. Rouviere O, Vitry T, Lyonnet D. Imaging of prostate cancer local recurrences: why and how? *Eur Radiol* 2010;**20**:1254.
54. De Visschere PJ, De Meerleer GO, Futterer JJ, et al. Role of MRI in follow-up after focal therapy for prostate carcinoma. *AJR Am J Roentgenol* 2010;**194**:1427.
55. Del Vescovo R, Pisanti F, Russo V, et al. Dynamic contrast-enhanced MR evaluation of prostate cancer before and after endorectal high-intensity focused ultrasound. *Radiol Med* 2013;**118**:851.
56. Punwani S, Emberton M, Walkden M, et al. Prostatic cancer surveillance following whole-gland high-intensity focused ultrasound: comparison of MRI and prostate-specific antigen for detection of residual or recurrent disease. *Br J Radiol* 2012;**85**:720.
57. Kim CK, Park BK, Lee HM, et al. MRI techniques for prediction of local tumor progression after high-intensity focused ultrasonic ablation of prostate cancer. *AJR Am J Roentgenol* 2008;**190**:1180.
58. Martino P, Scattoni V, Galosi AB, et al. Role of imaging and biopsy to assess local recurrence after definitive treatment for prostate carcinoma (surgery, radiotherapy, cryotherapy, HIFU). *World J Urol* 2011;**29**:595.
59. Chalasani V, Martinez CH, Lim D, et al. Salvage HIFU for recurrent prostate cancer after radiotherapy. *Prostate Cancer Prostatic Dis* 2009;**12**:124.
60. Cordeiro ER, Cathelineau X, Thuroff S, et al. High-intensity focused ultrasound (HIFU) for definitive treatment of prostate cancer. *BJU Int* 2012;**110**:1228.
61. Thuroff S, Chaussy C, Vallancien G, et al. High-intensity focused ultrasound and localized prostate cancer: efficacy results from the European multicentric study. *J Endourol* 2003;**17**:673.
62. Elterman DS, Barkin J, Radomski SB, et al. Results of high intensity focused ultrasound treatment of prostate cancer: early Canadian experience at a single center. *Can J Urol* 2011;**18**:6037.
63. Challacombe BJ, Murphy DG, Zakri R, et al. High-intensity focused ultrasound for localized prostate cancer: initial experience with a 2-year follow-up. *BJU Int* 2009;**104**:200.
64. Komura K, Inamoto T, Black PC, et al. Clinically significant urethral stricture and/or subclinical urethral stricture after high-intensity focused ultrasound correlates with disease-free survival in patients with localized prostate cancer. *Urol Int* 2011;**87**:276.
65. Maestroni U, Dinale F, Minari R, et al. High-intensity focused ultrasound for prostate cancer: long-term followup and complications rate. *Adv Urol* 2012;**2012**:960835.
66. Uchida T, Sanghvi NT, Gardner TA, et al. Transrectal high-intensity focused ultrasound for treatment of patients with stage T1b-2n0m0 localized prostate cancer: a preliminary report. *Urology* 2002;**59**:394.
67. Berge V, Dickinson L, McCartan N, et al. Morbidity associated with primary high intensity focused ultrasound and redo high intensity focused ultrasound for localized prostate cancer. *J Urol* 2014;191(6):1764–9.
68. Blana A, Rogenhofer S, Ganzer R, et al. Morbidity associated with repeated transrectal high-intensity focused ultrasound treatment of localized prostate cancer. *World J Urol* 2006;**24**:585.
69. Blana A, Walter B, Rogenhofer S, et al. High-intensity focused ultrasound for the treatment of localized prostate cancer: 5-year experience. *Urology* 2004;**63**:297.
70. Sung HH, Jeong BC, Seo SI, et al. Seven years of experience with high-intensity focused ultrasound for prostate cancer: advantages and limitations. *Prostate* 2012;**72**:1399.
71. Ahmed HU, Zacharakis E, Dudderidge T, et al. High-intensity-focused ultrasound in the treatment of primary prostate cancer: the first UK series. *Br J Cancer* 2009;**101**:19.
72. Netsch C, Bach T, Gross E, et al. Rectourethral fistula after high-intensity focused ultrasound therapy for prostate cancer and its surgical management. *Urology* 2011;**77**:999.
73. Topazio L, Perugia C, Finazzi-Agro E. Conservative treatment of a recto-urethral fistula due to salvage HIFU for local recurrence of prostate cancer, 5 years after radical prostatectomy and external beam radiotherapy. *BMJ Case Rep* 2012;2012. doi: 10.1136/bcr.03.2012.6115.
74. Andrews EJ, Royce P, Farmer KC. Transanal endoscopic microsurgery repair of rectourethral fistula after high-intensity focused ultrasound ablation of prostate cancer. *Colorectal Dis* 2011;**13**:342.
75. Bochove-Overgaauw DM, Beerlage HP, Bosscha K, et al. Transanal endoscopic microsurgery for correction of rectourethral fistulae. *J Endourol* 2006;**20**:1087.
76. Quinlan M, Cahill R, Keane F, et al. Transanal endoscopic microsurgical repair of iatrogenic recto-urethral fistula. *Surgeon* 2005;**3**:416.
77. Venkatesan K, Zacharakis E, Andrich DE, et al. Conservative management of urorectal fistulae. *Urology* 2013;**81**:1352.
78. Madersbacher S, Marberger M. High-energy shockwaves and extracorporeal high-intensity focused ultrasound. *J Endourol* 2003;**17**:667.
79. Poissonnier L, Chapelon JY, Rouviere O, et al. Control of prostate cancer by transrectal HIFU in 227 patients. *Eur Urol* 2007; **51**:381.
80. Gelet A, Chapelon JY, Bouvier R, et al. Local control of prostate cancer by transrectal high intensity focused ultrasound therapy: preliminary results. *J Urol* 1999;**161**:156.
81. Shoji S, Nakano M, Nagata Y, et al. Quality of life following high-intensity focused ultrasound for the treatment of localized prostate cancer: a prospective study. *Int J Urol* 2010;**17**:715.
82. Li LY, Lin Z, Yang M, et al. Comparison of penile size and erectile function after high-intensity focused ultrasound and targeted cryoablation for localized prostate cancer: a prospective pilot study. *J Sex Med* 2010;**7**:3135.
83. Ward JF, Pisters LL. Considerations for patient selection for focal therapy. *Ther Adv Urol* 2013;**5**:330.
84. Andreoiu M, Cheng L. Multifocal prostate cancer: biologic, prognostic, and therapeutic implications. *Hum Pathol* 2010;**41**:781.
85. Meiers I, Waters DJ, Bostwick DG. Preoperative prediction of multifocal prostate cancer and application of focal therapy: review 2007. *Urology* 2007;**70**:3.
86. Sartor AO, Hricak H, Wheeler TM, et al. Evaluating localized prostate cancer and identifying candidates for focal therapy. *Urology* 2008;**72**:S12.
87. de la Rosette JJ, Mouraviev SV, Polascik TJ. Focal targeted therapy will be a future treatment modality for early stage prostate cancer. *Eur Urol Suppl.* 2009;**8**:424.
88. Valerio M, Ahmed HU, Emberton M, et al. The Role of Focal Therapy in the Management of Localised Prostate Cancer: A Systematic Review. *Eur Urol* 2014;**66**(4):732–51.
89. Muller BG, Futterer JJ, Gupta RT, et al. The role of magnetic resonance imaging in focal therapy for prostate cancer: recommendations from a consensus panel. *BJU Int* 2014;**113**(2):218–27.
90. Bostwick DG, Waters DJ, Farley ER, et al. Group consensus reports from the Consensus Conference on Focal Treatment of Prostatic Carcinoma, Celebration, Florida, Urology, 70: 42, 2007.
91. de la Rosette JJ. Editorial comment on: mid-term results demonstrate salvage high-intensity focused ultrasound (HIFU) as an effective and acceptably morbid salvage treatment option for locally radiorecurrent prostate cancer. *Eur Urol* 2009;**55**:647.
92. Eggener SE, Scardino PT, Carroll PR, et al. Focal therapy for localized prostate cancer: a critical appraisal of rationale and modalities. *J Urol* 2007;**178**:2260.
93. Ahmed HU, Freeman A, Kirkham A, et al. Focal therapy for localized prostate cancer: a phase I/II trial. *J Urol* 2011;**185**:1246.
94. Ahmed HU, Hindley RG, Dickinson L, et al. Focal therapy for localised unifocal and multifocal prostate cancer: a prospective development study. *Lancet Oncol* 2012;**13**:622.

95. El Fegoun AB, Barret E, Prapotnich D, et al. Focal therapy with high-intensity focused ultrasound for prostate cancer in the elderly. A feasibility study with 10 years follow-up. *Int Braz J Urol* 2011;**37**:213.
96. Van Velthoven R, Aoun F, Limani K, et al. Primary zonal high intensity focused ultrasound for prostate cancer: results of a prospective phase iia feasibility study. *Prostate Cancer* 2014;**2014**: 756189.
97. Muto S, Yoshii T, Saito K, et al. Focal therapy with high-intensity-focused ultrasound in the treatment of localized prostate cancer. *Jpn J Clin Oncol* 2008;**38**:192.
98. Boris RS, Bhandari A, Krane LS, et al. Salvage robotic-assisted radical prostatectomy: initial results and early report of outcomes. *BJU Int* 2009;**103**:952.
99. Kaffenberger SD, Keegan KA, Bansal NK, et al. Salvage robotic assisted laparoscopic radical prostatectomy: a single institution, 5-year experience. *J Urol* 2013;**189**:507.
100. Nguyen PL, D'Amico AV, Lee AK, et al. Patient selection, cancer control, and complications after salvage local therapy for postradiation prostate-specific antigen failure: a systematic review of the literature. *Cancer* 2007;**110**:1417.
101. Vaidya A, Soloway MS. Salvage radical prostatectomy for radiorecurrent prostate cancer: morbidity revisited. *J Urol* 2000;**164**:1998.
102. Friedlander DF, Gu X, Prasad SM, et al. Population-based comparative effectiveness of salvage radical prostatectomy vs cryotherapy. *Urology* 2014;**83**:653.
103. Pisters LL, Leibovici D, Blute M, et al. Locally recurrent prostate cancer after initial radiation therapy: a comparison of salvage radical prostatectomy versus cryotherapy. *J Urol* 2009;**182**:517.
104. Berge V, Baco E, Karlsen SJ. A prospective study of salvage high-intensity focused ultrasound for locally radiorecurrent prostate cancer: early results. *Scand J Urol Nephrol* 2010;**44**:223.
105. Crouzet S, Murat FJ, Pommier P, et al. Locally recurrent prostate cancer after initial radiation therapy: early salvage high-intensity focused ultrasound improves oncologic outcomes. *Radiother Oncol* 2012;**105**:198.
106. Rouviere O, Sbihi L, Gelet A, et al. Salvage high-intensity focused ultrasound ablation for prostate cancer local recurrence after external-beam radiation therapy: prognostic value of prostate MRI. *Clin Radiol* 2013;**68**:661.
107. Song W, Jung US, Suh YS, et al. High-intensity focused ultrasound as salvage therapy for patients with recurrent prostate cancer after radiotherapy. *Korean J Urol* 2014;**55**:91.
108. Uchida T, Shoji S, Nakano M, et al. High-intensity focused ultrasound as salvage therapy for patients with recurrent prostate cancer after external beam radiation, brachytherapy or proton therapy. *BJU Int* 2011;**107**:378.
109. Uddin Ahmed H, Cathcart P, Chalasani V, et al. Whole-gland salvage high-intensity focused ultrasound therapy for localized prostate cancer recurrence after external beam radiation therapy. *Cancer* 2012;**118**:3071.
110. Zacharakis E, Ahmed HU, Ishaq A, et al. The feasibility and safety of high-intensity focused ultrasound as salvage therapy for recurrent prostate cancer following external beam radiotherapy. *BJU Int* 2008;**102**:786.
111. Heidenreich A, Bastian PJ, Bellmunt J, et al. EAU guidelines on prostate cancer. Part II: treatment of advanced, relapsing, and castration-resistant prostate cancer. *Eur Urol* 2014;**65**:467.
112. Yuh B, Ruel N, Muldrew S, et al. Complications and outcomes of salvage robot-assisted radical prostatectomy: a single-institution experience. *BJU Int* 2014;**113**:769.
113. Rosoff JS, Savage SJ, Prasad SM. Salvage radical prostatectomy as management of locally recurrent prostate cancer: outcomes and complications. *World J Urol* 2013;**31**:1347.
114. SonaCare Medical Progresses to FDA Panel Review for its Sonablate 450 Prostate Ablation System. Available from: http://www.prnewswire.com/news-releases/sonacare-medical-progresses-to-fda-panel-review-for-its-sonablate-450-prostate-ablation-system-258506861.html. 2014.
115. Berge V, Baco E, Dahl AA, et al. Health-related quality of life after salvage high-intensity focused ultrasound (HIFU) treatment for locally radiorecurrent prostate cancer. *Int J Urol* 2011;**18**:646.
116. Asimakopoulos AD, Miano R, Virgili G, et al. HIFU as salvage first-line treatment for palpable, TRUS-evidenced, biopsy-proven locally recurrent prostate cancer after radical prostatectomy: a pilot study. *Urol Oncol* 2012;**30**:577.
117. Murota-Kawano A, Nakano M, Hongo S, et al. Salvage high-intensity focused ultrasound for biopsy-confirmed local recurrence of prostate cancer after radical prostatectomy. *BJU Int* 2010; **105**:1642.
118. Roy C, Foudi F, Charton J, et al. Comparative sensitivities of functional MRI sequences in detection of local recurrence of prostate carcinoma after radical prostatectomy or external-beam radiotherapy. *AJR Am J Roentgenol* 2013;**200**:W361.
119. Pasticier G, Chapet O, Badet L, et al. Salvage radiotherapy after high-intensity focused ultrasound for localized prostate cancer: early clinical results. *Urology* 2008;**72**:1305.
120. Munoz F, Guarneri A, Botticella A, et al. Salvage external beam radiotherapy for recurrent prostate adenocarcinoma after high-intensity focused ultrasound as primary treatment. *Urol Int* 2013;**90**:288.
121. Riviere J, Bernhard JC, Robert G, et al. Salvage radiotherapy after high-intensity focussed ultrasound for recurrent localised prostate cancer. *Eur Urol* 2010;**58**:567.
122. Lawrentschuk N, Finelli A, Van der Kwast TH, et al. Salvage radical prostatectomy following primary high intensity focused ultrasound for treatment of prostate cancer. *J Urol* 2011; **185**:862.
123. Leonardo C, Franco G, De Nunzio C, et al. Salvage laparoscopic radical prostatectomy following high-intensity focused ultrasound for treatment of prostate cancer. *Urology* 2012;**80**:130.
124. Stolzenburg JU, Bynens B, Do M, et al. Salvage laparoscopic extraperitoneal radical prostatectomy after failed high-intensity focused ultrasound and radiotherapy for localized prostate cancer. *Urology* 2007;**70**:956.
125. ClinicalTrials.gov: Ablatherm Integrated Imaging High Intensity Focused Ultrasound for the Indication of Low Risk, Localized Prostate Cancer: US National Institutes of Health. Available from: http://clinicaltrials.gov/ct2/show/study/NCT00295802. 2014.
126. Dickinson L, Hu Y, Ahmed HU, et al. Image-directed, tissue-preserving focal therapy of prostate cancer: a feasibility study of a novel deformable magnetic resonance-ultrasound (MR-US) registration system. *BJU Int* 2013;**112**:594.

CHAPTER

# 60

# Focal Therapy for Prostate Cancer

*Neil Mendhiratta, BA*, Samir S. Taneja, MD***

*School of Medicine, New York University Langone Medical Center, New York, NY, USA

**Department of Urology and Radiology, New York University Langone Medical Center, New York, NY, USA

## INTRODUCTION

In contemporary clinical practice, multiple modalities of cancer treatment allow therapy to be tailored according to the continuum of disease severity. As trends in prostate cancer (PCa) presentation change, existing treatment options must be adapted to adjust risk and efficacy accordingly, or in some cases, new treatment modalities must be developed.

In recent years, PCa presentation and diagnosis has changed significantly with the identification of prostate-specific antigen (PSA) as a biomarker for prostate disease screening. In the "PSA era," a significant downward stage migration in detected PCa has been observed.[1] This trend has led to increased detection of early stage, localized cancers in the prostate,[2] which invites the opportunity to manage the disease at an earlier point in its natural history.

The rise in overall PCa detection includes diagnosis of many indolent cancers that are unlikely to cause harm. The consequent "overdiagnosis" of PCa has prompted the need for alternate diagnostic and treatment strategies to avoid side effects of conventional therapy. With such shifts in clinical characteristics of PCa at presentation, novel strategies must be explored to more accurately risk-stratify patients and provide alternatives to highly morbid whole-gland therapy.

One strategy that has evolved is active surveillance (AS) of low-risk PCa, a protocol utilized by clinicians to prevent overtreatment of potentially indolent disease. However, up to one-third of men under AS require definitive treatment, and nearly 30% of patients who fail AS are found to have locally advanced after radical prostatectomy.[3,4] Additionally, in its current form, AS requires frequent monitoring by serum testing, office visits, imaging, and biopsy. This results in anxiety and a reduction in quality of life for many patients. Therefore, there exists a need for management options that are more aggressive than surveillance but provide similar preservation of functional outcomes and quality of life.

Focal therapy may provide the capability to treat disease at an early stage while minimizing the risks associated with whole-gland therapy, thereby satisfying a middle-ground between AS strategies and radical therapy for men with localized PCa. In this chapter, we aim to describe the present state of focal therapy, including the goals of therapy, challenges of successful implementation, development of multiple treatment modalities, and contemporary clinical trials that have explored the potential of focal therapy of PCa.

## GOALS OF FOCAL THERAPY

The primary advantage of focal therapy is the selective eradication of PCa and preservation of healthy prostate tissue. The intent of this approach is to mitigate the sexual and urinary side effects that are frequently associated with radical therapy. In this regard, focal therapy holds promise in multiple clinical settings in which the goals of therapy might differ.

In men with low-risk, unifocal disease, focal therapy may provide complete eradication of isolated PCa, leaving the remaining prostate gland disease-free with reduced morbidity compared to whole-gland therapy. Although the incidence of such disease states has risen, recent analysis of 1400 men undergoing radical prostatectomy showed that only 11% had unilateral, low-risk disease (PSA <10 ng/mL, Gleason score <7, <10% tumor involvement).[5] A similar study showed that unifocal and unilateral disease in men with low to intermediate preoperative risk features (PSA ≤20 ng/mL, Gleason score ≤7, ≤66% cores positive) existed in only 14 and 13%, respectively.[6] In addition, identification of men with

Prostate Cancer. http://dx.doi.org/10.1016/B978-0-12-800077-9.00060-8

truly unifocal cancer is difficult with current biopsy techniques.[5,7–9] As such, the use of focal therapy in this setting alone would result in relatively few candidates.

Broader applications of focal therapy may be envisioned in light of other observations on the natural history of PCa. The concept of an index lesion, or a single, high-risk focus as the nidus of most lethal PCa, opens the potential for effective treatment of multifocal disease by targeting only the dominant focus of cancer. In a study of radical prostatectomy specimens, Ohori et al.[10] found that 80% of total tumor volume and 92% of extracapsular extension arose from one dominant focus of cancer within the prostate. Additional studies suggest that nondominant cancer foci typically demonstrate features consistent with low-risk disease. Ohori et al. also concluded that the Gleason score was rarely higher in a secondary lesion than the primary or index lesion,[10] while Villers et al.[11] observed that 80% of secondary tumors are less than 0.5 $cm^3$. Rukstalis et al.[12] similarly reported that the median volume of secondary lesions was 0.3 $cm^3$ and suggested that 79% of men would likely have insignificant residual cancer after ablation of the dominant tumor focus. Subsequent investigations have determined that sites of metastatic PCa, when present, are typically derived from one genotypic clonal cell population from the prostate,[13] which supports the potential efficacy of focal therapy in preventing disease spread if directed to the correct tumor focus.[14]

Accordingly, in men with multifocal, intermediate-risk disease, it may be reasonable to treat the dominant cancer focus while allowing residual "indolent" cancer to be monitored on AS, thereby expanding the pool of eligible men who may benefit from focal therapy by aiming for disease control rather than cure. Such an approach would extend the acceptable period of AS, prolong the natural history of the disease to avoid cancer-specific mortality within the patient's natural longevity, minimize the need for radical therapy, and reduce patient anxiety due to surveillance in the absence of therapy.

"Subtotal" strategies of PCa treatment, whereby ablative techniques are applied to nearly the entire gland while preserving the posterolateral capsule on one side to avoid damage to one neurovascular bundle (NVB) and preserve potency, has been put forth as an alternative to prostatectomy in medium- to high-risk patients.[15] Initially captured under the umbrella of focal therapy, a recent consensus panel[16] determined that such treatment strategies should not be considered "focal," and thus will not be in focus in this chapter. Finally, focal therapy presents utility in multimodal approaches to PCa, as either an adjunct to primary therapy or use in salvage therapy for recurrent disease. There are limited data in this regard, and as such, they will not be presented in this chapter.

In light of the specific goals of treatment, considerations for focal therapy include candidate selection, disease mapping, choice of treatment modality, and methods for clinical follow-up.

# CANDIDATE SELECTION

## Prevalence of Candidates for Focal Therapy

Among contemporary cohorts undergoing surgery, approximately 20% of men demonstrate unilateral disease on final pathology.[5,17] Conceivably a significant number of these men may have been candidates for focal therapy prior to surgery. Additional candidates include the nearly 50% of men who develop unfavorable disease features while on AS, and up to one-third who ultimately undergo radical therapy, presumably having missed the temporal window for localized disease control with focal treatment.[3,18,19] Traditionally, focal therapy trials have including men with unilateral disease as a surrogate for unifocal or limited focality cancer. Recently, the emergence of MRI assessment of the prostate has opened the door for men with bilateral, limited focality disease as potential candidates for targeted approaches.

## Risk Stratification

There are no broadly accepted criteria for identifying appropriate candidates for focal therapy. However, while the boundaries of acceptable risk are not strictly defined, it is clear that focal therapy is inappropriate for low-risk disease, which may do well on AS, as well as for the treatment of high-risk disease in which the window of opportunity for local control may be missed if focal therapy fails (Figure 60.1). Initial general consensus accordingly held that investigational efforts to implement

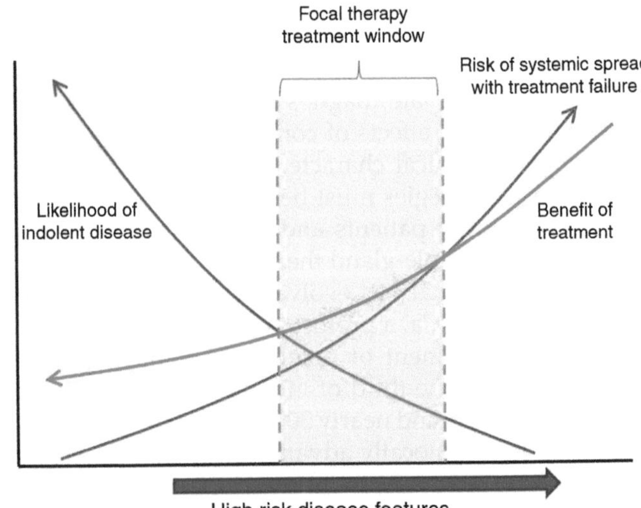


**FIGURE 60.1 The Window of Opportunity for Local Control Includes Low–Intermediate Risk Cancer.**

focal therapy protocols should be limited to men with low- to low-intermediate-risk disease in order to limit the likelihood of early systemic failure. In part this was also due to suboptimal methods of disease mapping, which meant that accurate selection of men with early localized intermediate- to high-risk disease was limited, and therefore inclusion of such men would risk early treatment failure.[20] Recently, however, there has been a proposed shift to focus on intermediate risk patients,[16] in light of growing clinical data with medium-term follow-up[21] as well as consideration for overtreatment of low-risk disease.[22]

Pretreatment cancer risk is often defined using features such as Gleason score, clinical stage, and PSA (among others), as in the risk classification schemes proposed by D'Amico[23] and Kattan.[24] Additional risk of treatment failure in focal therapy may be added by the presence of multifocal disease if either (1) the dominant lesion is not correctly identified or (2) ancillary lesions carry malignant potential. Few trials have explored the efficacy of focal therapy in multifocal disease. Ahmed et al.[25] demonstrated 85% biochemical disease-free survival (BDFS) and 77% negative biopsies at 6 months after focal therapy in a mixed cohort of men with unilateral unifocal, unilateral multifocal, and bilateral multifocal disease, in whom all detected cancer foci were targeted for therapy. Furthermore, potency and continence were preserved in these men (Table 60.1). Their results suggest that multifocal disease may be adequately treated with focal therapy and still provide superior functional outcomes to whole-gland therapy. In this study, seven of nine (89%) men with positive biopsies at 6 months had residual cancer evident on MRI in previously treated areas of the prostate, suggesting that improvements in lesion targeting and/or estimation of tissue ablation margin may improve the ability for focal therapy to treat multifocal cancer. Critically important in this assertion is the documentation of longer-term follow-up. In the study of Ahmed et al., men were staged prior to therapy with a transperineal template biopsy, but efficacy was assessed with a targeted biopsy of the treated area. The validity of this approach in the demonstration of disease control can only be assessed in longer-term follow-up.

Regardless of the specific combination of features chosen, a primary concern for reliable risk stratification is accurate PCa disease mapping, strategies and limitations of which are discussed next.

## METHODS FOR DISEASE MAPPING

The appropriateness and efficacy of focal therapy is heavily dependent on reliable determination of PCa risk, distribution, and extent, which underpins the ability to selectively target the foci of cancer. Two broad methods have been developed for determination of these factors: (1) prostate biopsy and (2) prostate imaging.

### Transrectal Ultrasound-Guided Biopsy

Standard transrectal ultrasound (TRUS)-guided biopsy has provided pretreatment diagnosis in the majority of patients in reported focal therapy trials to date.[30,31] However, the inherent insensitivity of this random biopsy technique[32,33] raises the potential for undetected disease dwelling in undersampled areas of the prostate, which may consequently be spared of treatment. Previous studies have demonstrated up to 30% false negative rates with sextant biopsy.[32] Increasing the number of biopsy cores obtained seems to improve the rate of detection to a certain point;[34] however, significant rates of cancer upgrading have been noted even after 10- to 12-core biopsy.[35,36] Tareen et al.[37] demonstrated that among 590 men undergoing radical prostatectomy, a unilaterally positive 12-core biopsy was able to accurately predict unilateral disease on final pathology in 26% of cases, with no improvements in the positive predictive value with increasing numbers of biopsy cores. Similarly, Gallina et al.[7] observed only 29% accuracy in predicting unilateral disease using extended transrectal biopsy. Other studies have confirmed equally poor prediction of unilateral disease with more extensive transrectal biopsy, including up to 24 cores.[8,9]

### Transperineal Saturation/Template Biopsy

Transperineal saturation biopsy (TPSB) presents another option for PCa mapping, which would appear to be favorable given the superior sensitivity for detecting cancer over standard transrectal biopsy, particularly in the anterior and apical regions of the prostate.[33] Other studies have additionally demonstrated upgrading of TRUS results with TPSB,[38] suggesting better capability for risk stratification. The accuracy of the transperineal biopsy relates in part to the number of cores utilized. TPSB is to be distinguished from transperineal template biopsy (TPTB) in which cores are taken every 5 mm across a standardized template. TPSB tends to utilize a fixed number of cores dispersed in quadrants, while TPTB utilizes as many cores as necessary to cover the whole gland. The latter is more accurate in mapping, although needle deflection can compromise localization, but incurs more morbidity. Onik and Barzell[38] provided evidence that TPTB is more discriminating of unilateral disease than TRUS biopsy. In a series of 180 men with unilateral disease detected on TRUS biopsy, TPTB detected bilateral cancer in 61% of men and upgraded the Gleason score in 23%. However, little evidence exists

**TABLE 60.1** Focal High-Intensity Focused Ultrasound (HIFU) Clinical Trials

| Authors | Pts | Clinical stage (%) | Ablation extent | Follow-up (months) | BDFS | Negative f/u biopsy | Urinary function | Potency preservation | Adverse events | Secondary conclusions |
|---|---|---|---|---|---|---|---|---|---|---|
| Barret et al.[26] | 21 | ≤T2a | Hemi | 12 | NR | NR | No incontinence | ↓IIEF-5 20–14 | 5 (24%) Retention | |
| Ahmed et al.[25] | 41 | T1c (90) T2a (10) | Focal(3–5 mm margin); 49% multifocal | 12 | 85% (ASTRO) | 30/39 (77%) at 6 months | No incontinence | No difference in IIEF-5 at 12 months from baseline | 7 (17%) UTI; 1 (2%) acute urinary retention; 1 (2%) urethral stricture | mpMRI has a high-sensitivity for residual disease |
| Ahmed et al.[27] | 20 | ≤T2b | Hemi | 12 | NR | 17/19 (89%) at 6 months | 2 men (10%) using pads | 100% Potent at 12 months | 1 (5%) Urethral stricture | mpMRI was 100% sensitive and specific for identifying residual cancer |
| El Fegoun et al.[28] | 12 | ≤T2a | Hemi | 127 | NR | 91% at 1 year 90% at 5 years 38% at 10 years | NR | NR | 2 (17%) UTI; 1 (8%) epididimo-orchitis; 1 (8%) acute urinary retention | 100% Cancer free survival at 10 years, though five underwent salvage therapy (mean 8.6 years later) |
| Muto et al.[29] | 29 | T1c (78) T2a (10) T2b (12) | All PZ 1 lobe TZ | 32 | 83% (low risk) 54% (med. risk) 0% (high-risk) | 25/28 (89%) at 6 months; 13/17 (77%) at 12 months | Minimal incontinence (not specified) | NR | 1 (4%) UTI; 1 (4%) urethral stricture | Focal HIFU provided comparable cancer control to whole gland HIFU |

BDFS, biochemical disease free survival; IIEF, international index of erectile function; PZ, peripheral zone; TZ, transition zone.

examining the overall positive predictive value of TPSB for unilateral disease. Hoshi et al.[39] noted only 30% positive predictive value of 24-core transperineal biopsy in determining unilateral disease in 86 men undergoing radical prostatectomy.

### Imaging: Multiparametric MRI

Prostate imaging using MRI exists as a possible method of noninvasive detection of PCa. The integration of anatomic and functional MRI sequences of the prostate, including T2-weighted, diffusion-weighted, and dynamic contrast-enhanced (DCE) images, provides improved diagnostic sensitivity for PCa over any one sequence alone.[40] Evidence supporting the role of multiparametric MRI (mpMRI) in the detection and characterization of PCa is evolving, and is discussed in detail elsewhere in this chapter. In the context of disease mapping for focal therapy, it is of particular significance to note the ability of mpMRI to characterize the location and extent of disease in the prostate and identify potential targets for biopsy and/or treatment.

In a study of 24 men who underwent T2-weighted and DCE MRI prior to radical prostatectomy, Villers et al.[41] assessed the sensitivity of MRI in identifying PCa determined from final pathology. They reported 77 and 90% mpMRI sensitivity for the 0.2 and 0.5 $cm^3$ tumors, respectively, with associated negative predictive values of 85 and 95%, respectively. Additional evidence has supported the finding that thresholds for mpMRI cancer detection vary with the degree of high-risk features present.[42,43] Rosenkrantz and coworkers[44] found that tumor size, Gleason score, and degree of solid growth were each significant independent predictors of detection on mpMRI. As a result, tumors that remain undetected on mpMRI likely represent smaller, lower-grade cancers with morphologic suggestion of a less aggressive disease.

Targeted biopsy of mpMRI lesions, either under MRI guidance or ultrasound visualization with MRI-US image fusion, may offer the best combination of biopsy and imaging techniques for PCa detection and mapping. Targeted biopsy has been shown to identify clinically significant disease with comparable or better sensitivity, improved sampling efficiency, and reduced over-detection of low-risk cancers compared to TRUS biopsy[44,45] and TPSB.[46] The sensitivity of mpMRI for identifying clinically significant disease is strongly suggestive that targeted biopsy may provide a reliable characterization of disease distribution and extent. The correspondingly robust ability of mpMRI to rule out significant cancer in the remaining prostate also has the potential to indicate a lower risk of focal treatment failure, improving the reliability of disease mapping for successful candidate selection.

## MODALITIES FOR FOCAL TREATMENT

### Focal Cryoablation

Focal cryosurgical ablation of the prostate is the longest studied focal therapy modality to date, with over 1500 cases reported in the literature (Table 60.2) and a growing body of clinical outcomes recorded in the Cryo On-Line Data (COLD) Registry (Endocare, Irvine, California). Cryoablation of prostate tissue is achieved by repeated cycles of freezing and thawing, leading to dismantling of cellular membranes and cell death.[47] Detailed physiologic mechanisms of cryotherapy are covered elsewhere in this chapter.

Initially investigated as a method for whole-gland therapy in the 1970s,[57] cryotherapy was piloted as a method of focal treatment by Onik et al.[56] from 1995 to 2000. In nine patients treated with either lesion-targeted, hemi-, or subtotal-ablation (all with one NVB spared) with a mean follow-up of 36 months, all had stable PSA after therapy, and six of six negative follow-up biopsies. Although whole-gland cryotherapy was expected to cause high-levels of potency degeneration,[58] 7/9 (78%) men remained potent after focal cryoablation in this series.

Subsequent trials have demonstrated nearly comparable levels of potency preservation (53–100%) after nerve-sparing cryotherapy, with minimal incontinence (0–4%) and reported adverse events (0.1%) (Table 60.1).[59] After primary therapy with focal cryoablation, follow-up biopsies in these studies have also typically suggested high-levels of cancer control in treated lobes of the prostate, as the majority of recurrences on biopsy occur on the untreated side. Follow-up of biochemical recurrence using ASTRO or Phoenix[60] criteria has shown 0–27% rates of biochemical failure, though such outcomes have not been concordant with biopsy results. In addition, results are often presented without rigorous implementation of follow-up biopsy, making data interpretation difficult. Two studies of focal cryotherapy as salvage therapy demonstrated lower rates of BDFS (50–68%), though with comparable rates of postprocedure incontinence and potency.[61,62]

Methods of diagnosis and disease mapping have been mixed between 12-core TRUS biopsy and transperineal template mapping, and in some cases unreported. Results of biopsy may have implications for the outcomes of therapy, as demonstrated by Truesdale et al.[50] who showed increased risk of positive repeat biopsy and biochemical failure with increasing numbers of positive cores on initial diagnostic biopsy. Similarly, Dhar et al.[51] observed higher rates of biochemical failure associated with higher pretreatment risk assessed by D'Amico criteria.[23]

At present, focal cryotherapy is increasing in popularity while the use of cryo for whole-gland therapy

**TABLE 60.2** Focal Cryoablation Clinical Trials

| Authors | Pts | Clinical stage | Ablation extent | Follow-up (months) | BDFS (%) | Negative f/u biopsy (%) | Urinary function | Potency preservation | Adverse events | Secondary conclusions |
|---|---|---|---|---|---|---|---|---|---|---|
| Barqawi et al.[48] | 62 | T1–T2b | Lesion-targeted | 28 | 71 (PSA > baseline) | 81% | No incontinence | No difference in IIEF-5 at 1 year from baseline | None | |
| Durand et al.[49] | 48 | ≤T2a | Hemi | 13.2 | 98 (Phoenix) | 74% (86% ipsilaterally) | No incontinence | No difference in IIEF-5 at 6 months from baseline | 7 (15%) retention1 (2%) each: perineal fistula, cavernous corpus necrosis, urethral stenosis | PSA follow-up is not a sensitive indicator of disease recurrence |
| Barret et al.[26] | 50 | ≤T2a | Hemi | 9 | NR | NR | No incontinence↓IPSS 9–5 | ↓IIEF-5 19–14 | 4 (8%) retention1 (2%) rectal fistula1 (2%) stricture | |
| Bahn et al.[21] | 73 | T1c (56%) T2a (43%) T2b (1%) | Hemi | 44 | NR | 36/48 (75%) (98% ipsilaterally) | No incontinence | 86% | None | |
| Truesdale et al.[50] | 77 | T1c (87%) T2a (13%) | Hemi | 24 | 73 (Phoenix) | 12/22 (54%) (16% of cohort) | No incontinence | NR | None | |
| Dhar et al.[51] | 795 | <T2b (90%)≥T2b (10%) | NR | 12 | 81 (ASTRO) | 36 Positive (5% of cohort; 25% of biopsied men) | 11 (2.8%) incontinent men | 65% | 3 (0.4%) rectal fistula | D'Amico risk pretreatment is predictive of 12 months BDFS |
| Onik et al.[52] | 48 | NR | Lesion-targeted | 48 | 85 (ASTRO) | 90% (100% ipsilaterally) | No incontinence | 90% | NR | |
| Lambert et al.[53] | 25 | T1c | Hemi | 28 | 84 (PSA nadir ≥50) | 4/7 (57%) with BCR (6/7 (86%) ipsilaterally) | No incontinence | 71% | 1 (4%) retention | Potency is reasonably preserved with one NVB ablated |
| Ellis et al.[54] | 60 | ≤T2a (93%) >T2a (7%) | Hemi | 12 | 80 (ASTRO) | 14/35 (60%) (34/35 (97%) ipsilaterally) | 2 (3.6%) incontinent men | 71%(vacuum therapy device intervention) | NR | |
| Bhan et al.[55] | 31 | NR | Lesion-targeted | 70 | 93 (ASTRO) | 24/25 (96%) (100% ipsilaterally) | No incontinence | 89% | None | |
| Onik et al.[56] | 9 | ≤T2a | 1 Lesion-targeted 5 Hemi 3 Subtotal | 36 | 100 (2 consecutive PSAs with out rise) | 6/6 (100%) | 1 (11%) transient stress incontinence | 7/9 (78%) | None | PSA stability was predictive of negative biopsy results in this small sample |

BDFS, biochemical disease free survival; IIEF, international index of erectile function; IPSS, international prostate symptoms score.

appears to be diminishing.[63] Evidence suggests that it is a well tolerated primary or salvage treatment option with favorable outcomes of cancer control and quality of life in medium-term follow-up. Yet challenges remain in successfully implementing focal cryotherapy into common practice. Optimal cryoneedle placement is yet to be discerned. Most clinical trials of focal cryotherapy have used hemiablative techniques, so the question remains whether lesion-targeted therapy is a viable approach to ablate cancer while improving posttreatment potency rates. Additionally, while not a unique barrier to cryotherapy as a focal modality, determination of long-term outcomes after therapy must be evaluated before appropriate patient selection criteria can be delineated.

## High-Intensity Focused Ultrasound

High-intensity focused ultrasound (HIFU) is another modality for prostate gland ablation that uses directed ultrasound energy to induce thermal tissue damage and coagulative necrosis.[64] More information on the mechanisms of HIFU can be found elsewhere in this chapter.

In 1995, Madersbacher et al.[65] demonstrated the ability of HIFU to induce controlled ablation when they reported sharply demarcated areas of coagulative necrosis in radical prostatectomy specimens after *in vivo* transrectal HIFU. Yet it was not until 2008 that the first clinical trial of HIFU for treatment of PCa was published by Muto et al.[29] In a direct comparison of whole-gland to focal HIFU (ablation of entire peripheral zone and half transition zone), they reported comparable rates of negative repeat biopsy at 6 and 12 months and BDFS at a median follow-up of 34 months. Although outcomes of potency were not reported, the level of postoperative continence were consistent between groups, and the focal HIFU group tended to have lower incidence of urethral stricture and symptomatic urinary tract infection.

In 2011, Ahmed et al.[27] demonstrated similar oncologic outcomes in 20 men treated with prostate hemiablation who were followed for 12 months posttreatment. After 1 year of treatment, in 2012, the same group published the results of a prospective clinical trial in which 41 men with unifocal or multifocal PCa, including 15 men with bilaterally detected disease, were treated with HIFU targeted to all detected lesions with a 3–5 mm margin, and a maximum of 60% prostate ablation. Negative biopsies at 6 months were found in 30/39 men (77%). Of 31 men with baseline erectile function who underwent repeat biopsy, 26 (84%) achieved the "trifecta" of cancer control, leak-free and pad-free continence, and preservation of erections sufficient for penetration.[25] Focal HIFU has also demonstrated efficacy in salvage therapy.[66]

So far, few trials of focal HIFU targeted and hemiablation support its utility as a means of localized cancer control. Compared to other focal treatment modalities, the transrectal probe placement of HIFU is relatively noninvasive; however, short-term adverse events appear to be relatively common, with acute urinary retention, dysuria, and UTI observed most frequently.[67] As a treatment method still under regulatory scrutiny in the United States, a majority of clinical trials of HIFU are from Japan and Europe. Additional clinical data may be required before more widespread adoption is feasible.

## Laser Ablation

Laser ablation refers to the destruction of tissue using a focused beam of electromagnetic radiation emitted from a laser. Other terms for laser ablation include photothermal therapy, laser interstitial thermal therapy, and interstitial laser photocoagulation; in this chapter, we will refer to it as focal laser ablation (FLA).

Laser-mediated tissue destruction is induced by thermal conversion of focused electromagnetic energy, which raises tissue temperature to cause coagulative necrosis.[68] Coagulative necrosis of prostate tissue can be achieved rapidly (seconds) at tissue temperatures above 60°C, or over time (hours to days) after reaching supraphysiologic temperatures between 42°C and 60°C.[69] Advances in bioheat transfer models have allowed for precise prediction of ablation margins, which is critical in dosimetric planning.[70]

FLA has historically been amenable to real-time ablation visualization. In 2009, Atri et al.[71] demonstrated feasibility of FLA with contrast-enhanced ultrasound (CEUS) monitoring of the ablation zone and placement of fluorotopic temperature probes for temperature monitoring. More recent clinical trials have been done under MRI guidance. Intraoperative visualization of ablation zones can be achieved through proton-resonance frequency (PRF) shift MR thermometry, which allows near-real-time quantification of tissue temperature using changes in the phase of gradient-recalled echo images to estimate relative temperature changes.[72] The accuracy of MRI thermometry in outlining histologically proven ablation margins,[73] and correlation with postoperative T1-weighted imaging of necrotic tissue,[70] has proven this to be a viable technique for real-time treatment monitoring.

Contemporary clinical trial outcomes are summarized in Table 60.3. To date, all published investigations of FLA for PCa are small nonrandomized studies with only short-term follow-up. Most have demonstrated >75% negative repeat biopsy in treated areas of the prostate. The most recent trials have additionally incorporated follow-up with MRI-guided targeted biopsy, allowing for better determination of residual cancer in ablated regions of the prostate. Few adverse events have been noted, with even fewer reports of continence

**TABLE 60.3** FLA Clinical Trials

| Authors | Pts | Laser wavelength | Real-time monitoring | Follow-up (months) | Negative f/u biopsy | Urinary function | Potency preservation | Adverse events | Secondary conclusions |
|---|---|---|---|---|---|---|---|---|---|
| Lee et al.[75] | 13 | 980-nm diode | MR thermometry | 6 | 12/13 (92%) (MRTB) | No incontinence ↓AUASS (5.7–4) | No difference in IIEF-5 at 6 months from baseline | NR | |
| Oto et al.[74] | 9 | 980-nm diode | MR thermometry | 6 | 7/9 (78%) (MRTB) | No difference in IPSS at 6 months from baseline | No difference in IIEF-5 at 6 months from baseline | 1 (11%) Perineal abrasion; 1 (11%) Transient paresthesia of glans penis | |
| Lindner et al.[76] | 38 | NR | MR thermometry | 18 | 16/34 (47%) (74% in treated areas) | 1 (3%) Transient stress incontinence | 96% | 3 (8%) Retention | |
| Woodrum et al. 2011[77] | 1 | NR | MR thermometry | 2 | NR | No incontinence | NR | None | FLA is feasible as a method of salvage therapy |
| Raz et al.[78] | 2 | 980-nm diode | MR thermometry | 1 | NR | No incontinence | NR | None | First demonstration of MR-guided FLA and MR thermometry |
| Lindner et al.[79] | 12 | 830-nm diode | Transrectal CEUS | 6 | 6/12 (50%) (8/12 (67%) in treated areas) | No incontinence | No difference in IIEF-5 at 6 months from baseline | None | No correlation between PSA nadir and presence of recurrent disease |
| Atri et al. 2009[71] | 1 | 830-nm diode | Transrectal CEUS | <1 | NR | NR | NR | NR | Ablation area seen on CEUS corresponds well with lesion measured MRI at 7 days |
| Amin et al.[68] | 1 | 805-nm diode | TRUS | <1 | Confirmed necrosis | No incontinence | Potency preserved | None | Laser ablation for the treatment of PCa is feasible |

MRTB, MRI-targeted biopsy; IIEF, international index of erectile function; IPSS, international prostate symptoms score; AUASS, American Urological Association Symptom Score; ; TRUS, transrectal ultrasound.

or potency degeneration. In a recent phase I trial, Oto et al.[74] successfully performed FLA in nine men under MR visualization. Follow-up demonstrated no change in IPSS or SHIM scores at 6 months, and 7/9 men (78%) had negative MRI-guided biopsies targeted to the initial tumor location.

FLA provides the advantage of accurate real-time treatment zone monitoring with either CEUS or MR thermometry. Additionally, the extent of ablation zones has been demonstrated to be highly predictable according to bioheat transfer modeling, allowing for accurate pretreatment planning. Although these results suggest that FLA is a safe and effective modality for focal treatment, more clinical evidence is required before a final determination can be made.

## Photodynamic Therapy

Photodynamic therapy (PDT) is a method of inducing tissue necrosis or vascular obliteration by using laser light to activate photosensitizer agents that have localized to target tissue. Once exposed to light of a specific wavelength, the activated photosensitizer causes direct cellular damage by producing cell-death-inducing superoxide and hydroxyl radicals or converting tissue oxygen into singlet oxygen.[80] Photosensitizers are often delineated as tissue-activated or vascular-activated agents. Although most photosensitizer agents localize to malignant cells preferentially,[81] tissue-activated agents also accumulate in other tissues of the body, requiring light protection of the skin and eyes after systemic administration. In contrast, poor distribution of vascular-activated photosensitizers into interstitial tissue results in rapid clearance, obviating the need for extended light protection after administration.[82]

Formal clinical trials of PDT are summarized in Table 60.4. Several additional phase I studies have demonstrated low occurrences of toxicity and adverse events associated with PDT.[83–85] Although the first use of PDT in PCa dates back to 1990,[86] the first formal PDT trial in a primary treatment setting was only reported in 2006 by Moore et al.[87] in six patients using the tissue-activated photosensitizer meso-tetra-(*m*-hydroxyphenyl) chlorin (mTHPC). All patients had residual cancer in at least one biopsy taken after PDT. However, areas of necrosis and fibrosis were demonstrated on MRI and biopsy histology, indicating that PDT provided a viable method of focal tissue ablation. More recent trials to date have focused on other photosensitizers, with the largest studied set of patients undergoing PDT using the vascular-activated photosensitizers padoporfin (TOOKAD WST-09) and water-soluble padeliporfin (TOOKAD Soluble WST-11) (Steba Biotech, Paris, Fr).[26,88,89] Activation of these palladium pheophorbide photosensitizers results in the immediate generation of reactive oxygen and nitrogen species within the tumor blood vessels causing instantaneous tumor vascular occlusion.[90] Follow-up biopsies at 6 months have yielded negative results in 53–100% of men in treated areas, though mild degeneration of potency and urinary function were noted.

PDT has several distinct advantages over other methods of focal therapy. Highly selective destruction of malignant prostate tissue may be achieved through either increased photosensitizer affinity for tumor cells (such as has been demonstrated with 5-ALA[94]), or by carefully controlling margins of light delivery. Photosensitizer localization to tumor cells may be improved with conjugation to a monoclonal antibody directed at proteins found more prevalently in malignant prostate cells.[95] Because the level of photosensitizer activation, and thus degree of cell damage, is correlated with photosensitizer concentration and intensity of light energy, tissue ablation can be additionally modulated by varying the amount of light energy administered to each unit of prostate tissue. This may allow for a more controlled ablation zone by minimizing light energy delivered to areas near the urethra, rectum, and NVBs. Additionally, as with cryoablation and FLA, transperineal optic fiber placement allows better access to all regions of the prostate gland, particularly the anterior gland, which is difficult to treat with transrectal modalities such as HIFU.

However, certain limitations exist. Systemic distribution of the photosensitizer remains a risk if exposed to light, particularly in the skin and eyes. Methods for real-time monitoring of treatment effect have not been reliably demonstrated. Additionally, there is insufficient clinical data to deduce optimal dosing for each photosensitizer and corresponding optimal light intensity to provide maximum treatment effect with controlled margins of tissue ablation. However, evidence using TOOKAD Soluble agents have postulated optimum dosing based on maximal tissue necrosis observed posttreatment.[88,89]

## Investigational Techniques

New modalities for focal therapy are under investigation. Targeted brachytherapy has been explored in the contexts of primary[96] and salvage[97] therapy. In a pilot study by Cosset et al.,[96] free Iodine-125 ($^{125}I$) seeds were placed strategically in the prostates of 21 men with focal disease to maximize radiation exposure to tumor tissue and minimize radiation to the rest of the prostate. In comparison with historical controls of whole-gland brachytherapy, focal application yielded reduced urinary toxicity and better preserved sexual function.

Other techniques, such as radiofrequency ablation (RFA) and focal electroporation, have been tested in animal prostate models.[98,99] Initial phase I trials of RFA for prostate tissue ablation have been reported in humans,

**TABLE 60.4** Focal PDT Clinical Trials

| Authors | Pts | Clinical stage | Photosensitizer | Follow-up (months) | Negative f/u biopsy | Urinary function | Potency preservation | Adverse events | Secondary conclusions |
|---|---|---|---|---|---|---|---|---|---|
| Moore et al.[89] | 38 | T1c (90%) T2a/b (10%) | Padeliporfin (WST-11) | 6 | 20/37 (53%) 10/12 (83%) in optimal tx group | ↓QoL score (IPSS 2.1–1.3) | No difference in overall IEFF-5; 4 patients reported ED | 2 (5%) Extraprostatic necrosis<br>1 (3%) Pelvic pain | 7 Day MRI showed max treatment effect in 4 mg/kg PS and 200 J/cm light dose; biopsy results optimal when LDI > 1 |
| Azzouzi et al.[88] | 84 | ≤T2b | Padeliporfin (WST-11) | 6 | 61/83 (74%) | ↓IPSS at 6 months | 9 (11%) reported ED<br>IEFF-5 mildly ↓ at 6 months | 1 (1%) Each: prostatitis, hematuria, epididymo-orchitis, stricture, optic neuropathy, inflammatory prostatic cyst | Likely optimal treatment combination is 4 mg/kg PS and 200 J/cm light dose |
| Barret et al.[26] | 23 | ≤T2a | Padeliporfin (WST-11) | 12 | NR | No incontinence | ↓IIEF-5 20–14 | None | |
| Patel et al.[91] | 17 | Salvage (EBRT or brachy) | Motexafin Lutetium | 12 | NR | 1 (6%) urge incontinence | 1 (6%) Grade 1 impotence | Many Grade 1 toxicities; 1 (6%) Grade II urinary toxicity | Higher photosensitizer dose leads to a more durable PSA response |
| Trachtenberg et al.[92] | 24 | Salvage (EBRT) | Padoporfin (WST-09) | 6 | 24/24 (100%) in treated zone<br>0/24 (0%) outside zone | No incontinence | 100% | 12 (50%) Intraoperative hypotension (fluid responsive) | PDT is a safe, viable treatment option for recurrence after EBRT |
| Haider et al.[93] | 25 | Salvage (EBRT) | Padoporfin (WST-09) | 6 | 8/25 (32%) | NR | NR | NR | Volume of necrosis seen on MRI correlates with PSA at 4 and 12 weeks post-PDT |
| Moore et al.[87] | 6 | ≤T2a (33%)<br>>T2a (66%) | mTHPC (Temoporfin) | 1–2 | 0% | 1 (16%) transient stress/urge incontinence; 1 (16%) ↑AUA-7 score (3–11) | 2/3 (66%) men with pretreatment erectile function remained potent | 1 (16%) Sepsis 2 (33%) recatheterizaiton | |

QoL, Quality of life; IPSS, International Prostate Symptoms Score; IIEF, International Index of Erectile Function; LDI, Light density index; EBRT, External beam radiotherapy; mTHPC, meso-tetra-(*m*-hydroxyphenyl) chlorin.

demonstrating the potential for further clinical investigation.[100,101] Results of ongoing clinical trials are awaited with interest.

## FOLLOW-UP

Assessment of focal treatment efficacy during clinical follow-up presents unique challenges. Traditional strategies of surveillance for disease recurrence, such as measurement of PSA kinetics, must be adjusted to reflect that healthy prostate tissue still remains posttreatment. Several clinical trials tracked biochemical recurrence using ASTRO or Phoenix criteria in short- to medium-term follow-up. However, these criteria were not intended for use in this setting, and poor concordance between BDFS and biopsy results were noted in multiple studies.[49,79] The relationship of cancer to PSA level may be reflective of the utility of PSA in follow-up. In cases of low-risk disease, in which the cancer detected may have had little to do with the PSA elevation noted at the time of biopsy, one could expect a focal ablation to result in an inconsistent effect on PSA, more likely related to the volume of prostate destroyed. In higher-risk or larger-volume cancers, a more profound effect on PSA may be noted, and this may indicate greater utility in PSA follow-up among those men.

Similarly, repeat biopsy (TRUS or transperineal) as a method for assessment of disease control suffers from the same limitations as initial biopsy for cancer detection as previously described. Biopsies that are positive in the treated region reflect failure of treatment, while those that are positive elsewhere in the gland reflects failure of baseline staging. Biopsy after focal ablation can be difficult, largely owing to the deformation of the gland upon tissue retraction. It is critically important, in our opinion, to perform the same biopsy following focal ablation, as was performed prior to ablation, to confirm the efficacy of the procedure. Fusion strategies, which account for gland deformation, may offer promise in improving the quality of the biopsy in follow-up.

Thus, the role of imaging in clinical follow-up continues to be explored (Figure 60.2). Multiple trials have noted the ability of MRI to accurately depict ablation zones posttreatment. Lindner et al.[79] demonstrated accurate estimation of ablation zones after FLA using T2 MRI 1 week posttreatment as compared to areas of necrosis determined from radical prostatectomy specimens. De Visschere et al.[102] observed that ablation zones appear similarly hypointense on T2 weighted imaging as PCa. However, Ahmed et al.[25] found that 7/9 (78%) cancers detected 6 months after focal HIFU were identifiable on mpMRI, indicating that residual disease may still be captured from imaging despite

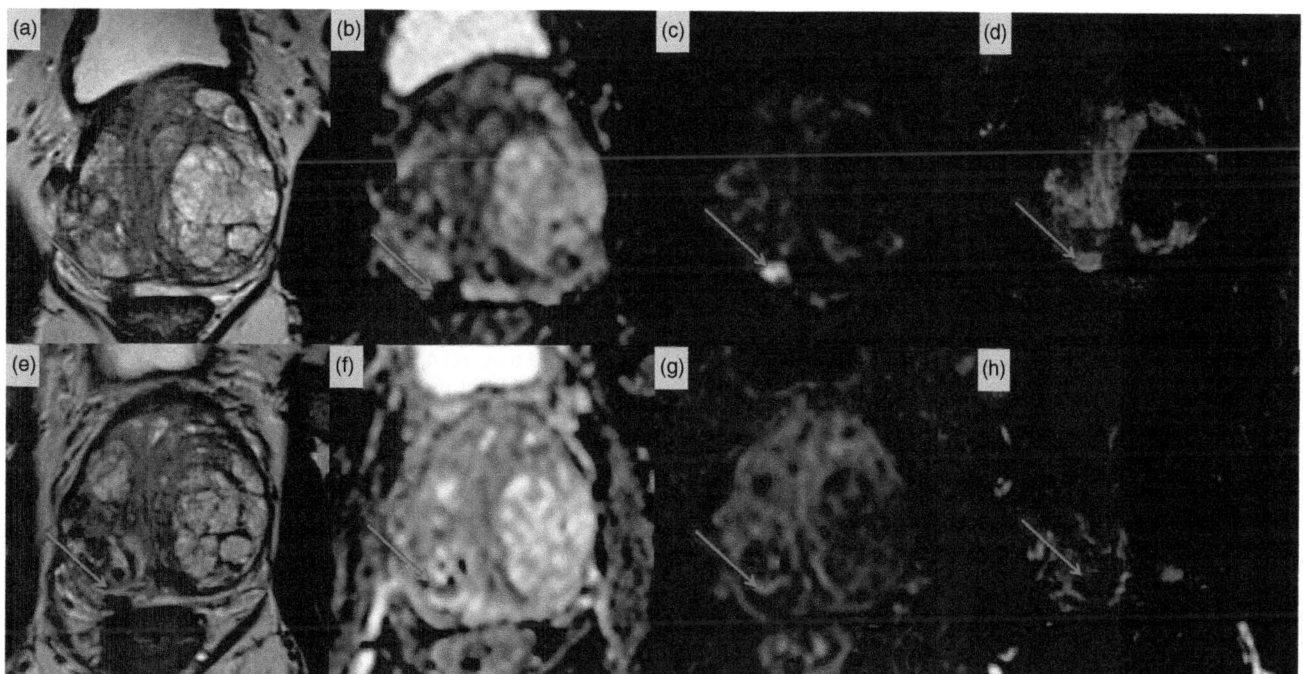


**FIGURE 60.2 Biopsy-Proven Right Peripheral Zone Lesion.** (a) Decreased T2 signal, (b) decreased ADC, (c) increased signal on calculated b1500 diffusion-weighted image, and (d) focal early postcontrast enhancement. On follow-up MRI obtained 10 days after cryoablation, T2WI shows marked decreased signal in the treatment zone (e). The lesion is no longer apparent on the ADC map (f) or high b DWI (g). A necrotic treatment cavity with no internal enhancement is visualized on the postcontrast image (h).

similarity of appearance and proximity to ablation zones. Although imaging alone may not provide optimal degree of sensitivity required to estimate complete oncological control, the selectivity of MRI in detecting clinically significant diseases may be well suited of all follow-up measures for the goals of focal therapy;[103] PSA and biopsy may pick up more diseases, but they are most likely to be insignificant diseases.

## LIMITATIONS OF THERAPY

Evidence has suggested that contemporary-focal therapy techniques are viable methods of selective tissue destruction. Many clinical trials to date have demonstrated the safety and efficacy of cryotherapy, HIFU, FLA, and PDT in the ability of each modality to reliably ablate target tissues with minimal incontinence, potency degeneration, and other adverse events. However, the main limitations of focal therapy remain independent of the technical challenges of applying the treatment itself, but rather lie in the difficulty of understanding long-term outcomes of treatment. Almost no data exists in this regard. Limitations in disease mapping and risk stratification, as previously described, have relegated focal therapy efforts largely to patients with early-stage, low-risk disease. Benefits of treatment in such men are unlikely to manifest in the short- to medium-term follow-up, which has been recorded in the literature to date. It is yet unclear what an effective trial design to assess long-term treatment outcomes and mortality might be.

Maximum efficacy of focal therapy can theoretically be achieved with appropriate patient selection, maximal disease mapping, accurate lesion targeting, real-time ablation monitoring, and follow-up protocols, which optimize monitoring of treatment efficacy and disease recurrence. Some of these challenges may soon be addressed with advances in mpMRI and image fusion platforms. Criteria for candidate selection may also be influenced by advances in focal ablation techniques, which allow for treatment of highgrade and multifocal disease with reliable and controlled tumor ablation.

## CONCLUSIONS

Focal therapy is a promising method of PCa treatment with the potential to bridge the gap between AS and whole-gland therapy. Emerging data regarding oncological and quality of life outcomes from clinical trials are encouraging. Few trials have demonstrated the "trifecta" of cancer control, continence, and potency, though the balance has typically been in favor of improved quality of life outcomes compared to radical therapy. Although the majority of contemporary-focal therapy options have yet to demonstrate comparable levels of cancer control as whole-gland or radical therapy, refinements in candidate selection, disease mapping, and accurate targeting of ablative energy to cancer foci may facilitate improvements in oncological outcomes. Some of the remaining challenges of focal therapy, including disease mapping, treatment monitoring, and evaluation at follow-up, may be addressed with advances in MRI sequencing and interpretation techniques. However, critical questions about long-term outcomes remain, which require further exploration before protocols for focal therapy in common clinical practice can be developed. Yet despite the uncertainty in long-term treatment efficacy, contemporary-focal therapy may still provide an appropriate and desirable option for men who may not tolerate radical treatment or who prioritize quality of life, understand the potential consequences of foregoing whole-gland treatment, and would accept the risk of reduced cancer control for preservation of potency and continence.

## References


1. Cooperberg MR, Lubeck DP, Meng MV, et al. The changing face of low-risk prostate cancer: trends in clinical presentation and primary management. *J Clin Oncol* 2004;**22**:2141–9.
2. Stattin P, Holmberg E, Johansson JE, et al. Outcomes in localized prostate cancer: national prostate cancer register of Sweden follow-up study. *J Natl Cancer Inst* 2010;**102**:950–8.
3. Dall'Era MA, Albertsen PC, Bangma C, et al. Active surveillance for prostate cancer: a systematic review of the literature. *Eur Urol* 2012;**62**:976–83.
4. Bul M, Zhu X, Rannikko A, et al. Radical prostatectomy for low-risk prostate cancer following initial active surveillance: results from a prospective observational study. *Eur Urol* 2012;**62**:195–200.
5. Tareen B, Sankin A, Godoy G, et al. Appropriate candidates for hemiablative focal therapy are infrequently encountered among men selected for radical prostatectomy in contemporary cohort. *Urology* 2009;**73**:351–4.
6. Bott SRJ, Ahmed HU, Hindley RG, et al. The index lesion and focal therapy: an analysis of the pathological characteristics of prostate cancer. *BJU Int* 2010;**106**:1607–11.
7. Gallina A, Maccagnano C, Suardi N, et al. Unilateral positive biopsies in low risk prostate cancer patients diagnosed with extended transrectal ultrasound-guided biopsy schemes do not predict unilateral prostate cancer at radical prostatectomy. *BJU Int* 2012;**110**:E64–68.
8. Falzarano SM, Zhou M, Hernandez AV, et al. Can saturation biopsy predict prostate cancer localization in radical prostatectomy specimens: a correlative study and implications for focal therapy. *Urology* 2010;**76**:682–7.
9. Abdollah F, Scattoni V, Raber M, et al. The role of transrectal saturation biopsy in tumour localization: pathological correlation after retropubic radical prostatectomy and implication for focal ablative therapy. *BJU Int* 2011;**108**:366–71.
10. Ohori M, Eastham J, Koh H, et al. Is focal therapy reasonable in patients with early stage prostate cancer (CaP): an analysis of radical prostatectomy (RP) specimens. *J Urol* 2006;**75**:507. Abstract 1574.
11. Villers A, McNeal JE, Freiha FS, et al. Multiple cancers in the prostate: morphologic features of clinically recognized versus incidental tumors. *Cancer* 1992;**70**:2313–8.

12. Rukstalis DB, Goldknopf JL, Crowley EM, et al. Prostate cryoablation: a scientific rationale for future modifications. *Urology* 2002;**60**:19–25.
13. Liu W, Laitinen S, Khan S, et al. Copy number analysis indicates monoclonal origin of lethal metastatic prostate cancer. *Nat Med* 2009;**15**:559–65.
14. Ahmed HU. The index lesion and the origin of prostate cancer. *N Engl J Med* 2009;**361**:1704–6.
15. Ward JF, Nakanishi H, Pisters L, et al. Cancer ablation with regional templates applied to prostatectomy specimens from men who were eligible for focal therapy. *BJU Int* 2009;**104**:490–7.
16. Donaldson IA, Alonzi R, Barratt D, et al. Focal therapy: patients, interventions, and outcomes – a report from a consensus meeting. *Eur Urol* 2015;**67**(4):771–7.
17. Mouraviev V, Mayes JM, Sun L, et al. Prostate cancer laterality as a rationale of focal ablative therapy for the treatment of clinically localized prostate cancer. *Cancer* 2007;**110**:906–10.
18. Klotz L, Zhang L, Lam A, et al. Clinical results of long-term follow-up of a large, active surveillance cohort with localized prostate cancer. *J Clin Oncol* 2010;**28**:126–31.
19. Barayan Ga, Aprikian AG, Hanley J, et al. Outcome of repeated prostatic biopsy during active surveillance: implications for focal therapy. *World J Urol* 2014; Nov 12 [Epub ahead of print].
20. Taneja SS, Mason M. Candidate selection for prostate cancer focal therapy. *J Endourol* 2010;**24**:835–41.
21. Bahn D, de Castro Abreu AL, Gill IS, et al. Focal cryotherapy for clinically unilateral, low-intermediate risk prostate cancer in 73 men with a median follow-up of 3.7 years. *Eur Urol* 2012;**62**: 55–63.
22. Wilt TJ, Ahmed HU. Prostate cancer screening and the management of clinically localized disease. *BMJ* 2013;**325**:1–9.
23. D'Amico AV, Whittington R, Malkowicz SB, et al. Biochemical outcome after radical prostatectomy, external beam radiation therapy, or interstitial radiation therapy for clinically localized prostate cancer. *JAMA* 1998;**280**:969–74.
24. Kattan MW, Eastham JA, Wheeler TM, et al. Counseling men with prostate cancer: a nomogram for predicting the presence of small, moderately differentiated, confined tumors. *J Urol* 2003;**170**: 1792–7.
25. Ahmed HU, Hindley RG, Dickinson L, et al. Focal therapy for localised unifocal and multifocal prostate cancer: a prospective development study. *Lancet Oncol* 2012;**13**:622–32.
26. Barret E, Ahallal Y, Sanchez-Salas R, et al. Morbidity of focal therapy in the treatment of localized prostate cancer. *Eur Urol* 2013;**63**:618–22.
27. Ahmed HU, Freeman A, Kirkham A, et al. Focal therapy for localized prostate cancer: a phase I/II trial. *J Urol* 2011;**185**(4):1246–54.
28. El Fegoun AB, Barret E, Prapotnich D, et al. Focal therapy with high-intensity focused ultrasound for prostate cancer in the elderly. A feasibility study with 10 years follow-up. *Int Braz J Urol* 2011;**37**:212–3.
29. Muto S, Yoshii T, Saito K, et al. Focal therapy with high-intensity-focused ultrasound in the treatment of localized prostate cancer. *Jpn J Clin Oncol* 2008;**38**:192–9.
30. Valerio M, Ahmed HU, Emberton M, et al. The role of focal therapy in the management of localised prostate cancer: a systematic review. *Eur Urol* 2014;**66**:732–51.
31. Marien A, Gill I, Ukimura O, et al. Target ablation – image-guided therapy in prostate cancer. *Urol Oncol* 2014;**32**:912–23.
32. Applewhite JC, Matlaga BR, McCullough DL. Results of the 5 region prostate biopsy method: the repeat biopsy population. *J Urol* 2002;**168**:500–3.
33. Taira AV, Merrick GS, Galbreath RW, et al. Performance of transperineal template-guided mapping biopsy in detecting prostate cancer in the initial and repeat biopsy setting. *Prostate Cancer Prostatic Dis* 2010;**13**:71–7.
34. Presti JC, O'Dowd GJ, Miller MC, et al. Extended peripheral zone biopsy schemes increase cancer detection rates and minimize variance in prostate specific antigen and age related cancer rates: results of a community multi-practice study. *J Urol* 2003;**169**: 125–9.
35. Walz J, Graefen M, Chun FKH, et al. High incidence of prostate cancer detected by saturation biopsy after previous negative biopsy series. *Eur Urol* 2006;**50**:498–505.
36. Keetch DW, Catalona WJ, Smith DS. Serial prostatic biopsies in men with persistently elevated serum prostate specific antigen values. *J Urol* 1994;**151**:1571–4.
37. Tareen B, Godoy G, Sankin A, et al. Can contemporary transrectal prostate biopsy accurately select candidates for hemi-ablative focal therapy of prostate cancer? *BJU Int* 2009;**104**:195–9.
38. Onik G, Barzell W. Transperineal 3D mapping biopsy of the prostate: an essential tool in selecting patients for focal prostate cancer therapy. *Urol Oncol Semin Orig Investig* 2008;**26**:506–10.
39. Hoshi S, Yamamuro T, Ogata Y, et al. 135 Trans-perineal saturation biopsy did not identify unilateral prostate cancer potentially amenable to focal treatment. *J Urol* 2014;**183**:e55.
40. Delongchamps NB, Rouanne M, Flam T, et al. Multiparametric magnetic resonance imaging for the detection and localization of prostate cancer: Combination of T2-weighted, dynamic contrast-enhanced and diffusion-weighted imaging. *BJU Int* 2011;**107**: 1411–8.
41. Villers A, Puech P, Mouton D, et al. Dynamic contrast enhanced, pelvic phased array magnetic resonance imaging of localized prostate cancer for predicting tumor volume: correlation with radical prostatectomy findings. *J Urol* 2006;**176**:2432–7.
42. Turkbey B, Mani H, Shah V, et al. Multiparametric 3T prostate magnetic resonance imaging to detect cancer: histopathological correlation using prostatectomy specimens processed in customized magnetic resonance imaging based molds. *J Urol* 2011;**186**: 1818–24.
43. Bjurlin MA, Meng X, Le Nobin J, et al. Optimization of prostate biopsy: the role of magnetic resonance imaging targeted biopsy in detection, localization and risk assessment. *J Urol* 2014;**192**:648–58.
44. Wysock JS, Rosenkrantz AB, Huang WC, et al. A prospective, blinded comparison of magnetic resonance (MR) imaging-ultrasound fusion and visual estimation in the performance of MR-targeted prostate biopsy: The PROFUS Trial. *Eur Urol* 2013;**66**:343–51.
45. Siddiqui MM, Rais-Bahrami S, Truong H, et al. Magnetic resonance imaging/ultrasound-fusion biopsy significantly upgrades prostate cancer versus systematic 12-core transrectal ultrasound biopsy. *Eur Urol* 2013;**64**:713–9.
46. Radtke JP, Kuru TH, Boxler S, et al. Comparative analysis of transperineal template saturation prostate biopsy versus magnetic resonance imaging targeted biopsy with magnetic resonance imaging-ultrasound fusion guidance. *J Urol* 2015;**193**(1):87–94.
47. Zenzes MT, Bielecki R, Casper RF, et al. Effects of chilling to 0°C on the morphology of meiotic spindles in human metaphase II oocytes. *Fertil Steril* 2001;**75**:769–77.
48. Barqawi AB, Stoimenova D, Krughoff K, et al. Targeted focal therapy for the management of organ confined prostate cancer. *J Urol* 2014;**192**:749–53.
49. Durand M, Barret E, Galiano M, et al. Focal cryoablation: a treatment option for unilateral low-risk prostate cancer. *BJU Int* 2014;**113**:56–64.
50. Truesdale MD, Cheetham PJ, Hruby GW, et al. An evaluation of patient selection criteria on predicting progression-free survival after primary focal unilateral nerve-sparing cryoablation for prostate cancer: recommendations for follow up. *Cancer J* 2010;**16**: 544–9.
51. Dhar N, Cher ML, Scionti SM, et al. Focal/partial gland prostate cryoablation: results of 795 patients from multiple centers tracked with the cold registry. *J Urol* 2009;**181**:715.

52. Onik G, Vaughan D, Lotenfoe R, et al. The "male lumpectomy": focal therapy for prostate cancer using cryoablation results in 48 patients with at least 2-year follow-up. *Urol Oncol Semin Orig Investig* 2008;**26**:500–5.
53. Lambert EH, Bolte K, Masson P, et al. Focal cryosurgery: encouraging health outcomes for unifocal prostate cancer. *Urology* 2007;**69**:1117–20.
54. Ellis DS, Manny TB, Rewcastle JC, et al. Focal cryosurgery followed by penile rehabilitation as primary treatment for localized prostate cancer: initial results. *Urology* 2007;**70**:9–15.
55. Bahn DK, Silverman P, Lee F, et al. Focal prostate cryoablation: initial results show cancer control and potency preservation. *J Endourol* 2006;**20**:688–92.
56. Onik G, Narayan P, Vaughan D, et al. Focal "nerve-sparing" cryosurgery for treatment of primary prostate cancer: a new approach to preserving potency. *Urology* 2002;**60**:109–14.
57. Bonney WW, Fallon B, Gerber WL, et al. Cryosurgery in prostatic cancer: survival. *Urology* 1982;**19**:37–42.
58. Wong WS, Chinn DO, Chinn M, et al. Cryosurgery as a treatment for prostate carcinoma: results and complications. *Cancer* 1997;**79**:963–74.
59. Shah TT, Ahmed H, Kanthabalan A, et al. Focal cryotherapy of localized prostate cancer: a systematic review of the literature. *Expert Rev Anticancer Ther* 2014;**14**:1337–47.
60. Roach M, Hanks G, Thames H, et al. Defining biochemical failure following radiotherapy with or without hormonal therapy in men with clinically localized prostate cancer: recommendations of the RTOG-ASTRO Phoenix Consensus Conference. *Int J Radiat Oncol Biol Phys* 2006;**65**:965–74.
61. De Castro Abreu AL, Bahn D, Leslie S, et al. Salvage focal and salvage total cryoablation for locally recurrent prostate cancer after primary radiation therapy. *BJU Int* 2013;**112**:298–307.
62. Eisenberg ML, Shinohara K. Partial salvage cryoablation of the prostate for recurrent prostate cancer after radiotherapy failure. *Urology* 2008;**72**:1315–8.
63. Ward JF, Jones JS. Focal cryotherapy for localized prostate cancer: a report from the national Cryo On-Line Database (COLD) Registry. *BJU Int* 2012;**109**:1648–54.
64. Beerlage HP, van Leenders GJ, Oosterhof GO, et al. High-intensity focused ultrasound (HIFU) followed after one to two weeks by radical retropubic prostatectomy: results of a prospective study. *Prostate* 1999;**39**(1):41–6.
65. Madersbacher S, Pedevilla M, Vingers L, et al. Effect of high-intensity focused ultrasound on human prostate cancer *in vivo*. *Cancer Res* 1995;**55**:3346–51.
66. Ahmed HU, Cathcart P, McCartan N, et al. Focal salvage therapy for localized prostate cancer recurrence after external beam radiotherapy: a pilot study. *Cancer* 2012;**118**:4148–55.
67. Crouzet S, Rouviere O, Martin X, et al. High-intensity focused ultrasound as focal therapy of prostate cancer. *Curr Opin Urol* 2014;**24**:225–30.
68. Amin Z, Lees WR, Bown SG. Interstitial laser photocoagulation for the treatment of prostatic cancer. *Br J Radiol* 1993;**66**:1044–7.
69. Van Nimwegen SA, L'Eplattenier HF, Rem AI, et al. Nd:YAG surgical laser effects in canine prostate tissue: temperature and damage distribution. *Phys Med Biol* 2009;**54**:29–44.
70. Stafford RJ, Shetty A, Elliott AM, et al. Magnetic resonance guided, focal laser induced interstitial thermal therapy in a canine prostate model. *J Urol* 2010;**184**:1514–20.
71. Atri M, Gertner MR, Haider MA, et al. Contrast-enhanced ultrasonography for real-time monitoring of interstitial laser thermal therapy in the focal treatment of prostate cancer. *Can Urol Assoc J* 2009;**3**:125–30.
72. Rieke V, Vigen KK, Sommer G, et al. Referenceless PRF shift thermometry. *Magn Reson Med* 2004;**51**:1223–31.
73. Peters RD, Chan E, Trachtenberg J, et al. Magnetic resonance thermometry for predicting thermal damage: an application of interstitial laser coagulation in an *in vivo* canine prostate model. *Magn Reson Med* 2000;**44**:873–83.
74. Oto A, Sethi I, Karczmar G, et al. MR imaging-guided focal laser ablation for prostate cancer: phase I trial. *Radiology* 2013;**267**:932–40.
75. Lee T, Mendhiratta N, Sperling D, et al. Focal laser ablation for localized trials, and our initial experience. *Rev Urol* 2014;**16**:55–66.
76. Lindner U, Davidson SRH, Fleshner NE, et al. 554 Initial results of MR guided laser focal therapy for prostate cancer. *J Urol* 2013;**189**:e227–8.
77. Woodrum DA, Mynderse LA, Gorny KR, et al. 3.0T MR-guided laser ablation of a prostate cancer recurrence in the postsurgical prostate bed. *J Vasc Interv Radiol* 2011;**22**:929–34.
78. Raz O, Haider MA, Davidson SRH, et al. Real-time magnetic resonance imaging-guided focal laser therapy in patients with low-risk prostate cancer. *Eur Urol* 2010;**58**:173–7.
79. Lindner U, Weersink Ra, Haider Ma, et al. Image guided photothermal focal therapy for localized prostate cancer: phase I trial. *Urology* 2009;**182**:1371–7.
80. Yoo J-O, Ha K-S. New insights into the mechanisms for photodynamic therapy-induced cancer cell death. *Int Rev Cell Mol Biol* 2012;**295**:139–74.
81. Zuluaga MF, Sekkat N, Gabriel D, et al. Selective photodetection and photodynamic therapy for prostate cancer through targeting of proteolytic activity. *Mol Cancer Ther* 2013;**12**:306–13.
82. Huang Z, Chen Q, Luck D, et al. Studies of a vascular-acting photosensitizer, Pd-bacteriopheophorbide (Tookad), in normal canine prostate and spontaneous canine prostate cancer. *Lasers Surg Med* 2005;**36**:390–7.
83. Verigos K, Stripp DCH, Mick R, et al. Updated results of a phase I trial of motexafin lutetium-mediated interstitial photodynamic therapy in patients with locally recurrent prostate cancer. *J Environ Pathol Toxicol Oncol* 2006;**25**:373–88.
84. Nathan TR, Whitelaw DE, Chang SC, et al. Photodynamic therapy for prostate cancer recurrence after radiotherapy: a phase I study. *J Urol* 2002;**168**:1427–32.
85. Zaak D, Sroka R, Höppner M, et al. Photodynamic therapy by means of 5-ALA induced PPIX in human prostate cancer – preliminary results. *Med Laser Appl* 2003;**18**:91–5.
86. Windahl T, Andersson S-O, Lofgren L. Photodynamic therapy of localised prostatic cancer. *Lancet* 1990;**336**:1139.
87. Moore CM, Nathan TR, Lees WR, et al. Photodynamic therapy using meso tetra hydroxy phenyl chlorin (mTHPC) in early prostate cancer. *Lasers Surg Med* 2006;**38**:356–63.
88. Azzouzi AR, Barret E, Moore CM, et al. TOOKAD® Soluble vascular-targeted photodynamic (VTP) therapy: determination of optimal treatment conditions and assessment of effects in patients with localised prostate cancer. *BJU Int* 2013;**112**:766–74.
89. Moore CM, Azzouzi AR, Barret E, et al. Determination of optimal drug dose and light dose index to achieve minimally invasive focal ablation of localized prostate cancer using WST11-Vascular Targeted Photodynamic (VTP) therapy. *BJU Int* 2014; May 19 [Epub ahead of print]. DOI: 10.1111/bju.12816.
90. Minimally Invasive Focal Treatment For Prostate Cancer. Available from: http://stebabiotech.com/index.php/TOOKAD-R-Soluble/TOOKAD-R-Soluble-Mechanism-of-Action/TOOKAD-R-Soluble-Mechanism-of-action-detailed-scientific-description.(accessed 12.08.14).
91. Patel H, Mick R, Finlay J, et al. Motexafin lutetium-photodynamic therapy of prostate cancer: short- and long-term effects on prostate-specific antigen. *Clin Cancer Res* 2008;**14**:4869–76.

92. Trachtenberg J, Bogaards a, Weersink Ra, et al. Vascular targeted photodynamic therapy with palladium-bacteriopheophorbide photosensitizer for recurrent prostate cancer following definitive radiation therapy: assessment of safety and treatment response. *J Urol* 2007;**178**:1974–9 discussion 1979.
93. Haider MA, Davidson SRH, Kale AV, et al. MR imaging appearance after vascular targeted photodynamic therapy with palladium-bacteriopheophorbide. *Radiology* 2007;**244**:196–204.
94. Sultan SM, El-Doray A-AM, Hofstetter A, et al. Photodynamic selectivity of 5-aminolevulinic acid to prostate cancer cells. *J Egypt Natl Canc Inst* 2006;**18**(4):382–6.
95. Sharman WM, Van Lier JE, Allen CM. Targeted photodynamic therapy via receptor mediated delivery systems. *Adv Drug Deliv Rev* 2004;**56**:53–76.
96. Cosset J-M, Cathelineau X, Wakil G, et al. Focal brachytherapy for selected low-risk prostate cancers: a pilot study. *Brachytherapy* 2013;**12**:331–7.
97. Hsu CC, Hsu H, Pickett B, et al. Feasibility of MR imaging/MR spectroscopy-planned focal partial salvage permanent prostate implant (PPI) for localized recurrence after initial PPI for prostate cancer. *Int J Radiat Oncol Biol Phys* 2013;**85**:370–7.
98. Onik G, Mikus P, Rubinsky B. Irreversible electroporation: implications for prostate ablation. *Technol Cancer Res Treat* 2007;**6**:295–300.
99. Richstone L, Ziegelbaum M, Okeke Z, et al. Ablation of bull prostate using novel bipolar radiofrequency ablation probe. *J Endourol* 2009;**23**:11–6.
100. Patriarca C, Bergamaschi F, Gazzano G, et al. Histopathological findings after radiofrequency (RITA) treatment for prostate cancer. *Prostate Cancer Prostatic Dis* 2006;**9**:266–9.
101. Shariat SF, Raptidis G, Masatoschi M, et al. Pilot study of radiofrequency interstitial tumor ablation (RITA) for the treatment of radio-recurrent prostate cancer. *Prostate* 2005;**65**:260–7.
102. De Visschere PJ, De Meerleer GO, Fütterer JJ, et al. Role of MRI in follow-up after focal therapy for prostate carcinoma. *Am J Roentgenol* 2010;**194**:1427–33.
103. Rosenkrantz AB, Scionti SM, Mendrinos S, et al. Role of MRI in minimally invasive focal ablative therapy for prostate cancer. *Am J Roentgenol* 2011;**197**(1):W90–W96.

CHAPTER

# 61

# Quality of Life: Impact of Prostate Cancer and its Treatment

*Simpa S. Salami, MD, MPH, Louis R. Kavoussi, MD, MBA*

**The Arthur Smith Institute for Urology, Hofstra North Shore LIJ School of Medicine, New Hyde Park, NY, USA**

## INTRODUCTION

Prostate cancer is the leading solid organ malignancy diagnosed in men in USA, with an estimated 238,590 cases in 2013.[1] The diagnosis of prostate cancer brings about a significant change in the health-related quality of life (HRQOL) of the patient, his sexual partner, and his family. Treatment effects and side effects present a significant socioeconomic burden, given the prevalence and incidence of prostate cancer. Options of treatment for clinically localized prostate cancer includes active surveillance, radical prostatectomy, and radiation therapy. Other measures that may be utilized in conjunction with the mentioned options or alone in the setting of locally advanced, recurrent, or metastatic diseases include hormonal therapy, chemotherapy, and cryotherapy. All of the aforementioned treatment options are associated with varying HRQOL. Prostate cancer thus presents a unique challenge to the physician and the patient when selecting a treatment modality.

## THE TRIFECTA, THE PENTAFECTA

The goal of prostate cancer care and treatment includes oncological control, maintenance of good sexual function, and urinary control. The definitions of these outcome measures (the trifecta) in published literature are highly variable.[2] Some authors have expanded this to include complication rates and margin status, termed the pentafecta.[3,4] In this chapter, we will focus on HRQOL outcomes, namely, sexual potency and urinary continence, and as indicated, bowel dysfunction.

One of the major issues with research into quality of life (QOL) after prostate cancer treatment is that most studies do not utilize a standardized instrument in the assessment of outcome measures. Moreover, most of the studies evaluating QOL outcomes of prostate cancer management are retrospective in nature or prospective cohort studies at best. With the absence of randomization to different treatment options, recall and selection bias are huge issues compounding the interpretation of the results of these studies. Patients without pretreatment assessment of QOL may be more likely to report a greater decrease in HRQOL posttreatment.[5] Furthermore, perception of QOL outcomes may be different between the doctor's and the patient's assessment.[6]

## INSTRUMENTS FOR QOL RESEARCH

Various validated instruments have been designed to standardize QOL assessment following the diagnosis of prostate cancer. Patients and/or their spouses typically complete these instruments as opposed to interviews conducted by physicians or other allied health professionals, since physician and patient estimates of QOL often disagree.[6] The instruments typically assess physical functions, emotions, symptoms, and social functioning. The length of time required to complete any such survey should be reasonable, no more than 20 min. Herein, we will review some of the most commonly used QOL instruments.

### Rand 36-Item Health Survey (SF-36)

The short-form health survey (SF-36) is a 36-item questionnaire focused on eight different domains: physical functioning, bodily pain, role limitations due to physical health problems, role limitations due to personal or emotional problems, general mental health, social functioning, energy/fatigue, and general health perceptions.[7] Each domain is scored on a scale of 0–100, with higher

Prostate Cancer. http://dx.doi.org/10.1016/B978-0-12-800077-9.00061-X

scores indicating higher levels of function. When two groups are compared using SF-36, any difference of 6.5–8.3 points is considered to be clinically significant.[8]

### UCLA Prostate Cancer Index (UCLA-PCI)

The UCLA-PCI is a survey used to assess prostate-cancer-specific HRQOL, focused on six different domains: urinary function, urinary bother, bowel function, bowel bother, sexual function, and sexual bother. Each domain is scored on a scale of 0–100, with higher scores indicating higher levels of function or less bother. This tool has been shown to be reliable and differences of 10 points between two comparison groups is deemed clinically significant.[9–11]

### Expanded Prostate Cancer Index Composite

The Expanded Prostate Cancer Index Composite (EPIC) prostate cancer HRQOL survey was developed to include questions targeted at the urinary side effects of brachytherapy, which were not included in the UCLA-PCI.[12] Although the AUA-SI (presented later) is routinely used in conjunction with UCLA-PCI, it was developed for benign prostatic hyperplasia (BPH) and thus, does not accurately assess treatment-related side effects of brachytherapy. Thus, the UCLA-PCI 20-item survey was expanded to the 50-item EPIC survey.

The EPIC survey also includes domain targeting hormone-related side effects, given that many patients undergoing radiation treatment are also on hormone therapy either in the short- or long-term. The EPIC survey utilizes the SF-12 to assess general QOL issues, instead of SF-36, in order to keep the survey short. Overall, EPIC measures outcomes related to urinary, bowel, sexual, and hormonal symptoms, a comprehensive assessment of HRQOL issues in the management of prostate cancer patients.[12] The 50-item EPIC survey has subsequently been shortened to a 26-item survey while retaining its ability to discern the HRQOL domains, which the original form was designed to measure.[13] A much shorter form containing 16-items, designed for use in clinical practice, was shown to be equally efficient and accurate.[14]

### AUA Symptom Index

The AUA Symptom Index (AUA-SI) was developed to assess obstructive and irritative urinary symptoms. This is a seven-item questionnaire focusing on frequency, nocturia, weak stream, hesitancy, intermittency, incomplete emptying, and urgency. Although it was developed to assess patients' complaints related to BPH,[15] the AUA-SI has been used to assess obstructive and irritating -voiding symptoms associated with prostate cancer treatment.

### Other HRQOL research instruments

Other HRQOL research instruments used in prostate cancer research include: The Functional Assessment of Cancer Therapy – General and Prostate Specific (FACT-G and FACT-P);[16,17] European Organization for Research and Treatment of Cancer (EORTC QLQ-C30);[18,19] and TAG Life/Family Scales.[20]

Although both EPIC and UCLA-PCI (discussed earlier) scored the highest in an assessment of HRQOL instruments with respect to validity and interpretability, the EPIC was recommended because it incorporates a hormonal domain and urinary subscales for incontinence, irritative, and obstructive voiding symptoms (UCLA-PCI mainly assessed incontinence).[21]

## PROSTATE CANCER HRQOL STUDIES

A number of HRQOL studies have critically examined functional outcomes experienced by patients and their spouses or partners following treatment for prostate cancer. These outcomes vary from one treatment modality to another. In this section, we will review three large cohort studies that have prospectively followed patients and/or their spouses/partners after undergoing treatment for prostate cancer or on active surveillance.

### Cancer of the Prostate Strategic Urologic Research Endeavor (CaPSURE)

The Cancer of the Prostate Strategic Urologic Research Endeavor (CaPSURE) is a national registry of men diagnosed with prostate cancer, enrolled from more than 40 urology practices across USA. Men with biopsy-proven diagnosis of prostate cancer were eligible for enrollment irrespective of the treatment modality or the prognosis. Participation in the study ends at death or study withdrawal. At study entry and after each visit, the treating physician collects and reports the registry clinical information including treatments, diagnostic tests, and any new diagnosis or conditions reported by the patients. In addition, a subgroup of patients complete regularly scheduled follow-up questionnaires every 6 months. However, QOL analysis evaluated a small subgroup of patients, and did not measure symptoms related to urinary irritation or hormonal treatment.[22,23]

### Prostate Cancer Outcomes Study

The Prostate Cancer Outcomes Study (PCOS) is a longitudinal, community-based multi-institutional study of patients with prostate cancer sponsored by the National Cancer Institute.[24] The SF-36 was used to assess HRQOL at 6, 12, and 24 months after treatment.

One limitation of the study is the inclusion of primarily patients treated with radical prostatectomy or external beam radiation therapy. Another limitation of this study was that the investigators relied on patients' recall of pretreatment HRQOL. Although recall of HRQOL issues may not be accurate,[25] a validation study in the PCOS showed that a high percentage of men reported accurately disease-related problems before diagnosis at 6 months later.

### Prostate Cancer Outcomes and Satisfaction with Treatment Quality Assessment

The Prostate Cancer Outcomes and Satisfaction with Treatment Quality Assessment (PROST-QA) consortium represents a prospective, multi-institutional effort designed to evaluate the HRQOL after brachytherapy, radiotherapy (including hormonal treatment), and prostatectomy in patients with clinically localized prostate cancer. Researchers also assessed the factors influencing QOL and how the latter relates to overall satisfaction with treatment outcome. The study cohort consisted of 1201 patients and 625 partners. The outcome measures were patients- (using the EPIC-26 and SCA) and partners- (using EPIC-partner and SCA-P) reported. These were collected via a phone survey conducted by a third-party before treatment, and at 2, 6, 12, and 24 months after the start of treatment. The median duration of follow up was 30 months.

Health-related quality of life (HRQOL) outcomes evaluated included urinary incontinence, urinary irritation or obstruction, urinary function, sexual function, bowel or rectal function, and vitality or hormonal function.

## HRQOL OUTCOMES OF PROSTATE CANCER TREATMENT

In this section, we will discuss the HRQOL outcomes of different treatment modalities for clinically localized and advanced prostate cancer.

### Active Surveillance

About 10% of those eligible for active surveillance will select this option of management, and age was the only demographic factor reported to influence this decision.[26] In patients who are eligible and elect active surveillance of prostate cancer, it is reasonable to expect that these men will be free from the toxicity associated with treatment of prostate cancer. Therefore, their QOL should be similar to those experienced by their age-matched controls. Although HRQOL domains decreased over time in men on active surveillance, these outcomes were reported to be similar or better than men without prostate cancer. Decrease in the HRQOL domains were attributable to the normal ageing process except sexual function, which decreased more than expected from the ageing process alone.[27] Does HRQOL outcome influence the decision of men to elect active surveillance, or to proceed to treatment while on active surveillance?

Significant level of anxiety is associated with the diagnosis of cancer, and the fear of the unknown, that is, patient's anxiety about the possibility that the cancer will progress to a point requiring treatment or that metastasis may occur during surveillance. In an analysis of the CaPSURE data, change in cancer anxiety was found to be a significant independent predictor of receiving treatment while on active surveillance.[28] Thus, psychosocial factors and availability of social support should be taken into consideration when counseling a patient for active surveillance.

### Radical Prostatectomy

Radical prostatectomy is a standard treatment option for clinically localized prostate cancer with or without pelvic lymphadenectomy. This modality of treatment is recommended for men with life expectancy of >10 years,[29] given that the adverse effects of surgery are more pronounced in older men.[30] With the introduction of the robotic surgical system and perfection of the nerve-sparing surgical technique, it is theorized that the robotic system may boost visibility thus improving nerve preservation. Prior to the development of the nerve-sparing technique, prostatectomy was associated with significant loss of sexual potency and urinary control. In patients undergoing prostatectomy, an improved recovery of sexual function with nerve-sparing procedures compared with those who did not have nerve-sparing procedures was reported.[31,32] Multi-institutional studies have shown that while improved sexual potency may be observed with the nerve-sparing technique, those with little baseline sexual function may not show significant sexual improvement with nerve-sparing procedures.[33–36] Factors identified to be associated with worse sexual function included older age, large prostate size, and high pretreatment PSA.[31,37]

Urinary incontinence after prostatectomy is worst in the immediate postoperative period, with improvements observed after 1–2 years. In the PROST-QA study at one year, only 7% of patients who underwent prostatectomy reported moderate or worse distress from overall urinary symptoms.[31] Similar trends were observed in the CaPSURE database, men who underwent radical prostatectomy experienced the worst urinary function but also had the greatest recovery compared to other treatment modalities.[37] Factors reported to be associated with worse urinary incontinence were older age, Negroidrace, and a high pretreatment PSA.[31]

In men with significant urinary irritation or obstructive symptoms prior to surgery, their symptoms tended to improve postsurgery, especially those with larger prostates as demonstrated in the PROST-QA study.[31] However, with regard to urinary incontinence, men with larger prostate size (>50 cm$^3$) were found to have lower rates of continence at 6 and 12 months after radical prostatectomy in an analysis of the CaPSURE database.[38] This difference in continence rate disappeared by 2 years after radical prostatectomy. It was hypothesized that the higher rate of incontinence in men with large prostate size after prostatectomy may be related to subclinical bladder dysfunction due to BPH that manifests after surgery.[38] Urethral stricture formation may occur in men after radical prostatectomy. The incidence is estimated at 8.7%, and tends to occur between 6 months and 24 months after surgery.[22] No significant bowel function-related QOL issues were encountered after radical prostatectomy, except when rectal injury occurs requiring bowel diversion.

Treatment satisfaction may play a role in the QOL of patients after radical prostatectomy. Higher treatment satisfaction was found to be associated with higher mental QOL scores, whereas lower fear of recurrence scores was associated with higher physical QOL scores.[39] Treatment satisfaction, however, is influenced by baseline health-related QOL, baseline fear of recurrence, and declines in urinary and bowel function.[40]

## Radiation Therapy

Radiation therapy (external beam or brachytherapy) is an acceptable form of treatment for clinically localized or locally advanced prostate cancer. Radiation therapy is associated with urinary, sexual, and bowel side effects.[31] All forms of radiation therapies are associated with worsened bowel function and bother after treatment. But eventual recovery to baseline can be expected.[37] Some short-term, commonly reported side effects include diarrhea, fecal incontinence, sexual dysfunction, and lower urinary tract symptoms; and in the long-term, urethral strictures, secondary malignancies, and radiation cystitis. Although the use of image-guided radiation therapy (IMRT) is associated with reduced radiation dose delivered to the rectum, bowel symptoms after radiation therapy may be seen in up to 11% of patients as reported in the PROST-QA study.[41] These symptoms persisted till 2 years after treatment.

Unlike radical prostatectomy, the occurrence of urethral strictures after external beam radiation therapy or brachytherapy is usually delayed and estimated to be 1.7 and 1.8%, respectively.[22] As reported in the PROST-QA study, patients receiving brachytherapy are at an increased risk of developing irritative voiding symptoms. Incontinence after brachytherapy was reported by 4–6% of patients at 1–2 years after treatment, with 18% reporting moderate or worse distress from overall urinary symptoms. Irritating-voiding symptoms were exacerbated by larger prostate size and androgen deprivation therapy (ADT) after brachytherapy.[31] Unlike sexual function that may not return to baseline after brachytherapy, urinary function may return to or approach baseline status.[42] Recovery of sexual function is worse in patients who received ADT in conjunction with external beam radiation therapy as opposed to radiation alone.

## Salvage Radical Prostatectomy

Performed mainly at specialized cancer centers, salvage radical prostatectomy after failure of radiation therapy is a highly morbid procedure with relatively high incontinence rates, erectile dysfunction, and rectal fistulae. Studies looking at the HRQOL outcomes of this procedure are limited. However, it has been shown to be associated with significantly lower scores for physical well-being and urinary incontinence compared with salvage radiation therapy.[43,44]

## Salvage Radiation Therapy

The administration of radiation for local control after radical prostatectomy is associated with worsening of incontinence and erectile dysfunction.[45] However, if considerable time has passed prior to external beam radiation therapy, urinary continence is often preserved. On the other hand, radiotherapy administered in the salvage settings results in less impairment of sexual function than primary radiotherapy.[46] Salvage radiation results in less impact on the HRQOL outcomes of patients with recurrence after primary radiation therapy compared with salvage prostatectomy.[43,44]

## Cryotherapy

Cryotherapy used as primary or salvage treatment is associated with significant urinary incontinence, urethral strictures, erectile dysfunction, and rectal fistulae formation. Lower urinary tract symptoms in the immediate postoperative period often result from sloughing of the urethral mucosa due to cooling. With improvement in technology and refinement in techniques including urethral warming, utilization of temperature probes to monitor the temperature around the rectum and neurovascular bundles, the incidence of these complications has reduced significantly. The incidence of urethral strictures after cryotherapy is estimated to be 2.5%.[22] In the salvage setting, cryotherapy was associated with urinary retention (6.8%), incontinence requiring pads (4.4%), and rectal fistulae formation (1.2%).[47]

### Androgen Deprivation Therapy (ADT)

ADT is associated with fatigue, weight change, gynecomastia, depression, and hot flashes. Compared with men on active surveillance for example, those receiving ADT were reported to have poorer urinary and sexual function in the CaPSURE study. Men on ADT were also more likely to be bothered by their sexual or urinary symptoms.[48] Also, hormonal therapy in addition to radical prostatectomy or radiation therapy worsen HRQOL outcomes in the urinary or sexual function domain.[49] ADT may result in a significant decline in mental, emotional, and physical wellbeing.[50,51]

Similarly in the PROST-QA study, ADT given in the adjuvant setting after brachytherapy or radiotherapy was associated with worse QOL outcomes. Patients and their partners may be affected, with 10–19% reporting distress by symptoms attributable to hormonal therapy. Not surprisingly, obesity was found to increase symptoms related to vitality and hormonal function after radiation or brachytherapy.[31]

## SPOUSAL/PARTNER ASSESSMENT OF QUALITY OF LIFE (QOL)

Just as the diagnosis of prostate cancer constitutes a life-changing event and distress to the patient and their partners, treatment for prostate cancer equally affects the patients and their spouses or partners.

In the study by Sanda et al., 44% of partners in the prostatectomy, 22% in the radiation therapy, and 13% in the brachytherapy groups reported distress related to patients' erectile dysfunction. At 1 year after treatment, 5% of partners reported being bothered by patient's urinary incontinence after prostatectomy or brachytherapy. On the other hand, 7% of partners of those who underwent brachytherapy and 3% each of those who underwent brachytherapy and prostatectomy reported being bothered by the patients' obstructive urinary symptoms, such as frequency. Bowel function related QOL issues reported by patients was associated with spousal distress, in up to 5 and 4% of radiation therapy and brachytherapy group, respectively.[31]

Chemotherapy for prostate cancer equally had a significant effect on the QOL of a patient's spouse or partner.[31]

## IMPACT OF QUALITY OF LIFE (QOL) CHANGES ON OVERALL OUTCOME OF TREATMENT

Satisfaction of patients and or their spouses/partners may affect their overall satisfaction with treatment outcome. Sanda et al.[31] reported that changes in QOL domains were associated with overall outcome satisfaction among patients and their partners. Symptoms related to sexual function, vitality, and urinary function were found to be independent predictors of overall outcome satisfaction.[31] Baseline HRQOL and fear of recurrence were also found to affect treatment satisfaction.[40]

## CONCLUSIONS

Treatment of prostate cancer results in significant HRQOL issues that the physician should be familiar with. Knowledge about these is important when counseling patients and their spouses/partners regarding treatment options for prostate cancer. As treatment modalities are modified or improved to reduce the morbidity resulting from treatment of prostate cancer, there will be a need for continued research into the HRQOL outcomes. Moreover, the instruments currently available for assessing these outcomes need to be continually refined, standardized, and tested for accuracy, efficiency, and reliability.

### References


1. *Cancer statistics*, 2013;63(1):11-30. Available from: http://eutils.ncbi.nlm.nih.gov/entrez/eutils/elink.fcgi?dbfrom=pubmed&id=23335087&retmode=ref&cmd=prlinks.
2. Borregales LD, Berg WT, Tal O, et al. "Trifecta" after radical prostatectomy: is there a standard definition? *BJU Int* 2013;**112**(1):60–7.
3. Patel VR, Sivaraman A, Coelho RF, et al. Pentafecta: a new concept for reporting outcomes of robot-assisted laparoscopic radical prostatectomy. *Eur Urol* 2011;**59**(5):702–7.
4. Patel VR, Abdul-Muhsin HM, Schatloff O, et al. Critical review of "pentafecta" outcomes after robot-assisted laparoscopic prostatectomy in high-volume centres. *BJU Int* 2011;**108**(6):1007–17.
5. Chamie K, Sadetsky N, Litwin MS. Physician assessment of pretreatment functional status: a process-outcomes link. *J Urol* 2011;**185**(4):1229–33.
6. Sonn GA, Sadetsky N, Presti JC, et al. Differing perceptions of quality of life in patients with prostate cancer and their doctors. *J Urol* 2009;**182**(5):2296–302.
7. Hays RD, Morales LS. The RAND-36 measure of health-related quality of life. *Ann Med* 2001;**33**(5):350–7.
8. Jaeschke R, Singer J, Guyatt GH. Measurement of health status. Ascertaining the minimal clinically important difference. *Control Clin Trials* 1989;**10**(4):407–15.
9. Litwin MS, Hays RD, Fink A, et al. Quality-of-life outcomes in men treated for localized prostate cancer. *JAMA* 1995;**273**(2):129–35.
10. Lubeck DP, Litwin MS, Henning JM, et al. Measurement of health-related quality of life in men with prostate cancer: the CaPSURE database. *Qual Life Res* 1997;**6**(5):385–92.
11. Brandeis JM, Litwin MS, Burnison CM, et al. Quality of life outcomes after brachytherapy for early stage prostate cancer. *J Urol* 2000;**163**(3):851–7.
12. Wei JT, Dunn RL, Litwin MS, et al. Development and validation of the expanded prostate cancer index composite (EPIC) for comprehensive assessment of health-related quality of life in men with prostate cancer. *Urology* 2000;**56**(6):899–905.
13. Szymanski KM, Wei JT, Dunn RL, et al. Development and validation of an abbreviated version of the expanded prostate cancer

index composite instrument for measuring health-related quality of life among prostate cancer survivors. *Urology* 2010;**76**(5):1245–50.

14. Chang P, Szymanski KM, Dunn RL, et al. Expanded prostate cancer index composite for clinical practice: development and validation of a practical health related quality of life instrument for use in the routine clinical care of patients with prostate cancer. *J Urol* 2011;**186**(3):865–72.
15. Barry MJ, Fowler FJ, O'Leary MP, et al. The American Urological Association symptom index for benign prostatic hyperplasia. The Measurement Committee of the American Urological Association. *J Urol* 1992;**148**(5):1549–57 discussion1564.
16. Cella DF, Tulsky DS, Gray G, et al. The functional assessment of cancer therapy scale: development and validation of the general measure. *J Clin Oncol* 1993;**11**(3):570–9.
17. Esper P, Mo F, Chodak G, et al. Measuring quality of life in men with prostate cancer using the functional assessment of cancer therapy-prostate instrument. *Urology* 1997;**50**(6):920–8.
18. Aaronson NK, Ahmedzai S, Bergman B, et al. The European Organization for Research and Treatment of Cancer QLQ-C30: a quality-of-life instrument for use in international clinical trials in oncology. *J Natl Cancer Inst* 1993;**85**(5):365–76.
19. Osoba D, Aaronson N, Zee B, et al. Modification of the EORTC QLQ-C30 (version 2.0) based on content validity and reliability testing in large samples of patients with cancer. The Study Group on Quality of Life of the EORTC and the Symptom Control and Quality of Life Committees of the NCI of Canada Clinical Trials Group. *Qual Life Res* 1997;**6**(2):103–8.
20. Lubeck DP, Litwin MS, Henning JM, et al. The CaPSURE database: a methodology for clinical practice and research in prostate cancer. CaPSURE research panel. Cancer of the prostate strategic urologic research endeavor. *Urology* 1996;**48**(5):773–7.
21. Schmidt S, Garin O, Pardo Y, et al. Assessing quality of life in patients with prostate cancer: a systematic and standardized comparison of available instruments. *Qual Life Res* 2014;**23**(8):2169–81.
22. Elliott SP, Meng MV, Elkin EP, et al. Incidence of urethral stricture after primary treatment for prostate cancer: data From CaPSURE. *J Urol* 2007;**178**(2):529–34.
23. Konety BR, Cowan JE, Carroll PR. CaPSURE Investigators. Patterns of primary and secondary therapy for prostate cancer in elderly men: analysis of data from CaPSURE. *J Urol* 2008;**179**(5):1797–803.
24. Potosky AL, Harlan LC, Stanford JL, et al. Prostate cancer practice patterns and quality of life: the prostate cancer outcomes study. *J Natl Cancer Inst* 1999;**91**(20):1719–24.
25. Litwin MS, McGuigan KA. Accuracy of recall in health-related quality-of-life assessment among men treated for prostate cancer. *J Clin Oncol* 1999;**17**(9):2882–8.
26. Barocas DA, Cowan JE, Smith JA, et al. CaPSURE investigators. What percentage of patients with newly diagnosed carcinoma of the prostate are candidates for surveillance? An analysis of the CaPSURE database. *J Urol* 2008;**180**(4):1330–4.
27. Arredondo SA, Downs TM, Lubeck DP, et al. Watchful waiting and health related quality of life for patients with localized prostate cancer: data from CaPSURE. *J Urol* 2004;**172**(5 Pt 1):1830–4.
28. Latini DM, Hart SL, Knight SJ, et al. The relationship between anxiety and time to treatment for patients with prostate cancer on surveillance. *J Urol* 2007;**178**(3 Pt 1):826–31.
29. Mohler JL, Kantoff PW, Armstrong AJ, et al. Prostate cancer, version 1.2014. *J Natl Compr Canc Netw* 2013;1471–9.
30. Wright JL, Lin DW, Cowan JE, et al. CaPSURE Investigators. Quality of life in young men after radical prostatectomy. *Prostate Cancer Prostatic Dis* 2008;**11**(1):67–73.
31. Sanda MG, Dunn RL, Michalski J, et al. Quality of life and satisfaction with outcome among prostate-cancer survivors. *N Engl J Med* 2008;**358**(12):1250–61.
32. Le JD, Cooperberg MR, Sadetsky N, et al. Changes in specific domains of sexual function and sexual bother after radical prostatectomy. *BJU Int* 2010;**106**(7):1022–9.
33. Quinlan DM, Epstein JI, Carter BS, et al. Sexual function following radical prostatectomy: influence of preservation of neurovascular bundles. *J Urol* 1991;**145**(5):998–1002.
34. Kundu SD, Roehl KA, Eggener SE, et al. Potency, continence and complications in 3,477 consecutive radical retropubic prostatectomies. *J Urol* 2004;**172**(6 Pt 1):2227–31.
35. Rabbani F, Stapleton AM, Kattan MW, et al. Factors predicting recovery of erections after radical prostatectomy. *J Urol* 2000;**164**(6):1929–34.
36. Harris CR, Punnen S, Carroll PR. Men with low preoperative sexual function may benefit from nerve sparing radical prostatectomy. *J Urol* 2013;**190**(3):981–6.
37. Huang GJ, Sadetsky N, Penson DF. Health related quality of life for men treated for localized prostate cancer with long-term followup. *J Urol* 2010;**183**(6):2206–12.
38. Konety BR, Sadetsky N, Carroll PR. CaPSURE Investigators. Recovery of urinary continence following radical prostatectomy: the impact of prostate volume – analysis of data from the CaPSURE Database. *J Urol* 2007;**177**(4):1423–5.
39. Hart SL, Latini DM, Cowan JE, et al. CaPSURE Investigators. Fear of recurrence, treatment satisfaction, and quality of life after radical prostatectomy for prostate cancer. *Support Care Cancer* 2008;**16**(2):161–9.
40. Resnick MJ, Guzzo TJ, Cowan JE, et al. Factors associated with satisfaction with prostate cancer care: results from Cancer of the Prostate Strategic Urologic Research Endeavor (CaPSURE). *BJU Int* 2013;**111**(2):213–20.
41. Hamstra DA, Conlon ASC, Daignault S, et al. Multi-institutional prospective evaluation of bowel quality of life after prostate external beam radiation therapy identifies patient and treatment factors associated with patient-reported outcomes: the PROSTQA experience. *Int J Radiat Oncol Biol Phys* 2013;**86**(3):546–53.
42. Downs TM, Sadetsky N, Pasta DJ, et al. Health related quality of life patterns in patients treated with interstitial prostate brachytherapy for localized prostate cancer – data from CaPSURE. *J Urol* 2003;**170**(5):1822–7.
43. Tefilli MV, Gheiler EL, Tiguert R, et al. Quality of life in patients undergoing salvage procedures for locally recurrent prostate cancer. *J Surg Oncol* 1998;**69**(3):156–61.
44. Tefilli MV, Gheiler EL, Tiguert R, et al. Salvage surgery or salvage radiotherapy for locally recurrent prostate cancer. *Urology* 1998;**52**(2):224–9.
45. Arredondo SA, Latini DM, Sadetsky N, et al. Quality of life for men receiving a second treatment for prostate cancer. *J Urol* 2007;**177**(1):273–8.
46. Hu JC, Elkin EP, Krupski TL, et al. The effect of postprostatectomy external beam radiotherapy on quality of life: results from the Cancer of the Prostate Strategic Urologic Research Endeavor. *Cancer* 2006;**107**(2):281–8.
47. Pisters LL, Rewcastle JC, Donnelly BJ. Salvage prostate cryoablation: initial results from the cryo on-line data registry. *J Urol* 2008;**180**(2):559–63 discussion 563-4.
48. Lubeck DP, Grossfeld GD, Carroll PR. The effect of androgen deprivation therapy on health-related quality of life in men with prostate cancer. *Urology* 2001;**58**(2 Suppl. 1):94–100.
49. Wu AK, Cooperberg MR, Sadetsky N, et al. Health related quality of life in patients treated with multimodal therapy for prostate cancer. *J Urol* 2008;**180**(6):2415–22.
50. Cary KC, Singla N, Cowan JE, et al. Impact of androgen deprivation therapy on mental and emotional well-being in men with prostate cancer: analysis from the CaPSURE™ registry. *J Urol* 2014;**191**(4):964–70.
51. Sadetsky N, Greene K, Cooperberg MR, et al. Impact of androgen deprivation on physical well-being in patients with prostate cancer: analysis from the CaPSURE (Cancer of the Prostate Strategic Urologic Research Endeavor) registry. *Cancer* 2011;**117**(19): 4406–13.

CHAPTER

# 62

# Impact of Prostate Cancer Treatments on Sexual Health

*Lawrence C. Jenkins, MD, MBA, John P. Mulhall, MD, MSc, FECSM, FACS*

**Sexual and Reproductive Medicine Program, Urology Service, Memorial Sloan-Kettering Cancer Center, New York, USA**

Erectile dysfunction (ED) is the most reported sexual dysfunction after prostate cancer treatments. However, there are other sexual dysfunctions that receive less attention and adversely impact quality of life after treatment, including absence of ejaculation, changes in orgasm or libido, sexual incontinence, and loss of penile length. Our understanding of the pathophysiological basis of ED and other sexual dysfunctions has evolved over the past decade. In this chapter, we will present the current knowledge of the impact and management of sexual dysfunctions after prostate cancer treatments.

## ERECTILE DYSFUNCTION

### Functional Anatomy of Erection

The penis consists of a paired corpora cavernosa dorsally and the corpora spongiosum ventrally, which contains the urethra. The tunica albuginea surrounds the corpora cavernosa. Blood flow is supplied by the internal iliac artery then the internal pudendal artery. The internal pudendal artery becomes the common penile artery and divides into the cavernosal, dorsal, and bulbourethral arteries. The cavernosal arteries supply the helicine arterioles, which open into the endothelial-lined lacunar spaces. The blood flow returns as these endothelial spaces drain into subtunical venules, which coalesce to form emissary veins. The emissary veins traverse the tunica to empty into the cavernosal, deep dorsal, or spongiosal veins where they drain into the prostatic venous plexus or the internal pudendal veins.[1] Accessory pudendal arteries (APA) arise from a source above the levator ani (iliac, femoral, vesical, or obturator arteries) and travel down to the perineum in the periprostatic region.

Prostate Cancer. http://dx.doi.org/10.1016/B978-0-12-800077-9.00062-1


### Physiology of Erection

In the flaccid state, cavernosal smooth muscle (SM) remains contracted under adrenergic control. With sexual stimulation, nerve impulses cause SM relaxation of the arteries and arterioles supplying the penis, facilitated by the nitric oxide (NO)/cyclic guanine monophosphate pathway (cGMP), and leading to an increased penile blood flow. Simultaneously, relaxation of the cavernosal SM allows the filling of lacunar spaces and compression of the subtunical venous plexuses, therefore blocking most of the venous drainage system and creating an erection.[2]

### Pathophysiology of ED after Radical Prostatectomy

The pathophysiology of ED after radical prostatectomy (RP) involves the interaction of three factors: neural injury, vascular injury, and corporal SM damage. Additionally, the extent and reversibility of these injuries ultimately will define the recoverability of erectile function (EF) (Table 62.1).

#### *Neural Trauma*

It is generally understood that transection of, or extensive thermal injury to, the cavernous nerves will result in permanent loss of EF after surgery. However, traction and/or percussive injury to the nerves may be just as harmful. In a recent study, Masterson et al. reported that alteration in technique whereby the urethral catheter is no longer used as a traction tool to apply tension to the lateral pedicles, resulted in a significant improvement in EF postprostatectomy.[3] It is indisputable that the nerve-sparing status of an RP is predictive about the recovery

TABLE 62.1 Predictors of Erectile Function Recovery

| Predictors of erectile function recovery |
|---|
| Nerve sparing status |
| Patient age |
| Baseline erectile function |
| Postoperative hemodynamics (arterial insufficiency) |
| Surgeon experience/volume |
| Vascular comorbidity profile |
| Preoperative serum testosterone |

of EF. Bilateral nerve sparing is associated with better spontaneous and oral therapy-assisted recovery of EF compared to unilateral nerve sparing, and is thus more likely to lead to functional erections than non-nerve-sparing surgeries.[4–6]

Postoperative factors such as edema and inflammation, around the bladder neck and the cavernous nerves, may play a role in the decline of EF. More recent evidence, Katz et al. have shown us that some men respond to phosphodiesterase type 5 (PDE5) inhibitors (PDE5i) within 4 weeks after surgery but by 12 weeks they do not respond. This is possibly due to ongoing postoperative Wallerian degeneration, or in part perhaps, due to perineural inflammation.[7]

Several animal models have recently focused on the pathophysiology of ED after RP.[8–12] The elucidation of the key mechanisms involved in the development of ED after cavernous nerve injury has created the potential application of penile rehabilitation protocols, aimed at reversing the complex consequences of cavernosal nerve injury during RP.

### *Arterial Injury*

During RP, APAs, which are supra-diaphragmatic arteries (lying above the levator ani), are predisposed to injury. The origin of these arteries is variable coming from femoral, obturator, vesicle, or iliac arteries. Rogers et al. have shown that APA preservation at the time of open RP results in an improvement in EF recovery and possibly even shortening of the time to recovery of erections.[13] Breza et al. studied 10 cadavers who underwent extensive pelvic dissection and in seven cadavers APA were present, in four they were the major source of arterial inflow, and in one patient they were the only source of inflow.[14]

### *Corporal Smooth Muscle Alterations*

In normal conditions, PGE1 inhibits collagen formation by inhibiting TGF-β1 that induces collagen synthesis. When PGE1 is inhibited, TGF-β1 is allowed to induce connective tissue synthesis.[15] The trabecular SM is then replaced with collagen, which alters the mechanical properties as well as the integrity of the corpora cavernosa.[9,11] Several studies have shown this process in the penile tissue of denervated animal models, identifying significant increases in collagen content and a decrease in the SM–collagen ratio compared to controls.[9,11,16–20]

An important mechanism associated with neural injury and exacerbated by absence of cavernosal oxygenation is apoptosis, or programmed cell death. In the penis, denervation has been shown to stimulate apoptosis, leading to increased deposition of connective tissue that may result in a decrease in penile elasticity, and in turn, venocclusive dysfunction (venous leak).[11,16,17] In a study by Mulhall et al., men who had partner-corroborated excellent erectile function prior to surgery, after undergoing duplex Doppler penile ultrasound after surgery, were found to have increased rates of venous leak (based on elevated end diastolic velocities) as time progressed after surgery.[21] In a different study, Mulhall et al. have shown in a cavernous nerve crush injury model that neural injury can cause apoptosis in SM and endothelium in a more delayed fashion compared to the neurectomy model.[10] These studies lead one to believe that after RP the neural injury induces pro-apoptotic (SM degeneration) and pro-fibrotic (increase in collagen) factors within the corpora cavernosa.

## Pathophysiology of ED after RT

The etiology of post radiation therapy (RT) ED is not completely understood and is likely multifactorial. Obayomi-Davies et al. state that the radiation may cause dose-dependent damage to the neurovascular bundles, the crura, and the penile bulb.[22] Zelefsky and Eid evaluated 98 patients for ED after RP or RT, with Duplex ultrasonography and classified them as having arteriogenic, cavernosal, mixed, or neurogenic impotence.[23] Thirty-eight patients were treated with RT and of these 24 (63%) were found to have arteriogenic dysfunction. This finding differed from post-RP patients who were more likely to have a cavernosal veno-occlusive dysfunction (CVOD). Endothelial cells are damaged in a time- and dose-dependent manner. These endothelial cells line the penile arteries and sinusoids of the corpora cavernosa. The capillaries and sinusoids are especially sensitive because endothelial cells are major cellular components. Damage to these cells leads to luminal stenosis and arterial insufficiency, which have a gradual progression over time. Pathologic evaluation of the small vessels shows hyalinization, fibrosis, and the deposition of lipid-laden macrophages.[24] The net effect of this endothelial damage is ischemia. Pisansky et al. theorized that ED after RT is related to cavernosal hypoxia and fibrosis causing penile endothelial dysfunction.[25]

## Pathophysiology of ED after ADT

The relationship of ED and androgen deprivation therapy (ADT) is primarily related to the lack of testosterone. In castrated men, erections are less frequent due to loss of libido and the absence of nocturnal erections.[26] Testosterone has effects on tumescence (positive effect on NO synthase and negative on Ras homolog gene family member A (RhoA) / Rho-associated coiled-coil containing protein kinase (ROCK)) and detumescense (positive effect on PDE5); however, the net effect is modest overall.[27] In a study by Suzuki et al., erectile response was measured in castrated male rats (model for venous leak) after electrical stimulation. The castrated rats had significantly decreased responses to stimulation compared to their noncastrate controls and after testosterone replacement, erectile responses were recovered.[28] This study showed that testosterone plays an important role in the neural pathways for tumescence. ADT also has indirect effects on the maintenance of endothelial and cavernosal SM. The absence of erections with ADT, leads to a decrease in cavernosal oxygenation creating an environment of prolonged absence of cavernosal oxygenation.[29] This leads to the inhibition of PGE, which allows TGF-β1 to promote connective tissue synthesis (primarily collagen), and the replacement of trabecular SM.[30,31] The change in SM/collagen ratio leads to a decreased ability for the SM to expand and occlude the subtunical venous plexuses, leading to venous leak.[32]

## Prevalence of ED after RP

A review of the existing medical literature demonstrates a large variance in reported rates of ED following RP (20–90%).[33–38] There are multiple reasons for these reported differences in outcomes including intrinsic patient factors, surgical factors, and reporting biases. Compared to postoperative complications like incontinence, ED is harder to characterize; the return of EF may be masked by the expected decline in function due to advancing age. Therefore, measuring EF prospectively has a moving baseline.[39,40] There is a significant psychological component for these men and their partners; however, it is poorly described in literature.[41,42]

Perhaps the gold standard recovery definition, the International Index of Erectile Function (IIEF) EF domain score of 26 is unequivocally normal, and in a study by Teloken et al., approximately 70% of men with EF domain scores of 22–25 agreed completely or somewhat with the statement that "they could have sexual intercourse whenever they wished."[43] Functional erectile ability likely lies somewhere between 22 and 26 on the EF domain score. Additionally, we believe that men who score 26 points or greater on the IIEF EF domain while using a PDE5i or other erectogenic agent are not truly normal.

The prevalence of ED following contemporary RP is poorly defined by the current peer-reviewed literature. The published studies have numerous methodological flaws, such as absence of consensus on a definition of success or failure to recover erections, small populations, incomplete data acquisition, limited data on quality of life and satisfaction with sexual life, and inadequate patient follow up. Mulhall et al. performed a comprehensive review of the literature, a composite prevalence rate among men following contemporary RP was 48% and where the nerve-sparing status was described, EF recovery was achieved in 50%.[35] It is safe to state that centers reporting EF recovery rates of 90% or higher are giving patients unrealistic expectations as they are unsubstantiated in the general RP population. Katz et al. showed that 20% of men with functional erections 3 months after RP would lose their erections by 6 months, and 90% of those who were functional at 3 months would retain or regain function at 12 months. In a study by Rabbani et al., they illustrate that younger age is associated with EF recovery beyond 2 years in patients who had a BNS RP.[44] In a more recent study by Sivarajan et al., EF showed significant improvements after RP up to 2 years, from 2 years to 10 years EF remained generally stable; however, men <60 were more likely to report improvements up to 7 years after RP when compared to older men.[45]

## Prevalence of ED after RT

The prevalence of ED after RT is likely underestimated, but has been reported as 36–59% after external beam radiotherapy and 24–50% after brachytherapy.[46] These numbers were based on several prospective studies examining the common side effects of radiation treatment. The Prostate Cancer Outcomes and Satisfaction with Treatment Quality Assessment (PROSTQA) reported that at 2 years after treatment in those who were potent at the beginning of treatment 42% (51/121) in the external radiotherapy group, 37% (59/158) in the brachytherapy group reported ED after treatment.[47] Just as with RP patients, methodological flaws exist, the most serious one being EF assessment too early after RT completion. Teloken et al. looked at the predictors of sildenafil response after RT and showed that EF post-RT has a progressive decline post treatment. On multivariable analysis, predictors included older age (OR 3.2), time after RT [time since RT >18 m (OR 4.1), >36 m (OR 9.3)], and ADT >4 m (OR 7.1),[48] thus illustrating that the nadir for EF deterioration is beyond 36 months.

## Prevalence of ED after ADT

Potosky et al. reported on sexual and erectile outcomes within a population-based sample of patients who were treated with primary ADT compared to a group with no treatment. At one year, of those who were potent at baseline, only 20% (20/88) of the ADT group versus 70% (163/223) of the no-treatment group were potent.[49] In a study by Mazzola et al., 38 patients who received ADT prior to RP were compared to 94 patients who did not using duplex Doppler penile ultrasound and the IIEF EF domain score.[50] Within 6 months of surgery, the incidence of venous leak in the ADT positive group compared to the ADT negative group was 60% versus 20%, respectively. The IIEF EF domain scores were also significantly different, showing worse outcomes for those who had ADT. Indeed, patients exposed to ADT pre-RP had outcomes similar to those not exposed to ADT who had non-nerve sparing RP.

## The Concept of Rehabilitation

### *PDE5 Inhibitors for Erectile Tissue Preservation*

There have been multiple experimental studies using animal models illustrating the benefits of PDE5 inhibitors (PDE5i) after cavernous nerve injury.[51,52] The PDE5i has been documented to be endothelium and SM protectants. Several studies in animal models maintained on PDE5 inhibitors have shown decreased amounts of collagen deposition resulting in decreased penile SM fibrosis.[16–20,53]

Several key human studies have shown significant benefits of treatment with PDE5i. Desouza et al. looked at brachial artery flow mediated dilation (FMD) in diabetic patients after treatment with sildenafil to show the effect on endothelial function.[54] They examined patients 1 h after a 25 mg dose of sildenafil and after 2 weeks of daily dosing, both groups showed increased FMD. They also found effects were still present in the 2-week dosing group 24 h after discontinuation, thus illustrating the beneficial effects on endothelial function. Rosano et al. showed that chronic therapy with tadalafil improved endothelial cell function in patients with increased cardiovascular risk regardless of EF, with a sustained benefit for at least 2 weeks after discontinuation of therapy.[55] In a study by Schwartz et al., they evaluated intracorporeal SM content pre-RP and 6 months post-RP.[56] In the group treated with sildenafil 50 mg every other night, there was no significantly different change in mean SM content; however, in the group treated with 100 mg every other night there was a statistically significant increase in mean SM content. The sildenafil not only preserved SM but in higher doses it increased SM content.

Sildenafil has shown to be beneficial in penile rehabilitation in multiple studies. Bannowsky et al. showed the benefit of nightly administration of sildenafil 25 mg after NS-RP and examined IIEF-5 scores serially for 52 weeks.[57] They found a significant difference in IIEF-5 scores and time to EF recovery between the sildenafil and control groups ($p < 0.001$) with potency rates of 86 and 66%, respectively. In a recent randomized double blind placebo-controlled trial by Pavlovich et al., they examined nightly versus on demand sildenafil in potent men after NS-RP and there was no difference in erectile recovery at 1 year.[58] This study shows that nightly sildenafil did not out perform on demand sildenafil in penile rehabilitation.

### *Intracavernosal Injections*

The first prospective randomized post-RP rehabilitation study involved penile injection therapy. Montorsi et al. demonstrated a benefit to using intracavernosal injections (ICI) using alprostadil (PGE1) monotherapy on the return of "spontaneous erections satisfactory for sexual intercourse."[59] The outcomes were patient-reported successful intercourse, nocturnal penile tumescence (NPT), and penile duplex Doppler studies. No preoperative NPT studies or Doppler studies were performed. In the treatment group, 8/12 (67%) completers recovered erections versus 3/15 (20%) in the observation group. This study provided the theory that early intervention after nerve-sparing prostate surgery may positively benefit EF. However, the study had a small overall population and was not repeated in a larger group.

In another study, Mulhall et al. allowed 132 patients to choose postoperative penile rehabilitation versus nil in a nonrandomized fashion.[60] Patients were given an initial challenge with sildenafil (on four occasions at 100 mg) and if an erection of "≥60% rigidity" (penetration hardness) was obtained, patients were instructed to obtain three erections per week using sildenafil. If sildenafil failed to achieve penetration hardness, patients were then encouraged to consider intracavernosal injection therapy. Again, patients were instructed to obtain three injection-induced erections per week. Sildenafil failures were instructed to rechallenge themselves every 4 months after surgery to determine if they had become responders. If patients progressed to responders, they were allowed to switch from injection therapy to sildenafil therapy yet still maintaining the protocol for three erections per week. Patients were followed at 4-month intervals until they were 18 months postoperative. Those in the rehabilitation group averaged 1.9 erections/week. The patients opting for rehabilitation had significant improvements in natural response (52% vs. 19%), sildenafil response (64% vs. 24%), and ICI response (95% vs. 76%). Although the results were positive, drawbacks include the nonrandomized nature and the possibility of investigation or participant biases.

### *PDE5 Inhibitors*

Sildenafil at night has been shown to enhance NPT.[61] This is significant because of the concept that nocturnal erections improve penile oxygenation and therefore minimize the hypoxic damage to the cavernous tissue after RP.

There are two large prospective, randomized, multicentered controlled studies after bilateral nerve-sparing radical prostatectomy (BNSRP), one assessing sildenafil and the other vardenafil. The sildenafil trial was designed shortly after the drug's approval and it consisted of a three-arm parallel design (placebo, sildenafil 50 mg, sildenafil 100 mg) with treatment starting 1 month after surgery and lasting for 9 months.[62] Enrollment stopped early, when a blinded review revealed only a 25% response rate, which was much lower than that reported in contemporary surgical series from centers of excellence. The two US centers carrying out NPT testing remained open and were able to conduct a subanalysis. At 48 weeks, 27% (14/51) of men on sildenafil versus 4% (1/25) in the placebo arm were considered responders ($p < 0.01$). Mean IIEF-EF domain scores in responders versus nonresponders off sildenafil were 26.8 versus 11.5 (sildenafil 100 mg), 26.3 versus 8 (sildenafil 50 mg), 23 versus 7.6 (placebo).

More recently, the vardenafil (REINVENT) trial was published, which also looked at EF recovery after BNSRP.[63] The study was designed to examine the differences in dosing intervals: nightly (N), on demand (OD), or placebo (P). Intervention was for 9 months commencing 14 days after RP. Like the sildenafil study, the primary efficacy variable was not met. There was no difference between the P group and the vardenafil (N or OD) groups at the end of the single-blind placebo washout phase. The proportions of patients with IIEF scores ≥22 were 28.9, 24.1, and 29.1% for the P, N, and OD, respectively.

### *Intraurethral Alprostadil Suppository*

In 1998, the intraurethral alprostadil suppository (IUA) was approved for the management of ED by the Food and Drug Administration (FDA). Several studies have shown that IUA is effective in treating ED after RP regardless of nerve-sparing status.[64,65] Raina et al., performed a nonrandomized, prospective observational study, examining three times weekly IUA versus no treatment for 9 months after RP.[65] They showed a benefit of IUA over no treatment in "successful vaginal intercourse with or without erectile aids" (40% vs. 11%). McCullough et al. conducted a randomized trial, aimed to determine if early post-BNSRP intervention with IUA nightly accelerated the return of EF.[66] The protocol consisted of two arms comparing nightly IUA administration to nightly sildenafil 50 mg. At the 3 and 6 month visits there was a slight insignificant difference in IIEF EF domain scores in favor of IUA.

Despite the above data, studies were small and not powerful enough to define the role of this strategy in rehabilitation. Furthermore, intraurethral PGE1 is notorious for inducing penile ache/pain. This is a major limitation in its clinical application. Fulgham et al. looked at in-office efficacy and evaluated pain scores with IUA use.[67] At 15 min after insertion, 16.8% rated the pain as very uncomfortable and 41.3% rated it as somewhat uncomfortable; however, these scores decreased to 9 and 25.4% at 60 min, respectively.

### *Vacuum Device Therapy*

The vacuum erection device (VED) has a long history of being effective in the treatment of ED; however, its efficacy in penile rehabilitation is unproven. Bosshardt et al. demonstrated that the use of the VED with a proximal constriction band results in only mixed venous oxygenation during use, with mean $pO_2$ levels of 34.1 mmHg, 30 min after applying the constriction band.[68] In a randomized prospective study by Raina et al., including nerve-sparing and non-nerve-sparing surgeries, 74 patients were instructed to apply the VED daily for 9 months compared to 35 men with no treatment.[69] The results showed no significant difference in spontaneous erections, 19/60 (32%) of the VED group and 13/35 (37%) in the no treatment.

Kohler et al. reported a randomized study of early intervention with VED compared to no treatment after RP.[70] Thirty-three men undergoing RP were randomized to early intervention (6 months of treatment) or no early treatment control group. The results showed there was no significant difference between the groups in SHIM score or in the percentage of men with moderate to severe ED, and neither group achieved spontaneous erections adequate for penetration. Interestingly, the VED group did actually gain penile length while the no-treatment group lost length.

## Rehabilitation in RP Patients

In a longitudinal penile biopsy study after RP, significant histologic changes were found in the cavernosal tissue of men as early as 2 months after surgery.[71] Trabecular elastic and SM fibers were decreased, and there was a significant increase in the deposition of collagen. In a survey conducted by Tal et al., among 618 physicians, approximately 86% performed some form of penile rehabilitation, and 87% of respondents used a PDE5i as their first choice regimen.[72] The second most preferred treatment was VED, followed by ICI and IUA. Slightly more than half of the respondents (65%) started penile rehabilitation immediately or at the time of catheter removal and 97% initiated their rehabilitative strategy within the first 4 months after RP. Although penile rehabilitation is widely accepted, a consensus or optimal strategy has yet to be determined. Figure 62.1 illustrates the penile rehabilitation algorithm

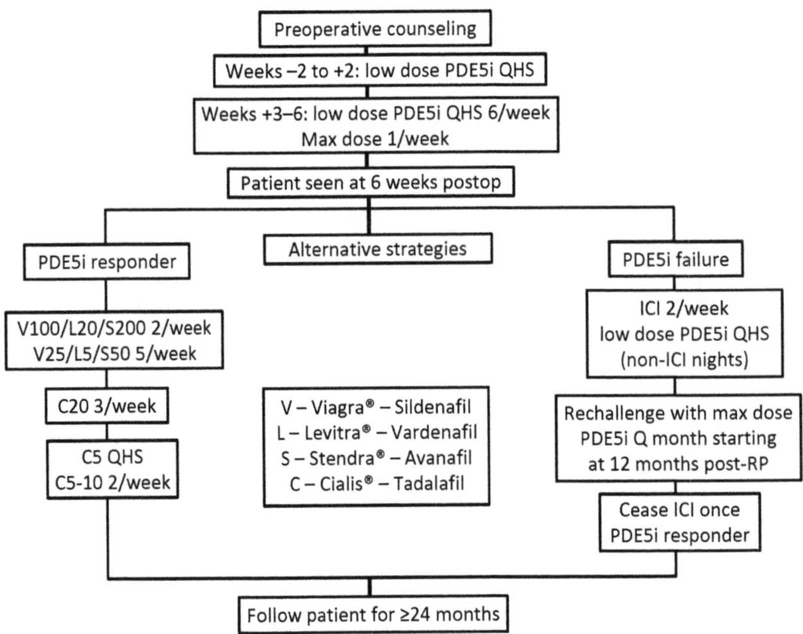


FIGURE 62.1 **MSKCC Penile Rehabilitation Algorithm.**

we currently use at Memorial Sloan-Kettering Cancer Center (MSKCC).

## Rehabilitation in RT Patients

There have been several recent studies examining rehabilitation after radiotherapy. Pisansky et al. performed a placebo-controlled, double-blinded, parallel group study to examine the impact of daily tadalafil for preserving spontaneous erectile function in patients treated with radiotherapy.[25] Patients were treated for 24 weeks and evaluated with the IIEF questionnaire to determine retention of erectile function. There were no significant differences identified at 30 and 52 weeks. Zelefsky et al. performed a randomized, double blind, placebo-controlled trial, to examine if sildenafil citrate would preserve sexual function 24 months after the start of radiotherapy.[73] They found that patients receiving radiotherapy without ADT treated with daily sildenafil had statistically significant improvements in various erectile function domains at 6, 12, and 24 months. At MSKCC, we use a penile rehabilitation algorithm similar to that for RP patients (Figure 62.1); however, the low- dose PDE5i is continued for 24 months.

## Rehabilitation in ADT Patients

There have been no human clinical studies examining the use of a penile rehabilitation program for patients who have had androgen deprivation treatment. However, we use a penile rehabilitation algorithm similar to that for RP patients (Figure 62.1) and the low-dose PDE5i is continued for 24 months.

## Future Strategies

### *Neuromodulatory Agents*

Although there have been many advances in the understanding of the basic science behind nerve injury associated with surgery and responses on a cellular level, the areas of neuroprotection and nerve regeneration are still early in their development. The theory of neuromodulatory therapy post-RP comprises the basic science of neurotrophic growth factors, neuroprotection, nerve regeneration and establishment of signaling continuity, neuronal stem cells, and the prevention of neuronal cell death regarding the nerve supply of the penis.[74]

A recent clinical trial (phase II, multicenter, randomized, double-blind, placebo-controlled, three-arm, 12-month study) conducted in men undergoing nerve-sparing RP and treated with GPI-1485, a neuroimmunophilin ligand aimed at protecting nerves from damage and promoting nerve regrowth, did not reveal significant efficacy.[75] The Johns Hopkins Group has presented data on the potential clinical use of erythropoietin as a neuromodulatory agent to preserve erectile function in men undergoing RP.[76]

There are multiple studies underway examining the utility of stem cells in the treatment of ED. Some advances have been made in preclinical studies using *in vitro* and *in vivo* multipotent stromal cells. Bone marrow-derived mesenchymal stromal cells and adipose-derived stromal cells are the two stem cell sources most likely to make their way to clinical practice in the treatment of ED.[77] More work is needed to assess side effects, define optimal treatment parameters, and human studies.

Other areas of research include the use of gene therapy to upregulate specific growth factors, and tissue engineering of various tissues (tunica albuginea, corpus cavernosum, and nerve conduits).[78]

## ANEJACULATION

The ejaculatory mechanism is an integral part of the prostate and is lost with its removal. Surgeons may assume patients do not care about their ability to ejaculate and may not discuss that anejaculation is an expected consequence of surgery. There is little medical literature focused on the true psychosocial consequences related to anejaculation. Anejaculation may lead to several new issues; patients may be "surprised" by the unexpected failure to ejaculate, it may interfere with their self-image, it may lead to reduced orgasmic satisfaction, and it makes them infertile. With the advent of early prostate cancer detection, men are diagnosed younger and have great possibilities for long-term survival.[79] In addition, a subset of men may still be interested in parenting postsurgery and should be advised about anejaculation and to cryopreserve semen prior to surgery.[80,81]

## ORGASM CHANGES

### Physiology of Orgasm

Despite many advances in the understanding of erectile function, the orgasmic process remains poorly understood. The orgasmic process is a combination of physiological and psychogenic elements. Ejaculation includes SM contraction of the accessory sex organs, buildup and release of pressure in the posterior urethra, sensation of ejaculatory inevitability, contraction of the urethral bulb and perineum, rhythmic contractions of the pelvic floor muscles, and semen emission. Also, ejaculation contributes to the sensation of orgasm as well as the reversal of generalized physiological changes and release of sexual tension.[82] Normally, orgasm is tied with ejaculation; however, after RP orgasm occurs without semen emission, ejaculatory muscular activity persists.

### Change in Orgasm Nature/Intensity and Orgasmic pain (Dysorgasmia)

Orgasmic changes receive less attention than common topic of erectile function; however, it is not less important. These changes can be categorized into three areas: (1) changes in orgasm intensity, (2) orgasmic pain (dysorgasmia), and (3) orgasm-associated incontinence. Schover et al., in a study of 1236 men treated for localized prostate cancer (52% RP, 48% RT), found that 65% of the sample reported a problem with their orgasms including 31% who no longer tried to reach orgasm, 17% who tried but were unable to reach orgasm, and 28% with orgasms that were disappointingly weak.[83] Dubbelman et al. examined the incidence and possible predictors for delayed orgasm.[84] In their study, at 2 years post-RP, 67% of men were able to achieve orgasm, and predictors included non nerve-sparing surgery, age older ≥60 years and urinary incontinence (necessitating more than two pads a day or urinary sphincter prosthesis). Barnas et al. used a proprietary questionnaire to further characterize the types of orgasmic changes in men after RP. Within the group, 22% of patients reported no change in orgasm intensity, 37% reported a complete absence of orgasm, 37% had decreased orgasm intensity, and 4% reported a more intense orgasm. Dysorgasmia occurred in 14% of the patients, most commonly located in the penis (63%), followed by the rectum (24%). Pain was reported to occur always (with every orgasm) in 33%, frequently in 13%, occasionally in 35%, and rarely in 19%. Most patients (55%) had orgasm-associated pain for less than a minute.[85]

Alterations in orgasmic intensity or anorgasmia may benefit from psychosexual counseling. Dysorgasmia may be uncommon but it can be a distressing problem for patients to endure. There is no consensus regarding the cause of pain but some believe bladder neck/pelvic floor spasms may play a role. Alpha-blockers (e.g., tamsulosin) may help reduce the intensity of pain.[86] However, the key to proper treatment is patient and partner education before surgery and supportive care afterward.

## SEXUAL INCONTINENCE

### Climacturia

Urinary incontinence during orgasm (climacturia) (Table 62.2) is another phenomenon that may adversely affect patient and partner sexual satisfaction following RP. Choi et al. reported that climacturia occurs in 20% following radical pelvic surgery, more often after RP than with radical cystectomy and it is unrelated to the type of prostatectomy performed (open or laparoscopic). It was more likely to be reported within the first year and associated with men who complain of orgasmic pain and/or penile shortening.[87] Lee et al. reported a higher prevalence of climacturia after RP of 45%.[88] In 68% of the cases reported it happened rarely or only occasionally, while in 21% it occurred most of the time or always. Urine leakage quantity was only a few drops in 58% of the subjects but 16% reported a loss of more than 1 ounce and bother was none or minimal in 52% and significant

TABLE 62.2 Climacturia

| Author | Year | Op | Patient number | Climacturia (%) | Outcomes |
|---|---|---|---|---|---|
| Manassero et al.[90] | 2013 | RP | 84 | 29 | Functional urethral length < control |
| Choi et al.[87] | 2007 | RP | 475 | 20 | No predictors ORP > LRP |
| Lee et al.[88] | 2006 | RP | 42 | 45 | No predictors 50% men bothered 25% partners bothered |
| Abouassaly et al.[91] | 2006 | RP | 26 | N/A | No predictors Large volume variability |
| Koeman et al.[92] | 1996 | RP | 17 | 64 | |

ORP, open radical prostatectomy; LRP, laparoscopic radical prostatectomy.

in 48%. Mitchell et al. reported that patients state they have a major bother from sexual incontinence 22.4 and 12.1%, at 3 and 24 months, respectively.[89] Additionally, bother from sexual incontinence and stress urinary incontinence were strongly associated at all times.

Manassero et al. investigated patients with climacturia using video-urodynamics after bladder neck sparing RP.[90] They found that functional length of urethra was significantly lower in the climacturia group compared to controls and postulated that this could more specifically be related to the membranous urethral length. Climacturia may be managed behaviorally (fluid intake restriction and bladder emptying prior to sexual activity) or mechanically (using a rubber constriction ring or condoms, if the leakage amount is small). The use of a constriction ring (penile variable tension loop) has been shown to significantly reduce the degree and frequency of orgasm-associated incontinence, and the associated distress experienced by patients and partners.[93]

### Arousal Incontinence

Another disorder not fully characterized is arousal incontinence, or incontinence with erection. Anecdotally, this is more common in the first few months after surgery and improves with continence.

## CHANGES IN SEXUAL DESIRE

Desire is dependent on stressors in the patient's life. The diagnosis of cancer and recovering from surgery can impact the patient's quality of life, sexual desire, and sexual function.[94] Reasons for a decline in the libido can be multifactorial including medication related (e.g., ADT), hormonal deficiency (e.g., hypogonadism), or psychological (e.g., PSA anxiety, fear of harming themselves during their recovery, or sexual/stress incontinence). These stressors may be manifested as avoidant behavior and not wanting to initiate sexual activity with their partner.

TABLE 62.3 Penile Length Loss

| Author | Year | Patient number | Time post-RP (months) | Main outcomes |
|---|---|---|---|---|
| Berookhim et al.[95] | 2013 | 65 | 6 | 41% no length loss; 11% ≥10 mm length loss;<br>Predictor of no length loss = men using daily PDE5i |
| Engel et al.[96] | 2010 | 127 | 11 | SFPL loss: 1 month – 0.64 cm; 3 months – 0.32 cm; 6 months – 0.26 cm<br>No significant SFPL loss at 9–11 months |
| Gontero et al.[97] | 2007 | 126 | 12 | 1.34 cm shortening – flaccid length<br>2.30 cm shortening – stretched length |
| Briganti et al.[98] | 2007 | 33 | 6 | No significant SFPL changes flaccid and erect states |
| Savoie et al.[99] | 2003 | 63 | 3 | 1.1 cm SFPL shortening<br>19% had ≥15% SFPL shortening |
| Munding et al.[100] | 2001 | 31 | 3 | 13% increased SFPL<br>16% no change<br>71% decreased:23% ≤0.5 cm; 35% 1–2 cm; 13% >2 cm |
| Fraiman et al.[101] | 1999 | 100 | 6 | 9% reduction in erect length<br>22% change in volume<br>Maximum changes seen 4–8 months post-RP |

SFPL, stretched flaccid penile length.

## PENILE LENGTH LOSS

Penile shortening, even if it does not preclude sexual intercourse, may have adverse effects on the patient's body image, self-esteem, sense of manhood, and self-confidence (Table 62.3). In a prospective study of men after RP, Munding et al. found a measured decrease in penile length at 3 months postoperatively in 22/31 (71%) and a loss of greater than 1 cm in 15/31 (48%) of the cases.[100] Similar findings were obtained in a study by Savoie et al. with a decrease in the stretched penile length in 43/63 (68%) of the cases and a greater than 15% decrease was found in 12/63 (19%).[99] Penile length loss has been shown to be independently associated with nerve preservation status and with postoperative erectile function outcome.[97] Engel et al. recently reported on a subset analysis of a penile rehabilitation study population (IUA vs. sildenafil) illustrating the time course of penile length changes over 11 months of follow-up after robotic-assisted BNSRP.[96] They showed that penile length was initially shorter at 1 month postsurgery, but gradually returned to the approximate presurgical length by 9 months, and penile length was not significantly different compared to baseline length. In a study by Dalkin and Christopher, they examined the ability of the VED to preserve penile length post-RP with early utilization, beginning treatment immediately after catheter removal and followed for 90 days. In patients with >50% compliance rates, 35/36 (97%) preserved their stretched penile length (SPL).[102]

Penile volume changes can be explained by the aforementioned penile structural physiological and functional changes. Fibrotic changes have been shown as early as 2 months postoperatively characterized by significantly decreased elastic and SM fiber content and increased collagen content.[71] One theory is that the prolonged absence of erectile activity leads to a state of cavernosal hypoxia. The low $pO_2$ tension favors the secretion of fibrogenic cytokines (TGF-$\beta_1$) that would normally be inhibited during erection, by the secretion of endogenous prostanoids ($PGE_1$).[103]

Penile length and girth changes can be divided into early and delayed. Early changes occur in response to the neural injury during RP; the cavernosal nerves undergo Wallerian degeneration, and in the early phase, sympathetic nerve activity dominates and the penis becomes a hypertonic organ. In response to the dominant adrenergic control, the penile SM contracts and this results in a penis that patients often refer to as "being drawn back in the body." This hypertonic state is most pronounced within the first 3–6 months after surgery. The delayed structural changes result from true irreversible structural alterations in the corporal SM. There is likely a multifactorial cause for these structural changes including neural injury-associated apoptosis and cavernosal hypoxia-induced collagenization. The difference between these irreversible structural changes and early hypertonicity is shown by the reduced or absent penile stretch in the group with irreversible SM alterations.

## References


1. Dean RC, Lue TF. Physiology of penile erection and pathophysiology of erectile dysfunction. *Urol Clin North Am* 2005;**32**(4):379–95.
2. Lue TF. Erectile dysfunction. *N Engl J Med* 2000;**342**(24):1802–13.
3. Masterson TA, Serio AM, Mulhall JP, et al. Modified technique for neurovascular bundle preservation during radical prostatectomy: association between technique and recovery of erectile function. *BJU Int* 2008;**101**(10):1217–22.
4. Catalona WJ, Carvalhal GF, Mager DE, et al. Potency, continence and complication rates in 1,870 consecutive radical retropubic prostatectomies. *J Urol* 1999;**162**(2):433–8.
5. Walsh PC, Partin AW, Epstein JI. Cancer control and quality-of-life following anatomical radical retropubic prostatectomy – results at 10 years. *J Urol* 1994;**152**(5):1831–6.
6. Zippe CD, Kedia AW, Kedia K, et al. Treatment of erectile dysfunction after radical prostatectomy with sildenafil citrate (Viagra). *Urology* 1998;**52**(6):963–6.
7. Katz D, Bennett NE, Stasi J, et al. Chronology of erectile function in patients with early functional erections following radical prostatectomy. *J Sex Med* 2010;**7**(2 Pt 1):803–9.
8. Klein LT, Miller MI, Buttyan R, et al. Apoptosis in the rat penis after penile denervation. *J Urol* 1997;**158**(2):626–30.
9. Leungwattanakij S, Bivalacqua TJ, Usta MF, et al. Cavernous neurotomy causes hypoxia and fibrosis in rat corpus cavernosum. *J Androl* 2003;**24**(2):239–45.
10. Mullerad M, Donohue JF, Li PS, et al. Functional sequelae of cavernous nerve injury in the rat: is there model dependency. *J Sex Med* 2006;**3**(1):77–83.
11. User HM, Hairston JH, Zelner DJ, et al. Penile weight and cell subtype specific changes in a post-radical prostatectomy model of erectile dysfunction. *J Urol* 2003;**169**(3):1175–9.
12. User HM, Zelner DJ, McKenna KE, et al. Microarray analysis and description of SMR1 gene in rat penis in a post-radical prostatectomy model of erectile dysfunction. *J Urol* 2003;**170**(1):298–301.
13. Rogers CG, Trock BP, Walsh PC. Preservation of accessory pudendal arteries during radical retropubic prostatectomy: surgical technique and results. *Urology* 2004;**64**(1):148–51.
14. Breza J, Aboseif SR, Orvis BR, et al. Detailed anatomy of penile neurovascular structures: surgical significance. *J Urol* 1989;**141**(2): 437–43.
15. Moreland RB, Traish A, McMillin MA, et al. PGE1 suppresses the induction of collagen synthesis by transforming growth factor-beta 1 in human corpus cavernosum smooth muscle. *J Urol* 1995;**153**(3 Pt 1):826–34.
16. Ferrini MG, Davila HH, Kovanecz I, et al. Vardenafil prevents fibrosis and loss of corporal smooth muscle that occurs after bilateral cavernosal nerve resection in the rat. *Urology* 2006;**68**(2): 429–35.
17. Ferrini MG, Kovanecz I, Sanchez S, et al. Fibrosis and loss of smooth muscle in the corpora cavernosa precede corporal veno-occlusive dysfunction (CVOD) induced by experimental cavernosal nerve damage in the rat. *J Sex Med* 2009;**6**(2):415–28.
18. Kovanecz I, Rambhatla A, Ferrini M, et al. Long-term continuous sildenafil treatment ameliorates corporal veno-occlusive dysfunction (CVOD) induced by cavernosal nerve resection in rats. *Int J Impot Res* 2008;**20**(2):202–12.
19. Kovanecz I, Rambhatla A, Ferrini MG, et al. Chronic daily tadalafil prevents the corporal fibrosis and veno-occlusive dysfunction that occurs after cavernosal nerve resection. *BJU Int* 2008;**101**(2):203–10.

20. Mulhall JP, Muller A, Donohue JF, et al. The functional and structural consequences of cavernous nerve injury are ameliorated by sildenafil citrate. *J Sex Med* 2008;**5**(5):1126–36.
21. Mulhall JP, Slovick R, Hotaling J, et al. Erectile dysfunction after radical prostatectomy: hemodynamic profiles and their correlation with the recovery of erectile function. *J Urol* 2002;**167**(3):1371–5.
22. Obayomi-Davies O, Chen LN, Bhagat A, et al. Potency preservation following stereotactic body radiation therapy for prostate cancer. *Radiat Oncol* 2013;**8**:256.
23. Zelefsky MJ, Eid JF. Elucidating the etiology of erectile dysfunction after definitive therapy for prostatic cancer. *Int J Radiat Oncol Biol Phys* 1998;**40**(1):129–33.
24. Fajardo LF. The pathology of ionizing radiation as defined by morphologic patterns. *Acta Oncol* 2005;**44**(1):13–22.
25. Pisansky TM, Pugh SL, Greenberg RE, et al. Tadalafil for prevention of erectile dysfunction after radiotherapy for prostate cancer: the radiation therapy oncology group [0831] randomized clinical trial. *JAMA* 2014;**311**(13):1300–7.
26. Corona G, Gacci M, Baldi E, et al. Androgen deprivation therapy in prostate cancer: focusing on sexual side effects. *J Sex Med* 2012;**9**(3):887–902.
27. Corona G, Maggi M. The role of testosterone in erectile dysfunction. *Nat Rev Urol* 2010;**7**(1):46–56.
28. Suzuki N, Sato Y, Hisasue S, et al. Effect of testosterone on intracavernous pressure elicited with electrical stimulation of the medial preoptic area and cavernous nerve in male rats. *J Androl* 2007;**28**(2):218–22.
29. Tal R, Mueller A, Mulhall JP. The correlation between intracavernosal pressure and cavernosal blood oxygenation. *J Sex Med* 2009;**6**(10):2722–7.
30. Moreland RB, Gupta S, Goldstein I, et al. Cyclic AMP modulates TGF-beta 1-induced fibrillar collagen synthesis in cultured human corpus cavernosum smooth muscle cells. *Int J Impot Res* 1998;**10**(3):159–63.
31. Moreland RB, Albadawi H, Bratton C, et al. O2-dependent prostanoid synthesis activates functional PGE receptors on corpus cavernosum smooth muscle. *Am J Physiol* 2001;**281**(2): H552–558.
32. Nehra A, Goldstein I, Pabby A, et al. Mechanisms of venous leakage: a prospective clinicopathological correlation of corporeal function and structure. *J Urol* 1996;**156**(4):1320–9.
33. Jaffe JS, Antell MR, Greenstein M, et al. Use of intraurethral alprostadil in patients not responding to sildenafil citrate. *Urology* 2004;**63**(5):951–4.
34. Kundu SD, Roehl KA, Eggener SE, et al. Potency, continence and complications in 3,477 consecutive radical retropubic prostatectomies. *J Urol* 2004;**172**(6 Pt 1):2227–31.
35. Mulhall JP. Defining and reporting erectile function outcomes after radical prostatectomy: challenges and misconceptions. *J Urol* 2009;**181**(2):462–71.
36. Penson DF, McLerran D, Feng Z, et al. 5-Year urinary and sexual outcomes after radical prostatectomy: results from the prostate cancer outcomes study. *J Urol* 2005;**173**(5):1701–5.
37. Rozet F, Galiano M, Cathelineau X, et al. Extraperitoneal laparoscopic radical prostatectomy: a prospective evaluation of 600 cases. *J Urol* 2005;**174**(3):908–11.
38. Stanford JL, Feng Z, Hamilton AS, et al. Urinary and sexual function after radical prostatectomy for clinically localized prostate cancer: the Prostate Cancer Outcomes Study. *JAMA* 2000;**283**(3):354–60.
39. Chew KK, Bremner A, Stuckey B, et al. Sex life after 65: how does erectile dysfunction affect ageing and elderly men? *Aging Male* 2009;**12**(2-3):41–6.
40. Laumann EO, Glasser DB, Neves RC, et al. A population-based survey of sexual activity, sexual problems and associated help-seeking behavior patterns in mature adults in the United States of America. *Int J Impot Res* 2009;**21**(3):171–8.
41. Zemishlany Z, Weizman A. The impact of mental illness on sexual dysfunction. *Adv Psychosom Med* 2008;**29**:89–106.
42. Molton IR, Siegel SD, Penedo FJ, et al. Promoting recovery of sexual functioning after radical prostatectomy with group-based stress management: the role of interpersonal sensitivity. *J Psychosom Res* 2008;**64**(5):527–36.
43. Teloken P, Valenzuela R, Parker M, et al. The correlation between erectile function and patient satisfaction. *J Sex Med* 2007;**4**(2): 472–6.
44. Rabbani F, Schiff J, Piecuch M, et al. Time course of recovery of erectile function after radical retropubic prostatectomy: does anyone recover after 2 years? *J Sex Med* 2010;**7**(12):3984–90.
45. Sivarajan G, Prabhu V, Taksler GB, et al. Ten-year outcomes of sexual function after radical prostatectomy: results of a prospective longitudinal study. *Eur Urol* 2014;**65**(1):58–65.
46. van der Wielen GJ, Mulhall JP, Incrocci L. Erectile dysfunction after radiotherapy for prostate cancer and radiation dose to the penile structures: a critical review. *Radiother Oncol* 2007;**84**(2):107–13.
47. Alemozaffar M, Regan MM, Cooperberg MR, et al. Prediction of erectile function following treatment for prostate cancer. *JAMA* 2011;**306**(11):1205–14.
48. Teloken PE, Parker M, Mohideen N, et al. Predictors of response to sildenafil citrate following radiation therapy for prostate cancer. *J Sex Med* 2009;**6**(4):1135–40.
49. Potosky AL, Reeve BB, Clegg LX, et al. Quality of life following localized prostate cancer treated initially with androgen deprivation therapy or no therapy. *J Natl Cancer Inst* 2002;**94**(6):430–7.
50. Mazzola CR, Deveci S, Heck M, et al. Androgen deprivation therapy before radical prostatectomy is associated with poorer postoperative erectile function outcomes. *BJU Int* 2012;**110**(1):112–6.
51. Magee TR, Ferrini M, Garban HJ, et al. Gene therapy of erectile dysfunction in the rat with penile neuronal nitric oxide synthase. *Biol Reprod* 2002;**67**(1):20–8.
52. Valente EG, Vernet D, Ferrini MG, et al. L-arginine and phosphodiesterase (PDE) inhibitors counteract fibrosis in the Peyronie's fibrotic plaque and related fibroblast cultures. *Nitric Oxide* 2003;**9**(4):229–44.
53. Vignozzi L, Filippi S, Morelli A, et al. Effect of chronic tadalafil administration on penile hypoxia induced by cavernous neurotomy in the rat. *J Sex Med* 2006;**3**(3):419–31.
54. Desouza C, Parulkar A, Lumpkin D, et al. Acute and prolonged effects of sildenafil on brachial artery flow-mediated dilatation in type 2 diabetes. *Diabetes Care* 2002;**25**(8):1336–9.
55. Rosano GM, Aversa A, Vitale C, et al. Chronic treatment with tadalafil improves endothelial function in men with increased cardiovascular risk. *Eur Urol* 2005;**47**(2):214–20.
56. Schwartz EJ, Wong P, Graydon RJ. Sildenafil preserves intracorporeal smooth muscle after radical retropubic prostatectomy. *J Urol* 2004;**171**(2 Pt 1):771–4.
57. Bannowsky A, Schulze H, van der Horst C, et al. Recovery of erectile function after nerve-sparing radical prostatectomy: improvement with nightly low-dose sildenafil. *BJU Int* 2008;**101**(10):1279–83.
58. Pavlovich CP, Levinson AW, Su LM, et al. Nightly vs on-demand sildenafil for penile rehabilitation after minimally invasive nerve-sparing radical prostatectomy: results of a randomized double-blind trial with placebo. *BJU Int* 2013;**112**(6):844–51.
59. Montorsi F, Guazzoni G, Strambi LF, et al. Recovery of spontaneous erectile function after nerve-sparing radical retropubic prostatectomy with and without early intracavernous injections of alprostadil: results of a prospective, randomized trial. *J Urol* 1997;**158**(4):1408–10.
60. Mulhall J, Land S, Parker M, et al. The use of an erectogenic pharmacotherapy regimen following radical prostatectomy improves recovery of spontaneous erectile function. *J Sex Med* 2005;**2**(4):532–40.
61. Montorsi F, Maga T, Strambi LF, et al. Sildenafil taken at bedtime significantly increases nocturnal erections: results of a placebo-controlled study. *Urology* 2000;**56**(6):906–11.

62. Padma-Nathan H, McCullough AR, Levine LA, et al. Randomized, double-blind, placebo-controlled study of postoperative nightly sildenafil citrate for the prevention of erectile dysfunction after bilateral nerve-sparing radical prostatectomy. *Int J Impot Res* 2008;**20**(5):479–86.
63. Montorsi F, Brock G, Lee J, et al. Effect of nightly versus on-demand vardenafil on recovery of erectile function in men following bilateral nerve-sparing radical prostatectomy. *Eur Urol* 2008;**54**(4):924–31.
64. Costabile RA, Spevak M, Fishman IJ, et al. Efficacy and safety of transurethral alprostadil in patients with erectile dysfunction following radical prostatectomy. *J Urol* 1998;**160**(4):1325–8.
65. Raina R, Agarwal A, Zaramo CE, et al. Long-term efficacy and compliance of MUSE for erectile dysfunction following radical prostatectomy: SHIM (IIEF-5) analysis. *Int J Impot Res* 2005;**17**(1):86–90.
66. McCullough AR, Hellstrom WG, Wang R, et al. Recovery of erectile function after nerve sparing radical prostatectomy and penile rehabilitation with nightly intraurethral alprostadil versus sildenafil citrate. *J Urol* 2010;**183**(6):2451–6.
67. Fulgham PF, Cochran JS, Denman JL, et al. Disappointing initial results with transurethral alprostadil for erectile dysfunction in a urology practice setting. *J Urol* 1998;**160**(6 Pt 1):2041–6.
68. Bosshardt RJ, Farwerk R, Sikora R, et al. Objective measurement of the effectiveness, therapeutic success and dynamic mechanisms of the vacuum device. *Br J Urol* 1995;**75**(6):786–91.
69. Raina R, Agarwal A, Ausmundson S, et al. Early use of vacuum constriction device following radical prostatectomy facilitates early sexual activity and potentially earlier return of erectile function. *Int J Impot Res* 2006;**18**(1):77–81.
70. Kohler TS, Pedro R, Hendlin K, et al. A pilot study on the early use of the vacuum erection device after radical retropubic prostatectomy. *BJU Int* 2007;**100**(4):858–62.
71. Iacono F, Giannella R, Somma P, et al. Histological alterations in cavernous tissue after radical prostatectomy. *J Urol* 2005;**173**(5):1673–6.
72. Tal R, Teloken P, Mulhall JP. Erectile function rehabilitation after radical prostatectomy: practice patterns among AUA members. *J Sex Med* 2011;**8**(8):2370–6.
73. Zelefsky MJ, Shasha D, Branco RD, et al. Prophylactic sildenafil citrate improves select aspects of sexual function in men treated with radiotherapy for prostate cancer. *J Urol* 2014;**192**(3):868–74.
74. Burnett AL, Lue TF. Neuromodulatory therapy to improve erectile function recovery outcomes after pelvic surgery. *J Urol* 2006;**176**(3): 882–7.
75. Mulhall JP. Exploring the potential role of neuromodulatory drugs in radical prostatectomy patients. *J Androl* 2009;**30**(4):377–83.
76. Burnett AL, Allaf ME, Bivalacqua TJ. Erythropoietin promotes erection recovery after nerve-sparing radical retropubic prostatectomy: a retrospective analysis. *J Sex Med* 2008;**5**(10):2392–8.
77. Albersen M, Kendirci M, Van der Aa F, et al. Multipotent stromal cell therapy for cavernous nerve injury-induced erectile dysfunction. *J Sex Med* 2012;**9**(2):385–403.
78. Hakim L, Van der Aa F, Bivalacqua TJ, et al. Emerging tools for erectile dysfunction: a role for regenerative medicine. *Nat Rev Urol* 2012;**9**(9):520–36.
79. Ung JO, Richie JP, Chen MH, et al. Evolution of the presentation and pathologic and biochemical outcomes after radical prostatectomy for patients with clinically localized prostate cancer diagnosed during the PSA era. *Urology* 2002;**60**(3):458–63.
80. Schover LR. Motivation for parenthood after cancer: a review. *J Natl Cancer Inst* 2005;(34):2–5.
81. Knoester PA, Leonard M, Wood DP, et al. Fertility issues for men with newly diagnosed prostate cancer. *Urology* 2007;**69**(1):123–5.
82. Kandeel FR, Koussa VK, Swerdloff RS. Male sexual function and its disorders: physiology, pathophysiology, clinical investigation, and treatment. *Endocr Rev* 2001;**22**(3):342–88.
83. Schover LR, Fouladi RT, Warneke CL, et al. Defining sexual outcomes after treatment for localized prostate carcinoma. *Cancer* 2002;**95**(8):1773–85.
84. Dubbelman Y, Wildhagen M, Schroder F, et al. Orgasmic dysfunction after open radical prostatectomy: clinical correlates and prognostic factors. *J Sex Med* 2010;**7**(3):1216–23.
85. Barnas JL, Pierpaoli S, Ladd P, et al. The prevalence and nature of orgasmic dysfunction after radical prostatectomy. *BJU Int* 2004;**94**(4):603–5.
86. Barnas J, Parker M, Guhring P, et al. The utility of tamsulosin in the management of orgasm-associated pain: a pilot analysis. *Eur Urol* 2005;**47**(3):361–5.
87. Choi JM, Nelson CJ, Stasi J, et al. Orgasm associated incontinence (climacturia) following radical pelvic surgery: rates of occurrence and predictors. *J Urol* 2007;**177**(6):2223–6.
88. Lee J, Hersey K, Lee CT, et al. Climacturia following radical prostatectomy: prevalence and risk factors. *J Urol* 2006;**176**(6 Pt 1): 2562–5.
89. Mitchell SA, Jain RK, Laze J, et al. Post-prostatectomy incontinence during sexual activity: a single center prevalence study. *J Urol* 2011;**186**(3):982–5.
90. Manassero F, Di Paola G, Paperini D, et al. Orgasm-associated incontinence (climacturia) after bladder neck-sparing radical prostatectomy: clinical and video-urodynamic evaluation. *J Sex Med* 2012;**9**(8):2150–6.
91. Abouassaly R, Lane BR, Lakin, et al. Ejaculatory urine incontinence after RP. *Urology* 2006;**68**:1248–52.
92. Koeman M, van Driel MF, Schultz WC, et al. Orgasm after RP. *Br J Urol* 1996;**77**:861–4.
93. Mehta A, Deveci S, Mulhall JP. Efficacy of a penile variable tension loop for improving climacturia after radical prostatectomy. *BJU Int* 2013;**111**(3):500–4.
94. Tsivian M, Mayes JM, Krupski TL, et al. Altered male physiologic function after surgery for prostate cancer: couple perspective. *Int Braz J Urol* 2009;**35**(6):673–82.
95. Berookhim BM, Nelson CJ, Kunzel B, et al. Prospective analysis of penile length changes after RP. *BJU Int* 2014;**113**:E131–136.
96. Engel JD, Sutherland DE, Williams SB, et al. Changes in penile length after robot-assisted laparoscopic radical prostatectomy. *J Endourol* 2011;**25**(1):65–9.
97. Gontero P, Galzerano M, Bartoletti R, et al. New insights into the pathogenesis of penile shortening after radical prostatectomy and the role of postoperative sexual function. *J Urol* 2007;**178**(2): 602–7.
98. Briganti A, Fabbri F, Salonia A, et al. Preserved postoperative penile size correlates well with maintained erectile function after bilateral nerve-sparing radical retropubic prostatectomy. *Eur Urol* 2007;**52**:702–7.
99. Savoie M, Kim SS, Soloway MS. A prospective study measuring penile length in men treated with radical prostatectomy for prostate cancer. *J Urol* 2003;**169**(4):1462–4.
100. Munding MD, Wessells HB, Dalkin BL. Pilot study of changes in stretched penile length 3 months after radical retropubic prostatectomy. *Urology* 2001;**58**(4):567–9.
101. Fraiman MC, Lepor H, McCullough AR. Changes in Penile Morphometrics in Men with Erectile Dysfunction after Nerve-Sparing Radical Retropubic Prostatectomy. *Mol Urol* 1999;**3**: 109–15.
102. Dalkin BL, Christopher BA. Preservation of penile length after radical prostatectomy: early intervention with a vacuum erection device. *Int J Impot Res* 2007;**19**(5):501–4.
103. Moreland RB. Is there a role of hypoxemia in penile fibrosis: a viewpoint presented to the Society for the Study of Impotence. *Int J Impot Res* 1998;**10**(2):113–20.

PART XI

# GOVERMENTAL POLICIES

CHAPTER

# 63

# Coding and Billing for Diagnosis and Treatment of Prostate Cancer

*Michael A. Ferragamo, MD, FACS*

**Department of Urology, State University of New York, University Hospital, Stony Brook, NY, USA**

## INTRODUCTION

Today, cancer of the prostate has become an everyday problem in most urological practices. In this chapter we will review the coding and billing of the various services provided by urologists for this clinical problem. The correct use of diagnostic and procedural codes will be presented so that entitled reimbursements may be collected for diagnostic and therapeutic services.

The International Classification of Disease, 9th and 10th revisions, Clinical Modification (ICD-9-CM and ICD-10-CM),[1,2] based on the official version of the World Health Organization's ICD-9 and ICD-10 manual, will be used for documenting the specific diagnostic codes. ICD-10-CM code set will begin October 1, 2015. Before that date one should only use the ICD-9-CM code set.

The 2014 and 2015 CPT – Current Procedural Terminology[3] will be used to document diagnostic and surgical procedures related to cancer of the prostate. The CPT Assistant[4] as well as the Internet Only Manual (IOM) will also be used to help maintain and ensure compliant and accurate coding.

## DIAGNOSES

ICD-9 code 185 and ICD-10 code C61 are the diagnostic codes used for malignant neoplasm of the prostate. Note that ICD-9 code 185 and ICD-10 code C61 contain three characters. 185 is all numeric and C61 is alpha-numeric. Most ICD-9 codes are numeric and all ICD-10 codes are alpha-numeric. Codes 185 and C61 include the various histological malignant tumors of the prostate namely, the common adenocarcinoma, transitional cell tumors, squamous cell tumors, and prostatic sarcoma. Other diagnoses often associated with neoplasms of the prostate include carcinoma *in situ* of the prostate, high grade prostatic intraepithelial neoplasm, high grade PIN, diagnostic codes 233.4 for ICD-9 and D07.5 for ICD-10, dysplasia of the prostate, PIN grades 1 and 2, diagnostic codes 602.3 for ICD-9 and N42.3 for ICD-10 (Table 63.1). ICD-9 code 790.93 and ICD-10 code R07.2 indicate an elevated prostate-specific antigen (PSA). A PSA level is considered elevated when determined as elevated by the attending urologist not only by the determined level itself. For example, an elevation of PSA above four may or may not represent a true elevation. However, a PSA level of 0.5 several months after a radical prostatectomy would represent an elevation. In either example the diagnosis would be 790.93, elevation of PSA.

Other diagnostic codes associated with carcinoma of the prostate are ICD-9 code 187.8 and ICD-10 code C63.7, malignant neoplasm of seminal vesicles, ICD-9 code 198.82 and ICD-10 code C79.82, secondary neoplasm of the seminal vesicles.

## PROSTATE CANCER SCREENING

In 2000, Medicare established specific coding guidelines and reimbursement fees for an annual prostate cancer screening. Two Healthcare Common Procedure Coding System (HCPCS)[5] codes have been assigned for this screening: (1) G0102 – prostate cancer screening; digital rectal examination and (2) G0103 – prostate cancer screening; prostate-specific antigen test. These HCPCS codes are payable for yearly screening (performed at least 11 1/2 months apart) in males 50 years of age and older with Medicare insurance and with no historical, clinical, or laboratory evidence of carcinoma of the prostate. The ICD-9 screening diagnoses for carcinoma of the prostate are V76.44 for ICD-9 and Z12.5 for ICD-10.

Prostate Cancer. http://dx.doi.org/10.1016/B978-0-12-800077-9.00063-3

**TABLE 63.1** Cancer of Prostate Gland ICD-9-CM and ICD-10-CM Coding

| ICD-9 | ICD-10 |
|---|---|
| • 185 Malignant neoplasm of prostate | • C61 Malignant neoplasm of prostate |
| • 198.82 Secondary malignant neoplasm of genital organs | • C79.82 Secondary malignant neoplasm of genital organs |
| • 233.4 CIS, prostate | • D07.5 CIS, prostate |
| • 602.3 PIN 1 and 2 | • N43.2 Pin 1 and 2 |
| • 790.93 Elevated PSA | • R97.2 Elevated PSA |

**TABLE 63.2** PSA Medicare Screening for CAP

- Code for CAP screening
  - No history, clinical or laboratory evidence of CAP
  - Annual
  - Have Medicare and be 50 years or older
- CPT
  - G0102 Screening DRE (99211 $18.82)
  - G0103 Screening PSA (84153 $25.70)
- ICD-9
  - V76.44

*Source: CMS, January 1, 2000.*

Reimbursements for G0102 and G0103 will be similar to payment for CPT E/M code 99211 and for G0103 similar to payment for 84153, PSAand total (Table 63.2).

Table 63.3 shows diagnoses supporting medical necessity for PSA determinations.

## OFFICE CONSULTATIVE SERVICES (UROLOGICAL CONSULTATIONS)

Often the urologist is requested by another physician or health care provider, (nonphysician provider, NPP), for his/her opinion (consultation) concerning an elevated PSA (790.93) or an abnormal rectal examination with a nodule (239.5). In 2010, Medicare deleted from their code set all office and hospital consultation codes, namely, 99241–99245 and 99251–99255, and replaced the office consult codes with 99201–99205, new patient office visits and hospital consultation codes with 99221–99223, initial hospital visits. Note that some commercial carriers continue to utilize the previous consultation codes 99241–99255, and one should check with their nonMedicare carriers as to which codes they prefer when billing for a urological consultation. The documentation of a consultation placed in the medical records should include a request from the primary care physician for the consultation or a statement thereof, as well as a consultation letter (a report) to the requesting physician with recommendations concerning the diagnosis and further care.

After an initial examination, a urologist may accept the patient for complete care. This constitutes a transfer of care from the requesting physician to the urologist. Contrary to the opinion of some carriers this initial encounter by the urologist does in fact constitute a consultation and a consultation code should be billed in this clinical scenario.

## DIAGNOSTIC PROCEDURE CODING FOR CARCINOMA OF THE PROSTATE

There are several diagnostic procedures a urologist may use in his/her investigation for carcinoma of the prostate (Table 63.4). Most often they are various biopsy techniques of the prostate gland. For most insurance carriers the transrectal ultrasonic guidance biopsy of the prostate gland is coded using the following CPT codes:

- 55700 biopsy, prostate; needle or punch, single or multiple, any approach,
- 76872 ultrasound, transrectal examination, and
- 76942 ultrasound guidance for needle placemen (e.g., biopsy).

Diagnoses indicating medical necessity for the biopsy include 790.93, elevated PSA, 236.5, prostatic neoplasm of uncertain behavior (used when a repeat biopsy is necessary for a final histological diagnosis), 239.5, nodule of prostate, 233.4, CIS of prostate, and 185, malignant neoplasm of prostate. These diagnoses may be used for all three of the mentioned CPT codes used in the biopsy code set. Occasionally, nonMedicare insurance carriers, such as Horizon, Cigna, and Oxford may deny payment of 76872, bundling (including) it within CPT code 76942. If this occurs, add modifier –59 to 76872 and resubmit the claim with diagnoses for 76872 including 222.2, 600.00, or 600.01, benign diseases of the prostate. Carriers may also limit the reimbursement of 76872 to twice per year per patient.

Other procedures often added to the above biopsy code set include 76376, three dimensional rendering, 76873, ultrasound, transrectal; prostate volume study, 64450, injection, anesthetic agent; other peripheral nerve or branch (used only for nonMedicare insurance carriers). Only a few insurance carriers still reimburse the anesthetic injection when performed by the operating surgeon. Some nonMedicare carriers may also reimburse the anesthetic agent used with HCPCS code S0020, Marcaine up to 30 $cm^3$.

TABLE 63.3 Medicare National Coverage Determination

| Diagnostic PSA, 84135: ICD-9 covered codes | |
|---|---|
| 185 | Malignant neoplasm, prostate |
| 188.5 | Malignant neoplasm of bladder neck |
| 196.5 | Secondary malignant neoplasm, lymph nodes |
| 196.6 | Secondary malignant pelvic lymph nodes |
| 196.8 | Secondary malignant neoplasm, nodes multiple sites |
| 198.5 | Secondary malignant neoplasm, bone/marrow |
| 198.82 | Secondary malignant neoplasm, genital organs |
| 233.4 | Carcinoma *in situ*, prostate |
| 236.5 | Neoplasm of uncertain behavior of prostate |
| 239.5 | Neoplasm of unspecified nature, other GU organs |
| 596 | Bladder neck obstruction |
| 599.6 | Urinary obstruction |
| 599.7 | Hematuria, unspecified |
| 599.71 | Gross hematuria |
| 599.72 | Microscopic hematuria |
| 600 | BPH without obstruction and/or LUTS |
| 600.1 | BPH with obstruction and/or LUTZ |
| 600.11 | Nodular prostate with obstruction and/or LUTZ |
| 600.21 | BPH, localized, with obstruction and/or LUTZ |
| 601.9 | Unspecified prostatitis |
| 602.9 | Unspecified disorder of prostate |
| 788.2 | Retention of urine, unspecified |
| 788.21 | Incomplete bladder emptying |
| 788.3 | Urinary incontinence, unspecified |
| 788.41 | Urinary frequency |
| 788.43 | Nocturia |
| 788.62 | Slowing of urinary stream |
| 788.63 | Urgency of urination |
| 788.64 | Urinary hesitancy |
| 788.65 | Straining on urination |
| 790.93 | Elevated PSA |
| 793.6–793.7 | Nonspecific abnormal result of radiological examination, evidence of malignancy |
| 794.9 | Bone scan evidence of malignancy |
| V10.46 | Personal history of malignant neoplasm, prostate |

For an endoscopic needle biopsy of the prostate gland, code only 55700, but for a transurethral prostatic cup or resection biopsy, use CPT code 52204, cystourethroscopy, with biopsy (s) for single or multiple biopsies. For a transurethral resection (TURP) of the prostate gland, and a transrectal needle biopsy performed at the same encounter, code 52601, TURP, diagnosis 600.01 and 55700-59 diagnosis 790.93 or 239.5.

TABLE 63.4 Malignant Neoplasm of Prostate

**Diagnostic procedures**

- 76942 Transrectal sonographic guided prostate biopsy
- 76872 Transrectal sonogram of the prostate, TRUS
- 55700 Endoscopic needle prostate biopsy
- 557006 Transperineal saturation prostate biopsy
- 52204 Transurethral prostate biopsy, cup or TURP biopsy

The urologist may perform extensive biopsies of the prostate gland, saturation biopsies, 35–60 biopsies, under general anesthesia in a hospital using a transperineal approach with a stereotactic template. Code 55706 is used for this procedure, which includes ultrasonic guidance. Unfortunately, several carriers (Blue Cross/Blue Shield, Aetna, and Cigna) consider this procedure investigational and will not reimburse for the saturation biopsies. In this case only bill the aforementioned biopsy code set. However, do not bill 55706 with 55700 or 76942. Before doing a saturation prostate gland biopsy in a hospital check with the insurance carrier whether the carrier recognizes this CPT code and will reimburse the procedure.

## CODING FOR SURGICAL PROCEDURES FOR CARCINOMA OF THE PROSTATE

As technology has improved, radical prostatectomies are being performed now more often via a laparoscopic/robotic approach. Table 63.5 includes the open procedure codes for radical prostatectomy, perineal, retropubic, and the popular laparoscopic robotic-assisted radical prostatectomy with CPT code 55866. Since the laparoscopic/robotic CPT code 55866 does not include a pelvic lymphadenectomy, the coding for a robotic retropubic radical prostatectomy with a bilateral pelvic node resection should include CPT codes 55866 and 38571, with bilateral total pelvic lymphadenectomy. When only a unilateral

TABLE 63.5 Malignant Neoplasm of Prostate

**Surgical procedures for prostatic carcinoma**

- 55810 Radical perineal prostatectomy
- 55815 With bilateral pelvic lymphadectomy
- 55840 Radical retropubic prostatectomy
- 55845 With bilateral pelvic lymphadectomy (append-52 for unilateral)
- 54250 Orchiectomy, for bilateral orchiectomy, add modifier -50
- 55866 Laparoscopic, radical retropubic prostatectomy including nerve sparing, includes robotic assistance when performed
  - 38571 Laparoscopy; with bilateral total pelvic lymphadenectomy
  - 38572 Extended bilateral total pelvic lymphadenectomy and sampling

**TABLE 63.6** Malignant Neoplasm of Prostate

| **Miscellaneous procedures for prostatic carcinoma** |
|---|
| • 55873 Cryosurgical ablation of prostate (includes ultrasonic guidance and monitoring) |
| • 55875 Placement of needles into prostate for interstitial radioelement application, with or without monitoring |
| • 55876 Placement of interstitial device(s) for radiation therapy guidance, prostate, single or multiple |
| • 52601 Transurethral prostatectomy Dx. 185 |
| • 52630 Repeat transurethral prostatectomy Dx. 185 |

pelvic node resection is performed, add modifier –52 (reduction modifier) to 38571. When a more extensive lymphadenectomy including nodes at and above the bifurcation of the large vessels is performed, code CPT code 38572 for this extended lymphadenectomy.

Other procedures (Table 63.6) for the treatment of carcinoma of the prostate include 55873, cryosurgical ablation of the prostate (includes ultrasonic guidance and monitoring), 55875, transperineal placement of needles or catheters into prostate for interstitial radioelement application with or without cystoscopy, and 55876, placement of interstitial device(s) for radiotherapy guidance (e.g., fiducial markers, dosimetry, prostate (via needle, any approach), single or multiple.

For the urologist, coding for prostate brachytherapy should include the following:

- 55875,
- 76965-26 ultrasonic guidance, and
- 51702 placement of a Foley catheter (if performed),

for nonMedicare carriers (Medicare does not reimburse for the Foley catheter placement).

The aforementioned coding is for a urologist especially when he/she is working in conjunction with a radiotherapist, who places the radioactive seeds after the needles have been passed by the urologist. Cystoscopic examination at the conclusion of the procedure by the urologist is included in CPT code 55875. In addition, cystoscopic removal of misplaced seeds at the time of the procedure is also included and not reimbursed separately by Medicare. However, some private/commercial carriers will pay for cystoscopic seed removal at the time of surgery using CPT codes 52310 for the removal of one misplaced seed and 52315 for the removal of more than one misplaced seed. The radiotherapist should not bill or code the above CPT codes, which should only be billed by the urologist.

In office coding for the endoscopic placement of fiducial markers should include:

- 55786 (Visicoil, Calyso),
- 76942 ultrasonic guidance,
- 77002 fluoroscopic guidance,
- 76872 transrectal ultrasound,
- A4648 supply cost for markers, and
- 64450 peri-prostatic anesthesia (not paid by Medicare; some private/commercial carriers will reimburse this).

For Medicare, enter the name of the supply and the number of units implanted, usually 1–3 units, in box 24G of the 1500 form. However, for private carriers only list 1 unit no matter how many units were used. In box 19 of the1500 form or its EMR electronic equivalent add "A4648 =3 tissue Markers" and the total acquisition cost. Many carriers, including Medicare will pay a percentage of the acquisition cost after requesting the payment invoice for the markers.

For the TURP of a carcinomatous prostate gland, code 52601 with diagnostic ICD-9, code 185 or C61 for ICD-10. For a repeat TURP use CPT code 52630 with modifier -78 if repeated within the 90-day global of the initial TURP or 52630 without a modifier when performed outside of the global period.

## CODING FOR DRUG MANAGEMENT OF PROSTATIC CARCINOMA

The coding of drug management of carcinoma of the prostate has become confusing and many times cost ineffective. The drugs often ordered by the urologist include Lupron/Eligard, (J9217), Zoladex, (J9209), and Trelstar, (J3315).

For most carriers the injection CPT code for these drugs is 96402, chemotherapy administration, subcutaneous or intramuscular; hormonal antineoplastic, which requires direct physician supervision when administered by a nurse or medical technician (urologist must be in the office suite but not necessarily in the examination room when the drug is being administered). One may also charge for an E/M service, 99212–99215, with modifier –25 added to the visit code. The administration of the drug may occur in the office, place of service (POS) 11, the home, POS 12, assisted living residence, POS 13, and in a group home, POS 14. Diagnoses may be 185, carcinoma of the prostate, 196.2–196.9, malignant lymph nodes, 233.4 carcinoma *in situ*, prostate, 198.5, bone and bone marrow neoplasm, and V10.46, history of carcinoma of the prostate. In box 19 of the 1500 form or an EMR's electronic equivalent space one must add the name of the drug, the amount given, the route of administration, and its National Drug Code, NDC. This information is always required to ensure payment of the drug when purchased by the office. For payment of the injection, 96402, this information must also be documented even when there is no drug cost to the office as when the drug is supplied by the patient or a specialty pharmacy.

## OTHER DRUGS USED IN THE TREATMENT OF PROSTATIC CARCINOMA

Zometa (zoledronic acid) 1 mg has a new HCPCS code J3489. This drug is administered intravenously, bill CPT code 96365, intravenous infusion for therapy, prophylaxis, or diagnosis up to 1 h. The standard dose is 4 mg. Therefore, code with 4 units in box 24G and adjust the fee charged to four times the reported fee for 1 mg. ICD-9 diagnoses should be 198.5, metastatic bone disease and 185. If an E/M visit at the same time is separate and distinct from the injection, bill for the E/M service (99212–99215). An E/M visit at level 99211 is included in the injection code and is not separately billable or payable. To ensure payment from Medicare the physician must document his presence in the office under direct supervision, when an injection is being administered by a nurse or other office personnel. In these cases, the billing for the injection and medication is made in the physician's name and numbers.

The coding for Zometa administration should be:

- 99213-25,
- J3489 (4 units),
- 96365.

Required diagnoses in this sequence are 198.5 and 185.

Histrelin implant (Vantas) 50 mg is billed with HCPCS code J9225. The specific coding for Histrelin 50 mg should be

- 11981 insertion, nonbiodegradable drug delivery implant, or
- 11983 removable with reinsertion, nonbiodegradable drug delivery implant, and
- J9225 Vantas, 50 mg.

Firmagon (Degarelix), 1 mg, is billed with HCPCS code J9155. Report 1 unit for every 1 mg. This drug is administered with a loading dose of 240 mg (two injections of 120 mg), (2 × 3 mL), NDC 55566-8401-01 and then a maintenance dose of 8 mg every 28 days (injection of 4 mL) NDC 55566-8301-01. Diagnoses are ICD-9 185 or ICD-10 C61.

The coding for Firmagon will include:

| | CPT | Box 24G/units |
|---|---|---|
| Starting dose | 96402 | 1 Unit |
| | 96402-59 | 1 Unit |
| | J9155 | 240 Units |
| or | | |
| | 96402 | 2 Units |
| | J9155 | 240 Units |
| Maintenance dose | 96402 | 1 Unit |
| | J9155 | 80 Units |

### Provenge (Sipuleucel-T) Dendreon

Per Medicare, a patient may receive this intravenous treatment once including three separate infusions within 2–3 weeks. The HCPCS code for this treatment is Q2043, minimum of 50 million autologous CD54+ cells activated with PAP-GM-CSF, including leukapheresis and all other preparatory procedures, per infusion billed once for each treatment.

Q2043 includes all preparatory procedures, for example, collection of cells from the patient, and all transportation services. Infusions are billed with CPT code 96413, chemotherapy administration, intravenous infusion technique; up to one hour, single or initial substance/drug and 96415, an add-on code, for each additional hour of infusion. One may also bill J1200, Benadryl HCl (up to 50 mg). Diagnoses must include the primary diagnosis, 185 and at least one secondary metastatic ICD-9 code (e.g., 196.1, 196.2, 196.5, 196.8, 197, 198.1, 198.5, 198.7, or 198.82).

### Denosumab (Xgeva) 1 mg

This drug is supplied as a 120 mg dose given every month by injection (96401, chemotherapy administration, subcutaneous, or intramuscular; non-hormonal antineoplastic). Place 120 units in box 24G (units given) of the 1500 form or in an electronic equivalent space of an EMR. Diagnoses in the following sequence should be 198.5, secondary neoplasm of bone and bone marrow, and 185. Prior authorization from insurance carriers such as Humana, UnitedHeathCare, Cigna, and Aetna may be required before administering this product. The National Drug Code, NDC, is 55513-0730-01 with Medicare reimbursement over $1732 per injection.

### Denosumab (Prolia), HCPCS code J0897, 1 mg

This product is supplied as a 60 mg dose given twice per year by injection (96401). Place 60 units in box 24G of the 1500 form or its electronic equivalent. For accurate payment diagnoses must be submitted in the following sequence: 733.90, disorder of bone and cartilage or 733.01, senile osteoporosis, male, V10.46, history of malignant neoplasm, prostate, V58.69, long-term (current) use of other medications. For Medicaid use diagnoses 733.09, other osteoporosis for example, drug-induced osteoporosis, and 185. For many carriers prior authorization may be required (Humana, UnitedHealthCare, Cigna, Aetna, and Coventry) before using this drug. Its National Drug Code, NDC, is 55513-0730-01 with Medicare reimbursing over $866 per injection.

## CODING FOR PATHOLOGICAL SERVICES

Many urology offices have added a pathologist to their staff. This pathologist, credentialed as part of the professional office staff, will perform the histological preparations (TC component) and interpretations (–26 component) of the prostatic biopsy cores. The office will bill in the pathologist's name and his/her NPI number.

Billing for these pathological services has recently changed. For Medicare, for the preparation and interpretation of any number of prostate needle core biopsies bill, only G0416, surgical pathology, gross and microscopic examinations for prostate needle biopsy, any method, once with one unit in box 24G. For nonMedicare, bill one unit 88305, surgical pathology, gross and microscopic examination, for each prostate needle core biopsy, with the total number of biopsies placed in box 24G. As some private carriers are beginning now to follow Medicare coding rules, check with each nonMedicare carrier as to how they wish these biopsies billed and coded. This testing requires a facility to have either a CLIA certificate of registration (certificate type code 9), a CLIA certificate of compliance (certificate type code 1), or a CLIA certificate of accreditation (certificate type code 3). A facility or office without one of these registered certificates will not be reimbursed for the above tests. If one does both the preparation and the interpretation of the slides, bill the global fees of either G0416 or 88305 without modifiers. If one only does the histological preparation of the slides, add the -TC modifier to each listed CPT code. If one only performs the slide interpretation, bill with modifier -26 appended to each of the pathology CPT codes. For Medicare and the typical 12 core biopsies, the pathology billing should be submitted with a single unit of G0416; for nonMedicare carriers, the pathology billing should be submitted with 12 units of 88305.

## MRI-ASSISTED TRANSRECTAL ULTRASOUND (FOR FUSION-GUIDED BIOPSY OF THE PROSTATE GLAND)

There are no specific CPT code(s) for a fusion-guided prostate biopsy, and many carriers consider this an investigational procedure and will not reimburse more than a standard transrectal needle prostate biopsy. If one uses this biopsy technique, be sure to obtain an advanced beneficiary notification, ABN, waiver of payment from the patient to indicate his possible financial responsibilities if an insurance carrier does not pay.

Suggested coding for a fusion-guided biopsy includes the following CPT codes:

- 55700 transrectal needle biopsy of the prostate gland,
- 76872 transrectal ultrasound examination of prostate,
- 76942 transrectal ultrasound guidance for biopsy,
- 76498 unlisted CPT code, MRI procedure, (diagnostic),
- 763763- 3 D rendering with interpretation and reporting of ultrasound with postprocessing.

For further information and questions on coding and billing contact Michael A. Ferragamo telephone 516-741-0118, fax 516-294-4736, and e-mail liqgold2@aol.com.

## References


1. 2014 International Classification of Disease, Clinical modification, Ninth Edition, ICD-9-CM, Professional Edition.
2. American Medical Association. *The Complete Official Draft Code Set, ICD-10-CM*. USA: AMA; 2014.
3. American Medical Association. *Current Procedural Terminology*. Professional ed. USA: AMA; 2014.
4. CPT Assistant 2014, Division Health.
5. Buck CJ. *Healthcare Common Procedure Coding System*. Professional ed. USA: AMA; 2014.

CHAPTER

# 64

# Health Policy for Prostate Cancer: PSA Screening as Case Study

*Shilpa Venkatachalam, PhD, MA*, Danil V. Makarov, MD, MHS***

***NYU Langone Medical Center, NYU School of Medicine, New York, NY, USA**

****Department of Urology, Population Health, and Health Policy, NYU School of Medicine, New York, NY, USA**

## PSA TESTING: AN ONGOING DIALOGUE

Prostate-specific antigen (PSA)-based screening for early detection of prostate cancer was approved in two stages by the US Food and Drug Administration (FDA).[1,2] In 1986, the blood test was approved mainly to monitor disease status in men already diagnosed with prostate cancer.[3] In 1992, it was approved for prostate cancer diagnosis.[4] Following its approval, the test was subsequently used to detect disease in asymptomatic men (screening) as well as in men presenting with urological symptoms, in spite of not having been approved for these indication.

## THE STATISTICS

Approximately one in six American males is diagnosed with invasive prostate cancer in his lifetime.[5] Prostate cancer is the second leading cause of death from cancer, following lung cancer;[6] a considerable public health burden.[7] However, in spite of its high incidence, only one out of every 36 American men over the age of 50 dies of prostate cancer.[6] In terms of mortality, it is true then, that more men die with, rather than of prostate cancer.[8] It is also true, that in the past 15 years, the risk of prostate cancer and incidence of the disease has increased with a corresponding decrease in mortality from the disease. For instance, it has been proposed that, "Prostate cancer-specific mortality has declined ... but several reasons for this trend have been suggested, such as improvements in medical treatment of advanced disease."[9]

What needs to be investigated is that the decrease in mortality from the disease is directly a cause of widespread PSA screening, or is a result of other factors such as aggressive surgery, radiation or hormonal therapy.[10] United States Preventative Services Task Force (USPSTF) chair, Virginia A. Moyer, has previously stated that there is no evidence regarding this:

> We actually don't have a solid explanation for the decline in mortality. PSA advocates believe that we should be paying attention to the Göteborg study, which shows that if there is benefit, it begins to accrue at about 9 years. However, the decline in prostate cancer mortality began around 1993, at the same time the widespread use of PSA began. If the decline were a result of screening, you would expect to see the curve drop around 2003, but it occurred in 1993. While there may be some benefit in PSA tests, the numbers don't bear out the theory supporting screening for the decline in mortality.[11]

## POLICY CONCERNS AROUND PSA SCREENING

Policy decisions, however, should not only be concerned about mortality, but must also assess the psychological, economic, and social burdens caused by the disease. Moreover, policy must consider what a decision to take away screening coverage, and hence an individual's autonomy, might entail. It has been suggested that ceasing to reimburse for PSA screening may have its own risks leading possibly to a "health care disparity problem" especially among financially disadvantaged populations and among men who may comprise high risk for aggressive disease, in particular, men of African-American, Latino, and/or Hispanic descent.[11]

Policy decisions around the risk benefit of prostate cancer screening have undergone significant change over time. Several decisions were made based on studies like the Prostate, Lung, Colorectal and Ovarian (PLCO; 1993 –2001) and ERSPC (1991–2003) randomized

**Prostate Cancer. http://dx.doi.org/10.1016/B978-0-12-800077-9.00064-5**

studies,[12] and modeling studies, since the introduction of PSA-based prostate screening in US in the early 1980s.[13] Politics and presidential campaigns began to push for health care reform based on preventative care. Then-Senator Barak Obama, for instance, stated:

> "Simply put, in the absence of a radical shift towards prevention and public health, we will not be successful in containing medical costs or improving the health of the American people." His opponent in the same campaign, Senator John McCain, argued that the "best care is preventative care." The logic behind this, for both candidates, was that it would help cut costs while keeping people well.[14]

This logic forms the core of debates around screening programs in general, and in this case specifically. Detractors of the argument emphasize that promoting screening is not the same as promoting good health habits and lifestyles and hence, would actually raise, not lower, healthcare costs. An effective screening modality, generally, increases the incidence of a given disease by increasing patients at risk of death from the disease and also by increasing overdiagnosis, consequently increasing health care costs, often when no treatment is warranted.[15] On the other hand, by not screening, we may miss detecting disease at an early stage when it is most curable, increasing mortality among those who might otherwise have been cured.[16] In the face of this dilemma, what do policy makers need to take into consideration?

According to data from the CDC, following the incidence of nonmelanoma skin cancer, prostate cancer is not only the leading type of cancer among males in the United States but is also the leading cause of mortality among men in the United States, second only to mortality from lung cancer (Figure 64.1).[17]

With the current recommendations put forth by the USPSTF against screening, D'Amico has argued that the USPSTF's recommendation against prostate cancer screening is not justified because benefits of screening only become clear after 15–20 years. The USPSTF, he argues, base their conclusions on studies that have a too short follow-up period. What are the risks of not screening for prostate cancer? How would it affect catching potentially aggressive prostate cancers that could be arrested by using baseline PSAs as a measure? Using this framework, the argument begins to alter and take a different course. PSA-based tests can be used as risk assessment tools. Hence the risk stratified recommendation by the AUA. The director of the Duke Prostate Cancer Center has argued that in his view, "the new American Urological Association (AUA) guidelines that, for example, recommend a risk-stratified PSA test at age 40, make sense, because there's evidence that baseline PSA measurements in certain age groups can predict aggressive prostate cancers, metastases, and disease-specific mortality years down the road. Over the past few years, we've learned how to better use PSA as a risk-assessment tool, which the Task Force didn't consider. So, I'm concerned that this policy might harm at-risk groups of men."[11]

Studies have shown a significantly greater reduction in mortality rates in US compared to UK.[18,19] Herein lies the conundrum: unarguably, there have been benefits and lives have been saved through PSA based screening tests. Figure 64.2 shows incidence and mortality age adjusted rates by all races. What is unclear is whether this trend was a result of screening and detection tools or aggressive treatment. Yet, following this trend, there was a subsequent trend toward overdiagnosis. As

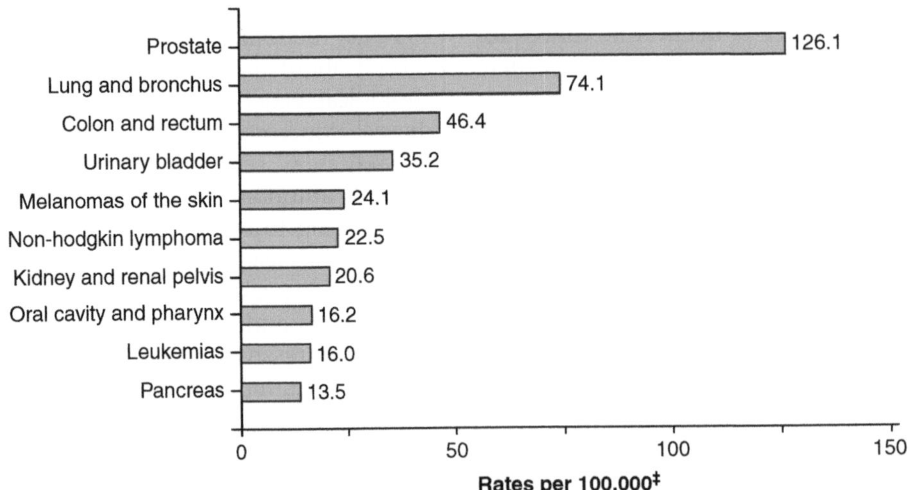


**FIGURE 64.1 Top 10 Cancer Sites: 2010, Male, United States – All Races.** *Source: CDC US cancer statistics.*

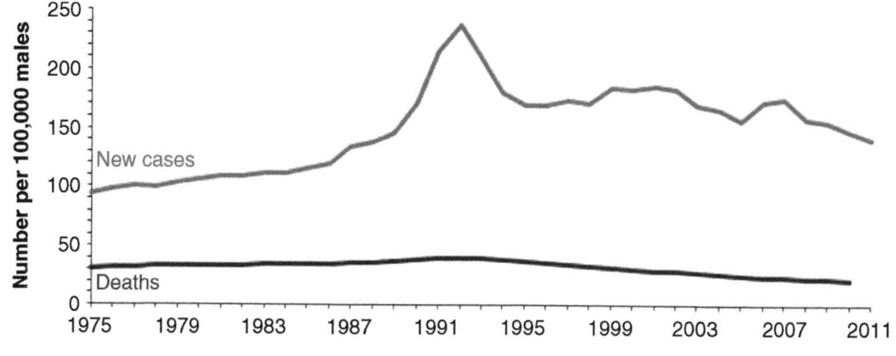


| Year | 1975 | 1980 | 1985 | 1990 | 1994 | 1998 | 2002 | 2006 |
|---|---|---|---|---|---|---|---|---|
| 5-Year relative survival (%) | 66.0 | 70.2 | 75.0 | 88.5 | 94.6 | 98.2 | 99.8 | 99.6 |

SEER 9 Incidence 1975–2011 & U.S. Mortality 1975–2010, All Races, Males. Rates are Age-Adjusted.

FIGURE 64.2 **New Cases, Deaths, and 5-Year Relative Survival.** *Source: Seer.cancer.gov.*

Welch, notes that; "Early diagnosis may help some, but it undoubtedly leads others to be treated for 'diseases' that would never have bothered them. That is called over diagnosis.... When it comes to cancer, for example, there is a very broad spectrum of diseases. Some kill rapidly, some progress slowly, and some do not progress at all."[15] In addition to this, in the United States, 6/10 cases are diagnosed in men who are 60 years or older and is generally rare in men below the age of 40 years. Average age at the time of a diagnosis is typically 66 years. In terms of race and ethnicity, incidence and mortality rates of, and from, prostate cancer are higher in males of African-American descent.[20] Framing policy around screening needs to necessarily take this too in consideration.[21] In order to avoid problems of overdiagnosis, policies in screening need to carefully consider what groups may be associated with a higher risk for the disease, what age groups need to be screened and of the detected cases which can be observed and which warrant treatment.[22]

## WHAT CONSIDERATIONS SHOULD DRIVE PROSTATE CANCER POLICY?

Should health policy take into account deaths averted per say 1000 men over a given period of time? Or should it be more concerned with the relative reduction in percent of deaths averted in a said population? Policy must always confront this issue: the conflict between individual interests and population level interests.[23] It must necessarily incorporate what follows after a diagnosis of such elevated levels of biomarkers and from diagnosis of a range of prostate cancers in terms of aggressive disease and prognosis following screening and biopsy. Furthermore, screening for prostate cancer may identify localized cancers but may not be quite so well equipped in identifying which may ultimately become metastatic.

One of the values of a screening test lies in not merely identifying the presence of a disease, but also the appropriate treatment and intervention strategy, in order to reduce morbidity, or mortality.[24] If appropriate treatment is not available, then there is no benefit to screening.[24] This has been clearly seen in the case of cervical cancer[25] where following screening and detection, treatment options have proven to be highly effective. However, in prostate cancer, there is a discrepancy for several reasons: is intervention required in individuals among the aged? Is it warranted for men with <10 years of life expectancy? As a result of all these issues, various organizations take varied stances on the subject of PSA screening, inherently contributing to the confusion. Should prostate cancer screening be integrated into routine preventative health care maintenance for American males?

## VARIATIONS IN GUIDELINES AND RECOMMENDATIONS

Guidelines proposed by various bodies, such as the USPSTF, AUA, and American Cancer Society (ACS), are constructed on an evidence-based approach and evaluate the risk–benefit ratio of a screening protocol.[26]

The USPSTF is comprised of individuals from diverse backgrounds such as primary care and preventive medicine, internal medicine, pediatrics, obstetrics, and gynecology, behavioral health, and nursing. However, no specialists such as oncologists or urologists sit on the board. Furthermore, the USPSTF is not armed to make any decisions based on cost-effectiveness. They state this explicitly in their Recommendation Statement:"[the USPSTF] bases its recommendations on the evidence of both the benefits and harms of the service, and an assessment of the balance. The USPSTF does not consider the costs of providing a service in this assessment."[27] In May 2012, based on results from The Prostate, Lung, Colorectal and Ovarian (PLCO) Cancer Screening Trial[28] and European Randomized Study of Screening for Prostate Cancer (ERSPC),[13] the USPSTF recommended against prostate screening, assigning it a grade D.[27] The USPSTF did not take into consideration modeling studies or the results of the Göteborg trial[29] on prostate cancer screening. These studies mainly conclude beneficial results of screening, quite in opposition to the PLCO that were considered by the USPSTF.[30]

The implications of a "D" grade are that the preventative service does not have to be covered by Medicare, or that if it is covered , a copayment must be made by the beneficiary.[31] The ramifications around such a decision involve access to health care, cost cuts for the government but increased costs for the individual, and can lead to individuals not benefitting from a service that they may need. To complicate the debate further, integrating routine PSA screenings for prostate cancer increases the economic burden carried by Medicare and hence recommending against it outright, as the USPSTF has suggested, can combat this problem and perhaps even serve certain financial interests.[32] As opposed to the "D" grade, questions were asked about why the USPSTF did not instead assign an "I" (insufficient evidence to make a decision) or a "C" (selective offering based on patient-clinical shared decision) grade, which would have been understood as more logical by many.[33,34] This would indicate that, "the USPSTF recommends selectively offering or providing this service to individual patients based on professional judgment and patient preferences. There is at least moderate certainty that the net benefit is small."[30]

It would be left to the discretion of the patient, along with input from the clinician, whether or not the patient was willing or wanting to be screened, despite the potential risks associated with the screening test. By assigning a grade D to the screening for prostate cancer, the USPSTF has in essence excluded the input and concerns of the patient from the decision-making process, almost entirely. But perhaps, this may have to do with clinicians routinely ordering PSA tests and the consequent overtreatment that may follow from elevated PSA levels.

None of the various bodies, governmental agencies, and federal agencies including scientific and professional bodies, recommend screening for all or mass screenings for early detection of prostate cancer. The AUA, for instance, explicitly states this in their revised guidelines in line with evidence-based studies.[35] A similar approach is followed by the American College of Preventive Medicine[34,36] which, while not recommending routine screenings for individuals, does emphasize informing men about the potential benefits and harmful effects associated with screening. They also reassert the importance of the specific individual's preferences in the decision-making process.

Screening recommendations need to integrate an understanding of the psychological dimensions associated with the framing of policies.[37,38] The potential harms associated with PSA screenings include increased anxiety, downstream effects of unnecessary biopsies that can carry risks of pain and infection, and treatment-related side effects such as bowel problems, sexual dysfunction, and incontinence.[39] Guidelines and recommendations need to necessarily take into account the psychological and behavioral impacts that may be impacted from the effects of screening as well. For instance, there has been considerable discomfort by men with regard to the USPSTF recommendations to not screen males above the age of 75 years. A national online panel survey illustrated that while 62% of men agreed with the USPSTF recommendations against prostate cancer screening, only 13% intended to actually follow these recommendations.[40] Older males and males worried about the disease, were more likely to get screened, as were men with a family history of prostate cancer and men of African-American descent. Policy makers must somehow strike a balance between individual demands and population-based needs in order to frame successful policies. Beyond the patient, implementing policy recommendations have to take into consideration clinicians as well. Clinicians may not implement recommendations, and may refer patients for routine and sometimes unnecessary screenings due to fear of potential malpractice suits and liability issues, or may not have enough time to weigh options through with patients.[41]

To address these disparate concerns, the American College of Physicians, The European Association of Urology, the National Comprehensive Cancer Center Network, and the American Urological Association, have increasingly been advocating a shared decision making (SDM) approach between patient and doctor, and have been working against the screening recommendations proposed by the USPSTF.[42] For instance the position of the AUA on this issue is as follows:

> The AUA feels strongly that any attempt to broadly reduce access to PSA testing would be a disservice to men, especially those with risk factors for prostate cancer (such as African American race or positive family history). Instead of instructing primary care physicians to discourage men from having a PSA

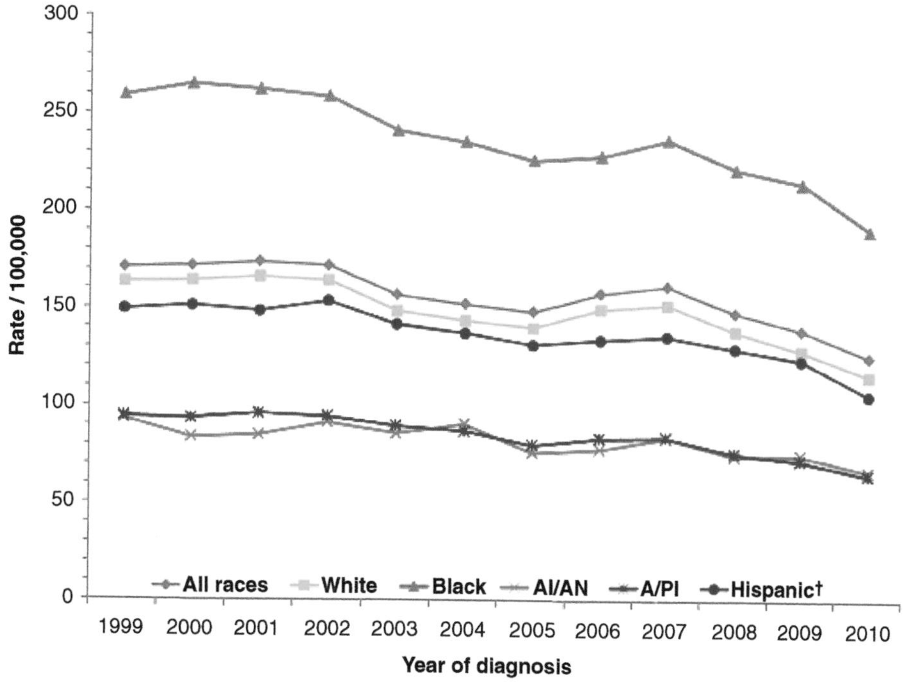


FIGURE 64.3 **Incidence of Prostate Cancer from 1999 to 2010 by Race.** Such studies may assist identifying persons of a high risk as opposed to those who are at average risk, allowing the development of targeted screenings in males across the country.

> test, the Task Force should focus on how best to educate primary care physicians regarding targeted screening and how to counsel patients about their prostate cancer risk. The decision to use or to forego the PSA test is a choice that should be individualized, made by informed patients in conjunction with their providers.[35]

The American Urological Association, for instance, does not support screening for men aged between 40 years and 55 years that are only at an average risk for disease. Defining average risk remains another important factor in decisions related to screening policies and indicates males who do not have a family history of prostate cancer, do not have a profile of early onset of prostate cancer in the family and are not of African-American descent. Identifying, through further studies, profiles of males who present a high risk for the disease may lead to targeted screening approaches, which may also prove highly cost-effective from a policy point of view (Figure 64.3).[43]

Incidence of prostate cancer from 1999 to 2010 by race. Such studies may assist identifying persons of a high risk as opposed to those who are at average risk, allowing the development of targeted screenings in males across the country.

## THE ROLE OF SHARED DECISION MAKING (SDM)

How do we identify who should qualify for prostate cancer screening? Who should decide whether or not an individual should be screened? And how do we integrate these issues within the system and still have broad consensus among the various decision-making boards?[44] A possible solution may be SDM. For certain men, the risks of prostate cancer screening may outweigh its benefits, while the opposite may hold true for others. SDM is a process that integrates the best-known evidence and the individual's preferences to tailor the decision to the necessities of the individual patient. Such SDM can be implemented using patient decision aids (PDA). The International Patient Decision Aids Standards Collaboration, states that PDAs are: "...designed to help people participate in decision making about health care options. They provide information on the options and help patients clarify and communicate the personal value they associate with different features of the options."[45] PDAs are not synonymous with SDM but are aids to it.[46] PDAs are specifically designed technologies that can assist in the SDM process that occurs between patient and clinicians, and may even include additional individuals, such as family and friends.[47] These are decision aids that are designed and developed, keeping in mind the patients' values and concerns while making decisions. Decision aids can be in the form of surveys and questionnaires, can be graphical in representing and providing information regarding risks, they can be pictorial or video-based, and may even include experiences shared by other patients regarding biopsies and treatment.[48]

The pictorial representation given next by The American Society of Clinical Oncology is a form of a patient decision aid that encapsulates in a precise and comprehensive manner, the issues surrounding whether or not one should get tested and the reliability of the PSA-based screening test. Studies suggest that individuals with inadequate numeracy and literacy skills, prefer receiving information from their health care provider to whom they are able to direct questions and specific concerns.[49] This relies on a participatory model of healthcare and medicine, where health literacy is not merely defined by the ability to read instructions and literature around medicine and healthcare, but to understand the issue with which the patient is confronted and to make informed decisions with support from PDAs and clinicians and experts in the field. Tools to incorporate patient views and concerns and to disseminate information may include video resources such as patient testimonies, infographics, pamphlets, questionnaires and patient–clinician structured dialogues.[50] Increasingly, the move in policy decisions has been modeled on enhancing the dynamic communication between patient, clinician, and PDAs. This tends to be the case, most importantly, in issues such as PSA testing, where evidence does not go entirely in one direction, toward harm or benefit.

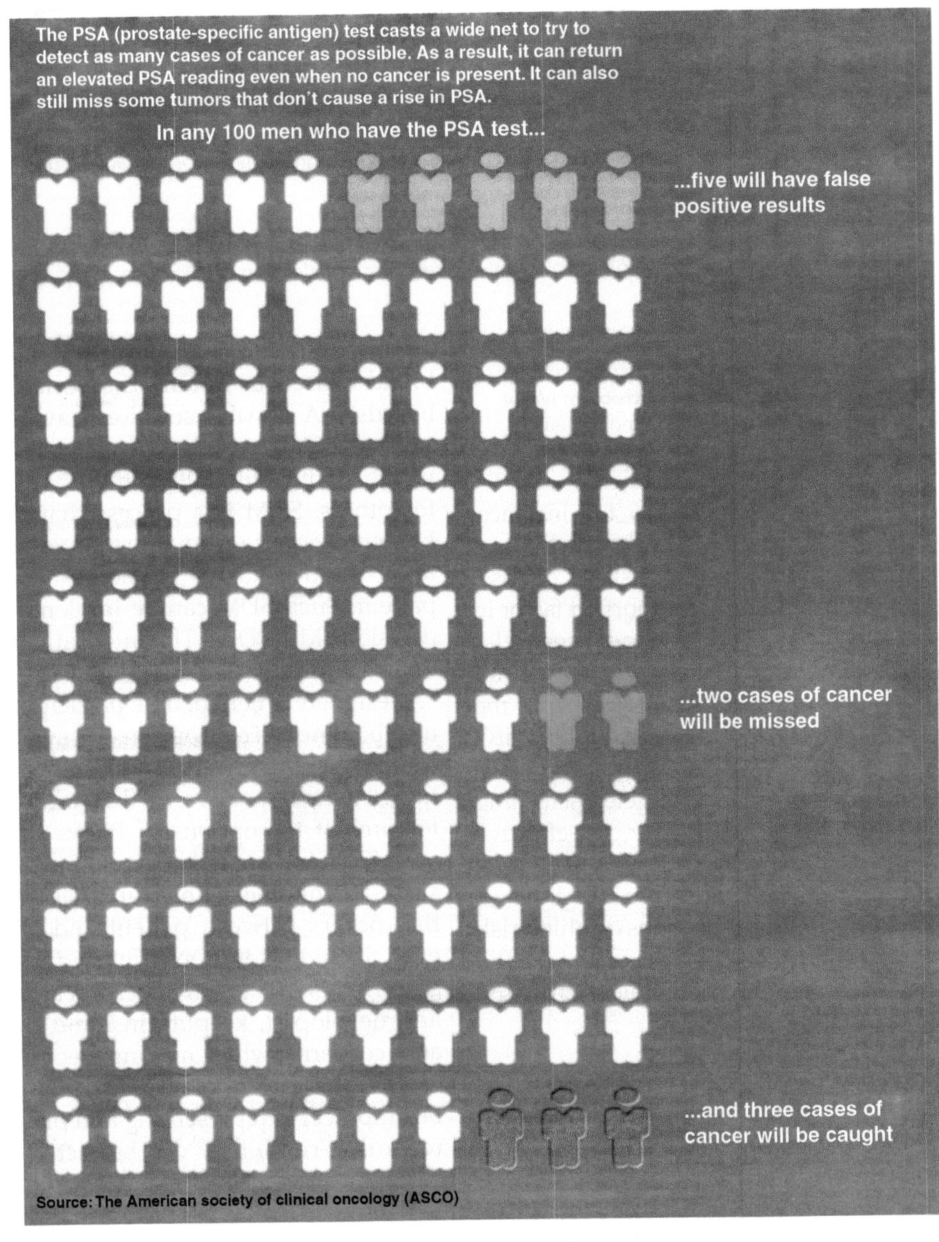

The PSA (prostate-specific antigen) test casts a wide net to try to detect as many cases of cancer as possible. As a result, it can return an elevated PSA reading even when no cancer is present. It can also still miss some tumors that don't cause a rise in PSA.
In any 100 men who have the PSA test...
...five will have false positive results
...two cases of cancer will be missed
...and three cases of cancer will be caught
Source: The American society of clinical oncology (ASCO)

In a recent national Health Interview Survey, 64.3% reported to not having any prior discussions with their physicians regarding PSA screening, including its benefits and risks. In the 8% who did report discussions, three of the main features of SDM that were discussed included the advantages, disadvantages, and uncertainties. Furthermore, these studies concluded that the lack or absence of SDM was linked to no screening rather than with screening. Policy recommendations must hence focus on such SDM procedures rather than focusing on an out-and-out policy that favors no routine screening.

SDM involves framing policies that can aid in disseminating tools to provide patients and their families with PDAs and information so that they are equipped to make informed decisions. At the time of proposing guidelines and recommendations, organizations such as the USPSTF must integrate views and representations from clinicians, specialists, public health personnel, and patients, as well as caregivers. Incorporating patients into decision-making boards is of utmost importance in implementing SDM. Patients must be those who have undergone PSA routine screenings as well as those who may not yet have been screened, or who may be consideringing screening. They must be individuals who represent the high-risk population as well as those who represent the average-risk population. Urologists and oncologists need to be consulted too. The process of arriving at a resolution regarding screening itself, must be seen as the ground for SDM, and it is through this that policy must be recommended.

SDM is predicated upon health literacy. The meaning of literacy expands beyond just the ability to read, for instance, instructions regarding treatment protocols.[51] When health literacy is combined with open dialogue between physician and patient in order to reach optimized decisions, patients may feel equipped to understand better their own conditions and their own bodies. This may give them a sense of agency; where they may feel that they have gained control over situations that may otherwise seem beyond any control.[52] Studies on SDM, however, have produced mixed results. There have been studies suggesting the SDM may be beneficial to more educated and literate groups, socioeconomically advantaged groups and may risk further marginalizing underrepresented, socioeconomically lower-bracket groups and those who are less literate. Some studies indicate they may work better for groups of individuals at high risk from the given disease. Studies are ongoing and extensive. However, many studies suggest SDM, and informed decision-making, lower overall expenditures in health care despite these differences.[53] With PSA testing and prostate cancer, it has been reported, based on a national study, that great improvement is necessary.[54]

## THE WAY FORWARD

The way forward requires substantial contribution and effort from all sectors of health care: patients, clinicians, policy makers, and health care professionals. This is possible only when channels of communication are opened between these different players in order to facilitate the exchange of questions, answers, and concerns from all involved. Educators need to disseminate information and awareness through appropriate media depending upon factors such as computer literacy, reading ability, and the range of information access by different individuals. Researchers need to study what may be the most appropriate population-specific media that clinicians need to address the concerns of their patients and guide them appropriately without being worried about malpractice litigations and patients need to feel empowered to make decisions. This can be facilitated through the joint efforts of all those involved including policy makers. Policy decisions around screening and other health issues finally need to be data driven. They must take into account qualitative data as well to account for the psychological costs of any decision-making in areas related to health. Keeping this framework in mind, the aim would include maximizing patient awareness and hence confidence and input in decision-making, minimizing anxiety by creating awareness regarding the disease and by empowering the individual to participate in decision-making and to do this in a cost-effective manner.

## References


1. American Society of Clinical Oncology. ASCO Daily News. <http://am.asco.org/although-better-defined-role-psa-prostate-cancer-screening-remains-controversial.>; 2013.
2. *National Cancer Institute*. Prostate Specific Antigen (PSA) Test. <http://www.cancer.gov/cancertopics/factsheet/detection/PSA.>; 2012.
3. Prensner JR, Rubin MA, Wei JT, Chinnaiyan AN. Beyond PSA: the next generation of prostate cancer biomarkers. *Sci Transl Med* 2012;**4**(127):127.
4. Shiels MS, Goedert JJ, Moore RD, Platz EA, Engles EA. Reduced risk of prostate cancer in U.S. men with AIDS. *Cancer Epidemiol Biomarkers Prev* 2010;**19**(11):2910–5.
5. Lim LS, Sherin K. ACPM Prevention Practice Committee. Screening for prostate cancer in U.S. men: ACPM position statement on preventive practice. *Am J Prev Med* 2008;**34**(2):164–70.
6. ACS.*Cancer.Org*. <http://www.cancer.org/cancer/prostatecancer/detailedguide/prostate-cancer-key-statistics.>; 2014.
7. Surveillance Research Program, NCI. *Surveillance, Epidemiology and End Results Program*. <http://seer.cancer.gov/statfacts/html/prost.html.>; 2014.
8. Whitmore FW. Localized prostatic cancer: management and detection issues. *Lancet* 1994;**343**(8908):1263–7.
9. La Rochelle J, Amling CL. Prostate cancer screening: what we have learned from the PLCO and ERSPC trials. *Curr Urol Rep* 2010;**11**(3):198–201.

10. Basch E, Oliver TK, Vickers A, et al. Screening for prostate cancer with prostate-specific antigen testing: American Society of Clinical Oncology Provisional Clinical Opinion. *J Clin Oncol* 2012;**30**(24):3020–5.
11. Piana, Ronald. *New PSA Recommendations: The Debate over Prostate Cancer Screening Continues.* <http://www.ascopost.com/issues/july-1-2012/new-psa-recommendations-the-debate-over-prostate-cancer-screening-continues.aspx.>; 2012.
12. Schröder FF H, Roobol MJ. ERSPC and PLCO prostate cancer screening studies: what are the differences? *Eur Urol* 2010;**58**(1):46–52.
13. Eckersberger F, Finkelstein J, Sadri H, et al. Screening for prostate cancer: a review of the ERSPC and PLCO trials. *Rev Urol* 2009;**11**(3):127–33.
14. Krauthammer C. Preventive care isn't the magic bullet for health care costs. The Washington Post. 2009.
15. Welch HG.Campaign Myth: Prevention as Cure-All. *New York Times*. 2008.
16. Carroll PR, Whitson JM, Cooperberg MR. Serum prostate-specific antigen for the early detection of prostate cancer: always, never, or only sometime. *J Clin Oncol* 2010;**29**(4):345–7.
17. Arcangeli S, Pinzi V, Arcangeli G. Epidemiology of prostate cancer and treatment remarks. *World J Radiol* 2012;**4**(6):241–6.
18. Collin SF M, Martin RM, Metcalfe C, et al. Prostate-cancer mortality in the USA and UK in 1975–2004: an ecological study. *Lancet Oncol* 2008;**9**(5):445–52.
19. Bryant RJ, Hamdy FC. Trends in Prostate Cancer Screening: Overview of the UK. In: Ankerst DP, editor. *Current Clinical Urology: Prostate Cancer Screening*. New York: Humana Press; 2009.
20. Wu I, Modlin CS. Disparities in prostate cancer in African American men: what primary care physicians can do. *Cleve Clin J Med* 2012;**79**(5):313–20.
21. Gann PH. Risk factors for prostate cancer. *Rev Urol* 2002; **4**(Suppl. 5):S3–S10.
22. Hoffman RM, Gilliland FD, Eley JW. Racial and ethnic differences in advanced-stage prostate cancer: the Prostate Cancer Outcomes Study. *J Natl Cancer Inst* 2002;**93**(5):388–95.
23. Arah OA. On the relationship between individual and population health. *Med Health Care Philos* 2009;**12**(3):235–44.
24. World Health Organisation. *Screening for various cancers.* <http://www.who.int/cancer/detection/variouscancer/en/.>; 2014.
25. Saslow, Debbie. *Cervical Cancer is an International Issue.* <http://www.cancer.org/cancer/news/expertvoices/post/2013/01/30/cervical-cancer-is-an-international-issue.aspx.>; 2013.
26. Gomella LG, Liu XS, Trabulsi EJ, et al. Screening for prostate cancer: the current evidence and guidelines controversy. *Can J Urol* 2011;**18**(5):5875–83.
27. Moyer VA. Screening for prostate cancer: US Preventive Services Task Force recommendation statement. *Ann Intern Med* 2012;**157**(2): 120–34.
28. National Cancer Institute. n.d. *Prostate, Lung, Colorectal, and Ovarian Cancer Screening Trial.* <http://prevention.cancer.gov/plco.>; 2014.
29. Hugosson J, Carlsson S, Aus G. Mortality results from the Göteborg randomised population-based prostate-cancer screening trial. *Lancet Oncol* 2010;**11**(8):725–32.
30. Kaffenberger SD, Penson DF. The politics of prostate cancer screening. *Urol Clin North Am* 2014;**41**(2):249–55.
31. U.S. Department of Health and Human Services. *Appendix A: How the U.S. Preventive Services Task Force Grades Its Recommendations: Guide to Clinical Preventive Services Task Force Grades Its Recommendations.* <http://www.ahrq.gov/professionals/clinicians-providers/guidelines-recommendations/guide/appendix-a.html.>; 2014.
32. Penson, D. "AUA Comments to MedPAC on USPSTF." American Urological Association. 2014.
33. U.S. Preventive Services Task Force. *Grade Definitions after May 2007.* <http://www.uspreventiveservicestaskforce.org/uspstf/gradespost.htm.>; 2008.
34. Knight SJ. Decision making and prostate cancer screening. *Urol Clin North Am* 2014;**41**(2):257–66.
35. AUA. *American Urological Association.* <www.auanet.org.>; 2014.
36. American Urological Association. *AUA releases new clinical guideline on prostate cancer screening*. <https://www.auanet.org/advnews/press_releases/article.cfm?articleNo=290>; 2013.
37. Kotwal AA, Schumm P, Mohile SG, Dale W. The influence of stress, depression, and anxiety on PSA screening rates in a nationally representative sample. *Med Care* 2012;**50**(12):1037–44.
38. Watts S, Leydon G, Birch B, et al. Depression and anxiety in prostate cancer: a systematic review and meta-analysis of prevalence rates. *BMJ Open* 2014;**4**(3):e003901.
39. Slatkoff S, Gamboa S, Zolotor AJ, Mounsey AL, Jones K. PSA testing: when it's useful, when it's not. *Clin Rev* 2011;**60**(6):357–60.
40. Abdi H, Black P. Prostate-specific Antigen Testing: Men's Responses to 2012 recommendation against screening: National Evidence on the Use of Shared Decision Making in Prostate-specific Antigen Screening. *Urology* 2014;**83**(1):4–5.
41. Hoffman RM, Barry MJ, Roberts RG, Sox HC. Reconciling primary care and specialist perspectives on prostate cancer screening. *Ann Fam Med* 2012;**10**(6):568–71.
42. Loeb S. Guideline of guidelines: prostate cancer screening. *BJU Int* 2014;**114**(3):323–5.
43. Jones BA, Liu WL, Araujo AB, et al. Explaining the race difference in prostate cancer stage at diagnosis. *Cancer Epidemiol Biomarkers Prev* 2008;**17**(10):2825–34.
44. McNaughton-Collin MF, Barry MJ. One man at a time-resolving the PSA controversy. *N Engl J Med* 2011;**365**(21):1951–3.
45. IPDAS. International Patient Decision Aids Collaboration. <http://ipdas.ohri.ca/what.html.>; 2013.
46. Oshima Lee E, Emanuel EJ. Shared decision making to improve care and reduce costs. *N Engl J Med* 2013;**368**(1):6–8.
47. Barry MJ, Edgman-Levitan S. Shared decision making — the pinnacle of patient-centered care. *N Engl J Med* 2012;**366**(9):780–1.
48. Informed Medical Decisions Foundation. *Informed Medical Decisions Foundation.* <http://www.informedmedicaldecisions.org/what-is-shared-decision-making/.>; 2014.
49. Gaglio B, Glasgow RE, Bull SS. Do patient preferences for health information vary by health literacy or numeracy? a qualitative assessment. *J Health Commun* 2012;**17**(Suppl. 3):109–21.
50. Robert Wood Johnson Foundation. "Shared Decision Making and Benefit Design: Engaging Employees and Reducing Costs for Preference-Sensitive Conditions." <http://www.rwjf.org/content/dam/farm/reports/reports/2013/rwjf405304.>; 2013.
51. Institute of Medicine. "Health Literacy: A prescription to End Confusion." 2004.
52. Wheeler DC, Szymanski KM, Black A, Nelson DE. Applying strategies from libertarian paternalism to decision making for prostate specific antigen (PSA) screening. *BMC Cancer* 2011;**11**:148.
53. Durand MA, Carpenter L, Dolan H, et al. Do interventions designed to support shared decision-making reduce health inequalities? A systematic review and meta-analysis. *PLoS One* 2014;**9**(4):e94670.
54. Han PK, Kobrin S, Breen N. National evidence on the use of shared decision making in prostate-specific antigen screening. *Ann Fam Med* 2013;**11**(4):306–14.

CHAPTER

# 65

# Legal Implications of Prostate Cancer Screening


C.J. Stimson, MD, JD

Department of Urologic Surgery, Vanderbilt University Medical Center, Nashville, TN, USA


## INTRODUCTION

The legal implications of prostate cancer screening flow from the formalization of expert opinion and scientific data into clinical practice guidelines (CPGs). These guidelines are a nexus between the daily clinical practice of prostate cancer care and the broader legal framework that permeates all medical practice. As public statements of appropriate care, they provide the lens through which clinical decisions between patients and their physicians can be judged, rightly and wrongly, by nonclinicians.

This chapter will explore how clinical decisions regarding prostate cancer screening are being scrutinized vis-a-vis CPGs, and highlight how these clinically oriented documents implicate myriad legal issues related to medical malpractice law. The content of the various prostate cancer screening CPGs are discussed elsewhere and will only be touched on briefly here.

## PROSTATE CANCER SCREENING RECOMMENDATIONS: A BRIEF REVIEW

There are currently four prostate cancer screening guidelines listed by the Agency for Healthcare Research and Quality's National Guideline Clearinghouse.[1] Additionally, the National Comprehensive Cancer Network (NCCN) published a CPG for prostate cancer early detection.[2] These guidelines all attempt to answer the same basic question: what group of asymptomatic men should be screened for prostate cancer? The answer to this question varies and has evolved over time.

The American Cancer Society (ACS) published an updated CPG on prostate cancer screening in 2010.[3] The guideline recommended that men with at least a 10-year life expectancy have an opportunity to make an informed decision about whether to be screened for prostate cancer. Specifically, the ACS recommended that a discussion regarding the risks and benefits of screening should occur based on a patient's risk of prostate cancer. Patients with the highest risk of prostate cancer should have this discussion at the age of 40 years, the next strata at 45 years, and average risk men at 50 years.

In contradistinction to the ACS CPG, the United States Preventive Services Task Force (USPSTF) recommended in 2012 that no men be screened for prostate cancer, regardless of prostate cancer risk, arguing that the risks of treatment outweighed the mortality benefit from screening.[4] Less than one year later, however, the American Urological Association (AUA)[5] and the American College of Physicians (ACP)[6] issued more nuanced recommendations contradicting the USPSTF CPG. Similar to the ACS, the AUA and ACP stratified their recommendations by prostate cancer risk. The AUA recommended no screening in men <40 years, average risk men aged 40–54 years, men ≥70 years, or men with a life-expectancy <10–15 years; however, individualized decisions and shared decision-making were recommended for higher risk patients <55 years and men 55–69 years, respectively. The ACP supported the 2010 ACS recommendations stratifying shared decision-making based on prostate cancer risk, but unlike the ACS the ACP did not endorse providers making decisions on behalf of men who are unable to decide whether to screen for prostate cancer.

Finally, in September 2014 the NCCN published a comprehensive CPG covering many facets of prostate cancer care, including when to initiate and stop screening.[2] There was uniform consensus among the panel members that baseline PSA testing be offered to healthy, well-informed patients aged 50–70 years, but nonuniform consensus regarding men outside of that age distribution.

As evidenced by the various CPG findings just outlined, the question of which men should be screened for prostate cancer remains controversial. Of course there


Prostate Cancer. http://dx.doi.org/10.1016/B978-0-12-800077-9.00065-7

is no novelty in the current debate over prostate cancer screening. Instead, what is new is the formalization of competing approaches to prostate cancer screening into contradictory CPGs. As will be discussed shortly, the implications of this formalization extend beyond the clinical exchange between patients and physicians.

## CLINICAL PRACTICE GUIDELINES: AN INTRODUCTION

Before addressing the legal implications of CPGs for prostate cancer screening, we will first explore the features of CPGs, how a CPG is developed, and the common criticisms leveled at these efforts to minimize practice variation while maximizing clinical value. The definition of CPGs varies across the medical literature,[7–12] although there are several common features.

### Features

CPGs are expert-driven, evidence-based statements defining appropriate diagnosis and treatment algorithms for particular clinical problems. Additionally, CPGs evolve over time in response to advancements in medical science and changes in professional and societal norms.

Whether disseminated by individual medical societies, physician consortiums, or networks of academic medical centers, the individuals charged with defining best or sound practices should have expertise relevant to the clinical question that the CPG is answering. For example, authorship of the NCCN, AUA, ACS, and ACP CPGs for prostate cancer screening included urologic surgeons, medical oncologists, and radiation oncologists with expertise in prostate cancer management.[13] A notable exception to this rule is the USPSTF CPG, authored by experts in primary care, public health, and epidemiology.[4]

The level of evidence available to support CPGs is variable and the strength of CPG recommendations should correspond to the strength of the evidence. Randomized, controlled trials (RCTs) are considered the highest level of evidence.[14] However, discrepancies between RCT patient populations and patient populations treated in the non-RCT setting make it difficult to project RCT evidence onto the general population. In these instances there is increased reliance on observational studies and simulation models.[15]

Finally, CPGs must be evolving. The contours of CPG recommendations must shift in response to advancements in medical science and changes in professional and societal norms regarding diagnostic and treatment options. This dynamic process may in fact be the greatest challenge in CPG science given the significant costs associated with CPG development.[16]

### Process

The process for developing CPGs was largely decentralized and without standardization until an Institute of Medicine (IOM) study published in 2011. The report, entitled "Clinical Practice Guidelines We Can Trust," established eight principles for CPG development. These principles set out national standards for establishing transparency, managing conflict of interest (COI), guideline development group composition, systematic evidence reviews, defining the strength of evidence for all recommendations, articulation of recommendations, external review, and CPG updates.[12] A detailed discussion of each standard is beyond the scope of this chapter; however, it is worth taking a closer look at the standards described for COI and guideline development group composition as these are often targeted by CPG critics.[17–19]

COI is a challenging issue to address in the CPG process, and it encompasses the individuals involved in the CPG process and those entities responsible for funding the effort. Regarding individual COI, it can be difficult to acquire the necessary expertise for CPGs with a zero tolerance COI policy.[20] To address this issue the IOM has called for a series of limits on COI, including divestment by CPG authors from financial interests related to the CPG topic, limits on the percentage of CPG authors with COI, and prohibitions against chairs or cochairs of CPG development groups from having any COI. Less concrete COI standards are prescribed by the IOM for organizations that fund CPGs, stating simply that such entities should have no role in CPG development.

The IOM makes particular recommendations regarding the composition of groups responsible for CPG development. In addition to a range of clinical experts, the report underscores the importance of diverse, multidisciplinary methodological expertise and patient participation. Specifically, individual patients and patient advocate organizations are identified as critical means for ensuring public involvement and support.

### Criticism

Despite the advancements and standardization in CPG science described earlier, there are many notable criticisms of CPGs aimed at the failure of CPGs to clarify best practices or standards. These criticisms can be broadly categorized into critiques of CPG content, evidence, and process.

#### *Content*

The publication of conflicting guidelines, lack of guideline specificity, and CPG obsolescence fuel the arguments of CPG content critics.[19,21,22] Exemplified by the CPGs for prostate cancer screening, particularly the 2012 USPSTF recommendations, the publication of conflicting guidelines works against the efforts of CPG authors to help patients and practitioners make informed medical decisions. Although debate over conflicting interpretations of medical evidence is an integral part of advancements in medical science, the publication of this debate to the laity is arguably counterproductive.[23] Furthermore, while CPG disclaimers and purposefully ambiguous language provide physicians with a space for discretionary clinical decisions outside of CPG recommendations, this lack of specificity undercuts the utility of CPGs as measures of adherence to best practices and standards.

Finally, CPG obsolescence in the setting of a rapidly evolving evidence base is more than a theoretical shortcoming. During the development and publication of a CPG there is a real risk that substantive changes to the evidence could render a CPG incomplete, irrelevant, or inaccurate. The evidence base for prostate cancer screening has not evolved rapidly enough to negatively impact the relevance of the associated contemporary CPGs, but additional longitudinal follow-up data may settle the current CPG discrepancy.[24] For a recent example of CPG obsolescence, see the 2013 American Urological Association guideline for castration-resistant prostate cancer and the role of enzalutamide in the prechemotherapy setting.[25,26]

#### *Evidence*

CPGs are built on a foundation of evidence culled from the medical literature, although the quality and interpretation of these data are variable and therefore a common source of CPG reproach. The most highly regarded level of evidence comes from RCTs, followed by cohort studies, case-controlled studies, and finally, case series/reports.[27] Systematic reviews of this evidence base are even more highly regarded, particularly those studies that synthesize multiple RCTs.

However, the utility of RCT data for CPG development is limited by its generalizability (i.e., projection of RCT cohort data to individual patients)[17,28–30] and variability in interpretation. To illustrate the latter point, consider the aforementioned prostate cancer screening CPGs. The bulk of the RCT data providing the foundation for these CPGs comes from the European Randomized Study of Screening for Prostate Cancer,[31,32] Goteborg,[33] and Prostate, Lung, Colorectal, and Ovarian[34,35] trials. That the same RCT data provide the evidentiary foundation for contradictory CPGs is per se evidence of the limitations of such data. Although there are compelling arguments that the USPSTF CPG represents an untenable, rather than an alternative, interpretation of the trial data,[13,36–39] the preventive services literature would argue otherwise.[40,41]

#### *Process*

COI and CPG authorship expertise represent the primary targets for CPG process critics. With CPGs carrying significant financial ramifications for individual physicians, medical societies, and private industry, these critics argue that the risk of contaminating the science of CPGs with financial gain could outweigh the potential benefits.[17,42–44] Contemporary examples supporting these concerns can be seen in the 2000 American Heart Association guidelines recommending alteplase for treatment of ischemic stroke[45] and the 2004 guidelines for statins.[46]

The concern with CPG authorship expertise is related to the COI described earlier, but it cuts both ways. While some critics argue that excessively homogenous expert panels (e.g., urologic surgeons, medical oncologists, etc.) lead to the formalization and propagation of clinical biases by those who stand to benefit financially from the recommendations,[30] others counter that there is a greater risk from CPG authors issuing clinical recommendations without any relevant clinical experience, informing data interpretation and guideline recommendations.[39]

As this introduction to CPGs highlights, the process and product are not a perfect mechanism for disseminating best practices and standards. The shortcomings of CPGs, however, do not provide sufficient justification for abandoning the process altogether. Like any scientific inquiry, CPGs are an iterative process that will improve over time as physician and policymaker buy-in continues to incentivize the promulgation of formal recommendations. The question that remains is what will be the legal implications of these dynamic prostate cancer-screening guidelines.

## THE LEGAL IMPLICATIONS OF PROSTATE CANCER SCREENING: CLINICAL PRACTICE GUIDELINES AND MEDICAL MALPRACTICE LAW

Prostate cancer screening CPGs are implicated throughout health care law. This section will focus primarily on the role of CPGs in medical malpractice tort law. Although there are no published court decisions directly addressing the discrepancy between current prostate cancer screening guideline recommendations, the court decisions analyzed in this review address analogous issues of fact and law.

### Medical Malpractice Tort Law

To understand the role of CPGs in medical malpractice tort law requires understanding the elements of the medical malpractice claim. To succeed in a medical malpractice claim a plaintiff must prove four elements: (1) duty, (2) breach of duty, (3) causation, and (4) injury.[47] Stated otherwise, a plaintiff must prove that a physician had a duty to care for the plaintiff, that the duty was breached by a deviation from the standard of care (SOC), and that this deviation was the proximate or direct cause of the plaintiff's injury.

The SOC determination creates a space for CPGs to influence the court's medical malpractice analysis, inasmuch as a CPG informs the evidentiary foundation for the SOC. Adversarial expert medical testimony provides this foundation by drawing evidence from multiple sources, including community practice patterns, peer-reviewed publications, and CPGs. Ultimately, if there is a question of fact (as opposed to a question of law) regarding the SOC, the jury is responsible for weighing the presented evidence and determining whether the SOC was followed. The question, then, is what role do prostate cancer screening CPGs play in the SOC determination.

### *Clinical Practice Guidelines Alone Do Not Define the Standard of Care*

The role of CPGs in establishing the SOC is limited by the depth of expert testimony regarding the guidelines. Without a discussion of the scientific evidence supporting the CPGs, and specific references to recommendations within the guidelines apropos the SOC analysis, the courts will not construe CPGs identified by expert testimony as the SOC. This provides further support for CPGs built on a sound evidentiary standard rather than expert opinion. It is not enough that experienced physicians come together to publish their collective practice patterns. Rather, there must be a rigorous, methodical, and scientifically sound process that incorporates the evidence base into CPG recommendations. As seen in the *Pearson* and *Conn* cases that follow, the judicial deference granted to the medical profession for self-regulation will not permit *ad hoc, ipse dixit* standard setting.

#### Linda Pearson-Heffner v. United States *(2006)*[48]

The facts of the case centered on Mr Waddell, a military veteran who was diagnosed and later died from metastatic prostate cancer. The patient established care with the Montgomery, AL, Veterans Administration (VA) hospital in December 1998. At that initial visit, he reported to his internist that he had a serum prostate-specific antigen (PSA) drawn at a civilian facility in August 1998 that was 1.4. A digital rectal exam (DRE) was not performed by the internist. He was seen on two subsequent occasions by the same primary care provider (PCP), March 1999 and September 2000. Between those clinic visits, he missed four scheduled appointments. During the September 2000 visit his internist ordered PSA testing for his next appointment in January 2001. The patient missed this January appointment and ultimately requested a different PCP. He was seen by a new PCP in March 2001, almost 32 months since his last PSA test, but did not have a PSA drawn or DRE performed. He presented to a non-VA emergency department complaining of headache and back pain 3 months later in June 2001. Head imaging demonstrated a brain lesion and his PSA was >1400. Prostate biopsy subsequently confirmed adenocarcinoma of the prostate. Mr Waddell succumbed to his disease 15 months following this diagnosis.

The executrix of Mr Waddell's estate filed a medical malpractice claim against the VA hospital under the Federal Tort Claims Act, claiming that the VA was negligent because it failed to meet the SOC for prostate cancer screening. The federal court entered judgment for the VA and against Mr Waddell's estate, a holding based largely on the failure of the plaintiff's expert witness to establish the SOC for prostate cancer screening. Despite submitting the AUA and ACS recommendations for annual prostate cancer screening, the court rejected the expert's testimony as insufficient evidence for establishing the SOC. Specifically, the court cited the plaintiff's expert's failure to provide "[p]ertinent evidence based on scientifically valid principles" or do more than "merely allude [...] to general policies espoused by... professional societies."

#### Conn v. United States *(2012)*[49]

The plaintiff is a veteran who filed a medical malpractice claim against the VA, arguing that the VA was negligent in their failure to diagnose and treat his acute myocardial infarction. The patient presented to the VA triage with chest pain. He had a cardiac workup, including cardiac enzymes and an EKG, which did not demonstrate an ischemic cardiac event. He was admitted overnight for observation and remained asymptomatic. He was discharged the next day without further workup. He presented again with chest pain 2 days later, this time to a civilian emergency department. There he had an acute myocardial infarction requiring advanced cardiac life support and electrical defibrillation. Ultimately, the plaintiff survived.

At trial, the plaintiff's expert witness implied that the VA failed to meet the SOC for management of acute myocardial infarction by referencing the American College of Cardiology and the American Heart Association guidelines on the subject. The court found that CPGs are "relevant to the question of the proper standard of care," and can be relied on "when enunciating a standard of care," but in this instance the expert did not identify specific CPG recommendations describing the actions that would be taken by a minimally competent physician. As a result, the court held that a "vague" reference to the CPGs did not establish the SOC.

### *Clinical Practice Guideline Consensus Helps Define the Standard of Care*

Consensus CPGs for cancer screening can be a powerful tool for defining the SOC for the courts, even in circumstances where the physician-defendant is not a member of the particular guideline-issuing organization. In *Daberkow v. United States* (2009),[50] the plaintiff's estate filed a medical malpractice claim alleging that the VA was negligent in screening Mr Daberkow for colorectal cancer. Specifically, the estate argued that the SOC for colorectal cancer screening required colonoscopy or combination flexible sigmoidoscopy and barium enema, and that the VA was negligent by relying only on a fecal occult blood test. The court disagreed with the plaintiff's SOC argument, based in large part on the consensus between colorectal cancer screening CPGs issued by the ACS, American Academy of Family Physicians, the American Gastroenterological Association, the American Society of Colon and Rectal Surgeons, and USPSTF. Judgment was therefore entered for the VA and against Mr Daberkow's estate.

It is important to highlight a difference between the holdings from *Pearson* and *Daberkow*. Specifically, the CPGs from the former were contradictory and this was part of the court's dismissive treatment of expert testimony defining the SOC with medical association-issued guidelines. In the latter case, however, the uniform consensus among the CPGs allowed the court to define the SOC for colorectal cancer screening with less scrutiny.

*Daberkow* shows that there is a practical benefit to uniformity in CPGs. This is not to say that disagreements over the evidence should be papered over or ignored in order to create a more legal-friendly document. On the contrary, the debate over prostate cancer screening should continue and be more inclusive. Rather than publish contradictory CPGs, the disparate parties could come together to develop a consensus statement to be published jointly. Not only would this allow for robust debate, it would also provide clarity to the medical community and the courts regarding the current consensus on the SOC for prostate cancer screening. Certainly such a consensus statement would be a resource-intensive endeavor, but these cases demonstrate that the cost of the status quo is high for patients and their physicians.

### Exception to the Rule: Clinical Practice Guidelines that Define the Standard of Care

Despite the previous discussion highlighting the court's general approach to CPGs and SOC, there is a notable exception to the rule that CPGs alone do not establish a SOC. To illustrate, consider the case of *Ellis v. Eng* (2010).[51]

The plaintiff's estate brought a medical malpractice suit against the plaintiff's surgeon and gastroenterologist for negligence in management of Stage II colon cancer. The patient had colon cancer resected by his surgeon that was Stage IIb on final pathology. The physician-defendants did not recommend adjuvant chemotherapy based on the American Society of Clinical Oncology (ASCO) guidelines. The patient died from progression of his disease.

The expert witnesses for the patient-plaintiff and physician-defendant both utilized the ASCO CPGs as evidence defining the SOC for Stage II colon cancer management in briefs before the court. Specifically, the plaintiff's expert claimed that adjuvant chemotherapy was the SOC according to ASCO CPG recommendations, while the defendant's expert claimed that the ASCO CPG did not require adjuvant chemotherapy for Stage II colon cancer. On appeal, the court analyzed the ASCO CPG document directly and determined that "there was no definite consensus that adjuvant therapy was warranted." The plaintiff's SOC claim was therefore dismissed and the case did not proceed to trial.

The court held that a CPG could be used to disprove a purported SOC where the same guideline was being put forth by plaintiff and defendant expert witnesses to support mutually exclusive SOCs. Although such circumstances are less common in medical malpractice law's adversarial setting, it may become more commonplace as CPG standardization and science progresses. Therefore, it is notable that the court is willing to engage CPGs as the final arbiters of the SOC in certain circumstances, particularly in the pretrial setting when a common CPG is cited by both parties.

### Treatment of Discordant CPGs: "Two Schools of Thought" or "Respectable Minority" Rule

Courts have long encountered SOC analyses requiring reconciliation of conflicting medical customs or practices. In certain circumstances, courts have held that physicians should be exempt from liability when medical authorities are divided over the SOC. Referred to as the "two schools of thought" or "respectable minority" rule, this line of jurisprudence developed from the courts' unwillingness to choose between conflicting practice patterns when physicians themselves cannot reach a consensus.[52]

The rule operates in a two-step fashion.[53] First, the court must decide as a matter of law whether a jury is allowed to consider "two schools of thought" as an affirmative defense for the physician-defendant. To reach this defense the defendant must demonstrate that there are a considerable number of professionals who agree with the treatment pursued. Then, if that burden is met, the second step is for the jury to determine as a question of fact whether (1) a considerable number of recognized and respected professionals advocate the same course of treatment, (2) the defendant was aware of the support for this course of treatment at the time the plaintiff-patient was treated, and (3) the defendant consciously chose to follow this recommended treatment. In the case that follows, we see how this rule is applied in the setting of CPGs, particularly as related to prostate cancer screening.

In *Mundis v. Neuburger* (2013), the Pennsylvania Superior Court held that the physician-defendant met the evidentiary burden required for an affirmative defense under the "two schools of thought" doctrine, upholding the jury verdict in favor of the physician.[54] The patient-plaintiff filed a claim against his primary care physician alleging negligence in prostate cancer screening. The plaintiff elected to undergo screening for prostate cancer at age 54 after his brother had been diagnosed with the condition. His DRE was unremarkable and his PSA was 3.2. The patient had repeat screening demonstrating an enlarged prostate and a PSA of 41, 2 years later. He was subsequently referred to a urologic surgeon and diagnosed with metastatic prostate cancer.

The patient's negligence claim alleged that the PCP did not meet the SOC for prostate cancer screening by failing to screen on an annual basis. The plaintiff argued for this SOC based on expert testimony. The defendant countered with expert testimony claiming that a "significant number" of physicians would consider less frequent prostate cancer screening as the SOC, and cited the 2003 USPSTF CPG on prostate cancer screening as an authority.

The appellate judges affirmed the trial court's support of an affirmative defense based on the "two schools of thought" rule. The court found that testimony from the defendant's experts in internal medicine and medical oncology met the first step in the evidentiary burden. Specifically, the testimony provided sufficient factual background to demonstrate "a differing school of thought" regarding annual prostate cancer screening based on the 2003 USPSTF CPG that (1) there was no medically recognized requirement for annual prostate cancer screening and (2) prostate cancer screening is an individualized decision. Additionally, the court was swayed by expert testimony that the majority of physicians do not follow rote PSA testing at prescribed intervals.

The "two schools of thought" doctrine could play an important role in future medical malpractice claims regarding prostate cancer screening, particularly given the current state of discordant CPGs. As demonstrated in *Mundis*, the doctrine allows the jury to determine whether they believe that there are two legitimate schools of thought such that a physician-defendant should be insulated from liability. Expert testimony presenting the science behind the CPGs for prostate cancer screening would certainly meet the defendant's two-step evidentiary burden showing that medical authorities are divided on this issue.

## CONCLUSIONS

Prostate cancer screening CPGs are a significant and worthwhile advancement in urologic surgery, although the utility of these published recommendations extends beyond the patient-physician relationship. As a formalization of expert opinion and data interpretation that is available to the public, these CPGs can play an important role in the negligence analysis for medical malpractice cases. A review of recent case law confirms that CPGs are informing SOC determinations and this role is likely to expand as the CPG process matures. The legal implications of discordant prostate cancer screening CPGs will similarly expand; however, current jurisprudence suggests that courts have several legal mechanisms for balancing these contradictory recommendations during the negligence analysis.

## References

1. National Guideline Clearinghouse; 2014. Available from: www.guideline.gov.
2. Carroll PR, Parsons JK, Andriole G, et al. Prostate cancer early detection, version 1.2014. *J Natl Compr Canc Netw* 2014;**12**(9):1211–9.
3. Wolf AM, Wender RC, Etzioni RB, et al. American Cancer Society guideline for the early detection of prostate cancer: update 2010. *CA Cancer J Clin* 2010;**60**(2):70–98.
4. Moyer VA. Screening for prostate cancer: U.S. Preventive Services Task Force recommendation statement. *Ann Intern Med* 2012;**157**(2):120–34.
5. Carter HB, Albertsen PC, Barry MJ, et al. Early detection of prostate cancer: AUA guideline. *J Urol* 2013;**190**(2):419–26.
6. Qaseem A, Barry MJ, Denberg TD, et al. Screening for prostate cancer: a guidance statement from the Clinical Guidelines Committee of the American College of Physicians. *Ann Intern Med* 2013;**158**(10): 761–9.
7. Woolf SH. Practice guidelines: a new reality in medicine. I. Recent developments. *Arch Intern Med* 1990;**150**(9):1811–8.
8. Field MJ, Lohr KN, Institute of Medicine (US). Committee to Advise the Public Health Service on Clinical Practice Guidelines, United States. Department of Health and Human Services. Clinical practice guidelines: directions for a new program. Washington, DC: National Academy Press; 1990.
9. Institute of Medicine (US). Committee on Clinical Practice Guidelines. Field MJ, Lohr KN, editors. Guidelines for clinical practice: from development to use. Washington, DC: National Academy Press; 1992.
10. Dahm P, Yeung LL, Gallucci M, et al. How to use a clinical practice guideline. *J Urol* 2009;**181**(2):472–9.
11. Lewis SZ, Diekemper R, Ornelas J, et al. Methodologies for the development of CHEST guidelines and expert panel reports. *Chest* 2014;**146**(1):182–92.
12. Institute of Medicine (US). Committee on Standards for Developing Trustworthy Clinical Practice Guidelines. Graham R, editor. Clinical practice guidelines we can trust. Washington, DC: National Academies Press; 2011.
13. Cooperberg MR. Implications of the new AUA guidelines on prostate cancer detection in the U.S. *Curr Urol Rep* 2014;**15**(7):420.
14. Greene KL, Punnen S, Carroll PR. Evolution and immediate future of US screening guidelines. *Urol Clin North Am* 2014;**41**(2):229–35.
15. Habbema JD, Wilt TJ, Etzioni R, et al. Models in the development of clinical practice guidelines. *Ann Intern Med* 2014;**161**(11):812–8.
16. Browman GP. Development and aftercare of clinical guidelines: the balance between rigor and pragmatism. *JAMA* 2001;**286**(12): 1509–11.
17. Sniderman AD, Furberg CD. Why guideline-making requires reform. *JAMA* 2009;**301**(4):429–31.
18. Jost TS. Oversight of marketing relationships between physicians and the drug and device industry: a comparative study. *Am J Law Med* 2010;**36**(2–3):326–42.

19. Mehlman MJ. Medical practice guidelines as malpractice safe harbors: illusion or deceit? *J Law Med Ethics* 2012;**40**(2):286–300.
20. Mendelson TB, Meltzer M, Campbell EG, et al. Conflicts of interest in cardiovascular clinical practice guidelines. *Arch Intern Med* 2011;**171**(6):577–84.
21. Sawka AM, Magalhaes L, Gafni A, et al. Competing interests in development of clinical practice guidelines for diabetes management: report from a multidisciplinary workshop. *J Multidiscip Healthc* 2008;**1**:29–34.
22. Gupta M, McCauley J, Farkas A, et al. Clinical practice guidelines on prostate cancer: a critical appraisal. *J Urol* 2014;**193**(4):1153–8.
23. Help or harm: the furious debate over screening for prostate cancer. The Economist. March 8, 2014.
24. Gomella LG, Liu XS, Trabulsi EJ, et al. Screening for prostate cancer: the current evidence and guidelines controversy. *Can J Urol* 2011;**18**(5):5875–83.
25. Cookson MS, Roth BJ, Dahm P, et al. Castration-resistant prostate cancer: AUA guideline. *J Urol* 2013;**190**(2):429–38.
26. Beer TM, Armstrong AJ, Sternberg CN, et al. Enzalutamide in men with chemotherapy-naive metastatic prostate cancer (mCRPC): results of phase III PREVAIL study. *Clin Adv Hematol Oncol* 2014;**12** (4 Suppl. 11):3–4.
27. Sackett DL. *Evidence-based medicine: how to practice and teach EBM.* 2nd ed. Edinburgh; New York: Churchill Livingstone; 2000.
28. Tricoci P, Allen JM, Kramer JM, et al. Scientific evidence underlying the ACC/AHA clinical practice guidelines. *JAMA* 2009;**301**(8):831–41.
29. Institute of Medicine (US). Committee on Comparative Effectiveness Research Prioritization. Initial national priorities for comparative effectiveness research. Washington, DC: Institute of Medicine of the National Academies; 2009.
30. Sniderman AD, Furberg CD. Pluralism of viewpoints as the antidote to intellectual conflict of interest in guidelines. *J Clin Epidemiol* 2012;**65**(7):705–7.
31. Schroder FH, Hugosson J, Roobol MJ, et al. Screening and prostate-cancer mortality in a randomized European study. *N Engl J Med* 2009;**360**(13):1320–8.
32. Schroder FH, Hugosson J, Roobol MJ, et al. Screening and prostate cancer mortality: results of the European Randomised Study of Screening for Prostate Cancer (ERSPC) at 13 years of follow-up. *Lancet* 2014;**384**(9959):2027–35.
33. Hugosson J, Carlsson S, Aus G, et al. Mortality results from the Goteborg randomised population-based prostate-cancer screening trial. *Lancet Oncol* 2010;**11**(8):725–32.
34. Andriole GL, Crawford ED, Grubb 3rd RL, et al. Mortality results from a randomized prostate-cancer screening trial. *N Engl J Med* 2009;**360**(13):1310–9.
35. Andriole GL, Crawford ED, Grubb 3rd RL, et al. Prostate cancer screening in the randomized Prostate, Lung, Colorectal, and Ovarian Cancer Screening Trial: mortality results after 13 years of follow-up. *J Natl Cancer Inst* 2012;**104**(2):125–32.
36. Carlsson S, Vickers AJ, Roobol M, et al. Prostate cancer screening: facts, statistics, and interpretation in response to the US Preventive Services Task Force Review. *J Clin Oncol* 2012;**30**(21):2581–4.
37. Catalona WJ, D'Amico AV, Fitzgibbons WF, et al. What the U.S. Preventive Services Task Force missed in its prostate cancer screening recommendation. *Ann Intern Med* 2012;**157**(2):137–8.
38. Scherger JE. PSA screening: the USPSTF got it wrong. *J Fam Pract* 2013;**62**(11):616, 618.
39. Kaffenberger SD, Penson DF. The politics of prostate cancer screening. *Urol Clin North Am* 2014;**41**(2):249–55.
40. Lefevre M. PSA screening: the USPSTF got it right. *J Fam Pract* 2013;**62**(11):617, 619.
41. Lin K, Croswell JM, Koenig H, Lam C, Maltz A. 2011 Prostate-Specific Antigen-Based Screening for Prostate Cancer: An Evidence Update for the U.S. Preventive Services Task Force. Rockville: Agency for Healthcare Research and Quality; 2011. Report No.: 12-05160-EF-1.
42. Lenzer J. Why we can't trust clinical guidelines. *BMJ* 2013;**346**: f3830.
43. Lenzer J. US stroke legislation is revised after BMJ airs controversy. *BMJ* 2004;**328**(7440):604.
44. Lenzer J. Controversial stroke trial is under review following BMJ report. *BMJ* 2002;**325**(7373):1131.
45. Lenzer J. Alteplase for stroke: money and optimistic claims buttress the "brain attack" campaign. *BMJ* 2002;**324**(7339):723–9.
46. Lenzer J. US consumer body calls for review of cholesterol guidelines. *BMJ* 2004;**329**(7469):759.
47. Stimson CJ, Dmochowski R, Penson DF. Health care reform 2010: a fresh view on tort reform. *J Urol* 2010;**184**(5):1840–6.
48. Pearson-Heffner v. United States, Dist. Court, MD Alabama 2006.
49. Conn v. US, 880 F. Supp. 2d 741, Dist. Court, SD Mississippi 2012.
50. Daberkow v. US, Dist. Court, D Colorado 2007.
51. Ellis v. Eng, 70 AD 3d 887, NY Appellate Div., 2nd Dept. 2010.
52. Peters Jr PG. The Quiet Demise of Deference to Custom: Malpractice Law at the Millenium. Wash & Lee L Rev 2000;**57**:163.
53. Jones v. Chidester, 610 A. 2d 964, PA Supreme Court 1992.
54. Mundis v. Neuburger, 91 A. 3d 1283, PA Superior Court 2013.

PART XII

# NEW HORIZONS FOR PROSTATE CANCER

CHAPTER

# 66

# New Markers for Prostate Cancer Detection and Prognosis

*E. David Crawford, MD*, Karen H. Ventii, PhD**, Neal D. Shore, MD, FACS*[†]

*University of Colorado Health Science Center, Aurora, CO, USA
**School of Medicine, Emory University, Atlanta, GA, USA
[†]Carolina Urologic Research Center, Atlantic Urology Clinics, Myrtle Beach, SC, USA

## INTRODUCTION

Prostate cancer (PCa) is the most commonly diagnosed solid tumor among men in the United States and Europe.[1] Even without treatment, PCa-specific mortality rates at 5 and 10 years remain low. Improvements in optimizing screening, diagnosis, and treatment have resulted in decreasing PCa mortality.[2] Nevertheless, certain challenges still remain for men facing important disease state decisions and there is a need for biomarkers that can help

- Increase the probability of an initial positive biopsy,
- Reduce the number of unnecessary repeat biopsies by better distinguishing benign from malignant disease,
- Stratify low risk from higher risk tumors,
- Stage the disease or classify the extent of disease, and
- Predict and monitor clinical response to an intervention.

Improved prognostic biomarkers that can provide individualized patient risk assessment are needed to assist informing treatment decisions for patients and physicians. Additionally, biomarkers, which can improve the precision of risk assessment, are needed to enhance decision-making for physicians and patients, especially when the traditional clinical parameters (prostate-specific antigen (PSA), digital rectal examination (DRE), pathology) do not provide an accurate assessment of risk.

Patients with early-stage PCa can benefit from a more precise, personalized assessment of their tumor biology given that current clinical risk assessment tools may be less than adequate.

For those with CRPC, the increasing complexity of treatment decisions has led to research efforts to define predictive biomarkers that identify men most likely to benefit from a given therapy.[3]

A new generation of biomarkers (Table 66.1) has emerged to help improve risk assessment, guide diagnostic strategies, and ultimately enhance treatment outcomes through more targeted screening, more accurate diagnosis, improved risk stratification, which should lead to improved treatment recommendations and subsequent selection of therapy (Figure 66.1).[4]

## WHO SHOULD BE BIOPSIED?

### Key Takeaways

- PSA testing became the cornerstone of early PCa detection after its approval over 20 years ago. However, due to the low disease mortality rate, controversies have emerged with early detection strategies.
- New biomarker assays have been developed to help reduce the burden of biopsies in men with a low probability of PCa.

### PSA

After its approval by the Food and Drug Administration (FDA) in 1986, the availability of PSA dramatically influenced PCa early diagnosis.[5,6] In the United States, approximately 19 million men receive annual PSA testing, which results in more than 1.3 million biopsy procedures

Prostate Cancer. http://dx.doi.org/10.1016/B978-0-12-800077-9.00066-9

**TABLE 66.1** Current Landscape of PCa Biomarkers

| **Clinical scenarios addressed by currently available biomarkers** | | | |
|---|---|---|---|
| **Who should be biopsied?** | **Who should be rebiopsied?** | **Who should be treated vs monitored?** | **Therapeutic response assessment** |
| • PSA<br>• Phi<br>• 4KScore | • PCA3<br>• ConfirmMDx<br>• PCMT<br>• PTEN<br>• PHI | • Onco*type* DX®<br>• Prolaris®<br>• Decipher™<br>• ProMark | • PSA<br>• CTC |

CTC, circulating tumor cells; PSA, prostate-specific antigen; PCA3, prostate cancer antigen 3; PCMT, Prostate Core Mitomic Test; PHI, Prostate Health Index; *PTEN*, phosphatase and tensin homolog; *TMPRSS2:ERG*, transmembrane protease, serine 2: ETS-related gene.
*Adapted from Crawford et al.*[4]

and a resultant 240,890 newly diagnosed findings of PCa cases.[1]

Nonetheless, reliance on PSA testing alone for the detection of PCa has inherent limitations. First, the test is prostate specific but not PCa specific, and it may give false-positive or false-negative results. Most men with an elevated PSA level (above 4.0 ng/mL)[7] are not found to have PCa; only approximately 25% of men undergoing biopsy for an elevated PSA level actually have PCa. Conversely, a negative result may give false assurances that PCa is not detected, when, in fact, a cancer may still exist. Second, the test does not always differentiate indolent from aggressive cancer and thus early detection of PCa may not impact eventual mortality from the disease[7] and potentially lead to overtreatment. This limitation of PSA testing was largely responsible for the recent recommendation of the USPTF against continued routine screening.[8]

The Prostate, Lung, Colorectal and Ovarian (PLCO) Cancer Screening Trial is a large population-based randomized trial designed and sponsored by the National Cancer Institute to determine the effects of screening on cancer-related mortality and secondary endpoints in men and women aged 55–74. Regarding the PCa arm of the trial, after 13 years of follow-up, there was no evidence of a survival benefit for planned annual PCa screening compared with mandated screening. Additionally, there was no clinical impact with benefit for scheduled versus unplanned screening as it related to age, baseline comorbidity,

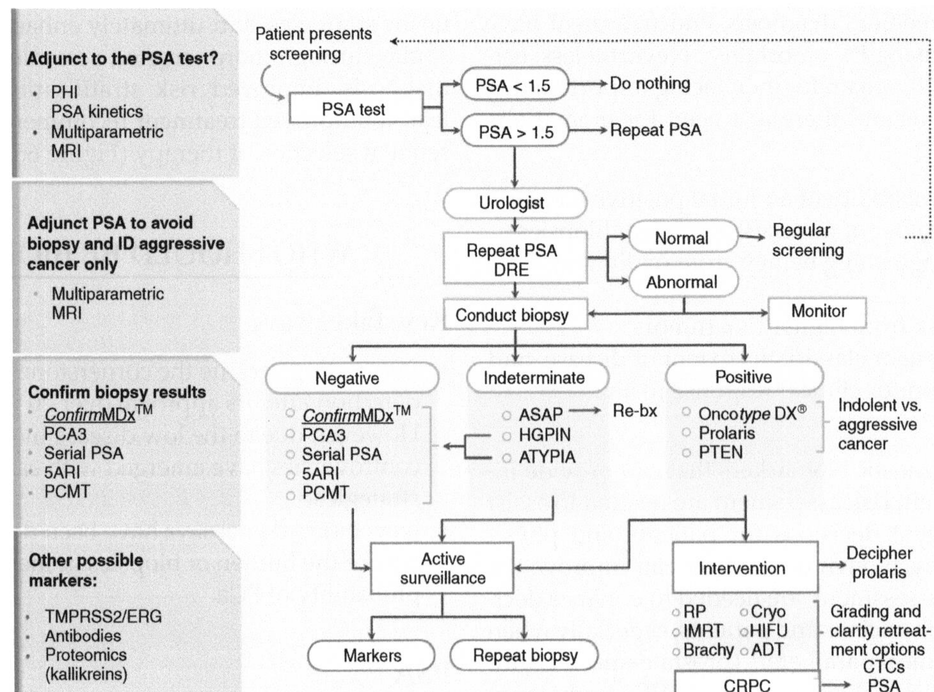


**FIGURE 66.1 The prostate cancer diagnostic pathway, with a focus on the role of biomarkers.** ADT, androgen deprivation therapy; 5ARI, 5-alpha-reductase inhibitors; ASAP, atypical small acinar proliferation; Brachy, brachytherapy; CRPC, castrate-resistant prostate cancer; Cryo, cryotherapy; CTCs, circulating tumor cells; DRE, digital rectal exam; HGPIN, high-grade prostatic intraepithelial neoplasia; HIFU, high intensity focused ultrasound; IMRT, intensity modulated radiotherapy; MRI, magnetic resonance imaging; phi, prostate health index; PCA3, prostate cancer antigen 3; PCMT, prostate core mitomic test; PSA, prostate serum antigen; *PTEN*, phosphatase and tensin homolog; RP, radical prostatectomy; TMPRSS2/ERG, transmembrane protease serine 2/ v-ets erythroblastosis virus E26 oncogene homolog.

or pretrial PSA testing.[9] PLCO had high rate of screening (~50%) in the control arm, thus limiting its conclusions. However, Crawford et al. have reported a survival benefit for screening in men without significant comorbidities.[10]

Follow-up results of 11 years from the European Randomized Study of Screening for Prostate Cancer study demonstrated that screening does significantly reduce death from PCa.[11] A potential reason for these differing results is that in the US-based PLCO Cancer Screening Trial, at least 44% of participants in the control arm were already PSA-tested prior to being randomized into the study,[9] confounding interpretation of the results.

Roobol et al.[12] stated that there was "poor compliance with biopsy recommendations" in PLCO, as the trial did not mandate biopsies. Screening test results were sent to the participant and his physician, and they decided upon subsequent biopsy per shared decision-making.

In order to improve the sensitivity and specificity of serum PSA, several PSA derivatives and isoforms (e.g., PSA Isoforms and PSA Density) have been used. The National Comprehensive Cancer Network (NCCN) recommends PSA density when assessing for very low risk PCa patients.[13]

Of note, the Goteborg trial, a prospective randomized trial of 20,000 men born between 1930 and 1944, showed that the benefit of PCa screening compared favorably to other cancer screening programs. PCa mortality was reduced by almost half over 14 years of follow-up.[14]

## Prostate Health Index (PHI)

Efforts have been made to reduce PSA-associated over-biopsying, which may lead to overtreatment in very low and low-risk patients. The PHI was approved by the FDA for use in 2012 in those with serum PSA values between 4 ng/mL and 10 ng/mL[15] in an effort to reduce the burden of biopsies in men with a low probability of PCa.

The PHI (PHI = [−2] proPSA/fPSA × PSA1/2; proPSA is a PSA subtype and fPSA is free PSA) was initially developed as an additional diagnostic biomarker in men with a serum PSA level of 2–10 ng/mL in European trials; an elevated proPSA/fPSA ratio is associated with PCa.[16] Percent-free PSA, PHI score, and PCA3 have been described as markers of specificity within the 2014 NCCN guidelines for early detection of prostate cancer.[15]

PHI score has a high diagnostic accuracy rate and can be used in PCa diagnosis. PHI score may be useful as a tumor marker in predicting patients harboring more aggressive disease and guiding biopsy decisions.[17]

PHI also predicts the likelihood of progression during active surveillance. Tosoian et al. showed that baseline and longitudinal values of PHI predicted which men would have reclassification to higher-risk disease on repeat biopsy during a median follow up of 4.3 years after diagnosis. Baseline and longitudinal measurements of PHI had C-indices of 0.788 and 0.820 for upgrading on repeat surveillance biopsy, respectively. In contrast, an earlier study in the Johns Hopkins active surveillance, PCA3 did not reliably predict short-term biopsy progression during active surveillance.[18]

In patients with persistent suspicion of PCa and a negative biopsy, testing with PCA3 and PHI has been proposed as a way to reduce the number of unnecessary repeat biopsies (Figure 66.2).[19]

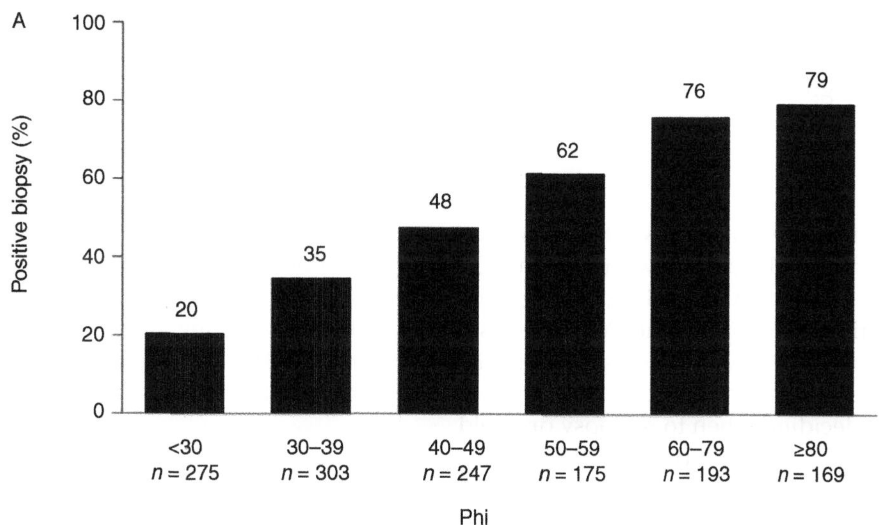


FIGURE 66.2 **The prostate health index (*phi*) score correlates with the percentage of positive prostate biopsies.** In the group of patients with a *phi* score ≥80, 79% had a positive prostate biopsy, whereas only 20% had a positive biopsy in the group with scores <30.[20]

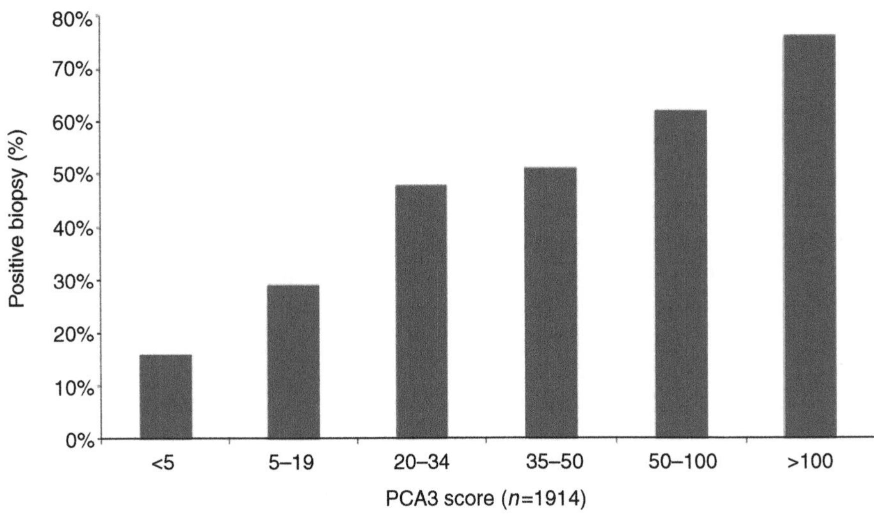


FIGURE 66.3 **Percentage of men with positive biopsy by PCA3 score.** In this prospective study, the PCA3 score was statistically significantly higher in men with a positive or suspicious prostate biopsy (Bx) versus a negative Bx.[23] *Adapted from Ref 23.*

## 4Kscore

4KScore is a newly available commercial assay panel that is designed to help predict which men with an elevated PSA will have high-grade disease upon tumor biopsy. By combining measures of total, free, and intact PSA with human kallikrein 2 (hK2) and other clinical parameters, the 4KScore was shown to be better than PCPT at predicting the occurrence of high-grade disease on biopsy.[21]

# WHO SHOULD BE REBIOPSIED?

## Key Takeaways

- For patients with negative biopsies who are believed to be at high risk for PCa, biomarker tests (such as PCA3 and PHI) should be consideration to improve specificity of the diagnosis.

## PCA3

PCA3 is a noncoding messenger RNA that has been shown to be elevated in >90% of men with known PCa, but not significantly elevated in normal prostatic glands or in benign prostatic hypertrophy. The PCA3 test is a urine-based assay approved by the FDA as a diagnostic test in the setting of a previous negative prostate biopsy. It may be helpful in deciding when to rebiopsy or avoid doing so, with its attendant morbidity and cost, and adds to the diagnostic information obtained from the PSA test.[22] A high PCA3 Score indicates a high probability of PCa, whereas a low score indicates a low likelihood. The mean PCA3 score was statistically significantly higher in men with a positive PCa biopsy, or those with atypical small acinar proliferation and/or high-grade prostatic intraepithelial neoplasia (HGPIN), compared with men who had a negative biopsy.[23] PCA3 testing may fail to identify transition zone cancers because the DRE does not elude cells into the urine (Figure 66.3).

## ConfirmMDx

ConfirmMDx is a tissue-based epigenetic assay to improve patient stratification on the decision for repeat biopsy. It is performed on the archived tissues from the previous negative biopsy and detects an epigenetic field effect resulting from increased hypermethylation of PCa-specific genes. This field effect around the cancer lesion can be detected despite the normal histologic appearance of cells, effectively extending the coverage of the biopsy. This test may help in the identification of high-risk men who require repeat biopsies and men without PCa who may avoid unnecessary repeat biopsies (Figure 66.4).[24,25]

## PCMT

PCMT is a tissue-based test that identifies a deletion in mitochondrial DNA that indicates cellular change associated with PCa. It detects presence of malignant cells in normal appearing tissue across an extended area.[26] Recent clinical data indicate that this test may be useful for identifying men who do not require a repeat biopsy.[27]

## PTEN

Dysregulation of *PTEN*, a tumor suppressor gene, has been associated with poor prognosis in PCa. Evidence suggests that loss of *PTEN* is associated with higher

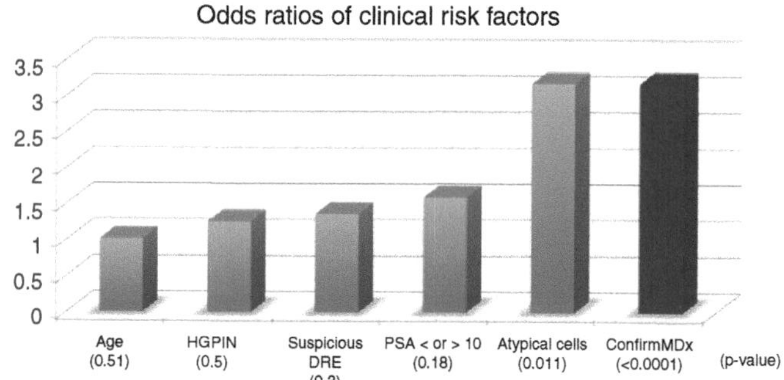


**FIGURE 66.4 The methylation analysis to locate occult cancer (MATLOC) validation study.** This study evaluated the clinical utility of an epigenetic quantitative methylation specific polymerase chain reaction assay to detect occult prostate cancer in histopathologically negative biopsies. The figure demonstrates a multivariate analysis of known risk factors and performance of the assay. Odds Ratios Reported in the MATLOC Study. *Adapted from Stewart et al.*[24]

Gleason grade, risk of progression, and recurrence after therapy.[28] Additionally, it is associated with advanced localized or metastatic disease and death.[29] The *PTEN* assay is a prognostic fluorescence *in situ* hybridization test typically ordered in conjunction with prostate biopsy tests, which will indicate partial or complete deletions of the gene.

# WHO SHOULD BE TREATED VS MONITORED?

## Key Takeaways

- Physicians and patients can consider disease monitoring as an alternative to treatment after careful consideration of the patient's PCa risk, general health, and age. Biomarkers can help with this decision-making.

## Onco*type* DX®

The Onco*type* DX® is a multigene RT-PCR expression assay that has been prospectively validated in several contemporary cohorts as an accurate predictor of adverse pathology in men with NCCN very low, low, and low–intermediate risk PCa.[30] Using very small biopsy tumor volumes, the assay measures expression of 17 genes (12 cancer-related genes from four relevant biological pathways and five reference genes). These are combined to calculate a genomic prostate score (GPS), which adds independent predictive information beyond standard clinical and pathologic parameters.[31–33] By predicting the likelihood of favorable pathology at the time of biopsy, Onco*type* DX® enables more confident selection of active surveillance or immediate therapy as an initial management strategy (Figure 66.5).

## Prolaris®

Prolaris® is a tissue-based cell cycle progression signature test that assesses 31 cell cycle progression genes to provide a risk assessment of PCa-specific progression and 10 year disease-specific mortality when combined with standard pathologic parameters.[34] It is designed as a risk stratification tool to help refine treatment/monitoring strategy for patients with PCa. Initially validated to predict prostate cancer specific survival (CSS) following radical prostatectomy, Prolaris has been subsequently validated in the biopsy setting as well.

## Decipher®

The Decipher® RNA assay directly measures the biological risk for metastatic PCa after radical prostatectomy. The test assesses the activity of 22 RNA markers associated with metastatic disease and has been demonstrated to

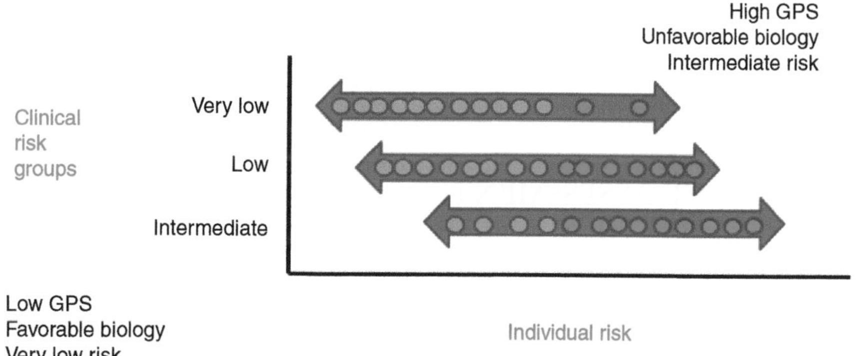


**FIGURE 66.5 Measuring individual tumor biology using the genomic prostate score (GPS).** The GPS method enables more accurate risk stratification. For example, for men with NCCN Low risk based on clinical criteria, a low GPS identifies a proportion (35%) who have likelihood of favorable pathology more closely matching patients with NCCN Very Low risk disease. On the other end, a higher GPS identifies 10% of men with likelihood of favorable pathology more closely aligned to that of men with NCCN Intermediate risk.

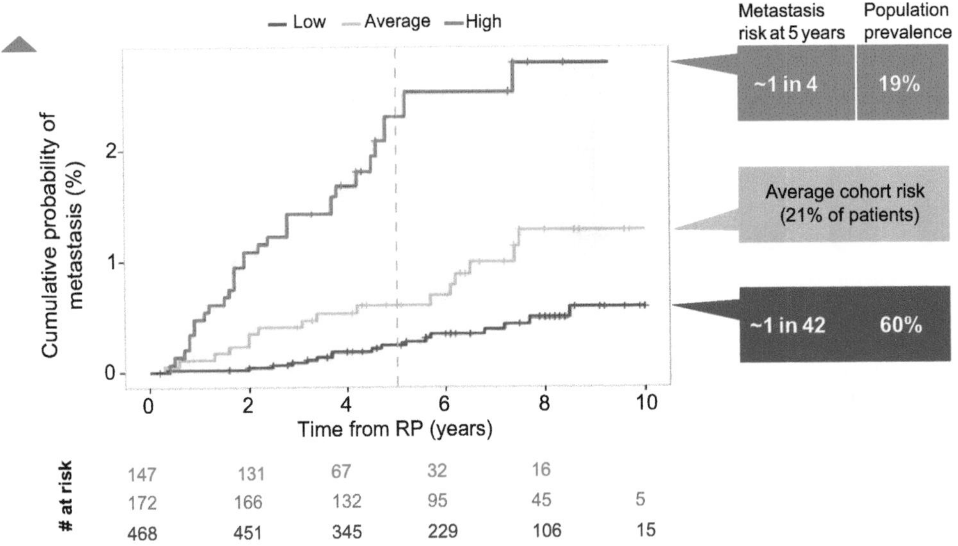


FIGURE 66.6 **Risk stratification of prostate cancer patients using decipher, a prognostic genomic marker test.** The Decipher test measures expression levels of multiple markers that are known to be associated with aggressive prostate cancer. Decipher scores are used to stratify an individual patient's risk of clinical metastasis following radical prostatectomy (RP). The blue, yellow, and red lines show men classified by Decipher as low-, moderate-, and high-risk; the risk of metastasis at 5 years post RP is approximately 1 in 4 for the high-risk patients. *Adapted from Karnes et al.*[36]

be independently prognostic of PCa death in a high-risk surgical cohort. In a validation study, over 70% of high-risk patients had low genomic classifier (GC) scores and good prognosis whereas patients with high GC scores had a cumulative incidence of metastasis over 25%.[35] This assay may better enable application of directed, multimodal, or adjuvant therapy for patients with high-risk PCa following radical prostatectomy (RP) (Figure 66.6).

### ProMark

ProMark is a prognostic biopsy-based PCa test. It uses immunofluorescent imaging analysis to quantify protein biomarker expression and classify patients' tumors. A clinical validation study demonstrated that ProMark can differentiate indolent from aggressive disease, based on data from standard formalin-fixed, paraffin-embedded tissue. The ability to monitor treatment effects and identify therapeutic targets at the time of treatment consideration are major unmet needs in PCa. PSA and CTCs are two markers designed to address this need.[37]

## THERAPEUTIC RESPONSE ASSESSMENT

### Key Takeaways

- Physicians who manage patients with CRPC are faced with complex decisions, given the numerous treatment options available. Beyond PSA, CTC testing offers an opportunity for predictive biomarkers to identify men most likely to benefit from a given therapy.

### PSA

In addition to its use in PCa screening and diagnosis, PSA has been used to monitor responses to therapy and has been investigated as a therapeutic target.[38] Newer therapies, such as sipuleucel-T and radium-223, may improve survival without decreasing PSA levels. For cytotoxic chemotherapies or newer oral targeted hormonal therapies, PSA responses may be indicative of clinical response to therapy. Radiographic progression and symptomatology are still key parameters for consideration of changing anti-neoplastic therapy.[39] Discordance between PSA kinetics and clinical response and progression of disease has been regularly observed. The need for biomarkers – beyond PSA – to predict response to treatment is well recognized. Other serologic tests that are helpful include hemoglobin, alkaline phosphatase, LDH, and others.

### CTCs

The CTC assay is intended for the enumeration of CTCs (CD45-, EpCAM+, and cytokeratins) in whole blood, which can be a biomarker for therapeutic response to antineoplastic regimens. An increased abundance of CTCs in the blood of CRPC patients can predict worse outcomes. Recently, evaluation of individual CTC cells has allowed further prognostication of PCa.[40] CTCs may be useful for predicting treatment response/survival with cytotoxic and hormonal therapies. However, ~50% of patients do not have detectable CTC levels by current detection methods. More sensitive CTC detection techniques are under investigation.

## Prostate Cancer Awareness and Education

The 1980s and 1990s marked the founding of several national education and advocacy groups focused on prostate cancer screening and treatment. These include Patient Advocates for Advanced Cancer Treatment, the Prostate Cancer Education Council, Man to Man, Us TOO! International, and other groups.

A number of these are notable for their efforts to promote PCa screening: The nonprofit Prostate Conditions Education Council (PCEC; www.prostateconditions.org), founded in 1989, is the leading resource for information on prostate health in the United States. A consortium of physicians and scientists, health educators, and prostate cancer advocates, the Council is the founder and coordinator of the national Prostate Cancer Awareness Week Program, and with its Screening Site Partners has screened about 5 million men for PCa across the United States.

In the summer of 2000, Arlene Mulder, former Mayor of Arlington Heights, Illinois, founded the Mayors' Coalition for Prostate Cancer Awareness and Education, to further promote PCa awareness and encourage screening. The same year, the United States Conference of Mayors designated September as Prostate Cancer Awareness Month, the first major national organization to do so.

In 2002, President George W. Bush officially designated September as National Prostate Cancer Awareness Month. To date, 154 mayors have joined the coalition, and over the past 7 years, mayors across the country have supported month-long promotions of PCa awareness.

## Cost vs Value

A substantial clinical and economic burden is associated with the diagnosis and treatment of PCa in the United States and worldwide. All men diagnosed with PCa and their physicians face the challenge of deciding whether to choose definitive therapy – which has a high cure rate but potentially significant complications – or monitor the disease through active surveillance or watchful waiting. Among physician, payer, and health care system stakeholders in the United States and globally there is widespread concern that the rate of treatment for PCa is too high, especially since prostate cancer in many men is often indolent, with little malignant potential. There is no question that aggressive therapy is indicated in a large subset of men with aggressive PCa. In our opinion, there is benefit with focusing our treatment on men with high-risk PCa.

When used appropriately, incorporation of biomarkers into clinical practice can have a positive economic impact. For example, a budget impact study of the ConfirmMDx assay identified men who may avoid unnecessary repeat prostate biopsies, thereby reducing overall healthcare spending.[41] By using the appropriate biomarkers, we can reduce unnecessary biopsies and help men make more effective clinical decisions about their individual plan of care, which should translate into economic savings.

## Future Perspective

One reason for the current controversy regarding screening for PCa is the lack of a test or a marker that can differentiate indolent from aggressive disease at the time of initial diagnosis. The ideal screening test should detect not just the presence of PCa but the aggressive PCa that needs to be treated, whereby its finding would impact progression and patient mortality.

Most urologists have experience treating men with apparent low-risk PCa who at RP, are found to have higher-grade and higher-stage cancers harboring a significant risk of progression and death. In addition to prognostic markers predicting likelihood of aggressive disease, we will need to develop markers that predict the response to therapies.

Molecular diagnostic researchers should ensure that the analytic validity of a biomarker test has been established prior to the evaluation of clinical utility. In planning clinical utility studies for biomarkers, protocols should specify the patient population intended to benefit from the decision guided by the test result. For validation studies of all types, prior evidence from early studies must be obtained from cohorts relevant to the intended use population. Another critical aspect is ascertaining that the samples studied are in fact those of the correct patient. To that end, another biomarker, Know Error (Strand Diagnostics) may be utilized to ensure that the specimen chain of custody is indeed accurate.

Ideally, clinical validation studies should use metrics that are clinically useful to physicians in order to assess the strength of association between the biomarker assay and PCa. Ideally, such studies should include outcome measures that assess the potential benefits and challenges from the patient perspective, recognizing that these outcomes may occur at different time points and are the result of clinical management decisions guided by test results. The development and use of reimbursement policy approaches to promote clinical utility must be evidence based.

Biomarker platforms that enable healthcare professionals to accurately interpret and communicate the results of biomarker diagnostic and predictive testing for patients and their caregivers must be prospectively validated and their contemporaneous use should be promoted. Such strategies should include Continuing Medical Education credits for biomarker-related training, ongoing clinical utility trials, as well as engaging professional societies to develop practice guidelines to assess the incorporation and utilization of biomarker test results when appropriate.

TABLE 66.2 Characteristics of Commonly Utilized Biomarkers in PCa

| Test | Indication | Science | Results | Cost |
|---|---|---|---|---|
| OncotypeDX Genomic Health | Biopsy tissue based test<br>NCCN Very Low, Low and Intermediate Risk<br>Provides Personalized Risk Assessment | Assay looks at 17 genes within 4 pathways (androgen signaling, stromal response, cellular organization, proliferation) to assess tumor aggressiveness | Genomic Prostate Score (GPS) from 0 to 100<br>Likelihood of freedom from Dominant 4 or higher GS and/or non-organ confined disease<br>GPS is reflective of the biology of the tumor at the time of biopsy | $3820<br>Medicare = No ABN required*<br>Other ins: If estimated out-of-pocket cost >$100, company will contact the patient to offer financial assistance program.<br>866-662-6897 |
| Prolaris Myriad Genetics | Biopsy tissue based test for patients who are active surveillance candidates<br>-or-<br>Postprostatectomy tissue based test to determine relative risk of BCR | 46-Gene expression signature includes cell cycle progression genes selected based upon correlation with prostate tumor cell proliferation | Prolaris score<br>Biopsy is < or = or > than AUA risk group and est. 10 years mortality risk<br>Postsurgical is similar but 10 years risk for BCR | $3400<br>Medicare = No ABN required*<br>Other ins: If estimated out-of-pocket cost>$375, company will contact patient to make arrangements…they have a financial assistance program.<br>800-469-7423 |
| Decipher® GenomeDx Biosciences | Postprostatectomy tissue based test used for patients who are candidates for secondary therapy post prostatectomy<br>pT2 with positive margins or pT3 or BCR | Analyzes the activity of 22 genetic markers in multiple pathways across the genome to measure the tumor's biological potential for metastasis after surgery | Decipher reports the probability of metastasis at 5 years after surgery and 3 years after PSA recurrence.<br>AUC 0.79<br>HR: 7.3 (Decipher high risk)<br>NPV 98.5% | $4250<br>Medicare =No ABN required*<br>Other ins: Financial assistance program available for out of pocket expenses.<br>888-792-1601 |
| ConfirmMDx MDxHealth | Biopsy tissue based test for patients who are repeat biopsy candidates<br>Provides risk stratification on decision for repeat biopsy<br>*Eligibility*: Prior negative or HGPIN biopsy result in past 24 months | Three-gene methylation assay to detect an epigenetic field effect associated with the cancerization process at the DNA level | Negative ConfirmMDx result: Avoid repeat biopsy and monitor with routine screening.<br>Positive ConfirmMDx result: Hypermethylated areas are marked as positive providing repeat biopsy guidance on a prostate map. | $2473 ($206 core/block)<br>Medicare = No ABN required*<br>Other ins: Financial assistance program is available for out of pocket expenses.<br>866-259-5644 |
| Know Error® Strand Diagnostics | Oral swab and biopsy tissue based test provides DNA tissue matching<br>Confirms pathology and/or confirms Biomarker is performed on correct patient<br>Increases diagnostic accuracy | Buccal swab in the clinic sent for DNA match to pathology specimen; may be used with all tissues. STR profiles assessed from multiplex panel of 16 genetic markers | DNA Match<br>DNA Nonmatch<br>Contamination | $1780 (out-of-network billed charge amount per test)<br>Medicare =$293/No ABN required*<br>Other ins: Patient is only responsible for "in-network" copays/deductibles. As of Mar/2014, only 2.4% of patients had any out-of-pocket costs and average is $65.<br>888-924-6779 ex. 2 |

** Without an Advanced Beneficiary Notification (ABN) the patient is not held responsible for any unreimbursed expenses.*
*Adapted from Shore et al.*[42]

## CONCLUSIONS

PCa biomarkers have the potential in assisting clinicians to improve decisions regarding who to biopsy, who to avoid a repeat biopsy, who to enhance risk assessment, and thereby reduce unnecessary biopsy strategies as well as overtreatment, and thus achieving more selective therapy for patients with high-risk disease. In effect, clinicians can strive for better outcomes and hopefully remain cost neutral or better yet achieve cost savings to the healthcare system. In the last few years, there has been rapid development of many new and novel biomarkers. These biomarkers should offer and assist clinicians with improved decision-making on when to biopsy, who to rebiopsy, and how to assist patients with treatment decisions. If third-party payers are to appropriately support reimbursement outcomes, prospective data are needed that will demonstrate beneficial and actionable utility metrics, which should positively impact patient outcomes and cost of care. Table 66.2 summarizes the characteristics of commercially available biomarkers in PCa.

## References


1. Siegel R, Ma J, Zou Z, et al. Cancer statistics. *CA Cancer J Clin* 2014;**64**(9).
2. Mortality, S.S. D.All COD, Aggregated with State, Total U.S. (1969-2010) <Katrina/Rita Population Adjustment>, National Cancer Institute, DCCPS, Surveillance Research Program, Surveillance Systems Branch, released April 2013. Underlying mortality data provided by NCHS (http://www.cdc.gov/nchs); 2013.
3. Armstrong AJ, Eisenberger MA, Halabi S, et al. Biomarkers in the management and treatment of men with metastatic castration-resistant prostate cancer. *Eur Urol* 2012;**61**:549.
4. Crawford ED, Ventii K, Shore ND. New biomarkers in prostate cancer. *Oncology* 2014;**28**(2):135–42.
5. Catalona WJ, Smith DS, Ratliff TL, et al. Measurement of prostate-specific antigen in serum as a screening test for prostate cancer. *N Engl J Med* 1991;**324**:1156.
6. Chou R, Croswell JM, Dana T, et al. Screening for prostate cancer: a review of the evidence for the U.S. Preventive Services Task Force. *Ann Intern Med* 2011;**155**:762.
7. Institute, N.C. Prostate-Specific Antigen (PSA) Test
8. Force, U.S. P. S. T.: Summaries for patients. Screening for prostate cancer: U.S. Preventive Services Task Force recommendation statement. Ann Intern Med, 157: I, 2012.
9. Andriole GL, Crawford ED, Grubb 3rd RL, et al. Prostate cancer screening in the randomized Prostate, Lung, Colorectal, and Ovarian Cancer Screening Trial: mortality results after 13 years of follow-up. *J Natl Cancer Inst* 2012;**104**:125.
10. Crawford ED, Grubb R, Black A, et al. Comorbidity and mortality results from a randomized prostate cancer screening trial. *J Clin Oncol* 2011;**29**:355.
11. Schroder FH, Hugosson J, Roobol MJ, et al. Prostate-cancer mortality at 11 years of follow-up. *N Engl J Med* 2012;**366**:981.
12. Roobol MJ, Bangma CH, Loeb S. Prostate-specific antigen screening can be beneficial to younger and at-risk men. *CMAJ* 2013;**185**:47.
13. NCCN: National Comprehensive Cancer Network. Clinical Practice Guidelines in Oncology. Prostate Cancer. Version 2.2014; 2014.
14. Hugosson J, Carlsson S, Aus G, et al. Mortality results from the Goteborg randomised population-based prostate-cancer screening trial. *Lancet Oncol* 2010;**11**:725.
15. NCCN: National Comprehensive Cancer Network. Prostate Cancer Early Detection. Version 1.2014; 2014.
16. Le BV, Griffin CR, Loeb S, et al. [-2]Proenzyme prostate specific antigen is more accurate than total and free prostate specific antigen in differentiating prostate cancer from benign disease in a prospective prostate cancer screening study. *J Urol* 2010;**183**:1355.
17. Wang W, Wang M, Wang L, et al. Diagnostic ability of %p2PSA and prostate health index for aggressive prostate cancer: a meta-analysis. *Sci Rep* 2014;4:5012.
18. Tosoian JJ, Loeb S, Feng Z, et al. Association of [-2]proPSA with biopsy reclassification during active surveillance for prostate cancer. *J Urol* 2012;**188**:1131.
19. Porpiglia F, Russo F, Manfredi M, et al. The roles of multiparametric magnetic resonance imaging, PCA3 and prostate health index-which is the best predictor of prostate cancer after a negative biopsy? *J Urol* 2014;**192**(1):60–6.
20. Stephan C, Vincendeau S, Houlgatte A, et al. Multicenter evaluation of [-2]proprostate-specific antigen and the prostate health index for detecting prostate cancer. *Clin Chem* 2013;**59**:306.
21. Daniel W, L.: The 4KscoreTM Test as a Predictor of High-Grade Prostate Cancer on Biopsy. In: American Urological Association Annual Meeting, 2014.
22. de la Taille A, Irani J, Graefen M, et al. Clinical evaluation of the PCA3 assay in guiding initial biopsy decisions. *J Urol* 2011;**185**:2119.
23. Crawford ED, Rove KO, Trabulsi EJ, et al. Diagnostic performance of PCA3 to detect prostate cancer in men with increased prostate specific antigen: a prospective study of 1,962 cases. *J Urol* 2012;**188**:1726.
24. Stewart G, Delvenne P, Delree P, et al. Clinical utility of a multiplexed epigenetic gene assay to detect cancer in histopathologically negative prostate biopsies: results of the multicenter MATLOC study. *J Urol* 2013;189(3):1110–6.
25. Trock BJ, Brotzman MJ, Mangold LA, et al. Evaluation of GSTP1 and APC methylation as indicators for repeat biopsy in a high-risk cohort of men with negative initial prostate biopsies. *BJU Int* 2012;**110**:56.
26. Robinson K, Creed J, Reguly B, et al. Accurate prediction of repeat prostate biopsy outcomes by a mitochondrial DNA deletion assay. *Prostate Cancer Prostatic Dis* 2010;**13**:126.
27. Robinson K, Creed J, Reguly B, et al. Accurate prediction of repeat prostate biopsy outcomes by a mitochondrial DNA deletion assay. *Prostate Cancer Prostatic Dis* 2013;**16**:398.
28. Lotan TL, Carvalho FL, Peskoe SB, et al. *PTEN* loss is associated with upgrading of prostate cancer from biopsy to radical prostatectomy. *Mod Pathol* 2014.
29. Chaux A, Peskoe SB, Gonzalez-Roibon N, et al. Loss of *PTEN* expression is associated with increased risk of recurrence after prostatectomy for clinically localized prostate cancer. *Mod Pathol* 2012;**25**:1543.
30. Knezevic D, Goddard AD, Natraj N, et al. Analytical validation of the Oncotype DX prostate cancer assay – a clinical RT-PCR assay optimized for prostate needle biopsies. *BMC Genomics* 2013;**14**:690.
31. Klein E.A, Maddala T, Millward C, et al. Development of a needle biopsy-based genomic test to improve discrimination of clinically aggressive from indolent prostate cancer. In: American Society of Clinical Oncology (ASCO) Annual Meeting, 2012.
32. Cooperberg MR, Simko S, Falzarano S, et al. Development and validation of the biopsy-based genomic prostate score (GPS) as a predictor of high grade or extracapsular prostate cancer to improve patient selection for active surveillance. *J Urol* 2013;**89**(4):e873.
33. Klein EA, Cooperberg MR, Magi-Galluzzi C, et al. A 17-gene assay to predict prostate cancer aggressiveness in the context of Gleason grade heterogeneity, tumor multifocality, and biopsy undersampling. *Eur Urol* 2014;66(3):550-60.
34. Ashok J, Kar MCS, Fegan JE, et al. The effect of cell cycle progression (CCP) score on treatment decisions in prostate cancer: results of an ongoing registry trial. *J Clin Oncol* 2014;**32**.
35. Badani K, Thompson DJ, Buerki C, et al. Impact of a genomic classifier of metastatic risk on postoperative treatment recommendations for prostate cancer patients: a report from the DECIDE study group. *Oncotarget* 2013;**4**:600.
36. Karnes RJ, Bergstralh EJ, Davicioni E, et al. Validation of a genomic classifier that predicts metastasis following radical prostatectomy in an at risk patient population. *J Urol* 2013;**190**:2047.
37. Saad F, Shipitsin M, Choudhury S, et al. Distinguishing aggressive versus nonaggressive prostate cancer using a novel prognostic

proteomics biopsy test, ProMark. In: 2014 ASCO Annual Meeting, vol. *J Clin Oncol* 325s, 2014 (suppl; abstr 5090), 2014.
38. Balk SP, Ko YJ, Bubley GJ. Biology of prostate-specific antigen. *J Clin Oncol* 2003;**21**:383.
39. Scher HI, Halabi S, Tannock I, et al. Design and end points of clinical trials for patients with progressive prostate cancer and castrate levels of testosterone: recommendations of the Prostate Cancer Clinical Trials Working Group. *J Clin Oncol* 2008;**26**:1148.
40. Chen CL, Mahalingam D, Osmulski P, et al. Single-cell analysis of circulating tumor cells identifies cumulative expression patterns of EMT-related genes in metastatic prostate cancer. *Prostate* 2013;**73**:813.
41. Aubry W, Lieberthal R, Willis A, et al. Budget impact model: epigenetic assay can help avoid unnecessary repeated prostate biopsies, reduce spending. *Am Health Drug Benefits* 2013;**6**:15.
42. Shore N, Ventii K, Crawford ED. A summary review of prostate cancer biomarkers OncologyLive, 2014 Published Online: Thursday, September 11, 2014.

CHAPTER

# 67

# Testosterone Therapy in Hypogonadal Men with Prostate Cancer

*Joshua R. Kaplan, MD*

**Department of Urology, Temple University School of Medicine, Philadelphia, PA, USA**

## BACKGROUND

### Prostate Cancer as a Contraindication for Testosterone Therapy

Numerous cross-sectional and longitudinal studies demonstrate that serum testosterone levels decrease with advancing age while the incidence of symptomatic male hypogonadism increases and may be as high as 500,000 new cases annually in men aged 40–69 years in the United States.[1–4] Similar data have been shown for men in Australia and Europe.[5,6] The implications and consequences of hypogonadism are increasingly diagnosed; they are wide-ranging and include decreased quality of life, energy and libido, loss of muscle mass, cognitive impairment, sexual and erectile dysfunction, poor attention span, depressed mood, and decreased cardiovascular and bone health. Treatment with testosterone replacement therapy can alleviate these signs and symptoms.[7–10]

Young men (<50 years of age) comprise approximately 2% of new prostate diagnoses in the United States.[11] The median age at prostate cancer diagnosis is 68 years of age, and 63% of patients are diagnosed at 66 years or older.[12] It is reasonable to conclude that there are a large percentage of hypogonadal men who have also undergone treatment for prostate cancer.[13] In addition, patients undergoing treatment for prostate cancer are at increased risk of depression, erectile and sexual dysfunction, and decreased libido. Men with low serum testosterone levels are thus at high risk of experiencing these symptoms, and may experience significant benefit from testosterone replacement therapy. Unfortunately, these are exactly the men in whom testosterone replacement therapy has been traditionally withheld.

The historical basis for withholding testosterone replacement therapy in patients with prostate cancer, even after definitive treatment for local disease, stems from the thinking that the prostate is an androgen-dependent organ and prostate cancer uses testosterone as a substrate or "fuel" for growth. Indeed, many currently practicing urologists were taught during training that administering testosterone to patients with prostate cancer is akin to "throwing gasoline on a fire" or "feeding a hungry tumor."[14] The original, oft-cited assertion that testosterone activates prostate cancer derives from Huggins' seminal studies more than 65 years ago, in which he showed that castration resulted in a rapid reduction in serum acid phosphatase and testosterone treatment caused a rapid increase in acid phosphatase.[15,16] These were conducted prior to discovery of the androgen receptor, accurate serum testosterone assays, and prostate-specific antigen (PSA). Subsequently, in 1967, Prout and Brewer reported that 5 of 10 men with recurrent disease after castration experienced progression or death when given several weeks of testosterone supplementation.[17] A 1981 study by Fowler and Whitmore demonstrated "unfavorable" outcomes in 45 of 52 men with metastatic disease administered exogenous testosterone.[18] The result of these trials was near-universal acceptance of the view that increased serum testosterone leads to prostate cancer growth and progression; accordingly, prostate cancer has long been considered a contraindication to testosterone supplementation, even after primary treatment for organ-confined disease.

### New Thinking: the Saturation Model

During the past decade, a number of authors began to question the validity of the link between serum testosterone level and prostate cancer activity, noting that the link between Huggins' castrated men and current hypogonadal men treated for prostate cancer is tenuous, with no clear evidence of increased risk of recurrence for men

Prostate Cancer. http://dx.doi.org/10.1016/B978-0-12-800077-9.00067-0

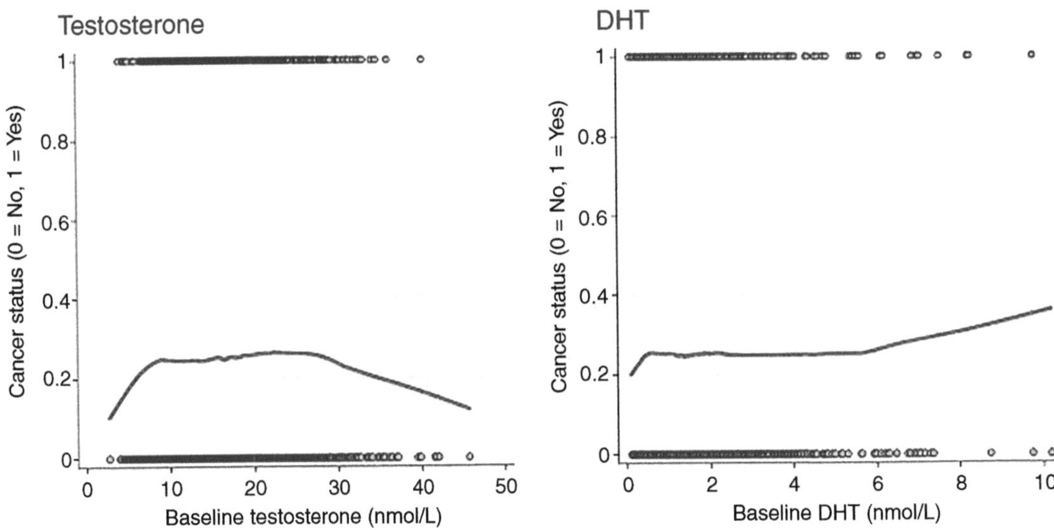


FIGURE 67.1 **Locally weighted scatterplot smoothing of serum levels of testosterone and dihydrotestosterone (DHT) at baseline and final cancer status after considering all biopsies during 4 years of the REDUCE trial.** The overlapping circles on the top and bottom of the chart represent each individual case. Men detected with prostate cancer were coded as 1, whereas men with no prostate cancer detected were coded as 0. *Adapted with permission from Muller et al.*[29]

successfully treated for primary prostate cancer.[13,19–21] Given the considerable overlap of patients with hypogonadism and prior treatment for prostate cancer, as well as increasing recognition of the efficacy and benefits of testosterone therapy, a limited yet significant body of literature has been published focusing on the safety and role for testosterone supplementation in men who have undergone treatment for prostate cancer.

In addition to questioning whether it is justifiable to withhold testosterone therapy in men with prostate cancer status post definitive treatment, Morgentaler and Traish proposed a revised view of the effect of testosterone on prostate cancer stimulation and growth: the saturation model.[22,23] Noting that androgens have a limited capacity to stimulate prostate cancer progression, they described a saturation point such that levels of testosterone at or below the near-castrate range are associated with a steep T-dependent rate of growth, while minimal or no growth is seen above the level of saturation. Morgentaler and Traish argue that administration of exogenous testosterone to hypogonadal (but not castrate-level) men would constitute serum fluctuation of testosterone at levels above the saturation point, which should not impact the rate of cancer recurrence or growth, yet may provide marked benefit to hypogonadal men who have undergone treatment for their prostate cancer.

In a rat model, Wright et al. demonstrated that intraprostatic androgen levels and prostate mass are sensitive to serum testosterone at near-castrate levels, but sensitivity decreases as levels increase.[24] A small, randomized controlled trial conducted by Marks et al. on 44 men with late-onset hypogonadism assessed the effect of androgen replacement therapy on prostate tissue.[25] Forty men underwent both randomization and follow-up biopsy at 6 months. Testosterone replacement increased serum levels but had no effect on intraprostatic androgen and biomarker levels, gene expression, or prostate cancer incidence (Figure 67.1). Other previously published studies of androgen binding capacity in animal and human prostate tissues have supported the possibility of a saturation model, demonstrating maximal binding (saturation) of androgen to androgen receptors in the 3nM or less range.[26–28] Viewed in consideration of these findings, the Huggins, Prout, and Fowler cohorts' unfavorable responses very well may have been experienced by patients who had undergone castration prior to testosterone treatment – and were thus functioning at levels of serum testosterone lower than the saturation point with unbound androgen receptors available to be activated by exogenous testosterone.

Muller et al. provided further compelling evidence in support of the saturation model in 2012 when they reported on the relationship of serum testosterone and dihydrotestosterone to prostate biopsy results in the reduction by dutasteride of prostate cancer events (REDUCE) trial.[29] Trial entry requirements included PSA between 2.5 ng/mL and 10.0 ng/mL and prior negative biopsy. The authors analyzed the 3255 men randomized to the placebo arm who underwent at least one of the planned biopsies at 2 and 4 years with regard to baseline serum testosterone and dihydrotestosterone values. No significant association between prostate cancer and serum androgens was identified. Cancer rates were similar between men with normal testosterone and men with hypogonadism defined as <10 nmol/L or 288 ng/dL (25.5% vs. 25.1%, respectively; $p = 0.831$). The authors included a LOWESS plot in their paper, presented later, stating "prostate cancer detection between men with low compared with normal baseline testosterone was similar, perhaps because the optimal threshold above which testosterone becomes saturated may be ≤10 nmol/L." The rates of prostate cancer are seen to plateau at a value below the normal range.

In an editorial in the same journal issue as the Muller study, Morgentaler wrote: "Muller et al. provide the final nail in the coffin for what had been a guiding principle of uro-oncology for >70 years: the androgen hypothesis."[30] Morgentaler went on to emphasize the importance of this study, as "the failure to find increased (prostate cancer) rates associated with higher serum androgens based on biopsies in a large at-risk population removes the last possible hope to those who wish to hold on to a disproved theoretical notion from a premodern era." The title of the editorial succinctly summed up the shift in thinking that has taken place regarding the use of testosterone therapy in patients with a diagnosis prostate cancer over the past decade: "Goodbye androgen hypothesis, hello saturation model."

## TESTOSTERONE SUPPLEMENTATION FOLLOWING DEFINITIVE THERAPY FOR LOCALIZED PROSTATE CANCER

Since the advent and diffusion of PSA testing and the resultant stage migration, prostate cancer is most-often diagnosed while still organ-confined. Surgical or radiation therapy often provides durable disease control or cure. The majority of these men are eugonadal before and after treatment, which suggests that serum testosterone levels above the saturation point does not adversely affect disease control. As previously noted, the subset of post treatment patients with symptomatic hypogonadism may experience substantial benefit with testosterone administration. The saturation model provides a possible explanation as to why exogenous testosterone therapy may be well-tolerated in symptomatic hypogonadal men, yet there is limited clinical evidence as to the safety and efficacy in men who have undergone such treatments, due to the fact that until recently prostate cancer was a contraindication to testosterone supplementation.

### The Evidence for Testosterone Therapy Following Radical Prostatectomy

In 2004, Kaufman and Graydon published their experience of seven patients treated with androgen replacement for symptomatic hypogonadism after prior curative radical prostatectomy.[31] All subjects had undetectable PSA and no recurrences were noted with follow-up for as long as 12 years with an average of 24 months. A year later, Agarwal and Oefelein reported no biochemical recurrences with testosterone supplementation following radical prostatectomy in 10 symptomatic hypogonadal men with undetectable PSAs during a mean follow-up period of 19 months.[32] In 2009, Khera et al. published the results of 57 men treated with prostatectomy and subsequent testosterone therapy.[33] Pre- and post treatment PSA levels remained stable at 0.005 ng/mL and average follow-up was 13 months. Nabulsi et al. reported the results of 22 men in their cohort with a follow-up mean of 24 months. They did have one case of biochemical recurrence in a patient with pathologic Gleason 8 disease.[34] In 2008, Davila et al. reported a series of 20 patients with prostate cancer who had undergone primary treatment and developed symptoms of hypogonadism subsequently treated with testosterone therapy.[35] Fourteen of these patients were treated with radical prostatectomy, followed for an average of 12 months after testosterone supplementation, and no patient experienced a biochemical recurrence.

The largest series to date was published by Pastuszak et al. in 2013, describing successful testosterone treatment for 103 men after radical prostatectomy, mean follow-up 27.5 months, and pretreatment PSA 0.004, post treatment 0.007.[36] Of special note, this study included 26 men with at least one of the following high-risk features: positive surgical margin, nodal disease, or Gleason score greater than or equal to 8. They compared their cohort to a group of 49 men with high-risk disease who did not receive testosterone and reported four PSA recurrences in the testosterone treatment group (4%) and eight recurrences in the untreated group (16%).

### The Evidence for Testosterone Therapy Following Radiotherapy

The decision to offer testosterone supplementation to hypogonadal men whose prostate cancer has been treated with primary radiation therapy warrants separate consideration from patients treated with extirpative surgery because the prostatic tissue remains *in situ* following radiotherapy.[37] According to the saturation model, testosterone supplementation for patients with serum testosterone levels above the castrate-level saturation point would not be expected to impact residual or untreated tumor behavior. The data for post radiation testosterone use is even more limited than in postprostatectomy patients.

#### *Brachytherapy*

Sarosdy's experience with testosterone supplementation in 36 men treated with brachytherapy for prostate cancer remains the only brachytherapy-only cohort to date.[38] Five men discontinued their testosterone due to perceived lack of benefit after 1–3 months, and the remaining 31 patients were treated for a median duration of 4.5 years with mean follow-up of 5 years. Median time to testosterone initiation after brachytherapy was 24 months. At last follow-up, 74% of patients' PSA was <0.1 ng/mL and all patients had PSA <1 ng/mL. Pastuszak et al. also published a study of 13 patients treated with radiation, three of whom were treated with brachytherapy.[39] All patients remained biochemical recurrence-free during a median follow-up of 29.7 months.

### *External Beam Radiation Therapy*

The remaining 10 patients in Pastuszak's study were successfully treated with external beam radiation therapy and did not recur during their testosterone treatment period or during nearly 30 months of follow-up. Davila's previously discussed cohort also included six patients who underwent external beam radiation therapy for prostate cancer and subsequently were treated with testosterone supplementation for hypogonadism a mean of 57 months after their radiation.[35] No patients experienced biochemical recurrence during a 9-month follow-up period. Morales et al. reported their experience with five men with severe hypogonadism treated with exogenous testosterone that had previously undergone external beam radiation for prostate cancer.[40] No biochemical recurrence was observed during a 14.6-month mean follow-up duration.

## Role of PSA

At final review of the available literature, the evidence for testosterone supplementation in patients with prostate cancer who have been successfully treated with radiation or surgery remains retrospective and focuses on small patient cohorts. Nonetheless, at this time, the mechanism outlined in the saturation model seems plausible in light of the efficacy and safety these few studies have demonstrated. Indeed, of the 210 men among the studies described who were treated with radical prostatectomy and the 55 men treated with primary radiotherapy, only five patients recurred (2.4% of patients treated with prostatectomy, 1.9% of all patients treated with testosterone after treatment of localized prostate cancer); the men who recurred all had high-risk disease and therefore were at increased risk of recurrence regardless of testosterone administration.

Regarding PSA cutoffs for institution of testosterone therapy in patients previously treated for prostate cancer, all 265 patients covered in the literature to date had PSA levels below 0.9 ng/mL (all prostatectomy cohorts were <0.1) at the time of testosterone administration. While the saturation model and studies to date support the use of testosterone in any patient with symptomatic hypogonadism who is without evidence of biochemical recurrence, there may be a role for testosterone therapy in patients with PSA levels above 0.9 ng/mL. Unfortunately, despite the likelihood that these patients are functioning at a testosterone level above the saturation point (unless on androgen deprivation therapy) and therefore may be able to be treated with exogenous testosterone, the paucity of evidence as to the safety of testosterone treatment in these patients mandates an extensive discussion of the available evidence as well as risks with patients prior to the initiation of treatment.

## Future Considerations

Despite the lack of harm demonstrated in the few studies to date, prostate cancer is still widely held to be a contraindication for testosterone therapy. The AndroGel® website cites "men with [...] known or suspected carcinoma of the prostate" as a contraindication for use.[41] At the time of submission, the European Association of Urology guidelines states that

> Testosterone therapy is clearly contraindicated in men with prostate cancer. A topic currently under debate involves the use of TRT in hypogonadal men with a history of prostate cancer and no evidence of active disease... Men who have been surgically treated for localized prostate cancer and who are currently without evidence of active disease (i.e., measurable PSA, abnormal rectal examination, evidence of bone/visceral metastasis) and showing symptoms of testosterone deficiency can be cautiously considered for TRT, although this approach is still an "off-label" treatment. In these patients, treatment should be restricted to patients with a low risk for recurrent prostate cancer (pre-surgery Gleason <8; pT1-2; PSA <10 ng/mL). Therapy should not start before 1 year of follow-up after surgery and there should be no PSA recurrence. Patients who have undergone brachytherapy or external-beam radiotherapy (EBRT) for low-risk prostate cancer can also be cautiously treated with TRT in case of hypogonadism, with close monitoring for prostate cancer recurrence.[42]

Despite increasing evidence supporting the safety of testosterone supplementation in these men, Kaplan et al. reported a SEER-Medicare population-based study that found decreasing utilization of this therapy from 1992 to 2006.[43] The earliest studies reporting on these types of patients was published in 2004,[31] and thus we may hope that utilization has increased since 2006, in what may be considered the post-Huggins era of testosterone and prostate cancer. The population-based sample also revealed that only 1181 patients of the 149,354 men total diagnosed with prostate cancer (0.79%) received testosterone. Notably, only ~102,000 of these men received primary treatment for their disease. Testosterone replacement use was also directly correlated to higher income, educational status, and inversely with age. Testosterone use was not associated with overall or cancer-specific mortality, or androgen deprivation therapy.

All currently available data support the concept that it may be safe to treat appropriate hypogonadal men

**TABLE 67.1** Considerations Prior to Offering Testosterone Therapy After Treatment of Localized Prostate Cancer

- The clinical picture is consistent with a diagnosis of testosterone deficiency.
- The patient must understand that safety data are limited and that there is an unknown degree of risk of (prostate cancer) progression or recurrence.
- The patient must be willing and able to provide informed consent.
- No medical contraindications to testosterone therapy (e.g., erythrocytosis) exist.
- There is an undetectable or stable PSA level.
- Clinicians must be prepared for the possibility of (prostate cancer) recurrence or progression, which will occur in some men regardless of testosterone therapy but may be attributed to testosterone therapy by patients, family, or other clinicians.
- Use testosterone therapy with extreme caution in men at high risk for (prostate cancer) recurrence or progression.
- Do not recommend testosterone therapy for men currently receiving any form of (androgen deprivation therapy).

*Adapted with permission from Khera et al.*[37]

with testosterone after they have undergone definitive treatment of their prostate cancer. Unfortunately, the available data are of low quality, retrospective or observational in nature, and definitive prospective studies are required. Until high-quality evidence is available, Khera et al. have offered a list of criteria to consider prior to offering testosterone supplementation to men with a history of treated prostate cancer (Table 67.1). The authors also call for the creation of an international registry for men treated with testosterone therapy after prostate cancer diagnosis. In light of the available evidence, their list seems prudent, and perhaps the most vital aspect of their criteria is a careful, frank discussion with each patient, to protect patient and provider.

## CAN PATIENTS ON ACTIVE SURVEILLANCE WITH SYMPTOMATIC HYPOGONADISM RECEIVE TESTOSTERONE THERAPY?

The question of whether to offer testosterone replacement therapy to men with untreated prostate cancer who are on an active surveillance regimen and symptomatic from hypogonadism is even more controversial than the decision to offer androgen replacement to men who have previously undergone treatment for prostate cancer. Concerns about utilization of exogenous testosterone by untreated cancer cells propel this debate, and the available studies addressing the safety of this approach are – as one might expect – even more limited. Like testosterone therapy in hypogonadal men treated for prostate cancer, the use of testosterone therapy in men with untreated prostate cancer on active surveillance relies on the prescribing physicians' belief in the saturation model.

One of the earliest reports in the literature of testosterone therapy in a man with prostate cancer came in 2009.[44] Morgentaler reported the effects of PSA on an 84-year-old man with Gleason 3 + 3 disease in two of six cores, constituting 30% and 5% of biopsies of the right and left base, respectively. The patient complained of erectile dysfunction refractory to oral sildenafil and anorgasmia. Labs were notable for testosterone of 400 ng/dL, free testosterone 7.4 pg/mL, and PSA 8.5 ng/mL. He noted improvement in erection quality, libido, energy, and sense of vigor with testosterone gel therapy. Additionally, despite a serum testosterone increase to a mean of 699 ng/dL and free testosterone to 17.1 pg/mL with therapy, serum PSA declined to a nadir of 5.2 ng/mL at 10 months after initiation of treatment. After 24 months of testosterone replacement the PSA did not increase beyond the level at time of presentation. The patient declined repeat biopsy.

In 2011, Morgentaler et al. reported the results of prostate biopsies, serum PSA, and prostate volume in 13 symptomatic testosterone-deficient men on an active surveillance for prostate cancer.[45] Twelve of the men had Gleason score 3 + 3 disease and one man had Gleason score 3 + 4. All men underwent at least one follow-up biopsy after at least 12 months of testosterone therapy, 26 biopsies in total. One patient had evidence of disease progression on repeat biopsy, from low volume Gleason 3 + 3 to Gleason 3 + 4 in 5% of one core. Two subsequent annual biopsies on this patient revealed only low volume Gleason 3 + 3 disease. Another man underwent radical prostatectomy when a follow-up biopsy showed Gleason 4_3 disease in 75% of one of 12 cores. This occurred 8 years after initial diagnosis of Gleason 3 + 3 disease. Final pathology revealed only Gleason 3 + 3 involving 5% of the prostate with negative margins and nodes. All patients experienced improvement in libido, sexual performance, mood, or energy during testosterone therapy. The results of this study led the authors to conclude that testosterone replacement did not lead to prostate cancer progression during modest follow-up duration, which was interpreted to constitute a validation of the saturation model of prostate cancer growth.

The results of Hult and Morgentaler's initial series were expanded upon in an abstract presented at the AUA annual meeting in 2013.[46] Of 33 men who had received testosterone while on active surveillance at the authors' practice, 28 had been treated for >6 months. No significant increase in PSA was seen during an average of

36.2 months of testosterone replacement, and no definite clinical progression (defined as an increase in positive cores by 3 or more, or persistent increase in Gleason score by 1) was noted during an average of 1.1 biopsies per patient. Seven men had upgrading of Gleason score on follow-up biopsy but all seven also had subsequent biopsies that revealed a lower Gleason score or no malignancy. Two men discontinued testosterone therapy due to lack of efficacy and two men proceeded to prostatectomy despite lack of clinical evidence of disease progression.

Morales reported more mixed results in terms of PSA behavior in his cohort of seven patients with a diagnosis of prostate cancer who underwent testosterone therapy for symptomatic hypogonadism.[47] Only six of these patients had a diagnosis of prostate cancer at the beginning of testosterone replacement. One patient had Gleason 4 + 4 disease and the remainder had Gleason 3 + 3 disease. The duration of therapy in this series ranged from 6 months to 96 months. Morales' patients had much more varied PSA responses to testosterone therapy than Morgentaler's, with one patient demonstrating stable PSA for 36 months and then a rapid rise that responded to cessation of treatment. This patient had a rapid rise yet again with reinitiation of therapy that resolved once therapy was again stopped. Another patient had a rapid rise at the time of testosterone administration and ultimately stopped therapy and underwent radical prostatectomy with good biochemical result. Finally, the patient without a prostate cancer diagnosis at initiation of therapy (and therefore not truly on active surveillance) underwent two negative biopsies during the first 4 years of treatment but was observed to have a rising PSA in the fifth year of treatment, and a third biopsy diagnosed prostate cancer when PSA was nearly 10 μg/L. This study highlighted the difficulty inherent in using PSA as a primary trigger for discontinuation of testosterone therapy in men on active surveillance. As Atin et al. noted in a 2013 review article on testosterone replacement treatment and prostate cancer after discussing the active surveillance series, "the present data are not mature enough to recommend (testosterone replacement therapy) routinely for symptomatic hypogonadism in patients with (untreated prostate cancer), but they are encouraging."[48]

## HOW IS THE NEW THINKING BEING EMPLOYED IN PRACTICE?

How are urologists incorporating the saturation model and the relatively scant but encouraging body of evidence for the safety of testosterone therapy in men with prostate cancer into practice? Unfortunately, there is no literature available to address this question. However, Kaplan et al. recently conducted a study employing Surveillance, Epidemiology, and End Results-Medicare data to exam population-based utilization and impact of testosterone replacement therapy in men with prostate cancer.[49] Using CPT-4 codes, the authors identified 348,372 men diagnosed with prostate cancer between 1991 and 2007 then excluded 113,844 men who were enrolled in an HMO or who were not enrolled in Medicare part A and B throughout the duration of the study period, as well as 20,060 subjects lacking 1 year of precancer diagnosis data, leaving 149,354 men with prostate cancer in the study. Of these men, only 1181 received injectable or depot testosterone replacement therapy (topical formulations were not captured). The testosterone replacement group included a mere 0.79% of the overall study population, and treatment utilization actually downtrended from 1.24% in 1992 to a nadir of 0.40% in 2006. Median follow-up was 6 years for the no-testosterone group and 8 years for the testosterone group. Testosterone use was associated with higher income census tract ($P = 0.009$), younger age ($P = 0.001$), and higher education ($P < 0.0001$). Men who underwent radical prostatectomy or who had well-differentiated tumors were more likely to receive testosterone replacement (24.7% vs. 18.3% and 11 vs. 6.6%, $P < 0.0001$, respectively). Overall and cancer-specific mortality were higher in the no-testosterone group ($P < 0.0001$ for both).

Notably, the Kaplan population study cohort only extended to 2007. Meanwhile, the saturation model papers and case series addressing the use of testosterone therapy in men with treated or untreated prostate cancer were published near or after 2007, and thus their impact on urologic practice cannot be assessed. With very few men with prostate cancer being treated with testosterone therapy in the years leading up to 2007, the logical conclusion, in the face of the evidence discussed in this chapter, must be that utilization of testosterone therapy can only increase. Prospective, randomized trials are needed to validate the findings of the primarily retrospective case series discussed. Barring definitive evidence to the contrary, it is only with high level data that the urologic community will move beyond historical teachings of testosterone as fuel for a hungry prostate cancer and universally begin to offer testosterone replacement therapy to the full complement of symptomatic hypogonadal patients who stand to benefit from treatment.

## References

1. Snyder PJ. Hypogonadism in elderly men – what to do until the evidence comes. *N Engl J Med* 2004;**350**:440–2.
2. Gray A, Feldman HA, McKinlay JB, Longcope C. Age, disease, and changing sex hormone levels in middle-aged men: results of the Massachusetts Male Aging Study. *J Clin Endocrinol Metab* 1991;**73**(5):1016–25.
3. Harman SM, Metter EJ, Tobin JD, Pearson J, Blackman MR. Longitudinal effects of aging on serum total and free testosterone levels in healthy men. *J Clin Endocrinol Metab* 2001;**86**(2):724–31.

4. Araujo AB, Esche GR, Kupelian V, et al. Prevalence of symptomatic androgen deficiency in men. *J Clin Endocrinol Metab* 2007;**92**(11):4241–7.
5. Liu PY, Beilin J, Meier C, et al. Age-related changes in serum testosterone and sex hormone binding globulin in Australian men: longitudinal analyses of two geographically separate regional cohorts. *J Clin Endocrinol Metab* 2007;**92**:3599–603.
6. Wu FC, Tajar A, Pye SR, et al. Hypothalamic-pituitary-testicular axis disruptions in older men are differentially linked to age and modifiable risk factors: the European male aging study. *J Clin Endocrinol Metab* 2008;**93**(7):2737–45.
7. Wang C, Nieschlag E, Swerdloff R, et al. Investigation, treatment, and monitoring of late onset-hypogonadism in males: ISA, ISSAM, EAU, EAA, and ASA recommendations. *J Androl* 2009;**30**:1–10.
8. Bassil N, Alkaade S, Morley JE, et al. The benefits and risks of testosterone replacement therapy: a review. *Ther Clin Risk Manag* 2009;**5**:427–48.
9. Hellstrom WJ, Paduch D, Donatucci CF, et al. Importance of hypogonadism and testosterone replacement therapy in current urologic practice: a review. *Int Urol Nephrol* 2012;**44**:61–70.
10. Traish AM, Miner MM, Morgentaler A, Zitzmann M. Testosterone deficiency. *Am J Med* 2011;**124**:578–87.
11. Jani AB, Johnstone PA, Liauw SL, Master VA, Brawley OW. Age and grade trends in prostate cancer (1974–2003): a Surveillance, Epidemiology, and End Results Registry Analysis. *Am J Clin Oncol* 2008;**31**:375–8.
12. Ries LA, Melbert D, Krapcho M, et al. *SEER cancer statistics review, 1975-2007*. http://www.seer.cancer.gov/archive/csr/1975_2007/index.html. [accessed 21.04.14].
13. Khera M. Androgen replacement therapy after prostate cancer treatment. *Curr Urol Rep* 2010;**11**:393–9.
14. Morgentaler A, Traish AM. Testosterone therapy: new concepts for a rapidly changing field webcast. Available at https://www.auanet.org/university/live-course.cfm?id=951. [accessed 8.08.14].
15. Huggins C, Hodges CV. Studies on Prostatic Cancer I. The effect of castration, of estrogen, and of androgen injection on serum phosphatases in metastatic carcinoma of the prostate. *Cancer Res* 1941;**1**:293–7.
16. Huggins C, Stevens RE, Hodges CV, et al. Studies on Prostatic Cancer II. The effects of castration on advanced carcinoma of the prostate gland. *Arch Surg* 1941;**43**:209–23.
17. Prout GR, Brewer WR. Response of men with advanced prostatic carcinoma to exogenous administration of testosterone. *Cancer* 1967;**20**:1871.
18. Fowler JE, Whitmore WF. The response of metastatic adenocarcinoma of the prostate to exogenous testosterone. *J Urol* 1981;**126**:372.
19. Morgentaler A. Testosterone and prostate cancer: an historical perspective on a modern myth. *Eur Urol* 2008;**53**:68–80.
20. Morgentaler A. Guilt by association: a historical perspective on Huggins, testosterone therapy, and prostate cancer. *J Sex Med* 2008;**5**:1834–40.
21. Isbarn H, Pinthus JH, Marks LS, et al. Testosterone and Prostate Cancer: Revisiting Old Paradigms. *Eur Urol* 2009;**56**:48–56.
22. Morgentaler A. Testosterone therapy in men with prostate cancer: scientific and ethical considerations. *J Urol* 2009;**181**:972–9.
23. Morgentaler A, Traish AM. Shifting the paradigm of testosterone and prostate cancer: the saturation model and the limits of androgen-dependent growth. *Eur Urol* 2009;**55**:310–21.
24. Wright AS, Douglas RC, Thomas LN, et al. Androgen-induced regrowth in the castrated rat ventral prostate: role of 5-alpha-reductase. *Endocrinology* 1999;**140**:4509–15.
25. Marks LS, Mazer NA, Mostaghel E, et al. Effect of testosterone replacement therapy on prostate tissue in men with late-onset hypogonadism: a randomized controlled trial. *JAMA* 2006;**296**:2351–61.
26. Ho SM, Damassa D, Kwan PW, et al. Androgen receptor levels and androgen contents in the prostate lobes of intact and testosterone-treated Noble rats. *J Androl* 1985;**6**:279–90.
27. Traish AM, Williams DF, Hoffman ND, et al. Validation of the Exchange Assay for the measurement of androgen receptors in human and dog prostates. *Prog Clin Biol Res* 1988;**262**:145–60.
28. Marks LS, Mostaghel EA, Nelson PS, et al. Prostate tissue androgens: history and current clinical relevance. *Urology* 2008;**72**:247–54.
29. Muller RL, Gerber L, Moreira DM, et al. Serum testosterone and dihydrotestosterone and prostate cancer risk in the placebo arm of the reduction of prostate cancer events trial. *Eur Urol* 2012;**62**:757–64.
30. Morgentaler A. Goodbye androgen hypothesis, hello saturation model. *Eur Urol* 2012;**62**:765–7.
31. Kaufman JM, Graydon RJ. Androgen replacement after curative radical prostatectomy for prostate cancer in hypogonadal men. *J Urol* 2004;**172**:920–2.
32. Agarwal PK, Oefelein MG. Testosterone replacement therapy after primary treatment for prostate cancer. *J Urol* 2005;**173**:533–6.
33. Khera M, Grober ED, Najari B, et al. Testosterone replacement therapy following radical prostatectomy. *J Sex Med* 2009;**6**:1165–70.
34. Nabulsi O, Tal R, Gotto G, Narus J, Goldenberg L, Mulhall JP. Outcomes analysis of testosterone supplementation in hypogonadal men following radical prostatectomy. *J Urol* 2008;**179**(Suppl.):406 abstract 1181.
35. Davila HH, Arison CN, Hall MK, et al. Analysis of the PSA response after testosterone supplementation in patients who previously received management for their localized prostate cancer. *J Urol* 2008;**179**:428 abstract 1247.
36. Pastuszak AW, Pearlman AM, Lai WS, et al. Testosterone replacement therapy in patients with prostate cancer after radical prostatectomy. *J Urol* 2013;**190**:639–44.
37. Khera M, Crawford D, Morales A, et al. A new era of testosterone and prostate cancer: from physiology to clinical implications. *Eur Urol* 2014;**65**:115–23.
38. Sarosdy MF. Testosterone replacement for hypogonadism after treatment of early prostate cancer with brachytherapy. *Cancer* 2007;**109**(3):536–41.
39. Pastuszak AW, Pearlman AM, Godoy G, et al. Testosterone replacement therapy in the setting of prostate cancer treated with radiation. *Int J Impot Res* 2013;**25**:24–8.
40. Morales A, Black AM, Emerson LE, et al. Testosterone administration to men with testosterone deficiency syndrome after external beam radiotherapy for localized prostate cancer: preliminary observations. *Br J Urol Int* 2009;**103**:62–4.
41. Http://www.androgel.com/questions-about-low-testosterone-and-androgel. [accessed 5.09.14].
42. Dohle GR, Arver S, Bettocchi C, et al. Guidelines on male hypogonadism. Uroweb 2014. Available from: http://www.uroweb.org/guidelines/online-guidelines. [accessed 25.04.14].
43. Kaplan AL, Trinh QD, Sun M, et al. Testosterone replacement therapy following the diagnosis of prostate cancer: outcomes and utilization trends. *J Sex Med* 2014;**11**:1063–70.
44. Morgentaler A. Two years of testosterone therapy associated with a decline in prostate-specific antigen in a man with untreated prostate cancer. *J Sex Med* 2009;**6**:574–7.
45. Morgentaler A, Lipshultz LI, Bennett R, et al. Testosterone therapy in men with untreated prostate cancer. *J Urol* 2011;**185**(4):1256–60.
46. Hult M, Morgentaler A. Testosterone therapy in men on active surveillance for prostate cancer. *J Urol* 2013;**189**(4):e271.
47. Morales A. Use of testosterone in men with prostate cancer and suggestions for an international registry. *BJU Int* 2011;**107**:1343–4.
48. Atin A, Altug T, Yesil S, et al. Serum testosterone level, testosterone replacement treatment, and prostate cancer. *Adv Urol* 2013;**2013**:275945.
49. Kaplan AL, Trinh Q, Sun M, et al. Testosterone replacement therapy following the diagnosis of prostate cancer: outcomes and utilization trends. *J Sex Med* 2014;**11**:1063–70.

# Subject Index

## C



## D

## Q



## R